Finite Element Simulations with ANSYS Workbench 19

Huei-Huang Lee

Department of Engineering Science
National Cheng Kung University, Taiwan

SDC Publications
P.O. Box 1334
Mission, KS 66222
913-262-2664
www.SDCpublications.com
Publisher: Stephen Schroff

ISBN-13: 978-1-63057-211-2

ISBN-10: 1-63057-211-X

Printed and bound in the United States of America.

Contents

Preface 4

Chapter 1 Introduction 9

1.1 Case Study: Pneumatically Actuated PDMS Fingers 10
1.2 Structural Mechanics: A Quick Review 24
1.3 Finite Element Methods: A Concise Introduction 35
1.4 Failure Criteria of Materials 40
1.5 Review 47
1.6 Appendix: An Unofficial History of ANSYS 51

Chapter 2 Sketching 56

2.1 W16x50 Beam 57
2.2 Triangular Plate 70
2.3 More Details 81
2.4 M20x2.5 Threaded Bolt 92
2.5 Spur Gears 97
2.6 Microgripper 103
2.7 Review 107

Chapter 3 2D Simulations 109

3.1 Triangular Plate 110
3.2 Threaded Bolt-and-Nut 125
3.3 More Details 140
3.4 Spur Gears 151
3.5 Structural Error, FE Convergence, and Stress Singularity 157
3.6 Review 170

Chapter 4 3D Solid Modeling 172

4.1 Beam Bracket 173
4.2 Cover of Pressure Cylinder 179
4.3 Lifting Fork 190
4.4 More Details 197
4.5 LCD Display Support 203
4.6 Review 207

Chapter 5 3D Simulations 209

5.1 Beam Bracket 210
5.2 Cover of Pressure Cylinder 219
5.3 More Details 227
5.4 LCD Display Support 231
5.5 Review 236

Chapter 6 Surface Models 238

6.1 Bellows Joints 239
6.2 Beam Bracket 249
6.3 Gearbox 256
6.4 Review 269

Chapter 7 Line Models 271

7.1 Flexible Gripper 272
7.2 3D Truss 283
7.3 Two-Story Building 295
7.4 Review 309

Chapter 8 Optimization 311

8.1 Flexible Gripper 312
8.2 Triangular Plate 323
8.3 Review 329

Chapter 9 Meshing 330

9.1 Pneumatic Fingers 331
9.2 Cover of Pressure Cylinder 346
9.3 Convergence Study of 3D Solid Elements 353
9.4 Review 365

Chapter 10 Buckling and Stress Stiffening 367

10.1 Stress Stiffening 368
10.2 3D Truss 379
10.3 Beam Bracket 383
10.4 Review 387

Chapter 11 Modal Analysis 389

11.1 Gearbox 390
11.2 Two-Story Building 395
11.3 Compact Disk 402
11.4 Guitar String 410
11.5 Review 417

Chapter 12 Transient Structural Simulations 419

12.1 Basics of Structural Dynamics 420
12.2 Lifting Fork 429
12.3 Harmonic Response Analysis: Two-Story Building 441
12.4 Disk and Block 448
12.5 Guitar String 456
12.6 Review 466

Chapter 13 Nonlinear Simulations 468

13.1 Basics of Nonlinear Simulations 469
13.2 Translational Joint 481
13.3 Microgripper 495
13.4 Snap Lock 508
13.5 Review 524

Chapter 14 Nonlinear Materials 526

14.1 Basics of Nonlinear Materials 527
14.2 Belleville Washer 536
14.3 Planar Seal 553
14.4 Review 568

Chapter 15 Explicit Dynamics 570

15.1 Basics of Explicit Dynamics 571
15.2 High-Speed Impact 577
15.3 Drop Test 587
15.4 Review 599

Index 601

Preface

Who is this Book for?

This book is developed mainly for graduate and senior undergraduate students. It may be used in courses such as Computer-Aided Engineering (CAE) or Finite Element Analysis (FEA).

Why ANSYS?

ANSYS is almost a synonym of finite element simulations. I've been using ANSYS as a teaching platform for more than 30 years. I prefer ANSYS to other CAE software because of its multiphysics capabilities, completeness of on-line documentation, and popularity among both academia and industry. Equipping engineering students with multiphysics capabilities is becoming a necessity. Complete documentation allow students to go steps further after taking an introductory CAE course. Popularity among academia and industry implies that an engineering student, after his graduation from college, can work with the software without any further training. In recent years, I have additional reasons to advocate this software: user-friendly **Workbench** and **free student products**.

Workbench

ANSYS Workbench has evolved for years but matured enough in recent years. For many years, I had been using ANSYS Classic, nowadays dubbed Mechanical APDL. The unfriendly APDL imposes unnecessary constraints, making the software difficult to use. As a result, students or engineers often limit themselves to certain applications; for example, working on component simulations rather than assembly simulations. Workbench adds friendliness on top of the power of APDL, releasing many unnecessary constraints.

Free Student Products

Starting from Release 17, ANSYS provides a free product license for students anywhere in the world. The only limitation is the problem size, which should be less than 32,000 nodes/elements (for shell elements, the limitation is 16,000 nodes/elements). All examples in this book are designed to meet this limitation and tested with the free student version. The free student product can be downloaded from the following ANSYS webpage:

http://www.ansys.com/Products/Academic/ANSYS-Student

Why a New Tutorial?

Preparing a tutorial for the Workbench needs much more effort than that for the APDL, due to the graphic nature of the interface. The most comprehensive tutorials, to my knowledge so far, are the training tutorials prepared by ANSYS Inc. However, they may not be suitable for use as a college tutorial for the following reasons. First, the cases used are either too complicated or too trivial. Many of the cases are geometrically too complicated for students to create from scratch. The students need to rely on the geometry files that come with the tutorials. Students usually obtain a better comprehension by working from scratch. Second, the tutorials cover too little on theories of finite element methods and solid mechanics while too much on software operation details. We try to provide a more suitable tutorial as a college textbook.

Organization of the Book

We'll describe the organization of the book in Section 1.1. Here is a quick overview of the book.

Using a case study, Section 1.1 walks through a typical Workbench simulation procedure. As more concepts or tools are needed, specific chapters or sections are pointed out, in which an in-depth discussion will be provided. The rest of Chapter 1 provides introductory background essential for the discussion in later chapters. This background includes a concise knowledge of structural mechanics: equations that govern the behavior of a mechanical or structural system (Section 1.2), the finite element methods that solve these governing equations (Section 1.3), and the failure criteria of materials (Section 1.4). Chapter 1 is the only chapter that doesn't have any hands-on exercises because, in the very beginning of a semester, students may not be ready to access the software facilities.

Starting from Chapter 2, a learning-by-doing approach is used throughout the book. Chapters 2 and 3 cover 2D geometric modeling and simulations. Chapters 4-7 introduce 3D geometric modeling and simulations. The next two chapters are dedicated to two useful topics: Chapter 8 to optimization and Chapter 9 to meshing. Chapter 10 deals with buckling and its related topic: stress stiffening. Chapters 11 and 12 discuss dynamic simulations: Chapter 11 is on modal analysis while Chapter 12 is on transient structural analysis. Up to this point, the discussions are mostly on linear problems. Although several nonlinear simulations have been performed, their nonlinear behaviors were not been discussed further. Chapters 13 and 14 discuss nonlinear simulations in a more in-depth way. Chapter 15 introduces an exciting topic: explicit dynamics, which is becoming a necessary discipline for a CAE engineer.

Features of the Book

To be **Comprehensive and friendly** is the ultimate goal of this book. To achieve this goal, the following features are incorporated in the book.

Real-World Cases. There are 45 step-by-step hands-on exercises in the book, each completed in a single section. These exercises are designed from 27 real-world cases, carefully chosen to ensure that they are neither too trivial nor too complicated. Many of them are industrial or research projects; pictures of real-world products are presented whenever available. The size of the problems are not too large so that they can be performed in a free ANSYS Student Product, which has a limitation of 32,000 nodes/elements. They are not too complicated so that the students can build each project step by step from scratch themselves. Throughout the book, the students don't need any supplementary files to work on these exercises.

Theoretical Background. Relevant background knowledge is provided whenever necessary, such as solid mechanics, finite element methods, structural dynamics, nonlinear solution methods (e.g., Newton-Raphson methods), nonlinear materials (e.g., plasticity, hyperelasticity), implicit and explicit integration methods, etc. To be efficient, the teaching methods are conceptual rather than mathematical; concise, yet comprehensive. The last four chapters (chapters 12-15) cover more advanced topics, each chapter having an opening section that gives basics of that topic in an efficient way to facilitate the subsequent learning.

How the Workbench internally solves a model is also illustrated throughout the book. Understanding these procedures, at least conceptually, is useful for a CAE engineer.

Key concepts are inserted whenever necessary. Must-know concepts, such as structural error, finite element convergence, and stress singularity, are taught by using designed hands-on exercises, rather than by abstract lecturing. For example, how finite element solutions converge to their analytical solutions, as the meshes become finer, is illustrated by guiding the students to plot convergence curves. That way, the students should have strong knowledge of the finite elements convergence behaviors. And, after hours of laborious working, they will not forget it for the rest of their life! Step-by-step procedures guiding the students to plot curves to illustrate important concepts are one of the featured teaching methods in this book.

Learning by Hands-on Exercises. A learning approach emphasizing hands-on exercises is used through the entire book. In my own experience, this is the best way to learn a complicated software such as ANSYS Workbench. A typical chapter, such as Chapter 3, consists of 6 sections, the first two sections providing two step-by-step examples, the third section giving a more systematic view of the chapter subjects, the following two sections providing more exercises in a not-so-step-by-step way, and the final section providing review problems.

Demo Videos. Each of the 45 step-by-step exercises has been screen-recorded. See Access Code or Author's Webpage, next page, for details.

ANSYS On-line References. One of the objectives of this book is to serve as a guide through the huge repository of ANSYS on-line documentation, a well of knowledge for many students and engineers. The on-line documentation includes a theory reference, an element reference, and many examples. Whenever helpful, we point to a location in the on-line documentation as further study for the students.

End-of-Chapter Keywords. Keywords are summarized at the ending section of each chapter in a quiz form. One goal of this book is to help the students comprehend the terminology and use it efficiently. That is not always easy for some students. For example, whenever asked "What are shape functions?" most of the students cannot satisfyingly define the terminology. Yes, many textbooks spend pages teaching students what the shape functions are, but the challenge is how to define or describe a term in less than two lines of words. This part of the book demonstrates how to define or describe a term in an efficient way; for example, "Shape functions serve as interpolating functions, to calculate continuous displacement fields from discrete nodal displacements."

To Instructors: How I Use the Book

I use this book in a 3-credit 18-week course, Computer-Aided Engineering. The progress is one chapter per week, except Chapter 1, which takes 2 weeks. Each week, after a classroom introduction of a chapter using lecture slides, I set up a discussion forum in an e-learning system maintained by the university. After completing the exercises of the chapter, the students are required to participate in the discussion forum.

In the forum, the students may post their questions, help or answer other students' questions, or share their comments. In addition to taking part in the discussion, I rate each posted article with 0-5 stars; the sum of the ratings becomes the grade of a student's performance for the week. The weekly discussion closes before next classroom hours.

The course load is not light. Nevertheless, most of the students were willing to spend hours working on these step-by-step exercises, because these exercises are tangible, rather than abstract. Students of this generation are usually better in picking up knowledge through tangible software exercises rather than abstract lecturing. Further, the students not only feel comfortable to post questions in the forum but also enjoy helping other students or sharing their comments with others.

At the end of the semester, each student is required to turn in a project. Students who are currently working as engineers usually choose topics related to their jobs. Students who are working on their theses usually choose topics related to their studies. They are also allowed to repeat a project found in any reference, as long as they go through all details by themselves. The purpose of the final project is to ensure that students are capable of carrying out a project independently, which is the goal of the course, not just following the step-by-step procedures in the book.

To Students: How My Students Use the Book

Many students, when following the steps in the book, often made mistakes and ended up with different results from those in the book. In many cases they cannot figure out in which steps the mistakes were made. In that case, they have to redo the exercise from the beginning. It is not uncommon that they redid the exercise several times and finally saw the beautiful, reasonable results.

What I want to say is that you may come across the same situation, but you are not wasting your time when you redo the exercises. You are learning from the mistakes. Each time you fix a mistake, you gain more insight. After you obtain the reasonable results, redo it and try to figure out if there are other ways to accomplish the same results. That's how I learned finite element simulations when I was a young engineer.

Finite element methods and solid mechanics are the foundation of mechanical or structural simulations. If you haven't taken these courses, plan to take them after you complete this course. If you've already taken them and still feel uncomfortable, review them.

Project files

Finished project files are available. See Access Code or Author's Webpage below for details.

If everything works smoothly, you don't need the finished project files at all. Every project can be built from scratch according to the steps described in the book. The author provides these project files just in case you need them. For example, when you run into troubles and you don't want to redo from the beginning, you may find these files useful. Or, when you have trouble following the geometry details in the book, you may need to look up the geometry details from the project files.

The most important reason we provide the finished project files is as follows. It is strongly suggested that, in the beginning of an exercise when previously saved project files are needed, you use the project files provided by the author rather than your own files so that you are able to obtain results that have minimum differences in numerical values from those in the textbook.

These finished project files are saved with the Workbench GUI command File/Archive; i.e., they are in WBPZ compressed format. To open a compressed project file, please use the Workbench GUI command File/Restore Archive.

Access Code

Each copy of this book includes an access code which gives you access to the project files, lecture slides, results animations, and video tutorials. Instructions to redeem your code are located on the inside of the front cover of this book.

Author's Webpage

A webpage dedicated to this book is maintained by the author. The project files, lecture slides, results animations, and video tutorials can also be downloaded from the webpage:

http://myweb.ncku.edu.tw/~hhlee/Myweb_at_NCKU/ANSYS19.html

Notations

To efficiently present the material, the writing of this book is not always done in a traditional format. Chapters and sections are numbered in a traditional way. Each section is further divided into subsections, for example, the 3rd subsection of the 2nd section of Chapter 1 is denoted as "1.2.3." Textboxes in a subsection are ordered with numbers enclosed by a pair of square brackets (e.g., [4]). We may refer to such a textbox as "1.2.3[4]." When referring to a textbox in the same subsection, we drop the subsection identifier; for the foregoing example, we simply write "[4]." Equations are numbered in a similar way, except that the equation number is enclosed by a pair of round brackets (parentheses) rather than square brackets. For example, "1.2.3(1)" refers to the 1st equation in the Subsection 1.2.3. Numbering notations are summarized as follows (some of these notations are reiterated in 1.1.1[6] (page 10), 1.1.2[9] (page 12), and 2.1.2[12] (page 58):

[1], [2], ...	A number enclosed by brackets is used to identify a textbox.
(1), (2), ...	A number enclosed by round brackets is used to identify an equation.
Reference[Ref 1]	Superscripts are used for references.
Mechanical	Boldface is used to highlight Workbench keywords.
Round-cornered textboxes	A round-cornered textbox indicates that mouse or keyboard actions are needed.
Sharp-cornered textboxes	A sharp-cornered textbox is for commentary; no mouse or keyboard needed.
→, ←, ↓, ↑, ↘, ↙, ↗, ↖	An arrow is used to point to the location of the next textbox.
↵	This symbol is used to indicate that the next textbox is on the next page.
#	This symbol is used to indicate that it is the last textbox of a section.

Acknowledgement

I feel thankful to students in my classroom listening to my lectures. It is my students, past and present, that motivated me to give birth to this book.

Many of the cases presented in this book are from my students' final projects. Some are industry cases while others are thesis-related research topics. Without these real-world cases, the book would never be so useful. The following is a list of students who contributed to the cases in this book.

"Pneumatic Finger" (1.1 and 9.1) is contributed by Che-Min Lin and Chen-Hsien Fan, ME, NCKU.
"Microgripper" (2.6 and 13.3) is contributed by P.W. Shih, ME, NCKU.
"Cover of Pressure Cylinder" (4.2 and 9.2) is contributed by M. H. Tsai, ME, NCKU.
"Lifting Fork" (4.3 and 12.2) is contributed by K.Y. Lee, ES, NCKU.
"LCD Display Support" (4.5 and 5.4) is contributed by Y.W. Lee, ES, NCKU.
"Bellows Tube" (6.1) is contributed by W. Z. Liu, ME, NCKU.
"Flexible Gripper" (7.1 and 8.1) is contributed by Shang-Yun Hsu, ME, NCKU.
"3D Truss" (7.2) is contributed by T. C. Hung, ME, NCKU.
"Snap Lock" (13.4) is contributed by C. N. Chen, ME, NCKU.

Many of the original ideas of these projects came from the academic advisors of the above students. I also owe them a debt of thanks. Specifically, the project "Pneumatic Finger" is a work led by Prof. Chao-Chieh Lan of the Department of ME, NCKU. The project "Microgripper" originates from a work led by Prof. Ren-Jung Chang of the Department of ME, NCKU.

Thanks to Dr. Shen-Yeh Chen, the CEO of FEA-Opt Technology, for letting me use his article "An Unofficial History of ANSYS," as an appendix at the end of Chapter 1.

S.Y. Kan, the chief structural engineer of the Taipei 101, has reviewed some of the text and corrected some mistakes.

Much of the information about the ANSYS Workbench was obtained from the training tutorials prepared by ANSYS Inc. I didn't specifically cite them in the text, but I appreciate these well-compiled training tutorials.

Thanks to Mrs. Lilly Lin, the CEO, and Mr. Nerow Yang, the general manager, of Taiwan Auto Design, Co., the partner of ANSYS, Inc. in Taiwan. The couple, my long-term friends, provided much substantial support.

Thanks to Professor Sheng-Jye Hwang, of the ME Department, NCKU, and Professor Durn-Yuan Huang, of Chung Hwa University of Medical Technology. They are my long-term research partners. Together, we have accomplished many projects, and, in carrying out these projects, I've learned much from them.

Lastly, thanks to my family, my wife and my son and the pets, for their patience and sharing the excitement with me during the writing of this book.

Huei-Huang Lee

Associate Professor
Department of Engineering Science
National Cheng Kung University, Tainan, Taiwan
e-mail: hhlee@mail.ncku.edu.tw
webpage: myweb.ncku.edu.tw/~hhlee

Chapter 1

Introduction

Purpose of This Chapter

ANSYS is a software implementation of finite element simulations on many types of problems, including structural, mechanical, fluid, electromagnetic, etc. This book discusses only structural and mechanical simulations. This chapter shows the procedure of a typical structural simulation using Workbench (Section 1.1), explains the organization of this book (Section 1.1), reviews solid mechanics (Sections 1.2 and 1.4), and gives a brief introduction of finite element methods (Section 1.3).

About Each Section

The procedure of structural simulations and the organization of this book is illustrated in Section 1.1 by using a case study. Section 1.1 also serves as a preamble for the topics in Sections 2, 3, and 4.

Section 1.2 introduces quantities such as displacements, stresses, and strains, which are used throughout the book. With these quantities, the equations that govern the behavior of a structural system are presented.

ANSYS solves these governing equations by finite element methods. Section 1.3 introduces the basic ideas and the procedure of the finite element methods. The introduction is conceptual rather than theoretical or mathematical. These concepts should be adequate for the purpose of understanding the topics in the later chapters.

One goal of structural simulations is to predict whether or not a system would fail. We usually compare the calculated stresses with certain critical values. If the calculated stresses are larger than the critical values, then the system is said to fail. Section 1.4 discusses the theories of failure criteria.

This is the Only Chapter without Hands-On Exercises

All chapters of this book use the learning-by-doing approach, except this chapter. There are no hands-on exercises in this chapter. The main reason is that an overall picture is usually helpful before any hands-on exercises. A secondary reason is that, in the first week of a semester, students may not be ready to access the software facility.

Section 1.1

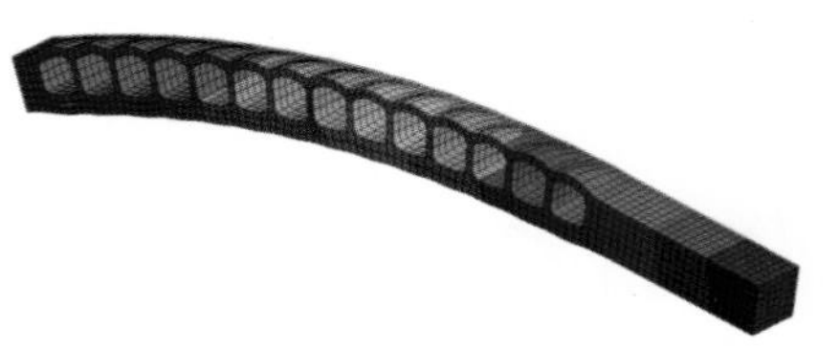

Case Study: Pneumatically Actuated PDMS Fingers[Ref 1]

This section demonstrates a typical ANSYS Simulation procedure and introduces the organization of this book. This is not a step-by-step exercise; you don't have to sit before ANSYS when reading this section. In Section 9.1, you will be guided to conduct this simulation.

1.1.1 Problem Description

About the Pneumatic Fingers

[1] The pneumatic fingers (see [2]) are designed as part of a surgical parallel robot system remotely controlled by a surgeon through the internet[Ref 2].

The fingers are made of a PDMS-based (polydimethylsiloxane) elastomer material. The geometry of a typical finger is shown in [3], in which 14 air chambers are cut from the material.

The air chambers locate close to the upper surface, so when the air pressure is applied, the finger bends downward [4-5]. →

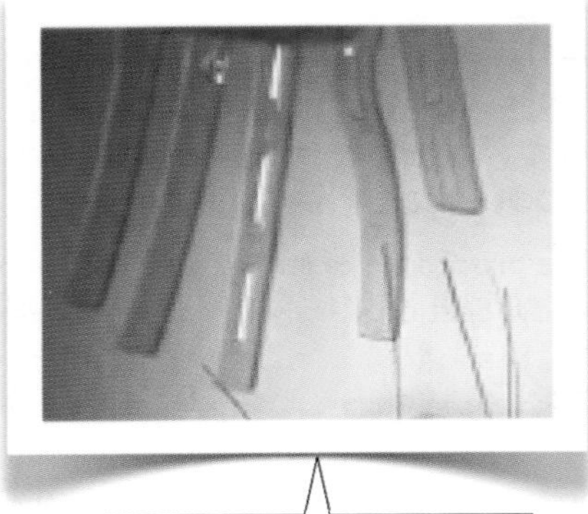

[2] A robotic hand has five fingers, remotely controlled by a surgeon through the internet. ←

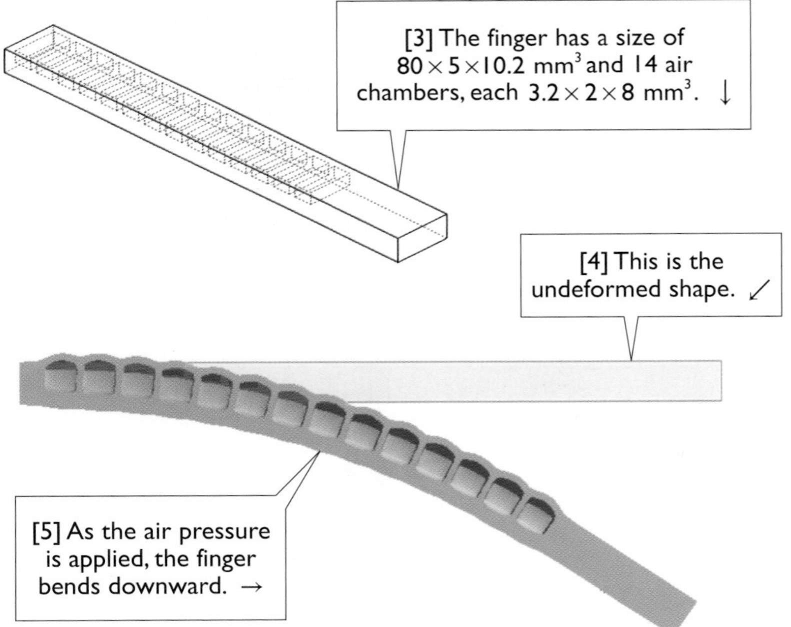

About Textboxes

[6] In this book, textboxes within a subsection (e.g., 1.1.1) are ordered with numbers enclosed by square brackets (e.g. [1]). When you read, please follow the order of the textboxes. An arrow is used at the end of a textbox to locate the next textbox (e.g., →, ←, ↓, ↙, and → in [1-5]). The symbol ↵ is used to indicate that the next textbox is on the next page (e.g., [6]). The symbol # is used to indicate that it is the last textbox of a subsection (e.g., [9], next page). ↵

About the PDMS Elastomer

[7] The mechanical properties of a PDMS elastomer are functions of the ingredients. The chart in [8] shows the stress-strain curve of the PDMS elastomer used in this case[Ref 3]. The curve exhibits a linear stress-strain relationship up to a strain of about 0.6. Within this range, the Young's modulus (the slope of the curve) is estimated to be 2.0 MPa (1.2 MPa divided by 0.6). Besides, the Poisson's ratio is 0.48, from another test. →

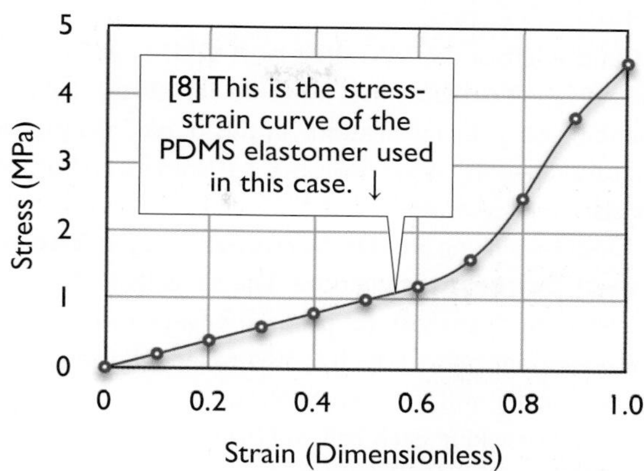

Purpose of the Simulation

[9] The purpose of this simulation is to assess the *efficiency* of the design, defined as the vertical deflection under an air pressure of 0.18 MPa. We also want to plot a deflection-versus-pressure chart (1.1.8[6], page 19; also see 1.1.12[3-4], page 23). #

1.1.2 Workbench GUI

[1] In a Windows system, you can launch Workbench from the **Start** menu or double-click a Workbench icon (if there is one) on the desktop [2]. A **Workbench GUI** (graphic user interface) then shows up [3]. The **Workbench GUI** is a gateway to Workbench applications. The Workbench applications that will be used in this book are **Project Schematic**, **Engineering Data**, **DesignModeler**, **Mechanical**, and **Design Exploration**. ↙

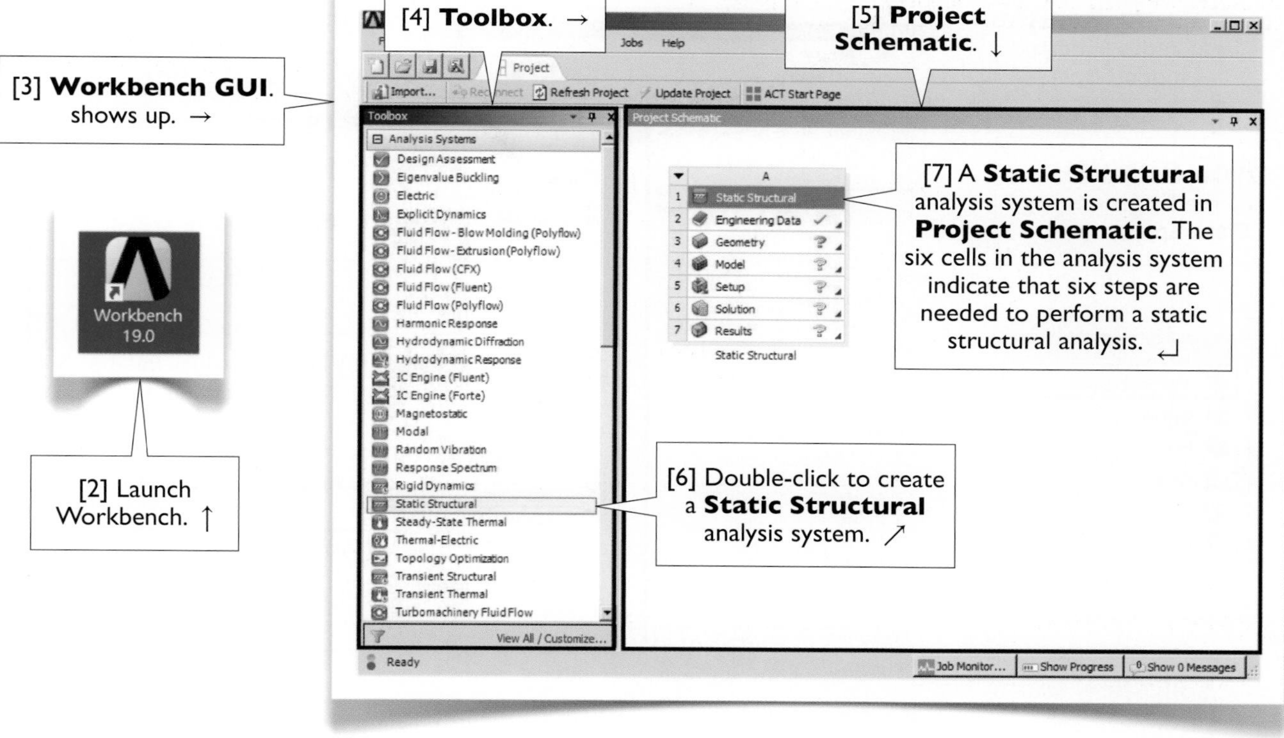

[8] On the left of the **Workbench GUI** is a **Toolbox** ([4], last page), and on the right is a **Project Schematic** window [5]. **Toolbox** lists available **Analysis Systems**, which may be different in your computer from here, depending on your installation. In this book, "analysis" and "simulation" are often interchangeable, for example, "static structural analysis" is synonymous to "static structural simulation." Here, we want to perform a **Static Structural** analysis.

Double-clicking **Static Structural** [6] in **Toolbox** creates a **Static Structural** analysis system [7] in the **Project Schematic** window. The six cells in the analysis system indicate that six steps are needed to perform a static structural analysis: (a) prepare engineering data, (b) create a geometric model, (c) divide the geometric model into a finite element mesh, which is also called a finite element model, (d) set up loads and supports, (e) solve the finite element model, and (f) view the results.

Double-clicking each cell will bring up a relevant application to process that step. ↓

[9] In this book, to facilitate the readability, a Workbench keyword is usually boldfaced, e.g., **Static Structural** in [8]. In cases where boldface does not help the readability, then the boldface may not be necessary, e.g., Workbench in [2]. #

1.1.3 Prepare Engineering Data

[1] Double-clicking **Engineering Data** cell [2] brings up **Engineering Data** [3]. Here, we want to specify material properties for the PDMS elastomer. The material is modeled as a linear **Isotropic Elasticity** material, for which we need to input a Young's modulus (2.0 MPa) and a Poisson's ratio (0.48) [4-6]. At completion, close the **Engineering Data** application [7] and return to **Project Schematic**. ↙

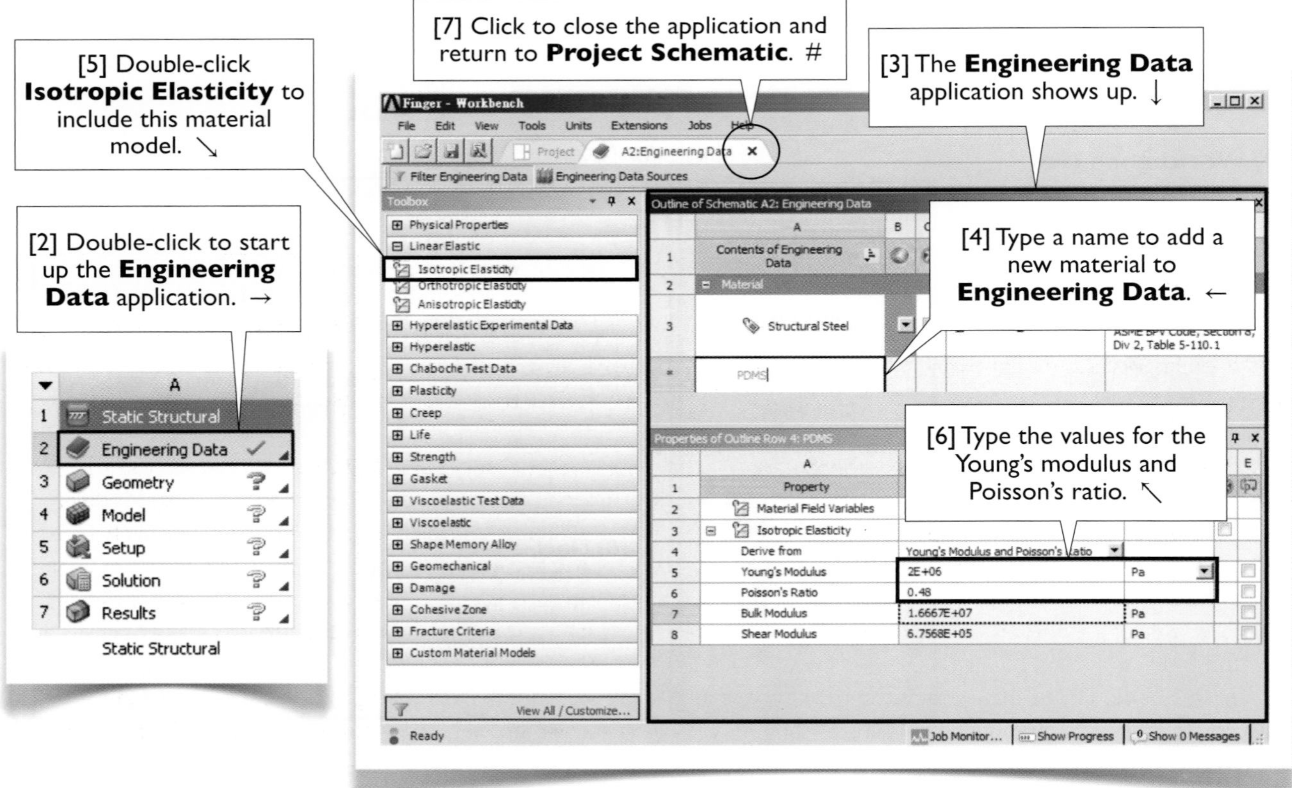

1.1.4 Create Geometric Model

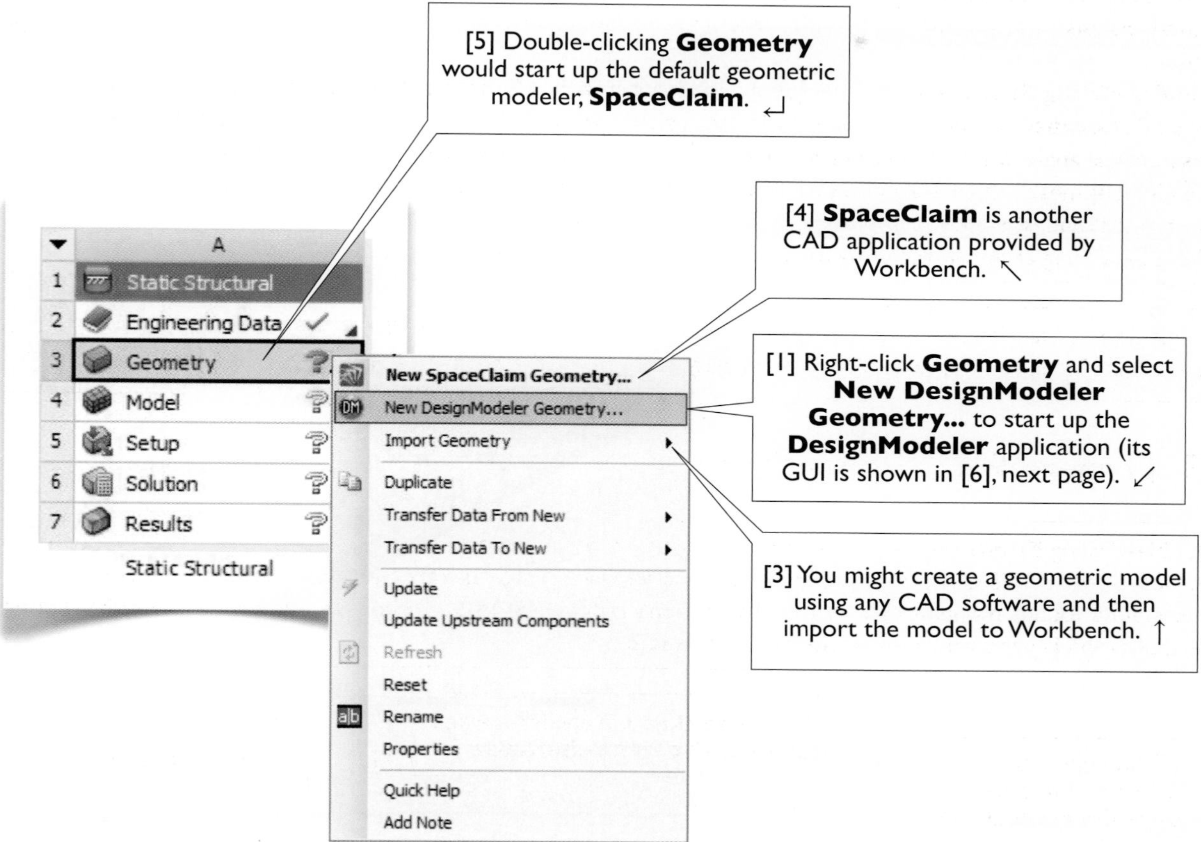

Ways of Creating Geometric Models

[2] You might create a geometric model using any CAD software (e.g., SOLIDWORKS, PTC Creo, Autodesk Inventor) and then import the model to Workbench [3]. However, I suggest that you create a geometric model using one of the two CAD applications provided by Workbench, since they are specifically designed for use in ANSYS Workbench simulations.

The two CAD applications in Workbench are **DesignModeler** [1] and **SpaceClaim** [4]. Until ANSYS 15, DesignModeler is the only built-in geometric modeler in Workbench. For simple and small models, DesignModeler serves well enough; but for complicated and large models, the engineers often resort to a CAD software such as SOLIDWORKS, PTC Creo, Autodesk Inventor, etc., and then import the model to Workbench [3]. In ANSYS 16 and 17, **SpaceClaim** was included in Workbench as a second modeler, and **DesignModeler** remained as the default modeler. Starting from ANSYS 18, **SpaceClaim** becomes the default modeler [5], and **DesignModeler** serves as an alternative modeler.

In this book, since all the geometric models are simple and small, we will use DesignModeler to create the geometric models. Remember, the focus of this book is the finite element simulations, not the geometric modeling.

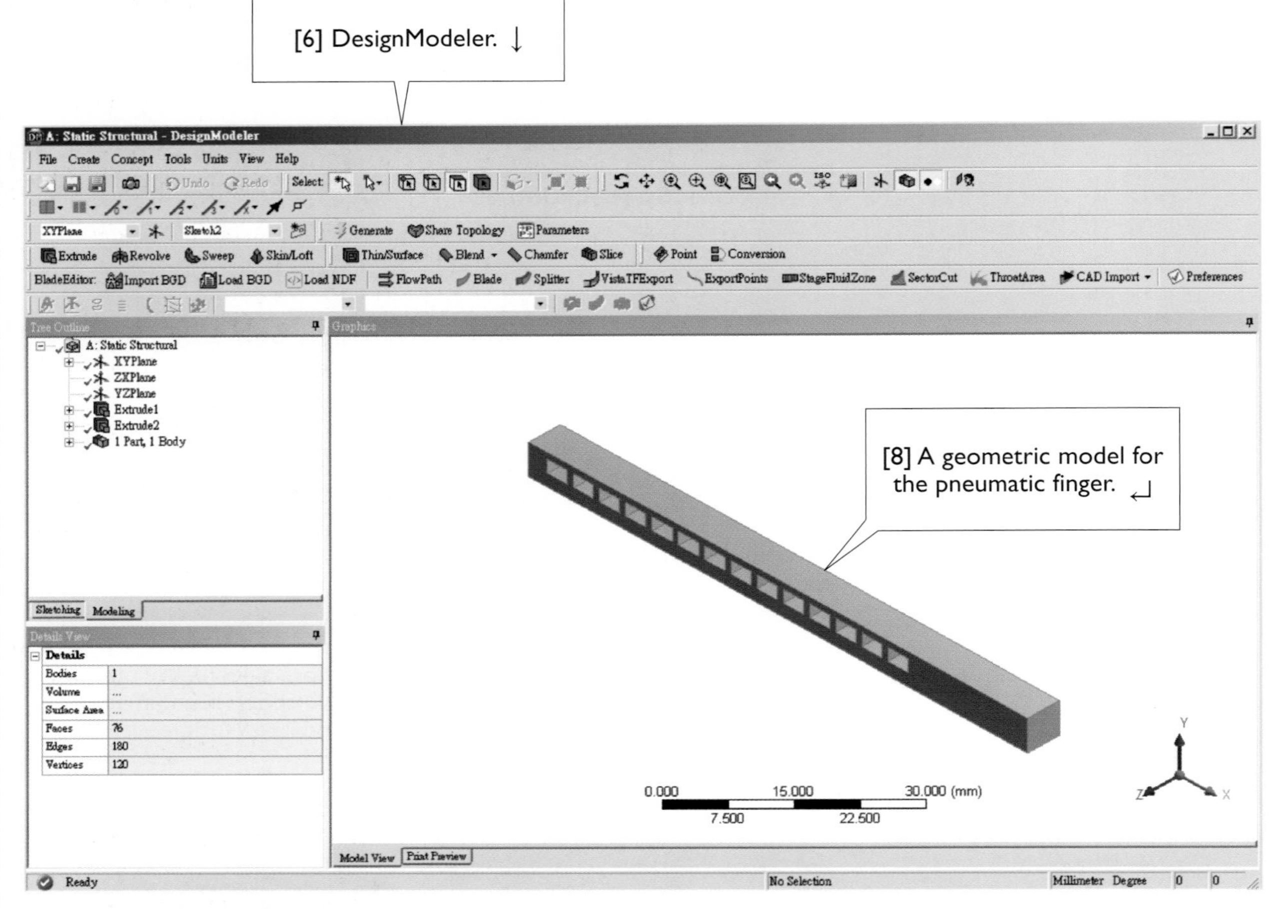

[7] The functions of DesignModeler are similar to any other CAD software except, as we mentioned, that DesignModeler is specifically designed to create geometric models for use in ANSYS Workbench simulations. Due to the symmetry, we will create only one half of the pneumatic finger, as shown in [8].

Creating the geometric model in our case is relatively simple. It can be viewed as a two-step task, and each step consists of a 2D **sketching** and a 3D **extrusion**. The procedure is summarized as follows.

Step 1. Create an 80 mm x 5 mm x 5.1 mm solid.
- Step 1.1. Sketch an 80 mm x 5 mm rectangle ([9], next page).
- Step 1.2. Extrude the sketch 5.1 mm to create a solid [10].

Step 2. Create fourteen 3.2 mm x 2 mm x 4 mm air chambers.
- Step 2.1. Sketch fourteen 3.2 mm x 2 mm rectangles [11].
- Step 2.2. Extrude the new sketch 4 mm deep to cut material from the existing solid [12]. ↑

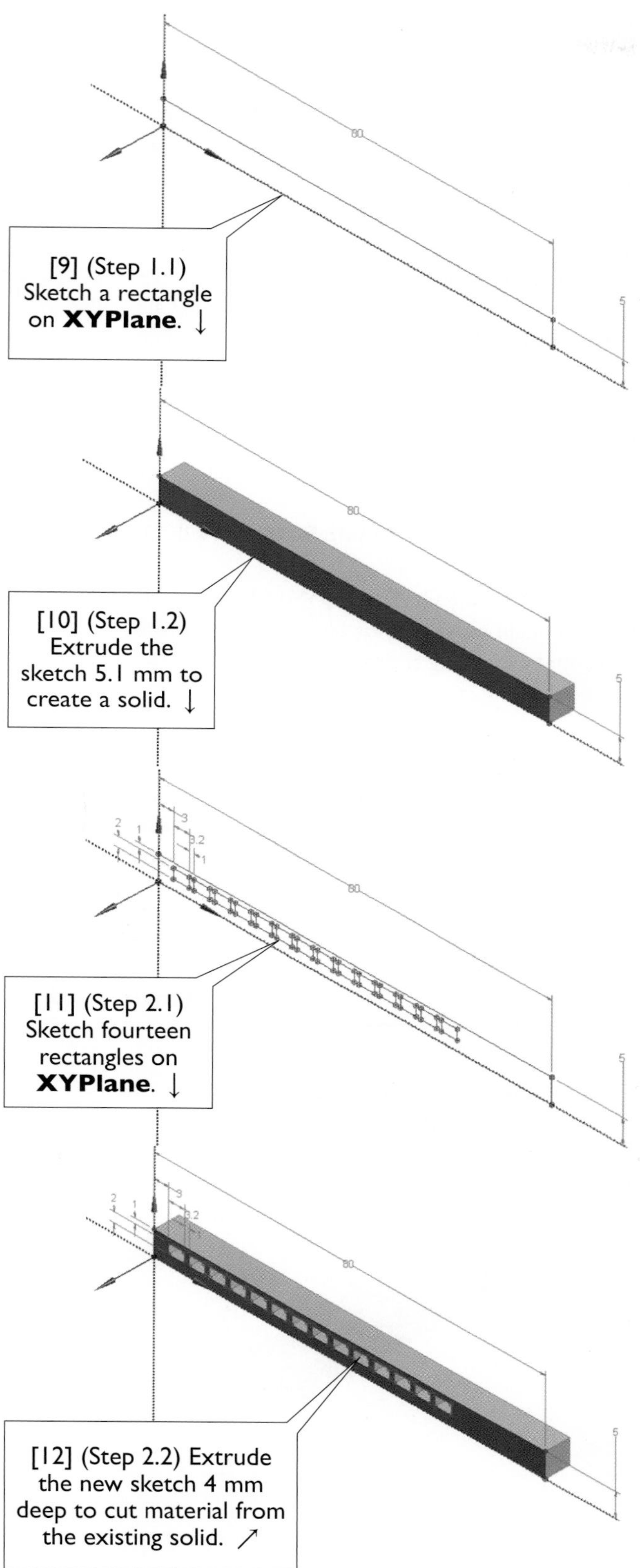

2D and 3D Simulations

[13] Workbench supports 2D and 3D simulations. For 3D simulations, Workbench supports three types of geometric bodies: *solid bodies* (which have volumes), *surface bodies* (which do not have volumes, but have surface areas), and *line bodies* (which do not have volumes or surface areas, but have length). Thin shell structures are often modeled as surface bodies (Chapter 6); Beam or frame structures are often modeled as line bodies (Chapter 7). Solid, surface, and line bodies can be mixed up in a 3D model. For 2D simulations, Workbench supports solid bodies only (Chapter 3). A 2D model must be created entirely on **XYPlane**. ↓

More on Geometric Modeling

[14] Creating a geometric model is usually more involved than that example; nonetheless, often, it still can be viewed as a series of two-step operations demonstrated in this case: drawing a sketch and then using the sketch to create a 3D body by one of the modeling tools, such as extrusion, revolution, sweeping, and skin/lofting (4.4.8[1], page 201).

Geometric modeling is the first step toward successful finite element simulations. Chapters 2 and 4 demonstrate some geometric modeling techniques using DesignModeler. Chapter 2 focuses on sketching methods. Some of the sketches created in Chapter 2 are reused in Chapter 3 to demonstrate the creation of 2D solid models. Simulations of these 2D solid models are then performed in Chapter 3.

Chapters 4-7 discuss 3D geometric modeling and linear static simulations (except 7.1, which involves geometric nonlinearities). Chapter 4 demonstrates the creation of 3D solid models and Chapter 5 demonstrates 3D simulations using the solid models created in Chapter 4. Chapter 6 demonstrates 3D surface modeling and simulations, and Chapter 7 demonstrates 3D line modeling and simulations.

In the real-world, there are no such things as surface/line geometries, therefore surface/line models (as well as 2D solid models) are called **conceptual models**. #

1.1.5 Divide the Geometric Model into Elements

[1] The procedure used by Workbench to solve a problem is called a **finite element method**, which can be viewed as two major steps: (a) establishing governing equations and (b) solving the governing equations. The geometric body of a real-world problem is often too complicated to write down the governing equations directly. A basic idea of the finite element method is to divide the body into many geometrically simpler shapes called *elements*. The elements are connected by *nodes*. The governing equations for each element then can be easily established, and the system of equations for all elements can be solved simultaneously. We will discuss this idea further in Section 1.3.

Since the elements have finite size (in contrast to the infinitesimal sizes of elements in Calculus), they are called **finite elements**. The collection of the elements is called a **finite element mesh**; in Workbench, a finite element mesh is also called a **finite element model**. In this book, we define the *finite element model as a finite element mesh plus its environment conditions* (loads and supports). The environment conditions are introduced in 1.1.6, next page.

Double-clicking the **Model** cell [2] in **Static Structural** brings up a **Mechanical GUI** [3]. The rest of the simulation will be done in **Mechanical**; that is, the functions of **Mechanical** include meshing, set up of loads and supports, solution, and viewing results [4].

Workbench can perform meshing task under your control [5]. In **Details of Mesh**, **Statistics** displays the number of nodes and elements [6].

The quality of a mesh cannot be overemphasized. Although it is possible to let ANSYS Workbench perform the meshing automatically, its quality is not guaranteed. Achieving a high quality mesh is not trivial; it needs much background knowledge and experience. Chapter 9 demonstrates many meshing techniques. ↓

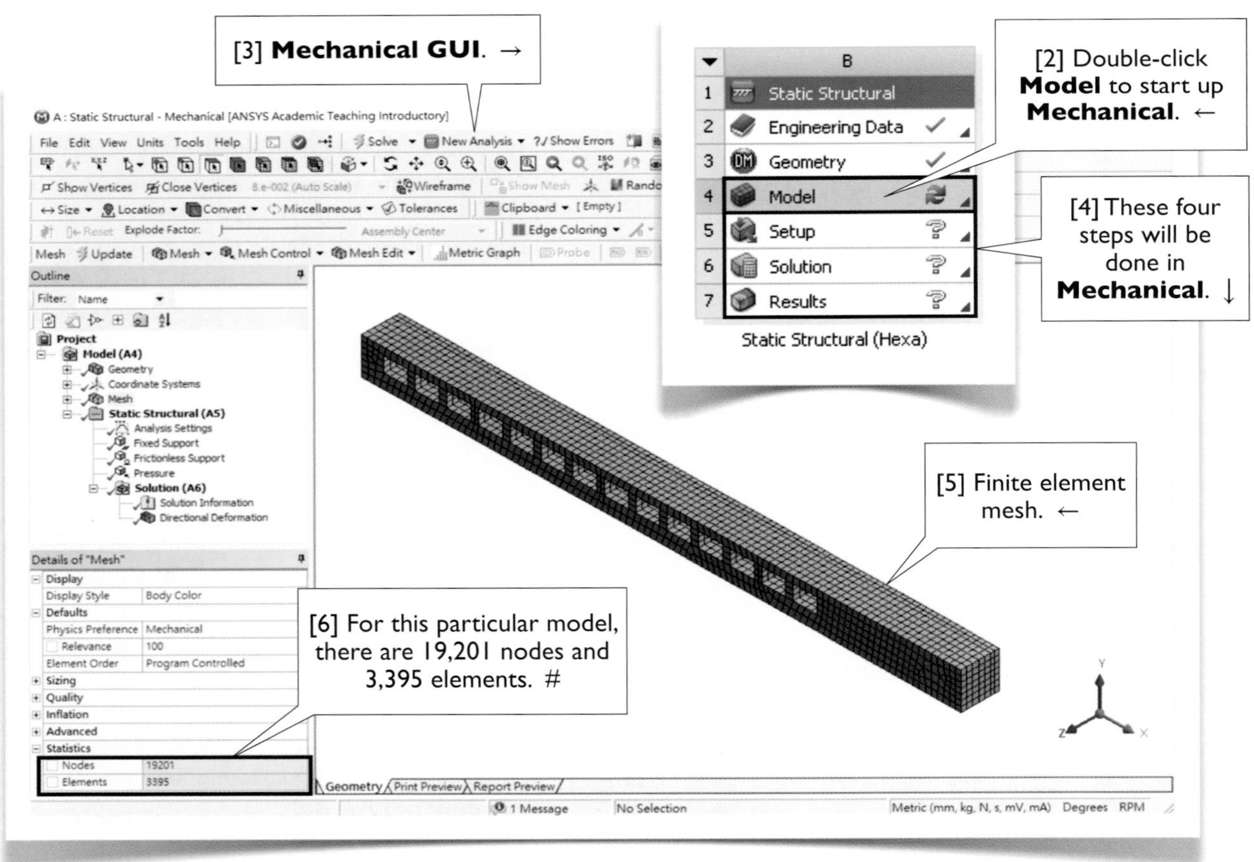

1.1.6 Set up Loads and Supports

[1] In the real-world, a body is a part of the world, interacting with other parts of the world. When we take a body apart for simulation, we are cutting it away from the rest of the world. The cutting surfaces are the boundary surfaces of the body. In theory, the choice of the boundary surfaces is arbitrary; however, we need to specify the *boundary conditions* on ALL of the boundary surfaces. In Workbench, conditions applying on the finite element mesh are called the *environment conditions*, including the boundary conditions and the conditions that are not specified on the boundaries. The temperature change INSIDE a body is an example of environment conditions that are not specified on the boundaries; another example is the gravitational forces.

In our case, things surrounding the half model of the pneumatic finger are pressurized air in the chambers [2], the material connecting the root of the finger [3], the material on the other side of the model [4], and the atmosphere air around the rest of the boundary surfaces.

Modeling the pressurized air in the chambers is straightforward: specify a pressure of 0.18 MPa for all the surfaces of the chambers [2]. The root of the finger is modeled as fixed support [3]. The plane of symmetry is modeled as frictionless support [4]. Note that a plane of symmetry is equivalent to a surface of frictionless support. Finally, assuming the atmospherical air has little effect on the model, we simply neglect it and model all the boundary surfaces surrounded by atmospherical air as free boundaries, i.e., boundaries with no forces acting on them.

As mentioned (1.1.5[1], last page), a finite element model is defined as a finite element mesh plus its environment conditions. We will stick to this definition throughout the book. Make sure that you can distinguish these three terms: *geometric model*, *finite element mesh*, and *finite element model*. ↓

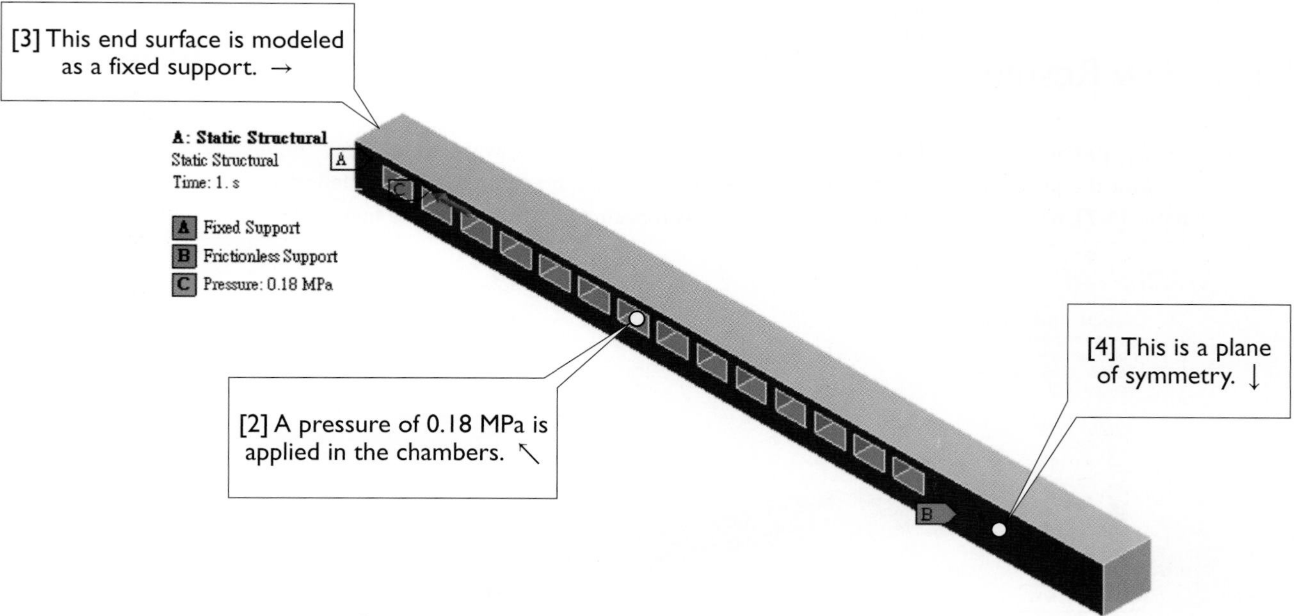

More on Environment Conditions

[5] Modeling environment conditions is sometimes not so easy as in this case. The challenge comes from the need of domain knowledge. It is not possible to perform a structural analysis if an engineer doesn't have enough domain knowledge of structural mechanics.

We will start to introduce environment conditions in Chapters 3. From then on, each chapter will involve some demonstrations of environment conditions. #

1.1.7 Solve Finite Element Model

[1] To solve a finite element model, simply click **Solve** in **Mechanical GUI** [2]. The time to complete a simulation depends on its problem size (number of nodes and number of elements), number of time steps, and nonlinearities.

As mentioned (1.1.5[1], page 16), the solution procedure can be viewed as two major steps: establishing governing equations and solving the governing equations. Section 1.2 overviews structural mechanics and summarizes the governing equations. How does ANSYS Workbench establish and solve these governing equations? The answer is *finite element methods*. Section 1.3 summarizes finite element methods, quickly equipping the students with enough concepts of the methods so they can proceed to learning the materials in this book. ↓

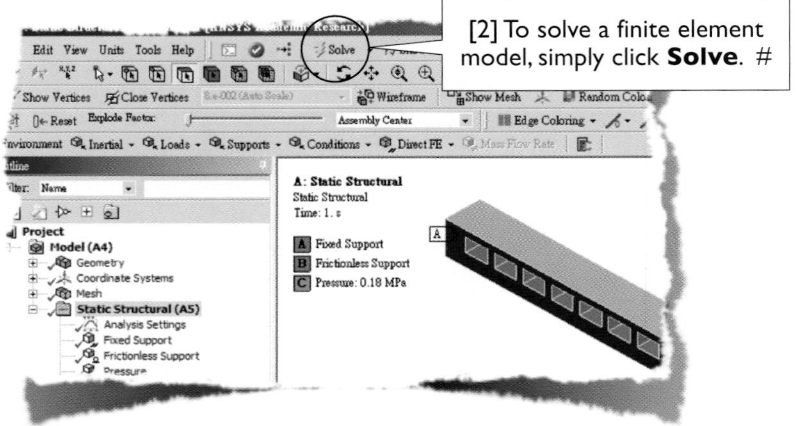

1.1.8 View Results

[1] After solving a problem, numerical results are stored in the databases, available for your request. In our case, we are concerned about the vertical deflection [2-4]. The deformation can be animated [5]. A deflection-versus-pressure chart is useful ([6-7], next page), in which the deflection is measured at the tip of the finger [4]. ↓

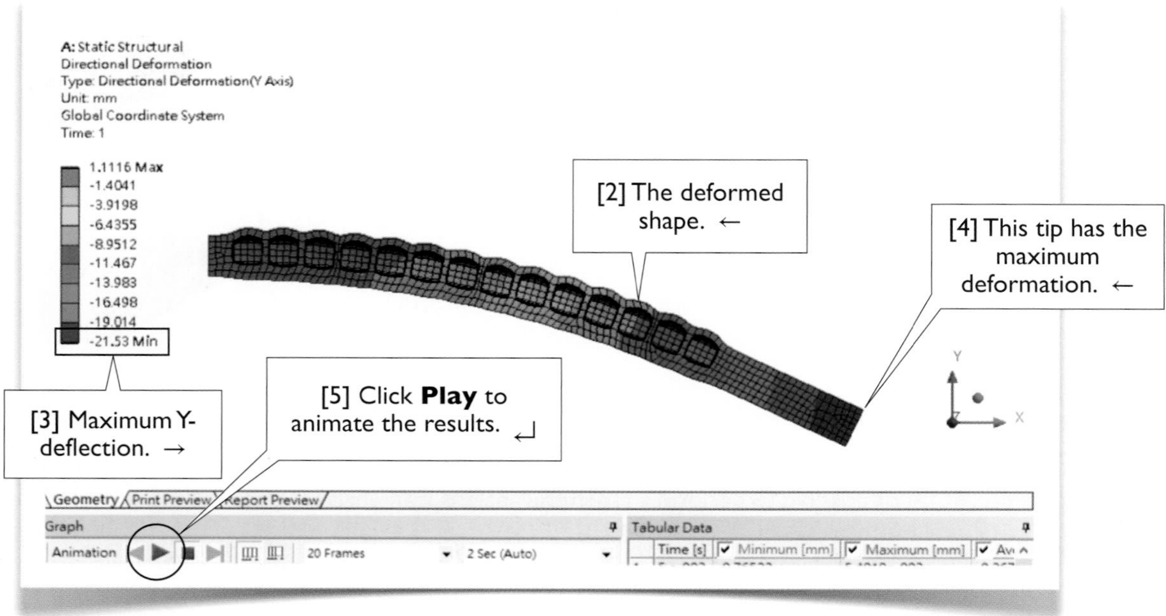

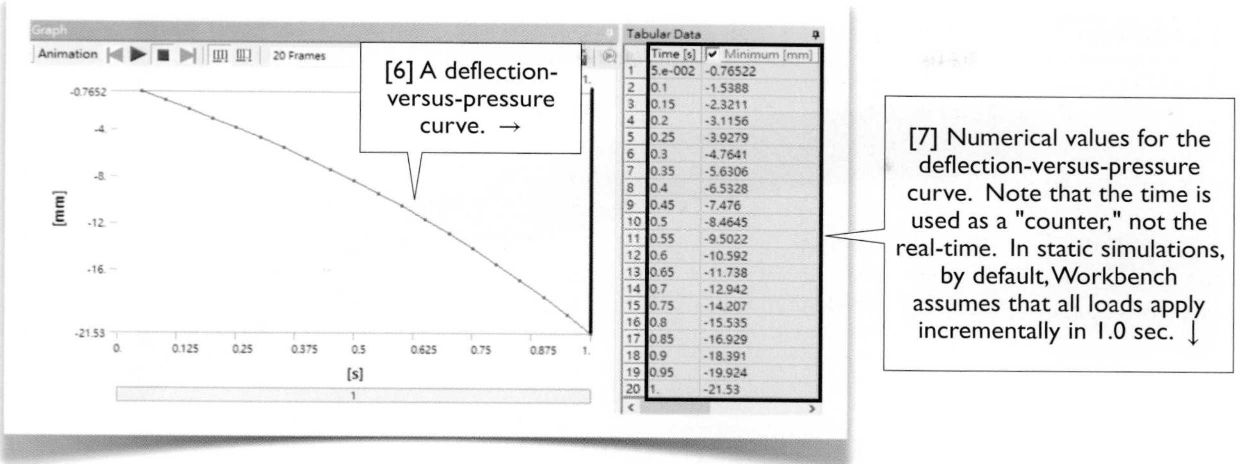

	Time [s]	Minimum [mm]
1	5.e-002	-0.76522
2	0.1	-1.5388
3	0.15	-2.3211
4	0.2	-3.1156
5	0.25	-3.9279
6	0.3	-4.7641
7	0.35	-5.6306
8	0.4	-6.5328
9	0.45	-7.476
10	0.5	-8.4645
11	0.55	-9.5022
12	0.6	-10.592
13	0.65	-11.738
14	0.7	-12.942
15	0.75	-14.207
16	0.8	-15.535
17	0.85	-16.929
18	0.9	-18.391
19	0.95	-19.924
20	1.	-21.53

[6] A deflection-versus-pressure curve. →

[7] Numerical values for the deflection-versus-pressure curve. Note that the time is used as a "counter," not the real-time. In static simulations, by default, Workbench assumes that all loads apply incrementally in 1.0 sec. ↓

[8] In this case, we assume the stress-strain relationship of the PDMS material is linear (1.1.3[5], page 12). This implies that the strains are less than 0.6 (1.1.1[7-8], page 11). This assumption has to be verified at this point. The results show that the maximum strain is about 0.47 [9], well below 0.6. Note that we also assumed the compressive behavior is the same as tensile behavior, but this is usually not true for an elastomer under such a large deformation. A more accurate material model, a hyperelasticity model, for the elastomer will be introduced in Chapter 14. ↓

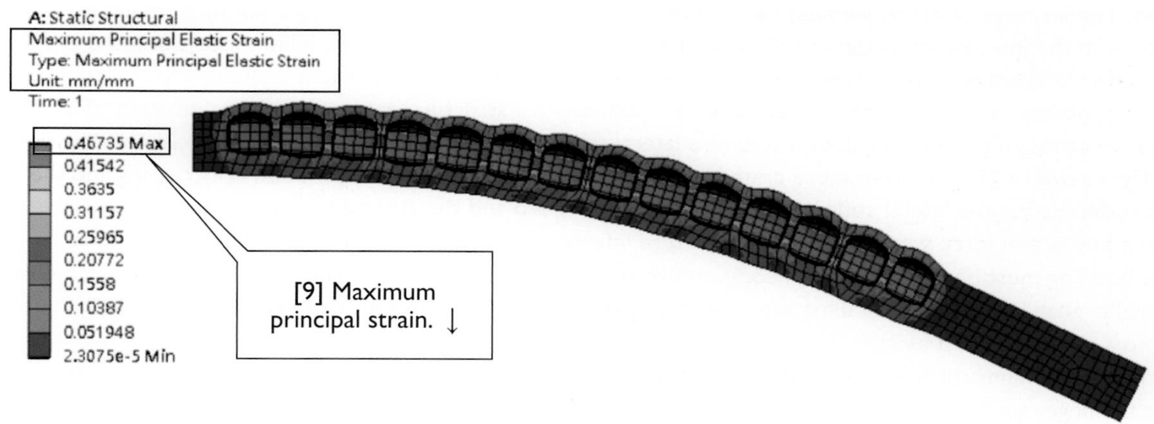

[9] Maximum principal strain. ↓

[10] Lastly, the stress is reviewed. It shows that the maximum von Mises stress, defined in Eq. 1.4.5(15) (page 46), is 1.06 MPa ([11], next page). The test data (1.1.1[8], page 11) show that the material can withstand up to 4.5 MPa without failure. The design is safe as far as the stress is concerned.

Failure Criteria

The main purpose of checking the stresses is to make sure the material doesn't fail under specified load. What is von Mises stress ([11], next page)? The test data (1.1.1[8], page 11) are produced according to a uniaxial tensile test, but the stress state in the pneumatic finger, as in any other real-world situations, is 3D by nature. How can we compare a 3D stress state with a uniaxial one, and judge the failure of the material? Section 1.4 will discuss failure criteria of materials. ↵

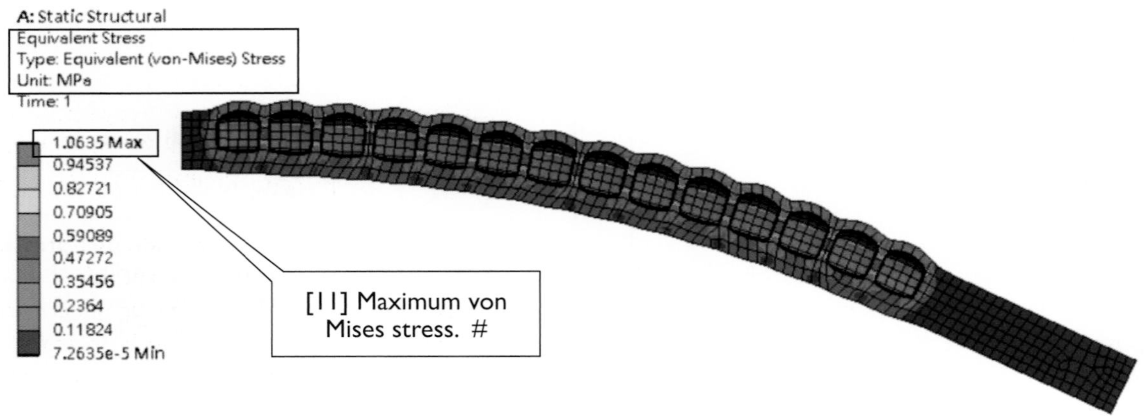

1.1.9 Buckling and Stress-Stiffening

[1] The more tension in a guitar string, the more force you need to deflect the string laterally. In technical words, the string's lateral stiffness increases as the longitudinal tensile stress increases; i.e., the longitudinal tensile stress causes the string to be stiffer in lateral direction. By lateral direction, we mean the direction orthogonal to the longitudinal direction. The increase of lateral stiffness can also be justified by the fact that the string's vibrating frequency (pitch) increases with the increase of its tension, since stiffer strings have higher frequencies. This effect is called the *stress-stiffening effect: a structure's lateral stiffness increases with the increase of its longitudinal tensile stress.*

Is the opposite also true? That is, does the lateral stiffness of a structure decrease with the increase of its axial compressive stress? For example, does a column's lateral stiffness decrease when subject to an axial compressive force? The answer is YES. An even more dramatic phenomenon is that, as the compression is increasing and the lateral stiffness is decreasing, the lateral stiffness will eventually reach zero and the structure is said to be in an unstable state, in which a tiny lateral force would deflect the structure infinitely. This phenomenon is called *buckling*.

The buckling must be considered in a compressive structural component, particularly when its lateral dimension is much smaller than the longitudinal dimension; for example, slender columns subject to axial loads, thin-walled pipes subject to a circumferential twist.

Back to the pneumatic finger. Instead of air pressure applied on the chambers' surfaces, we now apply an upward lifting at the finger tip [2]. The upper surface, which is essentially a layer of thin PDMS film, would undergo compressive stress [3]. Our concern then is to know the magnitude of the lifting force that will cause the thin film buckle.

Simulations of buckling and stress-stiffening will be covered in Chapter 10. ↓

1.1.10 Dynamic Simulations

[1] If the load (air pressure) is applied to the chambers very fast, the deformation would also occur very fast. When a body moves or deforms very fast, two effects must be taken into account: *inertia effect* and *damping effect*. Combination of these two effects is called *dynamic effects*. When the dynamic effects are considered in a simulation, it is called a *dynamic simulation*.

Imagine that the pressure in the chambers is increased from zero to 0.18 MPa in just 0.1 seconds. The pressure is applied so fast that the deformation must also be very fast and dynamic effects must play an important role in the structure's behavior. The figure and the chart below show the time-varying deflections of the finger tip under such a loading condition [2-5]. The curve [5] shows that the deflections in a dynamic simulation can be much larger than those obtained in a static simulation. Furthermore, the vibration lasts for several seconds. This, as a surgical application, is not desirable.

The foregoing simulation, a structure subject to dynamic loads, is called a *transient structural simulation*. Chapters 12 and 15 will cover transient structural simulations: Chapter 12 discusses *implicit methods* while Chapter 15 introduces *explicit methods*. →

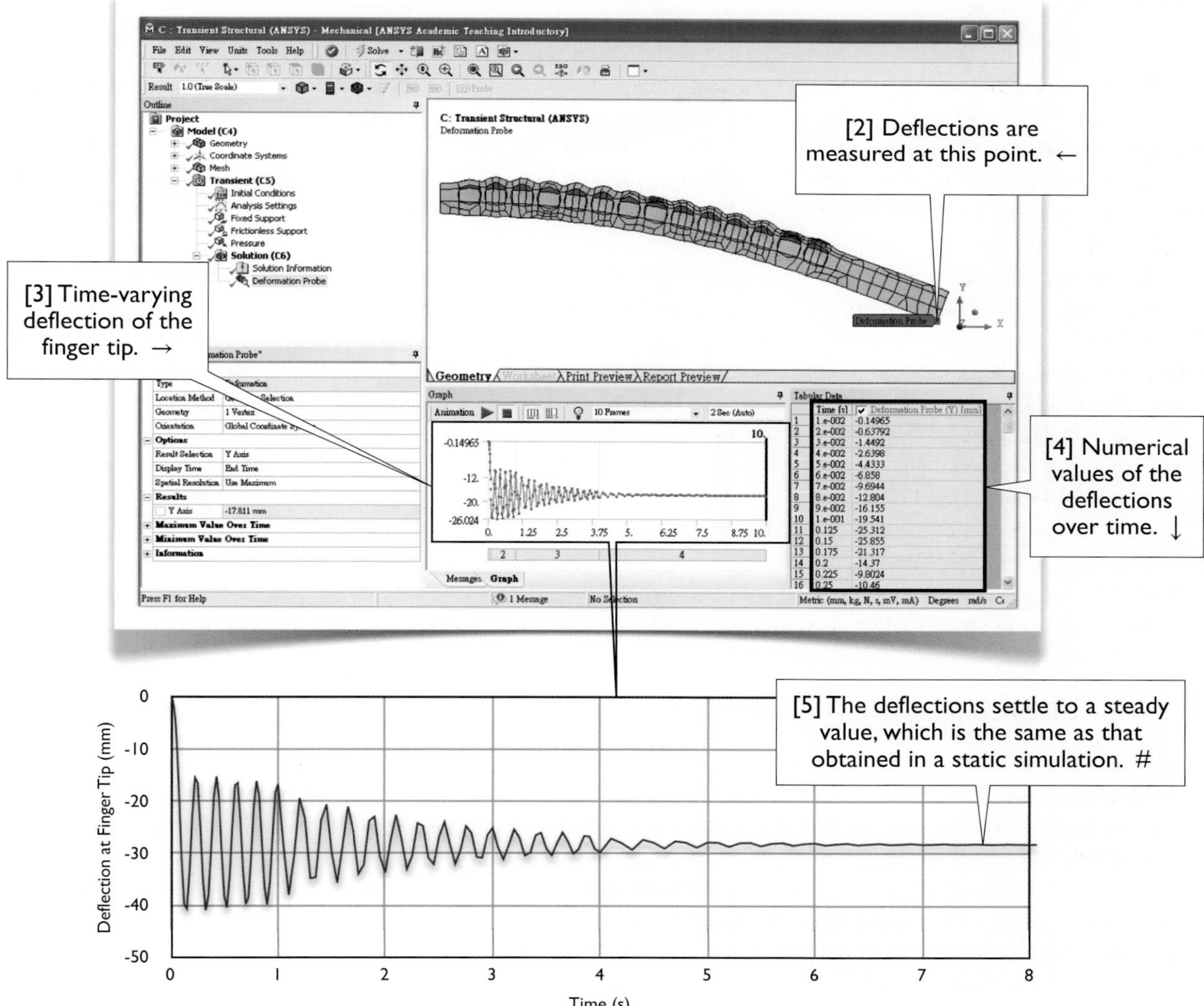

1.1.11 Modal Analysis

[1] A special case of dynamic simulations is the simulation of *free vibrations*, the vibrations of a structure without any external loading (but prestress is allowed). Consider that you deflect a structure and then release, causing the structure to vibrate without external forces. We want to know the behavior of this free vibration. The simulation is called a *modal analysis*. The results of modal analysis include the *natural frequencies* and the *vibration modes* of the structure. The figure below [2-5] shows the four lowest natural frequencies and the corresponding vibration modes.

A modal analysis is much less expensive (in terms of engineer's work hours and computing time) than a transient analysis and is often performed before a transient analysis to obtain preliminary dynamic characteristics of a structure. Chapter 11 will discuss modal analysis and its applications. ↓

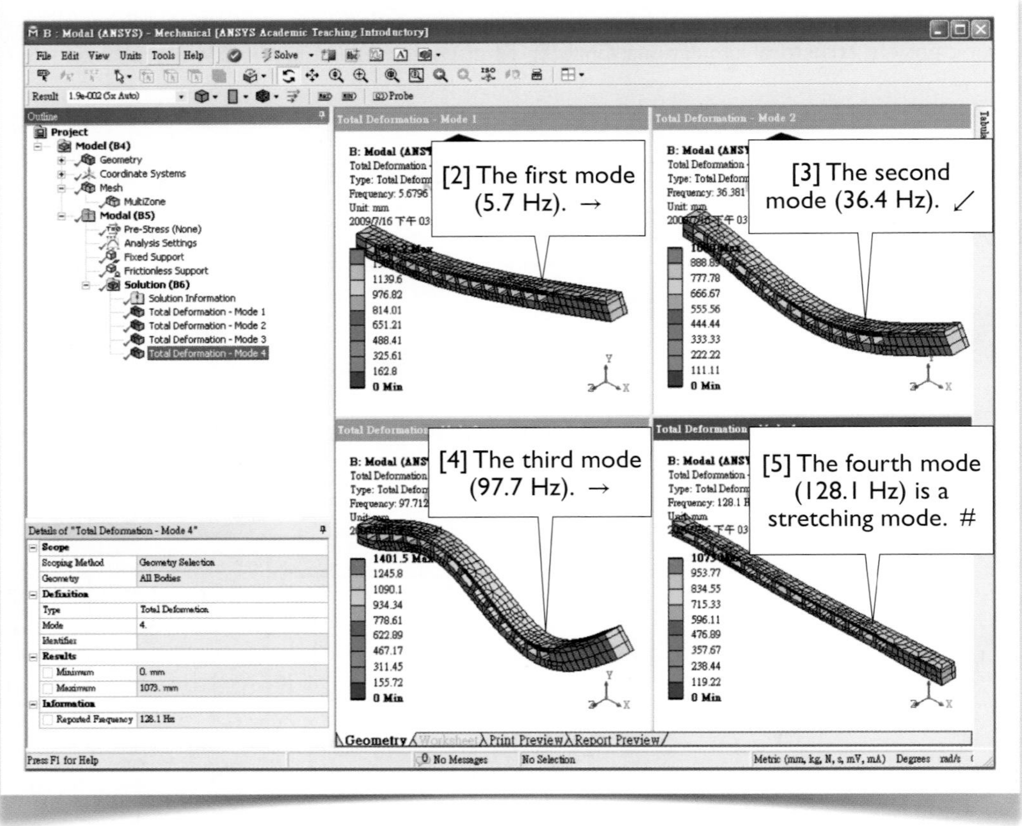

1.1.12 Structural Nonlinearities

[1] When the responses (deflection, stress, strain, etc.) of a structure are linearly proportional to the loads, the structure is called a *linear structure* and the simulation is called a *linear simulation*. Otherwise the structure is called a *nonlinear structure* and the simulation is called a *nonlinear simulation*.

Structural nonlinearities commonly come from three sources: (a) Due to large deformation; this is called *geometry nonlinearity*. (b) Due to the topological change of the structure; this is called *topology nonlinearity*. A common case of topology change is the change of contact status, and is called *contact nonlinearity*. (c) Due to the nonlinear stress-strain relationship of the material; this is called *material nonlinearity*. ↵

[2] In this case, the stress-strain relationship of the PDMS is almost linear within the range of operational air pressures, and there is no contact between any parts; therefore, there is no material nonlinearity or contact nonlinearity. It, however, has a deflection so large that it exhibits a certain degree of geometry nonlinearity.

The curve in [3] is reproduced from the curve in 1.1.8[6-7], page 19. It shows a nonlinear relationship between the deflection and the pressure. For comparison, we also include a linear solution [4]. The linear solution is obtained by turning **Large Deflection** off [5] and solving the model again.

A comparison between the nonlinear solution [3] and the linear solution [4] concludes that, in this case, the error would be significant if geometry nonlinearity were not taken into account.

Solving a nonlinear problem is often challenging and sometimes frustrating. Real-world problems often involve nonlinearities to some degree. We will experience nonlinear simulations as early as in Section 3.2, without detailed discussions of nonlinear solution controls. Chapters 13 and 14 are dedicated to the discussion of nonlinear simulations; Chapter 13 discusses general nonlinear solution methods and covers geometry nonlinearity and contact nonlinearity, while Chapter 14 discusses material nonlinearity. ↓

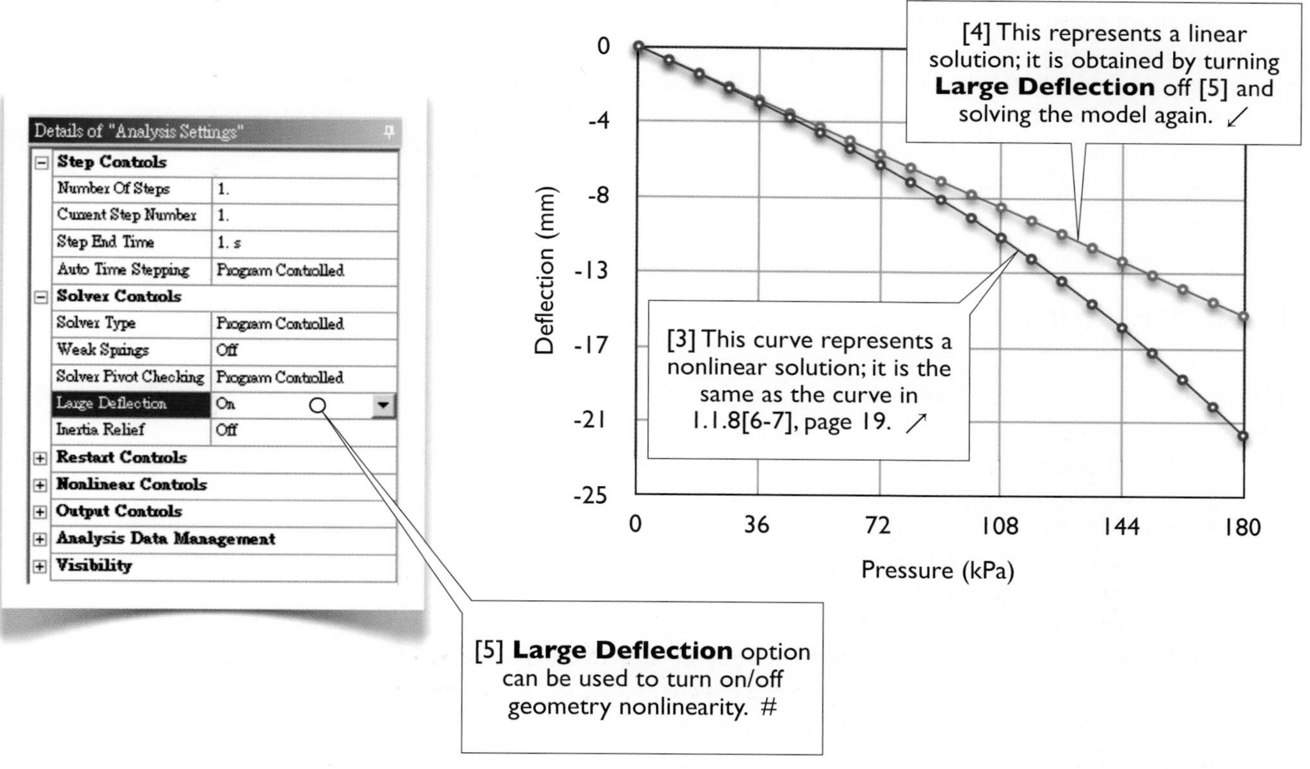

References

1. This case study is adapted from a work led by Prof. Chao-Chieh Lan of the Department of Mechanical Engineering, National Cheng Kung University, Taiwan.
2. Jeong, O. K and Konishi, S., "All PDMS Pneumatic Microfinger With Bidirectional Motion and Its Application," *Journal of Microelectomechanical Systems*, Vol. 15, No. 4, August 2006, pp. 896-903.
3. Draheim, J, Kamberger, R., and Wallrabe, U, "Process and material properties of polydimethylsiloxane (PDMS) for Optical MEMS," *Sensors and Actuators A: Physical*, Vol. 151, Issue 2, April 2009, pp. 95-99.

Section 1.2

Structural Mechanics: A Quick Review

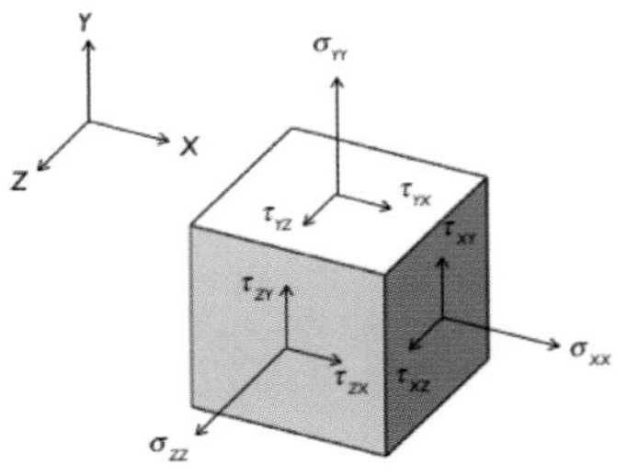

This section (a) defines a structural analysis problem, so the students have a clear picture about the input and the output of a structural analysis system; (b) introduces basic terminology that will be used throughout the book, such as displacements, stresses, and strains; (c) and summarizes equations that govern the behavior of the structure, so the students know what equations Workbench is solving.

To simplify the discussion, this review is limited to homogeneous, isotropic, linear static structural analyses. That is, (a) the material is assumed to be homogeneous, isotropic, and linearly elastic. Assumption of linear elasticity implies that Hooke's law is applicable; (b) the deformation is small enough so that we assume a linear relationship between the displacements and the strains; (c) there are no changes of topology; specifically, there are no changes of contact status during the deformation; (d) the deformation is slow enough so that the dynamic effects can be neglected.

The concepts introduced in this section can be generalized to include non-homogeneous, anisotropic, nonlinear, dynamic problems.

1.2.1 Structural Analysis Problems

[1] Engineering analysis (e.g., structural, mechanical, flow, electromagnetic) is to find the *responses* of a *problem domain* subject to *environmental conditions*.

In structural analyses, the problem domain consists of *solid bodies*; the environmental conditions include *loads* and *supports*; the responses can be described by the *displacements*, *stresses*, or *strains*.

For the pneumatic fingers case (Section 1.1), the problem domain is a body made of the PDMS elastomer. There are two support conditions: fixed support at one of the end faces and frictionless support at the face of symmetry. There is one load: the air pressure applied on the faces of the air chambers.

Note that these environmental conditions are applied on boundary faces, so these conditions are also called *boundary conditions*. Environmental conditions may not be applied on boundary surfaces. Common environmental conditions that do not apply on boundary surfaces include temperature changes and inertia forces; these loads distribute over the volumes (rather than boundary faces) of the problem domain. #

1.2.2 Displacement

[1] Deformation of a body can be described by a *displacement field* $\{u\}$ ([2-6], next page). Note that $\{u\}$ is a function of positions and, since it is a vector, we may express the displacement with three components,

$$\{u\} = \{ u_X \quad u_Y \quad u_Z \} \qquad (1)$$

The three components are, of course, functions of positions. The SI unit for displacements is meter (m). ↵

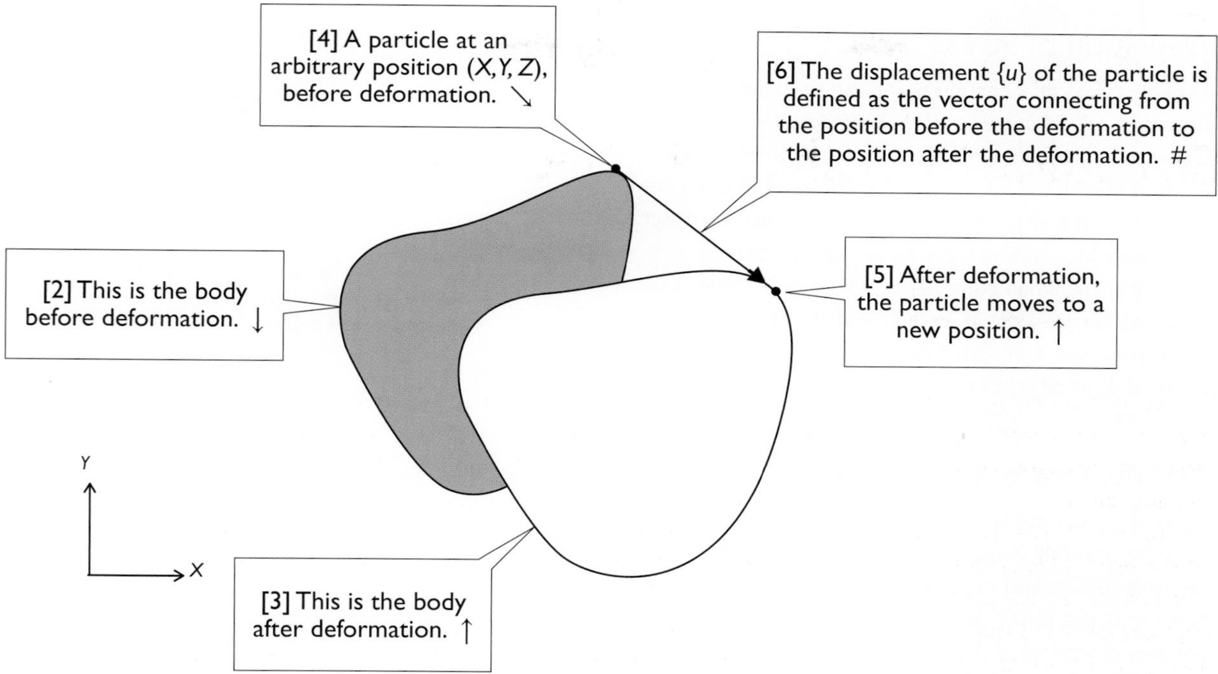

1.2.3 Stress

[1] The displacement is relatively easy to understand, since the displacement can be defined by a vector, and most college students are familiar with the mathematics of vectors. In contrast, the concepts of stress are not so obvious.

Stresses are quantities to describe the intensity of force in a body (either solid or fluid). Its unit is force per unit area (i.e., N/m^2 in SI). It is a position-dependent quantity.

Imagine that your arms are pulled by your friends with two forces of the same magnitude but opposite directions. What are the stresses in your arms? Assuming the magnitude of the forces is P and the cross-sectional area of your arms is A, then you may answer, "the stresses are P/A, everywhere in my arms." This case is simple and the answer is good enough. For a one-dimensional case like this, the stress σ may be defined as $\sigma = P/A$, where P is the applied force and A is the cross sectional area.

In 3D cases, things are much more complicated. Now, imagine that you are buried in the soil by your friends, and your head is deep below the ground surface. How do you describe the force intensity (i.e., stress) on your head (assuming your head is just like a particle in the sea of the soil)?

If the soil is replaced by still water, then the answer would be much simpler. The magnitude of the pressure (stress) on the top of your head would be the same as the pressure on your cheeks, and the direction of the pressure would always be perpendicular to the surface where the pressure applies. You've learned these in your high school. And you've learned that the magnitude of the pressure is $\sigma = \rho g h$, where ρ is the mass density of the water, g is the gravitational acceleration, and h is the depth of your head. In general, to describe the force intensity at a certain position in still water, we place an infinitesimally small body at that position, and measure the force per unit surface area on that body.

In the soil (which is a solid material rather than water), the behavior is quite different. First, the magnitude of the pressure on the top of your head may not be the same as that on your cheeks. Second, the direction of pressure is not necessarily perpendicular to the surface where the pressure applies. However, the above definition of stresses for water still holds (see [2]). ↵

Definition of Stress

[2] The stress at a certain point can be defined as *the force per unit area acting on the boundary faces of an infinitesimally small body centered at that point* [3]. In general, the stress values may be different at different faces. And the small body can be any shape. To describe the stress in a systematic way, we usually use an infinitesimally small cube [4] of which each edge is parallel to a coordinate axis. If we can find the stresses on a small cube, we then can calculate the stresses on any other shapes of small body. →

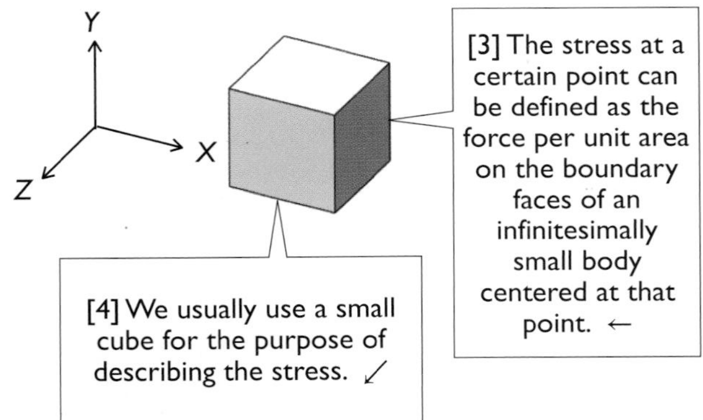

X-Face, Y-Face, and Z-Face

[5] Each of the six faces of the cube can be assigned an identifier, namely X-face, Y-face, Z-face, negative-X-face, negative-Y-face, and negative-Z-face, respectively [6-9]. Note that the outer normal of the X-face is in the positive direction of the X-axis, and so forth. →

Stress Components

[10] Let $\vec{p}_X$ be the force per unit area acting on the X-face. In general, $\vec{p}_X$ may not be normal to the X-face. We may decompose $\vec{p}_X$ into X-, Y-, and Z-component, and denote σ_{XX}, τ_{XY}, and τ_{XZ} respectively [11]. The first subscript (X) is used to indicate the **face** on which the stress components act, while the second subscript (X, Y, or Z) is used to indicate the **direction** of the stress components. Note that σ_{XX} is normal to the face, while τ_{XY}, and τ_{XZ} are parallel to the face. Therefore, σ_{XX} is called a normal stress, while τ_{XY}, and τ_{XZ} are called shear stresses. We usually use the symbol σ for a normal stress and τ for a shear stress.

Similarly, let $\vec{p}_Y$ be the force per unit area acting on the Y-face and we may decompose $\vec{p}_Y$ into a normal component (σ_{YY}) and two shear components (τ_{YX} and τ_{YZ}) [12]. Also, let $\vec{p}_Z$ be the force per unit area acting on the Z-face and we may decompose $\vec{p}_Z$ into a normal component (σ_{ZZ}) and two shear components (τ_{ZX} and τ_{ZY}) [13]. Organized in a matrix form, these stress components may be written as

$$\{\sigma\} = \begin{pmatrix} \sigma_{XX} & \tau_{XY} & \tau_{XZ} \\ \tau_{YX} & \sigma_{YY} & \tau_{YZ} \\ \tau_{ZX} & \tau_{ZY} & \sigma_{ZZ} \end{pmatrix} \quad (1)$$ →

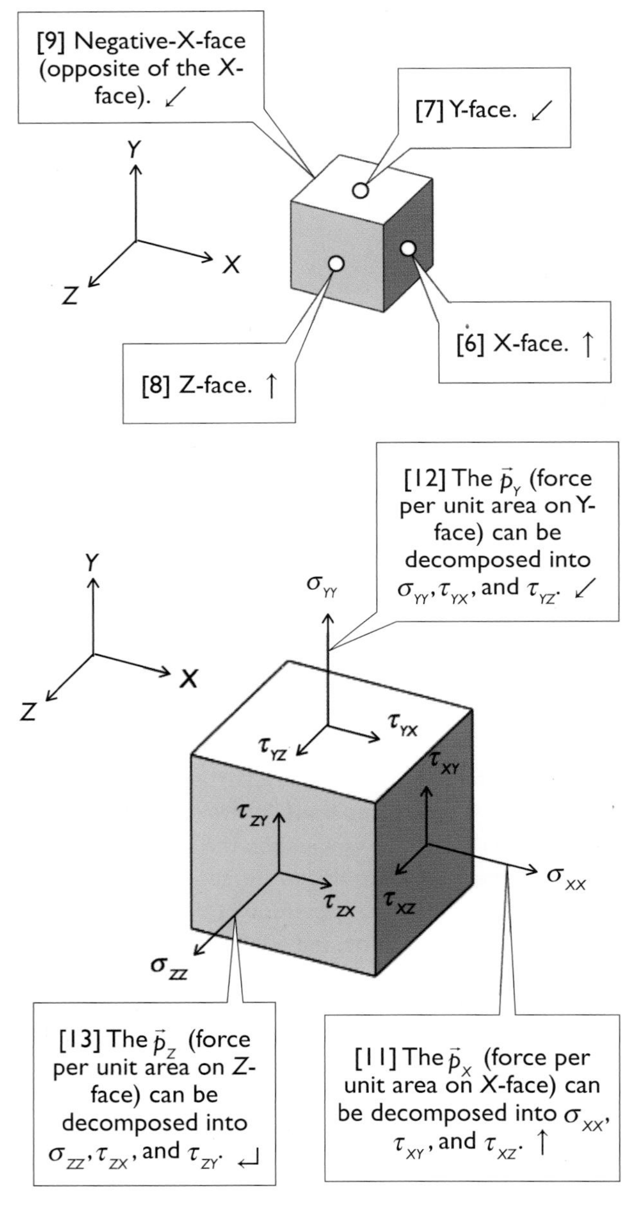

Stress Components on Other Faces

[14] It can be proven that the stress components on the negative-X-face, negative-Y-face, and negative-Z-face can be derived from the 9 stress components in Eq. (1), last page. For example, on the negative-X-face, the stress components have exactly the same stress values as those on the X-face but with opposite directions [15]. Similarly, the stress components on the negative-Y-face have the same stress values as those on the Y-face but with opposite directions [16], and the stress components on the negative-Z-face have the same stress values as those on the Z-face but with opposite directions [17].

These can be proved by taking the cube as free body and applying the force equilibria in X, Y, and Z directions respectively.

On an arbitrary face (which may not be parallel or perpendicular to an axis), the stress components also can be calculated from the 9 stress components in Eq. (1), last page. This can be done by using a Mohr's circle (1.4.2[6], page 41). →

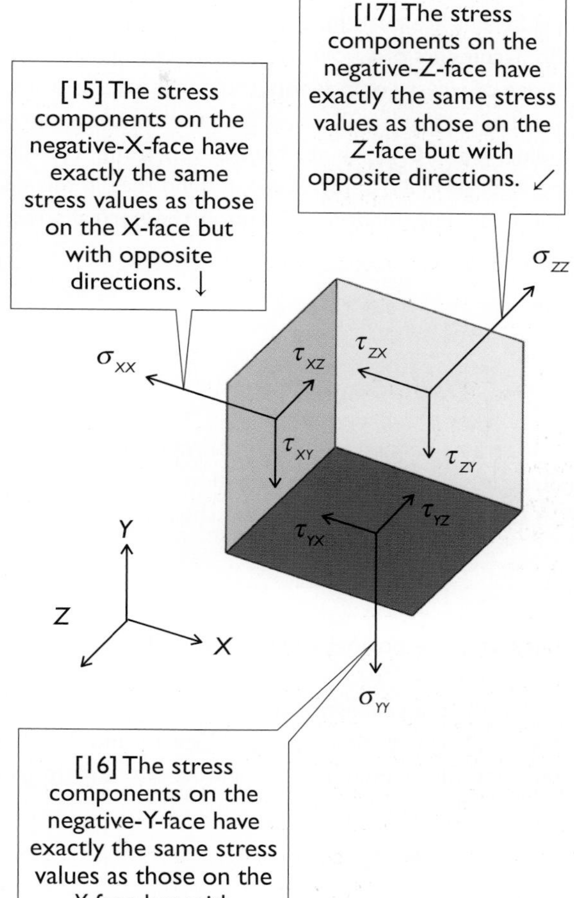

Symmetry of Shear Stresses

[18] It also can be proven that the shear stresses are symmetric; i.e.,

$$\tau_{XY} = \tau_{YX},\ \tau_{YZ} = \tau_{ZY},\ \tau_{ZX} = \tau_{XZ} \qquad (2)$$

These can be proved by taking the cube as free body and applying the moment equilibria in X, Y, and Z directions respectively.

Stress State

We now conclude that 3 normal stress components and 3 shear stress components are needed to describe the **stress state** at a certain point; therefore, a stress state may be described using a vector

$$\{\sigma\} = \{\ \sigma_X \quad \sigma_Y \quad \sigma_Z \quad \tau_{XY} \quad \tau_{YZ} \quad \tau_{ZX}\ \} \qquad (3)$$

Note that, to be more concise, we use σ_X in place of σ_{XX}, σ_Y in place of σ_{YY}, and σ_Z in place of σ_{ZZ}.

These 6 components are, of course, functions of position. #

1.2.4 Strain

[1] Strains are quantities to describe how the material in a body is stretched and distorted. Strains are defined as the displacements of a point relative to its neighboring points. Although the notations of strain components are similar to those of stress components, the concepts of strains are even more difficult to comprehend.

Let's consider 2D cases first; the concepts can be extended to 3D cases. Consider a point A and its neighboring points B and C, which are respectively along X-axis and Y-axis [2]. Suppose that, after deformation, ABC moves to a new configuration $A'B'C'$ [3]. Keep in mind that, in this discussion, we assume the deformation is infinitesimally small. Under the small deformation assumption, the normal strains in X-axis and Y-axis can be defined respectively as

$$\varepsilon_X = \frac{A'B' - AB}{AB} \text{ (dimensionless)} \tag{1}$$

$$\varepsilon_Y = \frac{A'C' - AC}{AC} \text{ (dimensionless)} \tag{2}$$

The strains defined in (1) and (2) represent the stretch at the point A in X-direction and Y-direction respectively. Stretch is not the only deformation modes; there are other deformation modes: changes of angles; e.g., from $\angle CAB$ to $\angle C'A'B'$, which is defined as the shear strain in XY-plane,

$$\gamma_{XY} = \angle CAB - \angle C'A'B' \text{ (rad)} \tag{3}$$

Note that the normal strains (1-2) and the shear strain (3) are all dimensionless, since the radian is also regarded as dimensionless.

In the above illustration, we consider only 2D cases. In general, the stretching may also occur in Z-direction and the shearing may also occur in YZ-plane and ZX-plane. Therefore, we need six strain components to completely describe the stretching and shearing of the material at a point:

$$\{\varepsilon\} = \{\ \varepsilon_X \quad \varepsilon_Y \quad \varepsilon_Z \quad \gamma_{XY} \quad \gamma_{YZ} \quad \gamma_{ZX}\ \} \tag{4}$$

These 6 components are, of course, functions of position. ↙

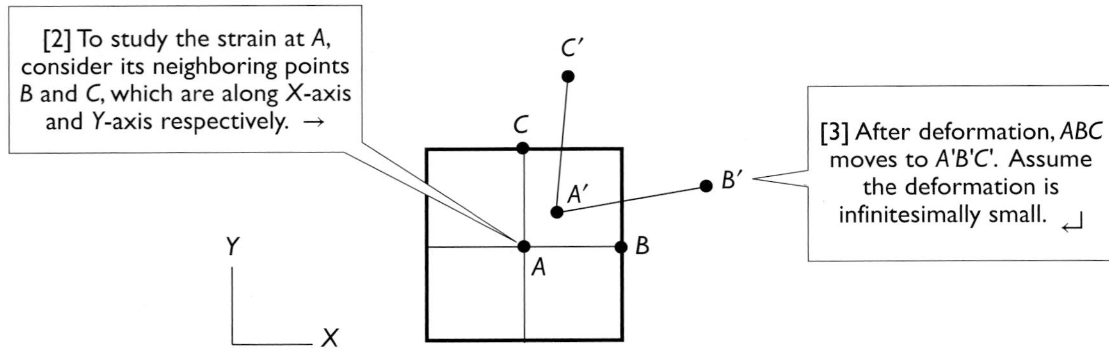

Why Are They Called Normal/Shear Strains?

[4] The definitions in Eqs. (1-3) do not explain why they are called "normal" strains and "shear" strain, respectively. To clarify this, let's redefine normal and shear strains using a different but equivalent way.

First, we translate and rotate $A'B'C'$ such that A' coincides with A and $A'C'$ aligns with AC [5]. Now the vector BB' is the "absolute" displacement (displacement excluding rigid body motion) of a neighboring point B which is on X-axis [6]. This displacement BB' can be decomposed into two components: BD and DB'; the former is called the normal component, while the latter is called the shear component. They are so named because BD is normal to the X-face and DB' is parallel to the X-face. The normal strain and shear strain on X-face are then defined respectively by dividing the components with the original length,

$$\varepsilon_X = \frac{BD}{AB} \text{ (dimensionless)} \tag{5}$$

$$\gamma_{XY} = \frac{DB'}{AB} \text{ (rad)} \tag{6}$$

Note that, under the assumption of small deformation, the definition in Eq. (5) is the same as that in Eq. (1), while the definition in Eq. (6) is the same as that in Eq. (3). Also note that there are two subscripts in the shear strain γ_{XY}. The first subscript X is the face where the shearing occurs, while the second subscript Y is the direction of the shearing.

Similarly, we may translate and rotate $A'B'C'$ such that A' coincides with A and $A'B'$ aligns with AB [7]. Now the vector CC' is the "absolute" displacement of a neighboring point C which is on Y-axis [8]. This displacement CC' can be decomposed into two components: CE and EC'; the former is the normal component, while the latter is the shear component. The normal strain and shear strain on Y-face is then defined by

$$\varepsilon_Y = \frac{CE}{AC} \text{ (dimensionless)} \tag{7}$$

$$\gamma_{YX} = \frac{EC'}{AC} \text{ (rad)} \tag{8}$$

Note that, under the assumption of small deformation, the definition in Eq. (7) is the same as that in Eq. (2), while the definition in Eq. (8) is the same as that in Eq. (3). From Eqs. (3, 6, 8), we may write

$$\gamma_{XY} = \gamma_{YX} = \text{change of a right angle in } XY\text{-plane (rad)} \tag{9}$$ ←

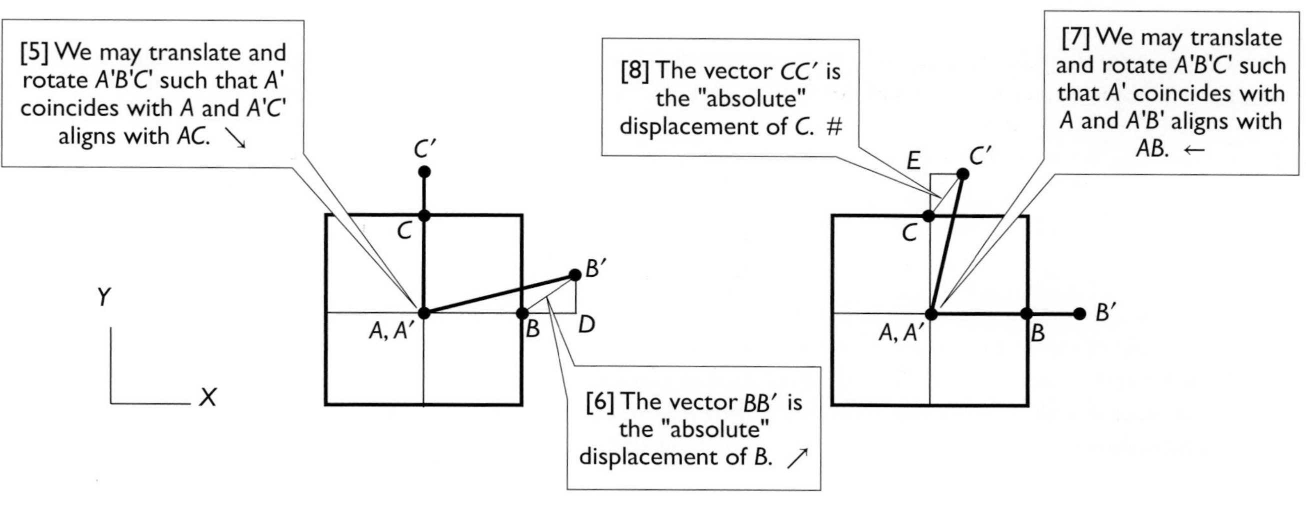

1.2.5 Governing Equations

[1] Let's summarize what we have concluded so far. In structural analysis, we can use the following quantities (all or part of them) to describe the response of a structure subject to environmental conditions:

$$\{u\} = \{\ u_X \quad u_Y \quad u_Z\ \}$$ Copy of 1.2.2(1), page 24

$$\{\sigma\} = \{\ \sigma_X \quad \sigma_Y \quad \sigma_Z \quad \tau_{XY} \quad \tau_{YZ} \quad \tau_{ZX}\ \}$$ Copy of 1.2.3(3), page 27

$$\{\varepsilon\} = \{\ \varepsilon_X \quad \varepsilon_Y \quad \varepsilon_Z \quad \gamma_{XY} \quad \gamma_{YZ} \quad \gamma_{ZX}\ \}$$ Copy of 1.2.4(4), page 28

To solve for these 15 quantities, we must establish 15 equations. These equations are called governing equations; they govern the structure's behaviors. These equations, introduced in the rest of this section, include 3 equilibrium equations, 6 strain-displacement relations, and 6 stress-strain relations. #

1.2.6 Equilibrium Equations

[1] The stress components in Eq. 1.2.3(3) (page 27) must satisfy the principle of force equilibrium:

$$\sum F_X = 0, \quad \sum F_Y = 0, \quad \sum F_Z = 0 \tag{1}$$

If we apply Eqs. (1) on a point INSIDE the structural body, we can obtain three equilibrium equations involving the stress components:

$$\begin{aligned} \frac{\partial \sigma_X}{\partial X} + \frac{\partial \tau_{XY}}{\partial Y} + \frac{\partial \tau_{XZ}}{\partial Z} + b_X &= 0 \\ \frac{\partial \tau_{YX}}{\partial X} + \frac{\partial \sigma_Y}{\partial Y} + \frac{\partial \tau_{YZ}}{\partial Z} + b_Y &= 0 \\ \frac{\partial \tau_{ZX}}{\partial X} + \frac{\partial \tau_{ZY}}{\partial Y} + \frac{\partial \sigma_Z}{\partial Z} + b_Z &= 0 \end{aligned} \tag{2}$$

where b_X, b_Y, b_Z are components of body forces (forces distributed in the body, with SI unit N/m^3). If we apply Eqs. (1) on a point ON the boundary surface of the structural body, the three equilibrium equations will have the form:

$$\begin{aligned} \sigma_X n_X + \tau_{XY} n_Y + \tau_{XZ} n_Z + S_X &= 0 \\ \tau_{YX} n_X + \sigma_Y n_Y + \tau_{YZ} n_Z + S_Y &= 0 \\ \tau_{ZX} n_X + \tau_{ZY} n_Y + \sigma_Z n_Z + S_Z &= 0 \end{aligned} \tag{3}$$

where S_X, S_Y, S_Z are components of surface forces (forces distributed on the boundary, with SI unit N/m^2), and n_X, n_Y, n_Z are components of the unit normal vector on the boundary surface.

We will not discuss the derivations of Eqs. (2-3); they can be found in any Solid Mechanics textbooks[Refs 1, 2]. Here, we just reiterate that the equilibrium equations originate from Eq. (1). Also note that, in order to derive Eq. (2), the stress components in 1.2.3[11-13, 15-17] (pages 26-27) must be expanded to include first differential terms. #

1.2.7 Strain-Displacement Relations

[1] There exist geometric relations between the displacement and the strain. Under the assumption of small deformation, the relations are linear:

$$\varepsilon_X = \frac{\partial u_X}{\partial X}, \quad \varepsilon_Y = \frac{\partial u_Y}{\partial Y}, \quad \varepsilon_Z = \frac{\partial u_Z}{\partial Z}$$

$$\gamma_{XY} = \frac{\partial u_X}{\partial Y} + \frac{\partial u_Y}{\partial X}, \quad \gamma_{YZ} = \frac{\partial u_Y}{\partial Z} + \frac{\partial u_Z}{\partial Y}, \quad \gamma_{ZX} = \frac{\partial u_Z}{\partial X} + \frac{\partial u_X}{\partial Z} \tag{1}$$

Again, we will not discuss the derivations of Eq. (1); they can be found in any Solid Mechanics textbooks[Refs 1, 2]. Here, we just want to reiterate that Eq. (1) is derived from mathematics (geometry) without applying any physical principles. Second, if the assumption of small deformation is removed, a strain-displacement relation still exists but is no longer linear; there will be some high-order differential terms in the equations. #

1.2.8 Stress-Strain Relations

[1] With 3 equilibrium equations (Eqs. 1.2.6(2 or 3), last page) and 6 strain-displacement relations (Eq. 1.2.7(1)), we need 6 additional equations to solve the governing equations. From an engineering point of view, it is practical to assume a relation between stress and strain. Experiment data show that a linear relation between stress and strain often can be adopted; it is called Hooke's law[Refs 1, 2, 3]:

$$\varepsilon_X = \frac{\sigma_X}{E} - \nu\frac{\sigma_Y}{E} - \nu\frac{\sigma_Z}{E}$$

$$\varepsilon_Y = \frac{\sigma_Y}{E} - \nu\frac{\sigma_Z}{E} - \nu\frac{\sigma_X}{E}$$

$$\varepsilon_Z = \frac{\sigma_Z}{E} - \nu\frac{\sigma_X}{E} - \nu\frac{\sigma_Y}{E} \tag{1}$$

$$\gamma_{XY} = \frac{\tau_{XY}}{G}, \quad \gamma_{YZ} = \frac{\tau_{YZ}}{G}, \quad \gamma_{ZX} = \frac{\tau_{ZX}}{G}$$

Although Eq. (1) is purely an assumption (no physical or mathematical principles applied), it is proved to be very useful and satisfactory in many cases, depending on the material and the application. Eq. (1) is called a *material model*; it characterizes the behavior of a material, independent of the geometry and environmental conditions.

There are three material parameters in Eq. (1): the Young's modulus E, the Poisson's ratio ν, and the shear modulus G. In SI, the Young's modulus and the shear modulus have a unit of pascal (Pa) and the Poisson's ratio is dimensionless. It can be shown that these three quantities are not independent to each other; they satisfy the relation[Ref 3]

$$G = \frac{E}{2(1+\nu)} \tag{2}$$

We conclude that, for an isotropic, linearly elastic material, any two of E, ν, and G can be used to describe the stress-strain relation. In Workbench, we usually input the Young's modulus and the Poisson's ratio to define an **Isotropic Elasticity** model (1.1.3[5-6], page 12). ↵

Thermal Effects

[2] Consider that the temperature changes over the structural body. Since a temperature change ΔT induces a strain $\alpha\Delta T$, in which α is the *coefficient of thermal expansion*, this *thermal strain* should be added to Eq. (1) (last page); i.e.,

$$\begin{aligned} \varepsilon_X &= \frac{\sigma_X}{E} - \nu\frac{\sigma_Y}{E} - \nu\frac{\sigma_Z}{E} + \alpha\Delta T \\ \varepsilon_Y &= \frac{\sigma_Y}{E} - \nu\frac{\sigma_Z}{E} - \nu\frac{\sigma_X}{E} + \alpha\Delta T \\ \varepsilon_Z &= \frac{\sigma_Z}{E} - \nu\frac{\sigma_X}{E} - \nu\frac{\sigma_Y}{E} + \alpha\Delta T \\ \gamma_{XY} &= \frac{\tau_{XY}}{G}, \quad \gamma_{YZ} = \frac{\tau_{YZ}}{G}, \quad \gamma_{ZX} = \frac{\tau_{ZX}}{G} \end{aligned} \tag{3}$$

Orthotropic Elasticity

For orthotropic materials (14.1.1[3], page 527), in which there exist three mutual orthogonal planes of material symmetry, Hooke's law can be generalized to[Refs 1, 2, 3]

$$\begin{aligned} \varepsilon_X &= \frac{\sigma_X}{E_X} - \nu_{YX}\frac{\sigma_Y}{E_Y} - \nu_{ZX}\frac{\sigma_Z}{E_Z} + \alpha_X\Delta T \\ \varepsilon_Y &= \frac{\sigma_Y}{E_Y} - \nu_{ZY}\frac{\sigma_Z}{E_Z} - \nu_{XY}\frac{\sigma_X}{E_X} + \alpha_Y\Delta T \\ \varepsilon_Z &= \frac{\sigma_Z}{E_Z} - \nu_{XZ}\frac{\sigma_X}{E_X} - \nu_{YZ}\frac{\sigma_Y}{E_Y} + \alpha_Z\Delta T \\ \gamma_{XY} &= \frac{\tau_{XY}}{G_{XY}}, \quad \gamma_{YZ} = \frac{\tau_{YZ}}{G_{YZ}}, \quad \gamma_{ZX} = \frac{\tau_{ZX}}{G_{ZX}} \end{aligned} \tag{4}$$

where E_X, E_Y, E_Z are Young's moduli in their respective directions, G_{XY}, G_{YZ}, G_{ZX} are the shear moduli in their respective planes, and $\nu_{XY}, \nu_{YZ}, \nu_{ZX}$ are the Poisson's ratios in their respective planes. The first subscript in each of the Poisson's ratios refers to the direction of the load, and the second to the direction of the contraction. For example, ν_{XY} represents the amount of contraction in *Y*-direction, when the material is stretched in *X*-direction. #

1.2.9 Summary

[1] We now have 15 equations, including three equilibrium equations, either (INSIDE the structural body)

$$\frac{\partial \sigma_X}{\partial X} + \frac{\partial \tau_{XY}}{\partial Y} + \frac{\partial \tau_{XZ}}{\partial Z} + b_X = 0$$
$$\frac{\partial \tau_{YX}}{\partial X} + \frac{\partial \sigma_Y}{\partial Y} + \frac{\partial \tau_{YZ}}{\partial Z} + b_Y = 0$$
$$\frac{\partial \tau_{ZX}}{\partial X} + \frac{\partial \tau_{ZY}}{\partial Y} + \frac{\partial \sigma_Z}{\partial Z} + b_Z = 0$$

Copy of 1.2.6(2), page 30

or (ON the boundary surface)

$$\sigma_X n_X + \tau_{XY} n_Y + \tau_{XZ} n_Z + S_X = 0$$
$$\tau_{YX} n_X + \sigma_Y n_Y + \tau_{YZ} n_Z + S_Y = 0$$
$$\tau_{ZX} n_X + \tau_{ZY} n_Y + \sigma_Z n_Z + S_Z = 0$$

Copy of 1.2.6(3), page 30

and six equations describing the strain-displacement relation

$$\varepsilon_X = \frac{\partial u_X}{\partial X},\quad \varepsilon_Y = \frac{\partial u_Y}{\partial Y},\quad \varepsilon_Z = \frac{\partial u_Z}{\partial Z}$$
$$\gamma_{XY} = \frac{\partial u_X}{\partial Y} + \frac{\partial u_Y}{\partial X},\quad \gamma_{YZ} = \frac{\partial u_Y}{\partial Z} + \frac{\partial u_Z}{\partial Y},\quad \gamma_{ZX} = \frac{\partial u_Z}{\partial X} + \frac{\partial u_X}{\partial Z}$$

Copy of 1.2.7(1), page 31

and six equations describing the stress-strain relation

$$\varepsilon_X = \frac{\sigma_X}{E} - \nu\frac{\sigma_Y}{E} - \nu\frac{\sigma_Z}{E} + \alpha\Delta T$$
$$\varepsilon_Y = \frac{\sigma_Y}{E} - \nu\frac{\sigma_Z}{E} - \nu\frac{\sigma_X}{E} + \alpha\Delta T$$
$$\varepsilon_Z = \frac{\sigma_Z}{E} - \nu\frac{\sigma_X}{E} - \nu\frac{\sigma_Y}{E} + \alpha\Delta T$$
$$\gamma_{XY} = \frac{\tau_{XY}}{G},\quad \gamma_{YZ} = \frac{\tau_{YZ}}{G},\quad \gamma_{ZX} = \frac{\tau_{ZX}}{G}$$

Copy of 1.2.8(3), page 32

In theory, these 15 equations can be solved for 15 quantities:

$$\{u\} = \{\, u_X \quad u_Y \quad u_Z \,\}$$

Copy of 1.2.2(1), page 24

$$\{\sigma\} = \{\, \sigma_X \quad \sigma_Y \quad \sigma_Z \quad \tau_{XY} \quad \tau_{YZ} \quad \tau_{ZX} \,\}$$

Copy of 1.2.3(3), page 27

$$\{\varepsilon\} = \{\, \varepsilon_X \quad \varepsilon_Y \quad \varepsilon_Z \quad \gamma_{XY} \quad \gamma_{YZ} \quad \gamma_{ZX} \,\}$$

Copy of 1.2.4(4), page 28

In practice, only a few simple "textbook problems" can be solved analytically. Most real-world problems are too complicated to solve analytically. The complexity mostly comes from the geometry and the environmental conditions. Numerical methods are usually the only feasible methods. The finite element methods, which have been the most successful numerical methods for *boundary-value problems* (such as the problems described in this section), are implemented in ANSYS Workbench to solve the governing equations. ↵

Remark

[2] By "governing equations" of a structural analysis problem, we sometimes mean the equilibrium equations: Eq. 1.2.6(2) (page 30) governs the behavior in the body and Eq. 1.2.6(3) (page 30) governs the behavior on the boundary. The two sets of equations constitute a boundary value problem.

It is possible to replace the stress components in the equilibrium equations by strain components using Eq. 1.2.8(3) (page 32), and in turn replace the strain components by displacement components using Eq. 1.2.7(1) (page 31). The result is a set of three equilibrium equations involving three displacement components, and we can solve the three differential equations for the three displacement components.

That is how ANSYS Workbench solves a structural problem (see Eq. 1.3.1(1), next page). #

References

1. Cook, R. D. and Young, W. C., *Advanced Mechanics of Materials, Macmillan*, 1985.
2. Haslach, H. W. Jr. and Armstrong, R. W., *Deformable Bodies and Their Material Behavior*, John Wiley & Sons, Inc., 2004; Chapter 9 Plasticity.
3. Beer, F. P., Johnston, E. R. Jr., and DeWolf, J. T., *Mechanics of Materials*, 3rd Ed., McGraw Hill, 2002.

Section 1.3

Finite Element Methods: A Concise Introduction

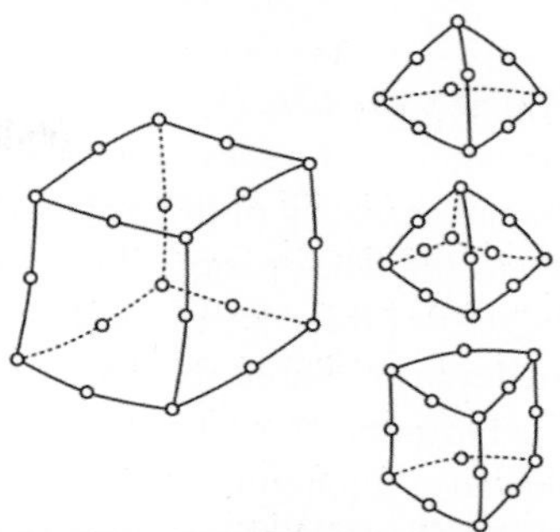

This section introduces (a) terminology relevant to finite element methods, such as degrees of freedom, shape functions, stiffness matrix, etc., that will be used throughout the book and (b) the basic procedure of finite element methods, so the students have a better understanding about how the Workbench performs a simulation.

1.3.1 Basic Procedure

[1] As mentioned (at the bottom of 1.2.9[1], page 33), most real-world problems are too complicated to be solved analytically, because of the complexity of geometry and/or environmental conditions. Further, when nonlinearity and dynamic effects are considered in the problems, then their analytical solutions are practically unreachable.

A basic idea of finite element methods is to divide the structural body into many small and geometrically simple bodies, called *elements*, so the equations of each element can be established, and all the equations are then solved simultaneously. The elements have finite sizes (in contrast to the infinitesimal sizes of elements in Calculus), thus the name *Finite Element Methods*.

The elements are assumed to be connected by *nodes* located at the elements' edges and vertices. An additional idea is to solve unknown discrete values (e.g., displacements at the nodes) rather than to solve unknown functions (e.g., displacement fields over the body). Since the displacement on each node is a vector and has three components (in 3D cases), the number of total unknown quantities to be solved is three times the number of nodes [2].

The types of elements available in the ANSYS Workbench and their specific configuration of nodes will be given in 1.3.3, pages 37-39. ↓

[2] In the case of the pneumatic finger, the structural body is divided into 3,395 elements (1.1.5[6], page 16). The elements are connected by 19,201 nodes. There are 3x19,201 unknown displacement values to be solved. ↓

[3] The nodal displacement components, collectively denoted by a vector $\{D\}$, are called the *degrees of freedom* (DOFs) of the structure. They are so called because these values fully define the response of a structure. In a static case, the system of equilibrium equations has the following form (also see 1.2.9[2], last page)

$$[K]\{D\} = \{F\} \quad (1)$$ ↵

[4] The size of Eq. (1) (last page) is determined by the number of degrees of freedom, which is, in 3D cases, three times the number of nodes. The vector $\{F\}$ is the external forces, which is calculated from the environmental conditions. Physical meaning of the matrix $[K]$ can be understood by thinking of the structure as a spring, $\{F\}$ as external force, and $\{D\}$ as the deformation of the spring. Then $[K]$ is the spring constant, or the stiffness of the spring. In general mutiple-degrees-of-freedom cases, $[K]$ is called the *stiffness matrix* of the structure, and the physical meaning of the ith column is the forces required to make the ith DOF a unit displacement while restraining the other DOFs from any displacements.

Note that, for a linear structure, $[K]$ is a constant matrix; while for nonlinear cases (1.1.12, pages 22-23), $[K]$ may be viewed as a function of $\{D\}$, which will be discussed in Chapters 13 and 14. For dynamic cases (1.1.10 and 1.1.11, pages 21-22), dynamic effects need to be added to Eq. (1) (last page), which will be discussed in Chapters 11 and 12.

After the discrete nodal displacements $\{D\}$ in Eq. (1) are solved, the displacement fields $\{u\}$ are calculated by interpolating the nodal displacements, either linearly or quadratically [5] (also see Eq. 1.3.2(2), this page). These interpolating functions are called *shape functions*. The concepts of shape functions are crucial in the finite element methods; they will be discussed further in 1.3.2.

As soon as the displacement fields $\{u\}$ become known, the strain fields can be calculated by Eq. 1.2.7(1) (page 31), the strain-displacement relation. The stress fields in turn can be calculated by Eq. 1.2.8(3) (page 32), the stress-strain relation. →

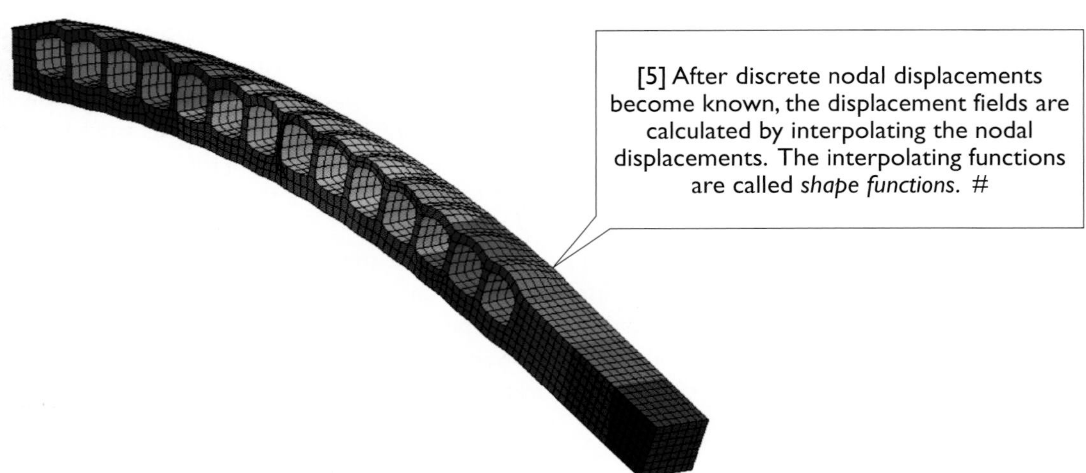

[5] After discrete nodal displacements become known, the displacement fields are calculated by interpolating the nodal displacements. The interpolating functions are called *shape functions*. #

1.3.2 Shape Functions and the Order of Element

[1] As mentioned (1.3.1[4-5], this page), the displacement fields $\{u\}$ are calculated by interpolating the nodal displacements. The interpolating functions are called shape functions; the shape functions establish a relation between the displacement fields and the nodal displacements.

As an example, consider a 2D 4-node quadrilateral element ([3], next page). The nodal displacements of the element, collectively denoted by a vector $\{d\}$, have 8 components

$$\{d\} = \{\, d_1 \;\; d_2 \;\; d_3 \;\; d_4 \;\; d_5 \;\; d_6 \;\; d_7 \;\; d_8 \,\} \tag{1}$$

The displacement fields $\{u\}$ can be calculated by interpolating the nodal displacements $\{d\}$

$$\{u\} = [N]\{d\} \tag{2}$$ ↵

[2] In Eq. (2), last page, [*N*] is the *matrix of shape functions*. The role of the shape functions is the interpolating functions from the nodal displacements {*d*} to the displacement fields {*u*}. Note that the components of nodal displacements {*d*} are discrete values and the components of displacement fields {*u*} are continuous functions of (*X*, *Y*, *Z*). The shape functions in [*N*] are the bridge between the continuous functions {*u*} and the discrete values {*d*}.

Now, let's examine the dimension of the [*N*] matrix using this example. Since {*d*} has 8 components and {*u*} has 3 components (Eq. 1.2.2(1), page 24), the matrix [*N*] must be of dimension 3x8.

Since {*u*} contains functions of (*X*, *Y*, *Z*) and {*d*} contains discrete values, [*N*] must contain functions of (*X*, *Y*, *Z*). Besides, for the element shown in [3], the interpolating points are at the vertices of the element, so the shape functions must be a linear form. When the shape functions are linear, the element is called a *linear element*, *first-order element*, or *lower-order element* [4].

Often, using quadratic polynomials as shape functions can be more efficient. In such cases, a node is added on the middle of each edge of the element; the added nodes are called the *midside nodes*; the element is called a *quadratic element*, *second-order element*, or *higher-order element*. In this book, we will not use the term *linear element*, to avoid confusing it with the terms such as linear material. ↓

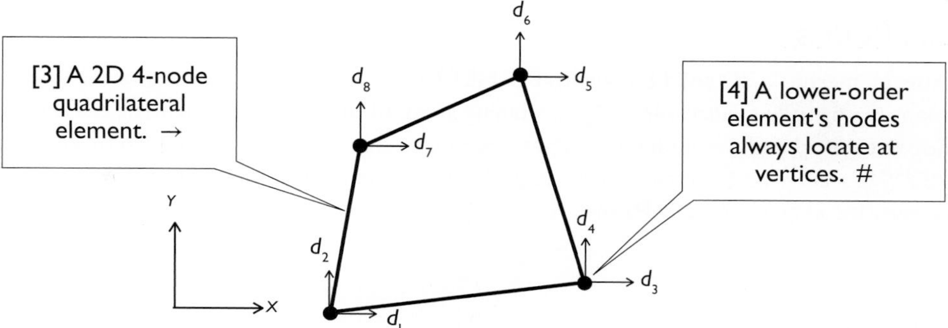

1.3.3 Workbench Element Types

[1] ANSYS literally provides hundreds of element types[Ref 1]. By default, Workbench automatically chooses appropriate element types during the meshing process[Ref 2]; to select a specific type of element, you need to use APDL. To identify these element types, each element type is assigned a code name (e.g., SOLID186, PLANE183, etc.).

For 2D solid-body cases, the element shapes available are quadrilateral (4-sided) and triangular (3-sided). For 3D solid-body cases, the available shapes are hexahedral (6-faced), triangle-based prism (5-faced), quadrilateral-based pyramid (5-faced), and tetrahedral (4-faced). As mentioned, Workbench chooses element types according to the types of the structural bodies. Currently Workbench supports 4 body types: 2D solid body, 3D solid body, 3D surface body, and 3D line body. 3D surface bodies are geometrically 2D but spatially 3D, while 3D line bodies are geometrically 1D but spatially 3D. How Workbench chooses an element type according to the body type is illustrated as follows.

3D Solid Bodies

Workbench meshes a 3D solid body with SOLID186[Ref 3], a 3D 20-node second-order structural solid element ([2], next page). The element is originally hexahedral. By combining some of the nodes, the element can degenerate to a triangle-based prism [3], quadrilateral-based pyramid [4], or tetrahedron [5]. Degeneration is useful since it allows different shapes of elements to be mixed up in a body. If a body is to be meshed with tetrahedral elements exclusively, Workbench meshes the body with SOLID187[Ref 4], which is a 3D 10-node tetrahedral second-order structural solid element and has a shape the same as that in [5]. Workbench allows an option to drop off elements' midside nodes; in that case the edges become straight and the element becomes first-order. ↵

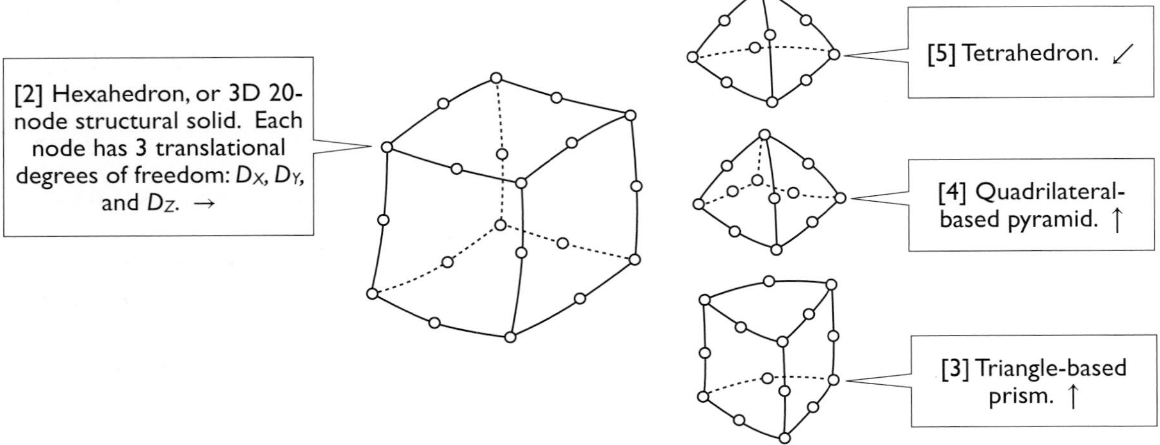

2D Solid Bodies

[6] Workbench meshes a 2D solid body with PLANE183[Ref 5], a 2D 8-node second-order structural solid element [7]. The element is originally quadrilateral. By combining some of the nodes, the element can degenerate to a triangle [8]. If you choose to drop midside nodes, the edges become straight, and the Workbench meshes the body with PLANE182[Ref 6], a 2D 4-node first-order structural solid element. It is important to remember that all 2D solid elements must be arranged on **XYPlane**. ↓

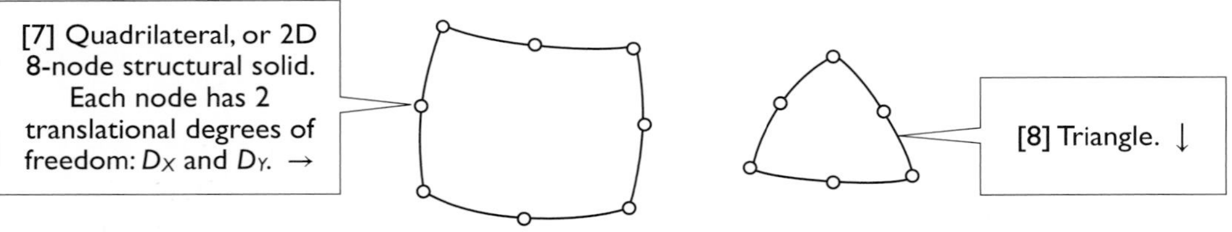

3D Surface Bodies

[9] Workbench meshes a 3D surface body with SHELL181[Ref 7], a 3D 4-node first-order structural shell element [10], or SHELL281[Ref 8], a 3D 8-node second-order structural shell. The element is originally quadrilateral but can degenerate to a triangle [11]. ↓

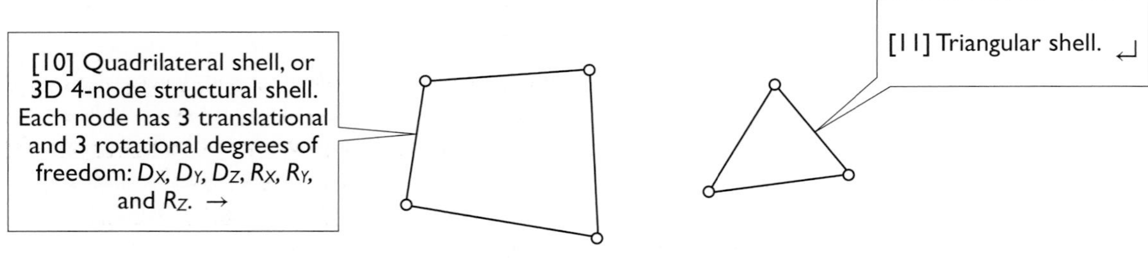

3D Line Bodies

[12] For a 3D line body, the Workbench meshes it with BEAM188[Ref 9], a 3D 2-node first-order beam element [13]. ↓

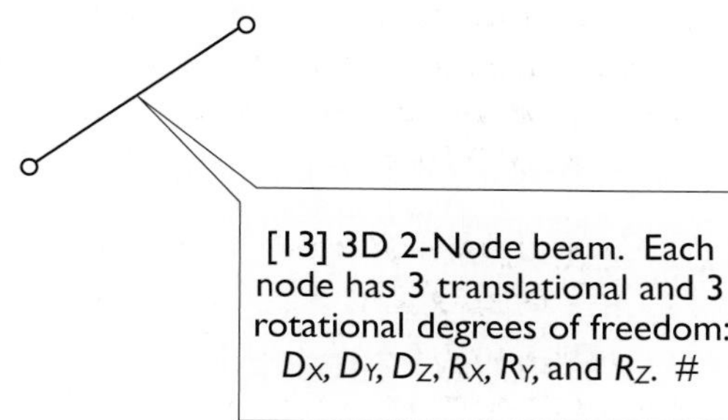

References

1. ANSYS Help//Mechanical APDL//Mechanical APDL Element Reference//I. Element Library
2. The following page is available in version 13 and earlier but removed since version 14.
   ANSYS Help System//Mechanical APDL//Theory Reference//1.4.1. Elements Used by the ANSYS Workbench Product
3. ANSYS Help//Mechanical APDL//Mechanical APDL Element Reference//I. Element Library//SOLID186
4. ANSYS Help//Mechanical APDL//Mechanical APDL Element Reference//I. Element Library//SOLID187
5. ANSYS Help//Mechanical APDL//Mechanical APDL Element Reference//I. Element Library//PLANE183
6. ANSYS Help//Mechanical APDL//Mechanical APDL Element Reference//I. Element Library//PLANE182
7. ANSYS Help//Mechanical APDL//Mechanical APDL Element Reference//I. Element Library//SHELL181
8. ANSYS Help//Mechanical APDL//Mechanical APDL Element Reference//I. Element Library//SHELL281
9. ANSYS Help//Mechanical APDL//Mechanical APDL Element Reference//I. Element Library//BEAM188

Section 1.4

Failure Criteria of Materials

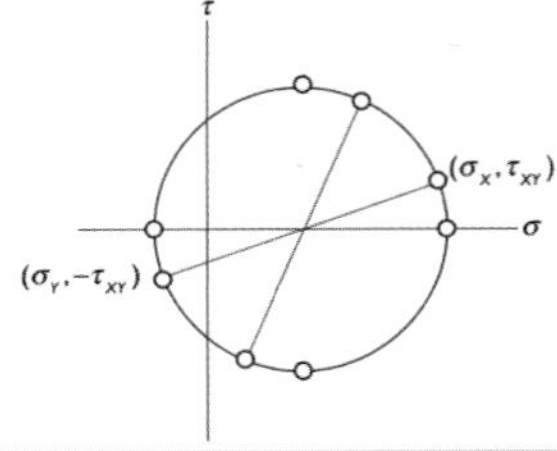

Achieving functionality, safety, and reliability are the main purposes of structural simulations. Deformation usually relates to the functionality, while stress to the safety and reliability. Which stress should we look into to ensure that a structure doesn't fail? Normal stress? Shear stress? Or something else? How can we say a stress value is too large? What critical values should the stresses be compared with? In short, what are the failure criteria of materials? This section intends to answer these questions.

1.4.1 Ductile versus Brittle Materials

[1] Stress-strain relation is the most important characteristic of a material. We usually obtain a stress-strain relation by conducting a uniaxial tensile test. Two examples of stress-strain relation from uniaxial tensile tests are shown in [2-3]. In [2], the material exhibits a large amount of strain before it fractures [4]; it is called a *ductile material*. In [3], the material's fracture strain is relatively small [5]; it is called a *brittle material*. Fracture strain is a measure of ductility. There are essential differences between these two types of materials.

Failure Points: Yield Point or Fracture Point?

Mild steel is a typical ductile material. For ductile materials, there often exists an obvious yield point [6], beyond which the deformation would be too large, and the material is no longer reliable or functional; the failure is accompanied by excess deformation. Therefore, for ductile materials, we are concerned about whether the material reaches the yield point [6]; its corresponding stress is called the yield stress σ_Y, which is the critical stress we want to compare with. But, with which stress to compare? σ_X? σ_Y? σ_Z? τ_{XY}? τ_{YZ}? τ_{ZX}? Or something else? 1.4.4 and 1.4.5 (pages 42-46) will answer this question.

Cast iron and ceramics are two examples of brittle materials. For a brittle material, there usually doesn't exist an obvious yield point, and we are concerned about its fracture point [5]; its corresponding stress is called the fracture stress σ_f, which is the critical stress we want to compare with. But, again, with which stress to compare? σ_X? σ_Y? σ_Z? τ_{XY}? τ_{YZ}? τ_{ZX}? Or something else? 1.4.3 (page 42) will answer this question.

Failure Modes: Tensile Failure or Shear Failure?

The fracture of brittle materials is mostly due to *tensile failure*; the yielding of ductile materials is mostly due to *shear failure*[Refs 1, 2]. The tensile failure of brittle materials is easy to observe: the failure always occurs after cracking, induced by tensile stresses. The shear failure of ductile materials can be observed in a standard uniaxial tensile test, in which the failure is accompanied by a necking phenomenon and a cone-shape breaking surface. It is important to remember that a material often fails due to a mix-up of these two mechanisms. ↓

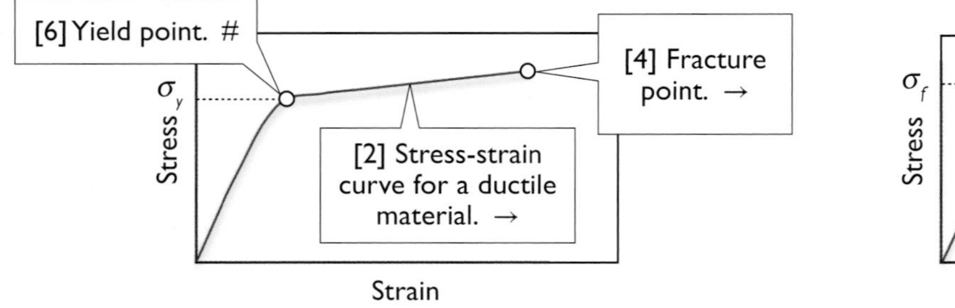

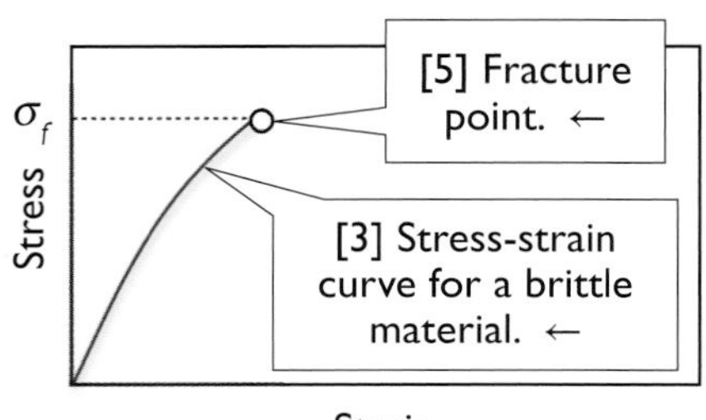

1.4.2 Principal Stresses

[1] We mentioned that, at a certain point, the stresses are different in different faces (or directions, 1.2.3[2], page 26). It then naturally raises a question: In which face does the normal stress reach its maximum? And in which face does the shear stress reach its maximum?

To present the concepts efficiently, consider a 2D case [2], and let's denote the X-direction the base direction. In the X-face, the stress can be expressed with a stress pair (σ_X, τ_{XY}). Likewise, in an arbitrary face, the stress can be expressed with a stress pair (σ, τ). Let's now try to find a relationship between normal stress σ and shear stress τ: how does σ vary with τ?

First, we mark the stress pair (σ_X, τ_{XY}) in the σ-τ space [3]. Second, noting that the stress $(\sigma_Y, -\tau_{XY})$ is also a stress pair, whose direction forms 90° (counter-clockwise) with the base direction, we mark the stress pair $(\sigma_Y, -\tau_{XY})$ in the σ-τ space [4]. Similarly, we could draw other stress pairs in the σ-τ space [5]. The collection of these points forms a circle in the σ-τ space [6], called a *Mohr's circle*. The details can be found in any textbook of Solid Mechanics[Refs 2, 3].

Here, we want to emphasize a concept: *a Mohr's circle represents a stress state*. With this useful concept, finding the maximum normal stress and maximum shear stress becomes straightforward. ↗

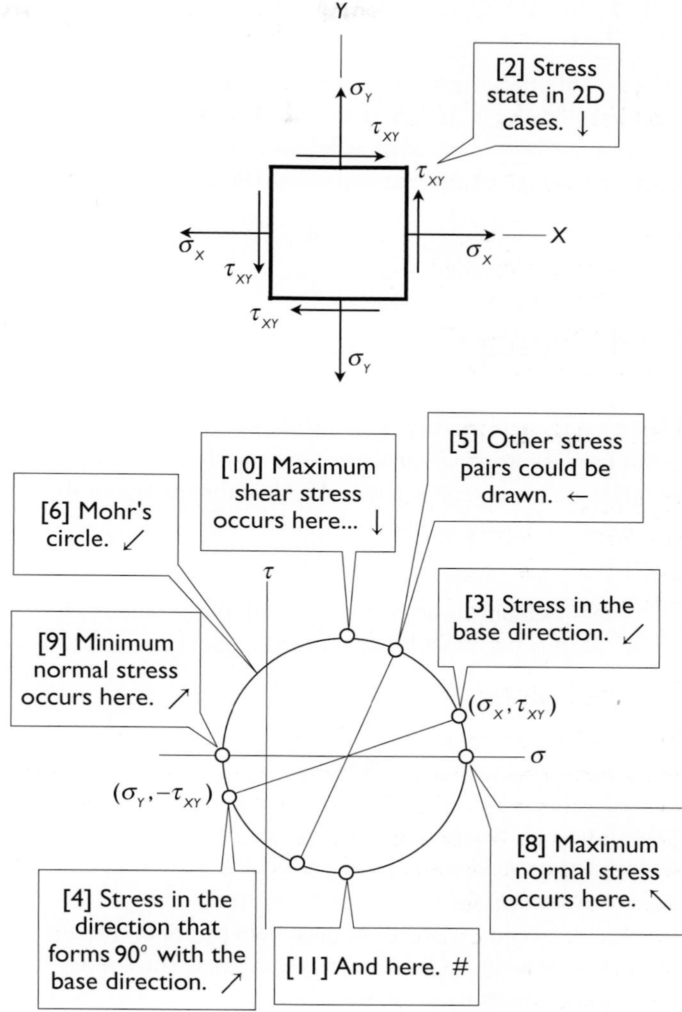

[7] A stress state defines a Mohr's circle, and vice versa. Further, the stress values, σ_X, σ_Y, and τ_{XY}, fully define a Mohr's circle. Once we have a Mohr's circle, the points of maximum normal stress, minimum normal stress, and maximum shear stress can be located. The maximum normal stress is located at the right quarter-point of the Mohr's circle [8]. The minimum normal stress is located at the left quarter-point of the Mohr's circle [9]. The *maximum shear stress* is located at either the upper quarter-point or the lower quarter-point of the Mohr's circle [10-11]. Note that at the points of maximum and minimum normal stresses [8-9], the shear stresses vanish.

The maximum normal stress [8] is called the *maximum principal stress* and is denoted by σ_1; The minimum normal stress [9] is called the *minimum principal stress* and is denoted by σ_3. Their corresponding directions are called *principal direction*. At a point of a 3D solid, there are three principal directions (and thus three principal stresses). The *medium principal stress* is denoted by σ_2. In our example, the maximum principal stress [8] is a positive value, a tension; the minimum principal stress [9] is a negative value, a compression.

Given three stress values, σ_X, σ_Y, and τ_{XY}, to define a Mohr's circle, we can easily calculate the values and their corresponding directions of the principal stresses and the maximum shear stress. We will not derive these formulas in this book; Workbench can report for you at your request. Finally, make sure you understand these concepts and can generalize the concepts to 3D cases. ↗

1.4.3 Failure Criterion for Brittle Materials

[1] As mentioned, the failure of brittle materials is a tensile failure (1.4.1[1], page 40). In other words, a brittle material fractures because its tensile stress reaches the fracture strength σ_f (1.4.1[5], page 40). Thus, we may state a failure criterion for brittle materials as follows: At a certain point of a body, if the maximum principal stress reaches the fracture strength of the material, it will fail. In short, a point of material fails if

$$\sigma_1 \geq \sigma_f \quad (1)$$

#

1.4.4 Tresca Criterion for Ductile Materials

[1] As mentioned, the failure of ductile materials is a shear failure (1.4.1[1], page 40). In other words, a ductile material yields because its shear stress reaches the shear strength τ_y. Thus, we may state a failure criterion for ductile materials as follows: At a certain point of a body, if the maximum shear stress reaches the shear strength of the material, it will fail. In short, a point of material fails if

$$\tau_{max} \geq \tau_y \quad (1)$$

The maximum shear stress τ_{max}, from the geometry of a Mohr's circle (1.4.2[6], page 41), is simply the radius of the circle. Noting that the diameter of the circle is $(\sigma_1 - \sigma_3)$, we may write down

$$\tau_{max} = \frac{\sigma_1 - \sigma_3}{2} \quad (2)$$

↙

[2] In a uniaxial tensile test, the material yields when undergoing its yield stress σ_y in the axial direction. Let the axial direction be X-direction, then the stress state $(\sigma_X, \sigma_Y, \tau_{XY})$ is $(\sigma_y, 0, 0)$ [3]; its maximum principal stress is σ_y while the minimum principal stress is zero. Thus, when the material yields, its shear stress is

$$\tau_y = \frac{\sigma_y}{2} \quad (3)$$

Substituting Eqs. (2-3) into (1), we have the criterion

$$\frac{\sigma_1 - \sigma_3}{2} \geq \frac{\sigma_y}{2} \quad (4)$$

→

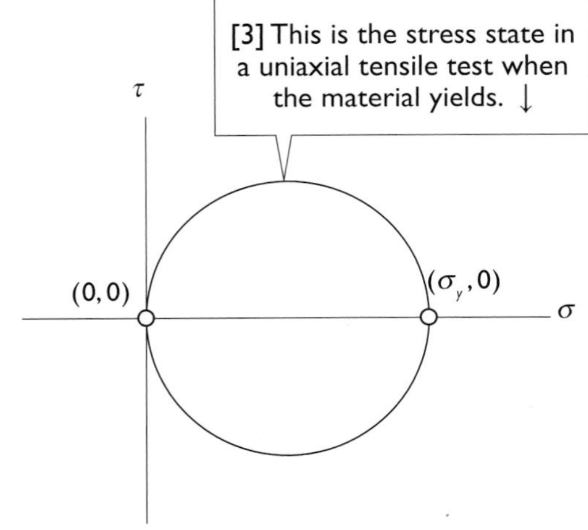

[4] Simplifying Eq. (4), we reach a conclusion that the material yields if

$$\sigma_1 - \sigma_3 \geq \sigma_y \quad (5)$$

In ANSYS, the quantity on the left-hand side, $(\sigma_1 - \sigma_3)$, is called the *stress intensity*, which the Workbench can report for you at your request. This failure criterion is called the *maximum shear stress criterion*, or *Tresca Criterion*, first proposed by Henri Tresca (1814-1885), a French mechanical engineer, in 1864. #

1.4.5 Von Mises Criterion for Ductile Materials

[1] The Tresca criterion (Eq. 1.4.4(5), last page) is simple but often not good enough for predicting the yielding of many ductile materials, particularly metals. The theory discussed in 1.4.4 (last page) is too simplified and may deviate from real-world situations. For example, the medium principal stress σ_2 plays no role in Eq. 1.4.4(5), but that is not always true. A more sophisticated theory, called the *von Mises criterion*, often predicts yielding more satisfactorily than Tresca criterion.

Hydrostatic Stress and Deviatoric Stress

Without loss of generality, a stress state

$$\{\sigma\} = \begin{Bmatrix} \sigma_X & \tau_{XY} & \tau_{XZ} \\ \tau_{YX} & \sigma_Y & \tau_{YZ} \\ \tau_{ZX} & \tau_{ZY} & \sigma_Z \end{Bmatrix}$$

can be expressed in a form of principal stresses if we choose the principal axes as coordinate axes,

$$\{\sigma\} = \begin{Bmatrix} \sigma_1 & 0 & 0 \\ 0 & \sigma_2 & 0 \\ 0 & 0 & \sigma_3 \end{Bmatrix} \tag{1}$$

Note that, as mentioned, in the principal directions, all the shear stresses vanish (1.4.2[7], page 41).

Define the hydrostatic stress as the average of the normal stresses,

$$p = \frac{\sigma_1 + \sigma_2 + \sigma_3}{3} \tag{2}$$

The stress state (1) can be decomposed into two parts

$$\{\sigma\} = \begin{Bmatrix} \sigma_1 & 0 & 0 \\ 0 & \sigma_2 & 0 \\ 0 & 0 & \sigma_3 \end{Bmatrix} = \begin{Bmatrix} p & 0 & 0 \\ 0 & p & 0 \\ 0 & 0 & p \end{Bmatrix} + \begin{Bmatrix} \sigma_1 - p & 0 & 0 \\ 0 & \sigma_2 - p & 0 \\ 0 & 0 & \sigma_3 - p \end{Bmatrix} \tag{3}$$

or, written in a more compact form,

$$\{\sigma\} = \{\sigma^p\} + \{\sigma^d\} \tag{4}$$

The first part $\{\sigma^p\}$ is called the *hydrostatic stress*, and the second part $\{\sigma^d\}$ is called the *deviatoric stress*, the stress deviating from the hydrostatic stress.

The deformation of a body can be thought of as a superposition of a dilation (volumetric change) and a distortion. The hydrostatic stress contributes exclusively to the dilation, while the deviatoric stress contributes exclusively to the distortion. The dilation plays no role in shear failure; it is the distortion that causes shear failure. In other words, the deviatoric stress

$$\{\sigma^d\} = \begin{Bmatrix} \sigma_1 - p & 0 & 0 \\ 0 & \sigma_2 - p & 0 \\ 0 & 0 & \sigma_3 - p \end{Bmatrix} \tag{5}$$

can be used to establish a shear failure criterion. However, it is not a scalar value. How can we use it to compare with a uniaxial yield stress σ_y? We need a more elaborate theory to derive a useful criterion. ↵

Von Mises Yield Criterion

[2] In 1913, Richard von Mises, an Austria-Hungary born scientist, proposed a theory for predicting the yielding of ductile materials. The theory states that the yielding occurs when the deviatoric strain energy density (or deviatoric energy for short) reaches a critical value.

For linearly elastic materials, the total strain energy *density* is half of the inner product of the stress and the strain,

$$w = \frac{1}{2}\{\sigma\}\cdot\{\varepsilon\} \tag{6}$$

This total strain energy can be decomposed into two parts: the energy w^p caused by the hydrostatic stress and the energy w^d caused by the deviatoric stress,

$$w = w^p + w^d \tag{7}$$

or

$$w^d = w - w^p \tag{8}$$

The von Mises criterion can be stated as follows: the yielding occurs when

$$w^d \ge w^{yd} \tag{9}$$

where w^{yd} is the deviatoric energy when the material yields in its uniaxial tension test. In the following discussion, we will express w^p and w^d in terms of stresses.

Deviatoric Energy w^{yd}

In a uniaxial tension test, when yielding occurs, the stress state (see 1.4.4[3], page 42) is

$$\begin{Bmatrix} \sigma_y & 0 & 0 \\ 0 & 0 & 0 \\ 0 & 0 & 0 \end{Bmatrix} = \begin{Bmatrix} \sigma_y/3 & 0 & 0 \\ 0 & \sigma_y/3 & 0 \\ 0 & 0 & \sigma_y/3 \end{Bmatrix} + \begin{Bmatrix} 2\sigma_y/3 & 0 & 0 \\ 0 & -\sigma_y/3 & 0 \\ 0 & 0 & -\sigma_y/3 \end{Bmatrix}$$

or, written in a more compact form,

$$\{\sigma^y\} = \{\sigma^{yp}\} + \{\sigma^{yd}\}$$

The first part $\{\sigma^{yp}\}$ is the hydrostatic stress and the second part $\{\sigma^{yd}\}$ is the deviatoric stress. Hooke's law, Eq. 1.2.8(1) (page 31), can be used to obtain the strains. The strains corresponding to the total stress $\{\sigma^y\}$ and the hydrostatic stress $\{\sigma^{yp}\}$ are respectively

$$\{\varepsilon^y\} = \frac{\sigma_y}{E}\begin{Bmatrix} 1 & 0 & 0 \\ 0 & -\nu & 0 \\ 0 & 0 & -\nu \end{Bmatrix} \text{ and } \{\varepsilon^{yp}\} = \frac{(1-2\nu)\sigma_y}{3E}\begin{Bmatrix} 1 & 0 & 0 \\ 0 & 1 & 0 \\ 0 & 0 & 1 \end{Bmatrix}$$

Calculated using Eq. (6), the energies corresponding to the total stress and the hydrostatic stress are respectively

$$w^y = \frac{1}{2}\{\sigma^y\}\cdot\{\varepsilon^y\} = \frac{\sigma_y^2}{2E} \text{ and } w^{yp} = \frac{1}{2}\{\sigma^{yp}\}\cdot\{\varepsilon^{yp}\} = \frac{(1-2\nu)\sigma_y^2}{6E}$$

By using Eq. (8), the deviatoric energy is

$$w^{yd} = w^y - w^{yp} = \frac{(1+\nu)\sigma_y^2}{3E} \tag{10}$$

↵

Deviatoric Energy w^d

[3] Now, we consider the general 3D stress state (see Eqs. (3-4), page 43). The strains corresponding to the total stress $\{\sigma\}$ and the hydrostatic stress $\{\sigma^p\}$ are, using Eq. 1.2.8(1) (page 31), respectively

$$\{\varepsilon\} = \frac{1}{E}\begin{Bmatrix} \sigma_1 - \nu(\sigma_2+\sigma_3) & 0 & 0 \\ 0 & \sigma_2 - \nu(\sigma_3+\sigma_1) & 0 \\ 0 & 0 & \sigma_3 - \nu(\sigma_1+\sigma_2) \end{Bmatrix} \text{ and } \{\varepsilon^p\} = \frac{(1-2\nu)p}{E}\begin{Bmatrix} 1 & 0 & 0 \\ 0 & 1 & 0 \\ 0 & 0 & 1 \end{Bmatrix}$$

Calculated using Eq. (6), the energies corresponding to the total stress $\{\sigma\}$ and the hydrostatic stress $\{\sigma^p\}$ are respectively

$$w = \frac{1}{2}\{\sigma\}\cdot\{\varepsilon\} = \frac{1}{2E}\left[\sigma_1(\sigma_1 - \nu\sigma_2 - \nu\sigma_3) + \sigma_2(\sigma_2 - \nu\sigma_3 - \nu\sigma_1) + \sigma_3(\sigma_3 - \nu\sigma_1 - \nu\sigma_2)\right]$$

and

$$w^p = \frac{1}{2}\{\sigma^p\}\cdot\{\varepsilon^p\} = \frac{3(1-2\nu)p^2}{2E}$$

From Eqs. (8, 2), the deviatoric energy is

$$\begin{aligned} w^d &= w - w^p \\ &= \frac{1}{2E}\left[\sigma_1(\sigma_1 - \nu\sigma_2 - \nu\sigma_3) + \sigma_2(\sigma_2 - \nu\sigma_3 - \nu\sigma_1) + \sigma_3(\sigma_3 - \nu\sigma_1 - \nu\sigma_2)\right] - \frac{3(1-2\nu)}{2E}\left(\frac{\sigma_1+\sigma_2+\sigma_3}{3}\right)^2 \end{aligned} \quad (11)$$

After some manipulations, the above equation can be simplified as

$$w^d = \frac{1+\nu}{6E}\left[(\sigma_1-\sigma_2)^2 + (\sigma_2-\sigma_3)^2 + (\sigma_3-\sigma_1)^2\right] \quad (12)$$

Derivation of Eq. (12)

$$\begin{aligned} w^d &= \frac{1}{2E}\left[\sigma_1(\sigma_1 - \nu\sigma_2 - \nu\sigma_3) + \sigma_2(\sigma_2 - \nu\sigma_3 - \nu\sigma_1) + \sigma_3(\sigma_3 - \nu\sigma_1 - \nu\sigma_2)\right] - \frac{3(1-2\nu)}{2E}\left(\frac{\sigma_1+\sigma_2+\sigma_3}{3}\right)^2 \\ &= \frac{1}{2E}\left[\sigma_1^2+\sigma_2^2+\sigma_3^2 - 2\nu(\sigma_1\sigma_2+\sigma_2\sigma_3+\sigma_3\sigma_1)\right] - \frac{1-2\nu}{6E}\left[\sigma_1^2+\sigma_2^2+\sigma_3^2 + 2(\sigma_1\sigma_2+\sigma_2\sigma_3+\sigma_3\sigma_1)\right] \\ &= \frac{1}{6E}\left[3(\sigma_1^2+\sigma_2^2+\sigma_3^2) - 6\nu(\sigma_1\sigma_2+\sigma_2\sigma_3+\sigma_3\sigma_1) - (1-2\nu)(\sigma_1^2+\sigma_2^2+\sigma_3^2) - (2-4\nu)(\sigma_1\sigma_2+\sigma_2\sigma_3+\sigma_3\sigma_1)\right] \\ &= \frac{1}{6E}\left[(2+2\nu)(\sigma_1^2+\sigma_2^2+\sigma_3^2) - (2+2\nu)(\sigma_1\sigma_2+\sigma_2\sigma_3+\sigma_3\sigma_1)\right] \\ &= \frac{1+\nu}{3E}\left[\sigma_1^2+\sigma_2^2+\sigma_3^2 - \sigma_1\sigma_2 - \sigma_2\sigma_3 - \sigma_3\sigma_1\right] \\ &= \frac{1+\nu}{6E}\left[(\sigma_1-\sigma_2)^2 + (\sigma_2-\sigma_3)^2 + (\sigma_3-\sigma_1)^2\right] \end{aligned}$$

↵

Von Mises Stress (Equivalent Stress)

[4] Substituting Eqs. (10, 12) into the von Mises Yield criterion, Eq. (9), we conclude that the material yields when

$$\frac{1+\nu}{6E}\left[(\sigma_1-\sigma_2)^2+(\sigma_2-\sigma_3)^2+(\sigma_3-\sigma_1)^2\right]\geq\frac{1+\nu}{3E}\sigma_y^2 \tag{13}$$

After simplification, we reach a more concise form

$$\sqrt{\frac{1}{2}\left[(\sigma_1-\sigma_2)^2+(\sigma_2-\sigma_3)^2+(\sigma_3-\sigma_1)^2\right]}\geq\sigma_y \tag{14}$$

The quantity on the left-hand-side is called *von Mises stress* or *effective stress* and is denoted by σ_e; in ANSYS, it is also called *equivalent stress*,

$$\sigma_e=\sqrt{\frac{1}{2}\left[(\sigma_1-\sigma_2)^2+(\sigma_2-\sigma_3)^2+(\sigma_3-\sigma_1)^2\right]} \tag{15}$$

To have more insight into Eq. (14), let's plot Eq. (14) in σ_1-σ_2-σ_3 space and consider only equal sign. It will be a cylindrical surface aligned with the axis $\sigma_1=\sigma_2=\sigma_3$ and with a radius of $\sqrt{2}\sigma_y$ (see 14.1.4[2], page 530). It is called the *von Mises yield surface*. Condition of Eq. (14) is equivalent to saying that the material fails when the stress state is on or outside the von Mises yield surface. When $\sigma_1=\sigma_2=\sigma_3$, the material is under hydrostatic pressure. It is the portion of stress that deviates from the axis $\sigma_1=\sigma_2=\sigma_3$ that contributes to the failure of the material.

Equivalent Strain

The *equivalent strain*, or von Mises strain, ε_e is defined by[Ref 4]

$$\varepsilon_e=\frac{1}{1+\nu'}\sqrt{\frac{1}{2}\left[(\varepsilon_1-\varepsilon_2)^2+(\varepsilon_2-\varepsilon_3)^2+(\varepsilon_3-\varepsilon_1)^2\right]} \tag{16}$$

Where ν', the effective Poisson's ratio, defaults to the Poisson's ratio of the material, 0.5, or 0, depending on various applications (for details, see Ref. 4). #

References

1. Haslach, H.W. Jr. and Armstrong, R.W., *Deformable Bodies and Their Material Behavior*, John Wiley & Sons, Inc., 2004; Chapter 9 Plasticity.
2. Cook, R. D. and Young, W. C., *Advanced Mechanics of Materials, Macmillan*, 1985.
3. Beer, F. P., Johnston, E. R. Jr., and DeWolf, J. T., *Mechanics of Materials*, 3rd Ed., McGraw Hill, 2002.
4. ANSYS Help//Mechanical APDL//Mechanical APDL Theory Reference//2.4.1. Combined Strains

Section 1.5

Review

1.5.1 Keywords (Part I)

Choose a letter for each keyword from the list of descriptions

1. ( ) APDL
2. ( ) Boundary Conditions
3. ( ) Brittle Materials
4. ( ) Buckling
5. ( ) Degenerated Element
6. ( ) Degree of Freedom
7. ( ) DesignModeler
8. ( ) Displacement
9. ( ) Ductile Materials
10. ( ) Dynamic Simulations
11. ( ) Engineering Data
12. ( ) Environment Conditions
13. ( ) Failure Criteria of Materials
14. ( ) Finite Element
15. ( ) Finite Element Mesh
16. ( ) Finite Element Model
17. ( ) First-Order Element
18. ( ) Free Boundaries
19. ( ) Governing Equations
20. ( ) Isotropic Elasticity

Answers:

1. (E) 2. (K) 3. (R) 4. (N) 5. (P) 6. (I) 7. (C) 8. (T) 9. (Q) 10. (S)
11. (A) 12. (L) 13. (O) 14. (B) 15. (G) 16. (H) 17. (J) 18. (M) 19. (D) 20. (F)

List of Descriptions

(A) An application of **Workbench GUI**. It can store material properties. Loads and boundary conditions also can be stored with the application.

(B) A small portion of a problem domain. Its geometry is so simple that the governing equations can be pre-formulated in terms of discrete nodal degrees of freedom.

(C) An application of the Workbench GUI. Its is similar to any other feature-based CAD software, except that it is specifically used to create geometric models for use in the ANSYS Workbench simulations.

(D) A set of equations governing the behavior of an engineering system (e.g., a structural system), usually in a form of differential equations.

(E) ANSYS Parametric Design Language. A set of text commands to drive ANSYS software.

(F) In **Engineering Data**, this term is used for materials whose stress-strain relation can be described by Hooke's law and characterized by two material parameters: Young's modulus and Poisson's ratio.

(G) A collection of elements and nodes.

(H) A finite element mesh plus its environment conditions.

(I) In the finite element methods, a term used for the discrete nodal values. In Workbench structural simulations, they are nodal displacements.

(J) Linear polynomials are used as shape functions; also called a linear element, or a lower-order element. Nodes are on the vertices of the element.

(K) Conditions applied on the boundaries of a finite element mesh.

(L) Conditions applied on the boundaries or interior of a finite element mesh.

(M) Boundaries with no boundary conditions specified. ANSYS assumes a zero pressure (stress normal to the surface) on a free boundary.

(N) When the compression in a structure is large enough such that its lateral stiffness vanishes, the structure becomes unstable.

(O) For a brittle material, it fails if the maximum principal stress reaches the fracture stress. For a ductile material, Tresca criterion or von Mises criterion may be used. The Tresca criterion states that the material fails if the stress intensity reaches the yield stress. The von Mises criterion states that the material fails if the von Mises stress reaches the yield stress.

(P) A 3D solid element, hexahedral in its natural shape, may combine some of its nodes to form a triangle-based prism, quadrilateral-based pyramid, or tetrahedron. A 2D solid element, quadrilateral in its natural shape, may combine some of its nodes to form a triangle.

(Q) The strain is large before it is stretched up to fracture. The fracture is mostly due to a shear failure. There is usually an obvious yield point in its stress-strain curve.

(R) The strain is small before it is stretched up to fracture. The fracture is mostly due to a tensile failure. There is usually no obvious yield point in its stress-strain curve.

(S) Structural simulations in which dynamic effects are included.

(T) In a deformed body, the vector connecting from its initial position to its final position.

1.5.2 Keywords (Part II)

Choose a letter for each keyword from the list of definitions

21. ( ) Linear Simulations
22. ( ) **Mechanical**
23. ( ) Modal Analysis
24. ( ) Node
25. ( ) Nonlinear Simulations
26. ( ) Procedure of Finite Element Method
27. ( ) Project Schematic
28. ( ) Principal Stress
29. ( ) Second-Order Element
30. ( ) Shape Functions
31. ( ) Stiffness
32. ( ) Stiffness Matrix
33. ( ) Strain
34. ( ) Strain State
35. ( ) Stress
36. ( ) Stress Intensity
37. ( ) Stress State
38. ( ) Stress Stiffening
39. ( ) Von Mises Stress
40. ( ) Workbench GUI

Answers:

21.(S) 22.(E) 23.(J) 24.(G) 25.(T) 26.(N) 27.(R) 28.(L) 29.(K) 30.(I)
31.(P) 32.(H) 33.(C) 34.(D) 35.(F) 36.(M) 37.(B) 38.(Q) 39.(O) 40.(A)

List of Definitions

(A) A gateway to ANSYS applications, including **Project Schematic**, **Engineering Data**, **DesignModeler**, **Mechanical**, **Design Exploration**, etc.

(B) A group of values describing the force intensity on the point in all directions of a point in a body. In 3D, three independent directions are needed to complete the description.

(C) A vector describing the stretch and twist in a direction in a point of a body. The vector can be decomposed into two components: one that is normal to the face and one that is parallel to the face.

(D) A group of values describing the stretch and twist in all directions of a point in a body. In 3D, three independent directions are needed to complete the description.

(E) An application of **Workbench GUI**. It performs structural and mechanical simulations, including meshing, setting up environment conditions, solving finite element models, and viewing results.

(F) The force per unit area acting on the boundary surfaces of an infinitesimally small body centered at a point of a body.

(G) Entities connecting elements. The elements share the same degrees of freedom values on the entities.

(H) The matrix that describes the linear relation between displacement vector and the force vector. Physical meaning of the ith column is the forces required on each of DOFs to maintain a unit displacement on the ith DOF and zero displacements on the other DOFs.

(I) In finite element methods, they are used as interpolating functions to calculate continuous displacement fields from discrete nodal displacements. Linear and quadratic polynomials are commonly used as the interpolating functions.

(J) Also called free vibration analysis. It is a special case of dynamic analysis, in which the structure is free of external forces. The results of the analysis include natural frequencies and their corresponding vibration modes.

(K) Quadratic polynomials are used as shape functions, also called a higher-order element. Nodes are on the vertices as well as on the middle of the edges.

(L) At a point in a body, different directions (faces) have different stress values. There exist directions in which the normal components are in their extremities and the shear components vanish; the directions and the corresponding normal stresses are called the principal directions and the principal stresses respectively. In 3D, there are 3 principal stresses; they are denoted as, in order starting from the largest, σ_1, σ_2, and σ_3 respectively. It is often used in a criterion for a brittle material: when the maximum principal stress is larger than the fracture strength, the material fails.

(M) Defined as the difference between the maximum principal stress and the minimum principal stress. It equals to twice the maximum shear stress. It is used in Tresca failure criterion for a ductile material: when it is larger than the yield strength, it is equivalent to saying that the shear stress is larger than the yielding shear strength and, thus, the material fails.

(N) (a) Calculate stiffness matrix for each element according to the element's geometry and its material properties. (b) Add up the element matrices to form a global stiffness matrix. The force vector is also calculated, according to the loading conditions. (c) Eq. 1.3.1(1) (page 35) is solved for the nodal displacements. (d) For each element, the displacement fields are calculated according to Eq. 1.3.2(2) (page 36); the strain fields are calculated according to Eq. 1.2.7(1) (page 31); the stress fields are calculated according to Eq. 1.2.8(3) (page 32).

(O) It is used in a failure criterion for a ductile material: when it is larger than the yield strength, it is equivalent to saying that the distortion strain energy density is larger than the yielding distortion strain energy density and, thus, the material fails.

(P) The forces required to deform the structure.

(Q) The phenomenon that, when a structure member is subject to a tensile stress, its lateral stiffness increases with the increase of the tensile stress.

(R) An application of **Workbench GUI**. Its function is to lay out simulation systems and their data flows.

(S) The responses of a system are linearly proportional to the loads.

(T) The responses of a system are not linearly proportional to the loads.

Section 1.6

Appendix: An Unofficial History of ANSYS

Shen-Yeh Chen, Ph. D.

About the Author

Dr. Shen-Yeh Chen earned his BS degree in civil engineering from National Chung Hsing University, Taiwan, in 1990, and his MS and Ph.D. degrees in structural mechanics from Arizona State University, USA, in December of 1997.

He began to work for Honeywell Engines and Systems in Phoenix, Arizona, in February of 1998. Shen-Yeh was a heavy user of ANSYS and LS-DYNA during his service at Honeywell. Soon he became a cross-department and cross-campus expert of ANSYS and LS-DYNA at Honeywell, and was very active on the internet (the XANSYS group) in the ANSYS internal users community. Dr. Chen published several popular ANSYS macros and articles during that time, and developed many in-house codes for Honeywell, including an optimizer to couple with ANSYS and LS-DYNA.

In August of 2002, Dr. Chen took a VP position offered from CADMEN, an ANSYS distributor in Taiwan. In 2005, Dr. Chen established his own company, FEA-Opt Technology (www.FEA-Optimization.com). About 80% of the revenue for FEA-Opt Technology is related to ANSYS and LS-DYNA. In 2006, he released a general purpose design optimization software, SmartDO. Today FEA-Opt Technology is well-known as a very successful CAE consulting firm in Taiwan, and its SmartDO is gaining more and more market share since 2006.

Dr. John Swanson holds B.S. and M.S. degrees in mechanical engineering from Cornell University. He holds a Ph.D. in applied mechanics from the University of Pittsburgh, obtained in night school with Westinghouse support.

In 1963, Dr. John Swanson worked at Westinghouse Astronuclear Labs in Pittsburgh, responsible for stress analysis of the components in NERVA nuclear reactor rockets. He used computer codes to model and predict transient stresses and displacements of the reactor system due to thermal and pressure loads. Swanson continued to develop 3D analysis, plate bending, nonlinear analysis for plasticity and creep, and transient dynamic analysis, in the next several years, using a finite element heat conduction program that was developed by Wilson at Aerojet. The old Westinghouse codes included a 2D/axisymmetric one also, possibly called FEATS (according to Kohnke). John wanted to combine these codes to remove the duplication, like equation solvers and some postprocessing.

Swanson believed an integrated, general-purpose FEA code could be used to do complex calculations that engineers typically did manually, such as heat transfer analysis. It would save money and time for Westinghouse and other companies.

Westinghouse didn't support the idea, and Swanson left the company in 1969. Before he left, he made sure that all code work had been sent to COSMIC, so that he could pick it up again from the outside.

Swanson Analysis Systems, Inc was incorporated in the middle of 1970 at Swanson's home. The offices were part of Swanson's home (there was no garage) in Pittsburgh. At the same time, Westinghouse realized that they needed John, so they hired him as a consultant. John said sure, but with the proviso that whatever he put into STASYS, the Westinghouse code, he could also put into ANSYS. Westinghouse had no trouble with this, as they just wanted to solve their problems. So this consulting kept bread on the table for the Swansons, and at the same time brought forth further improvements to ANSYS.

He developed his program using a keypunch and a time-shared mainframe at U.S. Steel. The first version of ANSYS was coded by the end of 1970, and the ANSYS program was first released soon after that. Westinghouse was the first customer, running as a data center. The data center was at the Telecommunications Center on Parkway East, on the east side of Pittsburgh. According to Dr. Swanson, the name ANSYS was used because the copyright lawyers assured Swanson that ANSYS was just a name, and did not stand for anything. This is understandable, because during that period all programs were "written" on punch card. When installing the program on the customer's computer, it meant carrying a relatively big case of punch cards to the customer's place, and feeding them into the machine.

Dr. Peter Kohnke met John Swanson first about early 1971. Swanson offered Peter a job in the fall of '71, but Peter did not accept. At that time Peter was a brand new father, and Westinghouse looked a lot more secure than SASI. Peter told John he was interested and would accept in the spring of '72, but then he had hired Gabe DeSalvo, and did not have the resources to hire Peter also. But John was finally able to hire Peter in the fall. Dr. Peter Kohnke's start date was 1/1/73.

When Peter started work, he asked what John wanted him to do. John told Peter that he developed code, did technical support, wrote manuals, gave seminars and did systems work so that the program would run on a variety of systems. John said he needed relief, so Peter should pick one or two of them. Ultimately, Peter did all of them, except systems work.

In around 1970, users ran ANSYS 2.x on a CDC 6600 machine over the Cybernet timesharing network. At the time, only fixed format input was available. The users would work up the input listing off-line, key it onto a tape cassette, log on, submit the run about quitting time for the best computer rates and stop by the CDC data center next morning to find out what went wrong. In 1973, ANSYS ran on three kinds of hardware: CDC, Univac, and IBM. And around 1973 the USS mainframe that they used to develop code was the US Steel CDC 6500.

The first minicomputer that ANSYS ran on was a MODCOMP 4 (or IV?). VAX came later. Being a small company, everyone did everything. When a "mini" computer was delivered, everyone helped wrestle it off of the truck. When printout paper was delivered, everyone helped unload the boxes.

In 1975, MITS began to build and sell the first PC ever in human history, the Altair. That, of course, did not have anything to do with ANSYS yet. The so-called PC was just a few switches and lights on the front board, and input had to be done in a binary fashion (no keyboard and monitor, of course). What was worse was that you had to assemble it by yourself, and it usually didn't work. Although Altair was rather popular, nobody really knew what to do with this machine. One former customer said that the most popular activity on Altair was to figure out what to do with this machine. At the same time, Microsoft built the BASIC language for Altair.

In 1977, Apple I was born.

In around 1979, Revision 3.0, ANSYS ran on a VAX 11-780 minicomputer. ANSYS evolved from fixed format input to purely command line driven and monocolor (green) on a Tektronix 4010 or 4014 vector graphics monitor. For a decent size model, the hidden lines plots could take 20-30 minutes. All of the nodes and elements were created separately without the benefit of importing CAD geometry. NGEN, EGEN, RPnnn, were used extensively. There was a geometry preprocessor, PREP7.

In 1980, we had Apple II.

In around 1980, John Swanson bought a Radio Shack TRS-80 machine, and planned to build a commercial version on it. However, later John returned the machine because Radio Shack left out (a socket for) a floating point processor. John decided that Finite Element Analysis probably should utilize a floating point processor, so he got his money back for that one.

Also around 1980, Rev 4 on a VAX 11-780 system was great, according to some old users. The chasm between batch and interactive running pretty much disappeared and file management was a very easy thing. No more element hard coding, the post processing got hugely better and you could mix batch and interactive running as you saw fit. Big dynamic transient runs or substructuring over night, post-processing and plotting next morning. Emag capabilities were first introduced at Rev 4.1.

Also in 1980, Microsoft signed a contract with IBM to provide the OS (PC DOS) for its up coming PC. This OS, however, was not created by Microsoft. Microsoft bought it from an engineer for 50K USD; and it was named the QDOS - the Quick and Dirty Operation System.

In 1981, IBM PC was born. This computer was created using the off the shelf technology, and an open architecture. The original reasons were to push the product to the market ASAP, so that IBM could catch up with the PC market. However, the BIOS was proprietary. Later Compaq reverse-engineered the BIOS and created a fully IBM PC compatible BIOS. This ignited the PC cloning market and war. The booming of the PC market directly changed the meaning of computing. The price of a PC dropped 30% in one month. And, it was the booming of the cloned IBM PC that really brought money into Microsoft.

In 1984, the revolutionary Macintosh was born. Macintosh was far more advanced then the IBM PC family at that time. The concept of GUI in the OS level and WYSIWYG was not possible on IBM PC until almost one decade later. However, the market of Macintosh did not pick up very soon, which caused Steve Jobs to leave Apple computer.

However, later the sales of Macintosh began to take off, which proved that Steve Jobs' vision had all been right. Macintosh saved Apple, and was directly responsible for the Apple phenomena.

A PC version of ANSYS was also available at around version 4.0 in about 1984. It was running on an Intel 286, with interactive command line input and limited graphics on the screens, like elements and nodes. No Motif GUI yet. In the first release of ANSYS on PCs, preprocessing, solution and post processing were performed in separate programs.

"Design Optimization" was introduced on Rev 4.2 (1985). This is also the release at which "Macro length is no longer limited to 400 characters."

FLOTRAN started as a graduate (PhD) project by Rita J. Schnipke at the University of Virginia circa 1986. After grad school Rita started (or helped start) Compuflo which was later sold to ANSYS in 1992. Rita later started her own shop which is in Charlottesville, VA, called Blue Ridge Numerics. They make CFDesign, a finite element based CFD code (www.cfdesign.com).

In 1988 at an ANSYS conference in California, IBM was there pushing their first unix machine, the "RT". It was slow. They asked Dr. Swanson if he would make a comment on it. He said "RT" must stand for Real Turkey.

SASI first started working with Compuflo (FLOTRAN) in 1989. In ANSYS Rev 5.0 and FLOTRAN V2.1A, SASI had what they called a "seamless interface" between the two programs (1993). FLOTRAN was "fully integrated" into ANSYS in Rev 5.1 (1994).

In 1993, Version 5.0 was released. And later the version 5.1 had a Motif GUI, which would have few changes to its layout up to 6.0.

Swanson Analysis Systems, Inc., was sold to TA Associates in 1994. The new company name, ANSYS, Inc., was announced at AUTOFACT '94 in Detroit.

According to many different people in the old SASI, John Swanson treated the people there pretty well. In contrast to the old "sandwich" jokes, John never passed up a chance to go to a restaurant. Indeed, for many years, John invited the entire staff to a restaurant regularly for the staff meetings.

At one time, Kohnke told Swanson that he ran the company like a benevolent dictatorship, and later Swanson told Kohnke that he liked that characterization.

Many people have said that John Swanson had an amazing overall understanding as well as detailed knowledge of the ANSYS code. Kohnke told a small story in an email to the author: "Sometime in the late '70s, a bug came my way. I wrestled with it for maybe half a day without making real progress. Then I went to John's office to ask him if he had any ideas. After I explained the bug, John thought about it for about 3 seconds (literally!) and said: 'Didn't you make a change in XXX about 6 months ago that would have a bearing on this?' In a nutshell, he was correct and I was then able to resolve the bug! John's knowledge and understanding of the code was always amazing to me."

In 1995, Windows 95 was published. Windows 95 was an important milestone for Microsoft. It bridged between the old DOS OS and the new NT technology. The birth of Windows 95 finally made it more and more acceptable for the engineering community to use the PC as a heavy duty calculation machine like a workstation.

In 1996, ANSYS 5.3 was published, with support for LS-DYNA. The feature of ANSYS/LS-DYNA in ANSYS 5.3 was still in the beginning stage.

On June 20, 1996, ANSYS Inc. common stock began trading on Nasdaq under ANSS after being a privately held company for 26 years. The IPO generated more than $41 million.

1998, ANSYS began to ship ANSYS/ed to university labs and paper reviewers. One of the copies arrived at the Structures Lab of Civil Engineering Department in Arizona State University, and that was the first time the author knew about ANSYS.

In the same year, on ANSYS's Annual report, it said "John is retiring from his direct role at ANSYS Inc., but will continue his association as a key consultant, mentoring all of us for many years to come."

On August 31, 2000, ANSYS acquired ICEM CFD.

January 2001, ANSYS announced the release of CADfix (International TechneGroup Incorporated) for ANSYS version 5.6.2 and 5.7. CADfix was to address the issue of importing CAD models into ANSYS with automatic geometric data repair.

In November 2001, ANSYS acquired CADOE S.A., an independent software vendor that specializes in the CAD/CAE market. In the same month, ANSYS announced a strategic OEM partnership with SAS LLC, a provider of NASTRAN simulation software and services. The alliance was focused on the joint development of a new NASTRAN computer-aided engineering solution that will be distributed exclusively by ANSYS Inc.

In November 2001, ANSYS announced AI*Environment. AI*Environment combines ICEM CFD Engineering's pre- and post-processor technologies.

In December 2001, ANSYS 6.0 was released. In this version, the Sparse solver was greatly improved. Efficient and reliable large scale model analysis (say, 1M DOF) finally became practical. The graphics screen of ANSYS was also painted blue in 6.0, which turned out to be a great disappointment to a lot of users.

In April of 2002, ANSYS 6.1 was released. The familiar Motif GUI was replaced by a Tcl/tk developed interface. It runs on 64-bit Intel Itanium architecture with Windows XP.

In February 26, 2003, ANSYS acquired CFX. ANSYS also announced that the functionality of Flotran would be "capped" at 8.1. That is, there will be no more development of Flotran after 8.1. Except for the Multiphysics platform, Flotran will be replaced by CFX.

In March 2004, ANSYS announces ParaMesh 2.3.

In May 2004, ANSYS 8.1 and CFX 5.7 was released. As previously mentioned, there will be no more revision of Flotran after 8.1.

In June 2004, ICEM CFD 5.0 was released.

On Jan 5, 2005, ANSYS announced that it acquired Century Dynamics. Century Dynamics' main product, AUTODYN, includes computational structural dynamics finite element solvers (FE), finite volume solvers for fluid dynamics (CFD), mesh-free particle solvers for high velocity, large deformation and fragmentation problems (SPH), and multi-solver coupling for multiphysics solutions including coupling between FE, CFD and SPH methods.

On February 16, 2006, ANSYS signed a definitive agreement to acquire Fluent.

As of March of 2012, Dr. Peter Kohnke is still working for ANSYS, but has cut back to 4 days a week. He turned 71 in December of 2012.

Dr. John Swanson is officially retired and living in The Villages, in Florida. But he still programs for ANSYS under contract on a varying schedule, on projects such as High performance mesh interpolation, Symmetric Multi Processing, 64 bit conversions, and APDL enhancements.

John is currently on the Board of Trustees of the University of Pittsburgh and the ASME Foundation and served two six-year terms as a Trustee of Washington and Jefferson College. He is a member of the Engineering College Council at Cornell University. His support of colleges and universities includes the donation of research laboratories to the Engineering Schools at Cornell, the University of Pittsburgh and (with Janet, his wife) the Veterinary School at Cornell. He gave the naming gift for the John A. Swanson Science Center at Washington and Jefferson College. The John A. Swanson School of Engineering at the University of Pittsburgh is named in his honor. Swanson recently invested in Applied Quantum Technology (AQT), a California startup company with the objective of reducing the cost of PV Solar Power by another factor of two. He serves on the AQT Board of Directors.

What is a data center?

I actually heard the word "data center" for the first time from Dr. Kohnke. For my age, I am more familiar with Workstation and PC and Xbox and PSP and so on. I used VMS when I was in college. And I remember later soon the whole room was occupied by PC and Macs. So I asked Dr. Kohnke what is a data center anyway? And this is what he told me.

Long ago (in the computer age), computers were relatively rare (and expensive). So, if you invested in one, you wanted to get the maximum use of it. You would run it day and night. This is unlike our PCs now (like our cars), which frankly just sit there most of the time, waiting for our command. Before it was very much a shared and continually used resource.

So, ANSYS got its start in that environment. Users at Westinghouse wanted to run ANSYS, so Westinghouse made a contract with SASI, where SASI would supply the code and support, and Westinghouse would pay to SASI so much per unit of time royalty for the time that the computer was actually being used to run ANSYS. As a result, many users were using the same machine (one after the other--parallel did not exist then). This was called a "data center."

The next step was external users. Knowing that even as a large company you might not need your computer full time, you would have your sales people out there selling time on it, often only at night, to external users.

Other companies sprang up that had no user base of their own, but only a computer(s), sales people, and external customers. The better ones also offered their own very good technical support. These were also called data centers.

And of course the whole concept of a data center disappeared as computers got cheaper and faster.

The "lease" jumped to the concept that the clock was not important; this "new" contract had SASI being paid so much per month, regardless of actual usage. This is closer to how things are now.

Acknowledgement

The author wants to acknowledge the help from many engineers and scientists in the XANSYS internet group. Some of the former employees of ANSYS also contributed greatly to this article, and many of them prefer not to be named. I also received emails from different people, and I usually tried to verify before I used them. Although I am trying to keep all the statements as accurate as possible, I really cannot guarantee the correctness of any information in this article.

Many of us, including the author in the XANSYS group, especially want to thank Dr. John Swanson, who invented ANSYS, and changed the life of many engineers forever in certain ways.

I have lived in the States for a total of 10 years. I lived there, was educated there, married there, had my son there, bought my first new car there, and had my first house there. It totally changed my life, my thinking (and my head) and everything. And I have to say most of my financial support was built on the existence of ANSYS and another program: LS-DYNA.

This article is also available on the web site www.FEA-Optimization.com. Anyone is welcome to distribute this article any way he or she wants, as long as the original article remains unchanged. Comments and suggestions should be forwarded to the authors directly. I will be glad to update this file continuously.

Shen-Yeh Chen

Chapter 2
Sketching

A 3D geometry can be viewed as a series of adding/removing material of simple solid bodies. Each solid body is often created by first drawing a 2D sketch, called a profile, and then extruding/revolving/sweeping the profile to generate the 3D solid body.

Purpose of This Chapter

This chapter provides exercises for the students so that they know how to draw 2D sketches using an ANSYS Workbench's geometry editor, DesignModeler. The profiles of several mechanical parts will be sketched in this chapter, and each sketch is then used to generate a mechanical part using a 3D modeling tool such as **Extrude** or **Revolve**. The use of these 3D modeling tools is trivial so that we may focus on 2D sketching techniques. More sophisticated use of 3D modeling tools will be introduced in Chapter 4.

About Each Section

Each mechanical part will be completed in a section. Section 2.1 sketches a cross section of W16x50; the cross section is then extruded to become a 3D beam. Section 2.2 sketches a triangular plate; the sketch is then extruded to become a 3D plate. Section 2.3 does not provide a hands-on case; rather, it overviews the sketching tools in a systematic way, attempting to complement what was missed in the first two sections. Sections 2.4, 2.5, and 2.6 provide three additional exercises, in which we purposely leave out some steps for the students to figure out the details themselves.

Section 2.1

W16x50 Beam

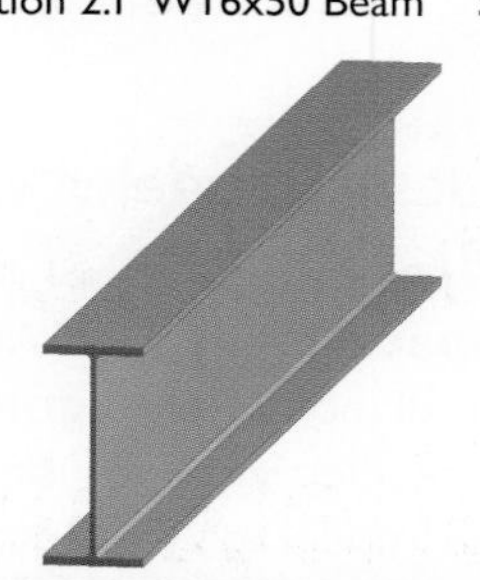

2.1.1 About the W16x50 Beam

[1] In this section, we will create a W16x50 steel beam (see [2-5]). The beam has a length of 10 ft. ↙

[2] Wide-flange I-shape section. →

[3] Nominal depth 16 in. →

[4] Weight 50 lb/ft. →

W16x50

[5] W16x50 cross section. #

2.1.2 Start Up DesignModeler

[1] Launch Workbench. ↗

[2] **Workbench GUI** appears. ↙

[3] Expand **Component Systems** by clicking the button with a plus sign; the plus sign becomes a minus sign (-). →

[4] Double-click **Geometry** to create a **Geometry** system in the **Project Schematic** window. If **Geometry** is not visible, see [5]. ↑

[5] If **Geometry** is not visible, scroll down to reveal it. ↑

[6] Right-click the **Geometry** cell and select **New DesignModeler Geometry...** to start up DesignModeler (see 1.1.4[1], page 13), a geometry modeler. ↲

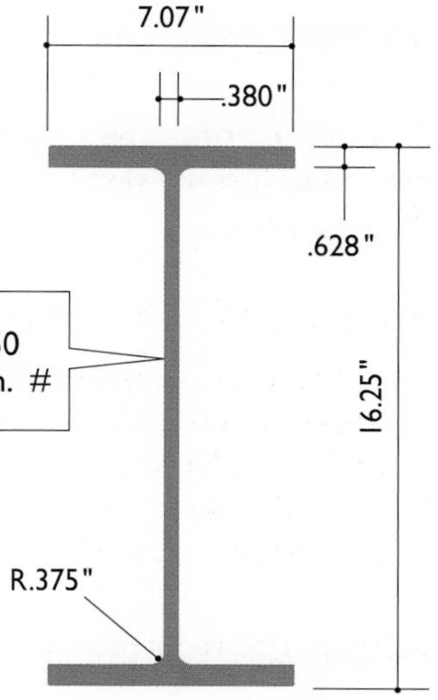

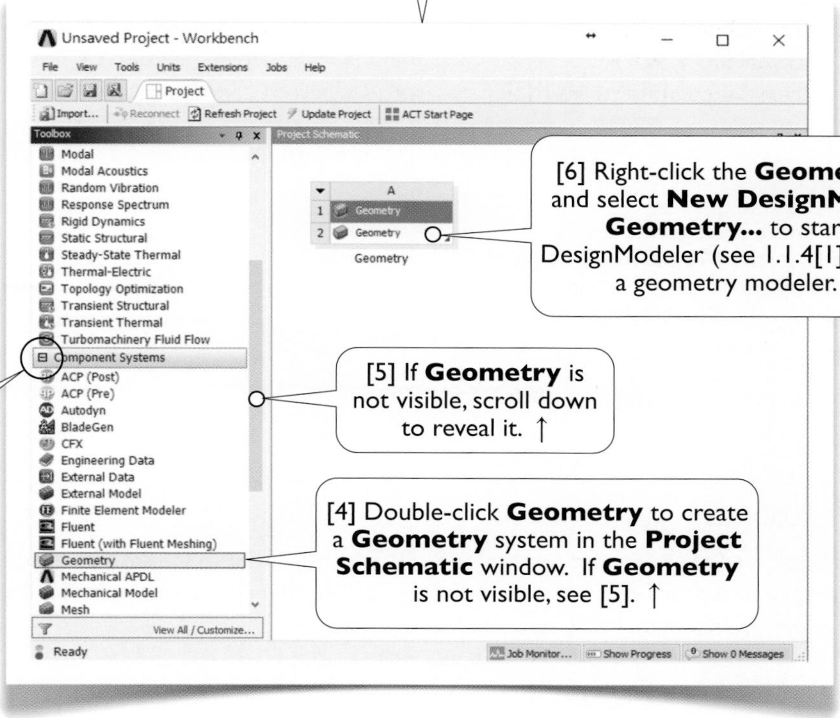

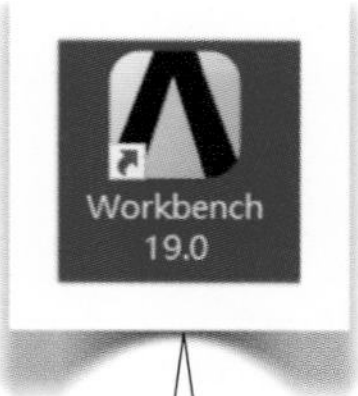

DesignModeler vs. SpaceClaim

[7] As mentioned in 1.1.4[2], page 13, Workbench provides two geometric modelers: **DesignModeler** and **SpaceClaim**. Until ANSYS 15, **DesignModeler** was the only modeler provided by Workbench. For simple and small models, **DesignModeler** serves well enough; but for complicated and large models, the engineers often create a geometric model using a CAD software such as SOLIDWORKS, PTC Creo, Autodesk Inventor, etc., and then import the model to Workbench. In ANSYS 16 and 17, **SpaceClaim** was included in Workbench as an alternative modeler, and **DesignModeler** remained as the default modeler. Starting from ANSYS 18, **SpaceClaim** becomes the default modeler, and **DesignModeler** serves as an alternative modeler.

In this book, since all the geometric models are simple and small, we will use **DesignModeler** to create the geometric models. Remember, the focus of this book is the finite element simulations, not the geometric modeling. ↙

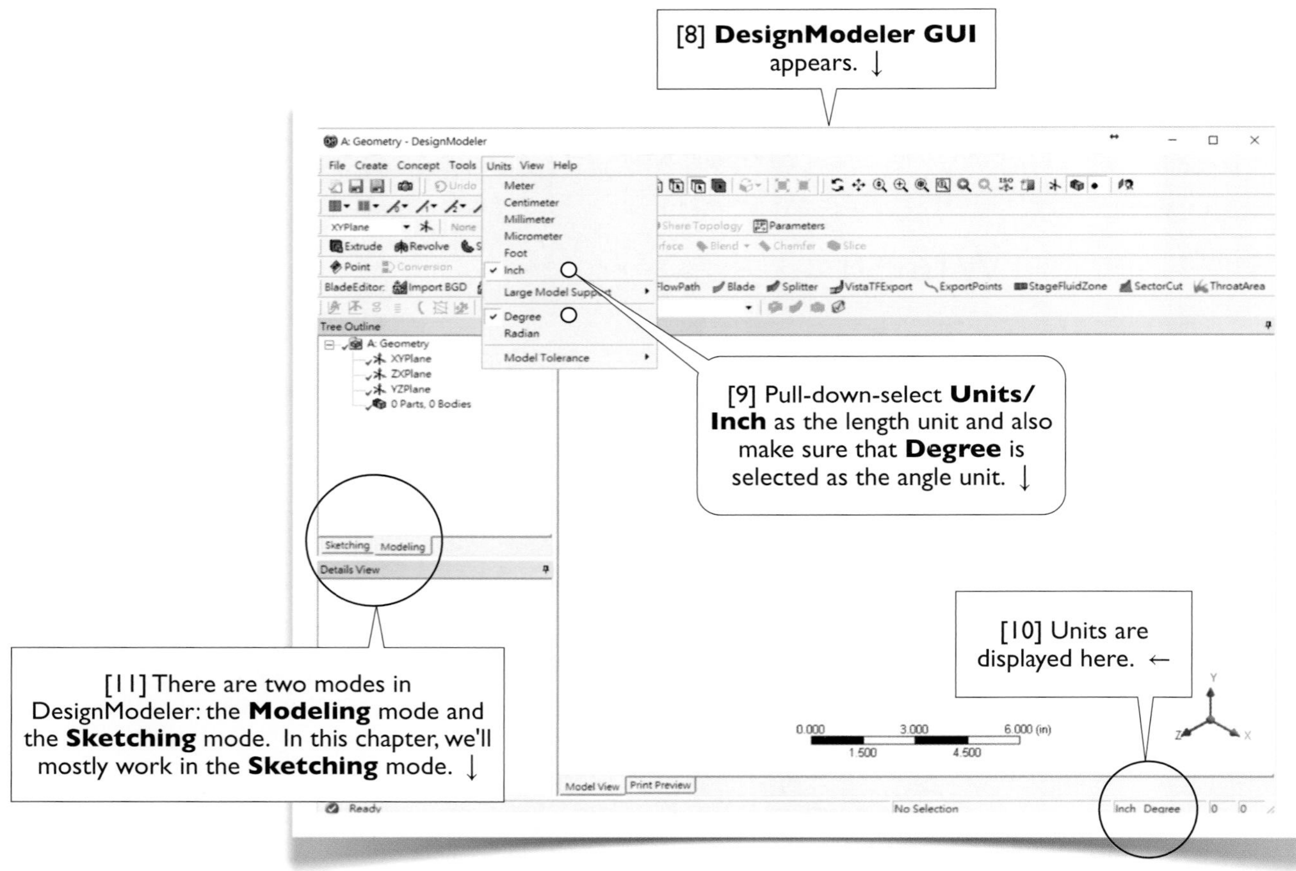

About Textboxes

[12] In this book, a round-cornered textbox (e.g., [1, 3-6, 9]) is used to indicate that mouse or keyboard ACTIONS are needed in that step. A sharp-cornered textbox (e.g., [2, 7-8, 10-11]) is used for commentary only; no mouse or keyboard actions are needed in that step. #

2.1.3 Draw a Rectangle on **XYPlane**

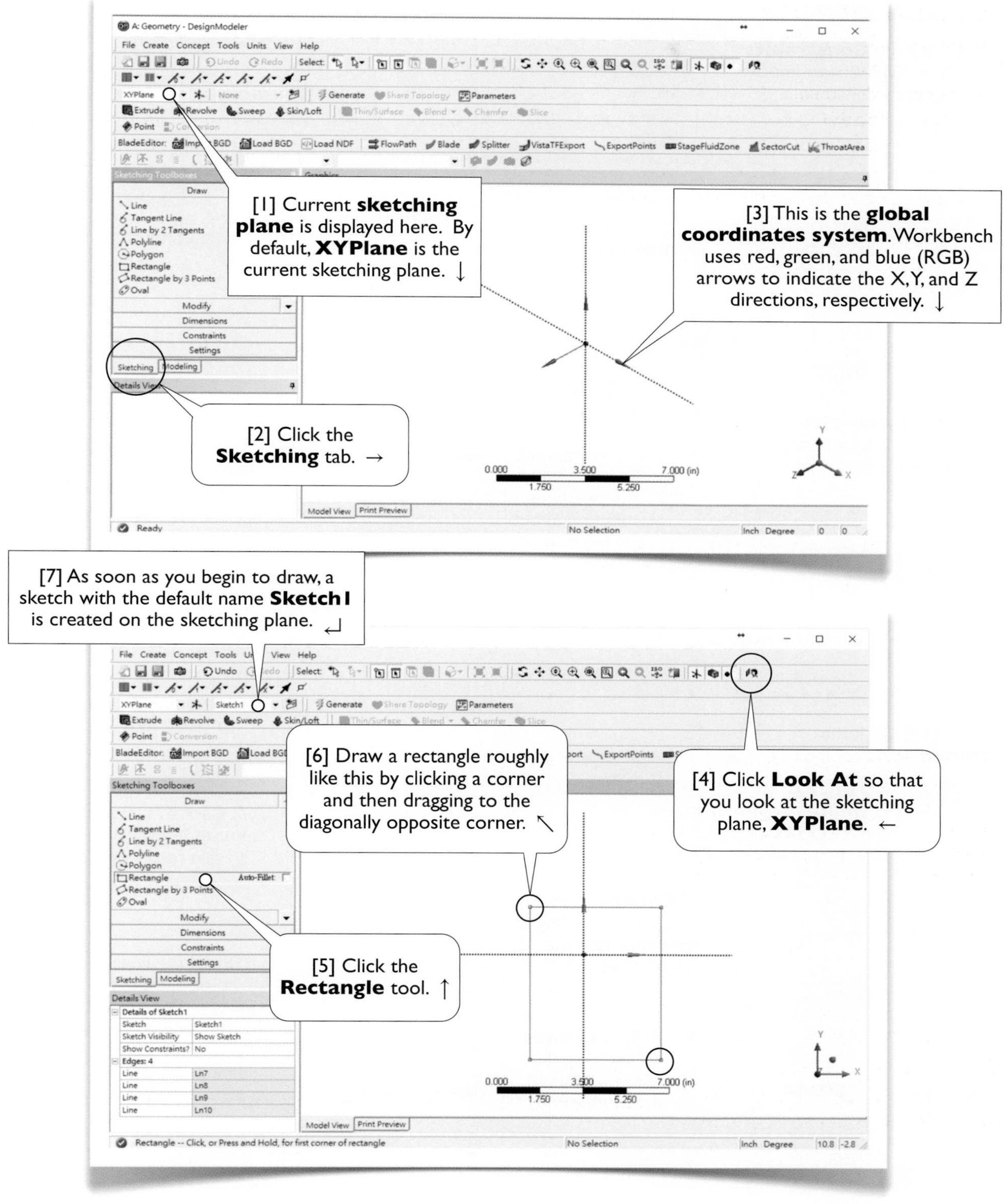

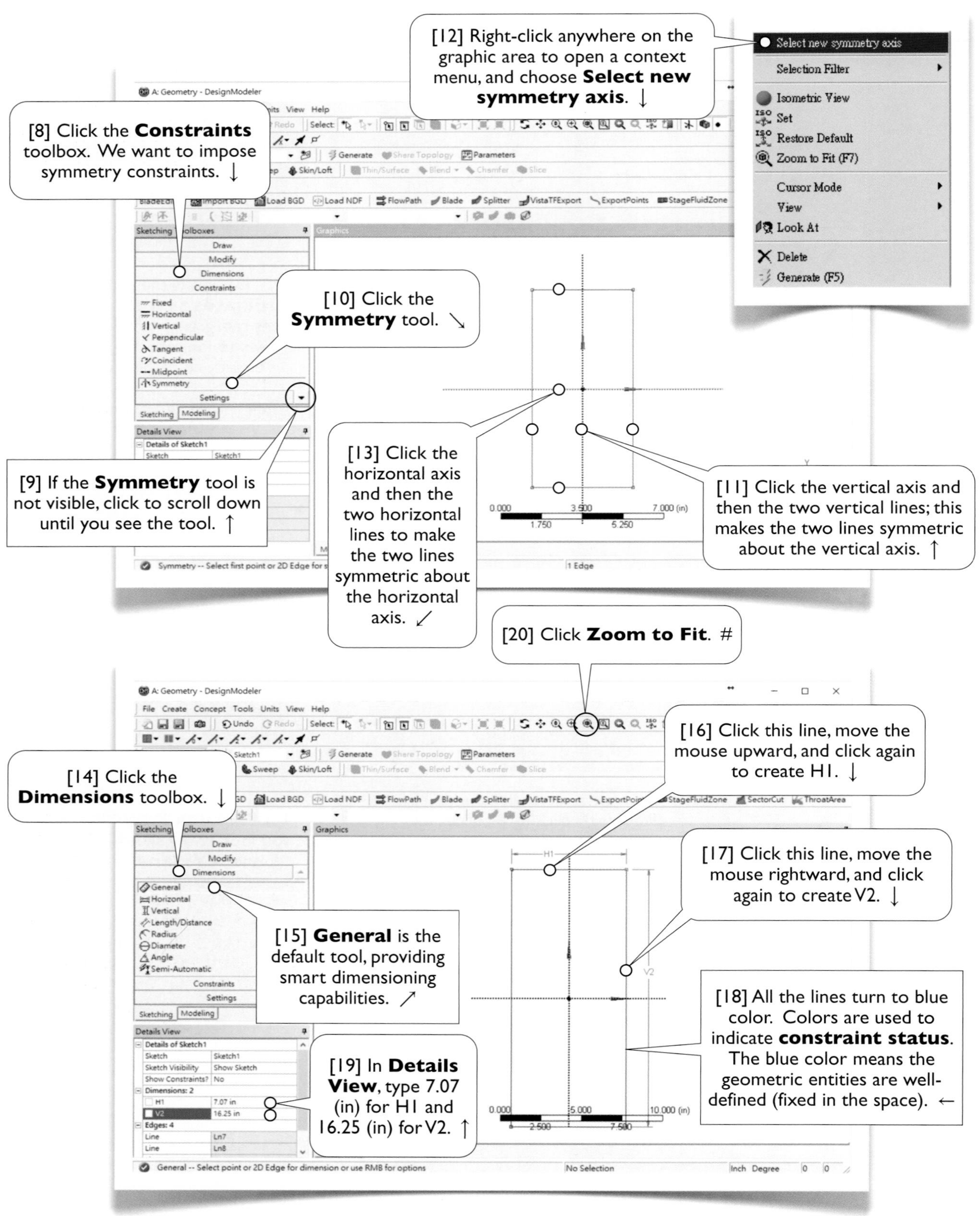

[12] Right-click anywhere on the graphic area to open a context menu, and choose **Select new symmetry axis**. ↓
Select new symmetry axis
Selection Filter
Isometric View
Set
Restore Default
Zoom to Fit (F7)
Cursor Mode
View
Look At
Delete
Generate (F5)
[8] Click the **Constraints** toolbox. We want to impose symmetry constraints. ↓
[10] Click the **Symmetry** tool. ↘
[9] If the **Symmetry** tool is not visible, click to scroll down until you see the tool. ↑
[13] Click the horizontal axis and then the two horizontal lines to make the two lines symmetric about the horizontal axis. ↙
[11] Click the vertical axis and then the two vertical lines; this makes the two lines symmetric about the vertical axis. ↑
[20] Click **Zoom to Fit**. #
[14] Click the **Dimensions** toolbox. ↓
[16] Click this line, move the mouse upward, and click again to create H1. ↓
[17] Click this line, move the mouse rightward, and click again to create V2. ↓
[15] **General** is the default tool, providing smart dimensioning capabilities. ↗
[18] All the lines turn to blue color. Colors are used to indicate **constraint status**. The blue color means the geometric entities are well-defined (fixed in the space). ←
[19] In **Details View**, type 7.07 (in) for H1 and 16.25 (in) for V2. ↑

2.1.4 Set Up Sketching Options

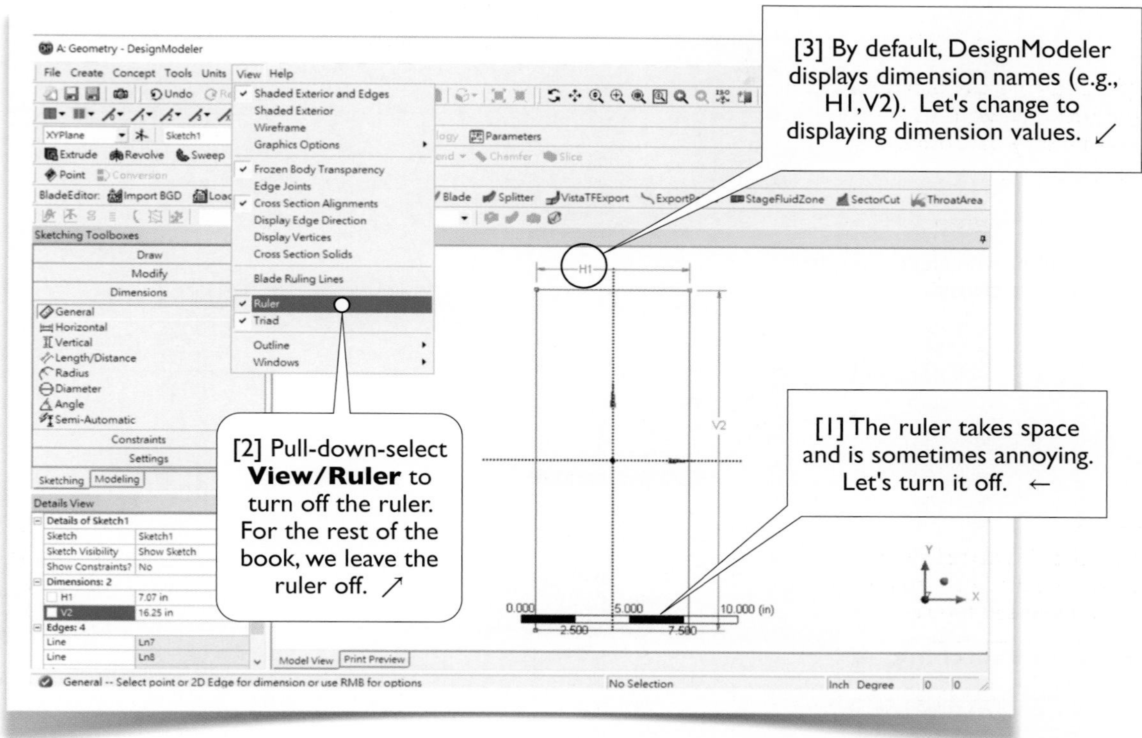

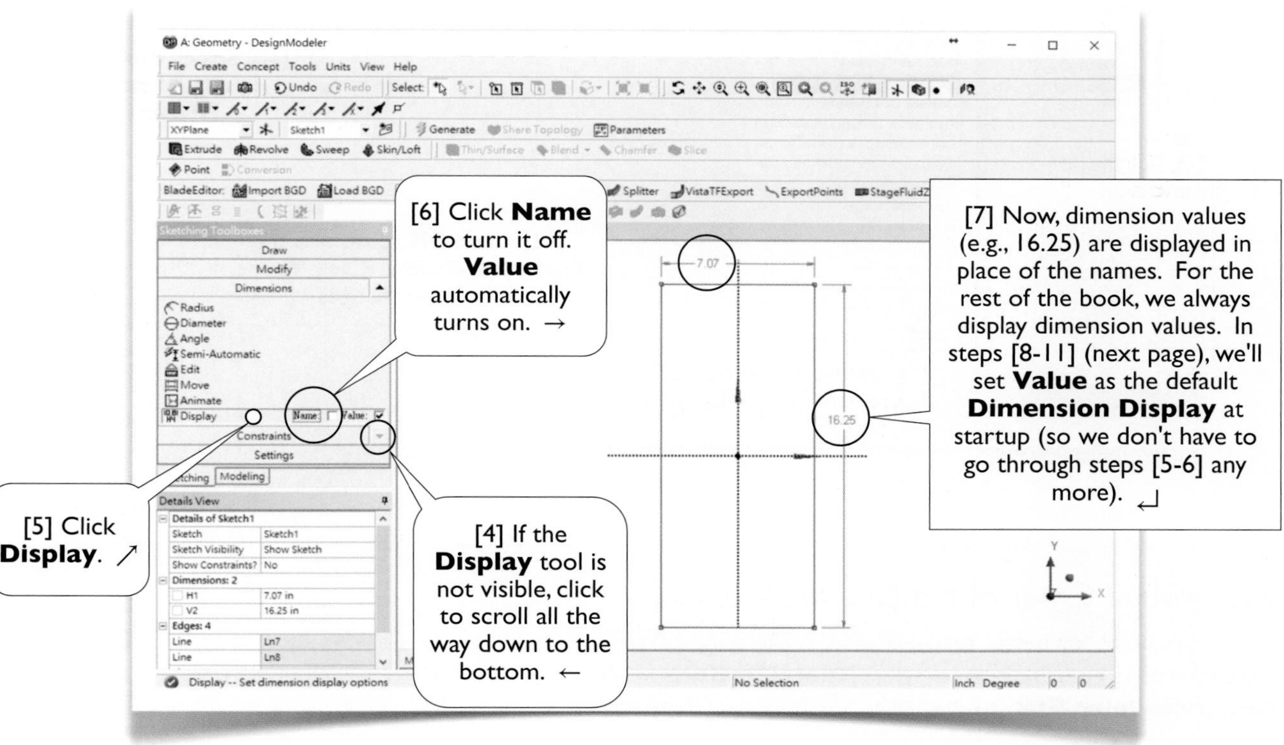

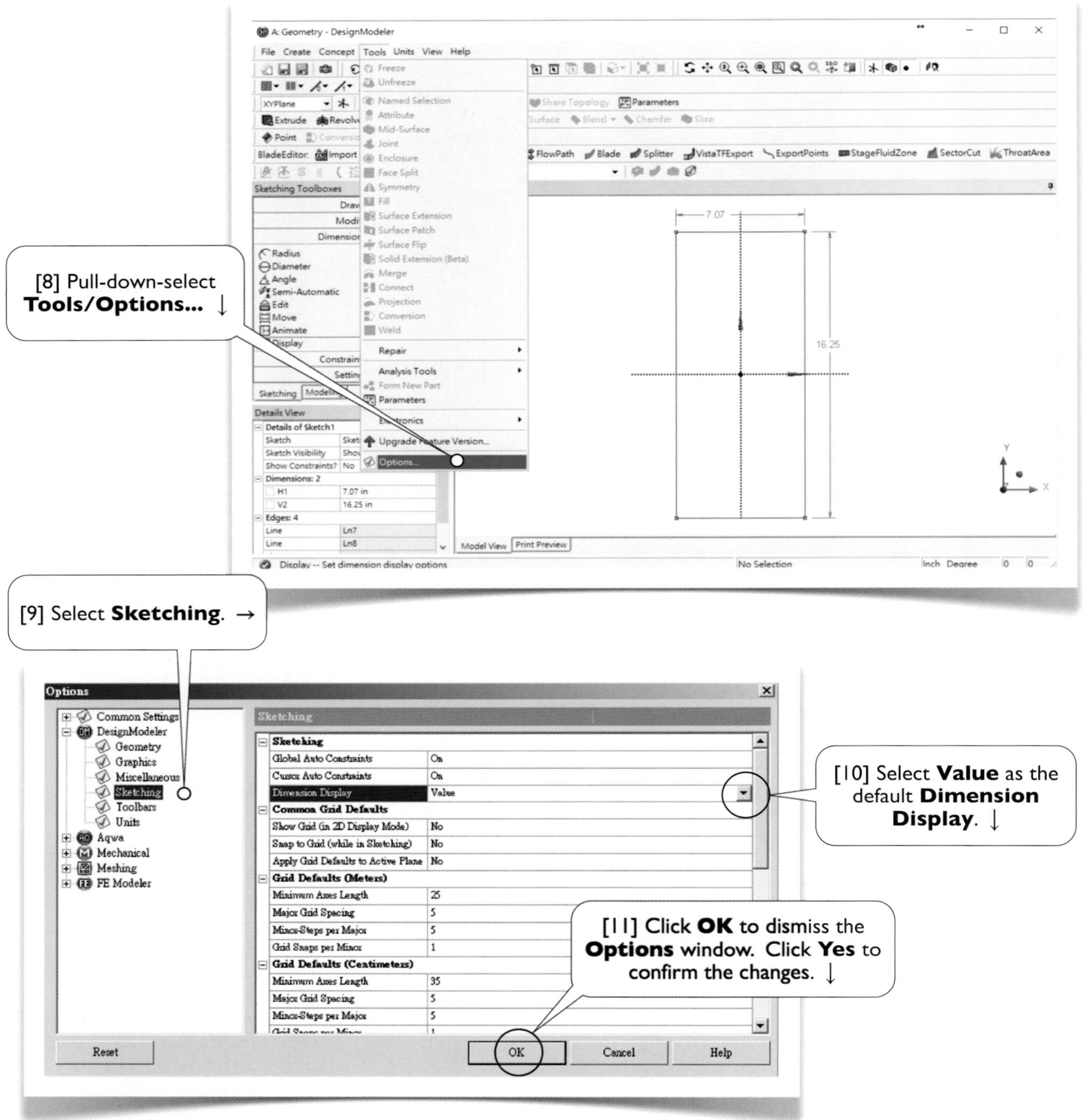

Background Color of the Graphic Area

[12] In this book, for better readability, the background color of the graphic area is always shown in white. To set up the background color, pull-down-select **Tools/Options** in **Workbench GUI** (2.1.2[2], page 57; not **DesignModeler GUI**, 2.1.2[8], page 58) and select **Appearance**. #

2.1.5 Draw a Polyline

2.1.6 Copy the Polyline

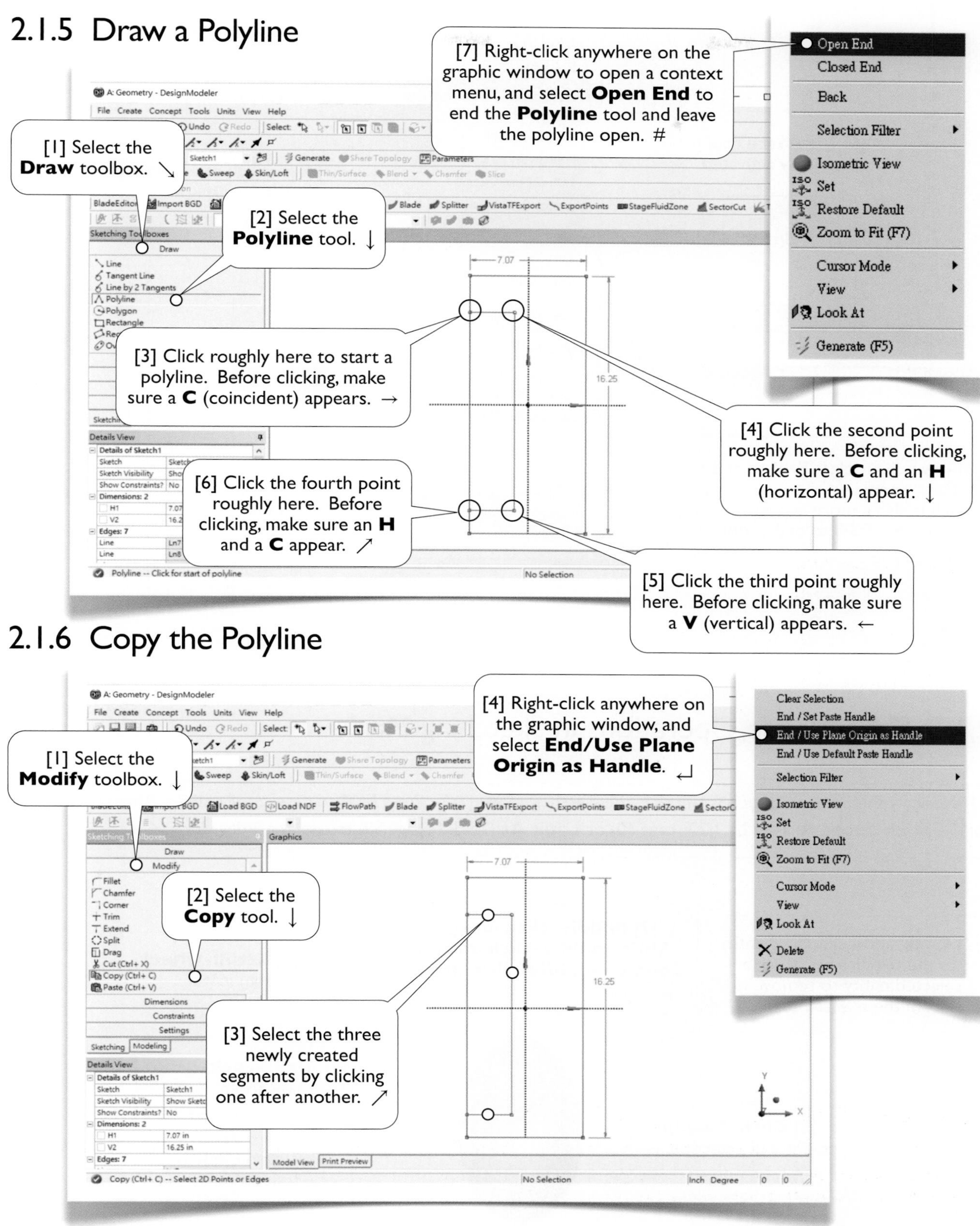

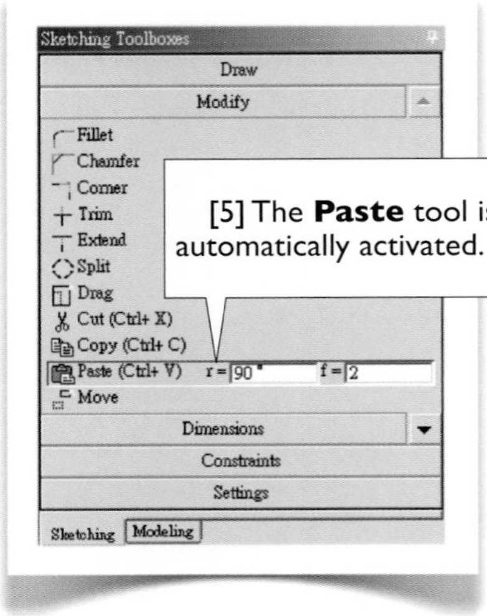

[5] The **Paste** tool is automatically activated. →

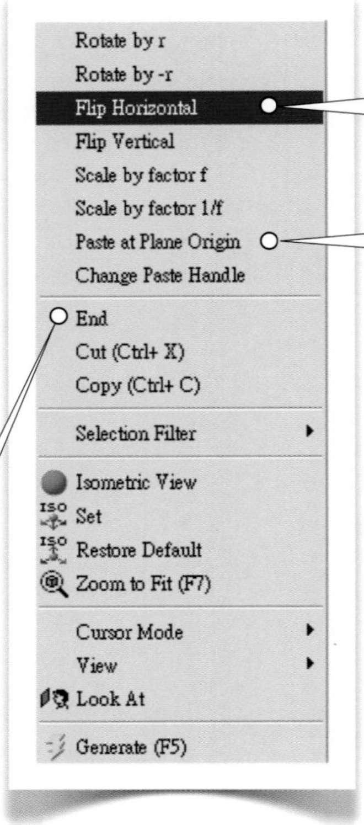

[6] Right-click-select **Flip Horizontal**. ↓

[7] Right-click-select **Paste at Plane Origin**. ↙

[8] Right-click-select **End** to end the **Paste** tool. An alternative way (which is also a more convenient way) is to press **ESC** to end a tool. #

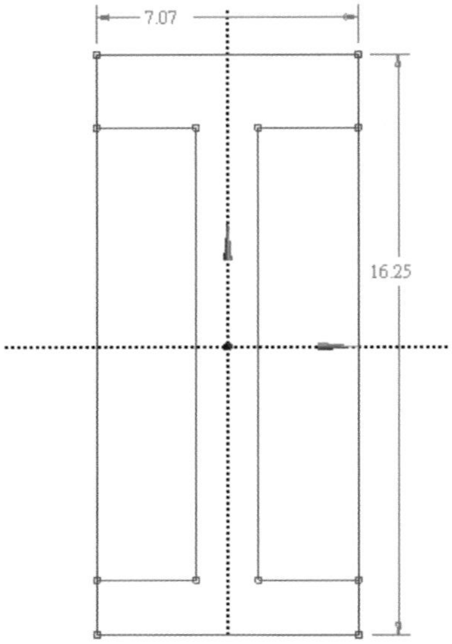

2.1.7 Basic Mouse Operations in Sketching Mode

[1] Now, try these basic mouse operations in the sketching mode [2-7]. Press **ESC** to deselect all entities. After trying any of [5-7], click **Zoom to Fit** (2.1.3[20], page 60) or **Look At** (2.1.3[4], page 59) to display a fitting view. ↙

[2] **Click**: add a sketching entity to the *selection set*. Click a selected entity to remove it from the *selection set*. ↓

[7] **Middle-click-drag**: rotate. **Shift-middle-click-drag**: zoom. **Control-middle-click-drag**: pan. #

[6] **Scroll-wheel**: zoom in/out. ←

[5] **Right-click-drag**: box zoom. ↑

[3] **Click-Sweep**: continuous selection. →

[4] **Right-click**: open a context menu. ↑

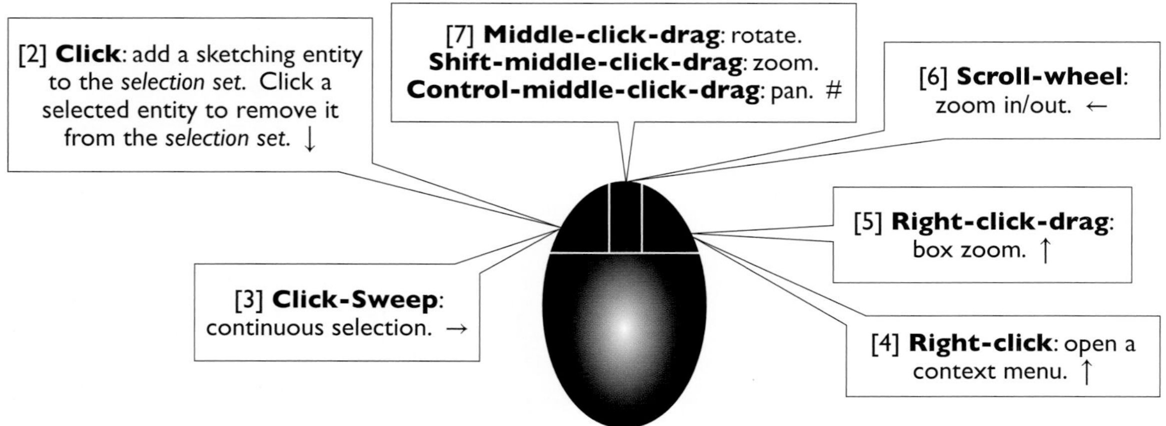

2.1.8 Trim Away Unwanted Segments

Sketching Toolboxes
Draw
Modify
Fillet
Chamfer
Corner
Trim
Ignore Axis
Extend
Split
Drag
Cut (Ctrl+ X)
Copy (Ctrl+ C)
Paste (Ctrl+ V)
Move
Replicate
Duplicate
Offset
Spline Edit
Dimensions
Constraints
Settings
Sketching
Modeling
[1] Select the **Trim** tool in the **Modify** toolbox. ↑
[2] Turn on **Ignore Axis**. Without turning it on, the axes would act as trimming tools. ↓
7.07
16.25
[3] Click to trim this segment. →
[4] And click to trim this segment. #

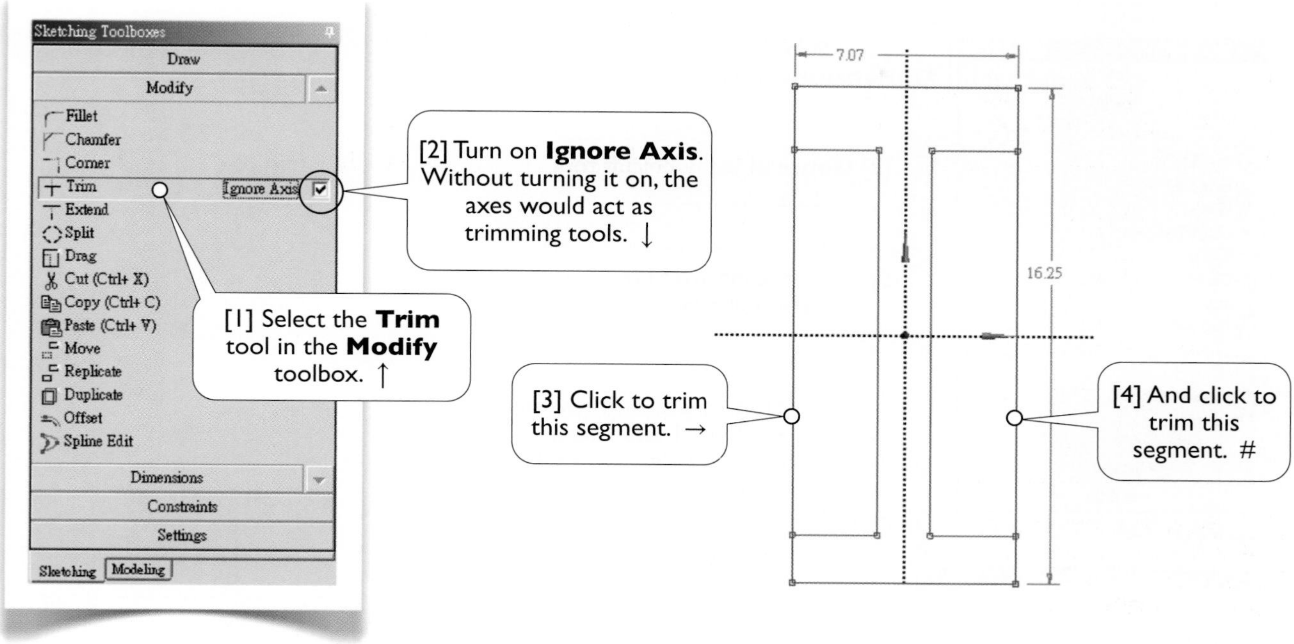

2.1.9 Impose Symmetry Constraints

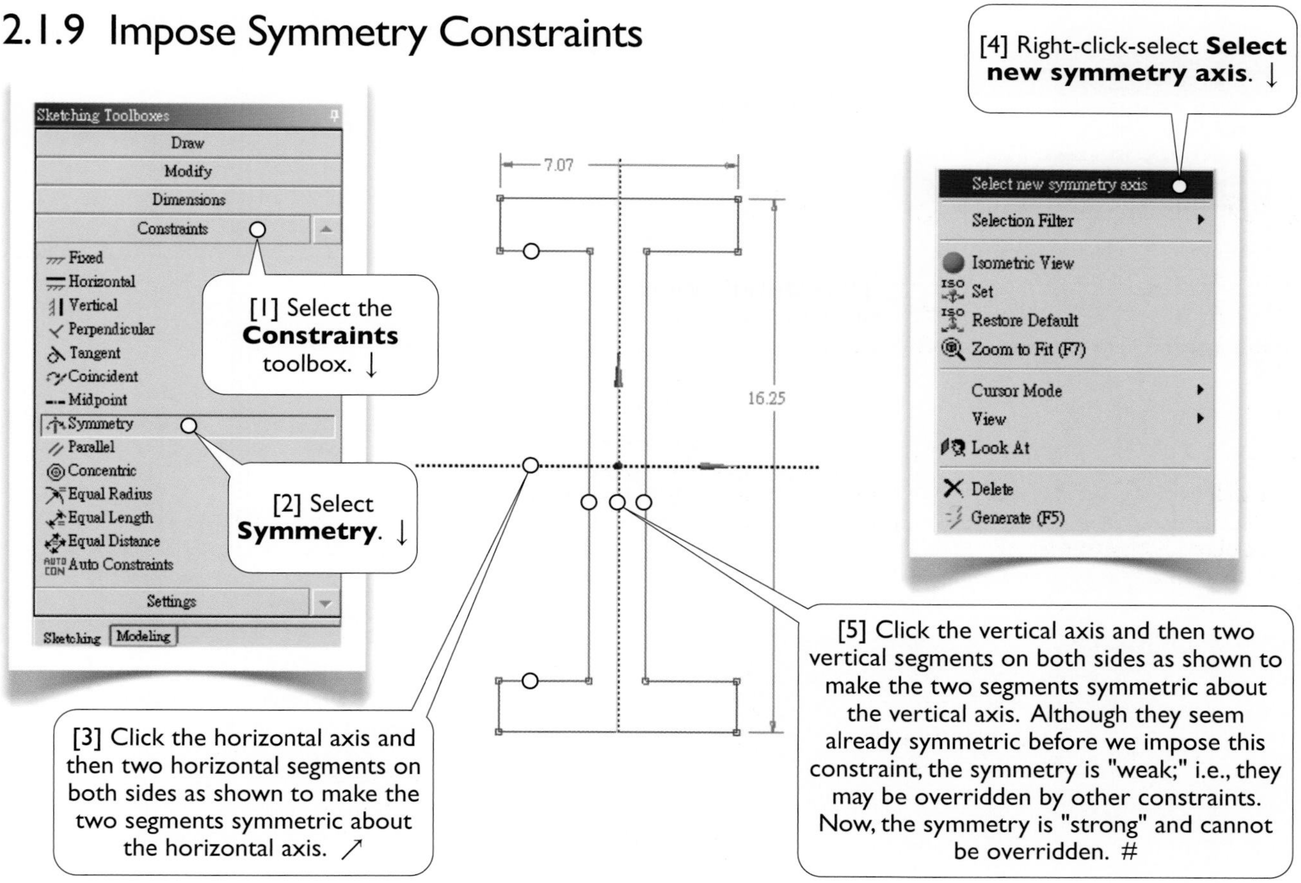

Sketching Toolboxes
Draw
Modify
Dimensions
Constraints
Fixed
Horizontal
Vertical
Perpendicular
Tangent
Coincident
Midpoint
Symmetry
Parallel
Concentric
Equal Radius
Equal Length
Equal Distance
Auto Constraints
Settings
Sketching
Modeling
[1] Select the **Constraints** toolbox. ↓
[2] Select **Symmetry**. ↓
7.07
16.25
[3] Click the horizontal axis and then two horizontal segments on both sides as shown to make the two segments symmetric about the horizontal axis. ↗
[4] Right-click-select **Select new symmetry axis**. ↓
Select new symmetry axis
Selection Filter
Isometric View
Set
Restore Default
Zoom to Fit (F7)
Cursor Mode
View
Look At
Delete
Generate (F5)
[5] Click the vertical axis and then two vertical segments on both sides as shown to make the two segments symmetric about the vertical axis. Although they seem already symmetric before we impose this constraint, the symmetry is "weak;" i.e., they may be overridden by other constraints. Now, the symmetry is "strong" and cannot be overridden. #

2.1.10 Specify Dimensions

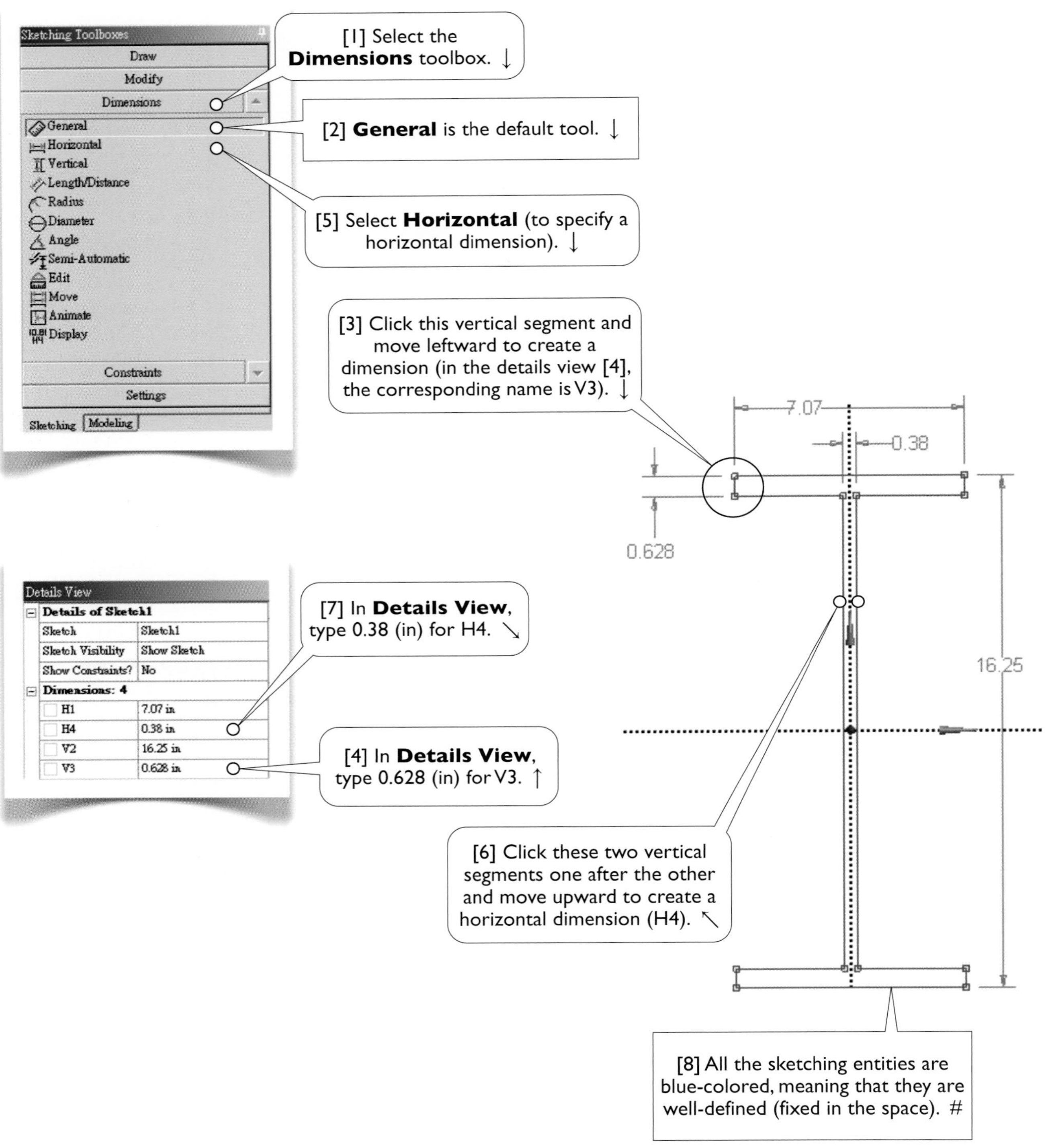

2.1.11 Add Fillets

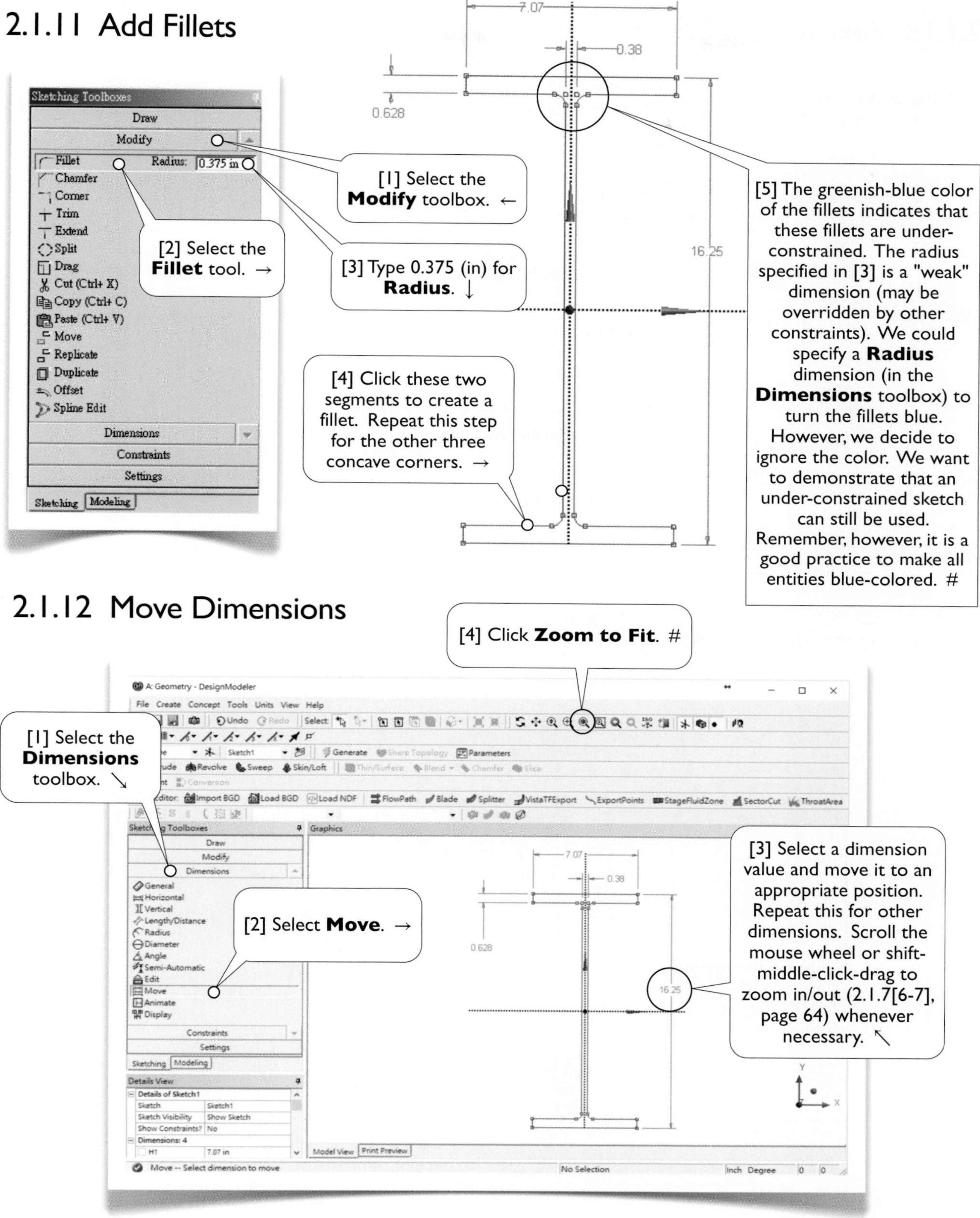

2.1.12 Move Dimensions

2.1.13 Generate a 3D Solid Body

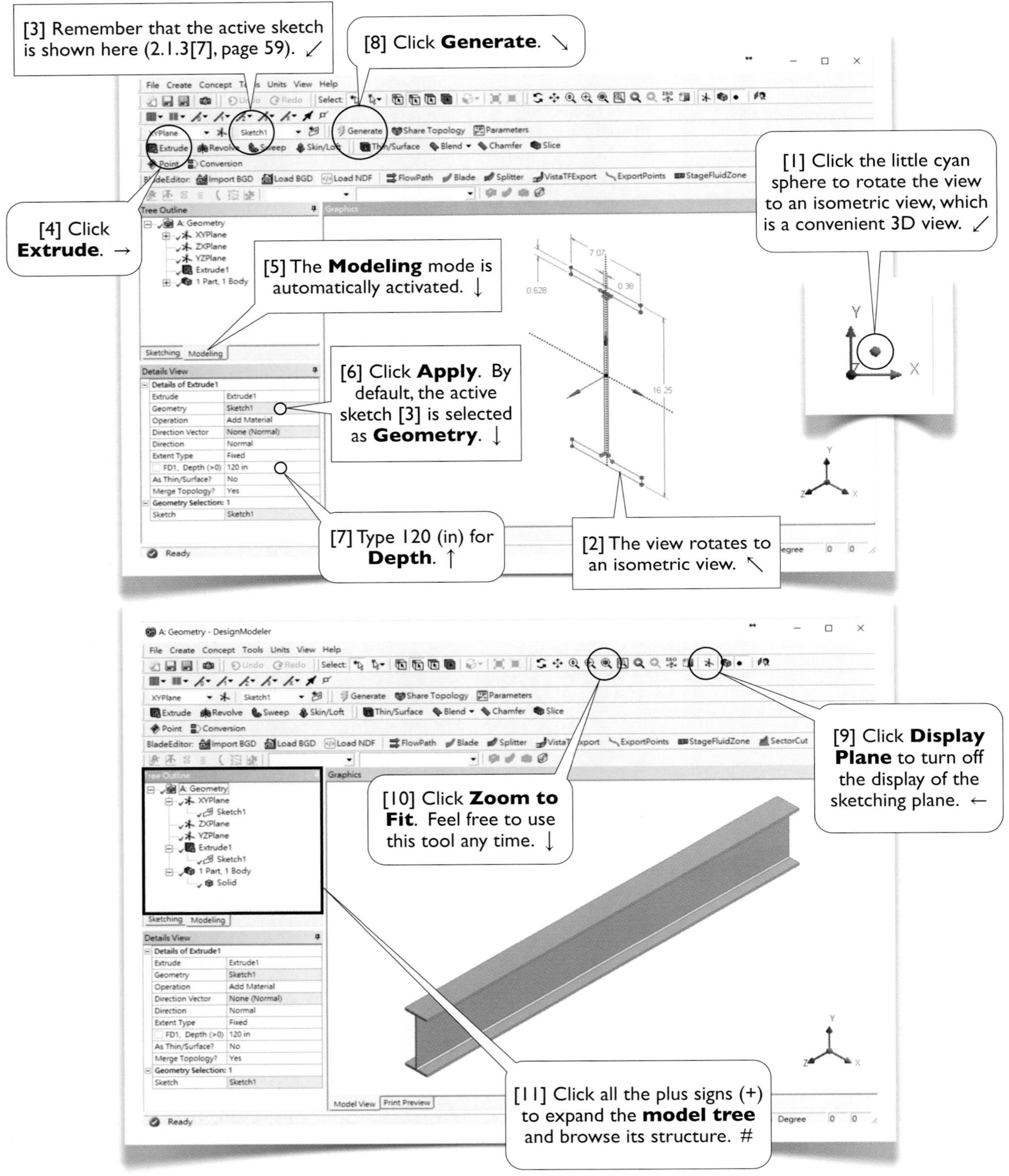

2.1.14 Save Project and Exit Workbench

Supporting Files

[5] To download the finished project files or view the demo videos, please visit SDC Publications' website or the author's webpage. See Access Code and Author's Webpage in **Preface** (page 7) for details. #

Section 2.2

Triangular Plate

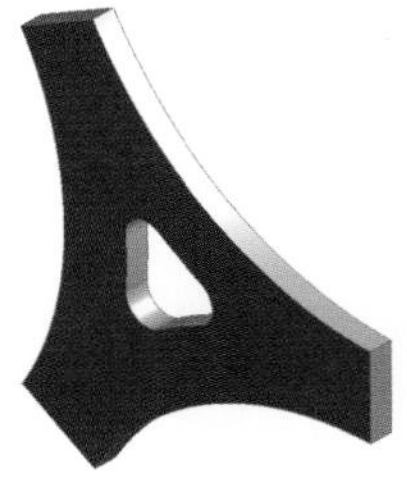

2.2.1 About the Triangular Plate

[1] A triangular plate [2-3], with a thickness of 10 mm, is used to withstand tensile forces on its three side faces [4].

In this section, we'll sketch a profile of the plate on **XYPlane** and then extrude the profile a thickness of 10 mm along *Z*-axis to generate a 3D solid body.

In Section 3.1, we will use this sketch again to generate a 2D solid model, which is then used for a static structural simulation to assess the stress under the tensile forces.

The 2D solid model will be used again in Section 8.2 to demonstrate a design optimization procedure. →

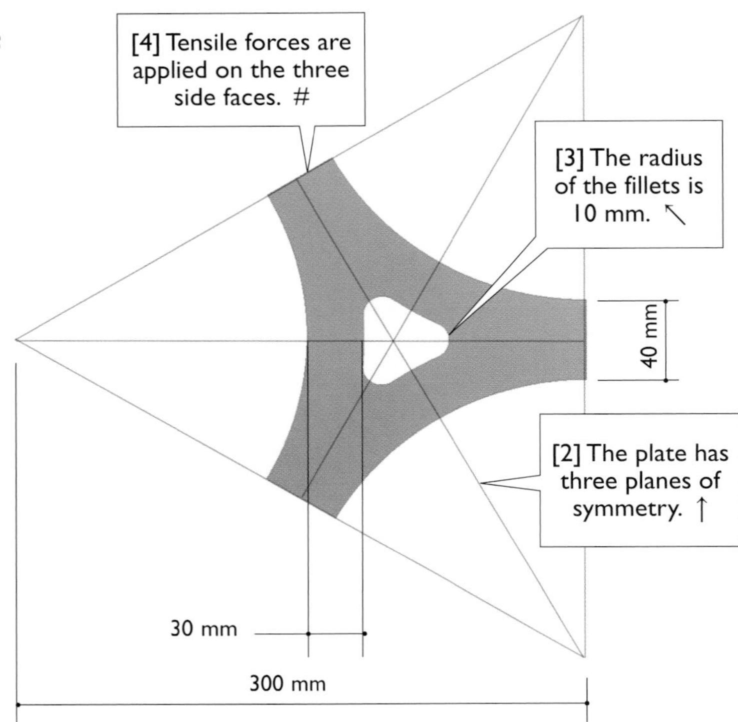

2.2.2 Start up DesignModeler

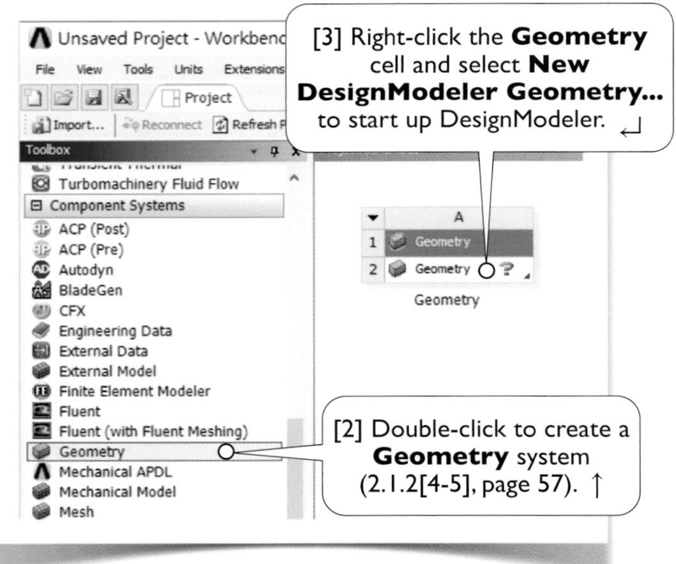

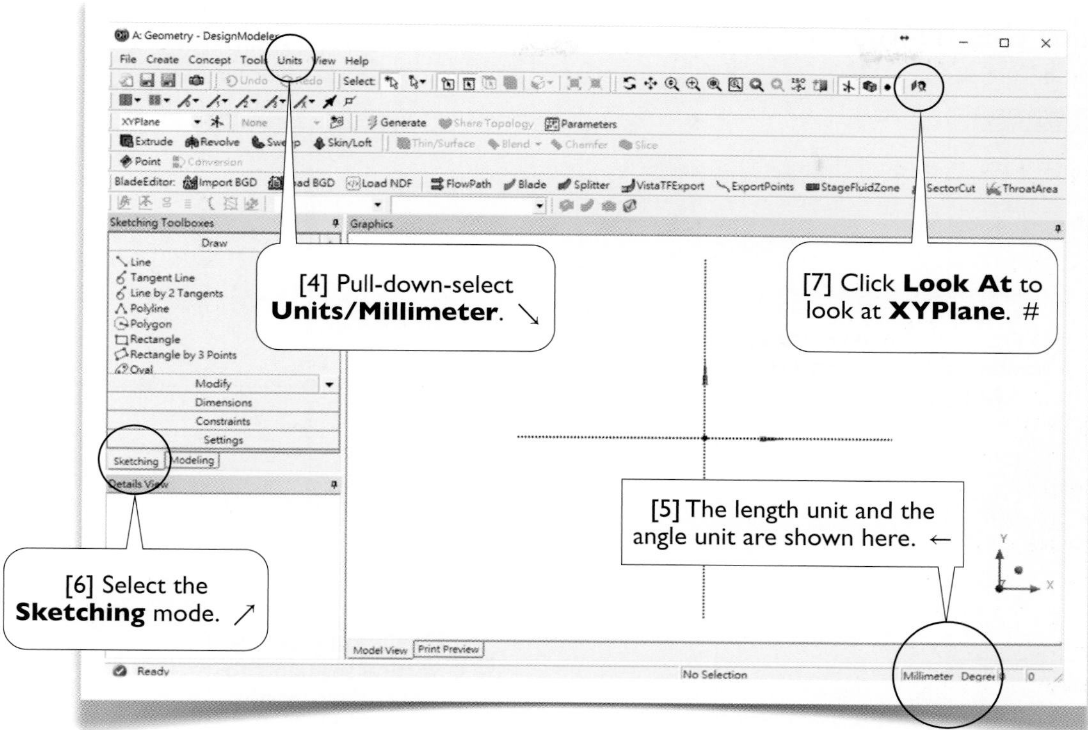

2.2.3 Draw a Triangle on XYPlane

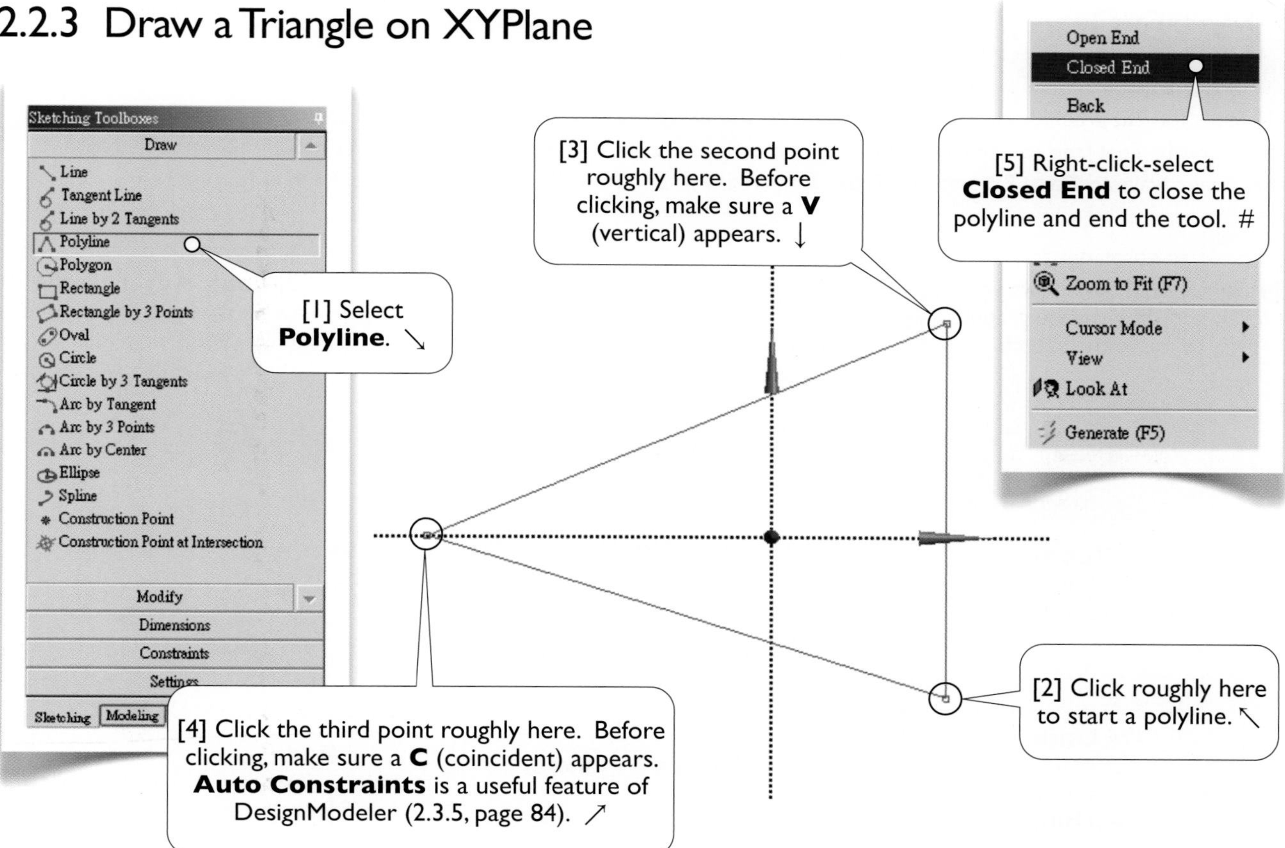

2.2.4 Make the Triangle Regular

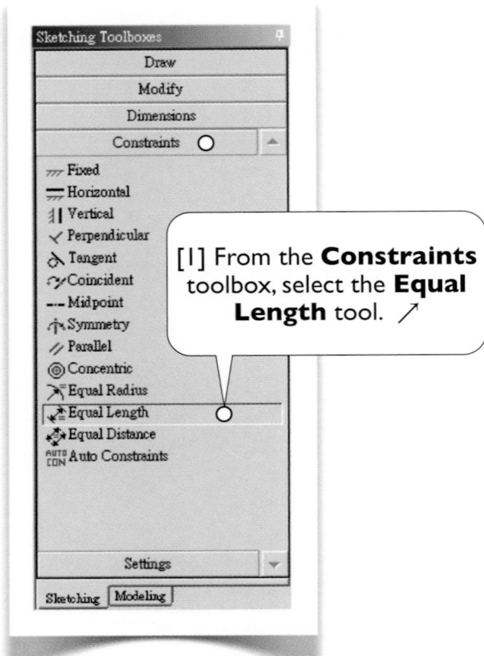

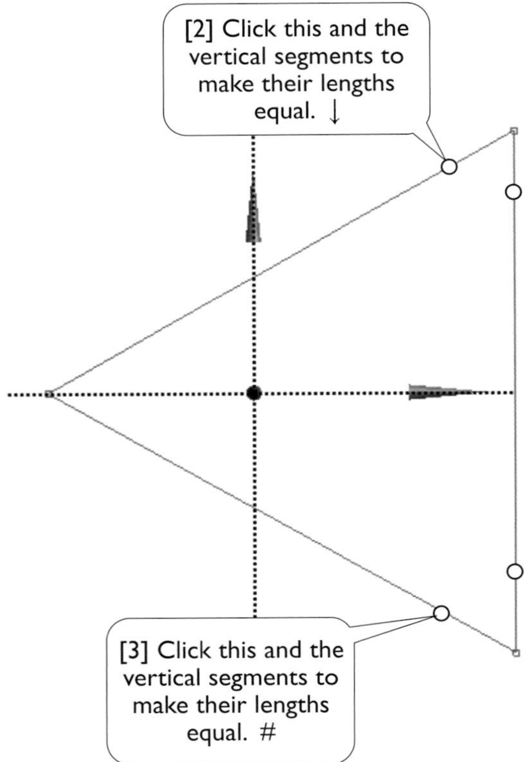

2.2.5 2D Graphics Controls

[1] Tools for 2D graphics controls are available in the **Display Toolbar** [2-10]. Click the tools in [4-6] to toggle them on/off. Feel free to use these tools any time. Try to click each tool now; they don't modify the model. Note that other ways to **Pan**, **Zoom**, and **Box Zoom** are given in 2.1.7[5-7] (page 64) and 2.3.4[1] (page 83). ↓

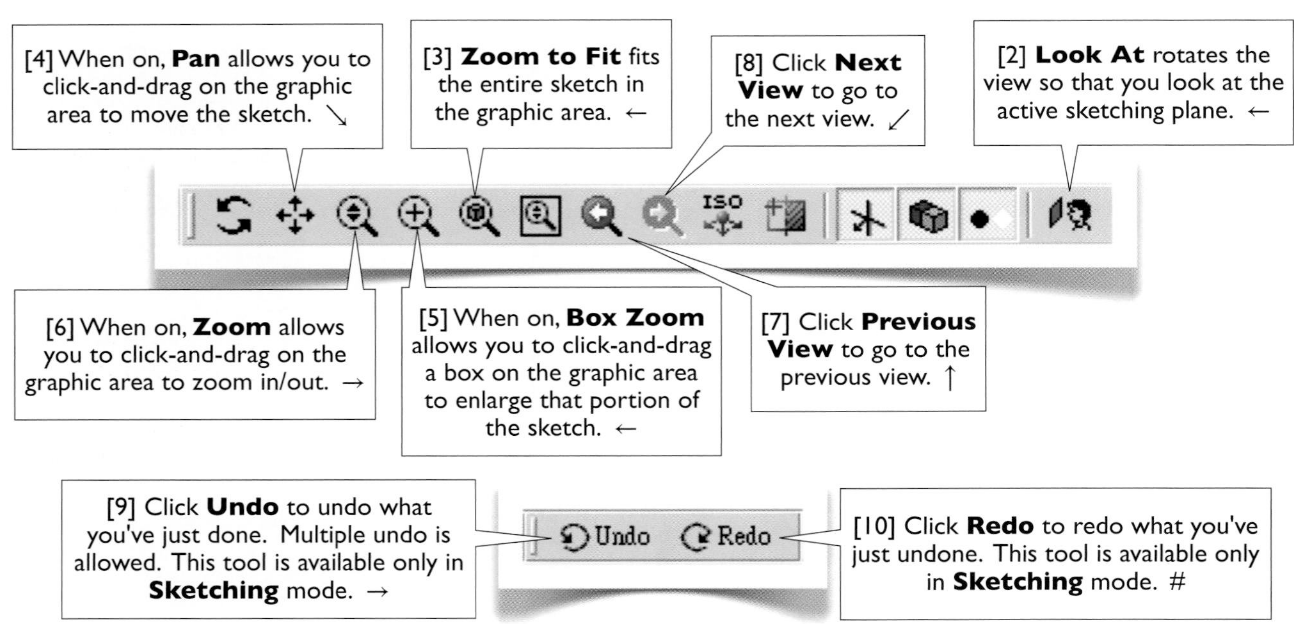

2.2.6 Specify Dimensions

2.2.7 Draw an Arc

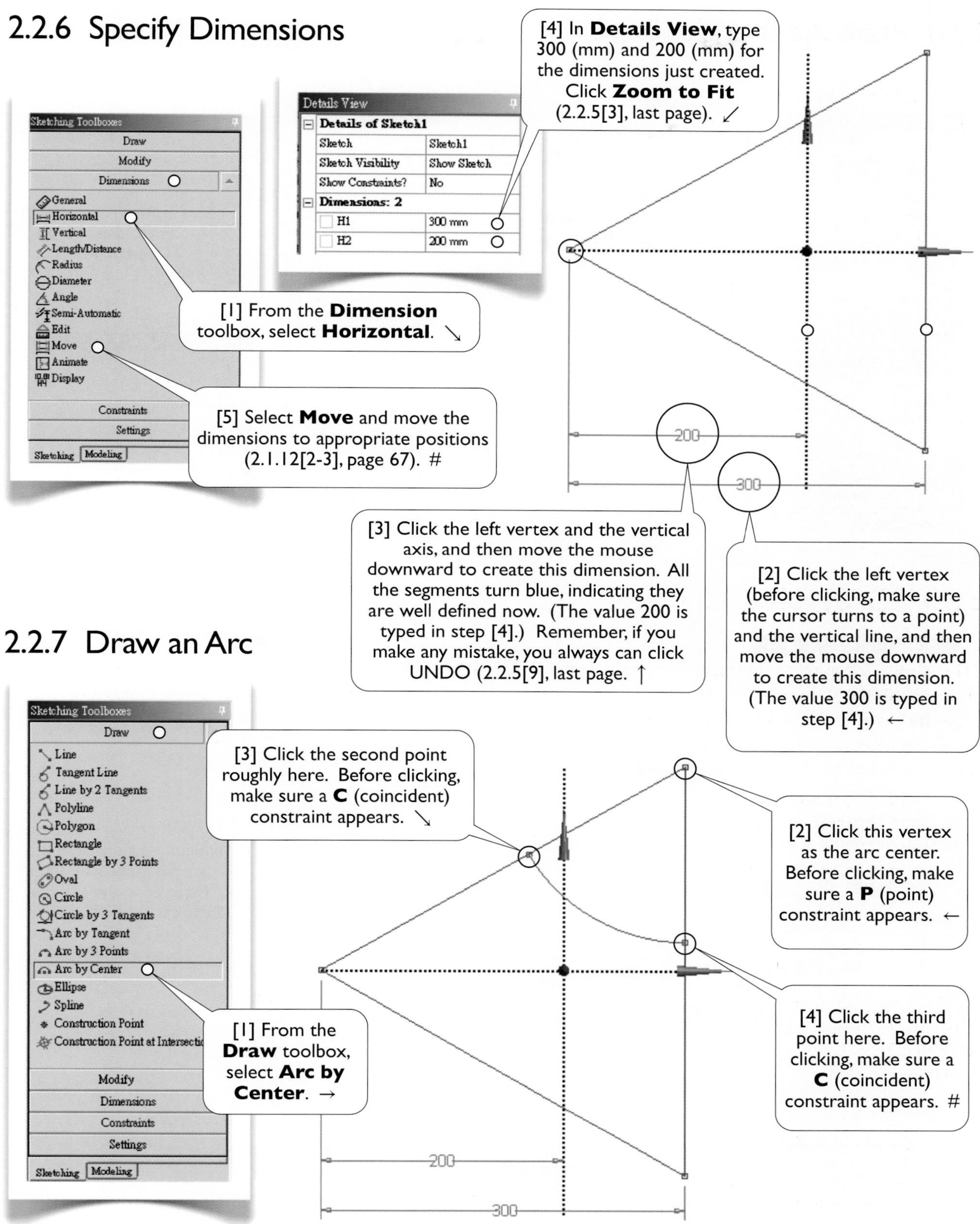

2.2.8 Replicate the Arc

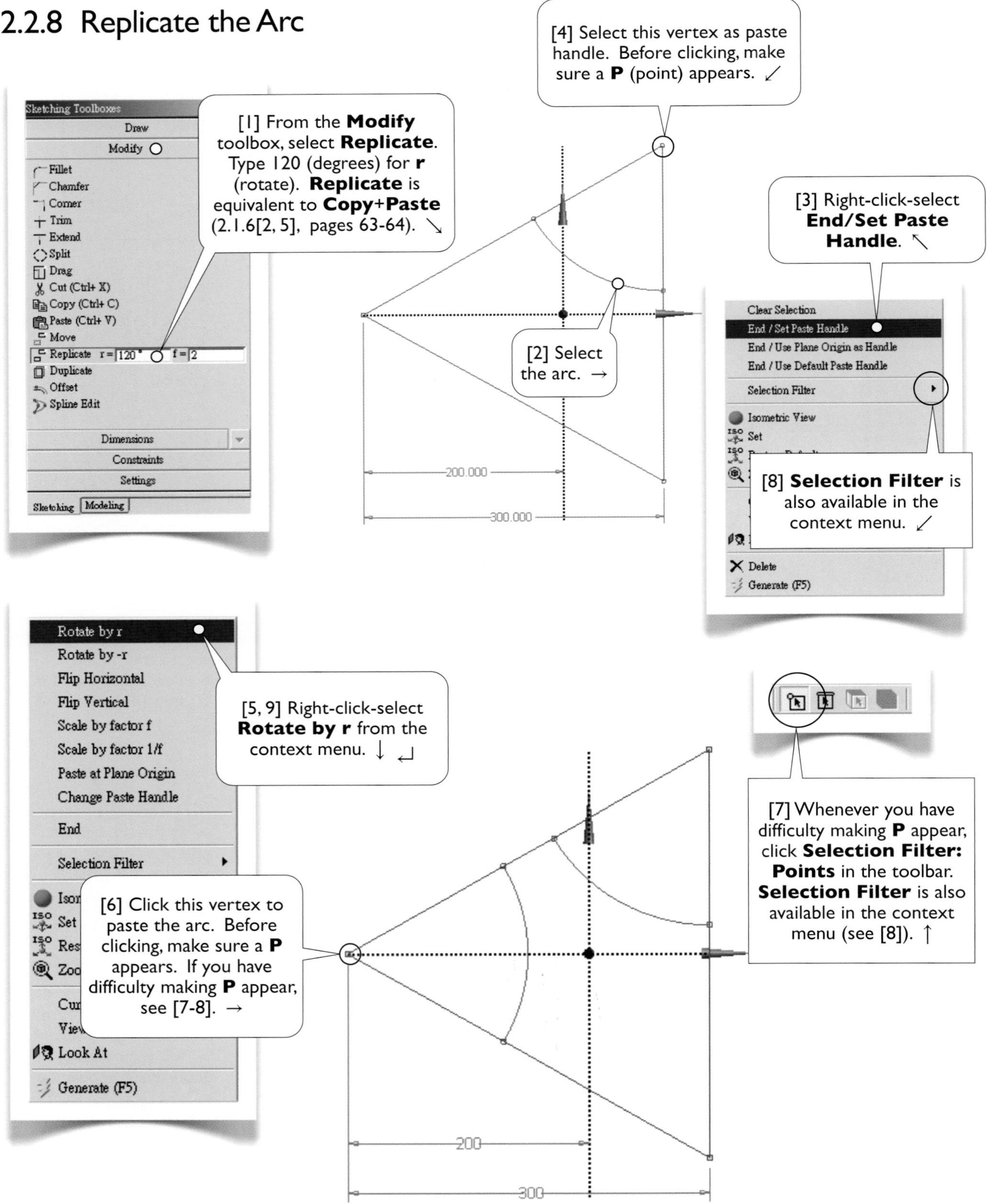

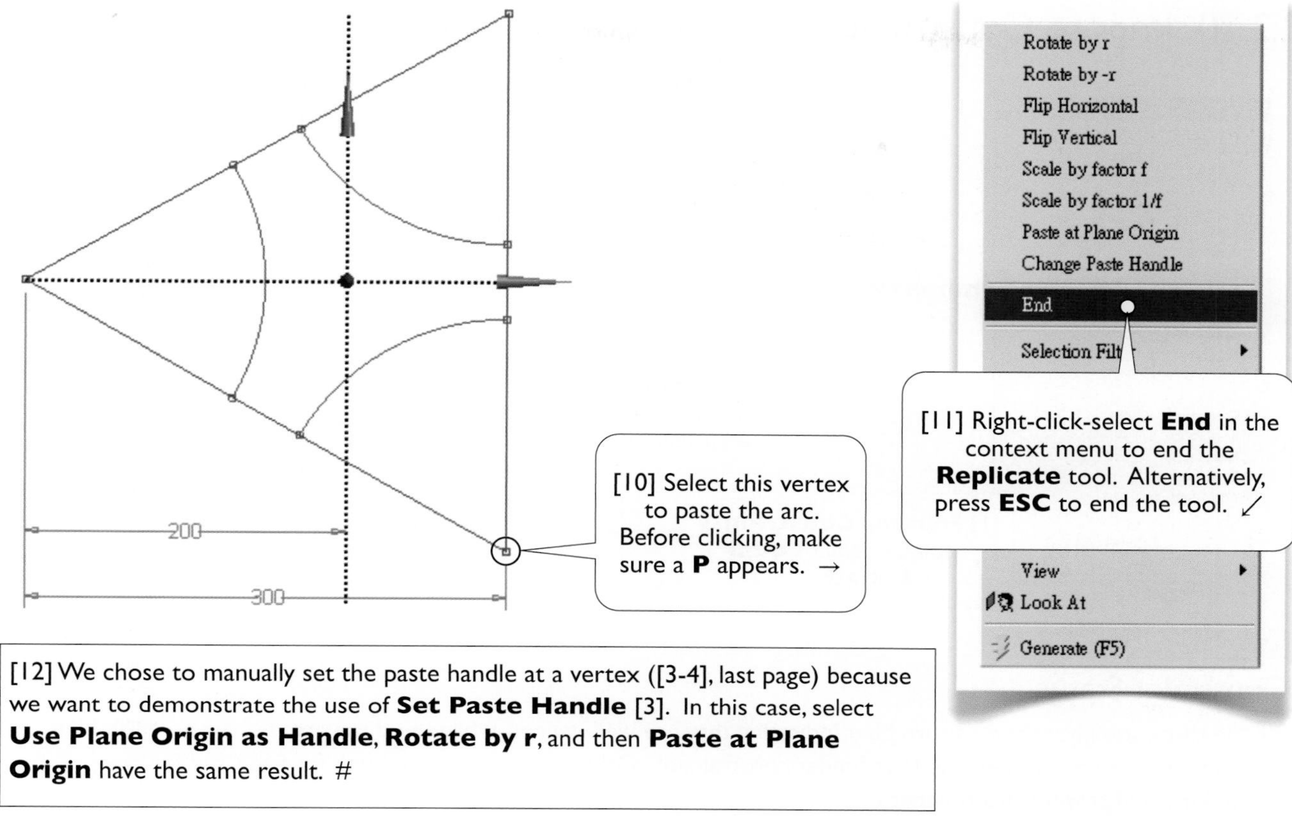

[12] We chose to manually set the paste handle at a vertex ([3-4], last page) because we want to demonstrate the use of **Set Paste Handle** [3]. In this case, select **Use Plane Origin as Handle**, **Rotate by r**, and then **Paste at Plane Origin** have the same result. #

2.2.9 Trim Away Unwanted Segments

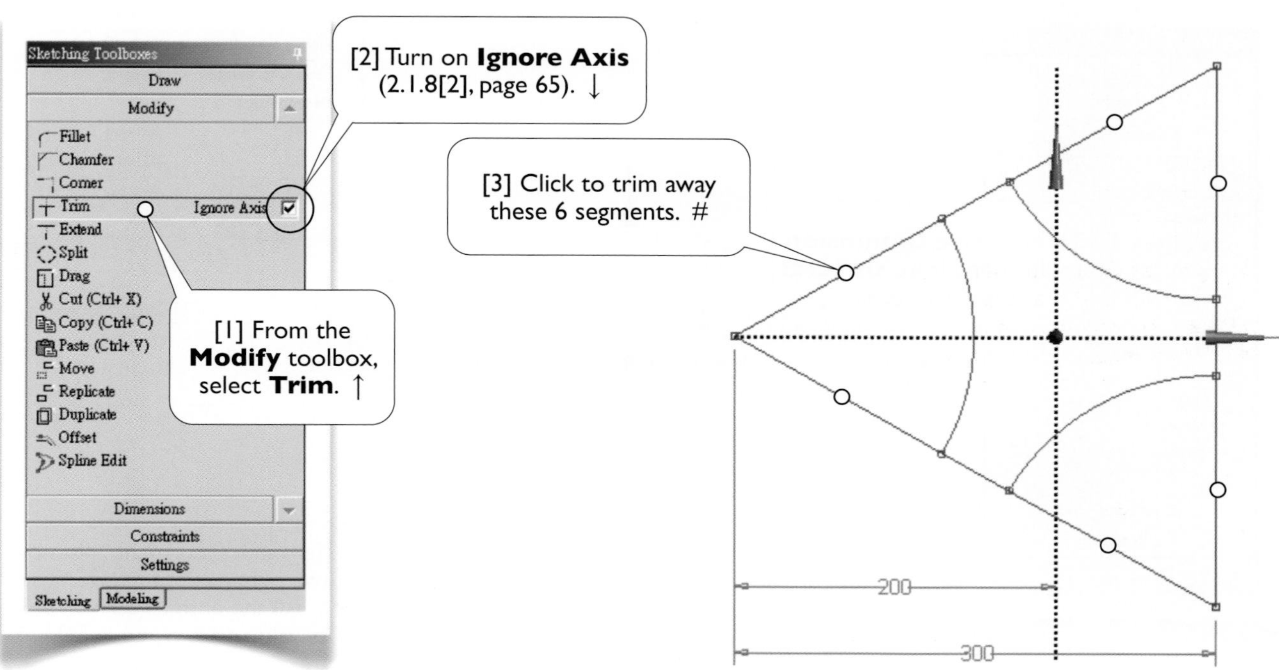

2.2.10 Impose Constraints

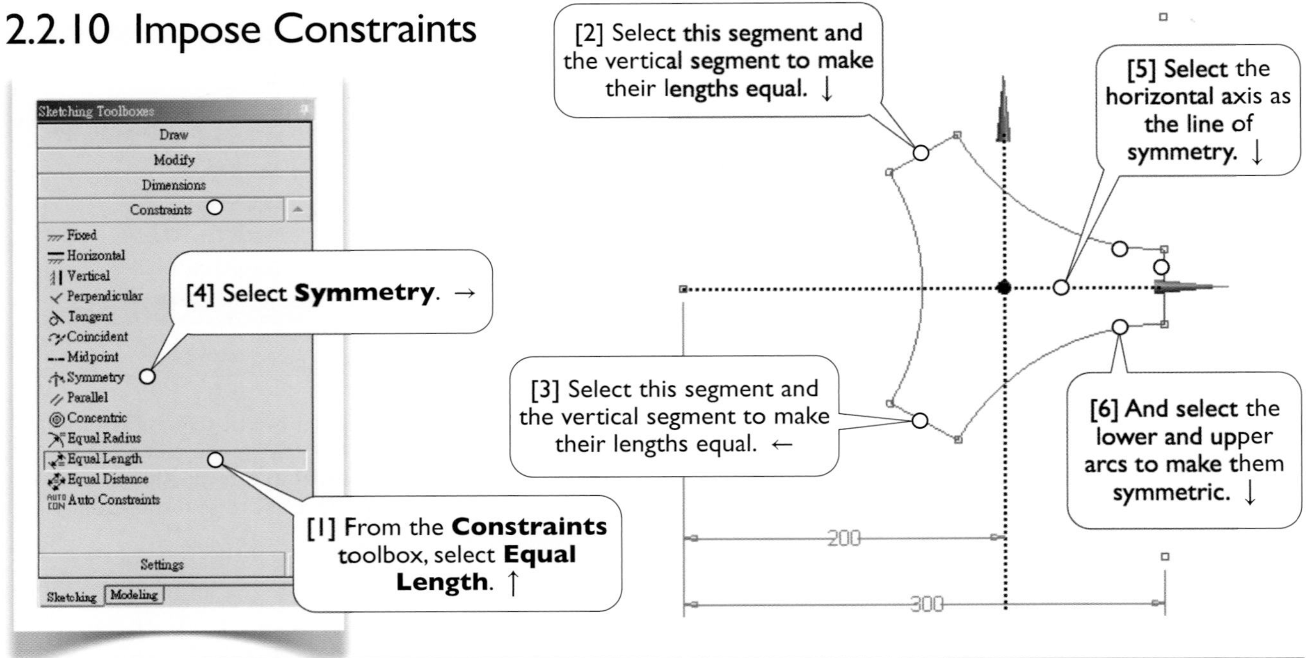

Constraint Status

[7] The three straight segments turn blue, indicating they are well-defined, while the three arcs remain greenish-blue, indicating they are not well-defined yet (under-constrained). Other color codes are black for fixed, red for over-constrained, and gray for inconsistency. #

2.2.11 Specify Dimension for Side Edges

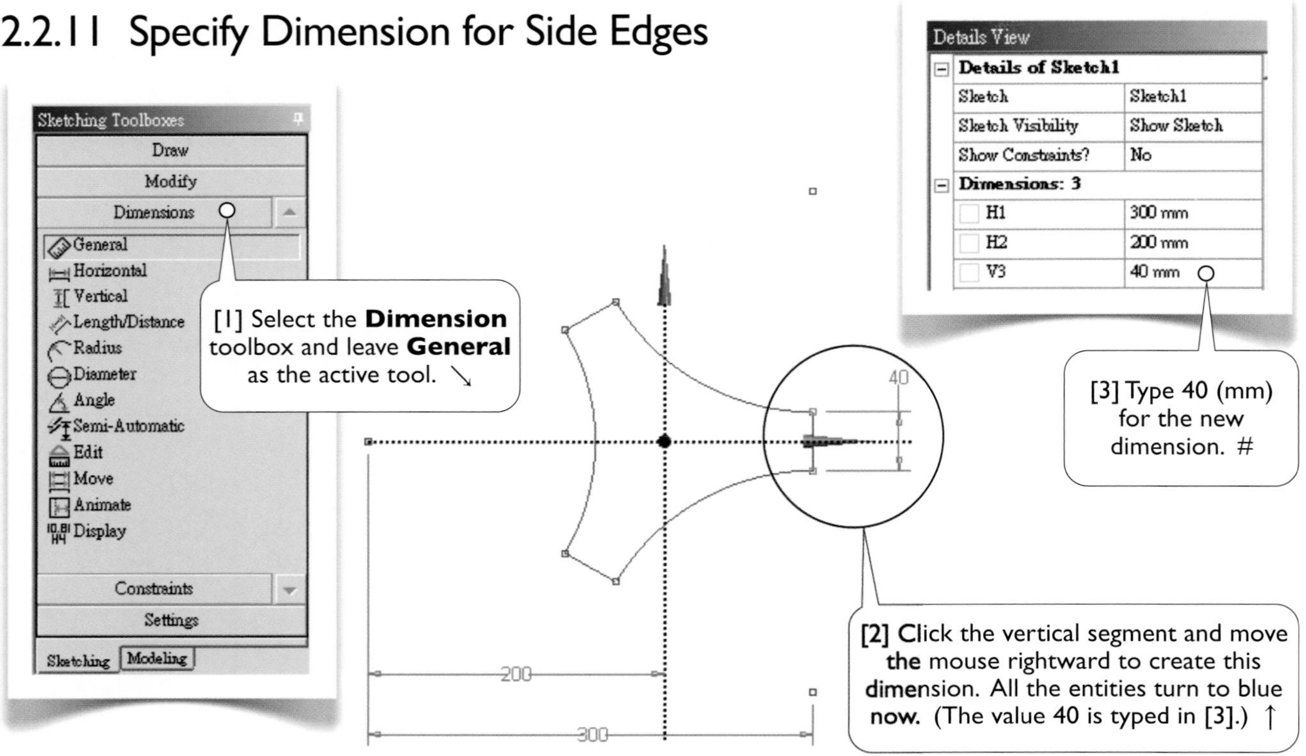

2.2.12 Create Offset

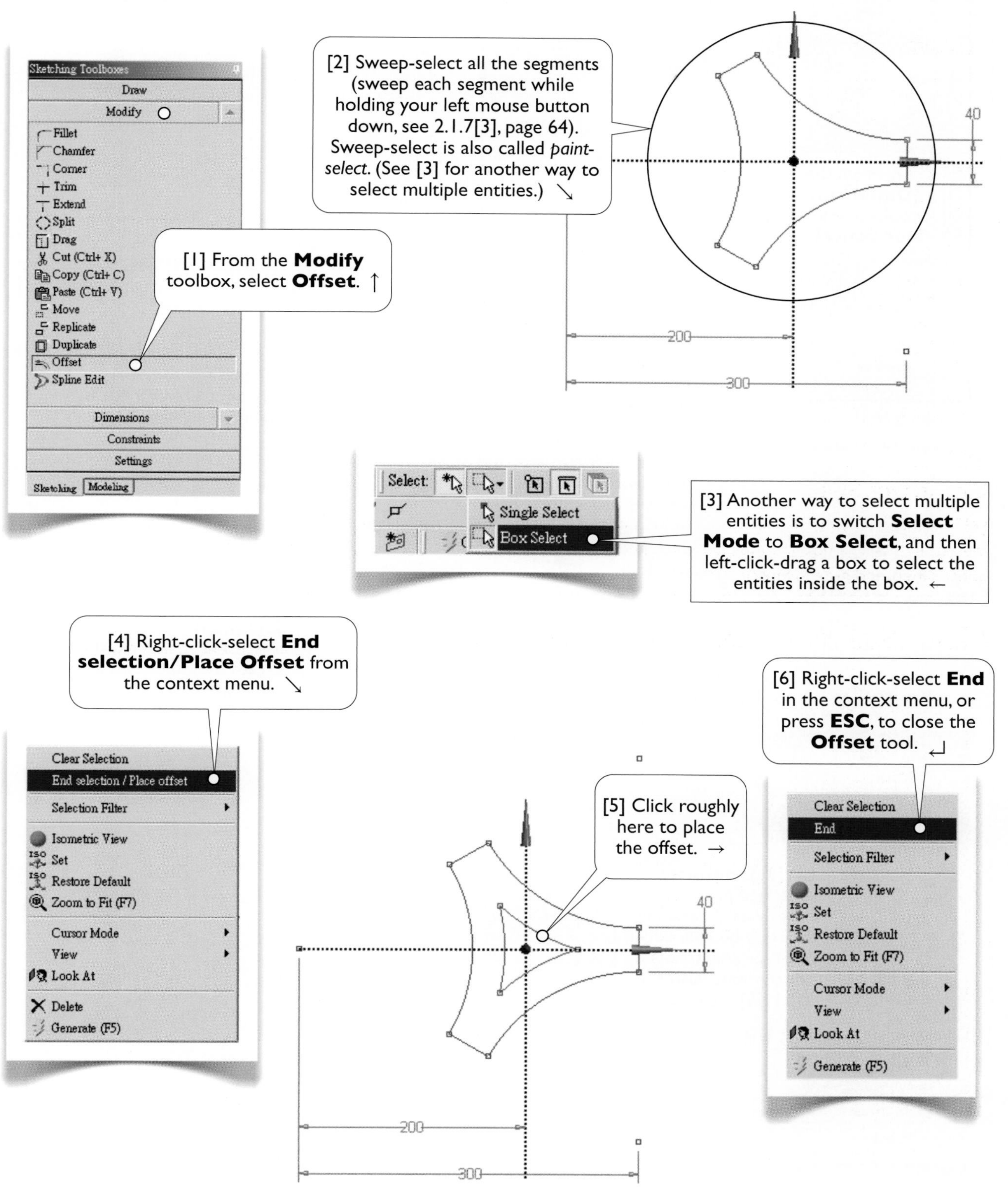

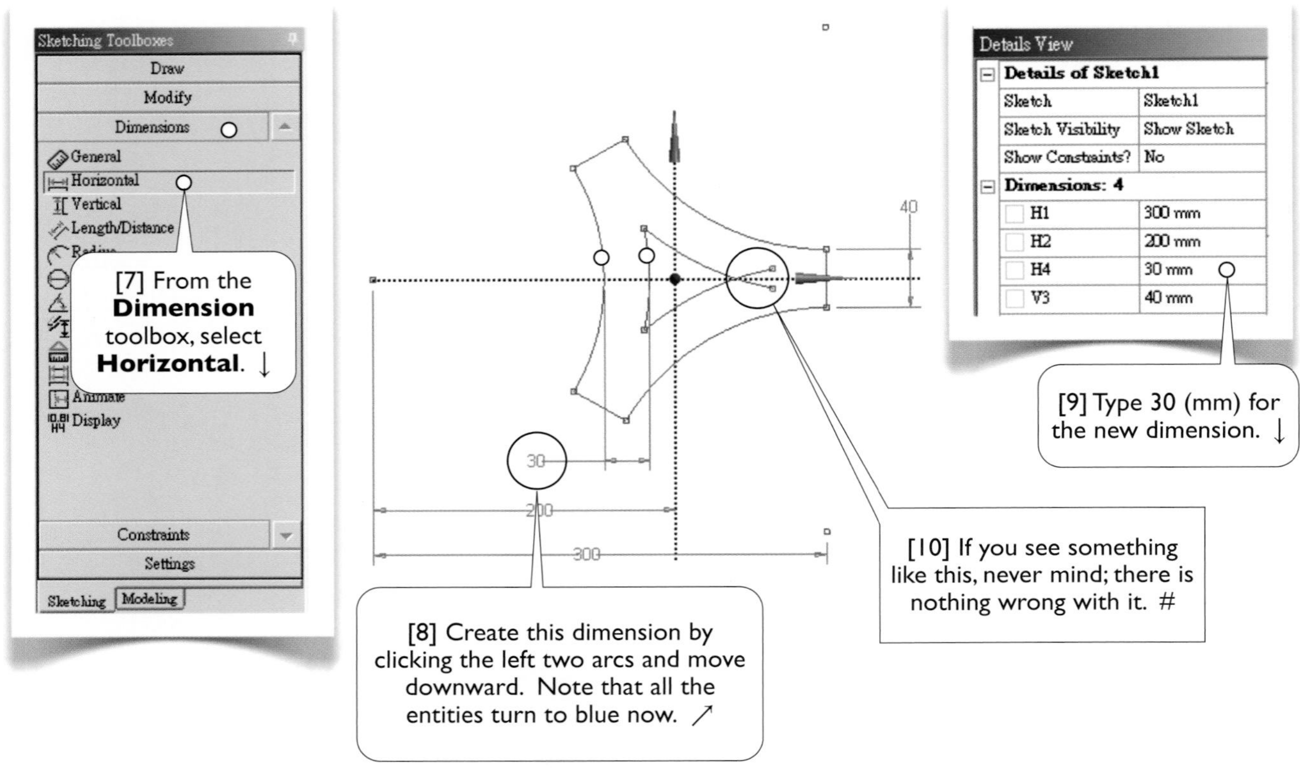

2.2.13 Create Fillets

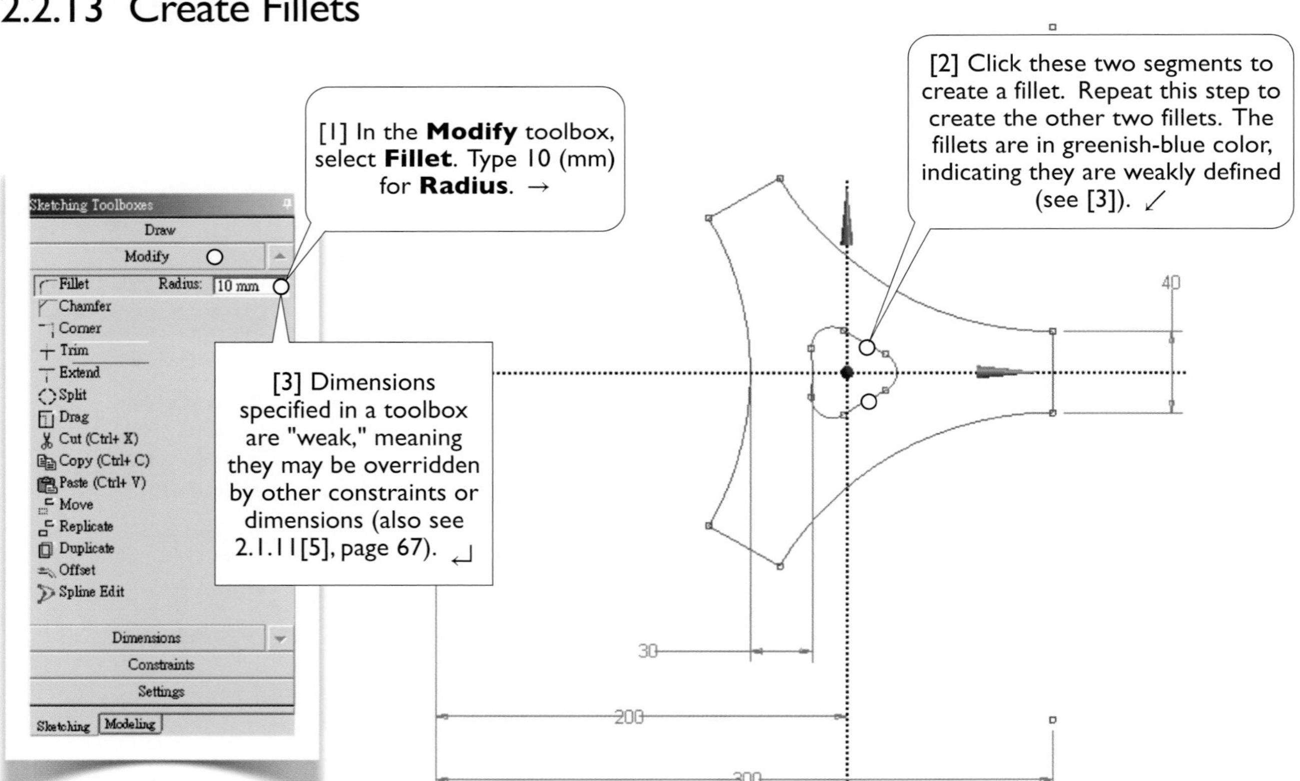

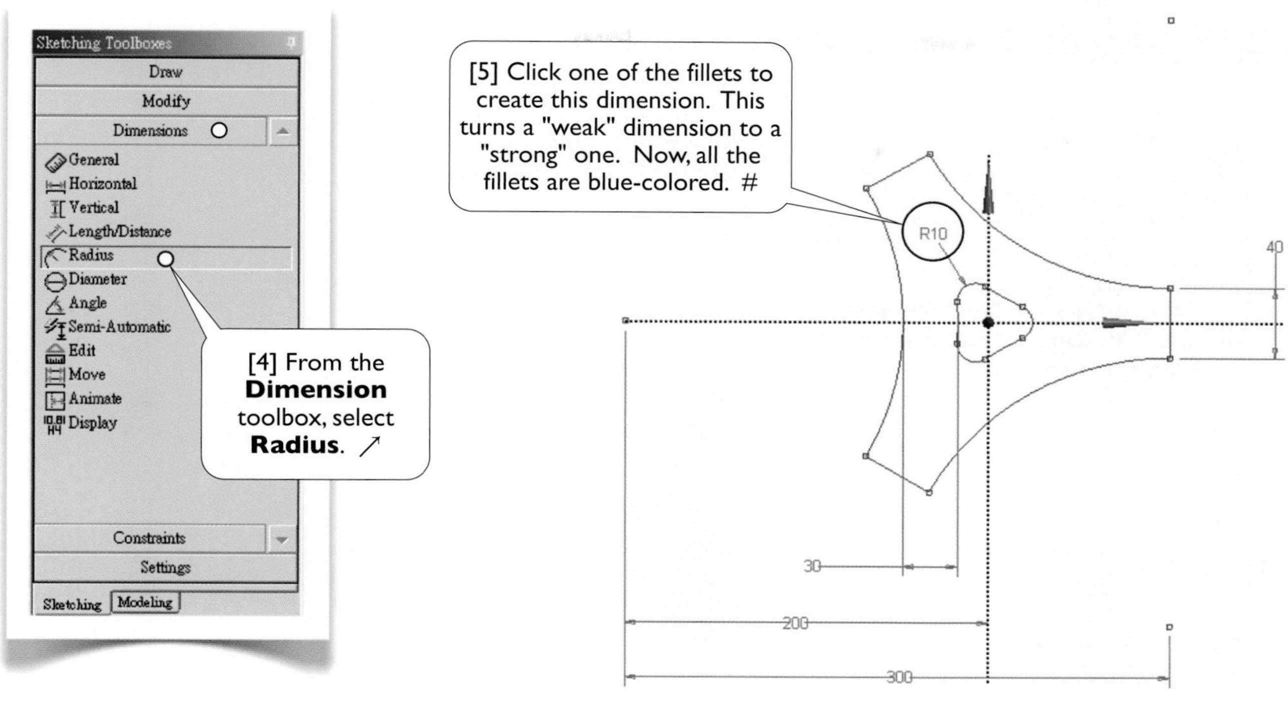

2.2.14 Extrude to Create 3D Solid

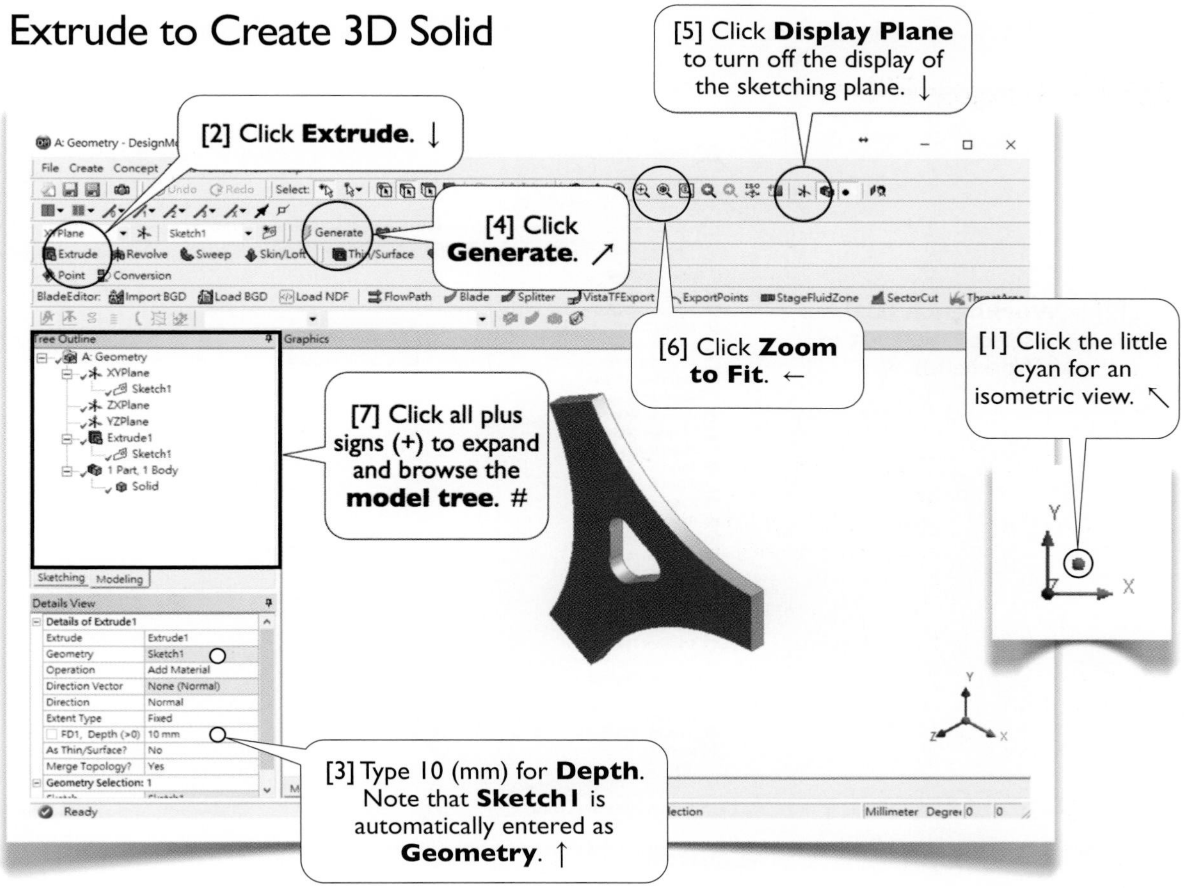

[1] Click the little cyan for an isometric view. ↖

[2] Click **Extrude**. ↓

[3] Type 10 (mm) for **Depth**. Note that **Sketch1** is automatically entered as **Geometry**. ↑

[4] Click **Generate**. ↗

[5] Click **Display Plane** to turn off the display of the sketching plane. ↓

[6] Click **Zoom to Fit**. ←

[7] Click all plus signs (+) to expand and browse the **model tree**. #

2.2.15 Save the Project and Exit Workbench

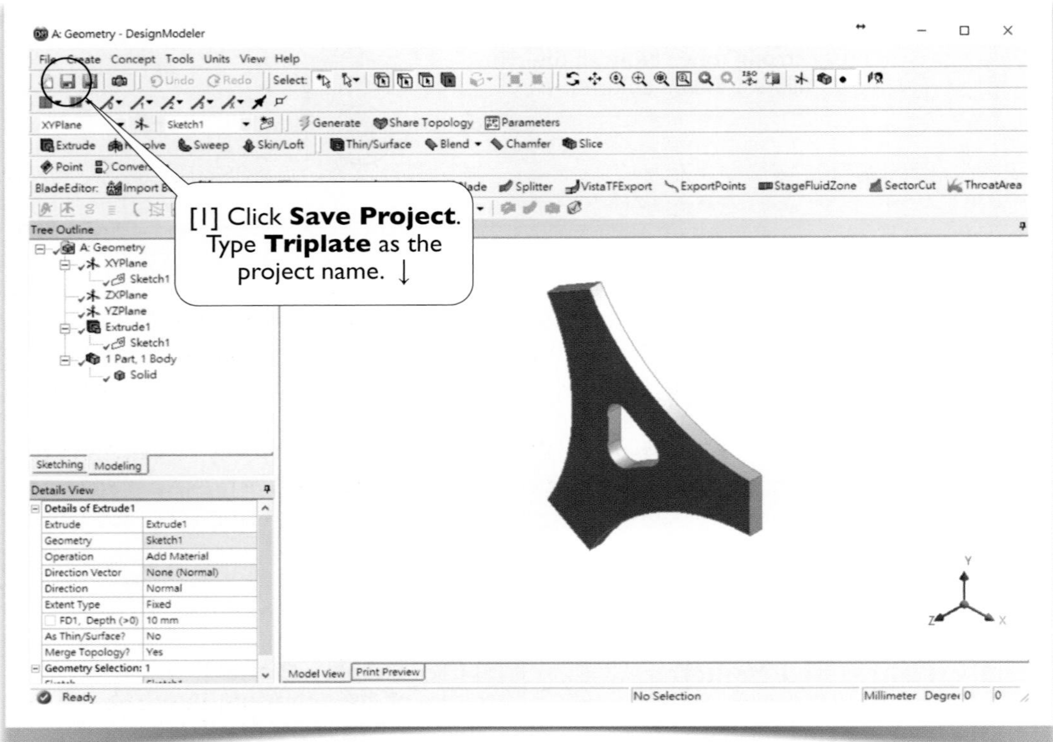

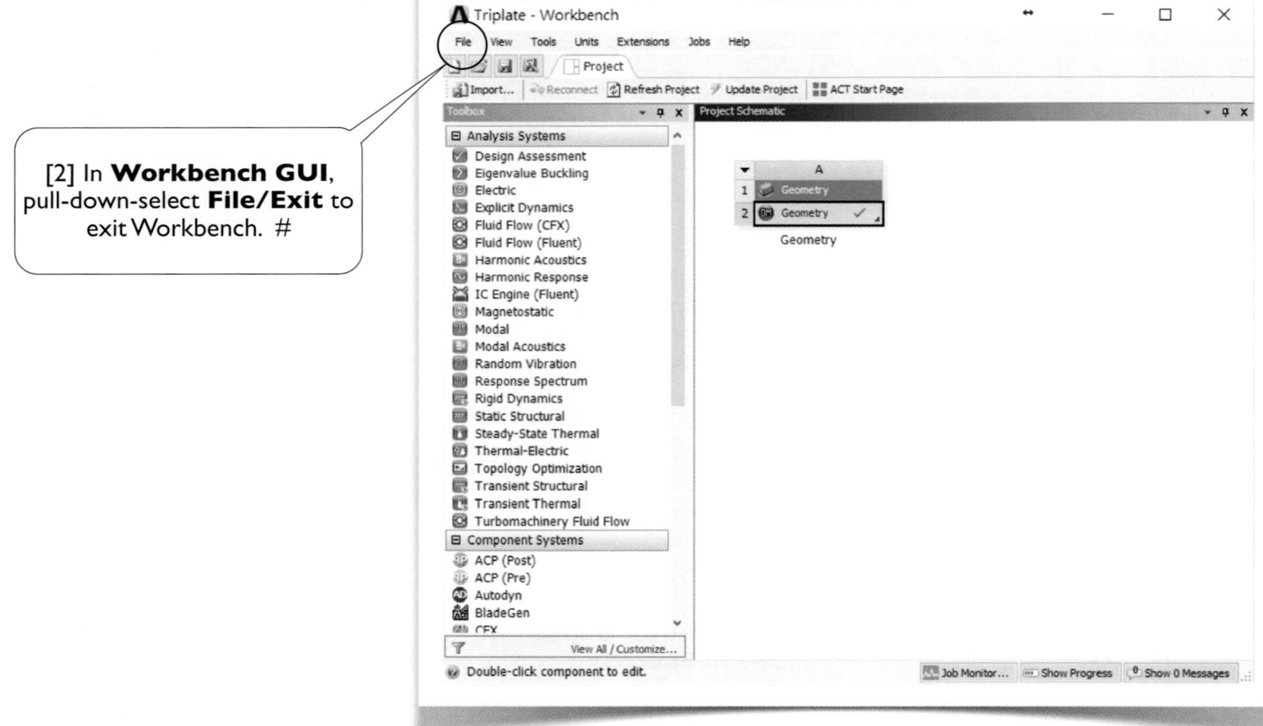

Section 2.3

More Details

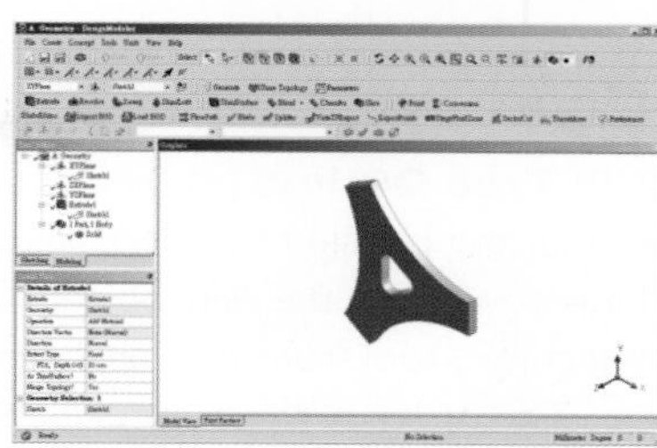

2.3.1 DesignModeler GUI

[1] **DesignModeler GUI** consists of several areas [2-8]. On the top are **pull-down menus** and **toolbars** [2]. On the bottom is a **status bar** [8]. In-between are several **window panes**. **Separators** [9] between window panes can be dragged to resize the window panes. You can make a window pane "float" by dragging or double-clicking its title bar. To return to its original position, simply double-click its title bar again.

Tree Outline [4] shares the same area with **Sketching Toolboxes** [5]. To switch between the **Modeling** mode and the **Sketching** mode, simply click a **mode tab** [3]. **Details View** [7] shows the detail information of the objects highlighted in **Tree Outline** [4] or **Graphics Window** [6]; the former displays a **Model Tree** (see [10], next page) while the latter displays a geometric model. Note that we discuss only the 2D functions of DesignModeler in this chapter and will discuss the 3D functions in Chapter 4. ↓

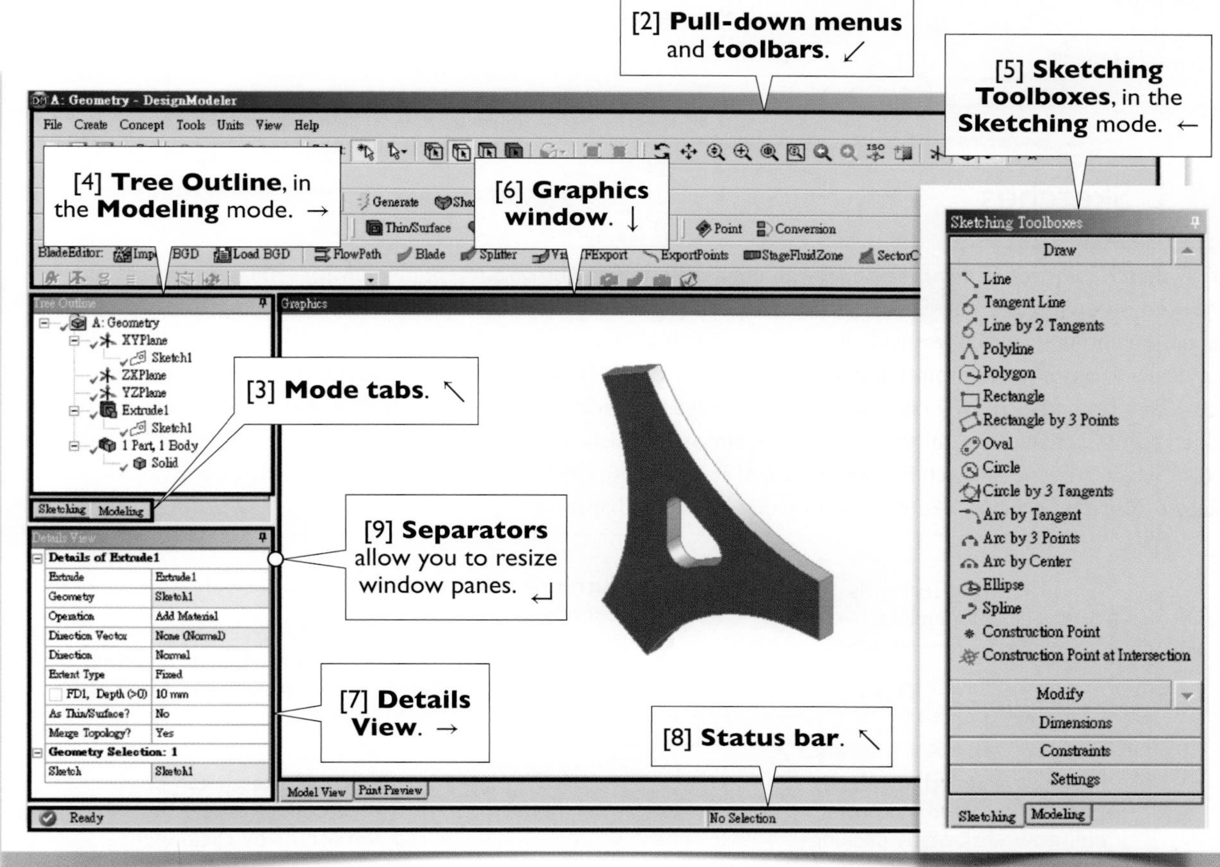

Model Tree

[10] **Tree Outline** [4] contains an outline of a **model tree**, the data structure of the geometric model. Each *branch* of the tree is called an *object*, which may contain one or more objects. At the bottom of the model tree is a **part** branch, which is the only object that will be exported to **Mechanical** for simulations. By right-clicking an object and selecting a tool from the *context menu*, you can operate on the object, such as delete, rename, duplicate, etc.

The order of the objects is relevant. DesignModeler renders the geometry according to the order of the objects in the model tree. New objects are normally added one after another. If you want to insert a new object BEFORE an existing object, right-click the existing object and select **Insert/...** from the context menu. After insertion, DesignModeler will re-render the geometry. #

2.3.2 Sketching Planes

[1] A sketch must be created on a *sketching plane*, or simply called a *plane*; each plane may contain multiple sketches. In the beginning of a DesignModeler session, three planes preexist: **XYPlane**, **YZPlane**, and **ZXPlane**. The currently active plane is shown on the toolbar [2]. You can create as many new planes as needed [3]. There are several ways to create a new plane [4]. In this chapter, since we always sketch on **XYPlane**, we will not discuss the creation of sketching planes now and will discuss it in Chapter 4. ↙

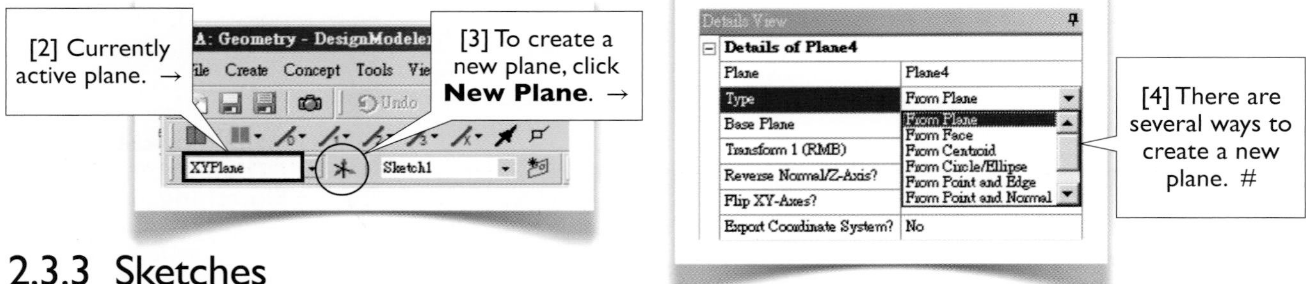

2.3.3 Sketches

[1] A sketch consists of *points* and *edges*; an edge may be straight or curved. Dimensions and constraints may be imposed on points and edges. As mentioned (2.3.2[1]), multiple sketches may be created on a plane. To create a new sketch on an empty plane, you simply switch to **Sketching** mode and draw any geometric entities on it. Later, if you want to add a new sketch on that plane, you have to click **New Sketch** [2]. Exactly one plane and one sketch is active at a time [3-6]; newly created points and edges are added to the active sketch, and newly created sketches are added to the active plane. In this chapter, we almost exclusively work with a single sketch; the only exception is Section 2.6, in which a second sketch is created (2.6.4[3], page 106). When a new sketch is created, it becomes the active sketch. More on creating sketches will be discussed in Chapter 4. ↘

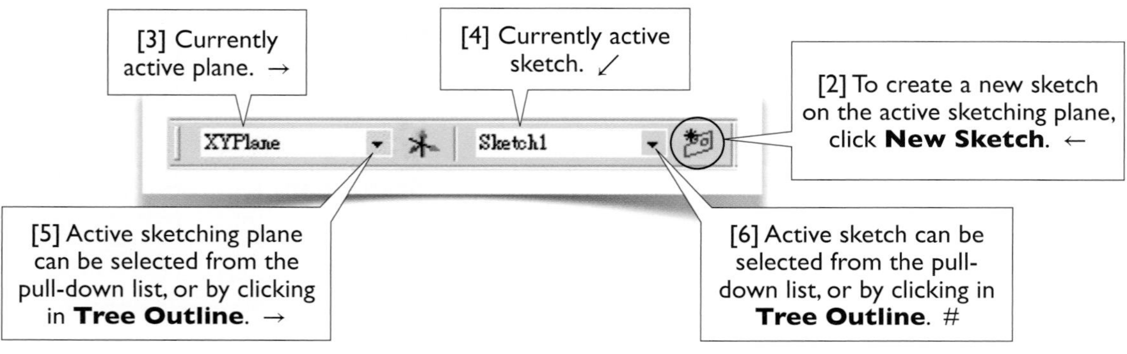

2.3.4 Sketching Toolboxes

[1] In the **Sketching** mode, five **Sketching Toolboxes** (2.3.1[5], page 81) are available: **Draw**, **Modify**, **Dimensions**, **Constraints**, and **Settings** [2-6]. Most of the tools in the toolboxes are self-explained. The best way to learn these tools is to try them out individually. During the tryout, whenever you want to clean up the graphics window, pull-down-select **File/Start Over**. These sketching tools will be briefly discussed, starting from 2.3.6. Before we discuss these sketching tools, let's reiterate some useful tips about sketching as follows.

Pan, Zoom, and Box Zoom

Besides the **Pan** tool in 2.2.5[4], page 72, a sketch can also be panned by dragging your mouse while holding down both the control key and the middle mouse button (2.1.7[7], page 64). Besides the **Zoom** tool in 2.2.5[6], page 72, a sketch can also be zoomed in/out by simply rolling your mouse wheel (2.1.7[6], page 64); the cursor position is the "zoom center." Besides the **Box Zoom** tool in 2.2.5[5], page 72, box zoom can also be done by dragging a rectangle in the graphics window using the right mouse button (2.1.7[5], page 64). After you are familiar with these mouse shortcuts, you usually don't need the **Pan**, **Zoom**, and **Box Zoom** tools in 2.2.5[4-6], page 72.

Context Menu

While most of the operations can be done by commands in pull-down menus or toolbars, many operations either require or are more efficient using a context menu. The context menu can be popped-up by right-clicking an entity in the graphics window or an object in the model tree. Try to explore whatever is available in the context menu.

Status Bar

The status bar (2.3.1[8], page 81) contains instructions on each operation. Look at the instructions whenever helpful. When a draw tool is in use, the coordinates of your mouse pointer are shown in the status bar. ✓

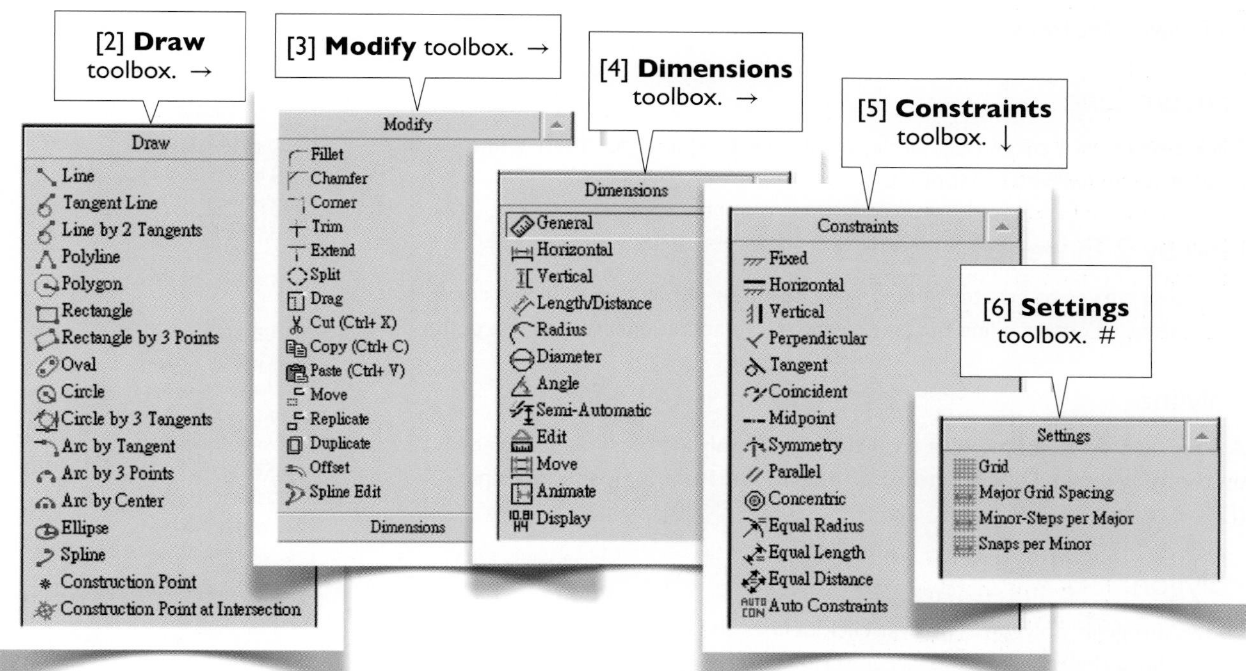

2.3.5 Auto Constraints[Refs 1, 2]

[1] By default, DesignModeler is in the **Auto Constraints** mode, both globally and locally. DesignModeler attempts to detect the user's intentions and tries to automatically impose constraints on sketching entities. The following cursor symbols indicate the kind of constraints that are applied:

C - The cursor is coincident with a line.
P - The cursor is coincident with a point.
T - The cursor is a tangent point.
⊥ - The cursor is a perpendicular foot.
H - The line is horizontal.
V - The line is vertical.
// - The line is parallel to another line.
R - The radius is equal to another radius.

Both **Global** and **Cursor** modes are based on all entities of the active plane (not just the active sketch). The difference is that **Cursor** mode only examines the entities nearby the cursor, while **Global** mode examines all the entities in the active plane. →

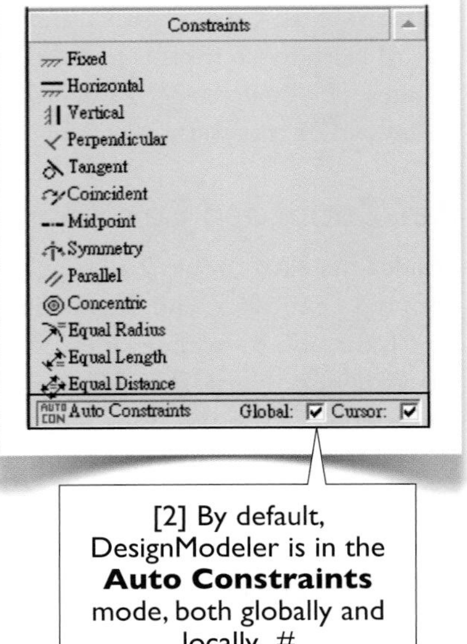

[2] By default, DesignModeler is in the **Auto Constraints** mode, both globally and locally. #

2.3.6 Draw Tools[Ref 3]

Line

[2] Draws a line by two clicks.

Tangent Line

Click a point on a curve (e.g., circle, arc, ellipse, or spline) to create a line tangent to the curve at that point.

Line by 2 Tangents

Click two curves to create a line tangent to these two curves. Click a curve and a point to create a line tangent to the curve and connecting to the point.

Polyline

A polyline consists of multiple straight lines. A polyline must be completed by choosing either **Open End** or **Closed End** from the context menu ([3], next page).

Polygon

Draws a regular polygon. The first click defines the center and the second click defines the radius of the circumscribing circle. ↵

[1] The **Draw** toolbox. ←

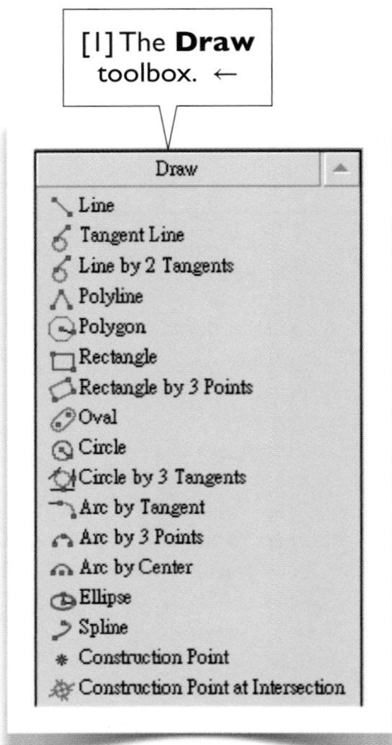

Rectangle by 3 Points

[4] The first two points define one side and the third point defines the other side.

Oval

The first two clicks define two centers, and the third click defines the radius.

Circle

The first click defines the center, and the second click defines the radius.

Circle by 3 Tangents

Select three edges (lines or curves) to create a circle tangent to these three edges.

Arc by Tangent

Click a point (usually an end point) on an edge to create an arc starting from that point and tangent to that edge; click a second point to define the other end and the radius of the arc.

Arc by 3 Points

The first two clicks define the two ends of the arc, and the third click defines a point between the ends.

Arc by Center

The first click defines the center, and two additional clicks define two ends.

Ellipse

The first click defines the center, the second click defines the major radius, and the third click defines the minor radius.

Spline

A spline is either rigid or flexible. A flexible spline can be edited or changed by imposing constraints, while a rigid spline cannot. After defining the last point, you must specify an ending condition [5]: either open end or closed end; either with fit points or without fit points.

Construction Point at Intersection

Select two edges; a construction point will be created at the intersection. ↗

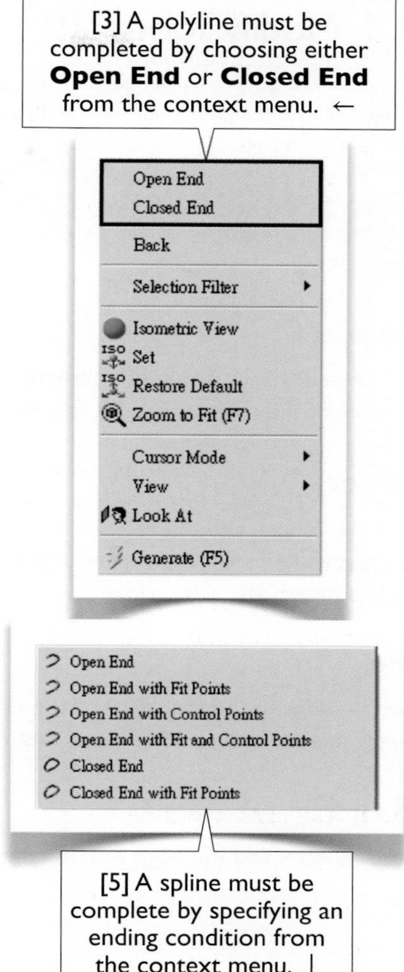

How to delete edges?

[6] To delete edges, select them and choose **Delete** or **Cut** from the context menu. Multiple selection methods (e.g., control-selection or sweep-selection) can be used to select the edges. To clean up the graphics window entirely, pull-down-select **File/ Start Over**. A general way of deleting any sketching entities (edges, dimensions, or constraints) is to right-click the entity in **Details View** and issue **Delete**. Also see 2.3.8[8-10], page 89, and 2.3.9[5-7], page 90.

How to abort a tool?

Simply press **ESC**. #

2.3.7 Modify Tools[Ref 4]

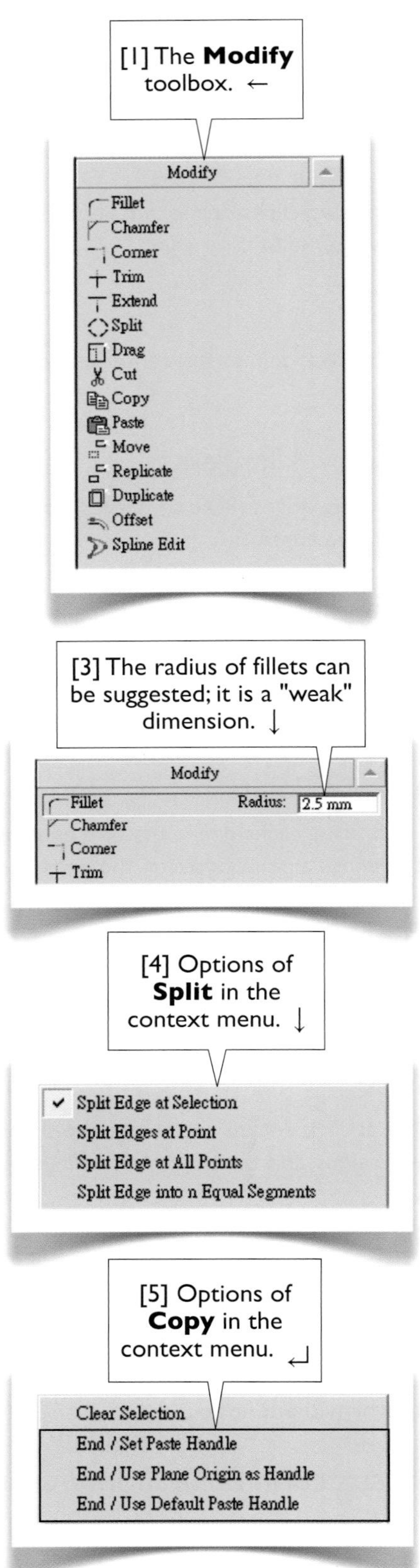

Fillet

[2] Select two edges or a vertex to create a fillet. The radius of the fillet can be specified in the toolbox [3]. Note that this radius value is a weak dimension; i.e., it can be changed by other dimensions or constraints.

Chamfer

Select two edges or a vertex to create an equal-length chamfer. The sizes of the chamfer can be specified in the toolbox.

Corner

Select two edges, and the edges will be trimmed or extended up to the intersection point and form a sharp corner. The clicking points decide which sides to be trimmed.

Trim

Select an edge, and the portion of the edge will be removed up to its intersection with another edge, axis, or point.

Extend

Select an edge, and the edge will be extended up to an edge or axis.

Split

This tool splits an edge into several segments depending on the options from the context menu [4]. **Split Edge at Selection**: Click an edge, and the edge will be split at the clicking point. **Split Edges at Point**: Click a point, and all the edges passing through that point will be split at that point. **Split Edge at All Points**: Click an edge, and the edge will be split at all points on the edge. **Split Edge into n Equal Segments**: Click an edge and specify a value *n*, and the edge will be split equally into *n* segments.

Drag

Drags a point or an edge to a new position. All the constraints and dimensions are preserved.

Copy

Copies the selected entities to a "clipboard." A **Paste Handle** must be specified using one of the methods in the context menu [5]. After completing this tool, **Paste** tool is automatically activated. ↗

Cut

[6] Similar to **Copy**, except that the copied entities are removed.

Paste

Pastes the entities in the "clipboard" to the graphics window. The click defines the point at which the **Paste Handle** positions. Many options can be chosen from the context menu [7], where the rotating angle **r** and the scaling factor **f** can be specified in the toolbox.

Move

Equivalent to a **Cut** followed by a **Paste**. (The original is removed.)

Replicate

Equivalent to a **Copy** followed by a **Paste**. (The original is preserved.)

Duplicate

Similar to **Replicate**. However, **Duplicate** copies entities to the same position in the active plane. **Duplicate** can be used to copy features of a solid body or plane boundaries.

Offset

Creates a set of edges that are offset by a distance from an existing set of edges.

Spline Edit

Used to modify flexible splines. You can insert, delete, drag the fit points, etc [8]. For details, see the reference[Ref 4]. ↗

[7] Options of **Paste** in the context menu. ↓

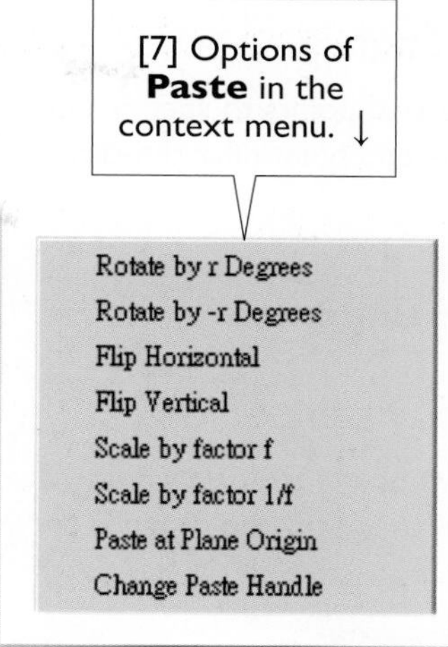

[8] Options of **Spline Edit** in the context menu. #

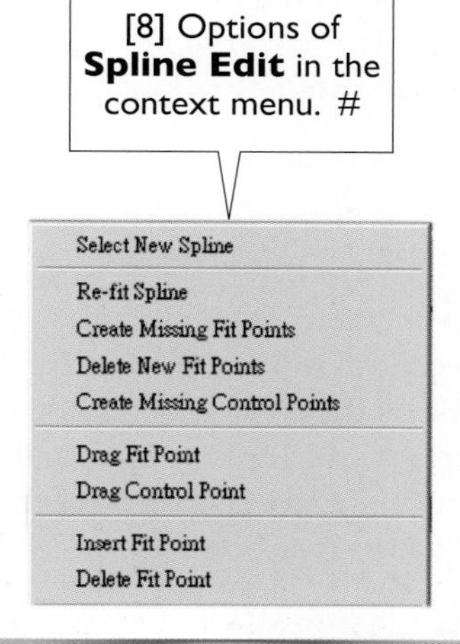

2.3.8 Dimensions Tools[Ref 5] [1]

[1] The **Dimension** toolbox. ↙

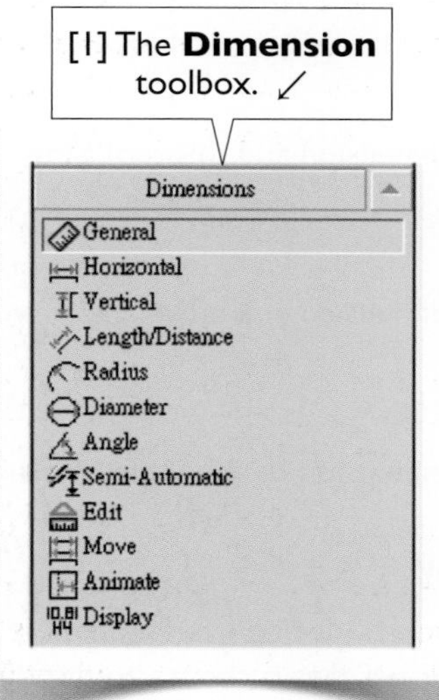

General

[2] Allows creation of any of the dimension types, depending on what edge and context-menu options are selected. If the selected edge is a straight line, the default dimension is its length ([3], next page.) If the selected edge is a circle or arc, the default dimension is its radius ([4], next page). ↵

Horizontal

[5] Select two points to specify a horizontal. If you select an edge (instead of a point), the end point near the click will be picked.

Vertical

Similar to **Horizontal**.

Length/Distance

Select two points to specify a distance dimension. You also can select a point and a line to specify the distance between the point and the line.

Radius

Select a circle or arc to specify a radius dimension. If you select an ellipse, the major (or minor) radius will be specified.

Diameter

Select a circle or arc to specify a diameter dimension.

Angle

Select two lines to specify an angle. By varying the selection order and location, you can control which angle you are dimensioning. The end of the lines that you select will be the arrow point of the hands, and the angle is measured counterclockwise from the first hand to the second. If the angle is not what you want, repeatedly choose **Alternate Angle** from the context menu until a correct angle is selected [6].

Semi-Automatic

This tool displays a series of dimensions automatically to help you fully dimension the sketch.

Edit

Click a dimension and this tool allows you to change its name or values.

Move

Click a dimension and move it to a new position.

Animate

Click a dimension to show the animated effects.

Display

Allows you to decide whether to display dimension names, values, or both. In this book, we always choose to display dimension values [7] rather than dimension names. ↗

[3] Options of **General** in the context menu if you select a line. ↓

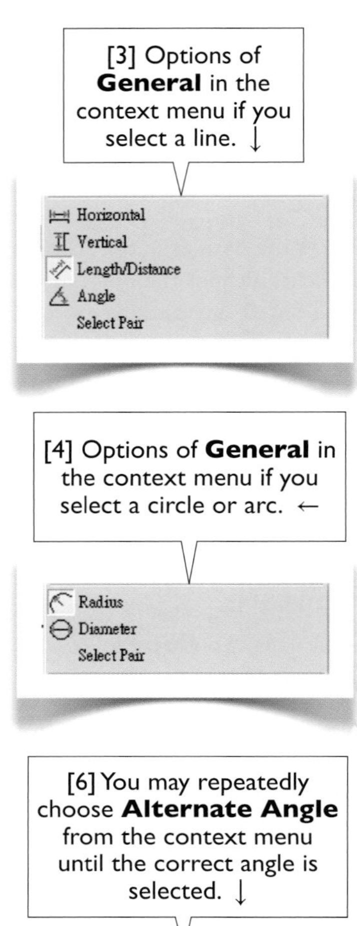

[4] Options of **General** in the context menu if you select a circle or arc. ←

[6] You may repeatedly choose **Alternate Angle** from the context menu until the correct angle is selected. ↓

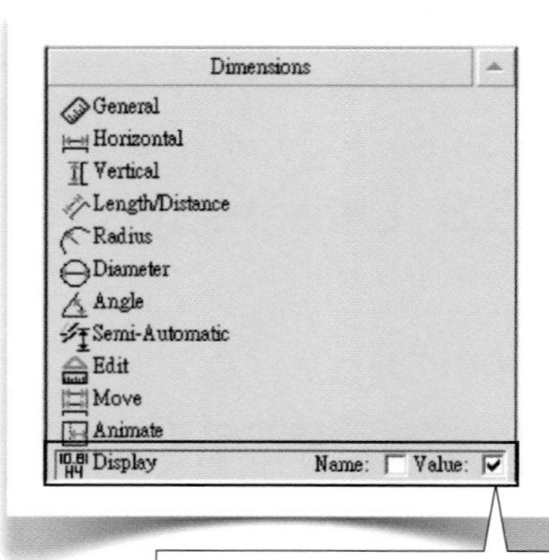

[7] In this book, we always display dimension values. ↲

How to delete dimensions?

[8] To delete a dimension, select the dimension in **Details View**, and choose **Delete** from the context menu [9]. You can delete ALL the dimensions by right-clicking **Dimensions** in **Details View** [10]. →

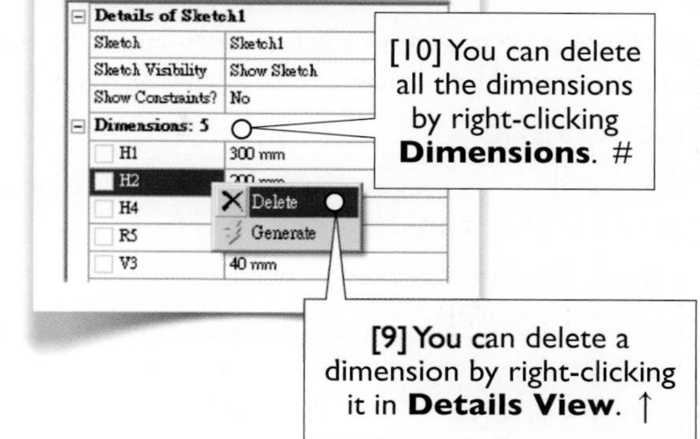

2.3.9 Constraints Tools[Ref 6]

Fixed

[2] Applies on an edge to make it fully constrained if **Fix Endpoints** is selected [3]. If **Fix Endpoints** is not selected, then the edge's endpoints can be changed, but not the edge's position and slope.

Horizontal

Applies to a line to make it horizontal.

Vertical

Applies to a line to make it vertical.

Perpendicular

Applies to two edges to make them perpendicular to each other.

Tangent

Applies to two edges, one of which must be a curve, to make them tangent to each other.

Coincident

Select two points to make them coincident. Or, select a point and an edge to make the edge or its extension pass through the point. There are other possibilities, depending on how you select the entities.

Midpoint

Select a line and a point to make the midpoint of the line coincide with the point.

Symmetry

Select a line or an axis, as the line of symmetry, and then select two entities to make them symmetric about the line of symmetry. →

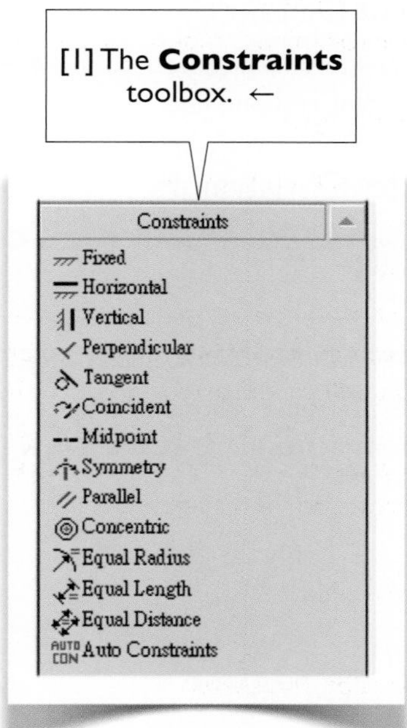

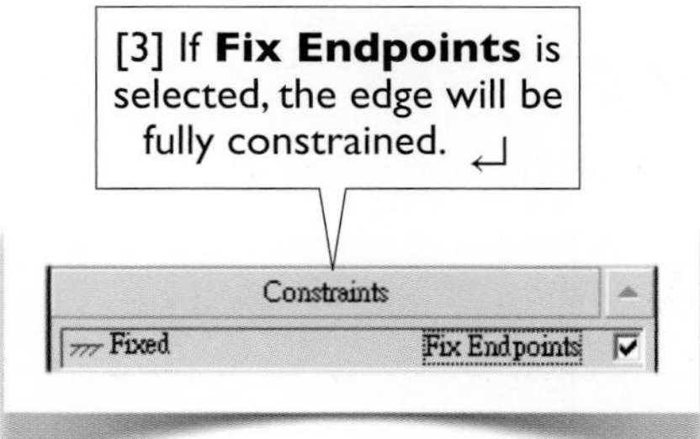

Parallel

[4] Applies to two lines to make them parallel to each other.

Concentric

Applies to two curves, which may be circle, arc, or ellipse, to make their centers coincident.

Equal Radius

Applies to two curves, which must be circle or arc, to make their radii equal.

Equal Length

Applies to two lines to make their lengths equal.

Equal Distance

Applies to two distances to make them equal. A distance can be defined by selecting two points, two parallel lines, or one point and one line.

Auto Constraints

Allows you to turn on/off **Auto Constraints** (2.3.5, page 84). ↓

How to delete constraints?

[5] By default, constraints are not displayed in **Details View**. To display constraints, select **Yes** for **Show Constraints?** in **Details View** [6]. To delete a constraint, right-click the constraint and issue **Delete** [7]. ↓

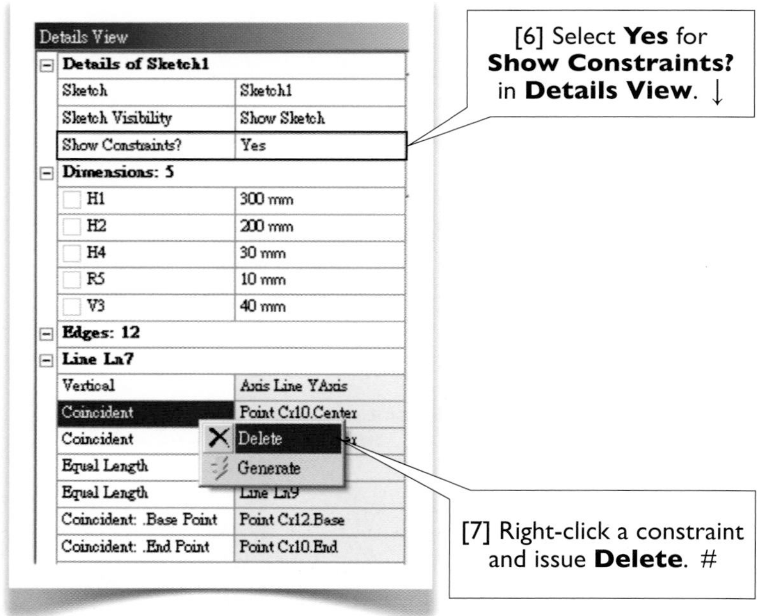

[6] Select **Yes** for **Show Constraints?** in **Details View**. ↓

[7] Right-click a constraint and issue **Delete**. #

2.3.10 Settings Tools[Ref 7]

Grid

[2] Allows you to turn on/off grid visibility and snap capability [3-4]. The grid is not required to enable snapping.

Major Grid Spacing

Allows you to specify **Major Grip Spacing** [5-6] if **Show in 2D** is turned on.

Minor-Steps per Major

Allows you to specify **Minor-Steps per Major** [7-8] if **Show in 2D** is turned on.

Snaps per Minor

Allows you to specify **Snaps per Minor** [9] if **Snap** is turned on. ↗

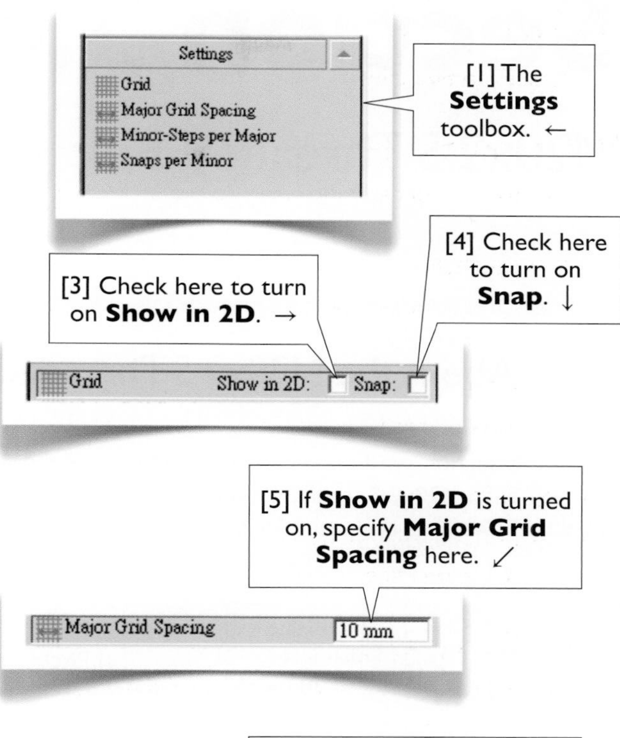

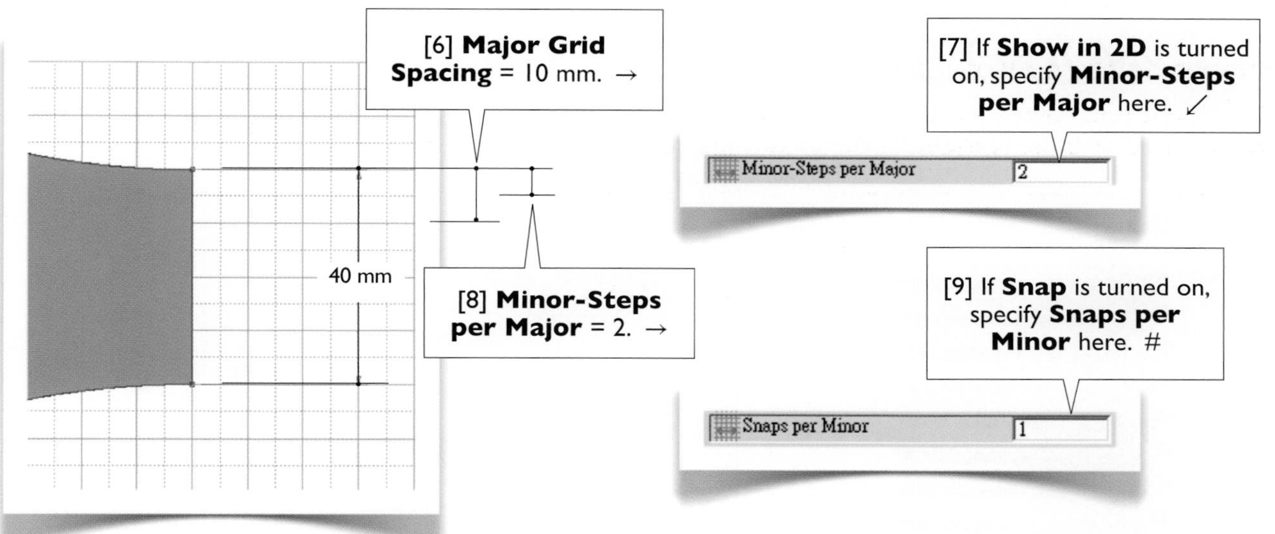

References

1. ANSYS Help//DesignModeler//ANSYS DesignModeler User's Guide//2D Sketching//Auto Constraints
2. ANSYS Help//DesignModeler//ANSYS DesignModeler User's Guide//2D Sketching//Constraints Toolbox//Auto Constraints
3. ANSYS Help//DesignModeler//ANSYS DesignModeler User's Guide//2D Sketching//Draw Toolbox
4. ANSYS Help//DesignModeler//ANSYS DesignModeler User's Guide//2D Sketching//Modify Toolbox
5. ANSYS Help//DesignModeler//ANSYS DesignModeler User's Guide//2D Sketching//Dimensions Toolbox
6. ANSYS Help//DesignModeler//ANSYS DesignModeler User's Guide//2D Sketching//Constraints Toolbox
7. ANSYS Help//DesignModeler//ANSYS DesignModeler User's Guide//2D Sketching//Settings Toolbox

Section 2.4

M20x2.5 Threaded Bolt

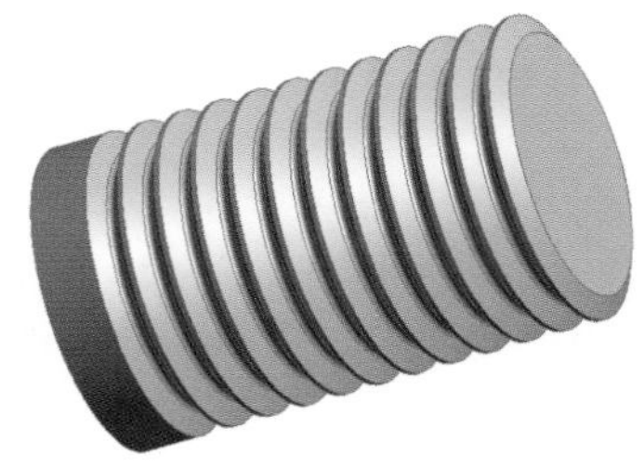

2.4.1 About the M20x2.5 Threaded Bolt[Refs 1, 2]

[1] In this section, we'll create a sketch and revolve the sketch 360° to generate a 3D solid body, a body representing a portion of an M20x2.5 threaded bolt as shown in [2-7]. We will use this sketch in Section 3.2 again to generate a 2D solid body, which is then used for a static structural simulation. ↓

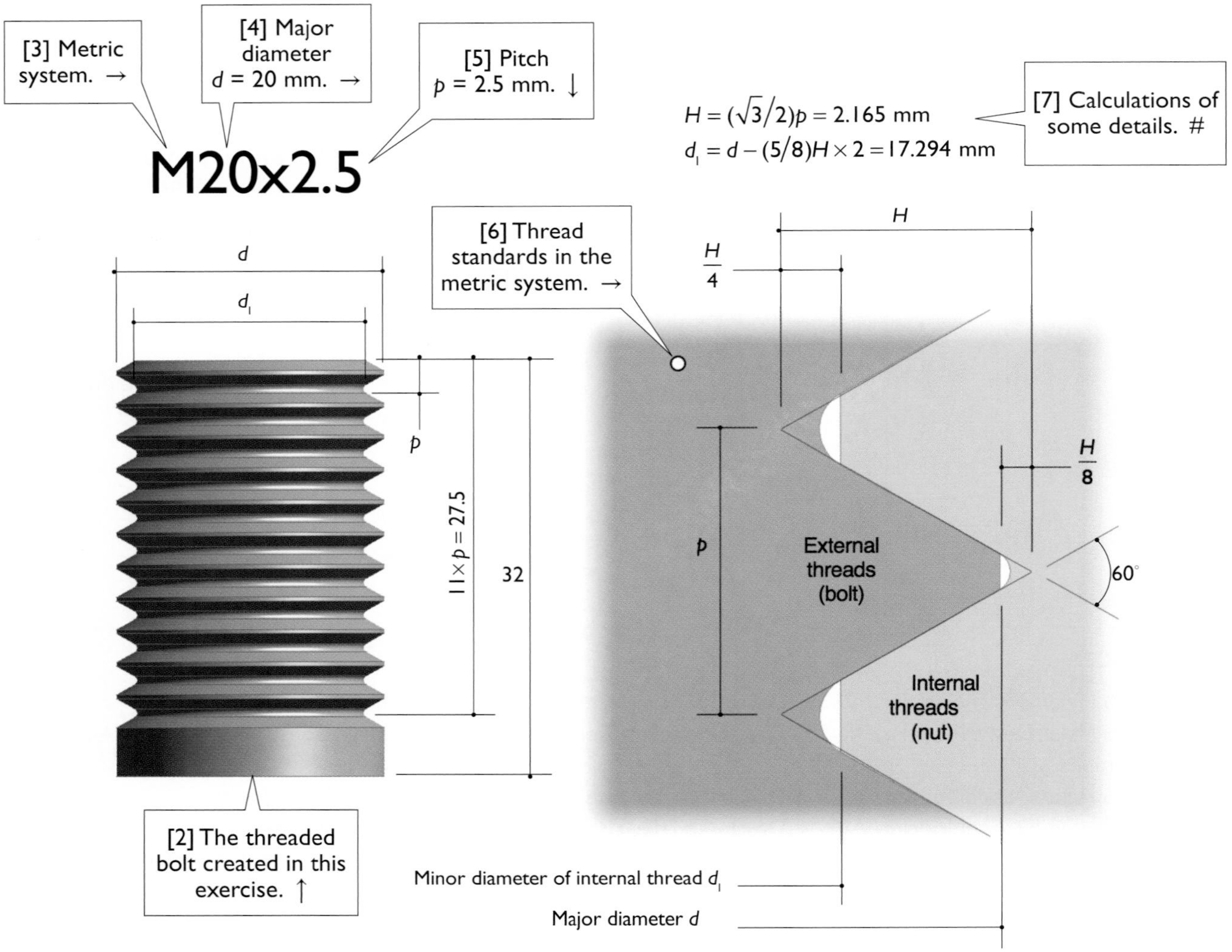

[3] Metric system. →

[4] Major diameter d = 20 mm. →

[5] Pitch p = 2.5 mm. ↓

M20x2.5

$H = (\sqrt{3}/2)p = 2.165$ mm

$d_1 = d - (5/8)H \times 2 = 17.294$ mm

[7] Calculations of some details. #

[6] Thread standards in the metric system. →

d

d_1

p

$11 \times p = 27.5$

32

[2] The threaded bolt created in this exercise. ↑

H

$\frac{H}{4}$

$\frac{H}{8}$

p

External threads (bolt)

Internal threads (nut)

60°

Minor diameter of internal thread d_1

Major diameter d

2.4.2 Draw a Horizontal Line

[1] Launch Workbench and create a **Geometry** system. Save the project as **Threads**. Start up DesignModeler. Select **Millimeter** as the length unit.

On **XYPlane**, draw a horizontal line and specify the dimensions (8.647 mm and 27.5 mm) as shown in [2]. →

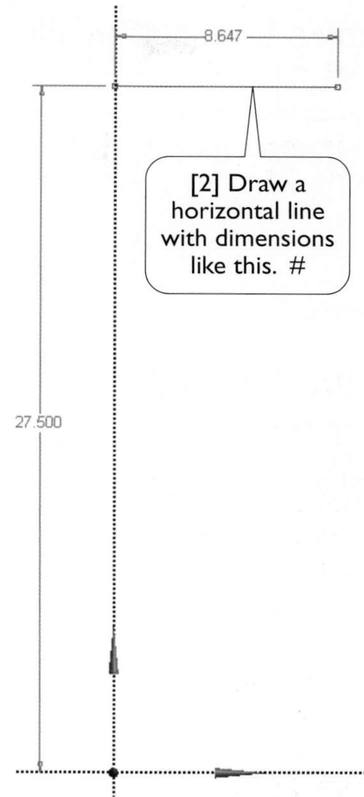

2.4.3 Draw a Polyline

[1] Draw a polyline of 3 segments [2-3] and specify the dimensions (30°, 60°, 60°, 0.541 mm, and 2.165 mm) as shown. To specify angle dimensions, please see **Angle**, 2.3.8[5], page 88. ↓

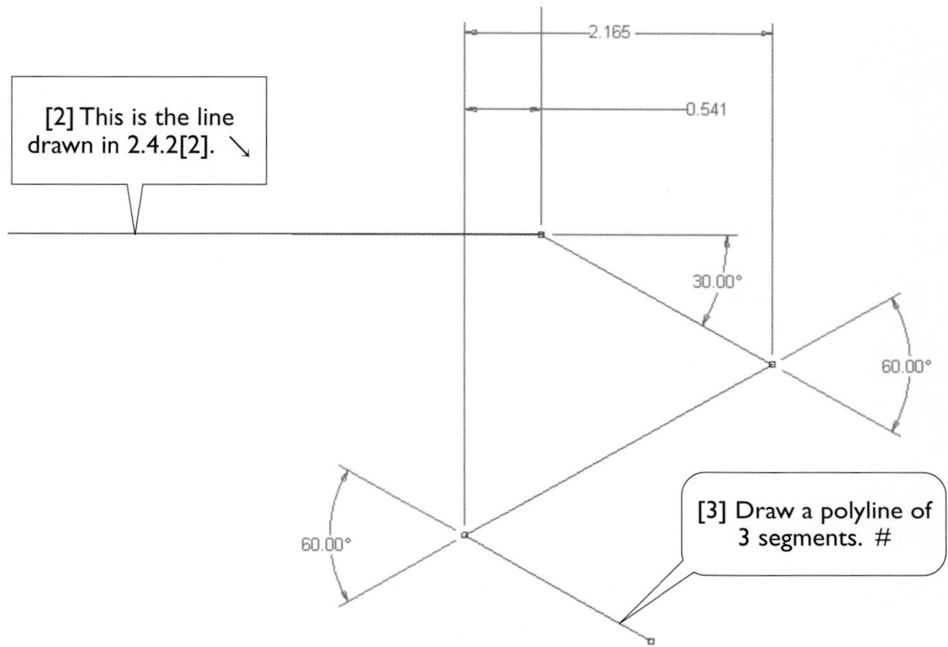

2.4.4 Draw a Line and a Fillet

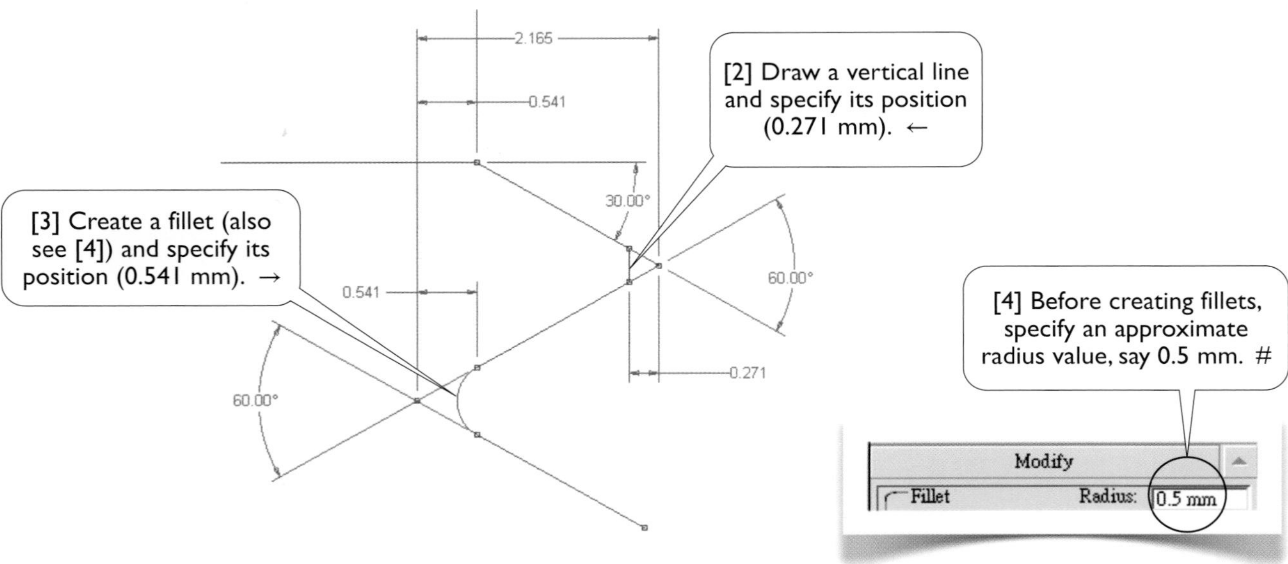

2.4.5 Trim Away Unwanted Segments

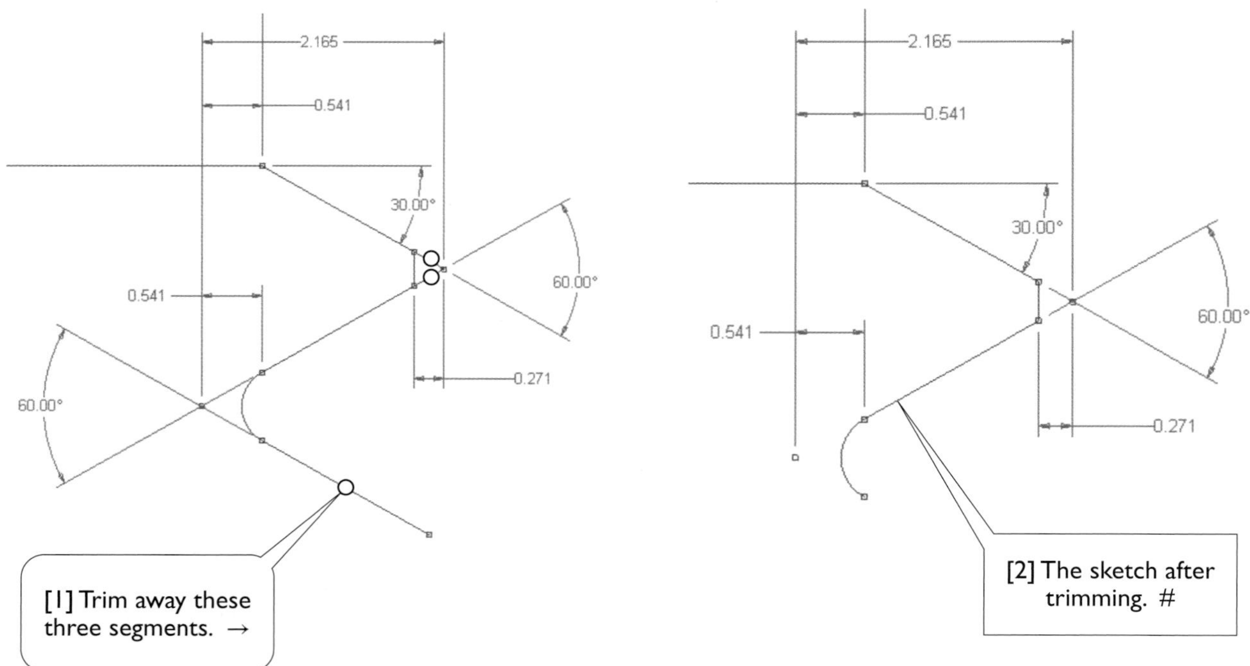

2.4.6 Replicate 10 Times

[1] Select the **Replicate** tool and select all segments except the horizontal line (4 segments in total), and replicate 10 times. Set the **Paste Handle** as shown in [2]. You may need to use **Selection Filter: Points** [3] (also see 2.2.8[7-8], page 74). ↓

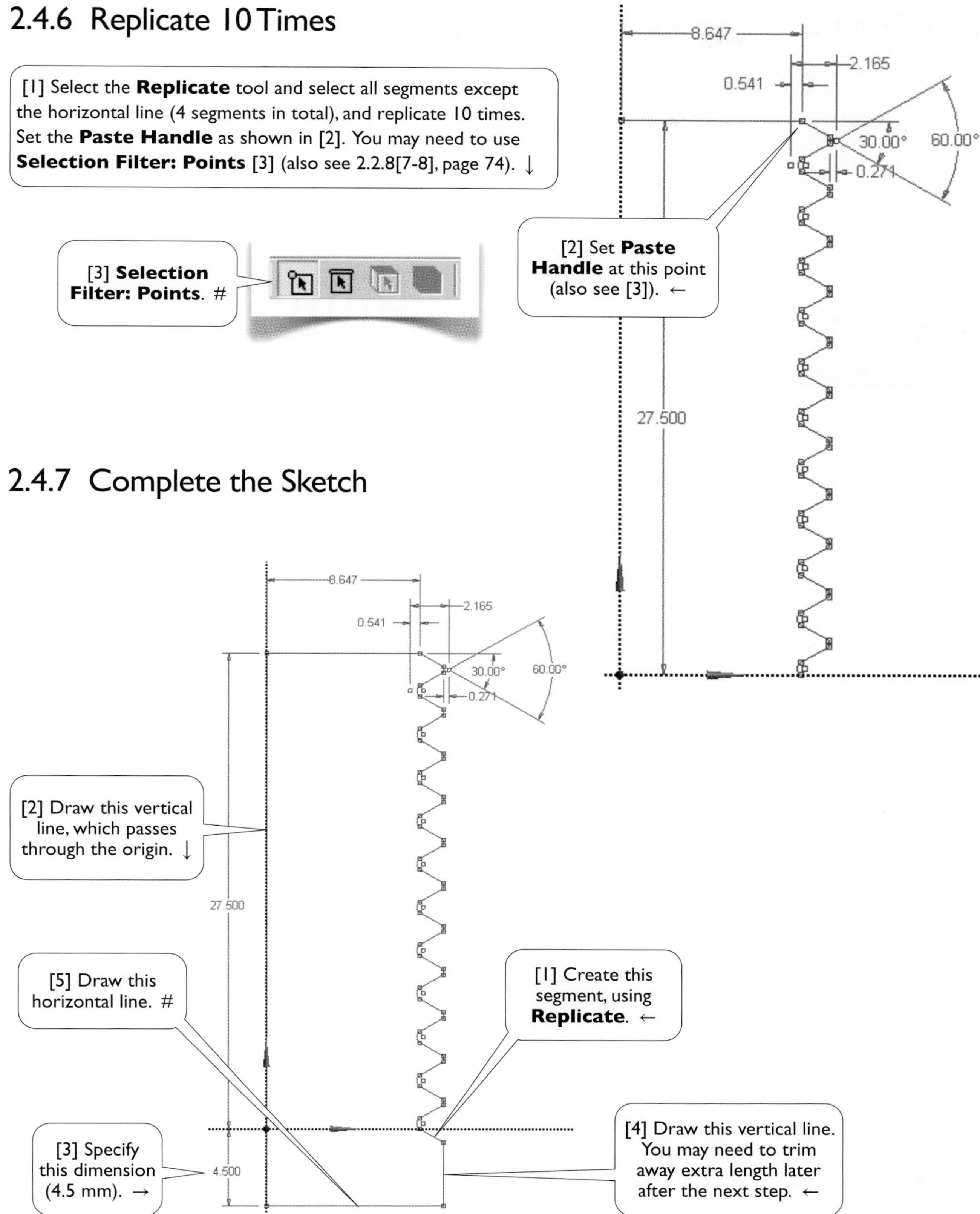

2.4.7 Complete the Sketch

2.4.8 Revolve to Create 3D Solid

[1] Click **Revolve** to generate a solid of revolution. Select the Y-axis as the axis of revolution [2]. Remember to click **Generate**. Save the project and exit Workbench. We will resume this project in Section 3.2. ↓

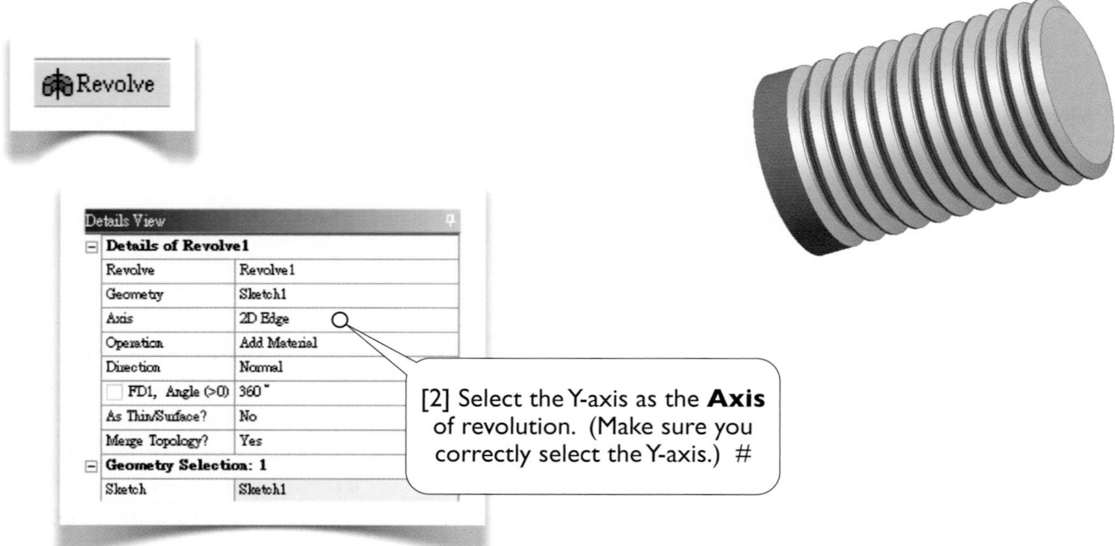

References

1. Zahavi, E., *The Finite Element Method in Machine Design*, Prentice-Hall, 1992; Chapter 7. Threaded Fasteners.
2. Deutschman, A. D., Michels, W. J., and Wilson, C. E., *Machine Design: Theory and Practice*, Macmillan Publishing Co., Inc., 1975; Section 16-6. Standard Screw Threads.

Section 2.5

Spur Gears

2.5.1 and 2.5.2 give the geometry of the spur gear used in this section. If you are not interested in these geometric details for now, you may skip them and jump directly to 2.5.3 (page 99).

2.5.1 About the Spur Gears[Refs 1, 2]

[1] The figure below shows a pair of identical spur gears in mesh [2-5]. Spur gears have their teeth cut parallel to the axis of the shaft on which the gears are mounted, transmitting power between the parallel shafts. To maintain a constant angular velocity ratio, two meshing gears must satisfy a fundamental law of gearing: the shape of the teeth must be such that the common normal [9] at the point of contact between two teeth must always pass through a fixed point on the line of centers[Ref 1] [6]. The contact point is called the *pitch point* [7].

The angle between the line of action [9] and the common tangent of the pitch circles [8] is known as the *pressure angle*. The spur gear is defined by its pitch radius (r_p = 2.5 in) [4], pressure angle (α = 20°) [9], and number of teeth (N = 20). The teeth are cut with a radius of addendum r_a = 2.75 in [10] and a radius of dedendum r_d = 2.2 in [11]. The shaft has a radius of 1.25 in [12]. All fillets have a radius of 0.1 in [13]. The thickness of the gear is 1.0 in. ✓

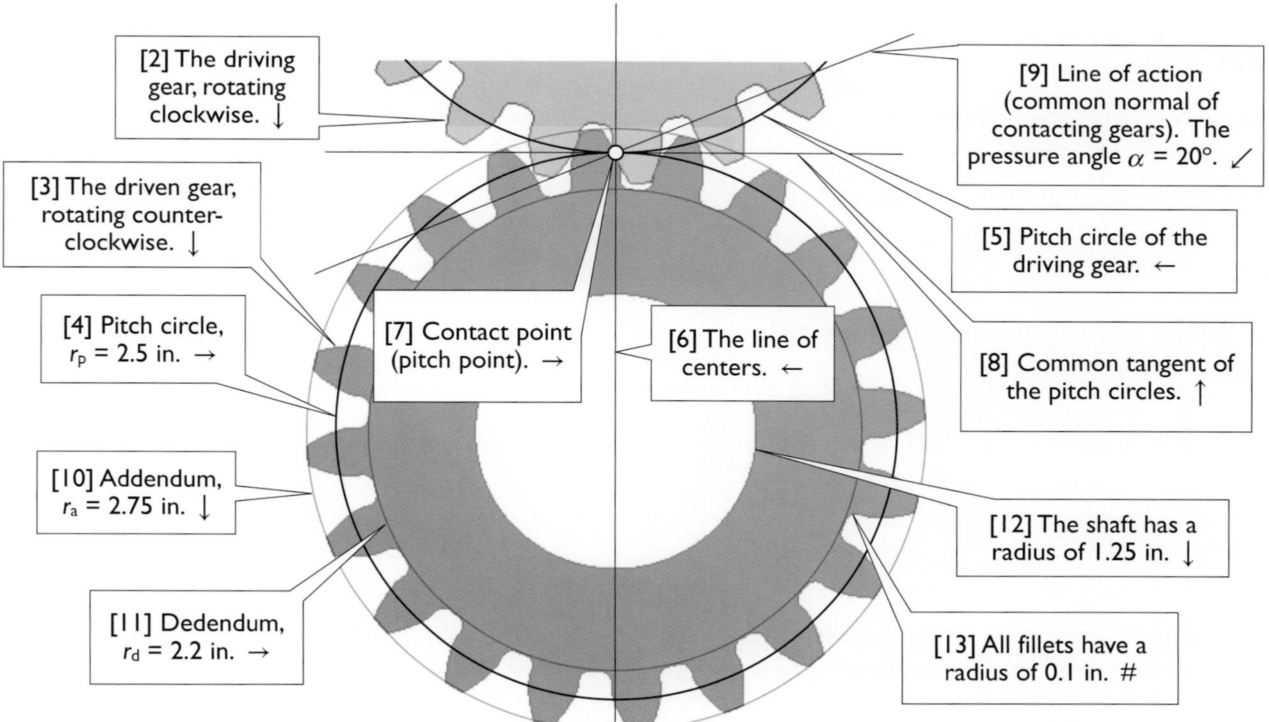

2.5.2 About Involute Curves[Refs 1, 2]

[1] To satisfy the fundamental law of gearing, gear profiles are usually cut to an *involute curve* [2], which may be constructed by wrapping a string (BA) around a *base circle* [3], and then tracing the path (A-P-F) of a point (A) on the string. Given the gear's pitch radius r_p and pressure angle α, we can calculate the coordinates of each point on the involute curve. ↘

[4] For example, let's calculate the polar coordinates (r,θ) of an arbitrary point A [5] on the involute curve. Note that BA and CP are tangent lines of the base circle, and F is a foot of perpendicular.

Since $\overset{\frown}{APF}$ is an involute curve and $BCDEF$ is the base circle, by the definition of involute curve,

$$\overline{BA} = \overset{\frown}{BCDEF} \quad (1)$$

$$\overline{CP} = \overset{\frown}{CDEF} \quad (2)$$

In ΔOCP,

$$r_b = r_p \cos\alpha \quad (3)$$

In ΔOBA,

$$r = \frac{r_b}{\cos\phi} \quad (4)$$

Or,

$$\phi = \cos^{-1}\frac{r_b}{r} \quad (5)$$ ↗

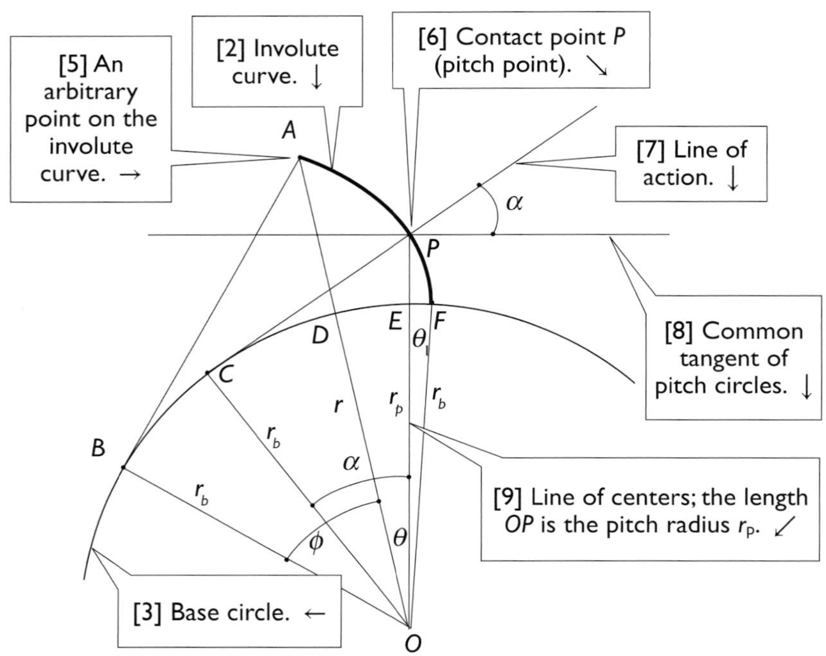

[10] To calculate θ, we notice that

$$\overset{\frown}{DE} = \overset{\frown}{BCDEF} - \overset{\frown}{BCD} - \overset{\frown}{EF}$$

Dividing the equation with r_b and using Eq. (1),

$$\frac{\overset{\frown}{DE}}{r_b} = \frac{\overline{BA}}{r_b} - \frac{\overset{\frown}{BCD}}{r_b} - \frac{\overset{\frown}{EF}}{r_b}$$

If radian is used, then the above equation can be written as

$$\theta = (\tan\phi) - \phi - \theta_1 \quad (6)$$

The last term θ_1 is the angle $\angle EOF$, which can be calculated by dividing Eq. (2) with r_b,

$$\frac{\overline{CP}}{r_b} = \frac{\overset{\frown}{CDEF}}{r_b}, \text{ or } \tan\alpha = \alpha + \theta_1, \text{ or}$$

$$\theta_1 = (\tan\alpha) - \alpha \quad (7)$$

We'll show how to calculate polar coordinates (r,θ) using Eqs. (3-7). The polar coordinates then can be easily transformed to rectangular coordinates, using O as origin and OP as y-axis,

$$x = -r\sin\theta, \quad y = r\cos\theta \quad (8)$$ ↵

Numerical Calculations of Coordinates

[11] In our case, the pitch radius $r_p = 2.5$ in, and pressure angle $\alpha = 20°$; from Eqs. (3) and (7) respectively,

$$r_b = 2.5\cos 20° = 2.349232 \text{ in}$$

$$\theta_1 = \tan 20° - \frac{20°}{180°}\pi = 0.01490438 \text{ (rad)}$$

The table below lists the calculated coordinates. The values in the first column (r) are chosen such that, except the pitch point (r = 2.5 in), the intermediate points are at the quarter points between r_b (r = 2.349232 in) and r_a (r = 2.75 in). Also note that, when using Eqs. (6) and (7), radians are used as the unit of angles; in the table below, however, degrees are used. #

r in.	ϕ Eq. (5), degrees	θ Eq. (6), degrees	$x = -r\sin\theta$ in.	$y = r\cos\theta$ in.
2.349232	0.000000	-0.853958	-0.03501	2.3490
2.449424	16.444249	-0.387049	-0.01655	2.4494
2.500000	20.000000	0.000000	0.00000	2.5000
2.549616	22.867481	0.442933	0.01971	2.5495
2.649808	27.555054	1.487291	0.06878	2.6489
2.750000	31.321258	2.690287	0.12908	2.7470

2.5.3 Draw an Involute Curve

[1] Launch Workbench. Create a **Geometry** system. Save the project as **Gear**. Start up DesignModeler. Select **Inch** as the length unit.

From the **Draw** toolbox, select the **Construction Point** tool, draw 6 points and specify dimensions as shown, where the vertical dimensions are measured from the X-axis and the horizontal dimensions are measured from the Y-axis. The pitch point [2] is coincident with the Y-axis.

Connect these six points using the **Spline** tool in the **Draw** toolbox, leaving **Flexible** option on, finishing the spline with **Open End**. →

[3] It is equally good that you draw the spline by using the **Spline** tool directly without first creating construction points. To do so, at the end of the **Spline** tool, select **Open End with Fit Points** from the context menu. After dimensioning each point, use the **Spline Edit** tool to edit the spline and select **Re-fit Spline** [4] from the context menu to smooth out the spline. ↗

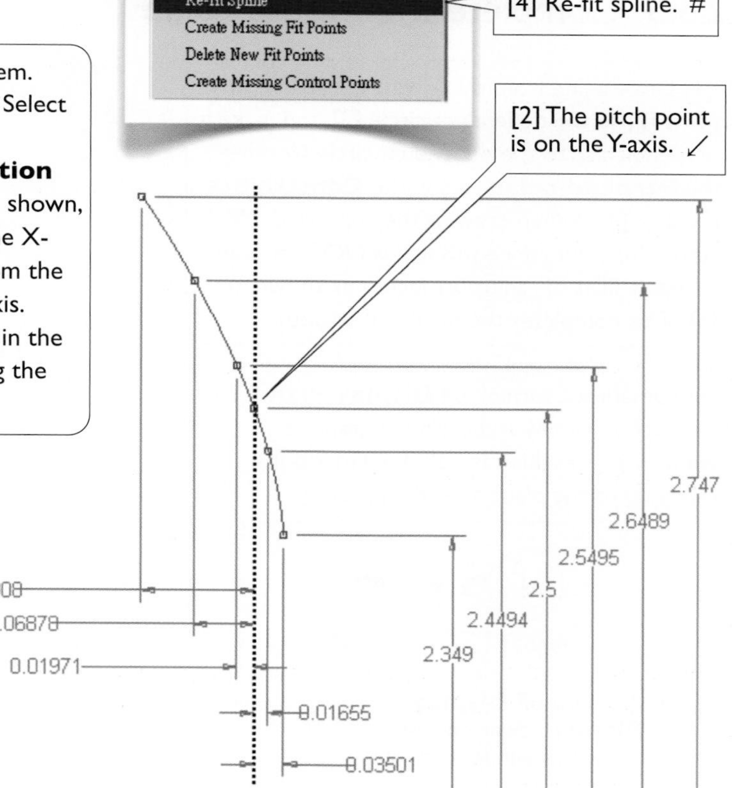

2.5.4 Draw Circles

[1] Using the **Circle** tool in the **Draw** toolbox, draw three circles as shown in [2-4]. Note that the radius of the addendum circle is defined by clicking the outermost construction point [4]. Specify radii for the shaft circle (1.25 in) and the dedendum circle (2.2 in). ↓

[4] The radius of the addendum circle is defined by clicking the outermost construction point. #

[3] The dedendum circle. ↗

[2] The shaft circle. ↑

R1.25

R2.2

2.5.5 Complete the Tooth Profile

[1] Draw a line from the lowest construction point [2] to the dedendum circle [3], and make it perpendicular to the dedendum circle by using the **Perpendicular** tool in the **Constraints** toolbox [4]. When drawing the line, avoid a **V** auto-constraint (since this line is NOT vertical). Create a fillet of radius 0.1 inches as shown in [5]. This completes the profile of a tooth. →

[6] Sometimes, turning off **Display Plane** may be helpful when working on the graphics window [7]. In this case, all the dimensions referring to the plane axes disappear. ↓

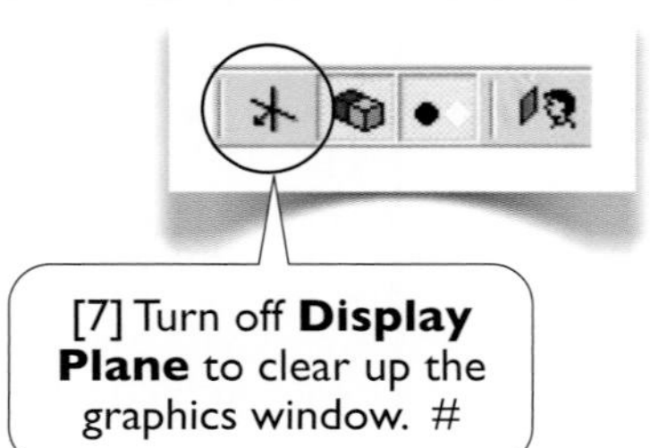

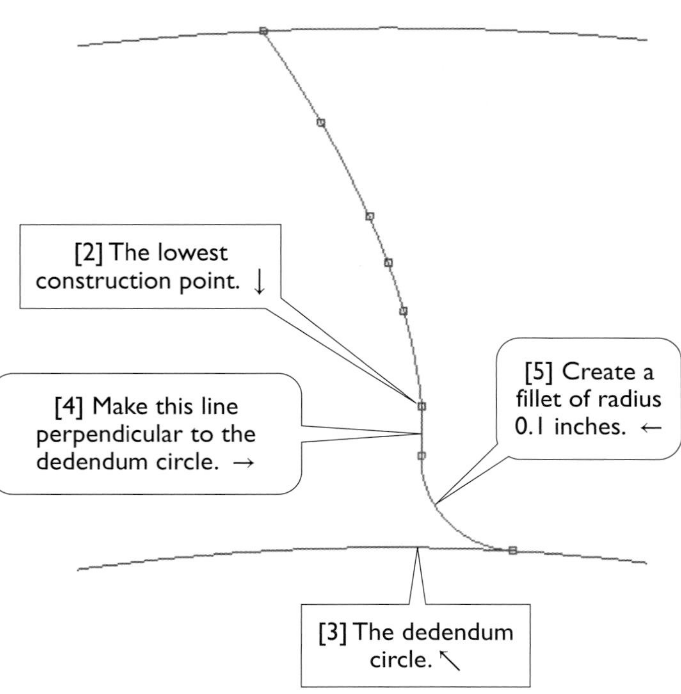

2.5.6 Replicate the Profile

[1] Select the **Replicate** tool, type 9 (degrees) for **r**. Select the profile (3 segments in total), **End/Use Plane Origin as Handle**, **Flip Horizontal**, **Rotate by r**, and **Paste at Plane Origin** [2]. End the **Replicate** tool by pressing **ESC**.

Note that the gear has 20 teeth, each spanning 18 degrees. The angle between the two pitch points [3] is 9 degrees. →

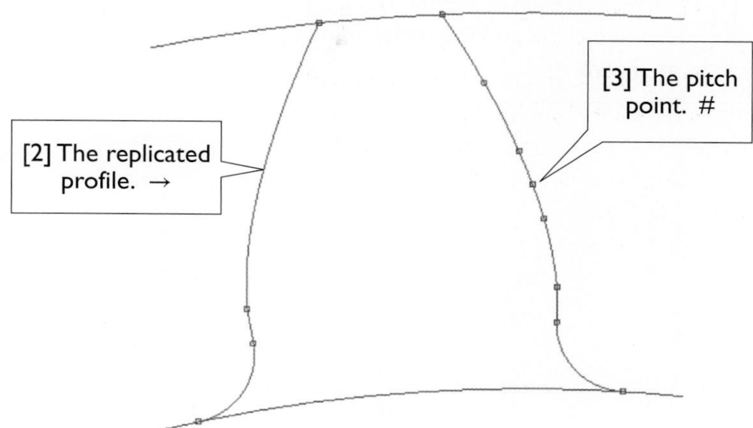

2.5.7 Replicate the Tooth 19 Times

[1] Click the **Replicate** tool again, type 18 (degrees) for **r**. Select both left and right profiles (6 segments in total), **End/Use Plane Origin as Handle**, **Rotate by r**, and **Paste at Plane Origin**. Repeat the last two steps (rotate and paste) until completing a full circle (20 teeth in total). #

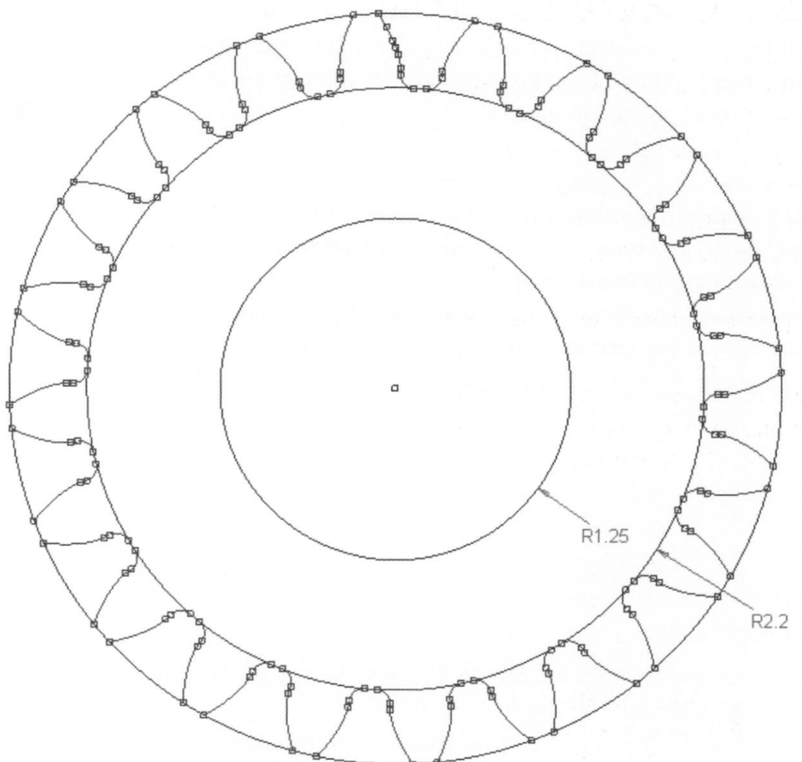

2.5.8 Trim Away Unwanted Segments

[1] Trim away unwanted segments in the addendum circle and the dedendum circle, as shown. #

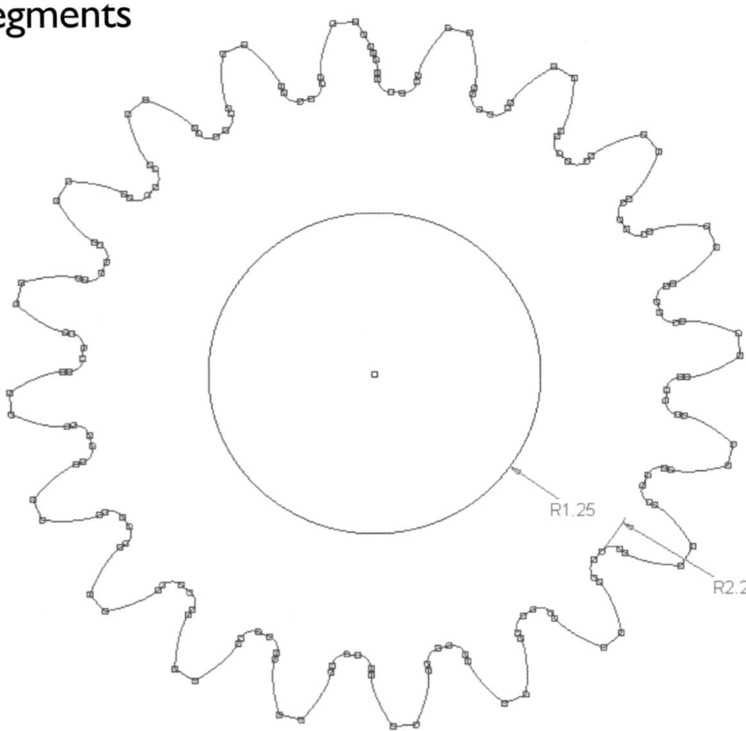

2.5.9 Extrude to Create 3D Solid

[1] Extrude the sketch 1.0 inch to create a 3D solid. Save the project and exit from Workbench. We will resume this project again in Section 3.4. ↓

[2] It is equally good that you create a single tooth (a 3D solid body) and then duplicate it by using **Create/ Pattern** in the **Modeling** mode. In this exercise, however, we use **Replicate** in **Sketching** mode because our focus in this chapter is to practice sketching techniques. #

References

1. Deutschman, A. D., Michels, W. J., and Wilson, C. E., *Machine Design: Theory and Practice*, Macmillan Publishing Co., Inc., 1975; Chapter 10. Spur Gears.
2. Zahavi, E., *The Finite Element Method in Machine Design*, Prentice-Hall, 1992; Chapter 9. Spur Gears.

Section 2.6

Microgripper

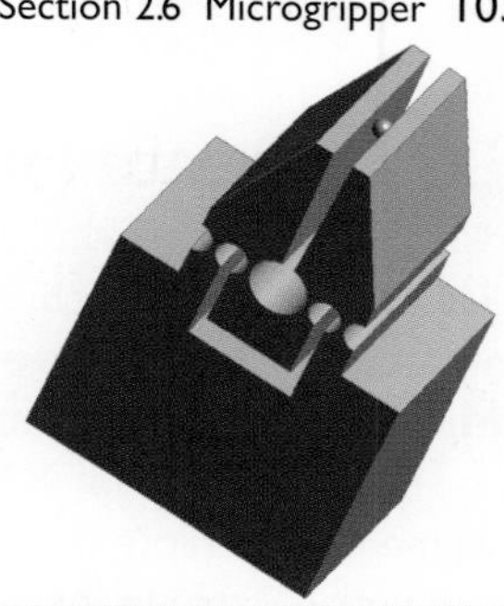

2.6.1 About the Microgripper[Refs 1, 2]

[1] The microgripper is made of a rubber-like polymer material and actuated by a shape memory alloy (SMA) actuator [2-4]. The motion of the SMA is caused by temperature change, which is controlled by electric current. In the lab, the microgripper is tested by gripping a steel bead of a diameter of 30 micrometers [5].

In this section, we will create a solid model for the microgripper. The model will be used for simulation in Section 13.3 to assess the gripping forces on the bead with an actuation force of the SMA actuator. ↓

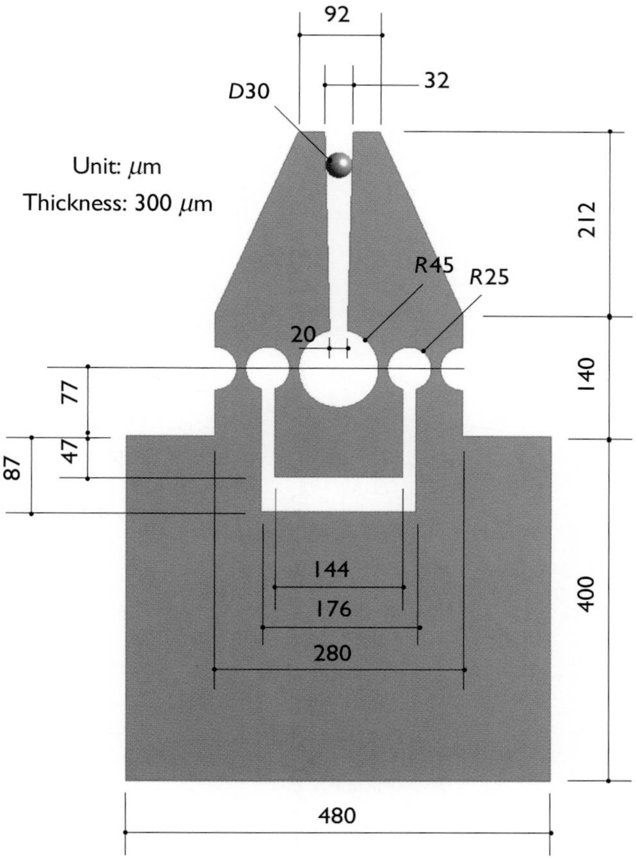

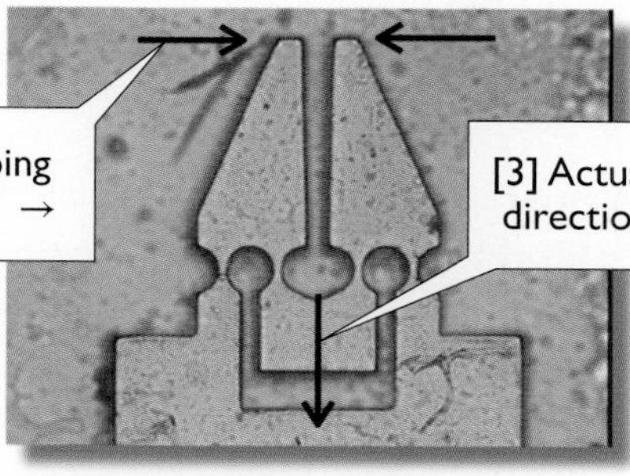

Unit: μm

Thickness: 300 μm

[2] Gripping direction. →

[3] Actuation direction. ↓

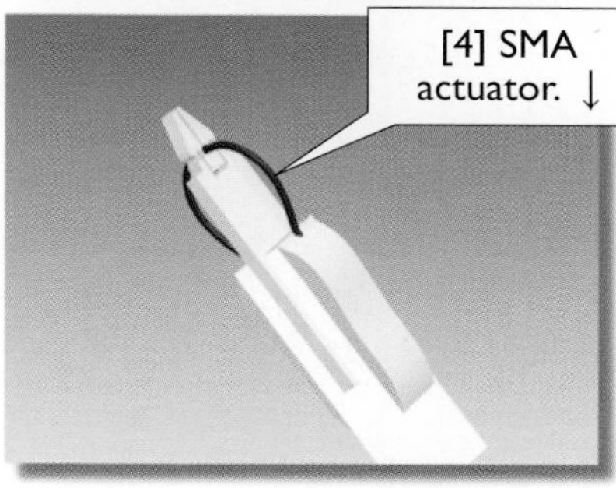

[4] SMA actuator. ↓

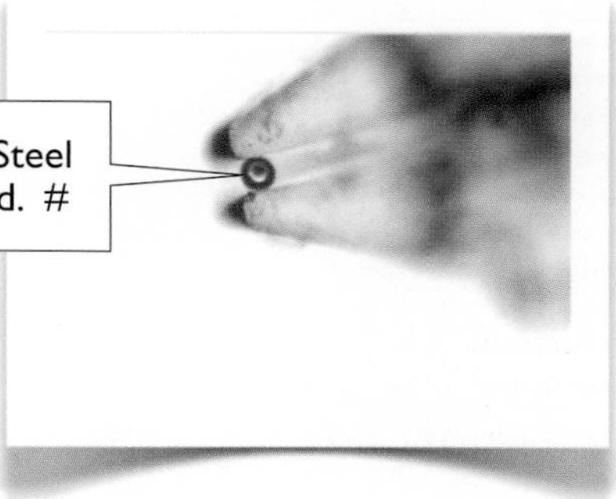

[5] Steel bead. #

2.6.2 Create Half of the Gripper

[1] Launch Workbench. Create a **Geometry** system. Save the project as **Microgripper**. Start up DesignModeler. Select **Micrometer** as the length unit.

Draw a sketch on **XYPlane** as shown in [2]. Trim away unwanted segments [3]. Note that we drew only half of the model because of the symmetry. Extrude the sketch 150 μm both sides symmetrically (the total depth is 300 μm) [4]. We now have a half of the gripper [5]. ↓

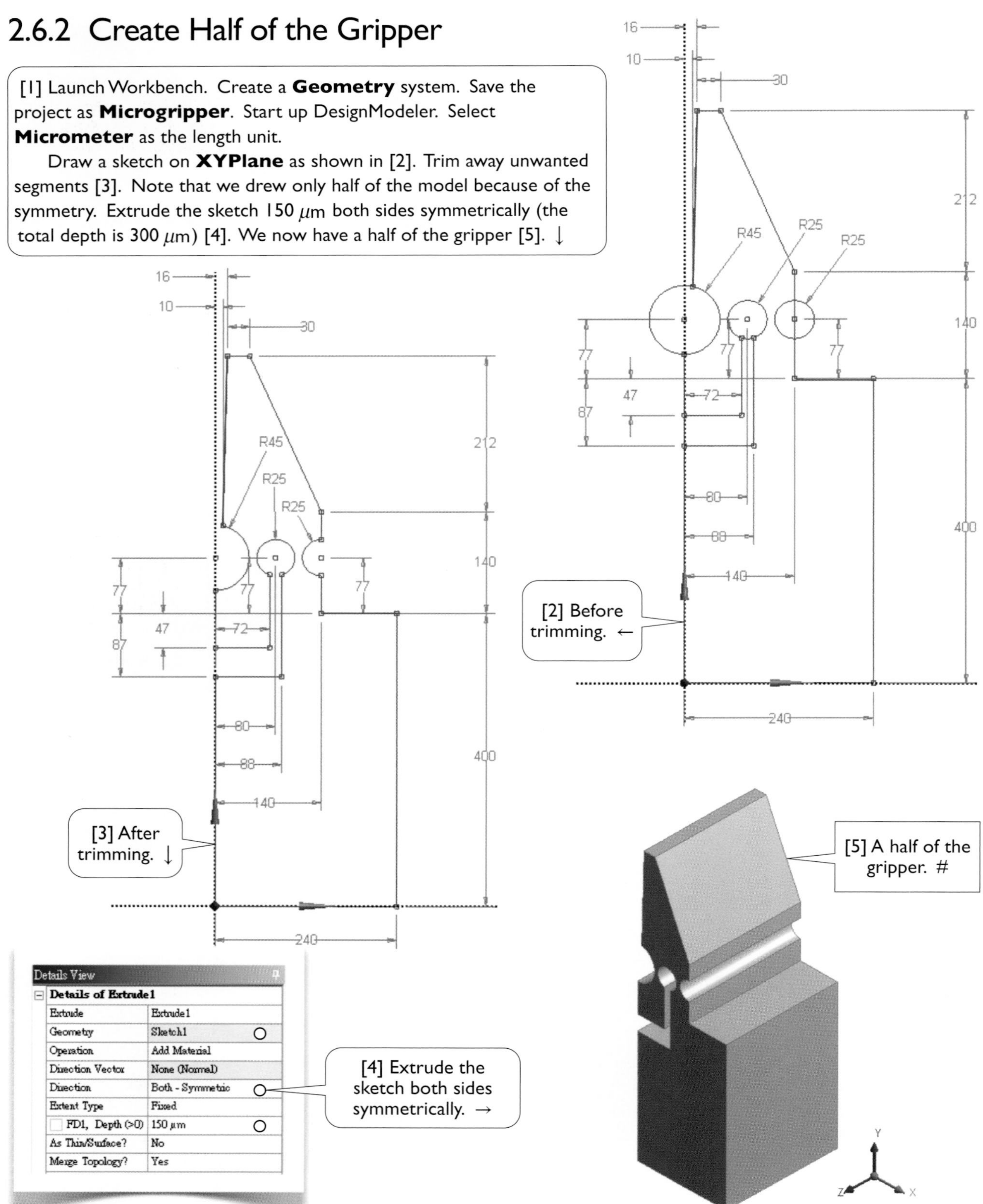

[2] Before trimming. ←

[3] After trimming. ↓

[4] Extrude the sketch both sides symmetrically. →

[5] A half of the gripper. #

2.6.3 Mirror Copy the Solid Body

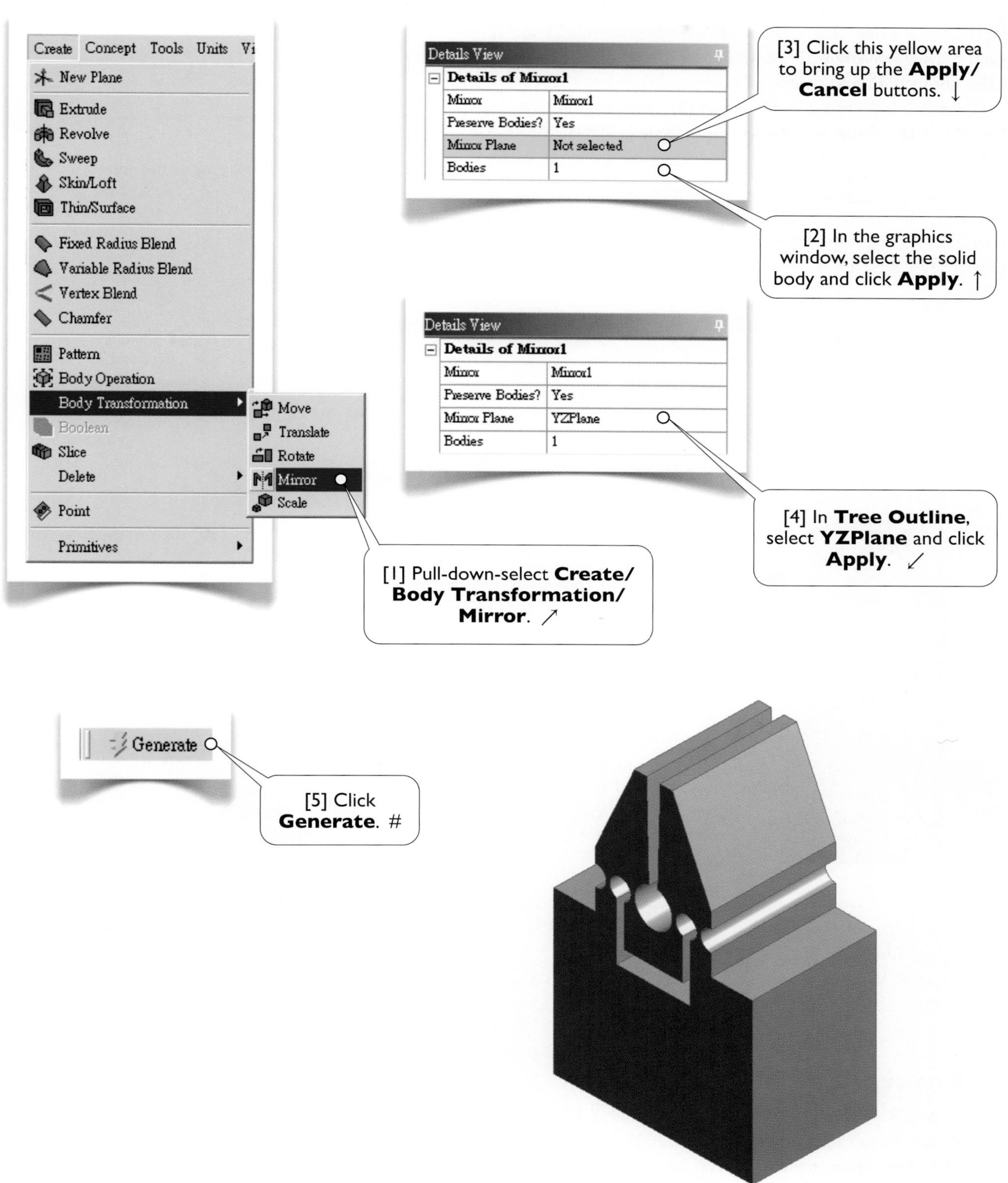

2.6.4 Create the Bead

[1] Create a new sketch on **XYPlane** as shown in [2-3] and draw a semicircle as shown in [4-7]. Revolve the sketch 360° about the Y-axis to create the bead [8]. Note that the two bodies are treated as two parts [9]. Rename the two bodies as **Gripper** and **Bead** respectively [10]. →

[2] Select **XYPlane**. →

[3] Click **New Sketch**. ↙

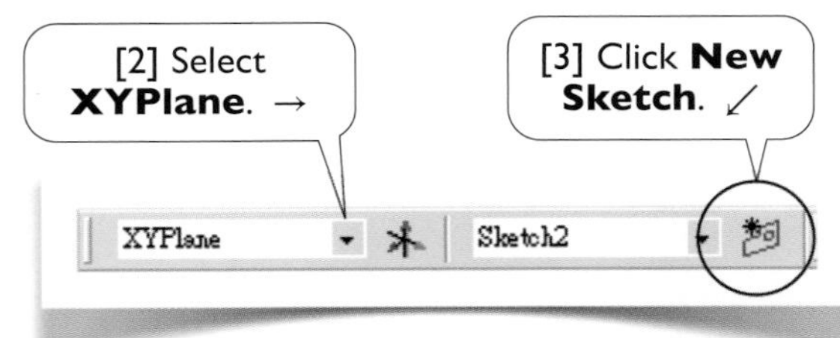

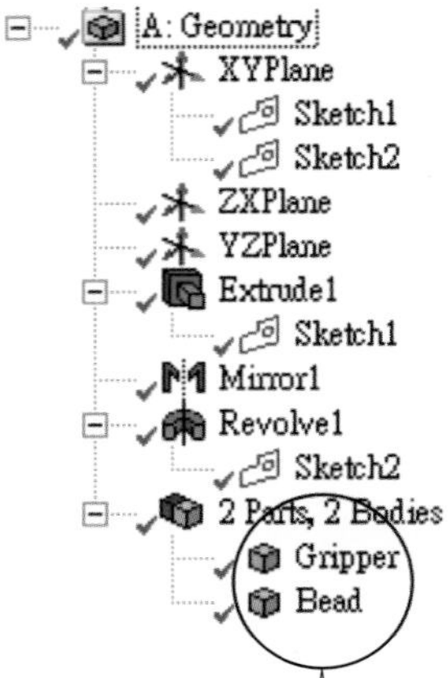

[4] The semicircle can be created by creating a full circle and then trimming it using the axis. ↓

[5] Close the sketch by drawing a vertical line. ↘

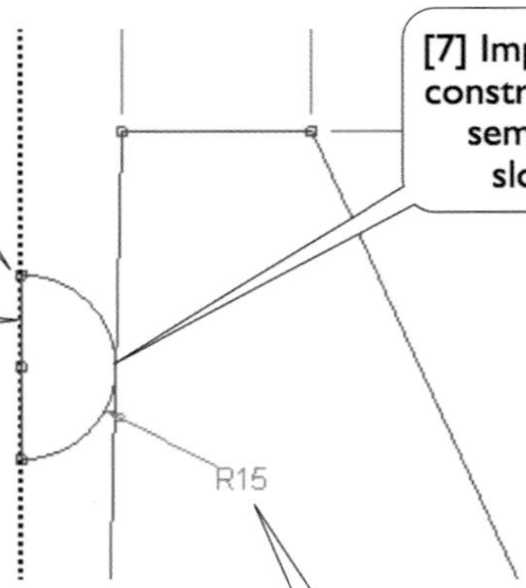

[6] Specify the radius (15 μm). ↑

[7] Impose a **Tangent constraint** between the **semicircle** and the **sloping** line. ↙

[8] **Revolve** the sketch about the Y-axis to create a sphere. ↓

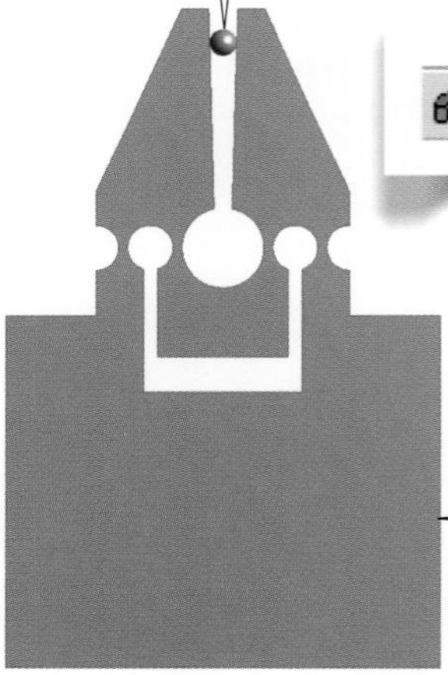

[9] The two bodies are treated as two parts (see [10]). ↗

[10] Right-click to rename each body like this. ↙

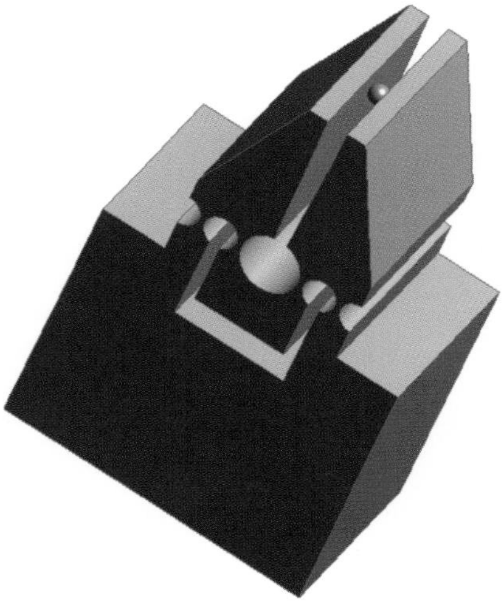

Wrap Up

[11] Close DesignModeler, save the project and exit Workbench. We will resume this project in Section 13.3. #

References

1. Chang, R. J., Lin ,Y. C., Shiu, C. C., and Hsieh,Y. T., "Development of SMA-Actuated Microgripper in Micro Assembly Applications," IECON, IEEE,Taiwan, 2007.
2. Shih, P.W., *Applications of SMA on Driving Micro-gripper*, MS Thesis, NCKU, ME,Taiwan, 2005.

Section 2.7

Review

2.7.1 Keywords

Choose a letter for each keyword from the list of descriptions

1. ( ) Auto Constraints
2. ( ) Branch
3. ( ) Constraint Status
4. ( ) Context Menu
5. ( ) Edge
6. ( ) Modeling Mode
7. ( ) Model Tree
8. ( ) Object
9. ( ) Paste Handle
10. ( ) Sketching Mode
11. ( ) Sketching Plane
12. ( ) Sketch
13. ( ) Selection Filter

Answers:

1. (J) 2. (G) 3. (M) 4. (I) 5. (D) 6. (B) 7. (F) 8. (H)
9. (L) 10. (A) 11. (C) 12. (E) 13. (K)

List of Descriptions

(A) An environment under DesignModeler; its function is to draw sketches on a plane.

(B) An environment under DesignModeler; its function is to create 3D or 2D bodies.

(C) The plane on which a sketch is created. Each sketch must be associated with a plane; each plane may have multiple sketches on it. Usage of planes is not limited for storing sketches.

(D) In **Sketching** mode, it may be a (straight) line or a curve. A curve may be a circle, ellipse, arc, or spline.

(E) It consists of points and edges. Dimensions and constraints may be imposed on these entities.

(F) It is the structured representation of a geometry and displayed on **Tree Outline** in **DesignModeler**. It consists of features and a part branch; their order is important. The parts are the only objects exported to **Mechanical**.

(G) An object of a model tree and consists of one or more objects under itself.

(H) A leaf or branch of a model tree.

(I) The menu that pops up when you right-click your mouse. The contents of the menu depend on what you click.

(J) While drawing in **Sketching** mode, by default, DesignModeler attempts to detect the user's intentions and tries to automatically impose constraints on points or edges. Detection is performed over entities on the active plane, not just active sketch. It can be switched on/off in the **Constraints** toolbox.

(K) It filters one type of geometric entity. When it is turned on/off, the corresponding type of entity becomes selectable/unselectable. In **Sketching** mode, there are two selection filters, namely points and edges filters. Along with these two filters, face and body selection filters are available in **Modeling** mode.

(L) A reference point used in a copy/paste operation. The point is defined during copying and will coincide with a specified location when pasting.

(M) In **Sketching** mode, entities are color coded to indicate their constraint status: greenish-blue for under-constrained; blue and black for well constrained (i.e., fixed in the space); red for over-constrained; gray for inconsistent.

2.7.2 Additional Workbench Exercises

Create Geometric Models with Your Own Way

After so many exercises, you should be able to figure out many alternative ways of creating the geometric models in this chapter. Try to re-create the models in this chapter using your own way.

Chapter 3

2D Simulations

All the real-world bodies are 3D bodies; there are no such things as 2D bodies. Some problems, however, can be simplified and simulated in a 2D space. As an example, consider an axisymmetric body subject to axisymmetric loads, in which all the particles with the same radial and axial coordinates (R and Y) share the same behaviors regardless of their tangential coordinate (θ). Thus, we can eliminate the tangential coordinate and reduce the problem to a 2D (in R-Y space) problem. Other 2D cases include *plane stress* problems and *plane strain* problems, which will be defined in Section 3.3.

Reducing a problem to 2D has many advantages over a 3D approach, and you should always do it whenever possible. These advantages include (a) simpler to build the geometry, (b) better mesh quality, (c) less computing time, (d) easier display and analysis of the results. In short, the simulation model becomes smaller and easier to handle. If the problem's nature is indeed 2D, it would not introduce inaccuracy for the solution.

Purpose of This Chapter

Since 2D simulations are usually easier to handle than 3D simulations, we start the learning of simulations by conducting 2D static structural simulations in this chapter, using some of the mechanical parts that we created in Chapter 2. This chapter also serves as a preliminary to 3D simulations, since most of the techniques and concepts in this chapter can be used in 3D simulations.

About Each Section

Sections 3.1 and 3.2 guide the students to perform a 2D simulation in a step-by-step fashion. Section 3.3 looks into more details and tries to provide what we are not able to cover in the first two sections. Section 3.4 provides an additional exercise. Problems in Sections 3.2 and 3.4 involve contact nonlinearities. In-depth discussion of contact nonlinearities will be postponed until Chapter 13. We introduce nonlinearities so early in this chapter to build motivation for learning nonlinear simulations in Chapters 13 and 14. Using a filleted bar subject to tension, Section 3.5 introduces some must-know concepts in finite element simulations, namely stress discontinuity, structural error, finite elements convergence, stress concentration, and stress singularity.

Section 3.1

Triangular Plate

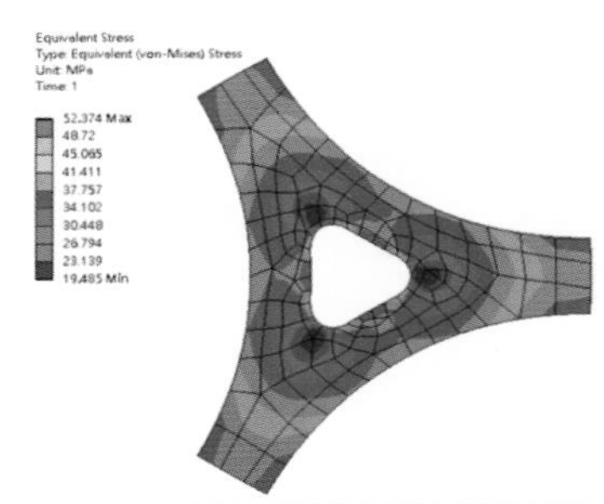

3.1.1 About the Triangular Plate

[1] In this section, we will perform a 2D static structural simulation using the triangular plate created in Section 2.2. The plate is made of steel and used to withstand a tensile force of 20,000 N on each of its three side faces. The size of the side faces is 40x10 mm^2; therefore, the applied tensile stress on the side faces is 50 MPa. The objective is to investigate the stresses in the plate.

We will model the problem as a 2D **plane stress** problem. Definition of the plane stress is given in 3.3.1, page 140; for now, what you need to know is that a thin plate subject to in-plane forces can be modeled as a plane stress problem.

There are two planes of symmetry in the model. In the first part of this section, we will analyze the full model without using the symmetries and then, in the second part of this section, reanalyze it by using the symmetries. #

3.1.2 Resume the Project Triplate

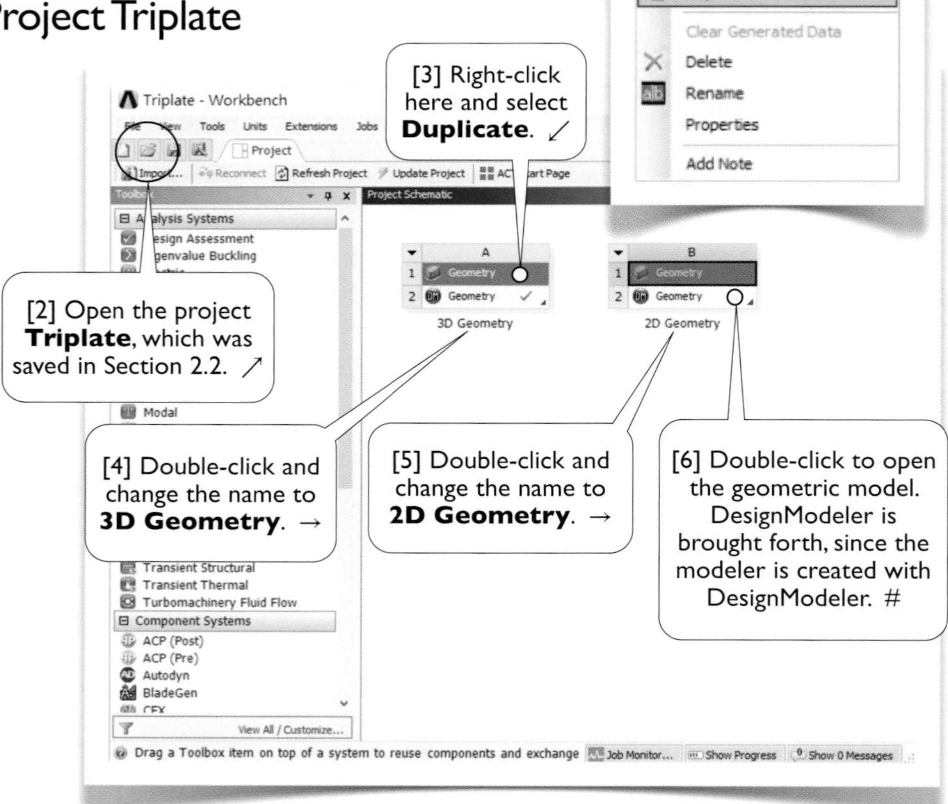

3.1.3 Delete the 3D Body and Create a 2D Body

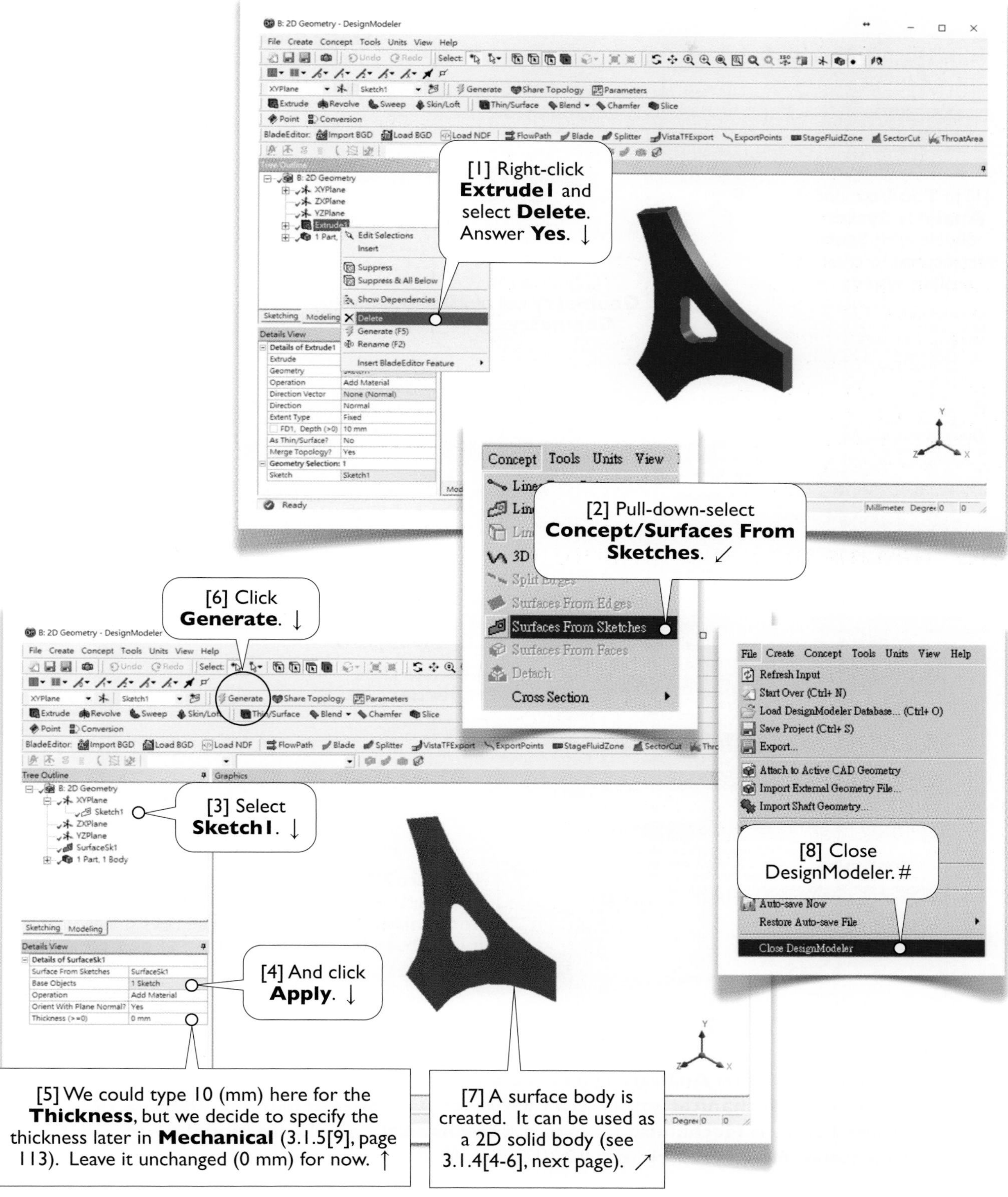

3.1.4 Create Analysis System and Specify Analysis Type

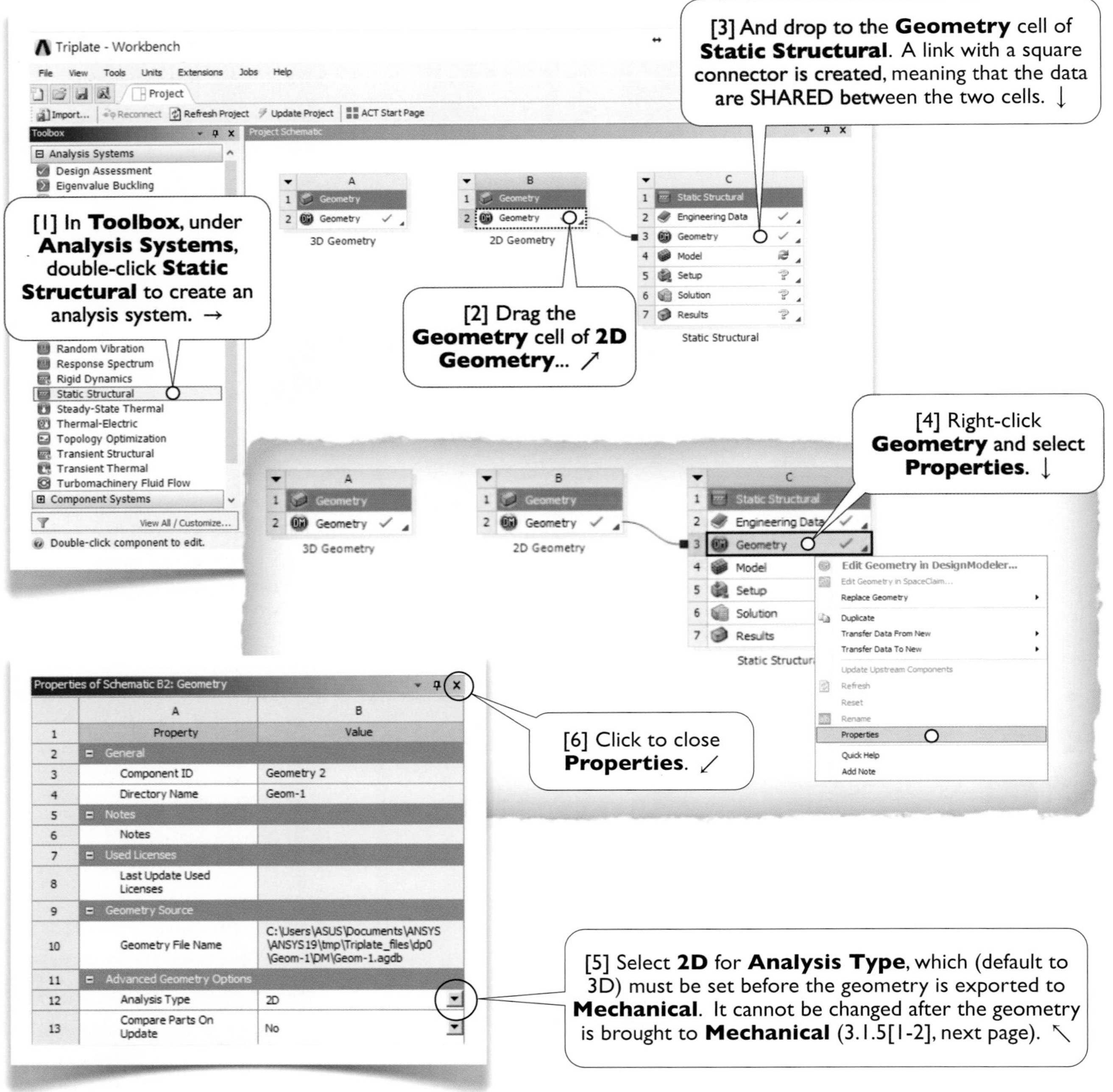

Analysis Type

[7] For a 2D simulation, always set **Analysis Type** to 2D before the geometry is brought to **Mechanical**. If you make a mistake (entering **Mechanical** without setting the property), the only way to change **Analysis Type** is to clear up the entire database in **Mechanical**. To do so, quit **Mechanical**, right-click the **Model** cell, and select **Reset** from the context menu, as exemplified in 9.2.2[1], page 346. #

3.1.5 Start Up Mechanical and Set Up Geometry

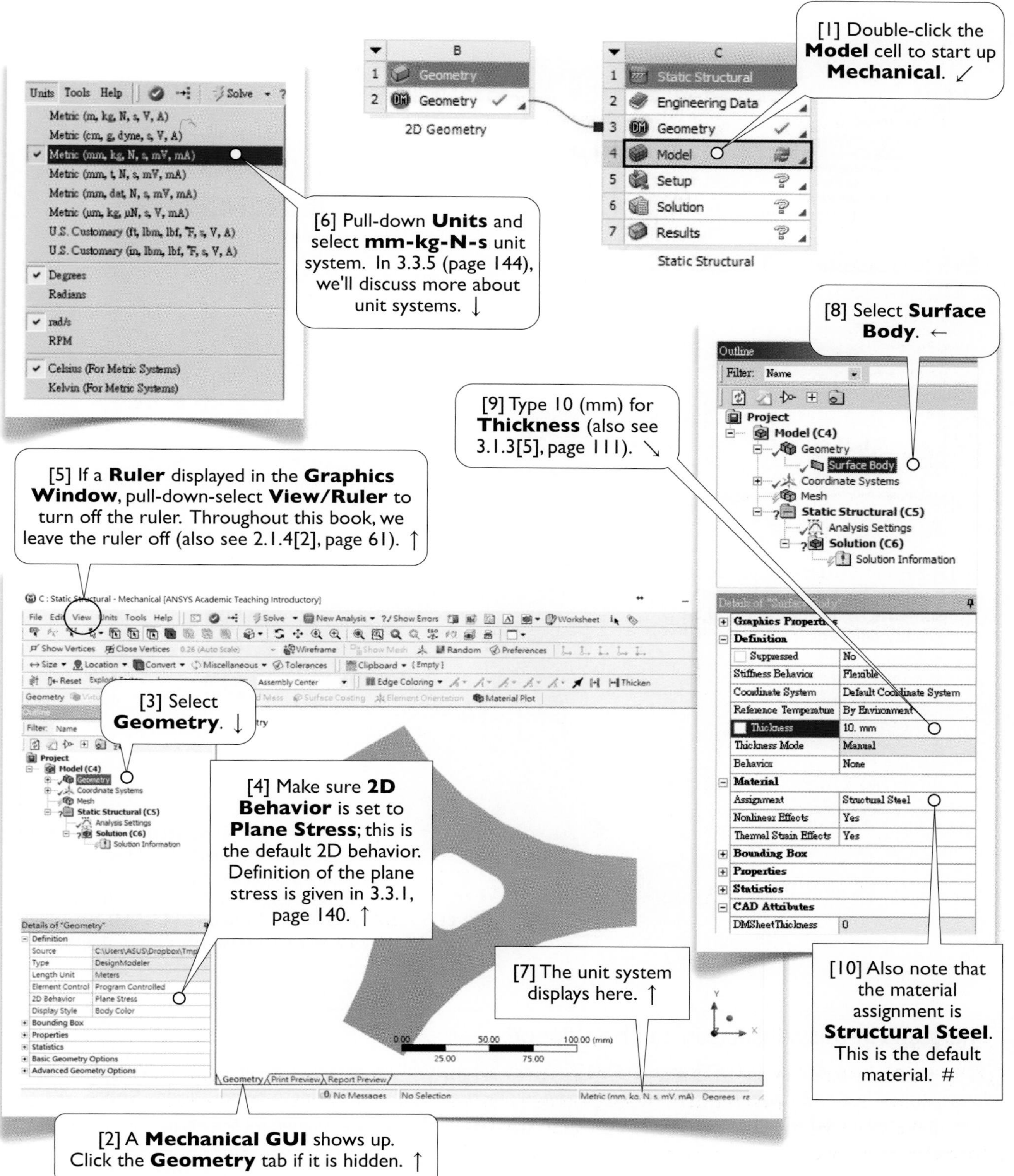

3.1.6 Generate Mesh

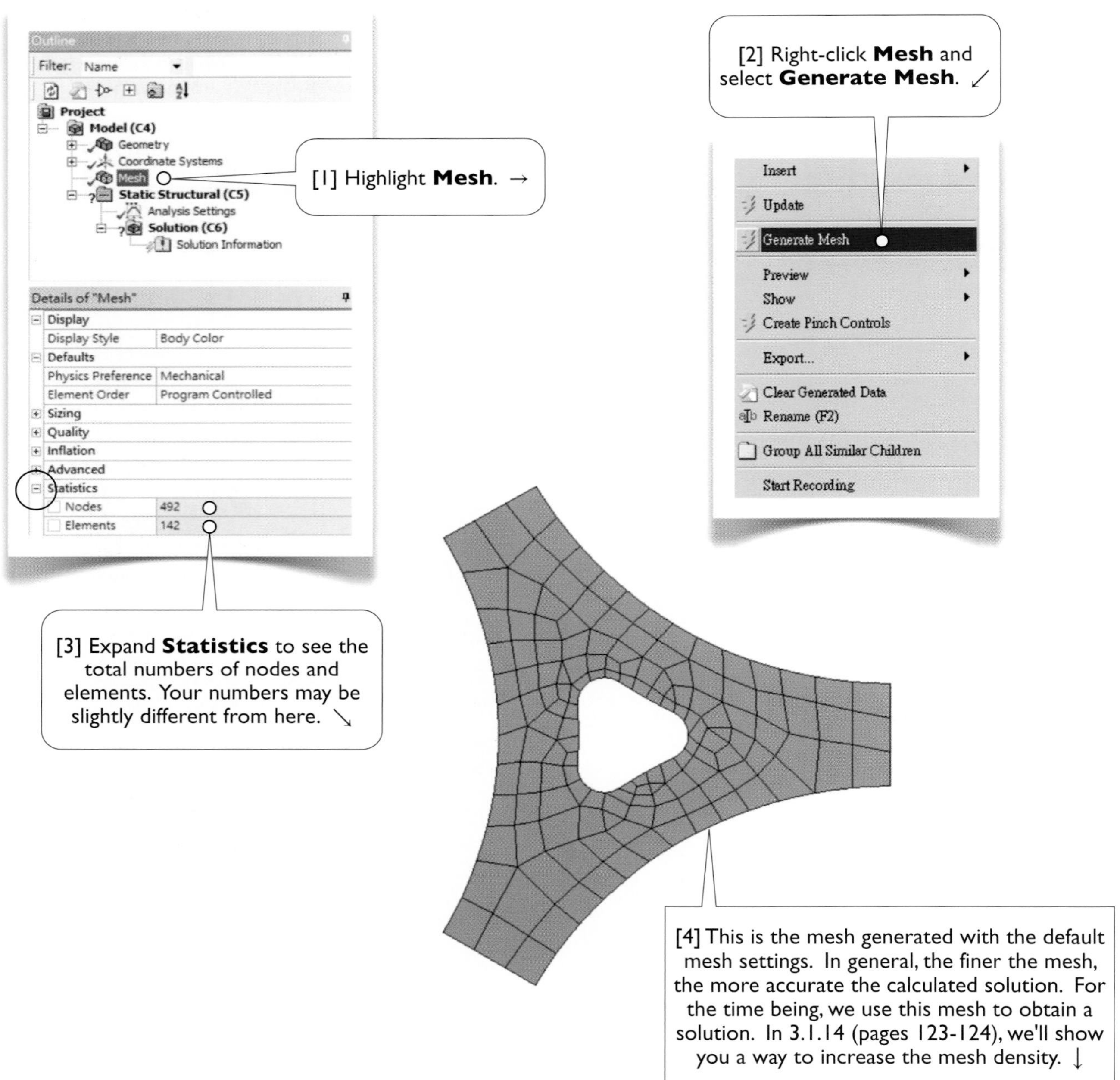

The steps in this subsection can be skipped.

[5] When you issue **Solve** (3.1.8[3], page 116), Workbench automatically generates a mesh if one does not exist, as demonstrated in 3.1.12 (page 121). However, it is a good practice to preview the mesh before clicking **Solve**. We often mesh the model with default settings first, and then adjust mesh controls to improve the mesh. #

3.1.7 Specify Loads

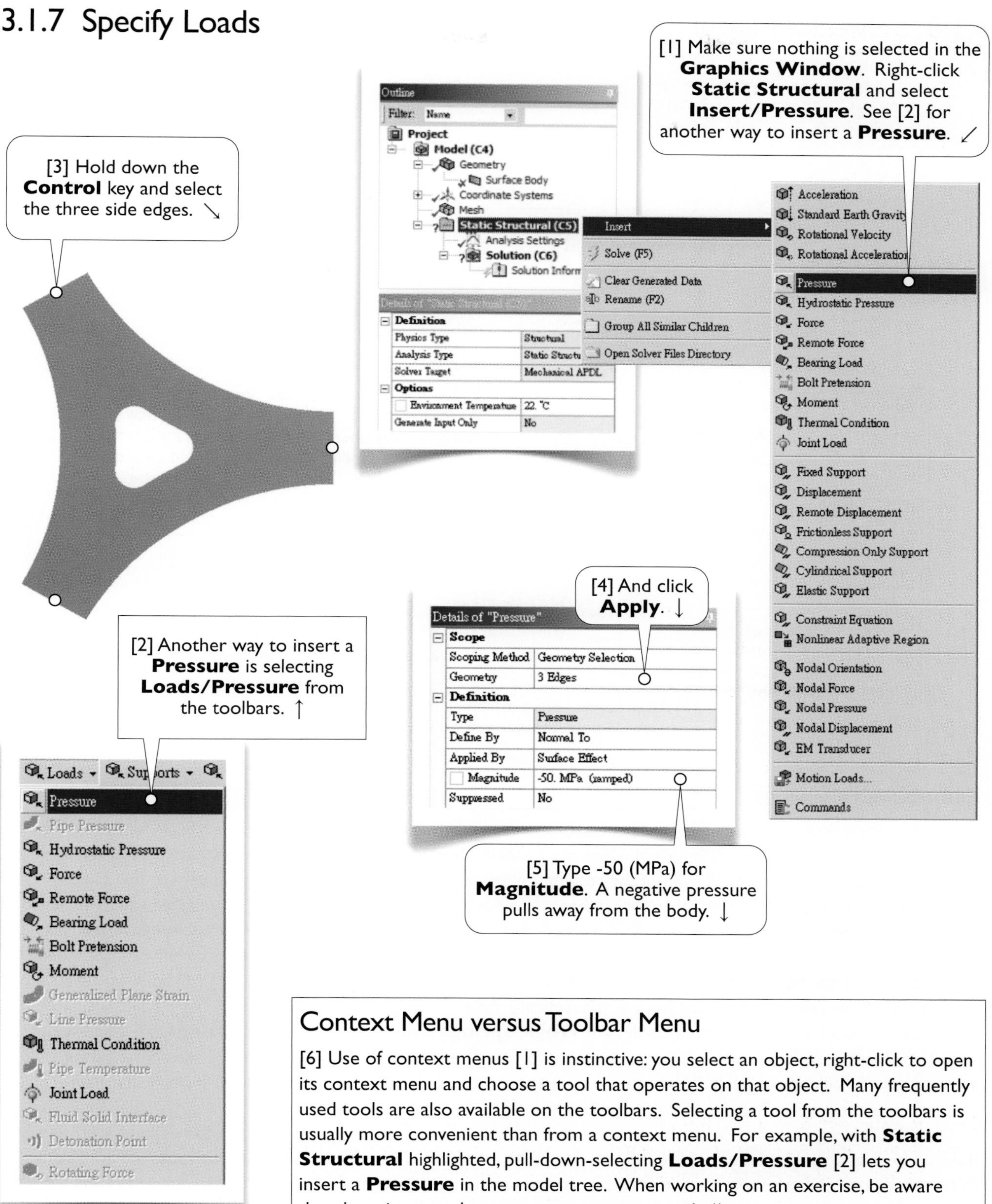

Context Menu versus Toolbar Menu

[6] Use of context menus [1] is instinctive: you select an object, right-click to open its context menu and choose a tool that operates on that object. Many frequently used tools are also available on the toolbars. Selecting a tool from the toolbars is usually more convenient than from a context menu. For example, with **Static Structural** highlighted, pull-down-selecting **Loads/Pressure** [2] lets you insert a **Pressure** in the model tree. When working on an exercise, be aware that there is more than one way to access a tool. #

3.1.8 Solve the Model

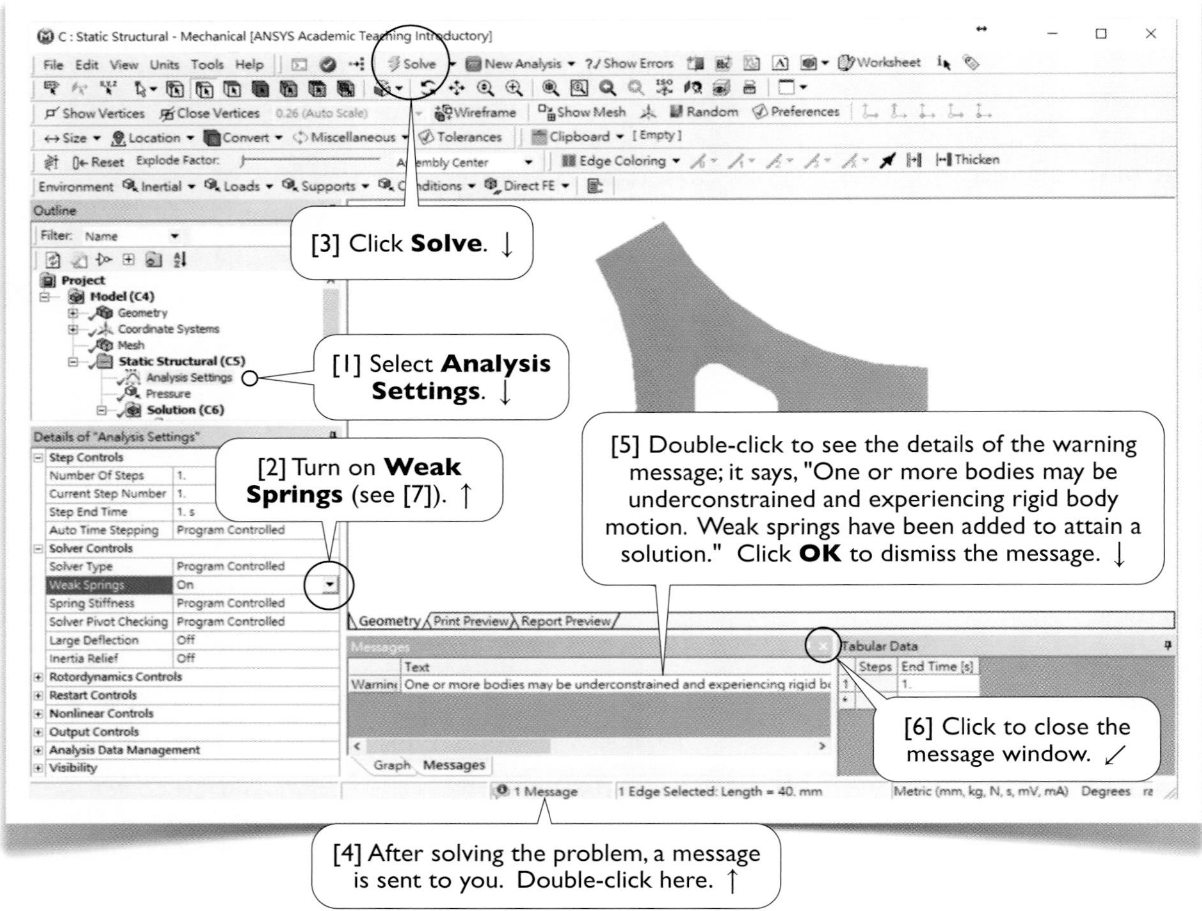

About Weak Springs

[7] If a structure does not have enough supports, the structure is unstable and any non-zero external forces would cause the structure to move indefinitely; the motion is called a *rigid body motion*. An unstable structure still can achieve static equilibrium (unstable equilibrium) if the resultant external force is zero. This is what happens in our case.

Traditionally, in a static structural analysis, whenever a finite element program detects an unstable structure, it stops and reports an error message. This has been practiced for decades, even if the resultant external force is zero. In a digital computer, the external forces rarely sum up to zero. There is usually a small residual force left when summing up the forces; this is the nature of numerical computations. In other words, an unstable structure will undergo rigid body motion even if the resultant external force is theoretically zero.

In a Workbench's static structural analysis, if **Weak Springs** is turned on [2], **weak springs** will be added to an unstable structure, so it can withstand small external forces. They are called **weak springs** because the spring constants are small and negligible. With **weak springs**, rigid body motions are small but still perceivable in an unstable structure even when the external force is zero.

Anyway, it is a good practice to provide enough supports. In our case, you might set up a fixed support on one of the side faces and apply pressure of -50 MPa on the other two faces. That way, all deformations reported are relative to the fixed edge. A better way to model this case is to use the symmetry conditions of the structure. We will present the procedure in the second part of this section, starting from 3.1.11. #

3.1.9 Insert Result Objects

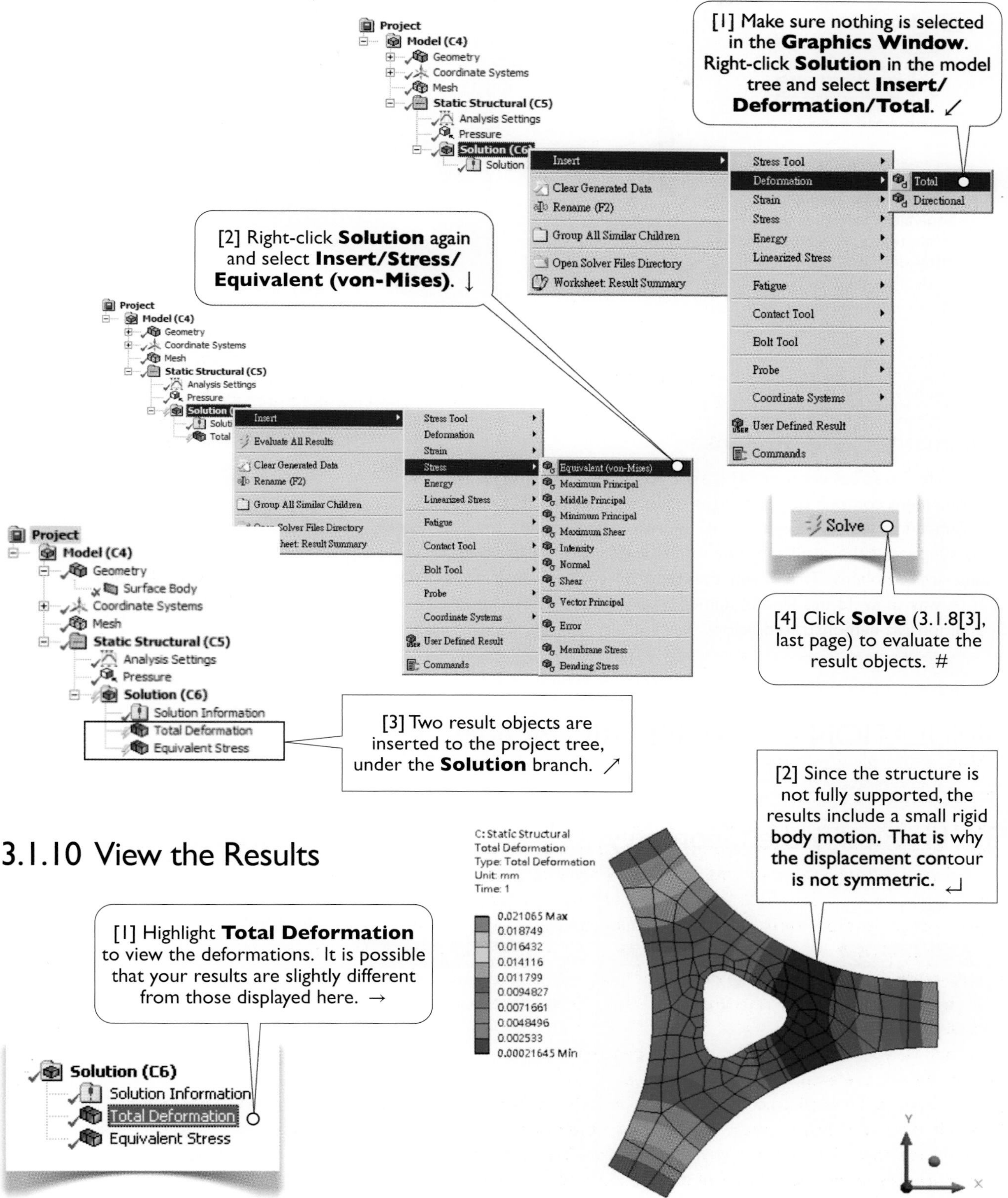

3.1.10 View the Results

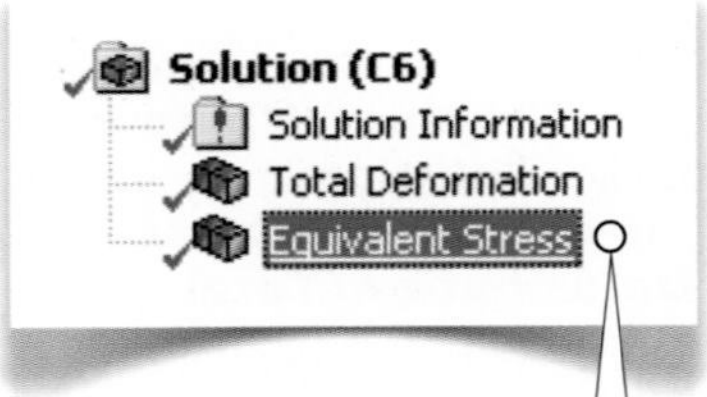

Solution (C6)
Solution Information
Total Deformation
Equivalent Stress

[3] Highlight **Equivalent Stress** to view the equivalent stresses. It is possible that your results are slightly different from those displayed here. →

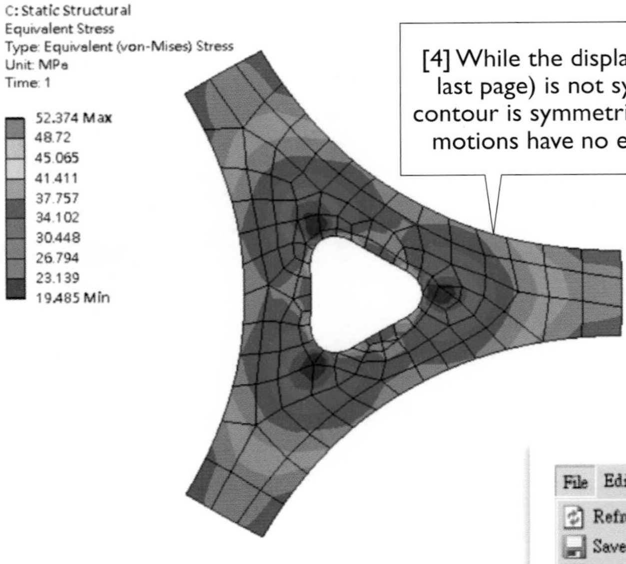

C: Static Structural
Equivalent Stress
Type: Equivalent (von-Mises) Stress
Unit: MPa
Time: 1
52.374 Max
48.72
45.065
41.411
37.757
34.102
30.448
26.794
23.139
19.485 Min

[4] While the displacement contour ([2], last page) is not symmetric, the stress contour is symmetric, since the rigid body motions have no effects on stresses. ↓

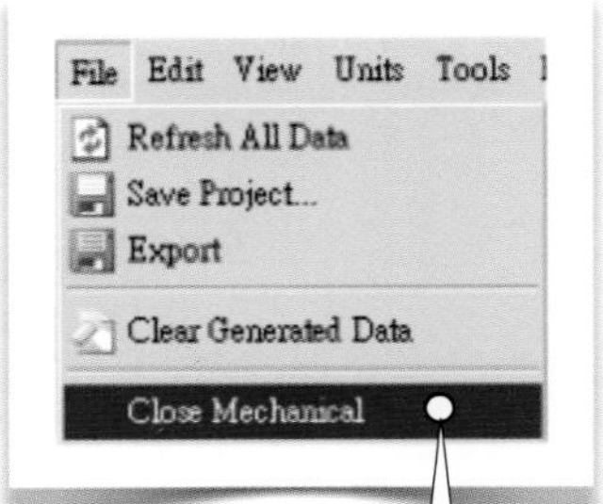

File
Edit
View
Units
Tools
Refresh All Data
Save Project...
Export
Clear Generated Data
Close Mechanical

[5] Select **File/Close Mechanical** from the pull-down menu. ←

Symmetry of the Response

[6] While the stress contour [4] is symmetric, you may wonder why the displacement contour ([2], last page) is not symmetric. Theoretically, the displacements should be symmetric. However, because the structure is not fully supported, the results include a small rigid body motion, due to numerical problems. That is why the displacement contour is not symmetric [2]. A way to avoid this kind of numerical problem is to use symmetry conditions, as demonstrated below. #

SIMULATION OF SYMMETRIC MODEL

3.1.11 Modify the Geometry

[1] Right-click here and select **Duplicate**. ↵

A
1 Geometry
2 Geometry
3D Geometry
B
1 Geometry
2 Geometry
2D Geometry
C
1 Static Structural
2 Engineering Data
3 Geometry
4 Model
5 Setup
6 Solution
7 Results
Static Structural
Refresh
Update
Duplicate
Replace With
Clear Generated Data
Delete
Rename
Properties
Add Note

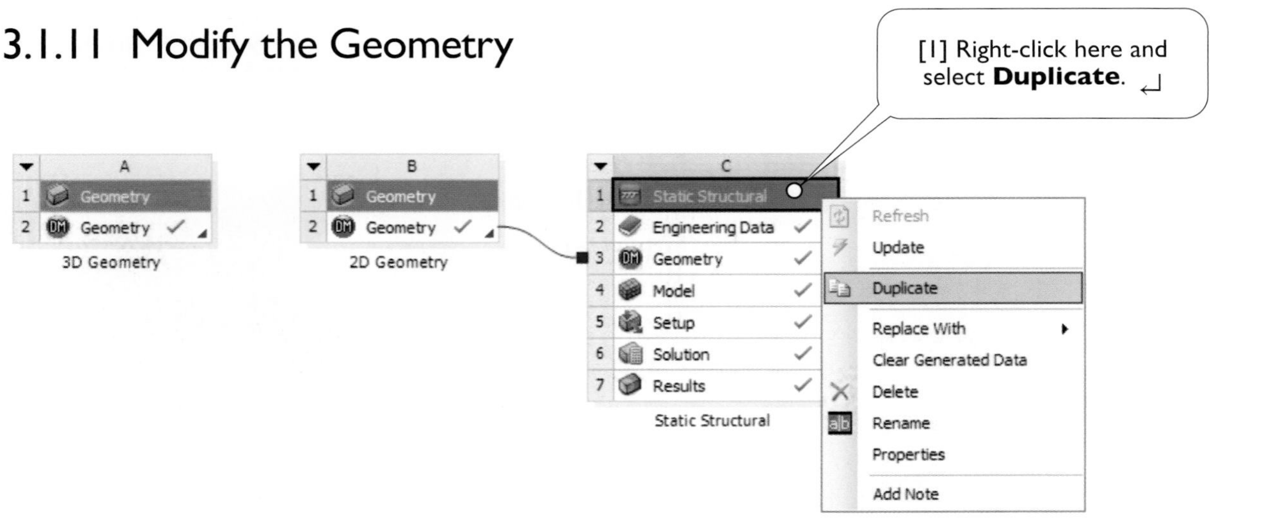

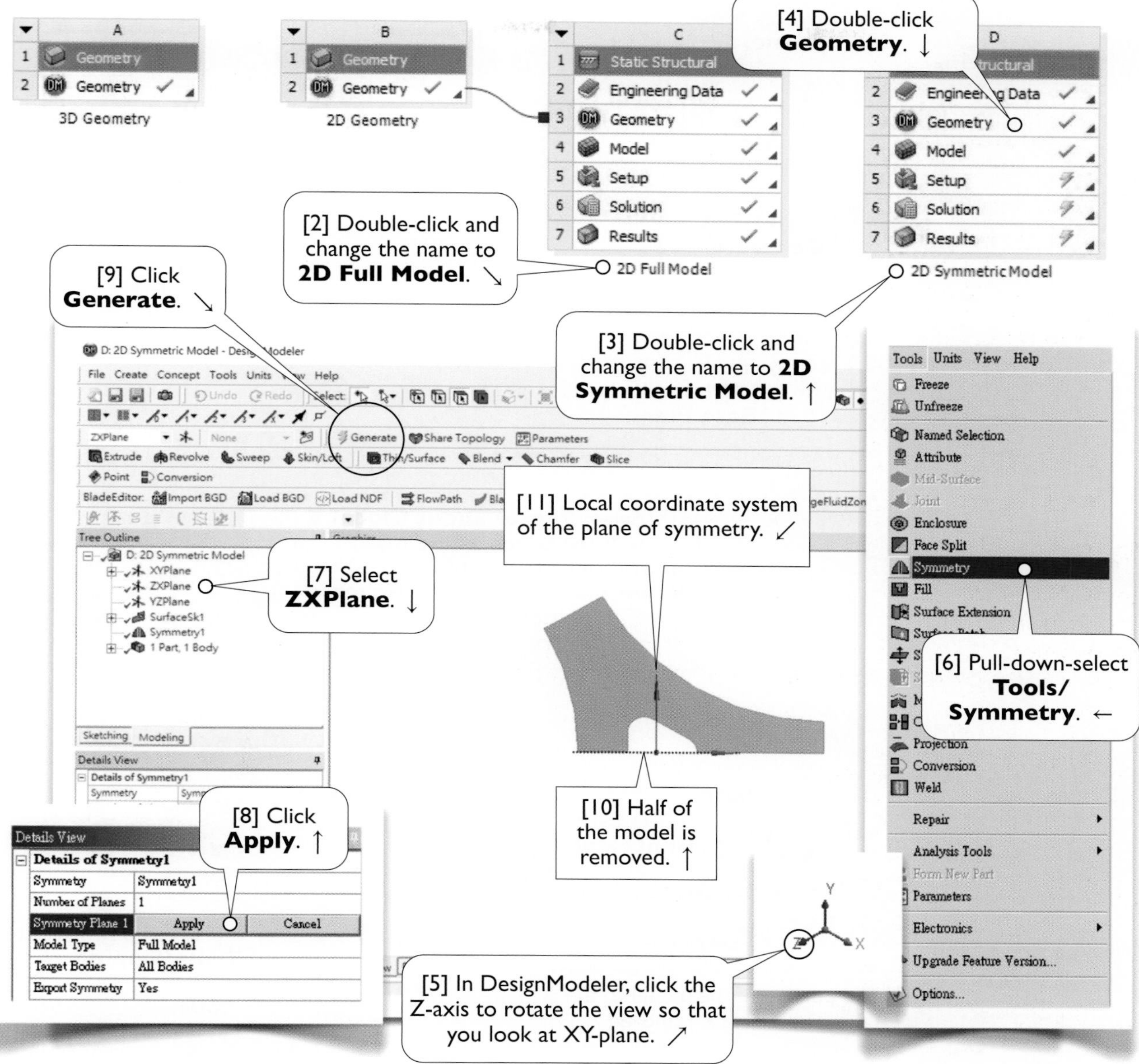

About Coordinate Systems

[12] There is a unique global coordinate system; its three axes are shown in the bottom-right corner of the graphics window [5]. Workbench uses R, G, and B colors to represent X-, Y-, and Z-axis respectively: red arrow for X-axis, green arrow for Y-axis, and blue arrow for Z-axis. In this book, we use upper-case (X, Y, Z) for both global and local coordinate systems, to be consistent with the notations used in Workbench.

Each plane has its own local coordinate system, using the same color codes. Let's take **ZXPlane** as an example [11]. Its local XY-plane coincides with global ZX-plane, and the local Z-axis points upward. When we specify a plane as the plane of symmetry, the plane's local XY-plane is used to cut away the portion of the model on the local negative-Z side. The portion of the model on the local positive-Z side remains [10].

The triangular plate has another plane of symmetry. None of the default planes can be used as the plane of symmetry; we need to create one. This plane can be derived from rotating **ZXPlane** by 60 degrees. ↵

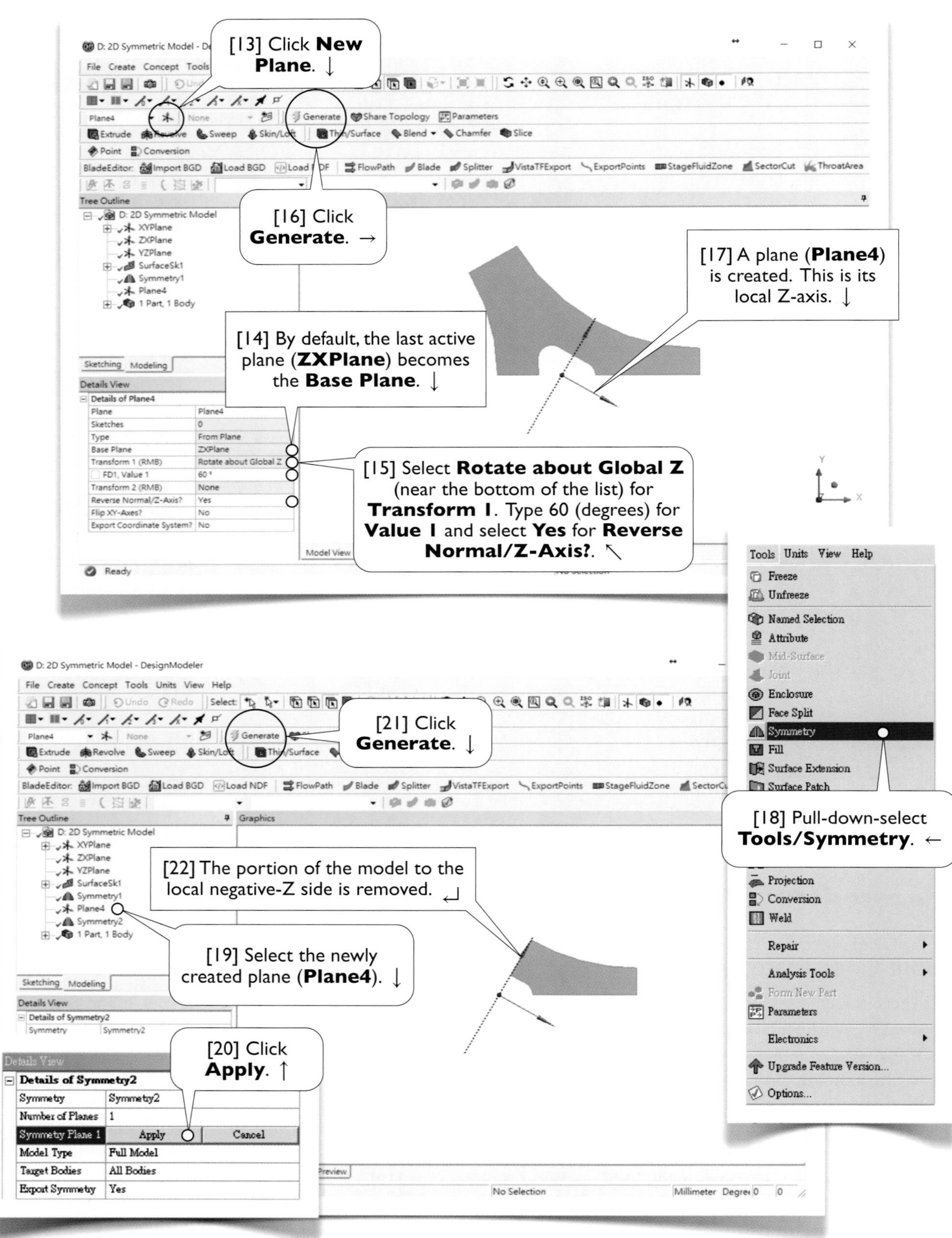
[13] Click **New Plane**. ↓
[14] By default, the last active plane (**ZXPlane**) becomes the **Base Plane**. ↓
[15] Select **Rotate about Global Z** (near the bottom of the list) for **Transform 1**. Type 60 (degrees) for **Value 1** and select **Yes** for **Reverse Normal/Z-Axis?**. ↖
[16] Click **Generate**. →
[17] A plane (**Plane4**) is created. This is its local Z-axis. ↓
[18] Pull-down-select **Tools/Symmetry**. ←
[19] Select the newly created plane (**Plane4**). ↓
[20] Click **Apply**. ↑
[21] Click **Generate**. ↓
[22] The portion of the model to the local negative-Z side is removed. ↵
D: 2D Symmetric Model - DesignModeler
File Create Concept Tools Units View Help
Tree Outline
D: 2D Symmetric Model
XYPlane
ZXPlane
YZPlane
SurfaceSk1
Symmetry1
Plane4
Symmetry2
1 Part, 1 Body
Details of Plane4
Plane Plane4
Sketches 0
Type From Plane
Base Plane ZXPlane
Transform 1 (RMB) Rotate about Global Z
FD1, Value 1 60 °
Transform 2 (RMB) None
Reverse Normal/Z-Axis? Yes
Flip XY-Axes? No
Export Coordinate System? No
Details of Symmetry2
Symmetry Symmetry2
Number of Planes 1
Symmetry Plane 1 Apply Cancel
Model Type Full Model
Target Bodies All Bodies
Export Symmetry Yes
Tools Units View Help
Freeze
Unfreeze
Named Selection
Attribute
Mid-Surface
Joint
Enclosure
Face Split
Symmetry
Fill
Surface Extension
Surface Patch
Projection
Conversion
Weld
Repair
Analysis Tools
Form New Part
Parameters
Electronics
Upgrade Feature Version...
Options...

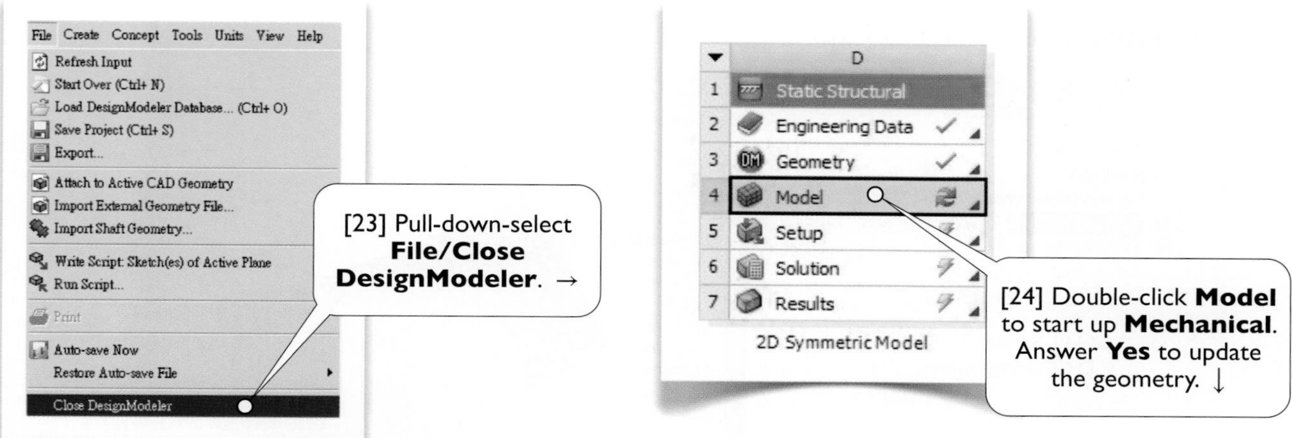

Do I need to close DesignModeler when working on **Mechanical**, or vice versa?

[25] When starting up, **Mechanical** automatically updates the geometry, including adding symmetry conditions on the model. I strongly suggest that the newcomers close one application while switching to another. That way, the update will be automatic; you won't mess up the work flow.

However, you don't have to close an application when switching to another application, but you have to "refresh" it by yourself. In **Project Schematic** you can right-click **Model** (of an analysis system) and select **Update Upstream Component**. Inside **Mechanical** you can right-click **Geometry** (of the project tree) and select **Update Geometry from Source**. Until you become an experienced user, I suggest that you always keep only one application open at a time to make life simpler. #

3.1.12 Solve the New Model

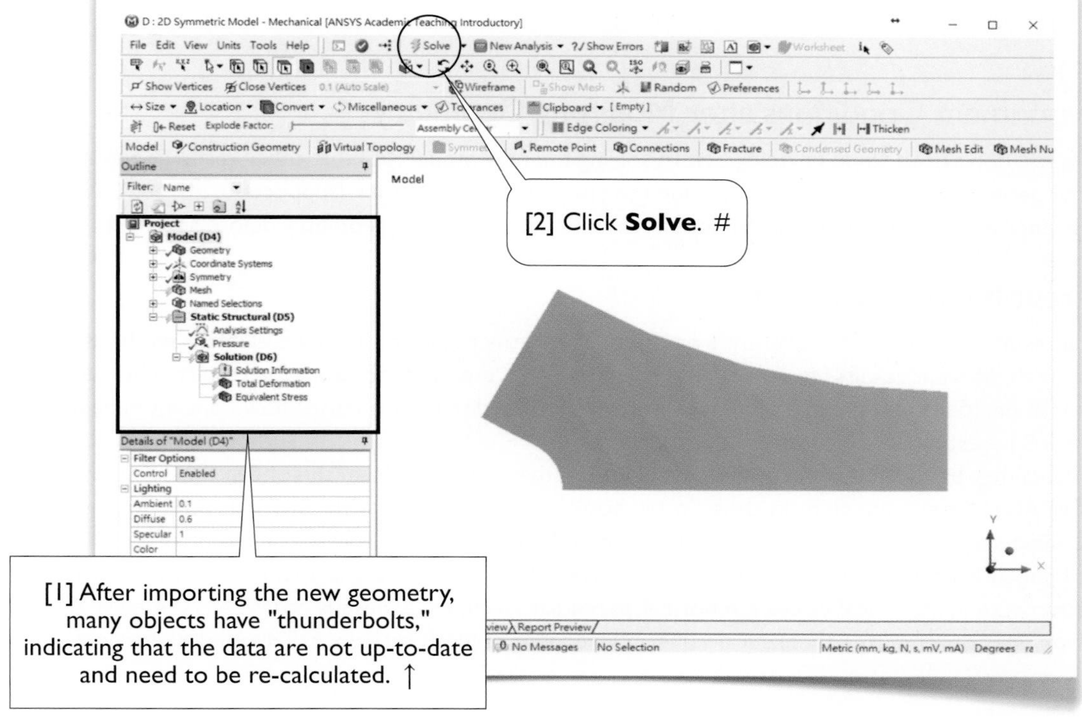

3.1.13 View the Results

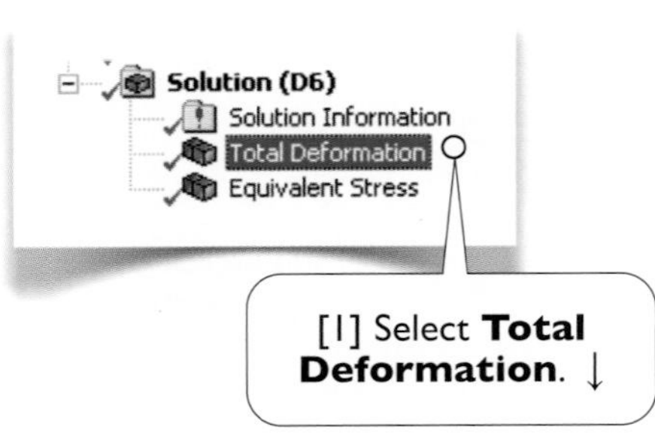

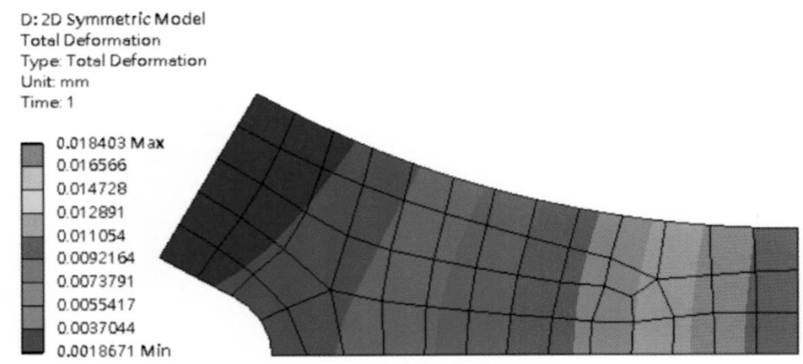

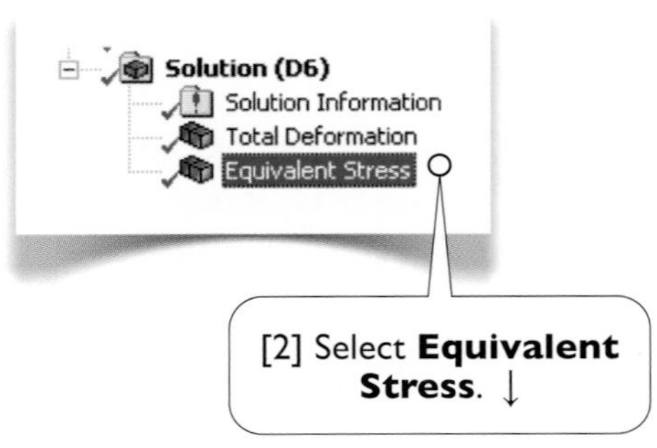

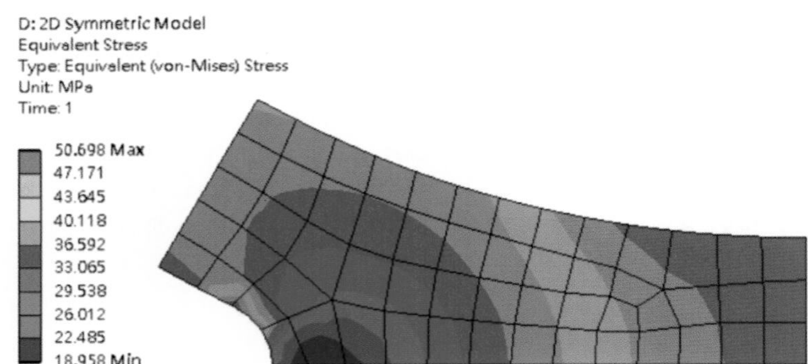

Check Environment Conditions after Modifying Geometry

[3] In this case, you don't need to do anything before solving the new model. In most other cases, however, whenever you modify your geometry, you may need to redefine the environment conditions (supports, loads, etc.). As a good practice, always check your environment conditions each time you modify the geometry before solving it.

Why Different Numerical Results?

Your numerical results may be slightly different from here. This is the nature of the finite element methods: different meshes end up with different results. In general, the finer the mesh, the more accurate the results. The question is how fine should a mesh be, to achieve enough accuracy? We will start to discuss this must-know concept in Section 3.5. The discussion will be extended to 3D cases in Section 9.3.

Some students may be puzzled about why they obtained a mesh different from the one in the book even though they followed EXACTLY the same steps as those in the book. The answer is that the students have no way to follow EXACTLY the same steps in the book. For example, a line in the book may be drawn from right to left while you drew it from left to right. It is possible that the direction of the line affects the meshing algorithm in the Workbench.

Limited differences in numerical values are normal, particularly when the mesh is coarse. As the mesh becomes finer, the solution will converge to a theoretical value, independent of mesh variations. This kind of puzzle would disappear. #

3.1.14 Improve Solution Accuracy

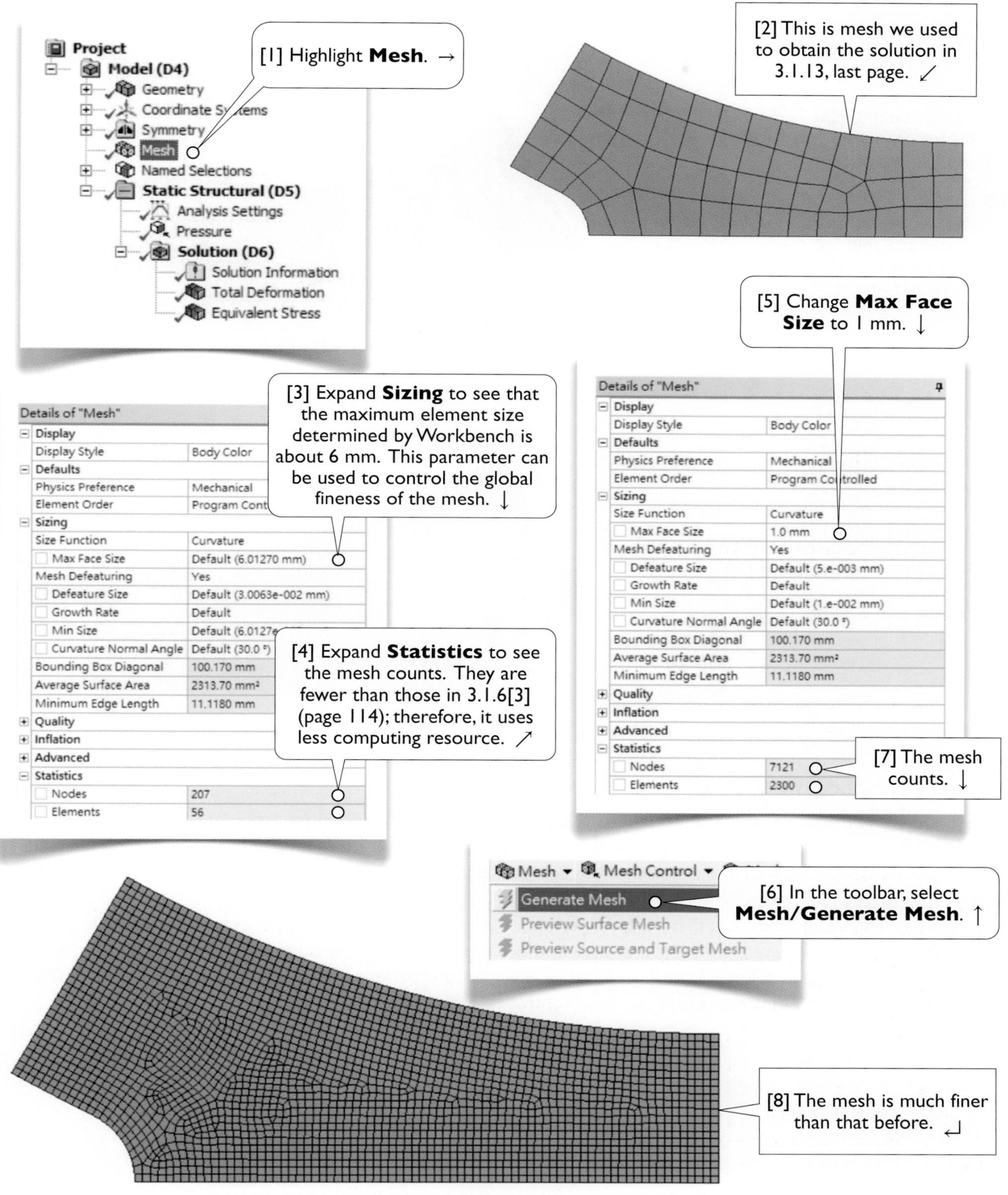

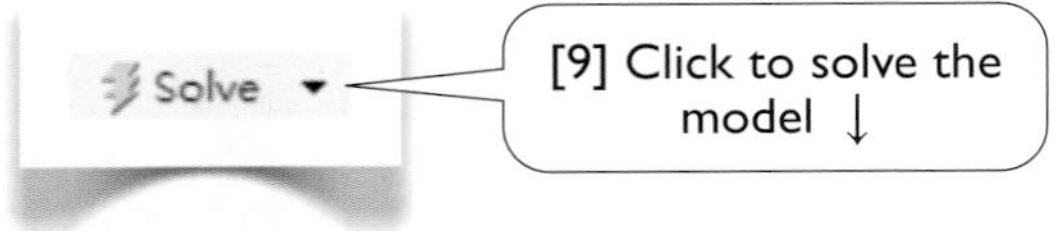

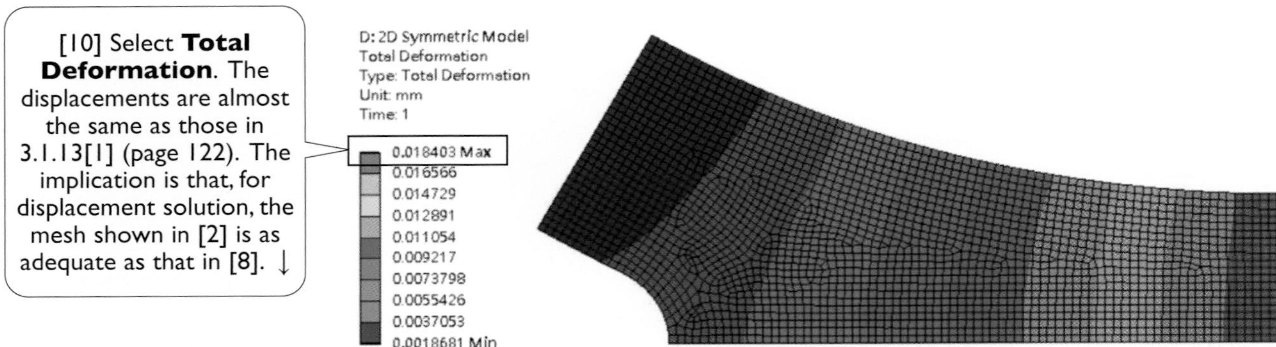

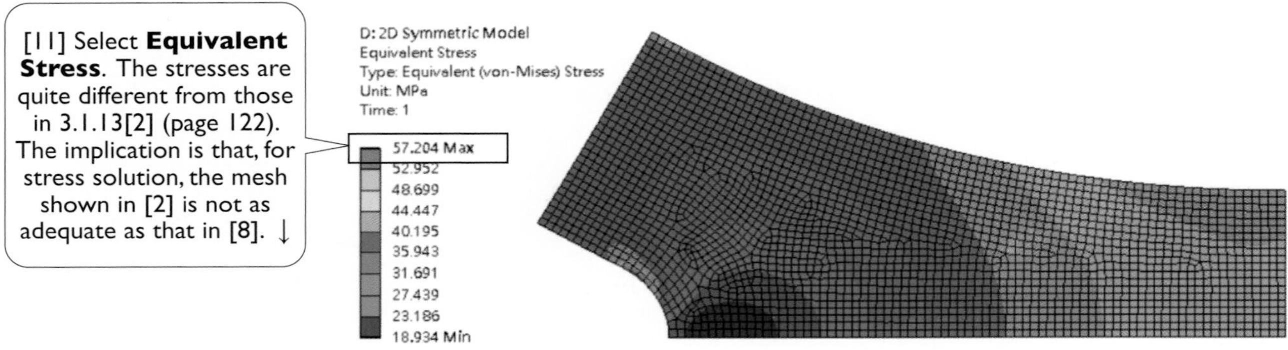

Accuracy of Solution

[12] Here, we've demonstrated an important behavior of finite element simulations: the finer the mesh, the more accurate the calculated solution. The big question is: how fine is adequate? There is no easy answer for this question. A comprehensive way to assure an adequate mesh is to conduct a "convergence study," demonstrated in PART C of Section 3.5 (pages 161-163).

Another important concept is that the element sizes usually need not be uniform; an example is shown in 3.2.8[18], page 135. We usually require finer mesh at an area in which the change of stress is more dramatic, i.e., at an area that has larger stress gradient. Of course, you need lots of engineering experiences to cultivate a sense of "stress gradient." In the next section, we will demonstrate the mesh control at local areas. ↓

Wrap Up

[13] Close **Mechanical**. Save the project and exit Workbench. #

Section 3.2

Threaded Bolt-and-Nut

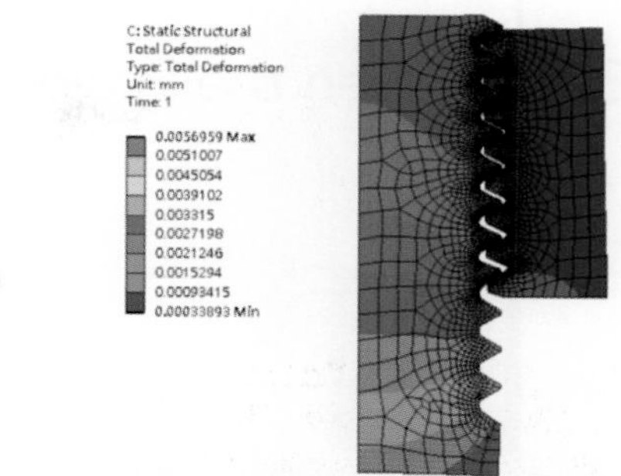

3.2.1 About the Threaded Bolt-and-Nut

[1] The threaded bolt created in Section 2.4 is a portion of a bolt-nut-plate assembly [2-5]. The bolt is preloaded with a tension of 10 kN, which is applied by tightening the nut with torque. We want to know the maximum stress at the threads under such a pretension.

Pretension is a built-in environment condition in Workbench 3D simulations, in which a pretension can be applied to a body or cylindrical surface. It is, however, not applicable for 2D simulations.

In this section, we will make some simplifications for a 2D simulation. Assuming a mirror-symmetry between the upper half and the lower half (which is not exactly true), we model only the upper part of the assembly [6-8]. To reduce the problem size further and alleviate contact nonlinearity, the plate is not included and its contact with the nut is modeled with a frictionless support [9].

The pretension is modeled using a uniform force applied on the lower face of the bolt [8]. The results will somewhat deviate from the reality, to be discussed at the end of this section, but the deviation has little effect on the stresses of the threads.

Assume that the coefficient of friction between the bolt and the nut is 0.3. ↗

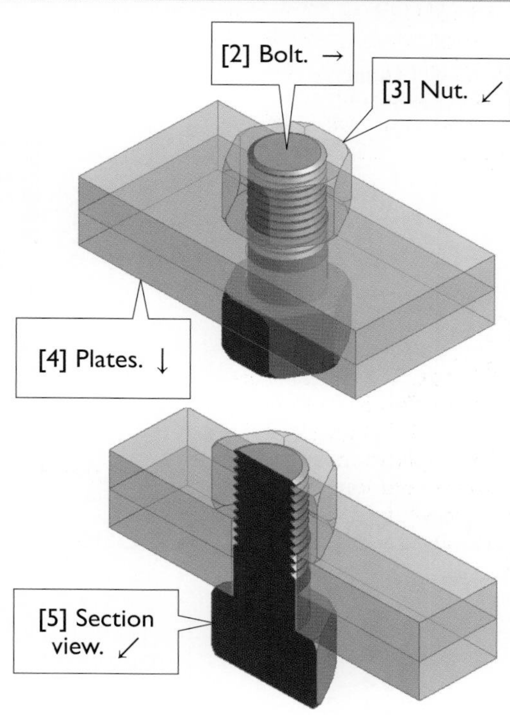

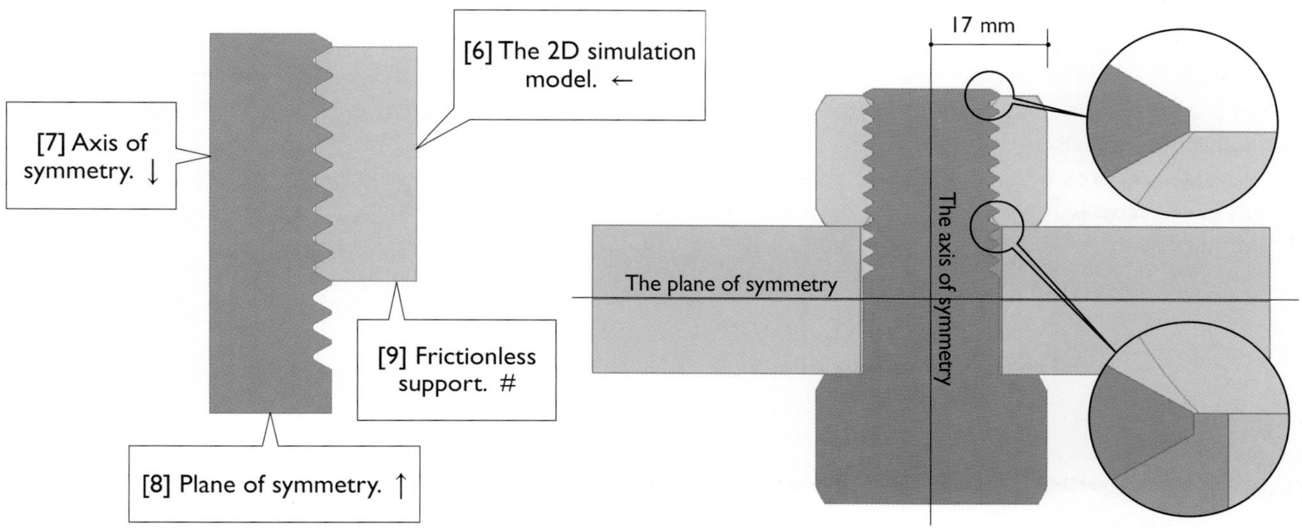

3.2.2 Open the Project **Threads**

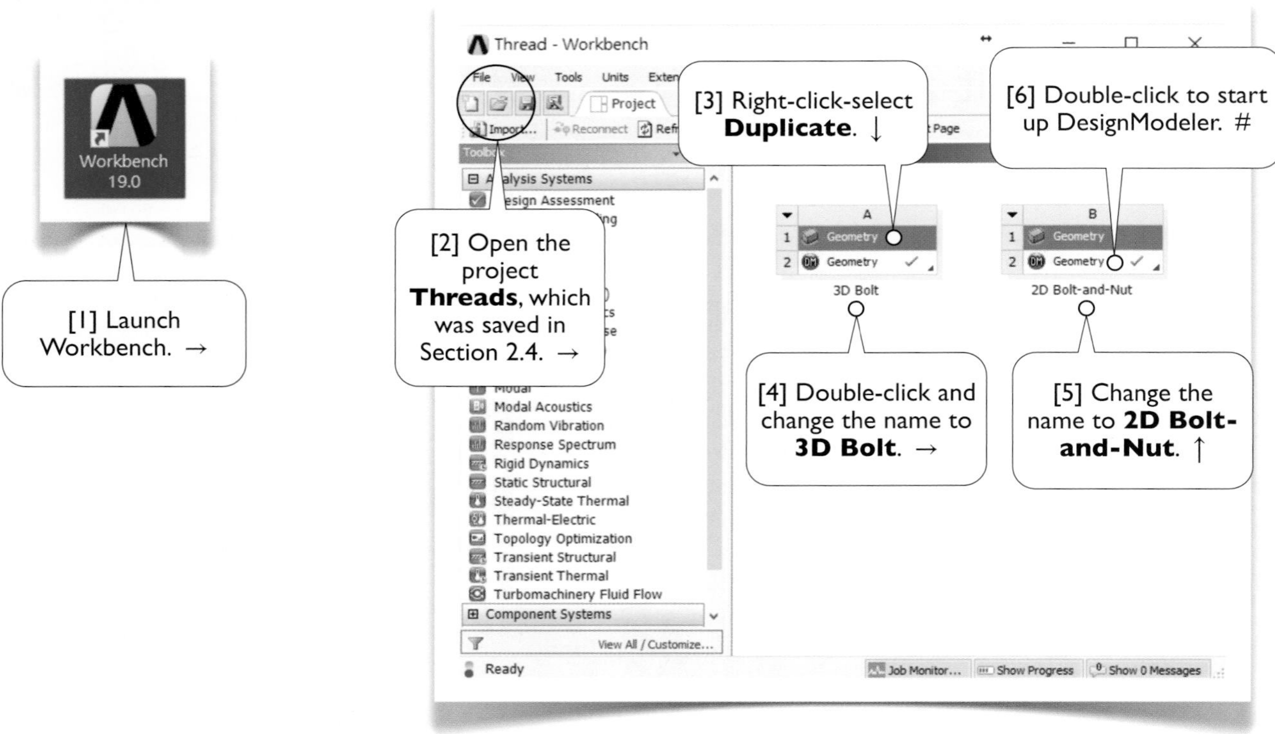

3.2.3 Delete the 3D Body and Create a 2D Body

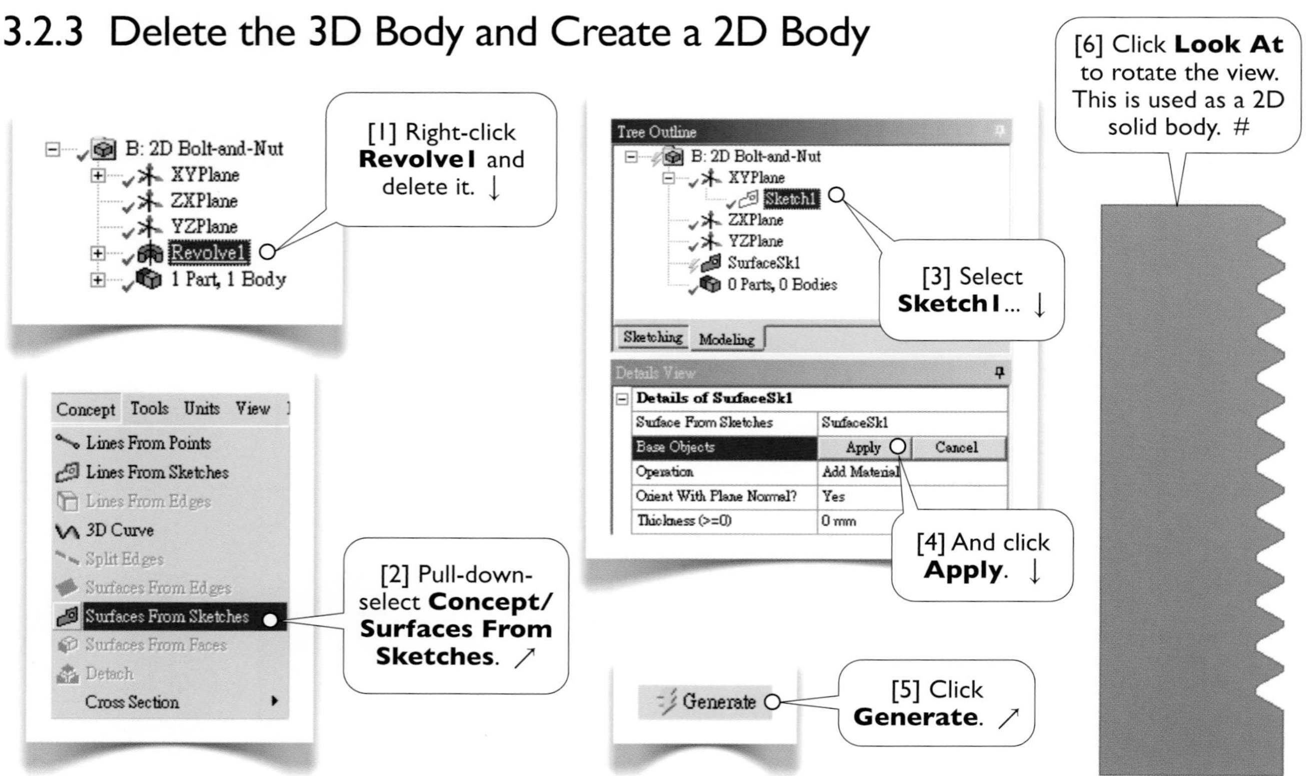

3.2.4 Create a 2D Body for the Nut

XYPlane
Sketch2

[1] Click **New Sketch**. ←

[2] **Sketch2** is created on **XYPlane** and becomes active. ↙

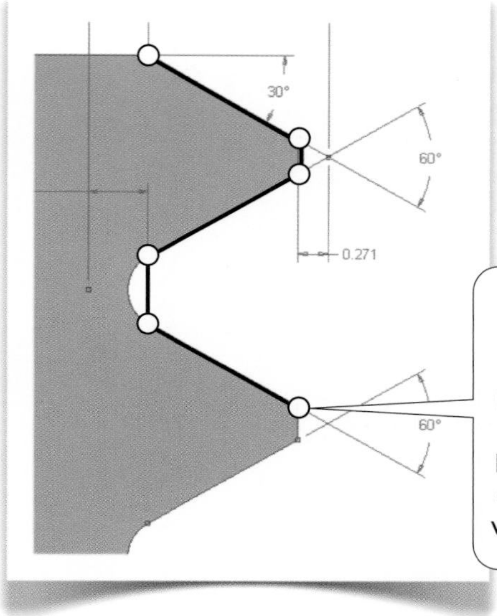

30°
60°
0.271
60°

[3] Switch to the **Sketching** mode. Draw a polyline connecting these 6 points. Make sure a **P** appears before you click a point. Note that both **Sketch1** and **Sketch2** are on **XYPlane** and we are working on **Sketch2** now. →

[4] Switch to **Modeling** mode, right-click **Sketch1** and select **Hide Sketch**. Click **Sketch2** to make it active. ↙

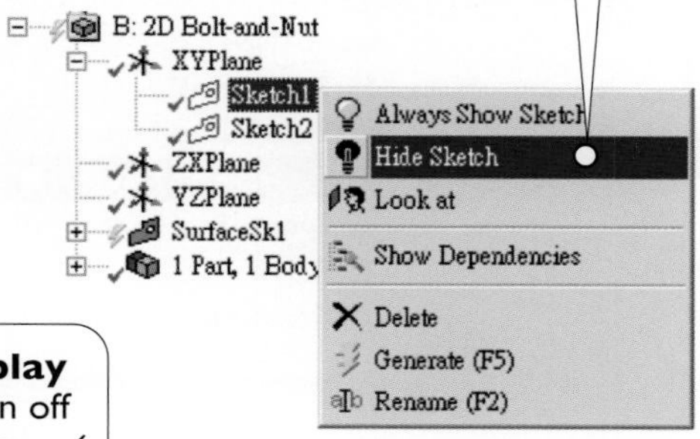

B: 2D Bolt-and-Nut
XYPlane
Sketch1
Sketch2
ZXPlane
YZPlane
SurfaceSk1
1 Part, 1 Body
Always Show Sketch
Hide Sketch
Look at
Show Dependencies
Delete
Generate (F5)
Rename (F2)

[5] Click **Display Model** to turn off the model display. ↙

[8] And trim away this segment. ↓

[6] Switch to the **Sketching** mode. Draw an arc like this using **Arc by 3 Points**. Before the third click to define the radius, make sure a **T** (tangent) appears, indicating that the arc is tangent to a line. ←

[7] Trim away this segment. ↑

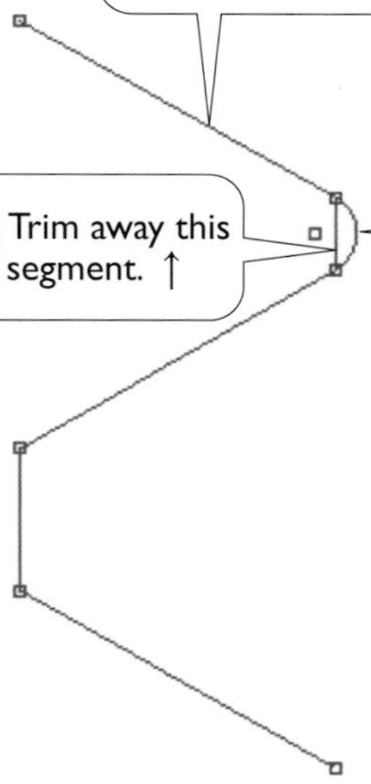

[10] Select the **Replicate** tool, select all the four segments, and set the **Paste Handle** at this point. Remember to use the selection filter (2.2.8[7-8], page 74). ↓

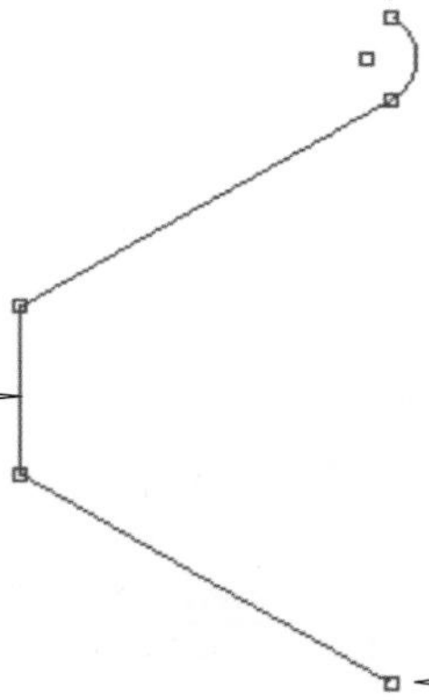

[9] This sketch represents one thread of the nut. Now, we want to replicate it 7 times. ↗

[11] And paste at this point. ↵

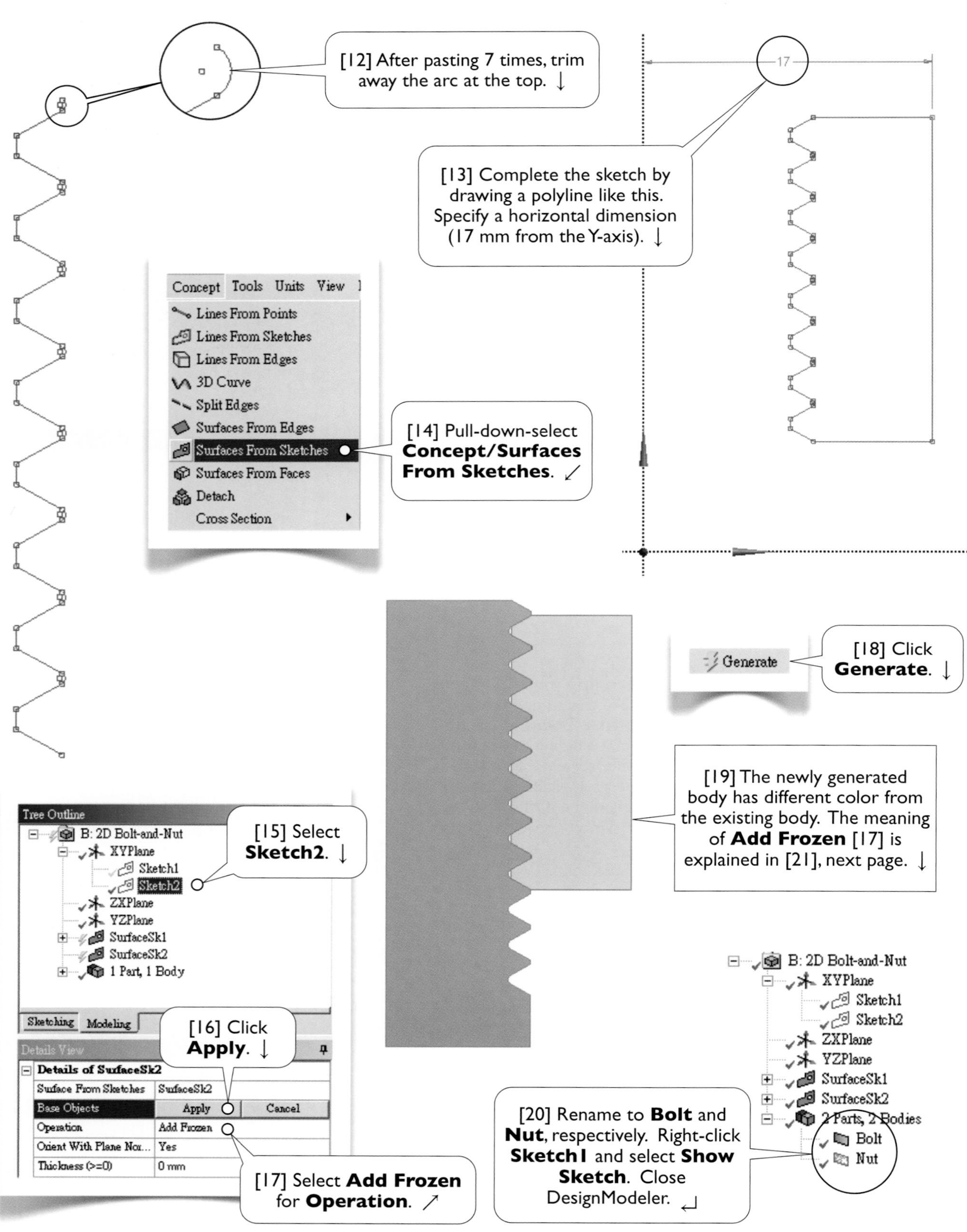
[12] After pasting 7 times, trim away the arc at the top. ↓
[13] Complete the sketch by drawing a polyline like this. Specify a horizontal dimension (17 mm from the Y-axis). ↓
17
Concept Tools Units View
Lines From Points
Lines From Sketches
Lines From Edges
3D Curve
Split Edges
Surfaces From Edges
Surfaces From Sketches
Surfaces From Faces
Detach
Cross Section
[14] Pull-down-select **Concept/Surfaces From Sketches**. ↙
Generate
[18] Click **Generate**. ↓
[19] The newly generated body has different color from the existing body. The meaning of **Add Frozen** [17] is explained in [21], next page. ↓
Tree Outline
B: 2D Bolt-and-Nut
XYPlane
Sketch1
Sketch2
ZXPlane
YZPlane
SurfaceSk1
SurfaceSk2
1 Part, 1 Body
[15] Select **Sketch2**. ↓
Sketching
Modeling
[16] Click **Apply**. ↓
Details View
Details of SurfaceSk2
Surface From Sketches
SurfaceSk2
Base Objects
Apply
Cancel
Operation
Add Frozen
Orient With Plane Nor...
Yes
Thickness (>=0)
0 mm
[17] Select **Add Frozen** for **Operation**. ↗
B: 2D Bolt-and-Nut
XYPlane
Sketch1
Sketch2
ZXPlane
YZPlane
SurfaceSk1
SurfaceSk2
2 Parts, 2 Bodies
Bolt
Nut
[20] Rename to **Bolt** and **Nut**, respectively. Right-click **Sketch1** and select **Show Sketch**. Close DesignModeler. ↵

Add Material versus **Add Frozen**

[21] With **Add Material** operation, the created material adds to the existing body and they become an integral part. On the other hand, if you choose **Add Frozen** ([17], last page), the created material does not add to the existing one; it becomes another part. This is what we intend: the bolt and nut are separate parts; they are not bonded to each other. In Workbench, each part is meshed independently. #

3.2.5 Create an Analysis System

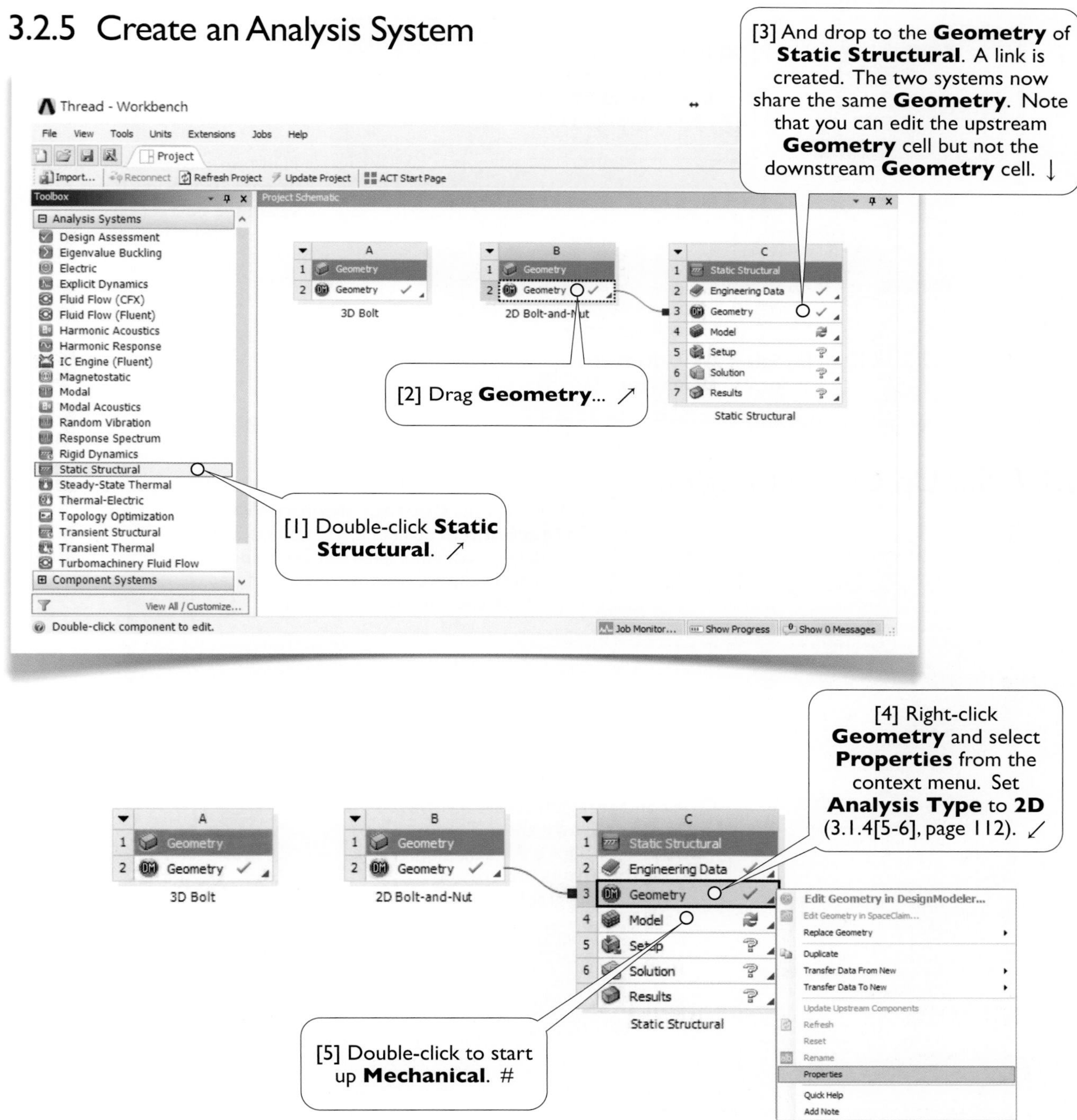

3.2.6 Set Up Geometry in Mechanical

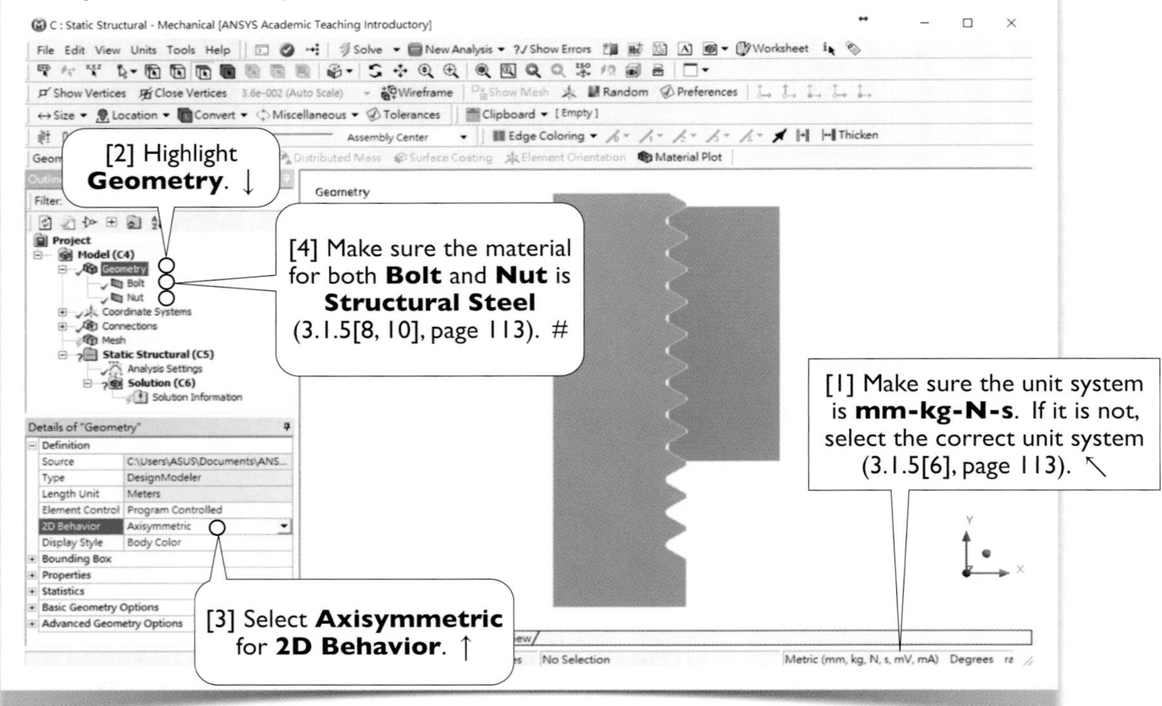

3.2.7 Set Up Contact Region

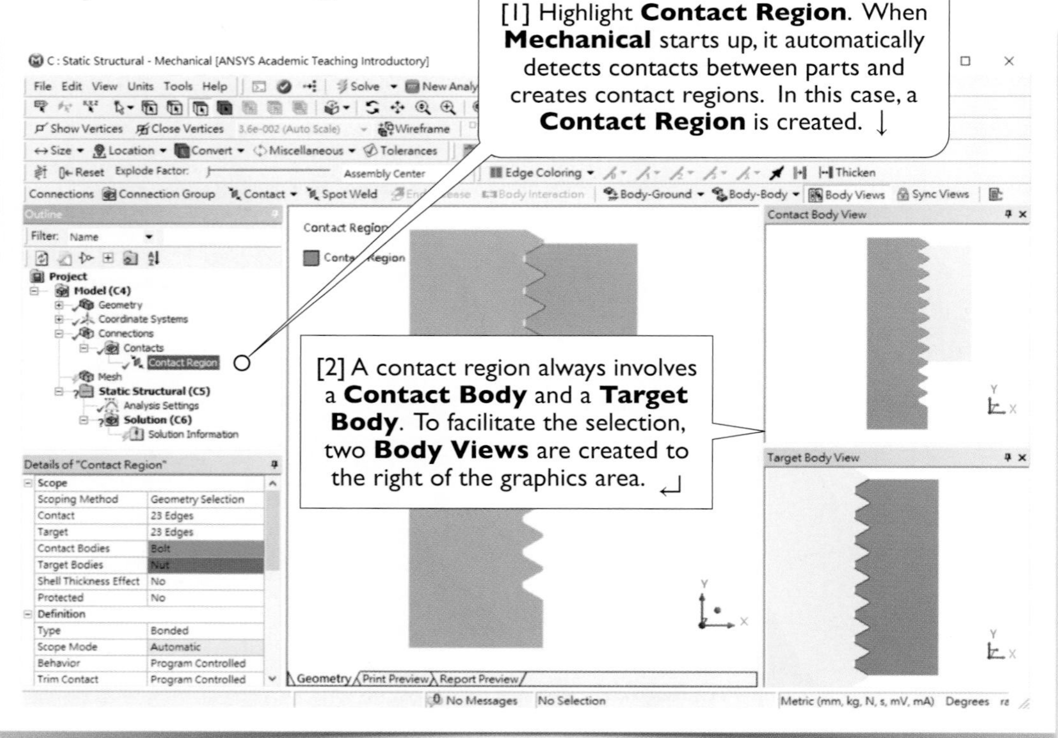

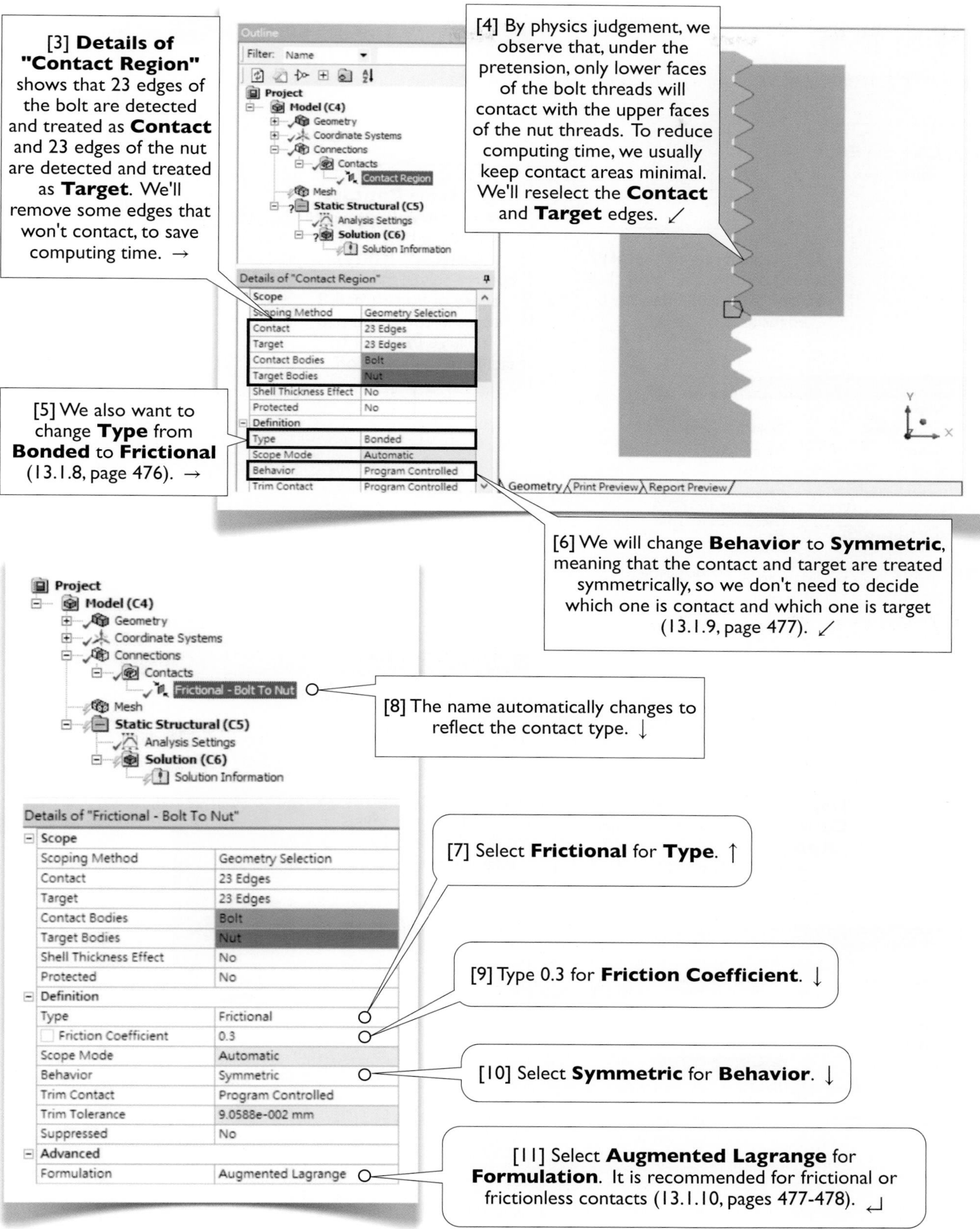

[3] **Details of "Contact Region"** shows that 23 edges of the bolt are detected and treated as **Contact** and 23 edges of the nut are detected and treated as **Target**. We'll remove some edges that won't contact, to save computing time. →
[4] By physics judgement, we observe that, under the pretension, only lower faces of the bolt threads will contact with the upper faces of the nut threads. To reduce computing time, we usually keep contact areas minimal. We'll reselect the **Contact** and **Target** edges. ↙
[5] We also want to change **Type** from **Bonded** to **Frictional** (13.1.8, page 476). →
[6] We will change **Behavior** to **Symmetric**, meaning that the contact and target are treated symmetrically, so we don't need to decide which one is contact and which one is target (13.1.9, page 477). ↙
[7] Select **Frictional** for **Type**. ↑
[8] The name automatically changes to reflect the contact type. ↓
[9] Type 0.3 for **Friction Coefficient**. ↓
[10] Select **Symmetric** for **Behavior**. ↓
[11] Select **Augmented Lagrange** for **Formulation**. It is recommended for frictional or frictionless contacts (13.1.10, pages 477-478). ↵
Outline
Filter: Name
Project
Model (C4)
Geometry
Coordinate Systems
Connections
Contacts
Contact Region
Mesh
Static Structural (C5)
Analysis Settings
Solution (C6)
Solution Information
Details of "Contact Region"
Scope
Scoping Method Geometry Selection
Contact 23 Edges
Target 23 Edges
Contact Bodies Bolt
Target Bodies Nut
Shell Thickness Effect No
Protected No
Definition
Type Bonded
Scope Mode Automatic
Behavior Program Controlled
Trim Contact Program Controlled
Geometry
Print Preview
Report Preview
Frictional - Bolt To Nut
Details of "Frictional - Bolt To Nut"
Scope
Scoping Method Geometry Selection
Contact 23 Edges
Target 23 Edges
Contact Bodies Bolt
Target Bodies Nut
Shell Thickness Effect No
Protected No
Definition
Type Frictional
Friction Coefficient 0.3
Scope Mode Automatic
Behavior Symmetric
Trim Contact Program Controlled
Trim Tolerance 9.0588e-002 mm
Suppressed No
Advanced
Formulation Augmented Lagrange

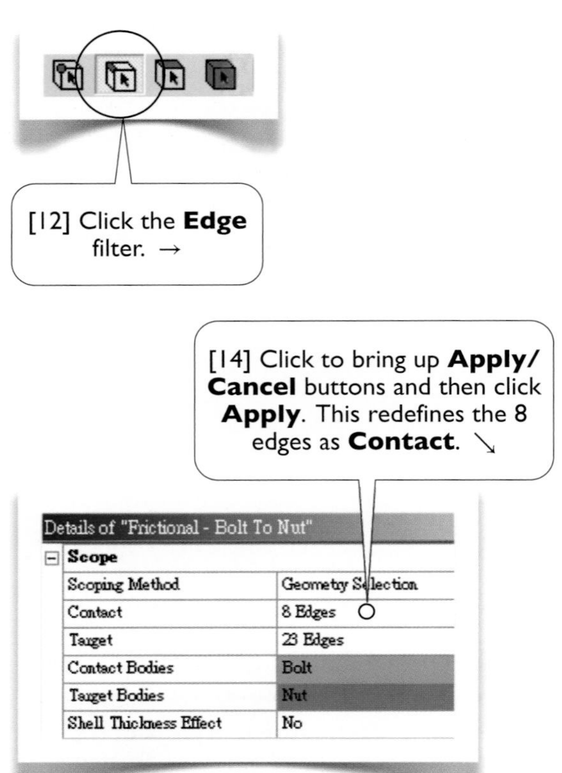
[12] Click the **Edge** filter. →
[14] Click to bring up **Apply/ Cancel** buttons and then click **Apply**. This redefines the 8 edges as **Contact**. ↘
Details of "Frictional - Bolt To Nut"
Scope
Scoping Method Geometry Selection
Contact 8 Edges
Target 23 Edges
Contact Bodies Bolt
Target Bodies Nut
Shell Thickness Effect No

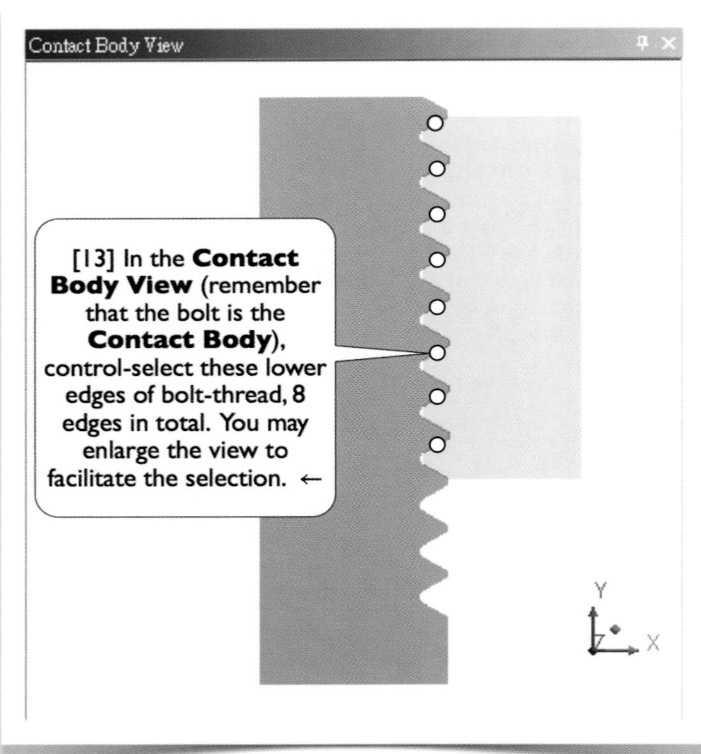
Contact Body View
[13] In the **Contact Body View** (remember that the bolt is the **Contact Body**), control-select these lower edges of bolt-thread, 8 edges in total. You may enlarge the view to facilitate the selection. ←
Y
X

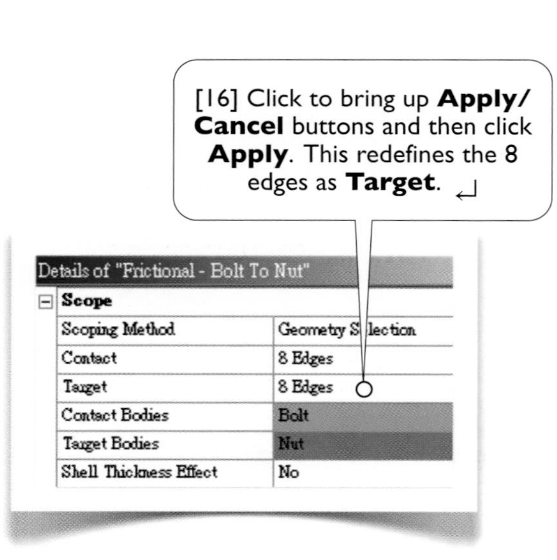
[16] Click to bring up **Apply/ Cancel** buttons and then click **Apply**. This redefines the 8 edges as **Target**. ↲
Details of "Frictional - Bolt To Nut"
Scope
Scoping Method Geometry Selection
Contact 8 Edges
Target 8 Edges
Contact Bodies Bolt
Target Bodies Nut
Shell Thickness Effect No

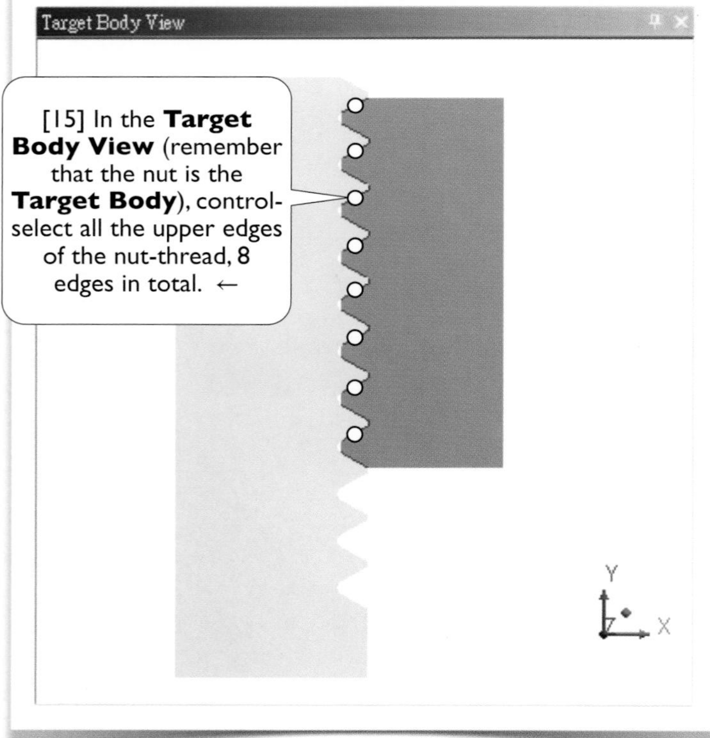
Target Body View
[15] In the **Target Body View** (remember that the nut is the **Target Body**), control-select all the upper edges of the nut-thread, 8 edges in total. ←
Y
X

Contact Elements

[17] During mesh generation (3.2.8, this page), Workbench creates contact elements between the contact edges and the target edges. Contact elements are used to prevent a *contact body* from penetrating into a target body. Therefore, you should set up contact regions wherever contacts may occur. As long as the behavior is **Symmetric** ([10], page 131), you may choose any one as contact body and the other as target body. PART B of Chapter 13 discusses contacts in detail (pages 476-479). #

3.2.8 Generate Mesh

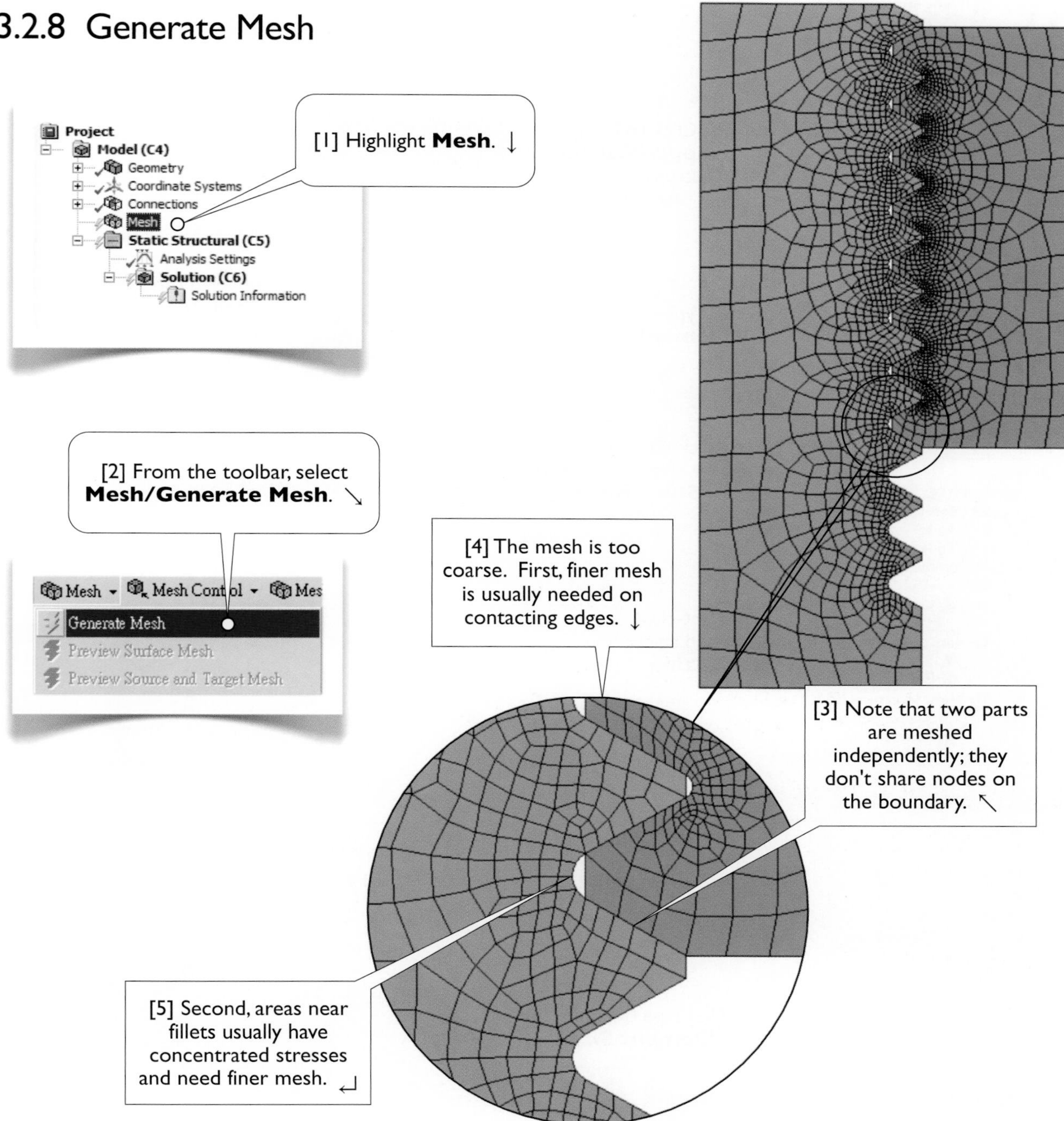

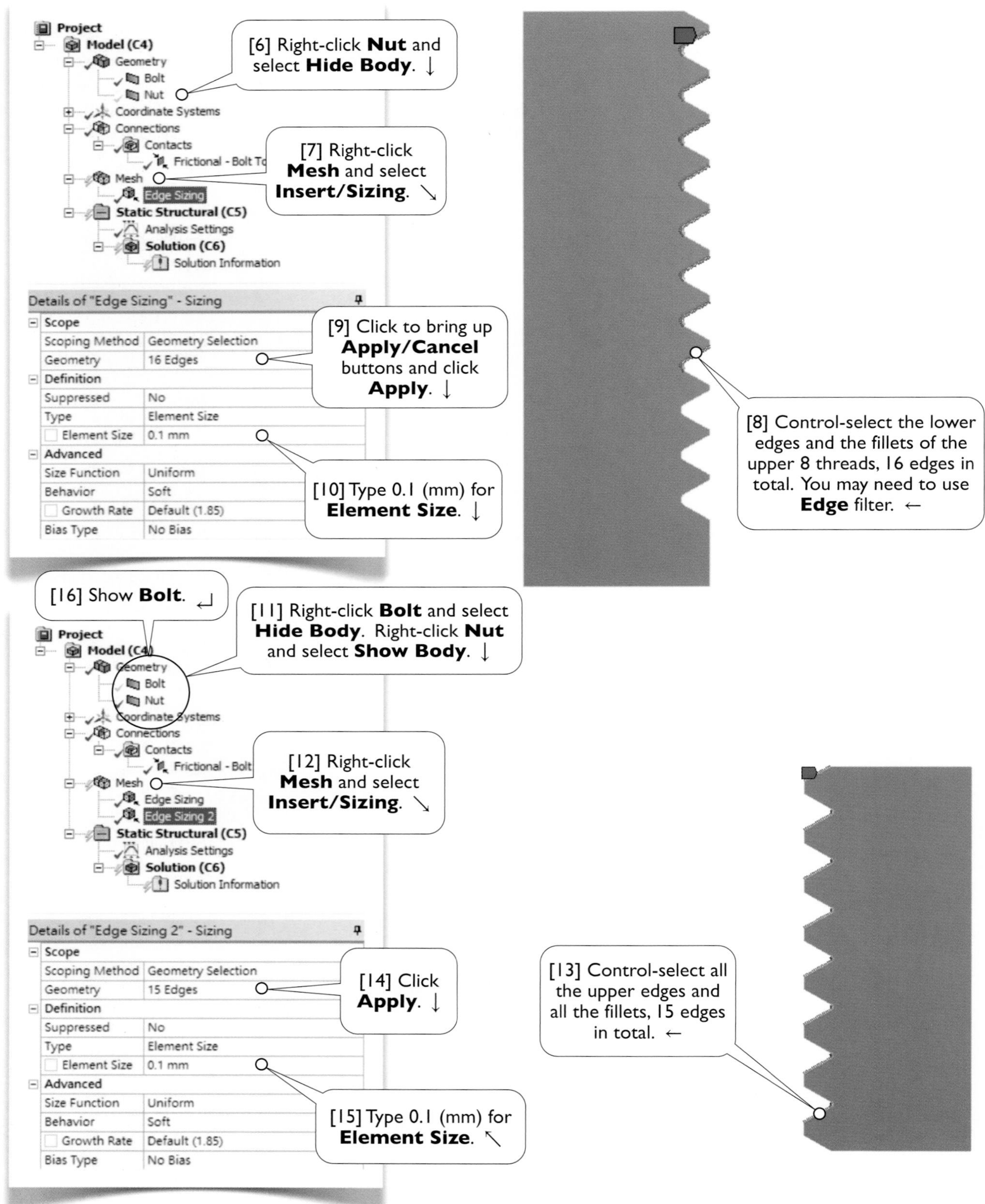
Project
Model (C4)
Geometry
Bolt
Nut
Coordinate Systems
Connections
Contacts
Frictional - Bolt To
Mesh
Edge Sizing
Static Structural (C5)
Analysis Settings
Solution (C6)
Solution Information
[6] Right-click **Nut** and select **Hide Body**. ↓
[7] Right-click **Mesh** and select **Insert/Sizing**. ↘
Details of "Edge Sizing" - Sizing
Scope
Scoping Method Geometry Selection
Geometry 16 Edges
Definition
Suppressed No
Type Element Size
Element Size 0.1 mm
Advanced
Size Function Uniform
Behavior Soft
Growth Rate Default (1.85)
Bias Type No Bias
[9] Click to bring up **Apply/Cancel** buttons and click **Apply**. ↓
[10] Type 0.1 (mm) for **Element Size**. ↓
[8] Control-select the lower edges and the fillets of the upper 8 threads, 16 edges in total. You may need to use **Edge** filter. ←
[16] Show **Bolt**. ↵
[11] Right-click **Bolt** and select **Hide Body**. Right-click **Nut** and select **Show Body**. ↓
Project
Model (C4)
Geometry
Bolt
Nut
Coordinate Systems
Connections
Contacts
Frictional - Bolt
Mesh
Edge Sizing
Edge Sizing 2
Static Structural (C5)
Analysis Settings
Solution (C6)
Solution Information
[12] Right-click **Mesh** and select **Insert/Sizing**. ↘
Details of "Edge Sizing 2" - Sizing
Scope
Scoping Method Geometry Selection
Geometry 15 Edges
Definition
Suppressed No
Type Element Size
Element Size 0.1 mm
Advanced
Size Function Uniform
Behavior Soft
Growth Rate Default (1.85)
Bias Type No Bias
[14] Click **Apply**. ↓
[15] Type 0.1 (mm) for **Element Size**. ↖
[13] Control-select all the upper edges and all the fillets, 15 edges in total. ←

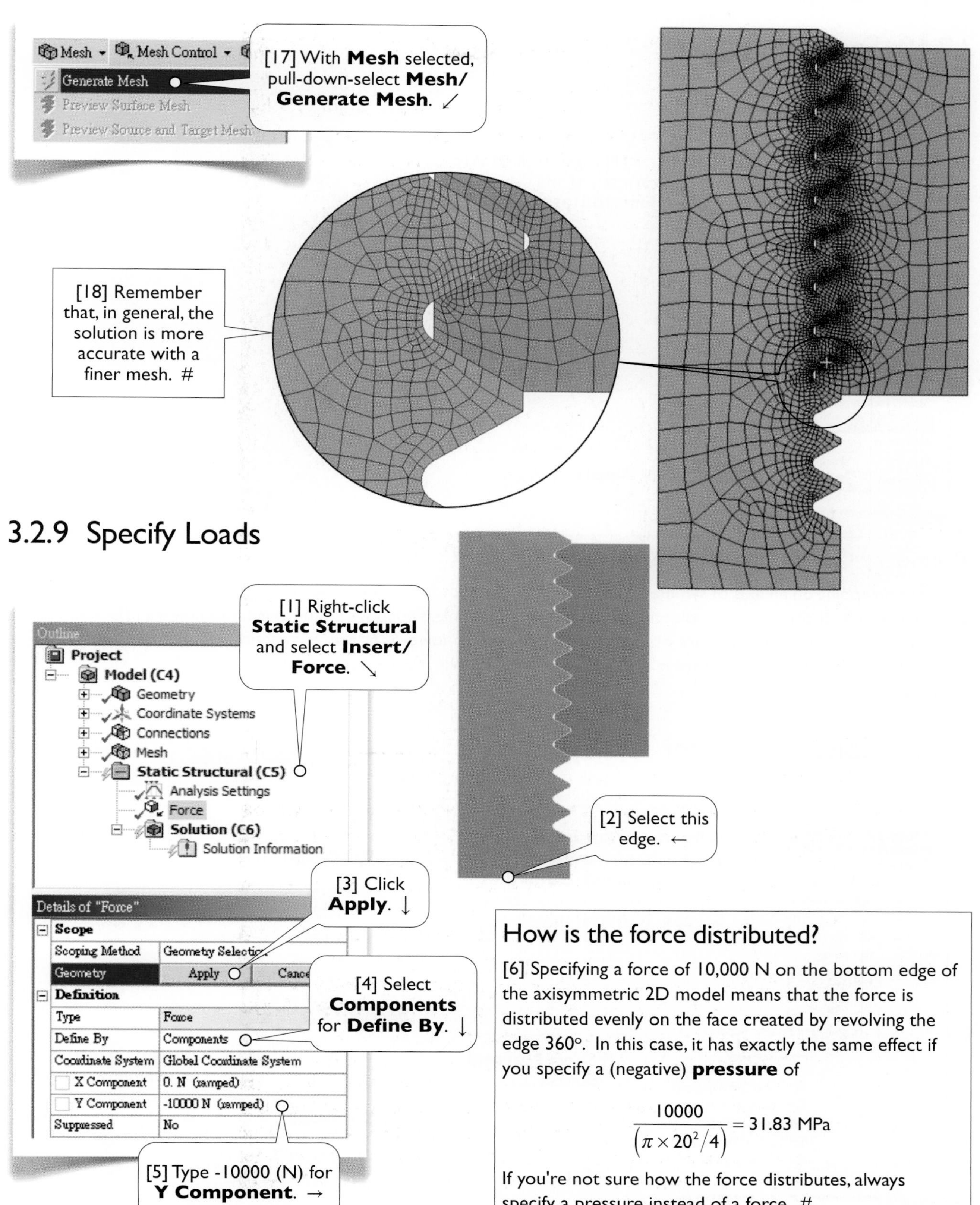

3.2.9 Specify Loads

How is the force distributed?

[6] Specifying a force of 10,000 N on the bottom edge of the axisymmetric 2D model means that the force is distributed evenly on the face created by revolving the edge 360°. In this case, it has exactly the same effect if you specify a (negative) **pressure** of

$$\frac{10000}{\left(\pi \times 20^2/4\right)} = 31.83 \text{ MPa}$$

If you're not sure how the force distributes, always specify a pressure instead of a force. #

3.2.10 Specify Supports

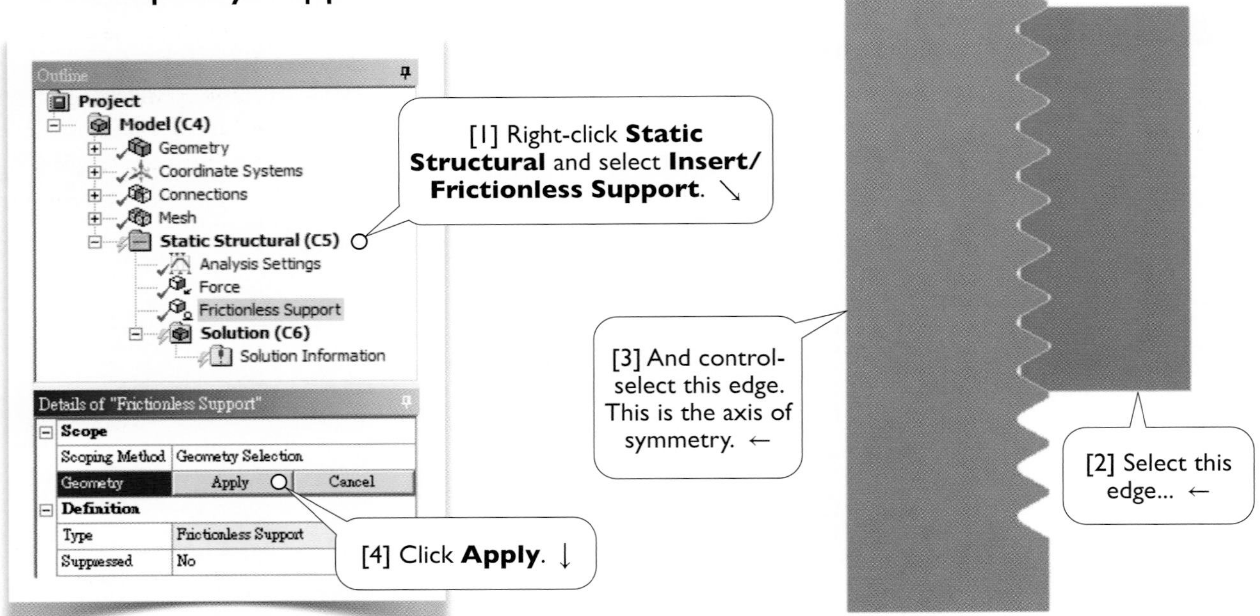

Boundary Conditions for the Axis of Symmetry

[5] Since any point on an axis of symmetry (here, *Y*-axis) does not move in the radial-direction (*X*-direction), you must specify a zero *X*-displacement condition or, equivalently, a frictionless support on the axis of symmetry [3]. Some FEA software can automatically take care of this boundary condition; however, as a good practice, always explicitly specify this boundary condition. If you leave it as a free boundary, the axis may become a small cylindrical "hole" after deformation. #

3.2.11 Solve the Model

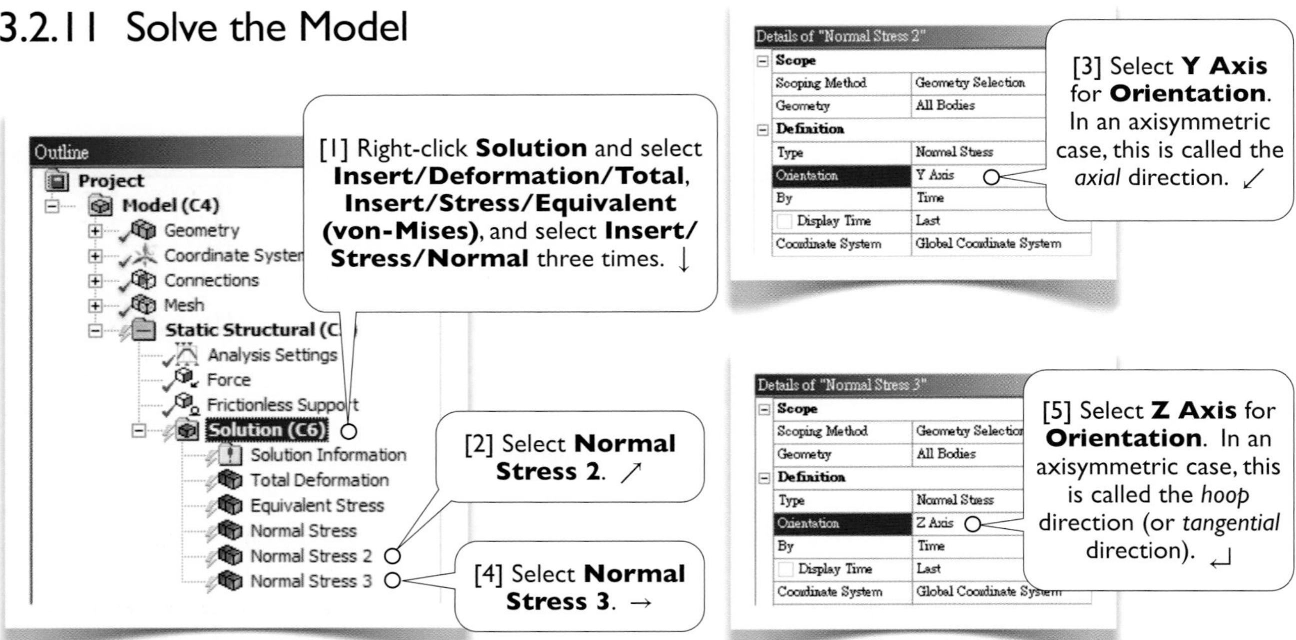

Nonlinear Simulations

[7] Due to the contact status, as the loads increase, the structure's stiffness also changes. The simulation is no longer linear. Workbench will solve the model using nonlinear solution methods. We will discuss nonlinear solution methods in Chapters 13-14. For now, to gain a feeling of nonlinear solution, let's watch how the solution proceeds [8-16]. ↙

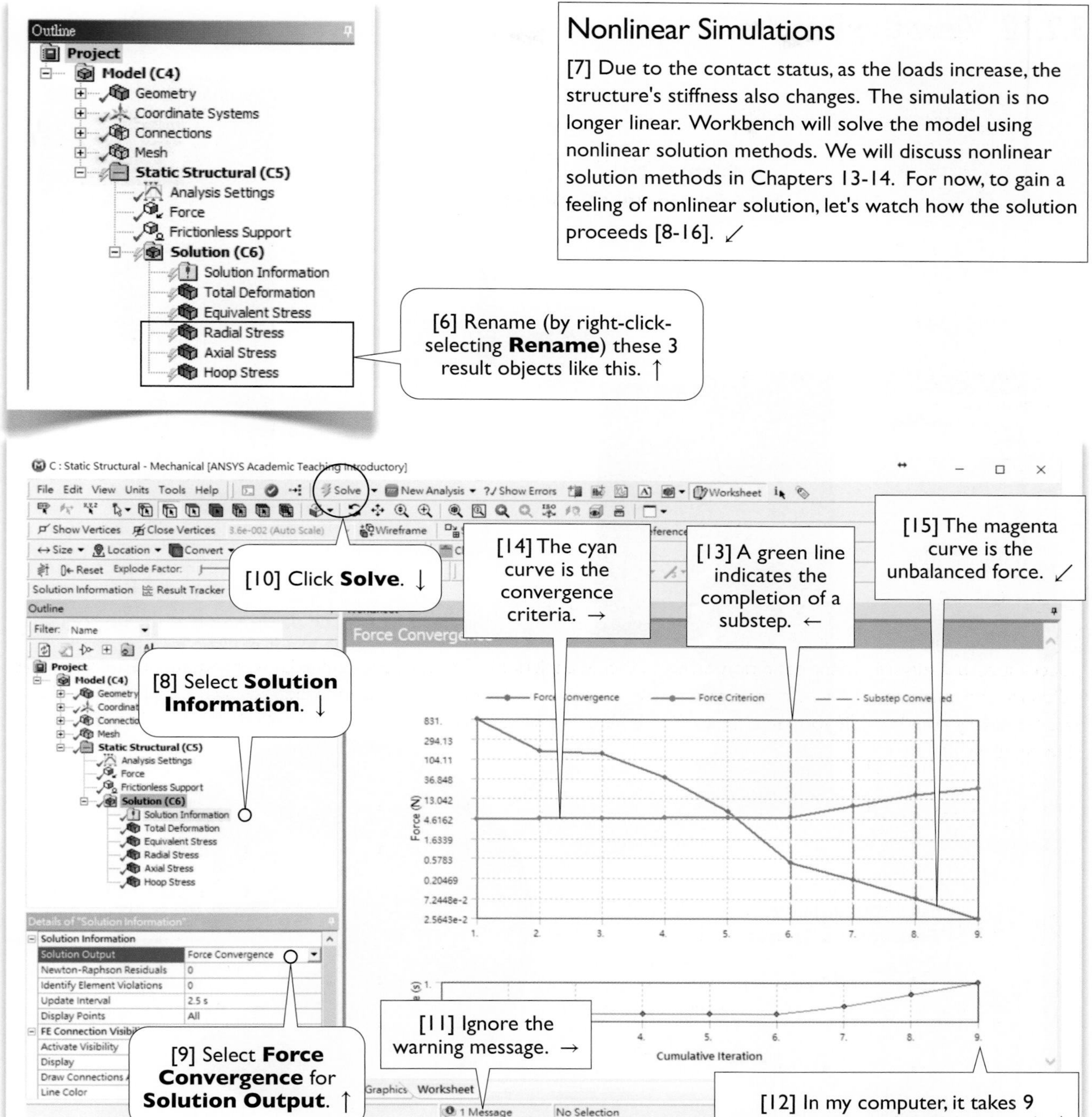

Nonlinear Solution Method: Force Convergence

[16] In this case, Workbench divides the load (10,000 N) into 4 substeps [13] (i.e., increasing 2,500 N each substep). The cyan curve is the convergence criteria [14] and the magenta curve is the "unbalanced" force [15]. A substep converges when the unbalanced force is less than the criterion [13-15]. PART A of Chapter 13 (pages 469-475) will discuss the details. #

3.2.12 View the Results

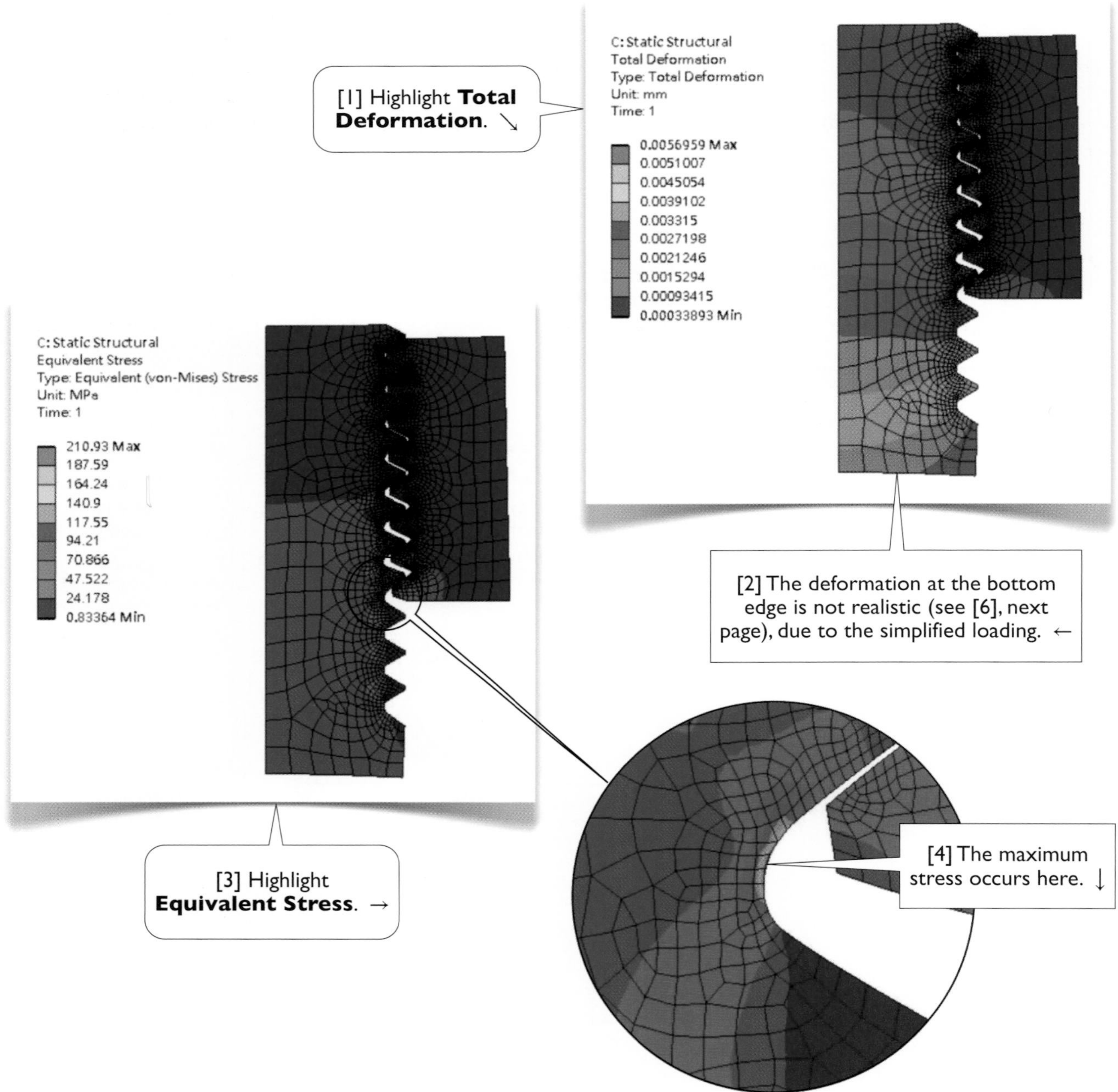

2D Models Must Be in *XY* Plane[Ref 1]

[5] ANSYS requires that, for a 2D problem (i.e., plane-stress, plane-strain, or axisymmetric problems), the 2D geometric model MUST lie in the global *XY* plane. Besides, for an axisymmetric problem, the global *Y*-axis MUST be the axis of symmetry and the model MUST be placed on the +X side.

A surface body created in DesignModeler is called a "2D solid body" if it is used in **Mechanical** for a 2D simulation, in which 2D solid elements (1.3.3[6-8], page 38) are used. ↵

Modeling Considerations

[6] The bottom edge ([2], last page) is actually a plane of symmetry; it must remain horizontal and must have no vertical displacements. This plane of symmetry might have been modeled as a frictionless support; however, a frictionless support cannot have an out-of-plane force acting on it. With this dilemma, we choose to apply a force, and expect an unrealistic deformation at the plane of symmetry [2]. Since we are only concerned about the stress at the threads, and the region of influence of this faulty boundary condition seems not so large as to reach the areas that concern us [4], we decide to accept this arrangement. ↓

Remark

[7] The quantities of bolt-and-nut used in daily industrial applications are huge. Their behavior should be carefully investigated. In our preliminary study, it shows that the stresses are distributed so unevenly that most of the stresses are taken by a few lower contacting threads. To improve the efficiency of the bolt-and-nut, one way is to allot some of the stresses to the upper contacting threads. In his books[Refs 2, 3], Zahavi has provided several alternatives to reduce the maximum stress. This case is adapted based on an example in his books. ↓

Wrap Up

[8] View other results. Close **Mechanical**, save the project, and exit Workbench. #

References

1. ANSYS Help//Mechanical APDL//Mechanical APDL Element Reference//I. Element Library//PLANE182.
2. Zahavi, E. and Barlam, E., *Nonlinear Problems in Machine Design*, CRC Press LLC, 2000; Chapter 10. Threaded Fasteners.
3. Zahavi, E., *The Finite Element Method in Machine Design*, Prentice-Hall, 1992; Chapter 7. Threaded Fasteners.

Section 3.3

More Details

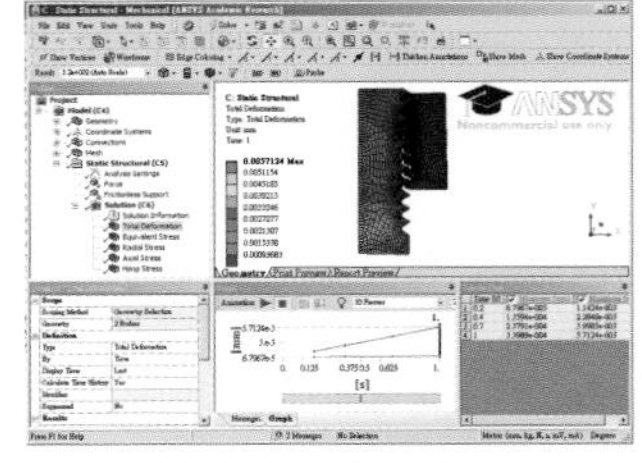

3.3.1 Plane-Stress Problems

Plane-Stress Condition

[1] Consider a plate of ZERO thickness on *XY* plane subject to in-plane forces. The stress state at any point can be depicted in [2]. Note that there are no stresses in *Z*-face; i.e.,

$$\sigma_Z = 0, \quad \tau_{ZY} = 0, \quad \tau_{ZX} = 0 \tag{1}$$

Eq. (1) is called a plane-stress condition. If the plane-stress condition holds everywhere, then it is called a plane-stress problem.

In the real world, there is no such thing as zero thickness. The triangular plate simulated in Section 3.1 is close to but not exactly a plane-stress problem; the triangular plate has finite thickness of 10 mm. However, since its stresses in *Z*-direction are negligible, we usually assume that the plane-stress condition holds for such a finite thickness plate.

In practice, a problem may assume the plane-stress condition if its thickness direction (*Z*-direction) is not restrained and thus free to expand or contract. As an example, a simply supported beam as shown in [3] is often solved by assuming the plane stress condition, even though its out-of-plane thickness is not zero.

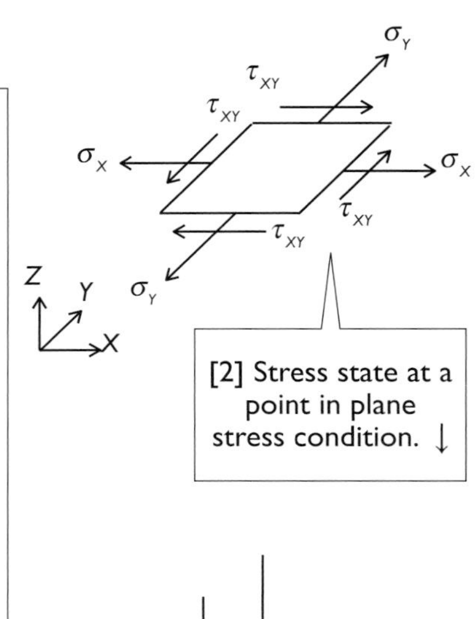

[2] Stress state at a point in plane stress condition. ↓

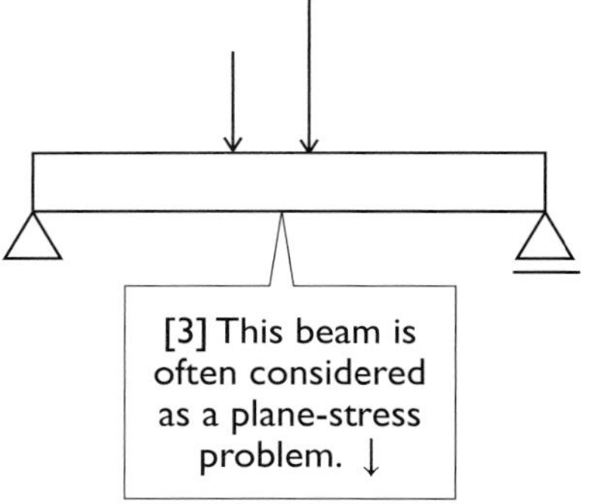

[3] This beam is often considered as a plane-stress problem. ↓

Hooke's Law for Plane-Stress Problems

Substituting the plane-stress condition, Eq. (1), into Eq. 1.2.8(1) (page 31), Hooke's law becomes

$$\begin{aligned}\varepsilon_X &= \frac{\sigma_X}{E} - \nu\frac{\sigma_Y}{E}\\ \varepsilon_Y &= \frac{\sigma_Y}{E} - \nu\frac{\sigma_X}{E}\\ \varepsilon_Z &= -\nu\frac{\sigma_X}{E} - \nu\frac{\sigma_Y}{E}\\ \gamma_{XY} &= \frac{\tau_{XY}}{G}, \quad \gamma_{YZ} = 0, \quad \gamma_{ZX} = 0\end{aligned} \tag{2}$$

Substitution of Eq. (1) into other governing equations (e.g., Eq. 1.2.6(2), page 30) will conclude that all quantities are independent of *Z*. That is, the particles with the same *X* and *Y* coordinates share the same behaviors regardless of their *Z* coordinate. Thus, we can eliminate the *Z* coordinate and reduce the problem to a two-dimensional problem, on *XY* space.

Note that, in Eq. (2), ε_Z is not zero; it can be calculated from σ_X and σ_Y. The nonzero ε_Z is easy to understand: since Z-direction is free to expand or contract, there must be strains in the Z-direction due to poisson's effect. ↗

2D models must be in *XY* plane[Ref 12]

[4] ANSYS requires that, for a 2D problem (i.e., plane-stress, plane-strain, or axisymmetric problems), the 2D geometric model MUST lie in a global *XY* plane. For an axisymmetric problem, the global *Y*-axis is always the axis of symmetry and the model MUST be placed on the +*X* side. (Also see 3.2.12[5], page 138.) #

3.3.2 Plane-Strain Problems

Plane-Strain Condition

[1] Consider a structure of INFINITE LENGTH in Z-direction. The Z-direction is restrained such that no particles can move in Z-direction. Further, all cross-sections perpendicular to the Z-direction have the same geometry, supports, and loads [2]. In such a case, the strain state at any point can be depicted in [3]. Note that there are no strains in the Z-face; i.e., $\varepsilon_Z = 0$ (otherwise the particle on the Z-face would move in Z-direction) and $\gamma_{ZX} = \gamma_{ZY} = 0$ (otherwise the cube would twist in ZX and ZY planes respectively, and that implies the particles would move in Z-direction),

$$\varepsilon_Z = 0, \quad \gamma_{ZX} = 0, \quad \gamma_{ZY} = 0 \qquad (1)$$

Eq. (1) is called a *plane-strain condition*. If the plane-strain condition holds everywhere, then it is called a plane-strain problem.

In the real world, there is no such thing as infinite length. In practice, a problem may assume the plane-strain condition if its Z-direction is restrained from expansion or contraction and all cross-sections perpendicular to the Z-direction have the same geometry, supports, and loads. As an example, a pressurized pipe buried under the earth is often considered as a plane-strain problem. Section 14.3 and the exercise problems in 3.6.2 (page 171) provide two examples for plane-strain problems.

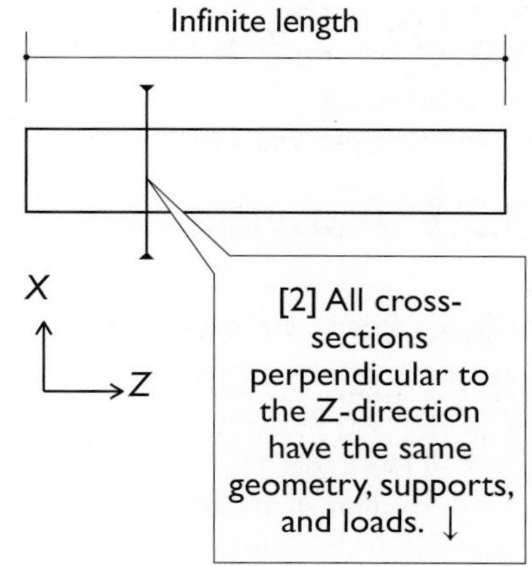

[2] All cross-sections perpendicular to the Z-direction have the same geometry, supports, and loads. ↓

Hooke's Law for Plane-Strain Problems

Eq. 1.2.8(1) (page 31), Hooke's law, can be inverted and rewritten as

$$\begin{aligned}
\sigma_X &= \frac{E}{(1+\nu)(1-2\nu)}\left[(1-\nu)\varepsilon_X + \nu\varepsilon_Y + \nu\varepsilon_Z\right] \\
\sigma_Y &= \frac{E}{(1+\nu)(1-2\nu)}\left[(1-\nu)\varepsilon_Y + \nu\varepsilon_Z + \nu\varepsilon_X\right] \\
\sigma_Z &= \frac{E}{(1+\nu)(1-2\nu)}\left[(1-\nu)\varepsilon_Z + \nu\varepsilon_X + \nu\varepsilon_Y\right] \\
\tau_{XY} &= G\gamma_{XY}, \quad \tau_{YZ} = G\gamma_{YZ}, \quad \tau_{ZX} = G\gamma_{ZX}
\end{aligned} \qquad (2)$$

The proof of Eq. (2) is in 3.3.14 (page 150).

Substitute the plane-strain condition Eq. (1) into Eq. (2), and Hooke's law becomes

$$\begin{aligned}
\sigma_X &= \frac{E}{(1+\nu)(1-2\nu)}\left[(1-\nu)\varepsilon_X + \nu\varepsilon_Y\right] \\
\sigma_Y &= \frac{E}{(1+\nu)(1-2\nu)}\left[(1-\nu)\varepsilon_Y + \nu\varepsilon_X\right] \\
\sigma_Z &= \frac{E}{(1+\nu)(1-2\nu)}\left[\nu\varepsilon_X + \nu\varepsilon_Y\right] \\
\tau_{XY} &= G\gamma_{XY}, \quad \tau_{YZ} = 0, \quad \tau_{ZX} = 0
\end{aligned} \qquad (3) \nearrow$$

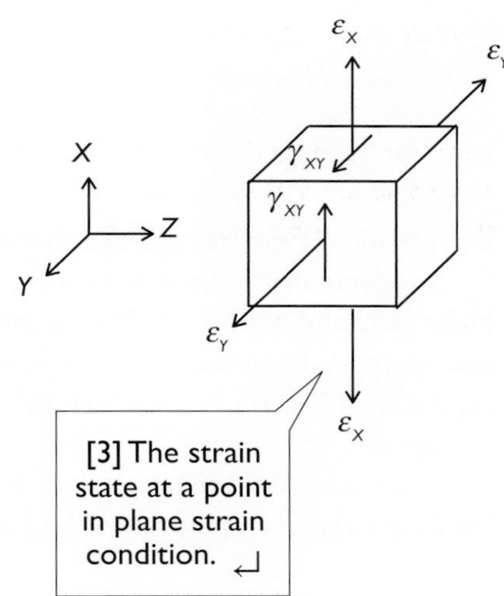

[3] The strain state at a point in plane strain condition. ↵

[4] Substitution of Eq. (1) into other governing equations (e.g., Eq. 1.2.6(2), page 30) will conclude that all quantities are independent of Z; i.e., the particles with the same X and Y coordinates share the same behaviors regardless of their Z coordinate. Thus, we can eliminate Z coordinate and reduce the problem to a two-dimensional problem, on XY space.

Note that, in Eq. (3), σ_Z is not zero; it can be calculated from ε_X and ε_Y. The nonzero σ_Z is easy to understand: since Z-direction is restrained from expansion or contraction, the material will develop stress to counteract the restriction. #

3.3.3 Axisymmetric Problems

[1] Consider a structure of which the geometry, supports, and loads are axisymmetric about the Y-axis. In such a case, all quantities are independent of θ coordinate; i.e., the particles with the same R and Y coordinates share the same behaviors regardless of their θ coordinate. Thus, we may eliminate θ coordinate and reduce the problem to a two-dimensional problem, on R-Y space.

The strain state at any point can be depicted in [2]. Note that there are no shear strains in θ-face (otherwise the θR-face and the θY-face would twist and the problem is no longer axisymmetric),

$$\gamma_{\theta R} = 0, \quad \gamma_{\theta Y} = 0 \qquad (1)$$

According to Hooke's law, Eq. (1) implies

$$\tau_{\theta R} = 0, \quad \tau_{\theta Y} = 0 \qquad (2)$$

Eqs. (1) and (2) can be regarded as the *axisymmetric condition*.

Axisymmetric problems are ubiquitous in engineering applications. Many problems are not strictly axisymmetric but can reasonably assume the axisymmetric condition, such as the bolt-and-nut problem simulated in Section 3.2, in which the threads are spiral and the nut is hexagonal.

In an axisymmetric problem, σ_r is called the *radial stress*, σ_θ is called the *hoop stress*, and σ_Y is called the *axial stress* [3]. ↗

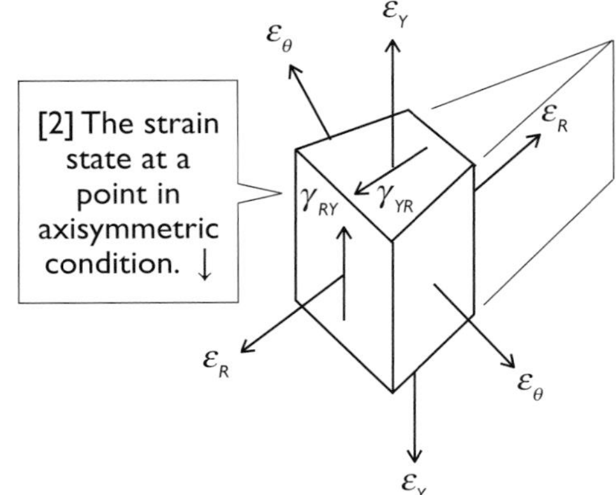

[2] The strain state at a point in axisymmetric condition. ↓

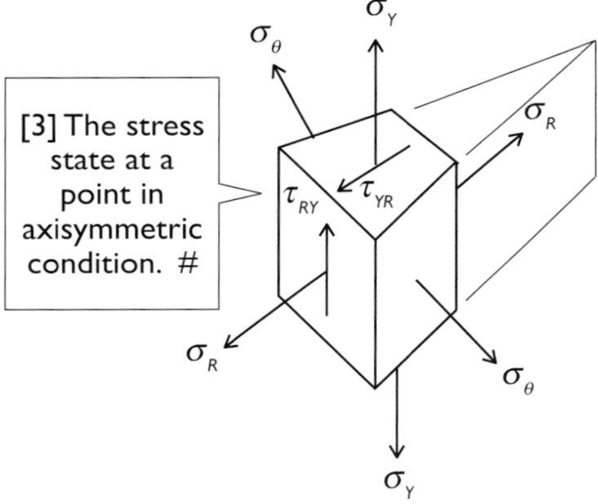

[3] The stress state at a point in axisymmetric condition. #

3.3.4 Mechanical GUI

[1] **Mechanical GUI** is composed of several areas ([2-8], next page); many of them are similar to those in **DesignModeler GUI** (2.3.1, page 81). On the top are pull-down menus and toolbars [2]; on the bottom is a status bar [8]. In-between are several "window panes" [3-7]. The separators [9] between window panes may be dragged to resize window panes. You even can move or dock a pane by dragging its title bar. Whenever you mess up the workspace, pull-down-select **View/Windows/Reset Layout** to reset to the default layout.

Outline [3] displays an outline of a *project tree*, which is a structured representation of the project (to be discussed). **Details** [4] shows the detail information of the object highlighted in the project tree or graphics window. The graphics window [5] displays a geometric model. **Graph** [6] typically shows a result-versus-time plot. **Tabular Data** [7] shows the numerical data of the result-versus-time plot. A set of animation tools are available in **Graph**; these tools allow you to play, stop, or save the animation. ↵

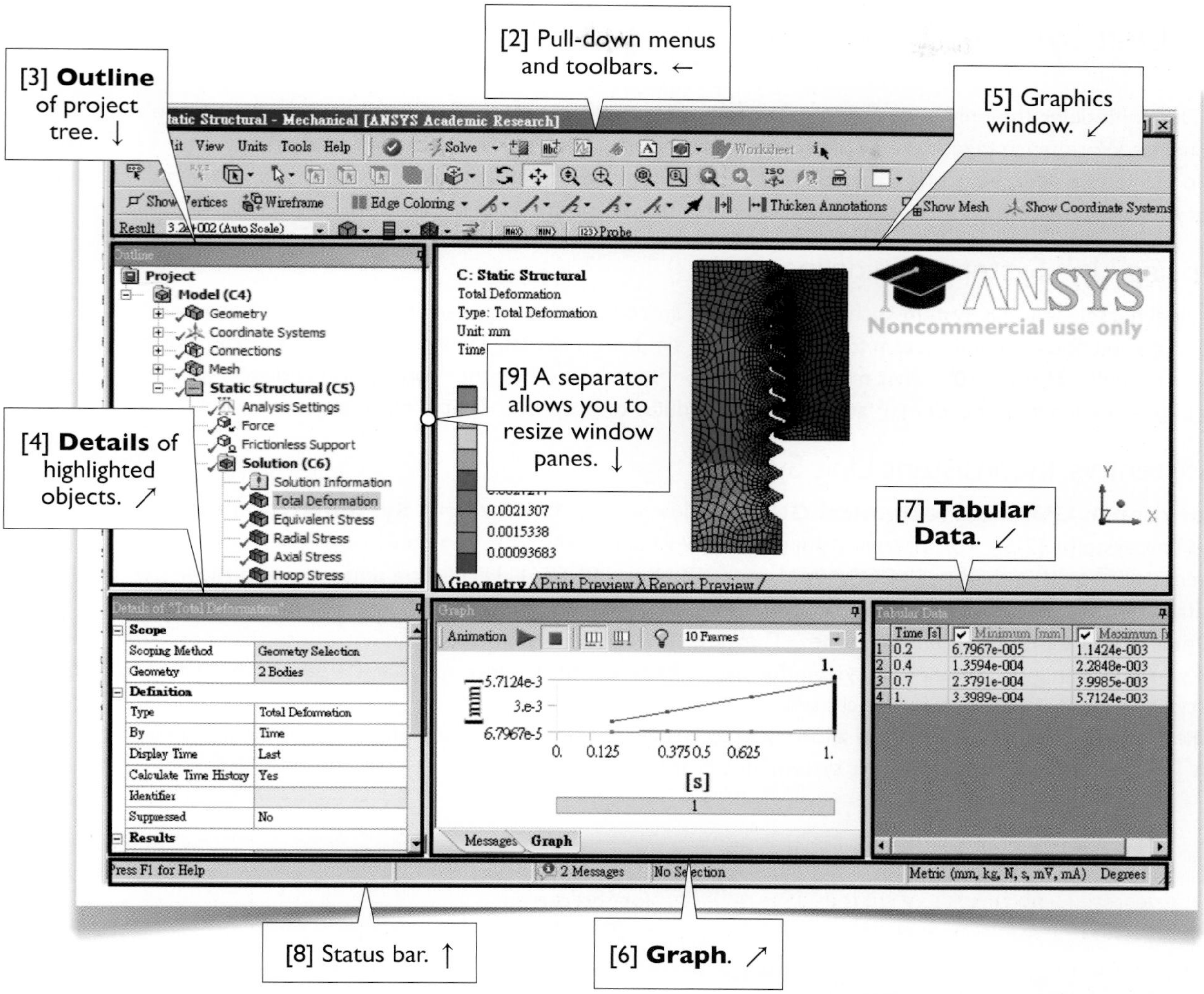

[2] Pull-down menus and toolbars. ←

[3] **Outline** of project tree. ↓

[5] Graphics window. ↙

[9] A separator allows you to resize window panes. ↓

[4] **Details** of highlighted objects. ↗

[7] **Tabular Data**. ↙

[8] Status bar. ↑

[6] **Graph**. ↗

Project Tree[Refs 1, 11]

[10] A project tree [11] is a structured representation of a project. A project tree may contain one or more **Models**; often, there is only one in a project tree. A **Model** may contain one or more **Environment** branches, along with other objects. Each can be renamed. Default name for an **Environment** branch is the name of the analysis system, for example: **Static Structural**. An **Environment** branch contains **Analysis Settings**, several objects that define the environment conditions, and a **Solution** branch, which contains a **Solution Information** and several results objects.

Right-clicking an object (or multiple objects) and selecting a tool from the context menu, you can operate on the object (or objects), such as delete, rename, duplicate, etc.

Unlike the objects of a model tree in DesignModeler, in which their order is important, the order of the objects in a project tree is not relevant. →

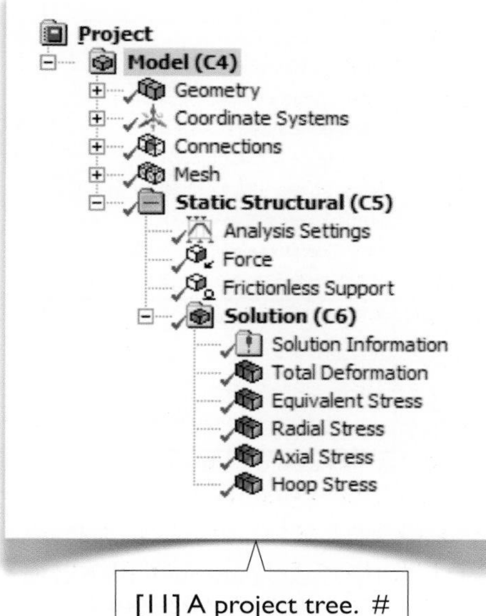

[11] A project tree. #

3.3.5 Unit Systems

[1] In DesignModeler, the only units used are length and angle. In **Mechanical**, units are much more involved; nevertheless, Workbench takes care of the consistency of unit systems, and all you have to do is to select a unit system suitable for your model. Selecting a suitable unit system for your model is crucial in many cases, in which the solution accuracy may deteriorate due to an accumulation of machine errors.

Choosing Unit Systems: Guideline

As a guideline, select a unit system such that the values stored in the computer have about the same order. For example, if you choose SI unit system for a micro-scale simulation model, you would have the lengths of order 10^{-6} and a Young's modulus of order 10^{11}. That may raise precision issues. On the other hand, if you choose a μMKS unit system, you will have the lengths of order 10^{0} and a Young's modulus of order 10^{5}. That is much better.

Consistent vs. Inconsistent Unit Systems

In **Workbench GUI** (not **Mechanical GUI**) pull-down-select **Units/Unit Systems...**, you will see a list of built-in unit systems [2-6]. For internal computation, Workbench always uses a consistent unit system. There are 6 *consistent unit systems* in the list: SI, CGS, NMM, μMKS, BIN, and BFT [6]. Highlight a unit system in the list; you will see the details of that unit system.

Other unit systems are *inconsistent* ones. They are, however, often more convenient to use than consistent ones. When you select an inconsistent unit system for use in **Mechanical**, it internally uses a consistent unit system that is closest to the inconsistent one you've chosen.

Like DesignModeler, **Mechanical** allows you to change the unit system any time, using the pull-down menu **Units** [7]. The internal consistent unit system also changes accordingly. ↙

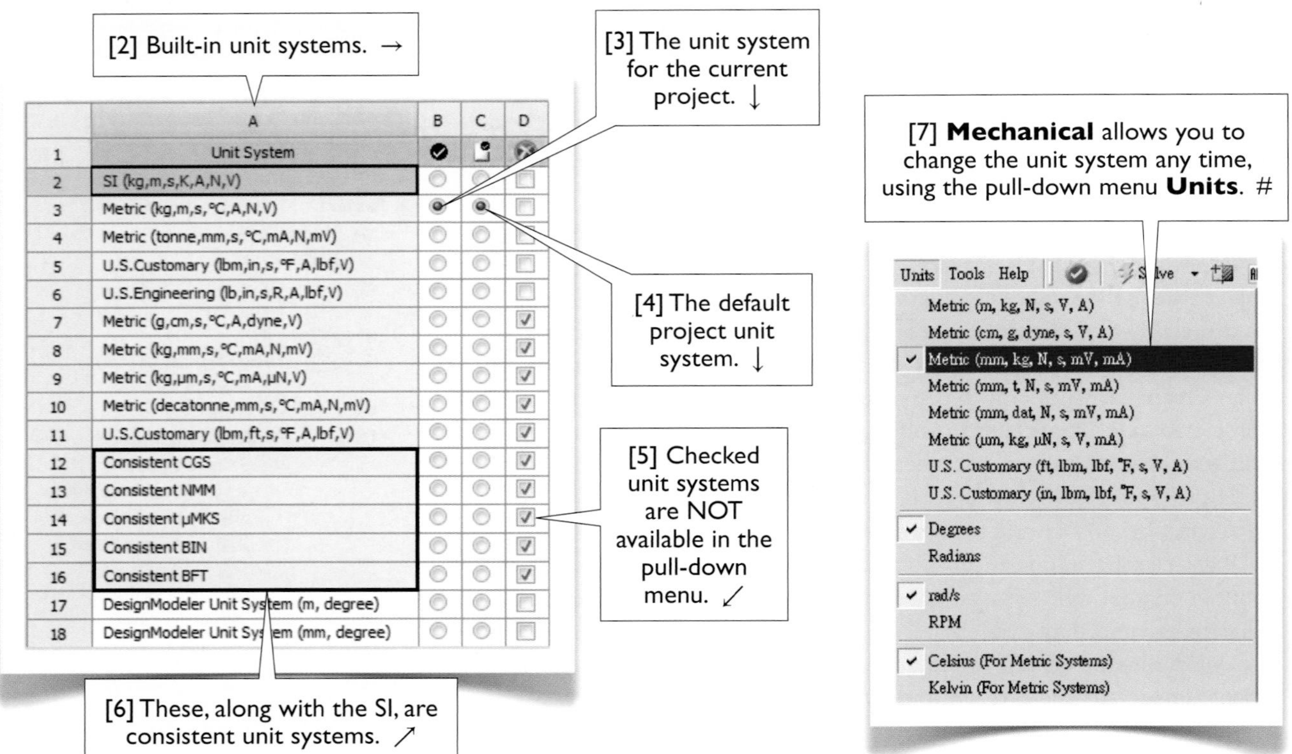

3.3.6 Environment Conditions

[1] Highlighting an **Environment** branch (e.g., **Static Structural**), you will see a row of environment conditions on the toolbar [2]. These environment conditions may also be accessed through the context menu.

Three groups of environment conditions are frequently used: **Loads** [3], **Supports** [4], and **Inertial** [5]. Environment conditions available in each group depend on the dimensionality (2D or 3D) as well as the type of analysis system (e.g., static or dynamic, structural or thermal).

Here, we will introduce the environment conditions available in 2D static structural simulations. Additional environment conditions will be introduced later, starting from Chapter 5. Many of the environment conditions are self-explanatory while others have many useful features. When going through each environment condition, we will point out its location in the ANSYS documentation system, in which many details can be found. You should consult this official documentation whenever needed.

Inside Workbench

Before jumping to individual environment conditions, let's describe how Workbench processes the environment conditions.

When an environment condition is applied, it will eventually be transferred to the NODES of the finite element model. For example, if you apply a pressure on a surface, the equivalent nodal forces are calculated and applied on nodes. The support conditions are processed in a similar way.

Consider Eq. 1.3.1(1) (page 35) again,

$$[K]\{D\}=\{F\} \qquad \text{Copy of Eq. 1.3.1(1)}$$

The vector $\{F\}$ is calculated according to the loads and inertia forces, while the vector $\{D\}$ is calculated according to the support conditions. Some of the nodal forces and nodal displacements are known values while the others are unknown. The unknown nodal forces to be solved are called *reaction forces*. For any degree of freedom, if the displacement is known then the corresponding force is unknown, and vice versa.

Magnitude of Environment Conditions[Ref 2]

The magnitude of most environment conditions can be specified in three ways: a constant value, a time-dependent tabular form, or a mathematical function with time, X, Y, or Z as the independent variable. ↗

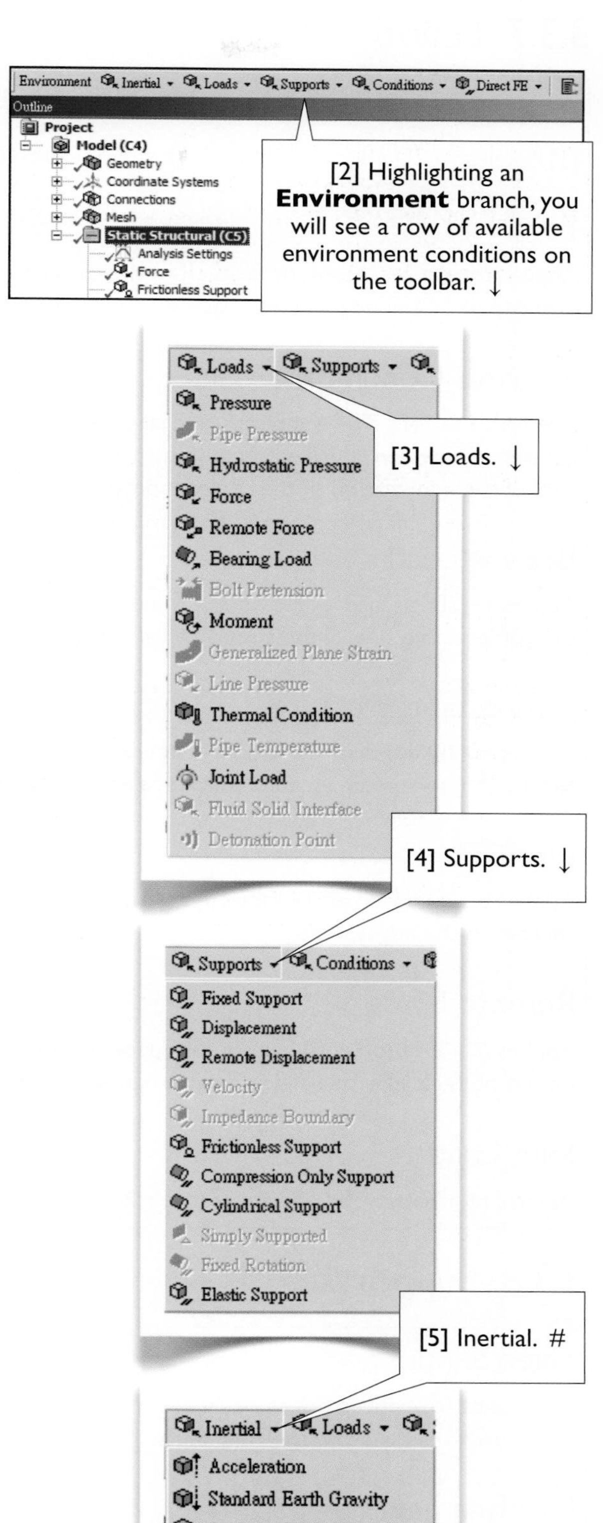

3.3.7 Loads[Ref 3]

Pressure

[1] Applies to 2D edges or 3D faces. It is possible to define a spatial varying pressure[Ref 4].

Force

Applies to vertices, edges, or faces. If it applies on edges/faces, the force is evenly distributed on the edges/faces.

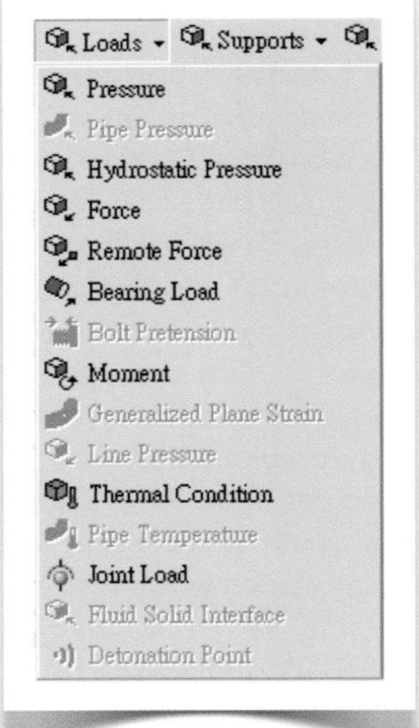

Thermal Condition

Applies to bodies. The temperature change ΔT (see Eq. 1.2.8(3), page 32) is the difference between specified temperature and the reference temperature, which is part of the information of the material properties, default to 22°C.

Bearing Load

Applies to 2D circular edges or 3D cylindrical faces. The total force is distributed on the compressive side of the circular edges or cylindrical faces.

Hydrostatic Pressure

Applies to 2D edges or 3D faces. It simulates pressure that occurs due to fluid weight. A free surface location may be specified, default to the surface at $X = 0$.

Moment

Applies to 2D edges or 3D faces. A statically equivalent pressure distributed on the edges/faces is calculated and applied on the edges/faces.

Remote Force

Applies at a location anywhere in the space. Workbench calculates the equivalent moment and force and applies them on the body. It may be used as an alternative way of building a rigid part and applying a force on it.

Joint Load

Applies to a **Joint**[Ref 5]. A joint load is a kinematic driving condition in a multi-body dynamic simulation. #

3.3.8 Supports[Ref 6]

Fixed Support

[1] Applies to vertices, edges, or faces. Prevents nodes from moving in X- Y- and Z-directions. It also prevents nodes from rotations for beam/shell elements.

Displacement

Applies to vertices, edges, or faces. Displacements in X- Y- and Z-directions can be specified. A zero value prevents nodes from moving in that direction. An unspecified value sets that direction free. ↵

Frictionless Support

[2] Applies to 2D edges or 3D faces. Prevents nodes from moving in the normal direction; allows nodes to freely move in the tangential direction.

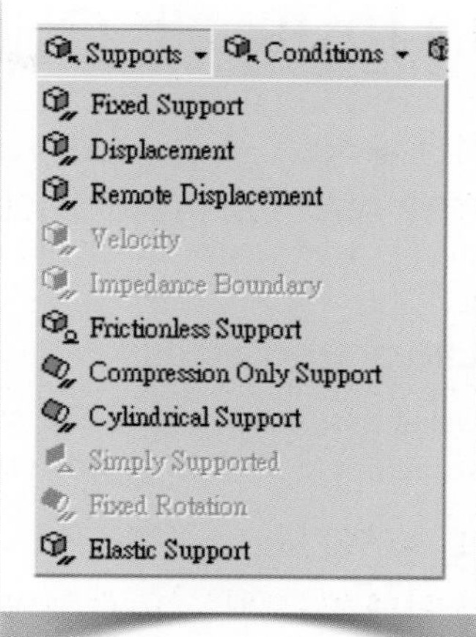

Compression Only Support

Applies to 2D edges or 3D faces. The associated body is free to depart from the edges or faces, but cannot penetrate them. It in effect sets up a frictionless contact region between the body and a rigid support; it introduces contact nonlinearity into the problem.

Cylindrical Support

Applies to 2D circular edges or 3D cylindrical faces. Each of the radial, tangential, and axial (3D only) directions can be set free or fixed.

Elastic Support

Applies to 2D edges or 3D faces. A foundation stiffness must be specified to establish the relation between the reaction pressure and the support displacement.

Remote Displacement

Applies at a location anywhere in the space. Workbench calculates the equivalent displacement and rotation and applies them on the body. It may be used as an alternative way of building a rigid part and applying a displacement to it. #

3.3.9 Inertial[Ref 3]

Standard Earth Gravity

[1] Applies to bodies. You must select a direction along which the gravitational force applies; it defaults to -Z direction.

Acceleration

Applies to bodies. You must specify the magnitude and direction of acceleration. The direction is where the bodies accelerate. An "Inertia force" will apply in the opposite direction.

Rotational Velocity

Applies to bodies. You must specify the magnitude and direction of the angular velocity of the bodies. A distributed "inertia force" will apply in the opposite direction of rotation. #

3.3.10 Results Objects[Ref 7]

[1] Highlighting **Solution** branch, you will see a row of results tools available on the toolbar [2]. These results tools may also be accessed through the context menu. Most of them are self-explanatory, but some of them need to be explained.

Linearized Stress

Using this tool, you can view stresses along a straight line path. You need to first define a straight line path using **Construction Geometry** under **Model**.

Probe[Ref 8]

It contains tools to explore the results of a point, or maximum/minimum values of the results of a scoped region, along the loading history. In other words, we are concerned about the results across the time domain instead of space domain.

Tools[Ref 7]

It contains some special results tools: **Stress Tool**, **Fatigue Tool**, **Contact Tool**, and **Beam Tool** [3].

User Defined Result[Ref 9]

You may define a result expression using available keywords, which can be listed by highlighting **Solution** in the project tree and clicking the **Worksheet** tab on the right side of the GUI (if you don't see the **Worksheet** tab, click the **Worksheet** toolbar button Worksheet). →

[3] Some special results tools. #

[2] Results toolbar. ↑

3.3.11 View Results[Ref 7]

[1] To view results, simply highlight a results object. Tools to control the visual effects of the results are shown on the toolbar [2-7]. ↘

[4] **Label**. ↓

[6] You can control how the contour displays. ↘

[2] Click to turn on/off the label of maximum/ minimum. →

[3] Click to turn on/off the probe. Results values will display along with your mouse pointer; click to label the value. To remove the label, activate **Label** [4] and press **Delete**. ←

[5] You may select the scale of deformation. ↑

[7] Some results can be displayed in a vector form. #

3.3.12 Insert APDL Commands[Ref 10]

[1] **Insert Commands** [2] allows you to insert APDL commands. For those who are familiar with APDL, this may be useful, since Workbench doesn't include all the functionalities provided by APDL. For the newcomers, my suggestion is that you do not worry about APDL for now; Workbench's functionalities are enough for most simulations.

After clicking **Insert Commands** tool, a text editor is opened with several lines of comments telling you WHEN the APDL commands will be executed. For example, an APDL commands object inserted under **Environment** will be executed just before **Solve** command. →

[2] Click **Insert Commands** to insert ANSYS APDL commands. ✓

What is APDL?

[3] In the old days, the users operated ANSYS using a set of text command language, called APDL (ANSYS Parametric Design Language). Compared with modern computer languages, the APDL is not a user-friendly language at all. For many users, use of APDL has been a painful experience.

Later, ANSYS started to provide a graphical user interface (GUI). The users operate ANSYS through pull-down menus, dialogs, etc. Basically, each APDL command has a corresponding operating path in the GUI. Using either APDL or the GUI, the users can use all the functionalities of ANSYS. Again, compared with **Workbench GUI**, the old ANSYS GUI is not efficient at all. Many experts and school teachers prefer APDL to the old GUI.

It is true that some capabilities of APDL are not directly supported in Workbench. It, however, provides two ways that you may access APDL commands: (a) You can insert APDL commands by clicking **Insert Commands** [2]. (b) You can create a **Mechanical APDL** system [4], which allows APDL files to be read into Workbench. →

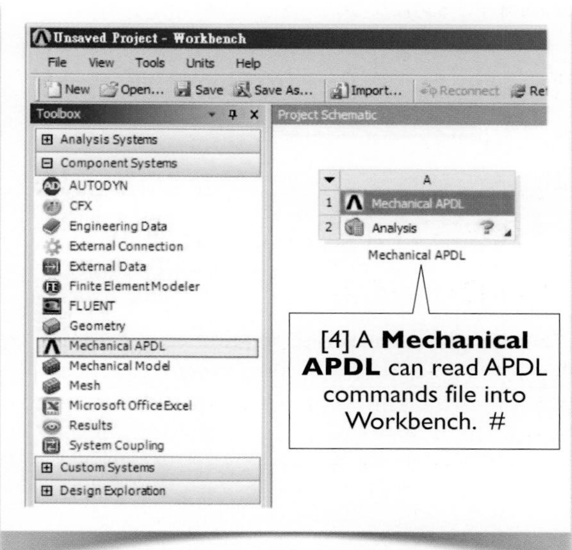

[4] A **Mechanical APDL** can read APDL commands file into Workbench. #

3.3.13 Status Symbols in Tree Outline[Ref 11]

[1] Each object of the project tree has a status symbol, explained below:

- Checkmark indicates branch is fully defined / OK.
- Question mark indicates item has incomplete data (need input).
- Lightning bolt indicates solving is required.
- Exclamation mark means a problem exists.
- "X" means item is suppressed (will not be solved).
- Transparent checkmark means body or part is hidden.
- Green lightning bolt indicates item is currently being evaluated.
- Minus sign means that mapped face meshing failed.
- Check mark with a slash indicates a meshed part/body.
- Red lightning bolt indicates a failed solution. #

3.3.14 Appendix: Proof of Eq. 3.3.2(2), page 141

[1] The first 3 equations and last 3 equations of Eq. 3.3.2(2) are decoupled; they can be proved independently. Proof of the last 3 equations from the last 3 equations in Eq. 1.2.8(1) (page 31) is trivial. Now, we prove the first 3 equations. The first 3 equations in Eq. 1.2.8(1) can be written in matrix form

$$\begin{Bmatrix} \varepsilon_X \\ \varepsilon_Y \\ \varepsilon_Z \end{Bmatrix} = \frac{1}{E} \begin{bmatrix} 1 & -\nu & -\nu \\ -\nu & 1 & -\nu \\ -\nu & -\nu & 1 \end{bmatrix} \begin{Bmatrix} \sigma_X \\ \sigma_Y \\ \sigma_Z \end{Bmatrix} \text{ or } \{\varepsilon\} = [D]\{\sigma\}$$

The first 3 equations in Eq. 3.3.2(2) can also be written in matrix form

$$\begin{Bmatrix} \sigma_X \\ \sigma_Y \\ \sigma_Z \end{Bmatrix} = \frac{E}{(1+\nu)(1-2\nu)} \begin{bmatrix} 1-\nu & \nu & \nu \\ \nu & 1-\nu & \nu \\ \nu & \nu & 1-\nu \end{bmatrix} \begin{Bmatrix} \varepsilon_X \\ \varepsilon_Y \\ \varepsilon_Z \end{Bmatrix} \text{ or } \{\sigma\} = [F]\{\varepsilon\}$$

Then

$$[D][F] = \frac{1}{E} \cdot \frac{E}{(1+\nu)(1-2\nu)} \begin{bmatrix} 1 & -\nu & -\nu \\ -\nu & 1 & -\nu \\ -\nu & -\nu & 1 \end{bmatrix} \begin{bmatrix} 1-\nu & \nu & \nu \\ \nu & 1-\nu & \nu \\ \nu & \nu & 1-\nu \end{bmatrix} = \begin{bmatrix} 1 & 0 & 0 \\ 0 & 1 & 0 \\ 0 & 0 & 1 \end{bmatrix}$$

This completes the proof. #

References

1. ANSYS Help//Mechanical Application//ANSYS Mechanical User's Guide//Objects Reference
2. ANSYS Help//Mechanical Application//ANSYS Mechanical User's Guide//Setting Up Boundary Conditions//Defining Boundary Condition Magnitude
3. ANSYS Help//Mechanical Application//ANSYS Mechanical User's Guide//Setting Up Boundary Conditions//Load Type Boundary Conditions
4. ANSYS Help//Mechanical Application//ANSYS Mechanical User's Guide//Setting Up Boundary Conditions//Spatial Varying Loads and Displacements
5. ANSYS Help//Mechanical Application//ANSYS Mechanical User's Guide// Setting Connections//Joints
6. ANSYS Help//Mechanical Application//ANSYS Mechanical User's Guide//Setting Up Boundary Conditions//Support Type Boundary Conditions
7. ANSYS Help//Mechanical Application//ANSYS Mechanical User's Guide//Using Results//Structural Results
8. ANSYS Help//Mechanical Application//ANSYS Mechanical User's Guide//Using Results//Result Output//Probe
9. ANSYS Help//Mechanical Application//ANSYS Mechanical User's Guide//Using Results/User Defined Results
10. ANSYS Help//Mechanical Application//ANSYS Mechanical User's Guide//Commands Objects
11. ANSYS Help//Mechanical Application//ANSYS Mechanical User's Guide//Application Interface//Tree Outline
12. ANSYS Help//Mechanical APDL//Mechanical APDL Element Reference//I. Element Library//PLANE182

Section 3.4

Spur Gears

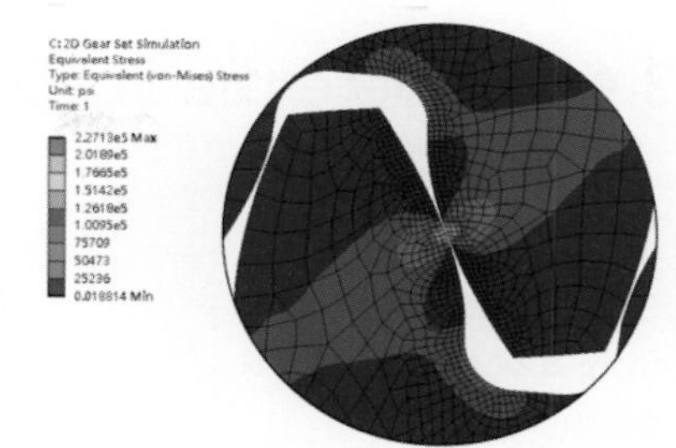

3.4.1 About the Spur Gears

[1] In this section, we'll perform a simulation for a pair of meshing spur gears introduced in Section 2.5. Consider a situation in which a clockwise torque of 15,000 lb-in is applied on the upper gear while both gears are at rest. Our goal is to assess the maximum stress during the transmission of the torque. By engineering judgement, the maximum stress occurs either at a contact point [2] or at the root of a tooth due to the bending of the tooth [3].

Since there is no restriction of deformation in the depth direction, i.e., the gears are free to expand (or contract) in the depth direction, so it is modeled as a plane stress problem (3.3.1, page 140). The U.S. customary unit system (**in-lbm-lbf-s**) is used in this exercise. →

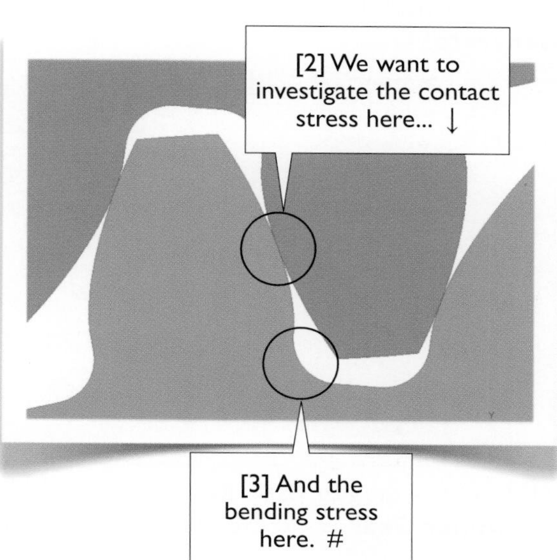

3.4.2 Set Up Project Schematic

[1] Launch Workbench. Open the project **Gear**, which was saved in Section 2.5 (see [2]). Duplicate the **Geometry** system [3-4]. Create a **Static Structural** system [5]. Create a link between the last two systems so that they share **Geometry** [6]. Double-click the **Geometry** cell in the **2D Gear Set** system to edit it [7]. ↙

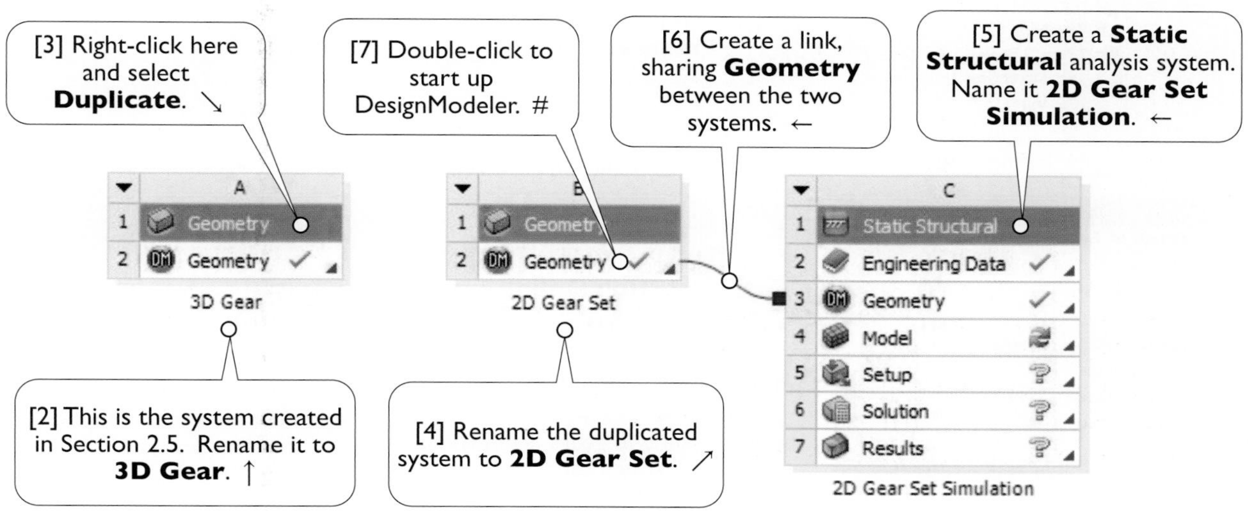

3.4.3 Delete the 3D Body and Create a Surface Body

[1] Delete **Extrude1**. Create a surface body from **Sketch1** by pull-down-selecting **Concept/Surfaces From Sketches**. Remember to click **Generate**. →

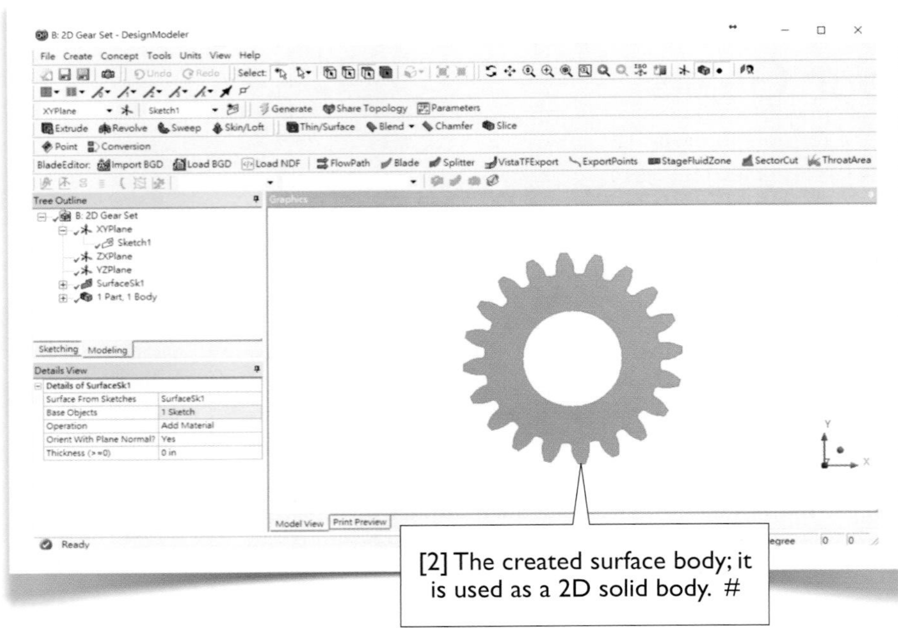

3.4.4 Duplicate the Gear

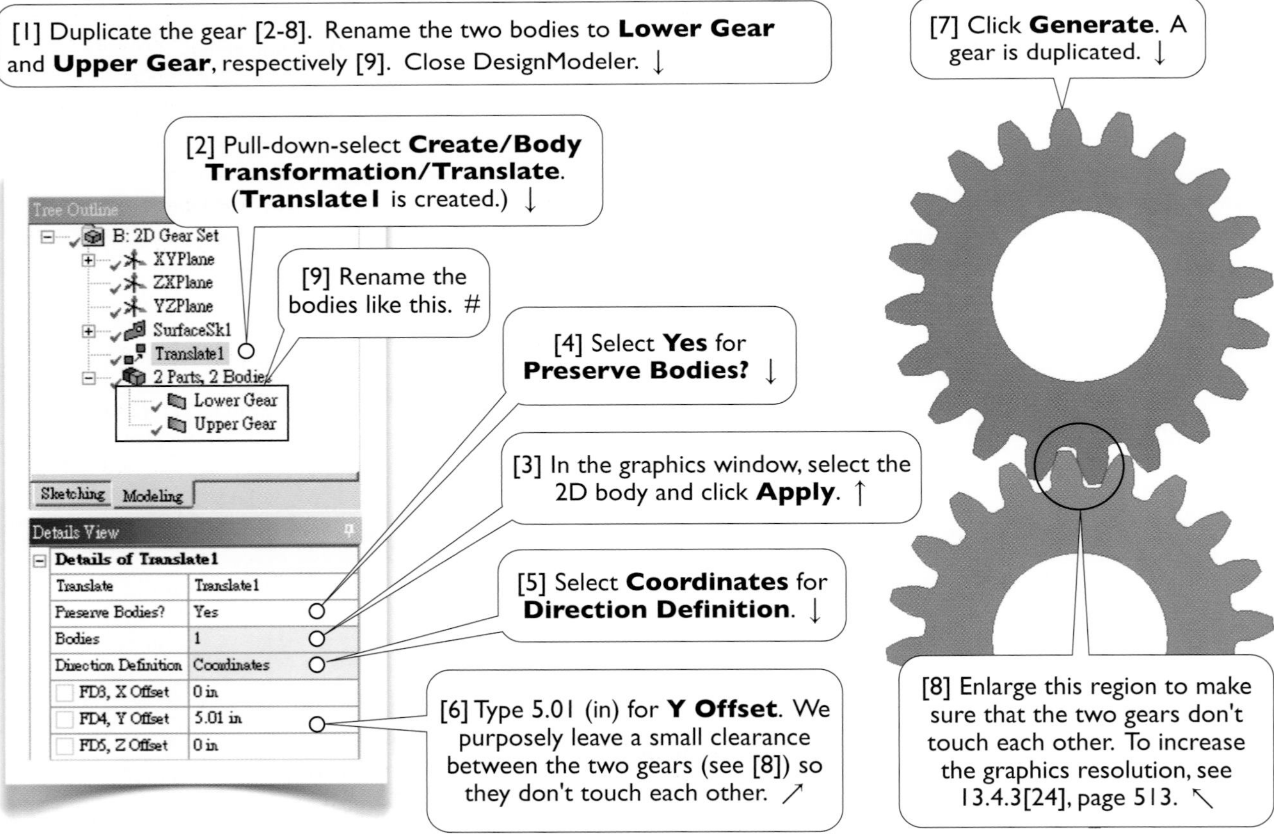

3.4.5 Set Up Geometry in Mechanical

[1] Before entering **Mechanical**, remember to specify **2D** for **Analysis Type** (3.1.4[4-6], page 112). This step is important since, after the geometry is brought to **Mechanical**, you cannot change it any more.

Start up **Mechanical** by double-clicking the **Model** cell. In **Mechanical**, change the units to **in-lbm-lbf-s** [2]. Make sure **2D Behavior** is set to **Plane Stress** [3-4]. Also make sure the material is **Structural Steel** (default) and the thickness is 1.0 inch (default) defaults [5]. →

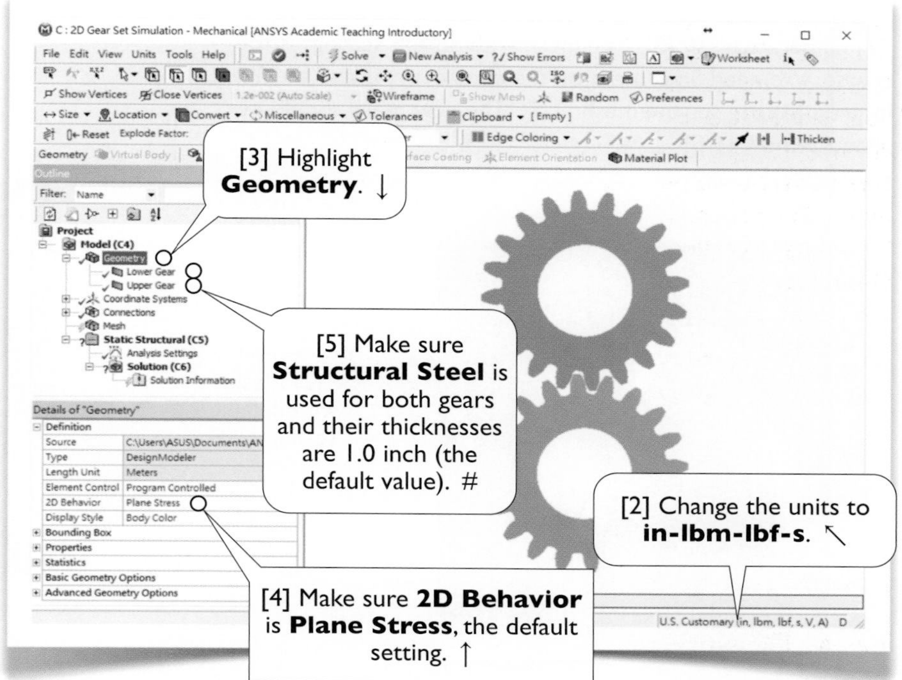

[3] Highlight **Geometry**. ↓

[5] Make sure **Structural Steel** is used for both gears and their thicknesses are 1.0 inch (the default value). #

[2] Change the units to **in-lbm-lbf-s**. ↖

[4] Make sure **2D Behavior** is **Plane Stress**, the default setting. ↑

3.4.6 Set Up Contact Region

[1] Redefine the contact region as follows. In the project tree, highlight **Connections/Contacts/Contact Region** and redefine **Contact** and **Target** [2-5]. Change the contact type to **Frictionless** [6]. Select **Augmented Lagrange** for **Formulation** [7]; this is generally recommended for frictionless contact (13.1.10, pages 477-478). Finally, select **Adjust to Touch** for **Interface Treatment** [8] (13.1.11, page 479). ↓

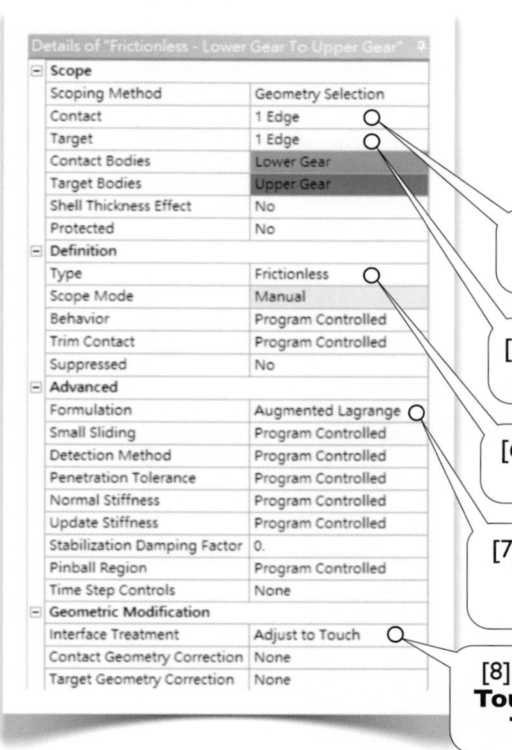

Details of "Frictionless - Lower Gear To Upper Gear"	
Scope	
Scoping Method	Geometry Selection
Contact	1 Edge
Target	1 Edge
Contact Bodies	Lower Gear
Target Bodies	Upper Gear
Shell Thickness Effect	No
Protected	No
Definition	
Type	Frictionless
Scope Mode	Manual
Behavior	Program Controlled
Trim Contact	Program Controlled
Suppressed	No
Advanced	
Formulation	Augmented Lagrange
Small Sliding	Program Controlled
Detection Method	Program Controlled
Penetration Tolerance	Program Controlled
Normal Stiffness	Program Controlled
Update Stiffness	Program Controlled
Stabilization Damping Factor	0.
Pinball Region	Program Controlled
Time Step Controls	None
Geometric Modification	
Interface Treatment	Adjust to Touch
Contact Geometry Correction	None
Target Geometry Correction	None

[3] Click to bring up buttons and apply as **Contact**. ↘

[5] Click to bring up buttons and apply as **Target**. ↓

[6] Select **Frictionless** for **Type**. ↓

[7] Select **Augmented Lagrange** for **Formulation**. ↓

[8] Select **Adjust to Touch** for **Interface Treatment**. #

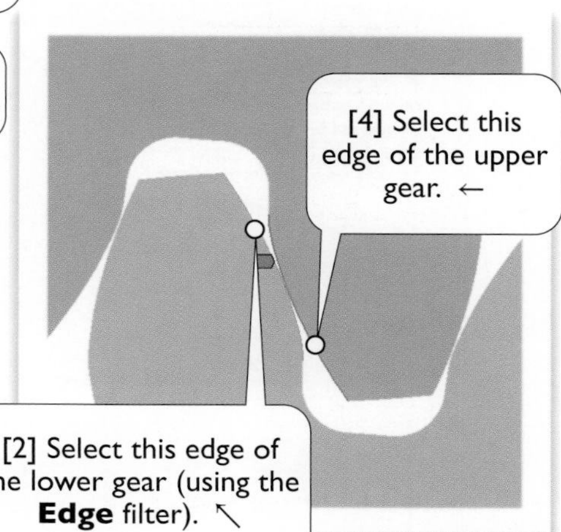

[4] Select this edge of the upper gear. ←

[2] Select this edge of the lower gear (using the **Edge** filter). ↖

3.4.7 Generate Mesh

[1] Generate mesh [2]. The mesh with default settings is too coarse. We need finer mesh on the contact areas and the fillets. Insert a **Sizing** for the mesh [3-5]. The new mesh should be adequate now [6-8]. ↓

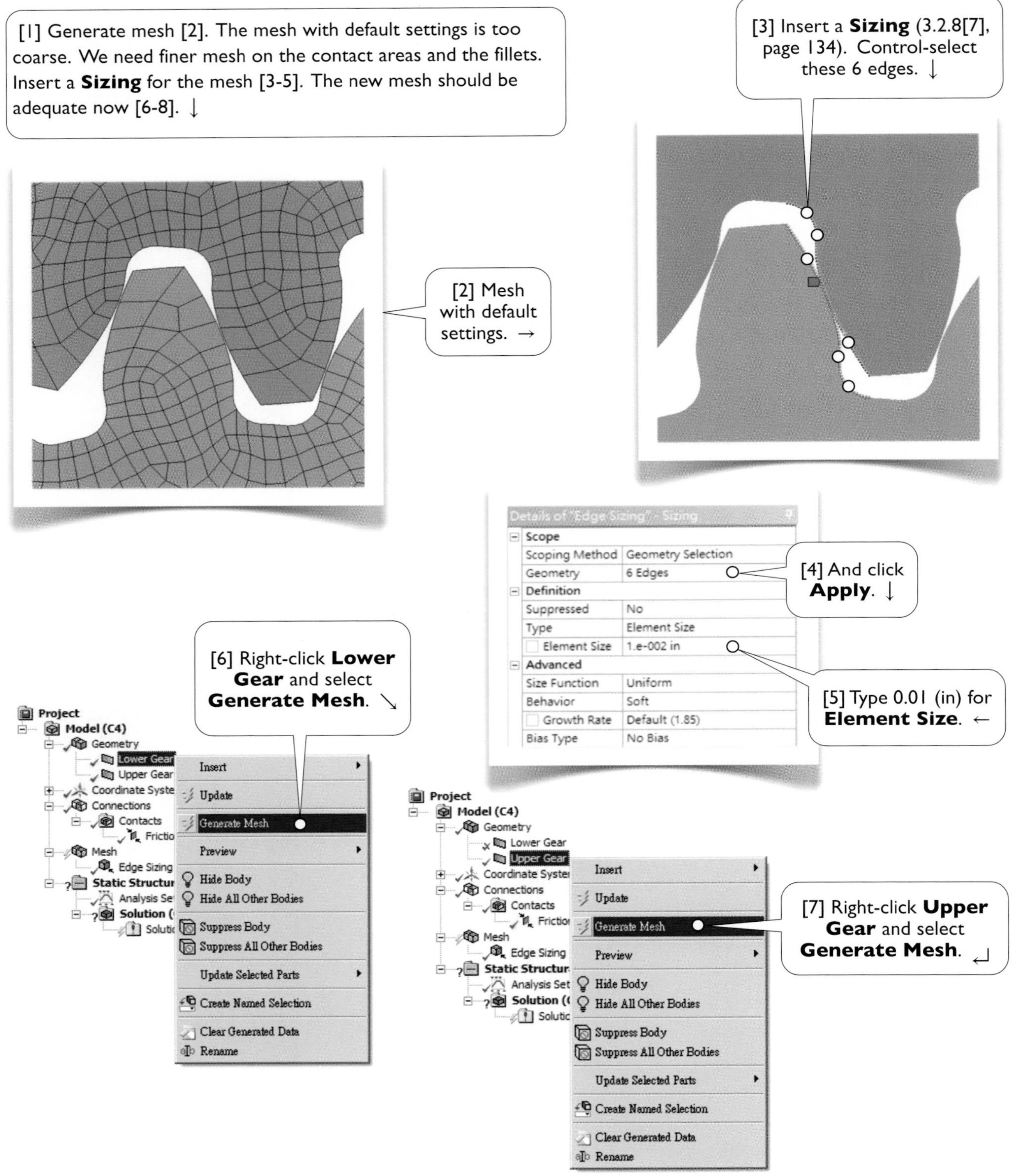

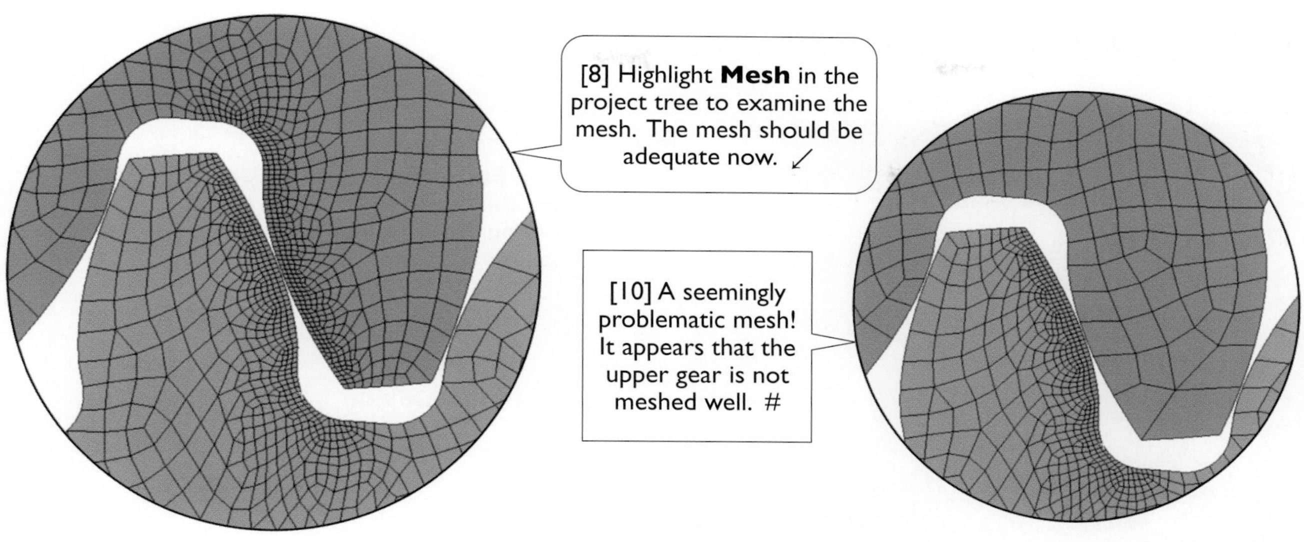

[9] Here, we've demonstrated that each body can be meshed separately. Another reason we meshed this way is to circumvent a small bug in Workbench. If you right-click **Mesh** and issue **Generate Mesh** as usual (instead of [6-7]), you would have a seemingly problematic mesh [10]. The mesh is actually fine; it is just a display problem. ↑

3.4.8 Specify Support and Load

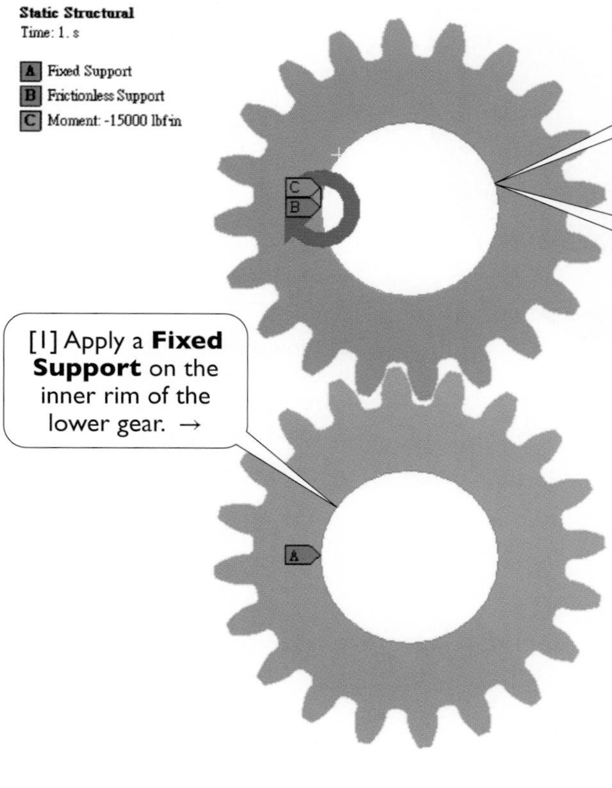

Static Structural
Time: 1. s

A Fixed Support
B Frictionless Support
C Moment: -15000 lbf·in

[3] Apply a **Loads/Moment** of 15,000 lb-in clockwise on the inner rim of the upper gear. (Type -15000 in the details view.) This is the driving torque. #

[2] Apply a **Frictionless Support** on the inner rim of the upper gear. This restricts the inner rim from radial translation, while allowing tangential rotation. ↑

[1] Apply a **Fixed Support** on the inner rim of the lower gear. →

Details of "Moment"

Scope	
Scoping Method	Geometry Selection
Geometry	1 Edge
Definition	
Type	Moment
Magnitude	-15000 lbf·in (ramped)
Suppressed	No
Behavior	Deformable
Advanced	

3.4.9 Solve the Model and View the Results

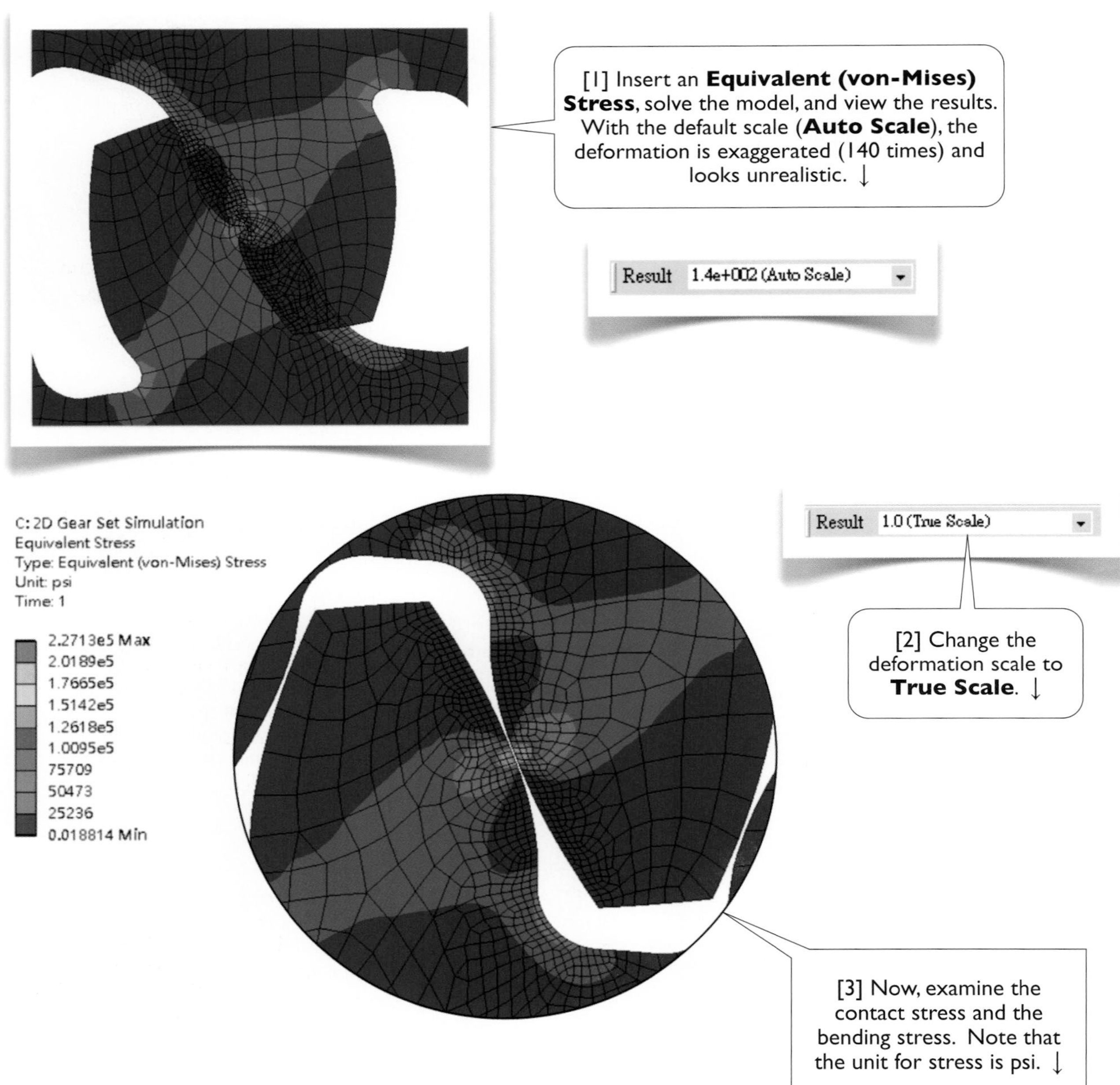

Wrap Up

[4] Close **Mechanical**, save the project, and exit Workbench. #

Section 3.5

Structural Error, FE Convergence, and Stress Singularity

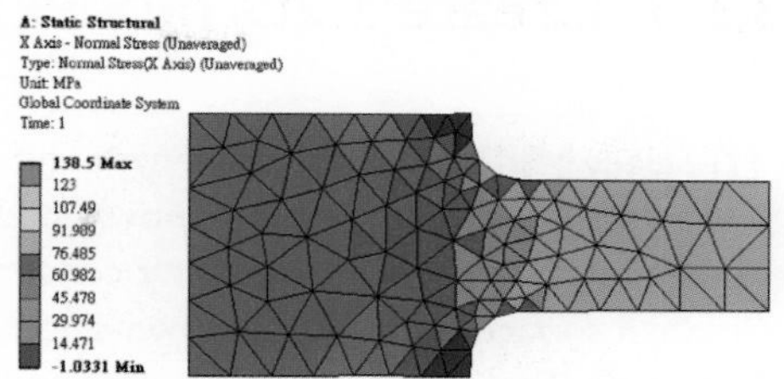

This exercise illustrates some must-know concepts in finite element simulations: (a) stress discontinuity, (b) structural error, (c) finite element convergence, (d) stress concentration, and (e) stress singularity. We use a filleted bar subject to tensile stresses to demonstrate these concepts.

3.5.1 About the Filleted Bar

[1] The filleted bar is made of steel and subject to a tension of 50 kN as shown in [2]. We want to investigate the maximum displacement and the maximum normal stress in horizontal direction. The maximum normal stress occurs near the fillets where the stress concentration occurs. The unit system used in this exercise is **mm-kg-N-s**. ↓

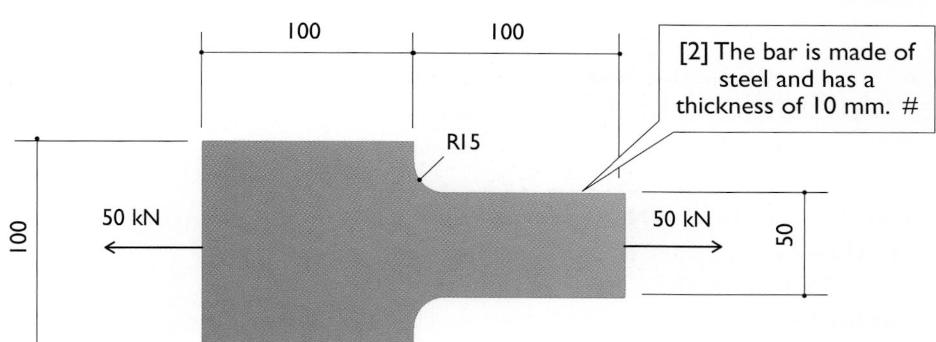

PART A. STRESS DISCONTINUITY

3.5.2 Start a New Project

[1] Launch Workbench. Create a **Static Structural** system. Save the project as **Bar**. Start up DesignModeler [2]. →

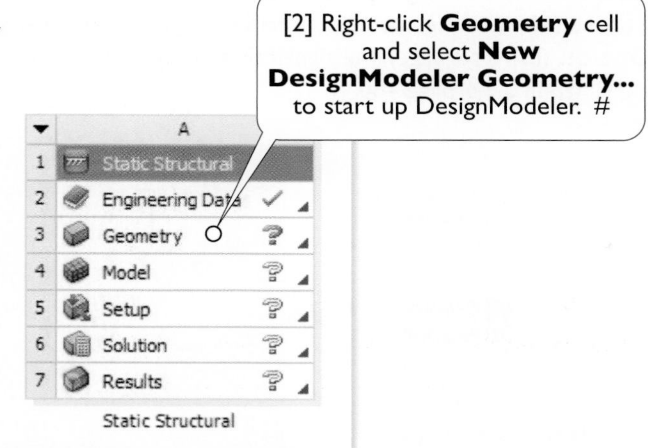

3.5.3 Create a 2D Model in DesignModeler

[1] Select **Millimeter** as the length unit. Create a sketch on **XYPlane** shown in [2]; remember to impose **Symmetry** constraints. Create a surface body by pull-down-selecting **Concept/Surfaces from Sketches** [3-4]. Close DesignModeler.

In the **Project Schematic**, specify **2D** for **Analysis Type** (3.1.4[4-6], page 112). Start up **Mechanical** by double-clicking **Model**. →

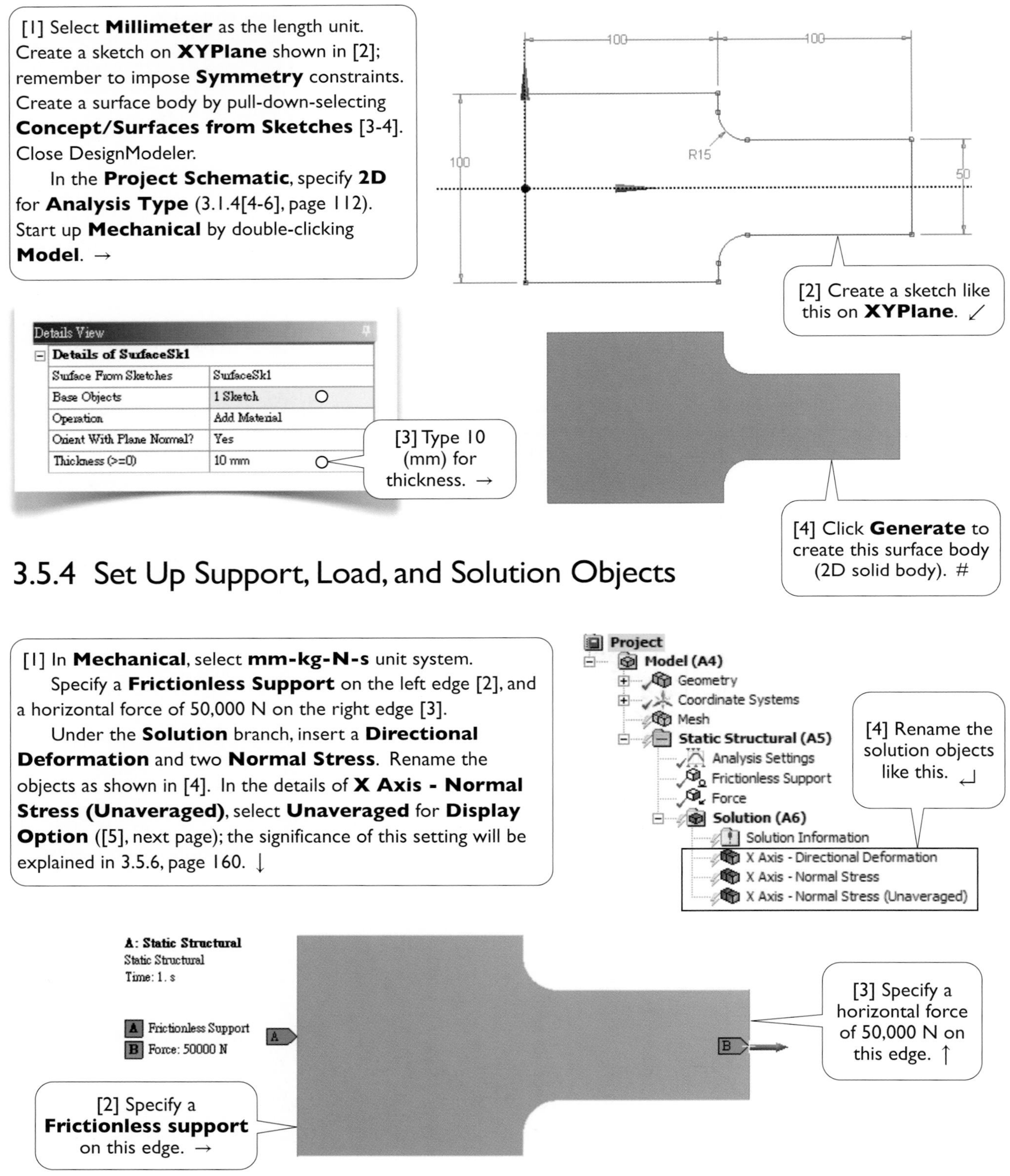

3.5.4 Set Up Support, Load, and Solution Objects

[1] In **Mechanical**, select **mm-kg-N-s** unit system.

Specify a **Frictionless Support** on the left edge [2], and a horizontal force of 50,000 N on the right edge [3].

Under the **Solution** branch, insert a **Directional Deformation** and two **Normal Stress**. Rename the objects as shown in [4]. In the details of **X Axis - Normal Stress (Unaveraged)**, select **Unaveraged** for **Display Option** ([5], next page); the significance of this setting will be explained in 3.5.6, page 160. ↓

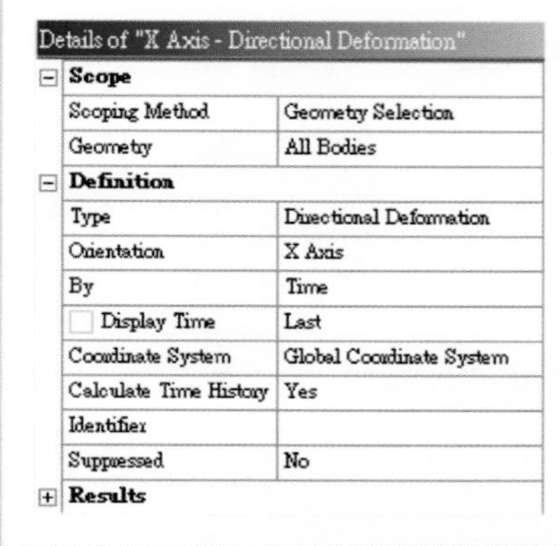

Details of "X Axis - Directional Deformation"	
Scope	
Scoping Method	Geometry Selection
Geometry	All Bodies
Definition	
Type	Directional Deformation
Orientation	X Axis
By	Time
Display Time	Last
Coordinate System	Global Coordinate System
Calculate Time History	Yes
Identifier	
Suppressed	No
Results	

Details of "X Axis - Normal Stress"	
Scope	
Scoping Method	Geometry Selection
Geometry	All Bodies
Definition	
Type	Normal Stress
Orientation	X Axis
By	Time
Display Time	Last
Coordinate System	Global Coordinate System
Calculate Time History	Yes
Identifier	
Suppressed	No
Integration Point Results	
Display Option	Averaged
Average Across Bodies	No
Results	

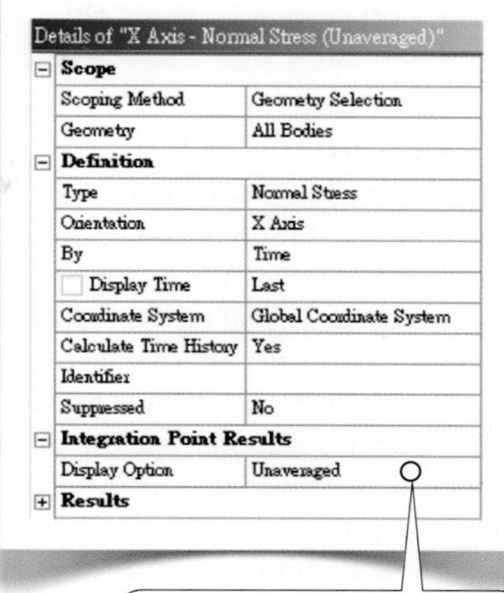

Details of "X Axis - Normal Stress (Unaveraged)"	
Scope	
Scoping Method	Geometry Selection
Geometry	All Bodies
Definition	
Type	Normal Stress
Orientation	X Axis
By	Time
Display Time	Last
Coordinate System	Global Coordinate System
Calculate Time History	Yes
Identifier	
Suppressed	No
Integration Point Results	
Display Option	Unaveraged
Results	

3.5.5 Set Up Mesh Controls

[1] Highlight **Mesh** in the project tree and, in the details view, select **Linear** for **Element Order** [2]. This sets to the use of lower-order elements (see 1.3.2[2-4], page 37).

Insert a **Mesh Control/Method** (select from the toolbar) and select the body as **Geometry** [3]. Select **Triangles** for **Method** [4]. This sets to the use of triangular elements (see 1.3.3[6], page 38).

Generate the mesh [5]. Turn on **Weak Springs** (3.1.8[1-2], page 116) and click **Solve** to solve the model.

Why Lower-Order Triangular Elements?

The purpose of **PART A** is to demonstrate *stress discontinuity*, a must-know in finite element solutions. We've set up a mesh that uses lower-order (linear) triangular elements. The purpose of this setup is to exaggerate the stress discontinuity behavior. Stress discontinuity is intrinsic in all the finite element software using displacements as degrees of freedom, and is not limited to triangular or lower-order elements. →

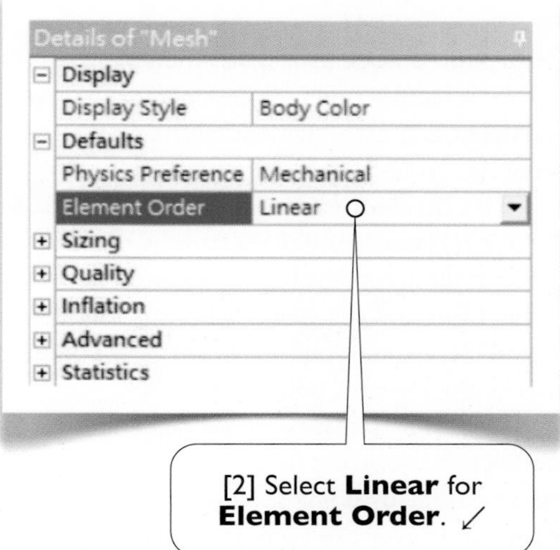

Details of "Mesh"	
Display	
Display Style	Body Color
Defaults	
Physics Preference	Mechanical
Element Order	Linear
Sizing	
Quality	
Inflation	
Advanced	
Statistics	

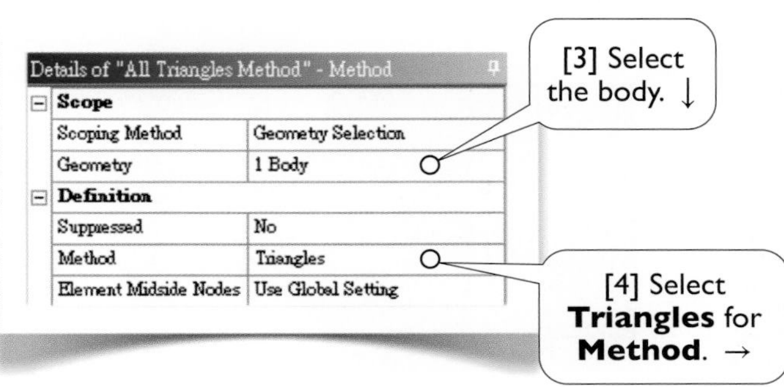

Details of "All Triangles Method" - Method	
Scope	
Scoping Method	Geometry Selection
Geometry	1 Body
Definition	
Suppressed	No
Method	Triangles
Element Midside Nodes	Use Global Setting

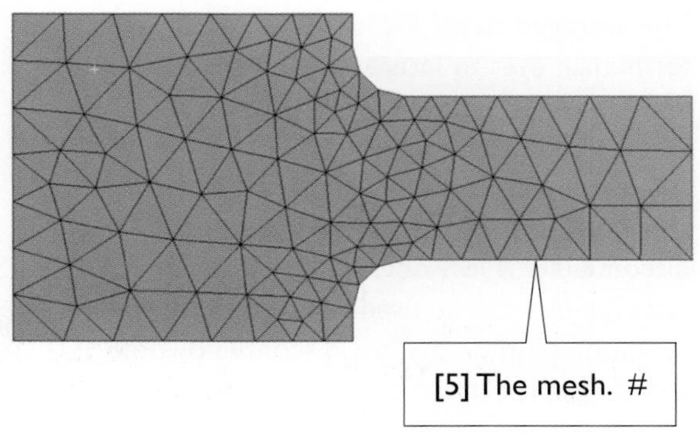

3.5.6 View the Results

Displacement Fields

[1] Nodal displacements, calculated by solving Eq. 1.3.1(1) (page 35), are single valued (each node has a single value). Therefore the displacement fields, calculated using Eq. 1.3.2(2) (page 36), are continuous over the entire structural body [2].

The displacement fields are continuous but not necessarily smooth. The use of continuous shape functions within an element guarantees the displacements field is piecewise smooth, but not necessarily smooth across the element boundaries.

Stress Fields

The strain fields are then calculated using Eq. 1.2.7(1) (page 31), and stress fields are calculated, element by element, using Eq. 1.2.8(1) (page 31). The figure in [3-4] is a typical result of stress calculation: a node usually has multiple stress values, since the node may connect to multiple elements, and each element has its own stress value.

This behavior can be easily understood. Since Eq. 1.2.7(1) involves differentiations of piecewise smooth displacement fields, this ensures the strain fields and the stress fields are continuous inside the element but not necessarily continuous across the element boundaries.

By default, stresses are averaged first on the nodes, and the stress fields are in turn recalculated. By doing so, the stress field is continuous over the body as shown in [5].

Usage of Unaveraged Stresses

The averaged stress fields [5] are visually efficient for human eyes to interpret the results, while unaveraged stress fields [3-4] provide a way of assessing the solution accuracy.

In general, as the mesh is getting finer, the solution is more accurate, and the stress discontinuity is less obvious. Thus, stress discontinuity can be used as an indicator for the solution accuracy: the less discontinuous the stress field, the more accurate the solution. ↗

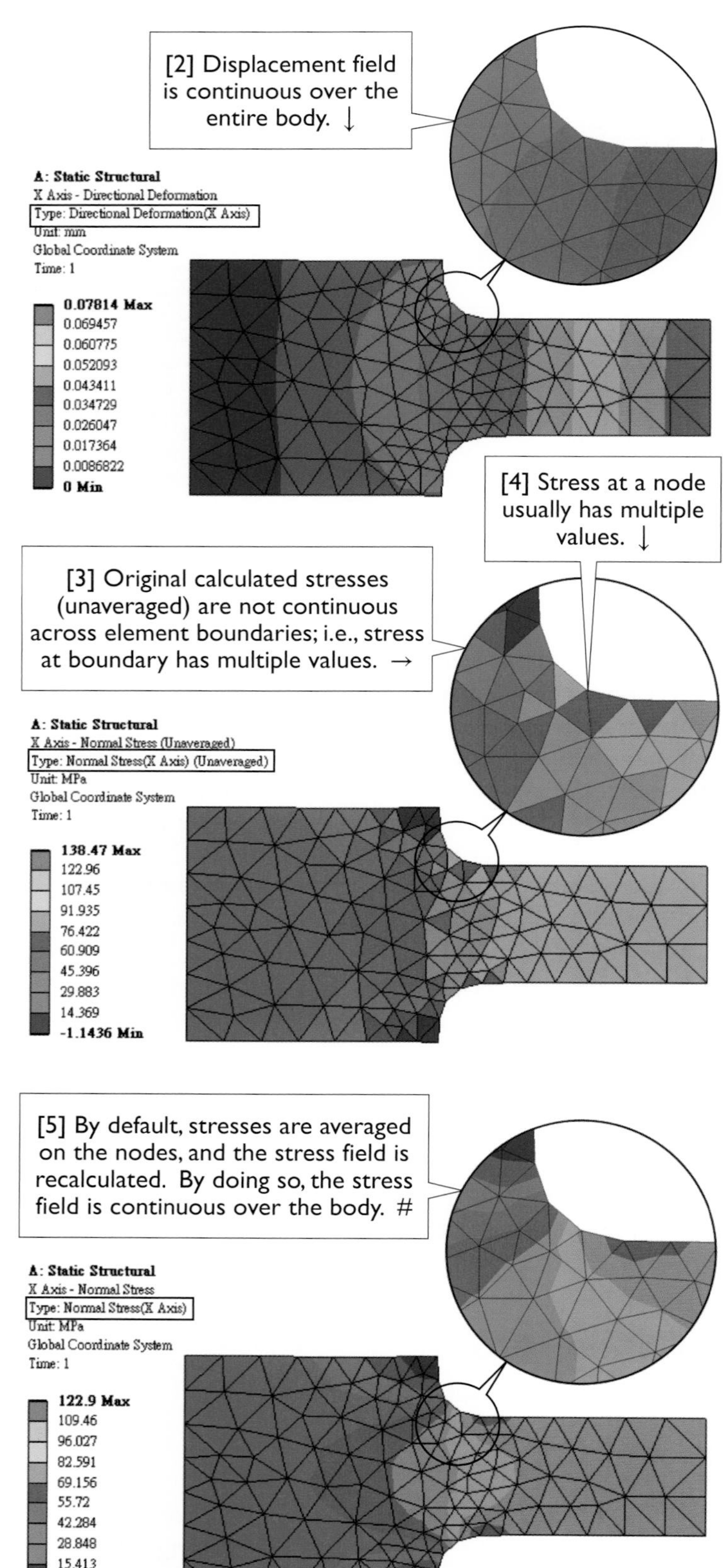

PART B. STRUCTURAL ERROR

3.5.7 Evaluate Structural Error

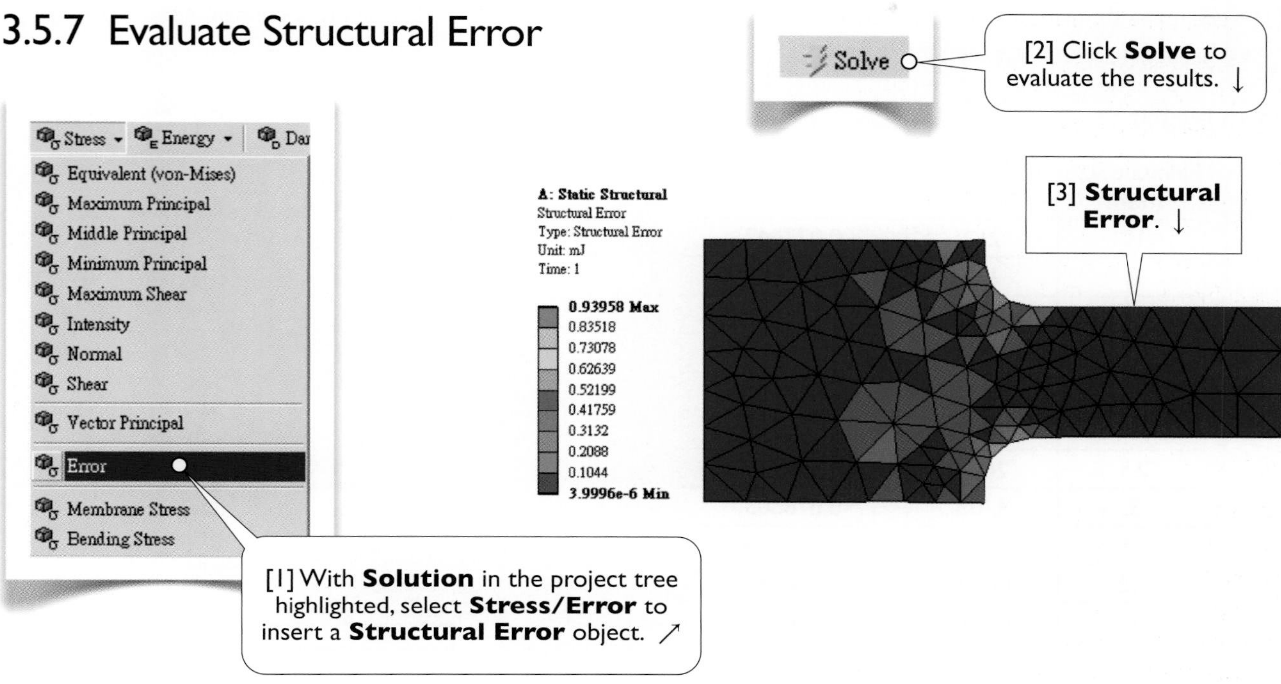

Structural Error[Ref 1]

[4] For an element, strain energy calculated using averaged stresses are different from that using unaveraged stresses. The difference between the two strain energy values is called **Structural Error** of the element. The finer the mesh, the smaller the structural error.

The structural error can be used for two purposes: (a) As an indicator of global mesh adequacy. In general, we want the values as small as possible. Refining the mesh globally is a way of reducing structural error. (b) As an indicator of the local mesh adequacy. In general, we want the structural error distribution as uniform as possible to optimize the efficiency of computation effort. That is, in the region of large values of structural error we should reduce the element size, while in the region of small values of structural error we should enlarge the element size. #

PART C. FINITE ELEMENT CONVERGENCE

One of the core concepts of the finite element methods is that *the finer the mesh, the more accurate the solution*. Ultimately, the solution will reach an analytical solution as the mesh is fine enough. But, how fast does it approach the analytical solution? This is what we want to answer in this part of the section. It is called the *element convergence behavior*.

The answer depends on what kind of element we are using. We'll draw a conclusion that, for 2D cases, *quadrilateral elements generally converge faster than triangular elements*.

We will compare lower-order triangular with lower-order quadrilateral. To be fair, the comparison is made under the same problem sizes for both types of elements. The problem size is reasonably defined by the number of nodes.

As a representative structural response, the maximum horizontal displacement is used for comparison.

3.5.8 Triangular Elements

[1] Repeat the simulation as follows: Change the element size for each run [2], record the number of nodes [3], and the maximum directional deformation (displacement). Tabulate the results like this: →

Element Size (mm)	Number of Nodes	Max Displacement (mm)
10	212	0.077647
8	290	0.077992
6	490	0.078262
5	698	0.078380
4	1066	0.078454
3	1857	0.078545
2.5	2687	0.078575
2	4108	0.078603
1.5	7238	0.078626
1.2	11050	0.078635
1	15962	0.078640

[2] Type the same element size for both **Min Size** and **Max Face Size**. This sets to a uniform-sized mesh. ↓

Details of "Mesh"

Display	
Display Style	Body Color
Defaults	
Physics Preference	Mechanical
Element Order	Linear
Sizing	
Size Function	Curvature
Max Face Size	10.0 mm
Mesh Defeaturing	Yes
Defeature Size	Default (5.e-002 mm)
Growth Rate	Default
Min Size	10.0 mm
Curvature Normal Angle	Default (30.0 °)
Bounding Box Diagonal	223.610 mm
Average Surface Area	15097 mm²
Minimum Edge Length	10.0 mm
Quality	
Inflation	
Advanced	
Statistics	
Nodes	212
Elements	363

[3] Number of nodes. #

3.5.9 Quadrilateral Elements

[1] Highlight **All Triangles Method** under the **Mesh** branch. In the details view, change the method to **Quadrilateral Dominant** [2], and select **All Quad** for **Free Face Mesh Type** [3]. Repeat the steps in 3.5.8, and tabulate the results like this: →

Element Size (mm)	Number of Nodes	Max Displacement (mm)
10	196	0.078111
8	299	0.078472
6	483	0.078470
5	691	0.078539
4	1045	0.078562
3	1806	0.078605
2.5	2619	0.078620
2	4009	0.078635
1.5	7056	0.078641
1.2	10960	0.078646
1	15686	0.078648

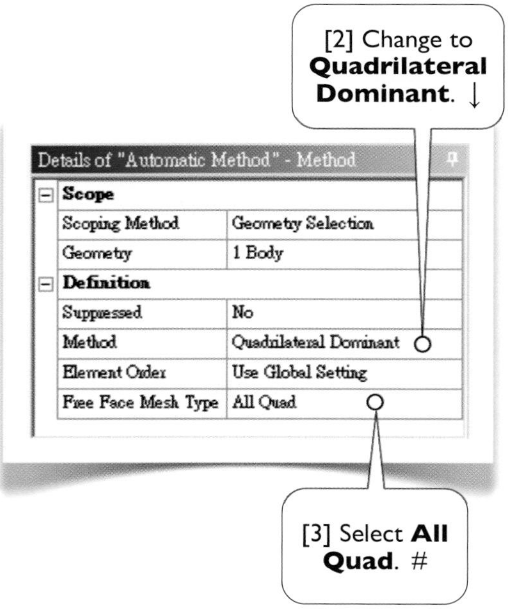

3.5.10 Comparison and Conclusions

[1] Plotting the results in 3.5.8 and 3.5.9 (last page) using a spreadsheet program such as Microsoft Excel, you should come up with a chart as shown below [2-4]. The two curves share the same horizontal asymptote (Displacement ≈ 0.07864 mm), which is the analytical solution. Additional behaviors can be observed as follows.

First, the quadrilateral element converges to the analytical solution faster than the triangular element. The difference seems undramatic in this particular case. However, in many other cases, the difference can be significant. In Section 9.3 we will investigate the convergence behavior of 3D elements, and we'll conclude that the convergence rate of the hexahedra (1.3.3[2], page 38) is faster than the prisms or the tetrahedra (1.3.3[3, 5], page 38).

Second, all the convergence curves approach the asymptote from below; i.e., displacements calculated by the finite element methods never exceed the analytical solution. In other words, the finite element solutions always underestimate the deformation, or in terms of stiffness (1.3.1[4], page 36), the stiffness is always overestimated. ↙

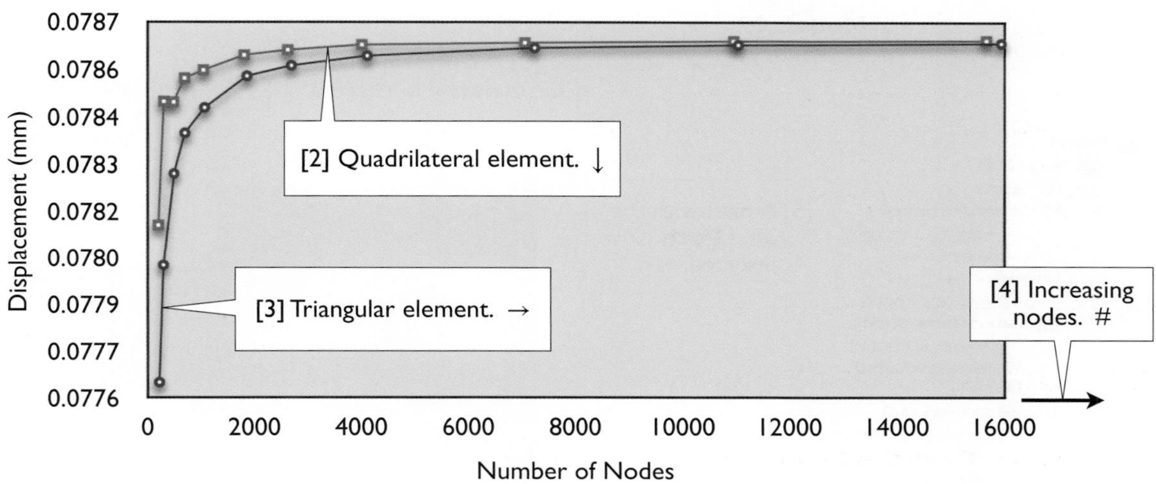

PART D. STRESS CONCENTRATION

3.5.11 View the Stress

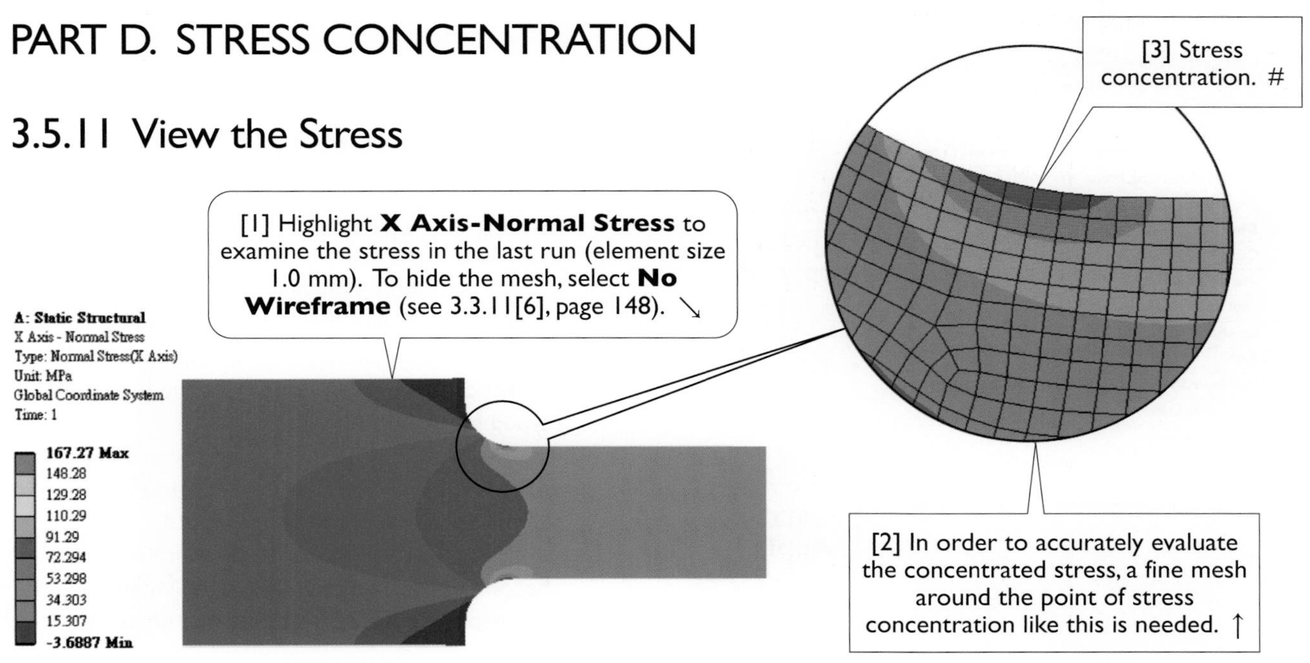

3.5.12 Define a Path

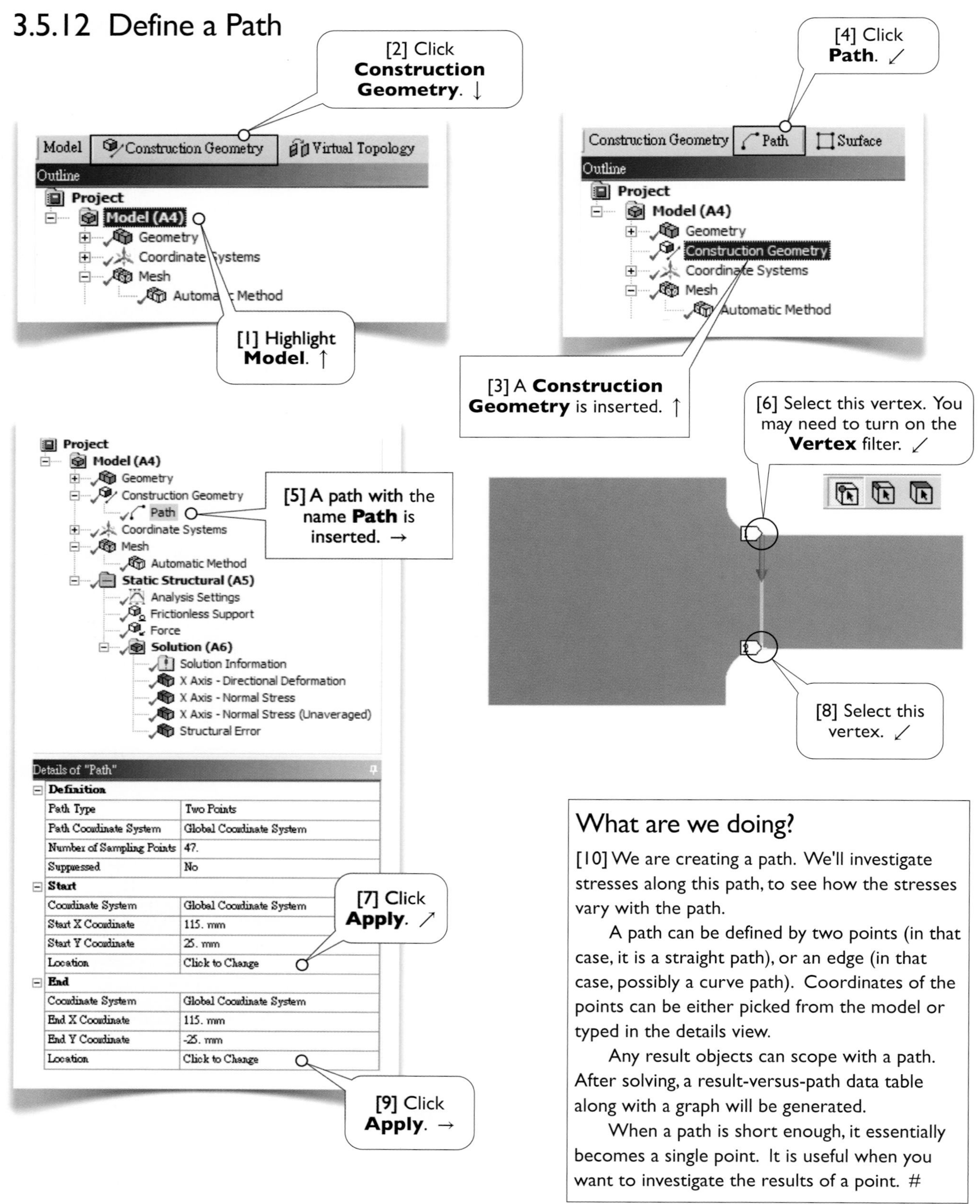

What are we doing?

[10] We are creating a path. We'll investigate stresses along this path, to see how the stresses vary with the path.

A path can be defined by two points (in that case, it is a straight path), or an edge (in that case, possibly a curve path). Coordinates of the points can be either picked from the model or typed in the details view.

Any result objects can scope with a path. After solving, a result-versus-path data table along with a graph will be generated.

When a path is short enough, it essentially becomes a single point. It is useful when you want to investigate the results of a point. #

3.5.13 View Stresses Along the Path

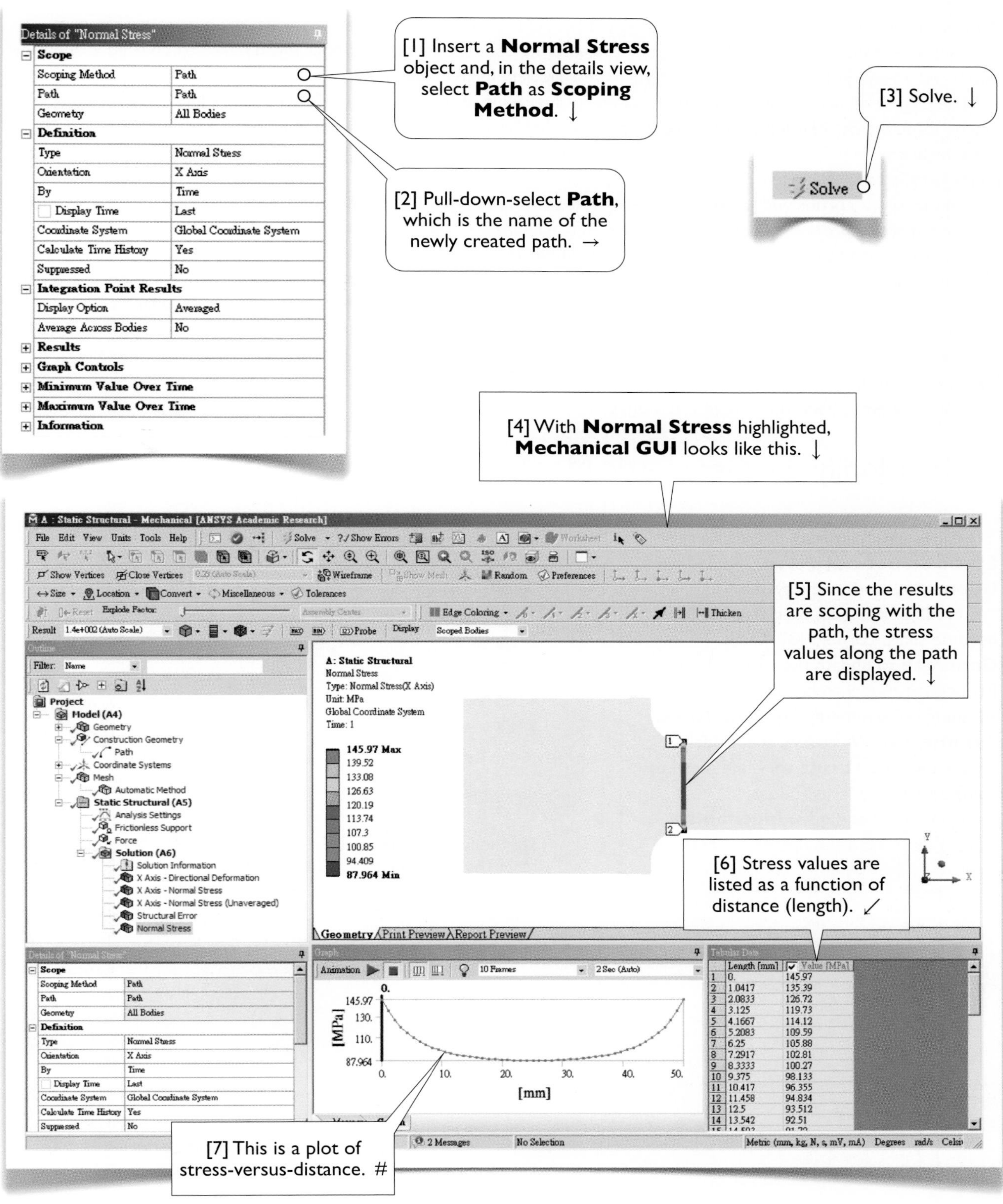

PART E. STRESS SINGULARITY

Many engineers are aware of the concept of stress concentration: stress is larger at a concave corner than stress at locations away from the corner. To accurately evaluate the concentrated stress, finer mesh needs to be used around the corner. As a general guideline, *an area of higher stress gradient requires finer mesh*.

The degree of stress concentration is related to the radius of the corner: the smaller the radius, the larger the concentrated stress. Naturally, a question comes up: what happens if the radius of the concave corner is zero (i.e., a sharp angle corner)? The elasticity theory predicts that, in that case, the stress at that sharp corner is infinity. A stress of infinity is called a *singular stress*.

Fortunately, a corner with zero radius never exists in the real-world. It is difficult to manufacture such a zero-radius fillet. It requires a process of infinite-high precision machining.

However, zero-radius fillets do exist in the virtual-world of simulations. Since many small features such as fillets do not significantly affect the global behavior (e.g., deformation) of a structure, these small features are often not modeled in the simulation model. The consequence is that zero-radius fillets exist everywhere in a simulation model.

It is important that the engineers be aware of the existence of stress singularities. Novice engineers often mistakenly take the maximum stress as the design stress, while, in fact, that is a singular stress—it doesn't exist in the real-world. Always check if the maximum stress occurs at a singular point. If so, the stress is meaningless. If a concentrated stress is important, always include the fillet in the model.

3.5.14 Modify the Geometric Model

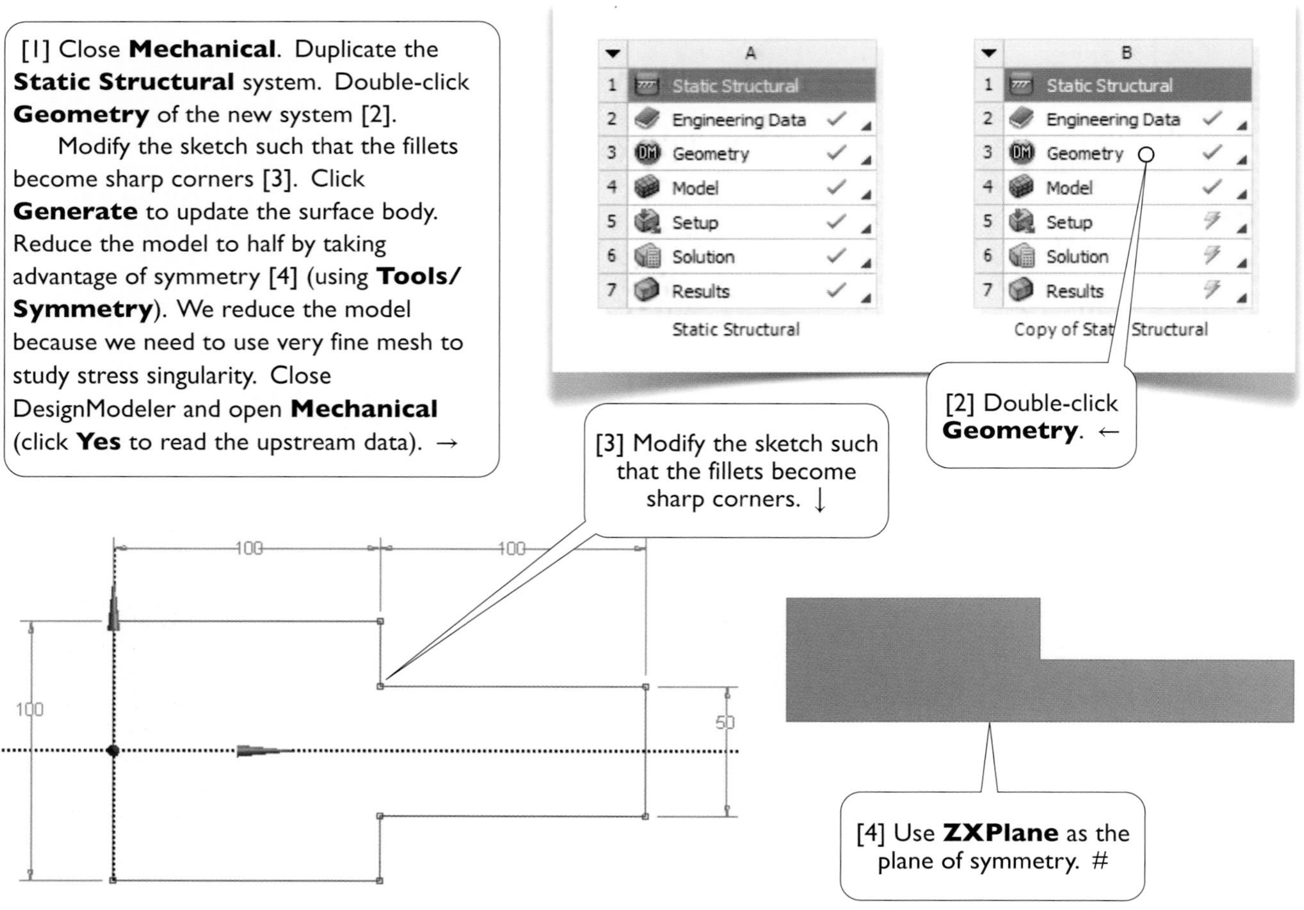

3.5.15 Modify the Environment Conditions in **Mechanical**

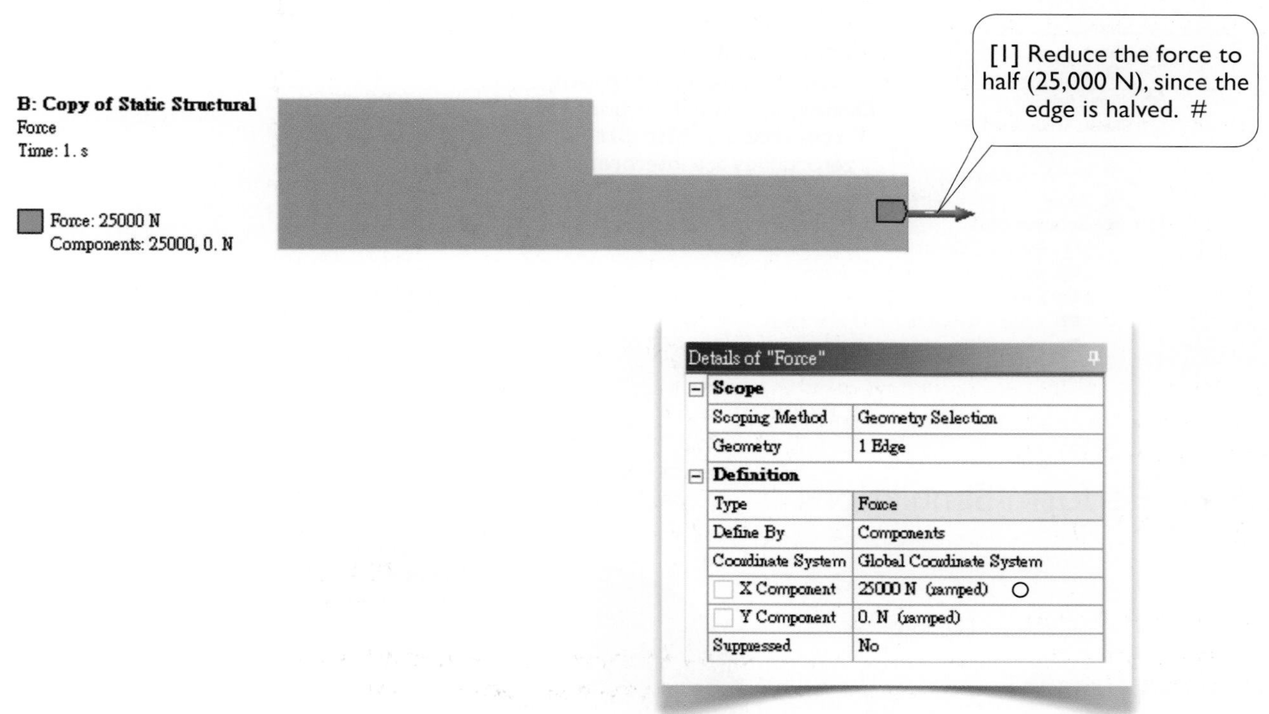

3.5.16 Set Up Mesh Controls

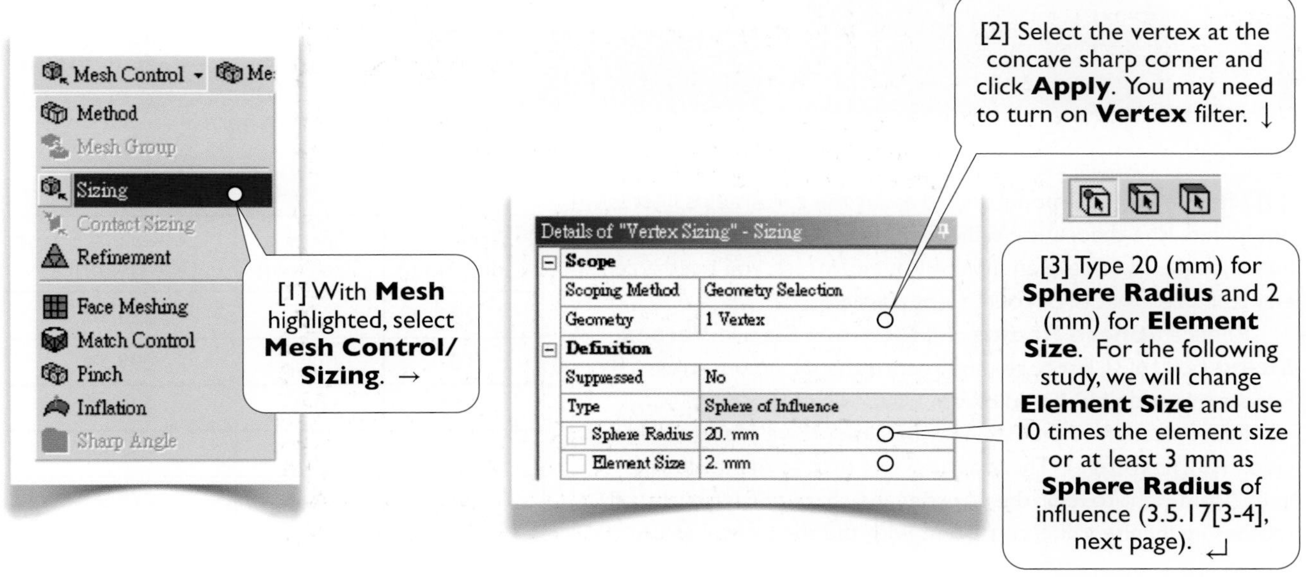

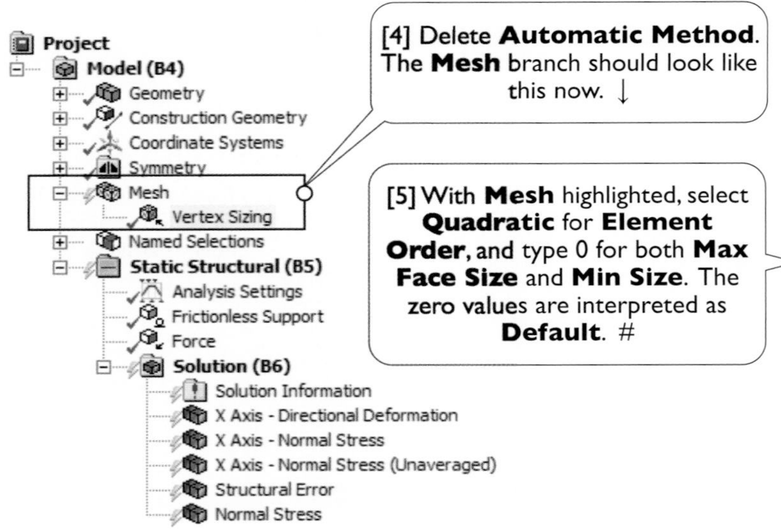

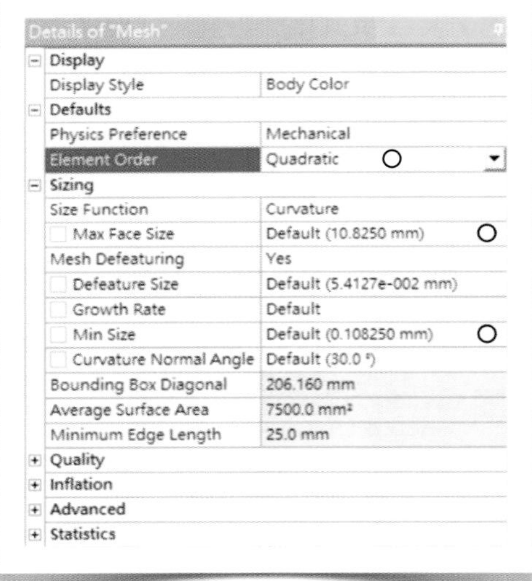

3.5.17 Perform Simulations

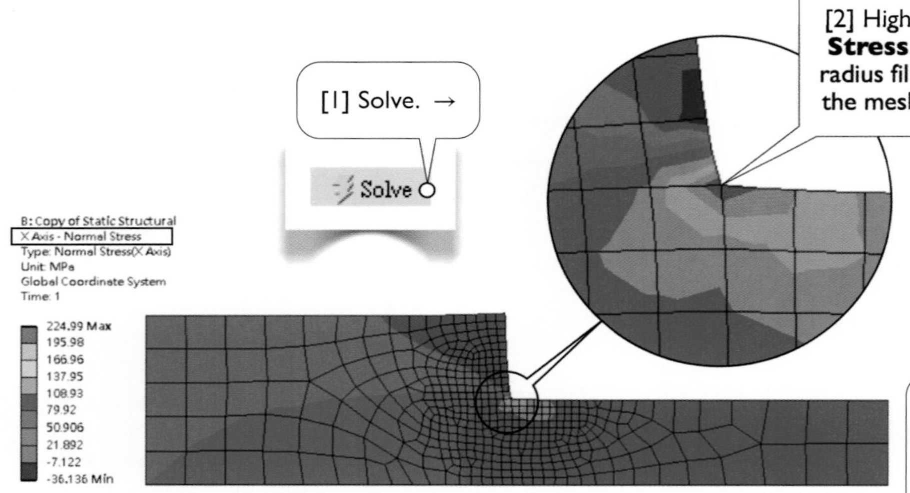

[3] After solving the model, the stress at the zero-radius fillet is not infinite, the theoretical value. This is because our mesh is not fine enough. To achieve an infinite value of stress, you need a zero element size. That is, of course, not possible.

Reduce **Sphere Radius** and **Element Size** in **Vertex Sizing** (3.5.16[3], last page) and record the maximum stress. Repeat this and you should obtain a table like [4].

Interpretation of the data may be easier if we plot them in a graph ([5], next page). The graph reveals that, as the mesh gets finer (from right to left), the stress at the sharp corner eventually reaches an infinite value, consistent with the theoretical value. ↗

[4] Change **Sphere Radius** and **Element Size** in **Vertex Sizing** (3.5.16[3], last page) and record the maximum normal stresses. ↵

Sphere Radius (mm)	Element Size (mm)	Max Normal Stress (X Axis) (MPa)
20	2	224.99
10	1	299.34
5	0.5	395.52
3	0.2	594.84
3	0.1	779.80
3	0.06	972.69

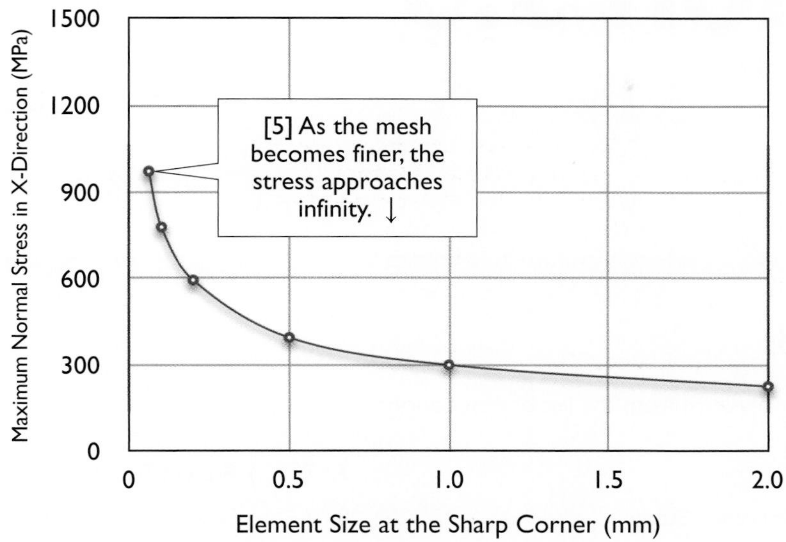

Remark

[6] Stress singularity is not limited to concave sharp corners. Any locations that have stress of infinity are called singular points. For example, a point of concentrated forces is also a singular point, since a point has zero area. ↓

Wrap Up

[7] Save the project. Close **Mechanical** and exit Workbench. #

Reference

1. ANSYS Help//Mechanical APDL//Mechanical APDL Theory Reference//17.6.1. Error Approximation Technique for Displacement-Based Problems

Section 3.6

Review

3.6.1 Keywords

Choose a letter for each keyword from the list of descriptions

1. (   ) APDL
2. (   ) Element Convergence Study
3. (   ) Environment Conditions
4. (   ) Inconsistent Unit System
5. (   ) Plane Strain Problem
6. (   ) Plane Stress Problem
7. (   ) Project Tree
8. (   ) Stress Discontinuity
9. (   ) Stress Singularity
10. (   ) Structural Error
11. (   ) Weak Spring

Answers:

1. (F) 2. (J) 3. (G) 4. (E) 5. (C) 6. (B) 7. (D) 8. (H) 9. (K) 10. (I) 11. (A)

List of Descriptions

(A) When Workbench detects a structure as unstable, it adds weak springs on the structure to make it capable of withstanding very small external forces.

(B) In a structural simulation problem, if all the stresses in a direction, say Z-direction, vanish, the problem can be reduced to a 2D problem.

(C) In a structural simulation problem, if all the strains in a direction, say Z-direction, vanish, the problem can be reduced to a 2D problem.

(D) A structured representation of an analysis system and displayed on the **Outline** in **Mechanical**. It contains one or more simulation models. A simulation model contains one or more **Environment** branches.

(E) A unit system in which at least one unit is not consistent with other units. For example, **mm-kg-N-s** is inconsistent, since if mm and kg is used, the force must be mN (milli-newton) instead of N.

(F) ANSYS Parametric Design Language. A set of text-based language that is used to drive ANSYS Classic program.

(G) Include loads, supports, and inertial effects, which apply on the simulation model.

(H) In the finite element methods, the shape functions are used to interpolate the displacement fields interior to the element. Across the element boundaries, the displacement fields are continuous but not smooth. The strains and stresses, which are calculated by differentiating the displacement fields, become discontinuous across the element boundaries.

(I) For an element, strain energies calculated using averaged stresses and unaveraged stresses are different. The difference between these two energy values is called the structural error of the element. The finer the mesh, the smaller the structural error. It is used as an indicator for mesh adequacy.

(J) Study of how the finite element solutions approach theoretical values as the mesh is getting finer. In 2D, quadrilateral elements converge faster than the triangular. In 3D, the order of convergence speeds are, from faster to slower, hexahedral, prism, pyramid, and tetrahedral.

(K) A stress that has infinitely large value. They are often found at concave fillets of zero radius, or at points subject to concentrated forces or displacement constraints.

3.6.2 Additional Workbench Exercises

Stress in a Long Cylinder

Consider the problem described in VM25 of the APDL verification manual[Ref 1]. Find the radial stress and the hoop stress at the inner and outer surfaces and at the middle wall thickness, for both loading cases. Model the problem as (a) an axisymmetric problem and (b) a plane strain problem.

Reference

1. ANSYS Help//Verification Manuals//ANSYS Mechanical APDL Verification Manual//I. Verification Test Case Description//1. Introduction//VM25

Chapter 4
3D Solid Modeling

Creating a 3D geometric model is usually much more elaborate than creating a 2D model. Most of the techniques you've learned in Chapters 2 and 3 for 2D cases can be used in 3D cases. Three types of 3D bodies supported by Workbench are *solid bodies*, *surface bodies*, and *line bodies*; they may coexist in a 3D model. In this chapter, we will focus on models consisting of solid bodies, except Section 4.3, in which a surface body and a solid body constitute a 3D model. Chapters 6 and 7 will discuss surface bodies and line bodies, respectively.

Purpose of This Chapter

This chapter guides students to familiarize themselves with 3D solid modeling using DesignModeler. Four mechanical parts are created in this chapter. These models will be used for simulations in the next and later chapters.

About Each Section

We start with a beam bracket model in Section 4.1 to introduce the basics of 3D bodies creation. Section 4.2 creates a more elaborated model, the cover of a pressure cylinder, to obtain some degree of proficiency at the 3D modeling techniques. Section 4.3 creates a model for a lifting fork, a combination of a solid body and a surface body. Section 4.4 overviews 3D solid bodies creation and manipulation tools in a systematic way, intending to cover what was missed in the first three sections. Section 4.5 provides an additional exercise.

Section 4.1

Beam Bracket

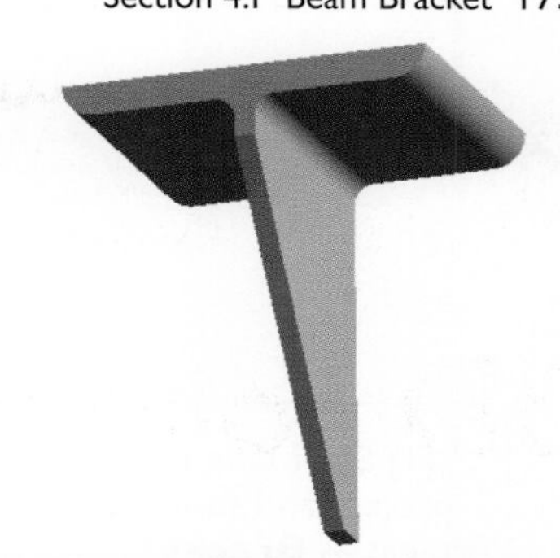

4.1.1 About the Beam Bracket

[1] When a steel beam-column structure, such as a high-rise building or a manufacturing plant, is constructed, its columns are erected before the beams can be elevated, positioned, and welded. The function of a beam bracket [2] is to precisely position a beam and safely transfer the loads from the beam to the column. The loads are determined by a thorough analysis of the entire structure subject to design loads, such as dead load, live load, earthquake, wind load, etc.

The beam bracket consists of a seat plate (the flange) [3] and a web plate [4]. The design considerations include: (a) Would the maximum stress exceed the allowable stress? (b) Would the web buckle under the load?

In this section, we will create a 3D solid model for the beam bracket. The model will be used for a static structural analysis in Section 5.1 and a buckling analysis in Section 10.3. The 3D solid model will be simplified to a surface model in Section 6.2 and the simulation results will be compared with those in Section 5.1. →

[2] Beam bracket. ↓

Global Coordinate System

[5] To create a geometric model, you need a global coordinate system [6]. For 2D, it is so trivial that we didn't even mention it. For 3D, you need to pay more attention to the global coordinate system. Many students are easily disoriented when working with 3D models on a 2D computer screen. In this section, the global coordinate system is set up as shown in [7-9]. ↘

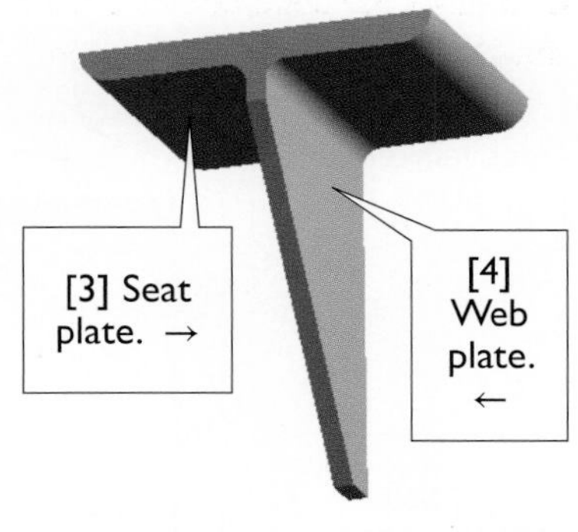

[3] Seat plate. →

[4] Web plate. ←

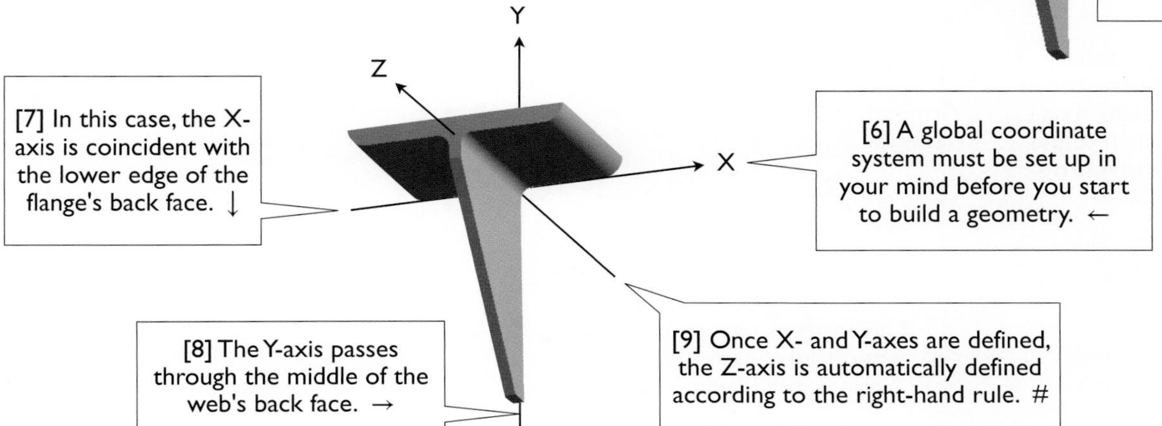

[7] In this case, the X-axis is coincident with the lower edge of the flange's back face. ↓

[6] A global coordinate system must be set up in your mind before you start to build a geometry. ←

[8] The Y-axis passes through the middle of the web's back face. →

[9] Once X- and Y-axes are defined, the Z-axis is automatically defined according to the right-hand rule. #

4.1.2 Start Up

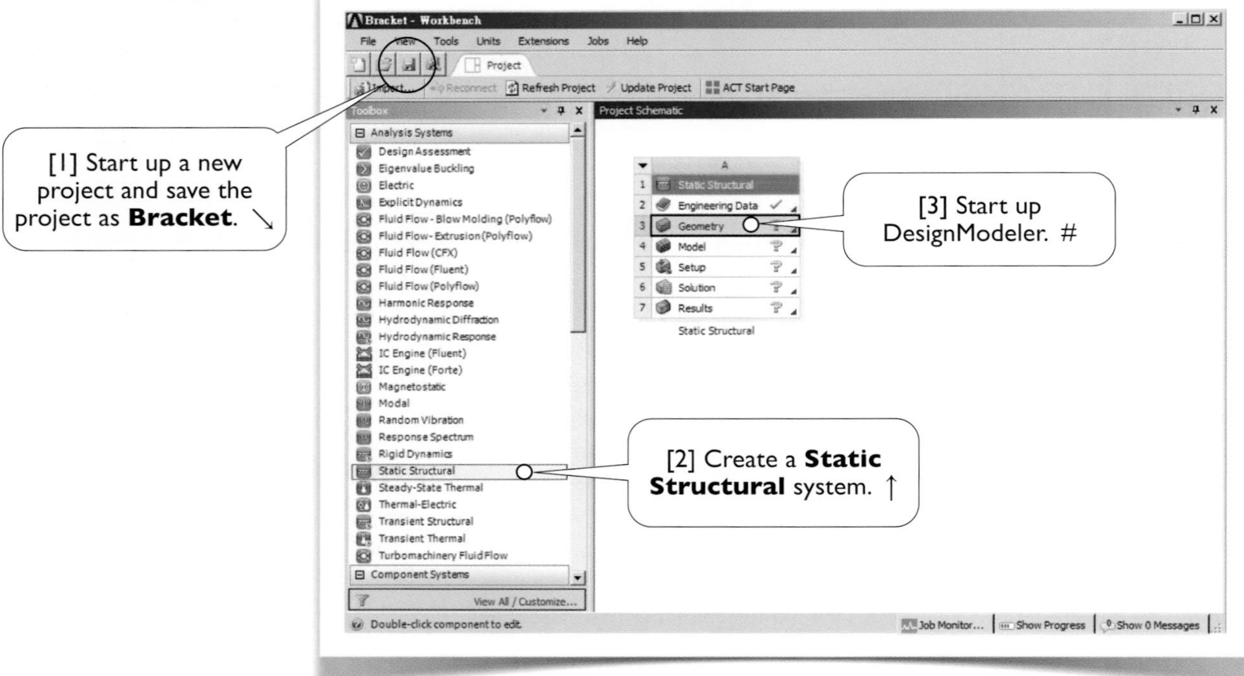

4.1.3 Sketch Seat Plate on ZXPlane

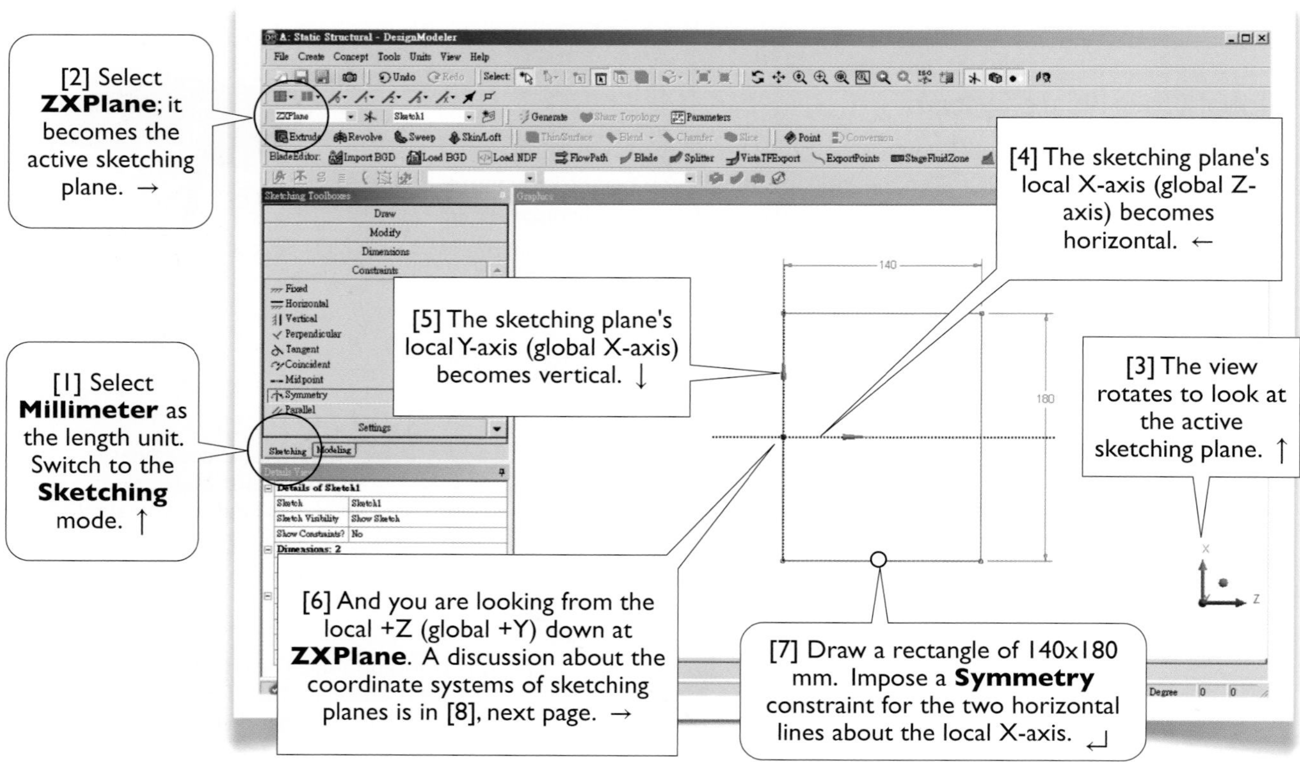

The Local Coordinate System of a Sketching Plane

[8] Each sketching plane has a local coordinate system (3.1.11[12], page 119). For the three pre-defined planes (**XYPlane**, **YZPlane**, and **ZXPlane**), their first two letters refer to two global axes, which become the local X-axis and Y-axis, respectively. The local Z-axis is then defined according to the right-hand rule. For example, **ZXPlane**'s local X-axis and local Y-axis coincide with the global Z-axis and global X-axis, respectively [9-11]. ↓

Plane View Sketching vs. 3D View Sketching

[12] When the view rotates to look at the sketching plane [3], the local X-axis and local Y-axis of the sketching plane becomes horizontal and vertical, respectively [4-5], and the local Z-axis points out toward you. It is a simple rule but may take time to get accustomed to. If you have trouble sketching with plane view [4-6], alternatively you may sketch in 3D view, as shown in [9-11]. The problem of sketching in 3D view is that all angles are distorted. For example, a right angle is no longer 90°. For most people, sketching with plane view should be more convenient. However, choose whichever is suitable for you. In this book, we will stick to the plane view sketching. #

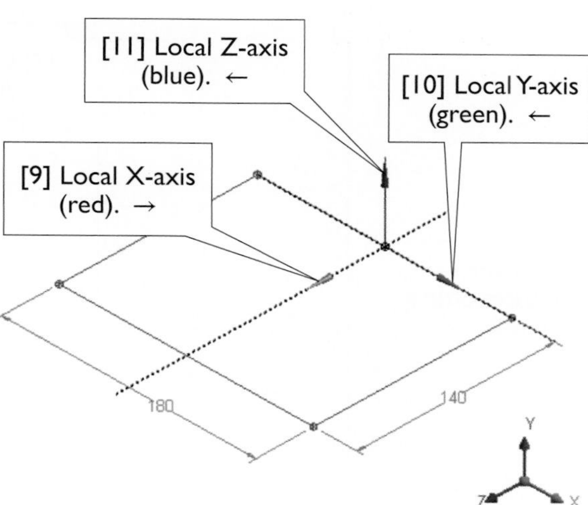

4.1.4 Extrude the Sketch to Create Seat Plate

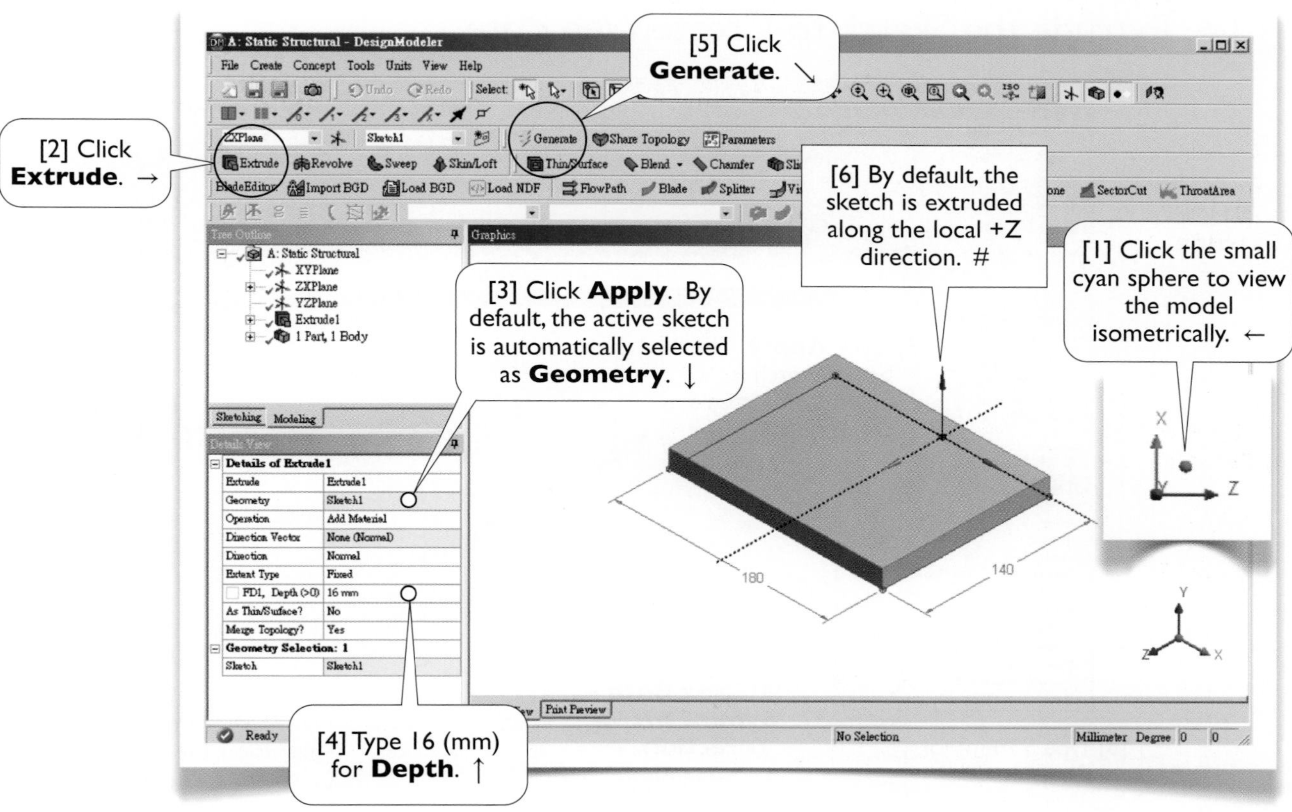

4.1.5 Sketch Web Plate on YZPlane

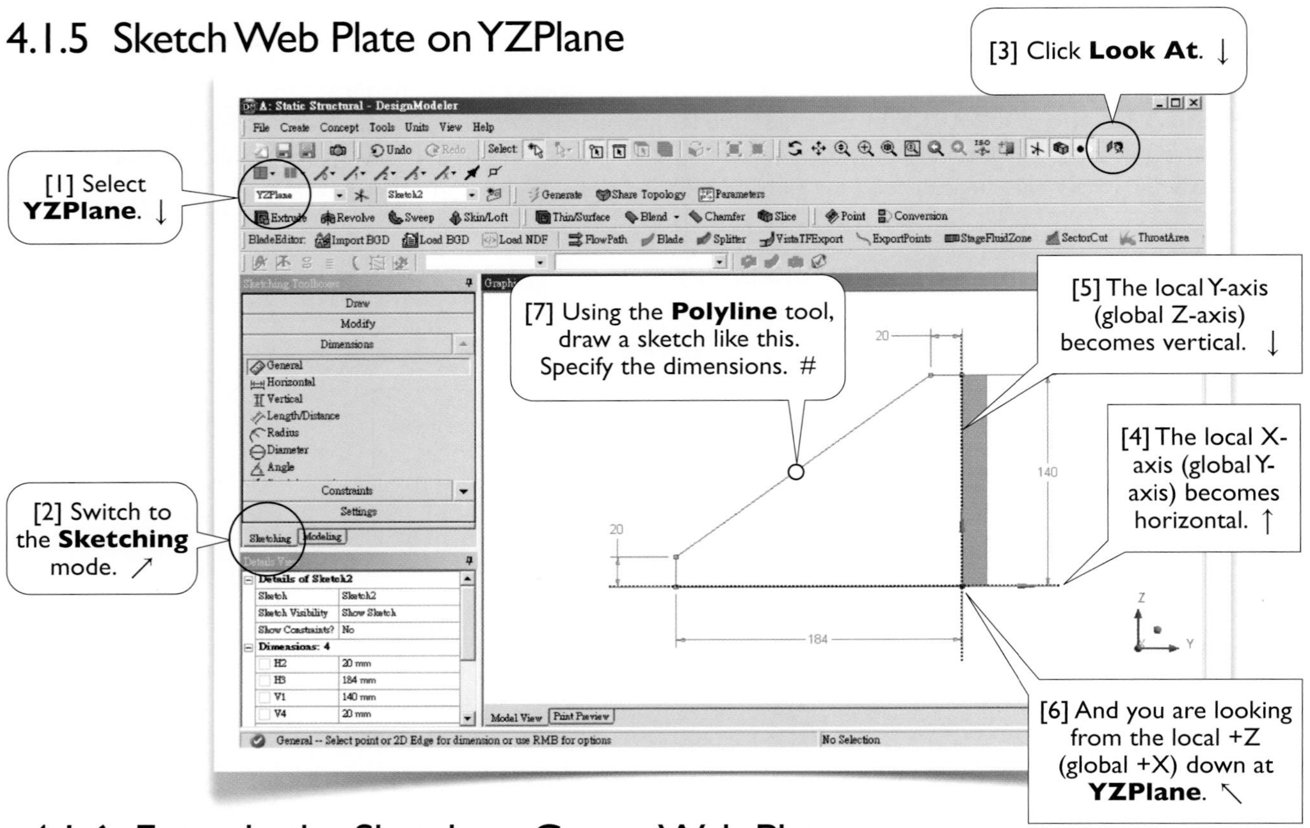

4.1.6 Extrude the Sketch to Create Web Plate

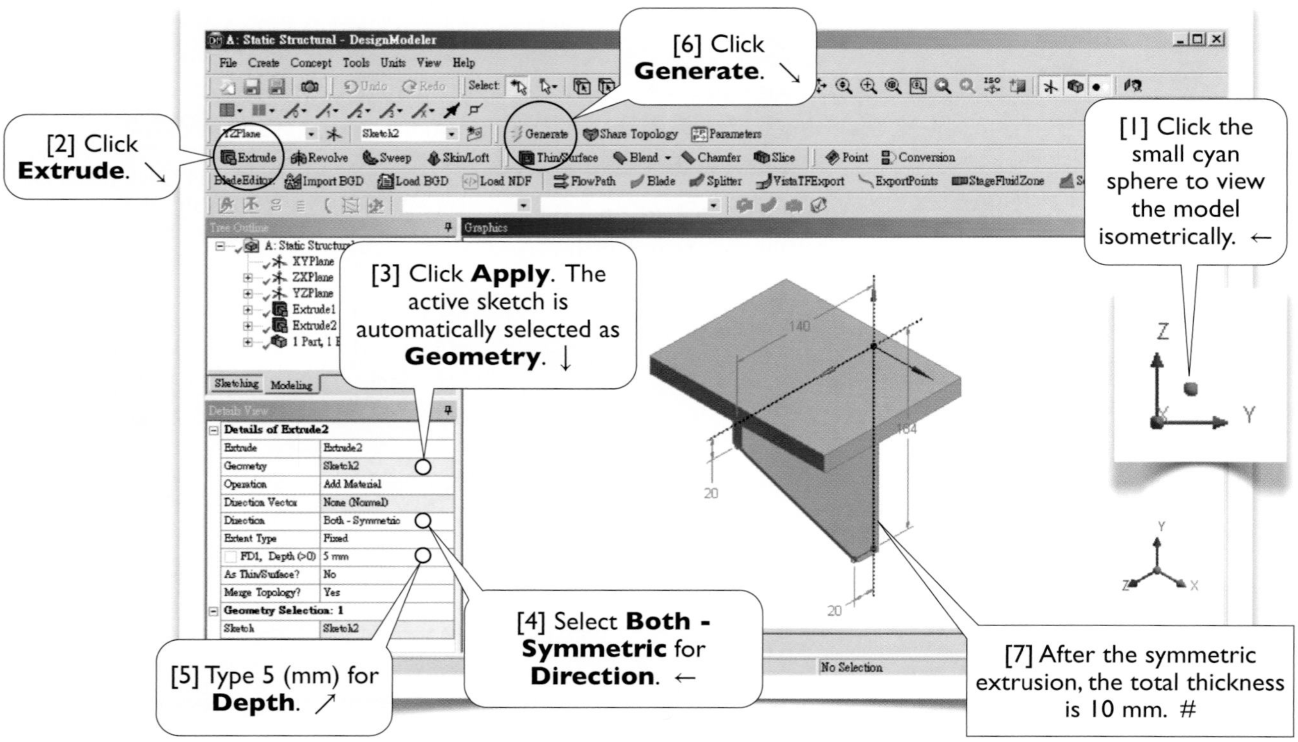

4.1.7 Create Rounds for the Flange

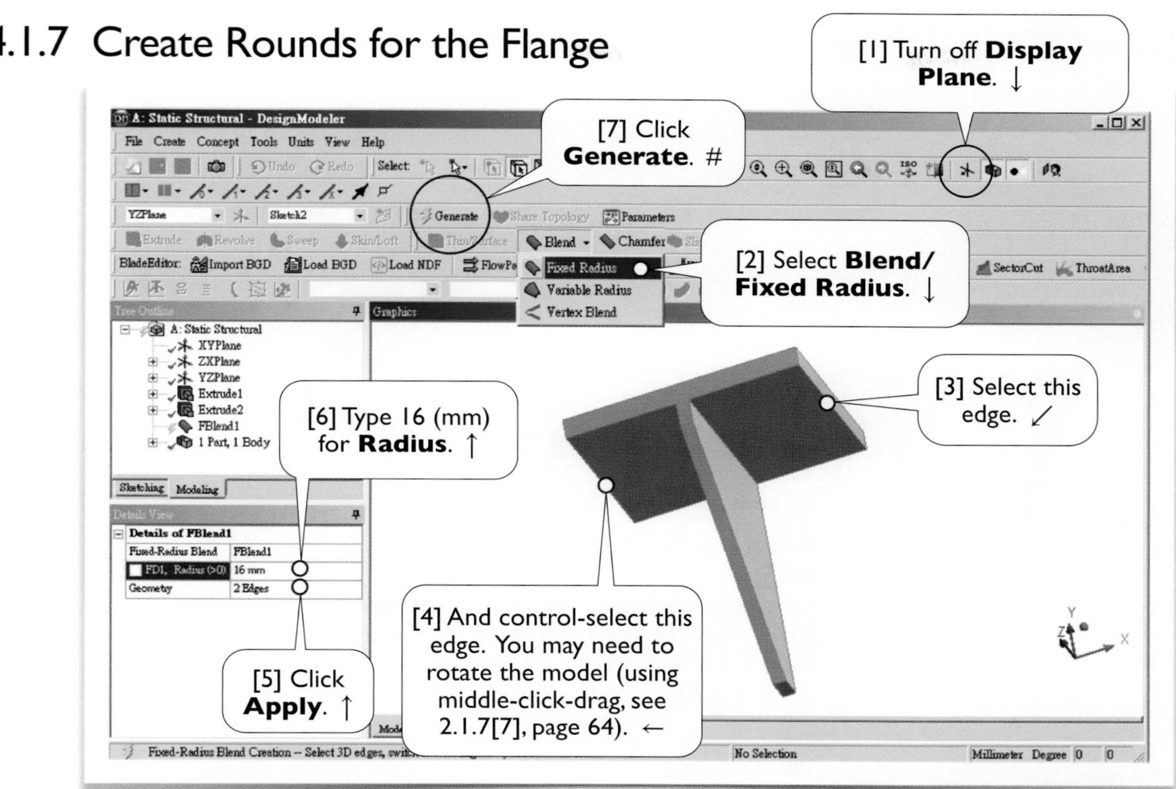

4.1.8 Create Fillets

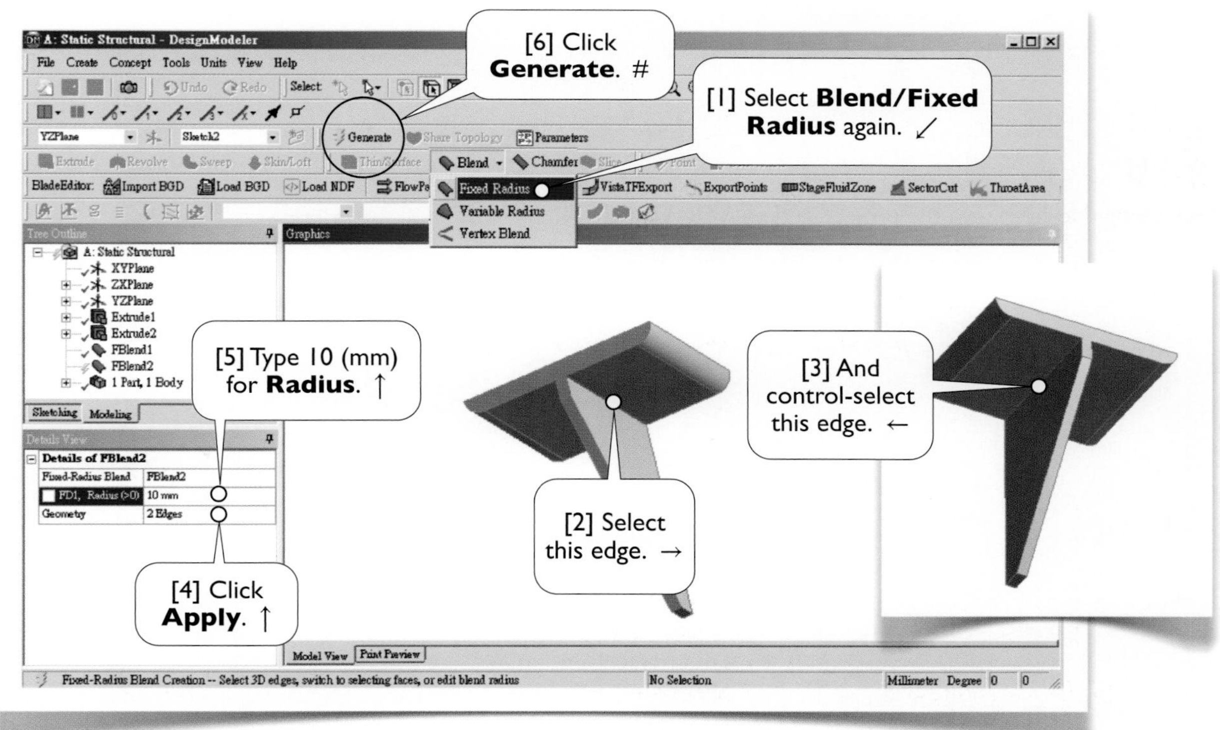

4.1.9 Turn Off Edges Display

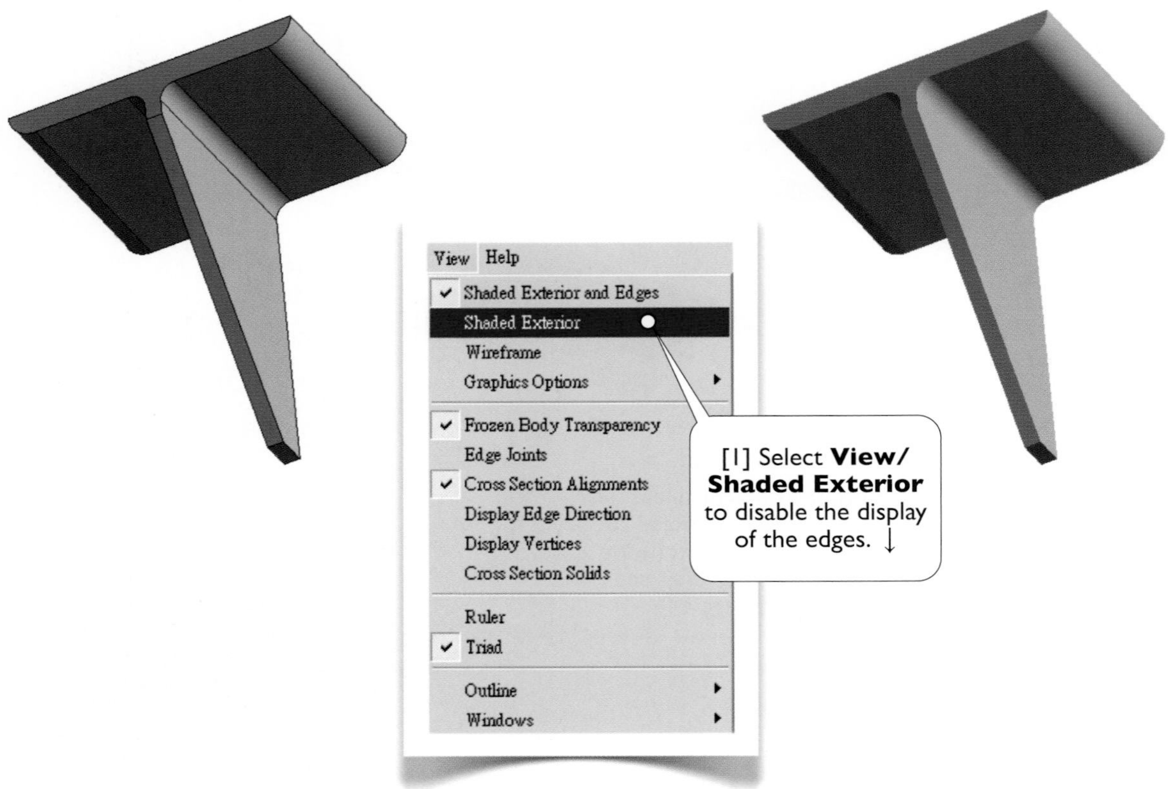

Wrap Up

[2] Close DesignModeler, save the project, and exit Workbench. #

Section 4.2

Cover of Pressure Cylinder

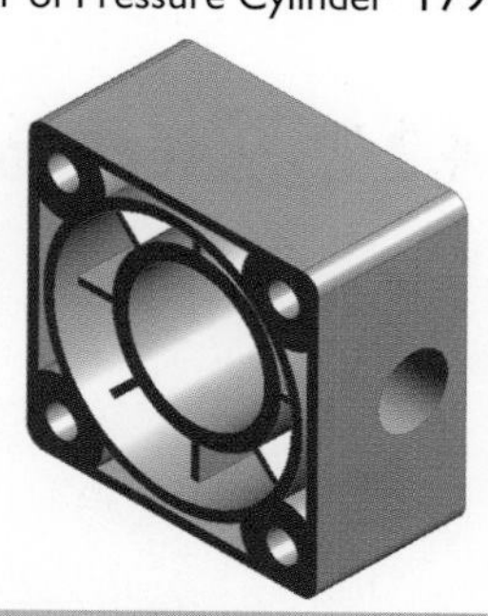

4.2.1 About the Cylinder Cover

[1] The pressure cylinder [2] contains a gas of 0.5 MPa. The cylinder cover [3-5] is made of a carbon-fiber reinforced plastic. In this section, we want to create a 3D solid model for the cover; the model will be used for a static structural simulation in Section 5.2 to investigate the deformation under pressure. ↙

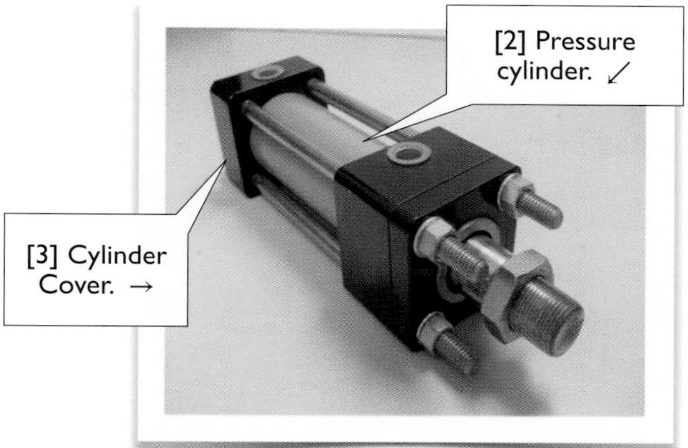

[2] Pressure cylinder. ↙

[3] Cylinder Cover. →

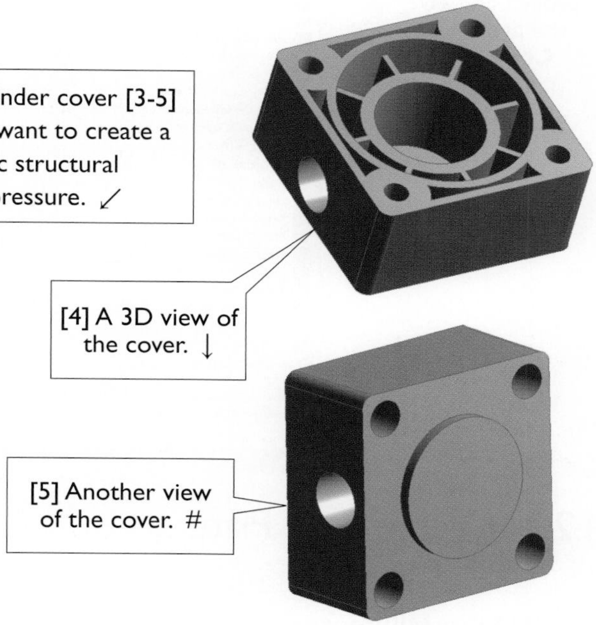

[4] A 3D view of the cover. ↓

[5] Another view of the cover. #

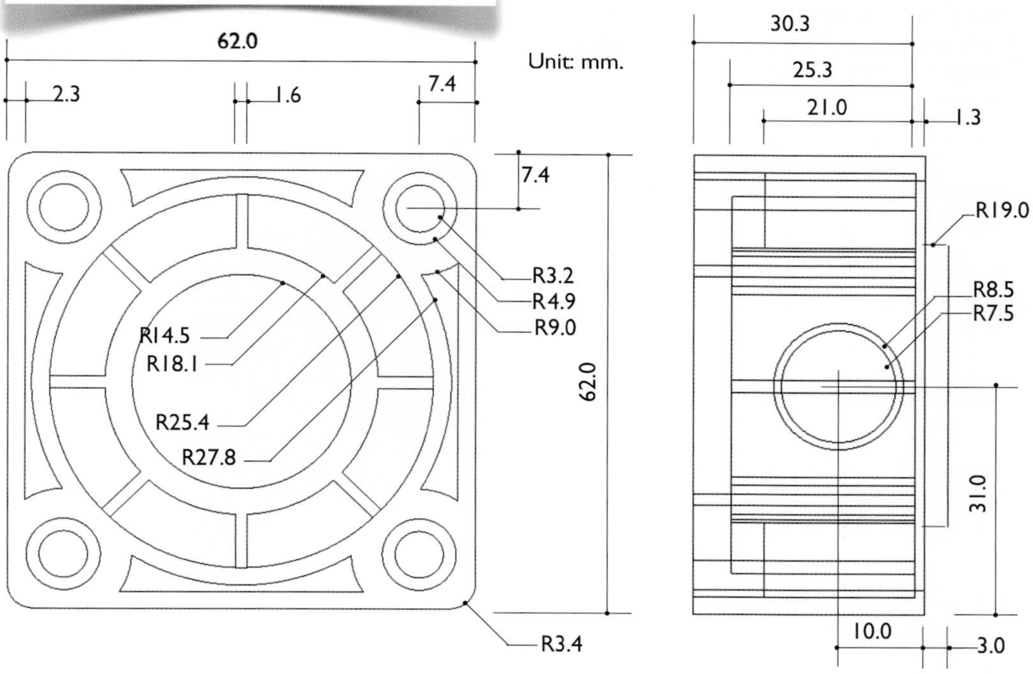

Unit: mm.

62.0 2.3 1.6 7.4 7.4 R3.2 R4.9 R9.0 R14.5 R18.1 R25.4 R27.8 62.0 R3.4

30.3 25.3 21.0 1.3 R19.0 R8.5 R7.5 31.0 10.0 3.0

4.2.2 Start Up

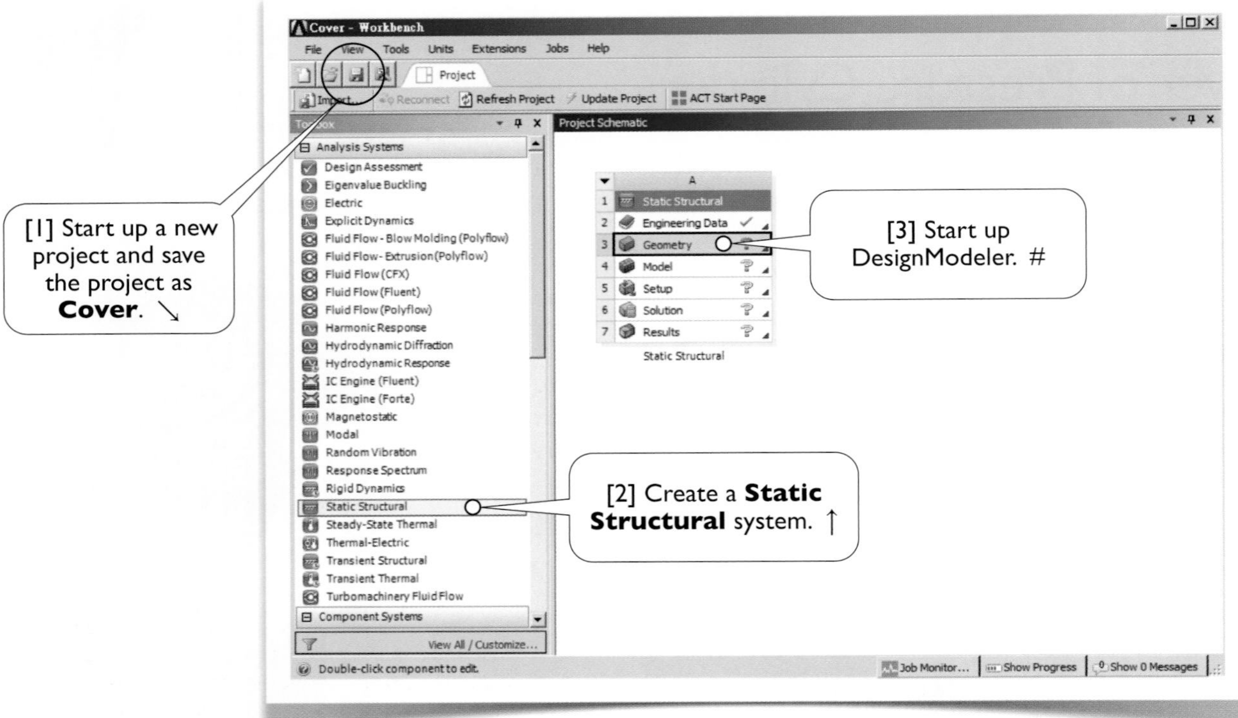

4.2.3 Create Base Plate

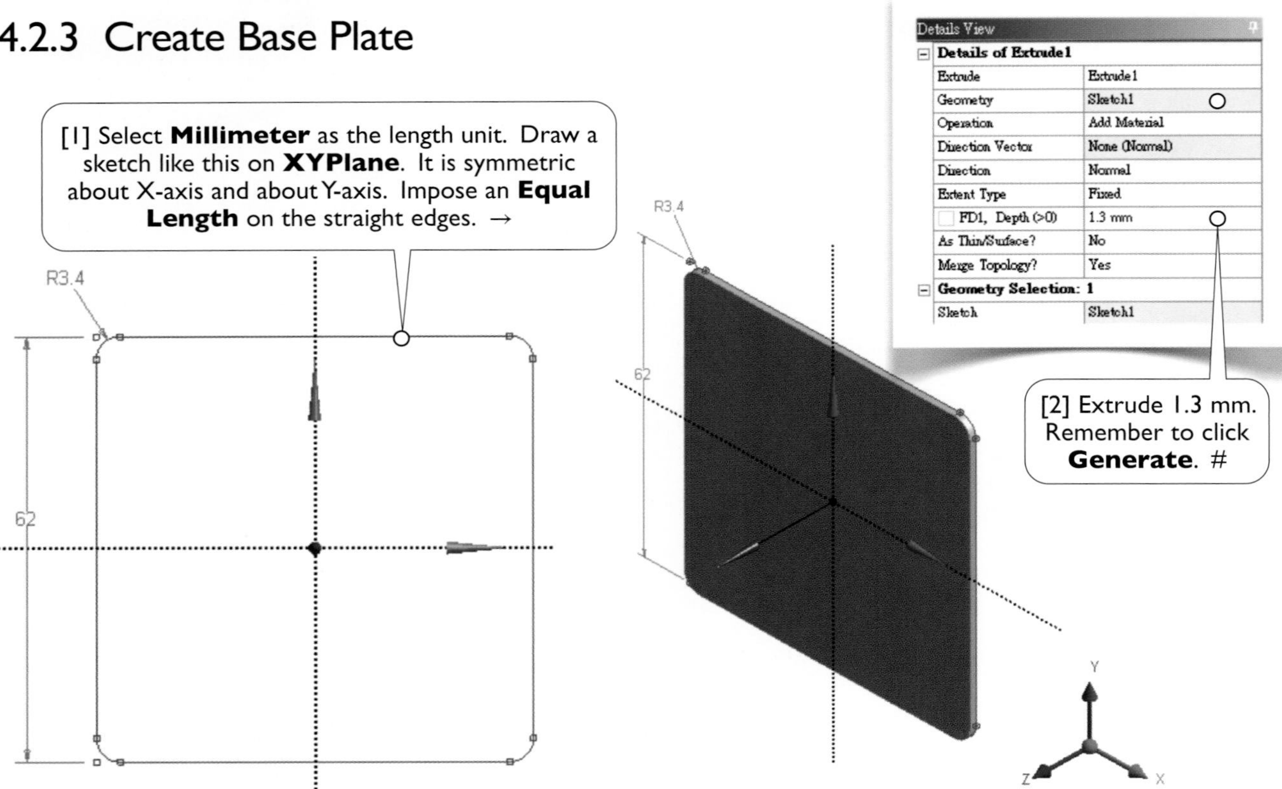

4.2.4 Create a New Sketching Plane

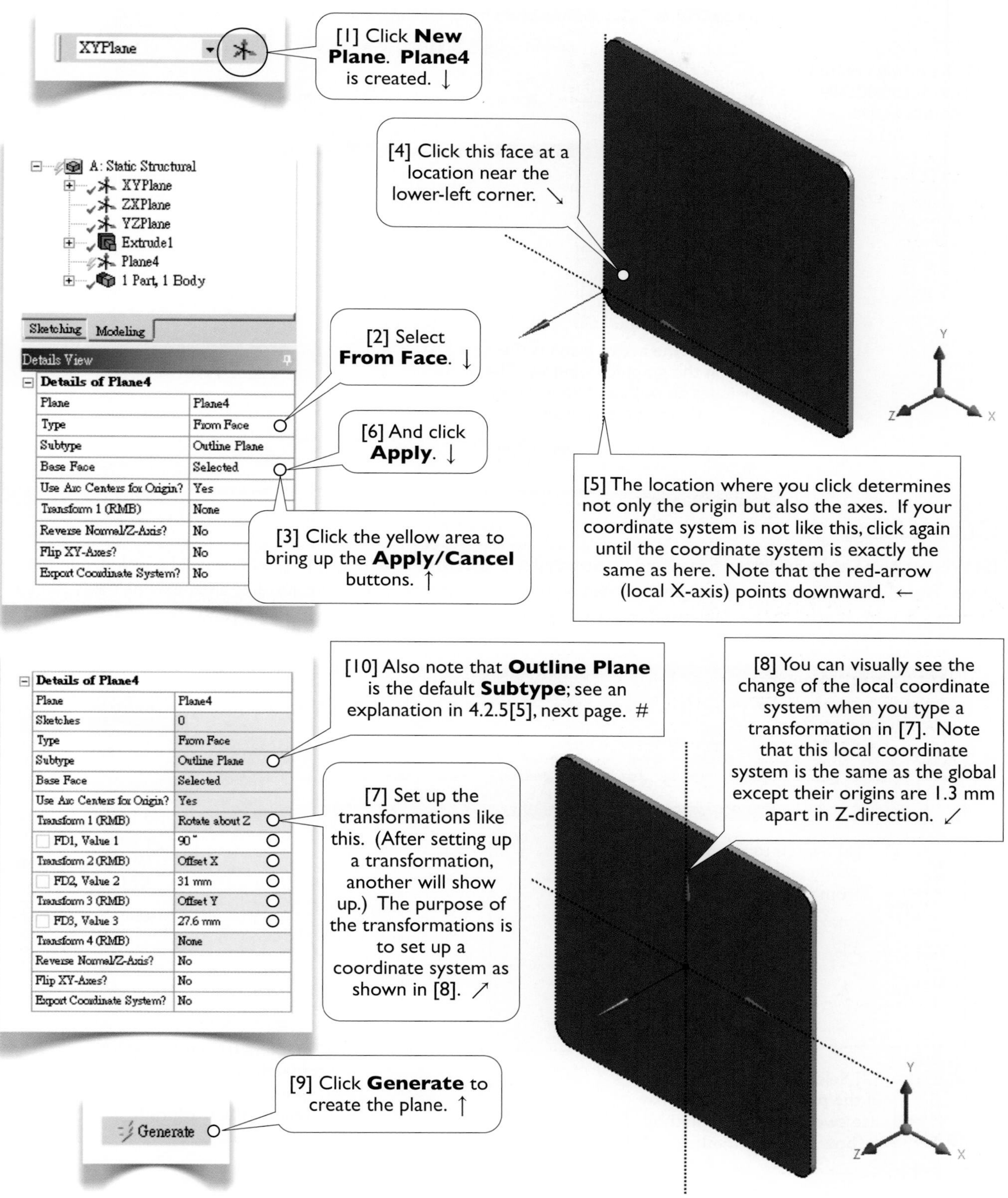

4.2.5 Duplicate the Plane Outline

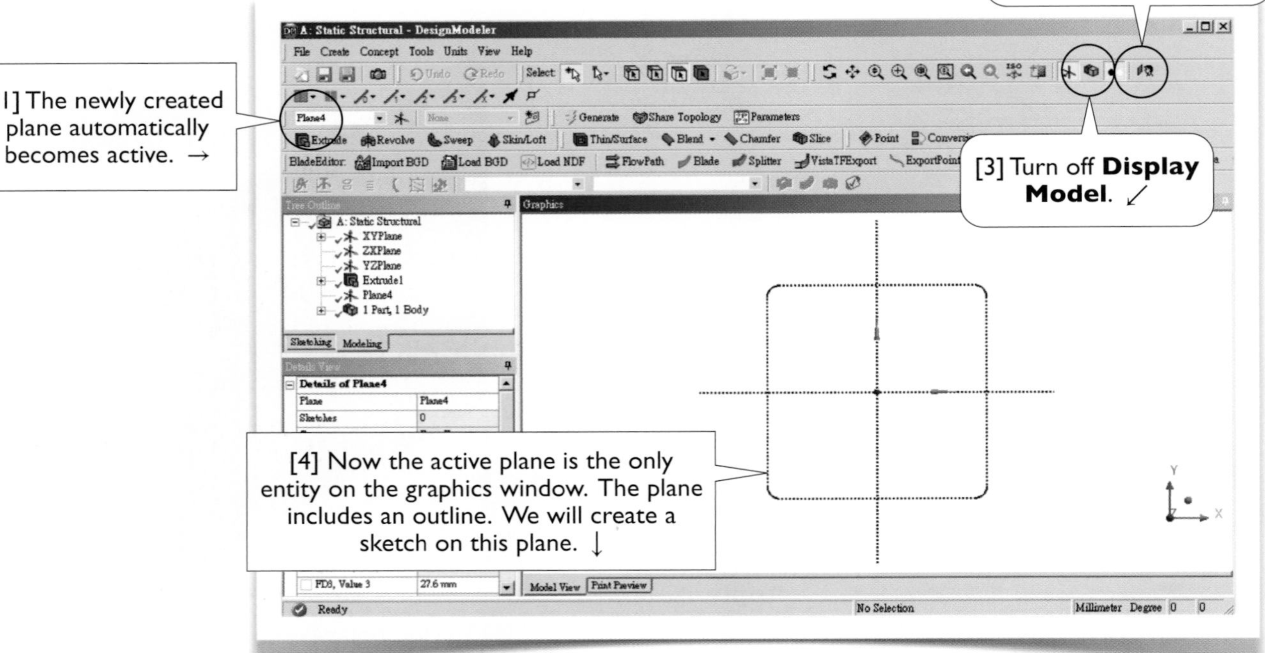

Outline Plane

[5] When a plane is derived from a face and its **Subtype** is **Outline Plane**, the face outline will be included in the plane. The outline is not a geometric entity; it can be used as a reference, or datum, just like an X-axis or Y-axis can be used as a datum. It also can be duplicated to create geometric entities, as demonstrated in [6-9]. ↙

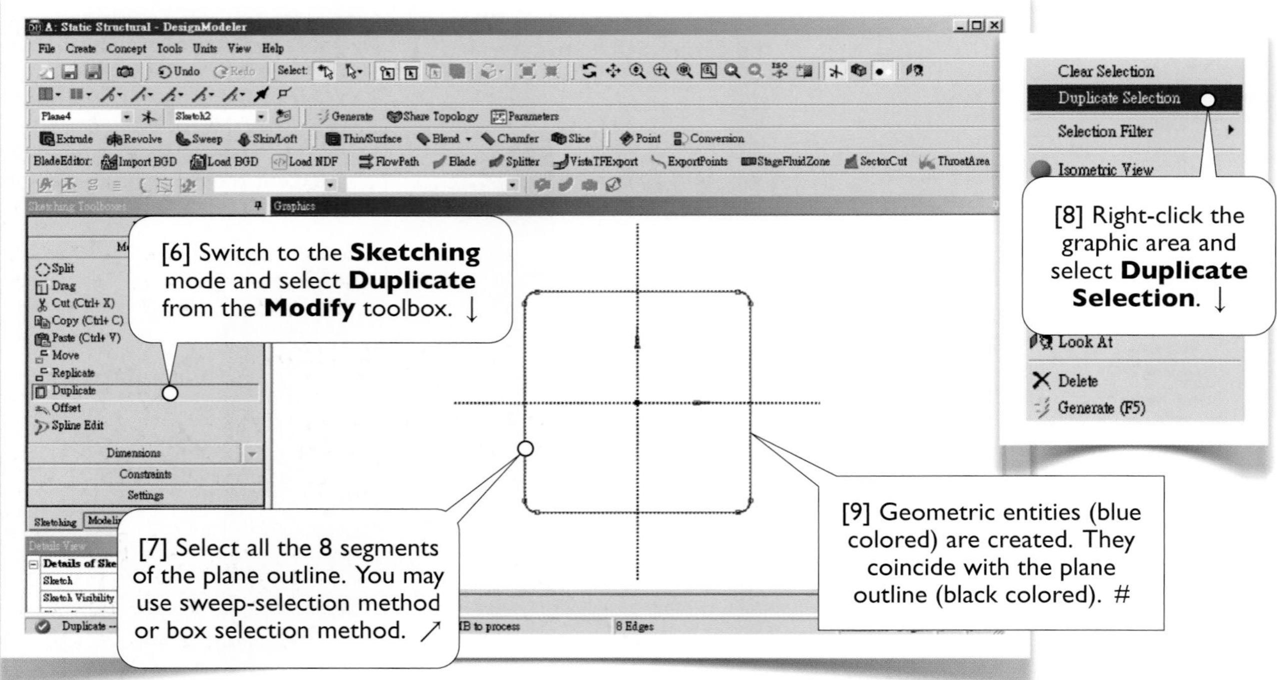

4.2.6 Finish the Sketch

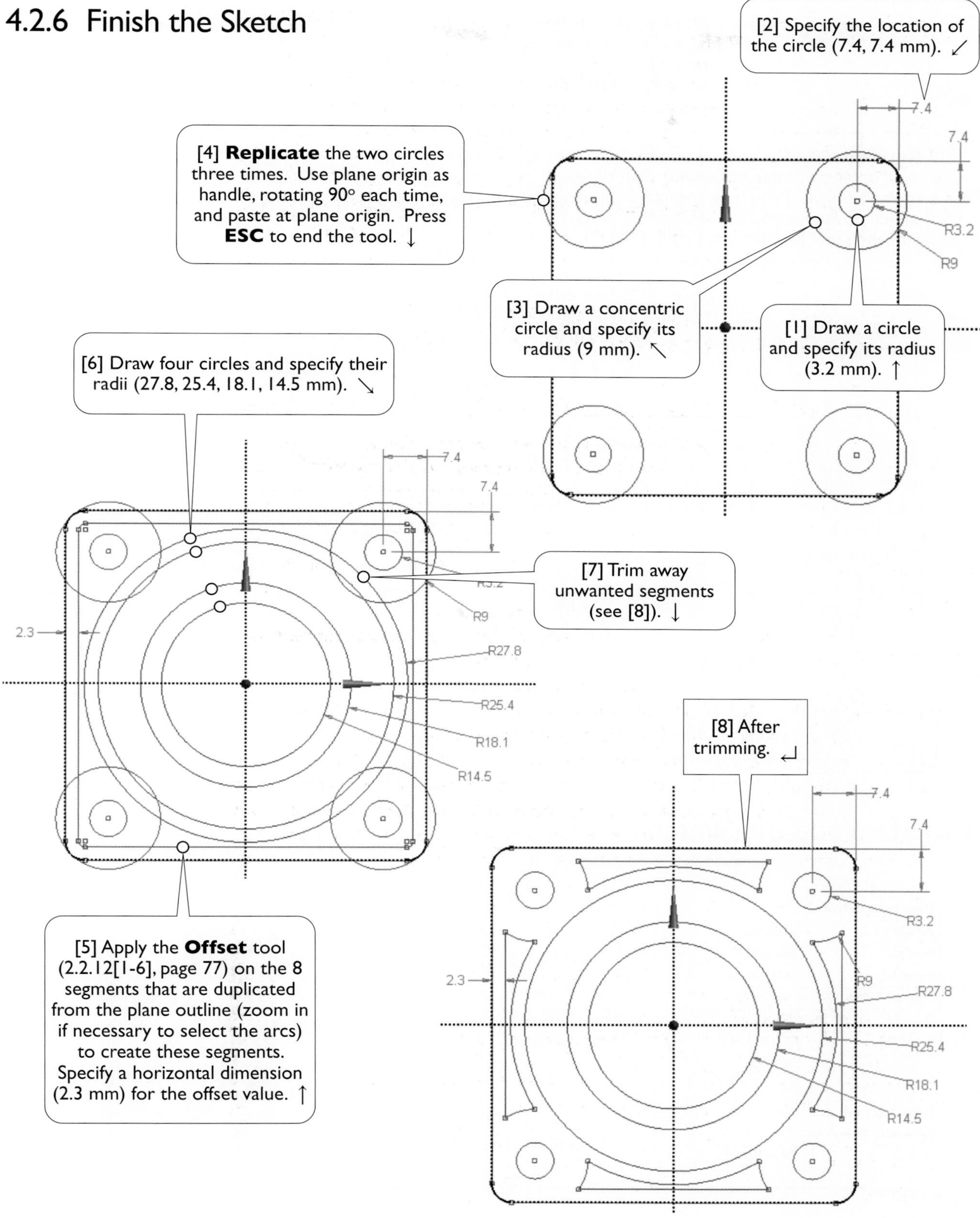

[9] Draw two vertical lines, impose **Symmetry**, and specify a horizontal dimension (1.6 mm). ↓

[10] **Replicate** the last two segments seven times. Use plane origin as handle, rotate 45° each time, and paste at plane origin. If you have trouble duplicating at a horizontal position, undo it and try other ways; you ran into the same problems as I did (see [11]). ↘

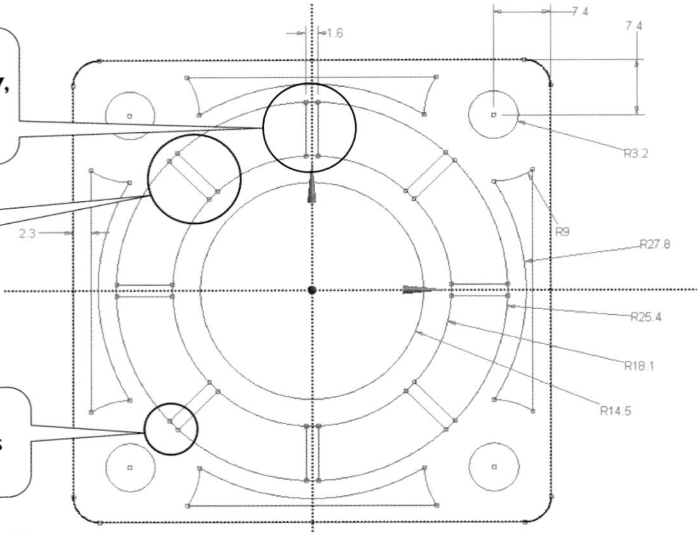

[12] Trim away unwanted segments (see [13]). ←

[13] After trimming. #

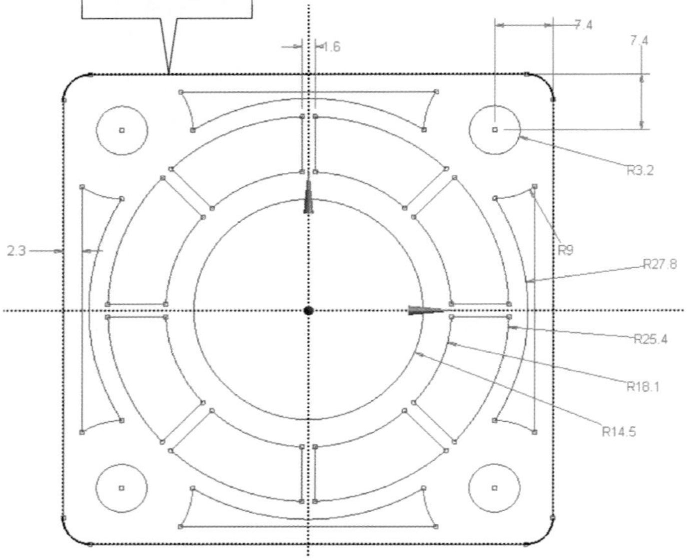

Working with Complicated Sketches

[11] When you are working with a complicated sketch like this one, things may not be smooth. You have to pan, zoom, turn on/off display modes frequently. Therefore, being familiar with the viewing control tools is crucial. The worst is that some peculiar behaviors may happen. For example, the replications [10] at horizontal positions may not be successful. These peculiar behaviors are possibly due to some unnecessary constraints automatically imposed (2.3.5, page 84). Fortunately these small problems usually can be easily circumvented by trying other methods. ↖

4.2.7 Extrude the Sketch

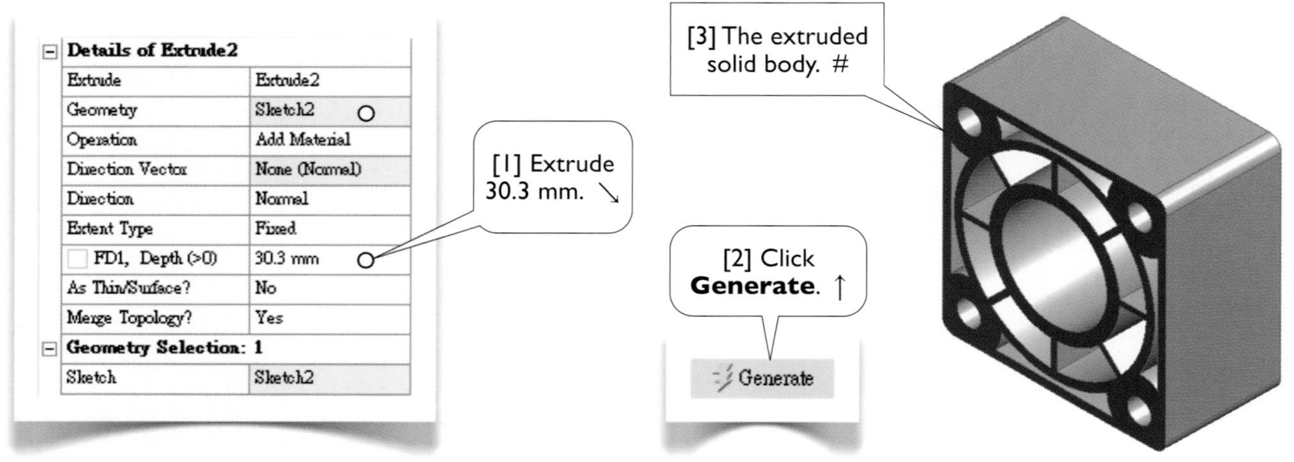

Details of Extrude2	
Extrude	Extrude2
Geometry	Sketch2
Operation	Add Material
Direction Vector	None (Normal)
Direction	Normal
Extent Type	Fixed
FD1, Depth (>0)	30.3 mm
As Thin/Surface?	No
Merge Topology?	Yes
Geometry Selection: 1	
Sketch	Sketch2

[1] Extrude 30.3 mm. ↘

[2] Click **Generate**. ↑

[3] The extruded solid body. #

4.2.8 Create a New Plane and Draw a Circle on the Plane

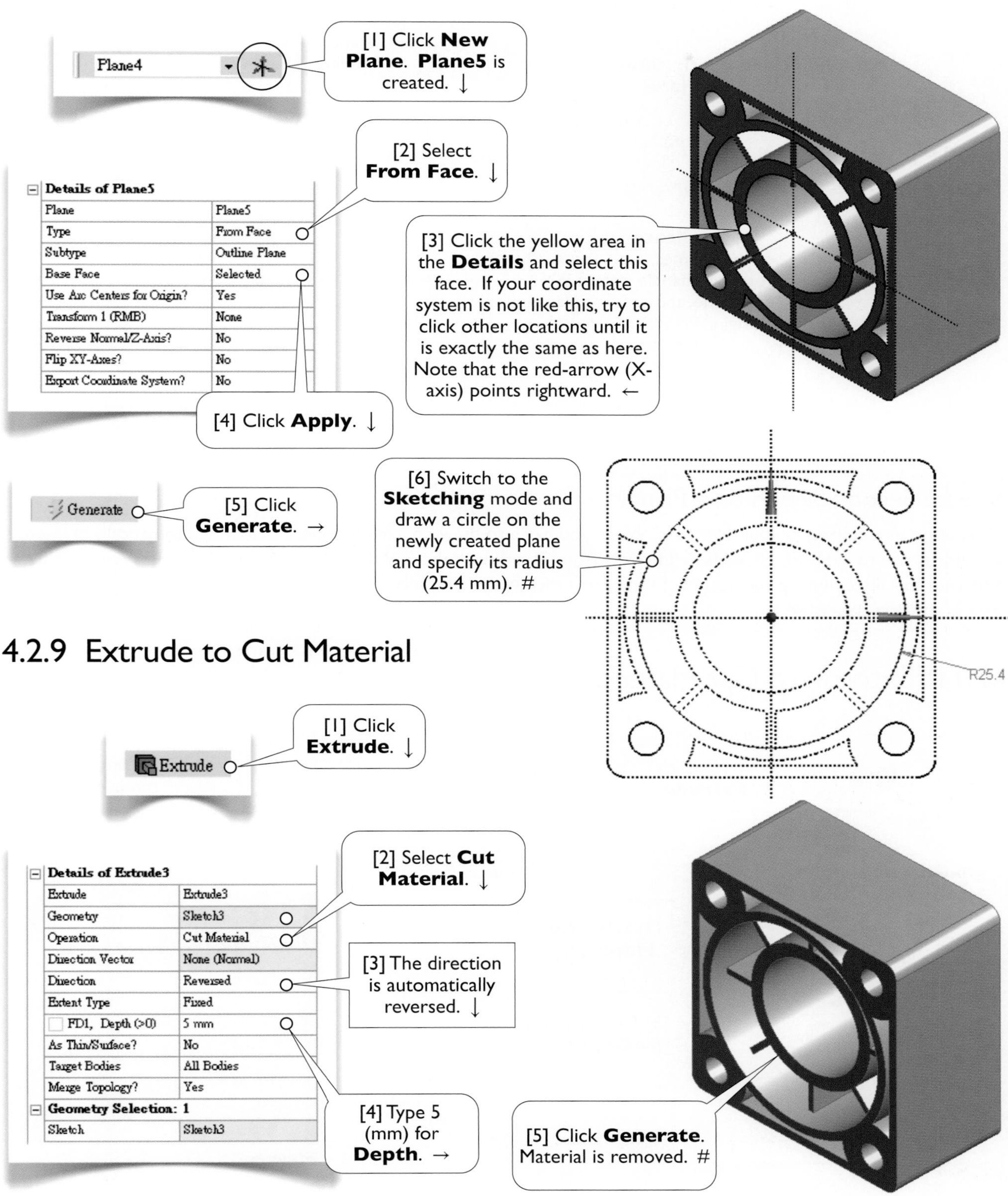

4.2.10 Create New Sketch on XYPlane and Draw Circles

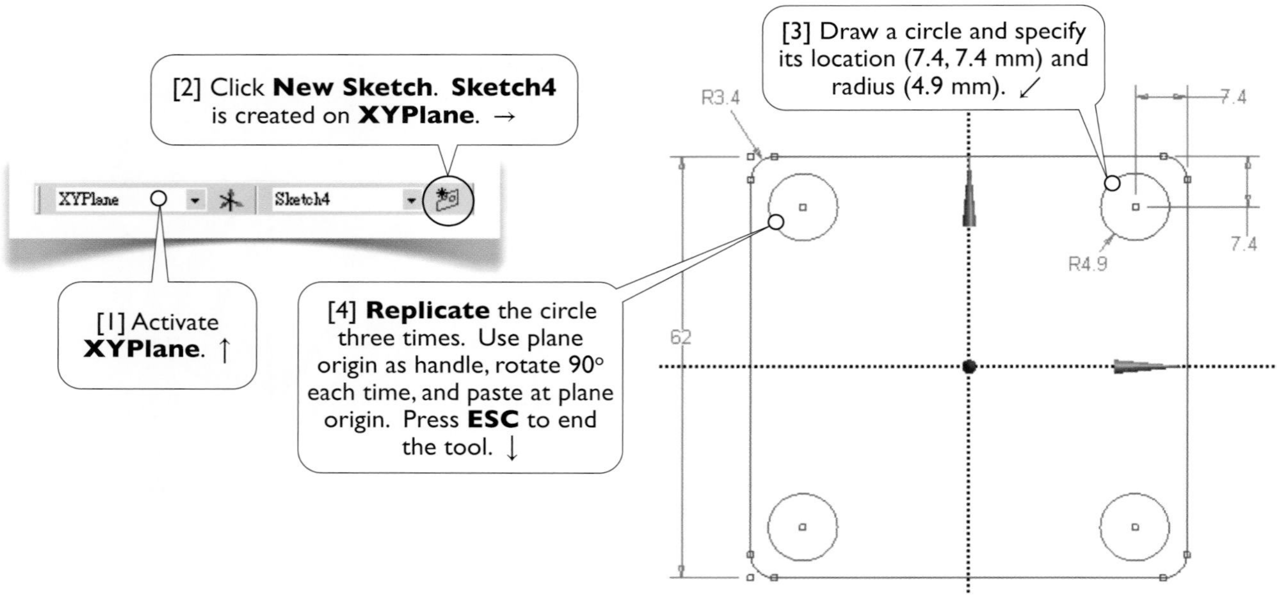

Create Second Sketch on a Plane

[5] A plane may contain multiple sketches. When you begin to draw on a fresh plane, a sketch is automatically created; you don't have to explicitly click **New Sketch** as in [2], although there is no harm in doing that. When you want to add a second sketch on a plane, you MUST explicitly click **New Sketch**, otherwise the entities you draw will become part of the first sketch. #

4.2.11 Extrude to Cut Material

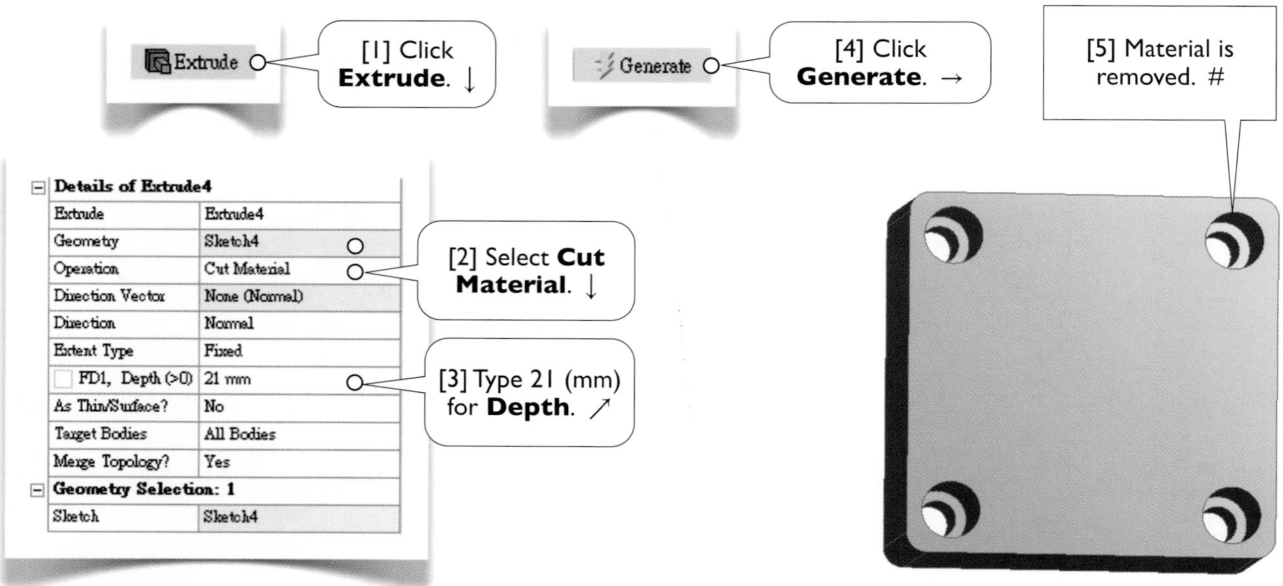

4.2.12 Create Another Sketch on XYPlane and Draw a Circle

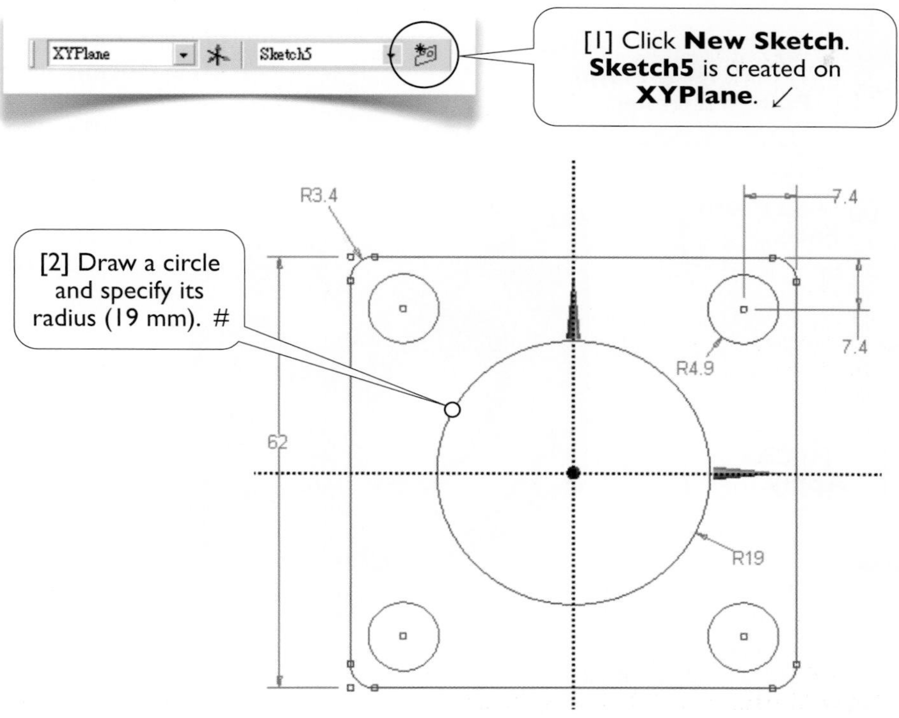

4.2.13 Extrude to Add Material

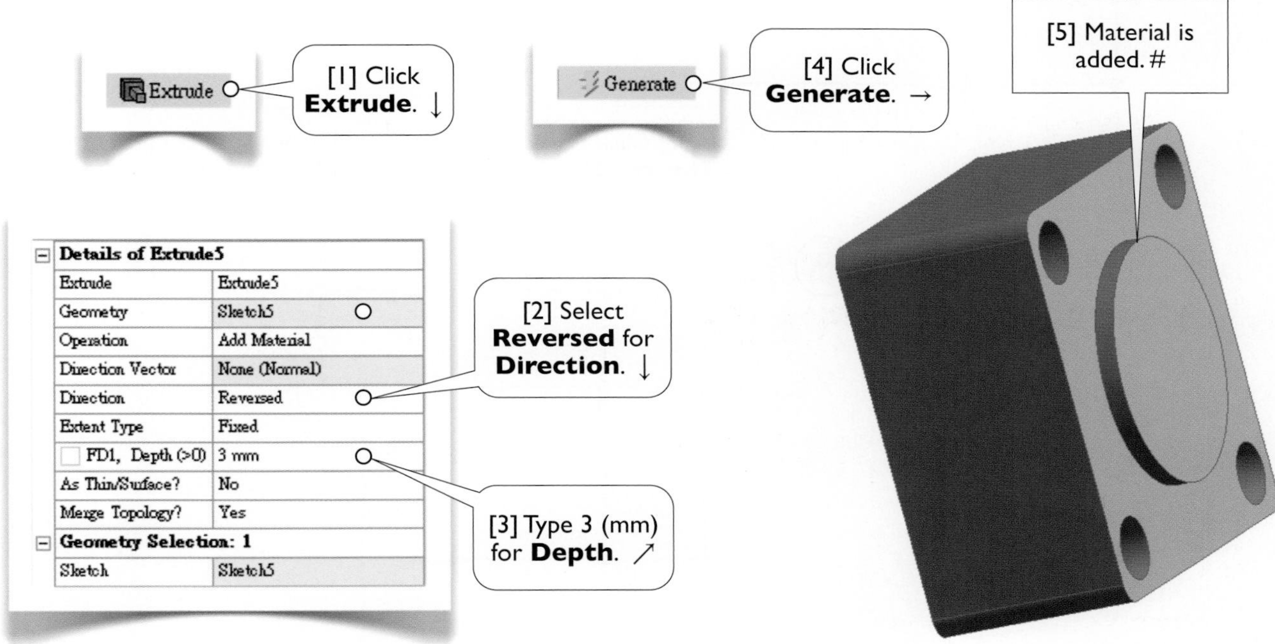

Details of Extrude5	
Extrude	Extrude5
Geometry	Sketch5
Operation	Add Material
Direction Vector	None (Normal)
Direction	Reversed
Extent Type	Fixed
FD1, Depth (>0)	3 mm
As Thin/Surface?	No
Merge Topology?	Yes
Geometry Selection: 1	
Sketch	Sketch5

4.2.14 Create a Plane on a Side Wall and Two Sketches on the Plane

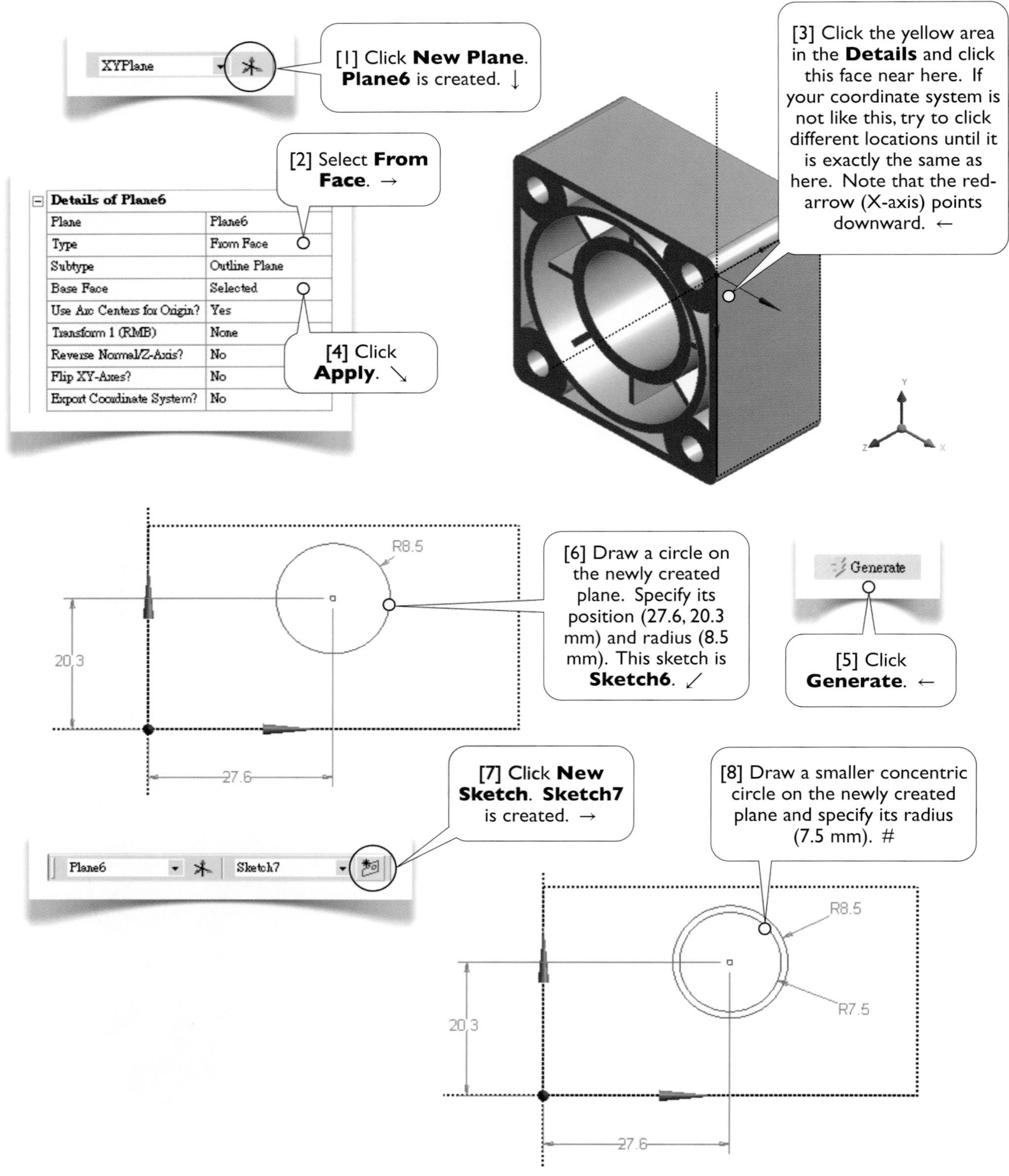

4.2.15 Complete the Model

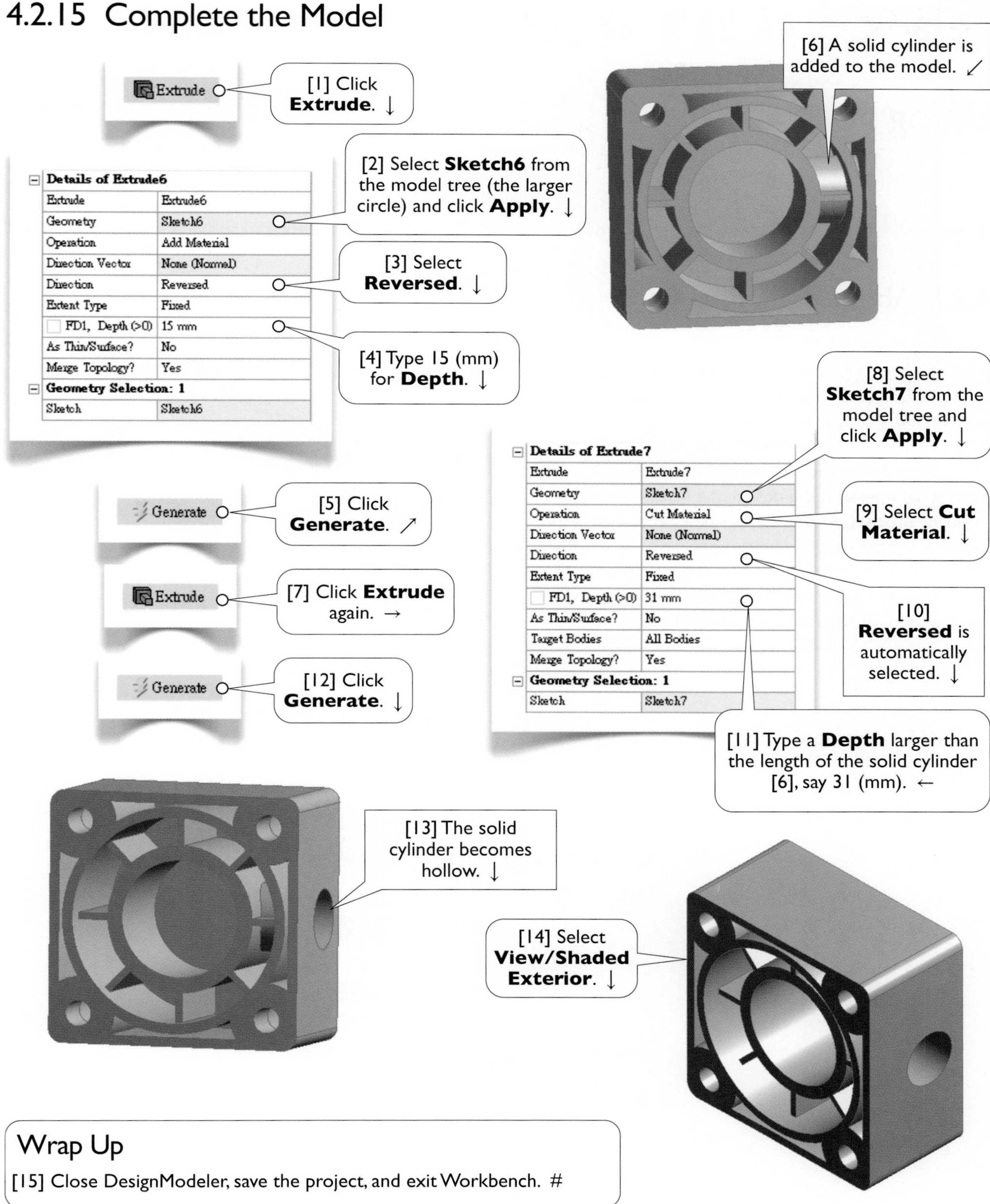

Wrap Up

[15] Close DesignModeler, save the project, and exit Workbench. #

Section 4.3

Lifting Fork

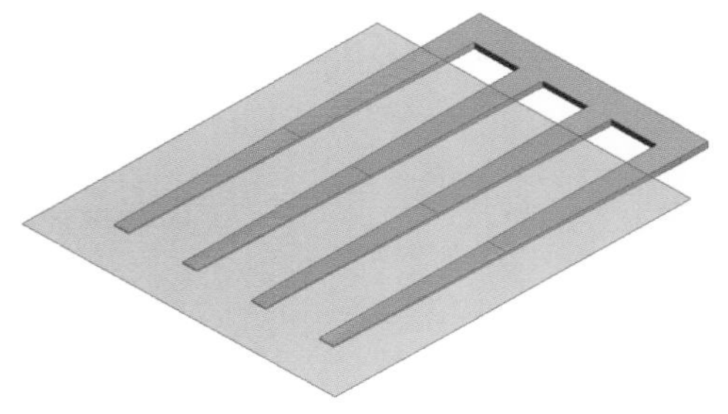

4.3.1 About the Lifting Fork

[1] The lifting fork [2-6] is used in an LCD (liquid crystal display) manufacturing factory to handle glass panels. The glass panel is large (2.5 m x 2.2 m) and thin (1.0 mm) and the engineers are concerned about its deflections during the dynamic handling, which is a critical parameter in the design of a precision machine handling process. Another important parameter is the time before the panel's vibrations cease.

In this section, we will model the fork as a solid body and the glass panel as a surface body. The model will be used for simulations in Section 12.2. →

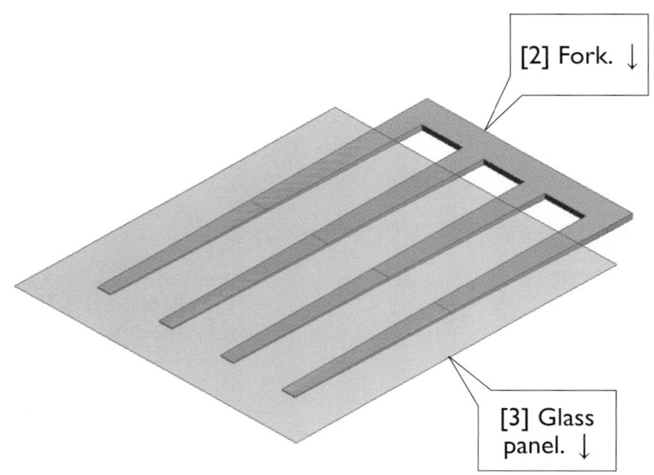

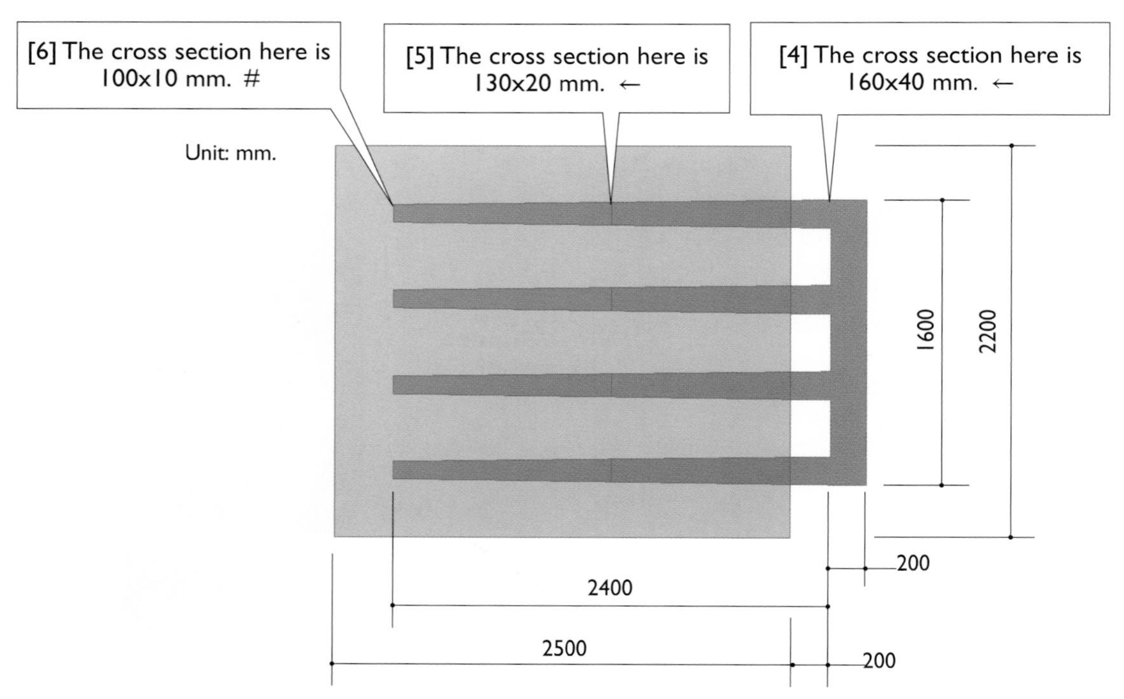

4.3.2 Start Up

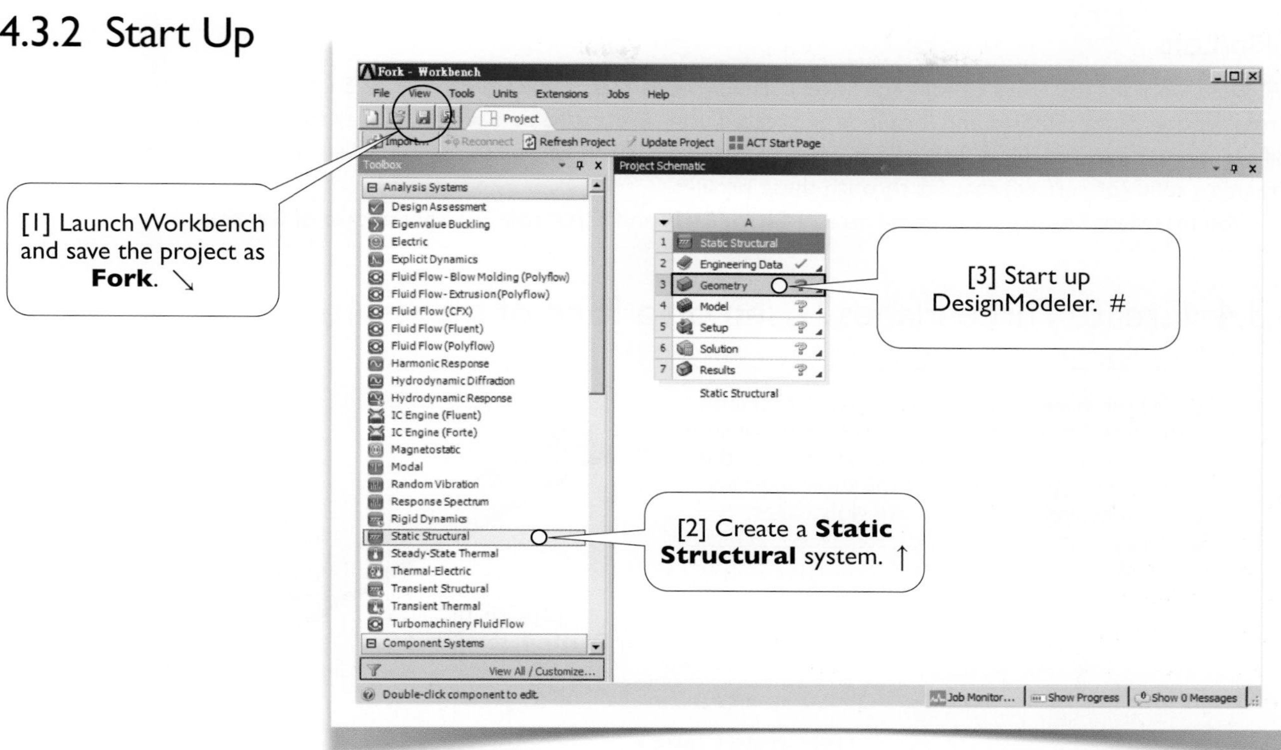

4.3.3 Create the Transversal Beam

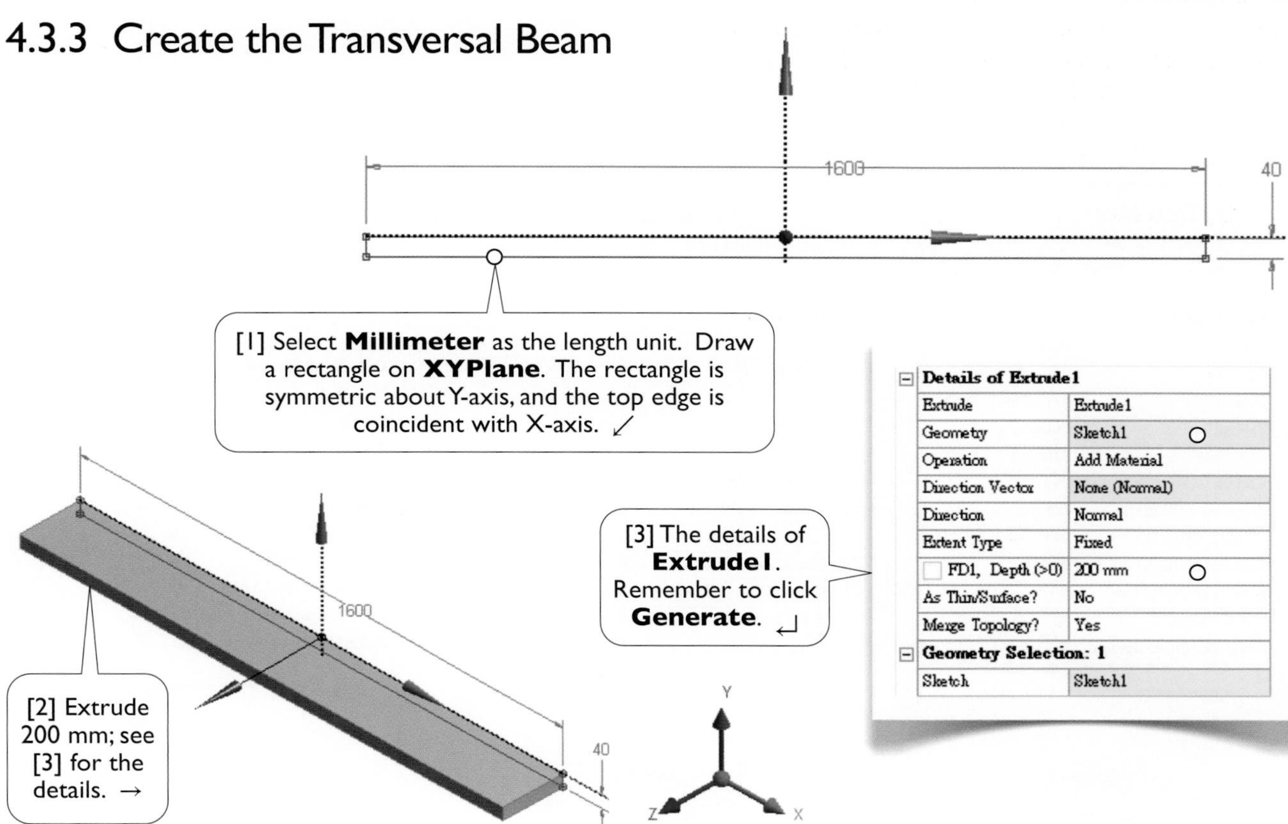

Details of Extrude1	
Extrude	Extrude1
Geometry	Sketch1
Operation	Add Material
Direction Vector	None (Normal)
Direction	Normal
Extent Type	Fixed
FD1, Depth (>0)	200 mm
As Thin/Surface?	No
Merge Topology?	Yes
Geometry Selection: 1	
Sketch	Sketch1

Skin/Loft

[4] Now we want to create a prong (or finger). The prong is then duplicated to create 3 additional prongs. The prong's cross section is not uniform, and it cannot be created using **Extrude** or **Sweep**. A way to create a solid body of different cross sections along its length is using **Skin/Loft**, which takes a series of profiles from different planes and creates a solid (or surface) that fits through these profiles.

You may view **Sweep** as a special case of **Skin/Loft** and **Extrude** as a special case of **Sweep**. #

4.3.4 Create Three Planes Based on a Face of the Beam

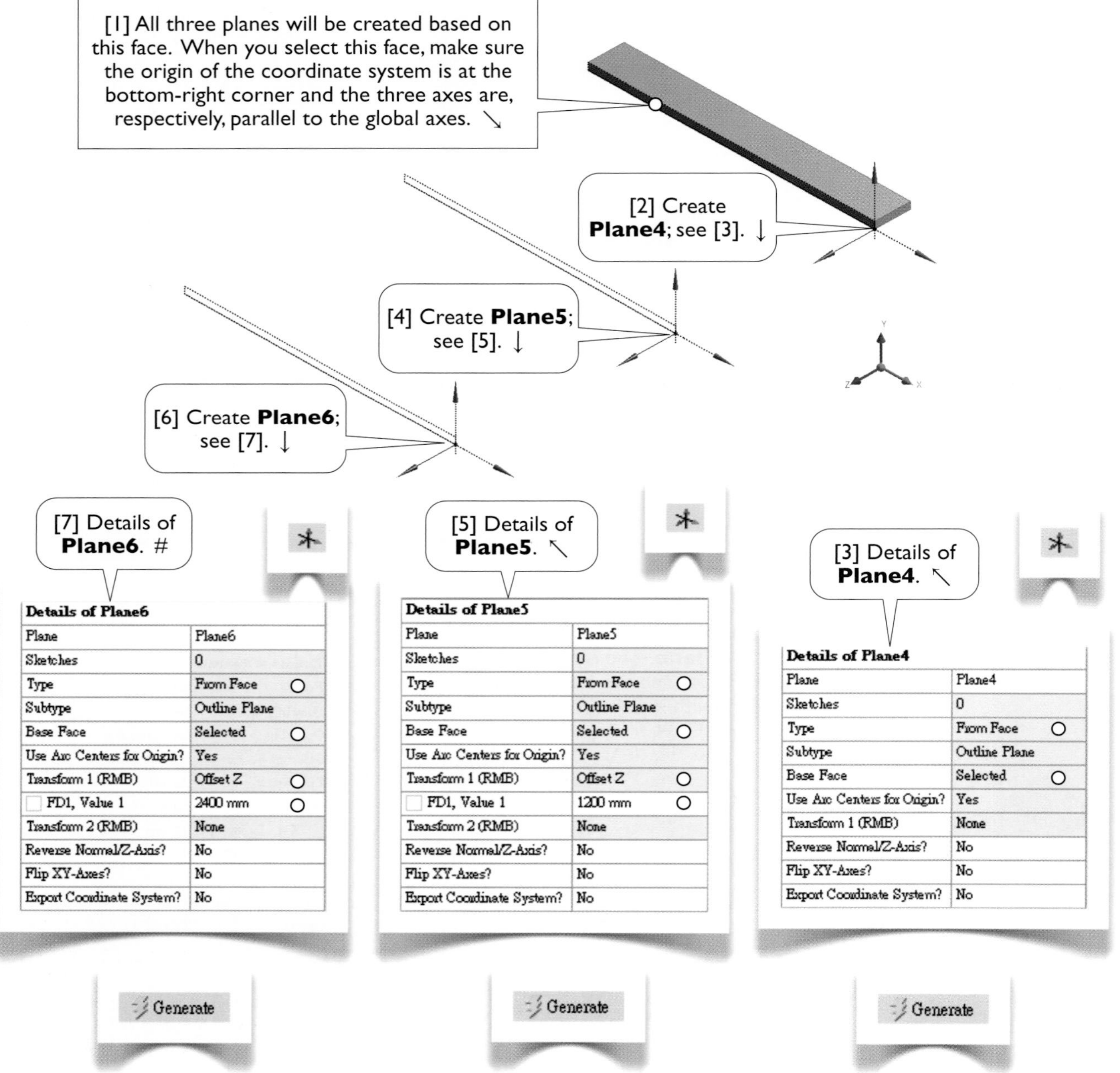

Details of Plane6	
Plane	Plane6
Sketches	0
Type	From Face
Subtype	Outline Plane
Base Face	Selected
Use Arc Centers for Origin?	Yes
Transform 1 (RMB)	Offset Z
FD1, Value 1	2400 mm
Transform 2 (RMB)	None
Reverse Normal/Z-Axis?	No
Flip XY-Axes?	No
Export Coordinate System?	No

Details of Plane5	
Plane	Plane5
Sketches	0
Type	From Face
Subtype	Outline Plane
Base Face	Selected
Use Arc Centers for Origin?	Yes
Transform 1 (RMB)	Offset Z
FD1, Value 1	1200 mm
Transform 2 (RMB)	None
Reverse Normal/Z-Axis?	No
Flip XY-Axes?	No
Export Coordinate System?	No

Details of Plane4	
Plane	Plane4
Sketches	0
Type	From Face
Subtype	Outline Plane
Base Face	Selected
Use Arc Centers for Origin?	Yes
Transform 1 (RMB)	None
Reverse Normal/Z-Axis?	No
Flip XY-Axes?	No
Export Coordinate System?	No

4.3.5 Create a Sketch on Each Plane

4.3.6 Create a Prong by Skin/Loft

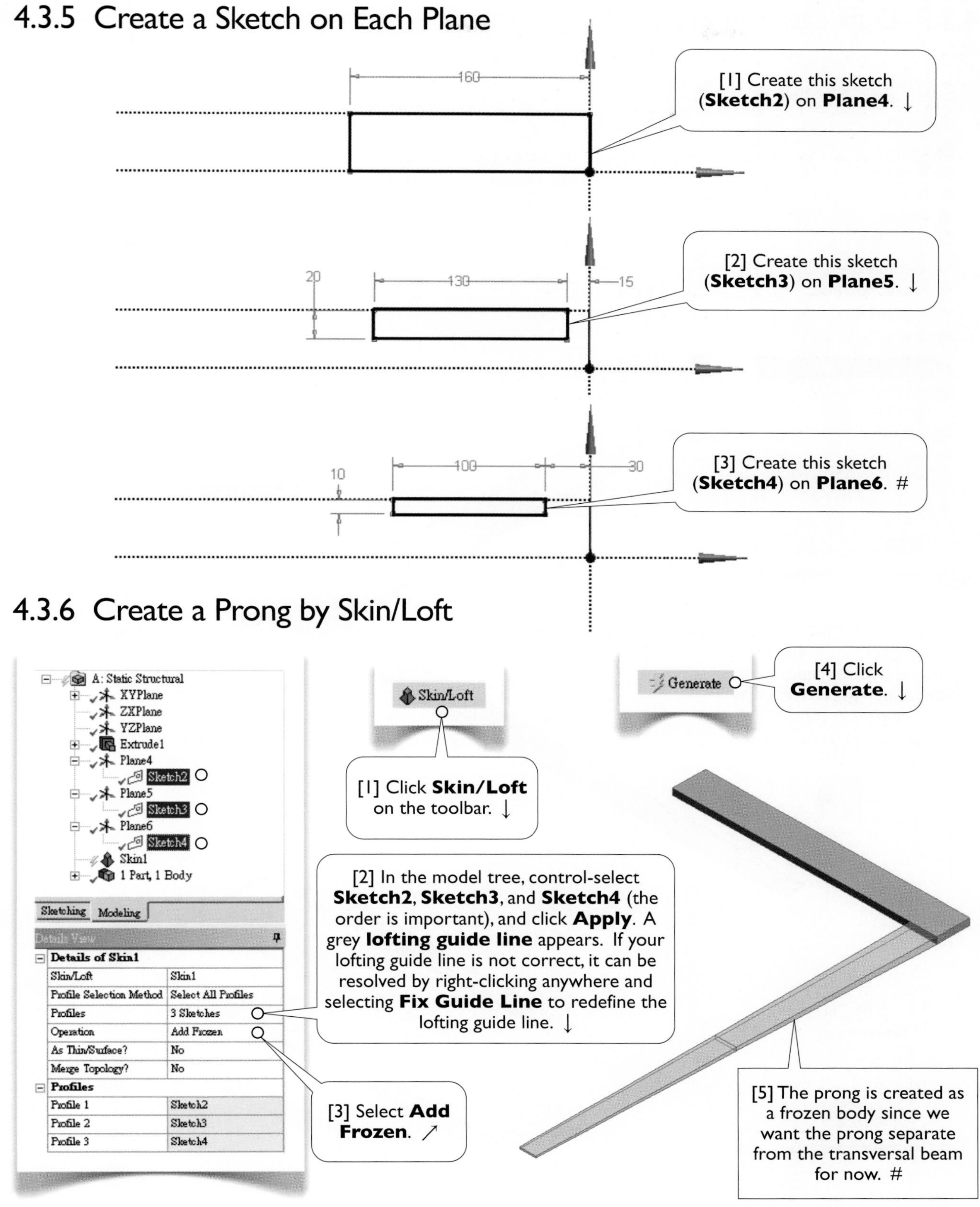

4.3.7 Duplicate the Prong Using Pattern

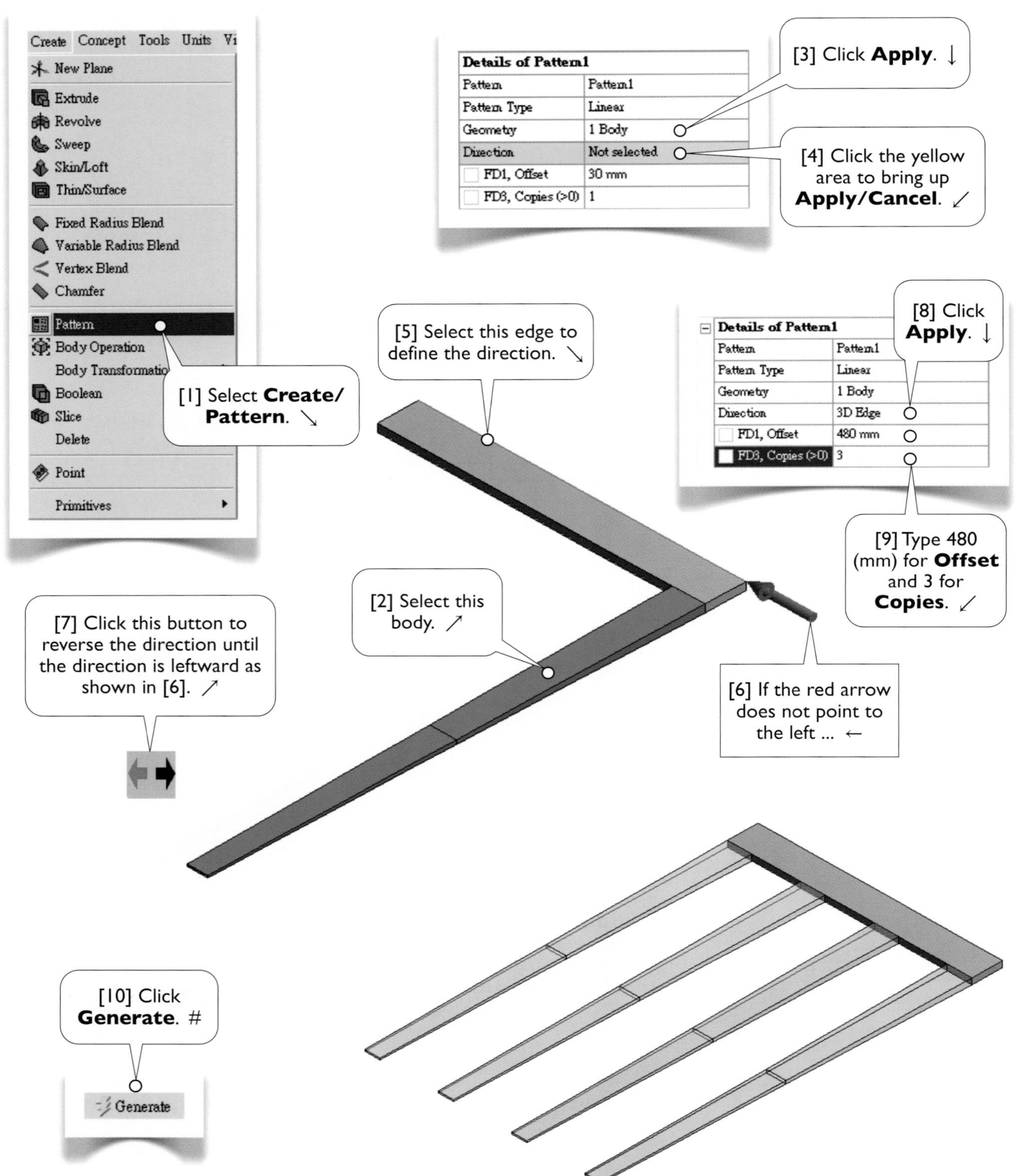

4.3.8 Combine the Bodies Using **Boolean**

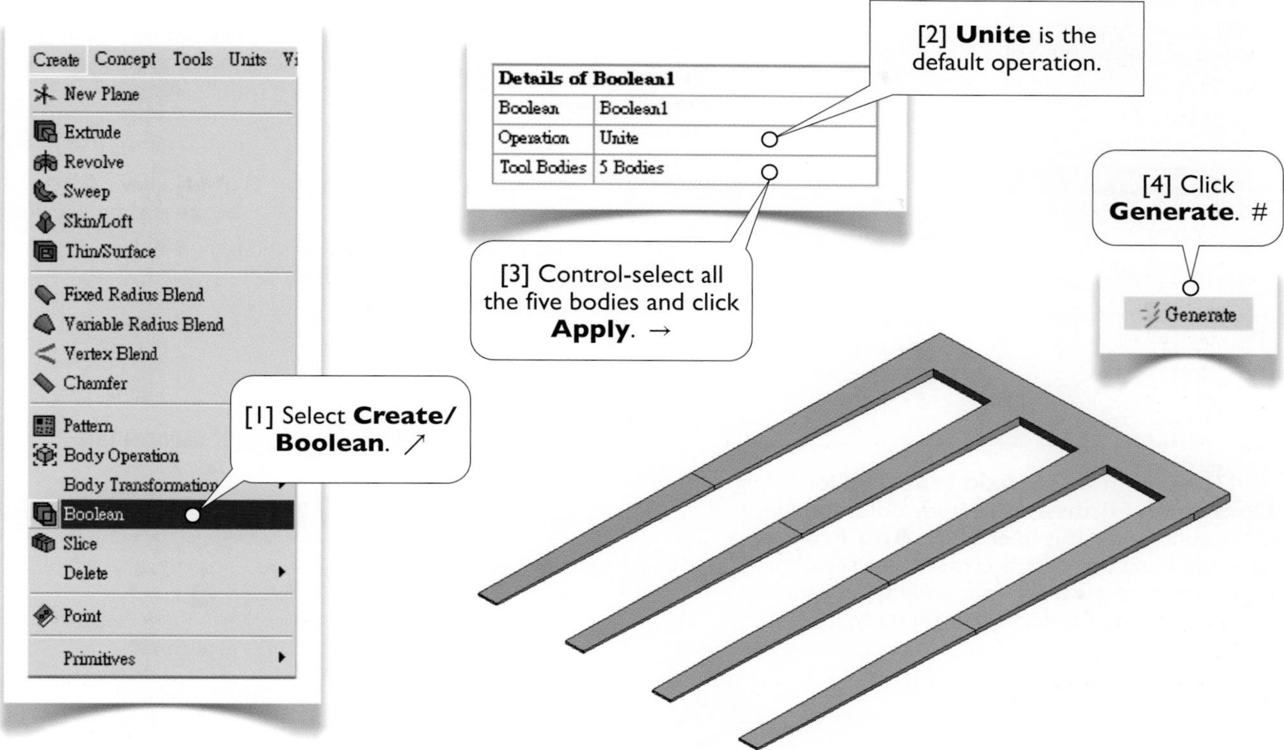

4.3.9 Create a Surface Body for the Glass Panel

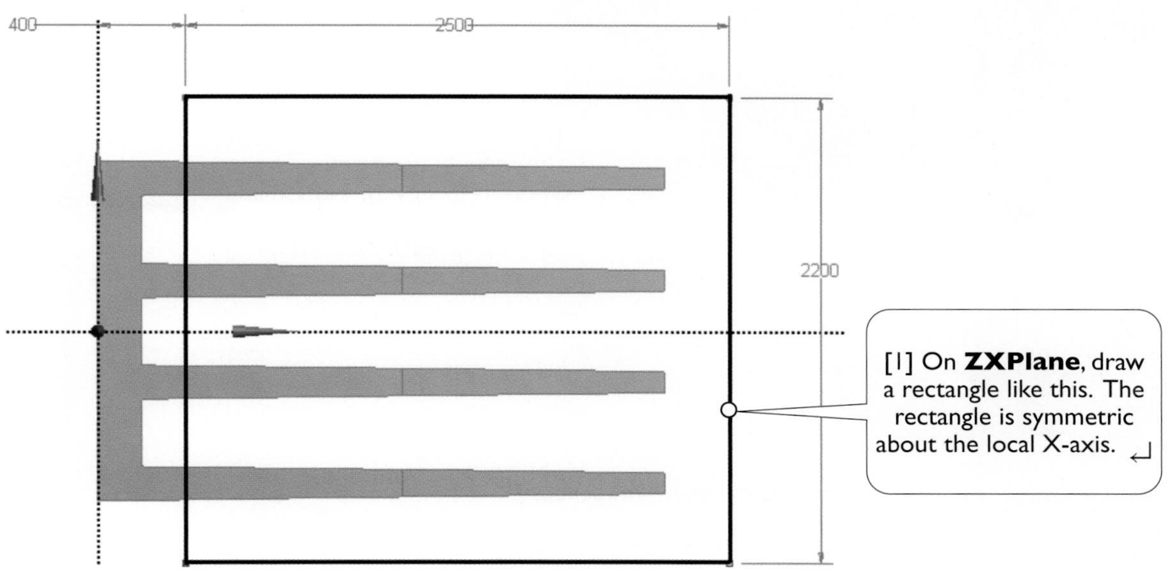

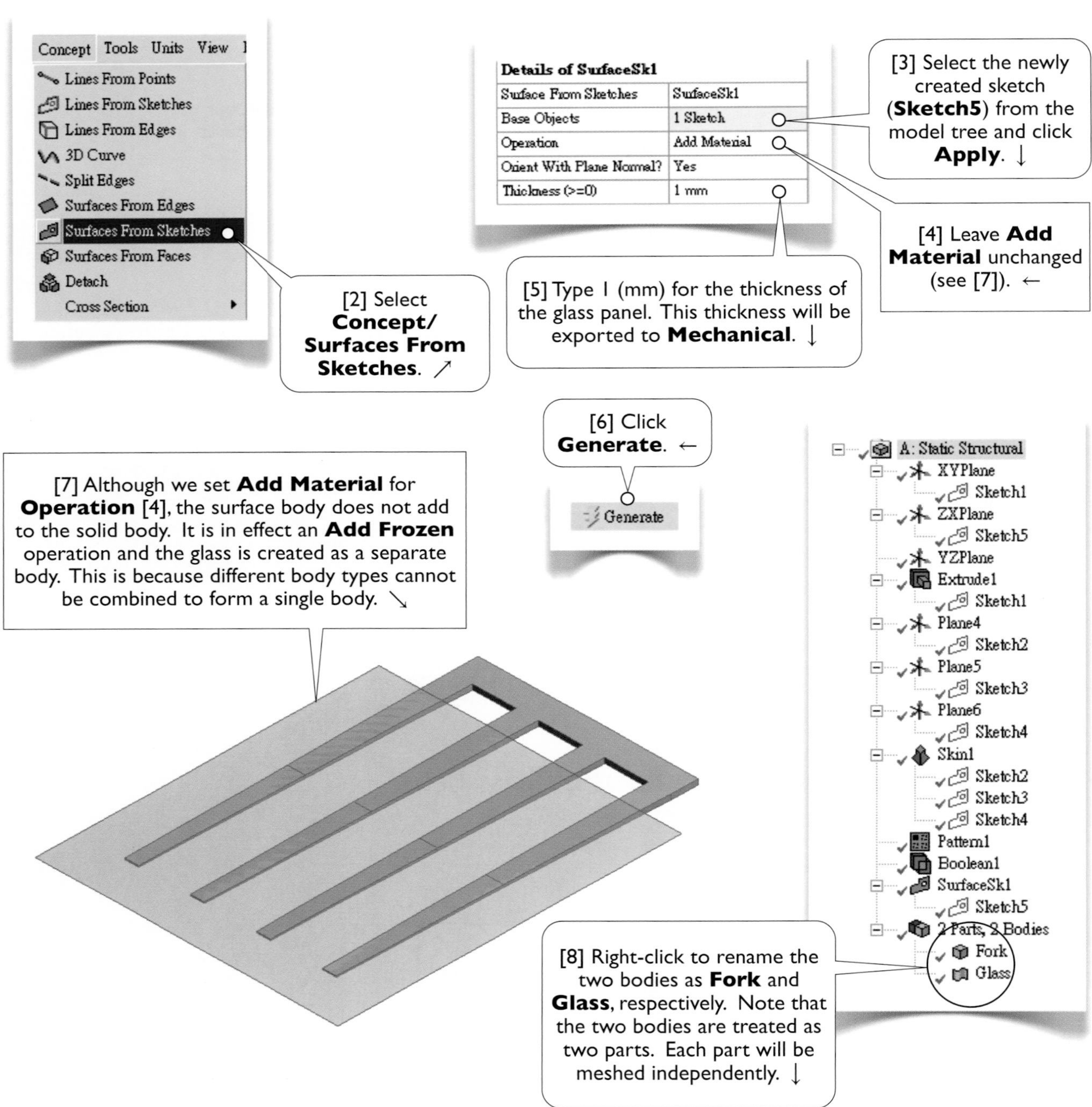

Wrap Up

[9] Close DesignModeler, save the project, and exit Workbench. #

Section 4.4

More Details

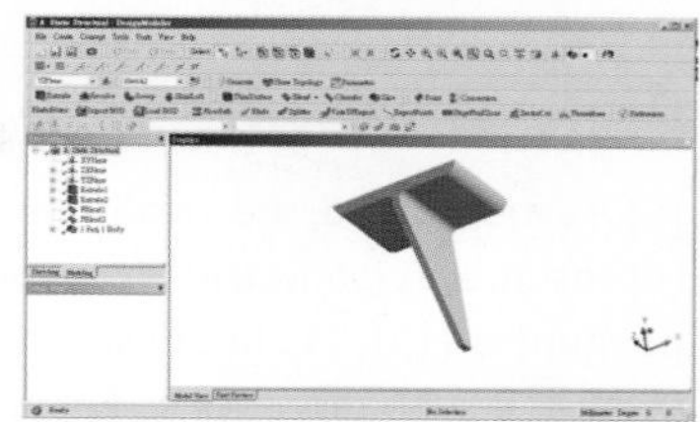

4.4.1 DesignModeler GUI Revisit

[1] In Section 2.3, we've overviewed **DesignModeler GUI**, yet skipped some 3D tools, such as view orientations [2-4], and selection tools [5-7]. Also on the toolbar are tools to create 3D features [8]. These tools will be covered in this section. ↘

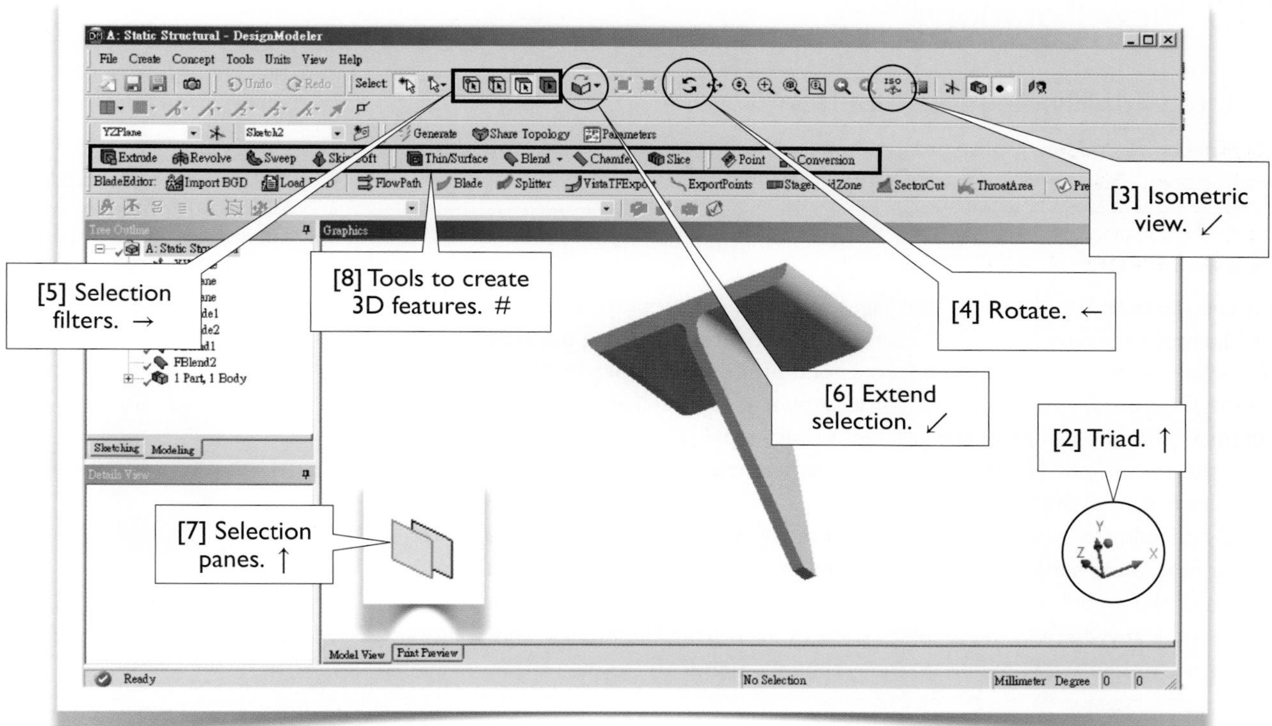

4.4.2 Principal Views and Isometric Views

Triad[Ref 1]

[1] On the bottom-right corner of the GUI is a triad (4.4.1[2]); clicking an arrow will rotate the view such that the view is normal to that arrow ([2], next page). Moving the mouse over the negative side of an arrow will reveal a black arrow ([3], next page), which represents the negative direction of the arrow. You may click a black arrow to rotate the view.

Accompanying the triad is a small cyan sphere. When you rotate the view, the triad and the small cyan sphere will rotate accordingly. The small sphere represents a point located at an "isometric axis," which is a collection of points having the same magnitude of coordinates (but could be different in signs) in all three axes. Its initial position is (1, 1, 1). Thus, if the sphere coincides with the origin, that means your view is an isometric view [4]. When the sphere does not coincide with the origin, clicking the sphere will rotate to an isometric view [5]. ↵

Isometric View[Ref 2]

[6] As mentioned, the small cyan sphere represents an isometric direction and its initial position is (1, 1, 1). In 3D space, there are a total of 8 such directions. For example: (-1, 1, 1), (1, -1, 1), etc. These are all isometric views. When you click **Isometric View** (4.4.1[3], last page), the view will rotate to the isometric view closest to the current view. #

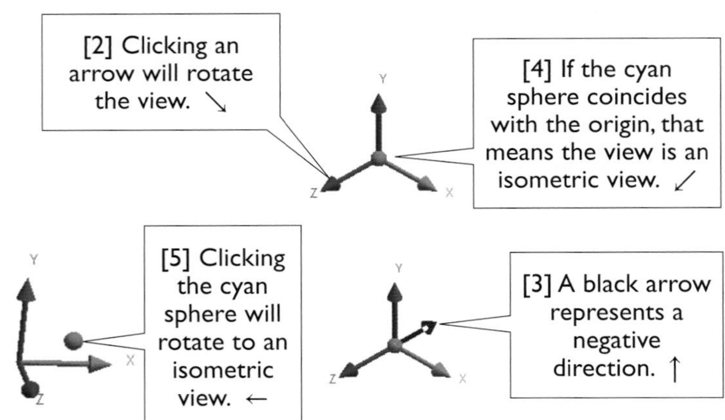

4.4.3 View Rotations[Ref 3]

Rotate with Middle Mouse Button

[1] Holding the middle mouse button down while moving the mouse around the graphic area, you can rotate the view [2]. It is simple and convenient. →

[2] Holding the middle mouse button down while moving the mouse around the graphic area, you can rotate the view. ←

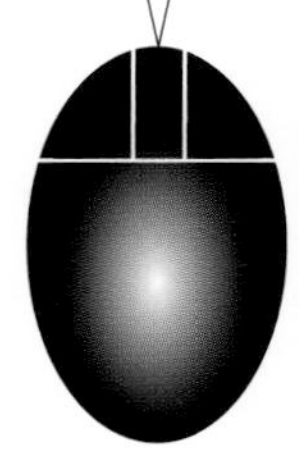

Rotate tool

[3] The **Rotate** tool (4.4.1[4], last page) gives you more controls than using the middle mouse button [1-2] for rotating the view. After activating the **Rotate** tool by clicking it, the mouse cursor becomes one of the four shapes [4-7], depending on the location of your mouse cursor as shown in [8]: free rotation [4] when the cursor is near the center of the graphics window; roll [5] when the cursor is away from the center; yaw [6] when the cursor is near the vertical edges; pitch [7] when the cursor is near the horizontal edges.

By default, the model center is the center of rotation. You can set the center of rotation (a red sphere) by clicking on the model. The red sphere will stay in the middle of the graphics window. To restore the center of rotation to the model center, click anywhere in the graphics window away from the model. This will re-center the model at the middle of the graphics window. ↓

[8] The type of rotation depends on the location of the cursor. #

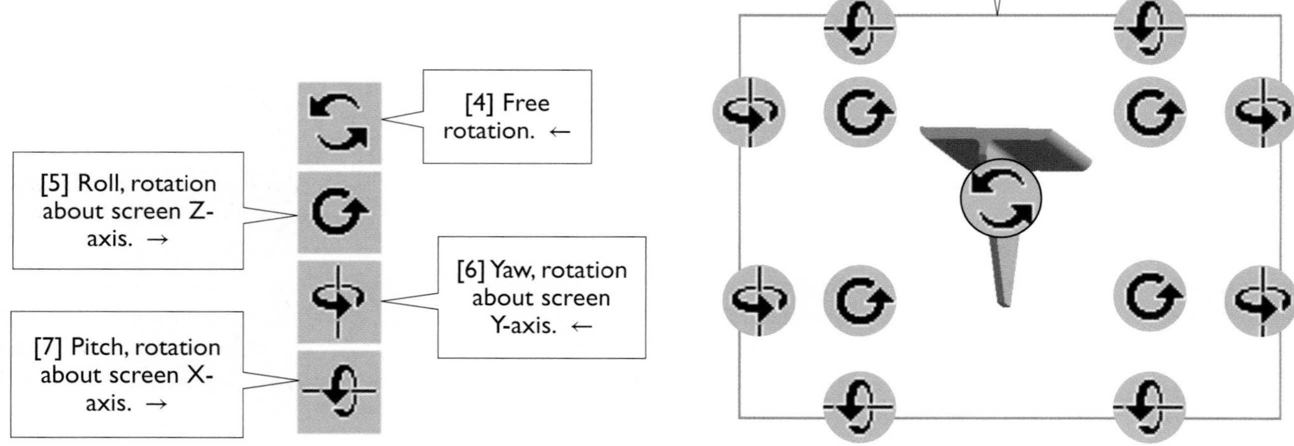

4.4.4 Mouse Cursor

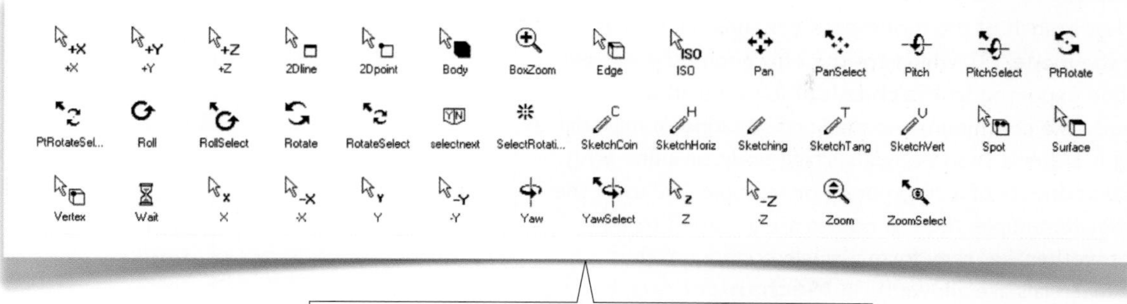

[1] The type of mouse cursor automatically changes to reflect the current operation. #

4.4.5 Selection

Selection Filters[Ref 4]

[1] By activating a selection filter (4.4.1[5], page 197), you can make one of four types of graphic entities (points, edges, faces, and bodies) selectable. Additional filters are available in the context menu (by right-clicking the graphic area) [2]. Multiple filters can be activated at the same time. →

Extend Selection[Ref 4]

[3] Using the current selection as a "seed", **Extend Selection** (4.4.1[6], page 197) allows you to include neighboring edges or faces into the selection set [4-7]. ↓

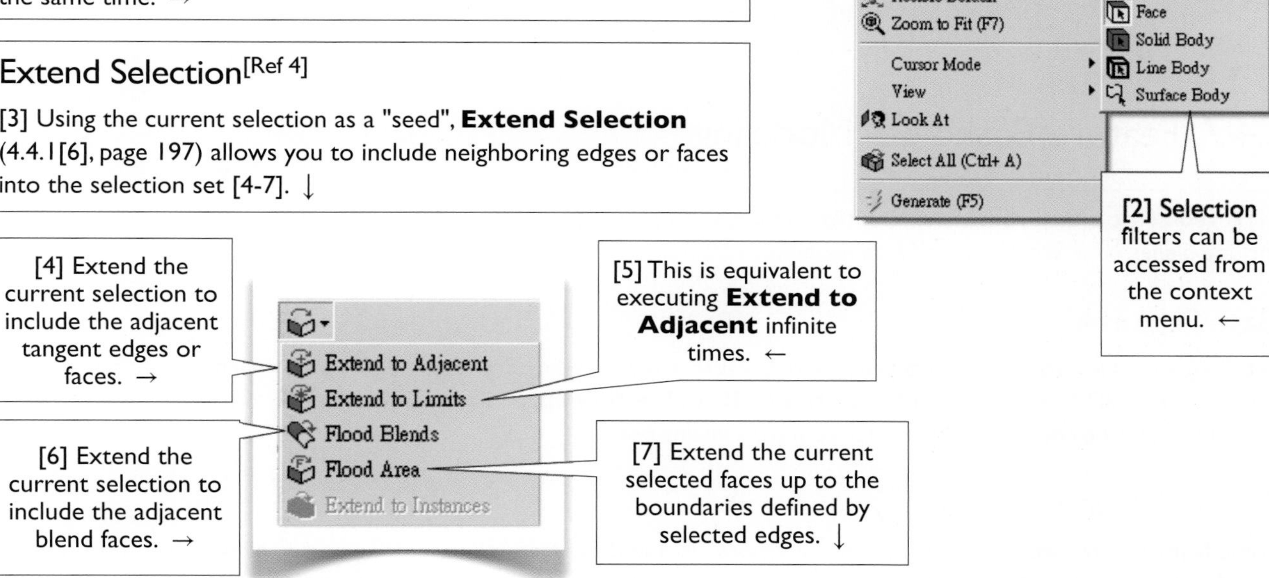

[2] **Selection** filters can be accessed from the context menu. ←

[4] Extend the current selection to include the adjacent tangent edges or faces. →

[5] This is equivalent to executing **Extend to Adjacent** infinite times. ←

[6] Extend the current selection to include the adjacent blend faces. →

[7] Extend the current selected faces up to the boundaries defined by selected edges. ↓

Selection Panes[Ref 5]

[8] When you select an entity by clicking your mouse on the model, and if more than one entity lies under the mouse cursor, the graphics window displays a stack of rectangles in the lower-left corner (4.4.1[7], page 197). The rectangles are stacked methodically with the topmost rectangle representing the most visible entity and subsequent rectangles representing entities underneath the mouse cursor, front to back. These rectangles are aliases of selectable entities, that is, highlighting and picking these rectangles are identical to that for the entities. When you move the mouse over these rectangles, the mouse cursor changes to show the type of the entity. #

4.4.6 Bodies and Parts[Ref 6]

[1] The last branch of the model tree contains the bodies and parts of the model [2], which are the only geometric entities that will be exported to **Mechanical** for simulations.

A body is a continuum and made of one kind of material. A 3D body is either a solid body, a surface body, or a line body.

A part consists of a single body or multiple bodies of the same type. If multiple, then all bodies are assumed to be bonded together; i.e., they form a single continuum (but multiple materials are allowed). In **Mechanical**, each body is meshed independently. If a part consists of multiple bodies, constraints are imposed on the boundaries of the contacting bodies to guarantee the compatibility between the boundaries.

A model may consist of one or more parts. Since each part is meshed independently, mesh at the boundaries between parts is not compatible. In **Mechanical**, connections[Ref 7] (e.g., contacts, joints) among parts must be established to complete a simulation model. →

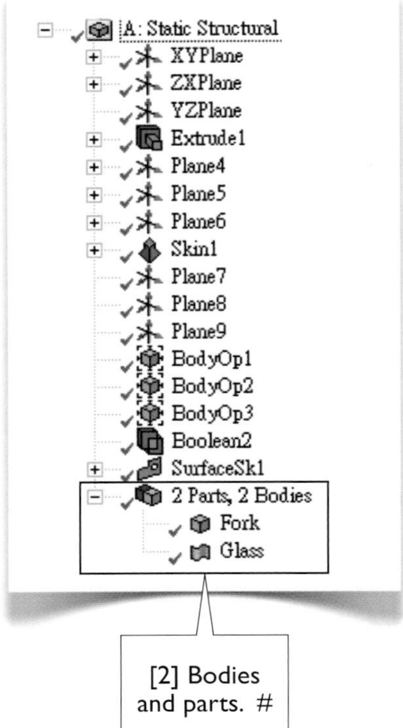

[2] Bodies and parts. #

4.4.7 Feature-Based 3D Modeling

[1] A geometric model consists of features; an object in a model tree is called a feature of the model. The features include *planes*, *base features*, *placed features*, etc.

Base Features

A base features is also called a *sketched feature* because it is created by drawing one or more sketches, and then "growing" to a 3D feature by means of **Extrude**, **Revolve**, **Sweep**, or **Skin/Loft**. A newly created base feature can add material to or subtract material from the existing bodies.

Placed Features

Some features have predefined shapes and behaviors. To add these features to existing bodies, all we have to do is to specify where we want to place these features, along with a few other settings. Therefore, these features are called *placed features*, e.g., **Blend**, **Chamfer**, **Thin/Surface**, **Slice**, etc. #

4.4.8 Base Features[Ref 8]

[1] A base feature is created by drawing one or more sketches, and then "growing" to a 3D feature by means of **Extrude**, **Revolve**, **Sweep**, or **Skin/Loft**.

Extrude

The tool is used to extrude a sketch along its normal direction to create a 3D body. The extrusion may be symmetric or asymmetric to the sketching plane. The extrusion depth may be a fixed value, through all bodies (used only for cutting the material), up to a face, or up to a surface. A *face* is a bounded region and has a finite area while a *surface* is an unbounded region and has infinite area. A surface is often the extension of a face.

Revolve

The tool is used to revolve a sketch about an axis to create a 3D body. An angle of revolution can be specified.

Sweep

Sweep can be thought of as a generalization of **Extrude**. The tool is used to sweep a profile along a path to create a 3D body. Both the profile and the path must be defined using sketches.

Skin/Loft

Skin/Loft can be thought of as a generalization of **Sweep**. It takes a series of profiles to create a 3D body by fitting through them. The profiles must be defined using sketches. #

4.4.9 Placed Features[Ref 8]

[1] A placed feature has predefined shape and behavior; it can be added to an existing body.

Thin/Surface

The tool is used to convert a solid into a thin solid body or a surface body. Typically, you will select one or more faces to remove, and then specify a thickness. If the thickness is a positive value, then a thin solid body is created. If the thickness is zero, then a surface body is created.

Blend

The tool is used to create *rounds* or *fillets* on edges, or on vertices. The radius of the rounds or fillets may be fixed or variable.

Chamfer

The tool is used to create chamfer faces on edges. #

References

1. ANSYS Help//DesignModeler//ANSYS DesignModeler User's Guide//Viewing//Model Appearance Controls//Triad
2. ANSYS Help//DesignModeler//ANSYS/DesignModeler User's Guide//Viewing//Rotation Modes Toolbar//Isometric View
3. ANSYS Help//Mechanical Application//ANSYS Mechanical User's Guide//Application Interface//Graphic Selection//Controlling the Viewing Orientation
4. ANSYS Help//DesignModeler//ANSYS DesignModeler User's Guide//Selection//Selection Toolbar
5. ANSYS Help//DesignModeler//ANSYS DesignModeler User's Guide//Selection//Graphical Selection//Depth Picking
6. ANSYS Help//DesignModeler//ANSYS DesignModeler User's Guide//3D Modeling//Bodies and Parts
7. ANSYS Help//Mechanical Application//ANSYS Mechanical User's Guide//Objects Reference//Connections
8. ANSYS Help//DesignModeler//ANSYS DesignModeler User's Guide//3D Modeling//3D Features

Section 4.5

LCD Display Support

4.5.1 About the LCD Display Support

[1] The LCD Display support is made of an ABS (acrylonitrile-butadiene-styrene) plastic. The thickness is 3 mm [2]. Details of the hinge [3] are not shown on this page but will be illustrated in 4.5.4 (pages 205-206).

This model will be used in Section 5.4 for a static structural simulation to assess the deformation and stress under a design load. →

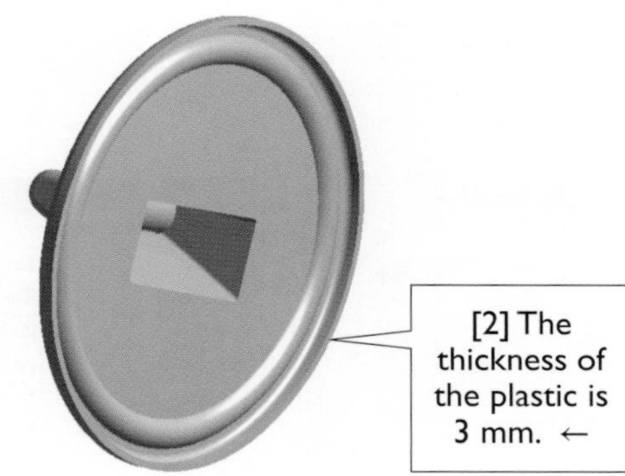

[2] The thickness of the plastic is 3 mm. ←

[3] Details of the hinge design will be illustrated in 4.5.4 (pages 205-206). #

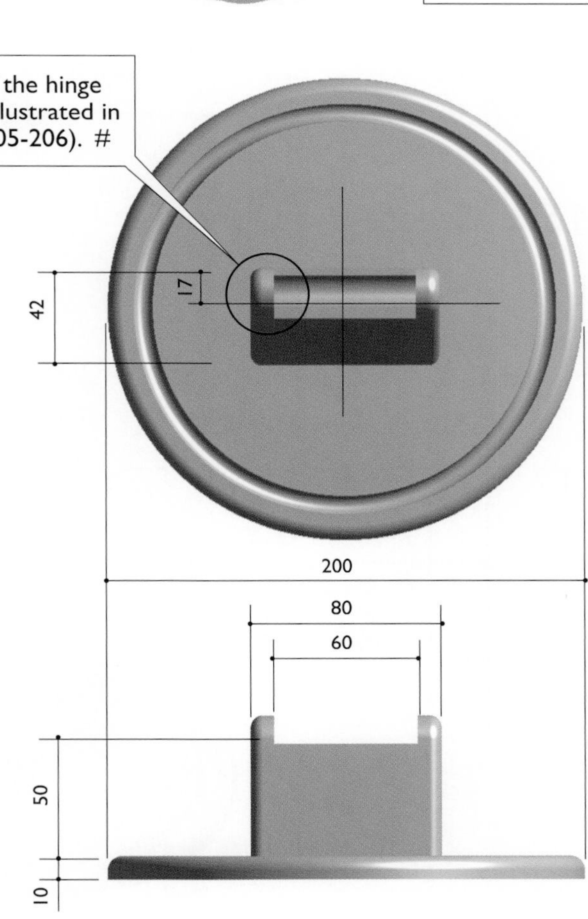

4.5.2 Create the Base

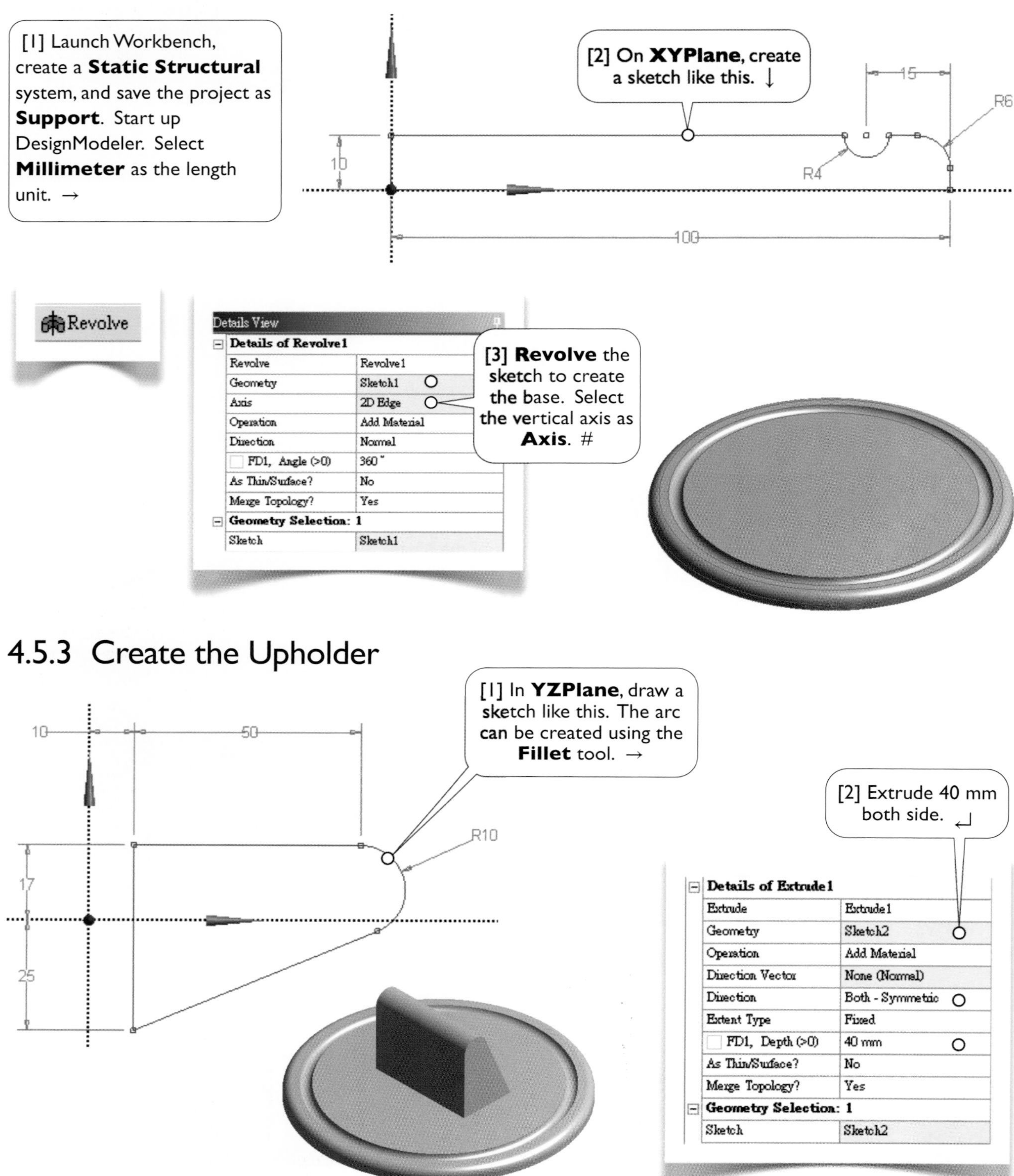

[1] Launch Workbench, create a **Static Structural** system, and save the project as **Support**. Start up DesignModeler. Select **Millimeter** as the length unit. →

[2] On **XYPlane**, create a sketch like this. ↓

[3] **Revolve** the sketch to create the base. Select the vertical axis as **Axis**. #

Details View

Details of Revolve1	
Revolve	Revolve1
Geometry	Sketch1
Axis	2D Edge
Operation	Add Material
Direction	Normal
FD1, Angle (>0)	360 °
As Thin/Surface?	No
Merge Topology?	Yes
Geometry Selection: 1	
Sketch	Sketch1

4.5.3 Create the Upholder

[1] In **YZPlane**, draw a sketch like this. The arc can be created using the **Fillet** tool. →

[2] Extrude 40 mm both side. ↵

Details of Extrude1	
Extrude	Extrude1
Geometry	Sketch2
Operation	Add Material
Direction Vector	None (Normal)
Direction	Both - Symmetric
Extent Type	Fixed
FD1, Depth (>0)	40 mm
As Thin/Surface?	No
Merge Topology?	Yes
Geometry Selection: 1	
Sketch	Sketch2

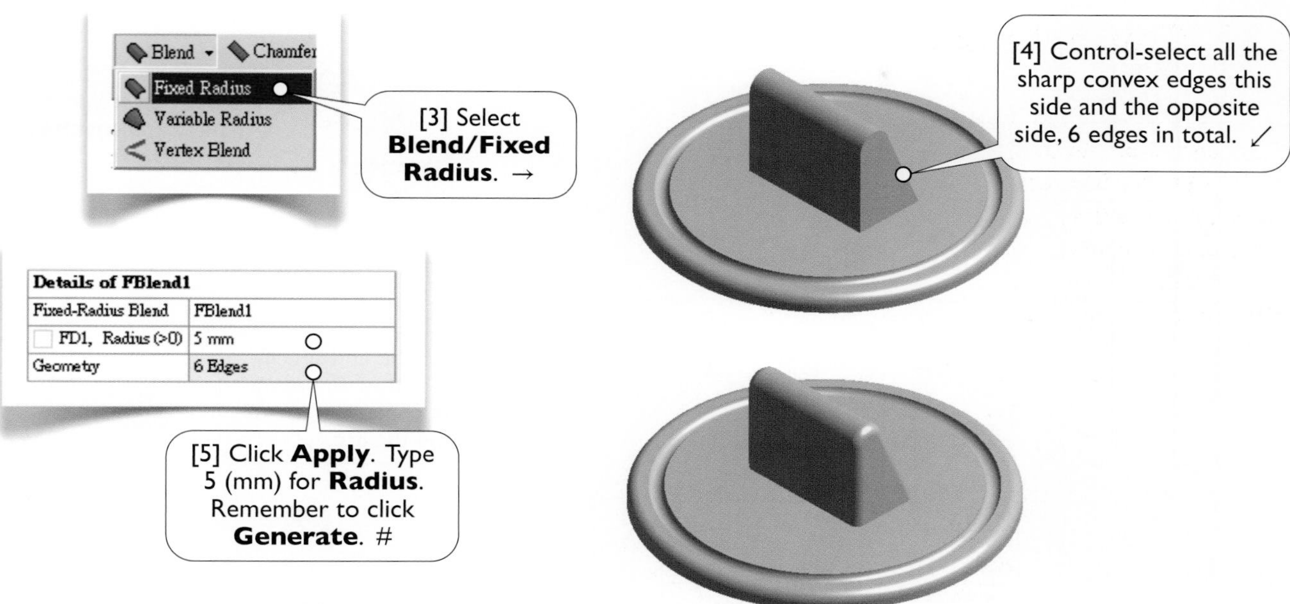

4.5.4 Create Hinge

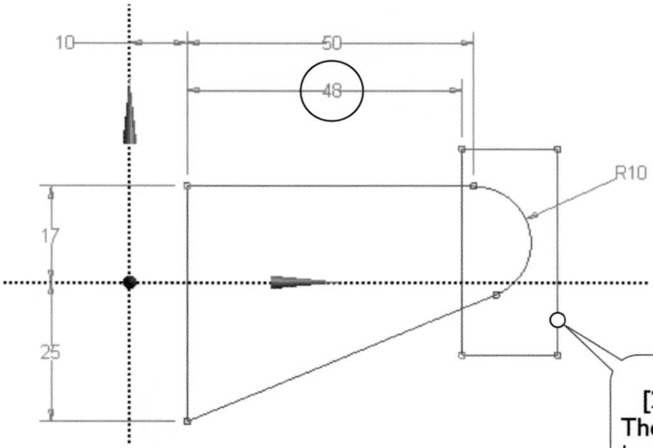

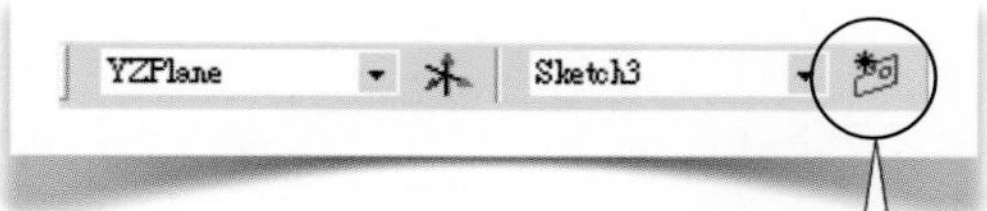

[1] Click to create a new sketch (**Sketch3**) in **YZPlane**. ↙

YZPlane | Sketch3

[2] Draw a rectangle like this. The size is not important but the location of the left edge (48 mm) must be specified. ↙

10 | 50 | 48 | R10 | 17 | 25

Details of Extrude2	
Extrude	Extrude2
Geometry	Sketch3
Operation	Cut Material
Direction Vector	None (Normal)
Direction	Both - Symmetric
Extent Type	Fixed
FD1, Depth (>0)	30 mm
As Thin/Surface?	No
Target Bodies	All Bodies
Merge Topology?	Yes
Geometry Selection: 1	
Sketch	Sketch3

[3] Extrude to cut material, 30 mm both sides. Remember to click **Generate**. ↵

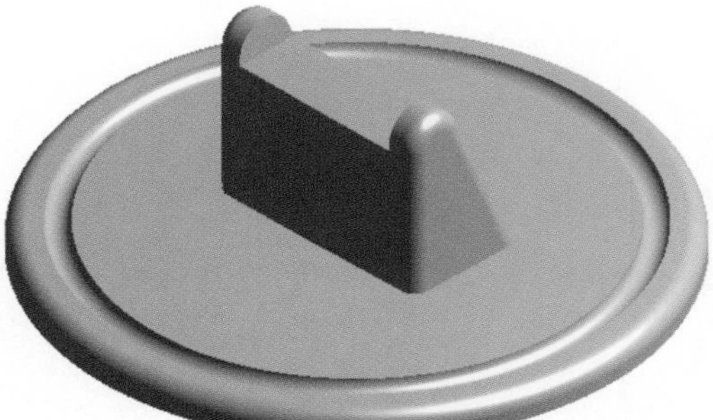

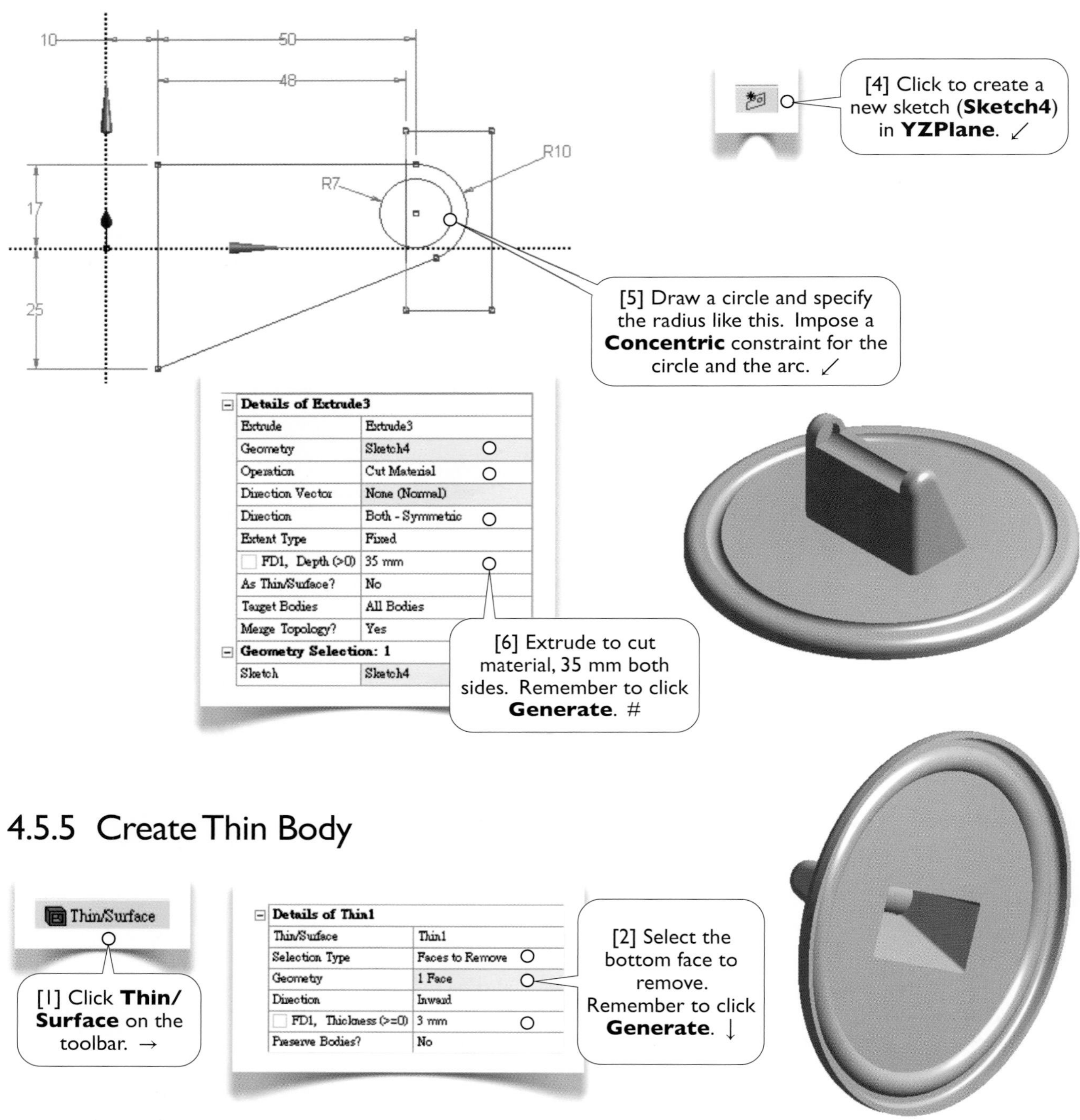

Details of Extrude3	
Extrude	Extrude3
Geometry	Sketch4
Operation	Cut Material
Direction Vector	None (Normal)
Direction	Both - Symmetric
Extent Type	Fixed
FD1, Depth (>0)	35 mm
As Thin/Surface?	No
Target Bodies	All Bodies
Merge Topology?	Yes
Geometry Selection: 1	
Sketch	Sketch4

4.5.5 Create Thin Body

Details of Thin1	
Thin/Surface	Thin1
Selection Type	Faces to Remove
Geometry	1 Face
Direction	Inward
FD1, Thickness (>=0)	3 mm
Preserve Bodies?	No

Wrap Up

[3] Close DesignModeler, save the project, and exit Workbench. #

Section 4.6

Review

4.6.1 Keywords

Choose a letter for each keyword from the list of descriptions

1. ( ) Base Features
2. ( ) Body
3. ( ) Connections
4. ( ) Extend Selection
5. ( ) Features
6. ( ) Isotropic View
7. ( ) Part
8. ( ) Pitch
9. ( ) Placed Features
10. ( ) Plane Outline
11. ( ) Plane Coordinate System
12. ( ) Roll
13. ( ) Selection Filters
14. ( ) Selection Panes
15. ( ) Yaw

Answers:

1. (N) 2. (J) 3. (L) 4. (H) 5. (M) 6. (C) 7. (K) 8. (F)
9. (O) 10. (B) 11. (A) 12. (D) 13. (G) 14. (I) 15. (E)

List of Descriptions

(A) The coordinate system attached to a plane. Its Z-axis always points out of the plane.

(B) A plane may be created by deriving from a face. In such cases, the plane may be designated as **Outline Plane** (i.e., a plane with boundary). You can draw entities beyond the boundary. The boundary itself is not a geometric entity. The plane boundary is used as datum edges, or as a source of replication/duplication.

(C) A model view in which the view direction follows an isometric axis, which forms the same angles with the three principal axes.

(D) Rotate the model about the screen Z-axis.

(E) Rotate the model about the screen Y-axis.

(F) Rotate the model about the screen X-axis.

(G) A tool used to make a specific type of entity selectable. Multiple filters can be activated at the same time.

(H) Using the current selection as seed, these tools allow you to extend the seed to include additional edges or faces into the selection set.

(I) Rectangles appear in the lower-left corner of the graphics window. These rectangles are aliases of selectable entities, that is, highlighting and picking these rectangles are identical and synchronized for the selectable entities.

(J) It is a continuum and made of one kind of material. A 3D body is either a solid body, a surface body, or a line body.

(K) A collection of same type of bodies. If multiple, then all bodies are assumed to be bonded together i.e., they form a single continuum (but multiple materials are allowed). In **Mechanical**, it is meshed independently.

(L) In **Mechanical**, they are kinematic relations between parts. Examples are contacts, joints, etc.

(M) They are the building block of a 3D body. Examples include planes, base features, and placed features.

(N) They are also called sketched features; they are created by first drawing one or more sketches, and then "growing" to 3D features by means of extrusion, revolution, sweeping, or lofting. A newly created one can add to or subtract material from the existing bodies.

(O) They have predefined shapes and behaviors. To add these features to existing bodies, all we have to do is to specify where we want to place these features, along with a few other settings.

Chapter 5

3D Simulations

Many concepts and techniques in 2D simulations (Chapter 3) are applicable to 3D simulations. The problem sizes (number of nodes and elements) in 3D cases are usually larger and the geometries are usually more complicated than those in 2D cases. That implies more computing and engineering time. On the other hand, because we've been accustomed to the 3D world, the 3D simulations are more intuitive to us. As a result, newcomers often stick to 3D ways of thinking and forget that a problem often can be modeled as 2D. Remember that, if a problem can be reduced to a 2D, you have no reason to go for a 3D simulation.

Purpose of This Chapter

This chapter focuses on 3D simulations with solid models. Problems involving surface models and line models are discussed in Chapters 6 and 7, respectively. Like the 2D simulations in Chapter 3, this chapter focuses on linear static structural simulations. Dynamic and nonlinear simulations will be discussed in later chapters.

About Each Section

We will conduct two step-by-step examples in the first two sections. Section 5.1 is a simple example, serving as an introductory tutorial. Section 5.2 is a more involved example. Section 5.3 provides a systematic discussion trying to complement what was missed in the first two sections. Section 5.4 is an additional exercise. All exercises in this chapter use models created in Chapter 4.

Section 5.1

Beam Bracket

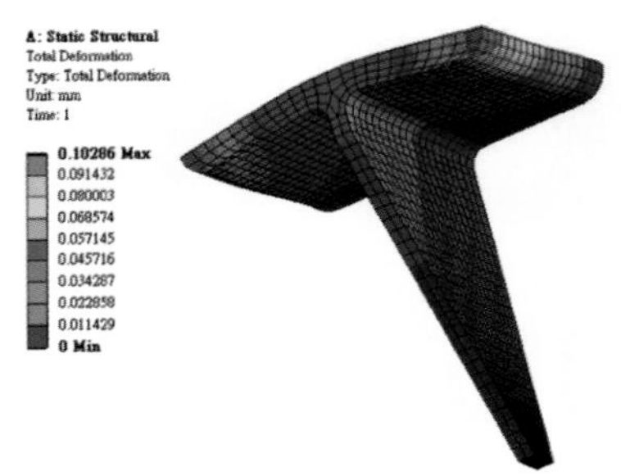

5.1.1 About the Beam Bracket

[1] In Section 4.1, we created a 3D solid model for a beam bracket. In this section, we will use the model for a static structural simulation. Besides the geometry, we need extra information for the simulation, which can be summarized into two categories: material properties and the environment conditions (i.e., loads and supports).

The beam bracket is made of structural steel [2] with a Young's modulus of 200 GPa, a Poisson's ratio of 0.3, and a yield strength of 250 MPa. The yield strength is used to assess safety factors.

The load is 27 kN uniformly distributed over the seat plate [3]. It was determined by an analysis of the entire structure.

As to the support conditions, we assume the beam bracket's back face is rigidly welded on a steel column; i.e., a fixed support [4]. →

[3] The bracket is designed to withstand a load of 27 kN uniformly distributed over the seat plate. ↓

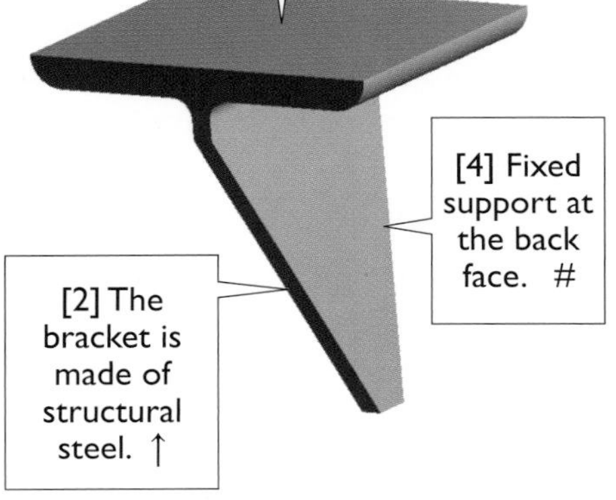

[4] Fixed support at the back face. #

[2] The bracket is made of structural steel. ↑

5.1.2 Open the Project **Bracket**

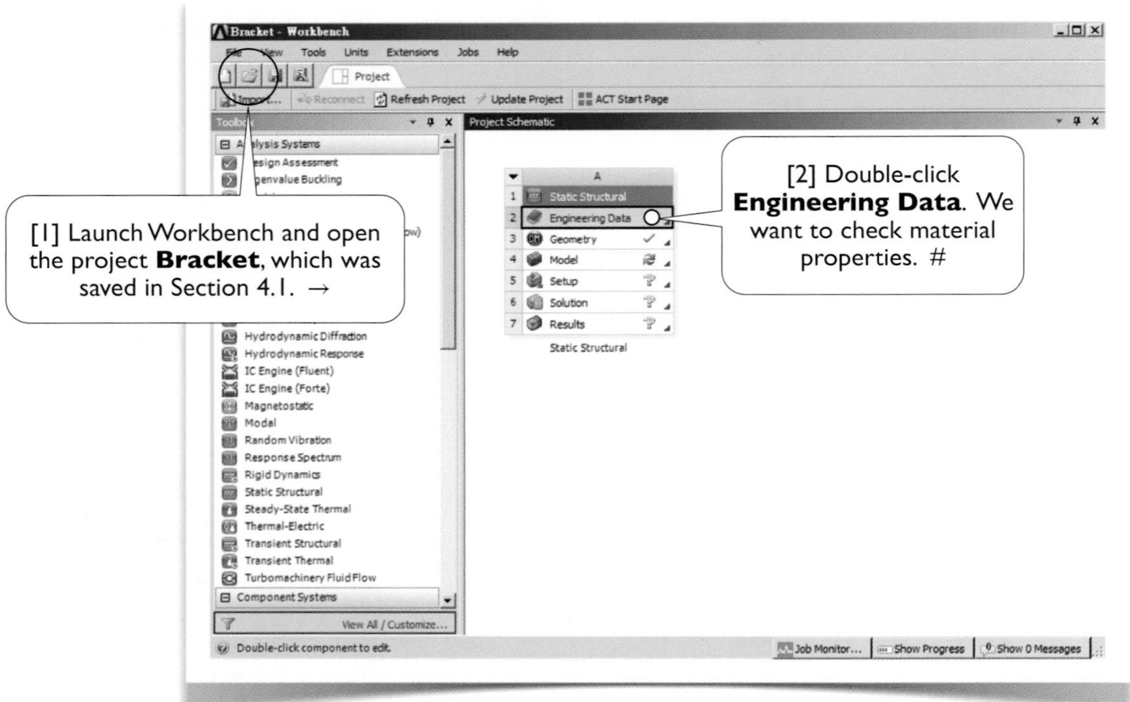

5.1.3 Check Material in **Engineering Data**

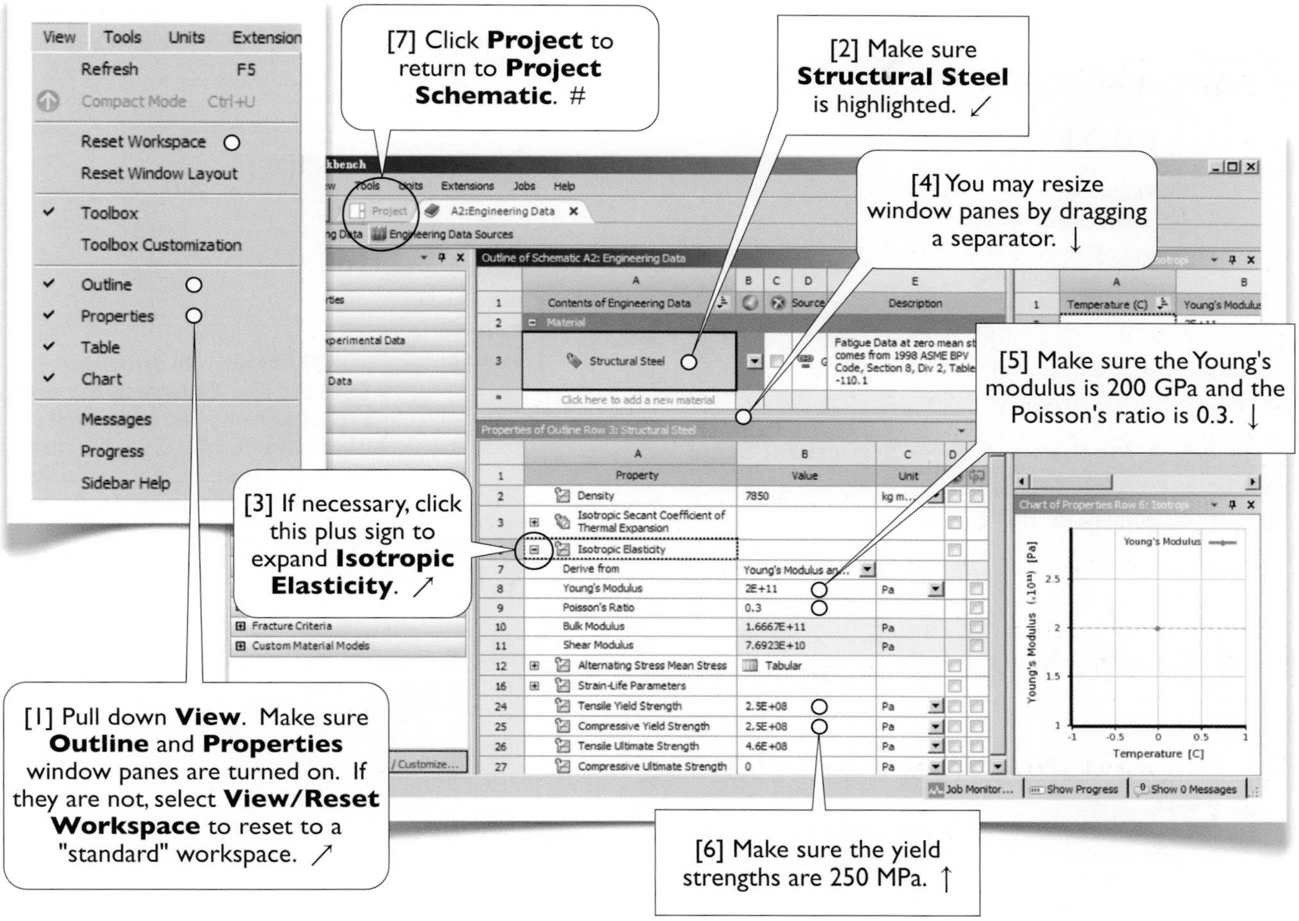

[7] Click **Project** to return to **Project Schematic**. #
[2] Make sure **Structural Steel** is highlighted.
[4] You may resize window panes by dragging a separator.
[5] Make sure the Young's modulus is 200 GPa and the Poisson's ratio is 0.3.
[3] If necessary, click this plus sign to expand **Isotropic Elasticity**.
[1] Pull down **View**. Make sure **Outline** and **Properties** window panes are turned on. If they are not, select **View/Reset Workspace** to reset to a "standard" workspace.
[6] Make sure the yield strengths are 250 MPa.
View Tools Units Extension
Refresh F5
Compact Mode Ctrl+U
Reset Workspace
Reset Window Layout
Toolbox
Toolbox Customization
Outline
Properties
Table
Chart
Messages
Progress
Sidebar Help
Outline of Schematic A2: Engineering Data
Contents of Engineering Data
Structural Steel
Properties of Outline Row 3: Structural Steel
Density 7850
Isotropic Elasticity
Young's Modulus 2E+11 Pa
Poisson's Ratio 0.3
Bulk Modulus 1.6667E+11 Pa
Shear Modulus 7.6923E+10 Pa
Tensile Yield Strength 2.5E+08 Pa
Compressive Yield Strength 2.5E+08 Pa
Tensile Ultimate Strength 4.6E+08 Pa
Compressive Ultimate Strength 0 Pa

5.1.4 Start Up **Mechanical**

[1] Double-click to start up **Mechanical**.
Static Structural
Engineering Data
Geometry
Model
Setup
Solution
Results
Static Structural

Units Tools Help Solve
Metric (m, kg, N, s, V, A)
Metric (cm, g, dyne, s, V, A)
Metric (mm, kg, N, s, mV, mA)
Metric (mm, t, N, s, mV, mA)
Metric (mm, dat, N, s, mV, mA)
Metric (μm, kg, μN, s, V, mA)
U.S. Customary (ft, lbm, lbf, °F, s, V, A)
U.S. Customary (in, lbm, lbf, °F, s, V, A)
Degrees
Radians
rad/s
RPM
Celsius (For Metric Systems)
Kelvin (For Metric Systems)
[2] In **Mechanical**, make sure the unit system is **mm-kg-N-s**. #

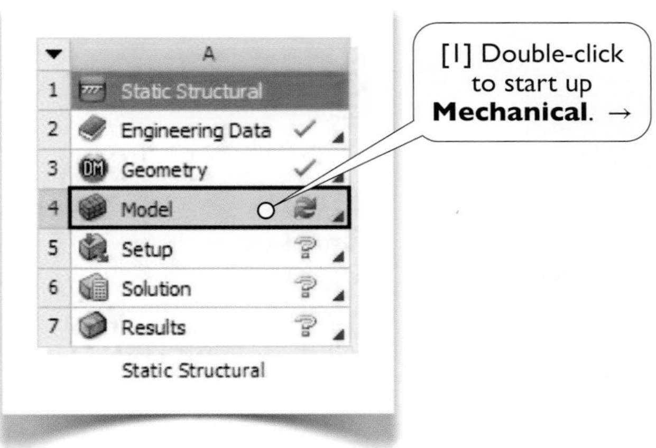

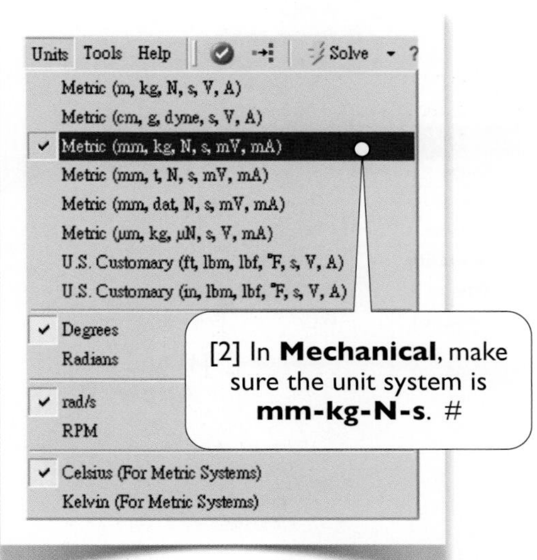

5.1.5 Check Material Assignment

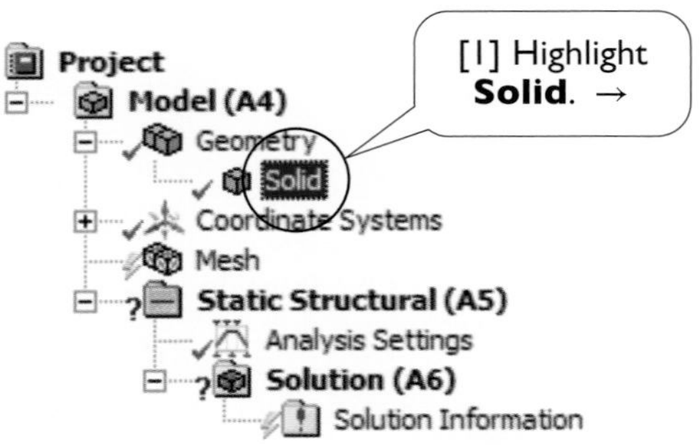

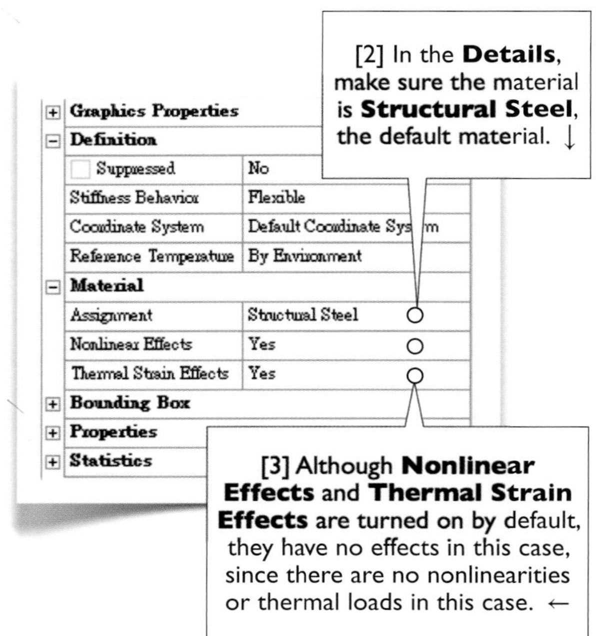

Material Assignment

[4] By default, **Mechanical** assigns **Structural Steel** for each body; therefore, we don't need to do anything about the material assignment in this case. #

5.1.6 Specify Support

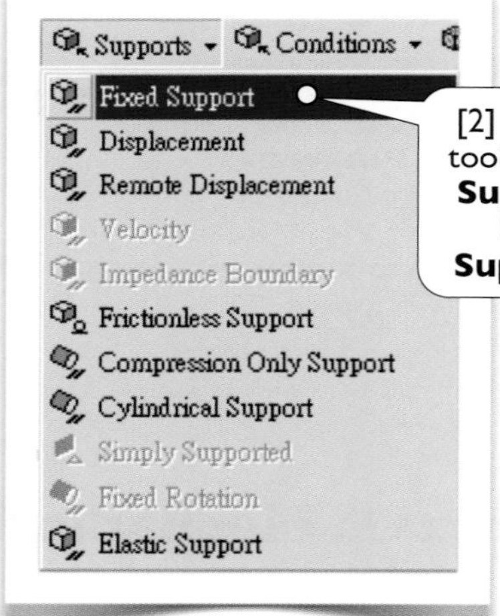

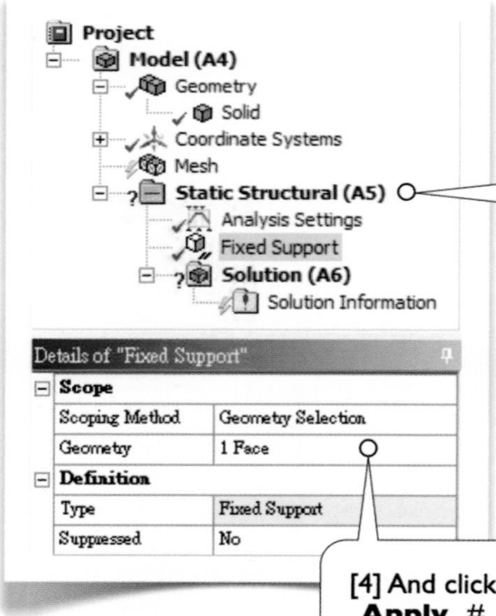

[1] Highlight **Static Structural**. →

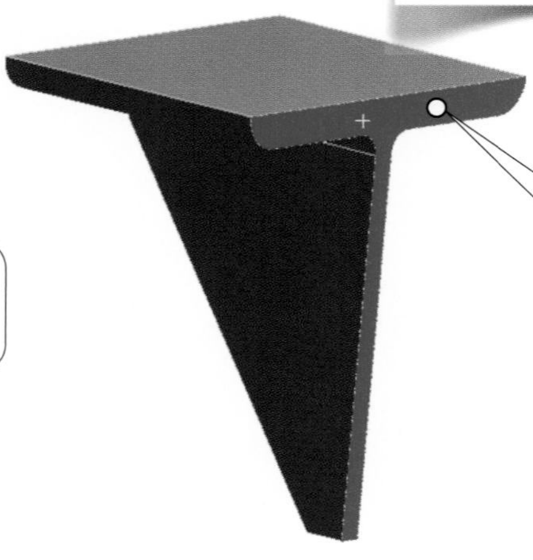

[2] From the toolbar, select **Supports/ Fixed Support**. ↓

[3] Rotate the model and select this face. ←

[4] And click **Apply**. #

5.1.7 Apply Force

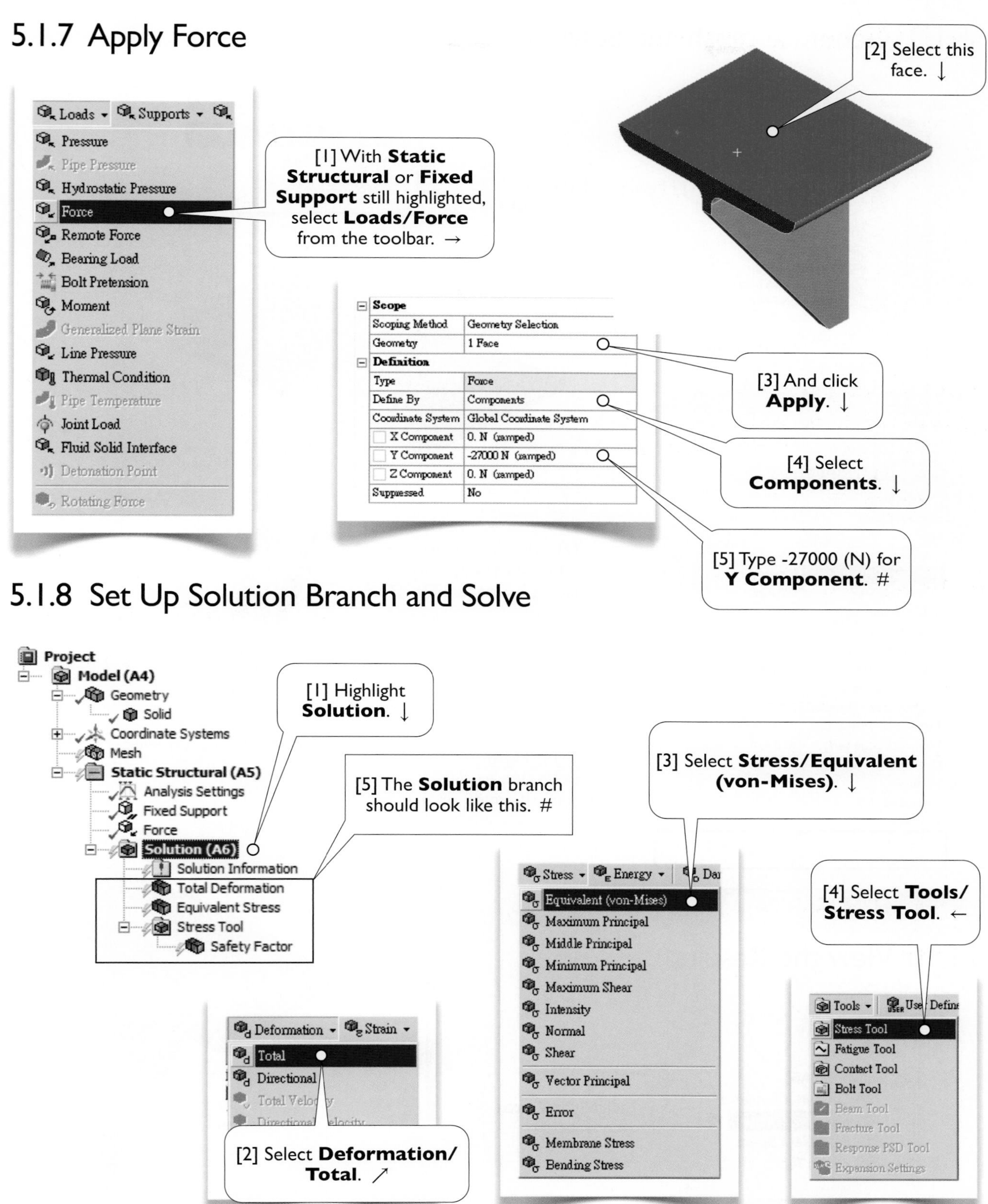

5.1.8 Set Up Solution Branch and Solve

5.1.9 Generate Mesh and Solve the Model

5.1.10 View the Results

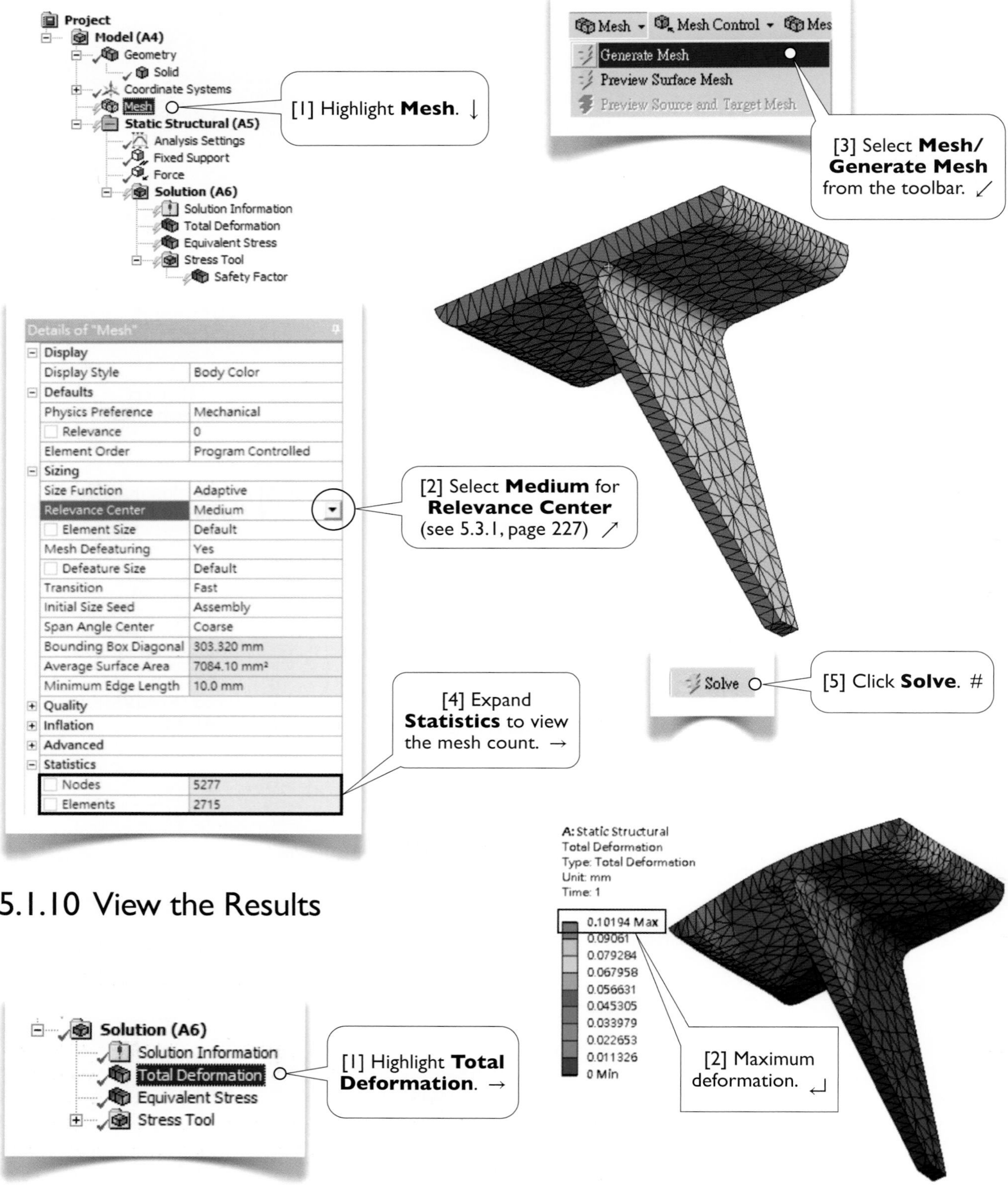

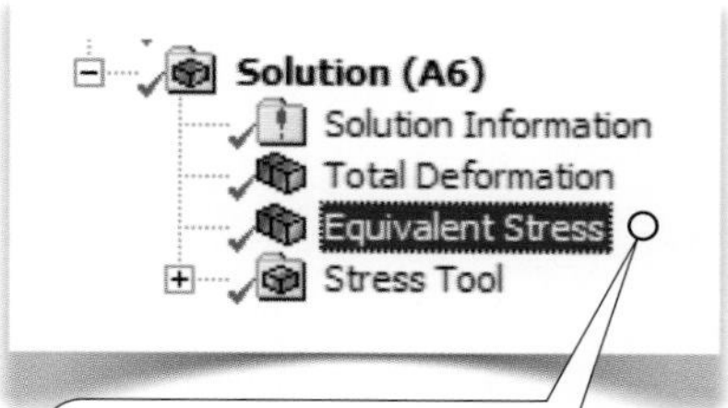

[3] Highlight **Equivalent Stress**. →

[4] Maximum von Mises stress. ↓

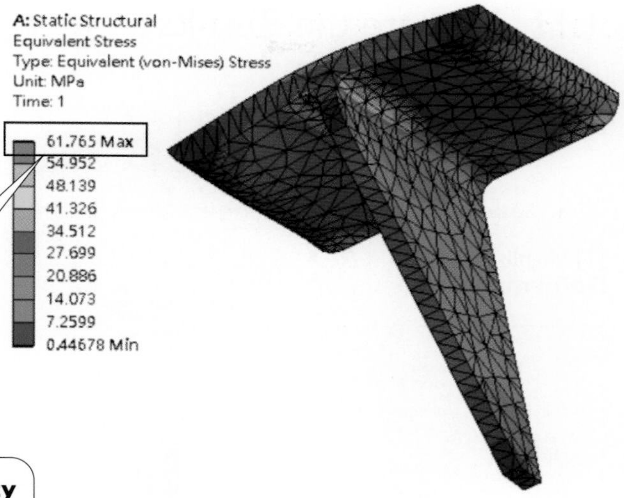

A: Static Structural
Equivalent Stress
Type: Equivalent (von-Mises) Stress
Unit: MPa
Time: 1

61.765 Max
54.952
48.139
41.326
34.512
27.699
20.886
14.073
7.2599
0.44678 Min

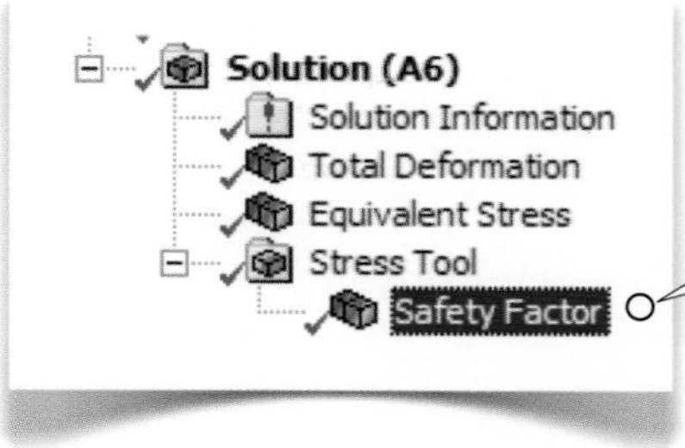

[5] Highlight **Safety Factor**. ↓

[6] Minimum safety factor. ↓

A: Static Structural
Safety Factor
Type: Safety Factor
Time: 1

15 Max
10
4.0476 Min
1
0

Safety Factors

[7] If you highlight **Stress Tool** in the project tree [8], the details view will show you how the safety factors are calculated [9]. In this case, the safety factor is the ratio between the tensile yield strength (250 MPa, see 5.1.3[6], page 211) and the calculated equivalent stress [3-4]. For example, at the most critical region, the equivalent stress is 61.765 MPa [4], therefore, the safety factor is 4.0476 (= 250/61.765) [6].

You can change the way safety factors are calculated by changing the settings in the details view [9]. ↓

[9] Details of **Stress Tool** shows how the safety factors are calculated. #

5.1.11 Animate the Results

5.1.12 Evaluate Structural Error

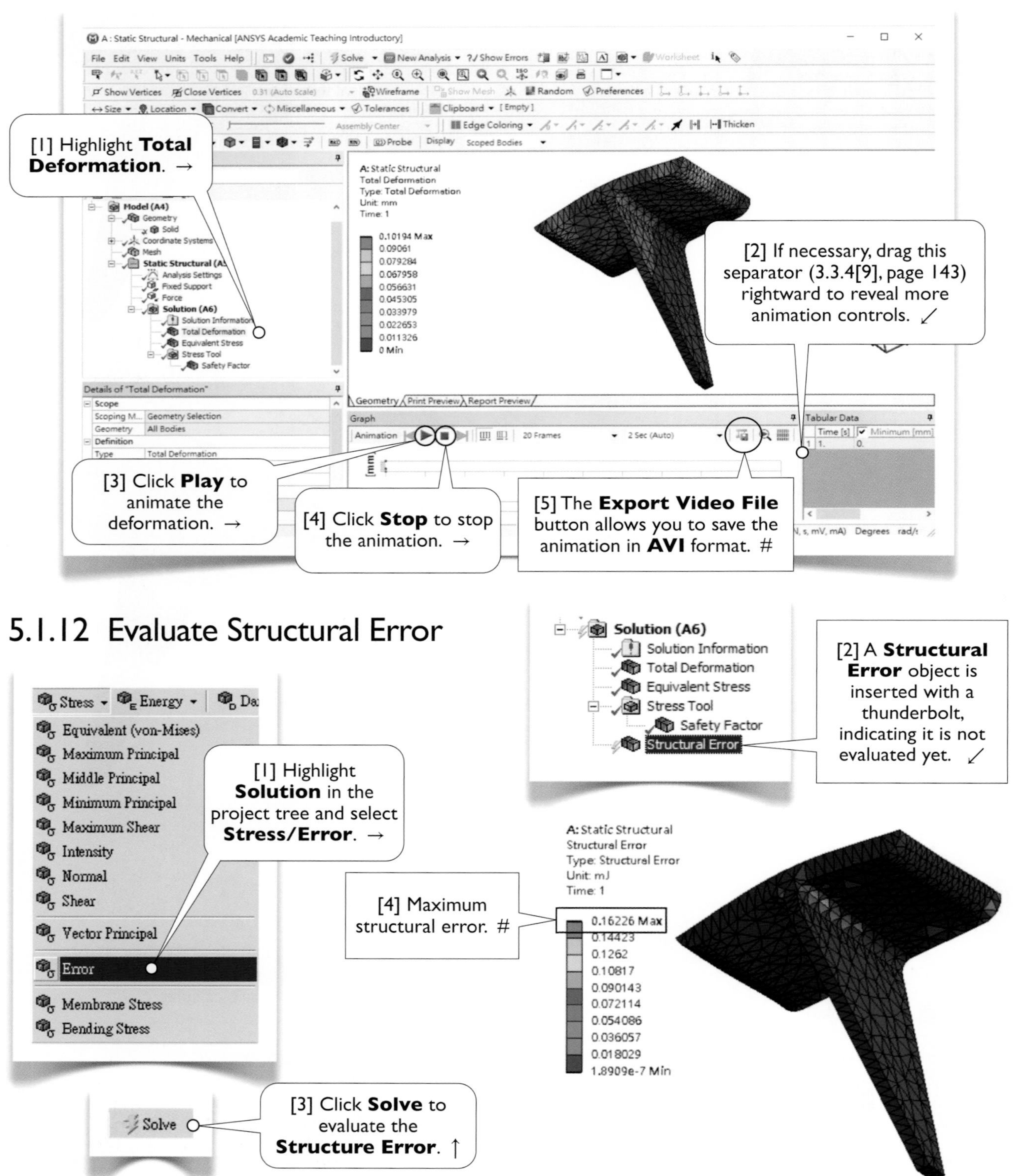

5.1.13 Improve Mesh Quality

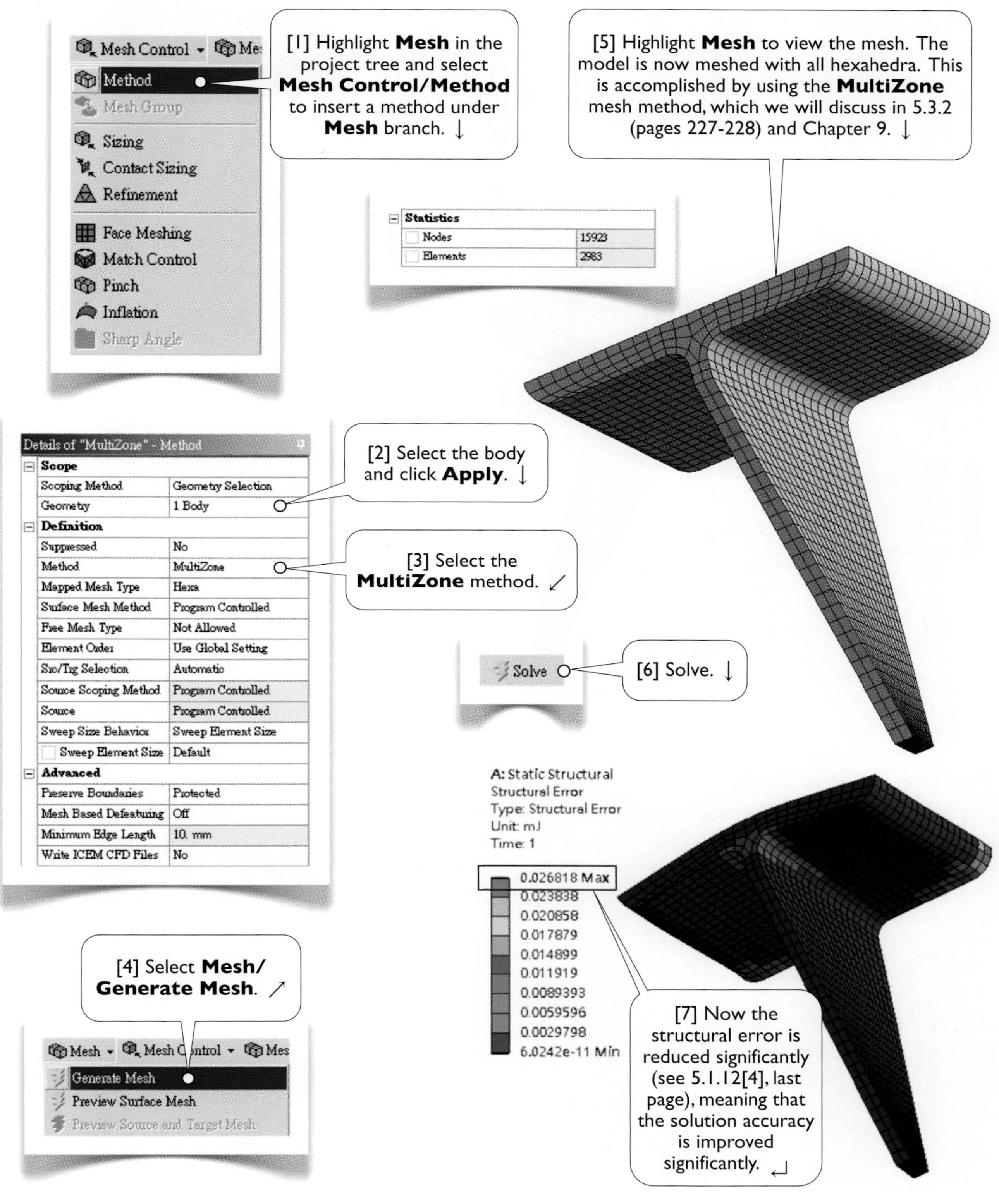

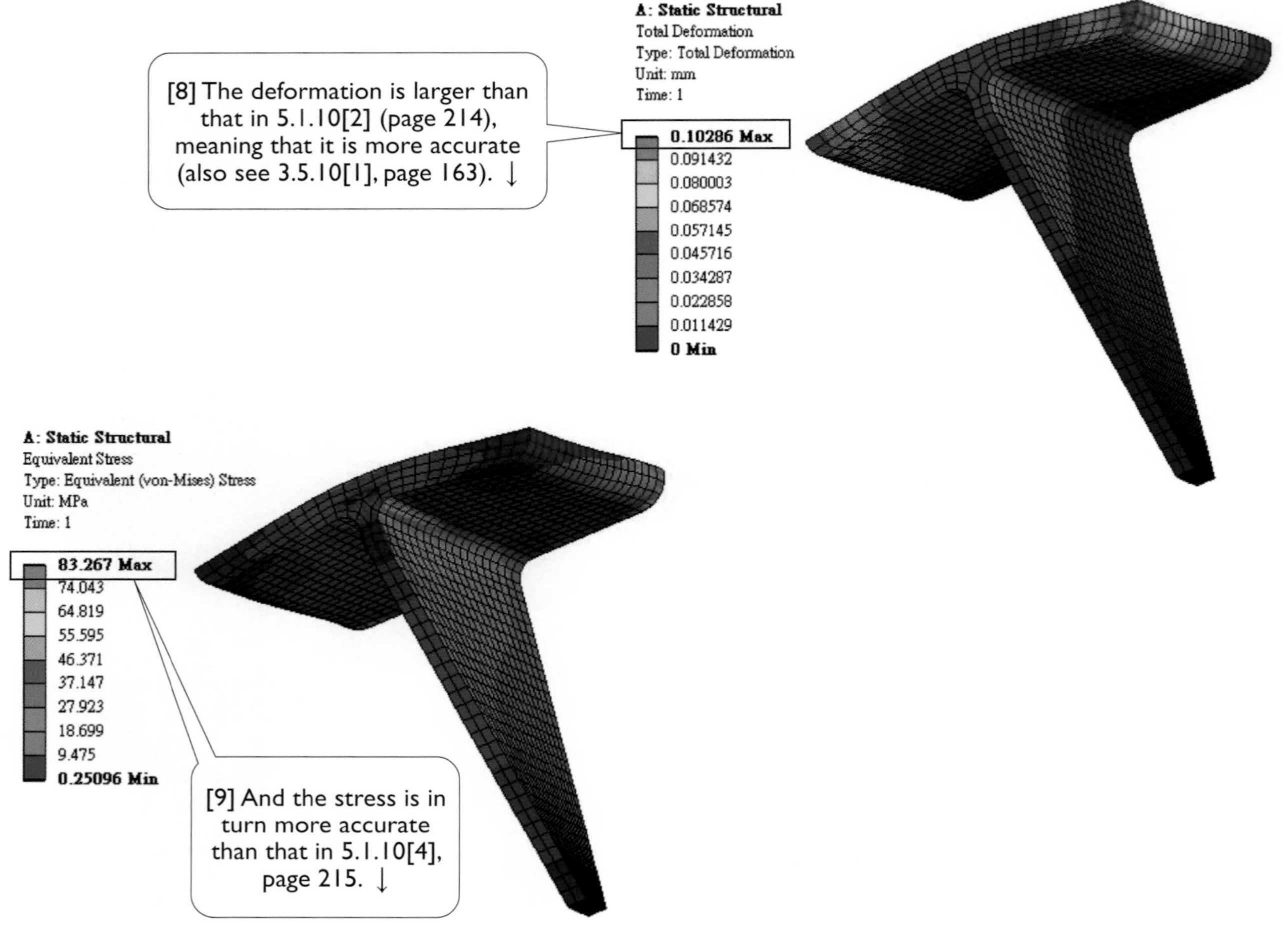

Meshing 3D Solid Bodies

[10] Solution accuracy depends not only on mesh density but also on mesh quality. In nonlinear problems, poorer mesh quality often leads to more computing time or even failure of finding a solution.

Achieving a high mesh quality is, however, not trivial. In this section, we've used a new meshing technology, namely the multi-zone method. It will be conceptually introduced in Section 5.3. Various meshing methods will be introduced in Chapter 9. ↓

Wrap Up

[11] Close **Mechanical**, save the project, and exit Workbench. #

Section 5.2

Cover of Pressure Cylinder

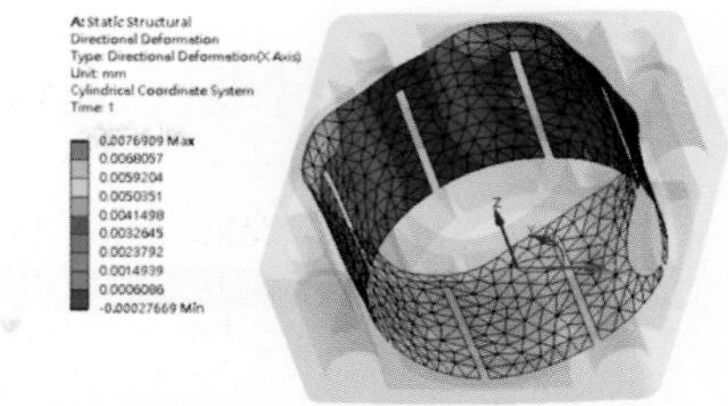

5.2.1 About the Cylinder Cover

[1] In Section 4.2, we created a 3D solid model for the cover of a pressure cylinder, which is designed to hold a pressure of 0.5 MPa. In this section, we will use this model for a static structural simulation.

The cover was originally made of aluminum alloy. The purpose of the simulation is to assess the possibility of replacing the aluminum alloy with a new type of engineering plastic. The engineering plastic has a Young's modulus of 22 GPa and a Poisson's ratio of 0.3.

The engineers are concerned about the deformation, specifically the circularity of the internal surface that encompasses the pressure cylinder [2]. The circularity of a cylindrical surface can be defined as the difference between the radii of its circumscribed circle and its inscribed circle. It is required that the circularity should be less than 10 micrometers; excess of circularity beyond this value may impair the tightness and cause a leakage of gas.

The unit system used in this exercise is **mm-kg-N-s**. →

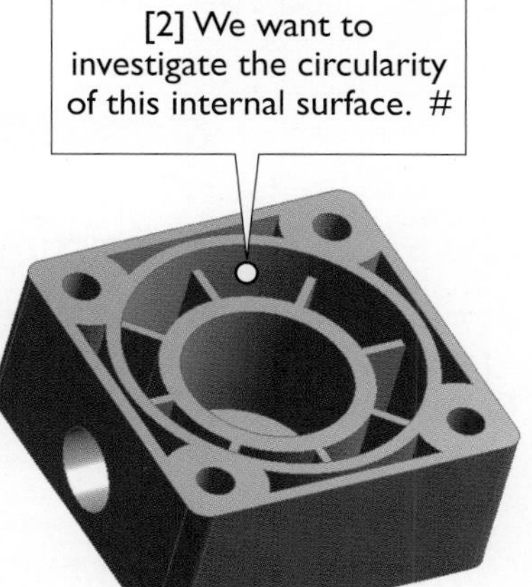

5.2.2 Open the Project **Cover**

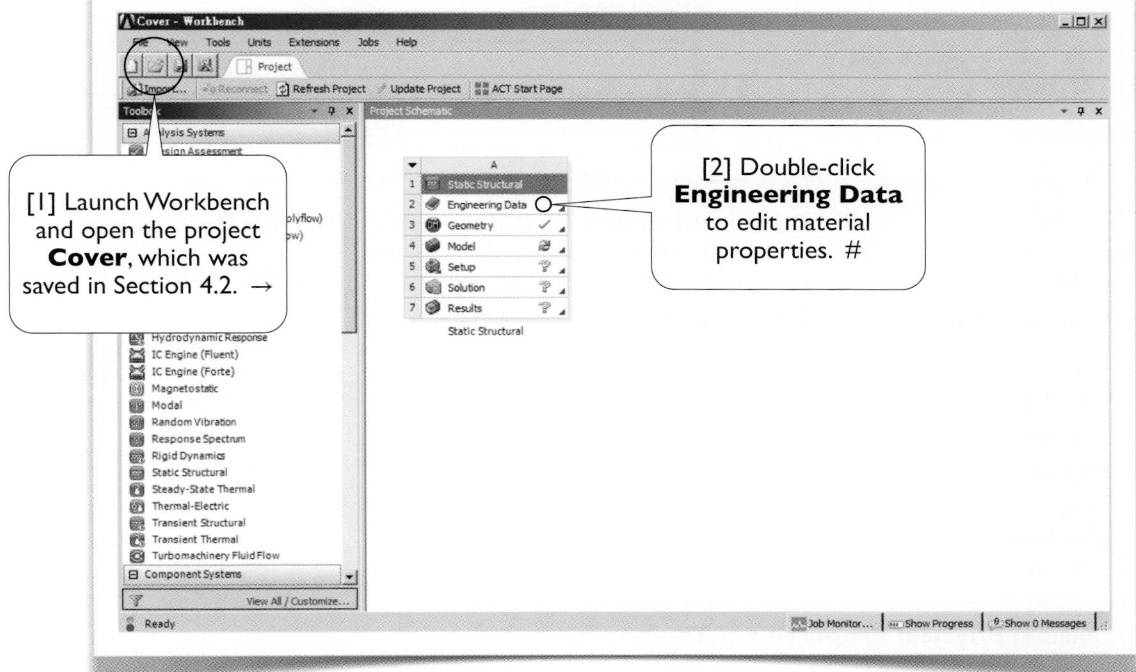

5.2.3 Prepare Material Properties

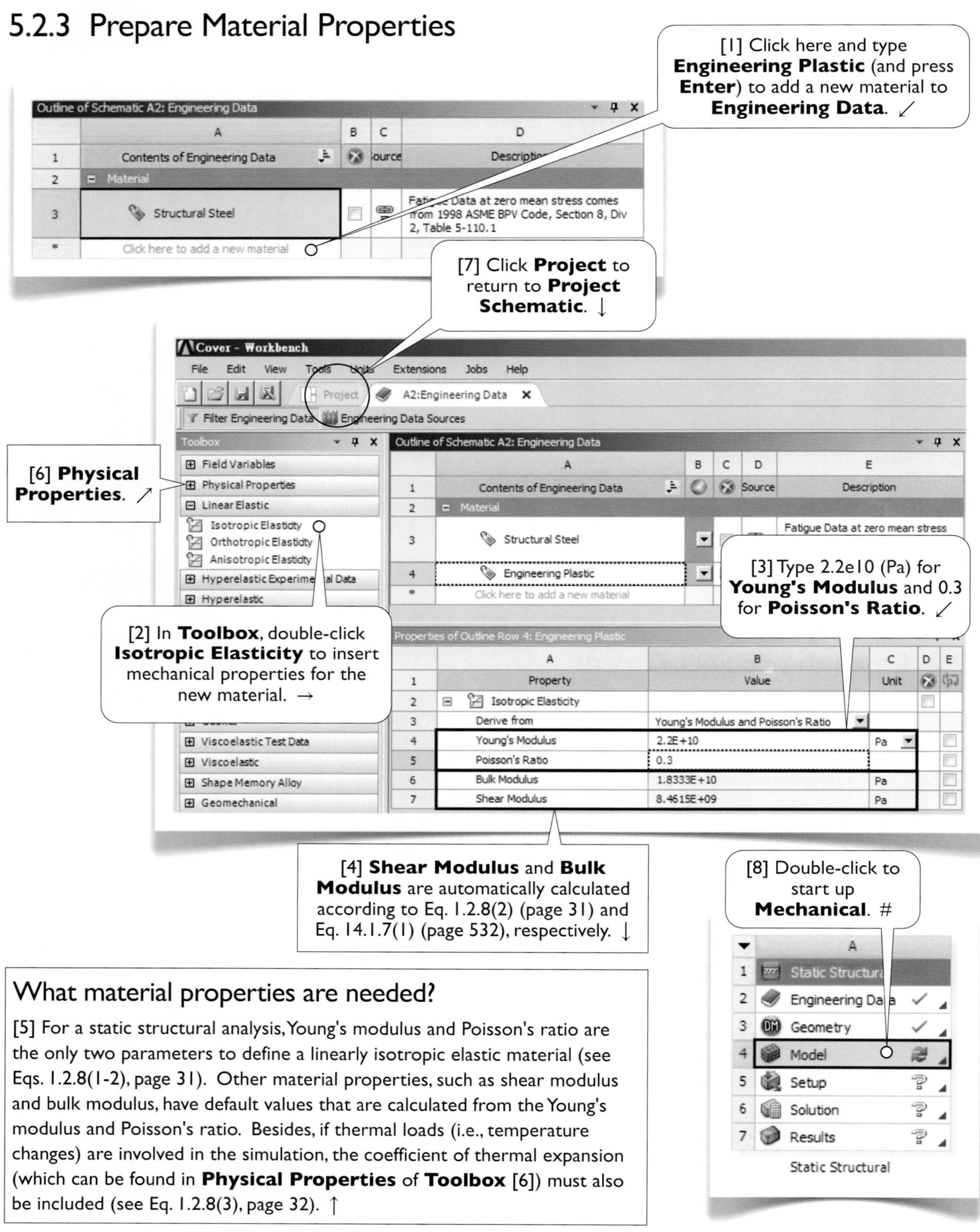

What material properties are needed?

[5] For a static structural analysis, Young's modulus and Poisson's ratio are the only two parameters to define a linearly isotropic elastic material (see Eqs. 1.2.8(1-2), page 31). Other material properties, such as shear modulus and bulk modulus, have default values that are calculated from the Young's modulus and Poisson's ratio. Besides, if thermal loads (i.e., temperature changes) are involved in the simulation, the coefficient of thermal expansion (which can be found in **Physical Properties** of **Toolbox** [6]) must also be included (see Eq. 1.2.8(3), page 32). ↑

5.2.4 Assign Material to the Body

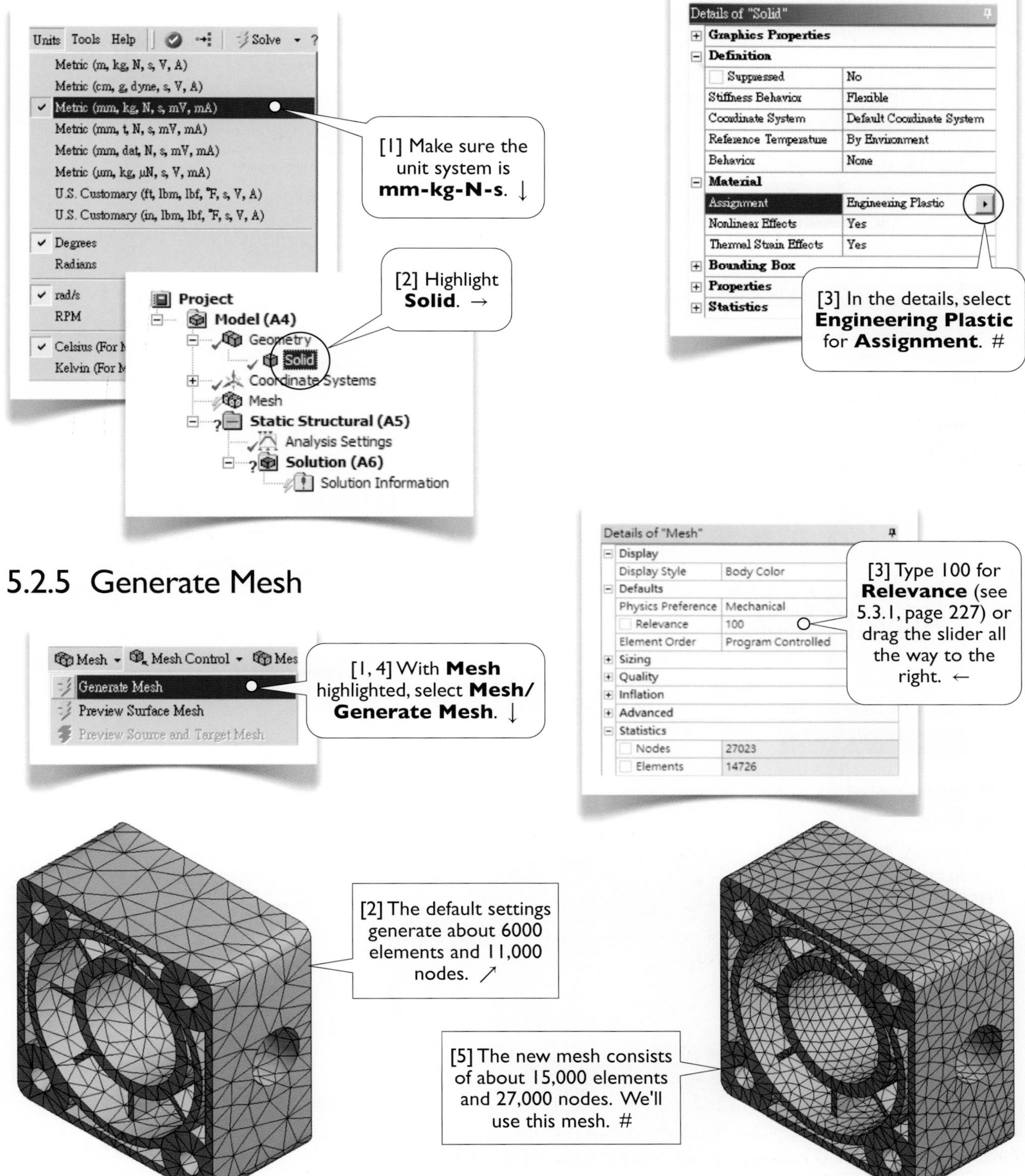

5.2.5 Generate Mesh

5.2.6 Specify Supports

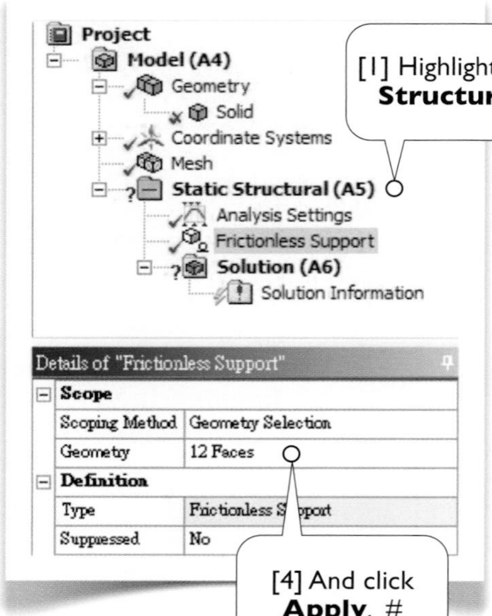

Project
Model (A4)
Geometry
Solid
Coordinate Systems
Mesh
Static Structural (A5)
Analysis Settings
Frictionless Support
Solution (A6)
Solution Information
Details of "Frictionless Support"
Scope
Scoping Method | Geometry Selection
Geometry | 12 Faces
Definition
Type | Frictionless Support
Suppressed | No

[1] Highlight **Static Structural**. →

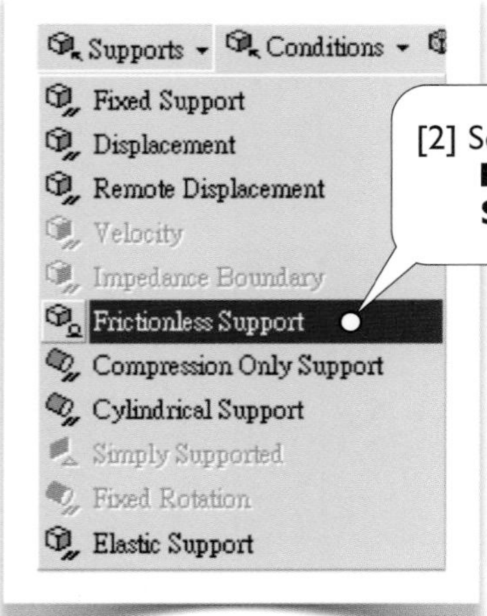

Supports
Conditions
Fixed Support
Displacement
Remote Displacement
Velocity
Impedance Boundary
Frictionless Support
Compression Only Support
Cylindrical Support
Simply Supported
Fixed Rotation
Elastic Support

[2] Select **Supports/ Frictionless Support**. ↙

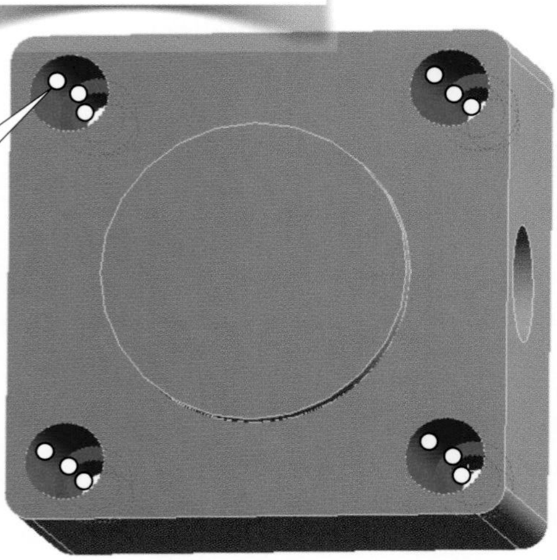

[3] Control-select all internal faces of the four holes, 12 faces in total. ←

[4] And click **Apply**. #

5.2.7 Apply Pressure

Loads
Supports
Pressure
Pipe Pressure
Hydrostatic Pressure
Force
Line Pressure
Thermal Condition
Pipe Temperature
Joint Load
Fluid Solid Interface
Detonation Point
Rotating Force

[1] With **Static Structural** or **Frictionless Support** still highlighted, select **Loads/Pressure**. →

[2] Control-select all internal faces like this. →

Details of "Pressure"
Scope
Scoping Method | Geometry Selection
Geometry | 39 Faces
Definition
Type | Pressure
Define By | Normal To
Applied By | Surface Effect
Magnitude | 0.5 MPa (ramped)
Suppressed | No

[3] And click **Apply**. There should be 39 faces selected. ↓

[4] Type 0.5 (MPa) for **Magnitude**. #

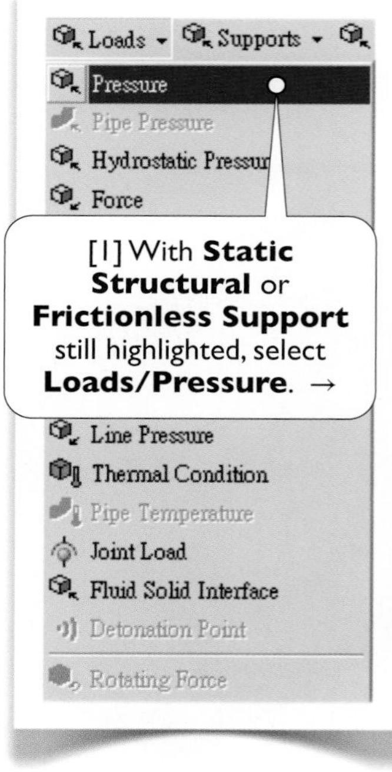

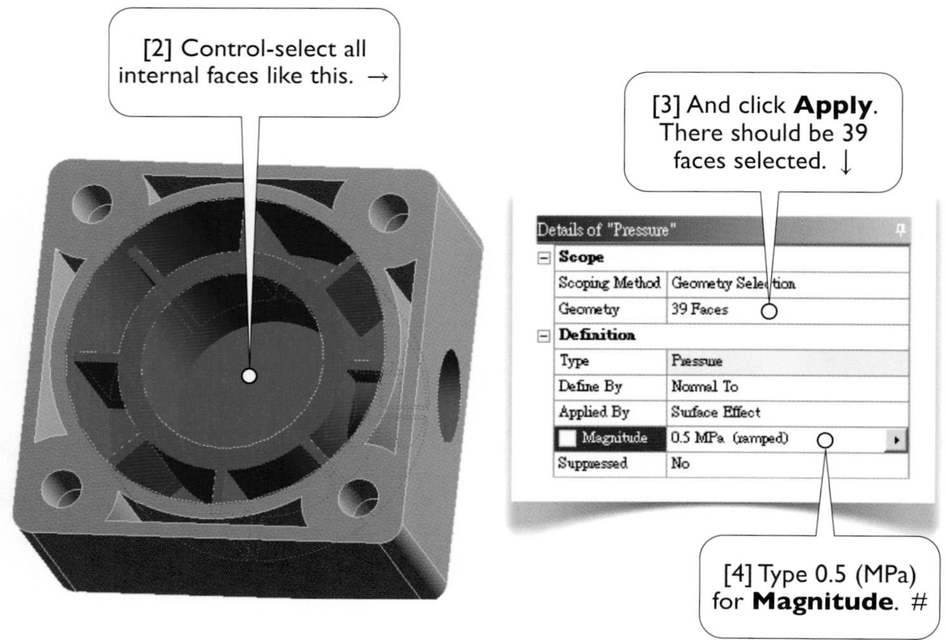

5.2.8 Set Up Solution Branch and Solve

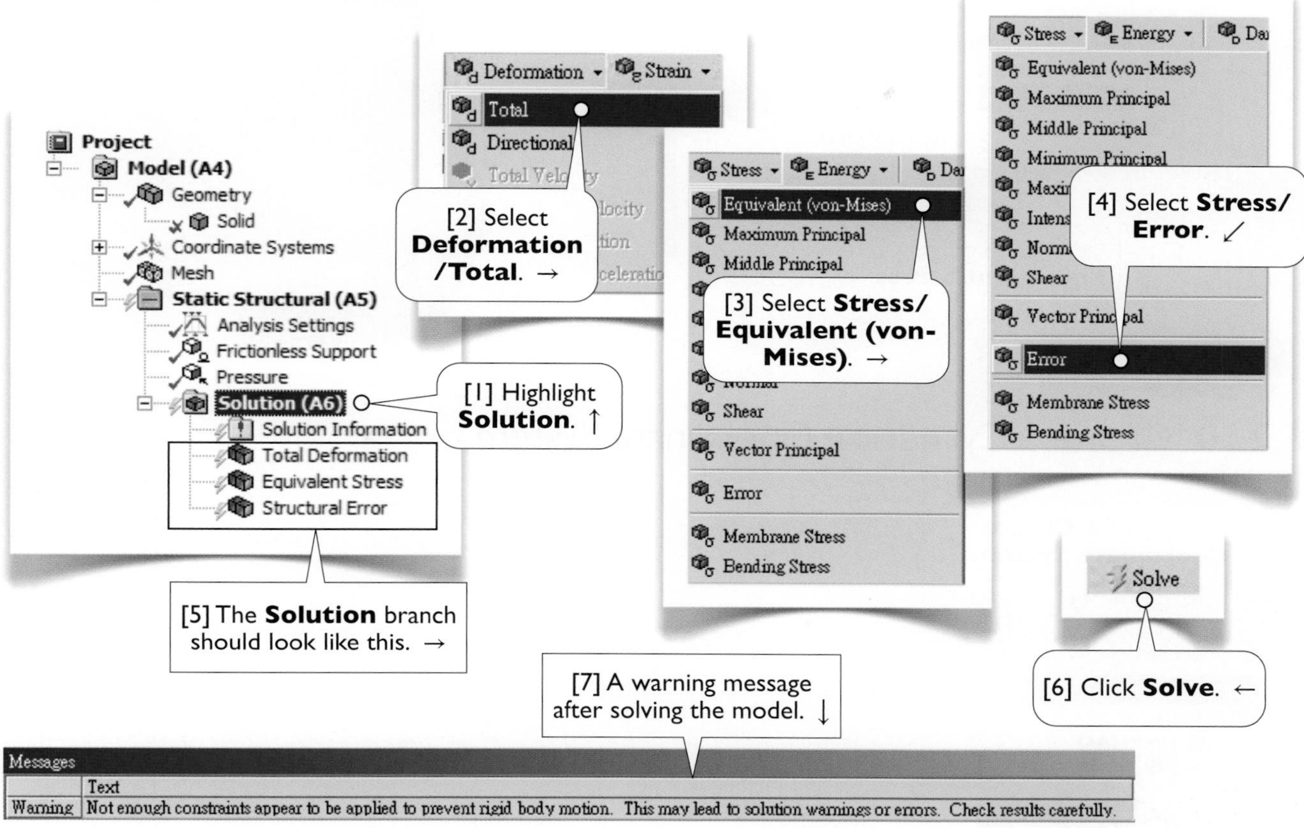

Not Enough Constraints?

[8] After solving the model, a warning message shows up [7]. If you carefully examine the support conditions, the rigid body motions in all directions have been constrained by the frictionless supports (5.2.6, last page). Ignore the warning message; it might be a bug. #

5.2.9 View the Results

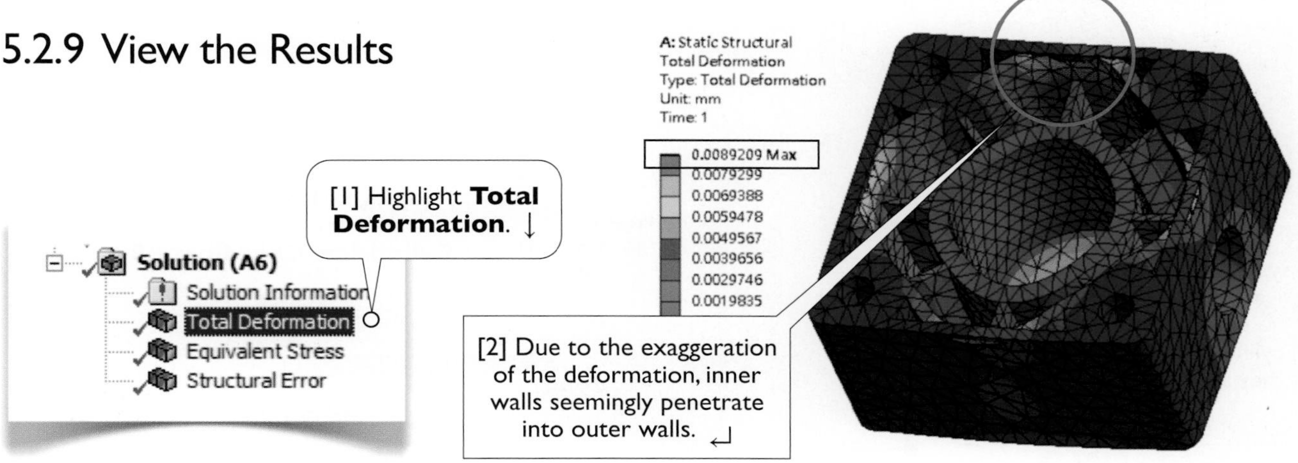

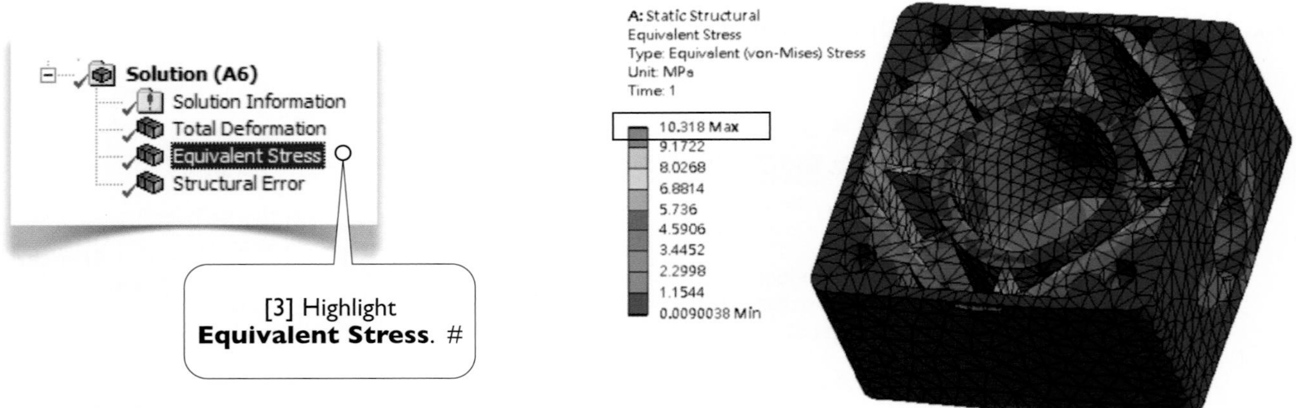

5.2.10 Create a Cylindrical Coordinate System

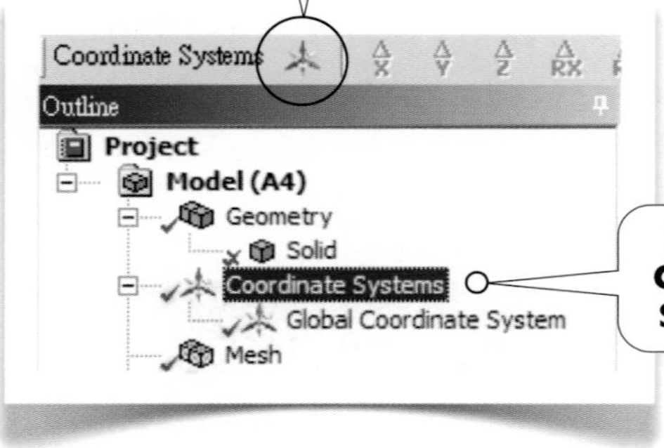

[1] Highlight **Coordinate Systems**. ↑

[2] Click **Create Coordinate System**. →

Coordinate Systems
Outline
Project
Model (A4)
Geometry
Solid
Coordinate Systems
Global Coordinate System
Mesh

[3] Select **Cylindrical**. ↓

[4] Select **Global Coordinates** for defining the origin. Leave other settings their default values. ↙

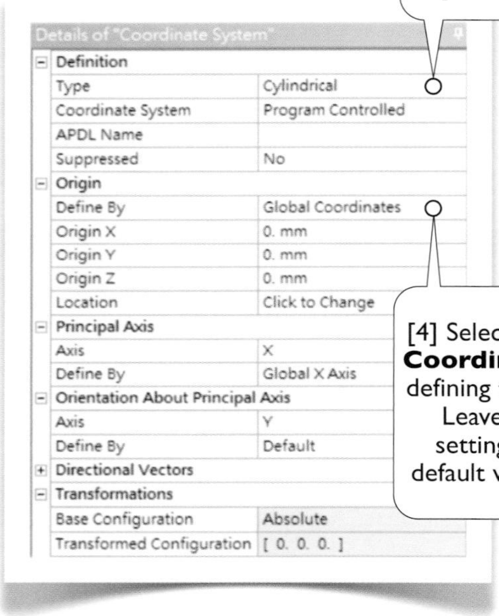

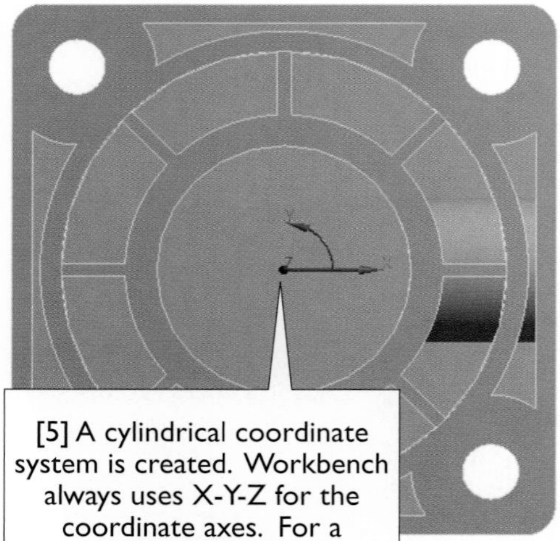

Details of "Coordinate System"	
Definition	
Type	Cylindrical
Coordinate System	Program Controlled
APDL Name	
Suppressed	No
Origin	
Define By	Global Coordinates
Origin X	0. mm
Origin Y	0. mm
Origin Z	0. mm
Location	Click to Change
Principal Axis	
Axis	X
Define By	Global X Axis
Orientation About Principal Axis	
Axis	Y
Define By	Default
Directional Vectors	
Transformations	
Base Configuration	Absolute
Transformed Configuration	[0. 0. 0.]

[5] A cylindrical coordinate system is created. Workbench always uses X-Y-Z for the coordinate axes. For a cylindrical coordinate system, they should be interpreted as R-Theta-Z. →

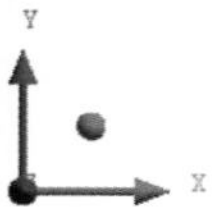

[6] Right-click and rename the newly created coordinate system as **Cylindrical Coordinate System**. #

Project
Model (A4)
Geometry
Solid
Coordinate Systems
Global Coordinate System
Cylindrical Coordinate System
Mesh

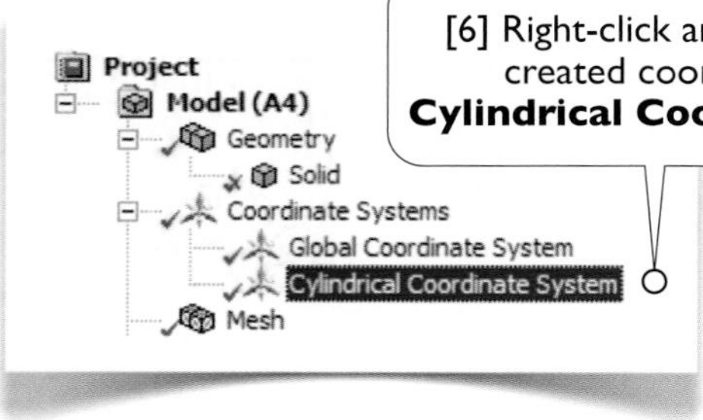

5.2.11 Assess Circularity

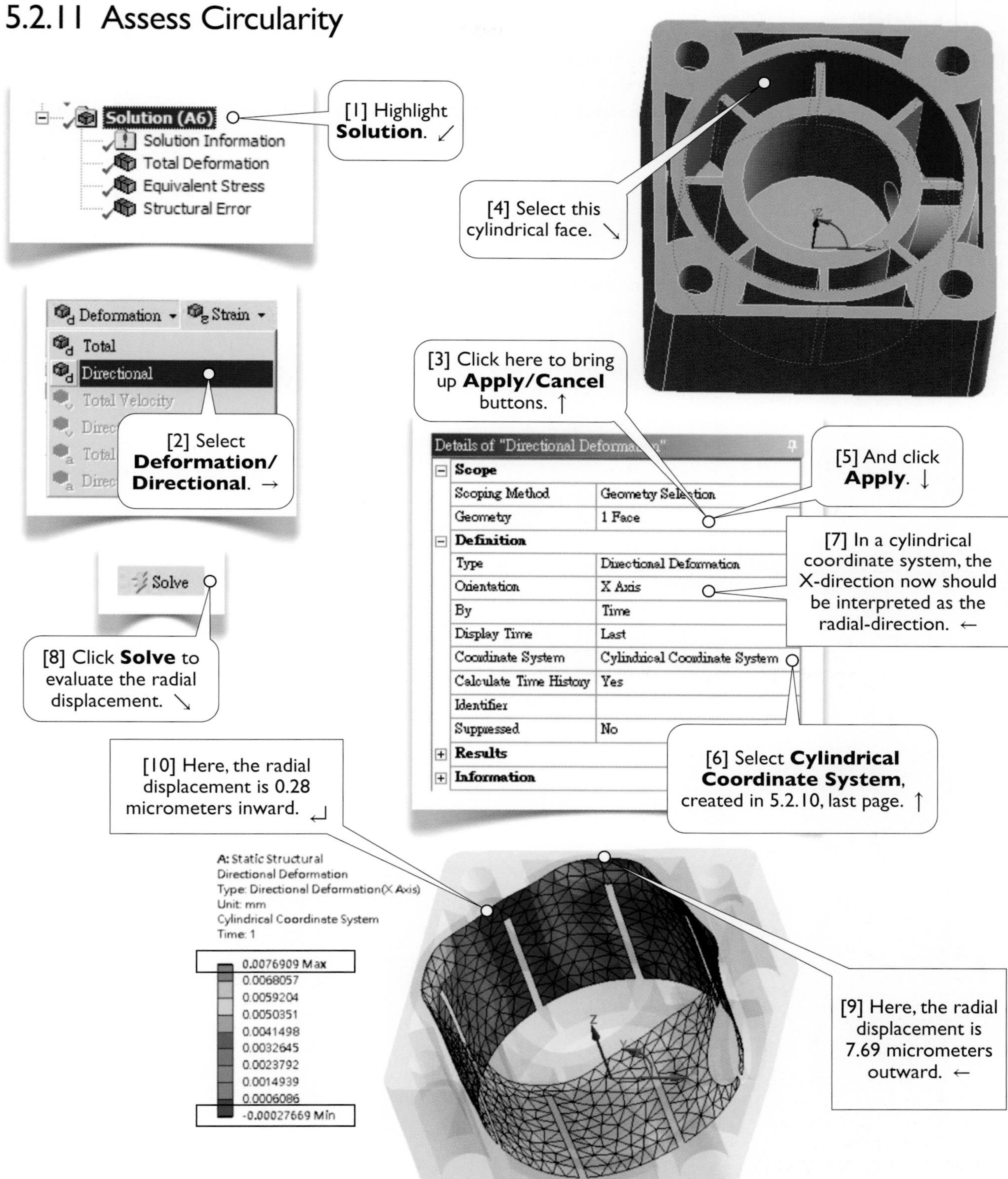

Conclusions

[11] The circularity of the cover under pressure, according to its definition, is 7.97 micrometers (7.69 + 0.28). It is within the spec requirement (10 micrometers).

The engineers need not worry about the stress, since the stress (10.318 MPa, 5.2.9[3], page 224) is well below the fracture stress (which is about 50+ MPa). ↓

Wrap Up

[12] Close **Mechanical**, save the project, and exit Workbench. #

Section 5.3

More Details

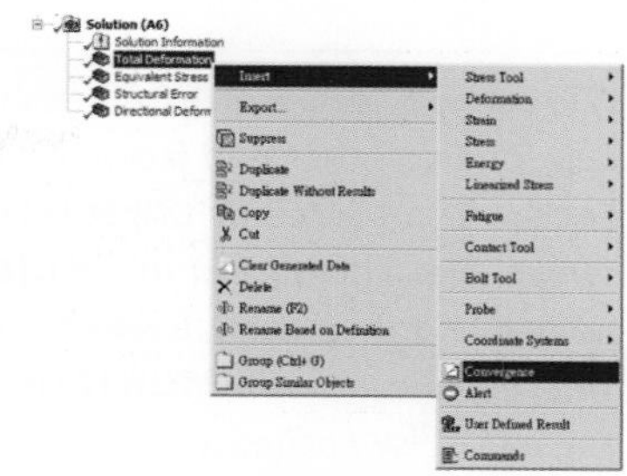

5.3.1 Global Mesh Controls[Ref 1]

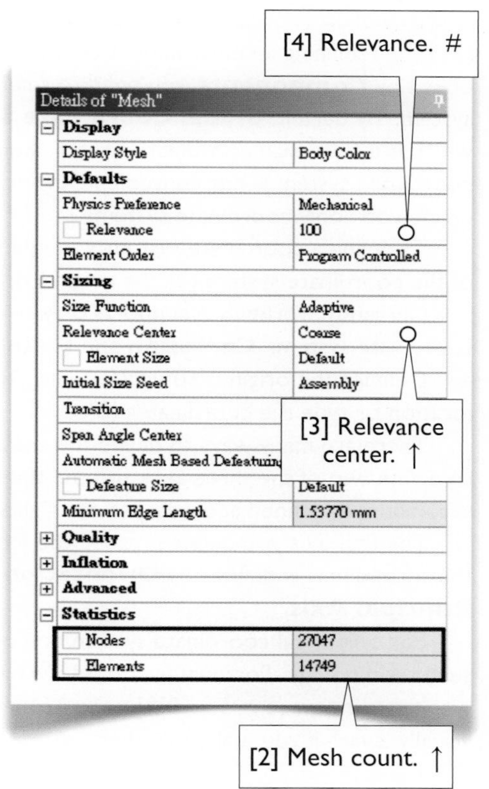

[1] By highlighting **Mesh** in a project tree, you can access the details of **Mesh** [2-4].

A statistics of the mesh count shows up at the bottom of the details [2]. The mesh count, number of nodes/elements, provides an estimation of the problem size. In 3D cases, as mentioned, the total degrees of freedom equals three times the number of nodes (1.3.1[2], page 35). In the case of Section 5.2, Workbench solves a system of equations of degrees of freedom 81141 (27047x3, see [2]). The matrix [*K*] has a size of 81141x81141!

Relevance Center [3] and **Relevance** [4] together provide a way of global mesh control. **Relevance Center** may be **Coarse**, **Medium**, or **Fine**. **Relevance** ranges from -100 to 100, the larger the finer. These two values are related roughly as follows:

(Coarse, 0) = (Medium, -100),
(Coarse, 100) = (Medium, 0) = (Fine, -100),
and (Medium, 100) = (Fine, 0). →

5.3.2 Mesh with MultiZone Method

[1] Generally, hexahedral elements are more desirable than other shapes, such as tetrahedral. The main reason is that hexahedral (or quadrilateral in 2D cases) has better convergence behavior (3.5.10, page 163; 9.3.13 and 9.3.14, page 361). That implies, with the same problem size, the hexahedral (or quadrilateral in 2D cases) gives more accurate results. Besides the shapes, mesh quality is also a key factor affecting convergence behavior. A mesh of hexahedral elements with poor mesh quality might be less desirable than tetrahedral with good mesh quality. Mesh quality will be discussed in Chapter 9.

For 2D models, Workbench usually does a good job and meshes them with all-quadrilateral elements. For example, the models in Sections 3.1 and 3.2 are meshed with quadrilateral elements without further mesh controls.

For 3D models, meshing is much more challenging. In 5.2.5 (page 221), the geometry is relatively complicated, so Workbench chooses to mesh with all-tetrahedra. ↵

[2] A simple idea to create hexahedral elements is to mesh faces of a body with quadrilaterals and then "sweep" along a depth up to other end faces of the body. The starting faces are called **source faces** and the ending faces are **target faces**. The source or target faces can be either manually or automatically selected. However, many bodies are not sweepable. The idea of the **MultiZone** method is to decompose a non-sweepable body into several sweepable bodies, and then apply **Sweep** method on each of the bodies. This is what we had done in 5.1.13[1-5] (page 217), where we inserted a **MultiZone** method and the Workbench decomposed the body into several sweepable bodies and easily "swept" each body with hexahedral elements. The result is an all-hexa mesh. #

5.3.3 Coordinate Systems[Ref 2]

[1] When defining an environment condition or a solution object by **Components**, you need to refer to a coordinate system. By default, **Global Coordinate System** is used, which is a Cartesian coordinate system. Sometimes this coordinate system is not convenient. In such cases, we may define additional coordinate systems.

To define a coordinate system, you need to define the type of the coordinate system [2], the origin [3], and the axes [4].

Currently, workbench supports only two types of coordinate systems: **Cartesian** and **Cylindrical** [2].

Defining the origin is straightforward. You can click a location or type the coordinates [3].

There are many ways to define the axes. Basically, you need to define two of the three axes and the third axis is automatically defined according to the right-hand rule.

The first axis you define is called the **Principal Axis** and the second axis is defined by **Orientation About Principal Axis**.

For cylindrical coordinate system, you may be confused by such terminology. Fortunately, a triad always appears on the graphics window (e.g., 5.2.10[5], page 224) and you should be visually aware if you make mistakes. ↗

[3] Definition of the origin. ↓

[2] Type of coordinate systems. ←

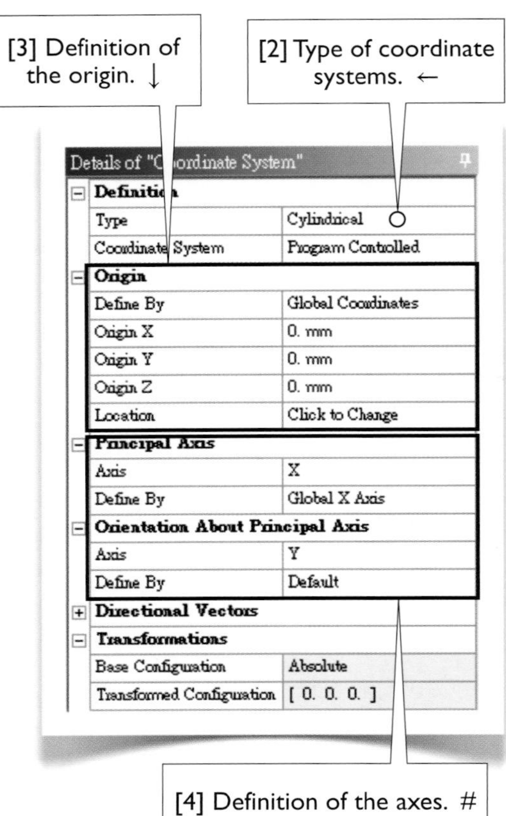

[4] Definition of the axes. #

5.3.4 Results View Controls[Ref 3]

[1] With a results object highlighted, the toolbar displays tools that can be used to control how the results are displayed [2-7]. ↓

[6] Vector display. ↘

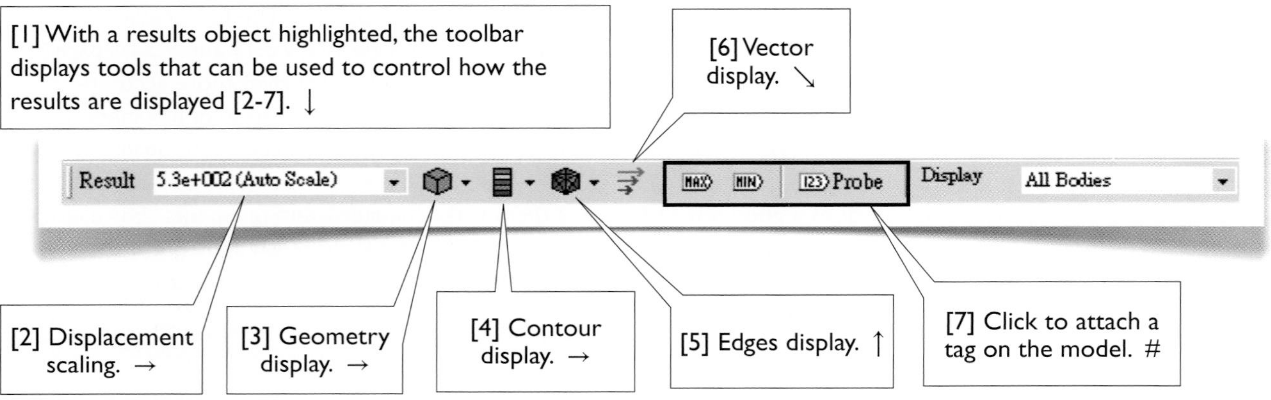

[2] Displacement scaling. →

[3] Geometry display. →

[4] Contour display. →

[5] Edges display. ↑

[7] Click to attach a tag on the model. #

5.3.5 Legend Controls

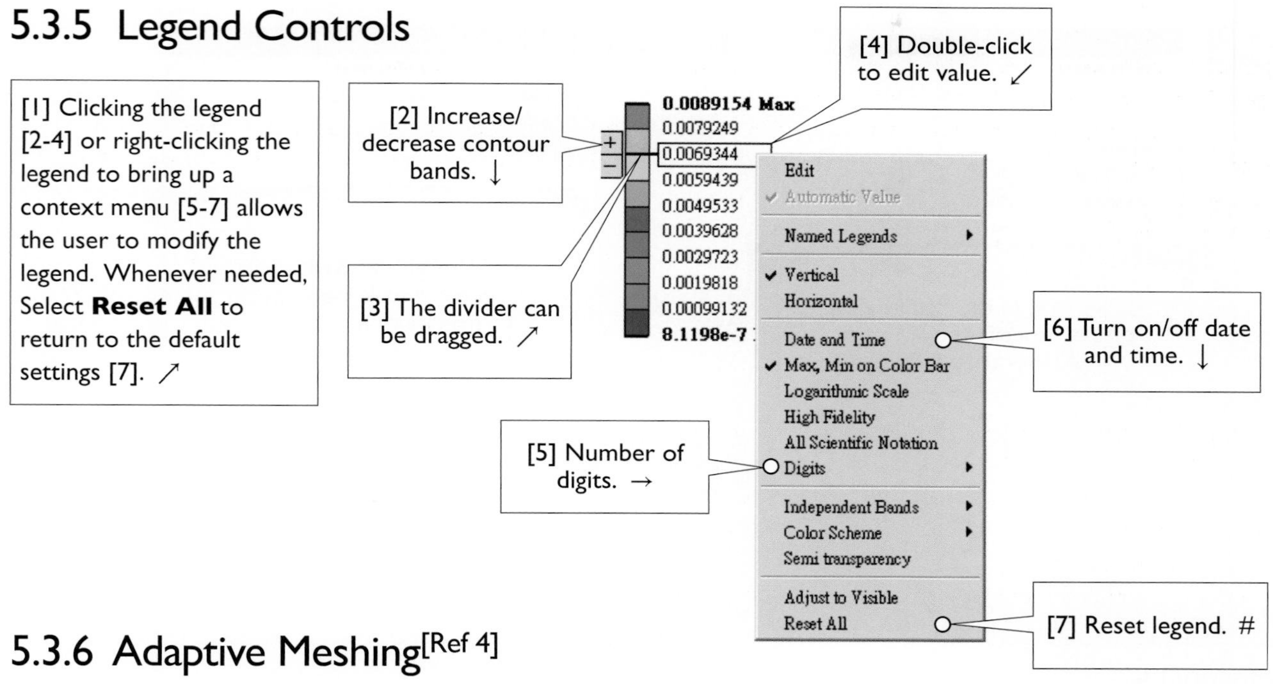

5.3.6 Adaptive Meshing[Ref 4]

[1] We demonstrated a procedure of finite element convergence study in PART C of Section 3.5 (pages 161-163). Given an error, say 5%, we may refine mesh until the accuracy reaches this level. Performing these tasks manually is cumbersome. Workbench provides a tool to automate the mesh refinement until a user-specified level of accuracy is reached. This idea is termed *adaptive meshing*. Internally, Workbench uses structural errors (3.5.7, page 161) to help adjust the mesh, that is, it refines the mesh size in the area of large structural errors.

To use this tool, right-click a results object and select **Insert/Convergence** [2]. In the details view, specify the accuracy, or **Allowable Change** ([3], next page). Also, highlight **Solution** in the project tree and, in the details view, specify the maximum number of mesh refinement loops ([4], next page). When solving the model, Workbench will iterate to refine the mesh until the difference between two iterations is less than the **Allowable Change** or the **Max Refinement Loops** is reached. ↓

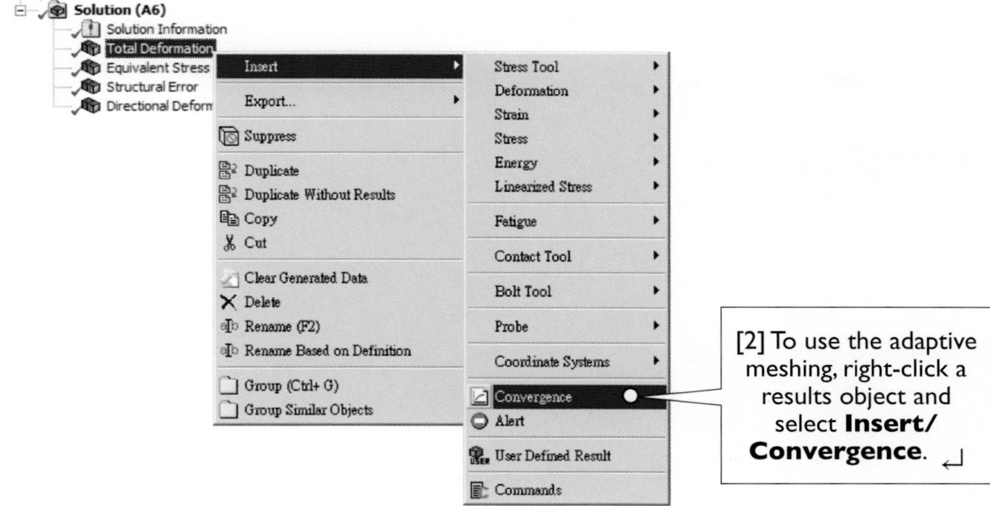

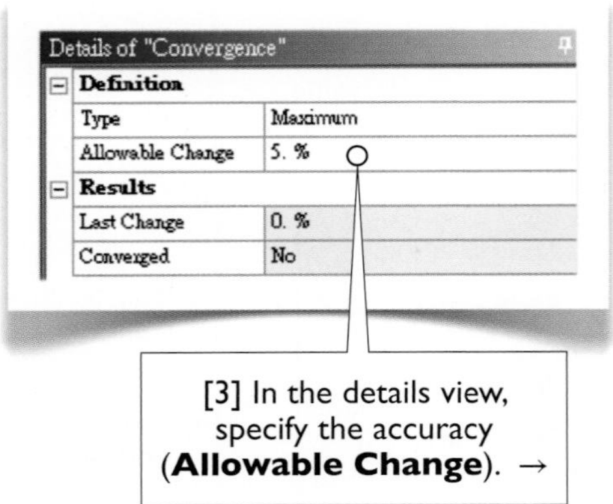

[3] In the details view, specify the accuracy (**Allowable Change**). →

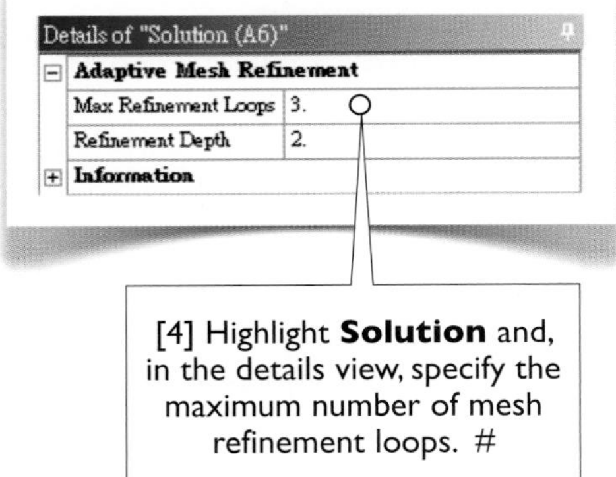

[4] Highlight **Solution** and, in the details view, specify the maximum number of mesh refinement loops. #

References

1. ANSYS Help//Meshing//Meshing User's Guide//Global Mesh Controls
2. ANSYS Help//Mechanical Application//ANSYS Mechanical User's Guide//Objects Reference//Coordinate System
3. ANSYS Help//Mechanical Application//ANSYS Mechanical User's Guide//Approach//Steps for Using the Application//Review Results
4. ANSYS Help//Mechanical Application//ANSYS Mechanical User's Guide//Objects Reference//Convergence

Section 5.4

LCD Display Support

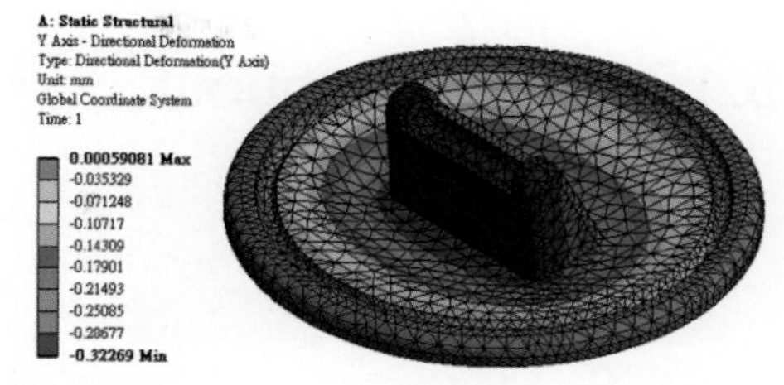

5.4.1 About the LCD Display Support

[1] The geometry created in Section 4.5 is used in this section for a static structural simulation to assess the deformation and stress under a design load [2].

The display support is made of an ABS (acrylonitrile-butadiene-styrene) plastic; its Young's modulus is 2.62 GPa and its Poisson's ratio is 0.34. In a tensile test, the material starts to develop fine cracks at 37 MPa, and fractures at 54 MPa.

The support is designed for a 17" LCD display, which weighs 40 N and is used as the design load. →

5.4.2 Prepare Material Properties

[1] Launch Workbench and open the project **Support**, which was saved in Section 4.5. Double-click **Engineering Data**. Add a material to **Engineering Data** [2] and specify the Young's modulus and Poisson's ratio [3-4].

Return to **Project Schematic** [5] and double-click **Model** to start up **Mechanical**. In **Mechanical**, make sure the unit system is **mm-kg-N-s**, and assign the newly created material to the solid body [6]. ↙

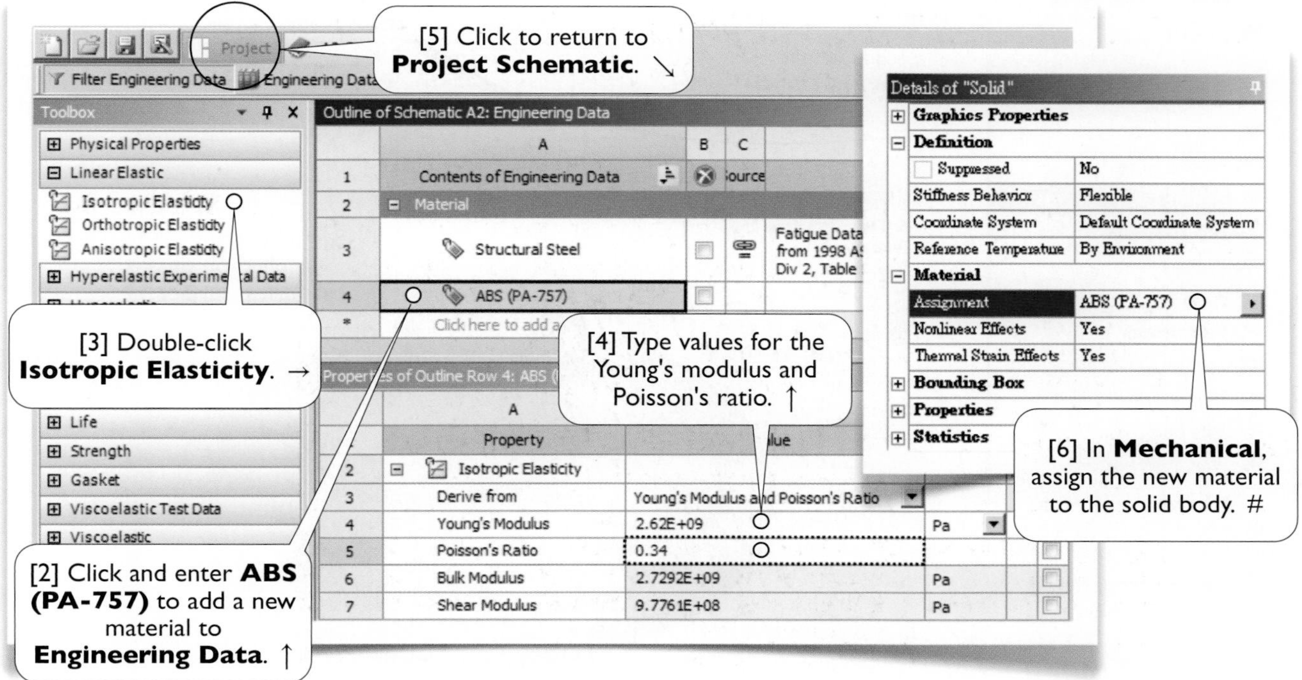

5.4.3 Specify Support

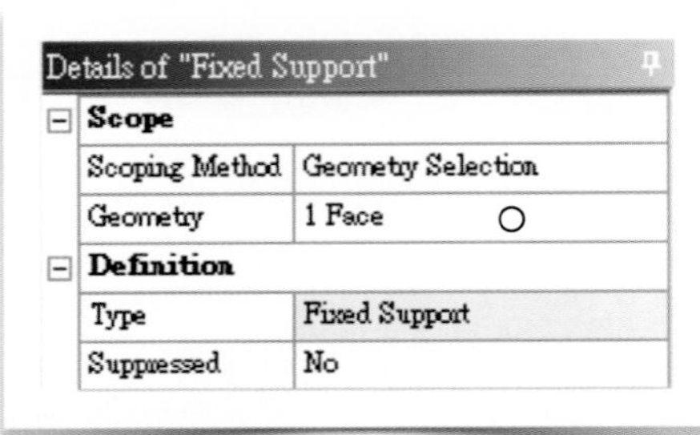

Details of "Fixed Support"	
Scope	
Scoping Method	Geometry Selection
Geometry	1 Face
Definition	
Type	Fixed Support
Suppressed	No

[1] Specify **Fixed Support** for the bottom rim face. Note that it would be more realistic if we specify **Compression Only Support** for the bottom rim face. For this case, however, the calculated results would be the same, since the rim face will not separate from the surface that supports the model. A **Compression Only Support** would introduce contact nonlinearity; that's why we avoid it. #

5.4.4 Specify Load

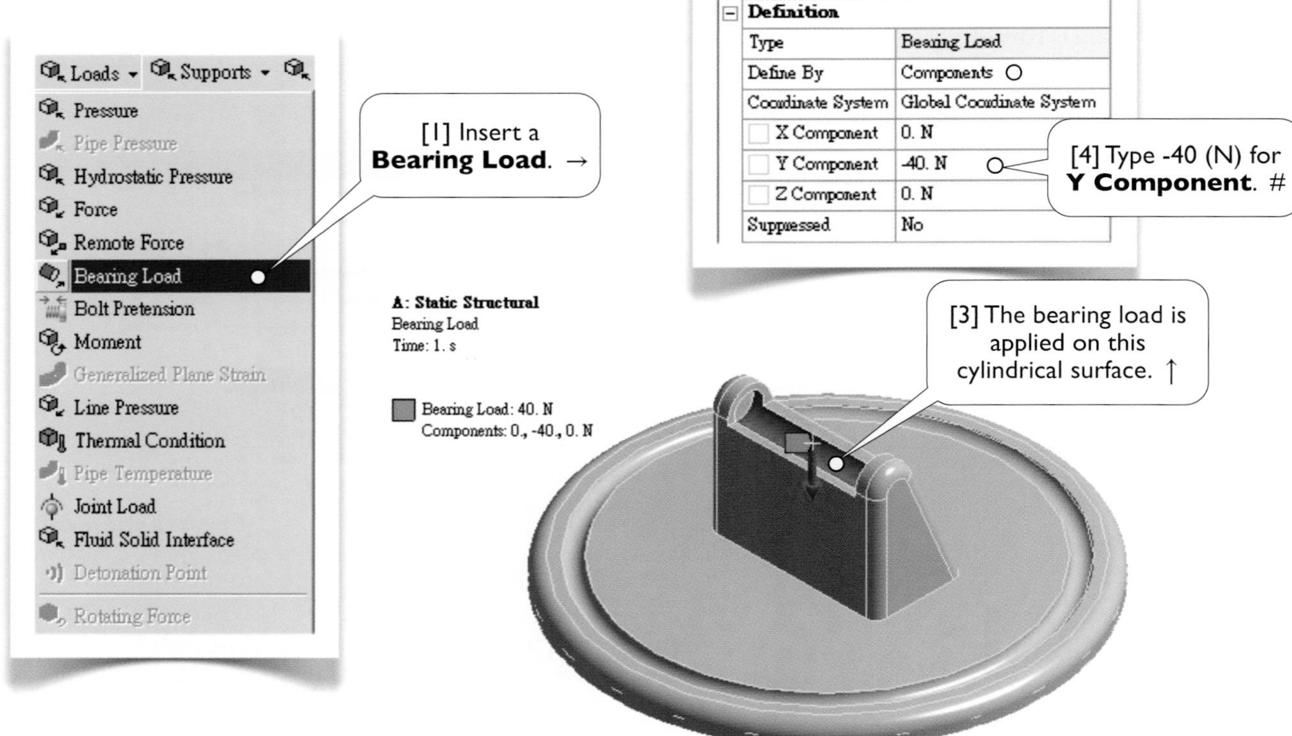

Details of "Bearing Load"	
Scope	
Scoping Method	Geometry Selection
Geometry	1 Face
Definition	
Type	Bearing Load
Define By	Components
Coordinate System	Global Coordinate System
X Component	0. N
Y Component	-40. N
Z Component	0. N
Suppressed	No

[1] Insert a **Bearing Load**. →

[2] Select the cylindrical surface as shown in [3]. ↓

[3] The bearing load is applied on this cylindrical surface. ↑

[4] Type -40 (N) for **Y Component**. #

5.4.5 Generate Mesh

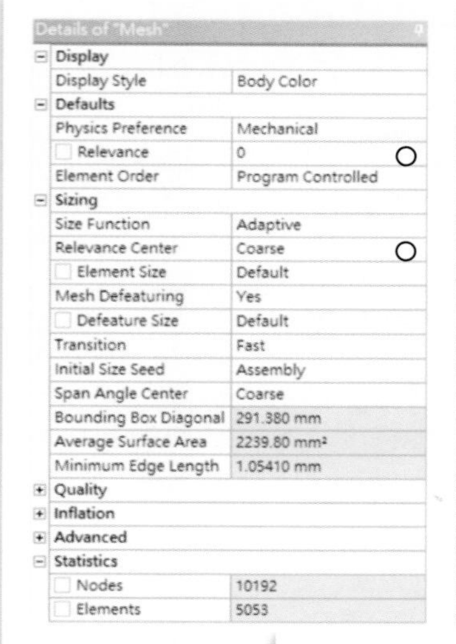

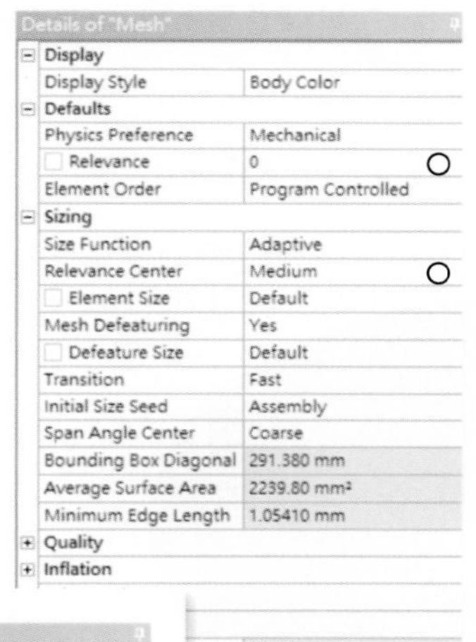

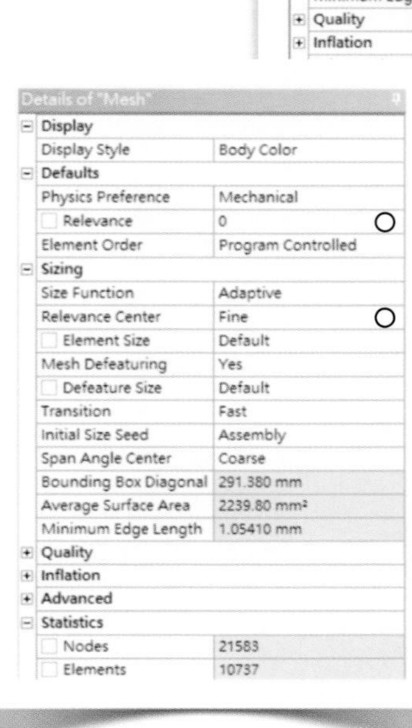

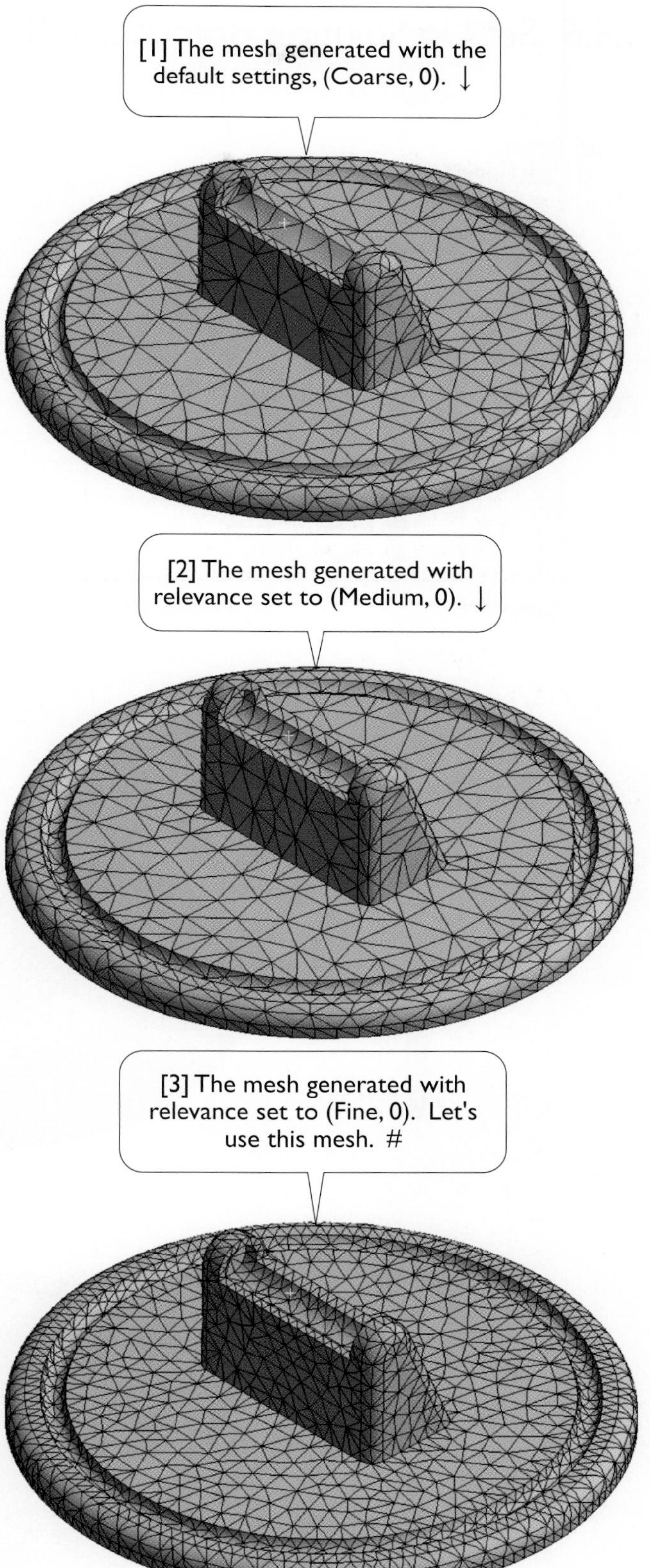

5.4.6 Set Up Solution Branch and Solve

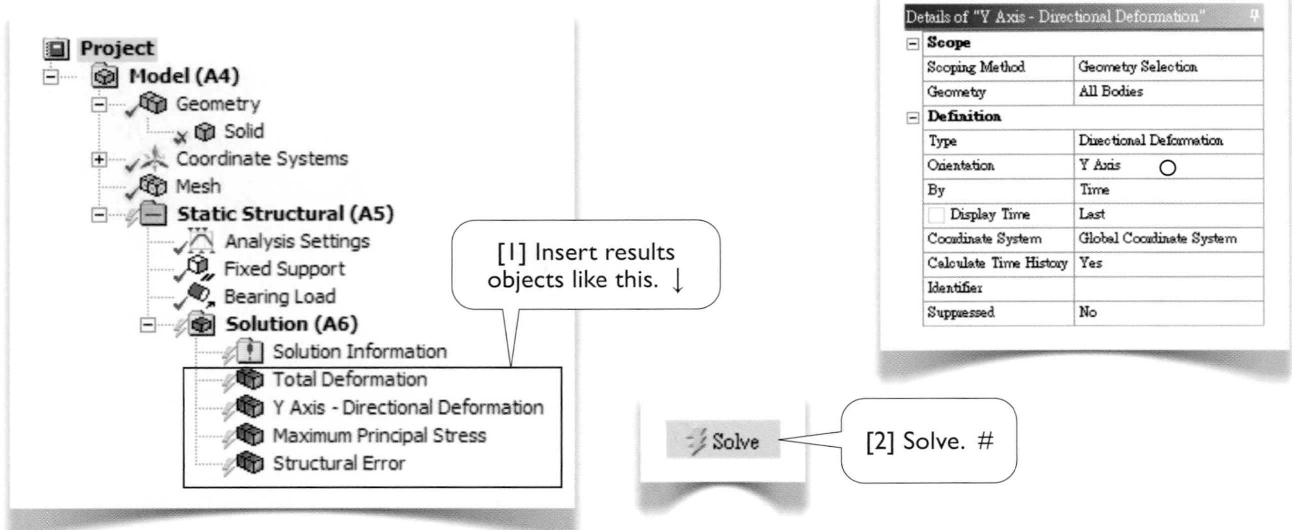

5.4.7 View the Results

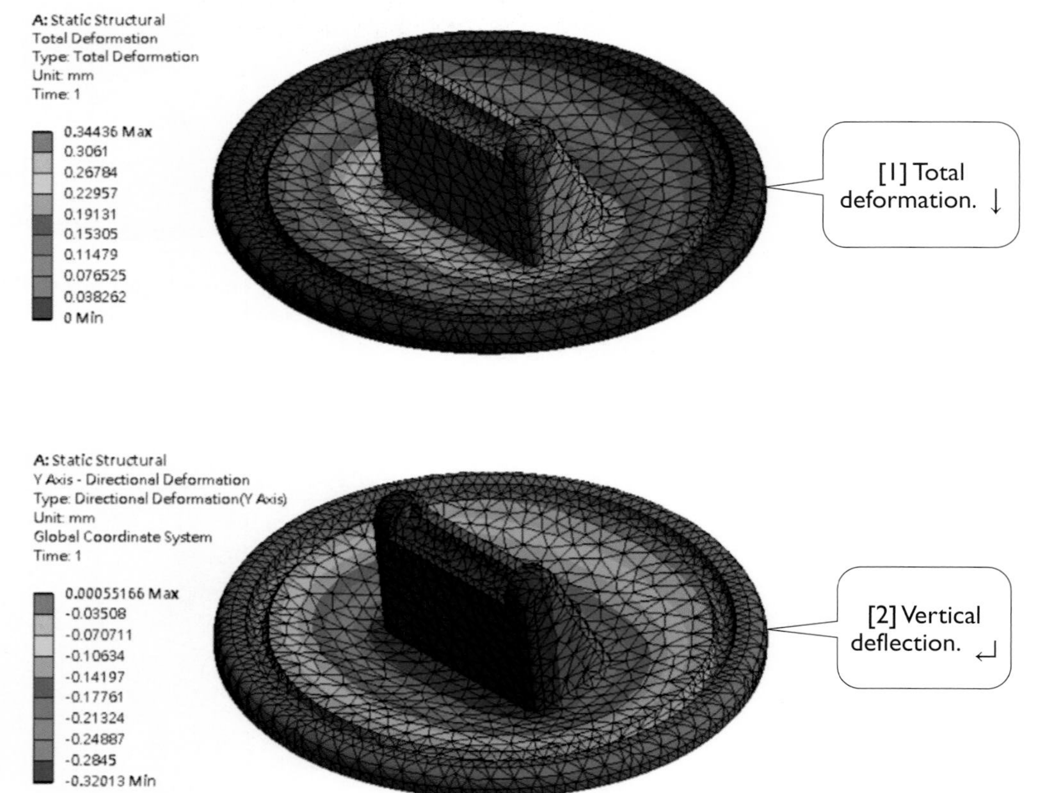

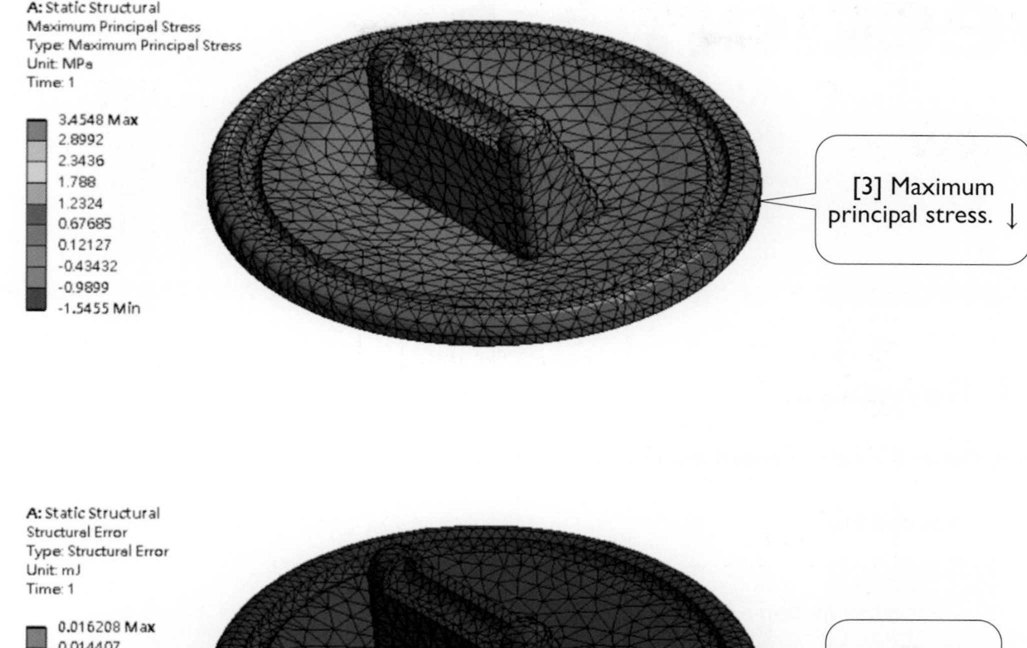

A: Static Structural
Structural Error
Type: Structural Error
Unit: mJ
Time: 1

0.016208 Max
0.014407
0.012606
0.010806
0.0090046
0.0072037
0.0054028
0.0036019
0.001801
3.8086e-8 Min

[4] Structural error. ↓

Wrap Up

[5] Close **Mechanical**, save the project, and exit Workbench. #

Section 5.5

Review

5.5.1 Keywords

Choose a letter for each keyword from the list of descriptions

1. ( ) Adaptive Meshing
2. ( ) Bearing Load
3. ( ) Coordinate System
4. ( ) MultiZone Method
5. ( ) Sweep Method

Answers:

1. (E) 2. (A) 3. (B) 4. (D) 5. (C)

List of Descriptions

(A) In 3D simulations, it applies on cylindrical faces. The total force is distributed on the compressive side of the cylindrical faces.

(B) To define a coordinate system, we need to specify the type of coordinate system, the location of origin, and the axes. The types of coordinate systems may be **Cartesian** or **Cylindrical** in the current version of the Workbench.

(C) A meshing method, in which the source faces, selected manually or automatically, are meshed, and the 2D mesh grows to become 3D elements by sweeping along a path up to target faces. Not all 3D solid bodies are sweepable. A sweepable 3D solid body often can be meshed with hexahedral elements.

(D) A meshing method. For a non-sweepable 3D solid body, the method tries to decompose the body into several sweepable bodies and then uses Sweep method to mesh each body.

(E) An automatic and iterative solution process to meet a user-specified solution accuracy. The basic idea is to refine the mesh size in the area of large structural errors until the specified accuracy is satisfied.

5.5.2 Questions

Symmetries

We didn't take any advantage of symmetries in all three simulation cases of this chapter. Point out the planes of symmetry in each model.

5.5.3 Additional Workbench Exercises

Simulation with Symmetric Models

It is a good exercise to redo the simulations in this chapter, taking advantage of the symmetries.

Additional Exercises

Additional Exercises of 3D simulations can be found in the Verification Manual for Workbench[Ref 1].

Reference

1. ANSYS Help//Verification Manuals//ANSYS Workbench Verification Manual

Chapter 6
Surface Models

Many real-world objects can be modeled as surface bodies. For example, in Section 4.3, we modeled the glass as a surface body. The 2D bodies in Chapter 3, although created as surfaces, are called 2D solid bodies in **Mechanical**. Remember that 2D solid bodies do not have out-of-plane bending. This section focuses on the simulations of 3D surface bodies.

When a real-world body is thin enough, it is usually a good candidate for a 3D surface body. Workbench will mesh surface bodies with shell elements (1.3.3[9-11], page 38). There are many advantages of using surface models over 3D solid models. First, creating surface models is usually easier. Second, the problem size is much smaller; that implies a much smaller computing time. Third, it often results in a more accurate solution due to the efficiency of shell elements. Therefore, engineers should consider using surface models instead of solid models whenever possible. In the old days, surface models were visually awkward since they had zero thickness and occupied zero volume in the space. Workbench avoids this awkwardness by allowing a rendering of thickness, so that the surface bodies can be visually the same as solid bodies.

Purpose of This Chapter

The main purpose of this chapter is to introduce the use of shell elements. This chapter guides the students to create surface models and perform simulations using surface models. The chapter uses three examples; two of them are purely surface models while the other one is a model mixed up with solid bodies and surface bodies.

About Each Section

Section 6.1 creates a bellows joint model and performs a simulation. In Section 6.2, the bracket, which we introduced in Sections 4.1 and 5.1, is used again to demonstrate how an existing solid model can be transformed to a surface model using the tool **Mid-Surface**. The solution is numerically comparable with that obtained using the solid model in Section 5.1, while using much less computing time. Section 6.3 demonstrates the simulation of a model that has a mix-up of solid bodies and surface bodies.

Section 6.1

Bellows Joints

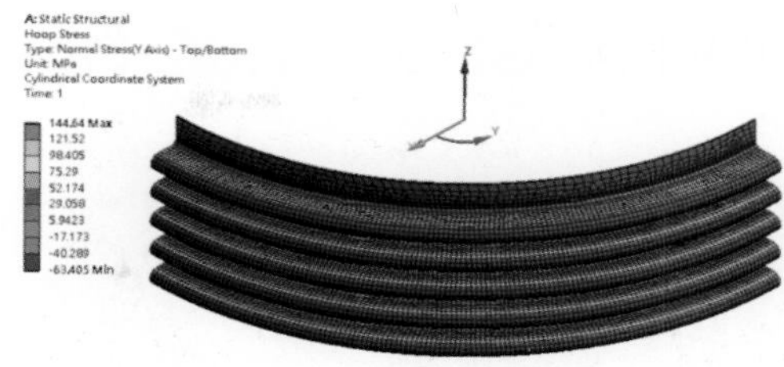

6.1.1 About Bellows Joints

[1] The bellows joints [2-3] are used to absorb thermal or vibrational movement in a pipeline system that transports high pressure gases; they are designed to sustain internal pressure as well as external pressure. The external pressure is considered when the piping system is used under the ocean. With the internal pressure, the engineers are concerned about its radial deformation (due to a tolerance consideration) and hoop stress (due to a safety consideration). With the external pressure, buckling is the major concern, which will be discussed in 10.4.2, page 388. ↓

[4] In this section, we will perform a static structural simulation under the internal pressure of 0.5 MPa.

The bellows joint is axisymmetric both in geometry and loading. It is also symmetric about a horizontal plane. We will take advantage of these symmetries and model 1/8 of the bellows joint as a 3D surface body.

We might model the bellows joint as an axisymmetric 2D solid body; however, it would result in a poorer solution than 3D surface body for this particular case, in which the bending dominates the structural behavior. #

[2] The bellows joints are made of an SU316 steel, which has a Young's modulus of 180 GPa and a Poisson's ratio of 0.28. ↓

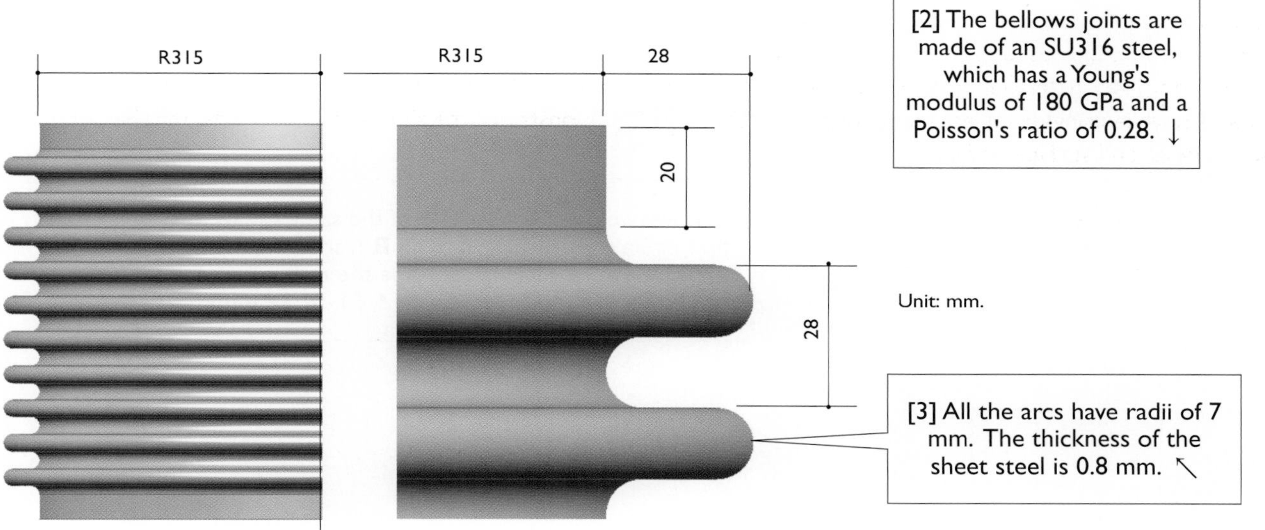

[3] All the arcs have radii of 7 mm. The thickness of the sheet steel is 0.8 mm. ↖

PART A. GEOMETRIC MODELING

6.1.2 Start Up

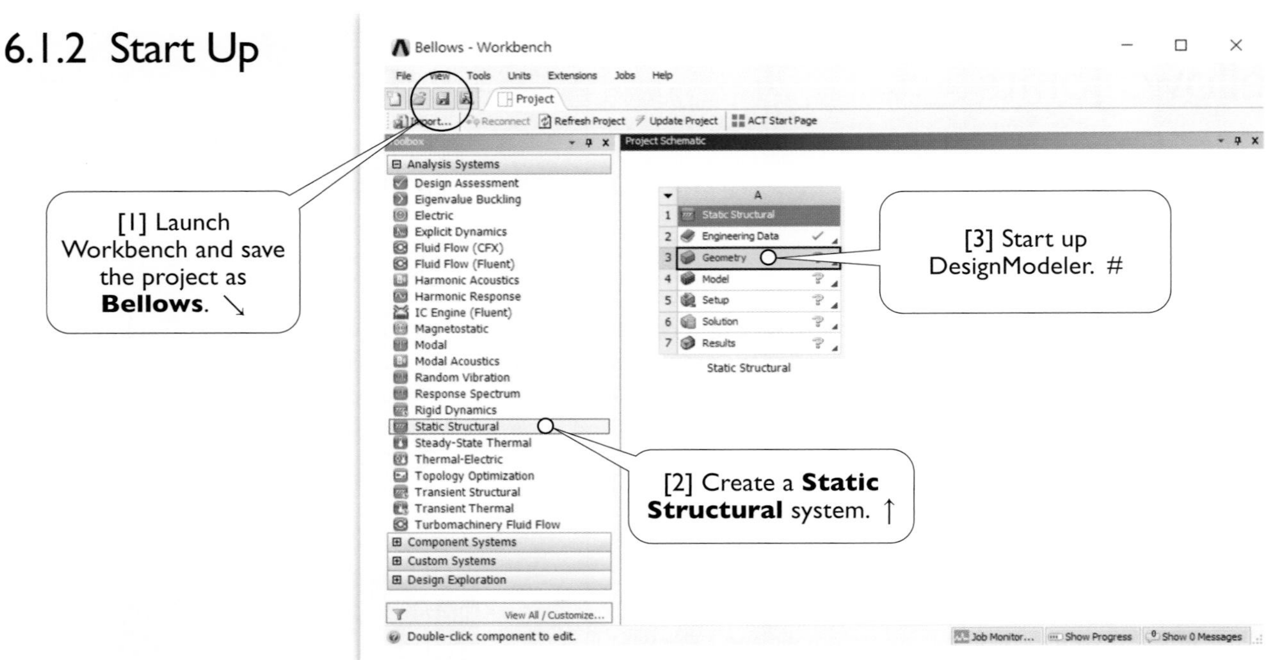

6.1.3 Create a Sketch

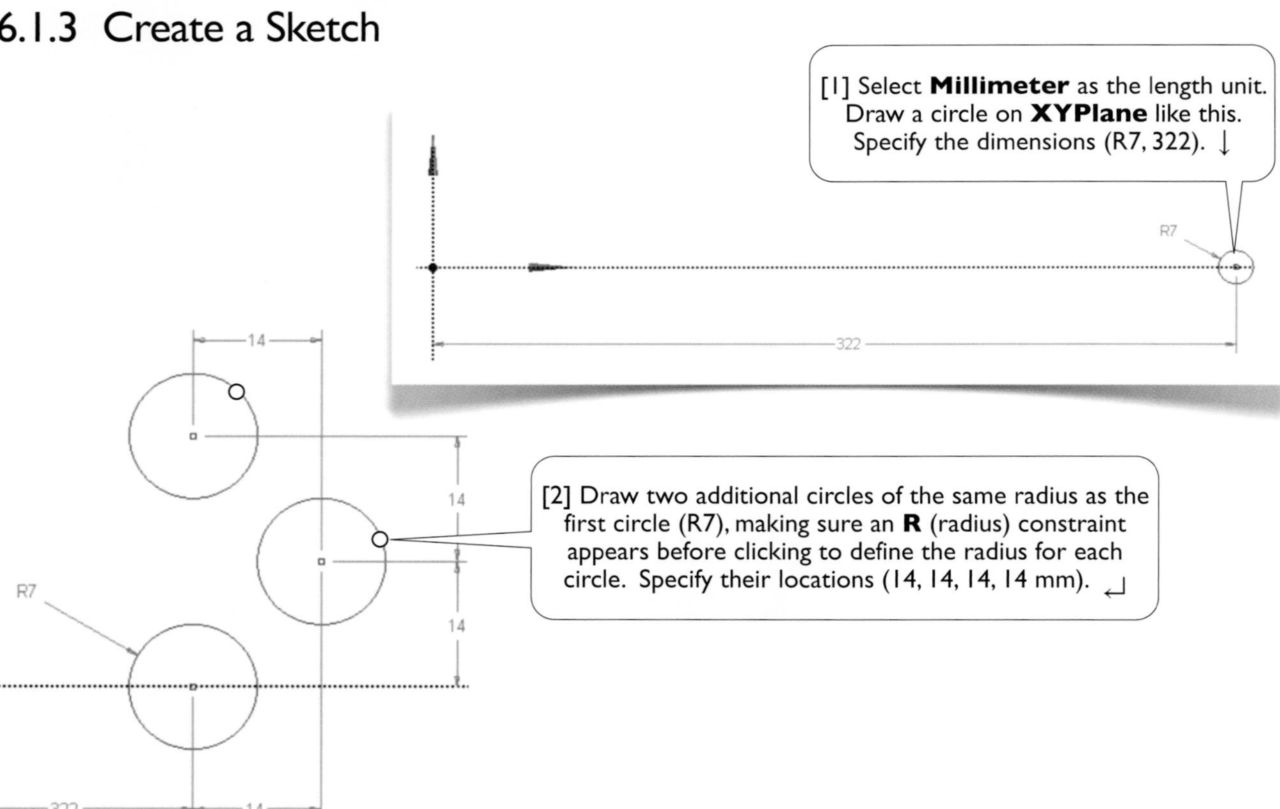

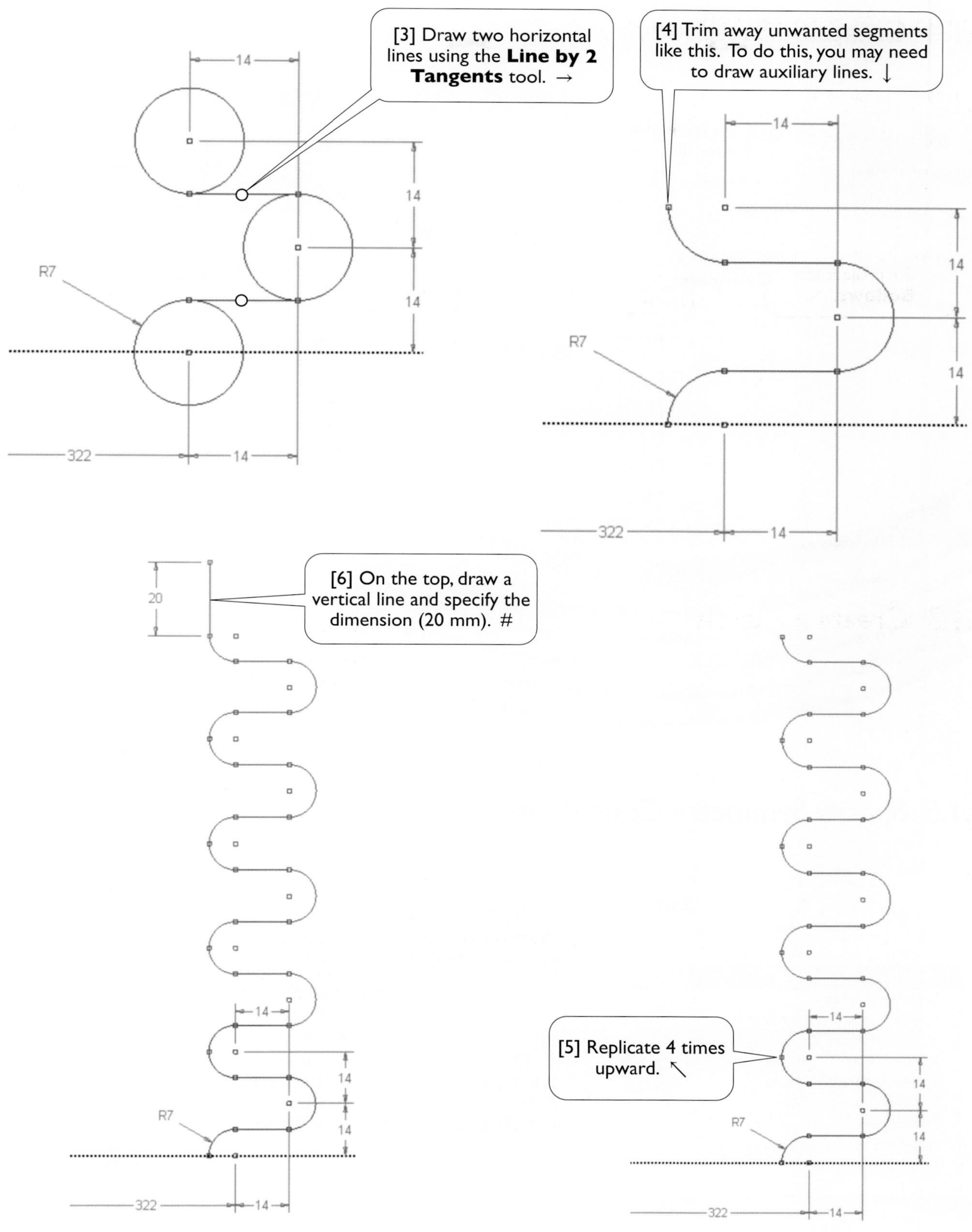
[3] Draw two horizontal lines using the **Line by 2 Tangents** tool. →
14
14
14
R7
322
14
[4] Trim away unwanted segments like this. To do this, you may need to draw auxiliary lines. ↓
14
14
14
R7
322
14
[6] On the top, draw a vertical line and specify the dimension (20 mm). #
20
14
14
14
R7
322
14
[5] Replicate 4 times upward. ↖
14
14
14
R7
322
14

6.1.4 Create Surface Body

Revolve

[1] Click **Revolve** on the toolbar. →

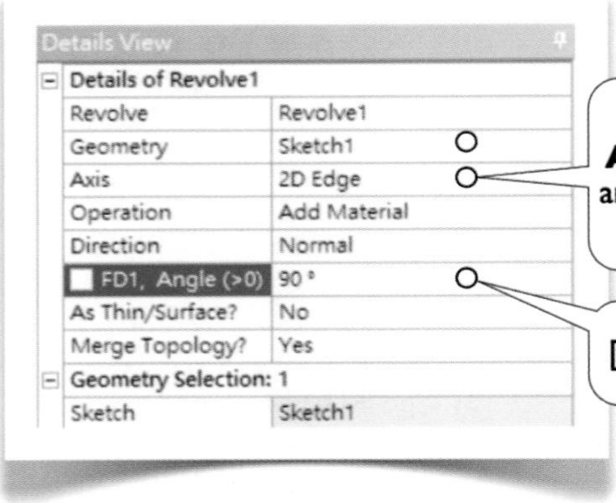

Details View	
Details of Revolve1	
Revolve	Revolve1
Geometry	Sketch1
Axis	2D Edge
Operation	Add Material
Direction	Normal
FD1, Angle (>0)	90 °
As Thin/Surface?	No
Merge Topology?	Yes
Geometry Selection: 1	
Sketch	Sketch1

[2] Click to bring forth **Apply/Cancel** buttons and select the Y-axis in the graphics window. ↓

[3] Type 90 (degrees). ←

Generate

[4] Click **Generate**. ↘

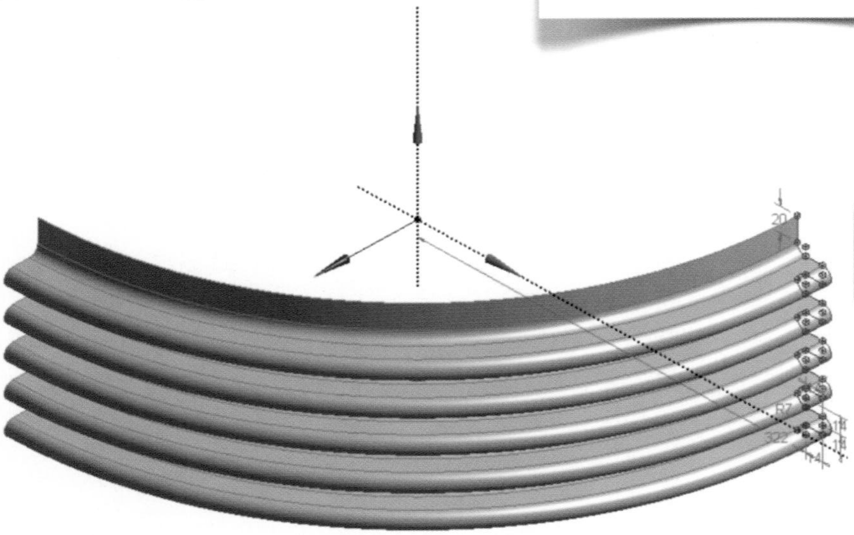

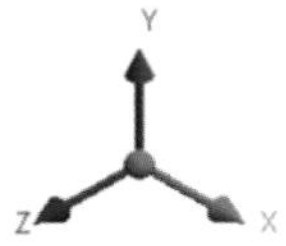

[5] A 3D surface body is created. Rotate to view the model isometrically. #

6.1.5 Specify Symmetry Conditions

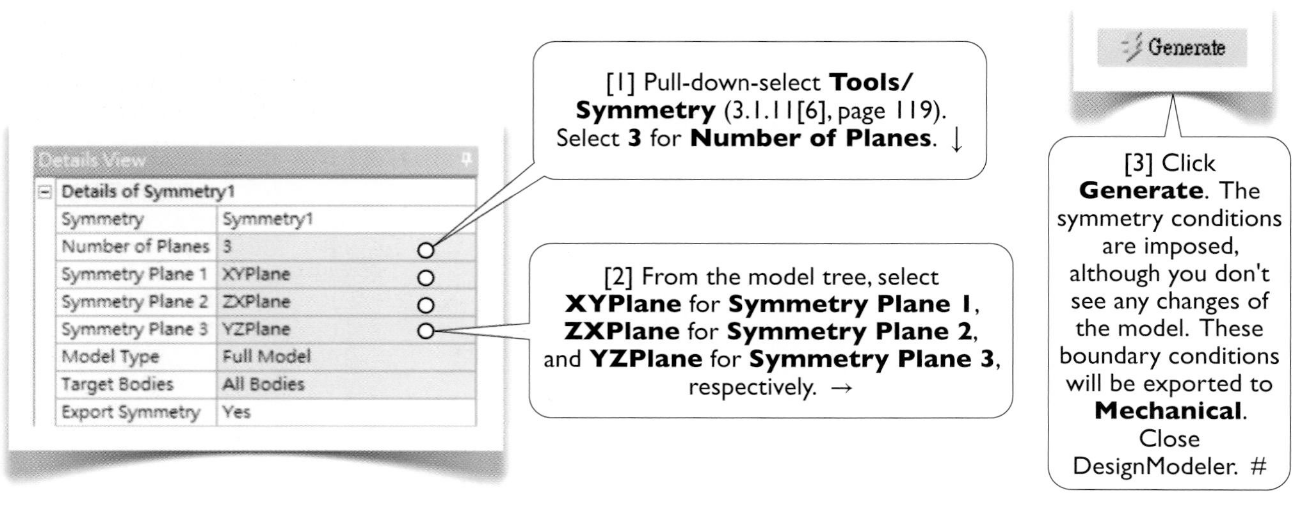

PART B. SIMULATION

6.1.6 Prepare Material Properties

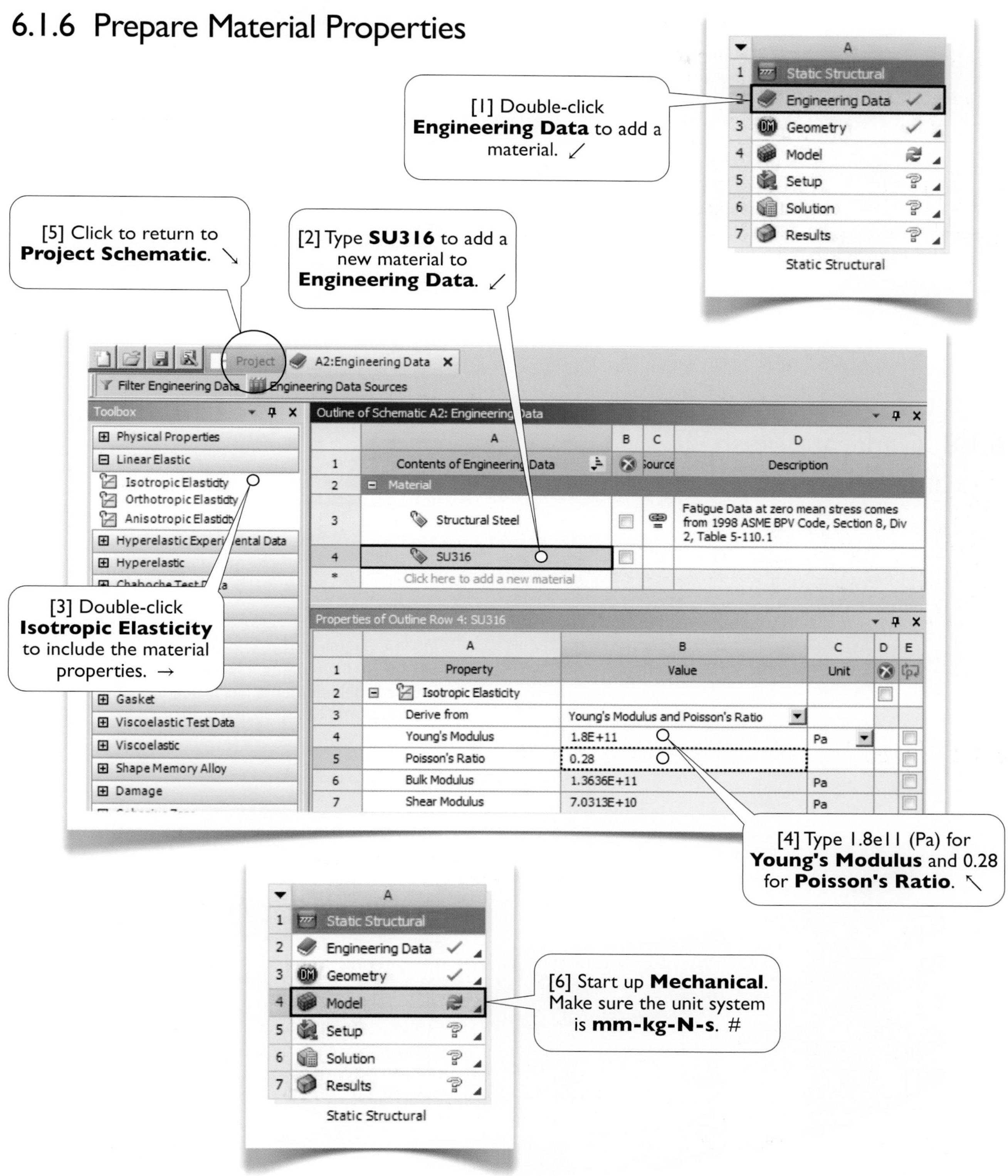

6.1.7 Assign Material and Thickness

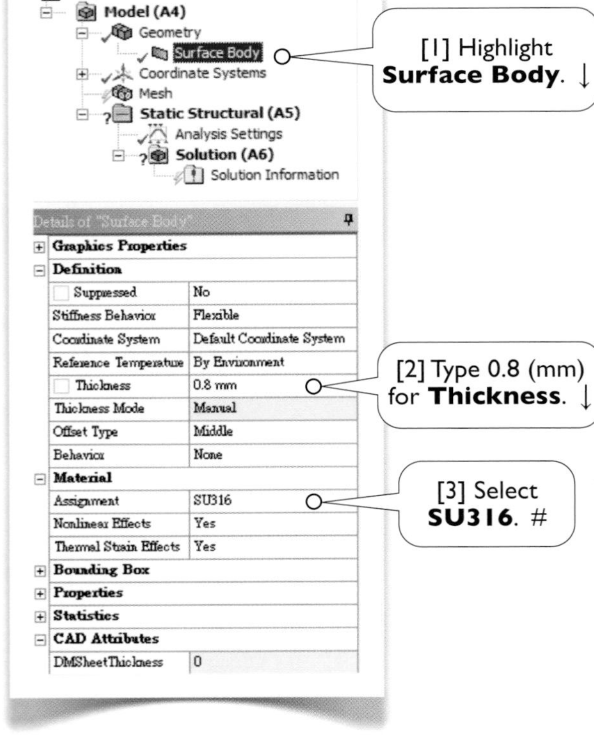

6.1.8 Generate Mesh

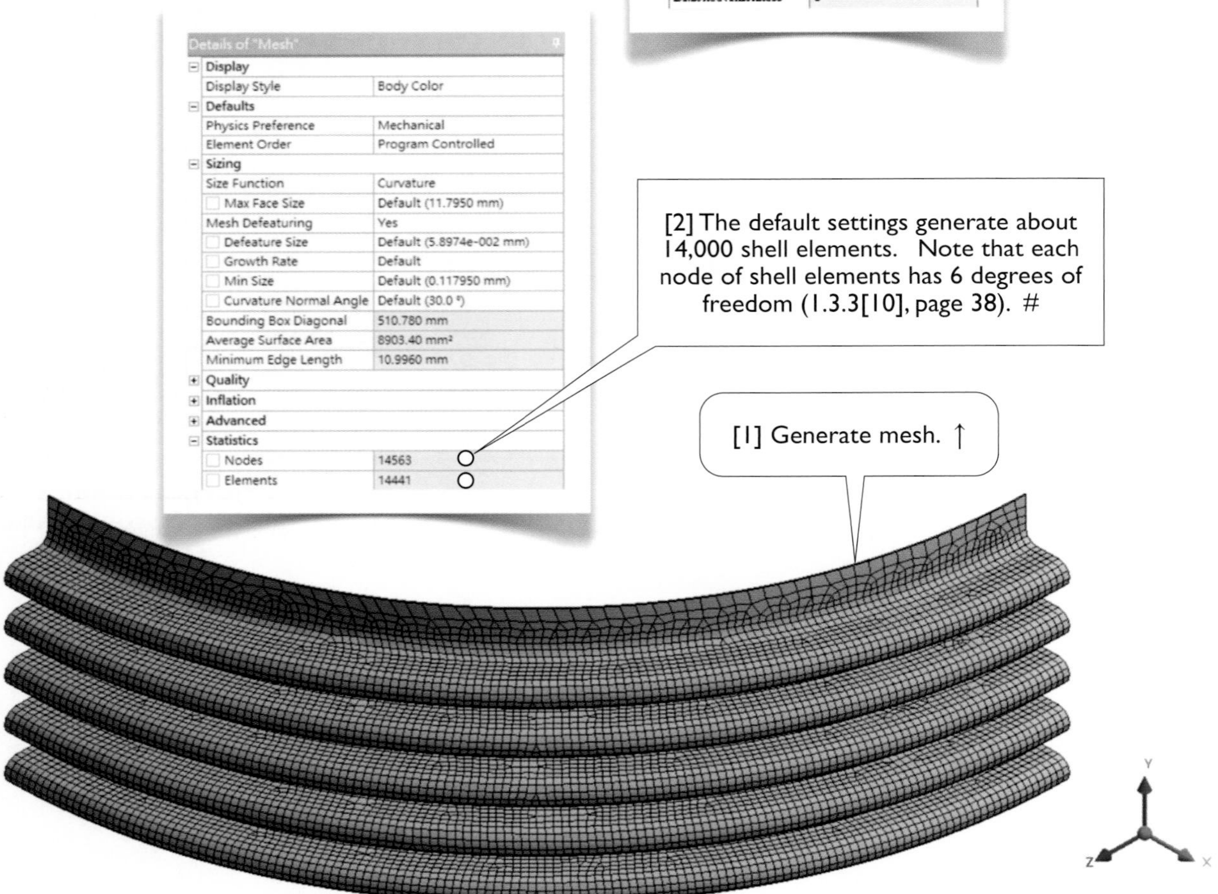

6.1.9 Specify Supports

6.1.10 Apply Pressure

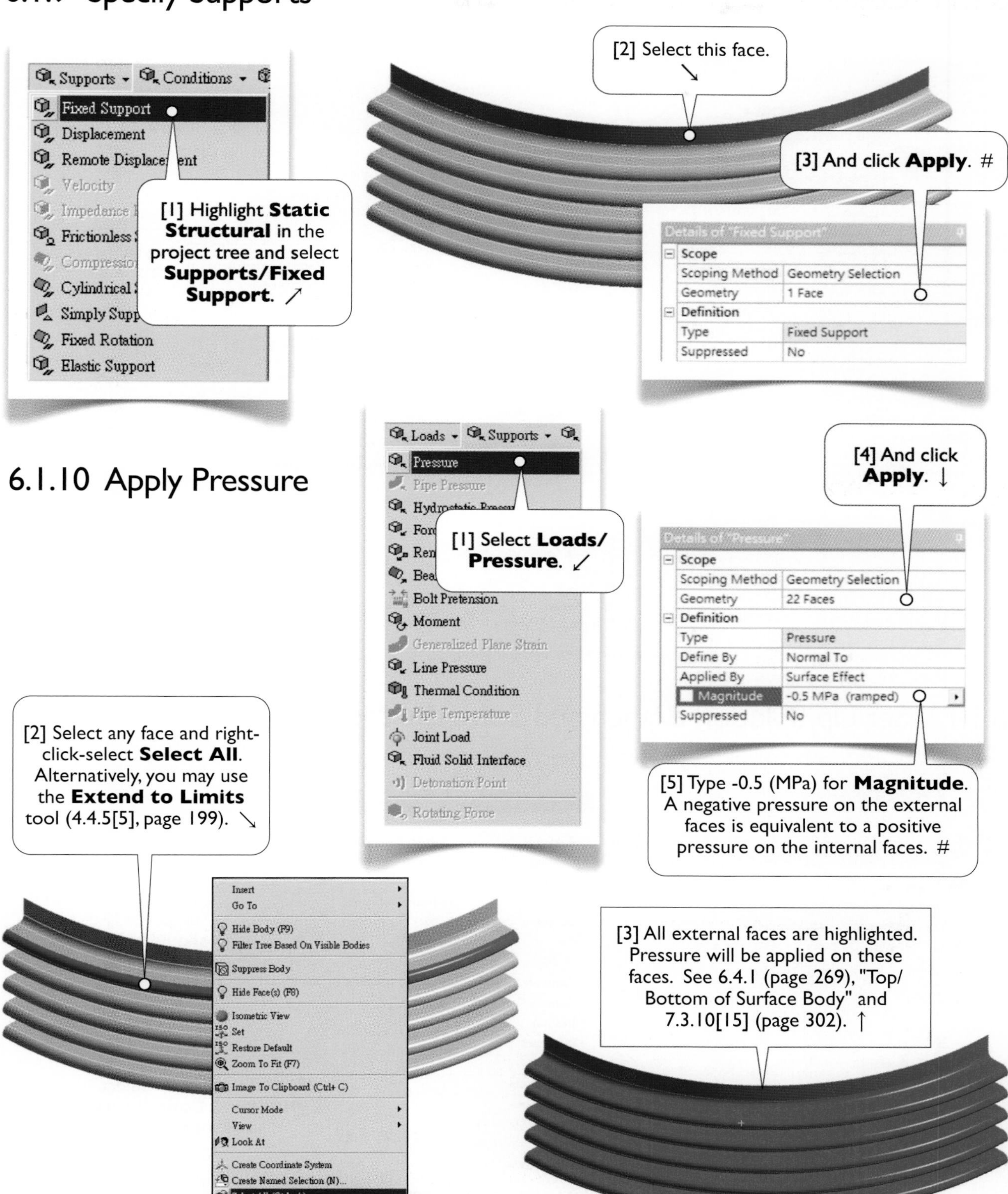

6.1.11 Create a Cylindrical Coordinate System

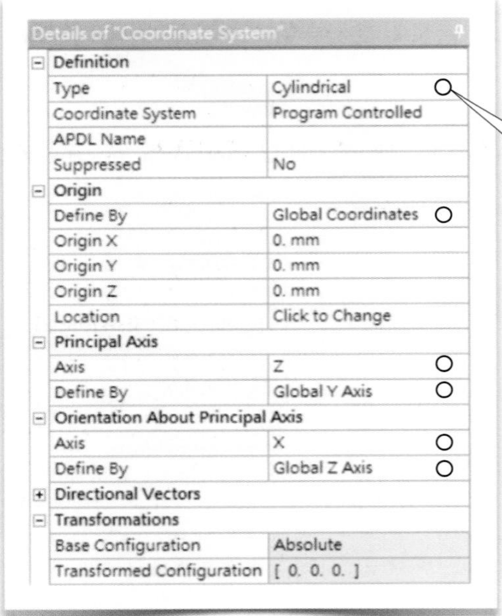

Details of "Coordinate System"	
Definition	
Type	Cylindrical
Coordinate System	Program Controlled
APDL Name	
Suppressed	No
Origin	
Define By	Global Coordinates
Origin X	0. mm
Origin Y	0. mm
Origin Z	0. mm
Location	Click to Change
Principal Axis	
Axis	Z
Define By	Global Y Axis
Orientation About Principal Axis	
Axis	X
Define By	Global Z Axis
Directional Vectors	
Transformations	
Base Configuration	Absolute
Transformed Configuration	[0. 0. 0.]

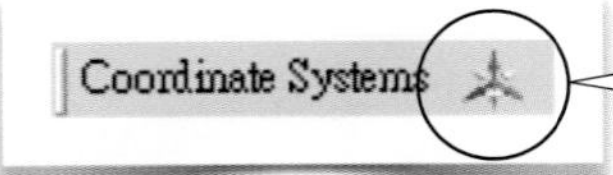

Coordinate Systems

[1] Highlight **Coordinate Systems** in the project tree, and click **Create Coordinate System** in the toolbar (5.2.10[2], page 224). ←

[2] Set up the coordinate system like this. →

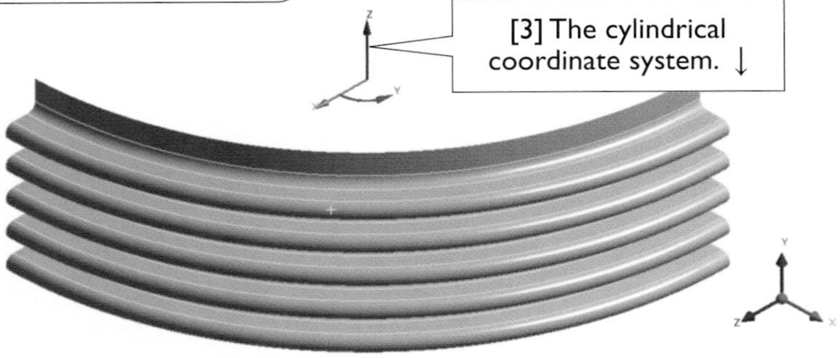

[3] The cylindrical coordinate system. ↓

Coordinate Systems
Global Coordinate System
XYPlane
ZXPlane
YZPlane
Cylindrical Coordinate System

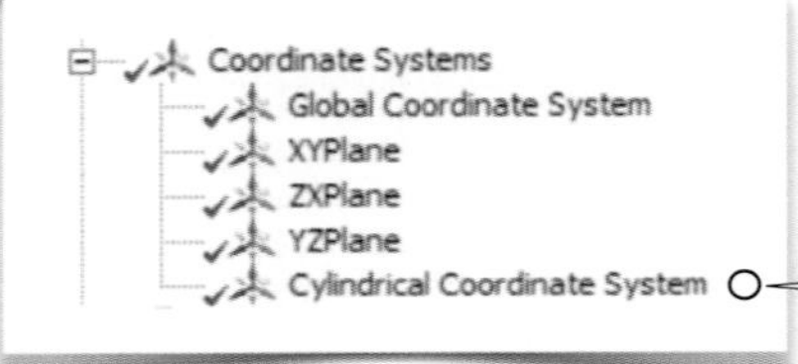

[4] In the project tree, rename the newly created coordinate system to **Cylindrical Coordinate System**. #

6.1.12 Set Up Solution Branch and Solve the Model

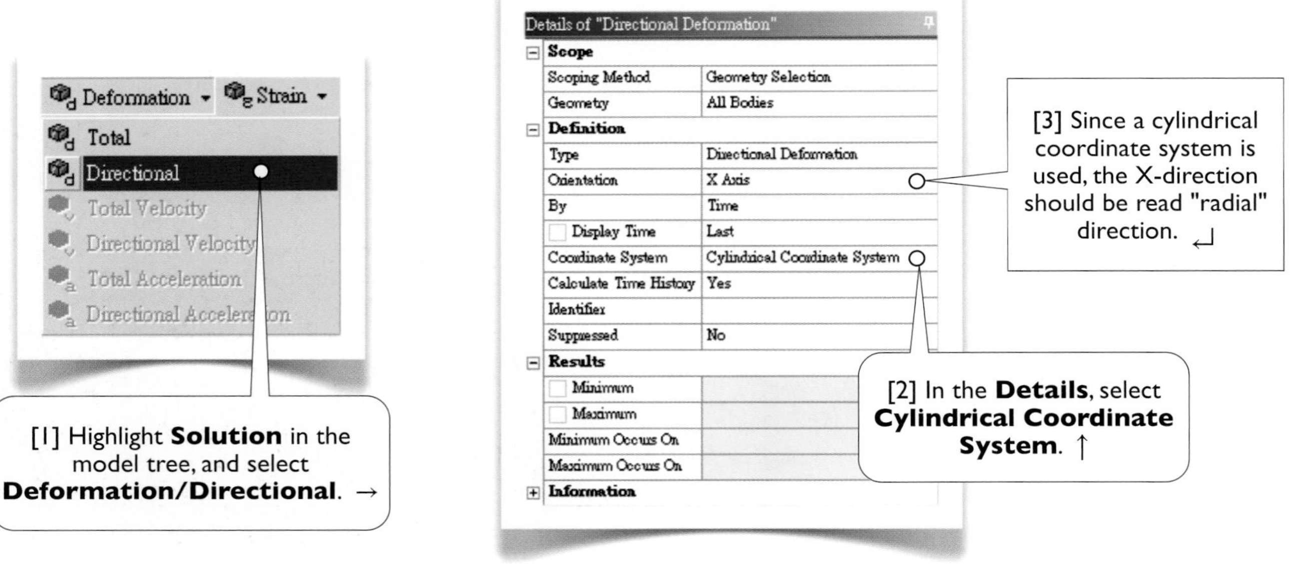

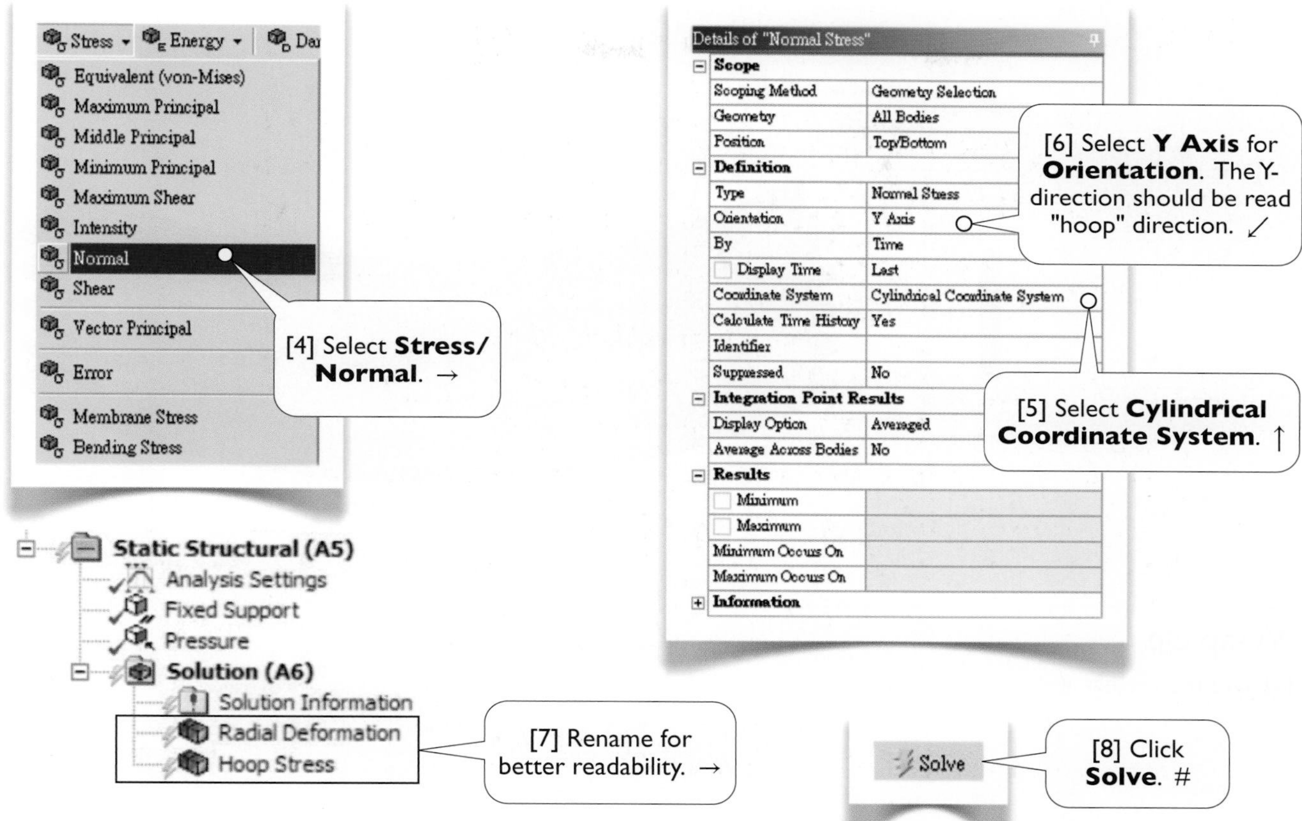

6.1.13 View the Results

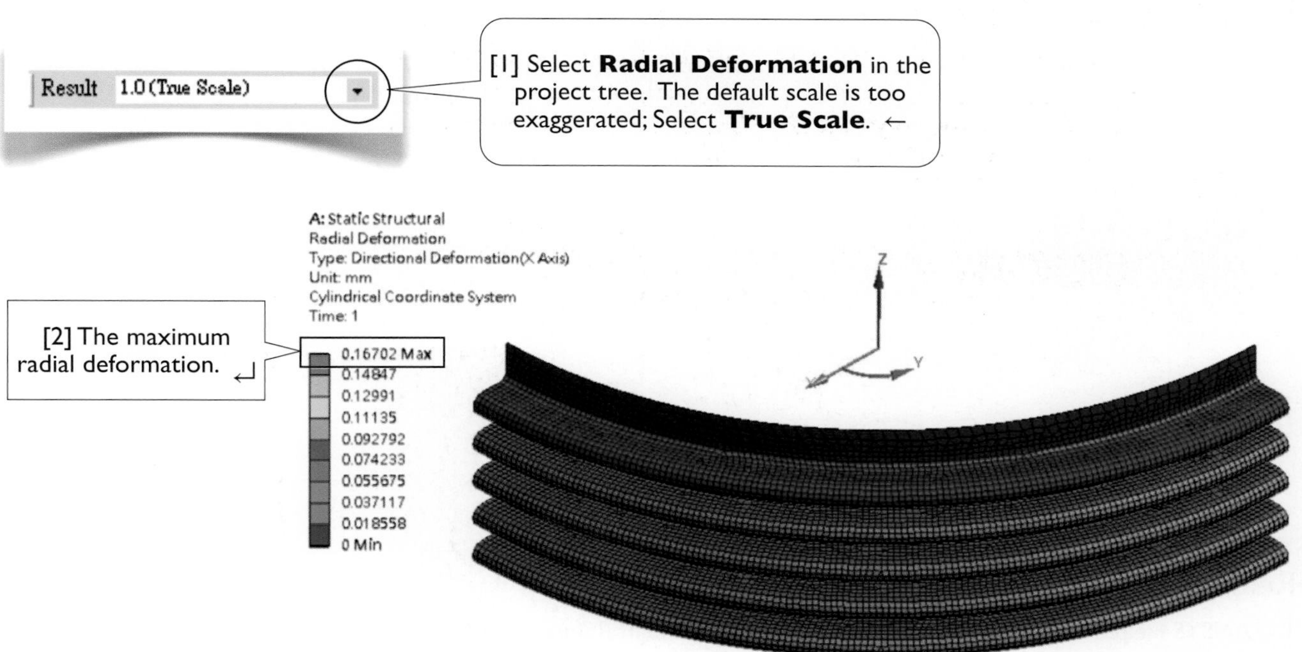

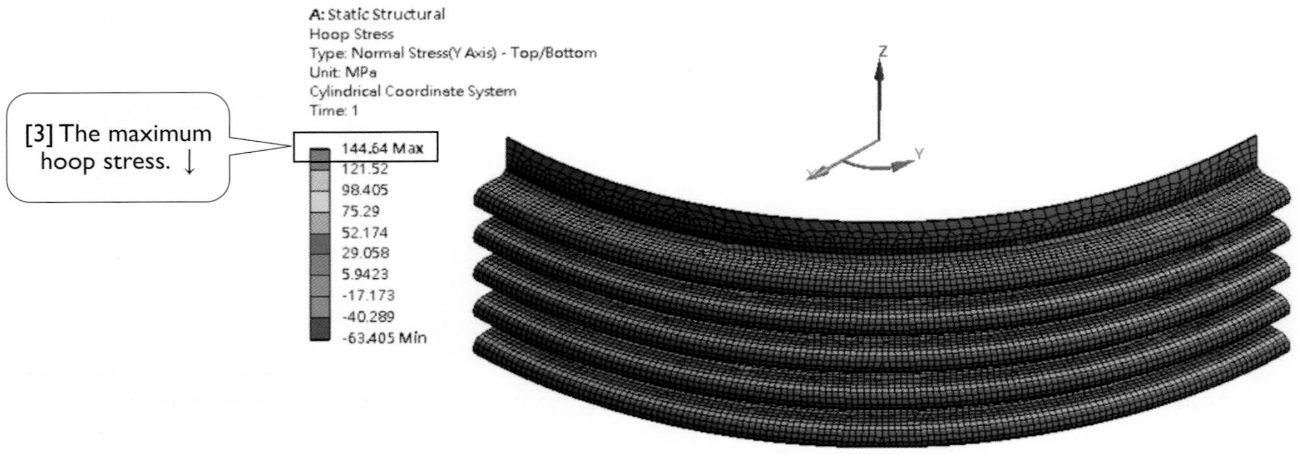

Wrap Up

[4] Save the project and exit Workbench. #

Reference

1. ANSYS Help//Mechanical APDL//Mechanical APDL Element Reference//I. Element Library//SHELL208

Section 6.2

Beam Bracket

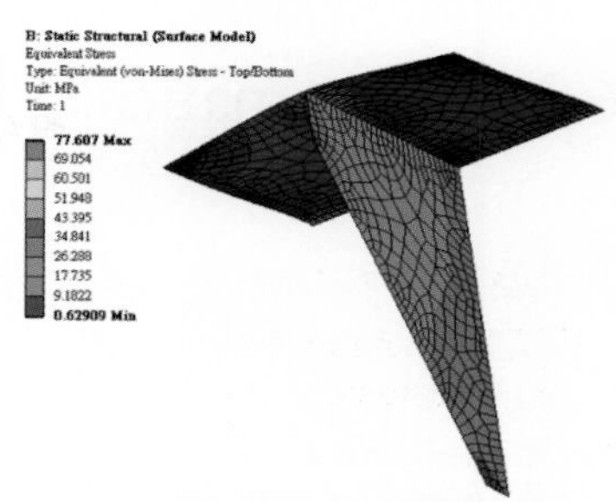

6.2.1 About the Beam Bracket

[1] In Section 4.1, we created a 3D solid model for the beam bracket, and the model was simulated in Section 5.1. Since the seat plate (flange) and the web plate are relatively thin and have uniform thicknesses, is it possible to model the beam bracket as a surface model and obtain a comparable result?

To create a surface model for the beam bracket, we don't have to start from scratch; we can use a tool in DesignModeler, called **Mid-Surface**. In fact, surface models are often created this way. CAD models are usually created as 3D solids, since they are created for multiple purposes, and simulation is only one of them. When a surface model is needed for simulation, **Mid-Surface** is a powerful tool to extract a surface model from a solid model. #

PART A. GEOMETRIC MODELING

6.2.2 Start Up

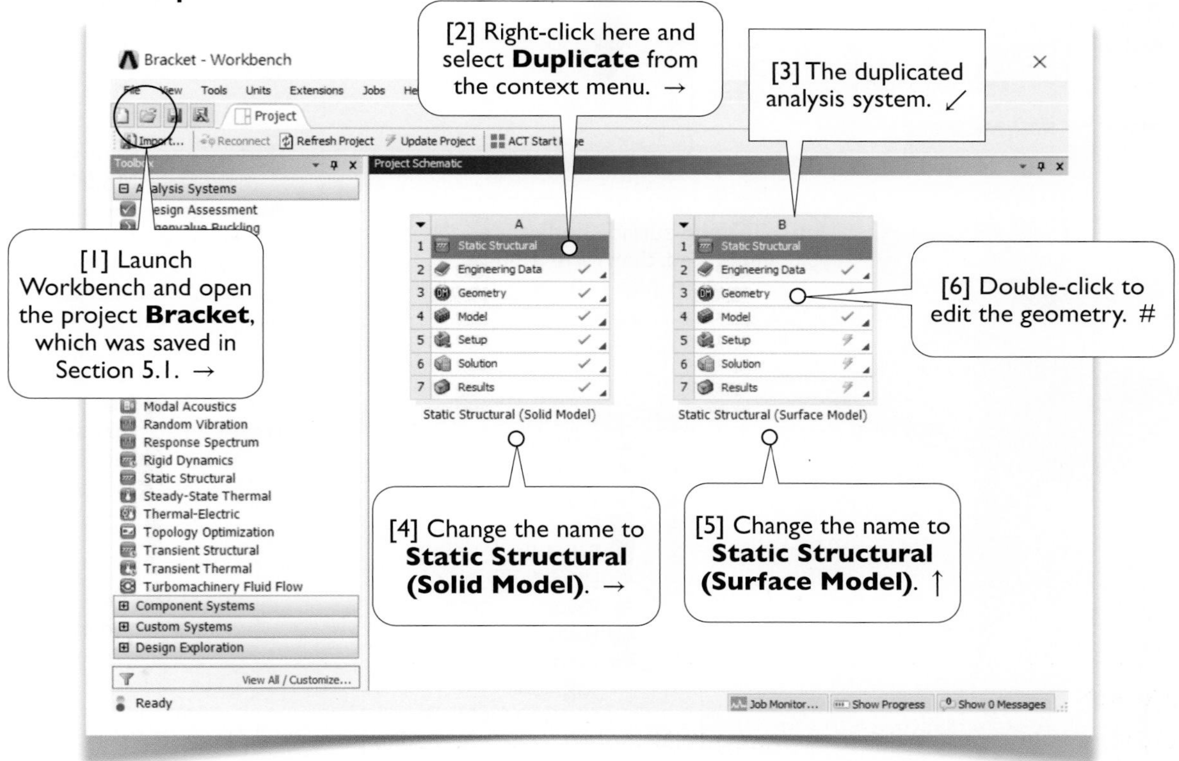

6.2.3 Automatically Find Face Pairs

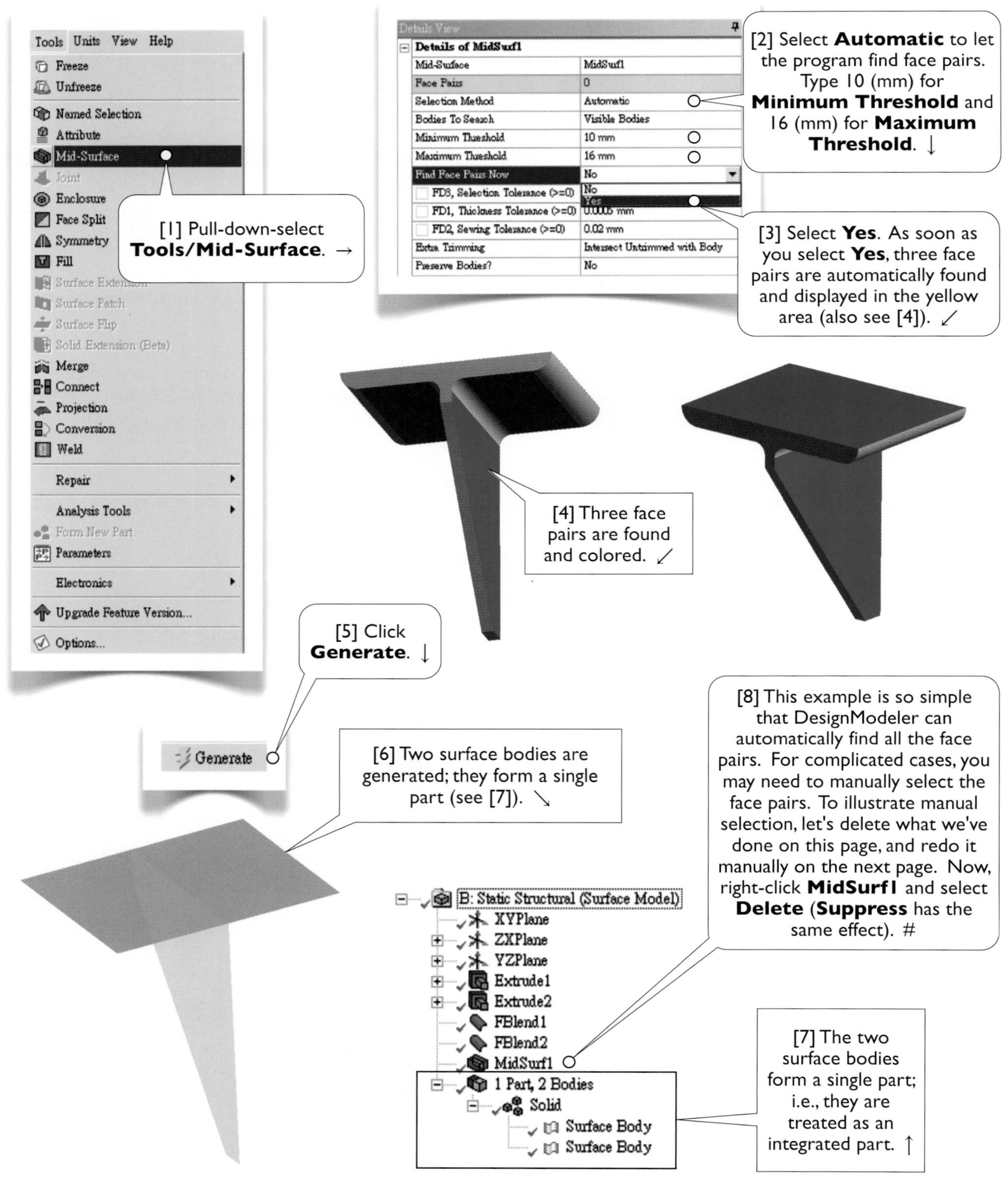

6.2.4 Manually Select Face Pairs

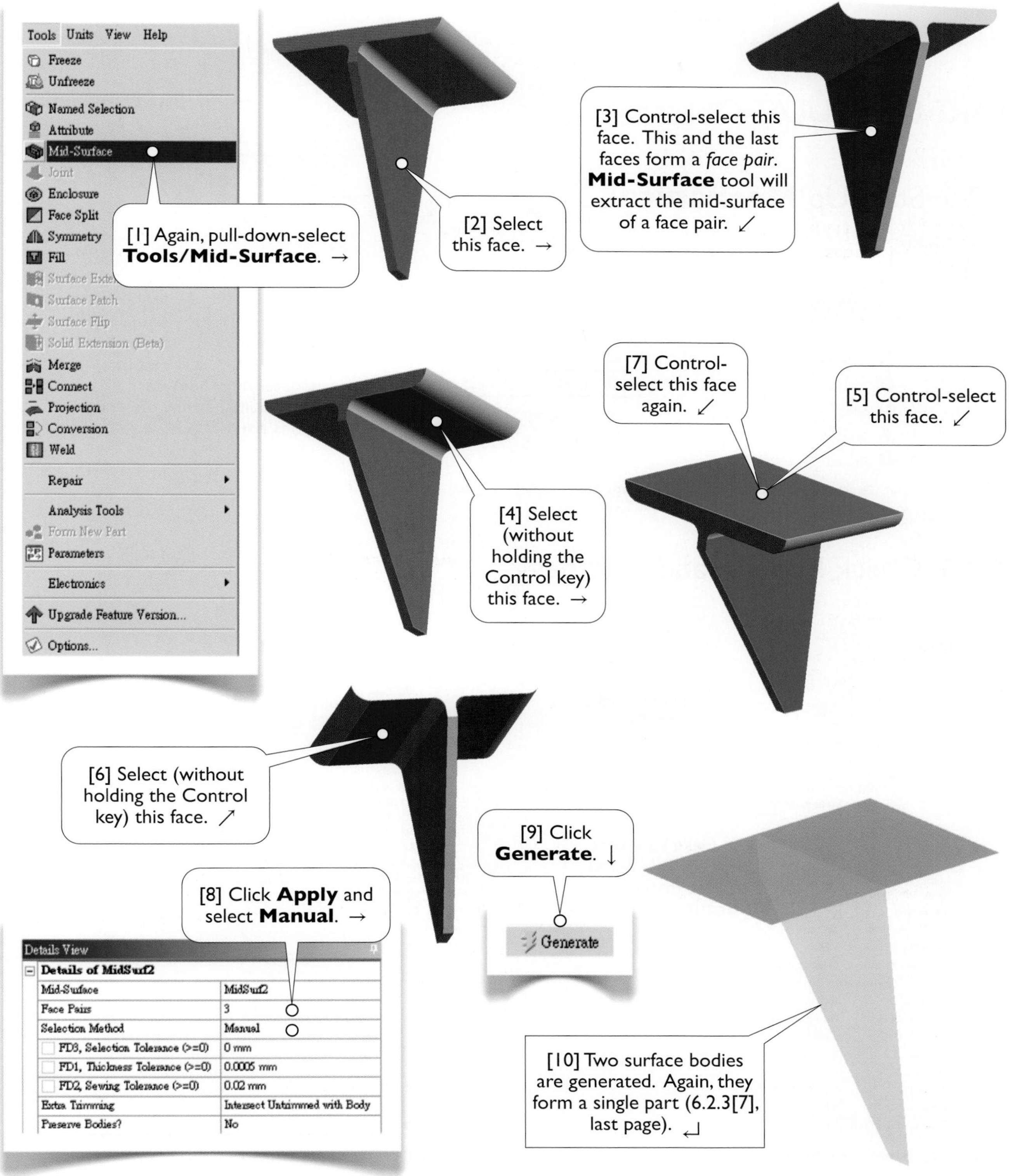

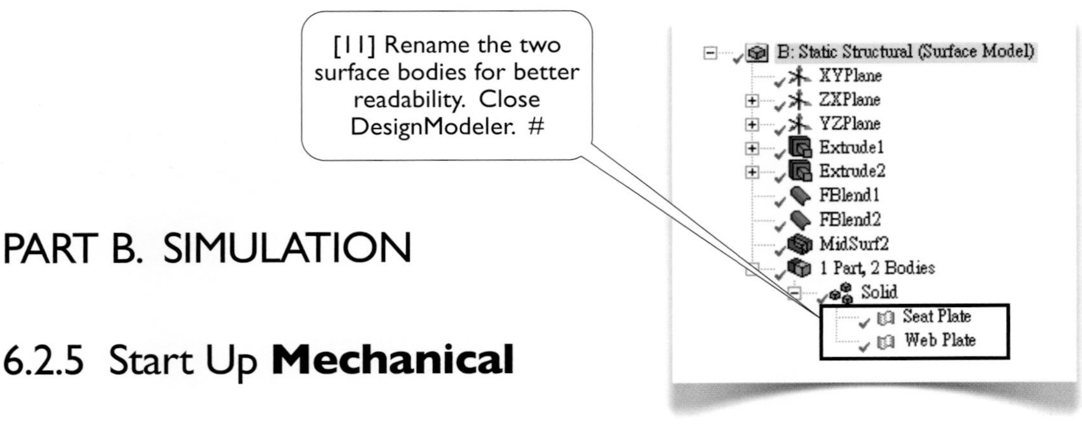

PART B. SIMULATION

6.2.5 Start Up **Mechanical**

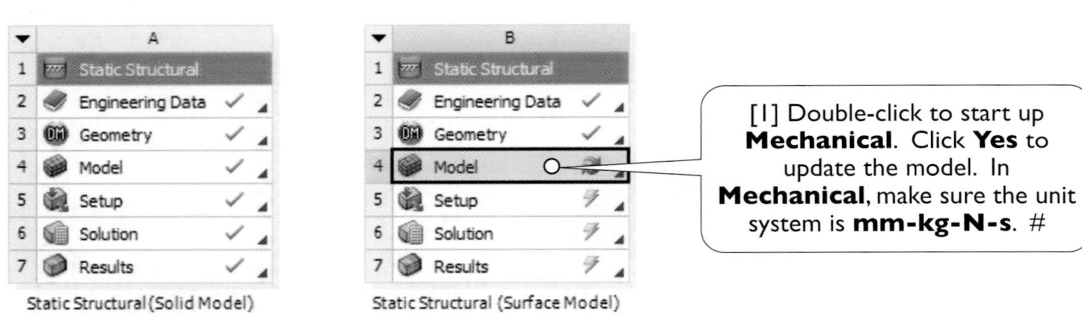

6.2.6 Check Material and Thickness

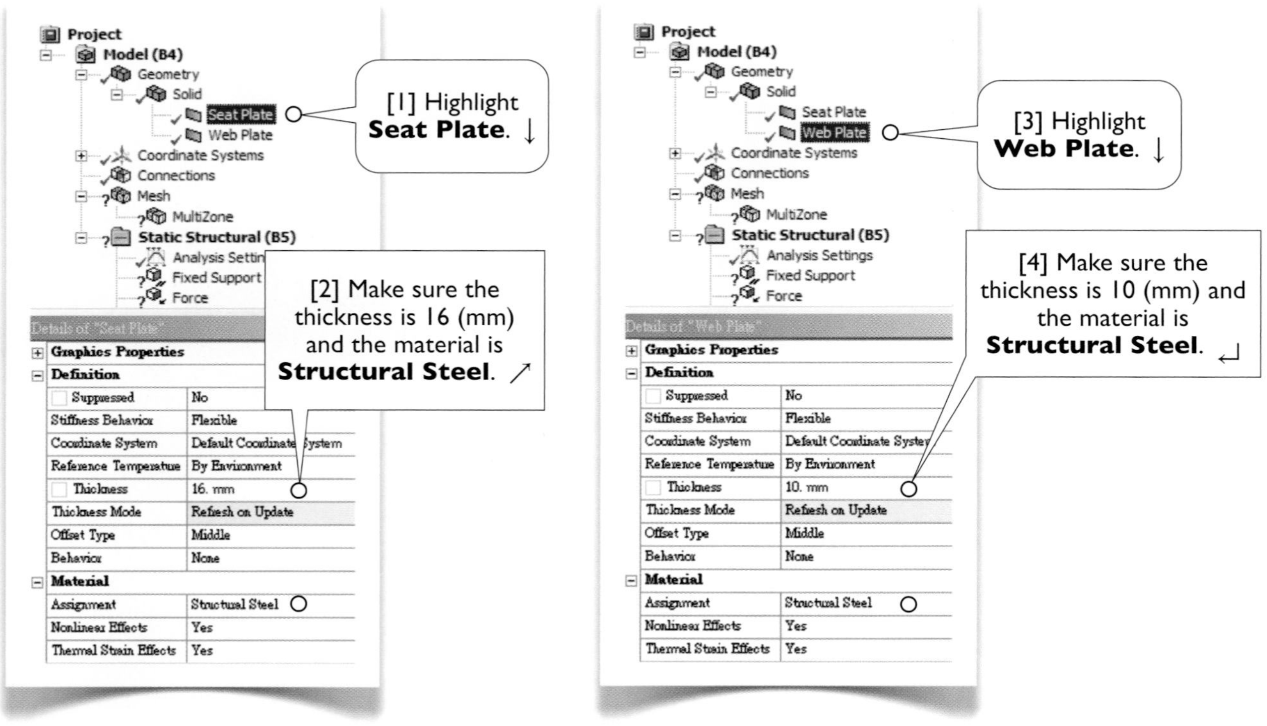

Thickness Assignment

[5] The surface bodies are created using the **Mid-Surface** tool. In most cases the thicknesses of bodies are correctly calculated and transferred to **Mechanical**. For some complicated cases, the thicknesses may not be correctly calculated and transferred. It is good practice to always check material and thickness assignments in **Mechanical**. #

6.2.7 Redefine Fixed Support

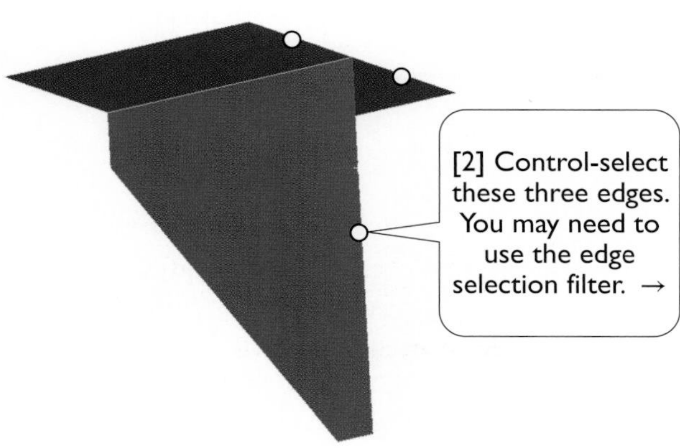

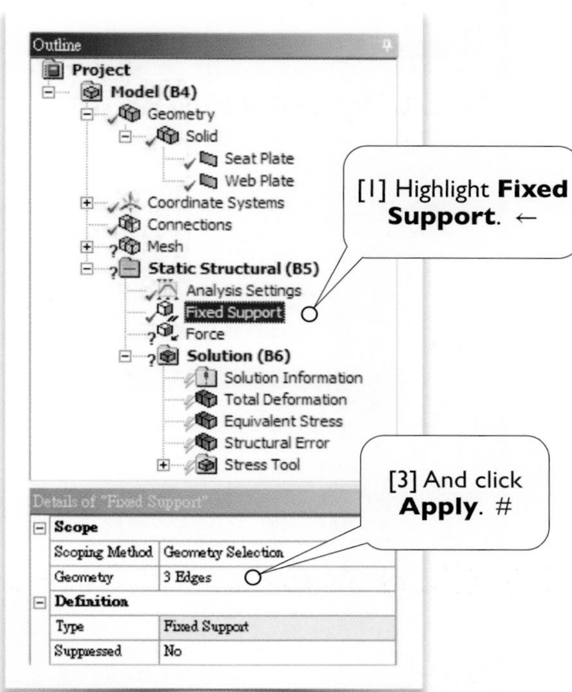

6.2.8 Redefine Force

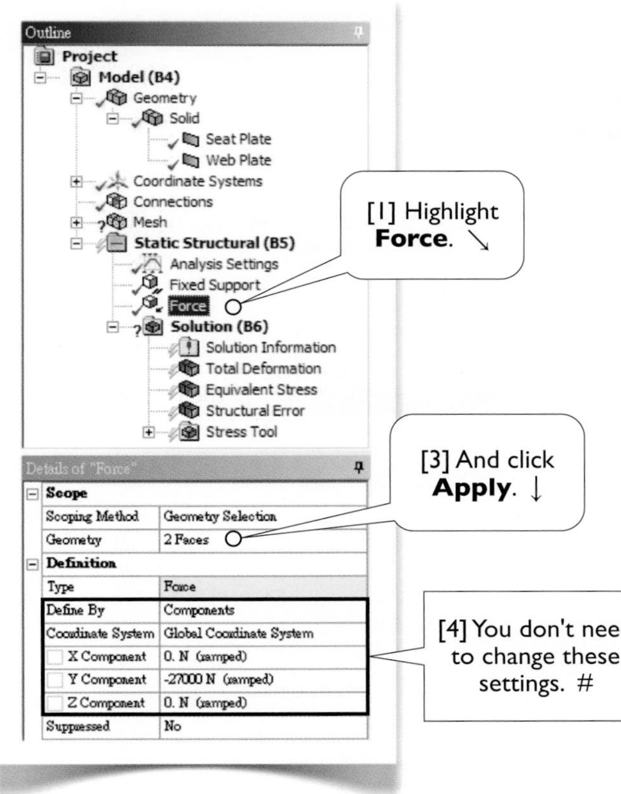

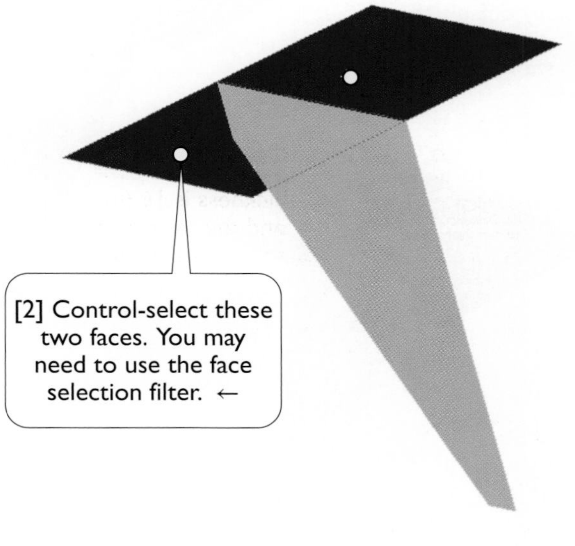

6.2.9 Generate Mesh

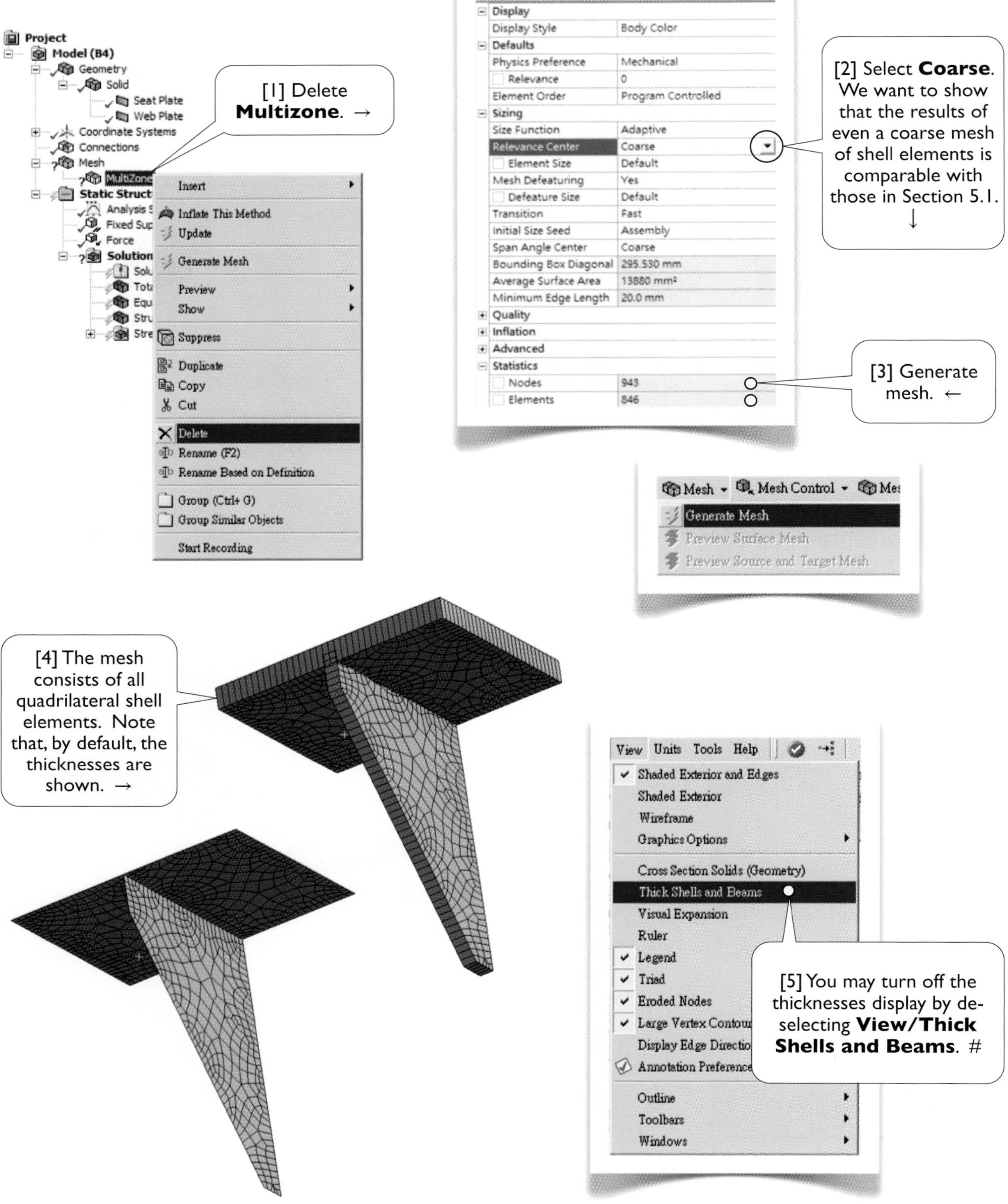

6.2.10 Solve the Model

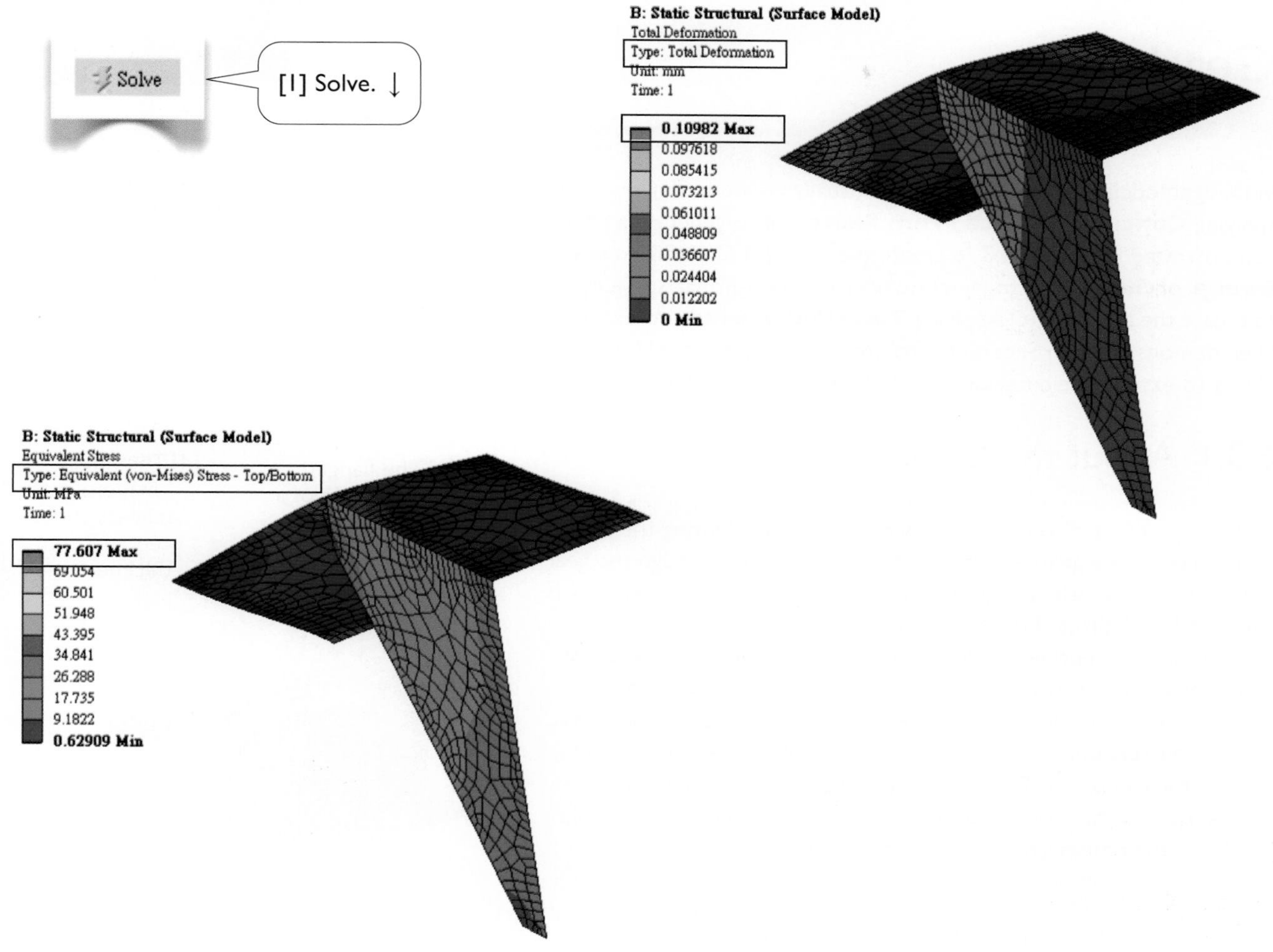

Surface Model versus Solid Model

[2] The results are quite comparable to those obtained in 5.1.13[8-9] (page 218). The surface model in this section consists of less than 1000 nodes (6.2.9[3], last page) while the solid model consists of about 16,000 nodes (page 217)! In general, before you decide to use a solid model, reconsider the possibility of using a surface (or line) model. ↓

Wrap Up

[3] Save the project and exit Workbench. #

Section 6.3

Gearbox

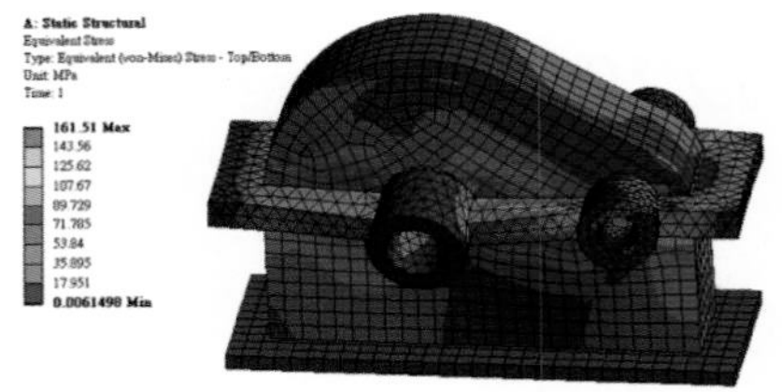

In DesignModeler, a 3D surface body is usually created using one of four techniques: (a) Creating a 2D sketch and applying **Concept/Surface From Sketches** to generate a planar surface body; this technique has been demonstrated in Section 4.3 to create the glass. (b) Creating an open sketch and then using **Extrude**, **Revolve**, **Sweep**, or **Skin/Loft** to generate planar or non-planar surfaces; this technique has been demonstrated in Section 6.1 to create the bellows. (c) Applying **Tool/Mid-Surface** to extract mid-surface from a solid body; this technique has been demonstrated in Section 6.2 to create a surface model for the beam bracket. (d) Using **Thin/Surface** tool; the idea is to extract the exterior "shell" of a solid body. We will demonstrate this technique in this section.

6.3.1 About the Gearbox[Ref 1]

[1] Gears are used to transmit mechanical power. During the power transmission, the gears exert forces on each other. Through the gear shafts, the forces eventually reach the bearing supports. A gearbox is designed to withstand the bearing forces.

In this case, engineers are concerned about the deformation of the gearbox, for it may cause a displacement of the rotating axes, which may reach a point where the transmission becomes defective.

In this section, we will create a 3D model for the gearbox and conduct a simulation. The model will consist of two solid bodies for the flange and the base, made of a gray cast iron [2-3], and a surface body for the housing, made of a stainless steel [4]. →

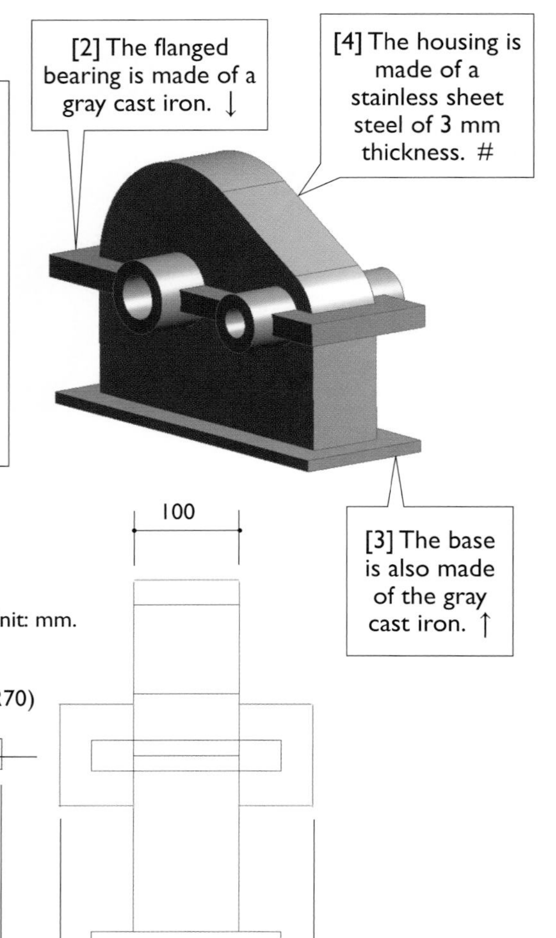

PART A. GEOMETRIC MODELING

6.3.2 Start Up

[1] Launch Workbench. Create a **Static Structural** system. Save the project as **Gearbox**. →

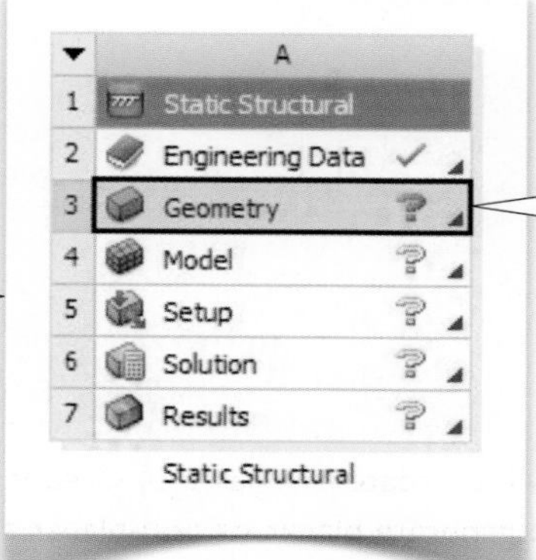

	A	
1	Static Structural	
2	Engineering Data	✓
3	Geometry	?
4	Model	?
5	Setup	?
6	Solution	?
7	Results	?

Static Structural

[2] Start up DesignModeler. #

6.3.3 Create the Housing

[1] Select **Millimeter** as the length unit. Create a sketch on **XYPlane** like this. Impose **Tangent** constraints wherever necessary. Trim away all internal segments. Leave only the outlines. →

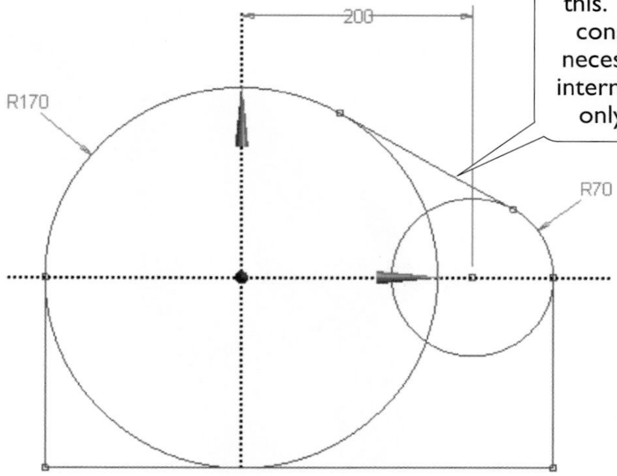

200
R170
R70

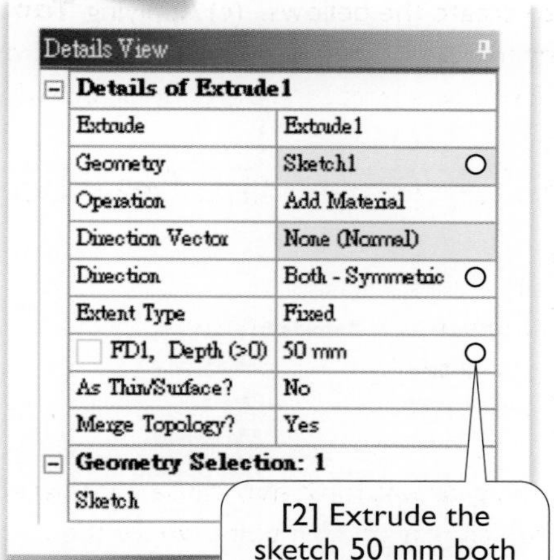

Details View

Details of Extrude1	
Extrude	Extrude1
Geometry	Sketch1
Operation	Add Material
Direction Vector	None (Normal)
Direction	Both - Symmetric
Extent Type	Fixed
FD1, Depth (>0)	50 mm
As Thin/Surface?	No
Merge Topology?	Yes
Geometry Selection: 1	
Sketch	

[2] Extrude the sketch 50 mm both sides. Remember to click **Generate**. ←

R170
200
R70

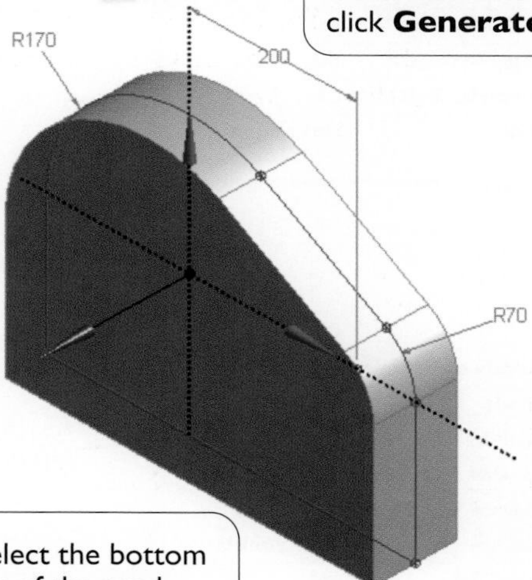

[3] Click **Thin/Surface** on the toolbar. ↓

Thin/Surface

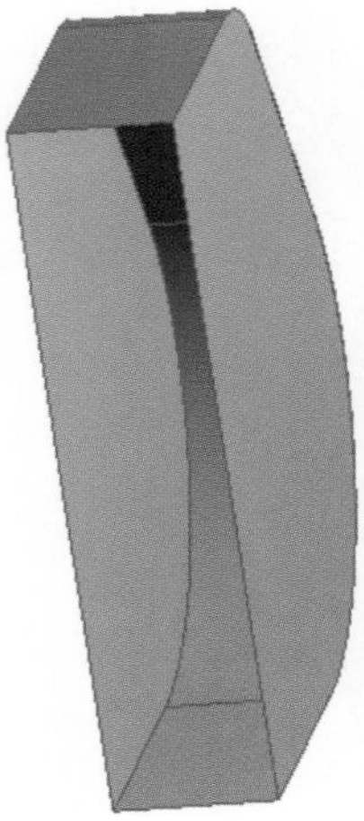

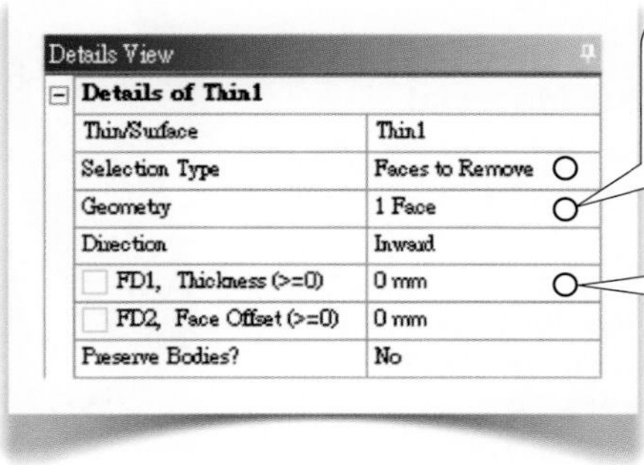

Details View

Details of Thin1	
Thin/Surface	Thin1
Selection Type	Faces to Remove
Geometry	1 Face
Direction	Inward
FD1, Thickness (>=0)	0 mm
FD2, Face Offset (>=0)	0 mm
Preserve Bodies?	No

[4] Select the bottom face of the newly created solid body. ↓

[5] To create a surface body, you must type a zero value for the thickness. If you type a nonzero value, it would create a "thin solid body" (see 4.5.5[1-2], page 206) instead of a surface body. Remember to click **Generate**. #

6.3.4 Create the Bearings

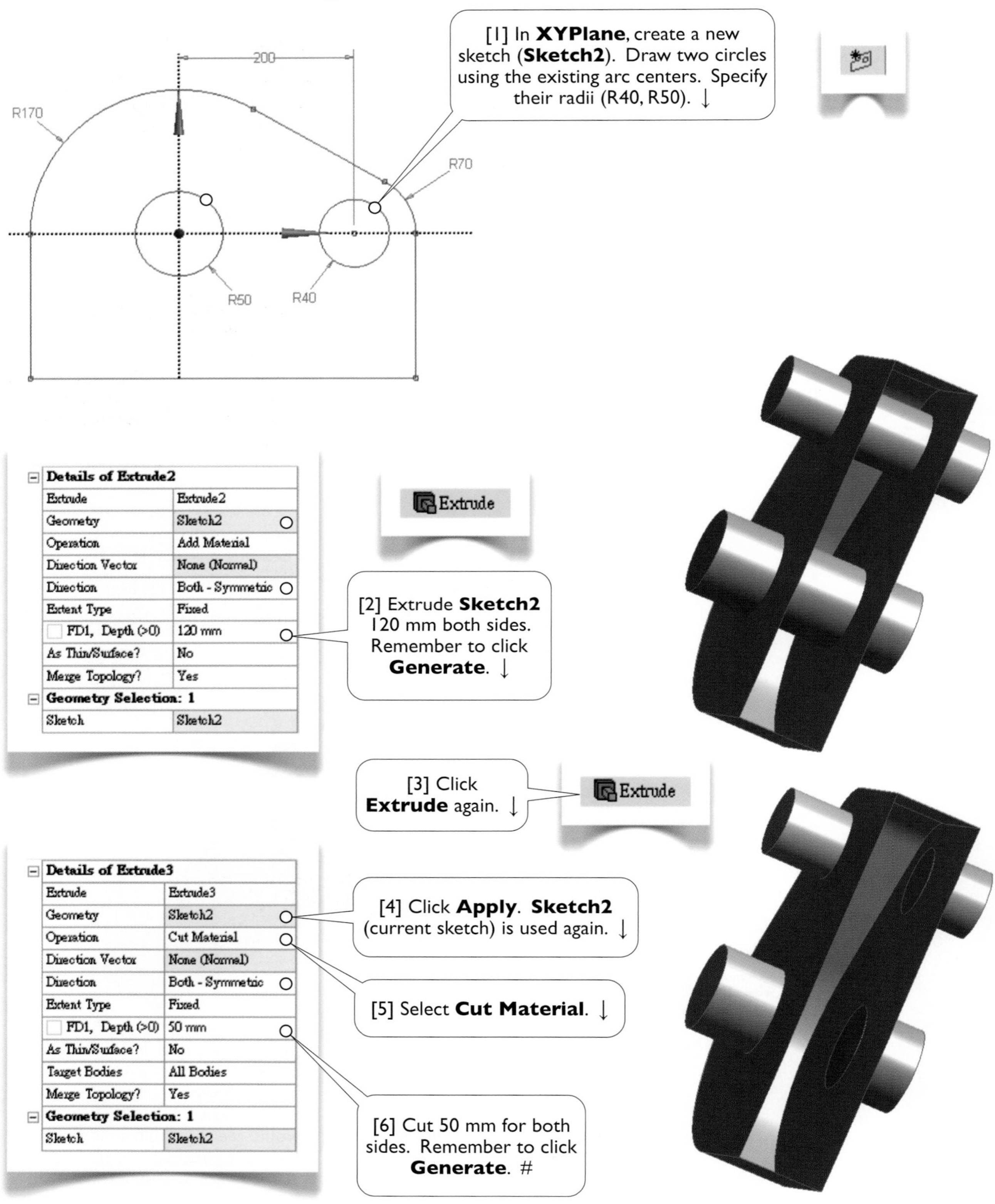

Details of Extrude2	
Extrude	Extrude2
Geometry	Sketch2
Operation	Add Material
Direction Vector	None (Normal)
Direction	Both - Symmetric
Extent Type	Fixed
FD1, Depth (>0)	120 mm
As Thin/Surface?	No
Merge Topology?	Yes
Geometry Selection: 1	
Sketch	Sketch2

Details of Extrude3	
Extrude	Extrude3
Geometry	Sketch2
Operation	Cut Material
Direction Vector	None (Normal)
Direction	Both - Symmetric
Extent Type	Fixed
FD1, Depth (>0)	50 mm
As Thin/Surface?	No
Target Bodies	All Bodies
Merge Topology?	Yes
Geometry Selection: 1	
Sketch	Sketch2

6.3.5 Create the Flange

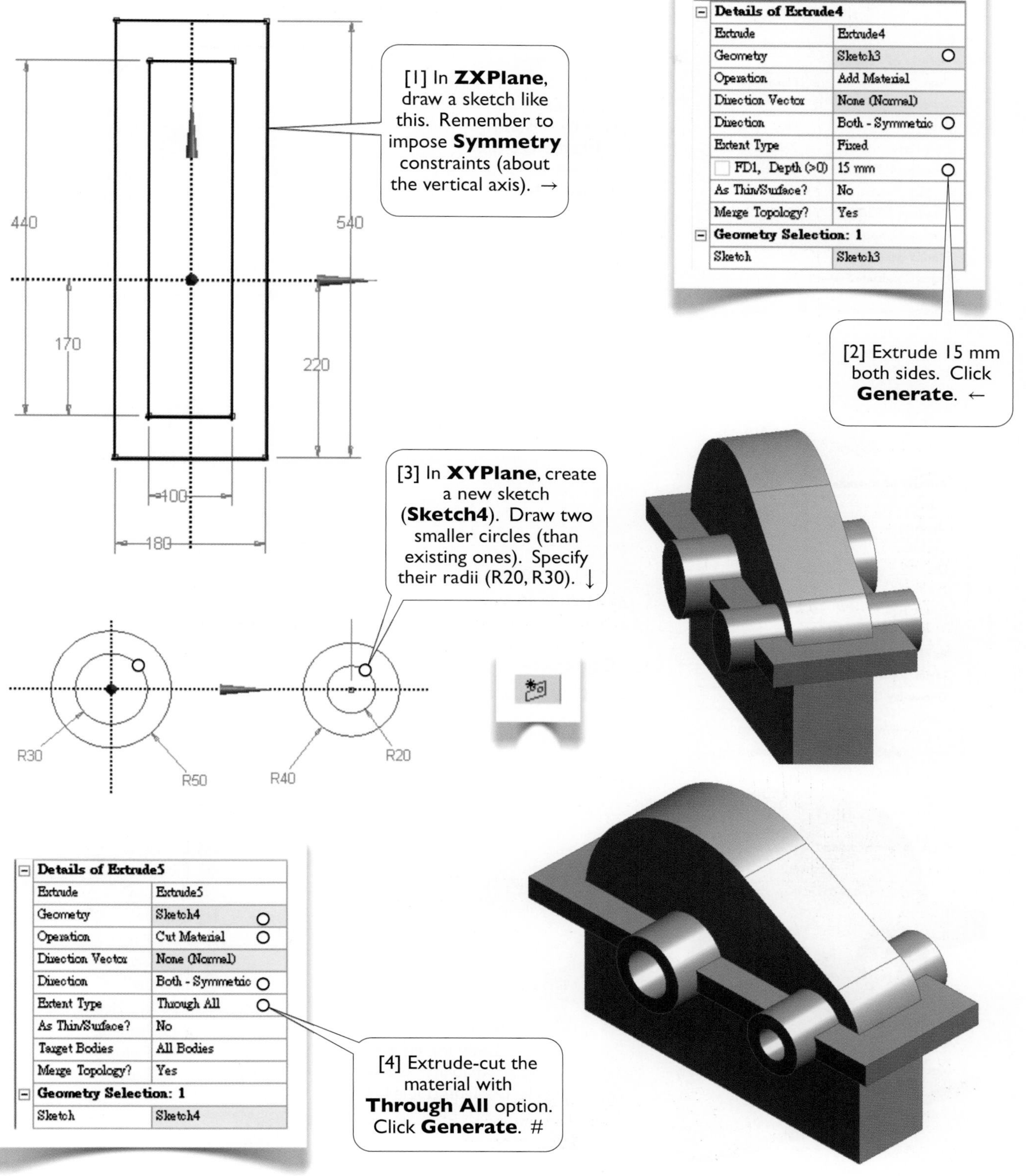

6.3.6 Create the Base

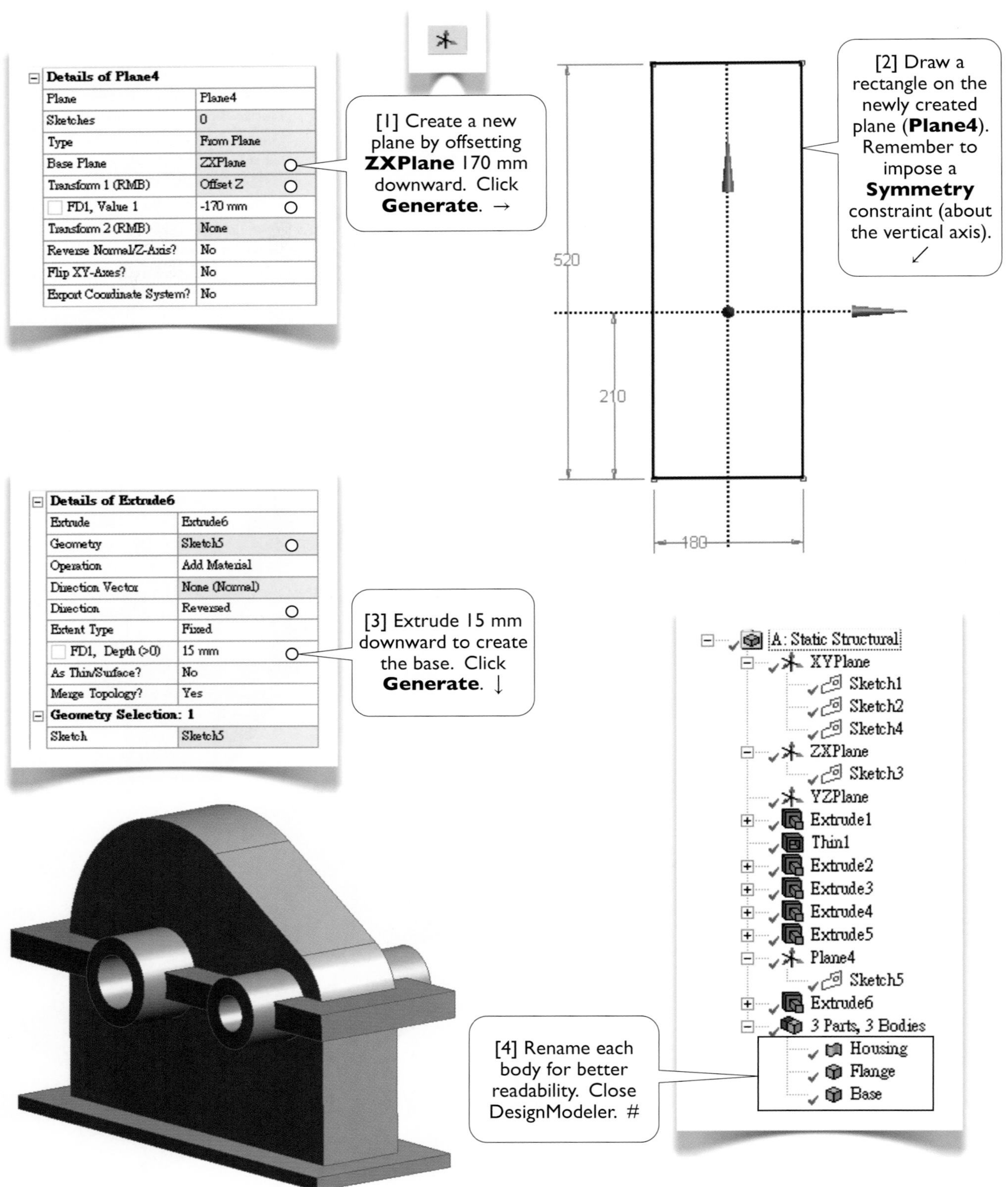

PART B. SIMULATION

6.3.7 Bearing Loads

[1] The bearing loads can be calculated according to the transmitted power, which is 175 hp in this case. Skipping the calculation details (if you are interested in the calculation details, please see the book by Zahavi[Ref 1]), we summarize the bearing loads as follows. Note that these forces sum up to zero. ↓

Bearing	F_X (N)	F_Y (N)	F_Z (N)
A	16000	30000	16000
B	6000	30000	0
C	3000	-30000	0
D	-25000	-30000	-16000

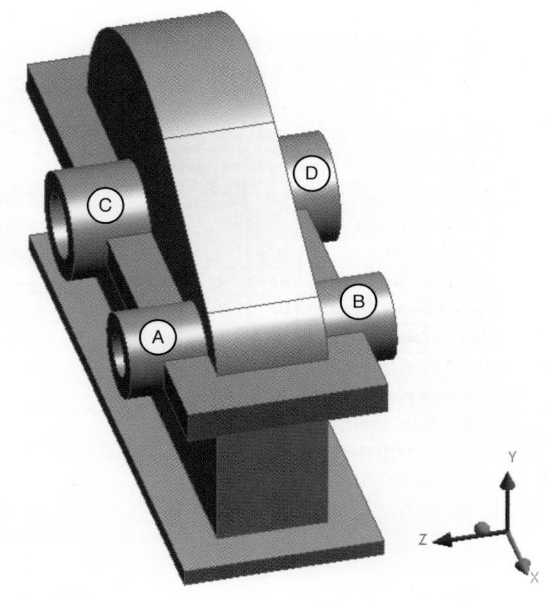

Specify Bearing Loads

[2] In **Mechanical**, a bearing load can be specified on a cylindrical surface. It is important to note that the bearing load is distributed on compressive side using projected area. This implies that axial components (here, the Z-components) will be zeros. In our case, the X- and Y-components (please refer to the global coordinates) of the bearing loads can be specified as bearing loads components, but the Z-components cannot be specified as bearing load components. The Z-components will be specified separately. #

6.3.8 Prepare Material Properties

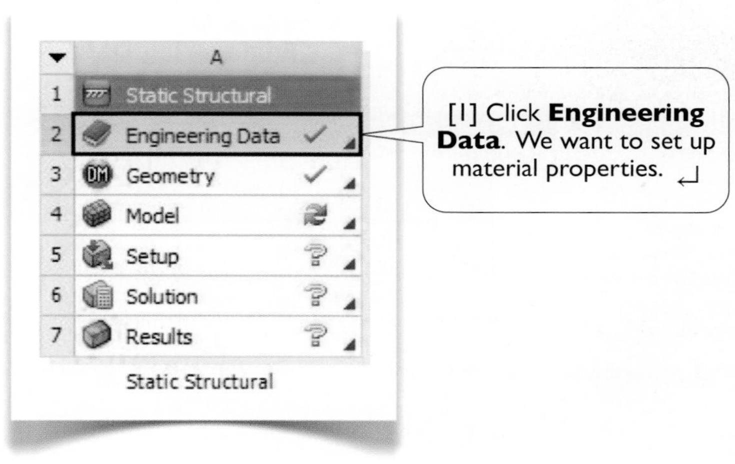

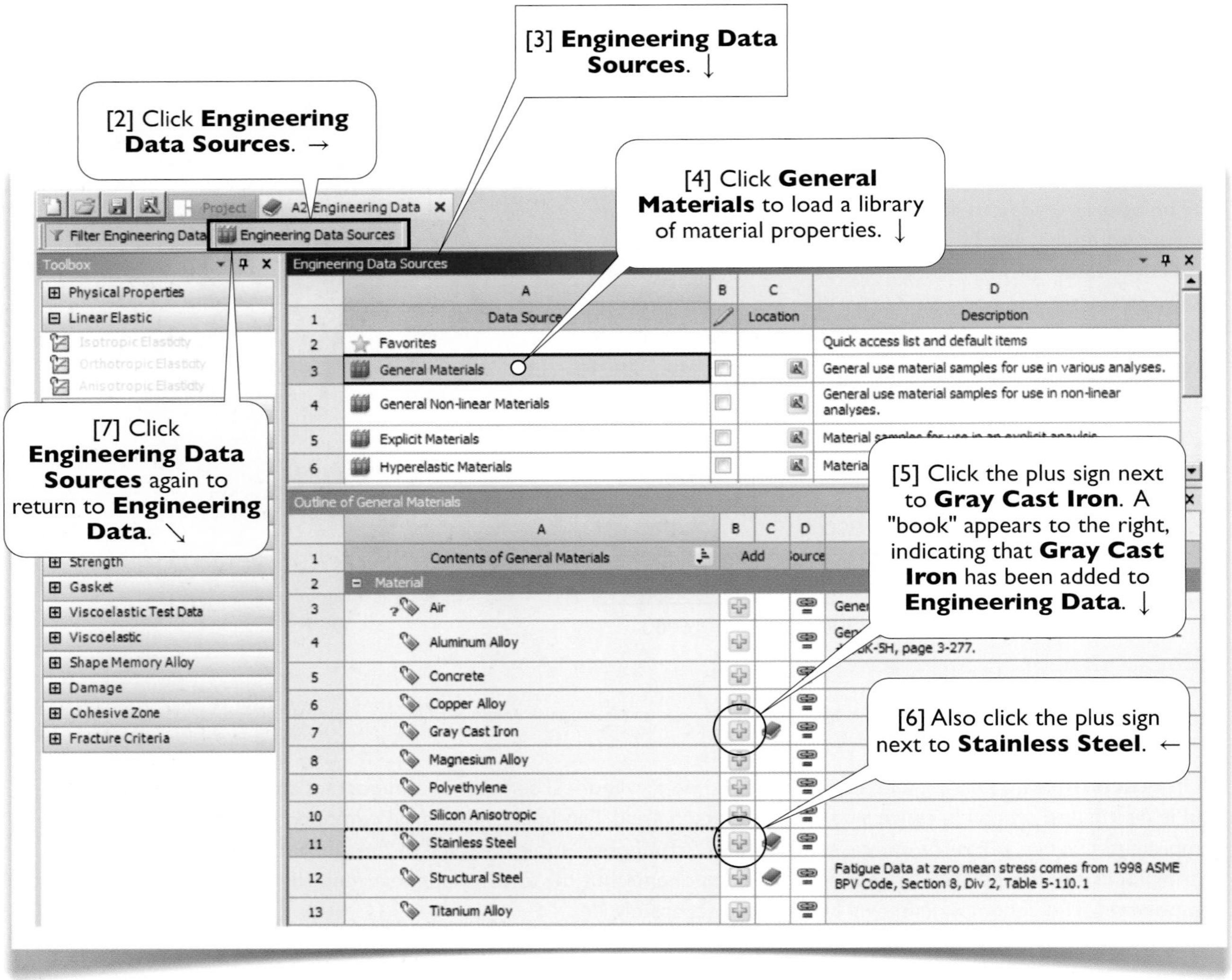

[2] Click **Engineering Data Sources**. →
[3] **Engineering Data Sources**. ↓
[4] Click **General Materials** to load a library of material properties. ↓
[5] Click the plus sign next to **Gray Cast Iron**. A "book" appears to the right, indicating that **Gray Cast Iron** has been added to **Engineering Data**. ↓
[6] Also click the plus sign next to **Stainless Steel**. ←
[7] Click **Engineering Data Sources** again to return to **Engineering Data**. ↘
Project
A2:Engineering Data
Filter Engineering Data
Engineering Data Sources
Toolbox
Physical Properties
Linear Elastic
Isotropic Elasticity
Orthotropic Elasticity
Anisotropic Elasticity
Strength
Gasket
Viscoelastic Test Data
Viscoelastic
Shape Memory Alloy
Damage
Cohesive Zone
Fracture Criteria
Data Source
Location
Description
Favorites
Quick access list and default items
General Materials
General use material samples for use in various analyses.
General Non-linear Materials
General use material samples for use in non-linear analyses.
Explicit Materials
Hyperelastic Materials
Outline of General Materials
Contents of General Materials
Add
Source
Material
Air
Aluminum Alloy
Concrete
Copper Alloy
Gray Cast Iron
Magnesium Alloy
Polyethylene
Silicon Anisotropic
Stainless Steel
Structural Steel
Titanium Alloy
Fatigue Data at zero mean stress comes from 1998 ASME BPV Code, Section 8, Div 2, Table 5-110.1

[9] Click to return to **Project Schematic**. #
[8] Now, **Gray Cast Iron** and **Stainless Steel** are included in **Engineering Data**. ←
Project
A2:Engineering Data
Filter Engineering Data
Engineering Data Sources
Toolbox
Physical Properties
Linear Elastic
Isotropic Elasticity
Orthotropic Elasticity
Anisotropic Elasticity
Hyperelastic Experimental Data
Hyperelastic
Chaboche Test Data
Outline of Schematic A2: Engineering Data
Contents of Engineering Data
Source
Description
Material
Gray Cast Iron
Stainless Steel
Structural Steel
Fatigue Data at zero mean stress comes from 1998 ASME BPV Code, Section 8, Div 2, Table 5-110.1
Click here to add a new material

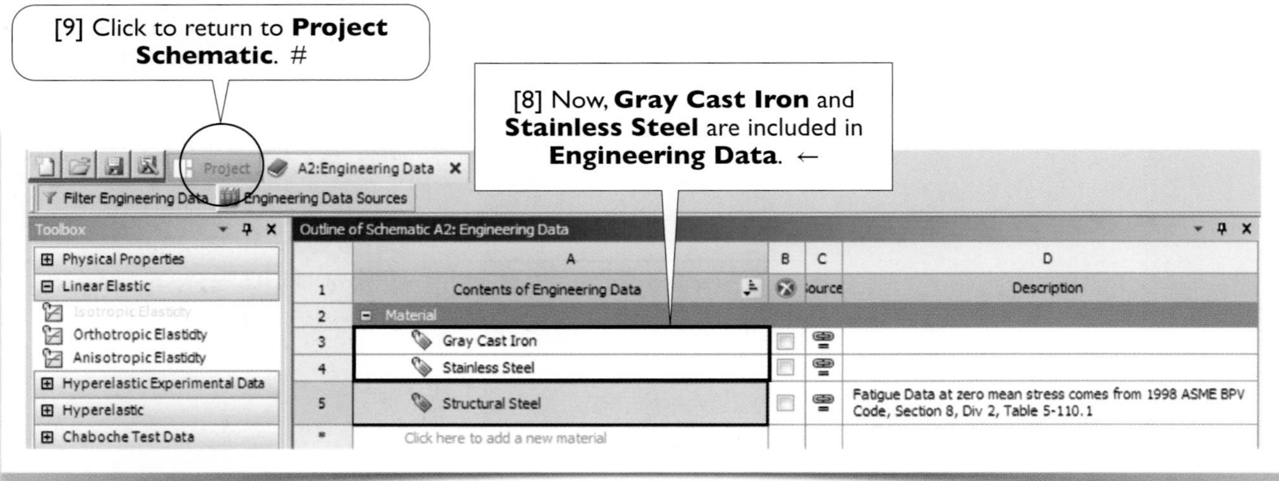

6.3.9 Assign Material and Thickness

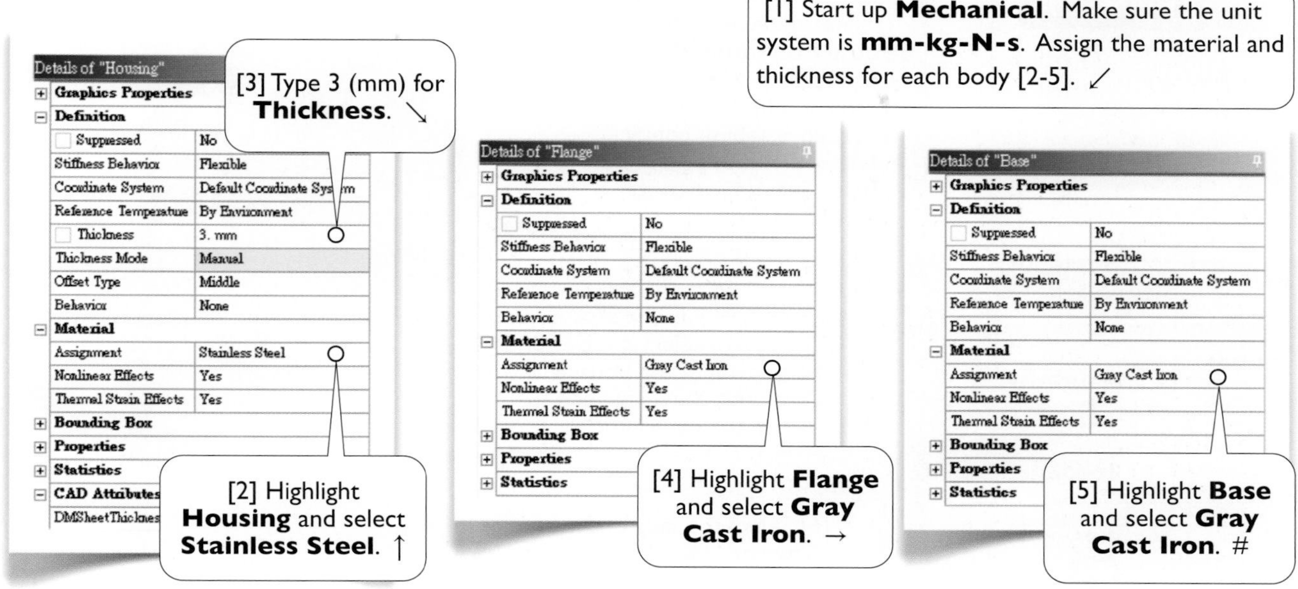

6.3.10 Set Up Connections between Parts

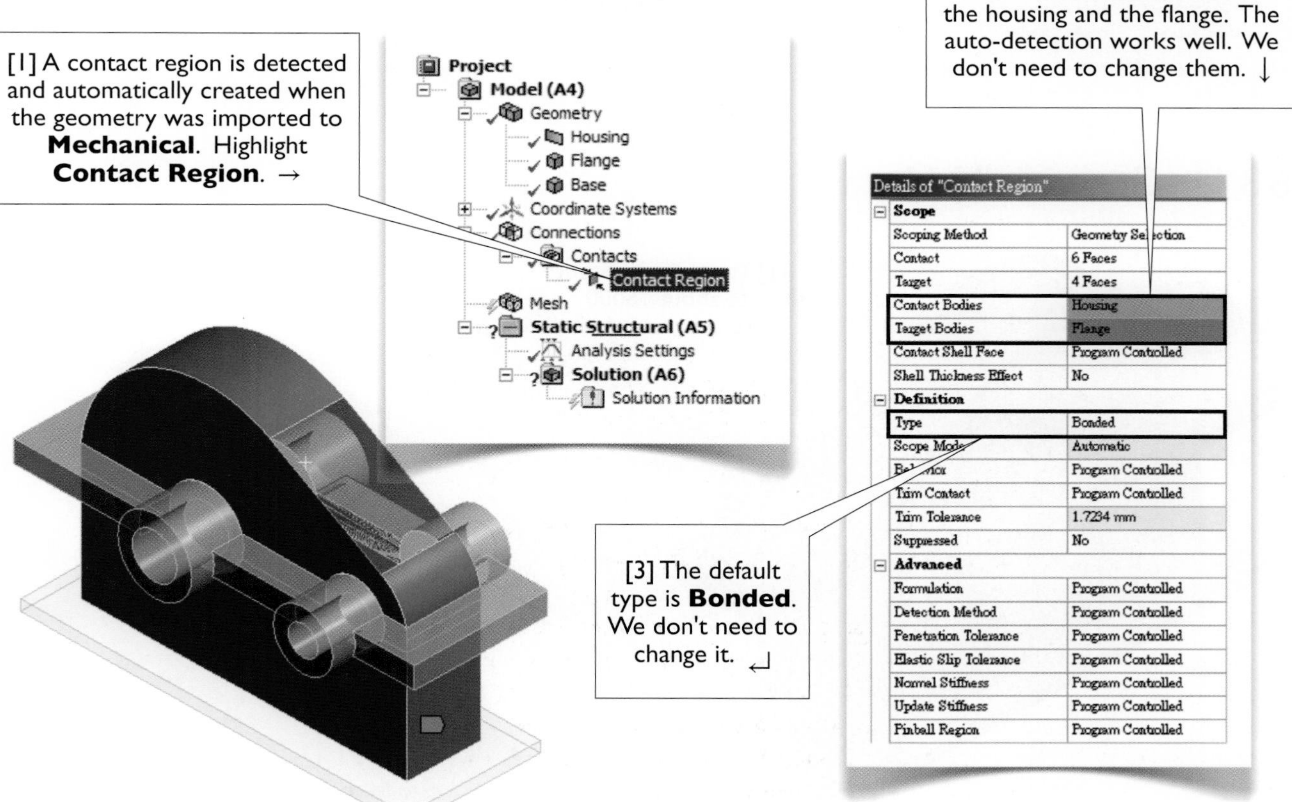

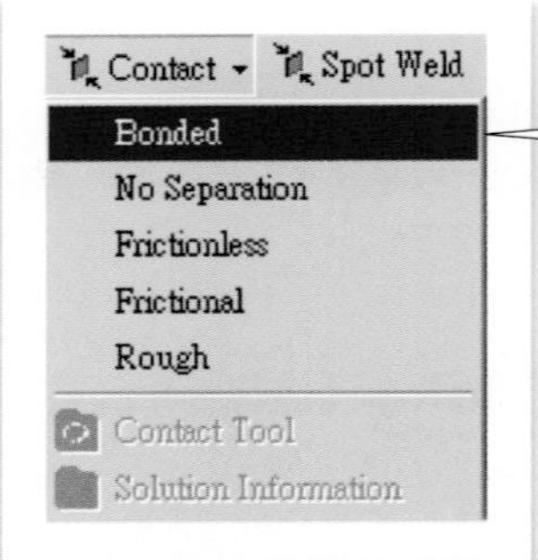

Contact ▾ Spot Weld
Bonded
No Separation
Frictionless
Frictional
Rough
Contact Tool
Solution Information

[4] Contact between the housing and the base, however, is not detected. This is because **Mechanical** doesn't detect edge-to-surface contact automatically. We need to manually establish contact between the housing and the base. Select **Contact/ Bonded**. ↓

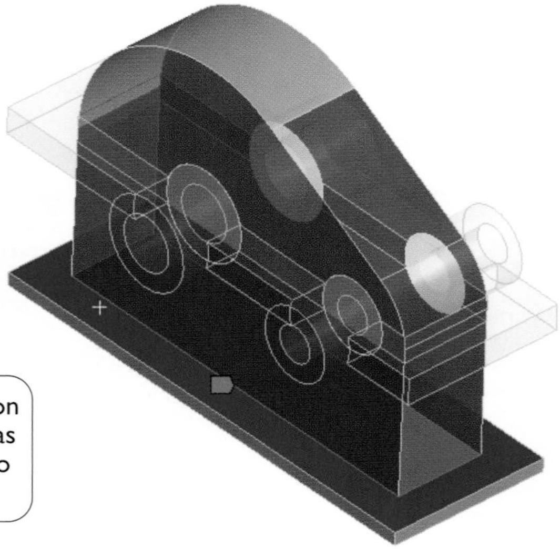

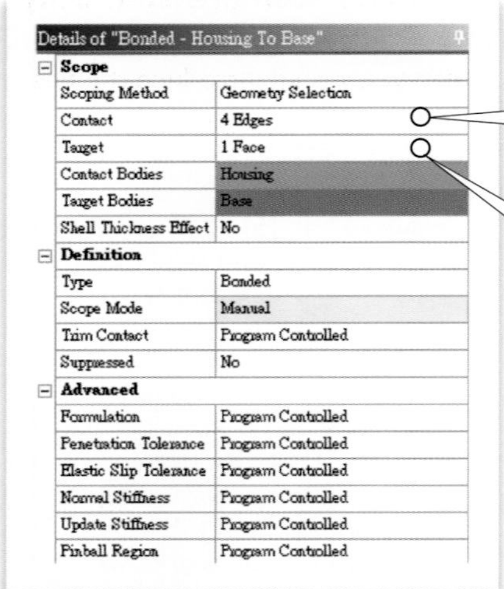

Details of "Bonded - Housing To Base"	
Scope	
Scoping Method	Geometry Selection
Contact	4 Edges
Target	1 Face
Contact Bodies	Housing
Target Bodies	Base
Shell Thickness Effect	No
Definition	
Type	Bonded
Scope Mode	Manual
Trim Contact	Program Controlled
Suppressed	No
Advanced	
Formulation	Program Controlled
Penetration Tolerance	Program Controlled
Elastic Slip Tolerance	Program Controlled
Normal Stiffness	Program Controlled
Update Stiffness	Program Controlled
Pinball Region	Program Controlled

[5] Control-select 4 edges on the bottom of the housing as **Contact**. You may need to use the edge filter. ↓

[6] Select top face of the base as **Target**. Use the face filter. ↓

Do contacts always introduce nonlinearity?

[7] It depends on the type of contacts. In this case, the **Bonded** contacts do not introduce any nonlinearities. For details, see 13.1.8, page 476. #

6.3.11 Generate Mesh

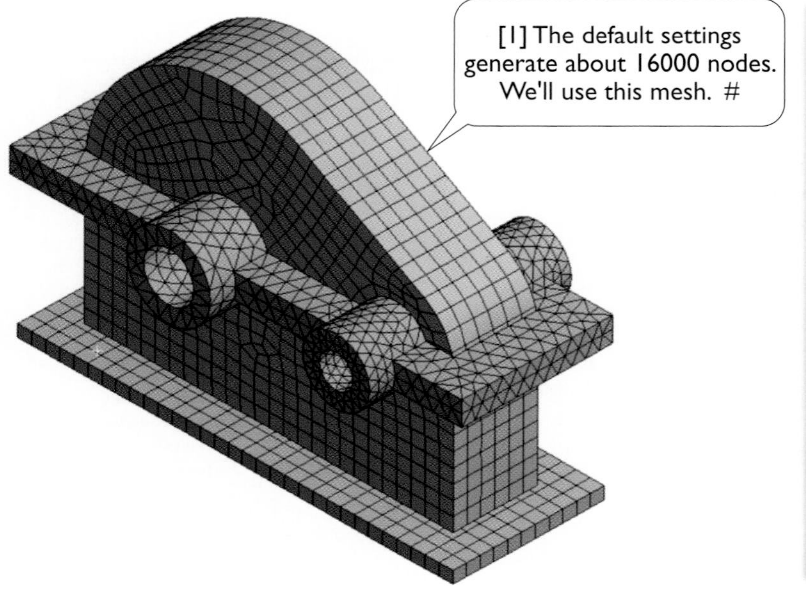

[1] The default settings generate about 16000 nodes. We'll use this mesh. #

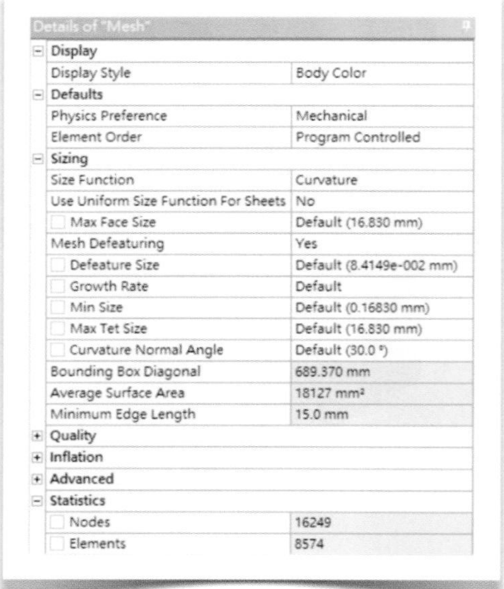

Details of "Mesh"	
Display	
Display Style	Body Color
Defaults	
Physics Preference	Mechanical
Element Order	Program Controlled
Sizing	
Size Function	Curvature
Use Uniform Size Function For Sheets	No
Max Face Size	Default (16.830 mm)
Mesh Defeaturing	Yes
Defeature Size	Default (8.4149e-002 mm)
Growth Rate	Default
Min Size	Default (0.16830 mm)
Max Tet Size	Default (16.830 mm)
Curvature Normal Angle	Default (30.0 °)
Bounding Box Diagonal	689.370 mm
Average Surface Area	18127 mm²
Minimum Edge Length	15.0 mm
Quality	
Inflation	
Advanced	
Statistics	
Nodes	16249
Elements	8574

6.3.12 Specify Support

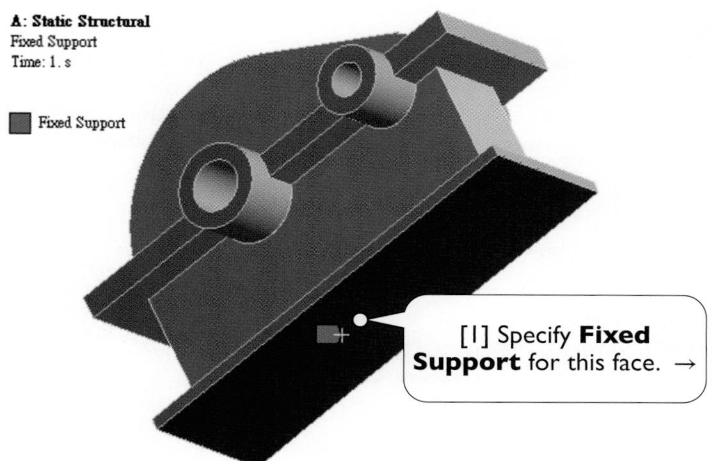

[1] Specify **Fixed Support** for this face. →

Support Conditions

[2] In this case, a **Compression Only Support** or a **Frictionless Support** would be more realistic than a **Fixed Support**. However, the results will be the same for these three support conditions. Since the external forces sum up to zero, the support reaction is zero. **Fixed Support** in this case has an advantage: it doesn't introduce nonlinearities or the need for weak springs (3.1.8[7], page 116). **Compression Only Support** would introduce nonlinearities, and **Frictionless Support** would introduce the need for weak springs. #

6.3.13 Apply Bearing Loads

[4] Apply a bearing load like this on the internal cylindrical surface of bearing B. ↘

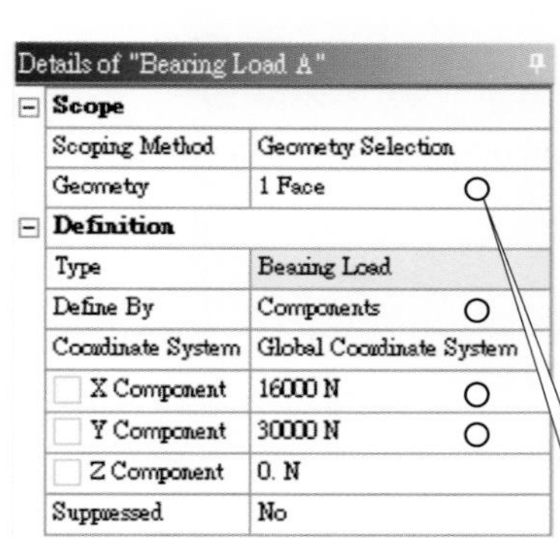

Details of "Bearing Load A"

Scope	
Scoping Method	Geometry Selection
Geometry	1 Face
Definition	
Type	Bearing Load
Define By	Components
Coordinate System	Global Coordinate System
X Component	16000 N
Y Component	30000 N
Z Component	0. N
Suppressed	No

[2] Apply a bearing load like this on the internal cylindrical surface of bearing A. →

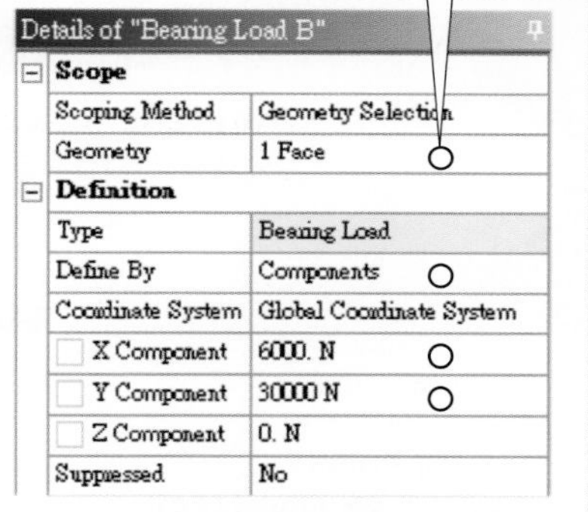

Details of "Bearing Load B"

Scope	
Scoping Method	Geometry Selection
Geometry	1 Face
Definition	
Type	Bearing Load
Define By	Components
Coordinate System	Global Coordinate System
X Component	6000. N
Y Component	30000 N
Z Component	0. N
Suppressed	No

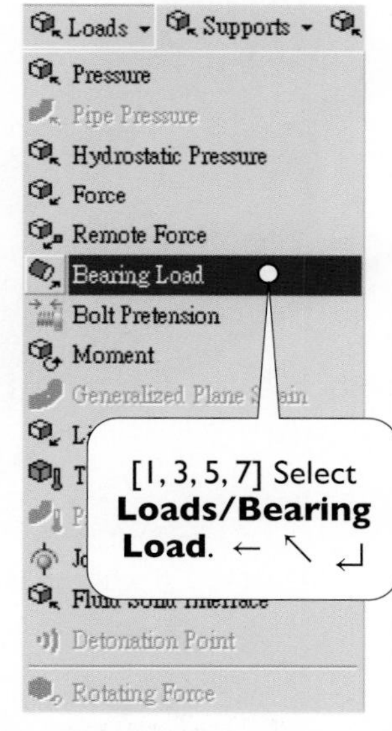

[1, 3, 5, 7] Select **Loads/Bearing Load**. ← ↖ ↵

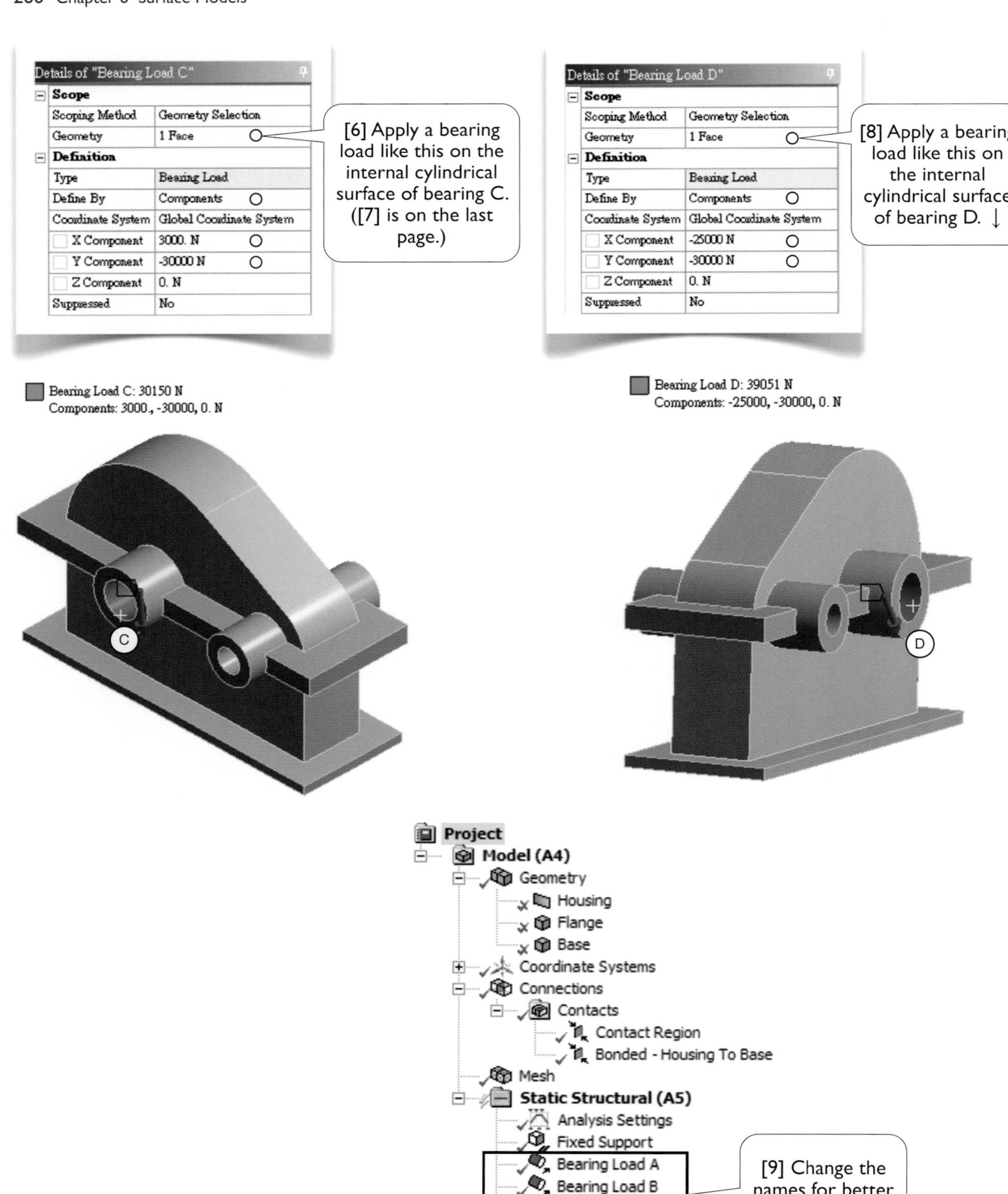
Details of "Bearing Load C"
Scope
Scoping Method
Geometry Selection
Geometry
1 Face
Definition
Type
Bearing Load
Define By
Components
Coordinate System
Global Coordinate System
X Component
3000. N
Y Component
-30000 N
Z Component
0. N
Suppressed
No
[6] Apply a bearing load like this on the internal cylindrical surface of bearing C. ([7] is on the last page.)
Details of "Bearing Load D"
Scope
Scoping Method
Geometry Selection
Geometry
1 Face
Definition
Type
Bearing Load
Define By
Components
Coordinate System
Global Coordinate System
X Component
-25000 N
Y Component
-30000 N
Z Component
0. N
Suppressed
No
[8] Apply a bearing load like this on the internal cylindrical surface of bearing D. ↓
Bearing Load C: 30150 N
Components: 3000., -30000, 0. N
C
Bearing Load D: 39051 N
Components: -25000, -30000, 0. N
D
Project
Model (A4)
Geometry
Housing
Flange
Base
Coordinate Systems
Connections
Contacts
Contact Region
Bonded - Housing To Base
Mesh
Static Structural (A5)
Analysis Settings
Fixed Support
Bearing Load A
Bearing Load B
Bearing Load C
Bearing Load D
Solution (A6)
Solution Information
[9] Change the names for better readability. #

6.3.14 Apply Axial Loads

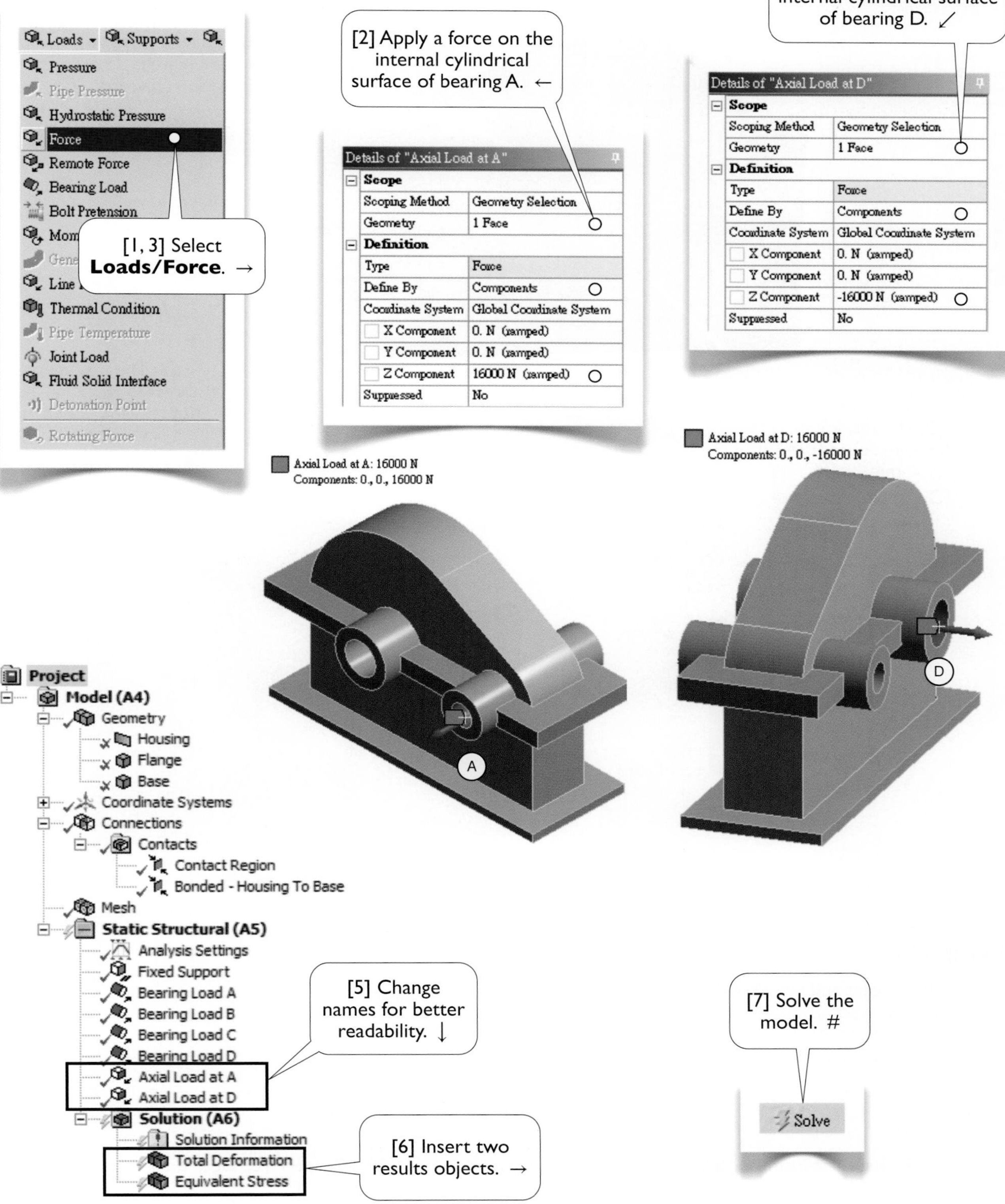

6.3.15 View the Results

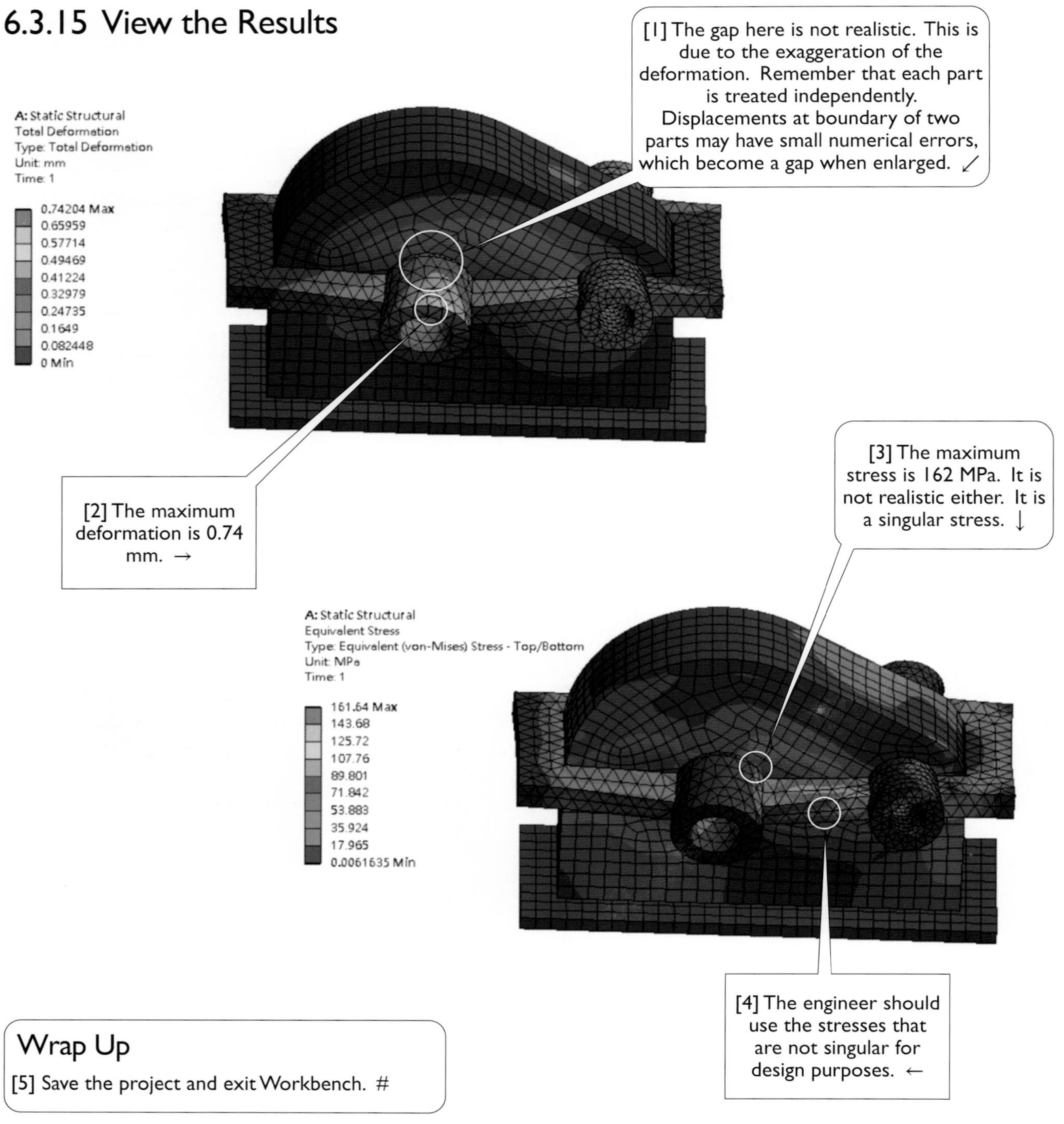

Wrap Up

[5] Save the project and exit Workbench. #

Reference

1. Zahavi, E., *The Finite Element Method in Machine Design*, Prentice-Hall, 1992; Chapter 10. Gear Box.

Section 6.4

Review

6.4.1 Keywords

Choose a letter for each keyword from the list of descriptions

1. ( ) Auto-Detection of Contact
2. ( ) Shell Elements
3. ( ) Top/Bottom of Surface Body

Answers:

1. (C) 2. (A) 3. (B)

List of Descriptions

(A) A planar (2D) element that can be arranged in the 3D space. It is used to mesh a body when one of its dimensions is much smaller than the other two dimensions. Each node has 6 degrees of freedom: 3 translational and 3 rotational. Due to the presence of rotational degrees of freedom, it is very efficient to model the problems dominated by the out-of-plane bending modes, contrasting to a solid element, which does not have rotational degrees of freedom.

(B) Each surface body has a top side and a bottom side. When you select a surface body, only the top side is highlighted. Loads are applied on the top side. By default, results are reported on both sides.

(C) When a geometry attaches to **Mechanical**, it automatically detects and establishes possible contacts between parts, wherever the gaps between parts are less than a tolerance. The contact type is **Bonded** by default.

6.4.2 Questions

Surface Body versus Thin Solid Body

Q: In 6.3.3[5] (page 257), we type a zero value for the thickness to create a surface body. If we provide a nonzero value, it would create a "thin" solid body instead of a surface body. What is an essential difference between a surface body and a thin solid body?
A: A surface body will be meshed with shell elements (1.3.3[10-11], page 38), while a thin solid body will be meshed with solid elements (1.3.3[2-5], page 38).

Shell Elements versus Solid Elements

Q: What is the essential difference between shell elements and solid elements?
A: The shell elements have rotational degrees of freedom (1.3.3[10-11], page 38), while solid elements do not (1.3.3[2-5], page 38).

Triangular Plate

Q: Can we perform a simulation with a 3D surface model for the triangular plate in Section 3.1? If positive, are there any advantages of doing that over the 2D solid model?
A: Yes, we could, but there is no advantage over the 2D solid model. Remember that 3D surface bodies are meshed with shell elements. The essential difference between shell elements and 2D solid elements is that shell elements can have out-of-plane deformation (warpage) while 2D solid elements cannot. In the case of triangular plate, there is no out-of-plane deformation. There is no need to use shell elements. The above discussions also apply on the spur gear of Section 3.4 and the filleted bar of Section 3.5.

Axisymmetric Bodies

In the beginning and the end of Section 6.1, we mentioned that we could model the bellows using axisymmetric 2D solid body or 2D line body. Make sure that you do understand the meaning of the statement.

References

1. ANSYS Help//Mechanical Application//ANSYS Mechanical User's Guide//Symmetry in the Mechanical Application

Chapter 7
Line Models

Many real-world objects can be modeled as line bodies. When a body has small lateral dimensions and has a uniform cross-section, it is often modeled as a line body. The most obvious applications of line models are frame, beam, and truss structures. Workbench meshes a line body with beam elements (1.3.3[13], page 39). Advantages of using line models over surface models or solid models include: (a) creating line models is usually easier, (b) the problem size is much smaller, and (c) the solution can be more accurate. Therefore, engineers should consider using line models instead of surface or solid models whenever possible. Workbench stores many built-in cross sections in the database to be chosen from by the users. Workbench allows the rendering of cross-sections when displaying the model, so that the line bodies visually look like solid bodies.

Purpose of This Chapter

The main purpose of this chapter is to introduce the use of beam elements (1.3.3[13], page 39). This chapter guides the students to create line models and perform simulations using these models. The chapter provides three examples; two of them are entirely line models, while the other one is a model mixed up with line bodies and surface bodies.

About Each Section

Section 7.1 creates a flexible gripper model and performs a simulation using the model. Section 7.2 demonstrates the creation of a truss structure and the simulation. Section 7.3 uses a two-story building as an example to demonstrate the creation of a building structure and the simulation. Another purpose of Section 7.3 is to demonstrate how surface bodies and line bodies can be mixed up in a simulation model.

Section 7.1

Flexible Gripper[Ref 1]

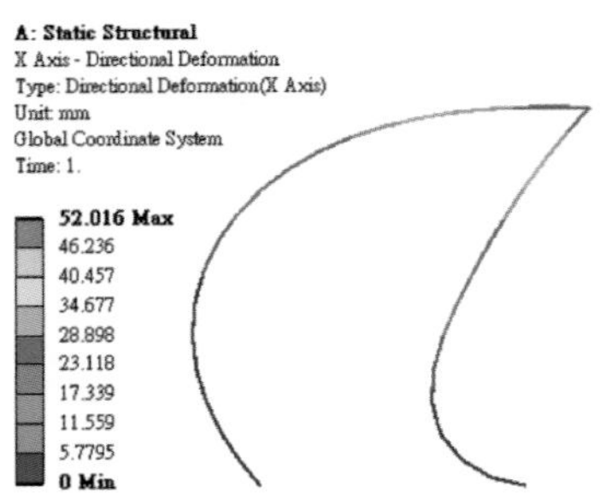

7.1.1 About the Flexible Gripper

[1] The gripper [2-5] is made of a POM (polyoxymethylene, a plastic polymer), which has a Young's modulus of 2 GPa and a Poisson's ratio of 0.35. It has a rectangular cross section of 1x5 mm^2.

A concern in designing this gripper is its geometric advantage (GA). The GA is defined as the ratio of the horizontal output displacement [5] to the input actuation [4]. The GA value is used to assess the efficiency of the gripper; the larger the better. The main purpose of this simulation is to assess the GA value of the current design as shown in the figure [6]. Note that only half of the gripper is modeled due to the symmetry.

The profile consists of two smooth spline curves defined by 7 key points, whose numbering and coordinates are shown in the figure [6-7].

The problem could be solved with a 2D simulation. With 2D simulations, however, the current version of **Mechanical** supports only solid models (it doesn't support 2D surface model or line model). Compared with a 3D line model, a 2D solid model is not efficient (in terms of accuracy and computing time.) We decide to go for a 3D line model although the geometry and the motion are entirely on a plane.

We will create a line model in the first part of this section. The model will be used in the second part of this section to simulate the motion of gripping and to assess the GA value. The model will be used again in Section 8.1 to demonstrate an optimization capability of Workbench, in which we want to relocate the positions of the key points to achieve an optimum GA value. ↗

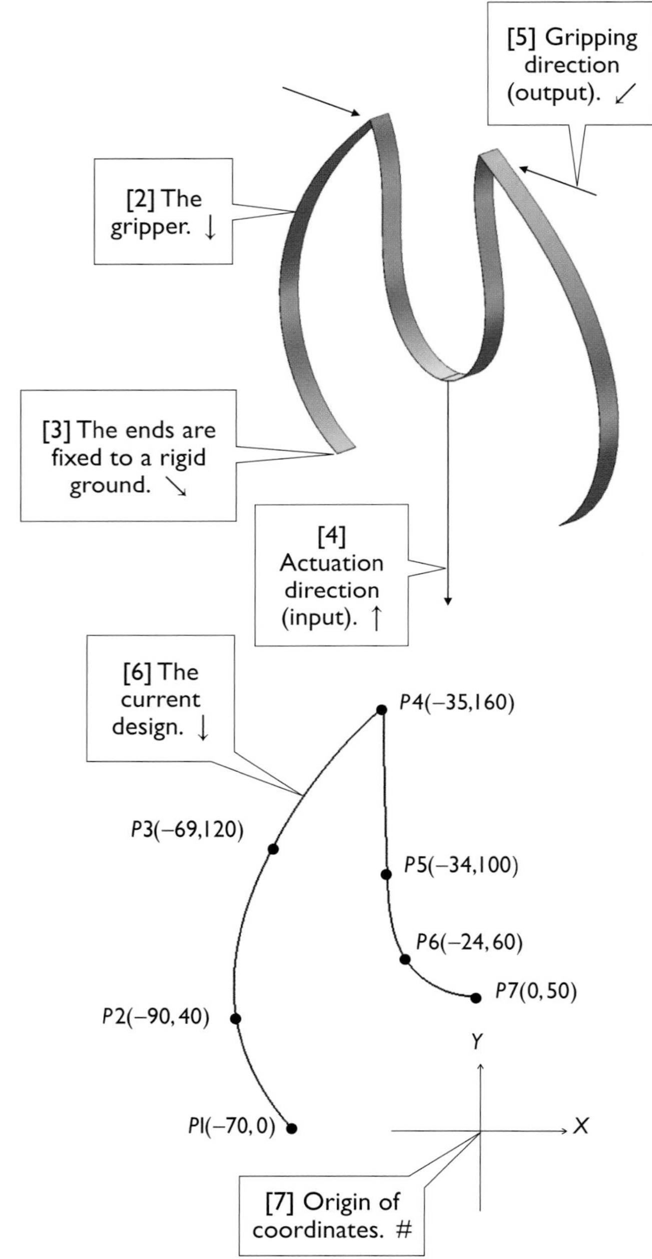

PART A. GEOMETRIC MODELING

7.1.2 Start Up

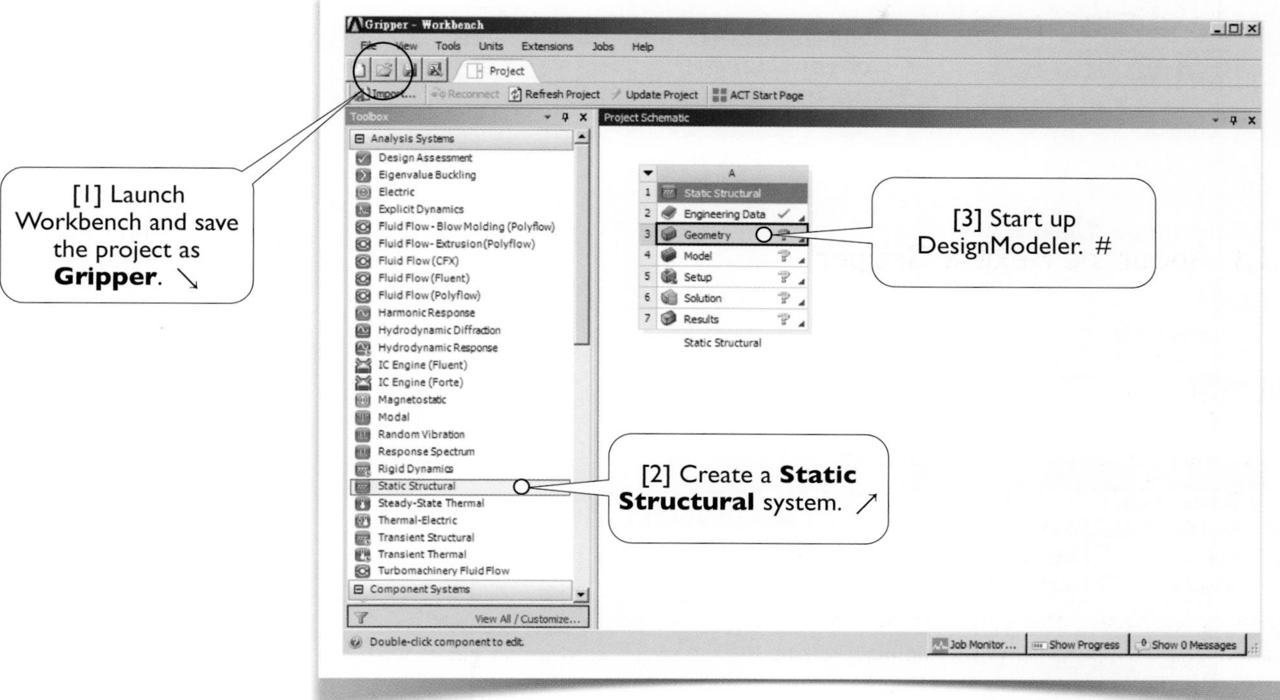

7.1.3 Create Sketch on **XYPlane**

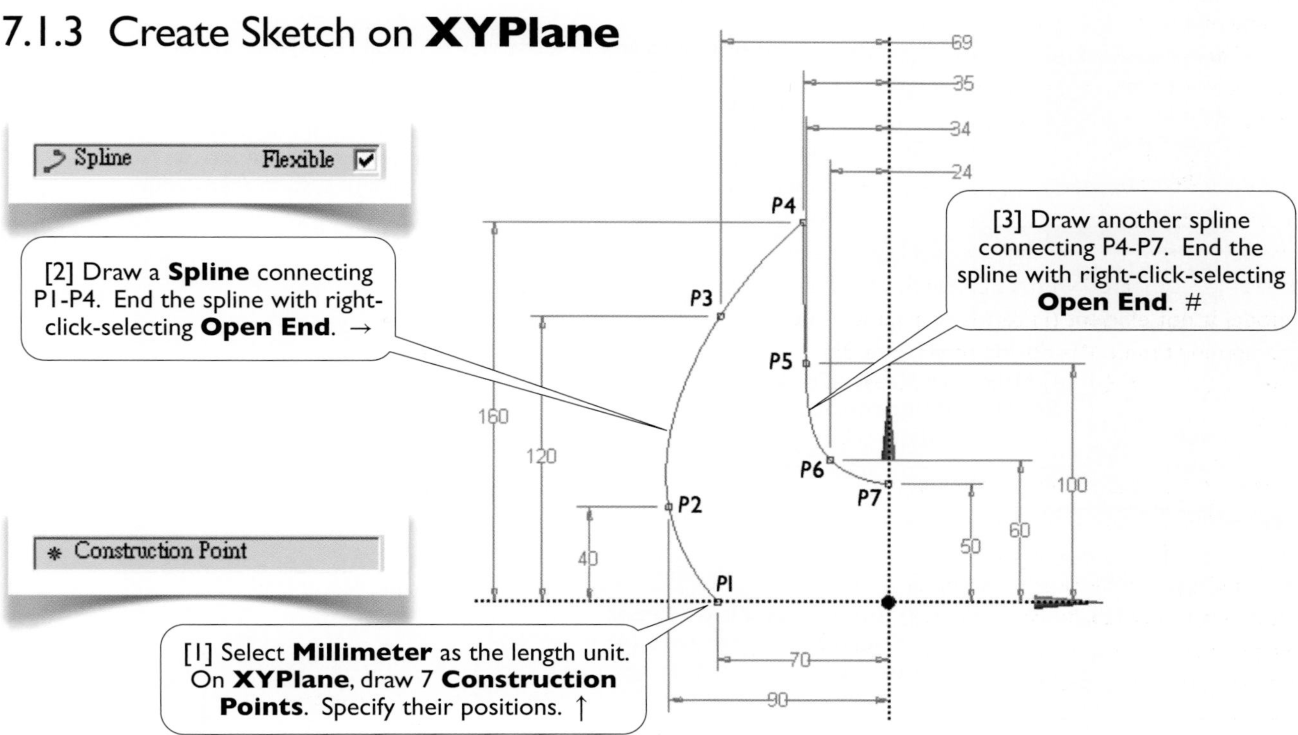

7.1.4 Create Line Body

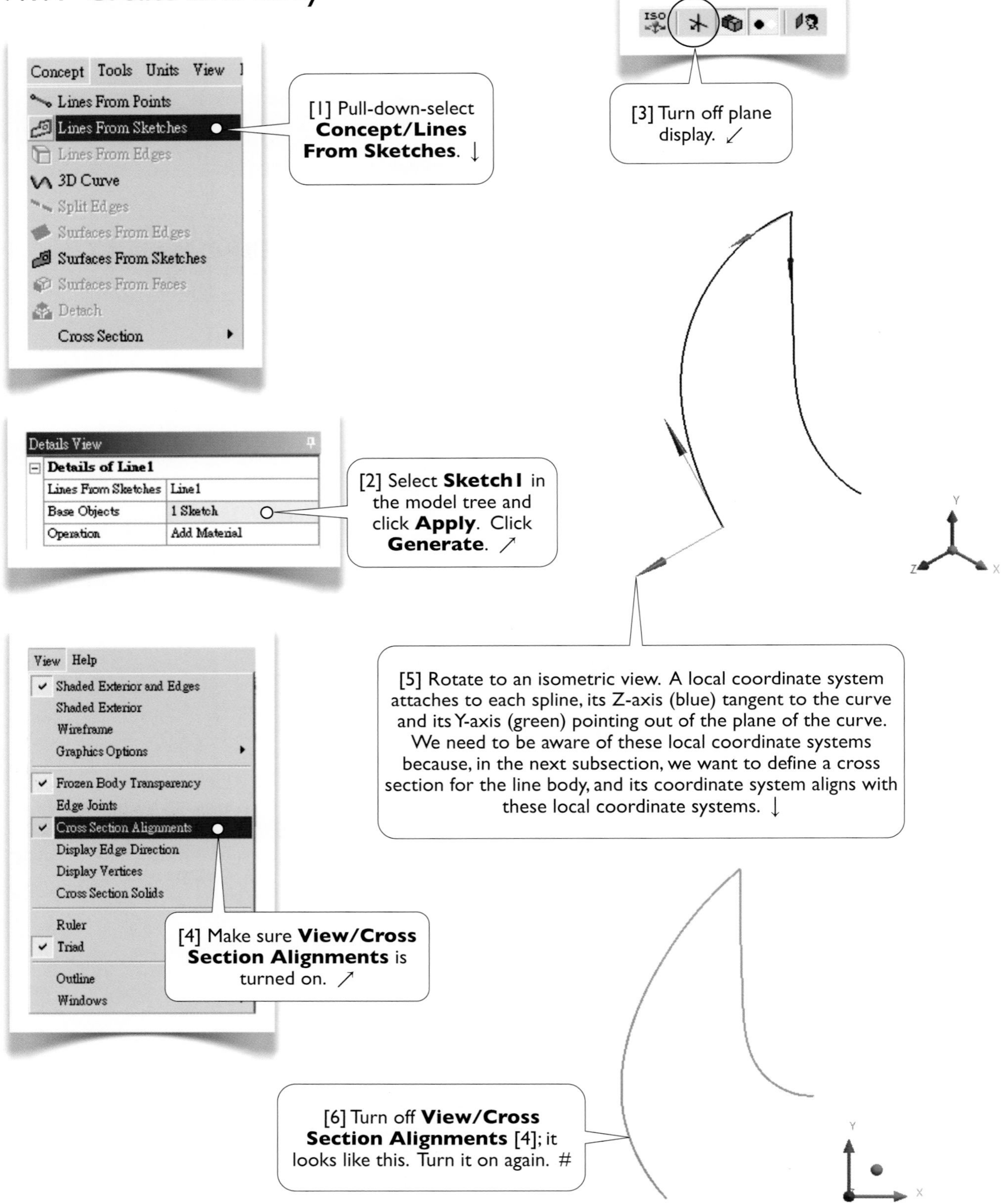

7.1.5 Create a Rectangular Cross Section

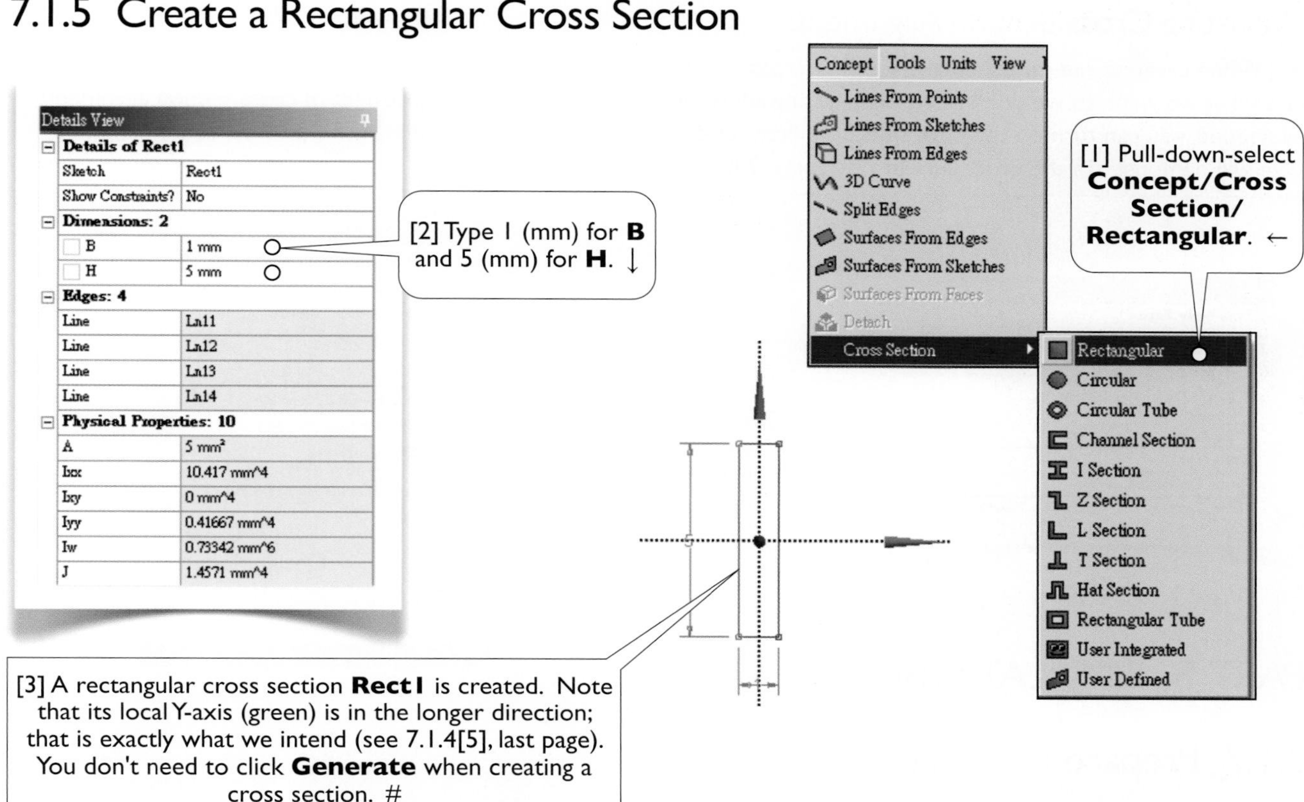

7.1.6 Assign the Cross Section to the Line Body

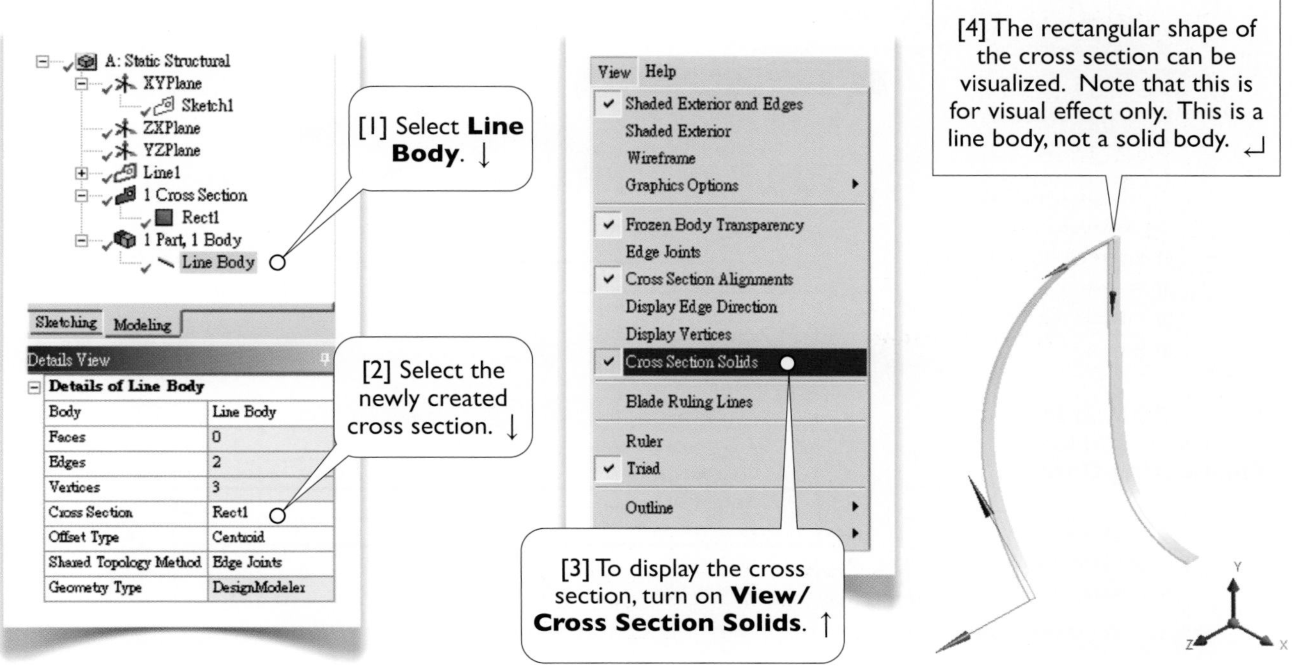

Adjusting Cross Section Alignments

[5] When creating line models, make sure the cross section alignments are correct. In our case, the default alignments are what we want, so we don't need to make any adjustment. In other cases, if adjustment of cross section alignments is needed, you can turn on the edge selection filter, select the edges of the line bodies, and type the rotation angle [6]. We will demonstrate the procedure in 7.3.9, page 300. ↓

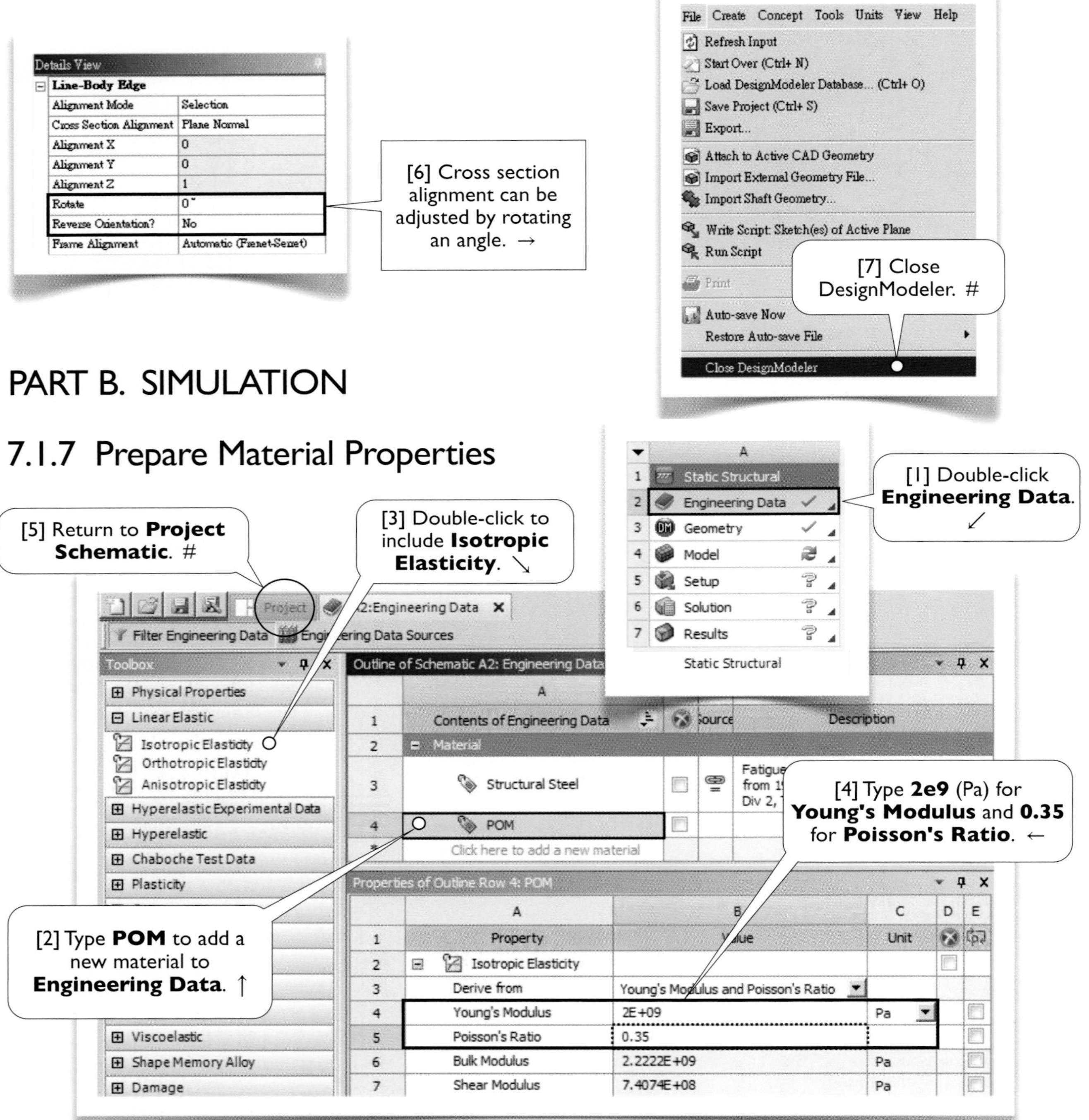

PART B. SIMULATION

7.1.7 Prepare Material Properties

7.1.8 Start Up **Mechanical** and Assign Material

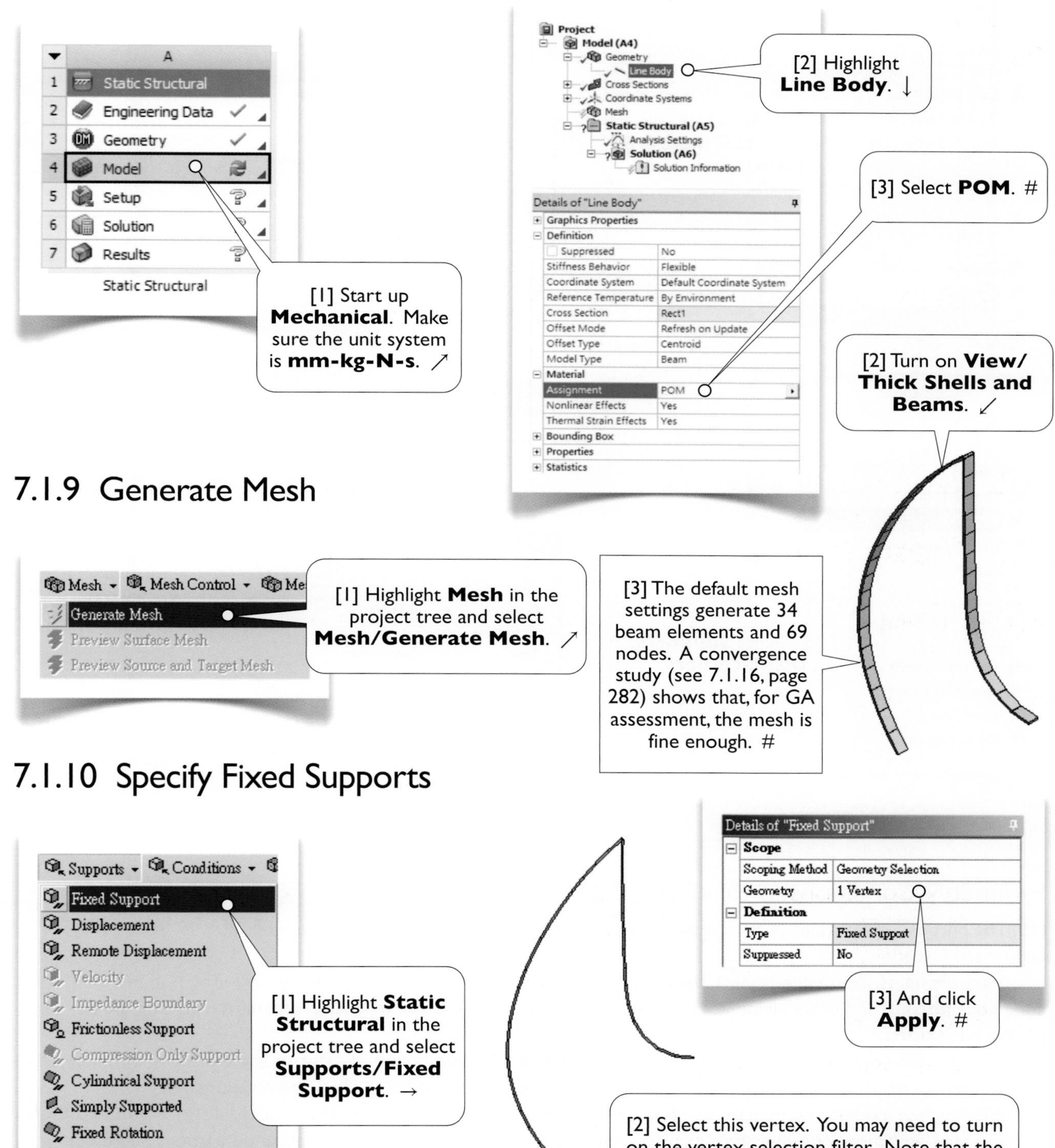

7.1.9 Generate Mesh

7.1.10 Specify Fixed Supports

7.1.11 Specify Symmetry Condition and the Actuation

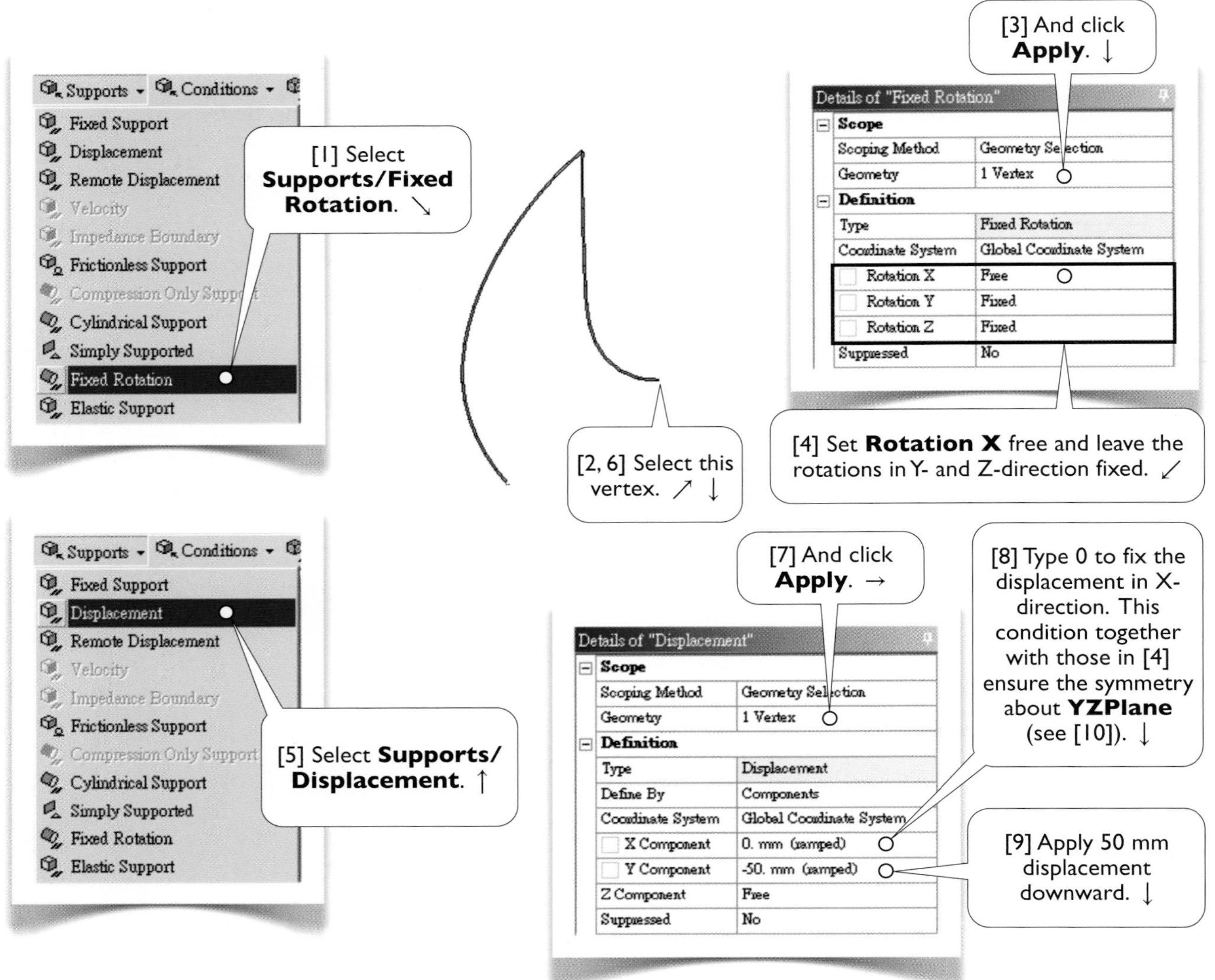

Symmetry Conditions for Shell and Beam Elements

[10] The purpose of the steps [1-8] is to set up a symmetry condition about **YZPlane**, while step [9] is to set up a downward actuation.

Beam elements, like shell elements, have rotational degrees of freedom (1.3.3[13], page 39). The rule of symmetry conditions for shell and beam elements is this: fixing (zero values) the **out-of-plane translations** and the **in-plane rotations**.

Consider the boundary conditions at the vertex shown in [2]. In this case, **YZPlane** is the plane of symmetry, X-displacement is the out-of-plane translation [8], Y-rotation and Z-rotation are the in-plane rotations [4].

With surface models, you may set up symmetry conditions using the **Symmetry** tool (3.1.11[6], page 119). Even so, make sure you know Workbench sets up symmetry conditions by fixing the out-of-plane translations and the in-plane rotations. As a good engineer, use software only when you know how it works.

The **Symmetry** tool (3.1.11[6], page 119) is not applicable for a line model. #

7.1.12 Set Up **Analysis Settings**

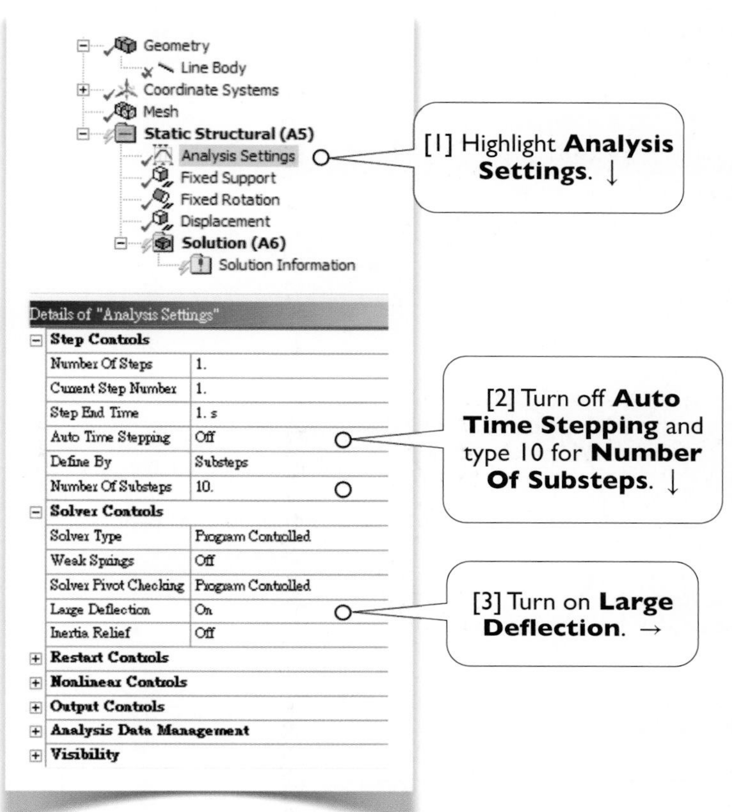

Why Large Deflection?

[4] Turning on large deflection is to include geometry nonlinearity, which always gives more accurate solutions but takes more computing time. To justify the inclusion of geometry nonlinearity, at the end of this section, turn off **Large Deflection** and rerun this case. A substantial difference in the results is an indication that the inclusion of geometry nonlinearity is necessary.

Auto Time Stepping

This model is simple enough that we actually don't need to change any default settings other than just turning on **Large Deflection**. The solution would be complete in just 3 substeps, with auto time stepping controlled by the program.

The reason we turn off **Auto Time Stepping** and specify 10 substeps is because we want to gather more data for plotting an input-displacement-versus-output-displacement chart (see 7.1.15[7], page 281). #

7.1.13 Set Up **Solution** Branch and Solve the Model

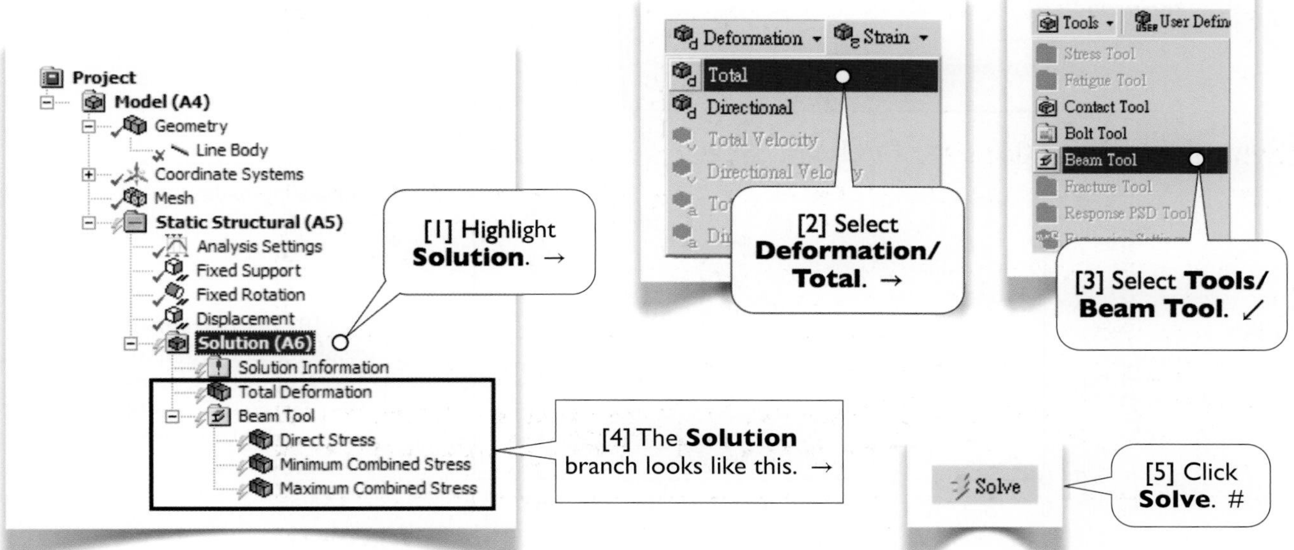

7.1.14 View Results

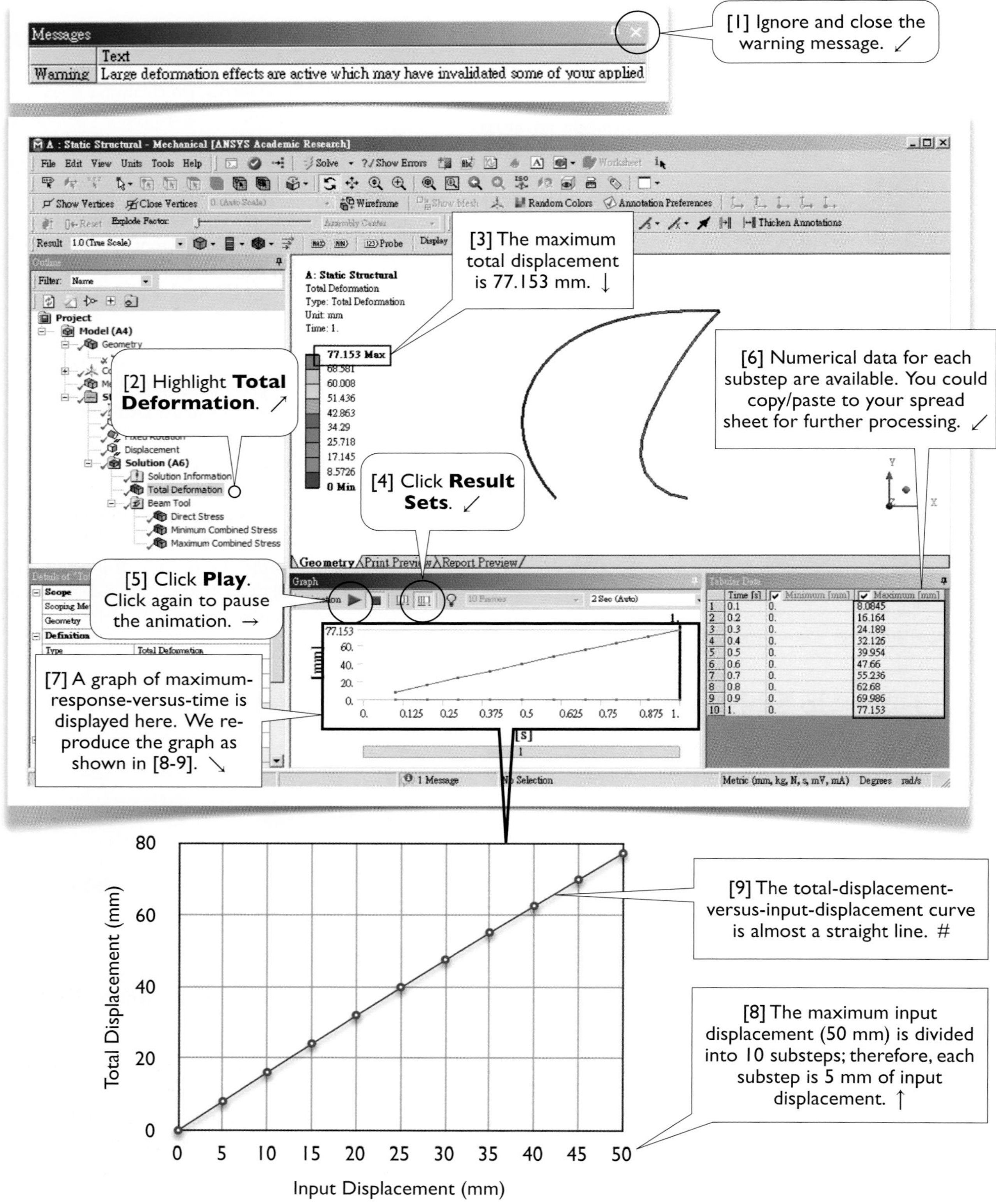

7.1.15 Assess GA Value

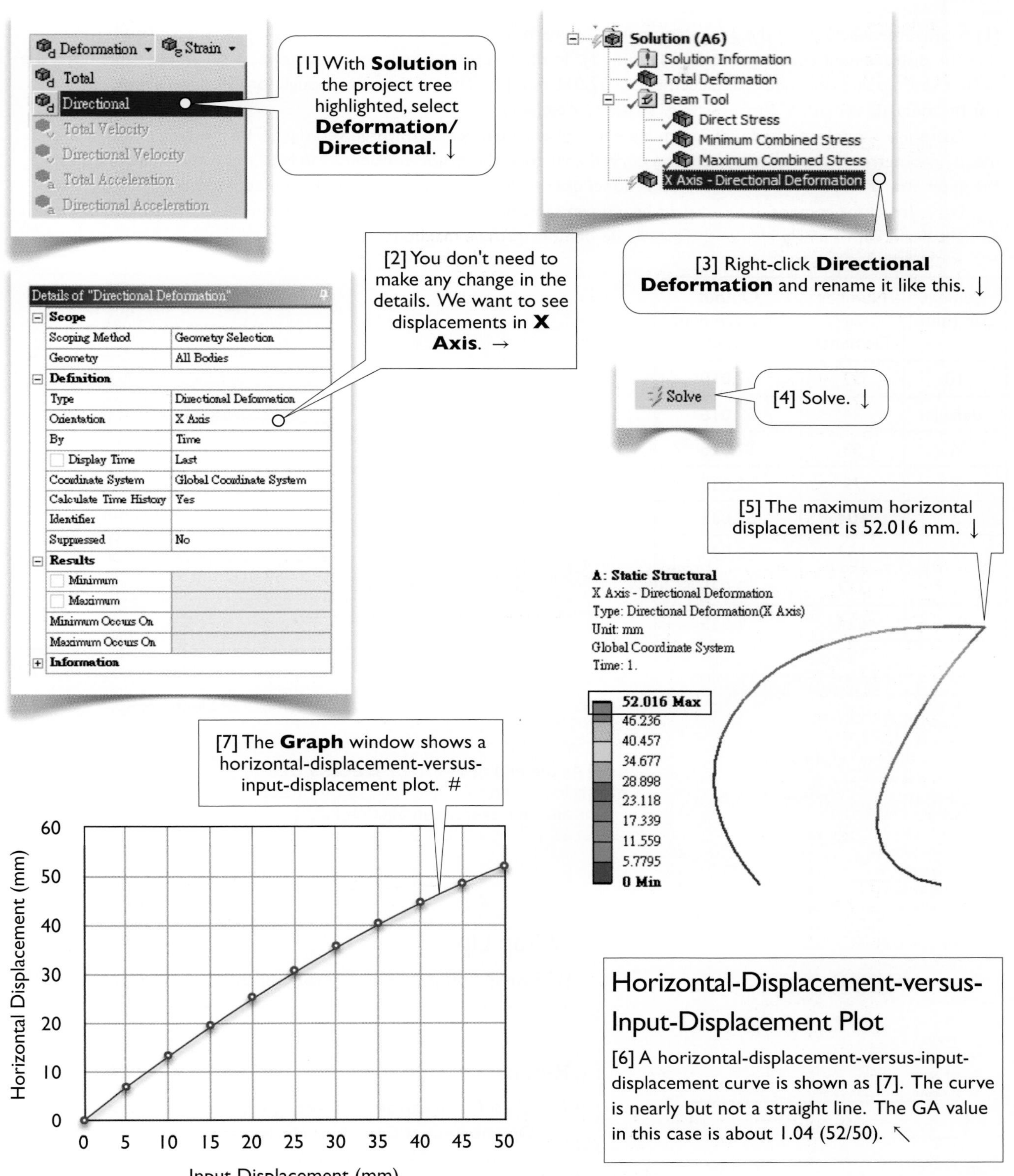

Horizontal-Displacement-versus-Input-Displacement Plot

[6] A horizontal-displacement-versus-input-displacement curve is shown as [7]. The curve is nearly but not a straight line. The GA value in this case is about 1.04 (52/50). ↖

7.1.16 Convergence Study of Beam Elements

[1] A convergence study of the gripper model is summarized in the table/figure below [2-3]. The table/figure shows that the displacement converges to 52.035 mm [2]. In the foregoing simulation, we meshed with 34 elements using the default settings and obtained a displacement of 52.016 mm [3]. That is accurate enough for GA assessment, but may not be adequate for other purposes (e.g., stress assessment).

Generally, models meshed with beam elements or shell elements converge very fast. The meshing consideration is mostly geometric. In our case, since it is a curved structure, the major consideration is to mesh the structure so that the geometry of the meshed finite element model doesn't deviate from the original geometry too much.

For a structure composed by straight beams, we can obtain a solution equal to theoretical values by meshing each straight beam with a single element! This will be demonstrated in Section 7.2. ↓

Element Size (mm)	Number of Elements	Output Displacement (mm)
10	32	52.010
default	34	52.016
6	53	52.028
4	79	52.032
3	104	52.033
2	157	52.034
1	312	52.035
0.5	624	52.035

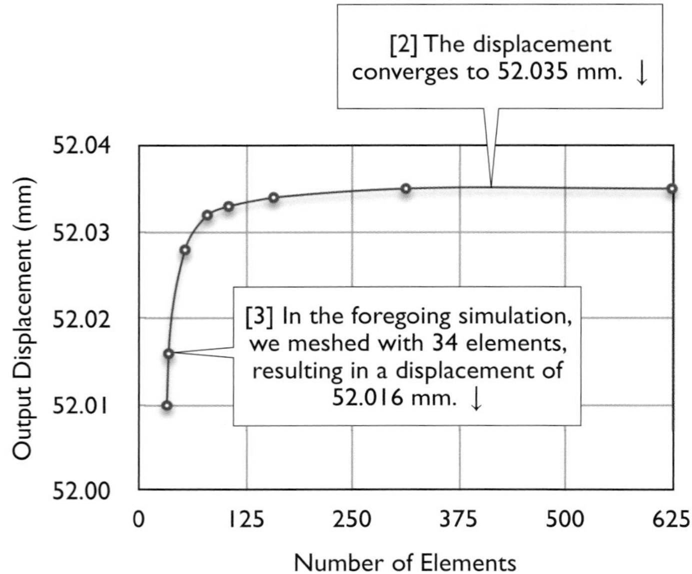

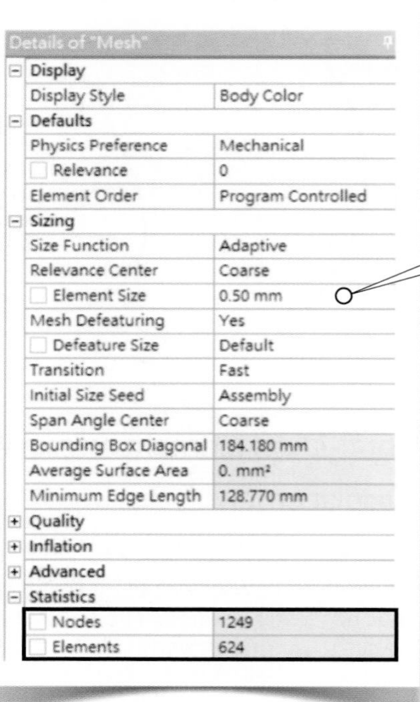

[4] At the end of this study, **Details of Mesh** looks like this. We will use this mesh for another exercise in Section 8.1. ↓

Wrap Up

[5] Save the project and exit Workbench. #

Reference

1. Chao-Chieh Lan and Yung-Jen Cheng, 2008, "Distributed Shape Optimization of Compliant Mechanisms Using Intrinsic Functions," *ASME Journal of Mechanical Design*, Vol. 130, 072304.

Section 7.2

3D Truss

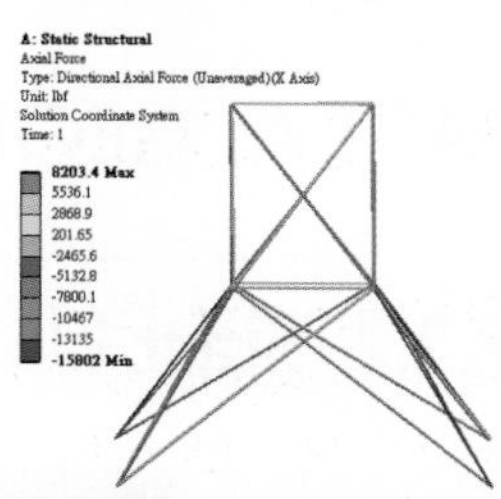

7.2.1 About the 3D Truss

[1] Traditionally, a truss is defined as a structure consisting of two-force members; i.e., the members are pin-jointed at ends and the loads are applied on the joints so that the members are either stretched or compressed but not bent. Two members connected by a pin-joint can rotate about the joint independently. In the real world, structural members are rarely connected by pin-joints. Modern structures are constructed using either welds or multiple bolt-and-nuts; i.e., the members are rigid jointed, not pin-jointed. Even in the old days, pin-jointed structures were not common. The main reason for pin-joint assumption was to ease the computational difficulty, in the days when computers were not widespread. Note that, due to the neglect of joint rigidity, pin-joint assumption leads to a conservative design: safer, but over-designed.

How much is the error caused by the pin-joint assumption? This is a good exercise problem for engineering students (7.4.2, page 310). The amount of error depends on the slenderness of the structural member. If the members are slender enough, there are no essential errors caused by the pin-joint assumption. On the other hand, if the members are not slender enough, then the pin-joints assumption may induce substantial errors.

The beam element **BEAM188**[Ref 1] (1.3.3[12-13], page 39) is the only element supported in Workbench for line bodies. The so called "truss elements" (such as LINK180[Ref 2]) are not supported in Workbench. To model a pin-jointed structure in Workbench, you need to either specify revolute joints between the members or utilize **End Release** feature (7.4.2, page 310).

In this section, we will create a line model for a power transmission tower as shown below. All members are made of structural steel angle of $1\frac{1}{2}\times1\frac{1}{2}\times\frac{1}{4}$ cross section. Note that we've assigned a number for each joint (*P1*-*P10*) and each member (1-25). The design loads are also tabulated below. #

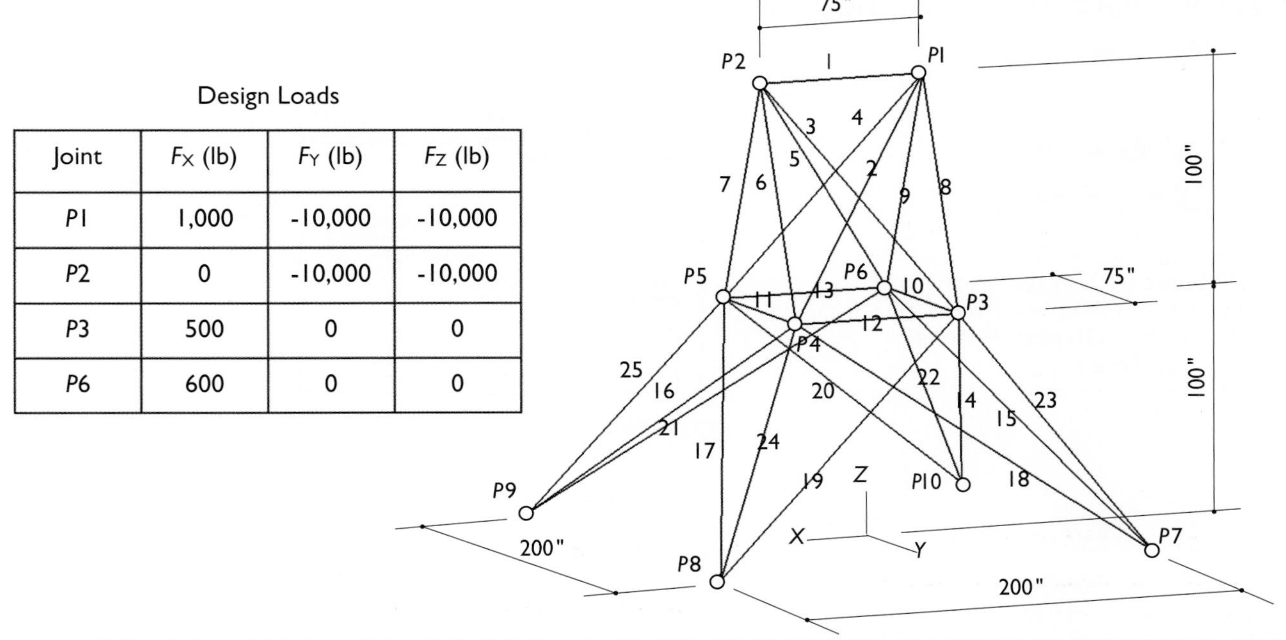

Design Loads

Joint	F_X (lb)	F_Y (lb)	F_Z (lb)
P1	1,000	-10,000	-10,000
P2	0	-10,000	-10,000
P3	500	0	0
P6	600	0	0

PART A. GEOMETRIC MODELING

7.2.2 Start Up

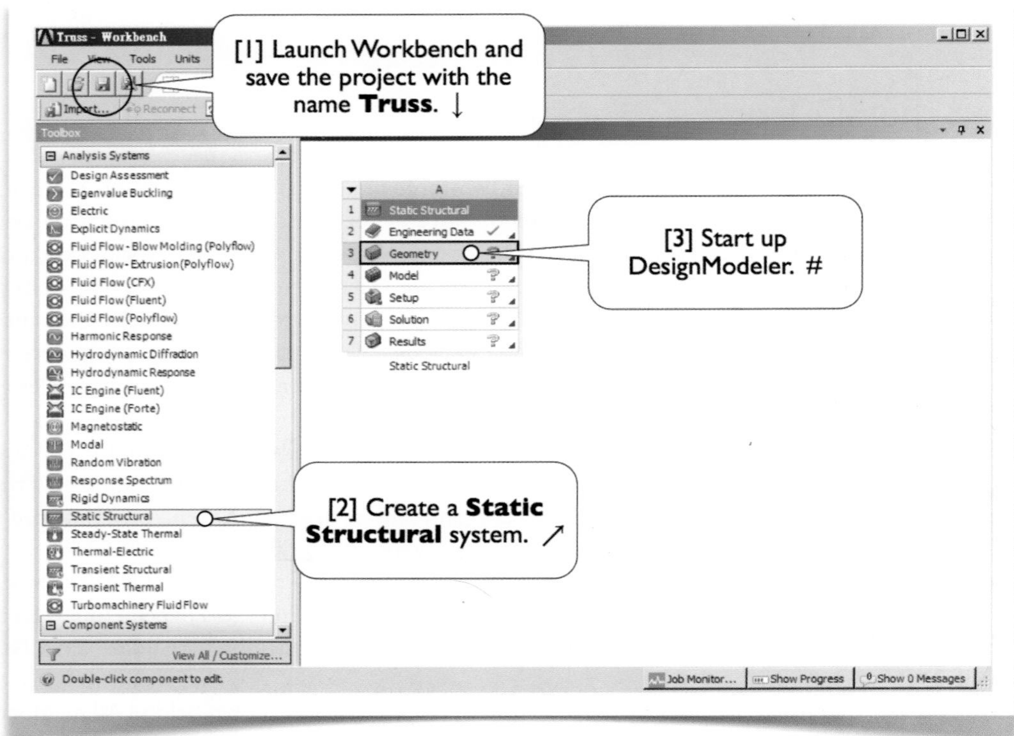

7.2.3 Create 10 Construction Points in 3D Space

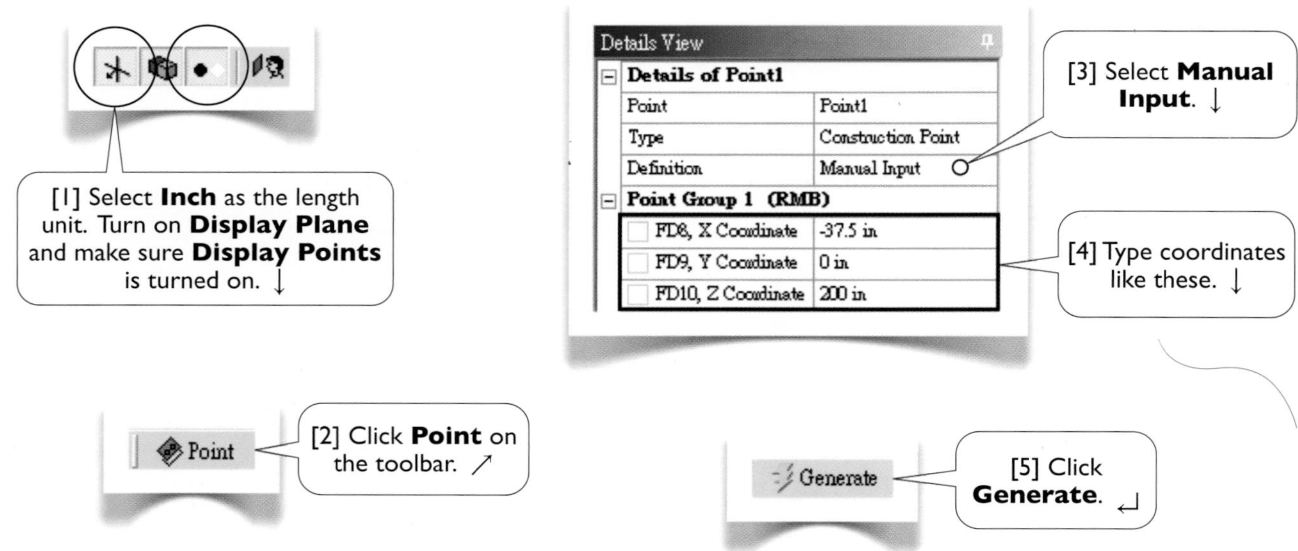

Point	X Coordinate (in)	Y Coordinate (in)	Z Coordinate (in)
1	-37.5	0	200
2	37.5	0	200
3	-37.5	37.5	100
4	37.5	37.5	100
5	37.5	-37.5	100
6	-37.5	-37.5	100
7	-100	100	0
8	100	100	0
9	100	-100	0
10	-100	-100	0

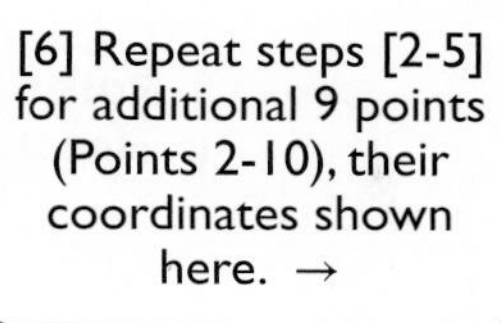

[7] The 10 points. (The numbers are added by the author for your convenience.) Note that the view is rotated such that Z-axis directs upward. ↓

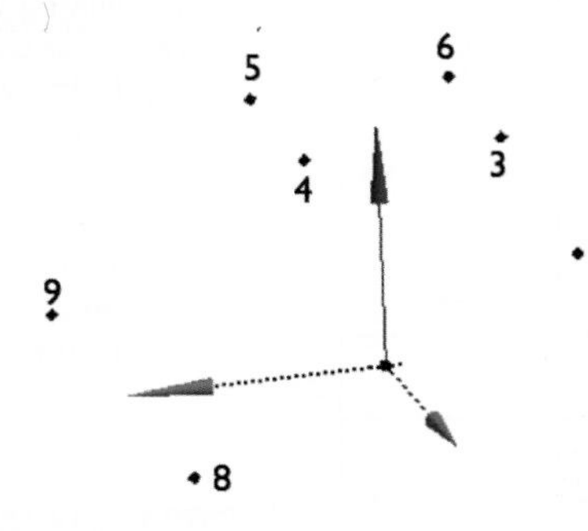

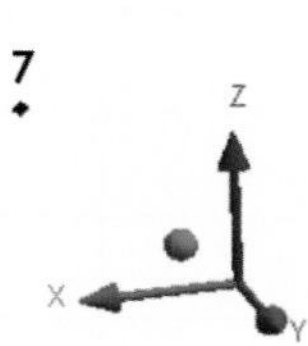

Using **Coordinates File** to Define Points

[8] An alternative way of defining points is to select **From Coordinates File** for **Definition** and read the coordinates from a file [9-10]. When the number of points is large, this is obviously a better way to input coordinates.

A **Coordinates file** [10] is a text file describing the coordinates of points. The file has 5 columns, or fields: (a) group number, (b) ID number, (c) X-coordinate, (d) Y-coordinate, and (e) Z-coordinate. The group number and the ID number can be arbitrarily chosen and they together uniquely identify a point from others.

The fields can be separated by spaces or TABs. You may prepare the text file using a text editor; another way is using a spread-sheet program such as Microsoft Excel and saving as a text file (e.g., a CSV file). ↙

[9] An alternative way of defining coordinates is to select **From Coordinates File** (default) for **Definition** and read the coordinates from a file. →

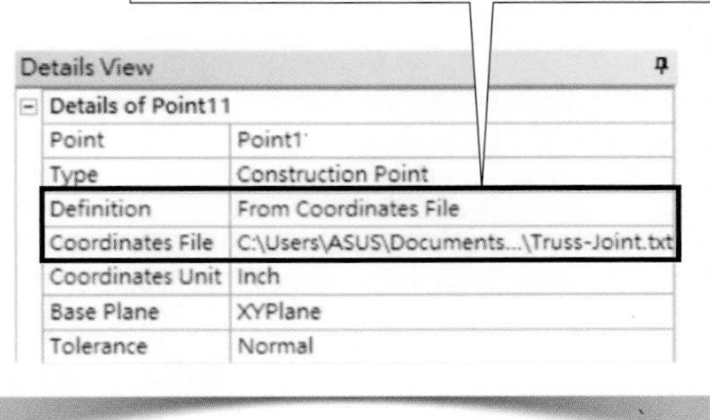

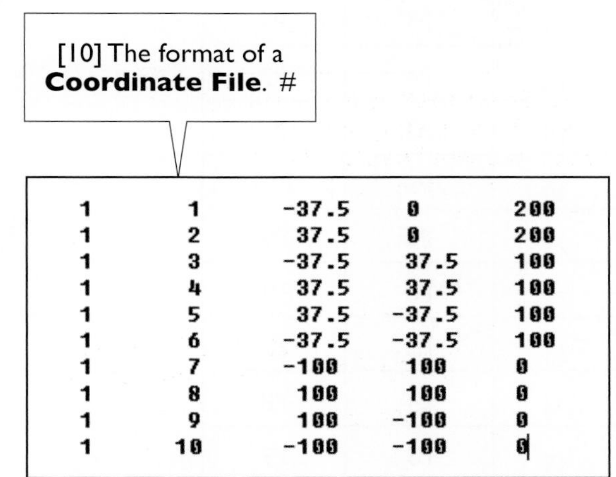
[10] The format of a **Coordinate File**. #

```
1    1    -37.5    0       200
1    2     37.5    0       200
1    3    -37.5    37.5    100
1    4     37.5    37.5    100
1    5     37.5   -37.5    100
1    6    -37.5   -37.5    100
1    7    -100     100     0
1    8     100     100     0
1    9     100    -100     0
1    10   -100    -100     0
```

7.2.4 Create Line Body

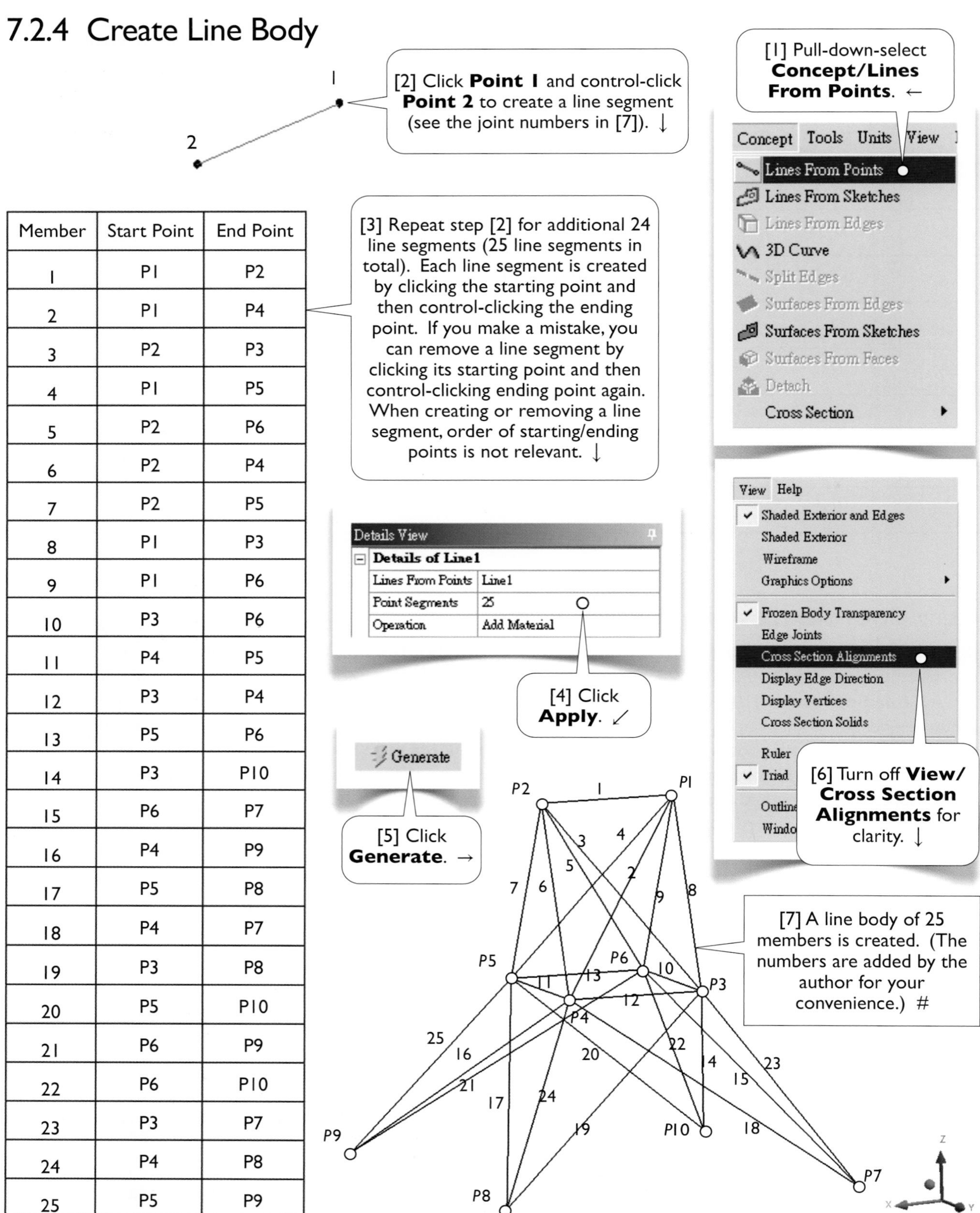

Member	Start Point	End Point
1	P1	P2
2	P1	P4
3	P2	P3
4	P1	P5
5	P2	P6
6	P2	P4
7	P2	P5
8	P1	P3
9	P1	P6
10	P3	P6
11	P4	P5
12	P3	P4
13	P5	P6
14	P3	P10
15	P6	P7
16	P4	P9
17	P5	P8
18	P4	P7
19	P3	P8
20	P5	P10
21	P6	P9
22	P6	P10
23	P3	P7
24	P4	P8
25	P5	P9

7.2.5 Create Cross Section

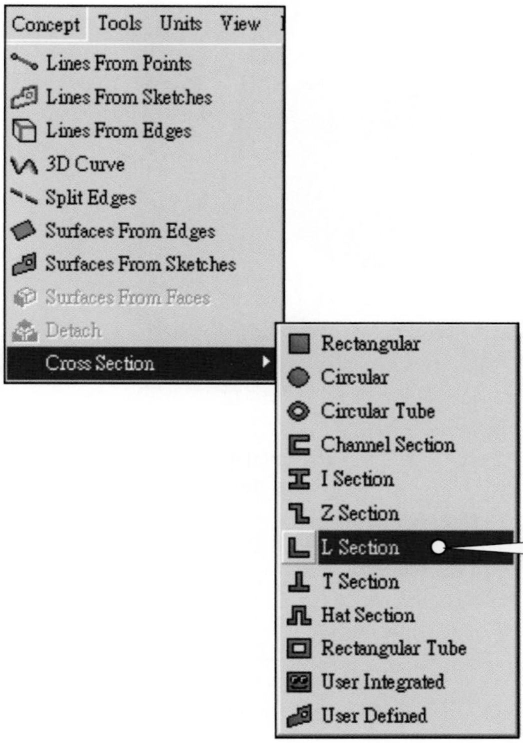

[1] Pull-down-select **Concept/Cross Section/L Section**. ↑

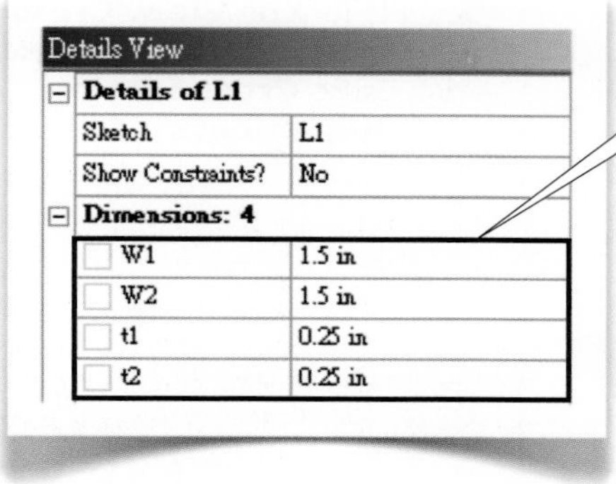

[2] Type dimensions for the steel angle. You don't have to click **Generate**. #

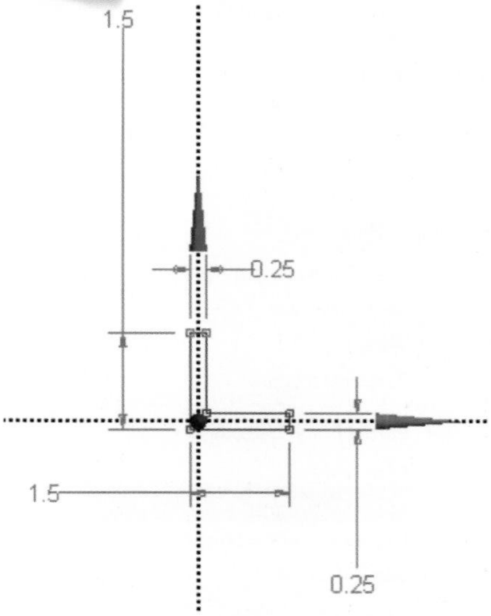

Details View	
Details of L1	
Sketch	L1
Show Constraints?	No
Dimensions: 4	
W1	1.5 in
W2	1.5 in
t1	0.25 in
t2	0.25 in

7.2.6 Assign Cross Section to the Line Body

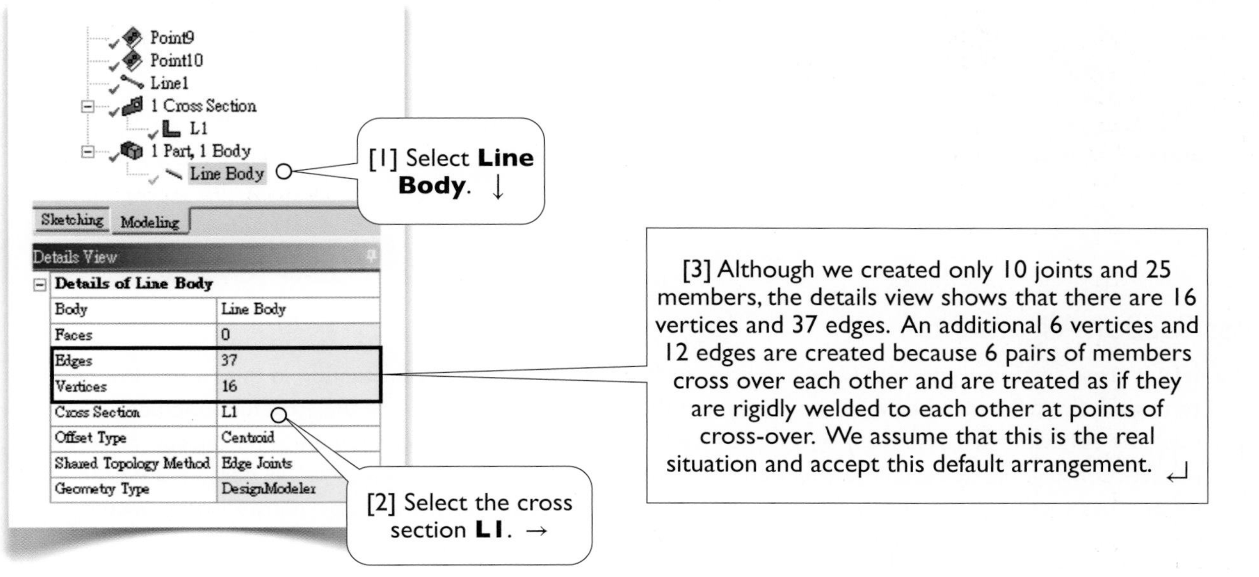

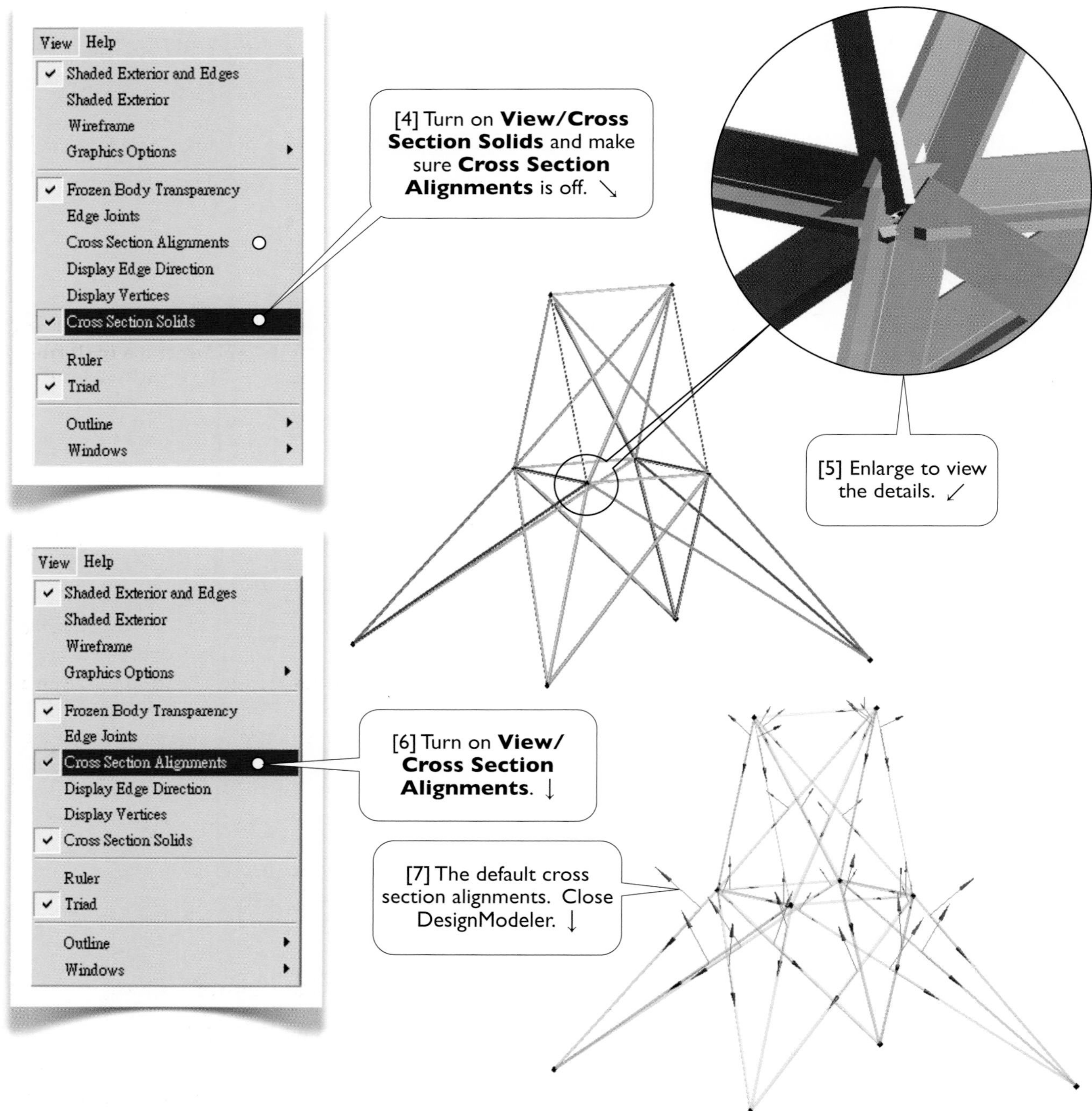

Cross Section Alignments

[8] The default cross section alignments [7] are not entirely consistent with the real situation in this case. However, we decided to neglect the difference between the reality and the default alignments. Since the structural members are quite slender, the behaviors are essentially two-force members, therefore alignments should not be critical. In other words, cross section alignments have little effect on the structural response in this case. We will demonstrate the adjustment of cross section alignments in 7.3.9 (page 300), in which the alignments must be adjusted, otherwise it would deviate from the reality too much. #

PART B. SIMULATION

7.2.7 Start Up **Mechanical**

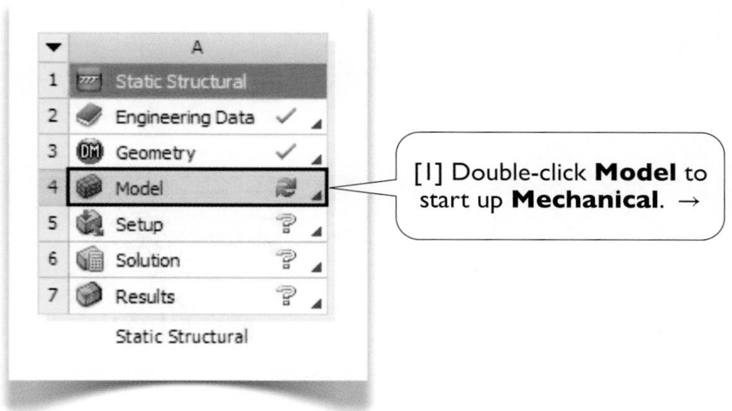

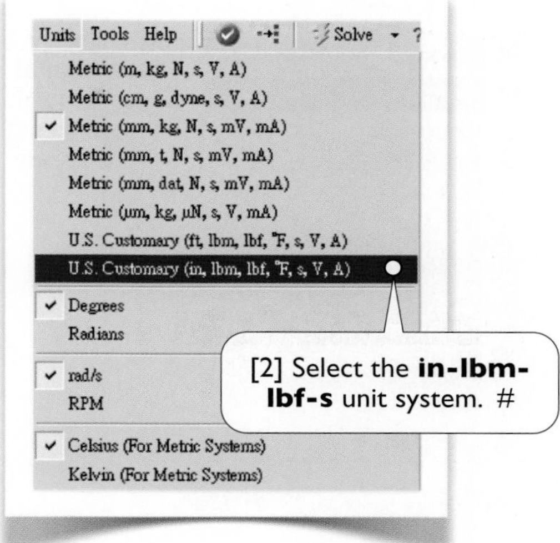

7.2.8 Generate Mesh

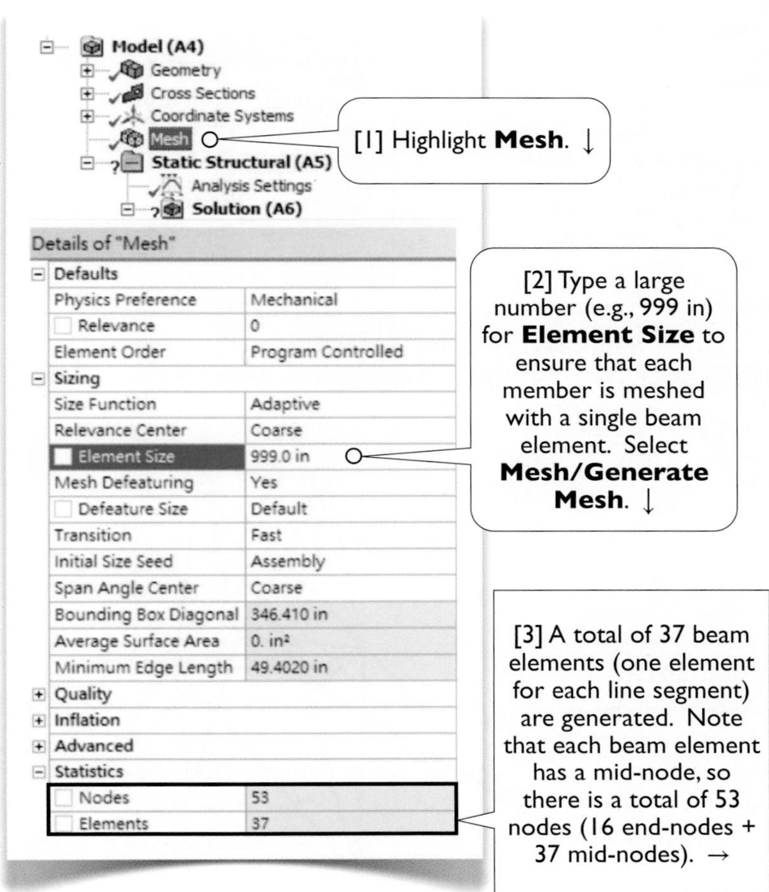

[3] A total of 37 beam elements (one element for each line segment) are generated. Note that each beam element has a mid-node, so there is a total of 53 nodes (16 end-nodes + 37 mid-nodes). →

Why mesh each member with a single element?

[4] We mentioned (7.1.16[1], page 282) that we can obtain a solution equal to analytical values by meshing each straight beam with a single element. The reason we meshed each member with a single beam element here is to demonstrate this fact.

The default settings of **Mesh** would mesh the model with 205 beam elements and would result in exactly the same solution as 37 elements. As an exercise (7.4.2, page 310), verify it yourself after the completion of this section.

Use Surface/Line Models Whenever Possible

Since the solution of a model meshed with beam elements or shell elements converges very fast (i.e., very accurate solution can be obtained with only a few elements), we should consider a line model or surface model whenever possible. This is particularly true for those problems requiring many iterations or substeps, such as nonlinear problems, dynamic problems, optimization problems, etc. #

7.2.9 Specify Supports

7.2.10 Specify Loads

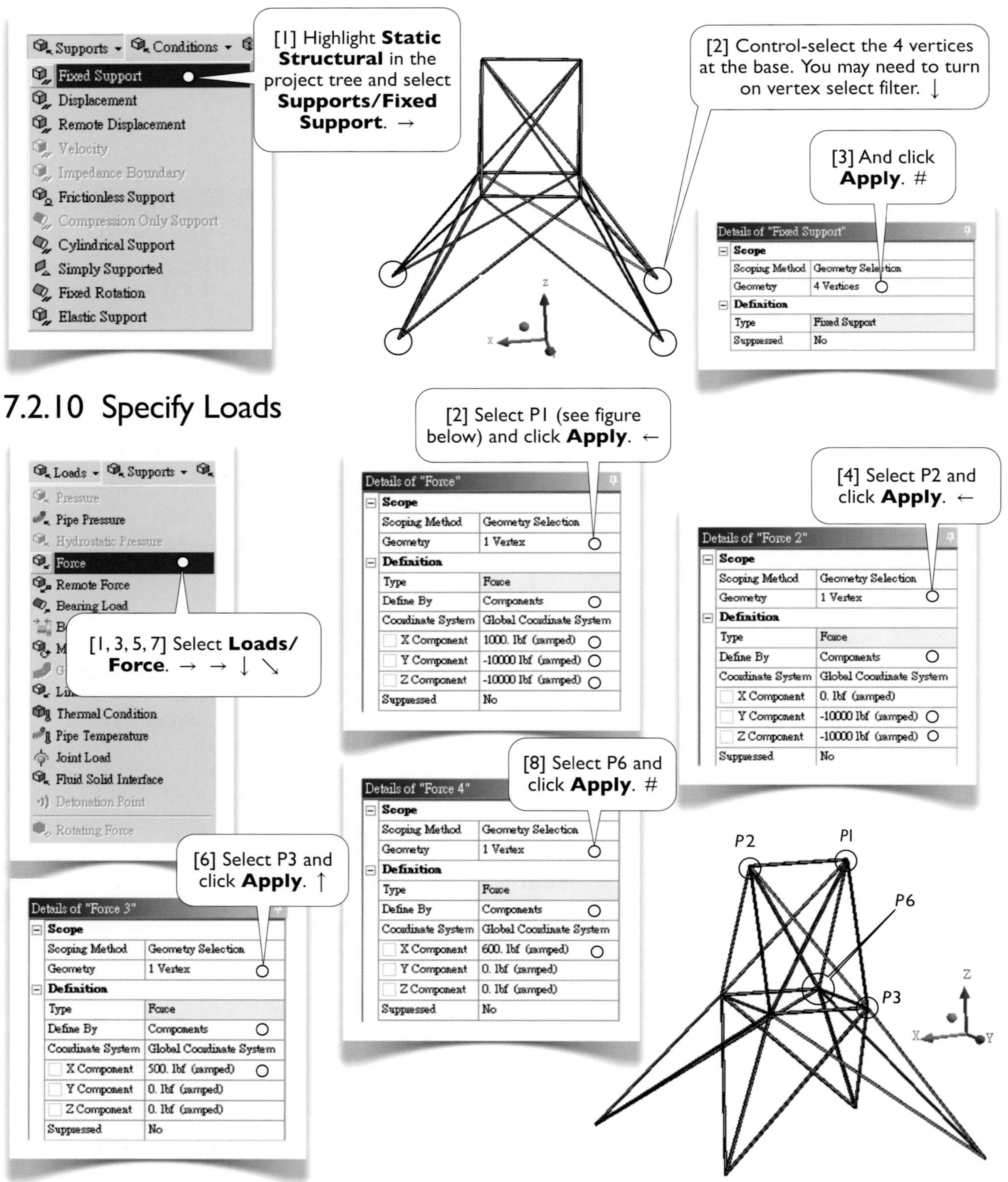

7.2.11 Set Up Solution Branch and Solve the Model

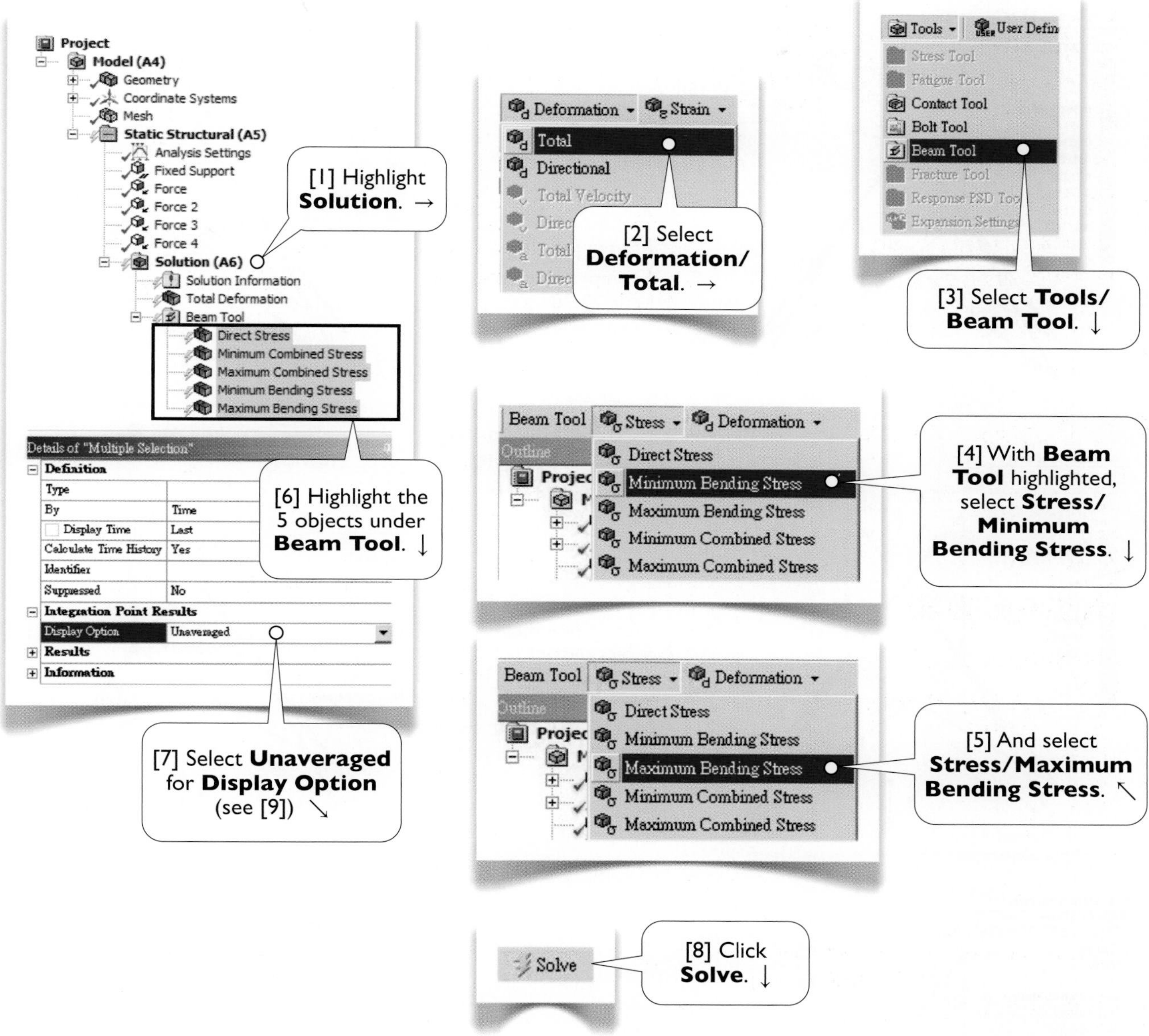

Why **Unaveraged** Stresses?

[9] We've introduced the notions of averaged and unaveraged stresses (3.5.6, page 160). Unaveraged stresses are also called *element stresses*, since they are calculated at points (usually the geometric center or the integration points) inside elements. On the other hand, averaged stresses are also called *nodal stresses*, since they are calculated at nodes, which are located at element boundaries. Since we mesh each member with a single beam element, and if we select to display averaged stresses, every two adjacent members' stresses would have been averaged and reported. The averaged stresses in this case would not have any meaning. #

7.2.12 View Results

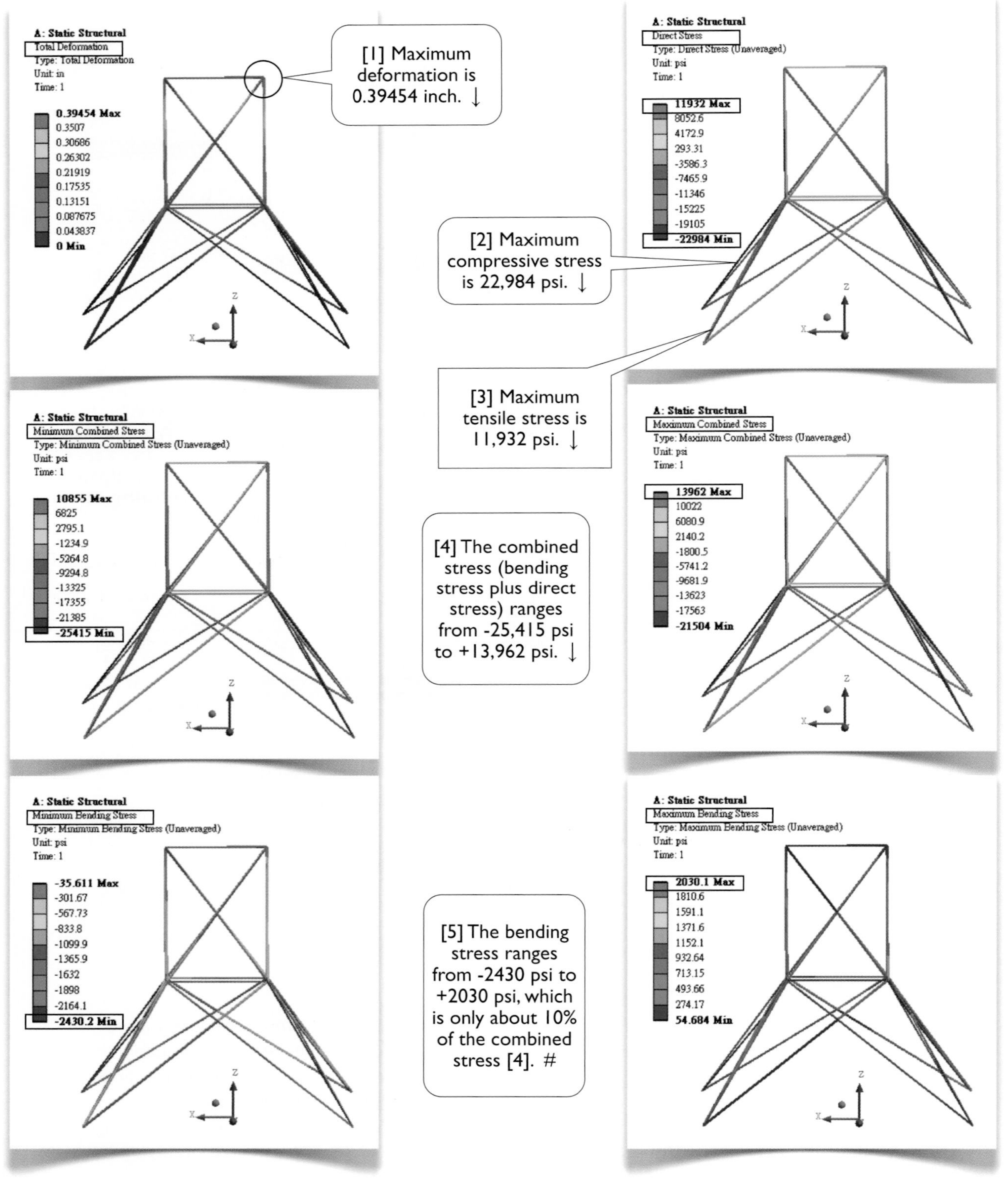

7.2.13 View Member Forces/Moments

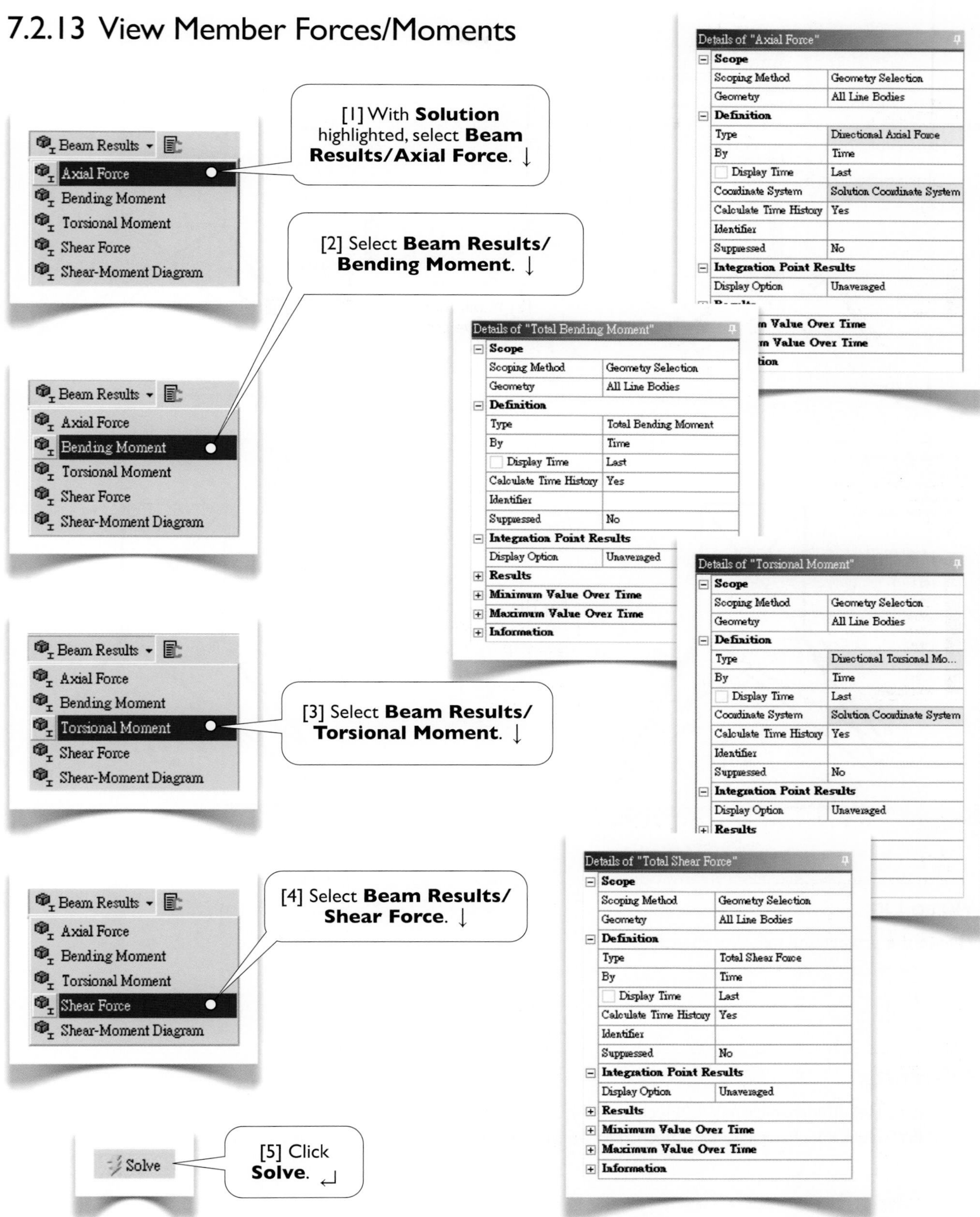

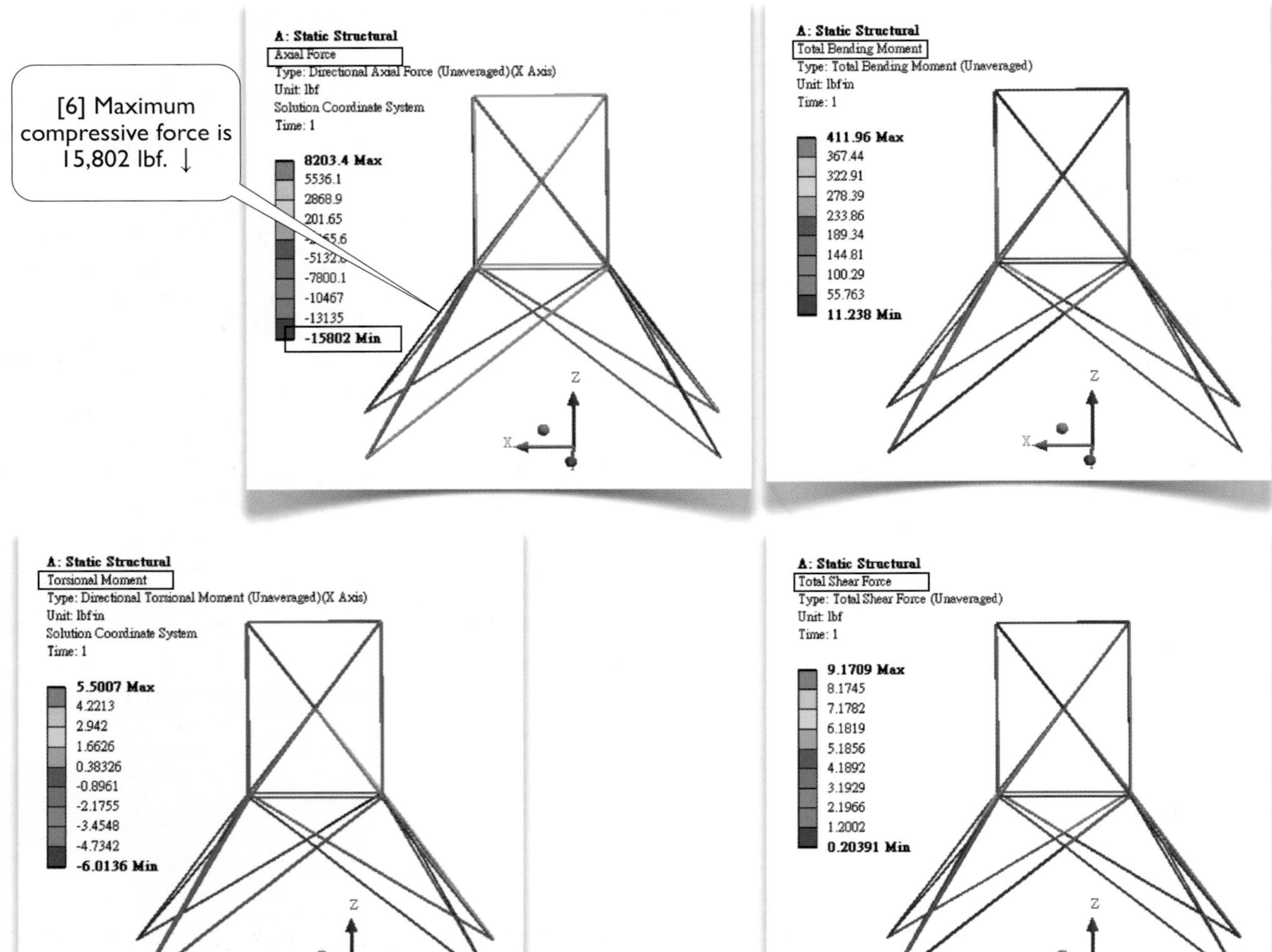

Wrap Up

[7] Save the project and exit Workbench. #

References

1. ANSYS Help//Mechanical APDL//Mechanical APDL Element Reference//I. Element Library//BEAM188
2. ANSYS Help//Mechanical APDL//Mechanical APDL Element Reference//I. Element Library//LINK180

Section 7.3

Two-Story Building

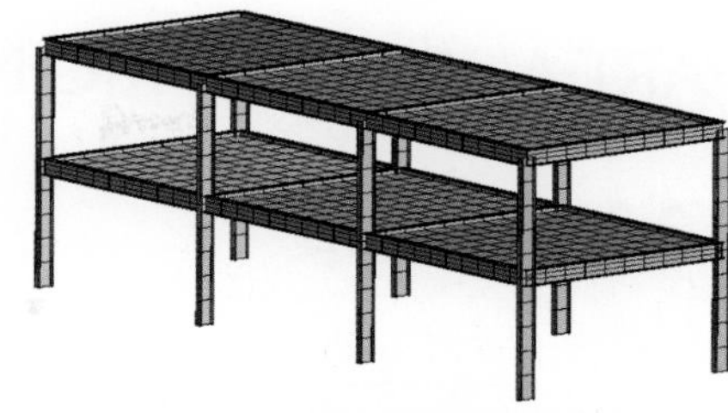

7.3.1 About the Two-Story Building

[1] The two-story building [2-7] is constructed for residential usage. The local building code requires that a live load of 50 lb/ft^2 be considered, along with its own weight. Since the building is in an earthquake zone, an earthquake load must be considered. For a low-rise building like this, the building code allows an equivalent static analysis instead of a dynamic analysis. Here we consider a static earthquake load, which is equivalent to 0.2 times gravitational acceleration, applying horizontally in the shorter direction of the building. In a practical design project, earthquake load applying in other directions should also be simulated.

The beams and columns will be modeled as line bodies and the floor slabs as surface bodies. We assume that all bodies are perfectly bonded together. The geometric model will be used for a static structural simulation in this section. The model will be used again in Section 11.2 for a modal analysis and Section 12.3 for a harmonic response analysis.

In this exercise, we will use **in-lbm-lbf-s** unit system most of the time. On some occasions, we will change the unit system to **ft-lbm-lbf-s** unit system. One feature of Workbench is the flexibility of using unit systems. When you switch to another unit system, Workbench will convert the units nicely for you. ↙

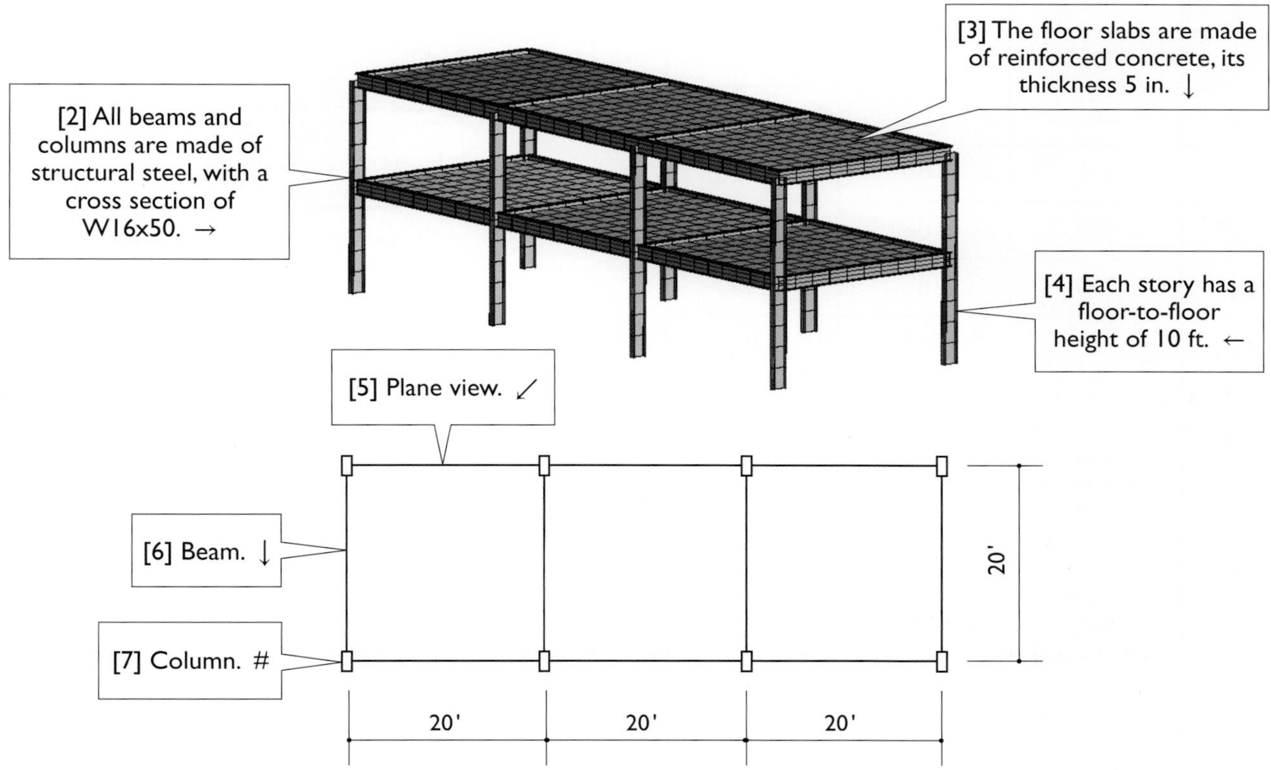

PART A. GEOMETRIC MODELING

7.3.2 Start Up

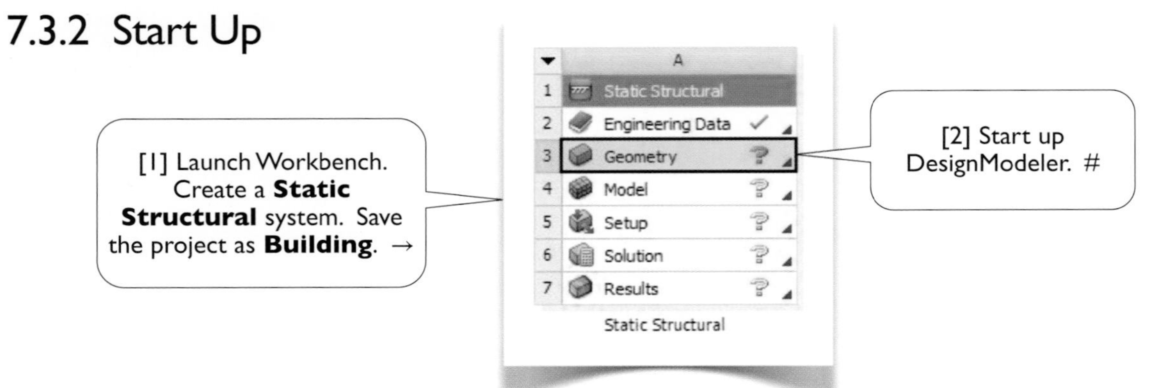

7.3.3 Create 16 Points in the Space

Point	X Coordinate (in)	Y Coordinate (in)	Z Coordinate (in)
1	0	0	0
2	240	0	0
3	480	0	0
4	720	0	0
5	0	0	240
6	240	0	240
7	480	0	240
8	720	0	240
9	0	120	0
10	240	120	0
11	480	120	0
12	720	120	0
13	0	120	240
14	240	120	240
15	480	120	240
16	720	120	240

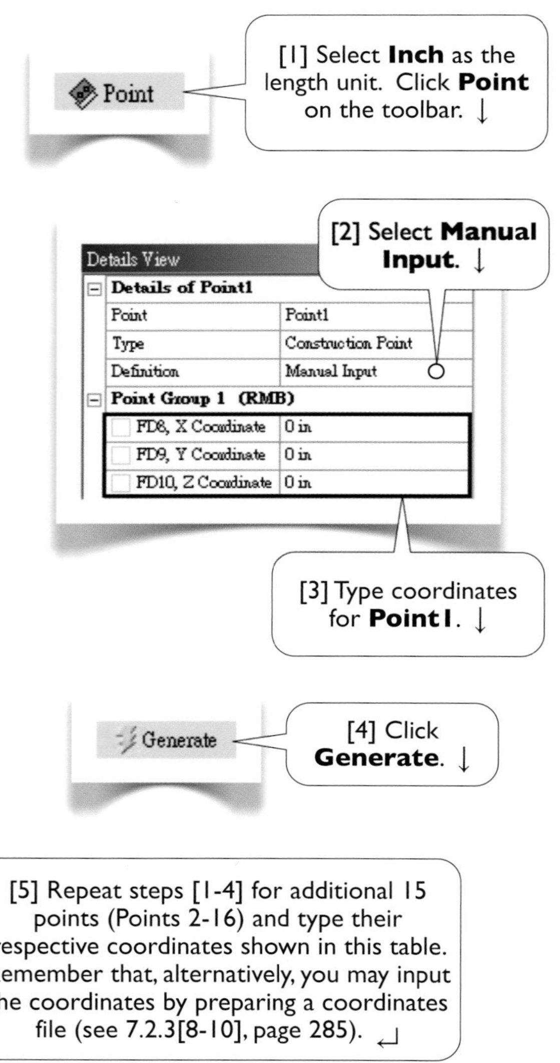

9
13
10
14
11
15
12
1
16
5
2
6
3
7
4
8

[6] These 16 points are starting and ending points of the 8 columns for the lower floor. (The numbers are added by the author for your convenience.) #

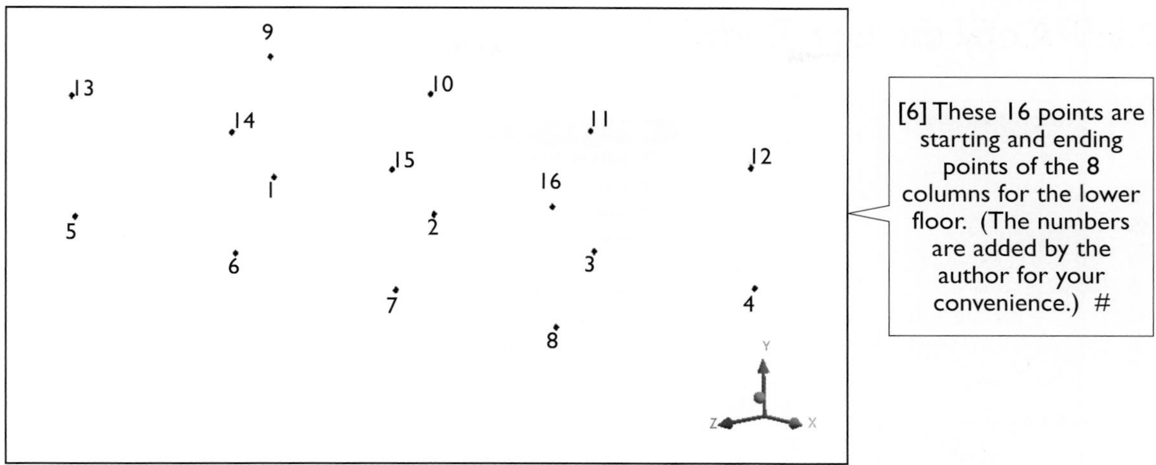

7.3.4 Create a Line Body of 10 Beams

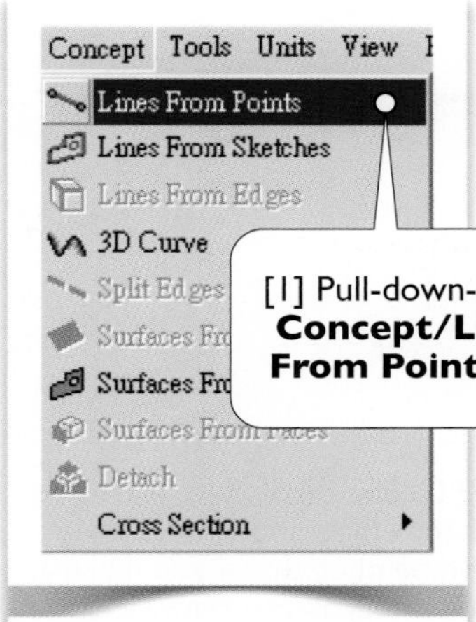

Concept
Tools
Units
View
Lines From Points
Lines From Sketches
Lines From Edges
3D Curve
Split Edges
Surfaces Fr
Surfaces Fr
Surfaces From
Detach
Cross Section

[1] Pull-down-select **Concept/Lines From Points**. →

[2] Define 6 line segments. Remember that each line segment is defined by first clicking its starting point and then control-clicking its ending point. ↓

Member	Start Point	End Point
1	P9	P12
2	P13	P16
3	P9	P13
4	P10	P14
5	P11	P15
6	P12	P16

[3] Click **Apply**. →

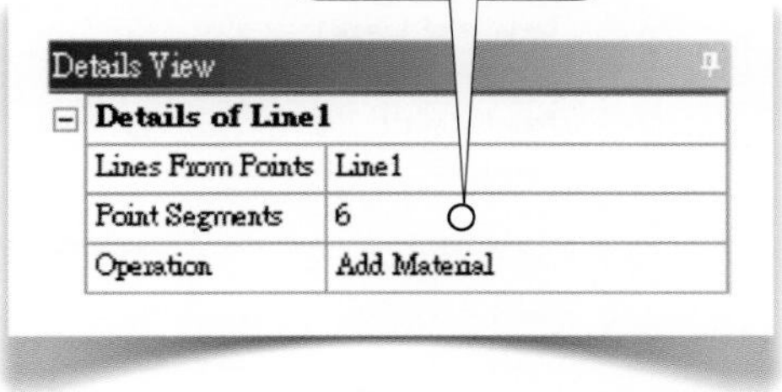

Details View
Details of Line1
Lines From Points
Line1
Point Segments
6
Operation
Add Material

[4] Click **Generate**. ↙

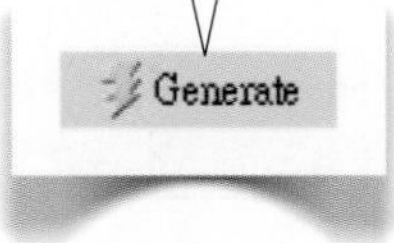

Generate

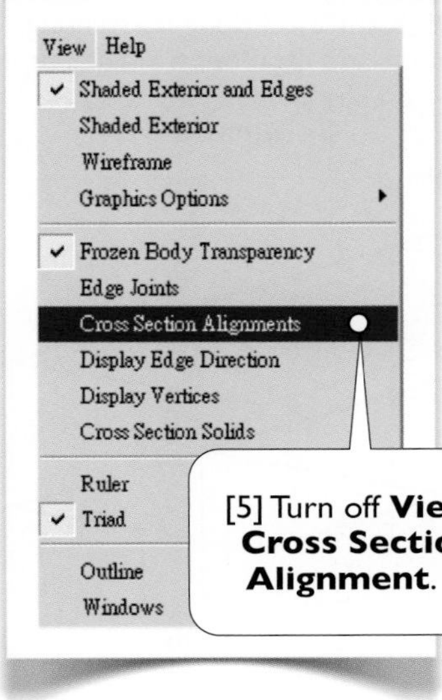

View
Help
Shaded Exterior and Edges
Shaded Exterior
Wireframe
Graphics Options
Frozen Body Transparency
Edge Joints
Cross Section Alignments
Display Edge Direction
Display Vertices
Cross Section Solids
Ruler
Triad
Outline
Windows

[5] Turn off **View/ Cross Section Alignment**. #

9
10
13
11
12
14
15
16

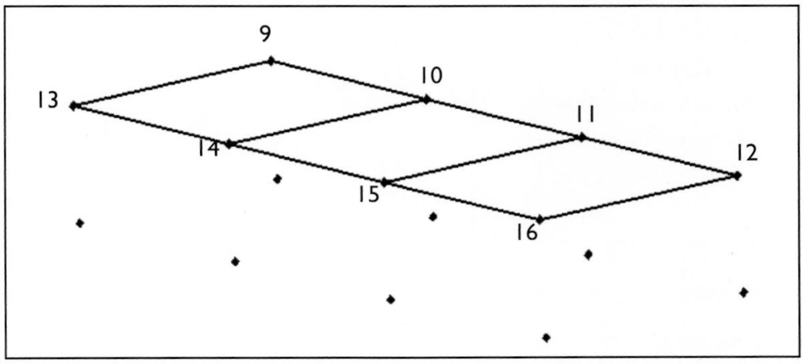

7.3.5 Copy the Line Body

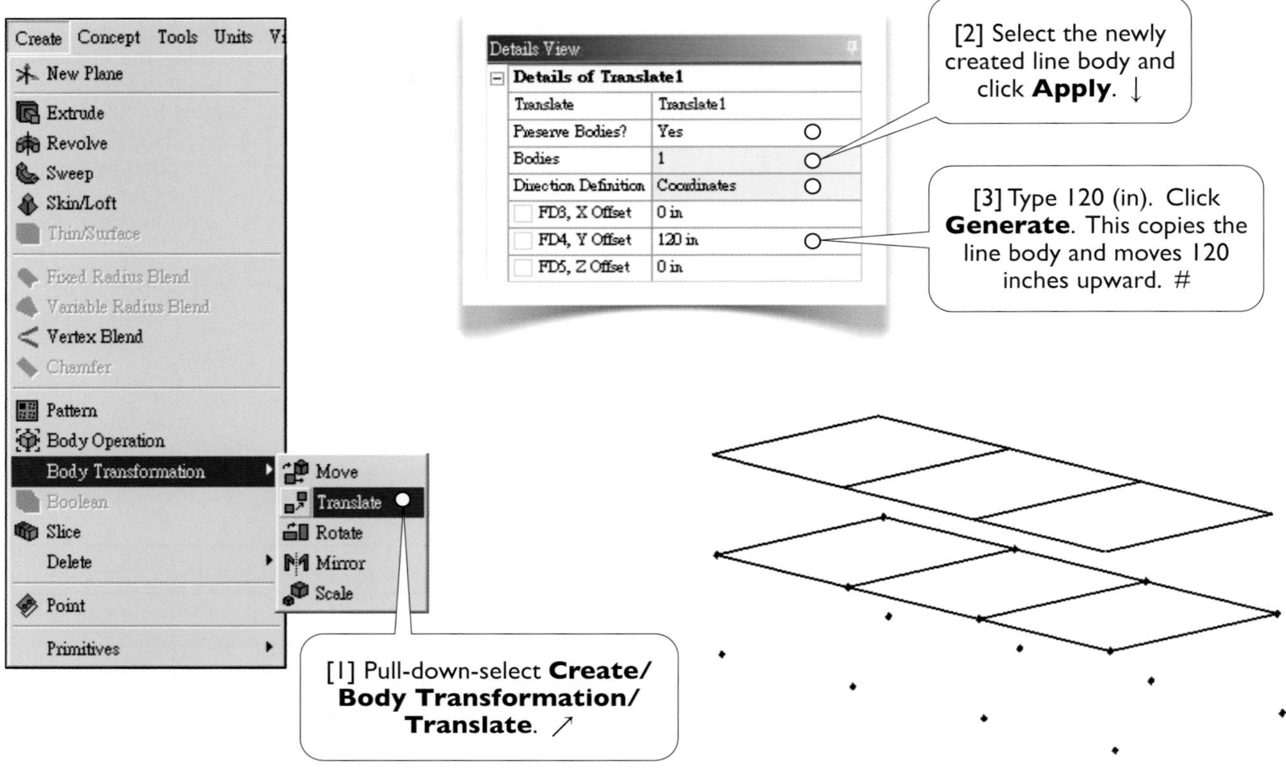

7.3.6 Create Line Body for Columns

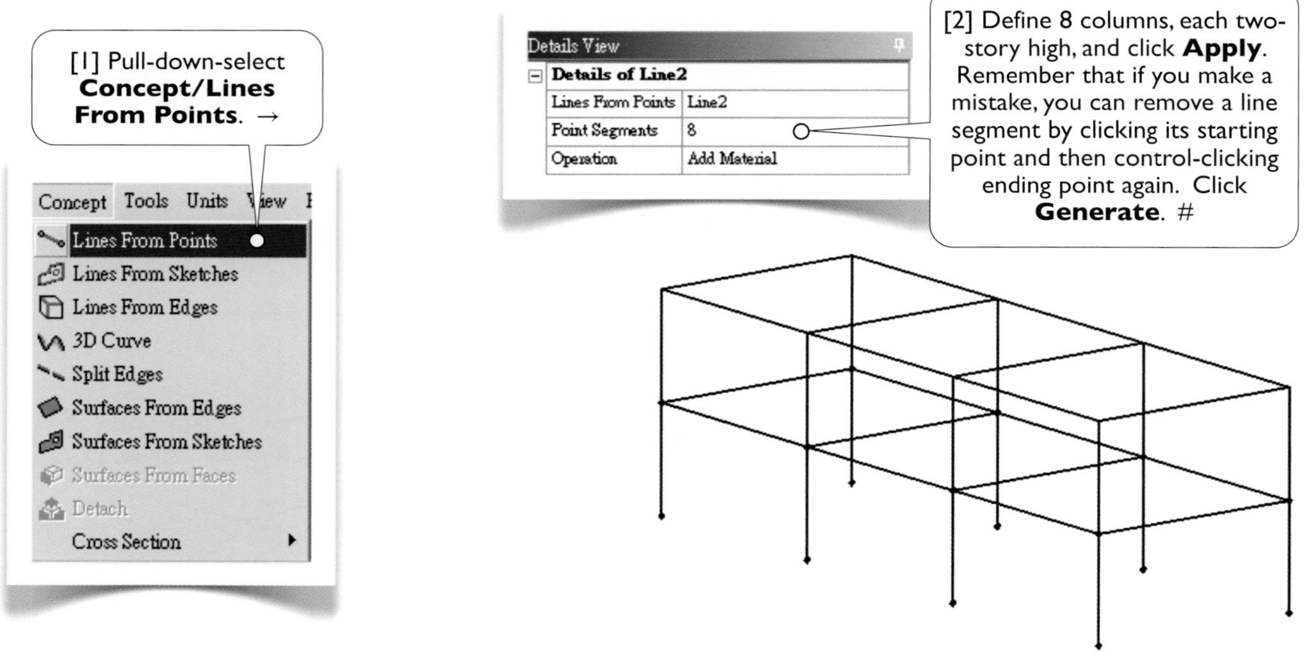

7.3.7 Create a Cross Section

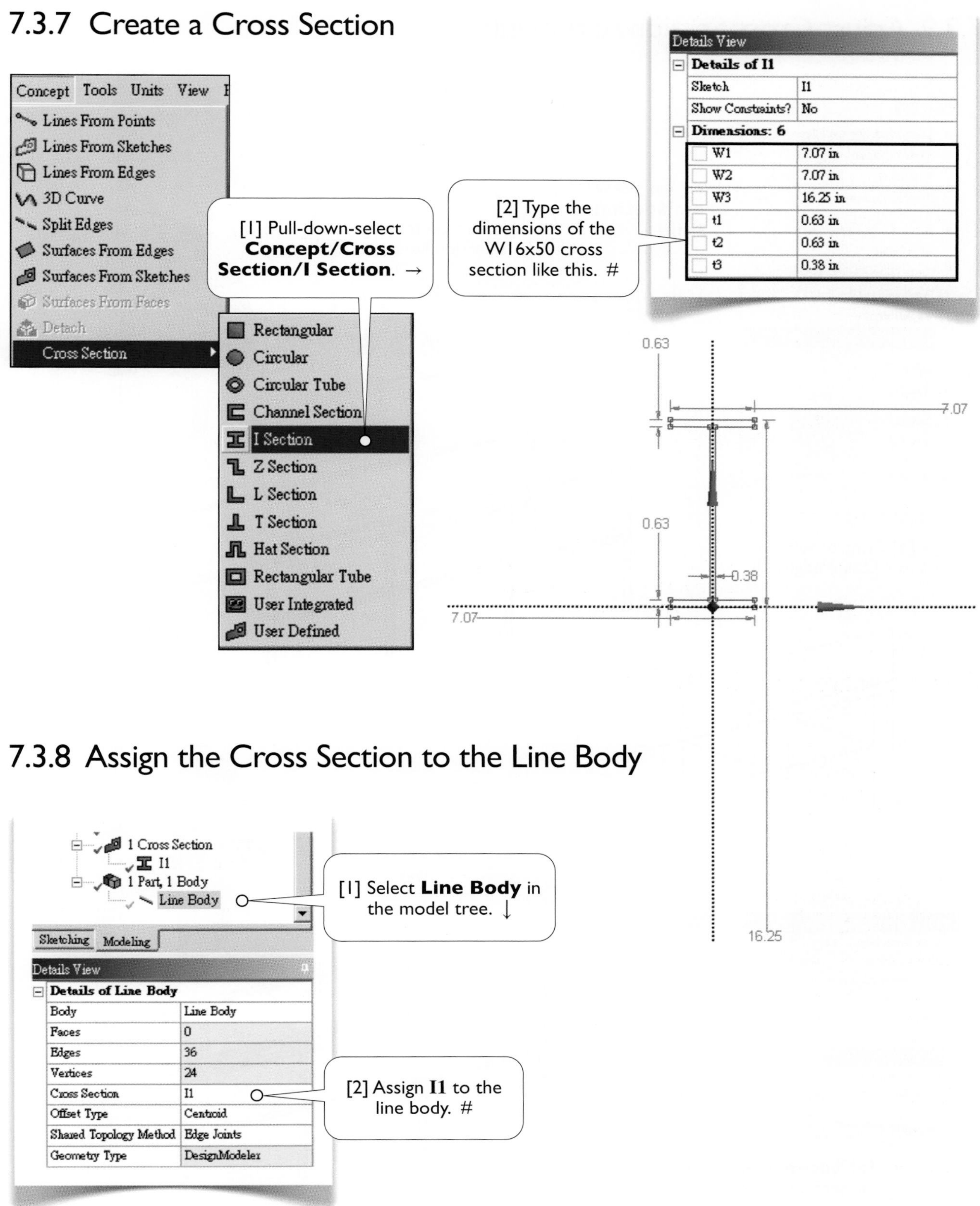

7.3.8 Assign the Cross Section to the Line Body

7.3.9 Adjust Cross Section Alignments

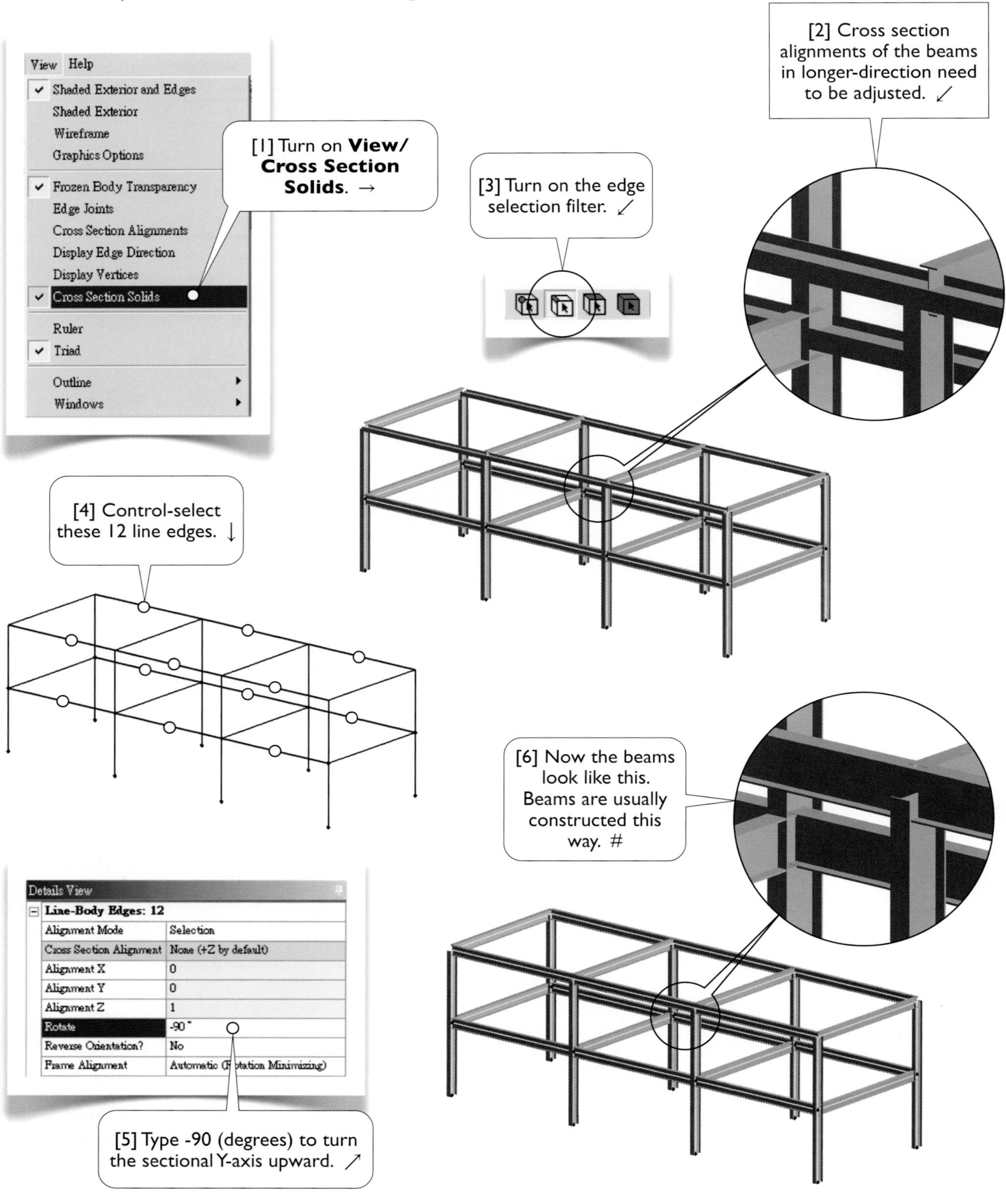

7.3.10 Create Surface Bodies

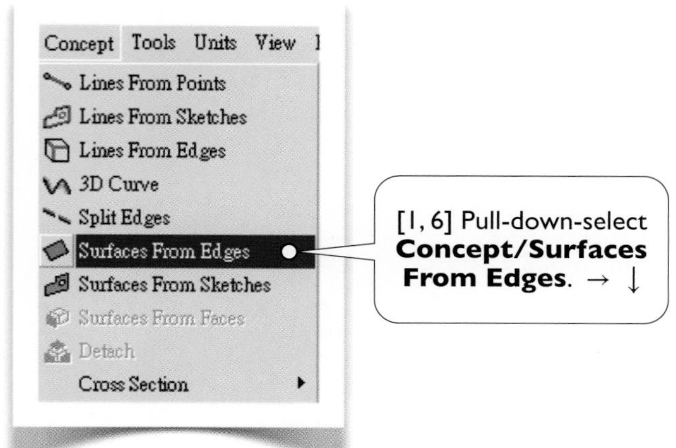

[1, 6] Pull-down-select **Concept/Surfaces From Edges**. → ↓

[2] Control-select these 4 edges. ↓

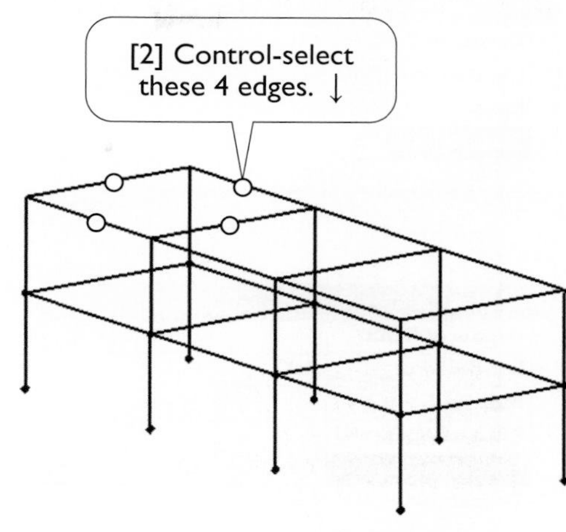

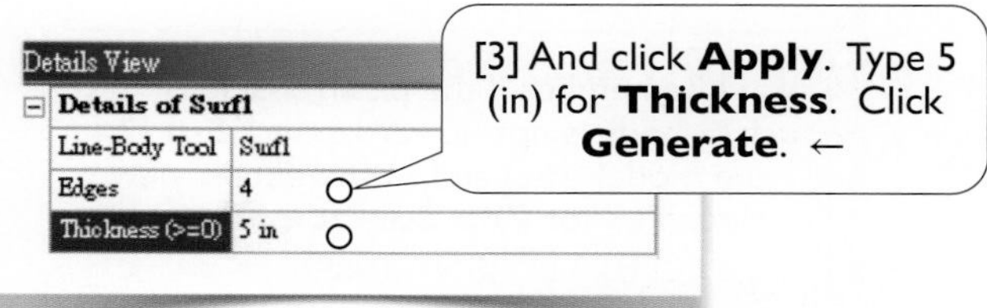

[3] And click **Apply**. Type 5 (in) for **Thickness**. Click **Generate**. ←

[4] Turn on the face selection filter and click the surface body. One of the surface faces is highlighted (with a green color), indicating that it is the positive-side of the surface body; the other side is the negative-side. ↘

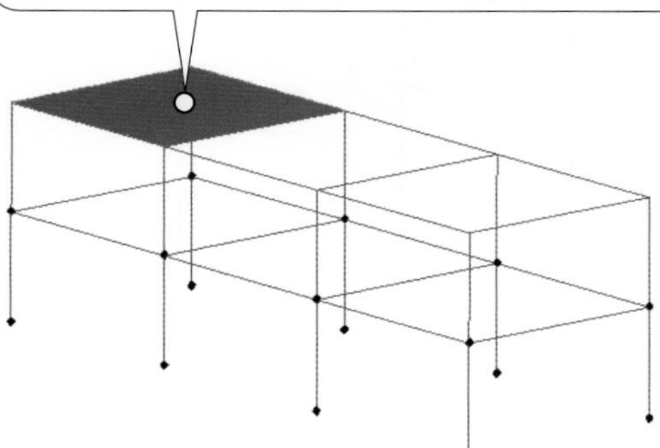

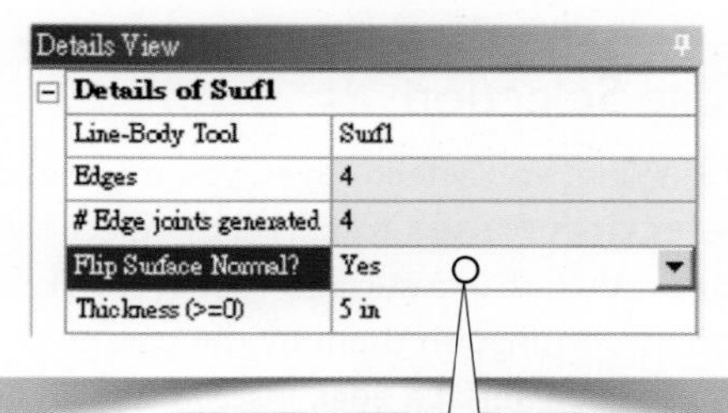

[5] If the positive-side of the surface body is not upward, select **Surf1** (in the model tree) and, in the details view, select **Yes** for **Flip Surface Normal?**. Click **Generate** and go back to [4] to make sure the positive-side is upward. ↖

[7] Select this edge. Use **Selection Panes** (see [8]) to correctly select a line edge rather than a surface edge. ↘

[9] Control-select the additional 3 edges. ↲

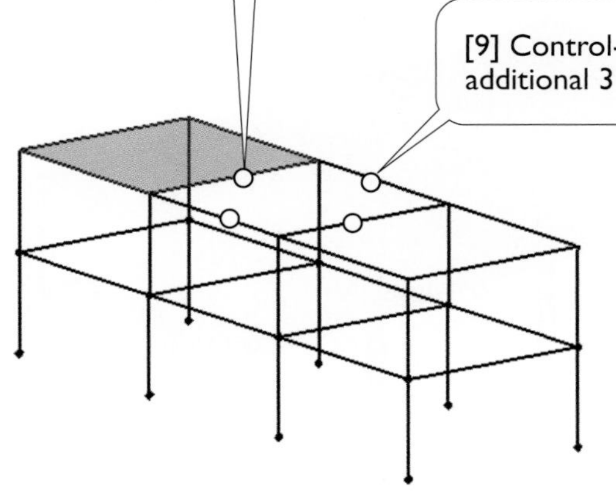

[8] **Selection Panes** can be used to select a line edge (instead of a surface edge). In this case one pane represents a line edge and the other pane represents a surface edge. You can distinguish them by the cursor shape. Also, you can deselect a pane by control-selecting it. ↖

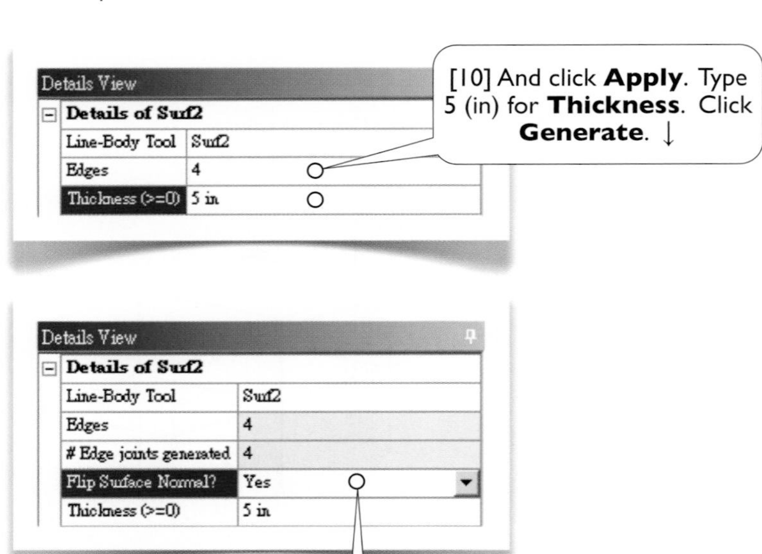

[11] Remember, if the positive-side of the surface body is not upward, select **Surf2** (in the model tree) and, in the details view, select **Yes** for **Flip Surface Normal?** (see [4-5], last page). →

[12] Repeat steps [6-11] to create 4 more surface bodies. ↙

Edge Selection

[13] When you select edges, you may need to use **Selection Panes** to correctly select the line edge rather than a surface edge ([7-8], last page). You can distinguish them by the cursor shape. If you select a surface edge instead of a line edge, the slab would not connect to the line body and would have a gap after deformation [14]. →

[14] If you select a surface edge instead of a line edge, the slab would not connect to the line body and would have a gap after deformation. ←

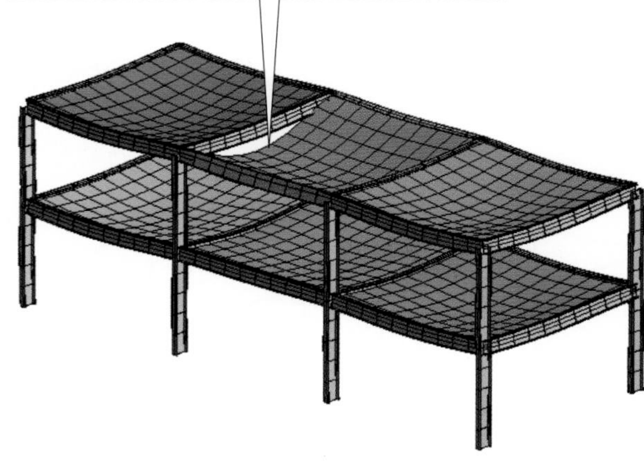

Direction of Surface Bodies

[15] Each surface body has a positive-side and a negative-side. In **Mechanical**, the pressure always applies on the positive-side. Therefore it is important to know which is the positive-side: turn on the face selection filter and click the surface body, and the positive-side will be highlighted with a green color. Make sure the positive-side of each surface body is upward in this case. If it is not, select **Yes** for **Flip Surface Normal?** in the details view (see [4-5], last page). #

7.3.11 Form a Single Part

PART B. SIMULATION

7.3.12 Prepare Material Properties

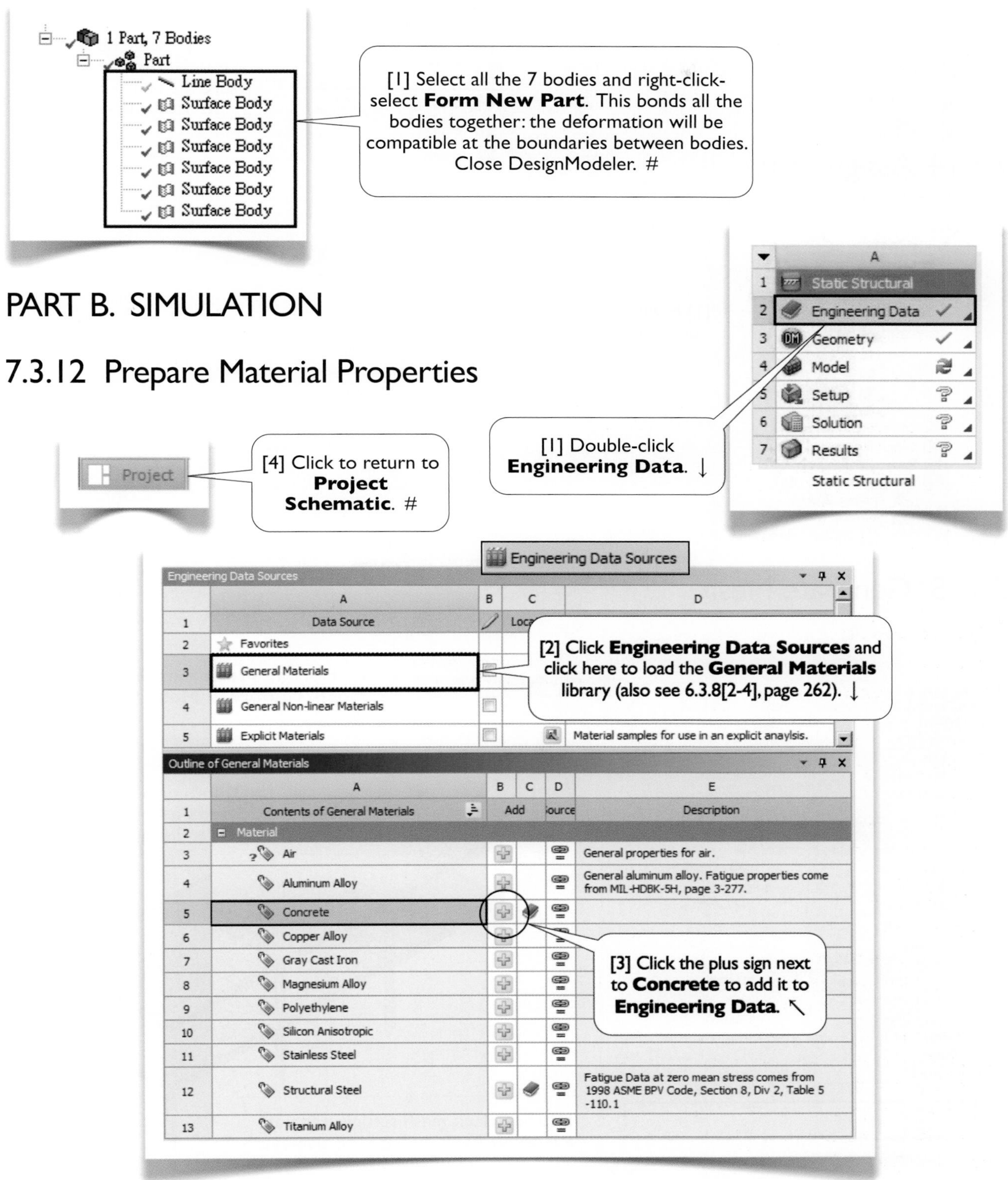

7.3.13 Start Up **Mechanical**

7.3.14 Assign Materials

7.3.15 Generate Mesh

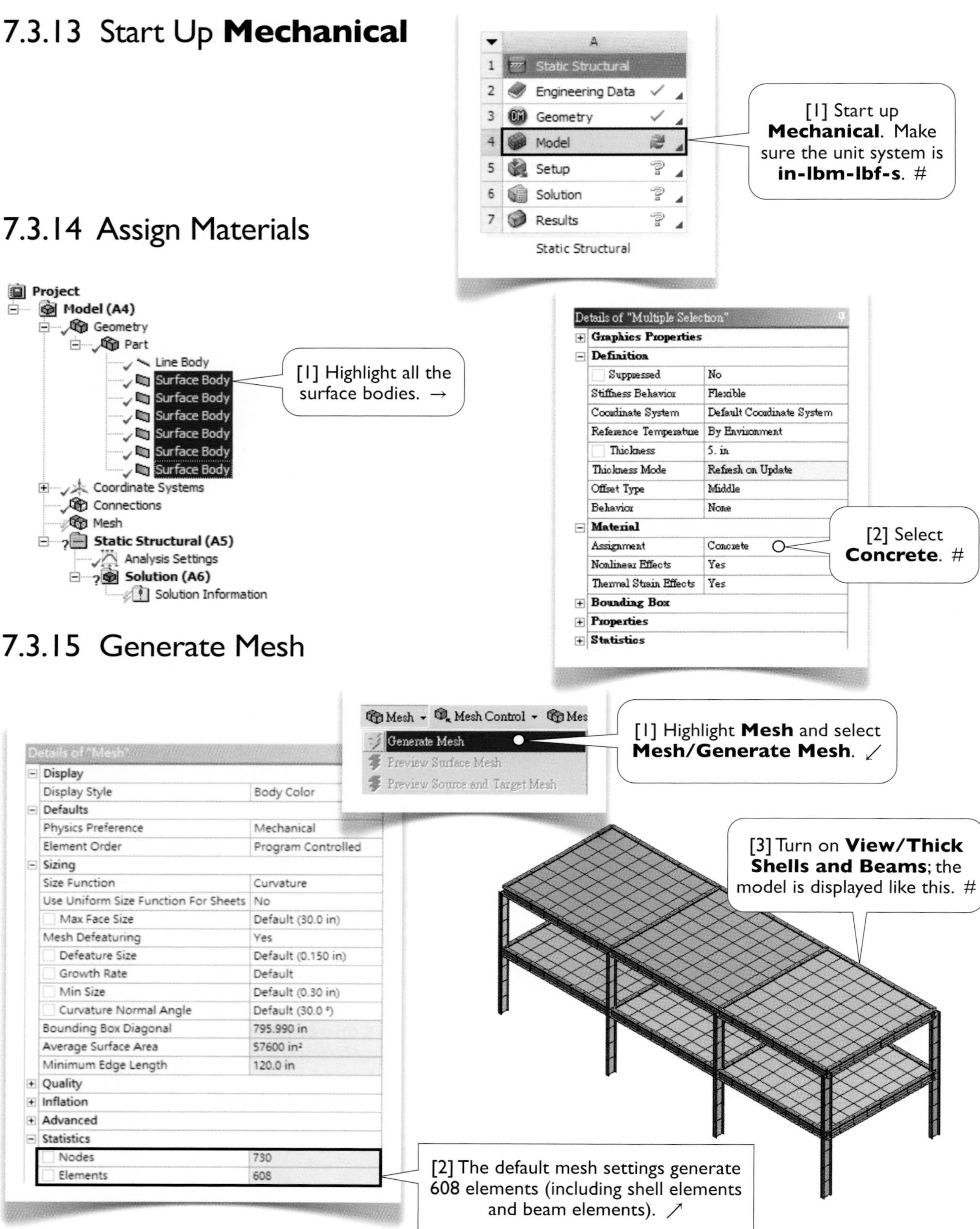

7.3.16 Specify Supports

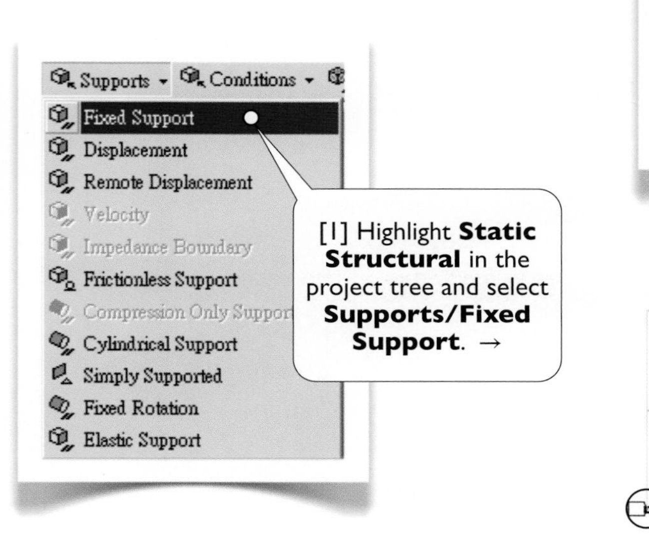

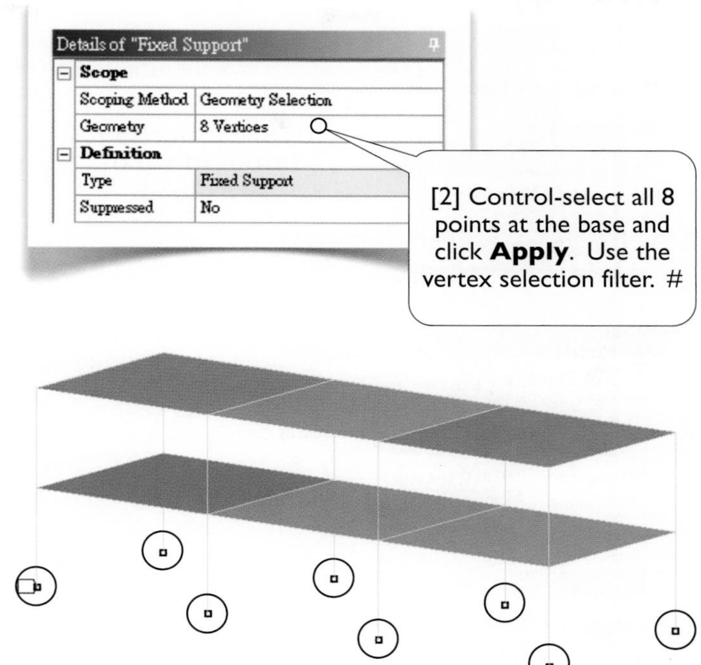

7.3.17 Apply the Live Load

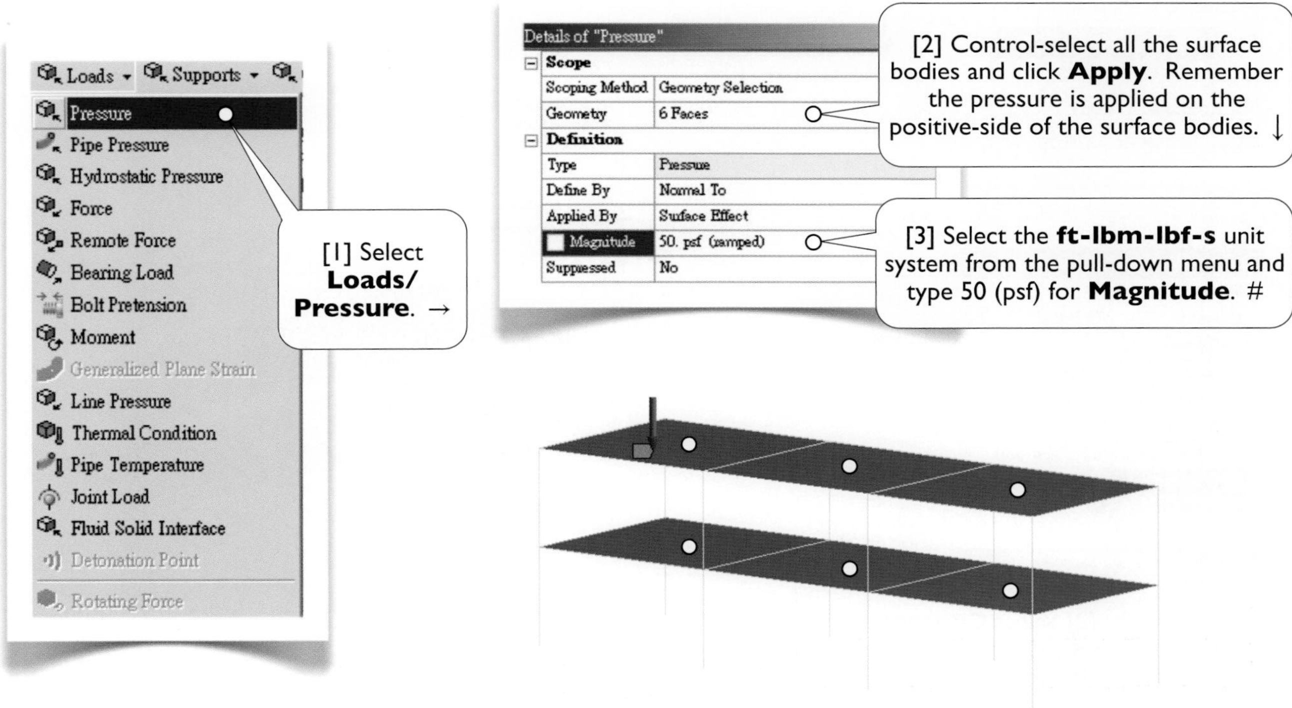

7.3.18 Apply Earth Gravity

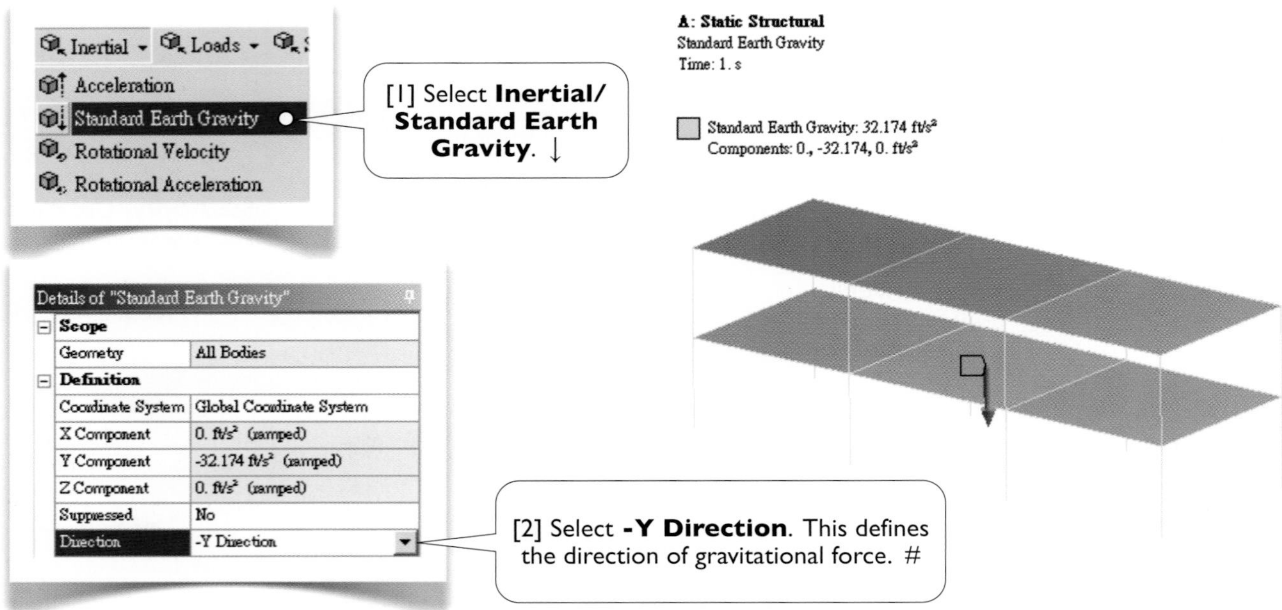

7.3.19 Apply the Earthquake Load

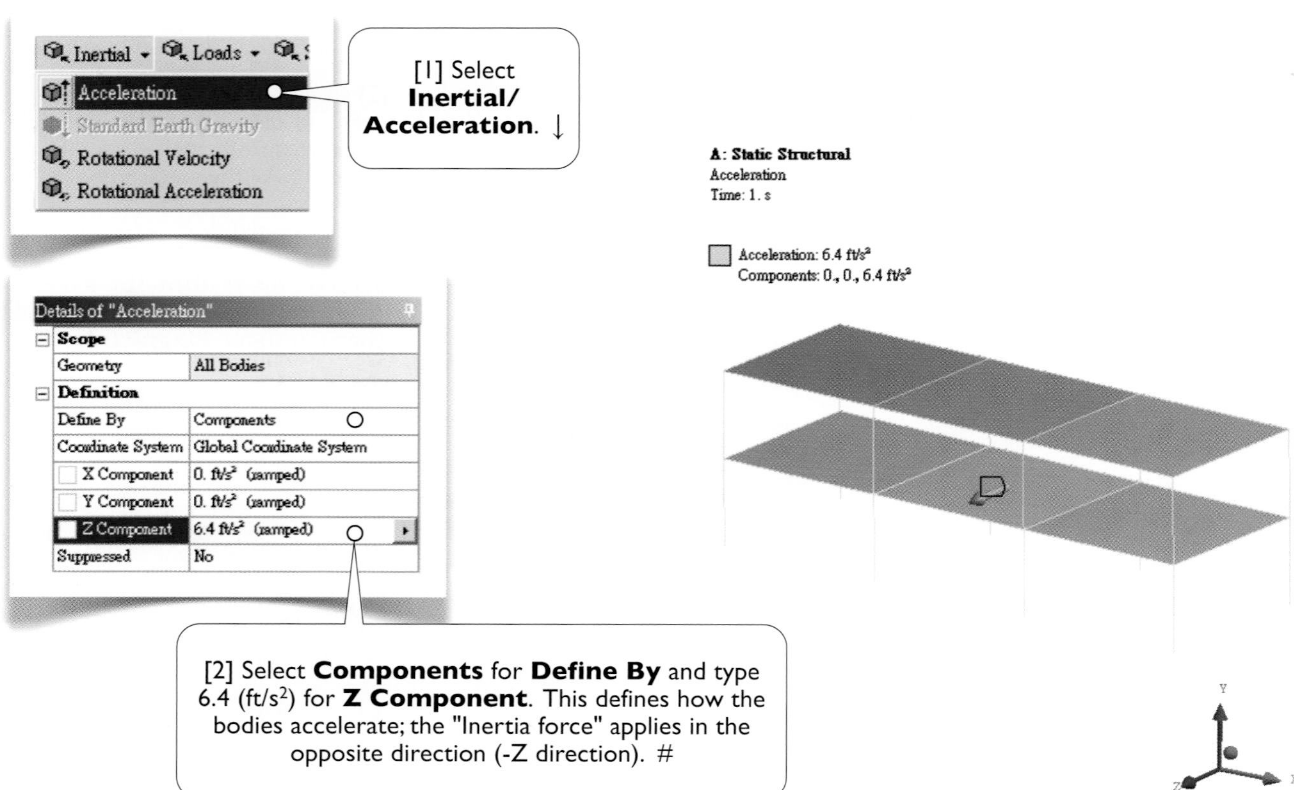

7.3.20 Solve the Model and View the Results

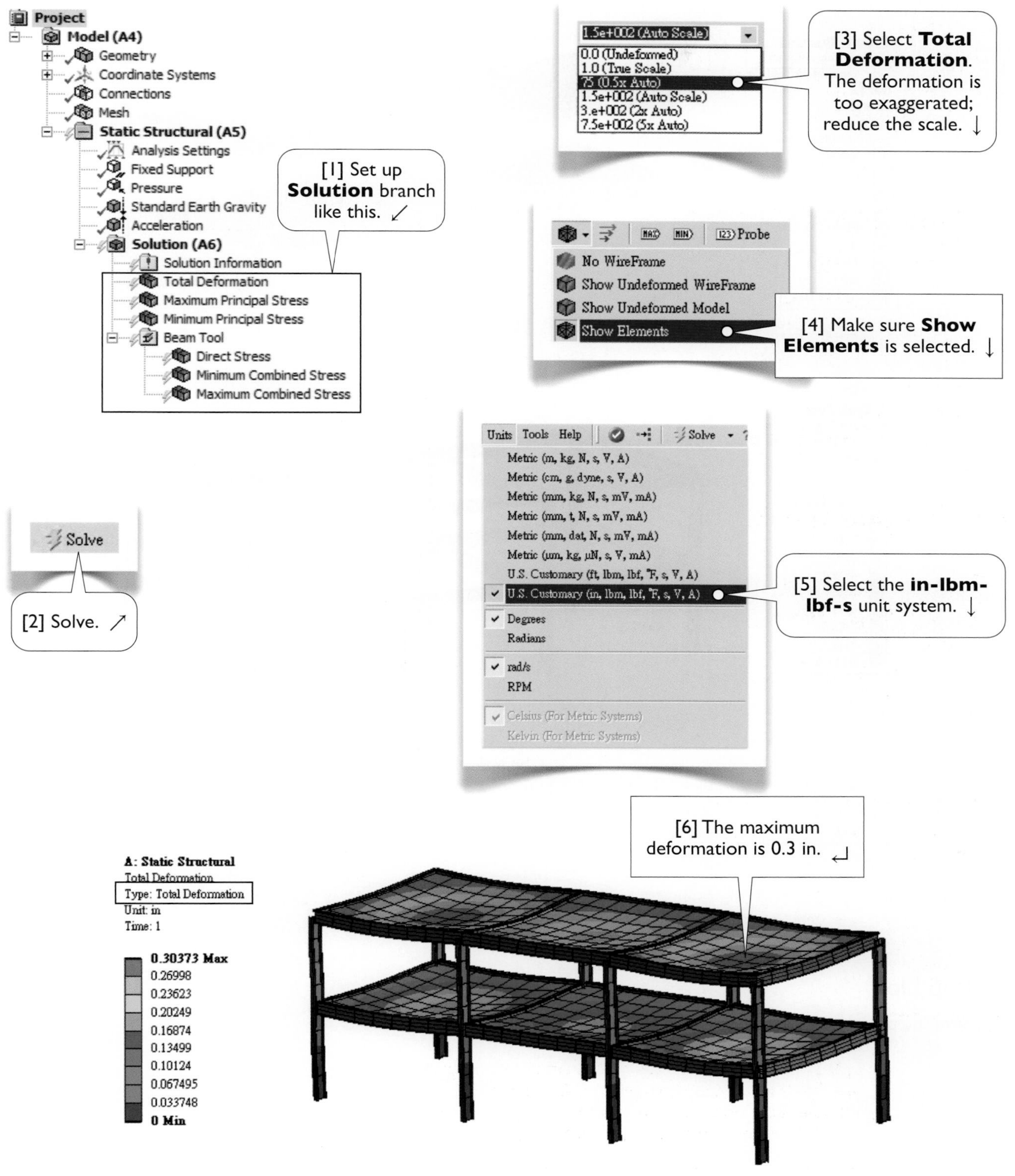

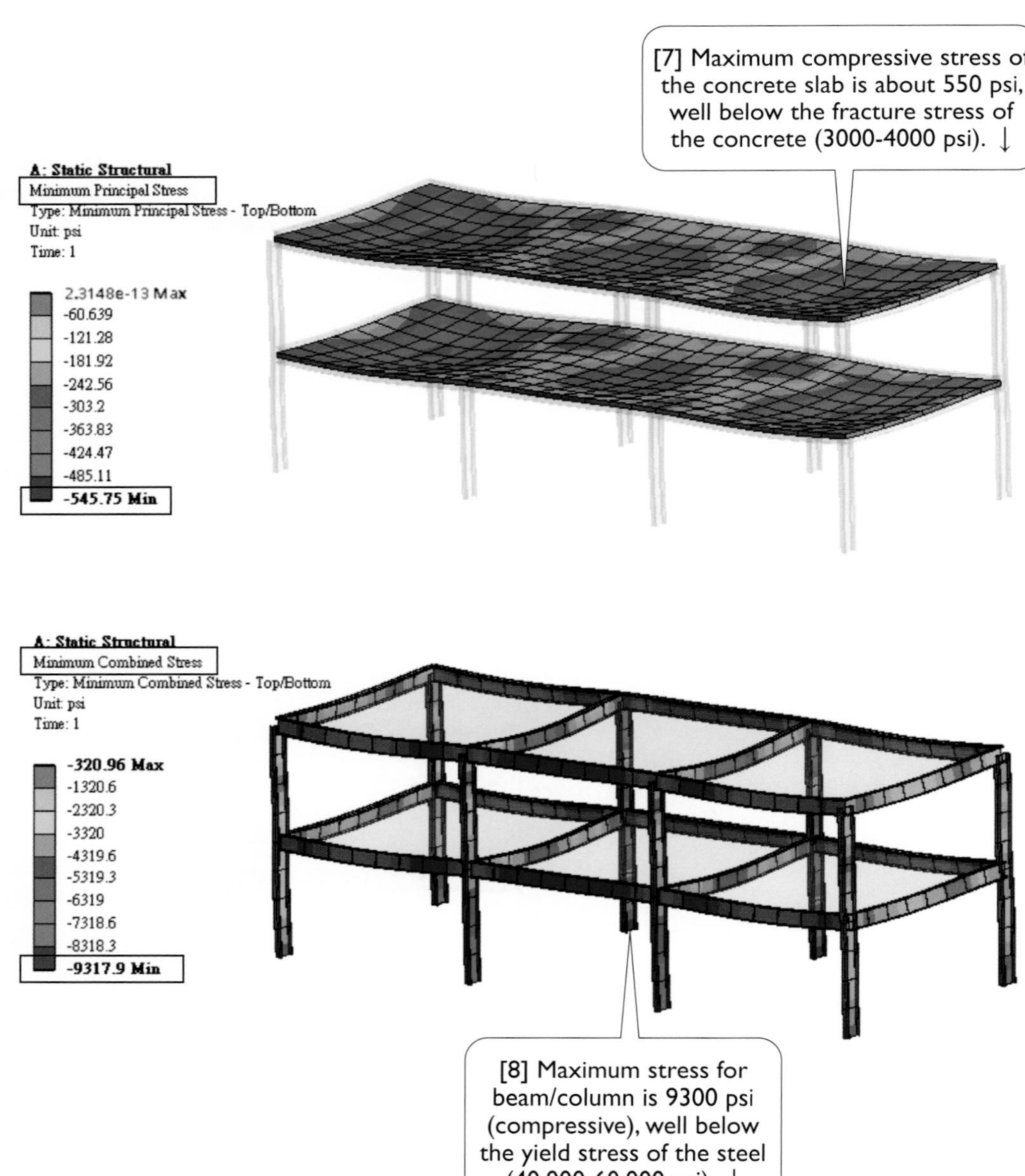

Wrap Up

[9] Save the project and exit Workbench. #

Section 7.4

Review

7.4.1 Keywords

Choose a letter for each keyword from the list of descriptions

1. (   ) Beam Elements
2. (   ) Coordinates File
3. (   ) Direct Stress and Bending Stress
4. (   ) Maximum/Minimum Bending Stress
5. (   ) Combined Stress
6. (   ) Truss

Answers:

1. (A) 2. (F) 3. (C) 4. (D) 5. (E) 6. (B)

List of Descriptions

(A) A line (1D) element that can be arranged in the 3D space. It is used to mesh a body when two of its dimensions are much smaller than the third dimension. Each node has 6 degrees of freedom: 3 translational and 3 rotational. Due to the presence of rotational degrees of freedom, it is very efficient to model the problems dominated by bending modes, contrasting to a solid element, which does not have rotational degrees of freedom.

(B) Defined as a structure consisting of two-force members. By two-force member, we mean that the members are pin-jointed at the ends, and the loads apply on the joints so that the members are either stretched or compressed but not bent.

(C) The resultant forces acting on a beam cross section can be summarized into three components: direct force *F*, bending moment *M*, and shear force *V*. Only the direct force and the bending moment contribute to the axial stress. The axial stress caused by the direct force is called the direct stress $\sigma = F/A$, where *A* is the cross-sectional area. The direct stress can be tensile or compressive. The axial stress caused by the bending moment is called the bending stress $\sigma_b = My/I$, where *y* is the distance from the neutral axis to the point of concern, and *I* is the moment of inertia of the cross section.

(D) Occurs at either the top or bottom beam edge, depending on the direction of the bending moment.

(E) The superposition of the direct stress and the bending stress.

(F) A text file describing the coordinates of points. The file has 5 columns, or fields: (a) group number, (b) ID number, (c) X-coordinate, (d) Y-coordinate, and (e) Z-coordinate. The group number and the ID number are arbitrary and they together uniquely identify a point from others.

7.4.2 Additional Workbench Exercises

Convergence Study of Beam Elements

Generate a convergence curve like the one in 7.1.16[2] (page 282) for the flexible gripper in Section 7.1.

Mesh Each Member with More Elements

We mentioned in 7.2.8[4] (page 289) that the default settings of **Mesh** would mesh the model with 205 beam elements and would result in exactly the same solution as 37 elements. Verify this. Type 0 for **Element Size** to set to the default element size.

Pin-Jointed or Rigid-Jointed?

We mentioned in 7.2.1[1] (page 283) that if the members of a truss structure are slender enough, there is no essential difference between a pin-jointed model and a rigid-jointed model. Verify this. You may use the 3D truss example (Section 7.2) or select a simple truss structure from any of your Engineering Mechanics textbooks. Solve the problem with the pin-joint assumption, and then solve the problem again using a rigid-jointed model. Compare the results and draw your conclusions.

To model a pin-jointed structure in Workbench, you need to either specify revolute joints between the members or utilize **End Release**[Ref 1] feature in **Mechanical** [1].

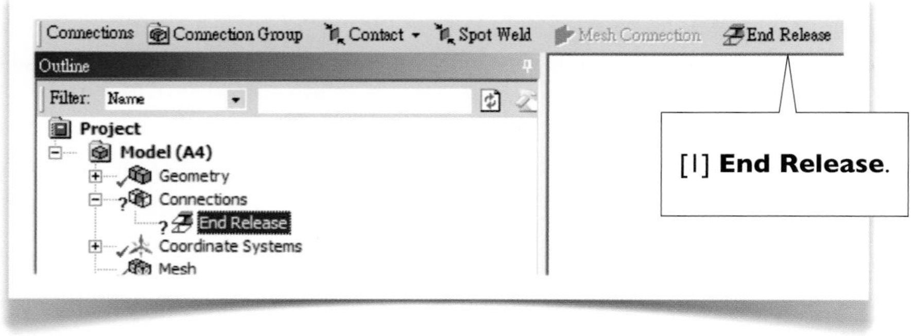

References

1. ANSYS Help//Mechanical Application//ANSYS Mechanical User's Guide//Setting Connections//End Releases

Chapter 8

Optimization

Design Process

A typical engineering design process involves several steps as shown in [1-6]. First, the engineer sets up an initial design [1]. The design is simulated [2] and the performance is evaluated [3]. If the performance is satisfied, the design is accepted [6], otherwise the design must be improved somehow [5]. It is an iterative process. After several iterations, the design hopefully converges to an optimal design.

Conventionally, the way of improving a design [5] relies on engineers' experience. The consequence is that the process is costly and often fails to find an optimal design.

This chapter focuses on a capability of Workbench, which automates the entire design process [1-6].

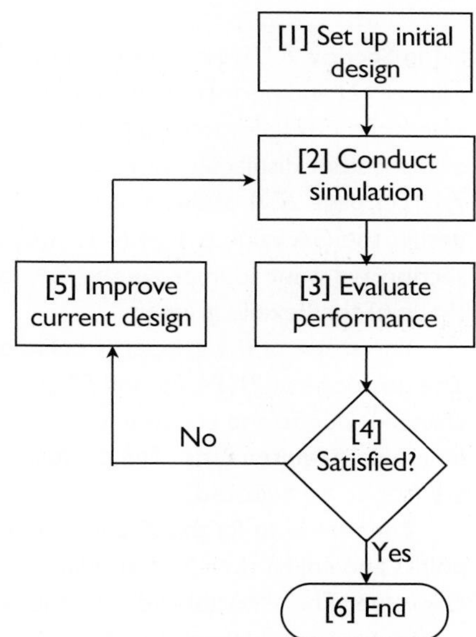

Purpose of This Chapter

This chapter is strategically arranged in the middle of the book because it is a suitable time to learn optimization techniques before we look into more advanced topics. The idea is that the students should be aware as early as possible of the optimization capabilities provided by Workbench so that they can use the features when doing their exercises.

This chapter provides two hands-on examples, which cover main ideas of optimization. After these exercises, the students should be able to use these capabilities on their own.

About Each Section

Section 8.1 revisits the flexible gripper, introduced in Section 7.1. This time, we want to improve the shape of the gripper to achieve an optimal geometric advantage. Section 8.2 provides an additional exercise, in which the triangular plate, introduced in Section 2.2 and simulated in Section 3.1, is to be improved to achieve a minimal weight design.

Section 8.1

Flexible Gripper[Ref 1]

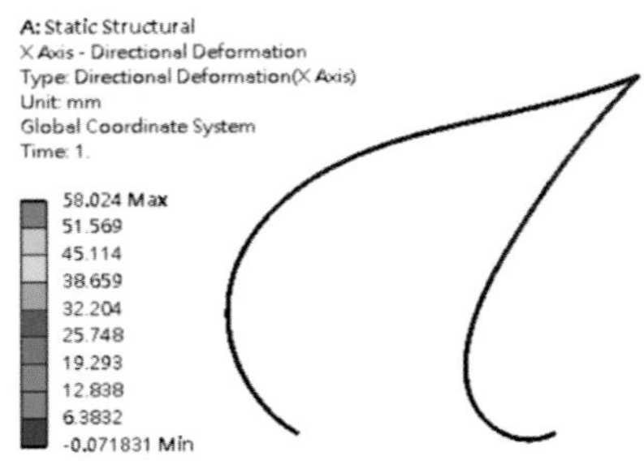

8.1.1 About the Flexible Gripper

[1] In Section 7.1, we created a model for the flexible gripper [2] and performed a static structural simulation to assess the GA (geometric advantage, defined as the ratio of the output displacement to the input displacement; see 7.1.1[1], page 272) of the design. For that particular design, the GA value is 1.04 (7.1.15[6], page 281). In this section, we want to improve the GA value by adjusting the shape of the flexible gripper.

The shape of the gripper is defined by 7 key points [3]. The positions of *P1*, *P4*, *P5*, and *P7* are fixed (cannot be changed) due to the constraints imposed by some functional requirements. The positions of *P2*, *P3*, and *P6* are free to be adjusted.

The idea is to fix the X-coordinates of these three points and adjust their Y-coordinates to achieve a better GA value. The allowable adjustment ranges for the Y-coordinates are ±10 mm for *P2*, ±20 mm for *P3* and ±5 mm for *P6*. Besides, there is a limitation on the stresses, either tensile or compressive, which should not exceed 15 MPa for a reliability consideration.

In Workbench, **DesignXplorer** is used to carry out the task. ↗

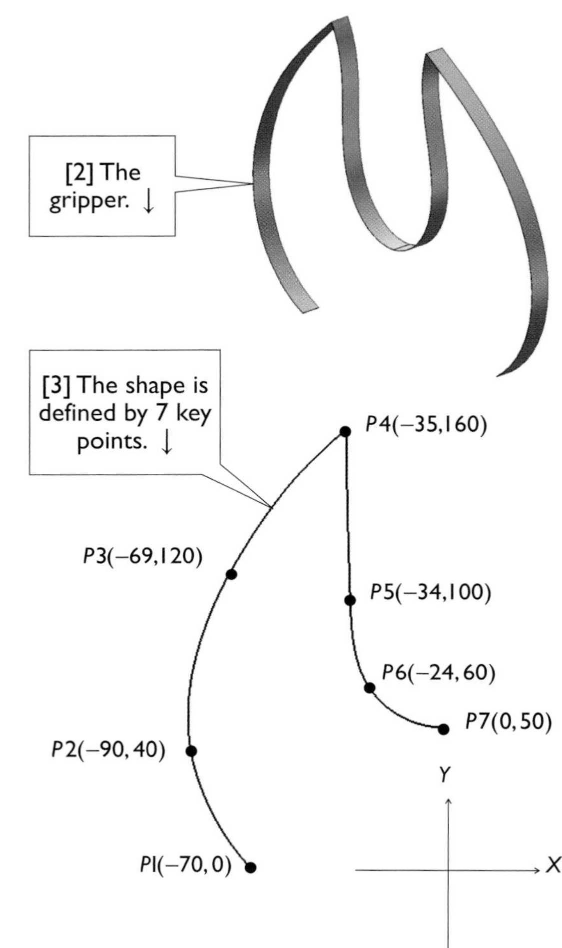

Keep the Number of Design Variables Minimum

[4] The (input) parameters that can be changed to improve a design are called the *design variables* (in this case, Y-coordinates of *P2*, *P3*, and *P6*). The overall computing time and complexity of an optimization process increases dramatically as the number of design variables increases; therefore, it is important to keep the number of design variables as low as possible.

In this case, we could have chosen both X-coordinates and Y-coordinates of *P2*, *P3*, *P6* as design variables. That would increase the number of design variables from 3 to 6 but without gaining any advantages. #

8.1.2 Resume the Project **Gripper**

8.1.3 Define Input Parameters in DesignModeler

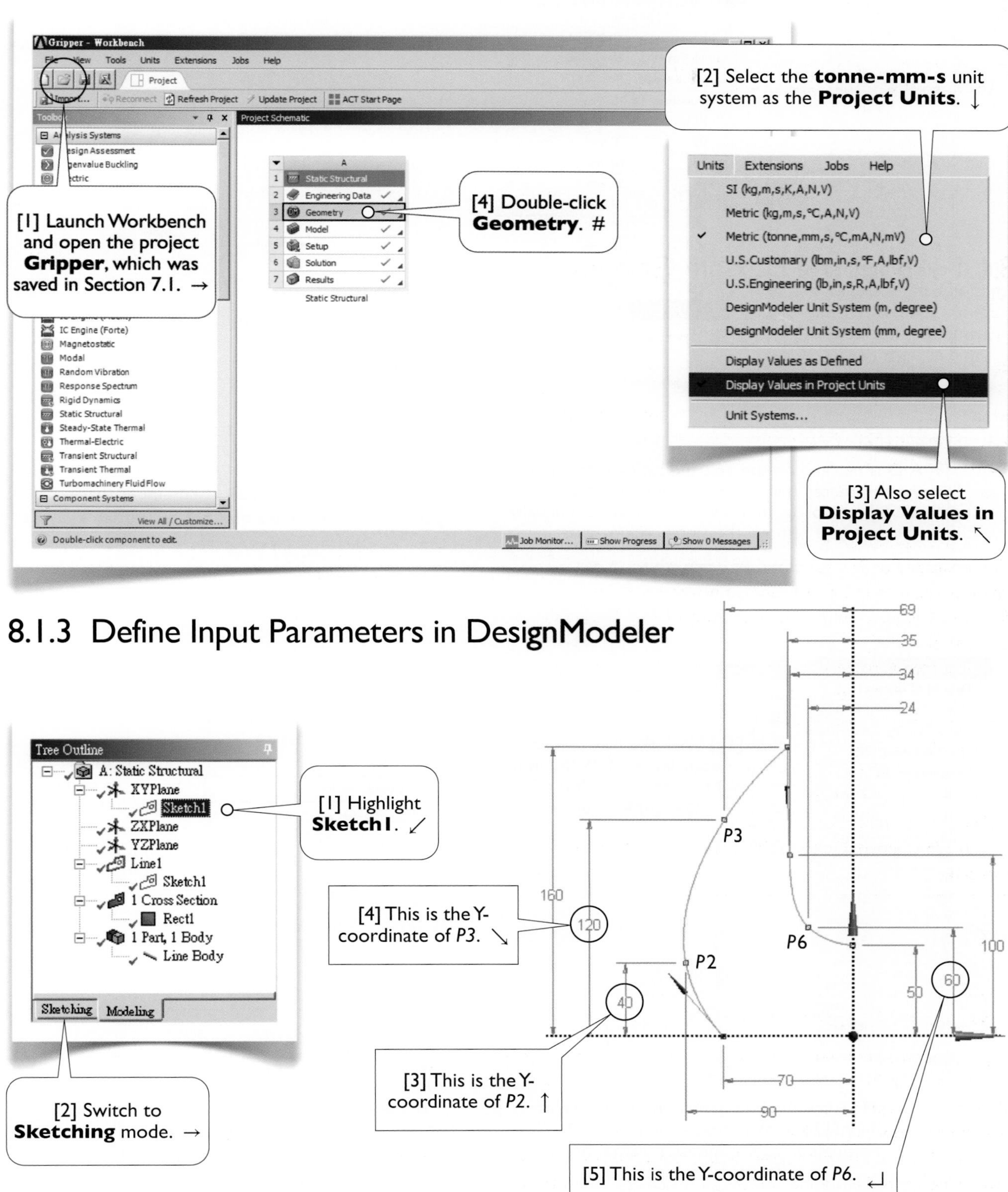

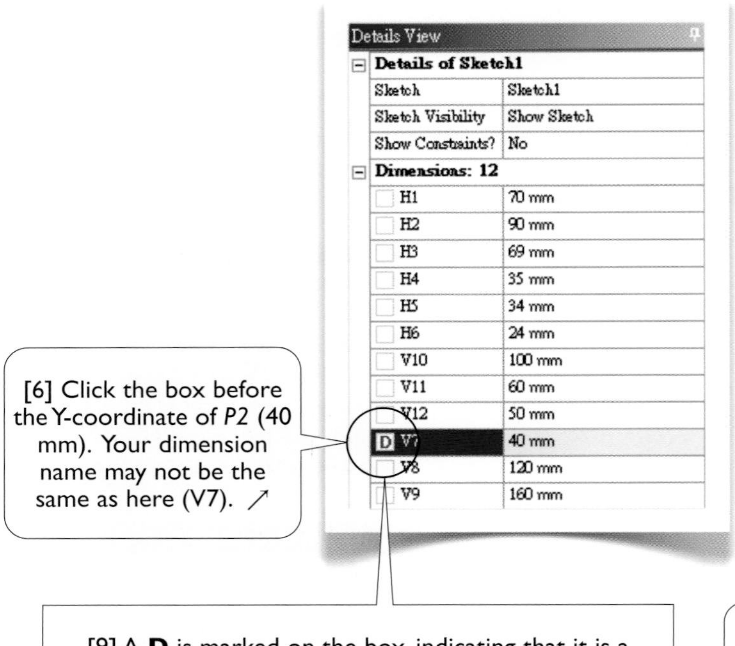

[6] Click the box before the Y-coordinate of *P2* (40 mm). Your dimension name may not be the same as here (V7). ↗

[7] Type **Y2** as the parameter name. ↓

[8] Click **OK**. ↙

[9] A **D** is marked on the box, indicating that it is a "driving parameter" (an input parameter). Its value is controlled by the program now and cannot be manually changed unless you remove the **D** by clicking it again. →

[10] Repeat steps [6-8] for the Y-coordinates of *P3* and *P6* (see [11-12]). Remember that your dimension names may not be the same as they appear here. ↙

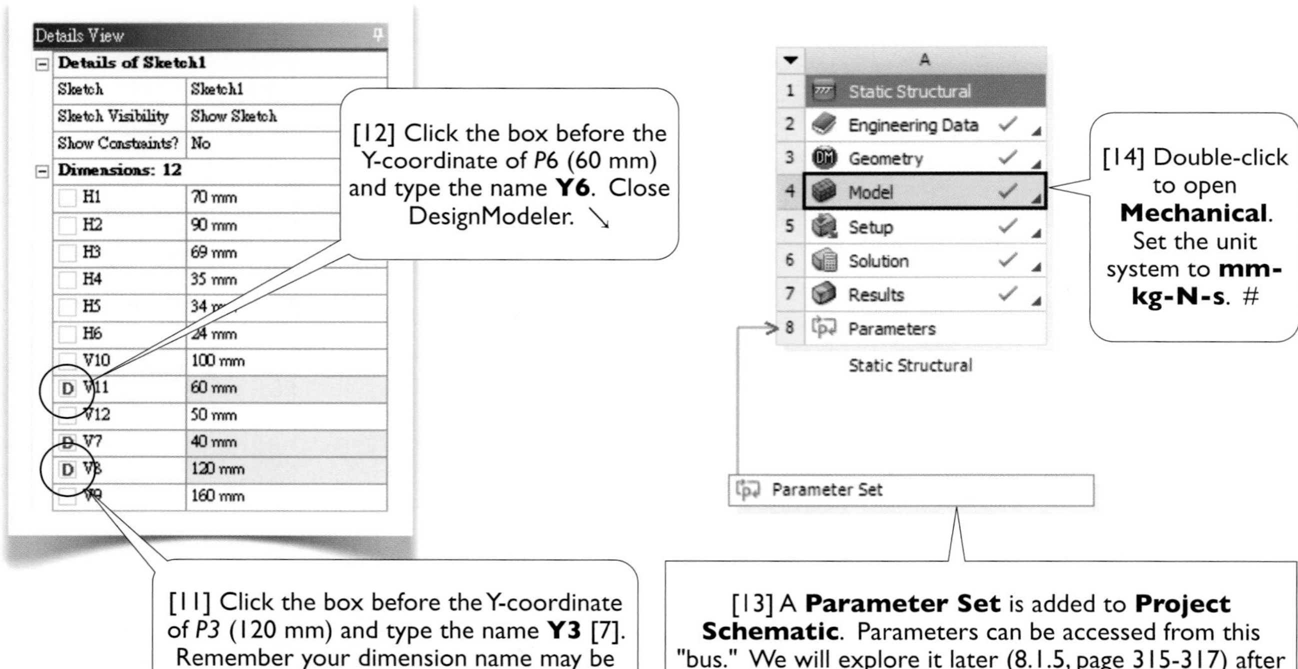

[12] Click the box before the Y-coordinate of *P6* (60 mm) and type the name **Y6**. Close DesignModeler. ↘

[14] Double-click to open **Mechanical**. Set the unit system to **mm-kg-N-s**. #

[11] Click the box before the Y-coordinate of *P3* (120 mm) and type the name **Y3** [7]. Remember your dimension name may be different from here. ↑

[13] A **Parameter Set** is added to **Project Schematic**. Parameters can be accessed from this "bus." We will explore it later (8.1.5, page 315-317) after we define additional parameters in **Mechanical**. ↑

8.1.4 Define Output Parameters in **Mechanical**

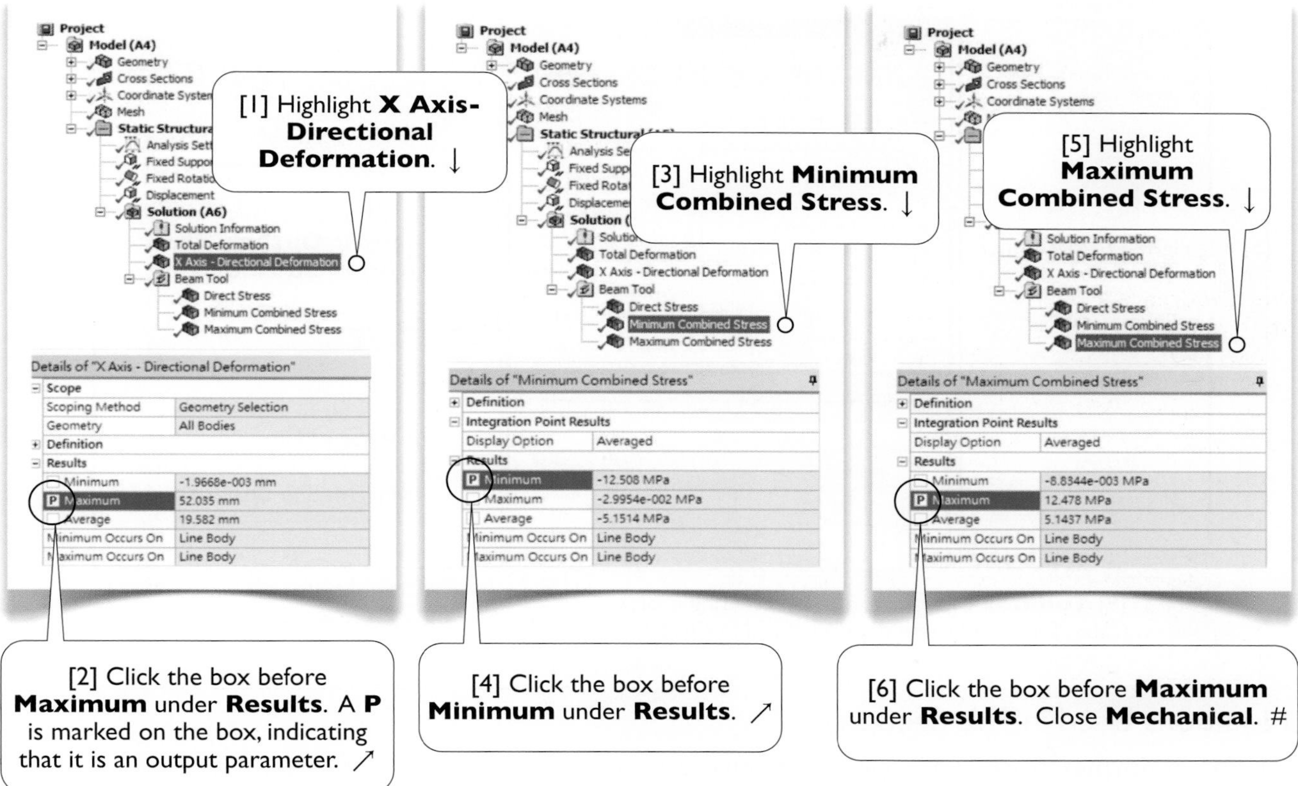

8.1.5 Explore the Parameter Set

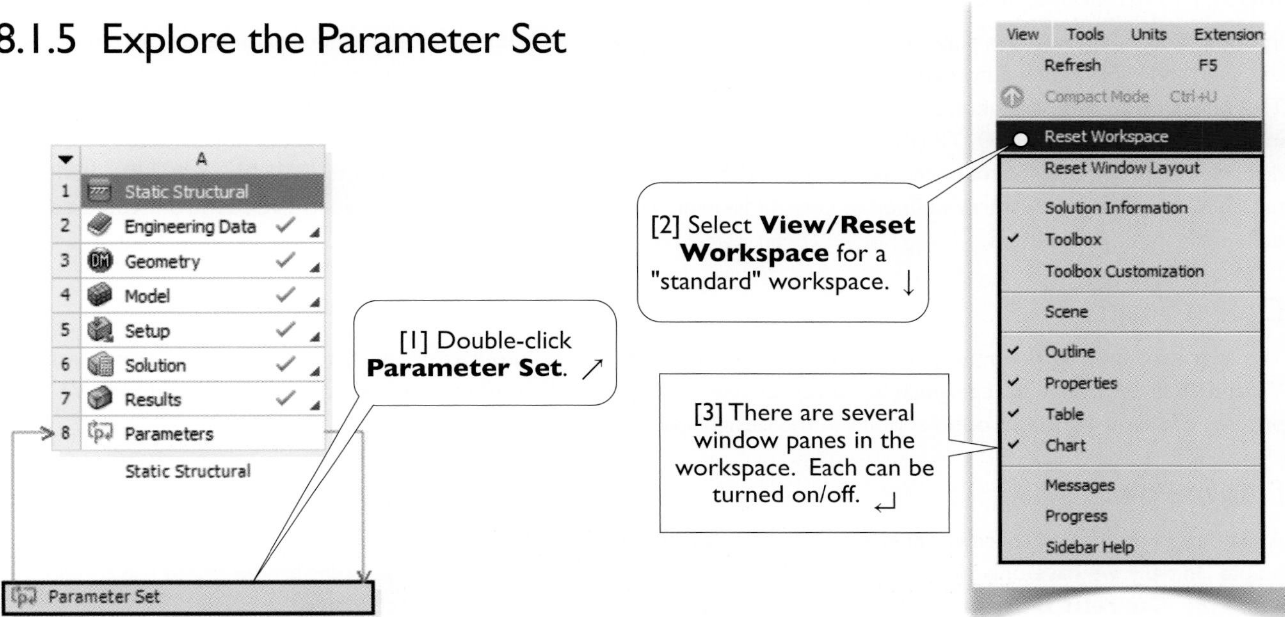

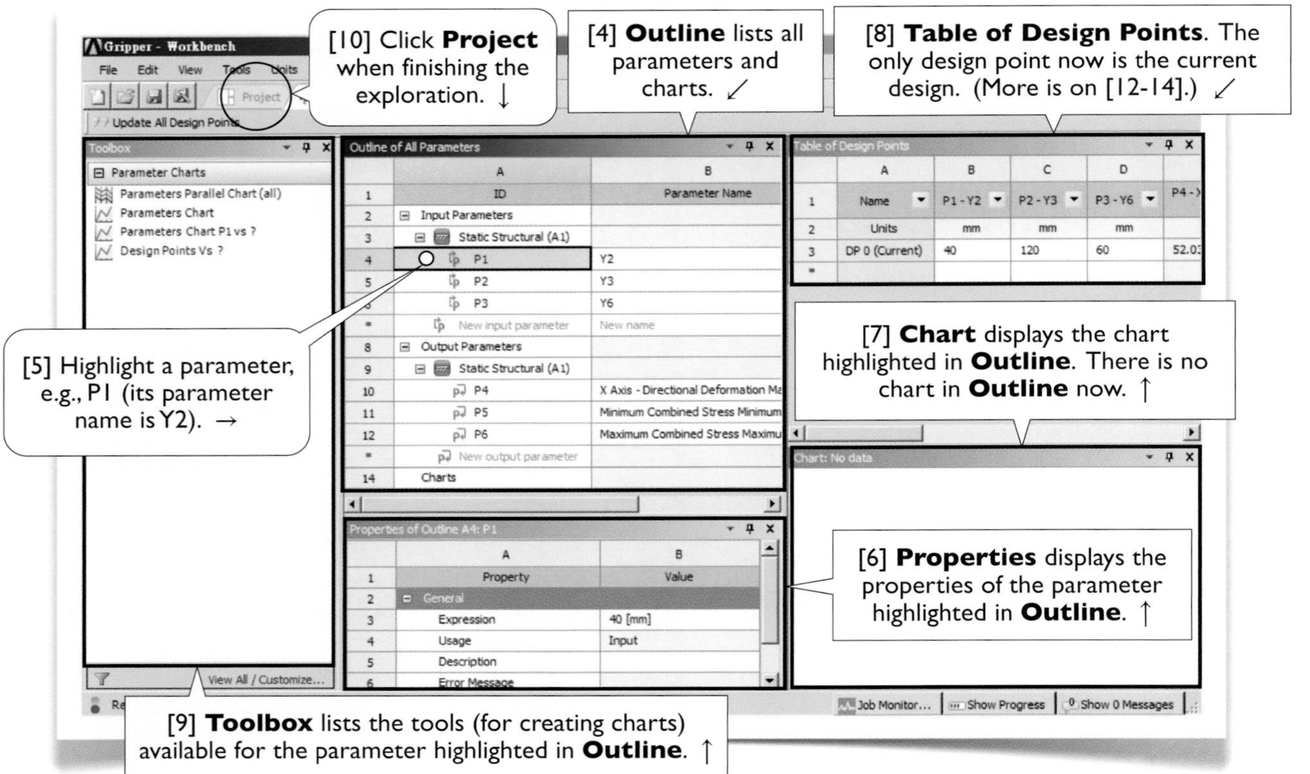

Notations Used in This Section

[11] Workbench uses P1, P2, P3(non-italic) and so on as parameter names [5]. In our case of the flexible gripper, we use *P2*, *P3*, and *P6* (italic) for three control points of the shape. Be careful not to confuse these two sets of notations.

Input Parameters and Output Parameters

In optimization jargon, input parameters are also called *design variables*. Initial values of input parameters are set up by the user, and Workbench automatically changes these values to improve the design according to the algorithm used. The values of output parameters are calculated from the simulation applications (e.g., **Mechanical**). In our case, the input parameters are Y2 (P1) ,Y3 (P2), and Y6 (P3), and the output parameters are the maximum horizontal displacement (P4), the minimum combined stress (P5), and the maximum combined stress (P6). Note that the input parameters are not necessarily defined in DesignModeler. For example, the loads defined in **Mechanical** could be defined as input parameters.

Design Space

The space spanned by the input parameters is called the *design space*. In our case, the space spanned by parameters Y2, Y3, and Y6 is the design space, which is a three-dimensional space. The optimization process can be thought of as a process of searching for an optimal point in the design space.

Design Points

Any point in the design space is called a *design point*. **Table of Design Points** [8] lists a series of calculated design points. Initially, we have only one design point: Y2 = 40, Y3 = 120, Y6 = 60. This design point is now called the *current design* [8]. **Current Design** is a design point of which the data are stored in **Mechanical** database. ↵

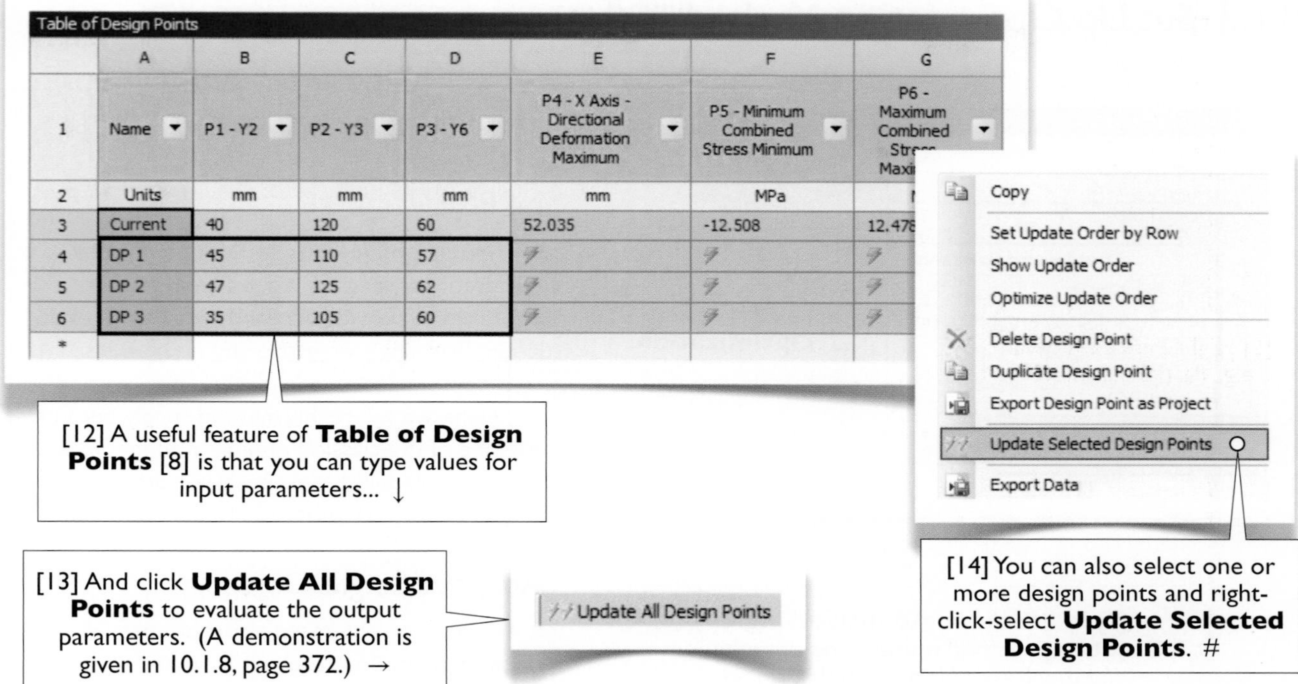

8.1.6 Create a Direct Optimization System

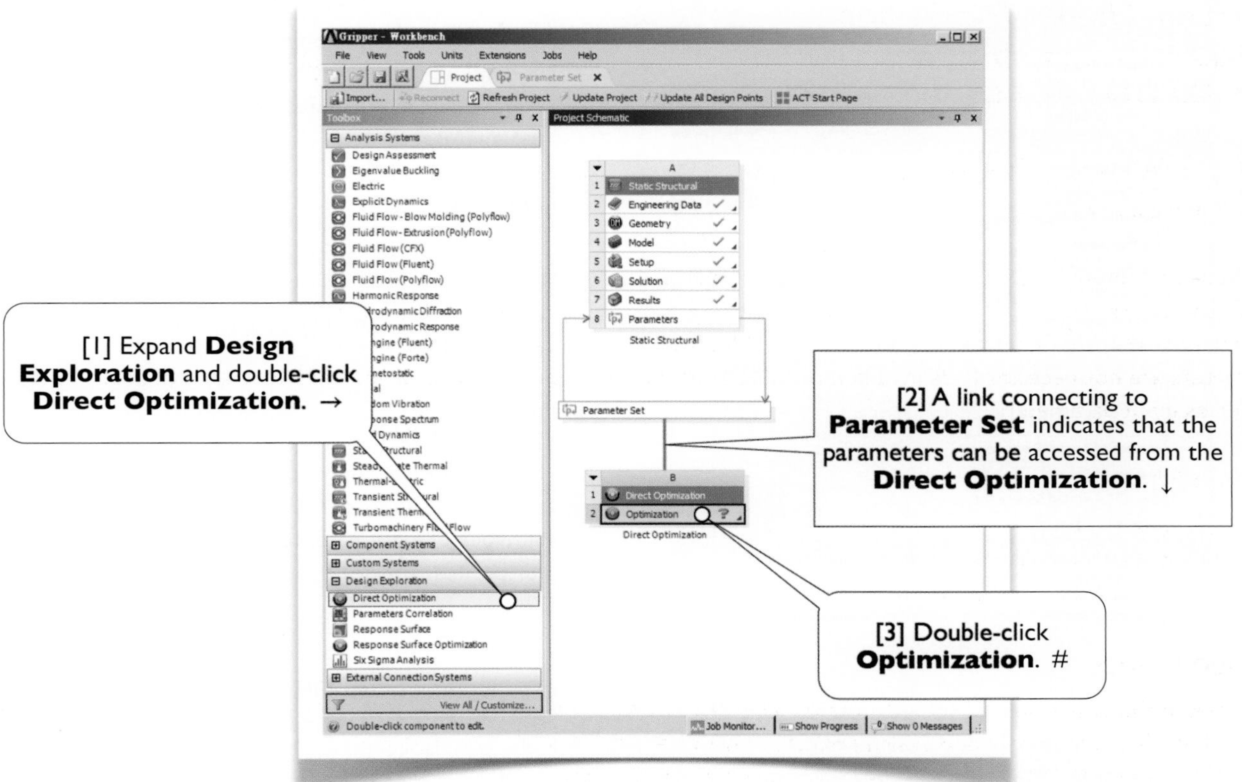

8.1.7 Set Up Optimization Method[Ref 2]

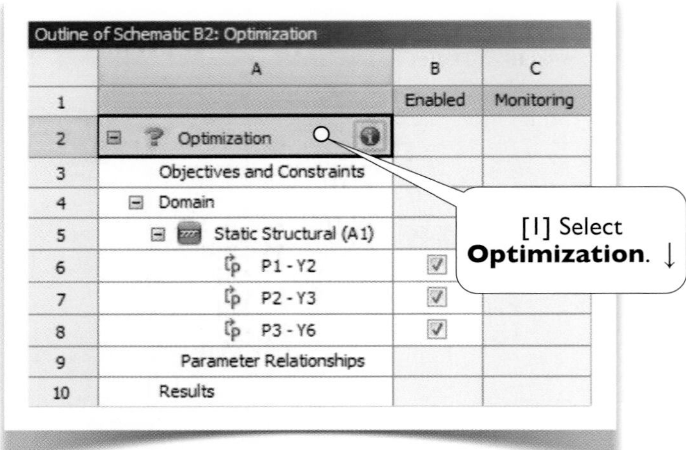

Outline of Schematic B2: Optimization

	A	B	C
1		Enabled	Monitoring
2	Optimization		
3	Objectives and Constraints		
4	Domain		
5	Static Structural (A1)		
6	P1 - Y2	☑	
7	P2 - Y3	☑	
8	P3 - Y6	☑	
9	Parameter Relationships		
10	Results		

[2] By default, the **Adaptive Single-Objective** method is used. The optimization methods available are listed in [3]. →

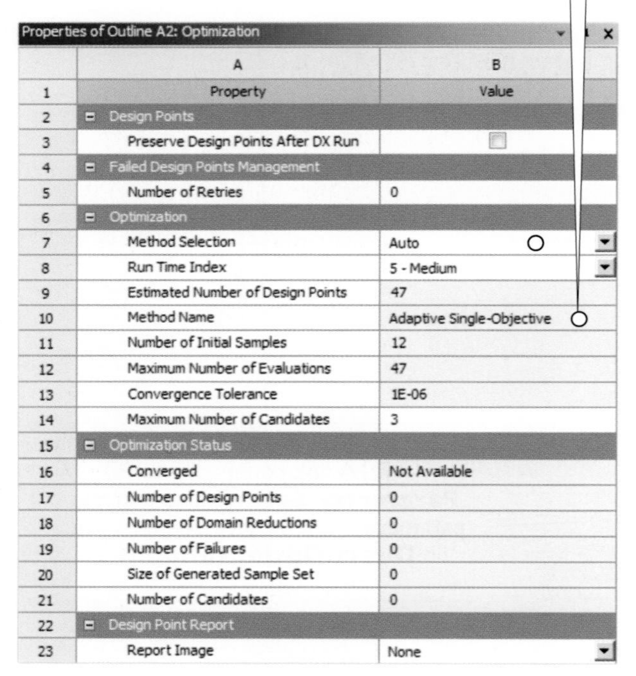

Properties of Outline A2: Optimization

	A	B
1	Property	Value
2	Design Points	
3	Preserve Design Points After DX Run	☐
4	Failed Design Points Management	
5	Number of Retries	0
6	Optimization	
7	Method Selection	Auto
8	Run Time Index	5 - Medium
9	Estimated Number of Design Points	47
10	Method Name	Adaptive Single-Objective
11	Number of Initial Samples	12
12	Maximum Number of Evaluations	47
13	Convergence Tolerance	1E-06
14	Maximum Number of Candidates	3
15	Optimization Status	
16	Converged	Not Available
17	Number of Design Points	0
18	Number of Domain Reductions	0
19	Number of Failures	0
20	Size of Generated Sample Set	0
21	Number of Candidates	0
22	Design Point Report	
23	Report Image	None

Optimization Methods

NLPQL

[3] The method (Nonlinear Programming by Quadratic Lagrangian) is a gradient-base algorithm to provide a local optimization result. It supports single objective and is limited to continuous input parameters.

Here, we select this method simply because it is one of the classical optimization methods. After completing this exercise, you are encouraged to try other methods.

Adaptive Single-Objective

It is also a gradient-based algorithm, providing a global optimization result. It is limited to single objective; it supports continuous and manufacturable input parameters.

Adaptive Multiple-Objective

It supports multiple objectives and aims at finding the global optimum. It supports continuous and manufacturable input parameters.

MISQP

The method (Mixed-Integer Sequential Quadratic Programing) solves mixed-integer nonlinear programming problems by a modified SQP method.

Screening

It uses a simple approach based on sampling and sorting. It supports multiple objectives as well as all types of input parameters. It is usually used for preliminary design, e.g., global behavior study.

MOGA

The method (Multi-Objective Genetic Algorithm) supports multiple objectives and aims at finding global optimum. #

8.1.8 Set Up Objective and Constraints

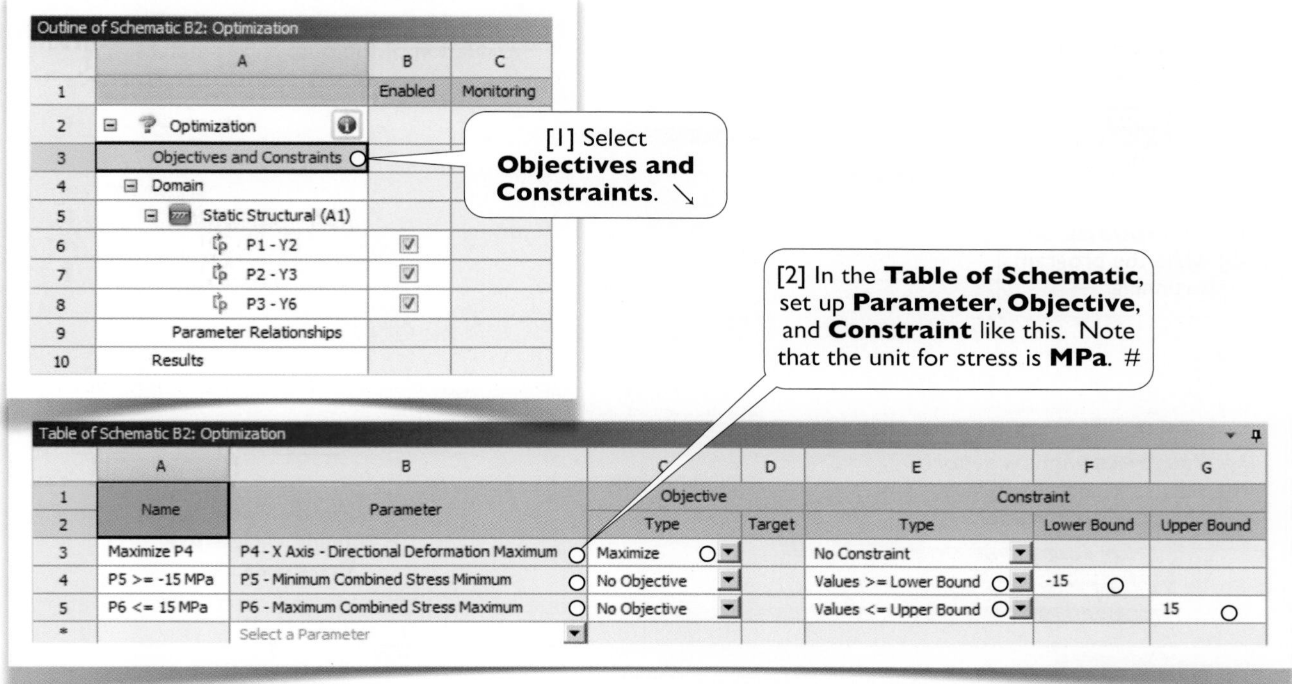

8.1.9 Set Up Lower/Upper Bounds for Input Parameters

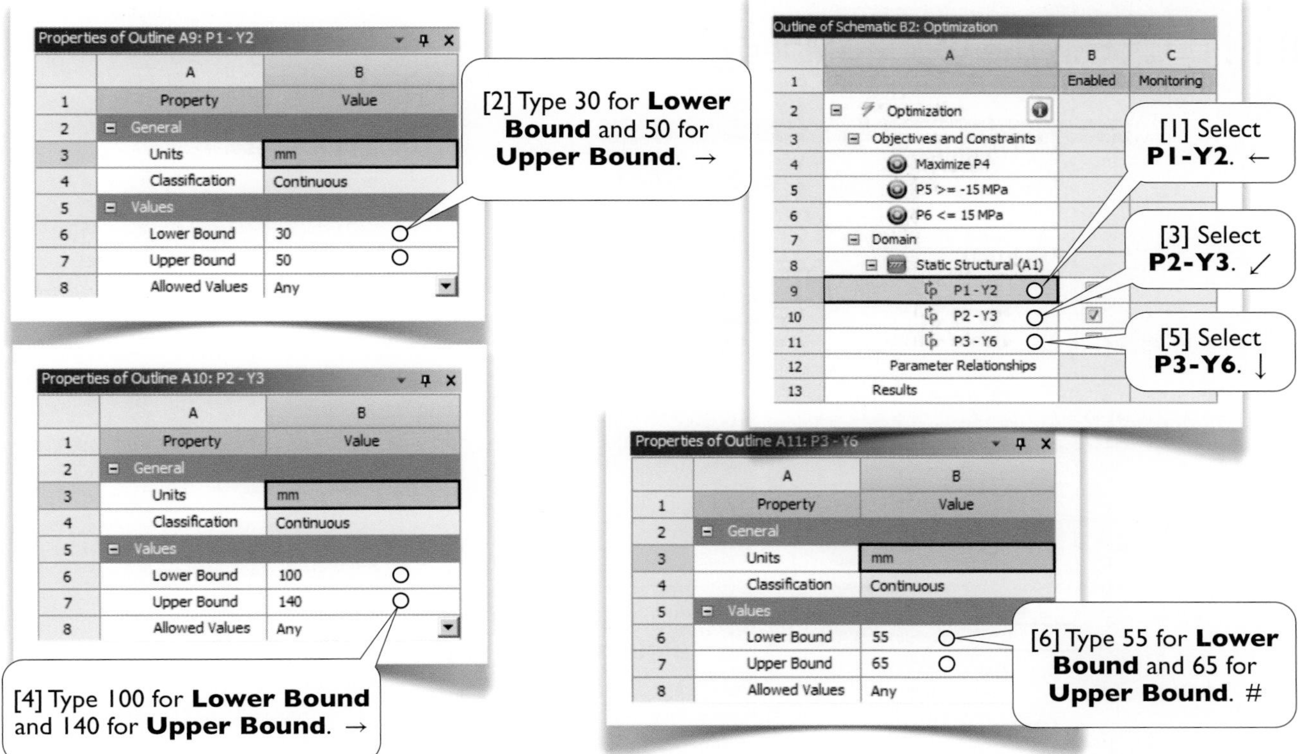

8.1.10 Run the Optimization

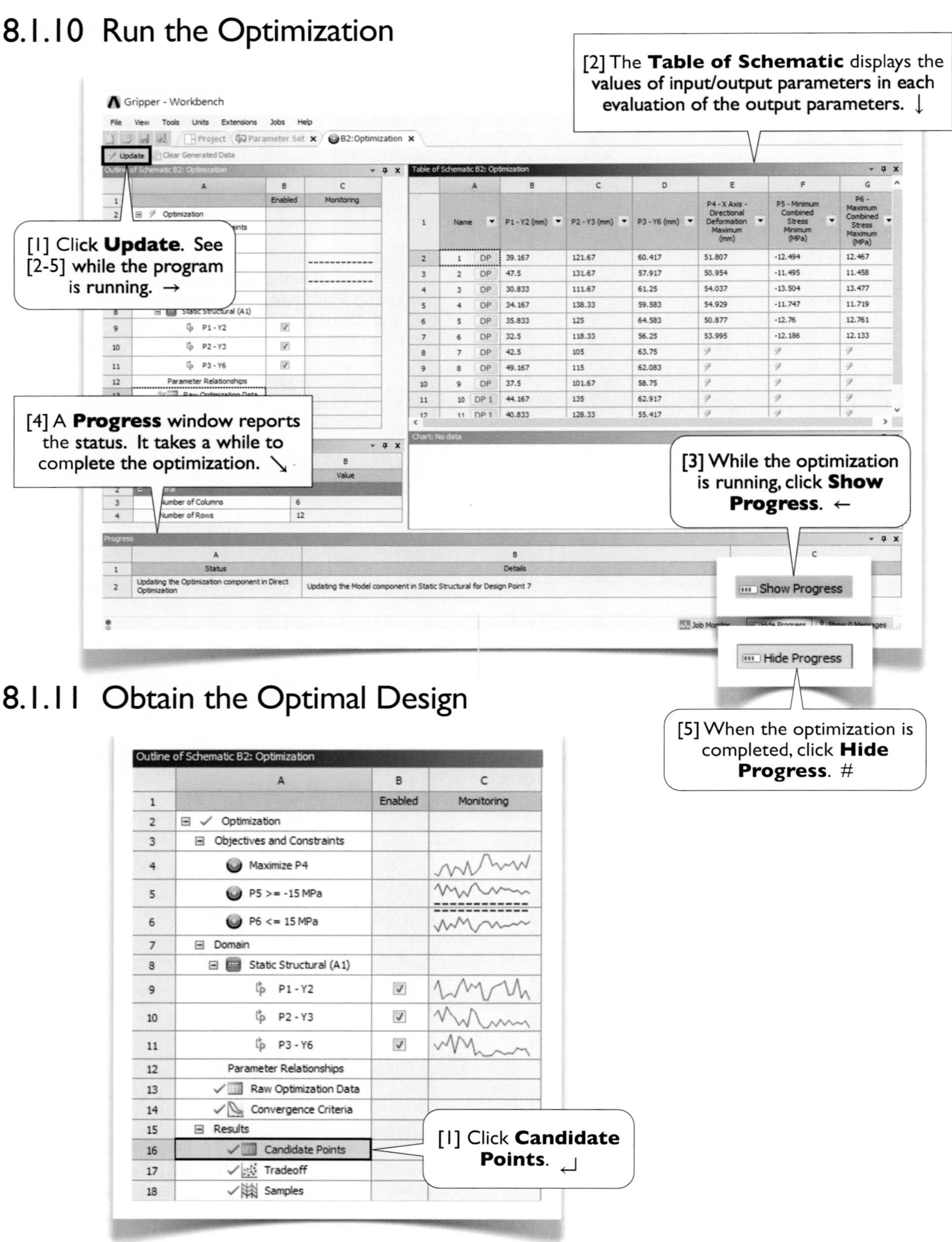

8.1.11 Obtain the Optimal Design

Table of Schematic B2: Optimization , Candidate Points

	A	B	C	D	E	F	G	H	I	J	K
1	Reference	Name	P1 - Y2 (mm)	P2 - Y3 (mm)	P3 - Y6 (mm)	P4 - X Axis - Directional Deformation Maximum (mm)		P5 - Minimum Combined Stress Minimum (MPa)		P6 - Maximum Combined Stress Maximum (MPa)	
2						Parameter Value	Variation from Reference	Parameter Value	Variation from Reference	Parameter Value	Variation from Reference
3	●	Candidate Point 1	30	100	55	58.024	0.00%	-13.211	0.00%	13.144	0.00%
4	○	Candidate Point 2	35.48	105.57	56.07	55.88	-3.69%	-12.863	2.63%	12.804	-2.59%
5	○	Candidate Point 3	42.855	102.77	57.475	55.279	-4.73%	-13.161	0.38%	13.107	-0.28%
*		New Custom Candidate Point	40	120	60						

[3] The values of the input parameters. ↓

[2] **Candidate Point 1** gives the best GA value (58.024/50 = 1.16). For a definition of GA value, see 7.1.1[1], page 272. ↖

[4] Now, we want to copy this design to the **Table of Design Points** (8.1.5[8], page 316). Right-click here and select **Insert as Design Point**. →

Paste
Insert as Design Point
Insert as Custom Candidate Point
Export Table Data as CSV

[5] Click **Project** to return to **Project Schematic**. #

Project

8.1.12 Set the Optimal Design as Current Design

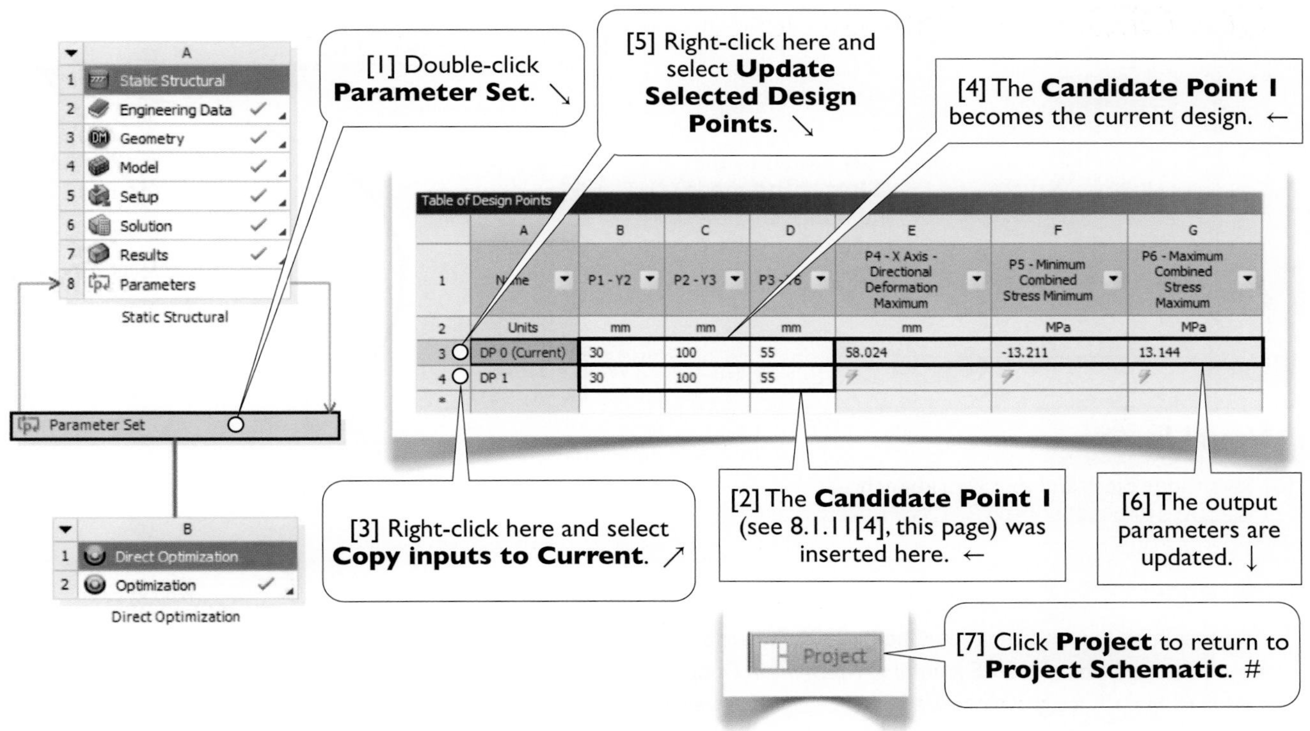

8.1.13 View the Current Design

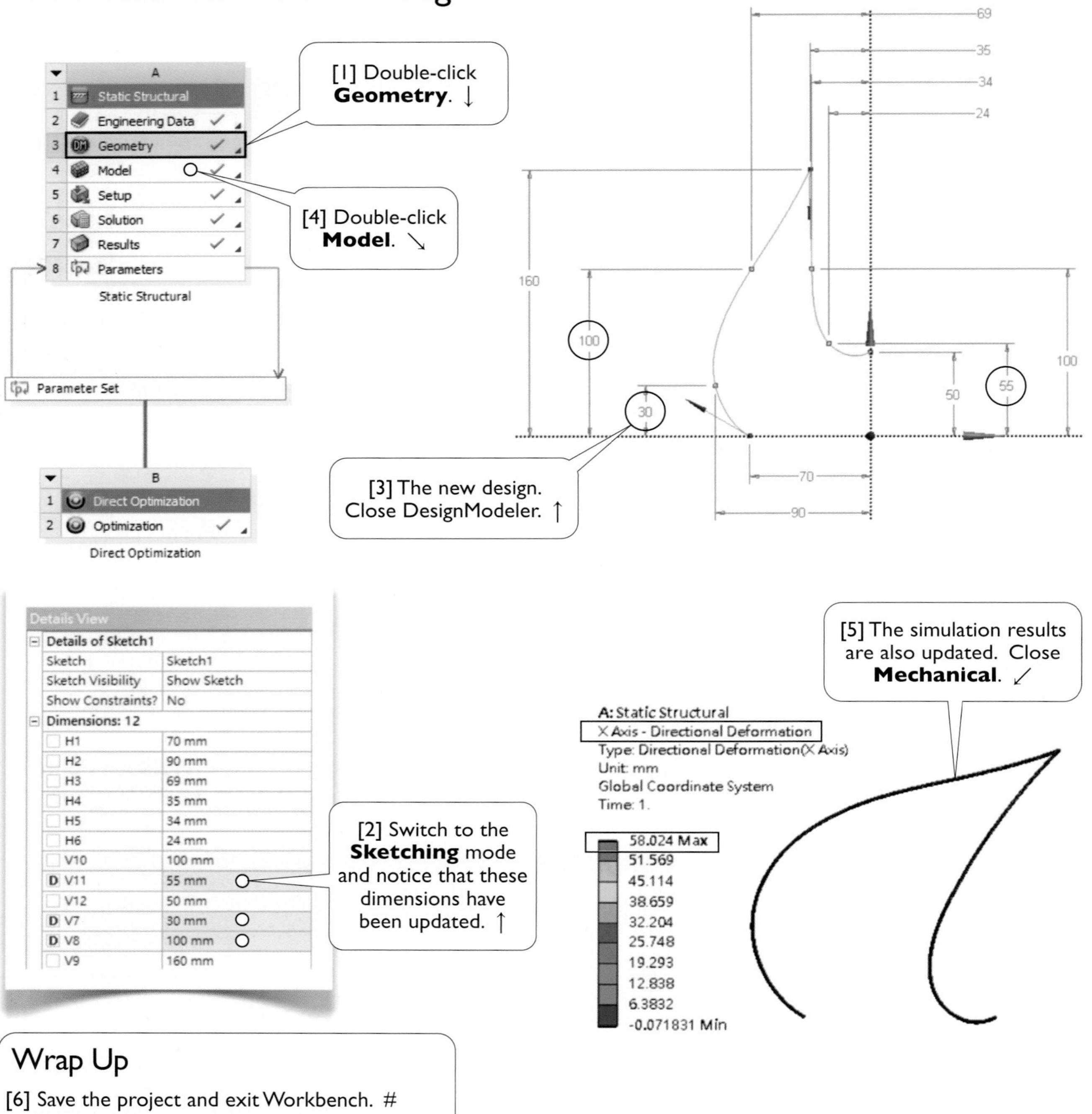

Wrap Up

[6] Save the project and exit Workbench. #

Reference

1. Chao-Chieh Lan and Yung-Jen Cheng, 2008, "Distributed Shape Optimization of Compliant Mechanisms Using Intrinsic Functions," *ASME Journal of Mechanical Design*, Vol. 130, 072304.
2. ANSYS Help//DesignXplorer//ANSYS DesignXplorer User's Guide//Using Goal Driven Optimization

Section 8.2

Triangular Plate

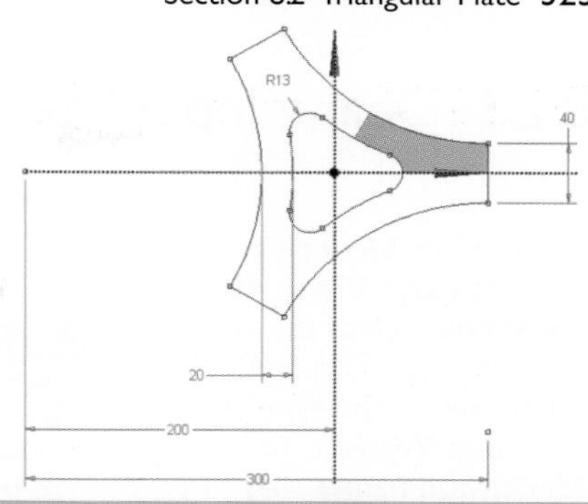

8.2.1 About the Triangular Plate

[1] In Section 3.1, we generated a 2D solid model and performed a 2D static simulation for the triangular plate [2-5] introduced in Section 2.2.

The plate is made of steel with an allowable stress of 100 MPa. According to the simulation results in Section 3.1, the initial design gives a maximum stress of about 51 MPa (3.1.13[2], page 122), well below the allowable stress. That means the initial design is over-designed: the material can be cut down somehow. In this section, we want to redesign the triangular plate to reduce the amount of material.

The design variables (input parameters) are the width of the bridges [4] and the radius of the fillets [5]. For the width *W*, the initial design is 30 mm and its allowable range is 20-30 mm. For the radius *R*, the initial design is 10 mm and its allowable range is 5-15 mm. The project unit system used in this exercise is **tonne-mm-s**. ↗

[2] The triangular plate. →

[3] One-sixth of the model will be analyzed due to the symmetries. ↙

[4] The initial design of the width of the bridges is 30 mm. ↓

W

R

[5] The initial design of the radius of the fillets is 10 mm. #

8.2.2 Resume Project **Triplate**

[1] Launch Workbench. Open the project **Triplate**, which was saved in Section 3.1. Select **tonne-mm-s** as the project units and select **Display values in Project Units** (see 8.1.2[2-3], page 313). →

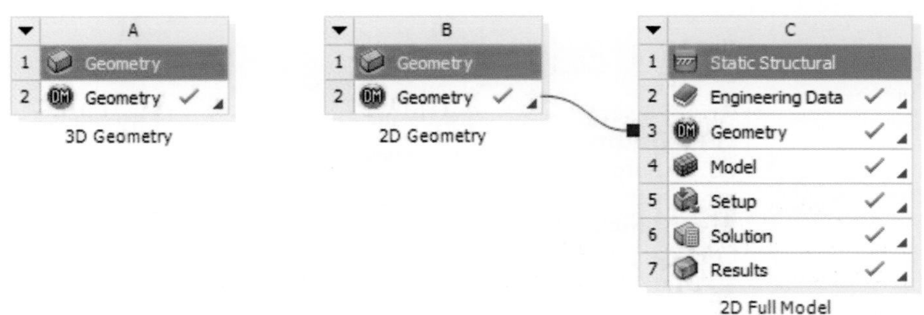

[2] Double-click to start up DesignModeler. #

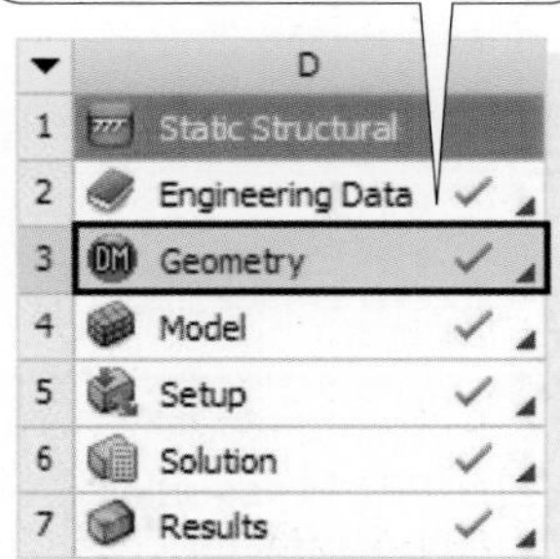

8.2.3 Define Input Parameters in DesignModeler

[1] Select **Sketch1**. In **Details View** of **Sketch1**, select the width (30 mm) as an input parameter and type the name **Width**. Your dimension names may not be the same as here. ↓

[2] Select the radius of the fillet (10 mm) as another input parameter and type the name **Radius**. Close DesignModeler. #

Details View
Details of Sketch1
Sketch | Sketch1
Sketch Visibility | Show Sketch
Show Constraints? | No
Dimensions: 5
H1 | 300 mm
H2 | 200 mm
D H4 | 30 mm
D R5 | 10 mm
V3 | 40 mm

R10
40
30
200
300

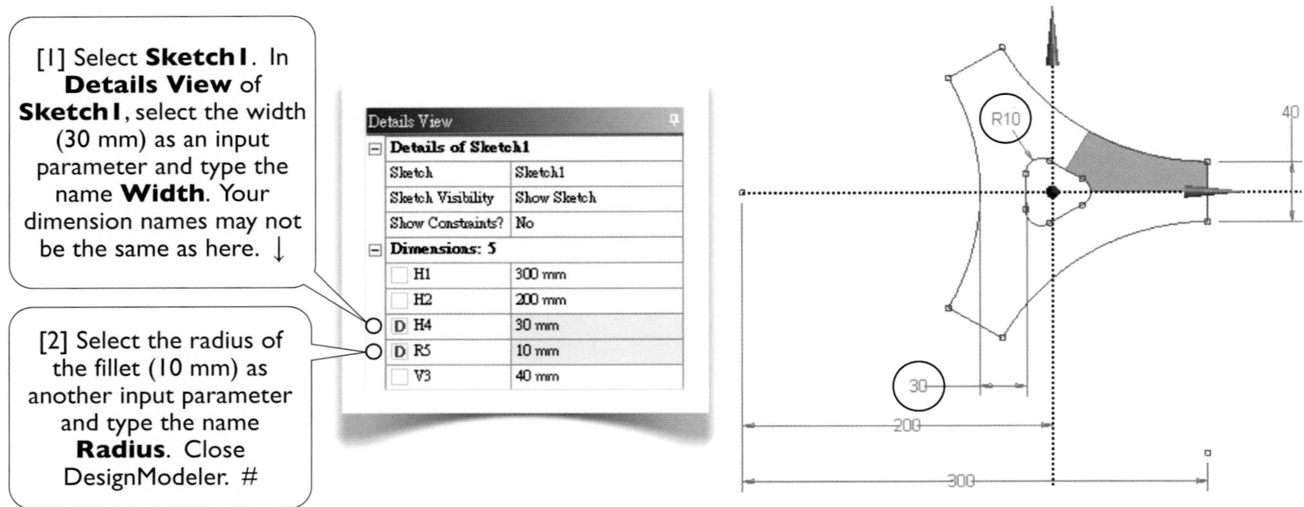

8.2.4 Define Output Parameters in **Mechanical**

[1] Double-click **Model** of **2D Symmetric Model** to start up **Mechanical**. Highlight **Geometry**. ↙

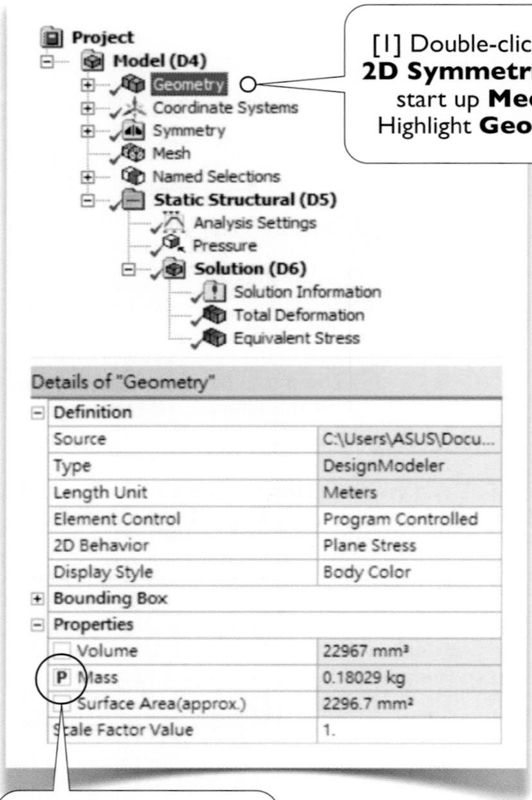

Project
Model (D4)
Geometry
Coordinate Systems
Symmetry
Mesh
Named Selections
Static Structural (D5)
Analysis Settings
Pressure
Solution (D6)
Solution Information
Total Deformation
Equivalent Stress
Details of "Geometry"
Definition
Source | C:\Users\ASUS\Docu...
Type | DesignModeler
Length Unit | Meters
Element Control | Program Controlled
2D Behavior | Plane Stress
Display Style | Body Color
Bounding Box
Properties
Volume | 22967 mm³
P Mass | 0.18029 kg
Surface Area(approx.) | 2296.7 mm²
Scale Factor Value | 1.

[2] Select **Mass** as an output parameter. ↗

[3] Highlight **Equivalent Stress**. ↙

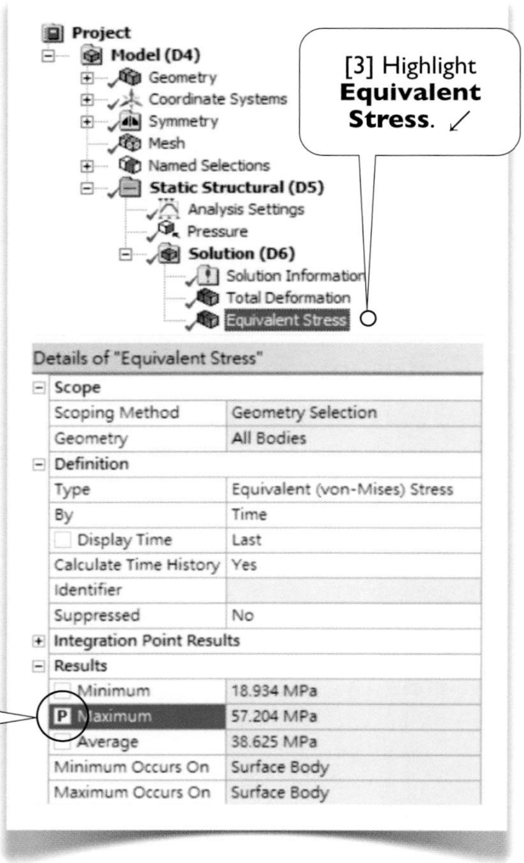

Project
Model (D4)
Geometry
Coordinate Systems
Symmetry
Mesh
Named Selections
Static Structural (D5)
Analysis Settings
Pressure
Solution (D6)
Solution Information
Total Deformation
Equivalent Stress
Details of "Equivalent Stress"
Scope
Scoping Method | Geometry Selection
Geometry | All Bodies
Definition
Type | Equivalent (von-Mises) Stress
By | Time
Display Time | Last
Calculate Time History | Yes
Identifier
Suppressed | No
Integration Point Results
Results
Minimum | 18.934 MPa
P Maximum | 57.204 MPa
Average | 38.625 MPa
Minimum Occurs On | Surface Body
Maximum Occurs On | Surface Body

[4] Select **Maximum** as an output parameter. Close **Mechanical**. #

8.2.5 Create a Direct Optimization System

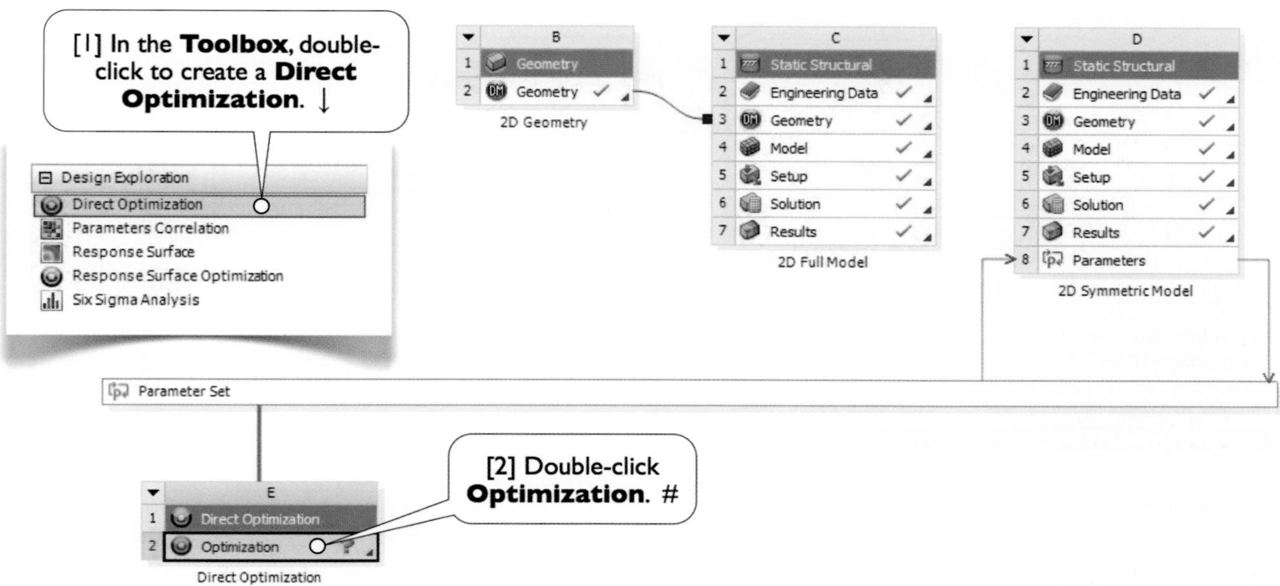

8.2.6 Define and Run the Optimization Problem

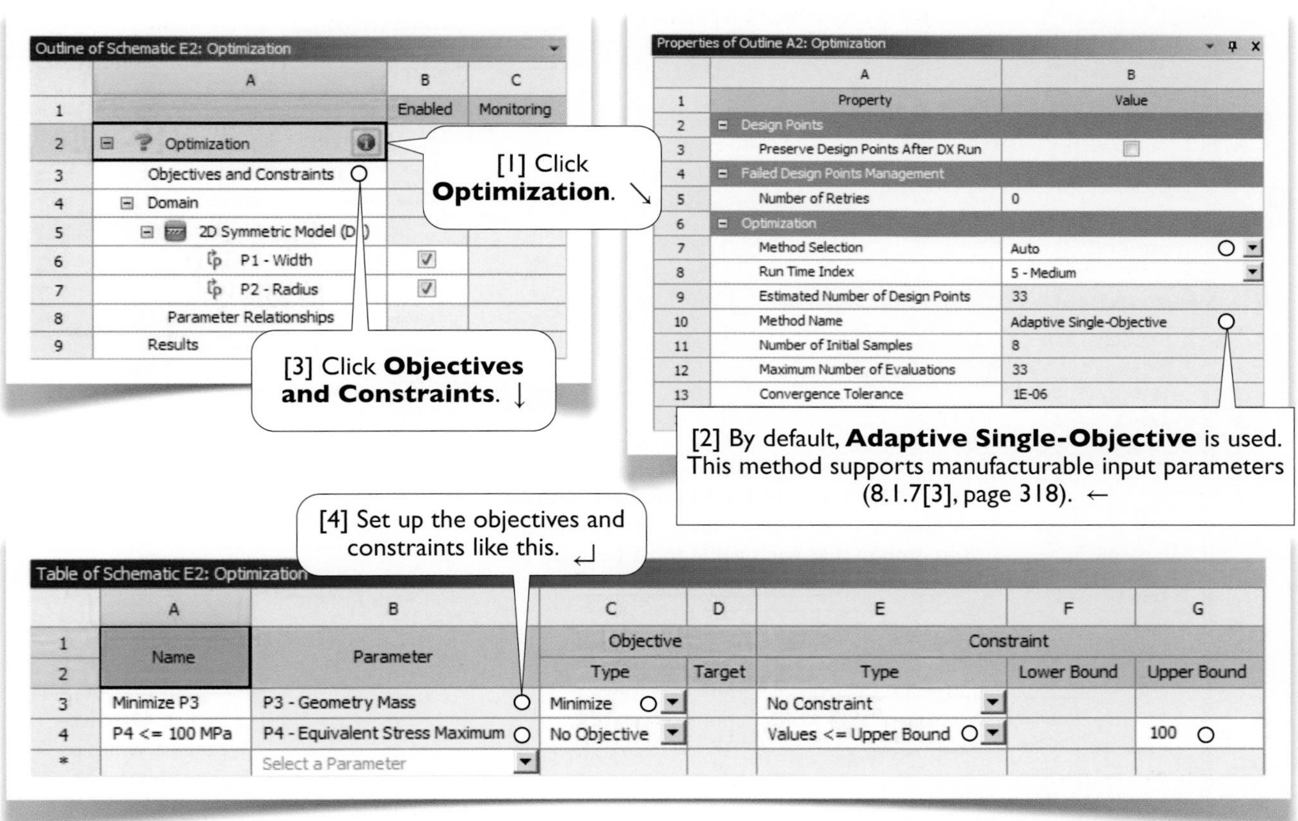

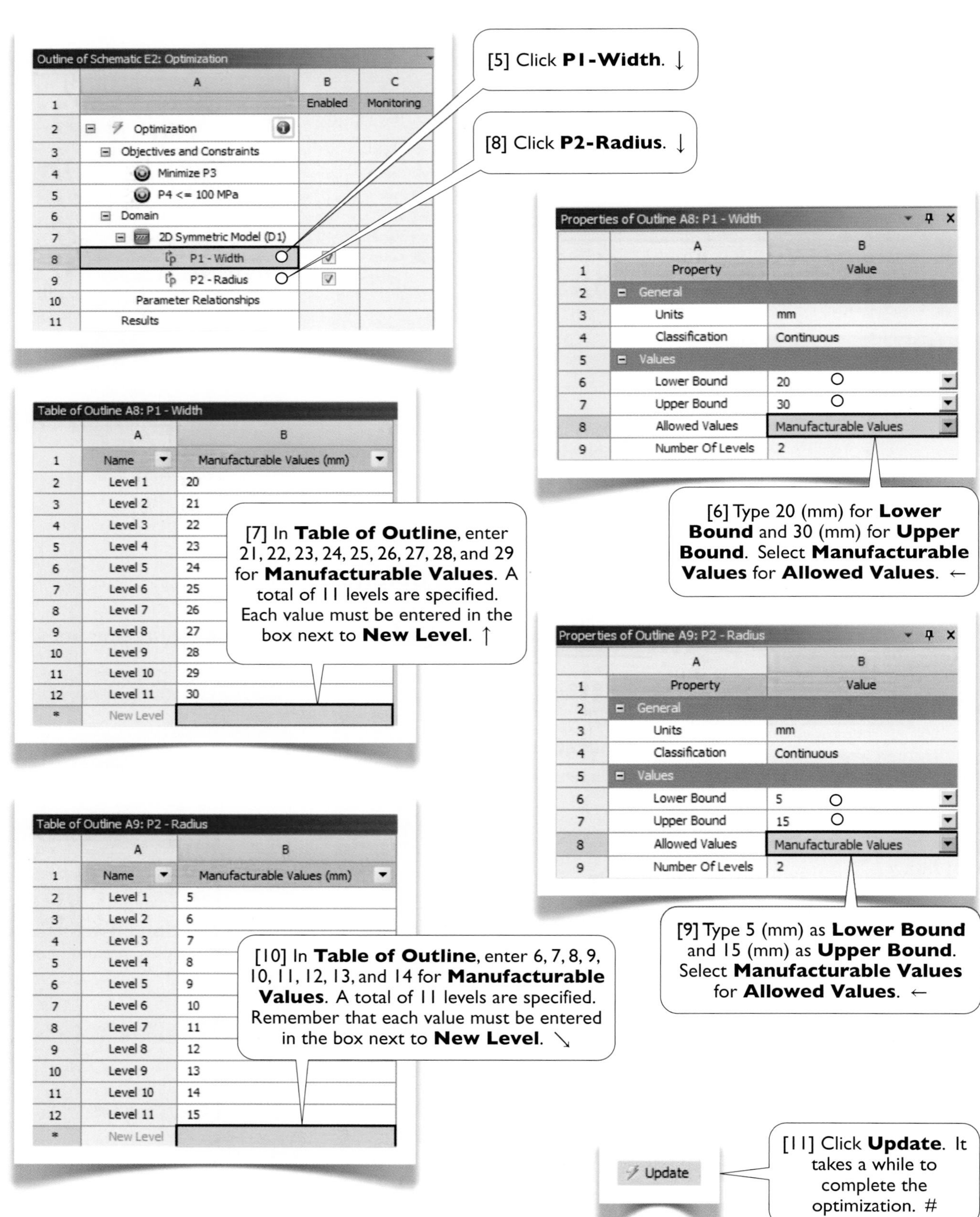
Outline of Schematic E2: Optimization
A
B
C
Enabled
Monitoring
Optimization
Objectives and Constraints
Minimize P3
P4 <= 100 MPa
Domain
2D Symmetric Model (D1)
P1 - Width
P2 - Radius
Parameter Relationships
Results
[5] Click **P1-Width**. ↓
[8] Click **P2-Radius**. ↓
Properties of Outline A8: P1 - Width
Property
Value
General
Units
mm
Classification
Continuous
Values
Lower Bound
20
Upper Bound
30
Allowed Values
Manufacturable Values
Number Of Levels
2
[6] Type 20 (mm) for **Lower Bound** and 30 (mm) for **Upper Bound**. Select **Manufacturable Values** for **Allowed Values**. ←
Table of Outline A8: P1 - Width
Name
Manufacturable Values (mm)
Level 1 20
Level 2 21
Level 3 22
Level 4 23
Level 5 24
Level 6 25
Level 7 26
Level 8 27
Level 9 28
Level 10 29
Level 11 30
New Level
[7] In **Table of Outline**, enter 21, 22, 23, 24, 25, 26, 27, 28, and 29 for **Manufacturable Values**. A total of 11 levels are specified. Each value must be entered in the box next to **New Level**. ↑
Properties of Outline A9: P2 - Radius
Property
Value
General
Units
mm
Classification
Continuous
Values
Lower Bound
5
Upper Bound
15
Allowed Values
Manufacturable Values
Number Of Levels
2
[9] Type 5 (mm) as **Lower Bound** and 15 (mm) as **Upper Bound**. Select **Manufacturable Values** for **Allowed Values**. ←
Table of Outline A9: P2 - Radius
Name
Manufacturable Values (mm)
Level 1 5
Level 2 6
Level 3 7
Level 4 8
Level 5 9
Level 6 10
Level 7 11
Level 8 12
Level 9 13
Level 10 14
Level 11 15
New Level
[10] In **Table of Outline**, enter 6, 7, 8, 9, 10, 11, 12, 13, and 14 for **Manufacturable Values**. A total of 11 levels are specified. Remember that each value must be entered in the box next to **New Level**. ↘
Update
[11] Click **Update**. It takes a while to complete the optimization. #

8.2.7 View the Optimal Design

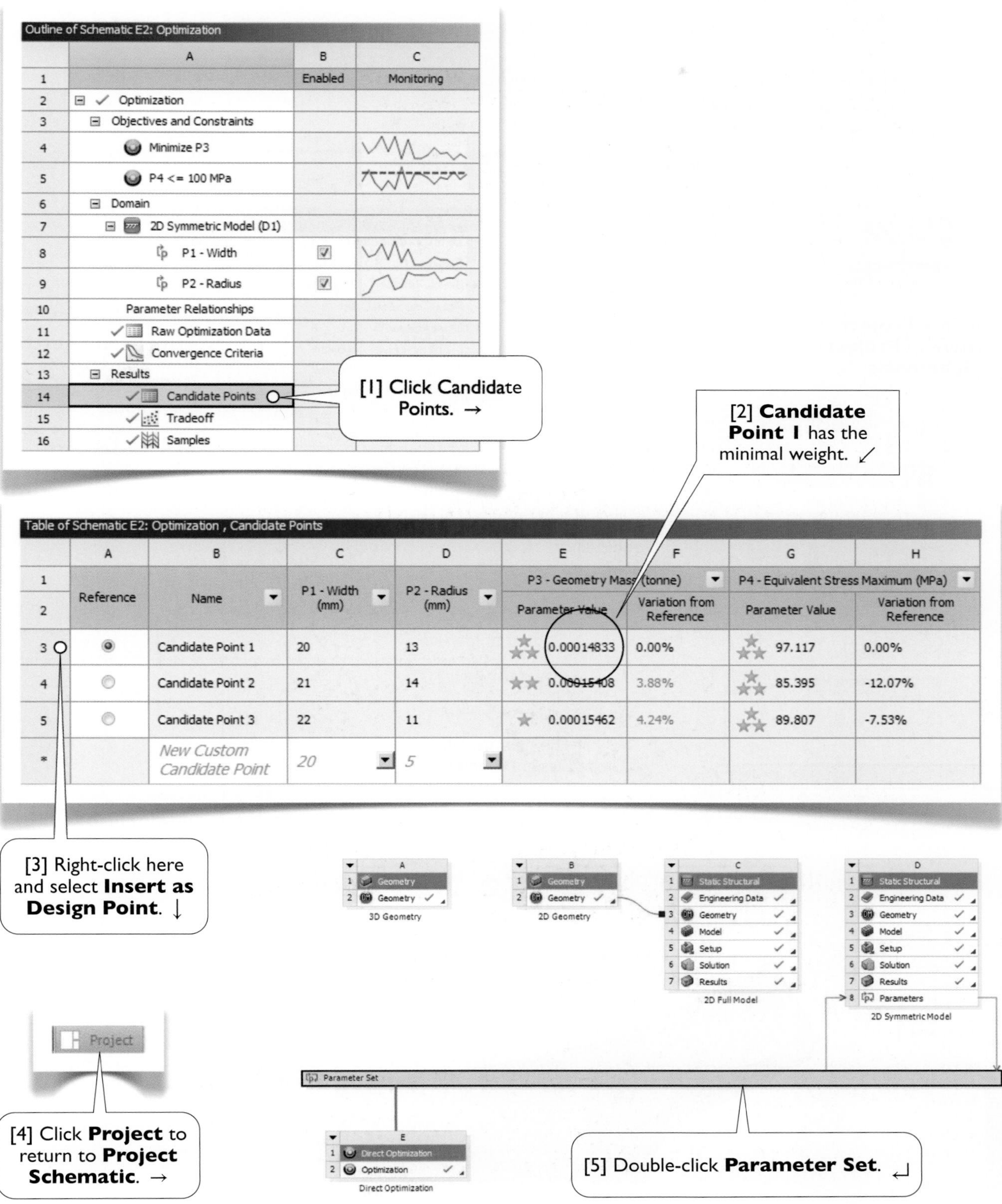

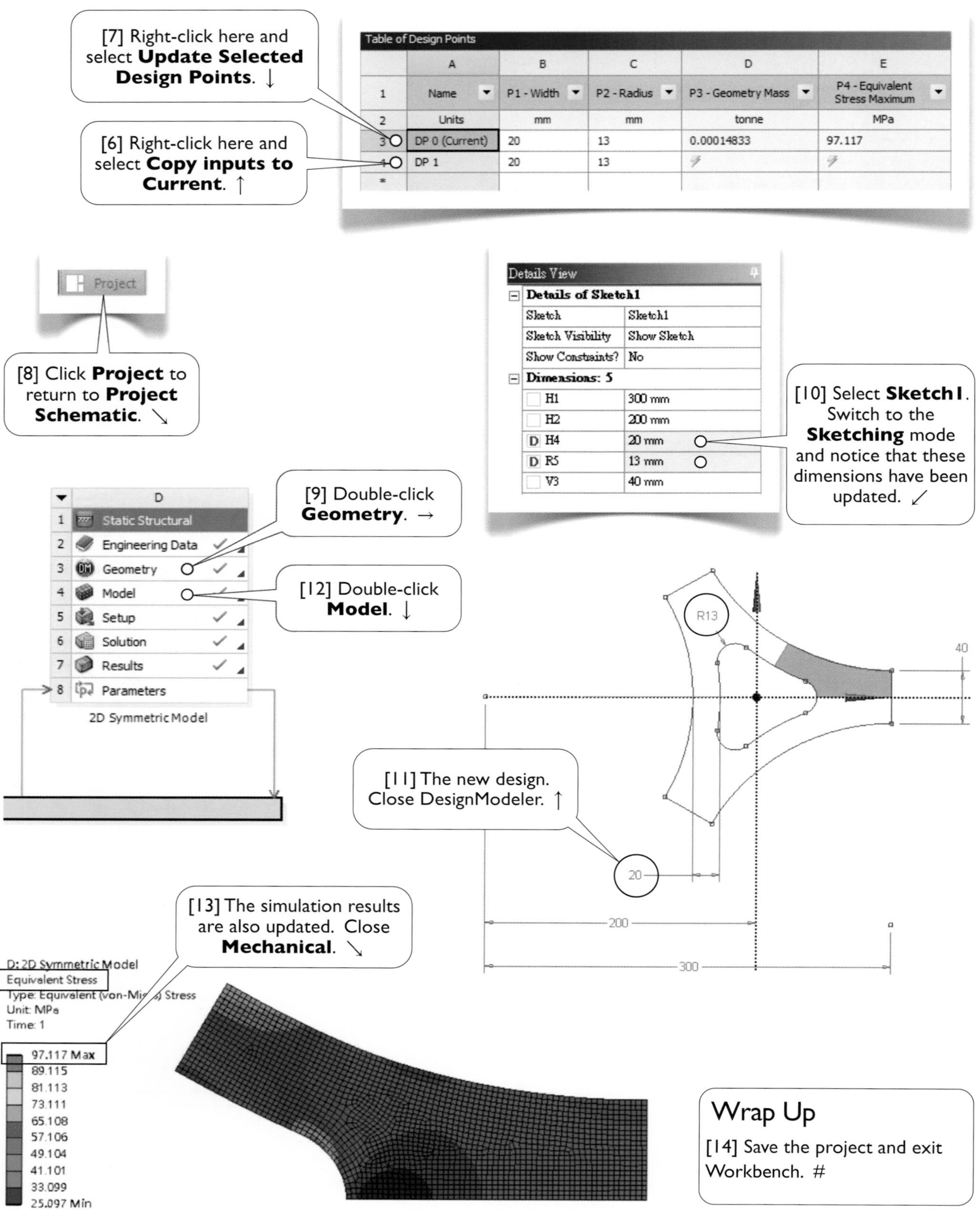

Wrap Up

[14] Save the project and exit Workbench. #

Section 8.3

Review

8.3.1 Keywords

Choose a letter for each keyword from the list of descriptions

1. ( ) Design Points
2. ( ) Design Space
3. ( ) Input Parameters and Output Parameters
4. ( ) NLPQL

Answers:

1. (C) 2. (B) 3. (A) 4. (D)

List of Descriptions

(A) Initial values of input parameters, also called design variables, are set up by the user and subsequently updated by Workbench. Values of output parameters, also called state variables, are calculated from the simulation applications.

(B) The space spanned by the input parameters, or design variables.

(C) Any points in the design space. In **DesignXplorer**, they are those with which the simulations have been carried out.

(D) Short for nonlinear programming by quadratic Lagrangian. A method of finding an optimal design. In each iteration, it approaches the problem using a quadratic polynomial and finds the optimal design in the subproblem described by the quadratic polynomial. The process is repeated until the optimal design is found. The method deals with only single-objective problems.

Chapter 9
Meshing

So far, we haven't discussed much on meshing because, for linear static problems, we usually can obtain solutions which are acceptable both in computing time and accuracy with global mesh controls (such as **Relevance Center**, **Relevance**, and **Element Size**) and some simple local mesh controls (such as **Sizing**). For the rest of the book, we will be dealing with dynamic and nonlinear problems, of which the solutions are sensitive to meshing quality. With poor mesh quality, a simulation may end up with solutions of poor accuracy or even run into convergence problems. Dynamic and nonlinear problems require many computational resources, and poor mesh quality may aggravate the situation and result in a lengthy computing time. For nonlinear problems, it is possible to reduce the runtime by improving the mesh quality (because they converge easier). In contrast, it is possible that a nonlinear solution fails to converge just because of the poor mesh quality.

Purpose of This Chapter

This chapter introduces meshing methods provided by Workbench and demonstrates how to use them. One of the mesh quality metrics, called *skewness*, is also introduced and used as a measure of mesh quality in this chapter.

About Each Section

Using the pneumatic finger example, Section 9.1 provides a step-by-step exercise to introduce some important concepts of meshing technologies. Section 9.2 uses a more involved model, the cover of pressure cylinder, to provide more exercises on meshing technologies. Section 9.3, a sequel of Section 3.5, studies 3D elements convergence behaviors. We postponed the study of 3D elements convergence until now because we need more meshing techniques to control the meshing density.

Section 9.1

Pneumatic Fingers[Ref 1]

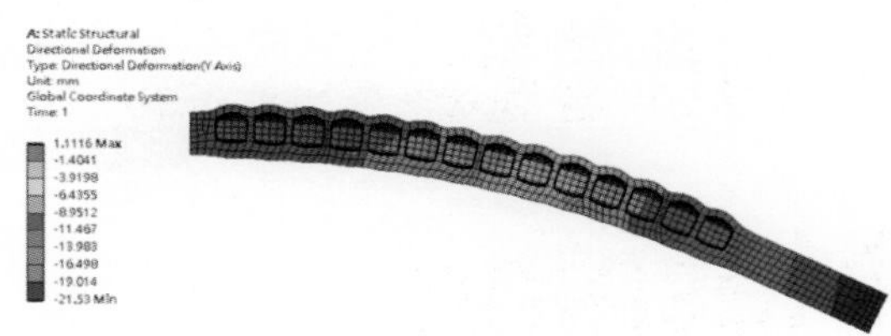

9.1.1 About the Pneumatic Fingers

[1] Simulation of the pneumatic finger was previewed in Section 1.1. In this section, we will walk through each step. Besides the information in 1.1.1 (pages 10-11), geometric details are given in the figure below. Due to the symmetry of the cross section [2], we model only half of the finger. The **mm-kg-N-s** unit system is used in this section. ↓

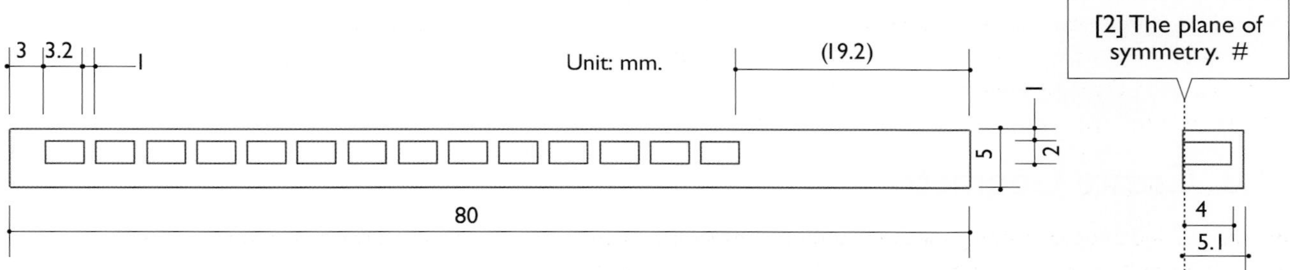

9.1.2 Start Up and Prepare Material Properties

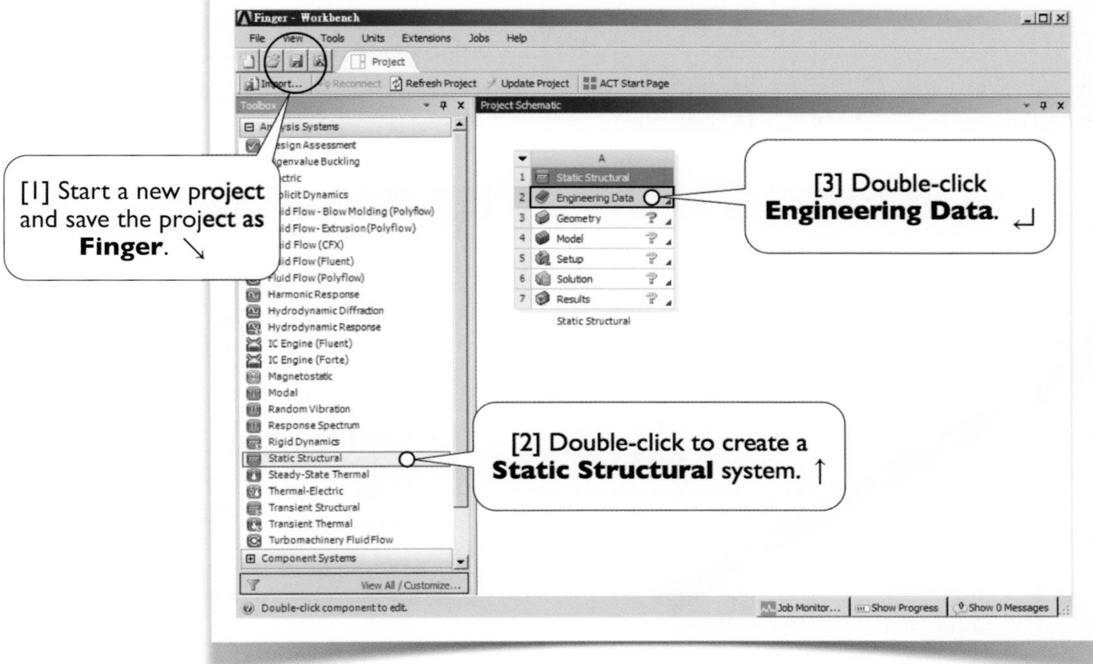

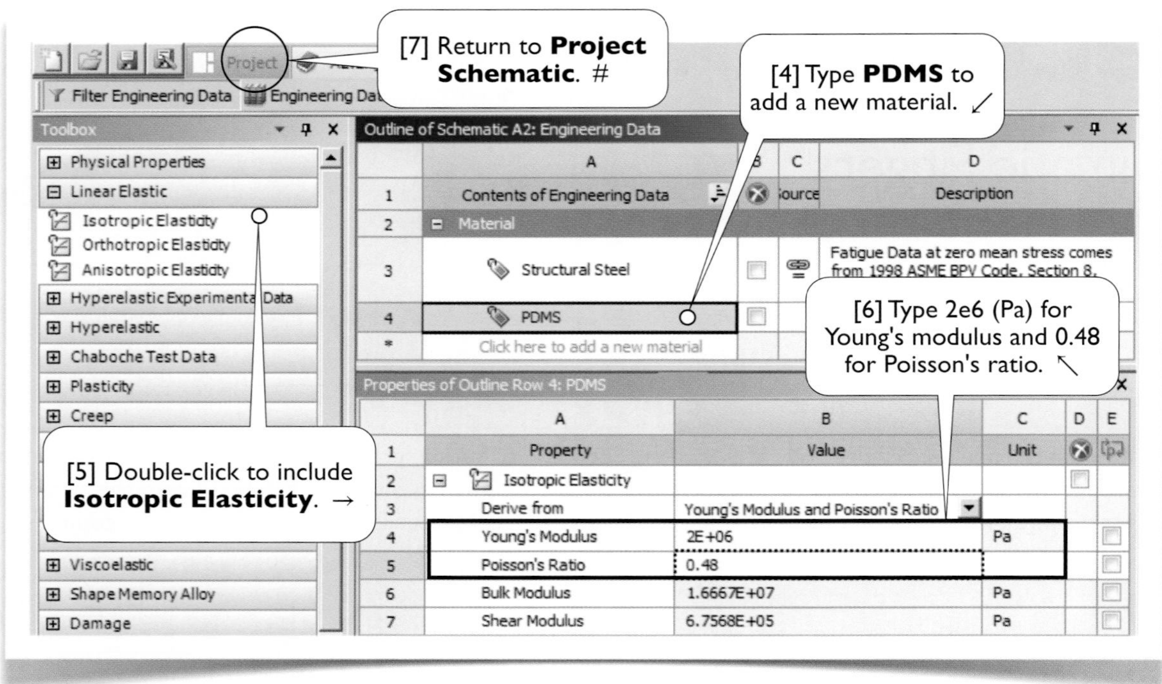

9.1.3 Create Geometry

[1] Start up DesignModeler. Select **Millimeter** as the length unit and create a geometric model as shown in [2-7]. →

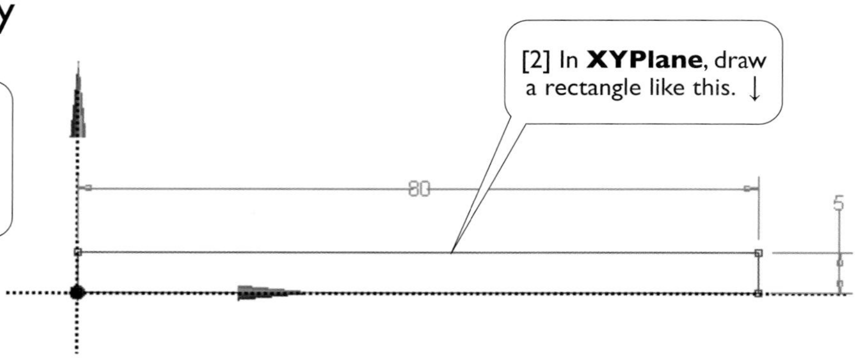

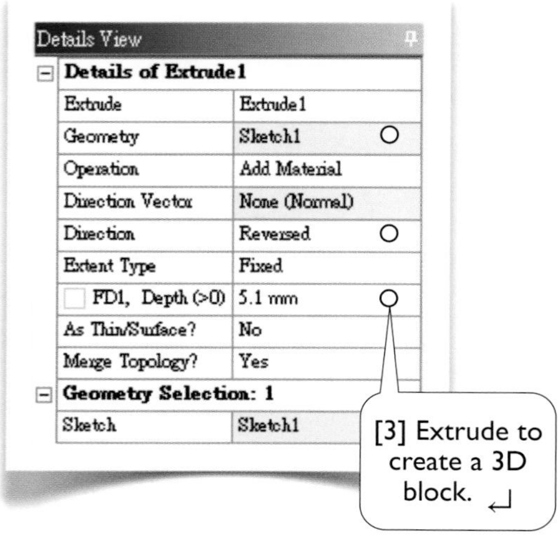

[2] In **XYPlane**, draw a rectangle like this. ↓

80

5

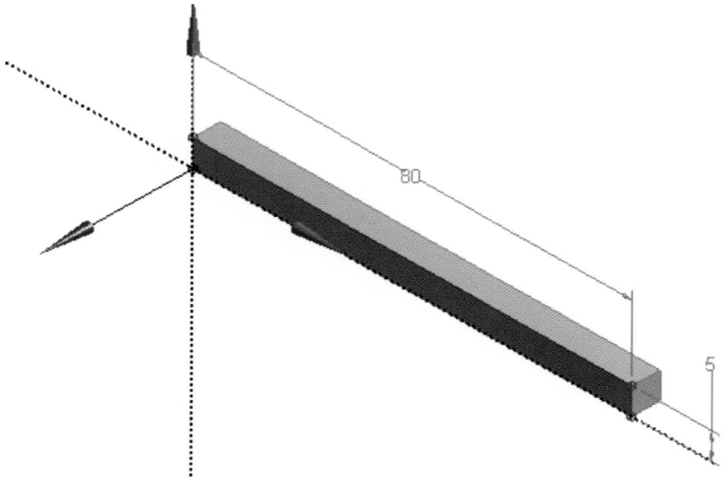

Details View	
Details of Extrude1	
Extrude	Extrude1
Geometry	Sketch1
Operation	Add Material
Direction Vector	None (Normal)
Direction	Reversed
Extent Type	Fixed
FD1, Depth (>0)	5.1 mm
As Thin/Surface?	No
Merge Topology?	Yes
Geometry Selection: 1	
Sketch	Sketch1

[3] Extrude to create a 3D block. ↵

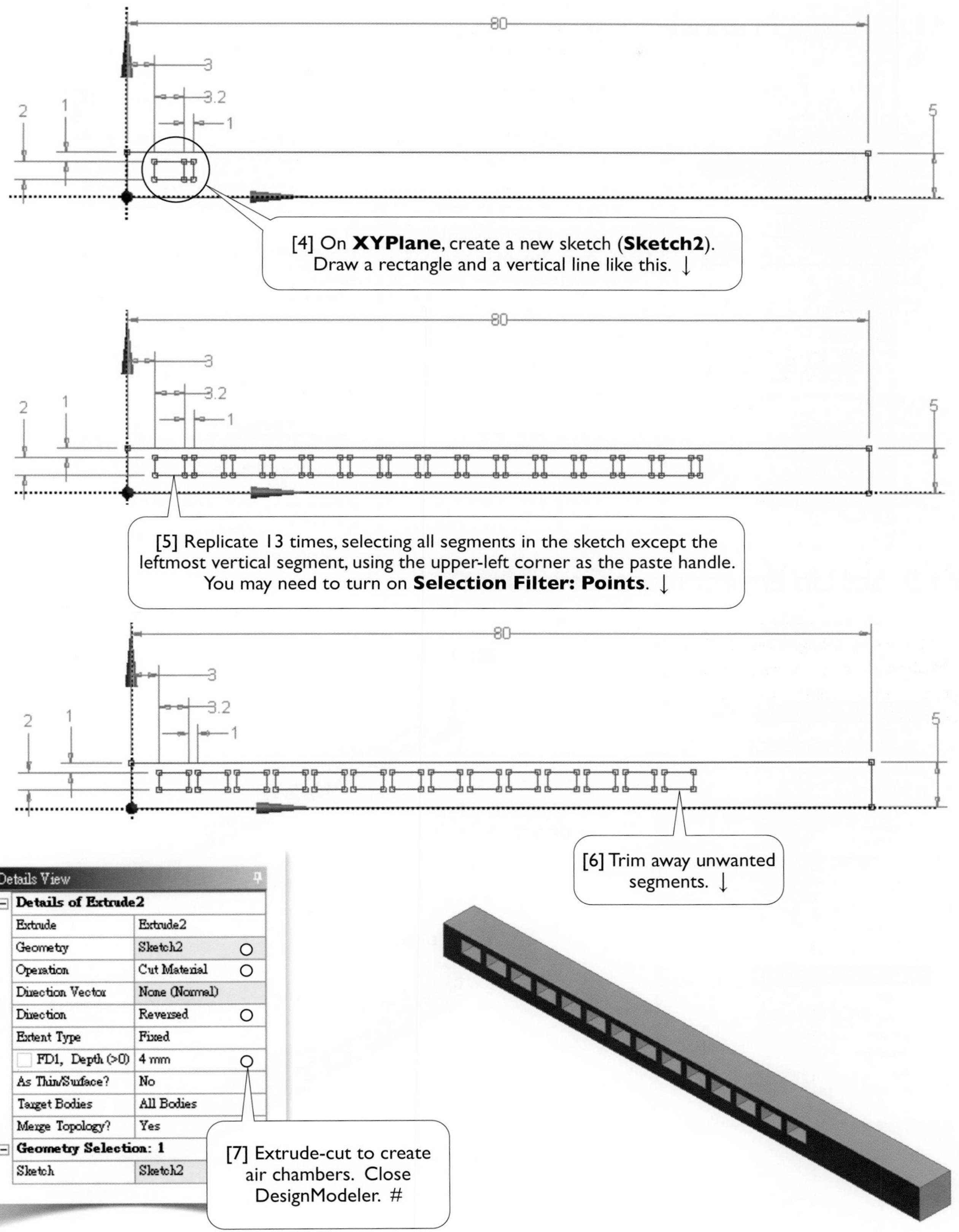
80
3
3.2
1
2
1
5
[4] On **XYPlane**, create a new sketch (**Sketch2**). Draw a rectangle and a vertical line like this. ↓
[5] Replicate 13 times, selecting all segments in the sketch except the leftmost vertical segment, using the upper-left corner as the paste handle. You may need to turn on **Selection Filter: Points**. ↓
[6] Trim away unwanted segments. ↓
Details View
Details of Extrude2
Extrude
Extrude2
Geometry
Sketch2
Operation
Cut Material
Direction Vector
None (Normal)
Direction
Reversed
Extent Type
Fixed
FD1, Depth (>0)
4 mm
As Thin/Surface?
No
Target Bodies
All Bodies
Merge Topology?
Yes
Geometry Selection: 1
Sketch
Sketch2
[7] Extrude-cut to create air chambers. Close DesignModeler. #

9.1.4 Assign Material

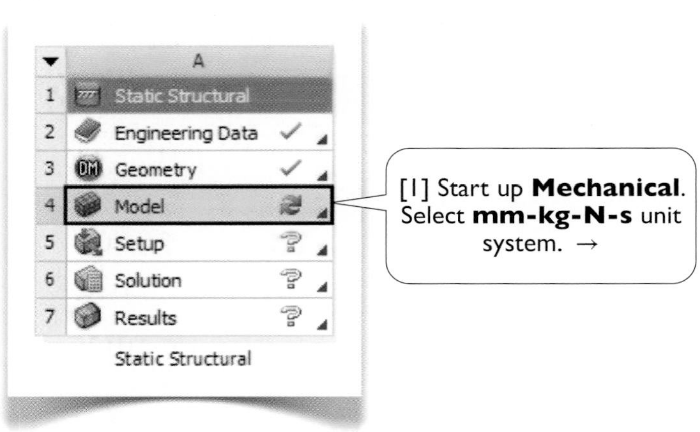

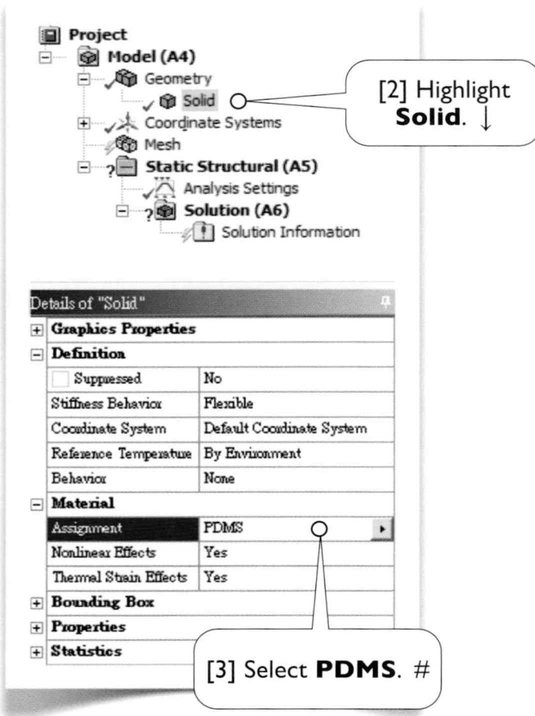

9.1.5 Set Up Environment Conditions

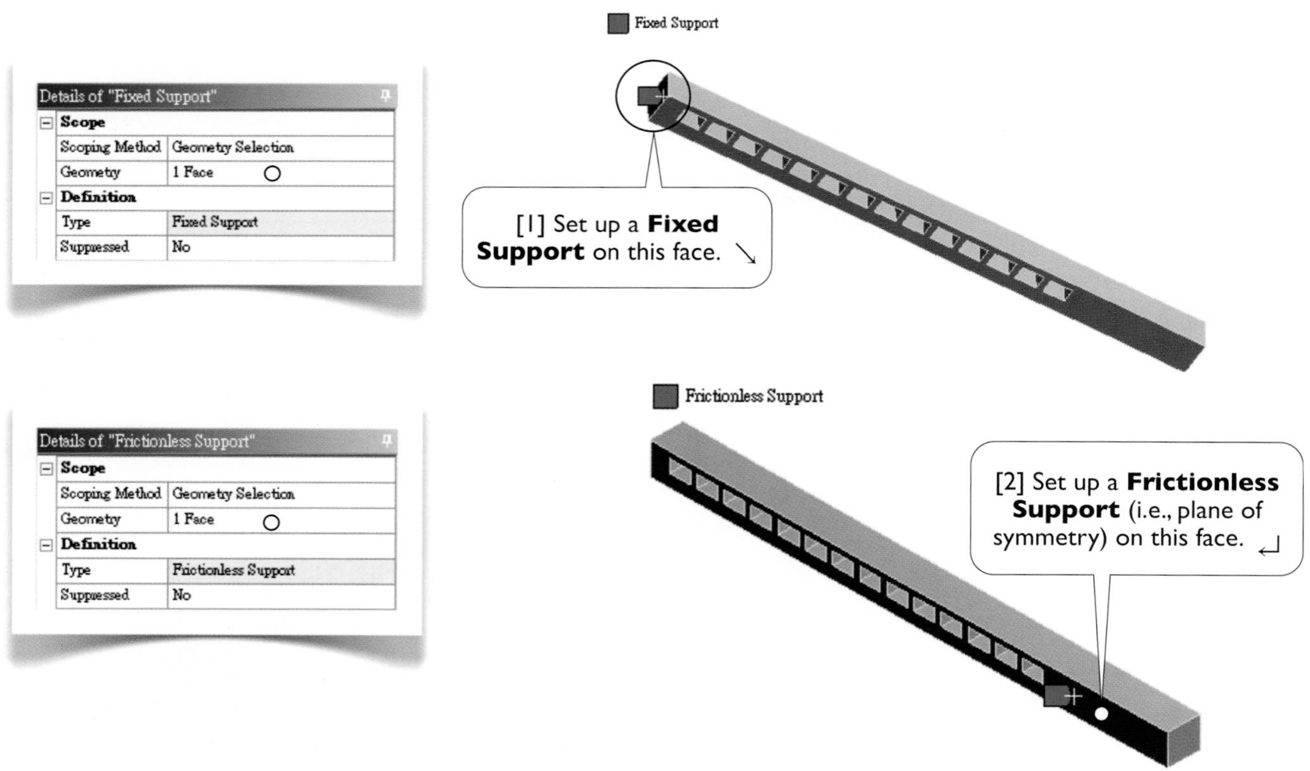

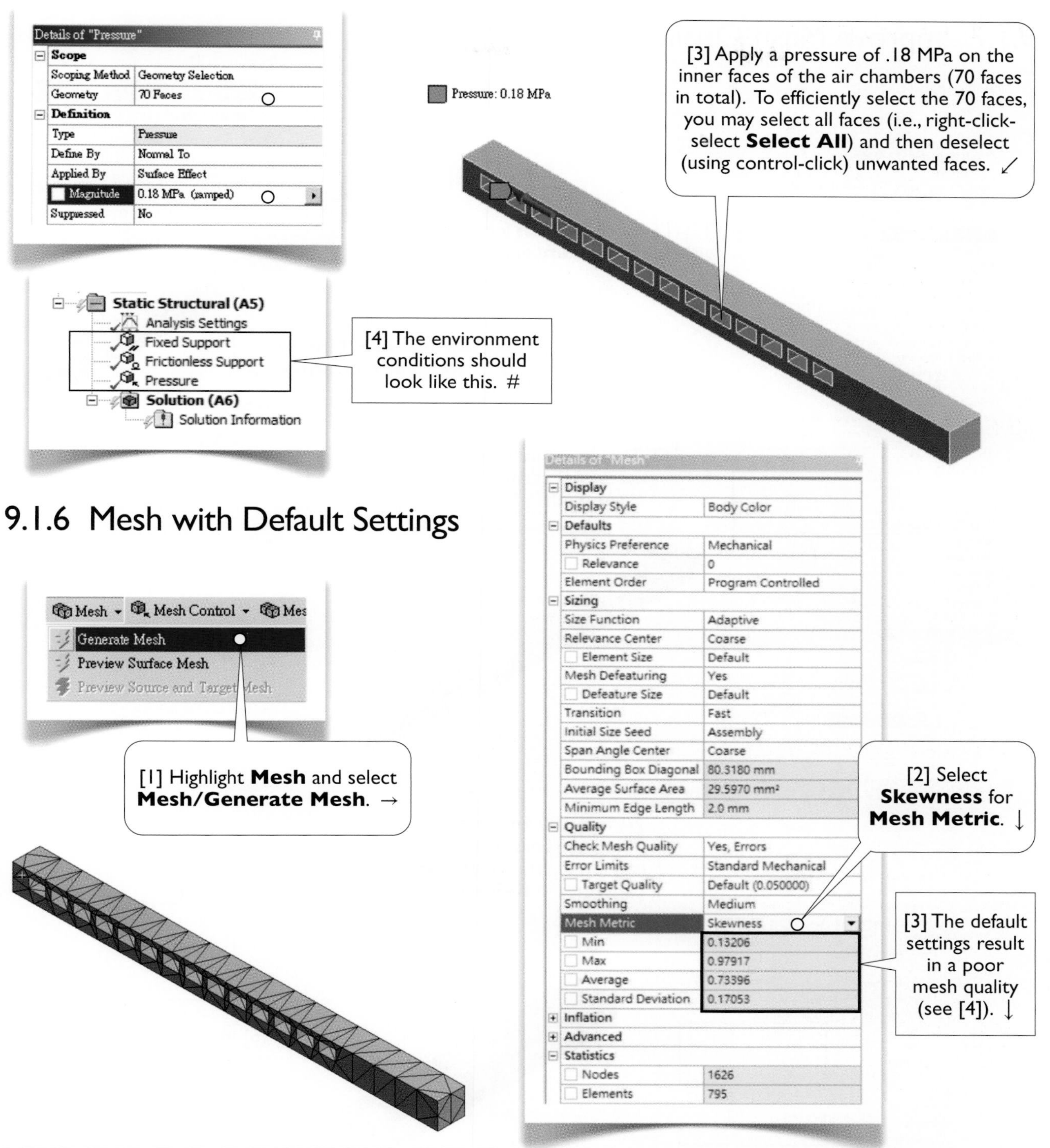

9.1.6 Mesh with Default Settings

Skewness[Refs 2, 3]

[4] Skewness, a measure of mesh quality, can be calculated for each element according to its geometry. Definition of skewness can be found in the on-line documentation[Refs 2, 3]. For now, all you need to know is that it is a value ranging from 0 to 1, the smaller the better, and, as a guideline, element skewness should not be larger than 0.95. #

9.1.7 Improve Mesh Quality

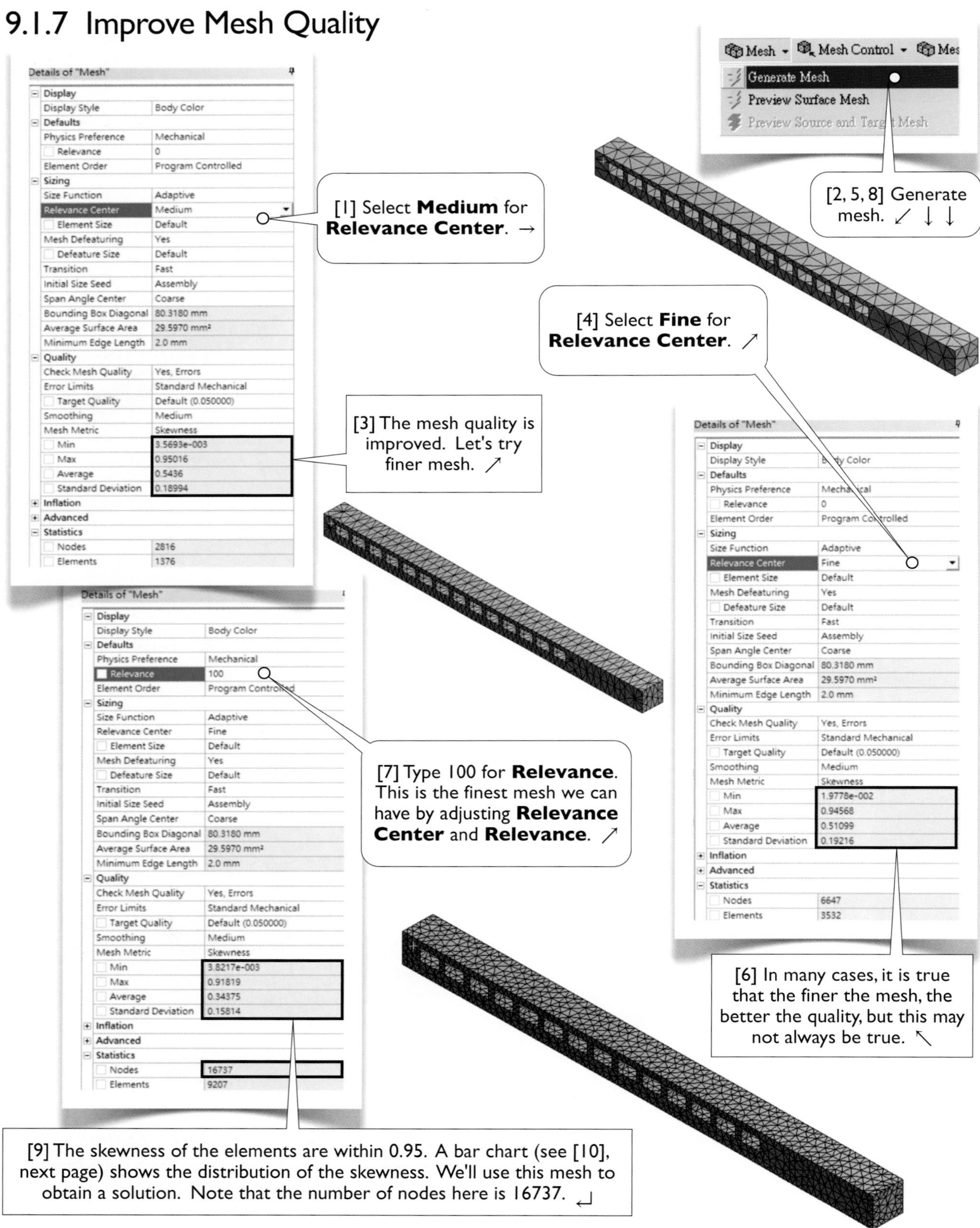

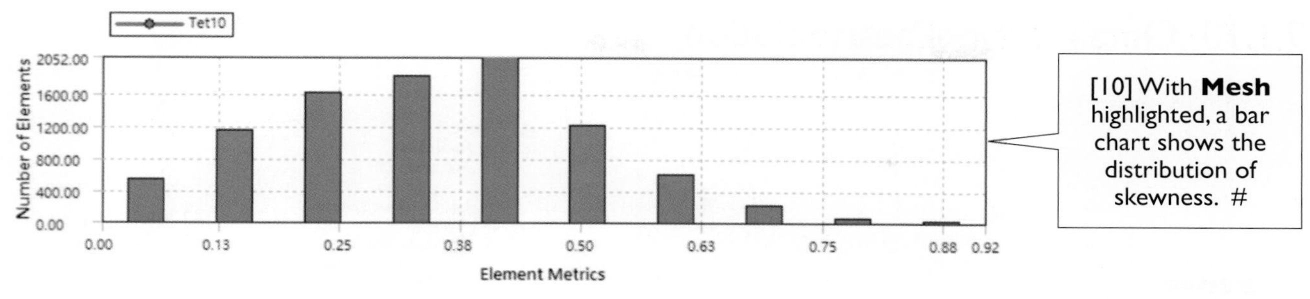

[10] With **Mesh** highlighted, a bar chart shows the distribution of skewness. #

9.1.8 Set Up Solution Branch

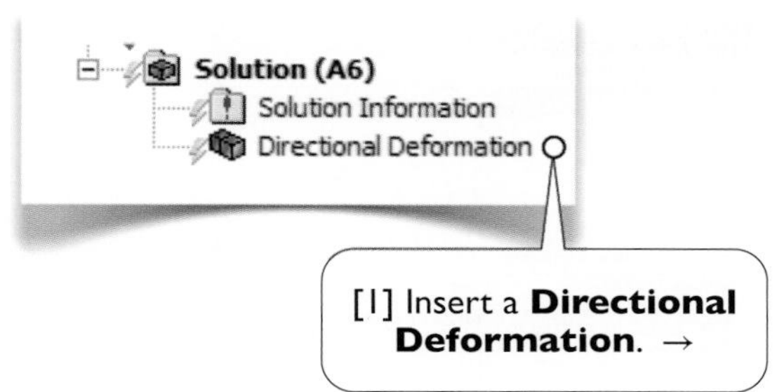

[1] Insert a **Directional Deformation.** →

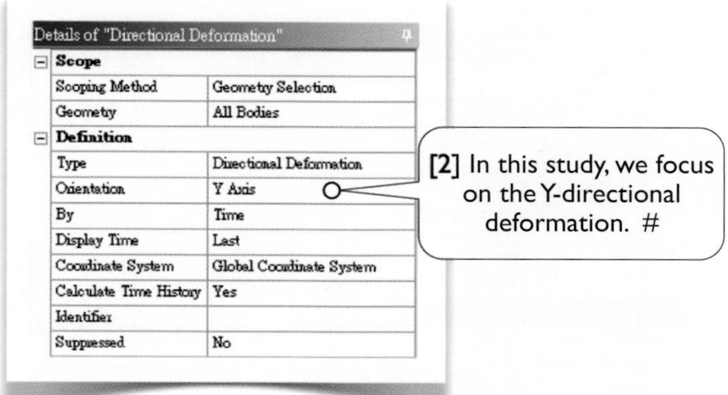

[2] In this study, we focus on the Y-directional deformation. #

9.1.9 Obtain a Linear Solution

[1] Solve. It takes only a few seconds to complete the linear solution. →

[2] Select **True Scale**. ↓

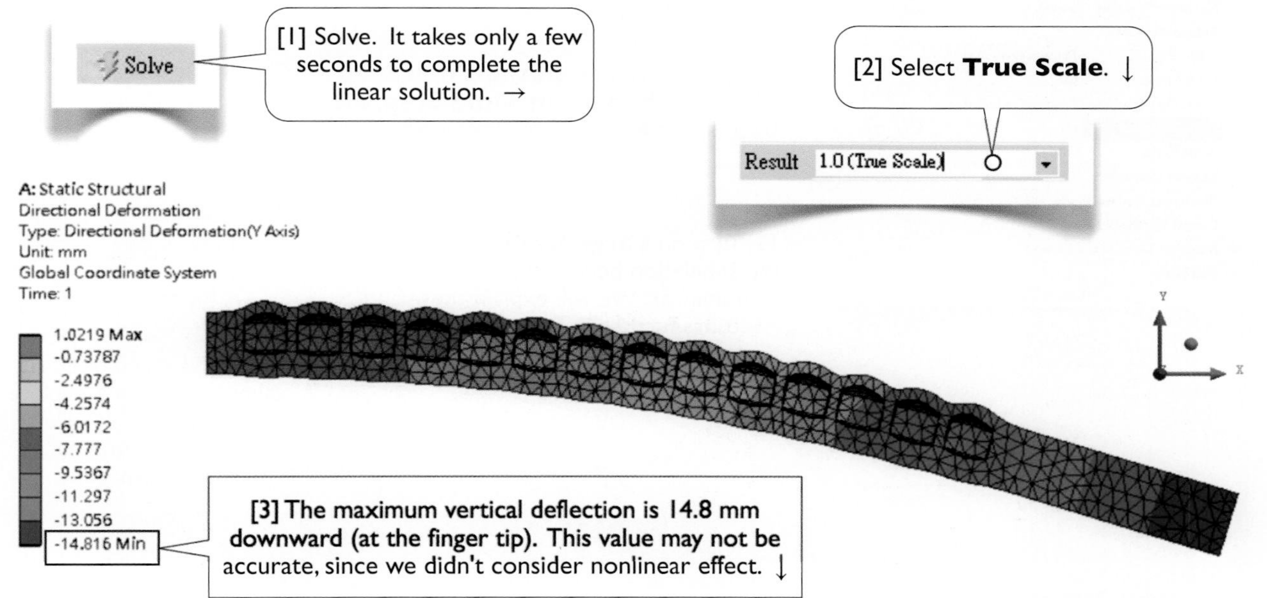

[3] The maximum vertical deflection is 14.8 mm downward (at the finger tip). This value may not be accurate, since we didn't consider nonlinear effect. ↓

Obtain a Linear Solution before Nonlinear Simulations

[4] It is a good practice to make sure a linear solution can be obtained before a nonlinear simulation is performed. A linear simulation takes much less computational time than a nonlinear one. Nonlinearity should be considered in this case, since the deflection is large. The linear solution, however, provides a way of model checking. #

9.1.10 Obtain a Nonlinear Solution

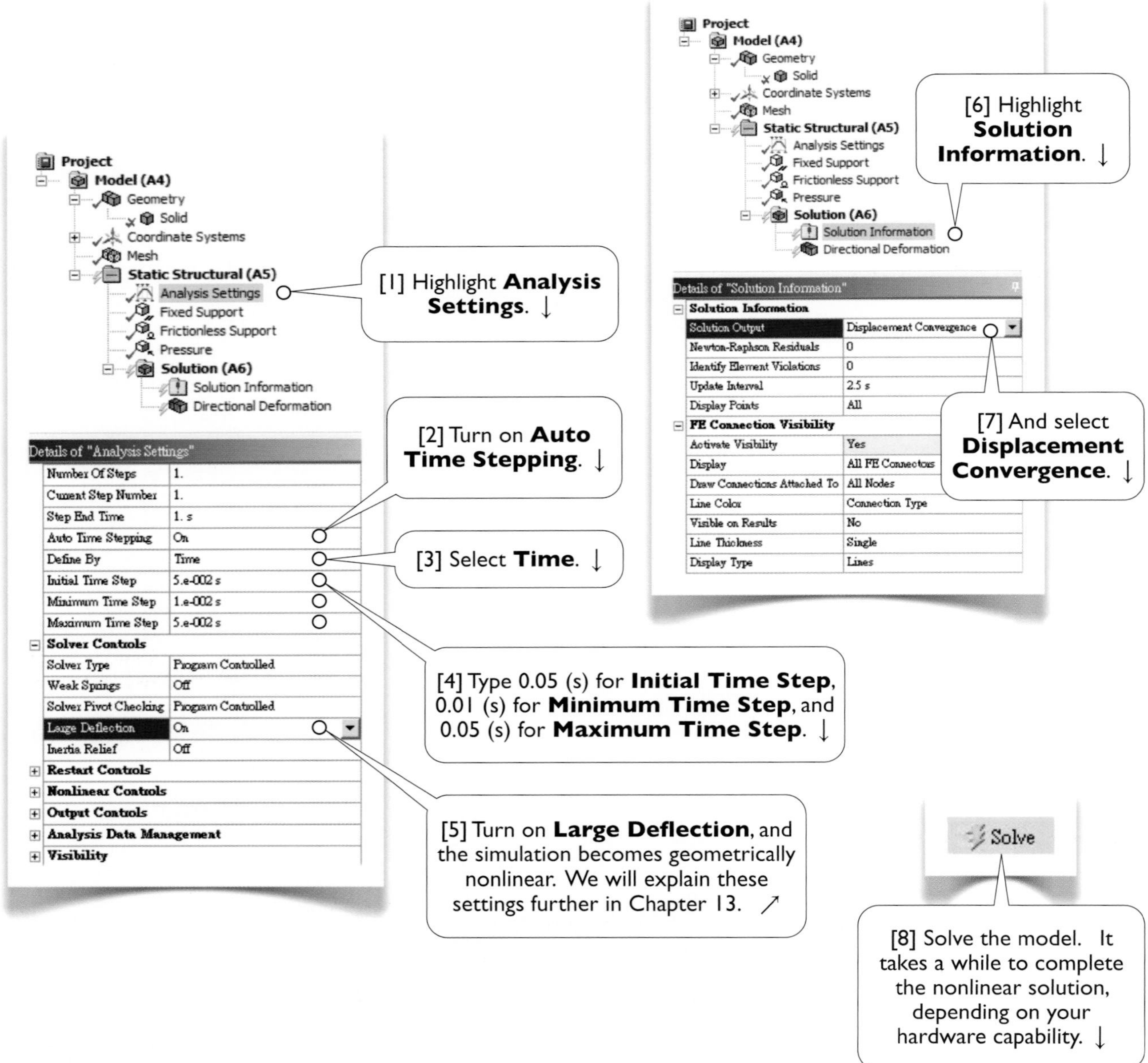

Displacement Convergence

[9] Each substep of a nonlinear simulation involves an iterative process. Force and displacement values are used as convergence criteria. These concepts will be further explained in 13.1.5 (page 473). ↵

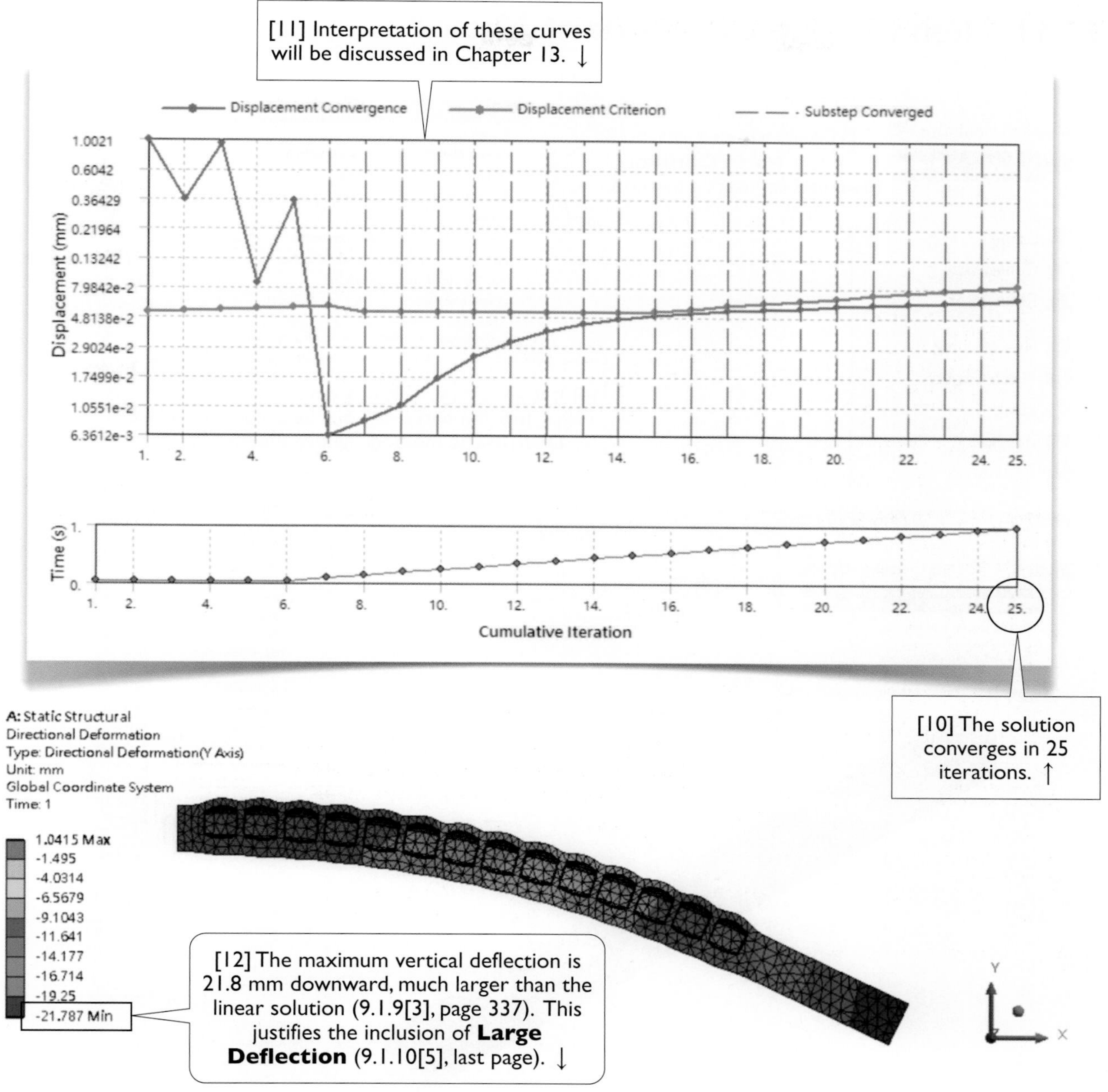

Element Shapes

[13] In many cases, nonlinear simulations can be challenging. Meshing quality plays an important role in the convergence of nonlinear solution. The mesh metric (e.g., skewness; see 9.1.7[9], page 336) is a measure of mesh quality. Skewness often can be improved by refining elements. It sometimes needs a large number of elements to achieve a mesh quality that is good enough to make the solution converge. In other cases, it may never achieve an acceptable mesh quality by simply refining elements.

Another factor affecting convergence is the shapes of elements. In general, hexahedra are more efficient than tetrahedra (see 9.3.13 and 9.3.14, page 361). In the following exercises, let's try to mesh the model with hexahedra. #

9.1.11 Mesh with **Hex Dominant** Method

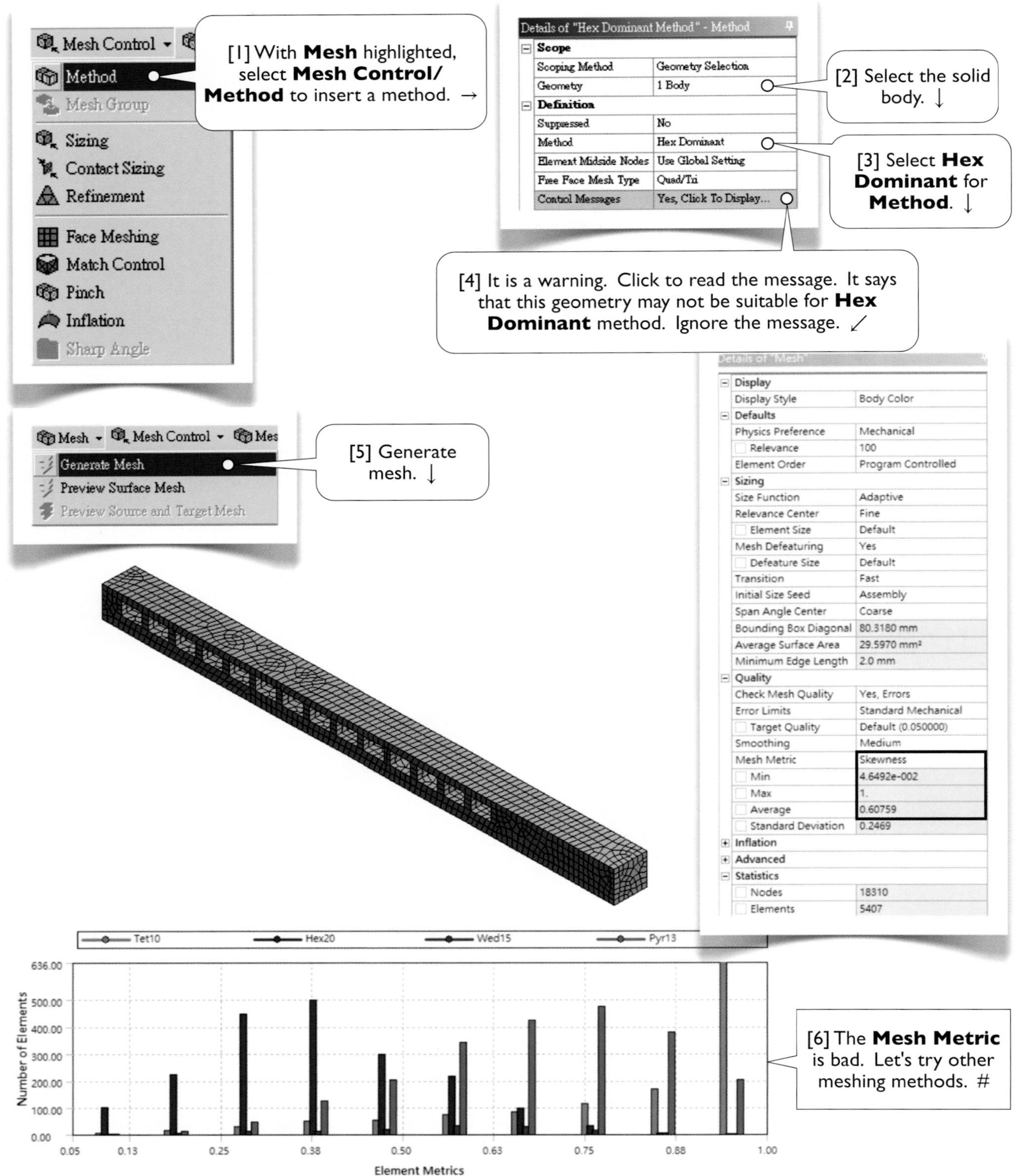

9.1.12 Mesh with **Sweep** Method

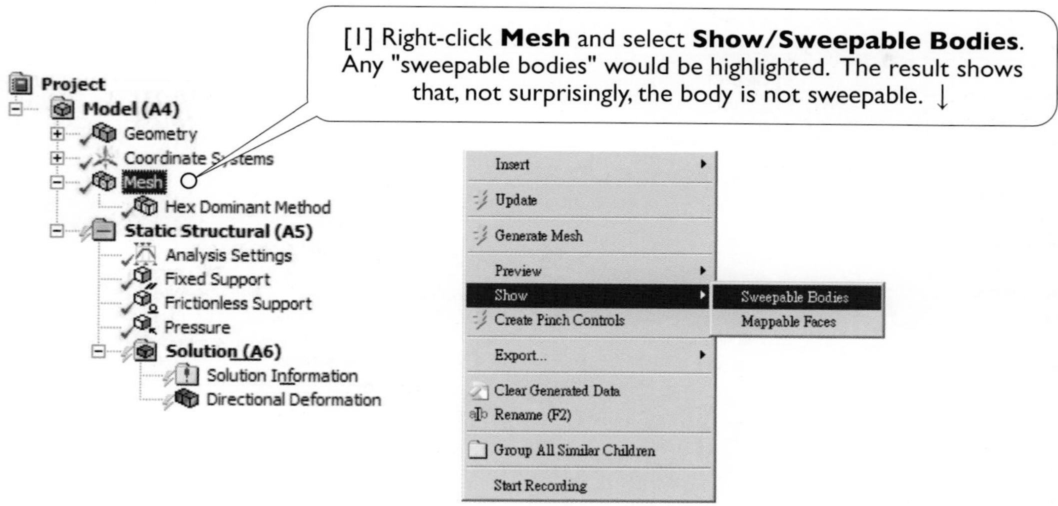

Sweepable Bodies

[2] As mentioned in 5.3.2[2] (page 228), a simple idea for generating hexahedral elements is to mesh a face (or faces) of a body with quadrilaterals and then "sweep" along its depth direction to the other end face (or faces) of the body. The starting faces are called the *source faces* and the ending faces are called the *target faces*. The source or target faces can be either manually or automatically selected.

Not all bodies are sweepable. In our case, there is only one body, and it is not sweepable.

Mesh with **MultiZone** Method

For non-sweepable bodies, Workbench provides a sophisticated method of generating hexahedral elements, called **MultiZone** method. The idea of **MultiZone** method is to divide a non-sweepable body into several sweepable bodies, and then apply **Sweep** method on each of the bodies. Actually, we already applied this method to mesh the beam bracket model in 5.1.13[1-5] (page 217). #

9.1.13 Mesh with **MultiZone** Method

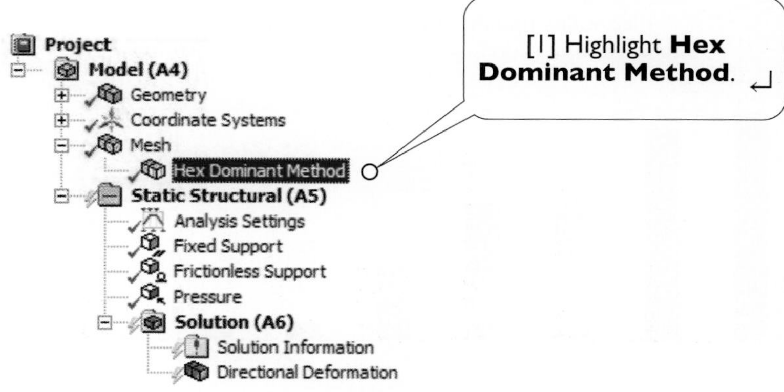

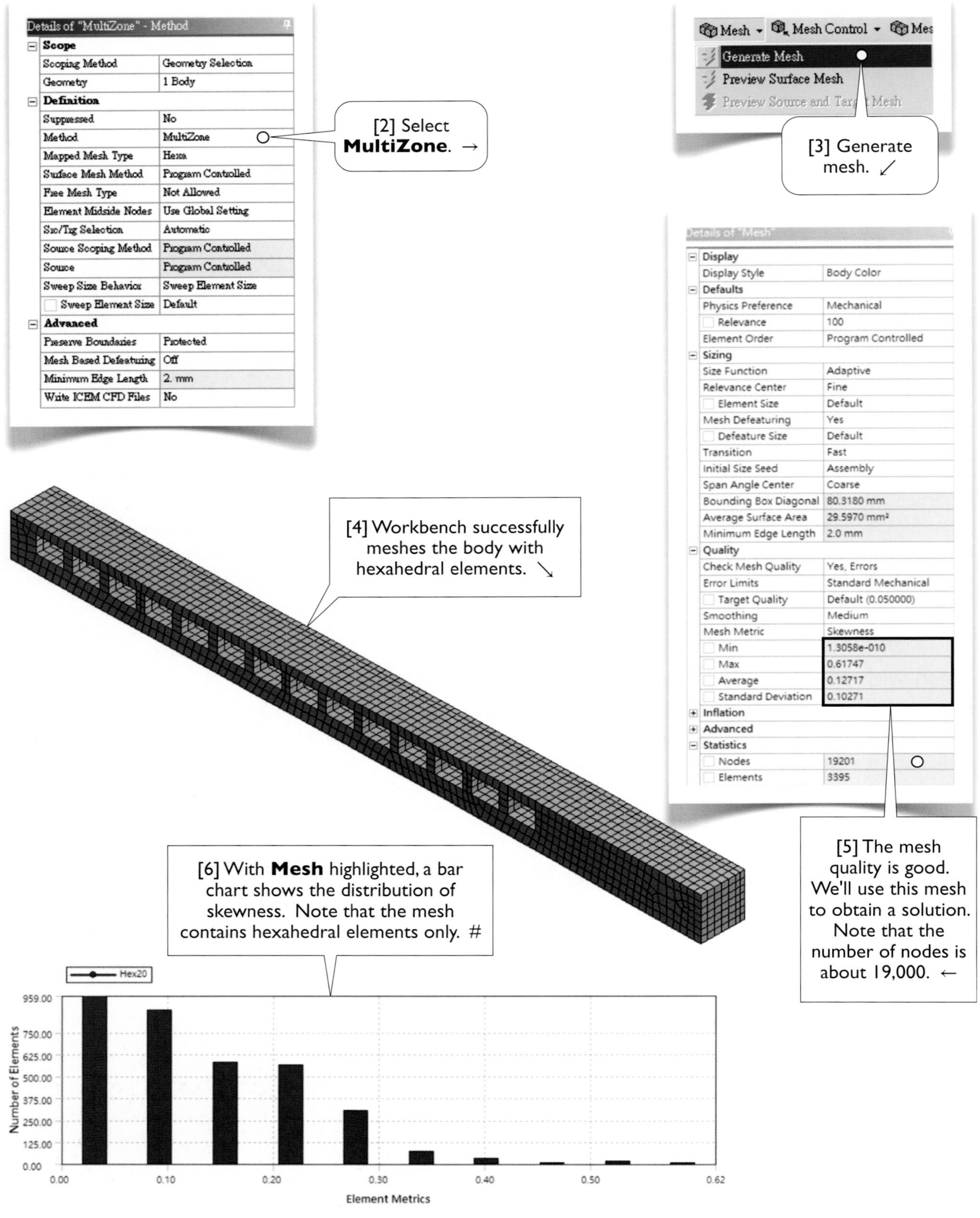

Details of "MultiZone" - Method
Scope
Scoping Method | Geometry Selection
Geometry | 1 Body
Definition
Suppressed | No
Method | MultiZone
Mapped Mesh Type | Hexa
Surface Mesh Method | Program Controlled
Free Mesh Type | Not Allowed
Element Midside Nodes | Use Global Setting
Src/Trg Selection | Automatic
Source Scoping Method | Program Controlled
Source | Program Controlled
Sweep Size Behavior | Sweep Element Size
Sweep Element Size | Default
Advanced
Preserve Boundaries | Protected
Mesh Based Defeaturing | Off
Minimum Edge Length | 2. mm
Write ICEM CFD Files | No
[2] Select **MultiZone**. →
Mesh
Mesh Control
Generate Mesh
Preview Surface Mesh
[3] Generate mesh. ↙
Details of "Mesh"
Display
Display Style | Body Color
Defaults
Physics Preference | Mechanical
Relevance | 100
Element Order | Program Controlled
Sizing
Size Function | Adaptive
Relevance Center | Fine
Element Size | Default
Mesh Defeaturing | Yes
Defeature Size | Default
Transition | Fast
Initial Size Seed | Assembly
Span Angle Center | Coarse
Bounding Box Diagonal | 80.3180 mm
Average Surface Area | 29.5970 mm²
Minimum Edge Length | 2.0 mm
Quality
Check Mesh Quality | Yes, Errors
Error Limits | Standard Mechanical
Target Quality | Default (0.050000)
Smoothing | Medium
Mesh Metric | Skewness
Min | 1.3058e-010
Max | 0.61747
Average | 0.12717
Standard Deviation | 0.10271
Inflation
Advanced
Statistics
Nodes | 19201
Elements | 3395
[4] Workbench successfully meshes the body with hexahedral elements. ↘
[5] The mesh quality is good. We'll use this mesh to obtain a solution. Note that the number of nodes is about 19,000. ←
[6] With **Mesh** highlighted, a bar chart shows the distribution of skewness. Note that the mesh contains hexahedral elements only. #
Hex20
959.00
750.00
625.00
500.00
375.00
250.00
125.00
0.00
Number of Elements
0.00
0.10
0.20
0.30
0.40
0.50
0.62
Element Metrics

9.1.14 Examine Mesh Using Section View

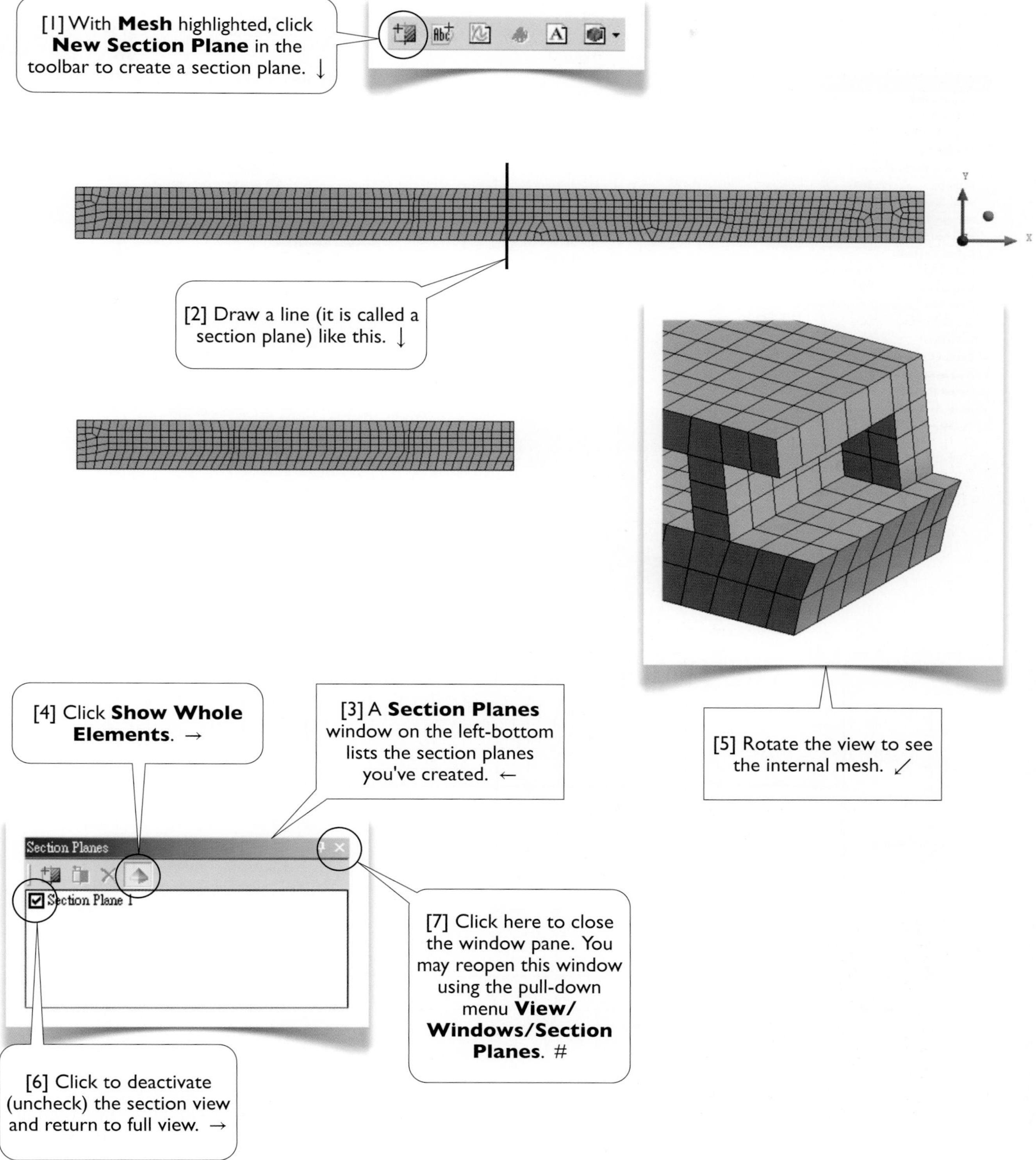

9.1.15 Obtain a Nonlinear Solution

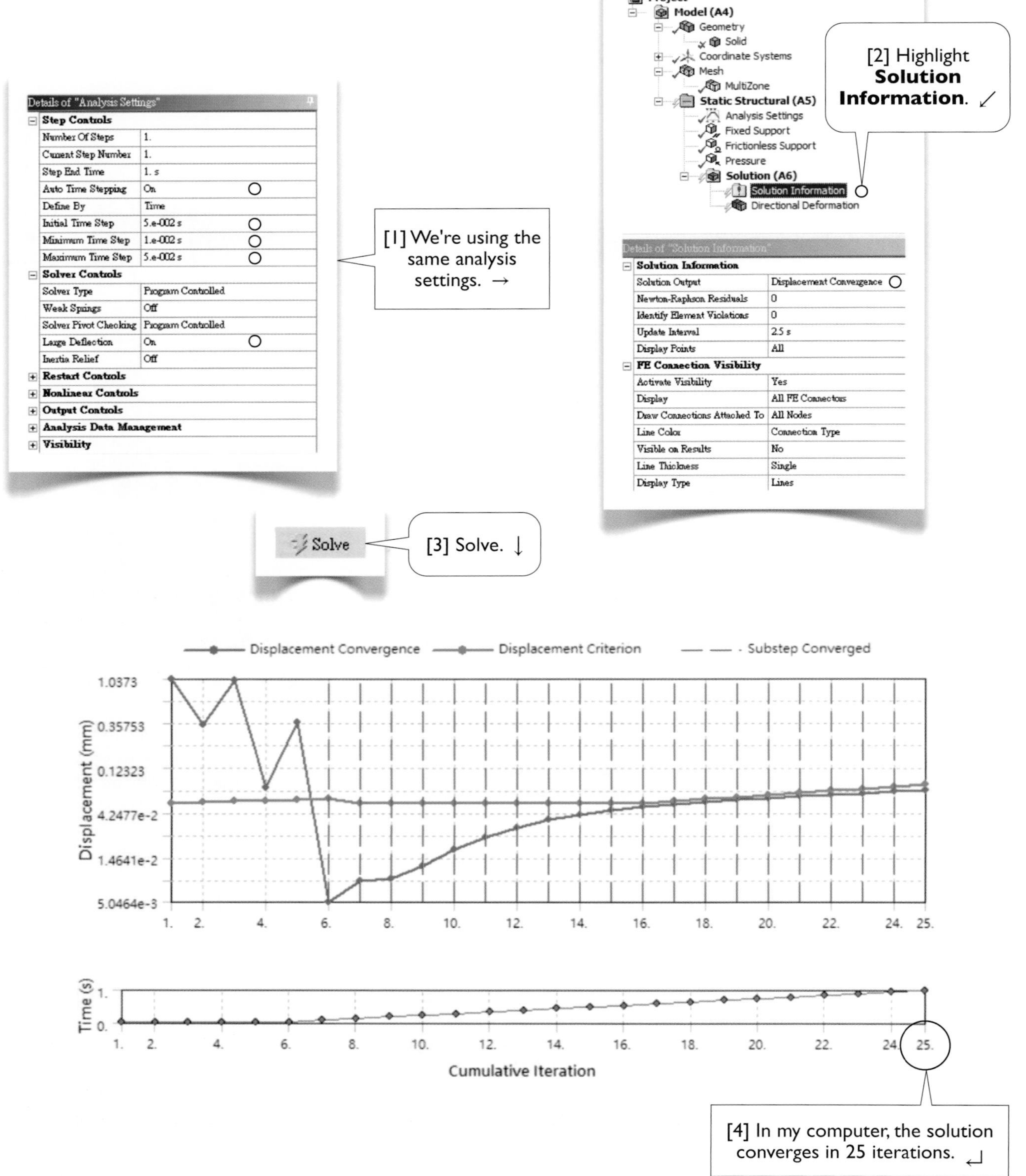

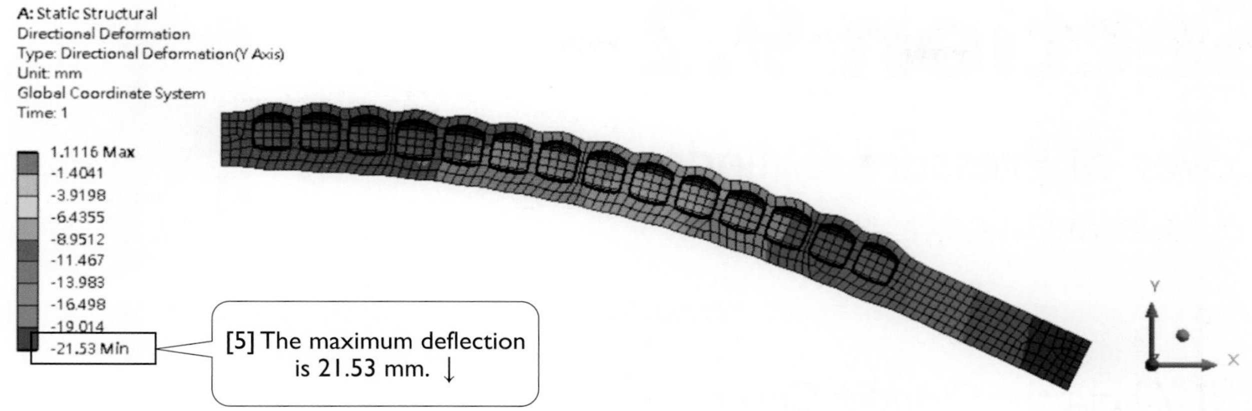

Wrap Up

[6] Save the project and exit Workbench. ↓

Remark

[7] As mentioned in 1.1.8[8] (page 19), when assuming a linear material, we are also assuming the compressive behavior is the same as tensile behavior, but this is usually not true for an elastomer under such a large deformation. (Note that the upper portion of the finger is subject to tension, while the lower portion is subject to compression.) Hyperelasticity, a more accurate material model for elastomer under large deformation, will be introduced in PART C of Section 14.1 (pages 532-535). #

References

1. This exercise is adapted from an unpublished work led by Prof. Chao-Chieh Lan of the Department of Mechanical Engineering, NCKU.
2. ANSYS Help//Meshing//Meshing User's Guide//Global Mesh Controls//Quality Group//Mesh Metric//Skewness
3. ANSYS Help//Mechanical APDL//Mechanical APDL Theory Reference//12. Element Tools//12.1. Element Shape Testing

Section 9.2

Cover of Pressure Cylinder

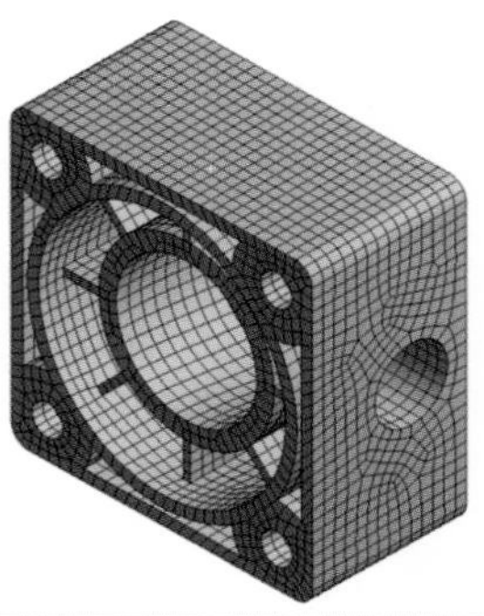

9.2.1 About the Cylinder Cover

[1] In this section, we will use the cylinder cover (Sections 4.2 and 5.2) to demonstrate some additional meshing techniques.

The geometry of the cover is relatively complicated. It seems that a tetrahedral mesh is the only feasible method. There is nothing wrong with a tetrahedral mesh as long as the mesh quality is good enough. Examining the mesh generated in 5.2.5[3-5] (page 221), we see that the mesh quality is bad (9.2.3[1-3], next page). The mesh quality needs to be improved. The simplest way is to adjust the relevance values. That sometimes works, although increasing the problem size, but sometimes fails. For a linear static simulation, problem size seems no big deal, but for a nonlinear or dynamic simulation, the problem size should be kept as small as possible, to maintain an acceptable computing time.

Note also that the purpose of this section is to demonstrate meshing techniques, rather than finding the best mesh for the cylinder cover. #

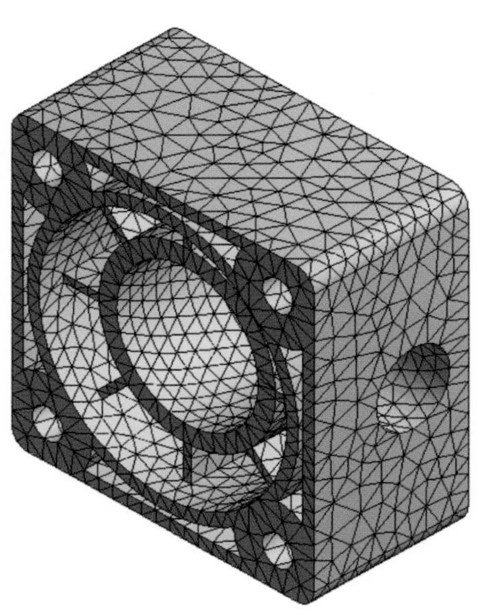

9.2.2 Open the Project **Cover**

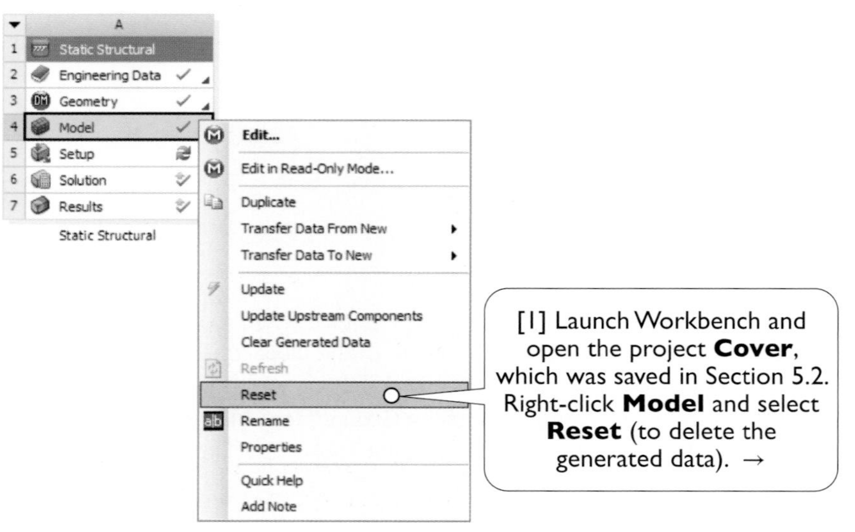

[1] Launch Workbench and open the project **Cover**, which was saved in Section 5.2. Right-click **Model** and select **Reset** (to delete the generated data). →

[2] Double-click **Model** to start up **Mechanical**. Make sure the unit system is **mm-kg-N-s**. #

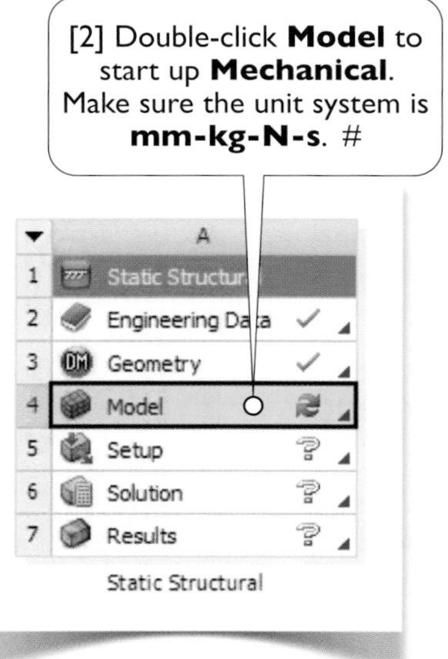

9.2.3 Increase Mesh Density Using Global Mesh Controls

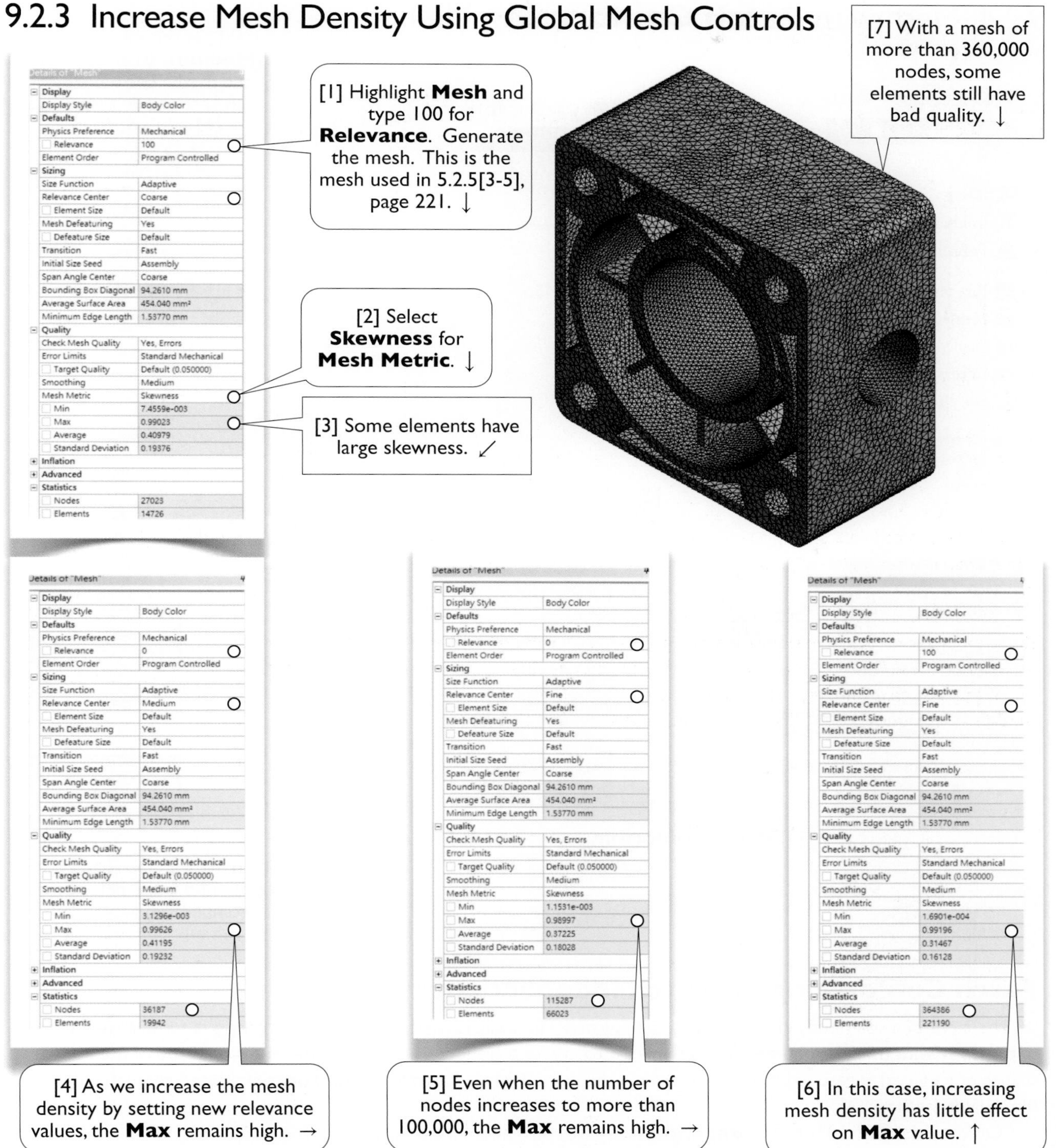

Increasing Mesh Density Is Not a Panacea

[8] The lesson we learned here is that increasing mesh density, although often reducing average skewness, is not a universal remedy for eliminating large skewness. We need to learn other meshing techniques. #

9.2.4 Mesh with **Patch Conforming** Method

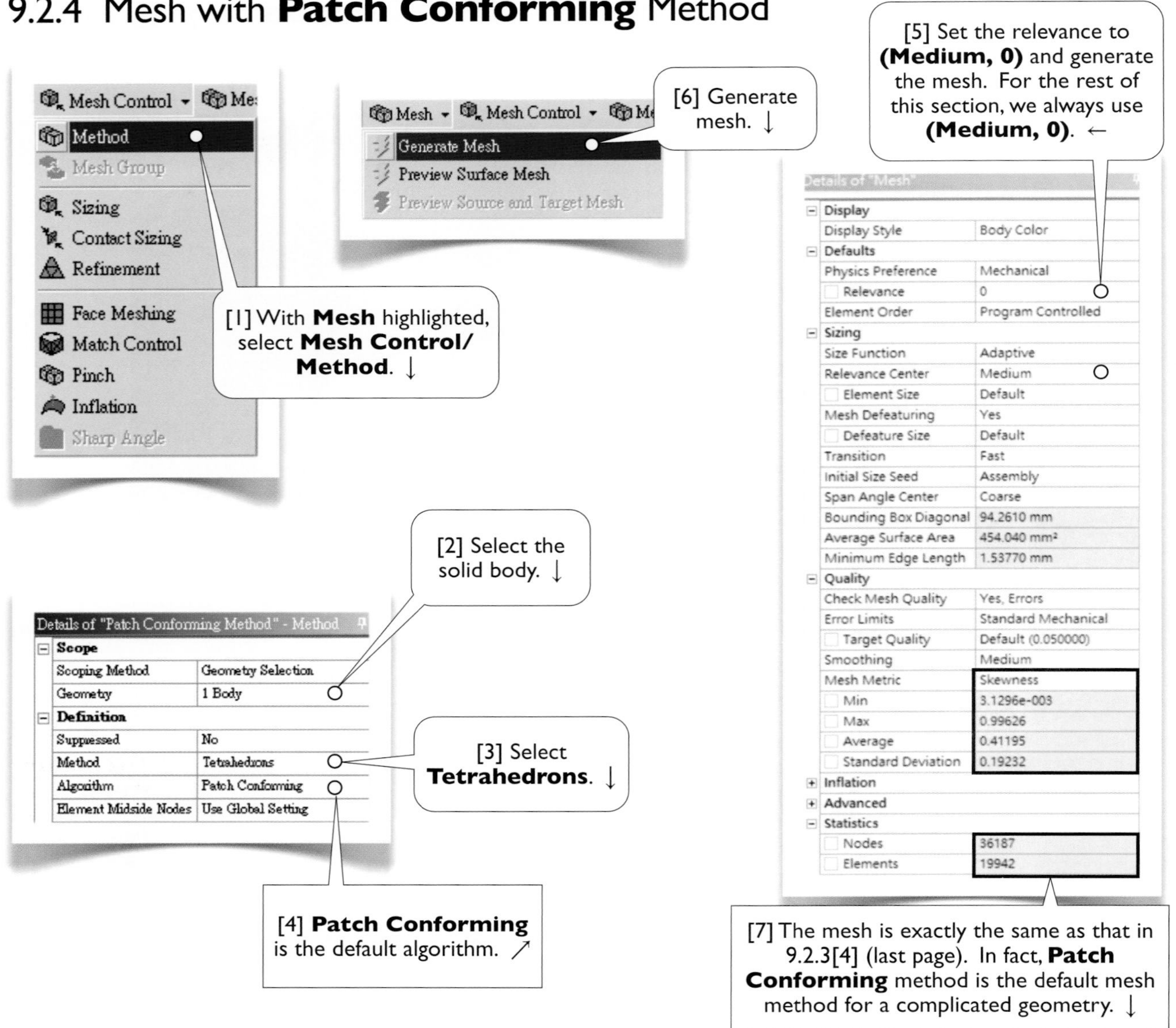

Patch Conforming and **Patch Independent** Methods

[8] The faces of a solid body are also called the *patches*. The basic idea of **Patch Conforming** is to mesh all the faces of the body with triangles and then "grow" inward to create tetrahedra. In this way, the exterior shape of the body (i.e., shapes of its faces) is respected (preserved), thus the name **Patch Conforming**. For complicated geometry, this is the default method.

On the other hand, **Patch Independent** creates tetrahedra from inside out. The outermost nodes are then projected onto the boundary faces and the element edges are created. In this way, the mesh's exterior shape may deviate from the original geometry, thus the name **Patch Independent**.

In some cases, when too many details exist that cause meshing difficulty, we may resort to **Patch Independent** algorithm and ignore these details. However, it is your responsibility to make sure that ignoring those details wouldn't distort the geometry too much. #

9.2.5 Mesh with **Patch Independent** Method

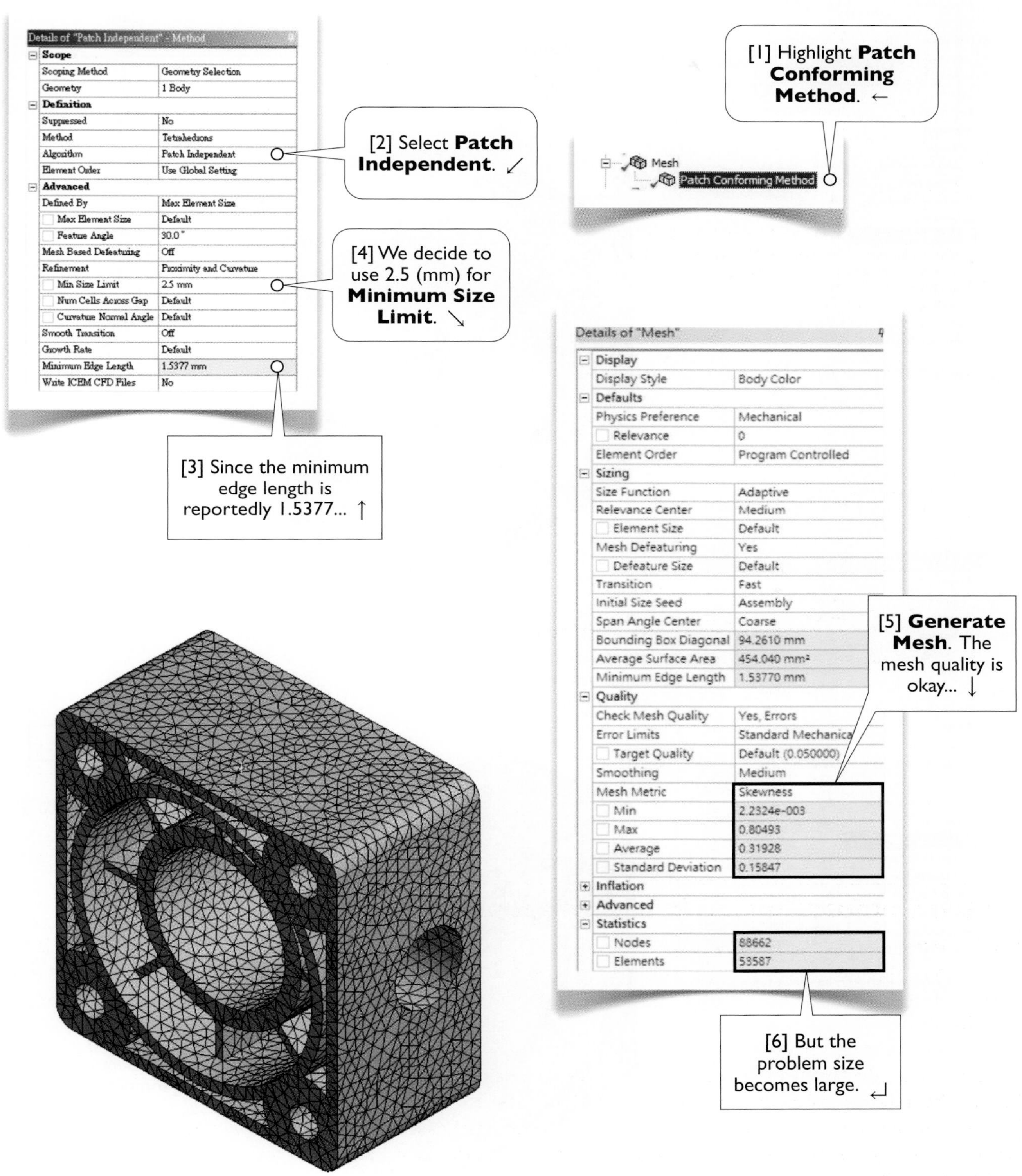

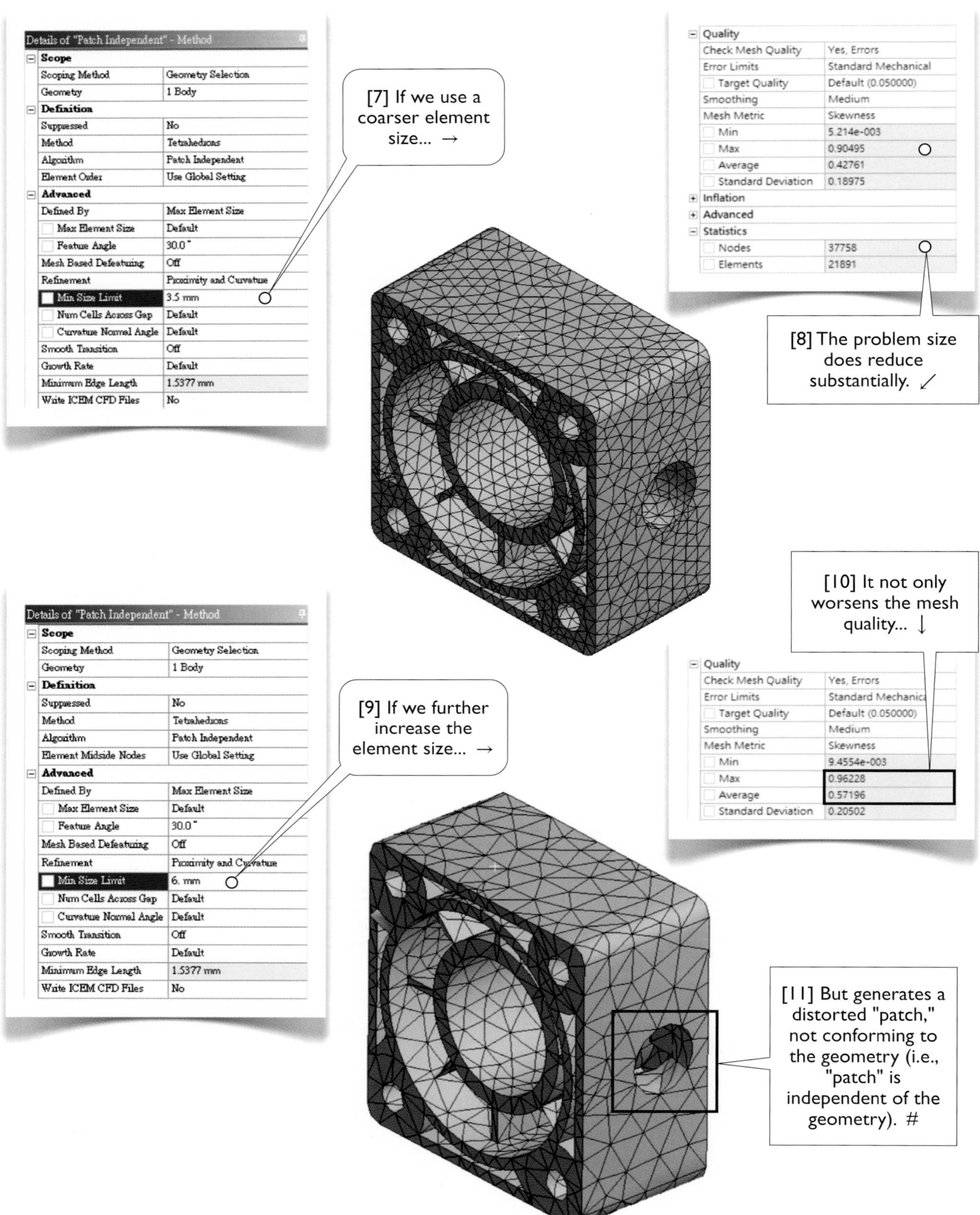
Details of "Patch Independent" - Method
Scope
Scoping Method | Geometry Selection
Geometry | 1 Body
Definition
Suppressed | No
Method | Tetrahedrons
Algorithm | Patch Independent
Element Order | Use Global Setting
Advanced
Defined By | Max Element Size
Max Element Size | Default
Feature Angle | 30.0°
Mesh Based Defeaturing | Off
Refinement | Proximity and Curvature
Min Size Limit | 3.5 mm
Num Cells Across Gap | Default
Curvature Normal Angle | Default
Smooth Transition | Off
Growth Rate | Default
Minimum Edge Length | 1.5377 mm
Write ICEM CFD Files | No
[7] If we use a coarser element size... →
Quality
Check Mesh Quality | Yes, Errors
Error Limits | Standard Mechanical
Target Quality | Default (0.050000)
Smoothing | Medium
Mesh Metric | Skewness
Min | 5.214e-003
Max | 0.90495
Average | 0.42761
Standard Deviation | 0.18975
Inflation
Advanced
Statistics
Nodes | 37758
Elements | 21891
[8] The problem size does reduce substantially. ↙
Details of "Patch Independent" - Method
Scope
Scoping Method | Geometry Selection
Geometry | 1 Body
Definition
Suppressed | No
Method | Tetrahedrons
Algorithm | Patch Independent
Element Midside Nodes | Use Global Setting
Advanced
Defined By | Max Element Size
Max Element Size | Default
Feature Angle | 30.0°
Mesh Based Defeaturing | Off
Refinement | Proximity and Curvature
Min Size Limit | 6. mm
Num Cells Across Gap | Default
Curvature Normal Angle | Default
Smooth Transition | Off
Growth Rate | Default
Minimum Edge Length | 1.5377 mm
Write ICEM CFD Files | No
[9] If we further increase the element size... →
[10] It not only worsens the mesh quality... ↓
Quality
Check Mesh Quality | Yes, Errors
Error Limits | Standard Mechanical
Target Quality | Default (0.050000)
Smoothing | Medium
Mesh Metric | Skewness
Min | 9.4554e-003
Max | 0.96228
Average | 0.57196
Standard Deviation | 0.20502
[11] But generates a distorted "patch," not conforming to the geometry (i.e., "patch" is independent of the geometry). #

9.2.6 Mesh with **Hex Dominant** Method

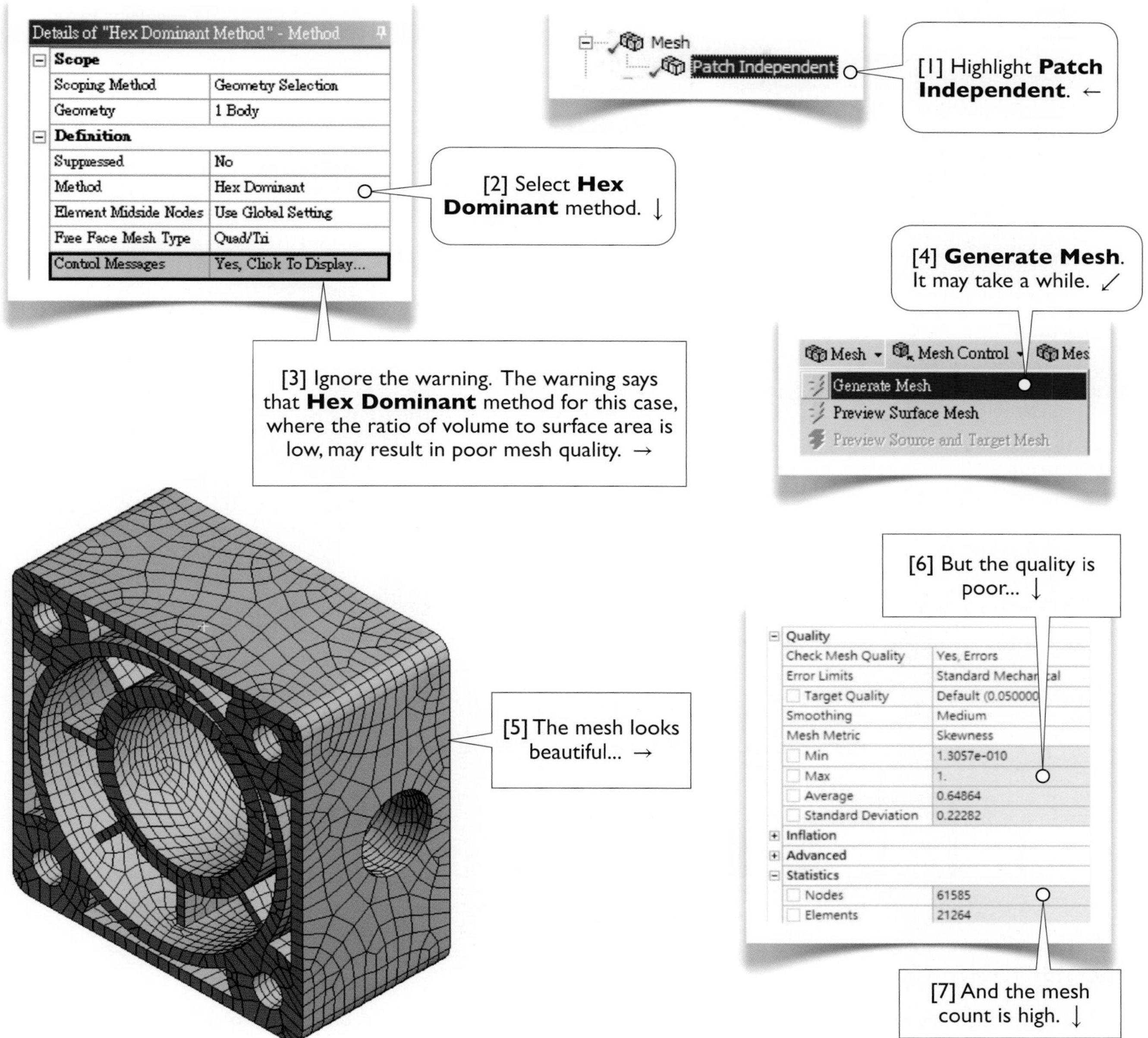

Hex Dominant Method

[8] An idea of **Hex Dominant** is to mesh the body with **Patch Conforming** first and then combine tetrahedra to form hexahedra: two tetrahedra form a hexahedron. It usually leaves some tetrahedra that cannot be combined to form hexahedra, thus the name **Hex Dominant**. After forming hexahedra, the algorithm tries to adjust the nodes to improve the mesh quality further.

Note that, **Hex Dominant** method, by its nature, is a method of patch conforming; that is, the faces are not distorted. In fact, all methods except **Patch Independent** are patch conforming. #

9.2.7 Mesh with **MultiZone** Method

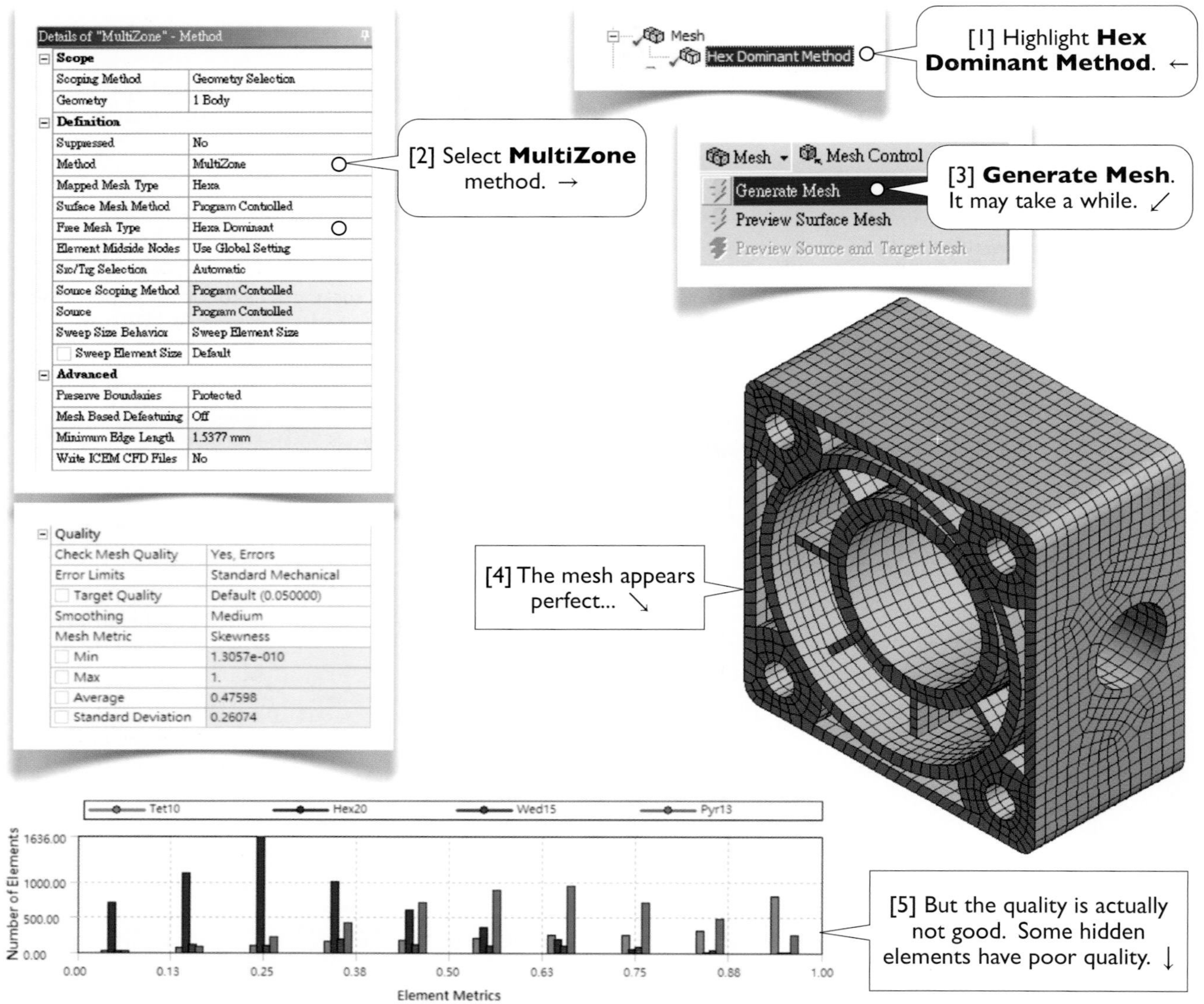

MultiZone Method

[6] As mentioned, the idea of **MultiZone** method is to decompose a non-sweepable body into several sweepable bodies, and then apply **Sweep** method on each body. The selection of source faces can be automatic or manual. In this case, you may try to select source faces manually.

Workbench is often smart enough to decompose the body into sweepable bodies. But, since these bodies are an integral part, the boundaries between the decomposed bodies must be conformal (i.e, the boundaries must have the same surface mesh); these constraints may complicate the meshing task. ↓

Wrap Up

[7] Save the project and exit Workbench. #

Section 9.3

Convergence Study of 3D Solid Elements

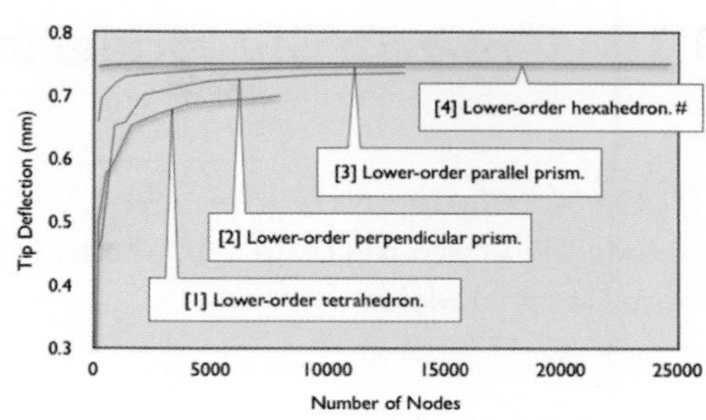

The main purpose of this section is to study 3D solid elements convergence behavior. A secondary purpose is to serve as an exercise for mesh controls techniques. A cantilever beam of rectangular cross section is used for these purposes. The conclusions drawn from the convergence study are crucial for CAE engineers. This section can be viewed as a sequel of PART C of Section 3.5 (pages 161-163).

9.3.1 About the Cantilever Beam

[1] The cantilever beam is made of steel and of size 100 mm x 10 mm x 10 mm; a uniform load of 1 MPa is applied on the upper face of the beam [2-4]. Convergence of three solid element shapes will be compared, namely hexahedron, prism, and tetrahedron (1.3.3[2-3, 5], page 38). ↘

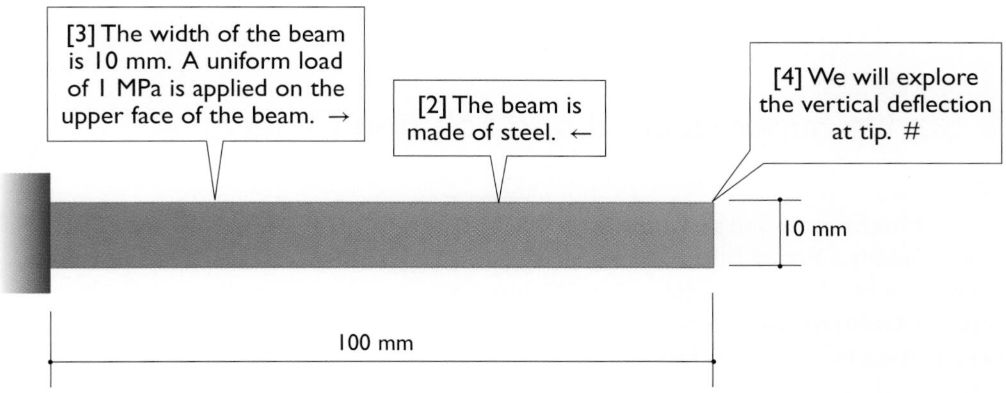

9.3.2 Start Up a New Project

[1] Launch Workbench. Create a **Static Structural** system. Save the project as **Cantilever**. Start up DesignModeler. Select **Millimeter** as the length unit. #

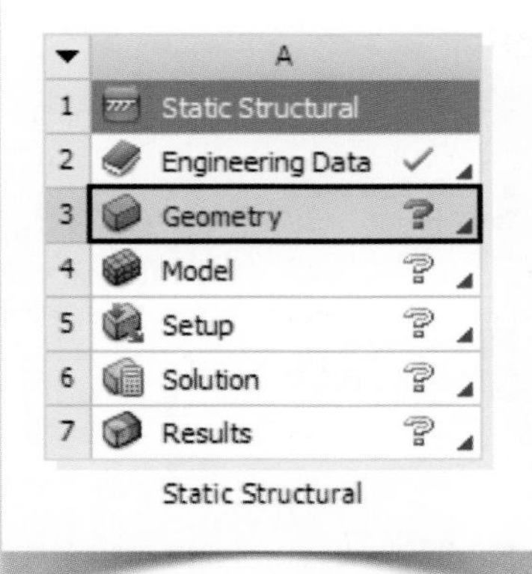

9.3.3 Create a 3D Model in DesignModeler

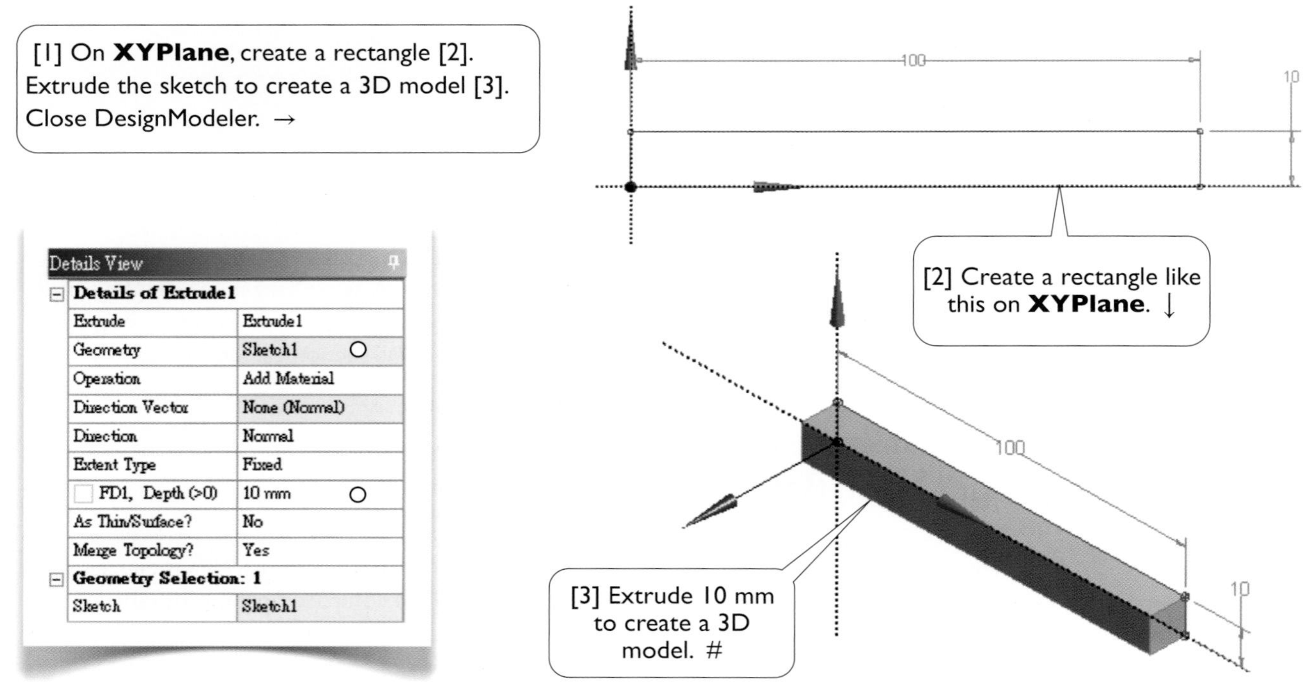

Details View	
Details of Extrude1	
Extrude	Extrude1
Geometry	Sketch1
Operation	Add Material
Direction Vector	None (Normal)
Direction	Normal
Extent Type	Fixed
FD1, Depth (>0)	10 mm
As Thin/Surface?	No
Merge Topology?	Yes
Geometry Selection: 1	
Sketch	Sketch1

9.3.4 Set Up Support, Load, and Solution Objects

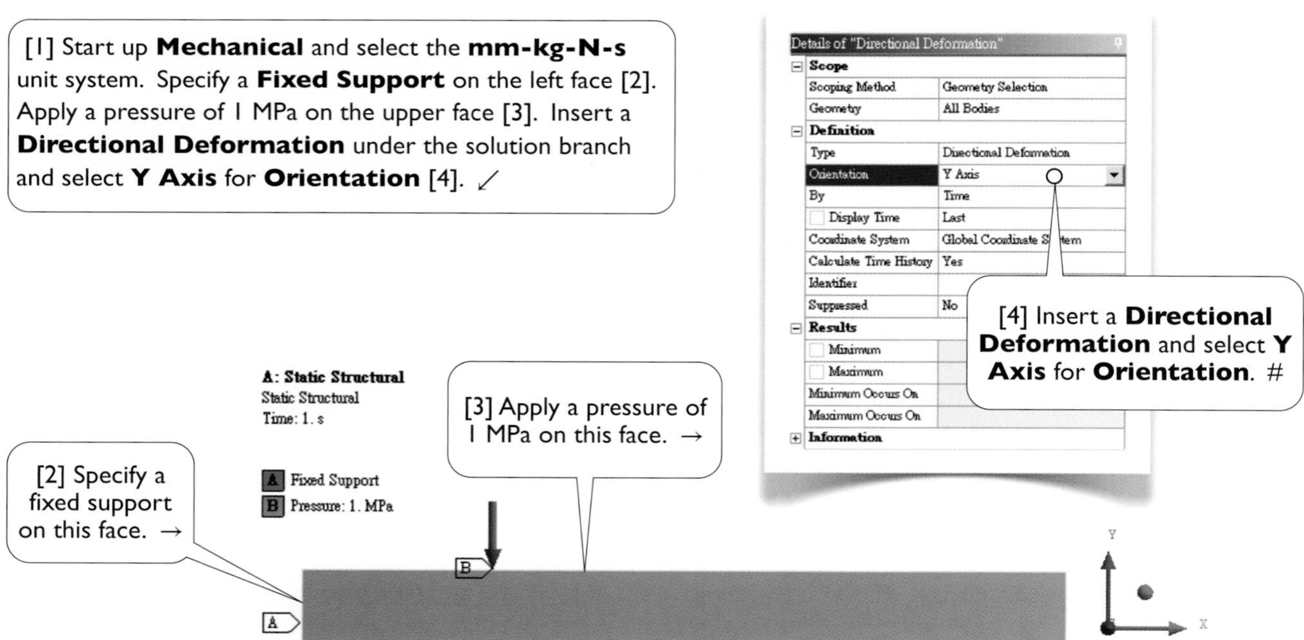

Details of "Directional Deformation"	
Scope	
Scoping Method	Geometry Selection
Geometry	All Bodies
Definition	
Type	Directional Deformation
Orientation	Y Axis
By	Time
Display Time	Last
Coordinate System	Global Coordinate System
Calculate Time History	Yes
Identifier	
Suppressed	No
Results	
Minimum	
Maximum	
Minimum Occurs On	
Maximum Occurs On	
Information	

9.3.5 Lower-Order Hexahedra

[1] For a model of such a regular geometry, the default mesh control settings will generate an all-hexahedra mesh of higher-order elements [2]. To generate lower-order hexahedra, select **Linear** for **Element Order** [3]. For each run, change the element size [4]. Resulting tip deflections are recorded in the table below. The convergence curve is shown in [5]. →

Element Size (mm)	Number of Nodes	Tip Deflection (mm)
5	189	0.74571
4	416	0.74693
3	875	0.74850
2	1836	0.74980
1.5	4352	0.75048
1.3	6318	0.75072
1.2	8500	0.75086
1	12221	0.75106
0.9	19097	0.75120
0.8	24696	0.75129

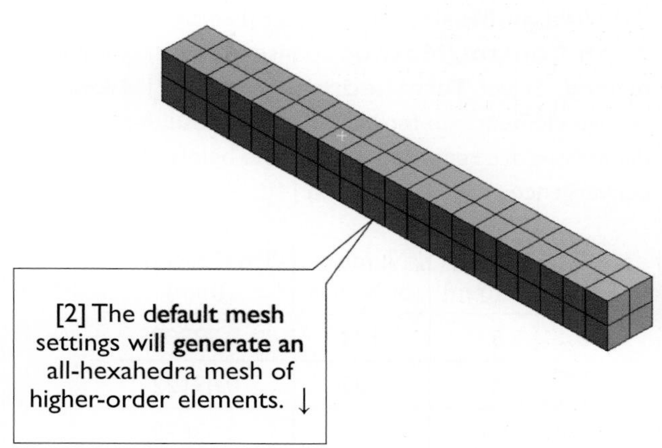

[2] The **default mesh** settings will **generate an** all-hexahedra mesh of higher-order elements. ↓

[3] Select **Linear** for **Element Order.** ↓

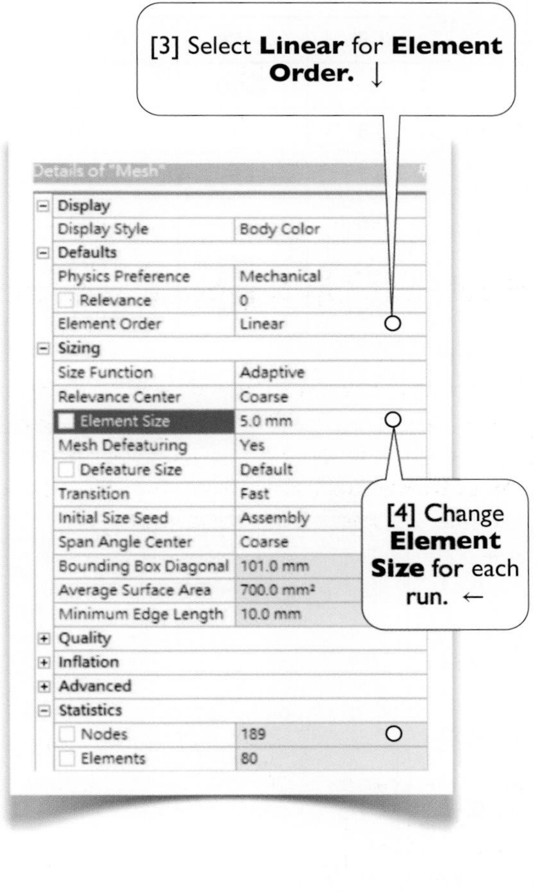

[4] Change **Element Size** for each run. ←

[5] Convergence curve for lower-order hexahedra. #

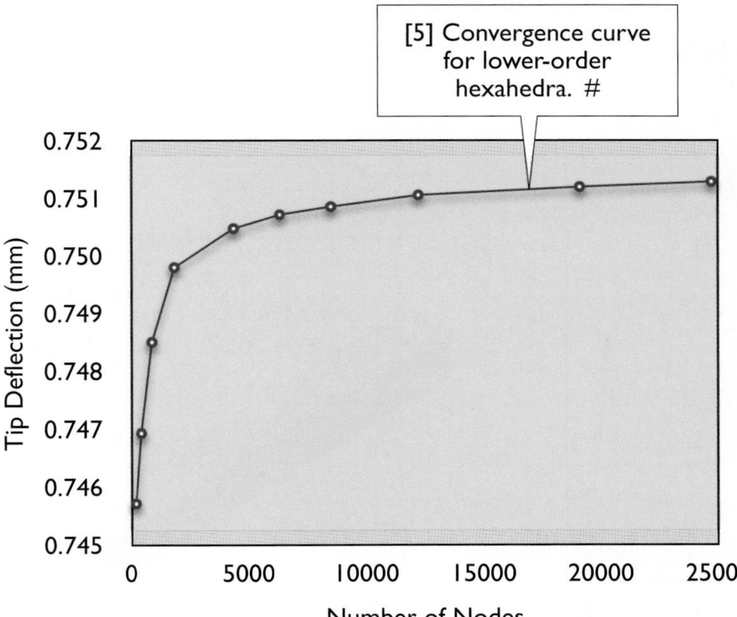

9.3.6 Lower-Order Tetrahedra

[1] Highlight **Mesh** in the project tree and select **Mesh Control/Method** to insert a mesh control method. Select **Tetrahedrons** method [2]. Now, change element size for each run [3]. Resulting tip deflections are recorded in the table below. The convergence curve is shown in [4]. →

Element Size (mm)	Number of Nodes	Tip Deflection (mm)
5	103	0.27809
4	183	0.49122
2.5	345	0.53698
2	527	0.58908
1.5	747	0.59064
1.2	1586	0.65228
1	2020	0.66045
0.8	2740	0.67014
0.7	4060	0.68632
0.6	5157	0.68912
0.55	6583	0.69492
0.5	7998	0.69706

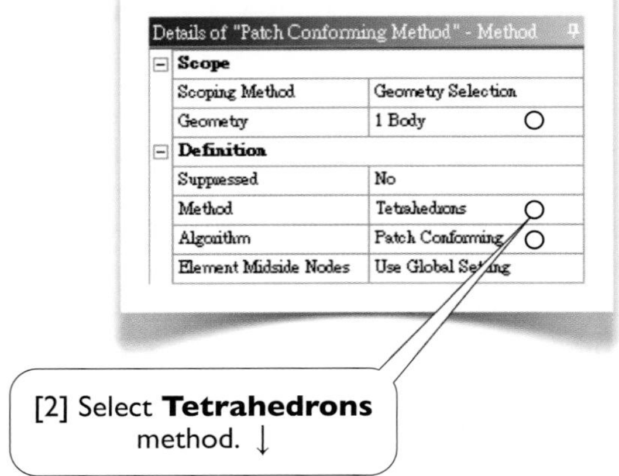
Details of "Patch Conforming Method" - Method

Scope	
Scoping Method	Geometry Selection
Geometry	1 Body
Definition	
Suppressed	No
Method	Tetrahedrons
Algorithm	Patch Conforming
Element Midside Nodes	Use Global Setting

[2] Select **Tetrahedrons** method. ↓

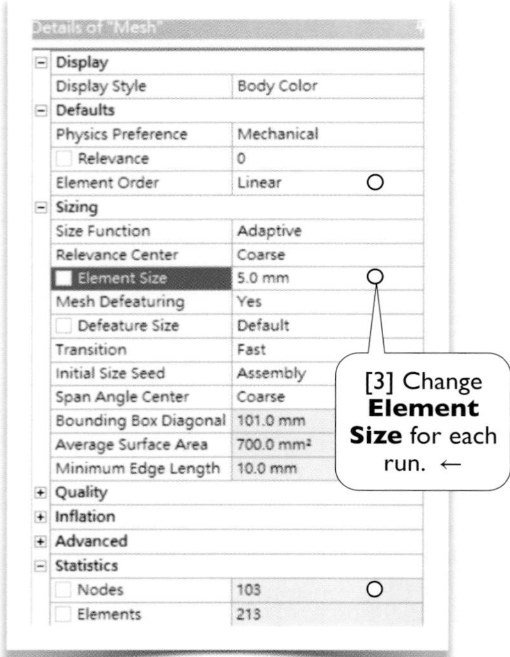
Details of "Mesh"

Display	
Display Style	Body Color
Defaults	
Physics Preference	Mechanical
Relevance	0
Element Order	Linear
Sizing	
Size Function	Adaptive
Relevance Center	Coarse
Element Size	5.0 mm
Mesh Defeaturing	Yes
Defeature Size	Default
Transition	Fast
Initial Size Seed	Assembly
Span Angle Center	Coarse
Bounding Box Diagonal	101.0 mm
Average Surface Area	700.0 mm²
Minimum Edge Length	10.0 mm
Quality	
Inflation	
Advanced	
Statistics	
Nodes	103
Elements	213

[3] Change **Element Size** for each run. ←

[4] Convergence curve for lower-order tetrahedra. #

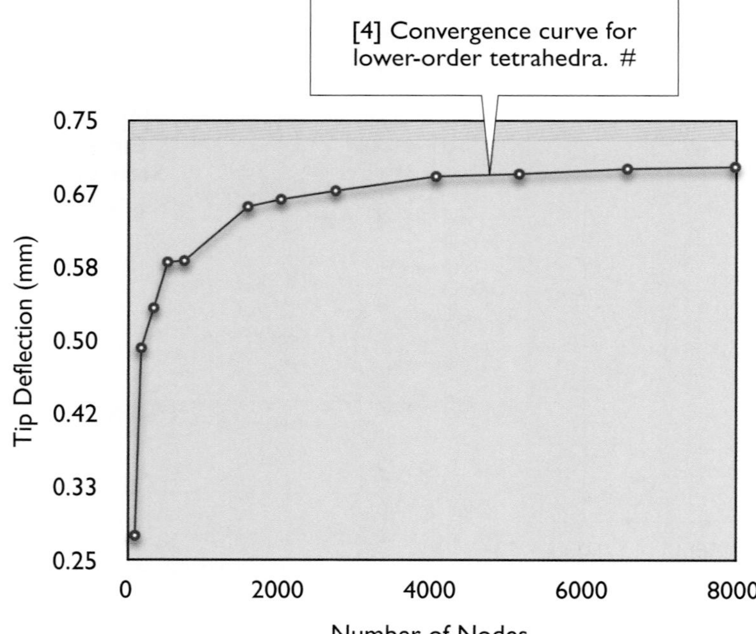

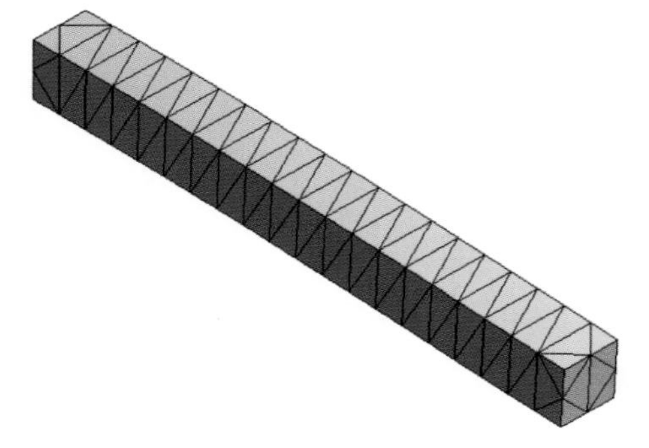

9.3.7 Lower-Order Prisms (Parallel to Loading Direction)

[1] Highlight **Patch Conforming Method**. In the details view, change to **Sweep** method [2]. Set up the source face [3-4]. For each run, change both sweep element size [5] and global element size [6]. Resulting tip deflections are recorded in the table below. The convergence curve is shown in [7].

Note that the prisms are oriented such that their heights are parallel to the loading (bending) direction. We will refer to the elements oriented in this way as "parallel prisms" for the rest of this section. We will show in 9.3.8 (next page) that the convergence curve will be different if the prisms are oriented differently. →

Element Size (mm)	Number of Nodes	Tip Deflection (mm)
5	186	0.65779
4	324	0.69584
3	875	0.71844
2.4	1326	0.72969
2	2112	0.73344
1.5	4864	0.74156
1.2	9300	0.74510
1	13288	0.74676

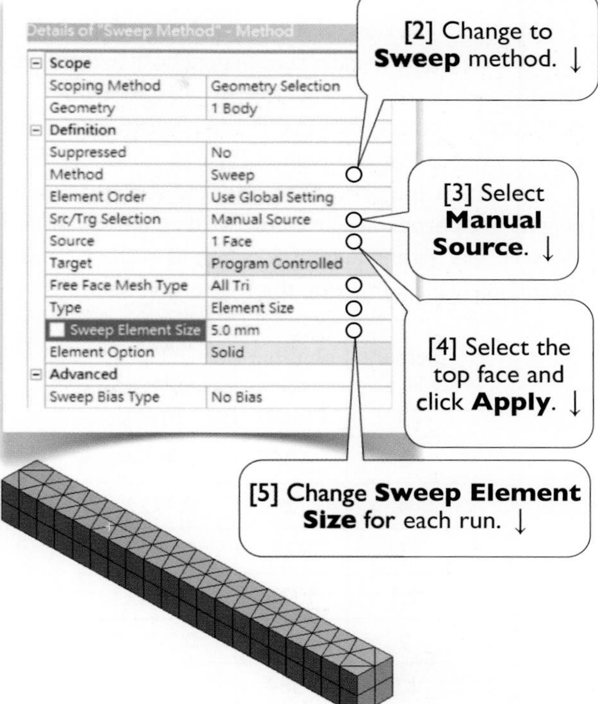

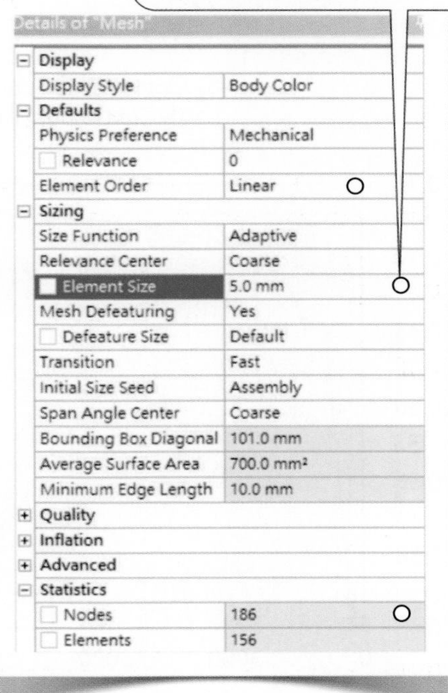

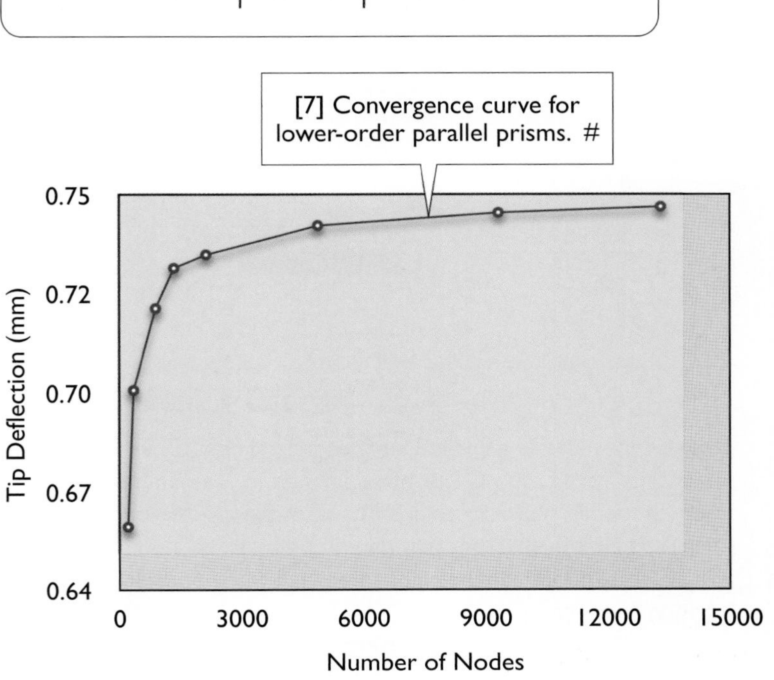

9.3.8 Lower-Order Prisms (Perpendicular to Loading Direction)

[1] In the details view of **Sweep Method**, re-select the source face [2]. For each run, change both sweep element size [3] and global element size [4]. Resulting tip deflections are recorded in the table below. The convergence curve is shown in [5].

Note that the prisms are oriented such that their heights are perpendicular to the loading (bending) direction. We will refer to the elements oriented in this way as "perpendicular prisms" for the rest of this section. →

Element Size (mm)	Number of Nodes	Tip Deflection (mm)
5	186	0.45546
4	324	0.47229
3	875	0.65078
2.4	1326	0.65622
2	2112	0.70027
1.5	4864	0.72260
1.2	9300	0.73279
1	13288	0.73582

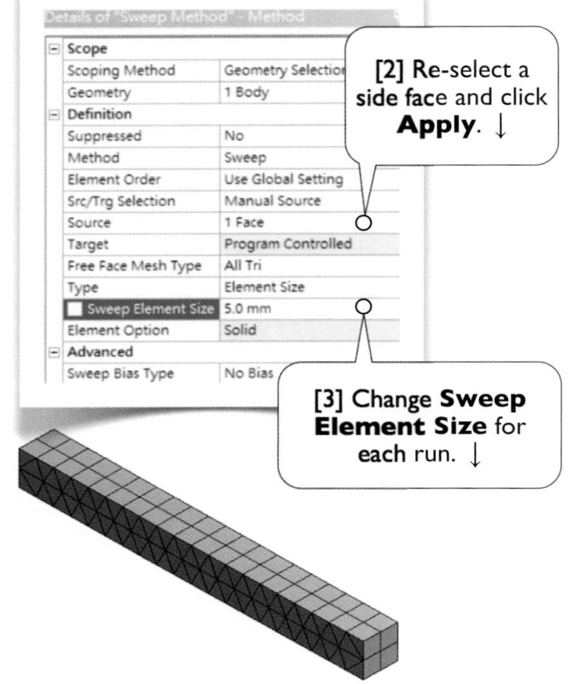

[4] Also change **Element Size** (using the same value as that in [3]) for each run. ←

Details of "Mesh"

Display	
Display Style	Body Color
Defaults	
Physics Preference	Mechanical
Relevance	0
Element Order	Linear
Sizing	
Size Function	Adaptive
Relevance Center	Coarse
Element Size	5.0 mm
Mesh Defeaturing	Yes
Defeature Size	Default
Transition	Fast
Initial Size Seed	Assembly
Span Angle Center	Coarse
Bounding Box Diagonal	101.0 mm
Average Surface Area	700.0 mm²
Minimum Edge Length	10.0 mm
Quality	
Inflation	
Advanced	
Statistics	
Nodes	186
Elements	156

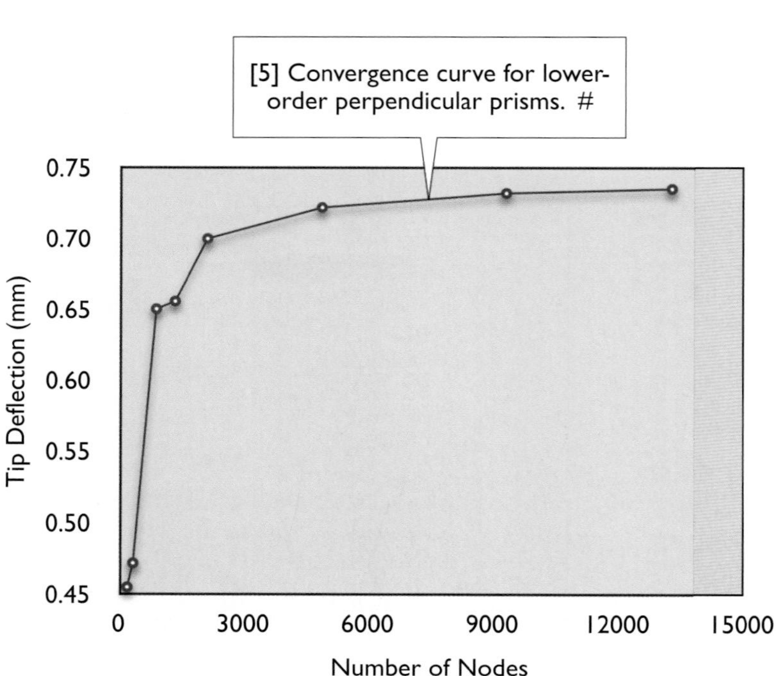

9.3.9 Higher-Order Hexahedra

[1] Delete **Sweep Method**. Repeat all the steps in 9.3.5 (page 355), using quadratic elements [2] and change the element sizes as shown below. →

Element Size (mm)	Number of Nodes	Tip Deflection (mm)
5	621	0.74899
4	1440	0.75011
3	3125	0.75071
2	6696	0.75108
1.5	16256	0.75136
1.3	23787	0.75145

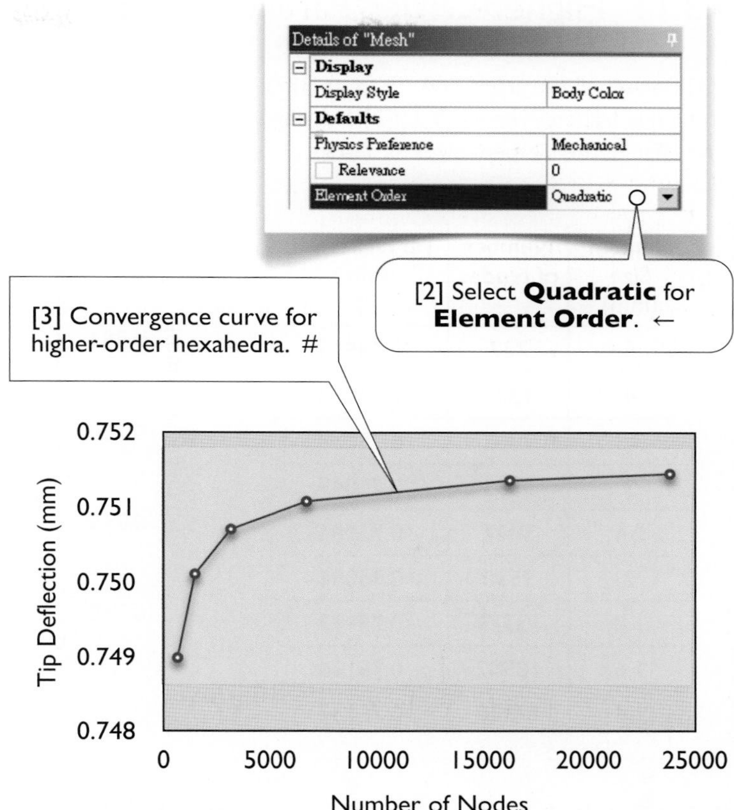

9.3.10 Higher-Order Tetrahedra

[1] Repeat all the steps in 9.3.6 (page 356), using quadratic elements (9.3.9[2], this page) and change the element sizes as shown below. →

Element Size (mm)	Number of Nodes	Tip Deflection (mm)
5	508	0.74377
4	934	0.74817
3	1393	0.74980
2.5	1912	0.75017
2	2935	0.75055
1.5	4171	0.75102
1.2	9536	0.75118
1	12027	0.75128
0.8	16430	0.75139
0.7	24877	0.75152

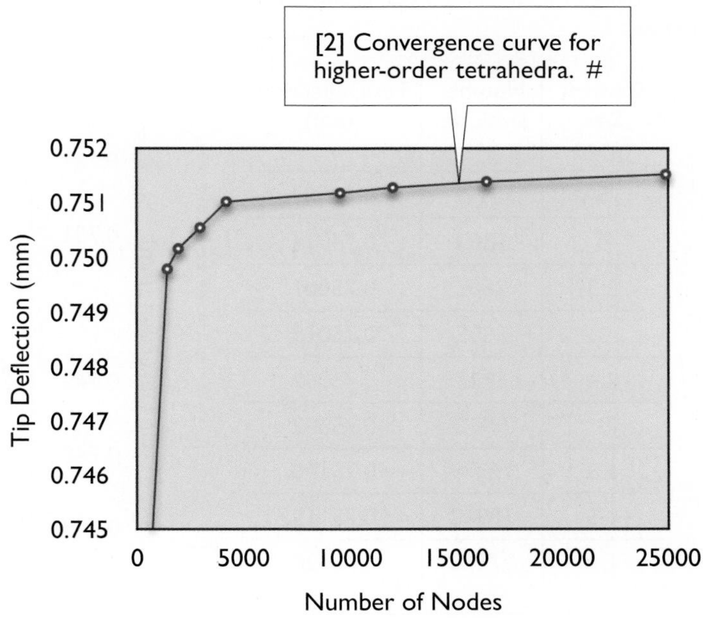

9.3.11 Higher-Order Parallel Prisms

[1] Repeat all the steps in 9.3.7 (page 357), using quadratic elements (9.3.9[2], last page) and change the element sizes as shown below. →

Element Size (mm)	Number of Nodes	Tip Deflection (mm)
5	727	0.74815
4	1303	0.74958
3.3	2803	0.75038
3	3805	0.75049
2.4	5827	0.75084
2	9530	0.75096
1.8	12779	0.75113
1.6	18902	0.75126
1.4	27529	0.75137

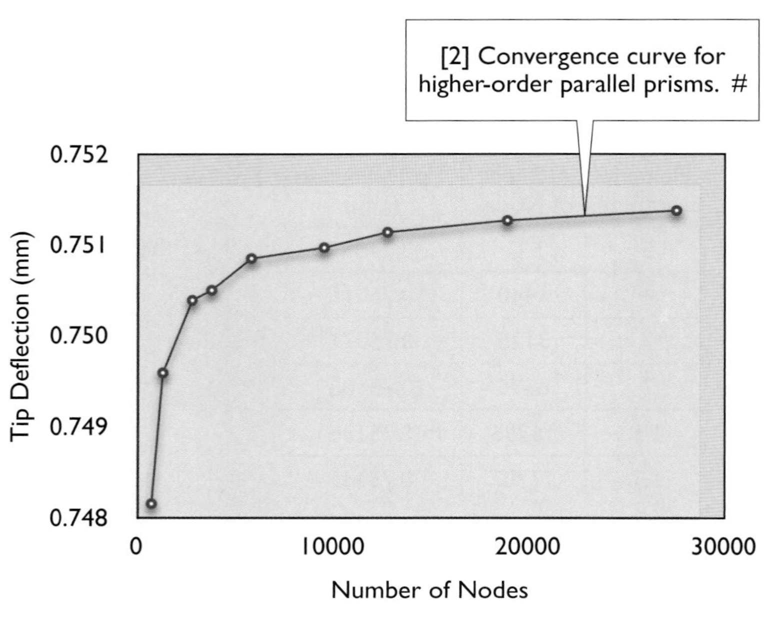

[2] Convergence curve for higher-order parallel prisms. #

9.3.12 Higher-Order Perpendicular Prisms

[1] Repeat all the steps in 9.3.8 (page 358), using quadratic elements (9.3.9[2], last page) and change the element sizes as shown below. →

Element Size (mm)	Number of Nodes	Tip Deflection (mm)
5	727	0.74767
4	1303	0.74914
3.3	2803	0.75001
3	3805	0.75018
2.4	5827	0.75062
2	9530	0.75082
1.8	12779	0.75100
1.6	18902	0.75115
1.4	27573	0.75128

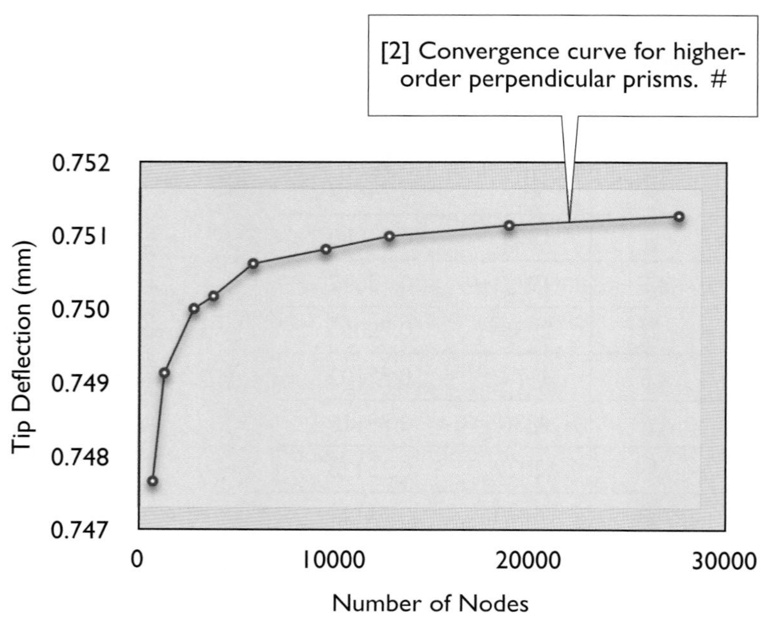

[2] Convergence curve for higher-order perpendicular prisms. #

9.3.13 Comparison: Lower-Order Elements

[1] The chart below is made by a collection of the convergence curves in 9.3.5 to 9.3.8 (pages 355-358) to compare the convergence behaviors of the lower-order elements. The order of the convergence speed is, from fast to slow, hexahedron, parallel prism, perpendicular prism, and tetrahedron. The differences among them are obvious. The lower-order tetrahedron converges so poorly that it is not practically useful. As a guideline, NEVER use lower-order tetrahedral elements. ↓

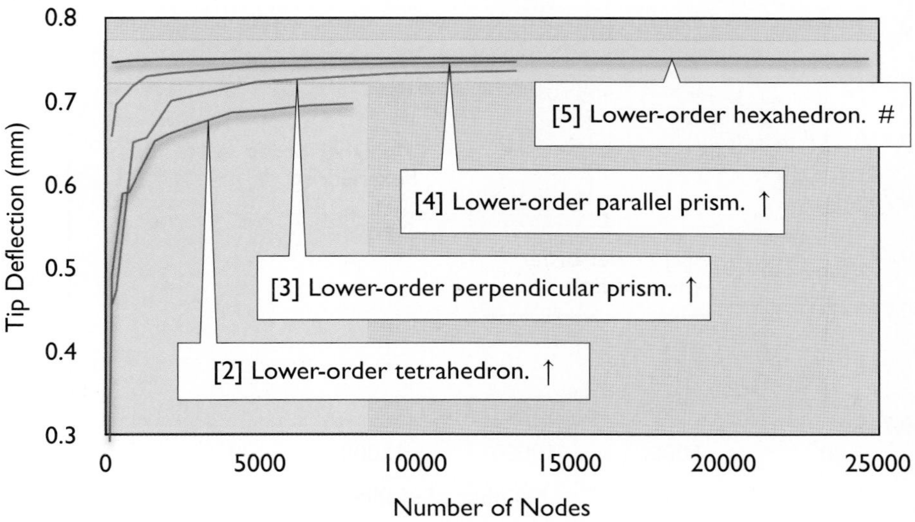

9.3.14 Comparison: Higher-Order Elements

[1] The chart below is made from a collection of the convergence curves in 9.3.9 to 9.3.12 (pages 359-360) to compare the convergence behaviors of the higher-order elements. The differences among them are not obvious but still distinguishable. In contrast to the lower-order tetrahedron, the higher-order tetrahedron is still practically useful as long as the mesh is fine enough. ↓

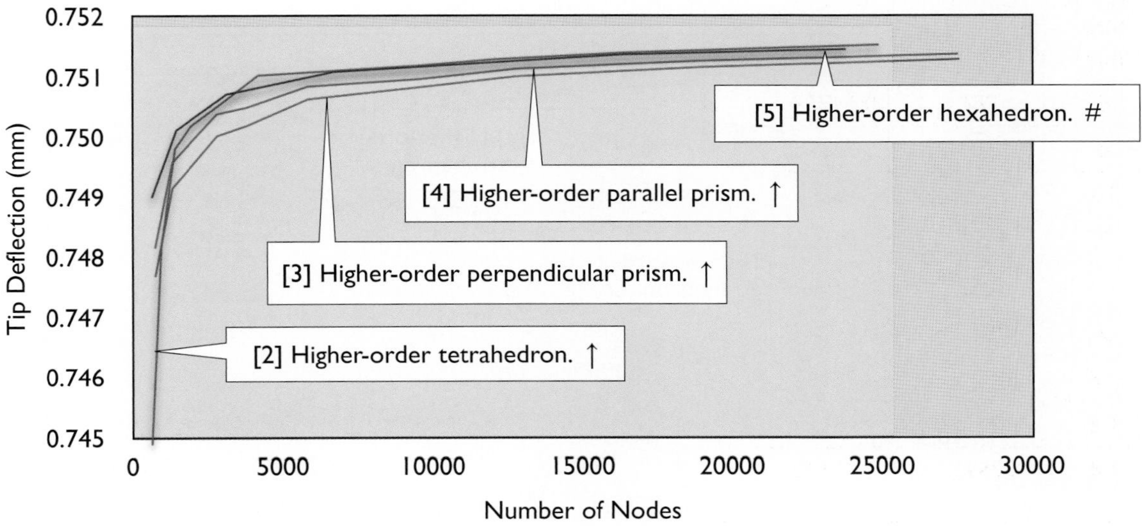

9.3.15 Comparison: Hexahedra

[1] The chart below is made by a collection of the convergence curves in 9.3.5 (page 355) and 9.3.9 (page 359) to compare the convergence behaviors between the lower- and higher-order hexahedra. It is obvious that the higher-order hexahedral element is better than the lower-order hexahedral, but the difference is not so dramatic as the tetrahedral element (see 9.3.16, this page). ↓

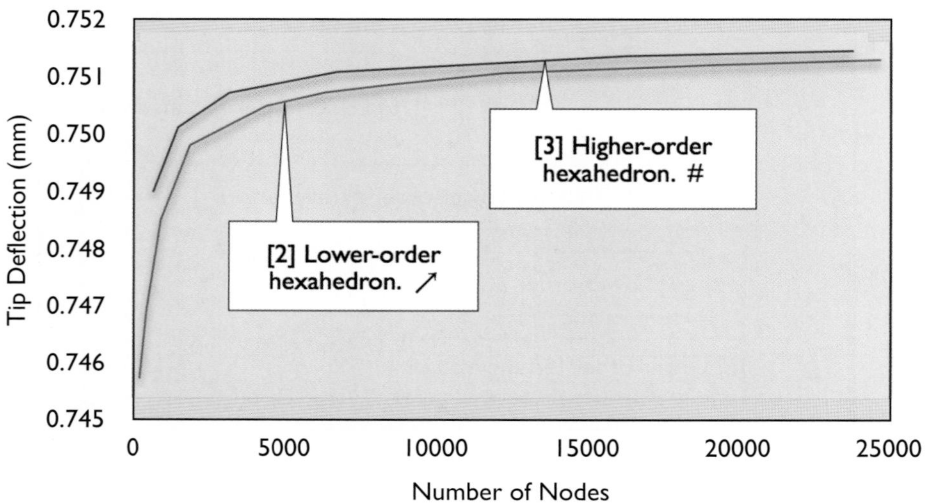

9.3.16 Comparison: Tetrahedra

[1] The chart below is made from a collection of the convergence curves in 9.3.6 (page 356) and 9.3.10 (page 359) to compare the convergence behaviors between the lower- and higher-order tetrahedra. It is obvious that higher-order element is much better than the lower-order one. Remember: NEVER use lower-order tetrahedral elements. ↓

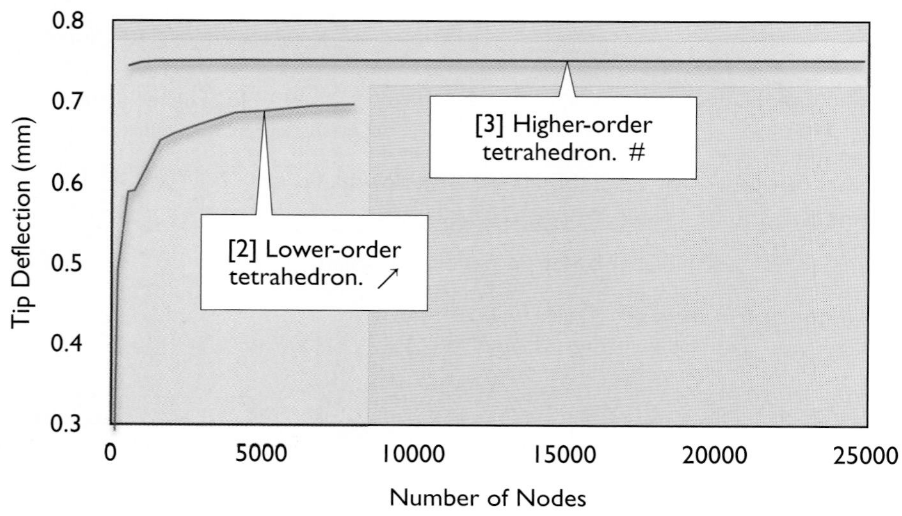

9.3.17 Comparison: Parallel Prisms

[1] The chart below is made by a collection of the convergence curves in 9.3.7 (page 357) and 9.3.11 (page 360) to compare the convergence behaviors between the lower- and higher-order parallel prisms. It is obvious that higher-order element is much better than the lower-order one. Like lower-order tetrahedral, lower-order prismatic elements are not recommended. ↓

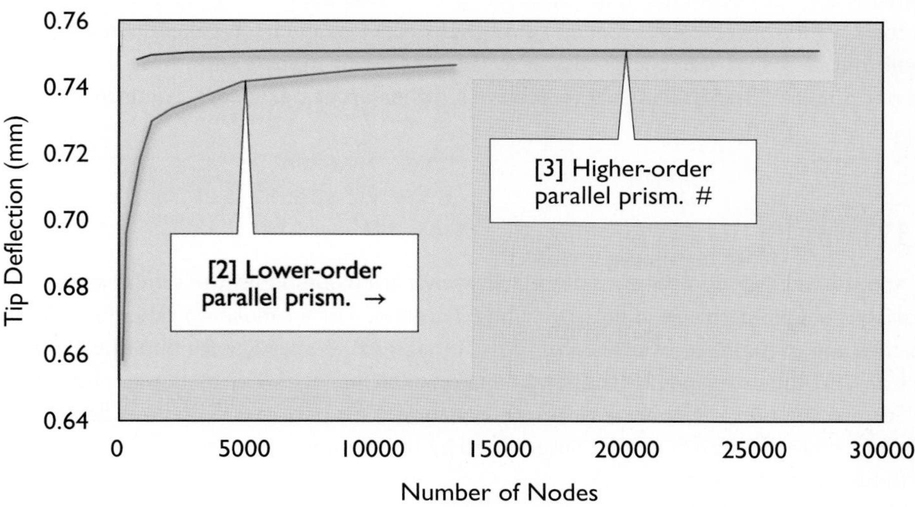

9.3.18 Comparison: Perpendicular Prisms

[1] The chart below is made by a collection of the convergence curves in 9.3.8 (page 358) and 9.3.12 (page 360) to compare the convergence behaviors between the lower- and higher-order perpendicular prisms. It is obvious that the higher-order element is much better than the lower-order one. Like lower-order tetrahedral, lower-order prismatic elements are not recommended. ↓

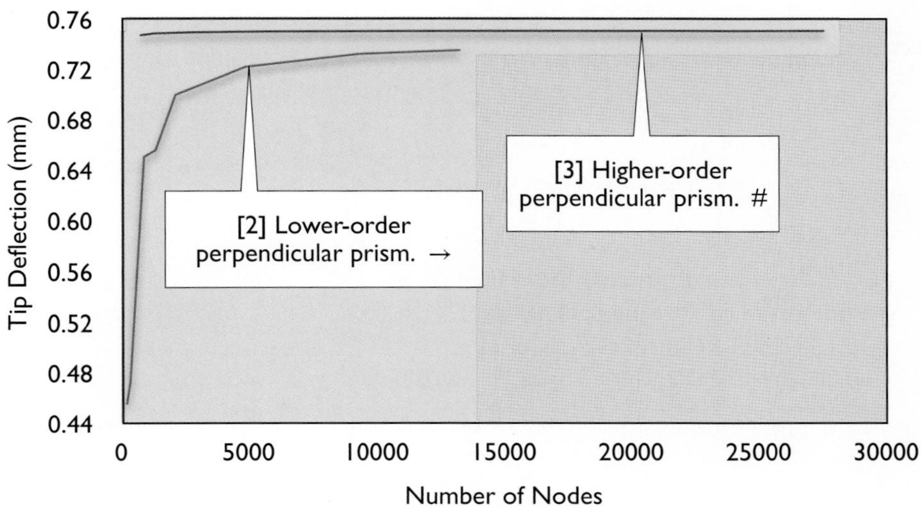

9.3.19 Summary and Guidelines

Summary and Guidelines

[1] Combining the observations in Section 3.5 and this section, we may summarize the conclusions as follows: (a) Never use lower-order tetrahedra or triangles. (b) Higher-order tetrahedra or triangles are as good as other elements as long as the mesh is fine enough. In cases of coarse mesh, however, they perform poorly and are not recommended. (c) Lower-order prisms are not recommended. (d) Lower-order hexahedra and quadrilaterals can be used, but they are not as efficient as their higher-order counterparts. (e) Higher-order hexahedra, parallel-prisms, and quadrilaterals are among the most efficient elements we have discussed so far. Mesh your models with these elements whenever possible. If that is not possible, then at least try to achieve a higher-order hexahedra-dominant or quadrilateral-dominant mesh. ↓

Remark: CPU Time

[2] In Section 3.5 and this section, comparisons among elements are made under the same number of nodes. More reasonable comparisons should be made under the same CPU time. For a simulation task, the CPU time consists of three parts. First, the time required to establish Eq. 1.3.1(1) (page 35). It may involve numerical integrations for each element. This part of CPU time depends on the total number of elements as well as the number of integration points of each element. Second, the time required to solve the equation. This part of CPU time is determined solely by the number of degrees of freedom, which is in turn determined by the number of nodes and the dimensionality (2D or 3D). Third, the others (housekeeping, overhead, etc).

For small problems, the overall CPU time is dominated by the third part. That is why we didn't use CPU time for comparison, since all cases are small when coarsely meshed. For large problems, the third part is negligible and the CPU time is essentially the sum of the first two parts.

Therefore, strictly speaking, our comparison was not perfectly accurate. Nevertheless, the discussions and conclusions in this section pretty much reflect the reality. These guidelines should be useful. ↓

Wrap Up

[3] Save the project and exit Workbench. #

Section 9.4

Review

9.4.1 Keywords

Choose a letter for each keyword from the list of descriptions

1. ( ) Convergence Criteria
2. ( ) Displacement Convergence Criterion
3. ( ) Force Convergence Criterion
4. ( ) Hex Dominant Method
5. ( ) Parallel Prisms
6. ( ) Patch Conforming Method
7. ( ) Patch Independent Method
8. ( ) Perpendicular Prisms
9. ( ) Skewness
10. ( ) Sweep Thin Method

Answers:

1. (B) 2. (C) 3. (D) 4. (H) 5. (I) 6. (F) 7. (G) 8. (J) 9. (A) 10. (E)

List of Descriptions

(A) A measure of mesh quality, calculated for each element according to its geometry. Its value ranges from 0 to 1, the smaller the better. Elements of skewness of more than 0.95 are considered unacceptable.

(B) In nonlinear simulation, the loading is divided into substeps and applied substep by substep. By default, a substep is said to be complete when both displacement convergence criterion and force convergence criterion are met during the iterations.

(C) During the iterations of a substep of a nonlinear simulation, the displacement convergence criterion is met when the increment of displacement is less than a criterion, which is, by default, 0.5% of maximum displacement.

(D) During the iterations of a substep of a nonlinear simulation, the force convergence criterion is met when the unbalanced force is less than a criterion, which is, by default, 0.5% of applied force.

(E) **Sweep** mesh control method can be classified into **Sweep** and **Sweep Thin**. **Sweep** allows a more complex sweeping path while **Sweep Thin** allows only a simple sweeping path. The advantage of **Sweep Thin** is that it allows multiple faces as source or target while **Sweep** allows only one face for both source and target.

(F) A mesh control method. It meshes all the faces of the body with triangles; the triangles then "grow" inward to create tetrahedra. In this way, the shapes of the faces are respected (preserved).

(G) A mesh control method. It creates tetrahedra from inside out. The outermost nodes are then projected onto the boundary faces and the element edges are created. In this way, the mesh's outline may be different from the original geometry.

(H) A mesh control method. It meshes a body with Patch Conforming method first and then combines tetrahedra to form hexahedra. It usually leaves some tetrahedra that cannot be combined to form hexahedra.

(I) When a body is meshed with prismatic elements and the prisms are oriented such that their heights are parallel to the bending direction, the prismatic elements oriented in this way are referred to as parallel prisms. (Note: this term is used only in this book.)

(J) When a body is meshed with prismatic elements and the prisms are oriented such that their heights are perpendicular to the bending direction, the prismatic elements oriented in this way are referred to as perpendicular prisms. (Note: this term is used only in this book.)

9.4.2 Additional Workbench Exercises

Convergence Study for Higher-Order 2D Elements

In Section 9.3, we study the convergence of 3D elements, both higher-order and lower-order elements. In Section 3.5, we study the convergence of 2D elements only for the lower-order elements. We haven't studied the higher-order 2D elements yet. Conduct a study of the higher-order 2D elements.

Chapter 10

Buckling and Stress Stiffening

Functionality, safety, and reliability are the main purposes of structural simulations. Stresses usually relate to safety and reliability. In the 3D truss example (Section 7.2), calculated stresses are well below the material's yield strength (7.2.12[4], page 292). Can we conclude that the design is safe? Not yet. For any structural members (particularly slender or thin members) subject to compressive stresses, we need to check their stability before concluding their safety. This chapter mainly discusses *stability analysis*, or *buckling analysis*.

Buckling can be viewed as an ultimate case of a more general effect, called *stress stiffening*: a structure member's bending stiffness increases with increasing axial tensile stress, and, on the other hand, the member's bending stiffness decreases with the increasing compressive stress. Buckling occurs when the compressive stress reaches a level such that the bending stiffness reduces to zero; in that situation, the applying load is called a *buckling load* and the corresponding deforming shape is called a *buckling mode*. The purpose of buckling analyses is to find the buckling loads and the corresponding *buckling mode*.

Purpose of This Chapter

The main purpose of this chapter is to introduce *linear buckling analysis* (also known as *eigenvalue buckling analysis*). Since buckling can be viewed as an ultimate case of stress stiffening, the discussion will start with a thorough understanding of stress stiffening. As usual, the concepts are introduced using step-by-step exercises.

About Each Section

Section 10.1 introduces the stress stiffening effects, using a simply supported beam as an example. The results of these nonlinear analyses can be used to predict the buckling load using an extrapolation method; this procedure is called a *nonlinear buckling analysis*. A linear buckling analysis is then carried out to find the buckling loads and buckling modes. Compared with nonlinear buckling analyses, linear buckling analysis tends to overestimate the buckling load.

Section 10.2 performs a linear buckling analysis on the 3D truss structure introduced in Section 7.2. Section 10.3 carries out a linear buckling analysis on the beam bracket which has been discussed in Sections 4.1, 5.1, and 6.2.

Section 10.1

Stress Stiffening

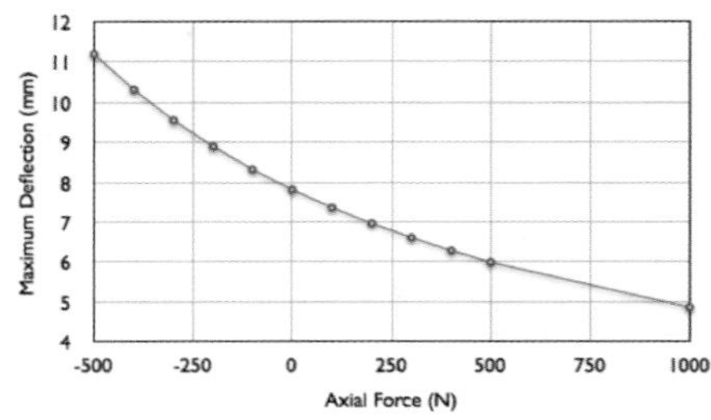

This section introduces *stress stiffening effect*, which is closely related to buckling. Stress stiffening effect is often observed in slender or thin structural members, such as cables, shells, columns, walls, towers, trusses, etc. In a slender structural member, such as a column, the bending stiffness is affected by its axial stress. In a thin structure member, such as a wall, the bending stiffness is affected by its in-plane stress. More specifically, when subject to axial (or in-plane) tension, the bending stiffness tends to increase, while when subject to axial (or in-plane) compression, the bending stiffness tends to decrease.

In this section, we will use a simply supported slender beam to demonstrate the stress stiffening effects.

10.1.1 About the Simply Supported Beam

[1] Consider a simply supported beam shown below [2-4]. The beam is made of steel and has a uniform cross section of 10 mm x 10 mm [2]. A uniformly distributed load of 0.1 N/mm is applied downward on the beam [3]. An axial force is applied at the beam's end which is free to move horizontally [4].

The vertical load, which causes bending, is a constant 0.1 N/mm, while the horizontal force *P* will change from -500 N to 1000 N. Note that the negative *P* produces a compressive axial stress and the positive *P* produces a tensile axial stress. We will examine the maximum vertical deflection occurring at the middle of the span.

Let δ_0 be the beam's middle-span deflection when $P = 0$. Then, when the beam is subject to a positive *P*, we will obtain a deflection less than δ_0. We can then conclude that the bending stiffness increases with the increasing tensile axial stress. On the other hand, when the beam is subject to a negative (compressive) *P*, we will obtain a deflection larger than δ_0. We can then conclude that the bending stiffness decreases with the increasing compressive axial stress. This effect is called the *stress stiffening*.

Since the bending stiffness decreases with the increasing compressive axial stress, you may raise a question: How large the compressive force *P* will cause the bending stiffness to reduce to zero? A zero bending stiffness implies an unstable structure, because a small lateral load would cause an infinitely large deflection. This phenomenon is called the *buckling*. ↓

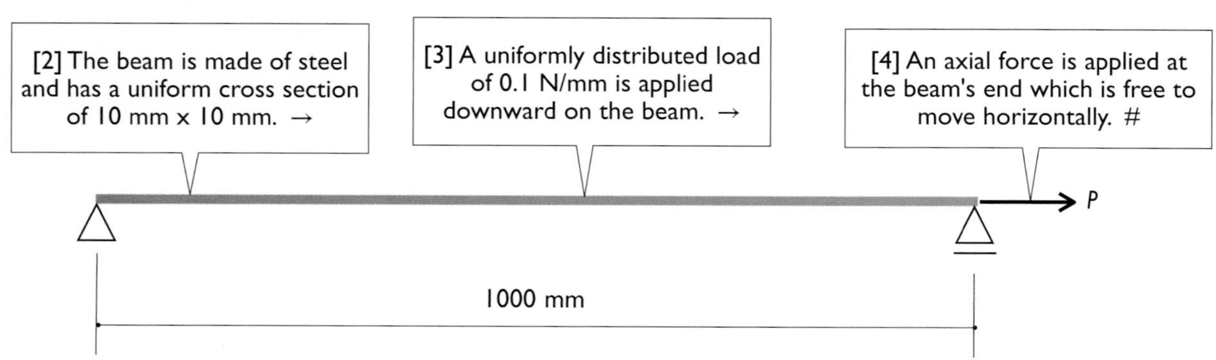

10.1.2 Start a New Project

[1] Launch Workbench. Create a **Static Structural** system. Save the project as **SimpleBeam**. Start up DesignModeler. #

A
1 Static Structural
2 Engineering Data
3 Geometry
4 Model
5 Setup
6 Solution
7 Results
Static Structural

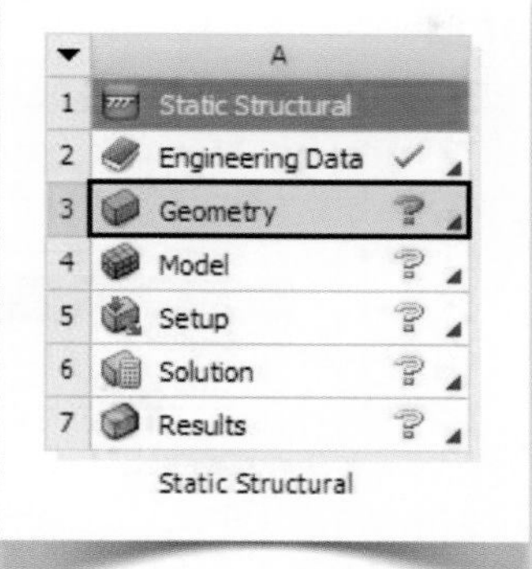

10.1.3 Create a Line in DesignModeler

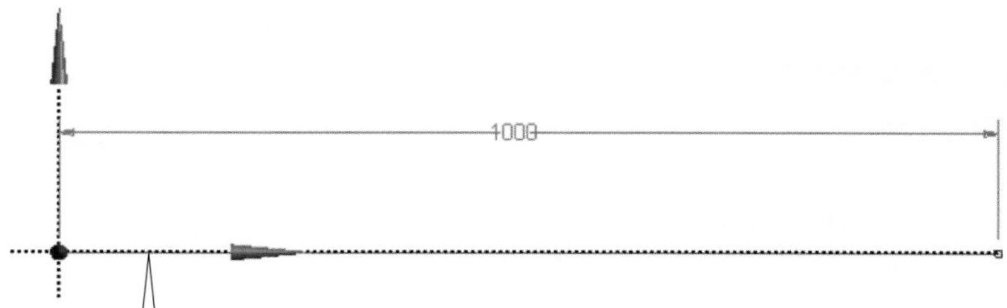

1000

[1] Select **Millimeter** as the length unit. On **XYPlane**, draw a straight line of length 1000 mm along X-axis. →

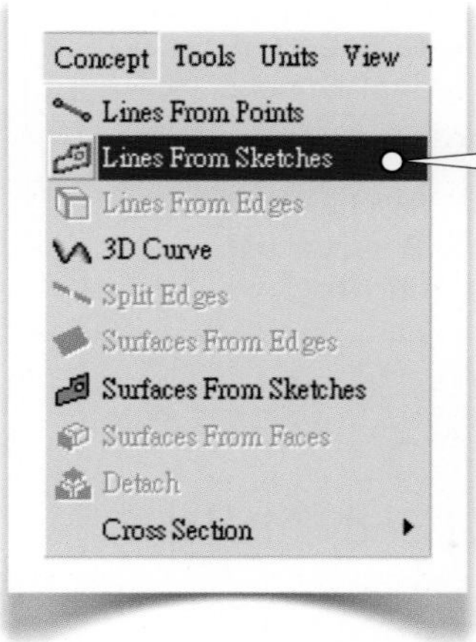

Concept Tools Units View
Lines From Points
Lines From Sketches
Lines From Edges
3D Curve
Split Edges
Surfaces From Edges
Surfaces From Sketches
Surfaces From Faces
Detach
Cross Section

[2] Select **Concept/Lines From Sketches** and create a line body using the newly created sketch (**Sketch1**). Remember to click **Generate**. ↓

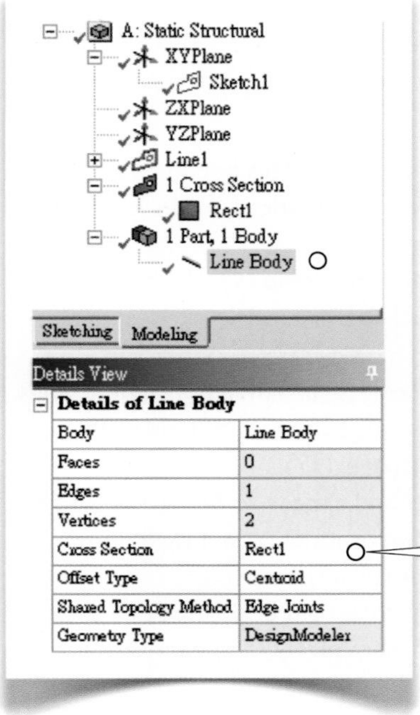

A: Static Structural
XYPlane
Sketch1
ZXPlane
YZPlane
Line1
1 Cross Section
Rect1
1 Part, 1 Body
Line Body
Sketching
Modeling
Details View
Details of Line Body
Body
Line Body
Faces
0
Edges
1
Vertices
2
Cross Section
Rect1
Offset Type
Centroid
Shared Topology Method
Edge Joints
Geometry Type
DesignModeler

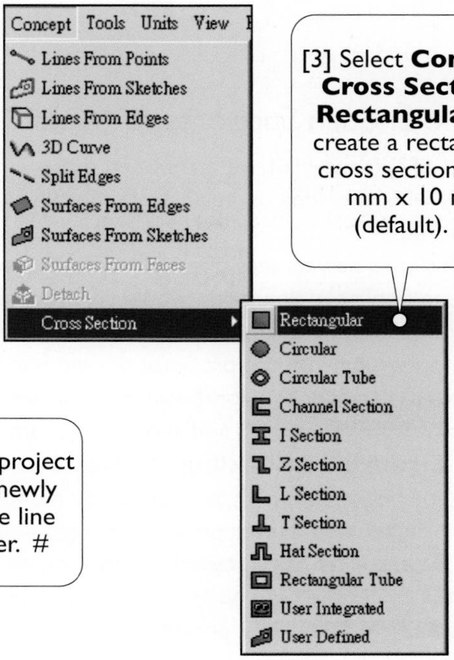

Concept Tools Units View
Lines From Points
Lines From Sketches
Lines From Edges
3D Curve
Split Edges
Surfaces From Edges
Surfaces From Sketches
Surfaces From Faces
Detach
Cross Section
Rectangular
Circular
Circular Tube
Channel Section
I Section
Z Section
L Section
T Section
Hat Section
Rectangular Tube
User Integrated
User Defined

[3] Select **Concept/Cross Section/Rectangular** and create a rectangular cross section of 10 mm x 10 mm (default). ↙

[4] With **Line Body** in the project tree highlighted, assign the newly created cross section to the line body. Close DesignModeler. #

10.1.4 Set Up Supports

[1] Start up **Mechanical** and select **mm-kg-N-s** unit system.

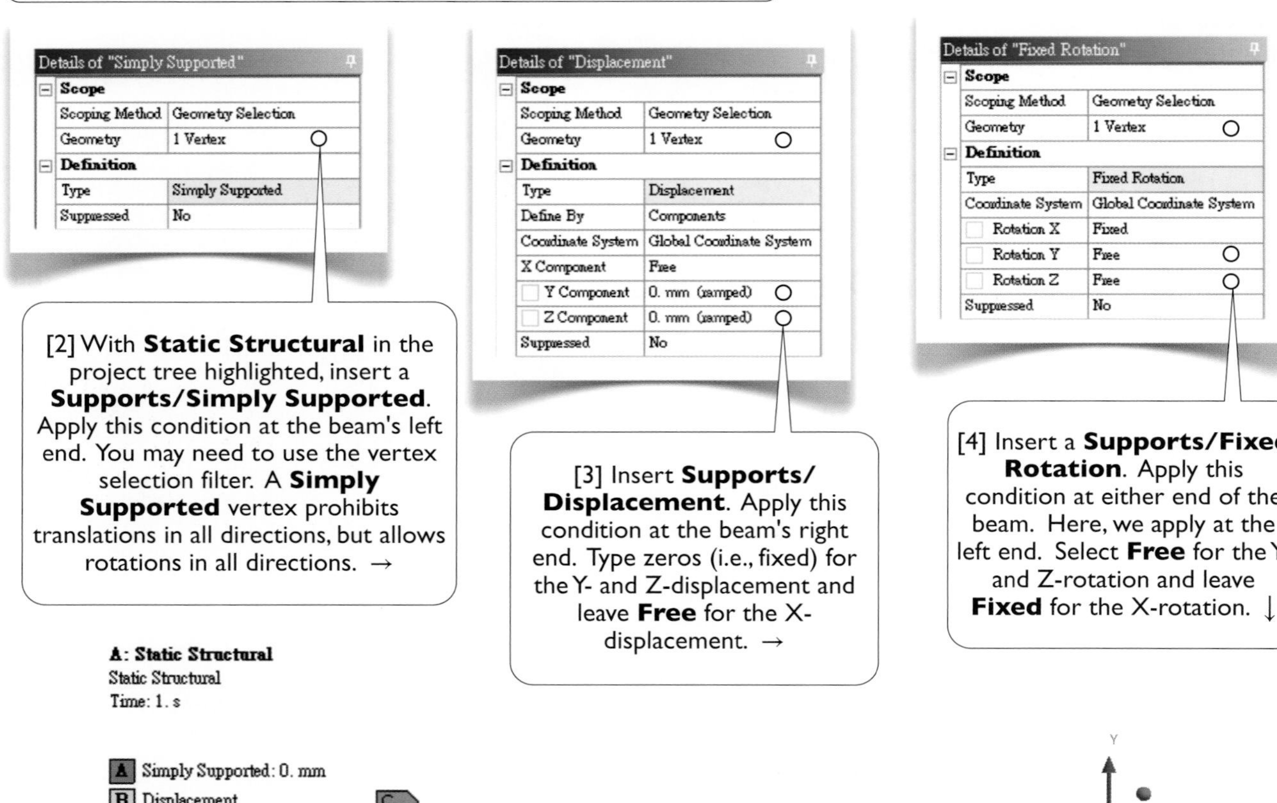

Provide Enough Supports to Avoid Rigid Body Modes

[5] In 3.1.8[7] (page 116), we mentioned that it is a good practice to provide enough supports. This becomes a necessity when working on buckling or modal analyses, where rigid body modes are not automatically eliminated with weak springs.

In this case, newcomers often fail to fix Z-displacements in step [3] and the X-rotation in step [4]. Without these supports, a nonlinear simulation would run into convergence difficulties. In case of buckling or modal analyses, rigid body modes would present.

What's wrong with the presence of rigid body modes? The answer depends on the type of simulation. For **Static Structural** simulations, Workbench can add weak springs to prevent an uncontrolled large amount of rigid body motions (3.1.8[7], page 116) and allow a small amount of rigid body motion (3.1.10[2], page 117).

For **Eigenvalue Buckling** and **Modal** simulations, rigid body modes are trivial and appear to be harmless. The buckling load corresponding to a rigid body mode is zero, and the natural frequency corresponding to a rigid body mode is also zero. However, presence of rigid body modes may deteriorate the numerical accuracy.

For **Transient Structural** simulations, we usually don't need to artificially eliminate rigid body modes if they exist naturally; let the rigid body modes be present, and the program will take care of them nicely. The exercises in Sections 12.4, 15.2, and 15.3 provide examples for these situations. #

10.1.5 Set Up Loads

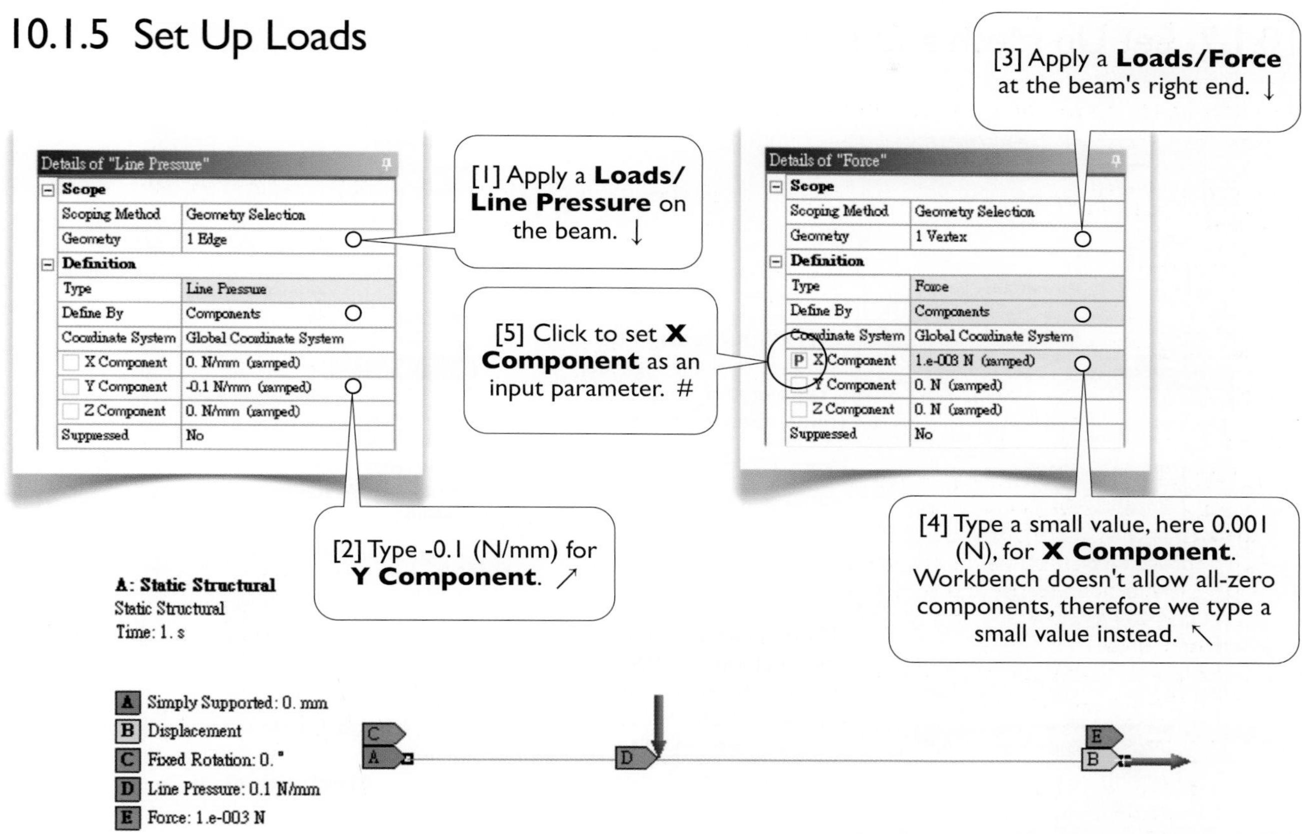

10.1.6 Set Up Solution Objects

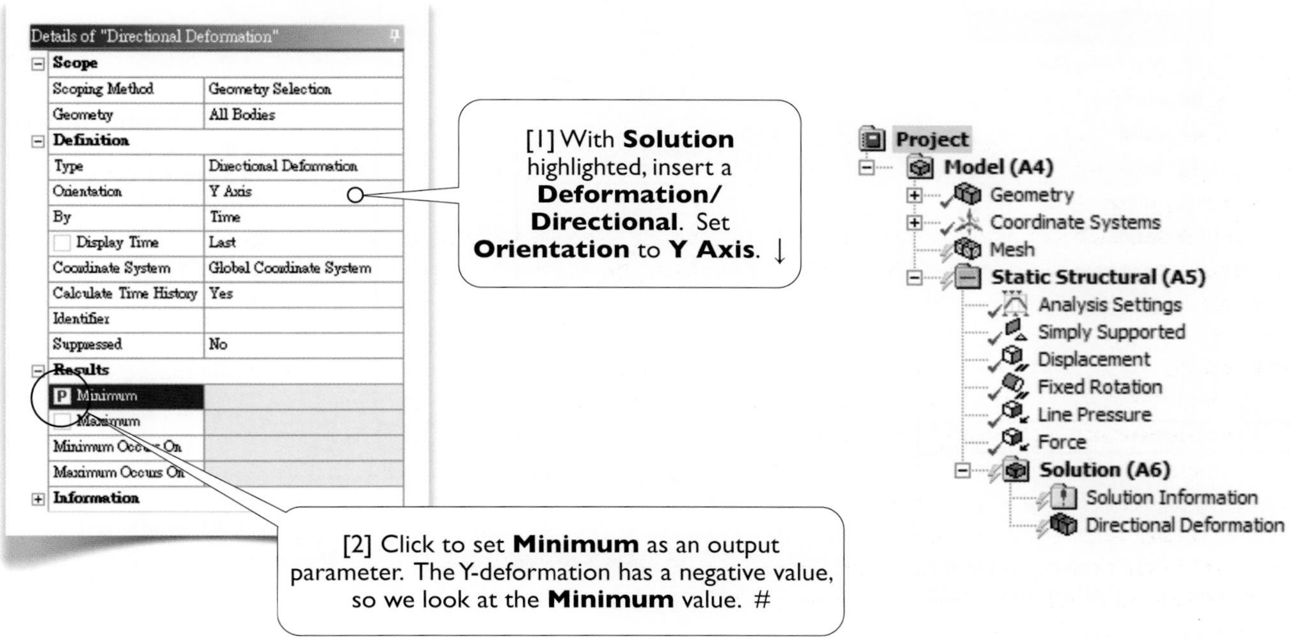

10.1.7 Set Up Mesh and Turn on Large Deflection Effect

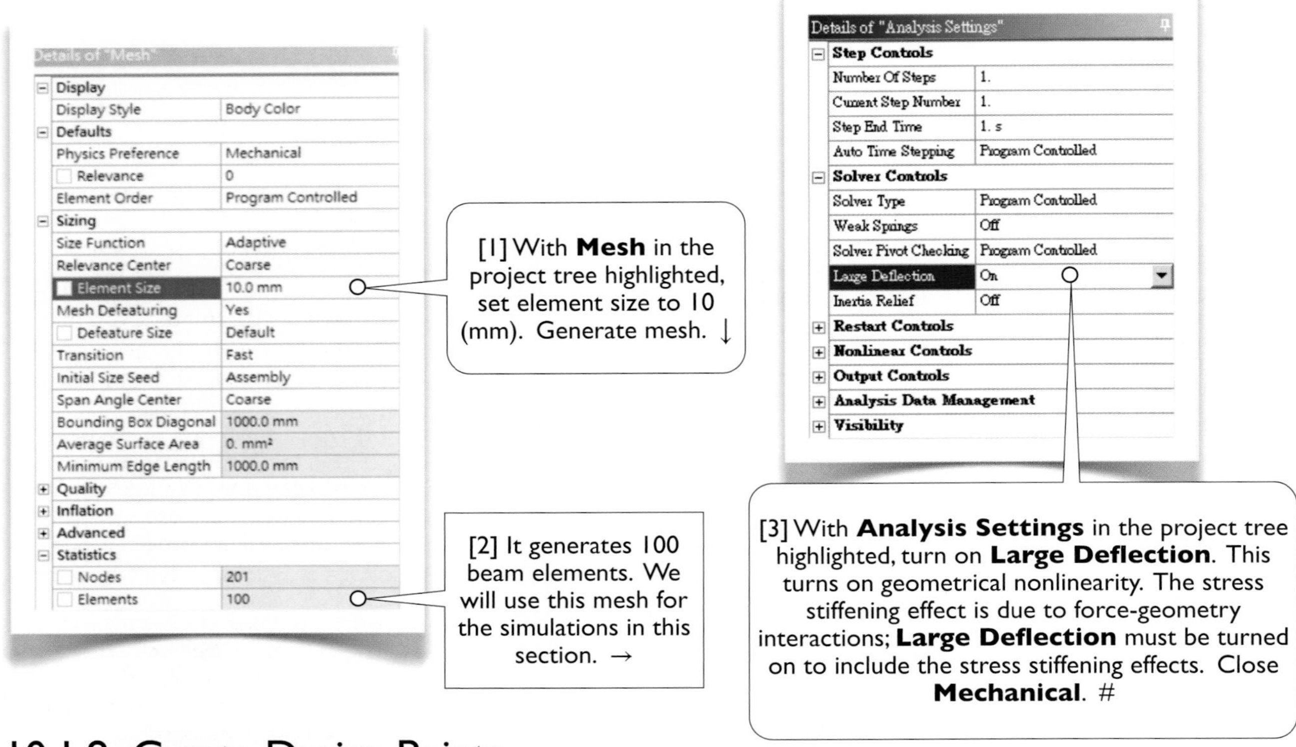

[1] With **Mesh** in the project tree highlighted, set element size to 10 (mm). Generate mesh. ↓

[2] It generates 100 beam elements. We will use this mesh for the simulations in this section. →

[3] With **Analysis Settings** in the project tree highlighted, turn on **Large Deflection**. This turns on geometrical nonlinearity. The stress stiffening effect is due to force-geometry interactions; **Large Deflection** must be turned on to include the stress stiffening effects. Close **Mechanical**. #

10.1.8 Create Design Points

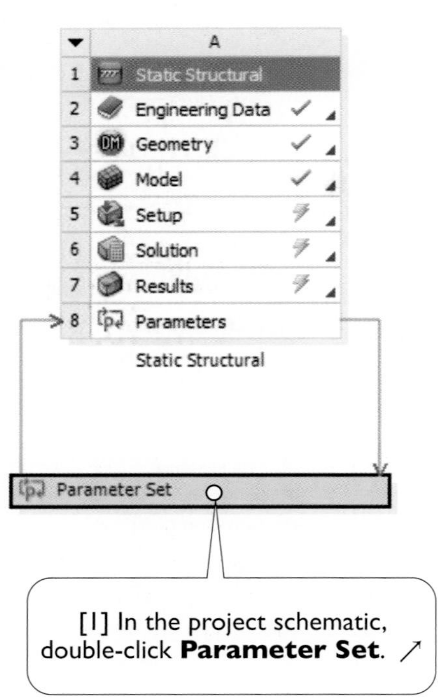

[1] In the project schematic, double-click **Parameter Set**. ↗

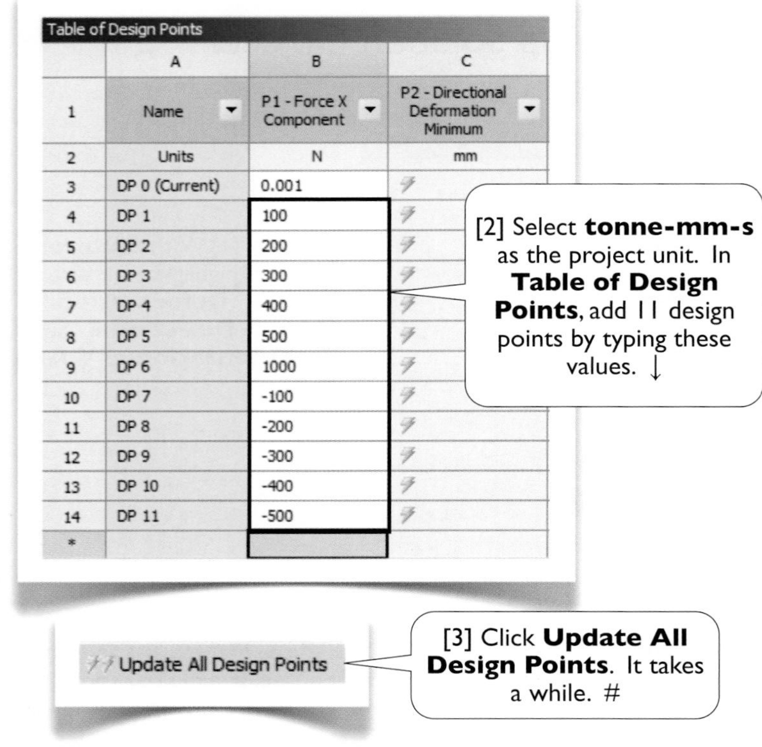

Table of Design Points

	A	B	C
1	Name	P1 - Force X Component	P2 - Directional Deformation Minimum
2	Units	N	mm
3	DP 0 (Current)	0.001	
4	DP 1	100	
5	DP 2	200	
6	DP 3	300	
7	DP 4	400	
8	DP 5	500	
9	DP 6	1000	
10	DP 7	-100	
11	DP 8	-200	
12	DP 9	-300	
13	DP 10	-400	
14	DP 11	-500	
*			

[2] Select **tonne-mm-s** as the project unit. In **Table of Design Points**, add 11 design points by typing these values. ↓

[3] Click **Update All Design Points**. It takes a while. #

10.1.9 The Results and Discussion

Table of Design Points

	A	B	C
1	Name	P1 - Force X Component	P2 - Directional Deformation Minimum
2	Units	N	mm
3	DP 0 (Current)	0.001	-7.8114
4	DP 1	100	-7.3626
5	DP 2	200	-6.9624
6	DP 3	300	-6.6034
7	DP 4	400	-6.2796
8	DP 5	500	-5.9863
9	DP 6	1000	-4.8551
10	DP 7	-100	-8.3184
11	DP 8	-200	-8.8953
12	DP 9	-300	-9.5571
13	DP 10	-400	-10.321
14	DP 11	-500	-11.205
*			

[2] And the deflections are used as the vertical axis. ↓

Parameters Studies with Workbench

[5] We've demonstrated a useful feature of Workbench: parameters study, mentioned in 8.1.5[12-14], page 317.

To study the relations among parameters, you may create various "design points" and then issue an **Update All Design Points** command. Workbench will do the rest of the work.

The resulting data can be copied and pasted to a spread sheet application, such as **Excel**, for plotting a chart. ↓

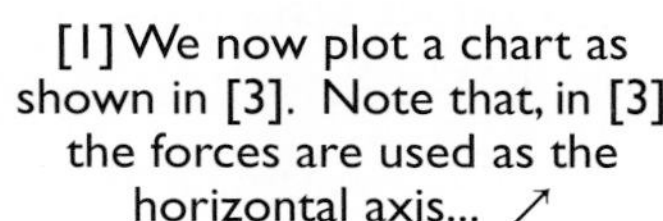

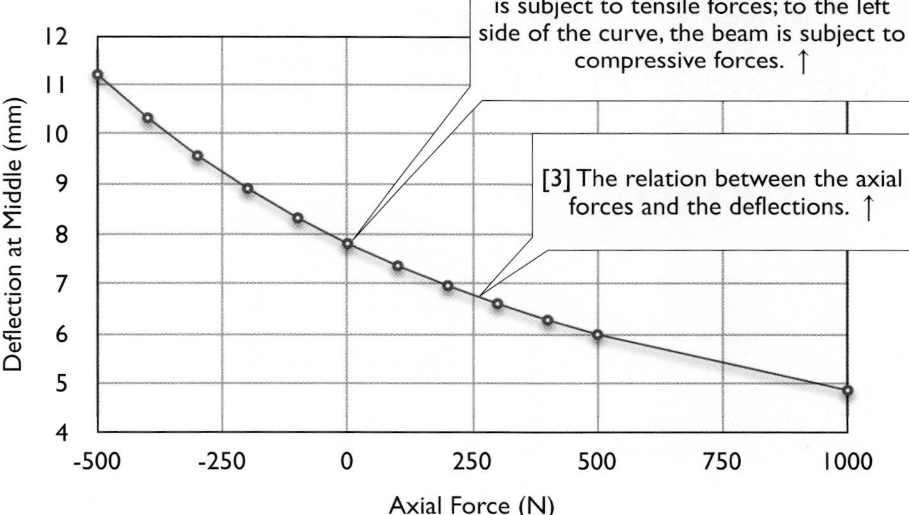

Stress Stiffening Effects

[6] The curve [3] manifests the stress stiffening effects. As the tensile axial force increases, the deflection decreases, indicating an increase of bending stiffness. On the other hand, as the compressive axial force increases, the deflection increases, indicating a decrease of bending stiffness. When the compressive axial force reaches a certain point, the bending stiffness decreases so much that the deflection is enlarged dramatically. You may raise a question: what is the compressive force such that the bending stiffness completely vanishes? A zero bending stiffness implies an unstable structure: a small vertical load would cause the beam to collapse. This phenomenon is called the *buckling* and the compressive force causing the structure to buckle is called the *buckling load*, or *critical load*. ↵

[7] We can continue the above process and extend the curve ([3], last page) leftward until a vertical asymptote can be drawn. The force value intercepted by the asymptote will be the buckling force. And this procedure is basically a *nonlinear buckling analysis*[Ref 1].

Workbench provides an **Eigenvalue Buckling** system to assess the linear buckling load. A linear buckling analysis usually takes much less computing time than a nonlinear buckling analysis but usually overestimates the buckling load.

The linear buckling theory predicts the buckling load as

$$P_{buckling} = \frac{\pi^2 EI}{L^2} = \frac{\pi^2 (200,000)(833.33)}{(1000)^2} = 1645 \text{ N} \qquad (1)$$

Note that the above calculation doesn't include the distributed beam load (0.1 N/mm). Although tending to overestimate the buckling load, the linear buckling analysis is useful for two reasons: (a) It is computationally much cheaper than a nonlinear buckling analysis and should be run as a first step to estimate the buckling load. (b) It can be used to determine the possible buckling mode.

We will proceed to demonstrate the linear buckling analysis for this case in the rest of the section. #

10.1.10 Set Up Project Schematic

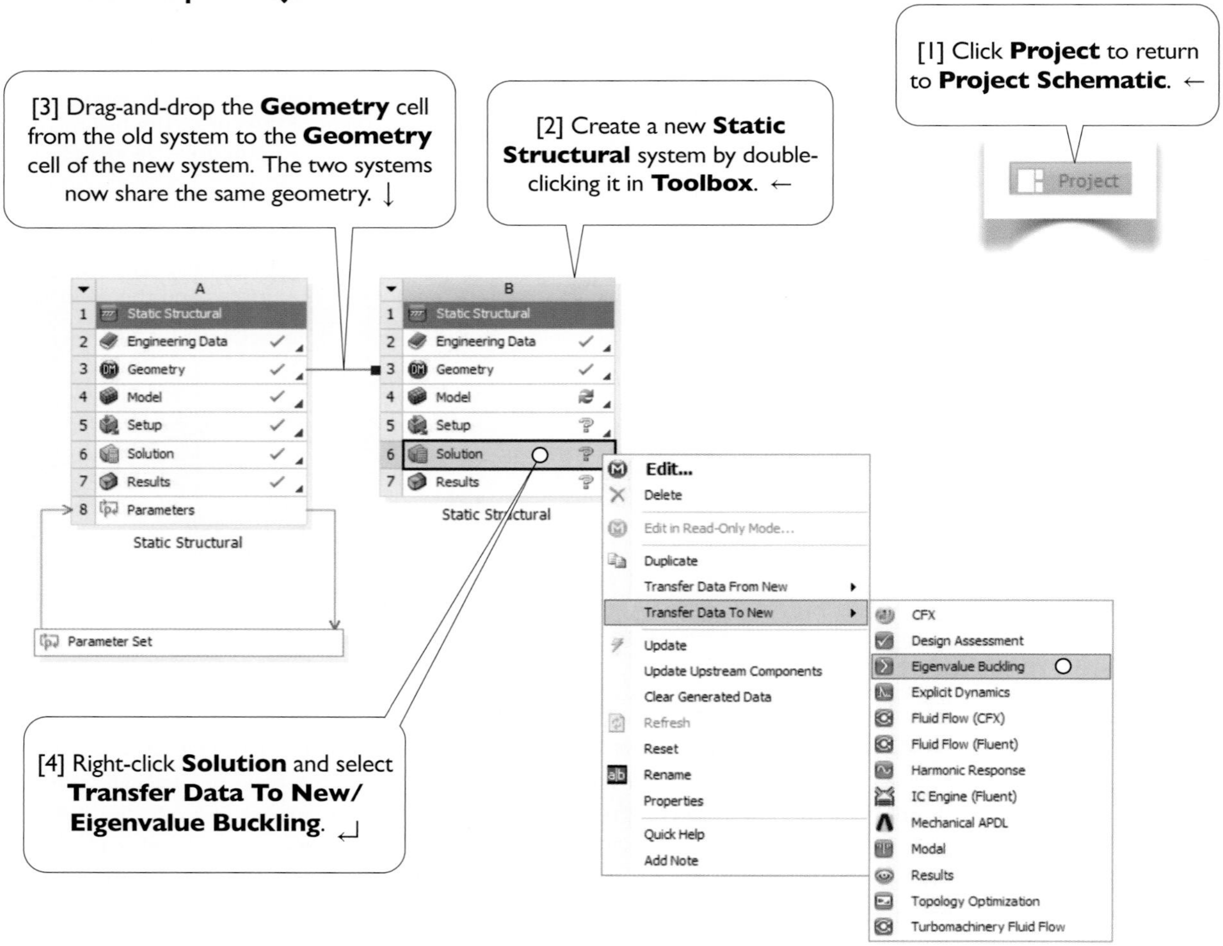

[1] Click **Project** to return to **Project Schematic**. ←

[2] Create a new **Static Structural** system by double-clicking it in **Toolbox**. ←

[3] Drag-and-drop the **Geometry** cell from the old system to the **Geometry** cell of the new system. The two systems now share the same geometry. ↓

[4] Right-click **Solution** and select **Transfer Data To New/ Eigenvalue Buckling**. ↵

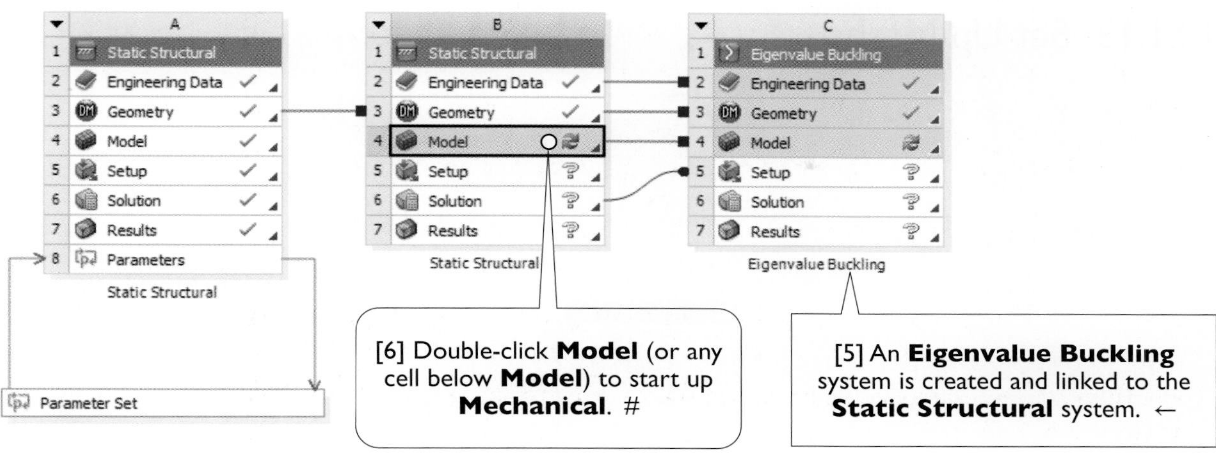

10.1.11 Set Up Supports

10.1.12 Set Up Load

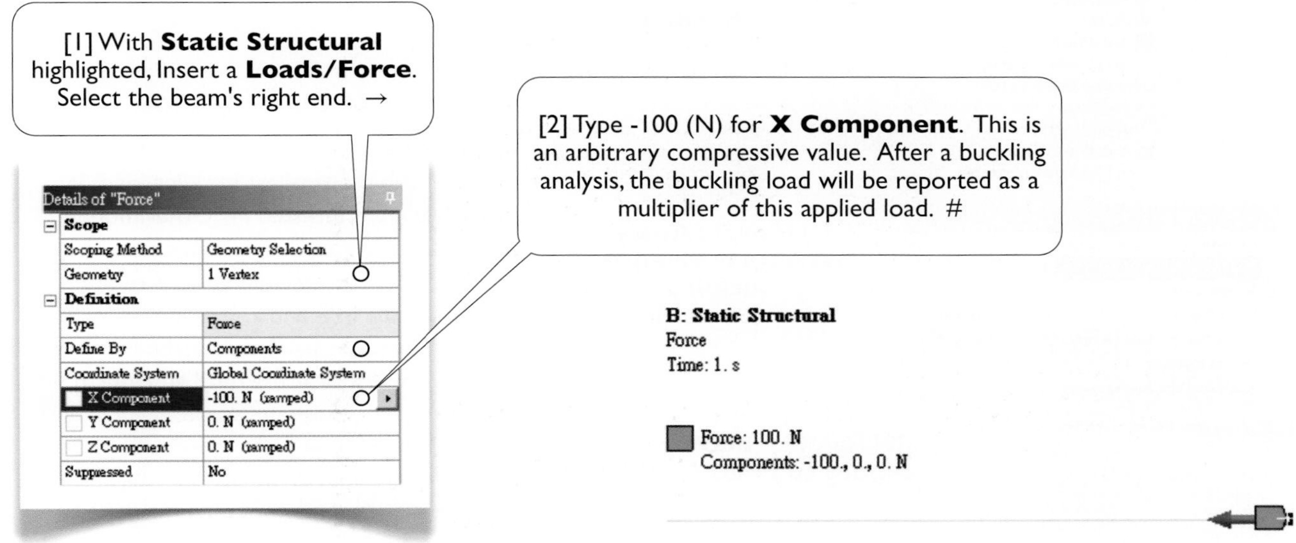

10.1.13 Set Up Mesh

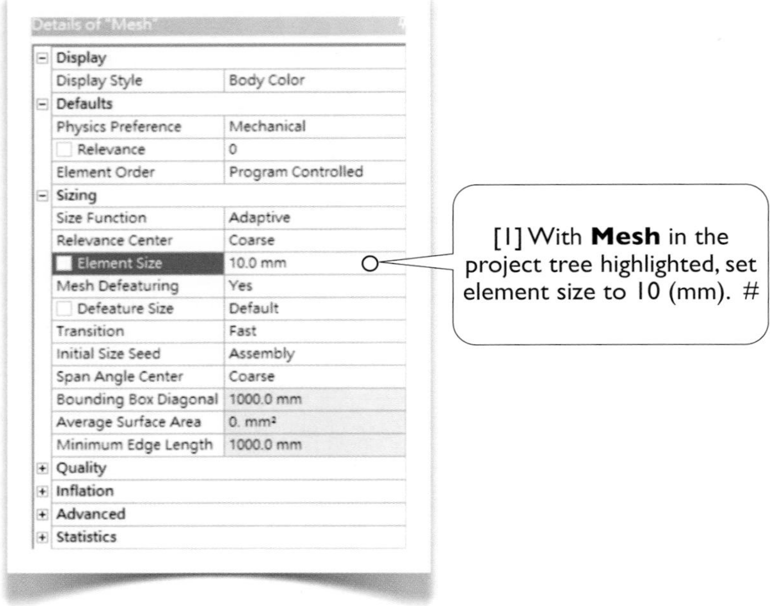

10.1.14 Specify Number of Buckling Modes and Solve

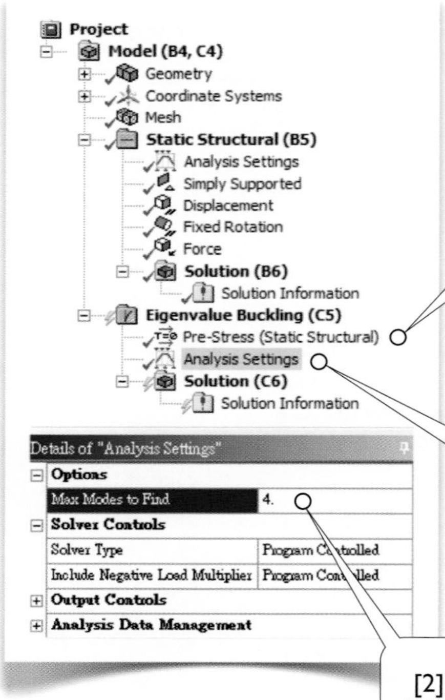

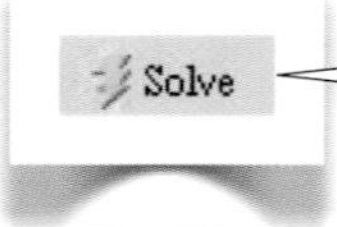

[1] Highlight **Analysis Settings** of **Eigenvalue Buckling**. ↓

[2] Type 4 for **Max Modes to Find**. ↑

[3] The name of this object indicates that the stress results of **Static Structural** are used as a pre-stress for **Eigenvalue Buckling**. ↗

[4] Highlight **Eigenvalue Buckling** and click **Solve**. This also triggers the solving of **Static Structural** (for calculating the pre-stress). If you highlight **Static Structural** when you click **Solve**, it would solve the **Static Structural** only. ↓

Number of Buckling Modes

[5] In this case, the first buckling mode (mode with the smallest buckling load) is what we are interested in. Other modes have little practical usage in this case.

Due to the same dimensions of the cross section in Y- and Z-direction, buckling modes appear as pairs, that is, the first two modes are identical, the third and fourth modes are identical, and so on.

The reason we want to evaluate the third and the fourth modes is to make sure they are far beyond our concern. #

10.1.15 View the Results

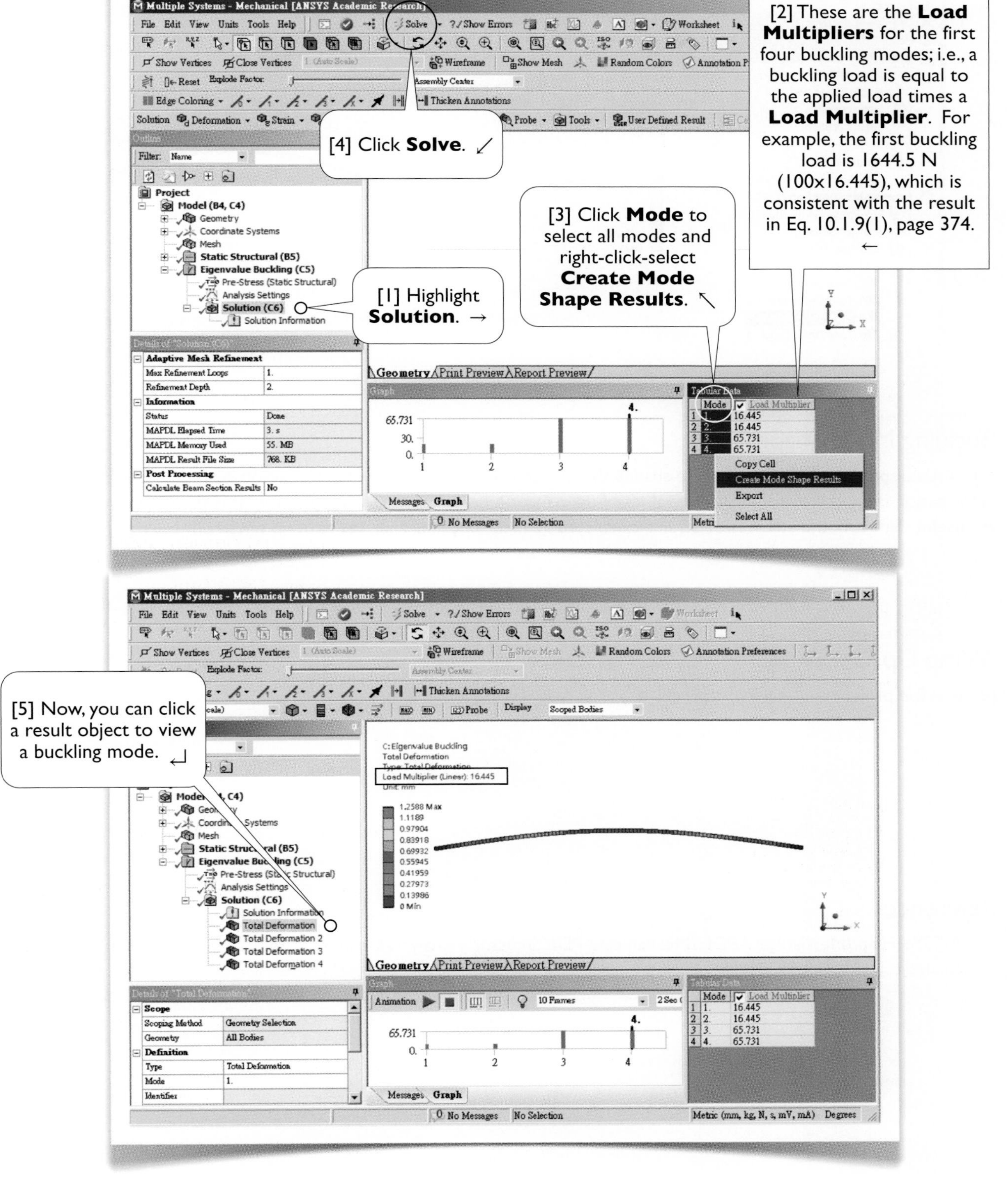

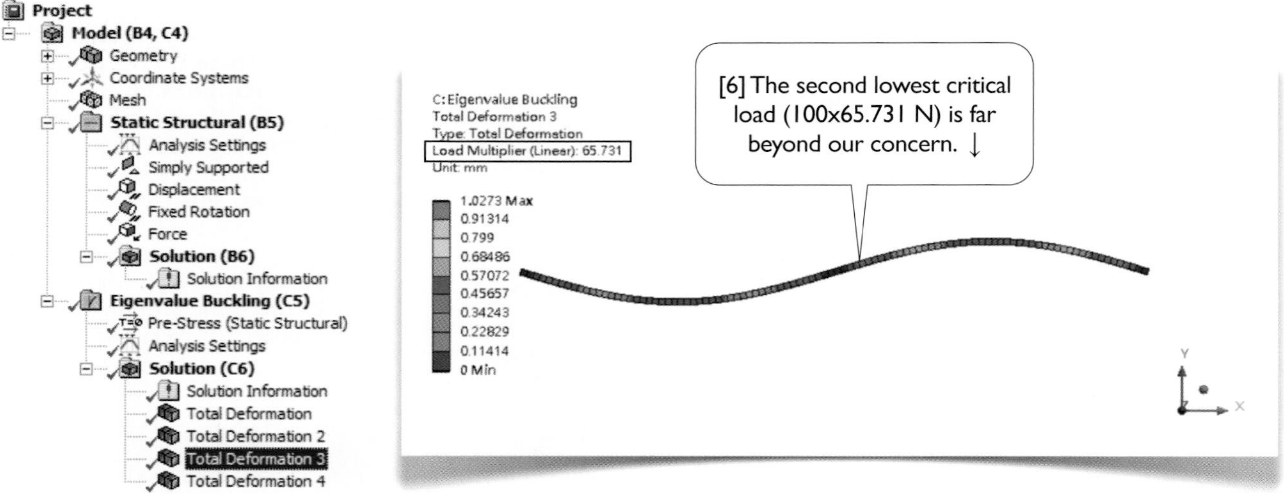

Buckling Mode Shapes

[7] When displaying the buckling mode shapes, Workbench scales the values of deformation such that the maximum deformation is approximately 1.0. The values of deformation have no physical significance. It is the mode shapes that are useful. Similarly, the stresses or strains calculated in linear buckling analyses have no physical meaning. ↓

Wrap Up

[8] Save the project and exit Workbench. #

Reference

1. ANSYS Help//Mechanical APDL//Mechanical APDL Structural Analysis Guide//7.3. Performing a Nonlinear Buckling Analysis

Section 10.2

3D Truss

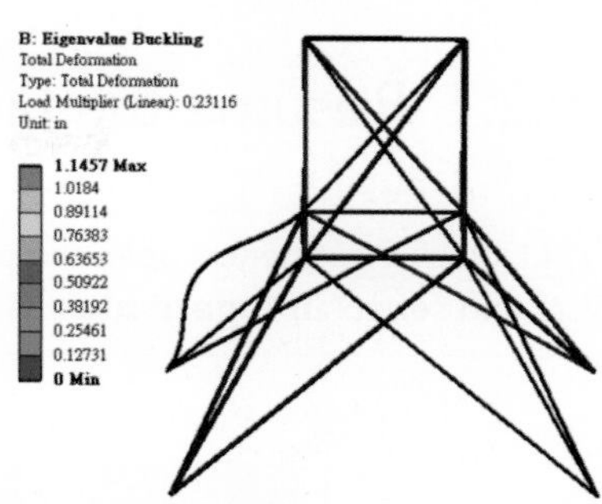

10.2.1 About the 3D Truss

[1] In Section 7.2, we analyzed a 3D truss; the stresses range from -25,415 psi to +13,962 psi (7.2.12[4], page 292). The structure seems safe, since the structural steel's yield strength can be as high as 40,000 psi for both tension and compression. However, stress is only one design consideration. Structural stability must be verified whenever compressive members are involved in the structure system.

Let's make some simple calculations to check whether structural stability should be an issue, by considering the member that has the maximum compressive force, which is P = 15,802 lb (see [2]). →

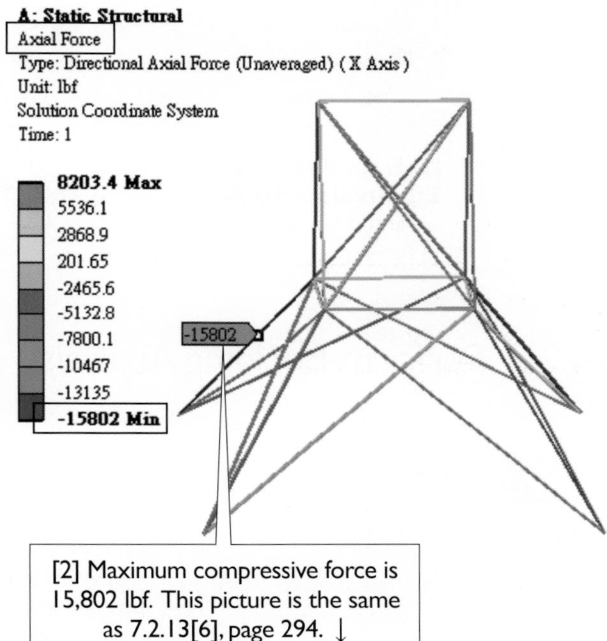

[2] Maximum compressive force is 15,802 lbf. This picture is the same as 7.2.13[6], page 294. ↓

[3] The member (of maximum compressive force [2]) has a length of 133.46 inches and a cross section ($L\ 1\frac{1}{2}\times 1\frac{1}{2}\times\frac{1}{4}$) of an area moment of inertia of 0.13852 in^4. The structural steel's Young's modulus is 29,000,000 psi. Its buckling load, according to a linear buckling theory, is estimated to be

$$P_{buckling} = \frac{\pi^2 EI}{L^2} = \frac{\pi^2(29{,}000{,}000)(0.13852)}{(133.46)^2} = 2{,}226 \text{ lb} = 0.14(15{,}802) = 0.14P$$

In other words, merely 14% of the design loads (see 7.2.1[1], page 283) would cause at least one of the structural members to buckle. It is now obvious that the structural stability might be a problem.

In this section, we want to perform a linear buckling analysis for the entire 3D truss structure, rather than a single member. The linear buckling analysis will result in a higher multiplier than 14% (10.2.4[6], page 382). Two factors cause the deviation from the above simple calculation. First, the rigid joints provide additional rigidity, increasing the buckling load. Second, even assuming pin-jointed, this truss is a statically indeterminate structure; buckling of one member does not necessarily cause a buckling of the entire structure.

The unit system used in this section is **in-lbm-lbf-s**. #

10.2.2 Resume the Project **Truss**

[1] Launch Workbench, open the project **Truss** (which was saved in Section 7.2), and set up an **Eigenvalue Buckling** system shown in [2-5]. ↙

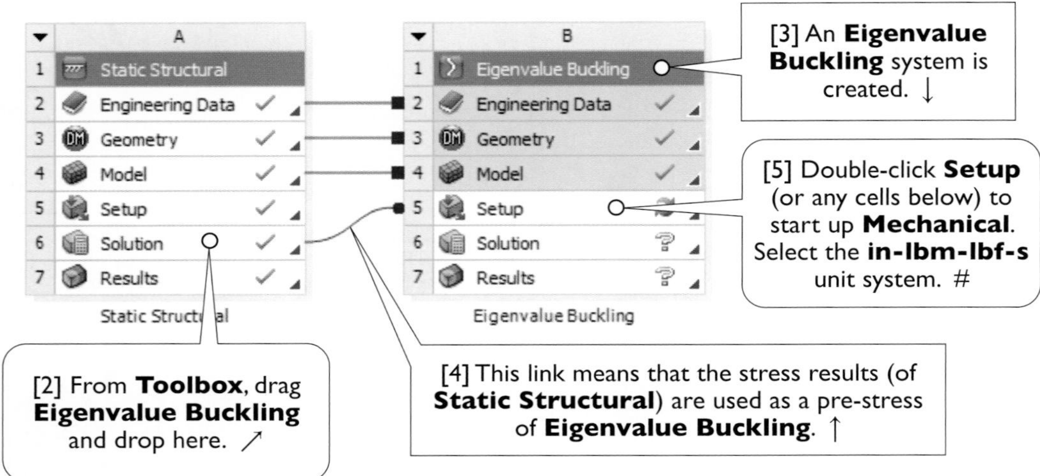

[3] An **Eigenvalue Buckling** system is created. ↓

[5] Double-click **Setup** (or any cells below) to start up **Mechanical**. Select the **in-lbm-lbf-s** unit system. #

[2] From **Toolbox**, drag **Eigenvalue Buckling** and drop here. ↗

[4] This link means that the stress results (of **Static Structural**) are used as a pre-stress of **Eigenvalue Buckling**. ↑

10.2.3 Perform Buckling Analysis

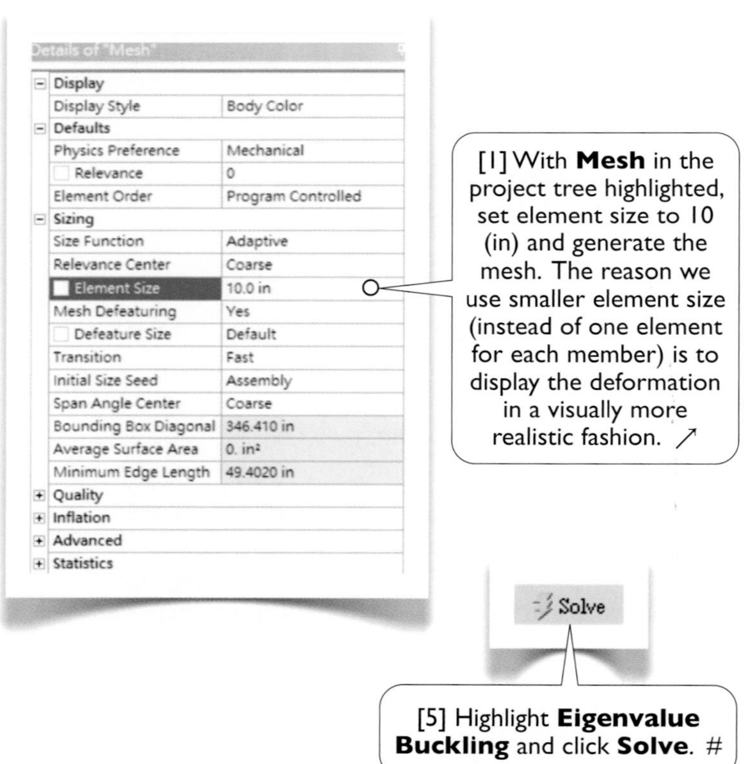

[1] With **Mesh** in the project tree highlighted, set element size to 10 (in) and generate the mesh. The reason we use smaller element size (instead of one element for each member) is to display the deformation in a visually more realistic fashion. ↗

[5] Highlight **Eigenvalue Buckling** and click **Solve**. #

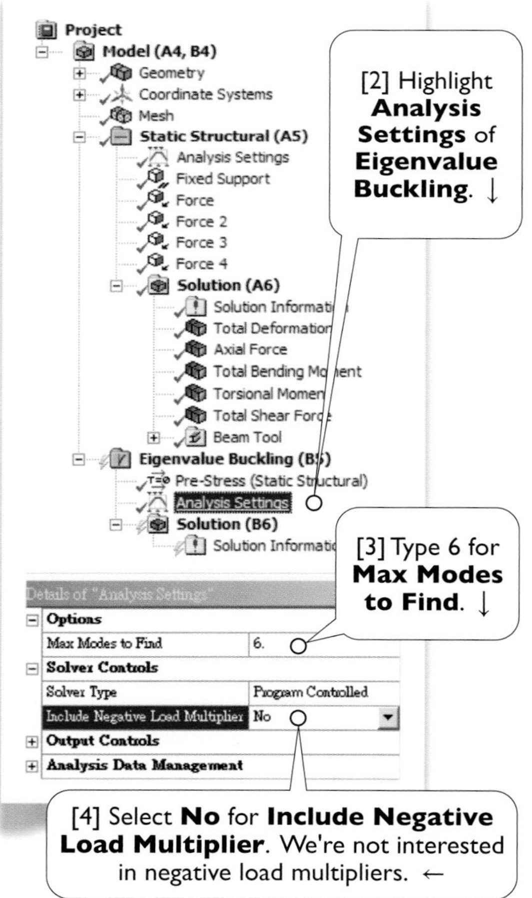

[2] Highlight **Analysis Settings** of **Eigenvalue Buckling**. ↓

[3] Type 6 for **Max Modes to Find**. ↓

[4] Select **No** for **Include Negative Load Multiplier**. We're not interested in negative load multipliers. ←

10.2.4 View the Results

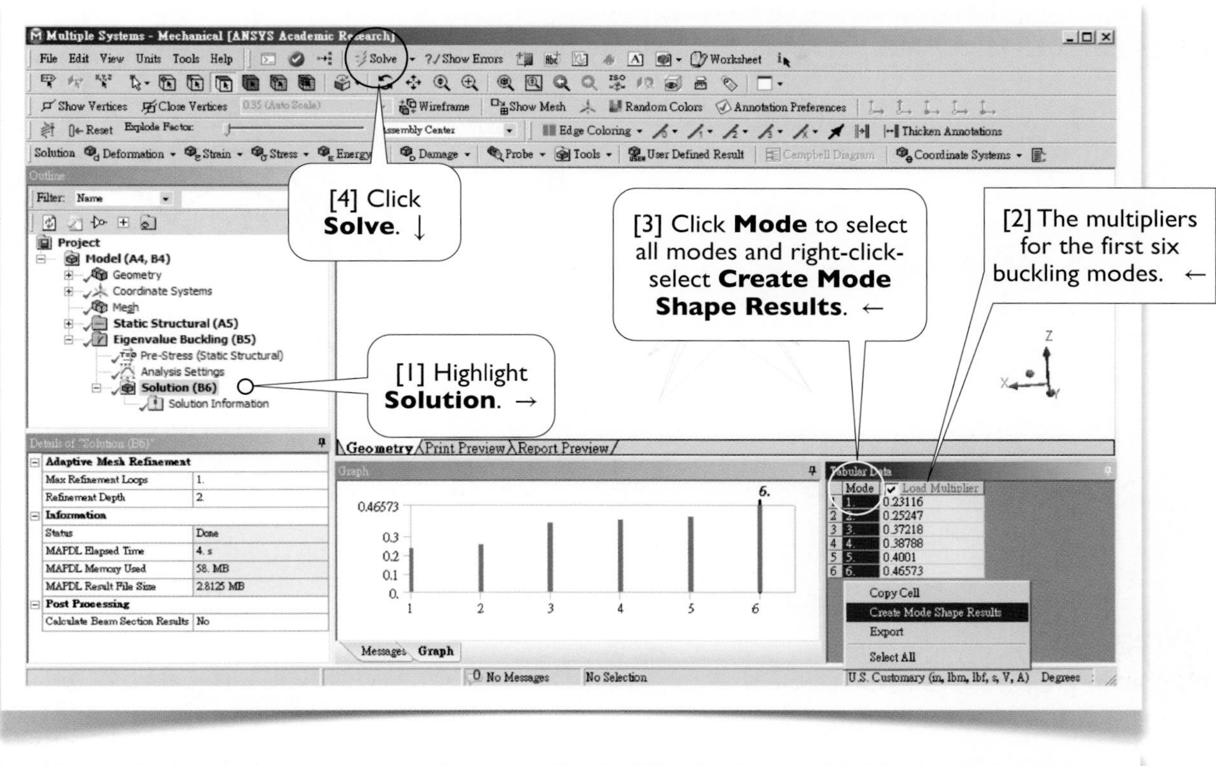

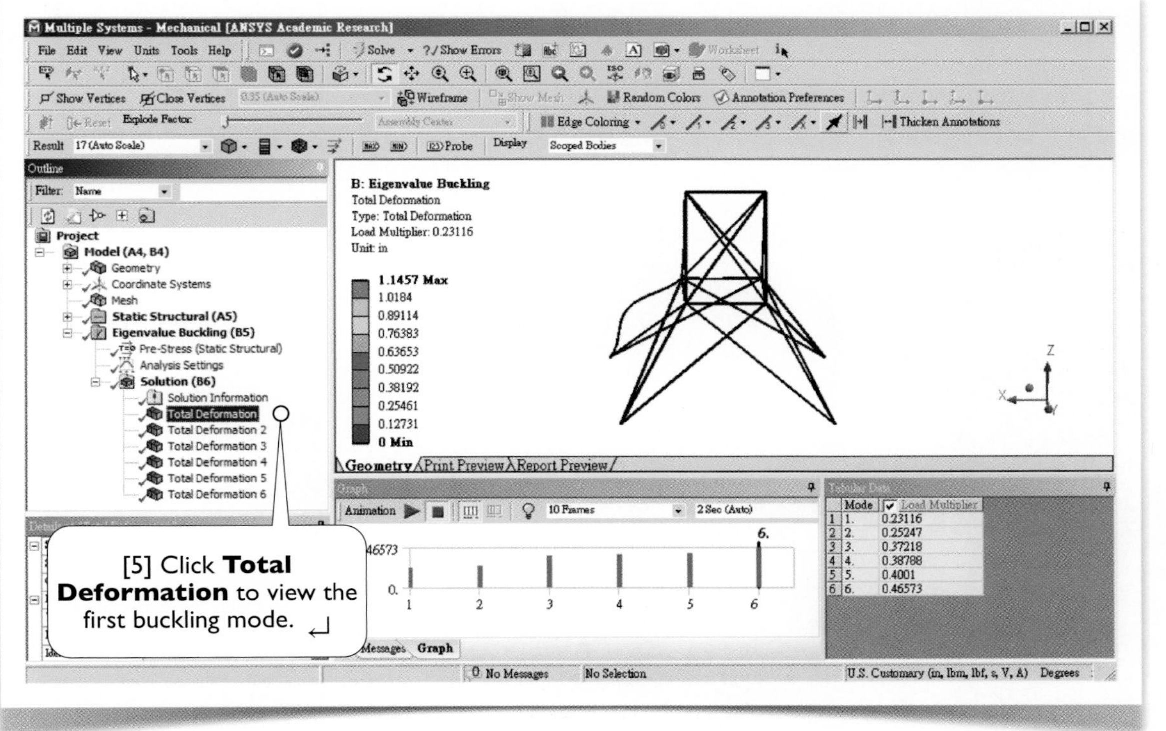

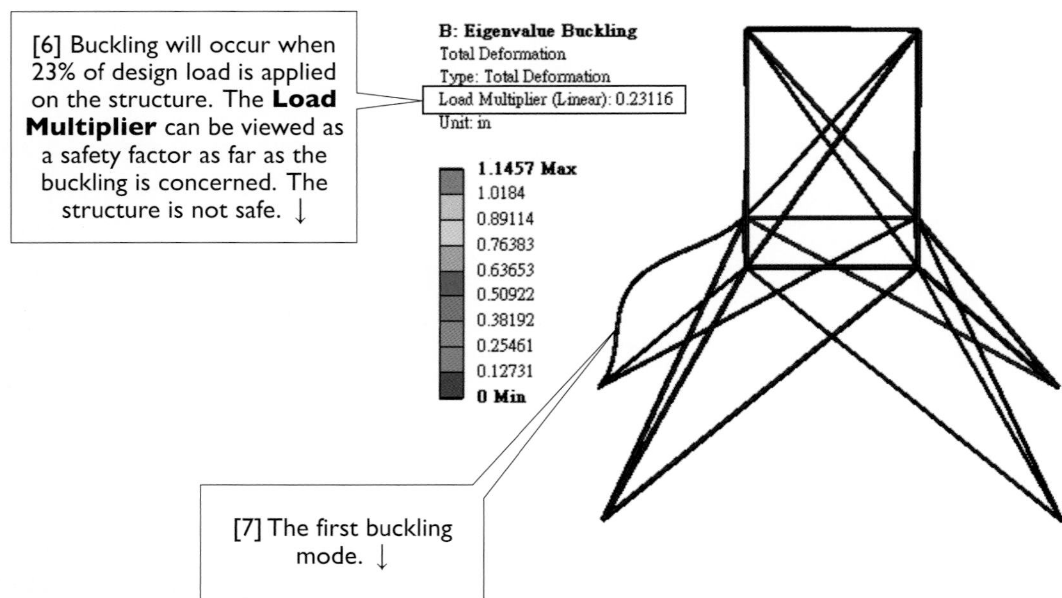

Wrap Up

[8] Save the project and exit Workbench. #

Section 10.3

Beam Bracket

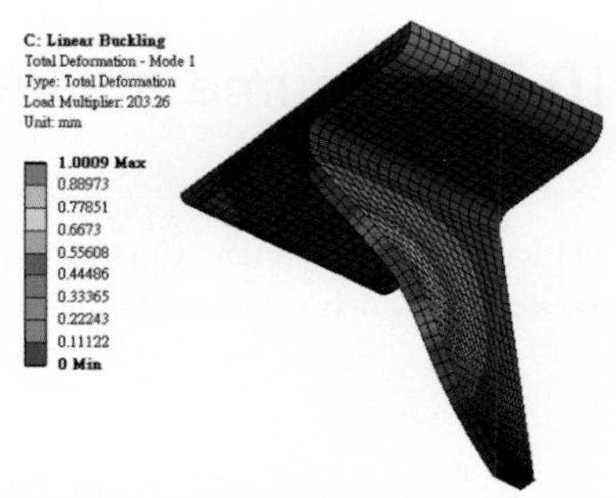

10.3.1 About the Beam Bracket

[1] In Section 5.1, we simulated a beam bracket using a 3D solid model, and in Section 6.2 using a surface model. The maximum von Mises stress is 83 MPa [2], well below the yield strength, 250 MPa.

By examining **Minimum Principal Stress** [3], we see that the web is subject to compressive stress; its magnitude is about 23 MPa [4]. It is a good practice that an engineer always checks the structural stability whenever compressive stresses exist, unless he has enough experience to judge that the stability is not an issue and therefore the checking is not necessary.

In this section, we want to make sure that, under the design load, the web does not buckle.

The unit system **mm-kg-N-s** is used in this section. ↗

[2] The maximum von Mises stress. This picture is a duplicate of 5.1.13[9], page 218. ↙

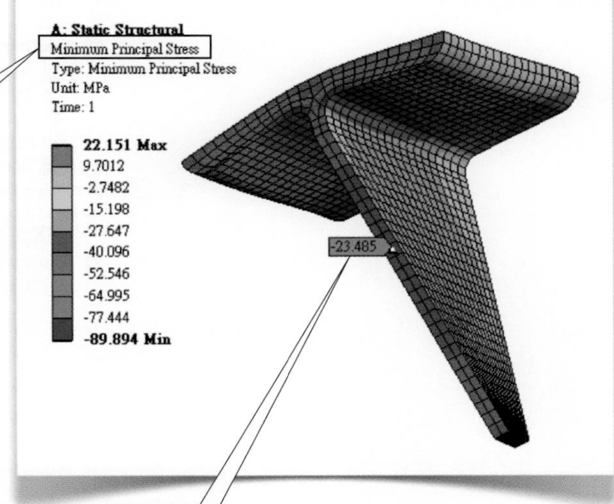

[3] **Minimum Principal Stress** is used to examine compressive stresses. ↓

[4] Compressive stress at the web. #

10.3.2 Resume the Project **Bracket**

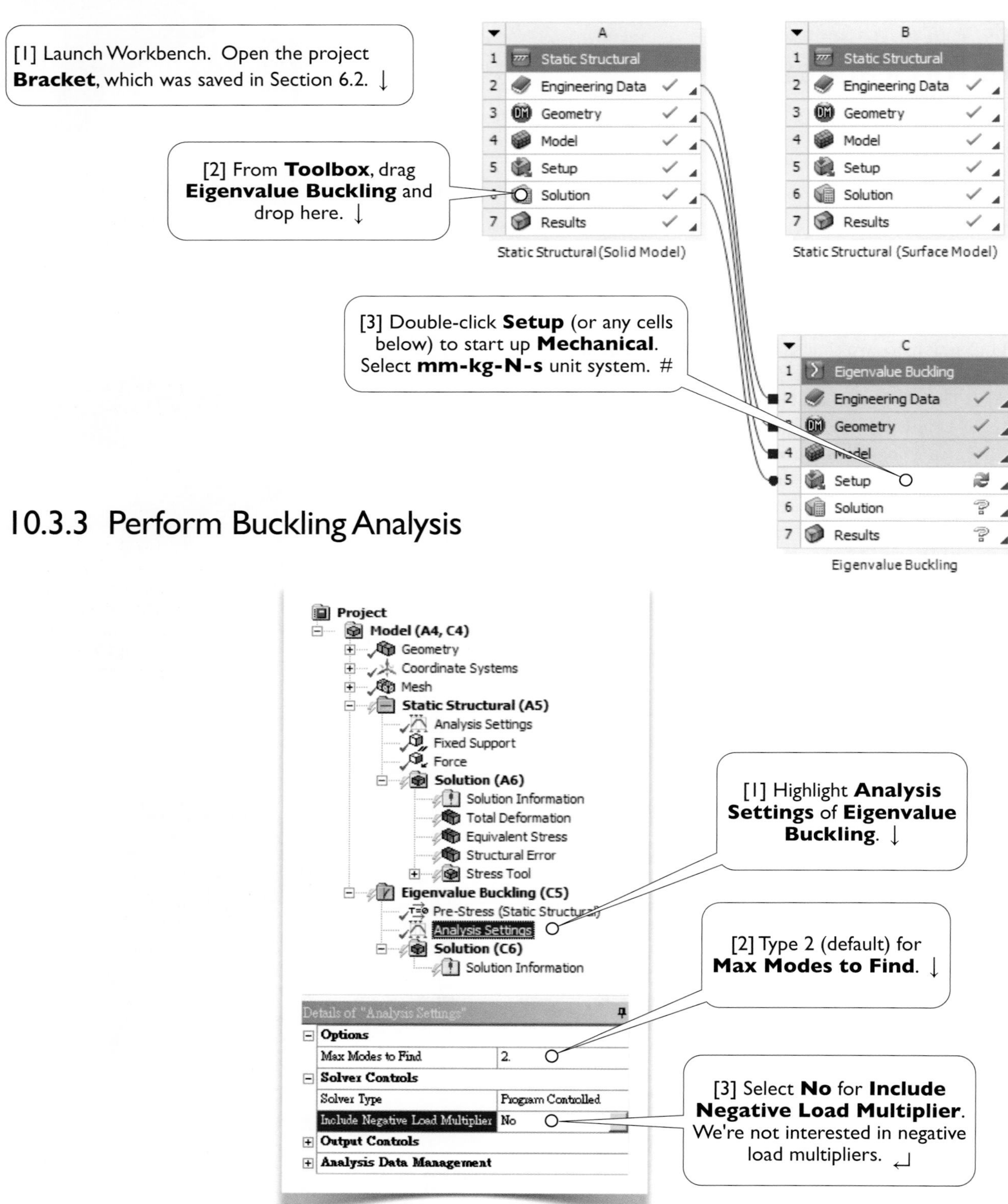

10.3.3 Perform Buckling Analysis

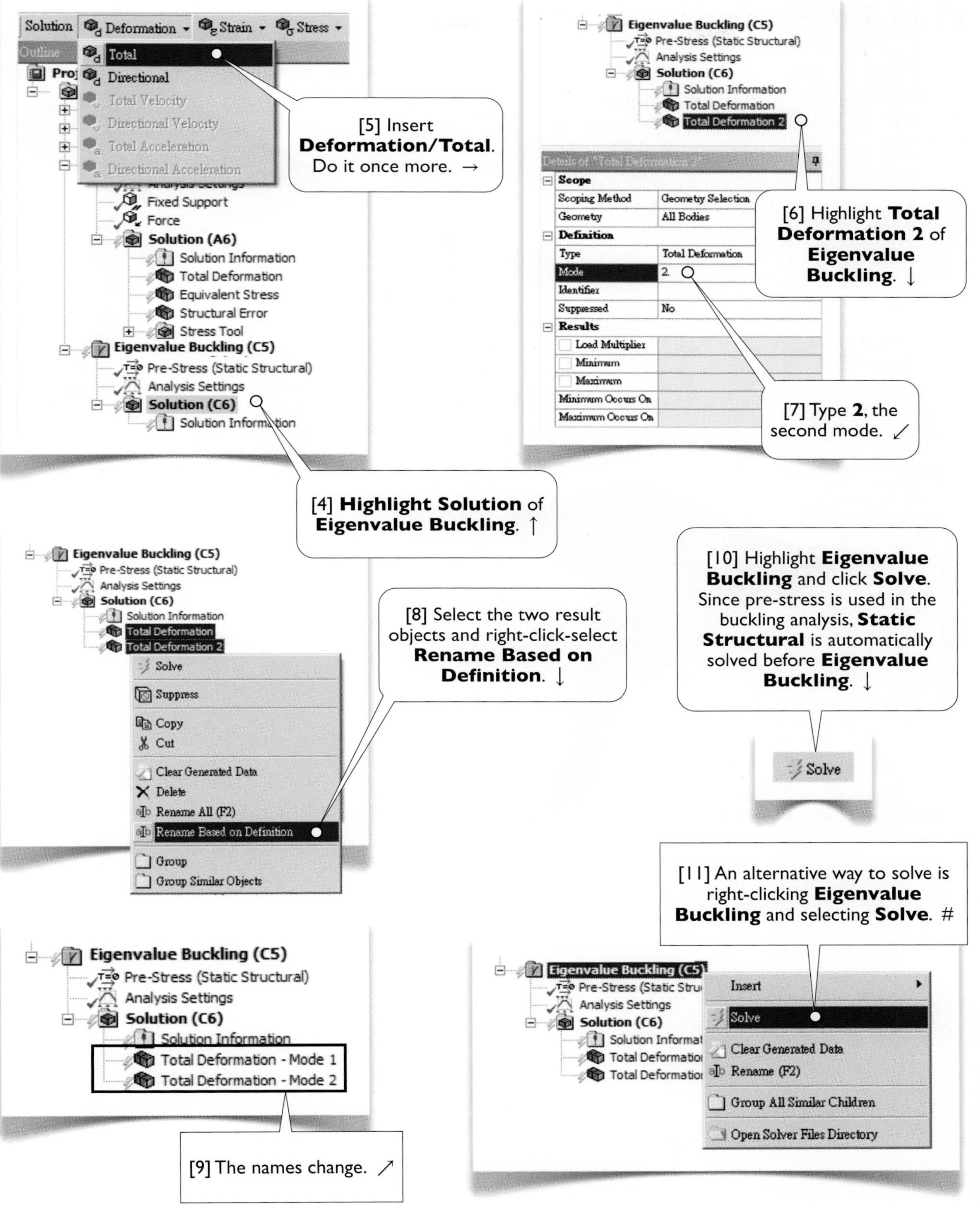
Solution
Deformation
Strain
Stress
Outline
Total
Directional
Total Velocity
Directional Velocity
Total Acceleration
Directional Acceleration
Fixed Support
Force
Solution (A6)
Solution Information
Total Deformation
Equivalent Stress
Structural Error
Stress Tool
Eigenvalue Buckling (C5)
Pre-Stress (Static Structural)
Analysis Settings
Solution (C6)
Solution Information
[5] Insert **Deformation/Total**. Do it once more. →
[4] **Highlight Solution** of **Eigenvalue Buckling**. ↑
Total Deformation
Total Deformation 2
Details of "Total Deformation 2"
Scope
Scoping Method
Geometry Selection
Geometry
All Bodies
Definition
Type
Total Deformation
Mode
2.
Identifier
Suppressed
No
Results
Load Multiplier
Minimum
Maximum
Minimum Occurs On
Maximum Occurs On
[6] Highlight **Total Deformation 2** of **Eigenvalue Buckling**. ↓
[7] Type **2**, the second mode. ↙
Solve
Suppress
Copy
Cut
Clear Generated Data
Delete
Rename All (F2)
Rename Based on Definition
Group
Group Similar Objects
[8] Select the two result objects and right-click-select **Rename Based on Definition**. ↓
[10] Highlight **Eigenvalue Buckling** and click **Solve**. Since pre-stress is used in the buckling analysis, **Static Structural** is automatically solved before **Eigenvalue Buckling**. ↓
Total Deformation - Mode 1
Total Deformation - Mode 2
[9] The names change. ↗
[11] An alternative way to solve is right-clicking **Eigenvalue Buckling** and selecting **Solve**. #
Insert
Rename (F2)
Group All Similar Children
Open Solver Files Directory

10.3.4 View the Results

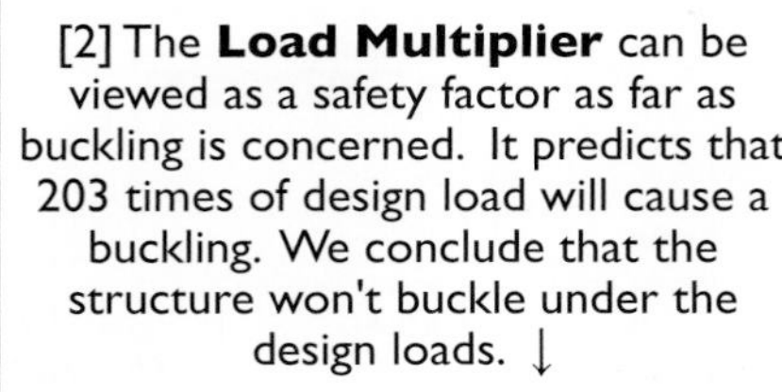

[2] The **Load Multiplier** can be viewed as a safety factor as far as buckling is concerned. It predicts that 203 times of design load will cause a buckling. We conclude that the structure won't buckle under the design loads. ↓

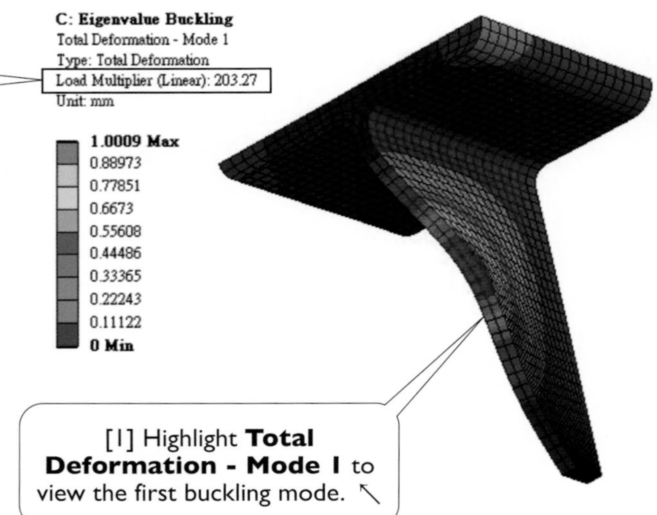

[1] Highlight **Total Deformation - Mode 1** to view the first buckling mode. ↖

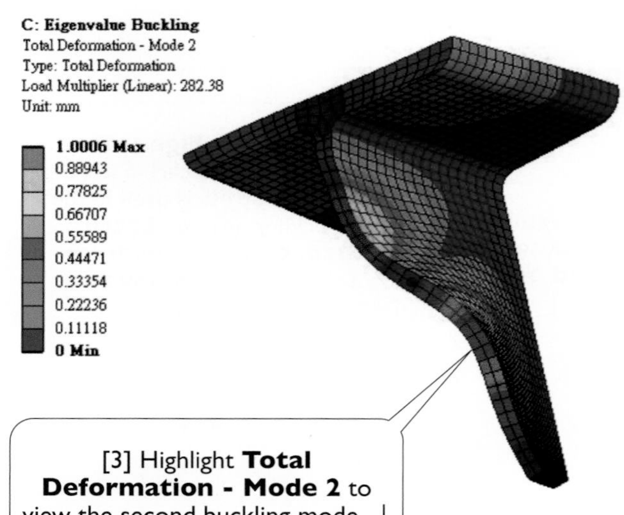

[3] Highlight **Total Deformation - Mode 2** to view the second buckling mode. ↓

Wrap Up

[4] Save the project and exit Workbench. #

Section 10.4

Review

10.4.1 Keywords

Choose a letter for each keyword from the list of descriptions

1. ( ) Eigenvalue Buckling Analysis
2. ( ) Nonlinear Buckling Analysis
3. ( ) Stress Stiffening Effects

Answers:

1. (B) 2. (C) 3. (A)

List of Descriptions

(A) A slender or thin structural member's bending stiffness increases with increasing axial tensile stress. Similarly, the member's bending stiffness decreases with increasing compressive stress. The extra (or deficient) stiffness is called the stress stiffness.

(B) Prediction of buckling loads and buckling modes based solely on initial stress stiffening effect. It doesn't account for large deformation effect.

(C) Prediction of buckling loads using a nonlinear analysis technique. It accounts for all nonlinear effects, including large deformation effect and nonlinear material effect.

10.4.2 Additional Workbench Exercises

Linear Buckling Analysis with Constant Loads

In the buckling analysis at the end of Section 10.1, we didn't include the lateral load (0.1 N/mm). You may wonder if the existence of the lateral load would alter the linear buckling load. Our engineering intuition tells us that a little lateral load would significantly decrease the buckling load. In linear buckling analysis, the large deformation effect is not considered. As a result, the lateral load has limited influence on the buckling loads. In our case, since the lateral load is so small, the buckling load predicted by a linear buckling analysis will be essentially the same regardless of the presence of the lateral load. This exercise requires you to verify this point.

To do this, you are performing a linear buckling analysis with constant loads. Remember that a linear buckling analysis reports a multiplier. Buckling loads are ALL the applied loads multiplied by the multiplier. If the model involves a constant load, then there is no sense to multiply that constant load unless the multiplier equals a unity. Iterating on buckling analysis and trying to obtain a multiplier of unity is exactly the approach you should use when performing a linear buckling analysis with constant loads.

Buckling Pressure of Bellows Joints

The bellows joint, simulated in Section 6.1, is subject to external pressure when used in the deep ocean. If the external pressure is larger than the internal pressure, then its stability must be checked. This exercise asks you to predict the net pressure (difference between external and internal pressures) that causes the bellows joint to buckle. Also study its buckling modes.

Buckling Torque of a Beverage Can

Applying a twist on a thin beverage can introduces tensile stress on a principal direction and compressive stress on another principal direction. Excess compressive stress may cause the skin to buckle. Model a beverage can as a cylindrical surface of length 122 mm, a diameter of 64 mm, and a thickness of 0.1 mm [1]. Assume that the can is made of AA3004, which has a Young's modulus of 68.9 MPa and a Poisson's ratio of 0.35. Predict the torque that causes the skin to buckle. Study the buckling modes [2].

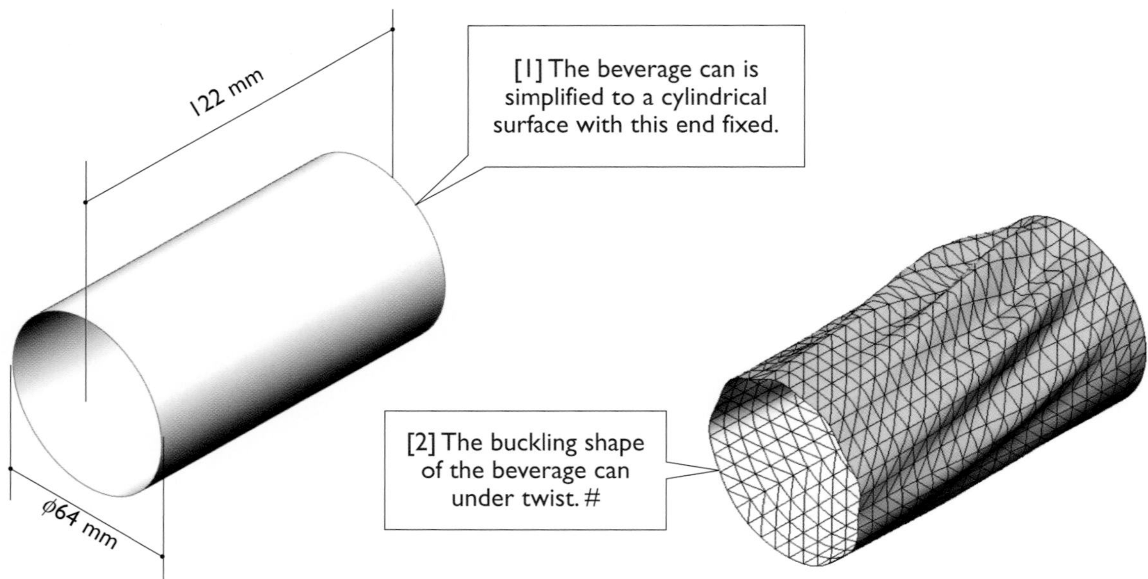

Chapter 11

Modal Analysis

When a structure deforms (e.g., vibrates) very fast, dynamic effects must be included in the governing equations. Modal analysis is a special type of dynamic simulation, which explores the behavior of free vibrations, vibrations without external forces. The most important characteristics of free vibrations are the natural frequencies and their corresponding vibration mode shapes.

Purpose of This Chapter

Why do we need to know the natural frequencies and the mode shapes? First, modal analysis has many usages in its own right; for example, to find the consonance frequencies of a structure in order to avoid or exploit them. This chapter provides several examples to demonstrate the usage of modal analyses. Second, these dynamic characteristics are important for further dynamic simulations; for example, transient dynamic simulations or harmonic response analyses. Chapter 12 will have several examples to demonstrate this.

About Each Section

Section 11.1 uses the gearbox, introduced in Section 6.3, as an example to illustrate a common consideration when designing a machine involving rotatory parts: avoiding resonance. Section 11.2 uses the two-story building, introduced in Section 7.3, to demonstrate the use of modal analysis to find the weakest direction of a structure and improve the stiffness of that direction. Section 11.3 discusses a case in the popular TV series *Mythbusters*, in which they succeeded in shattering CDs with a high rotational speed. Some people may believe the myth that the shattering is due to the excessive centrifugal stress. This section is designed to bust that myth, by proving that the shattering is due to resonant vibrations rather than the centrifugal stress. Section 11.4 discusses the physics of music, which is closely related to modal analysis, using a guitar string as an example.

Section 11.1

Gearbox

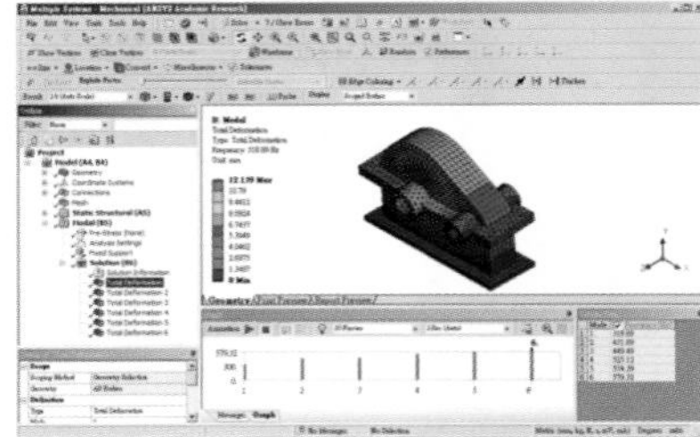

11.1.1 About the Gearbox

[1] In Section 6.3, we performed a static structural simulation for a gearbox. Deformation and stresses seem within safety margin under the static design loads. Dynamic behavior, however, should also be investigated.

The gearbox is designed for a speed reducer. The maximum speed at input is 630 rpm (10.5 Hz). The gearbox must be stiff enough such that, during the operation, resonance does not occur. Resonance, in this case, is harmful because it enlarges the deformation and the stresses. It also causes noises.

In this section, we want to perform a modal analysis to investigate the natural frequencies of the gearbox, to make sure these natural frequencies are much higher than the operational frequency (630 rpm).

You may raise a question: when evaluating the natural frequencies, whether the design bearing loads (6.3.7[1], page 261) should be applied or not. In general, the answer is yes, since the prestress will modify the stiffness (called the stress stiffening effect, see Section 10.1). In the old days, since a "prestressed modal analysis" was expensive to perform, engineers tended to neglect the prestress effect when, according to their experiences, they knew that the prestress effect was negligible.

In this section, we will perform modal analysis twice, one with prestress, another without, to show that the effect of the prestress in this case is negligible. Note that, in other cases, the effect of prestress may be significant. #

11.1.2 Resume the Project **GearBox**

[3] Double-click **Setup** (or any cells below) to start up **Mechanical**. Make sure the unit system is **mm-kg-N-s**. ↓

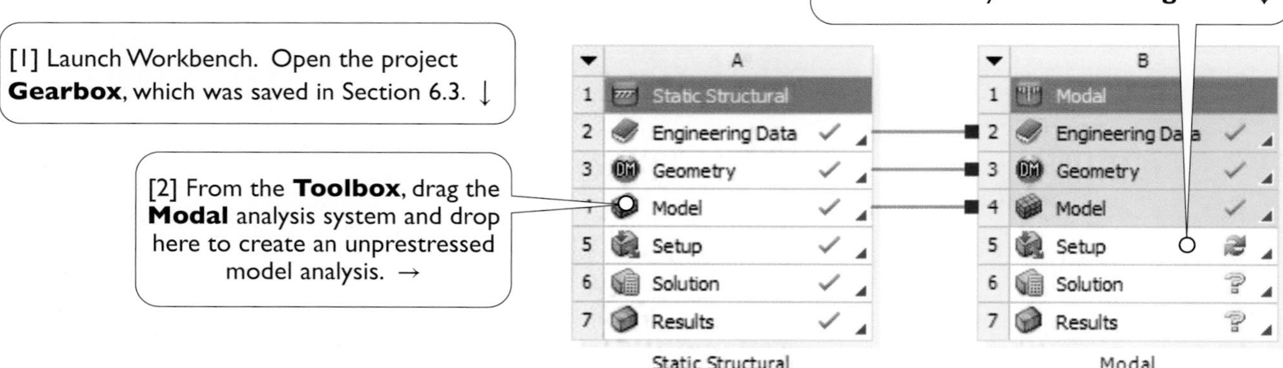

Static Structural analysis is not always needed for a modal analysis

[4] We are now performing an "unprestressed modal analysis." When performing an unprestressed modal analysis, a **Static Structural** analysis is not needed (i.e., you might delete it). We keep **Static Structural** system because we want to perform a prestressed modal analysis later (11.1.5, pages 393-394). #

11.1.3 Perform Modal Analysis

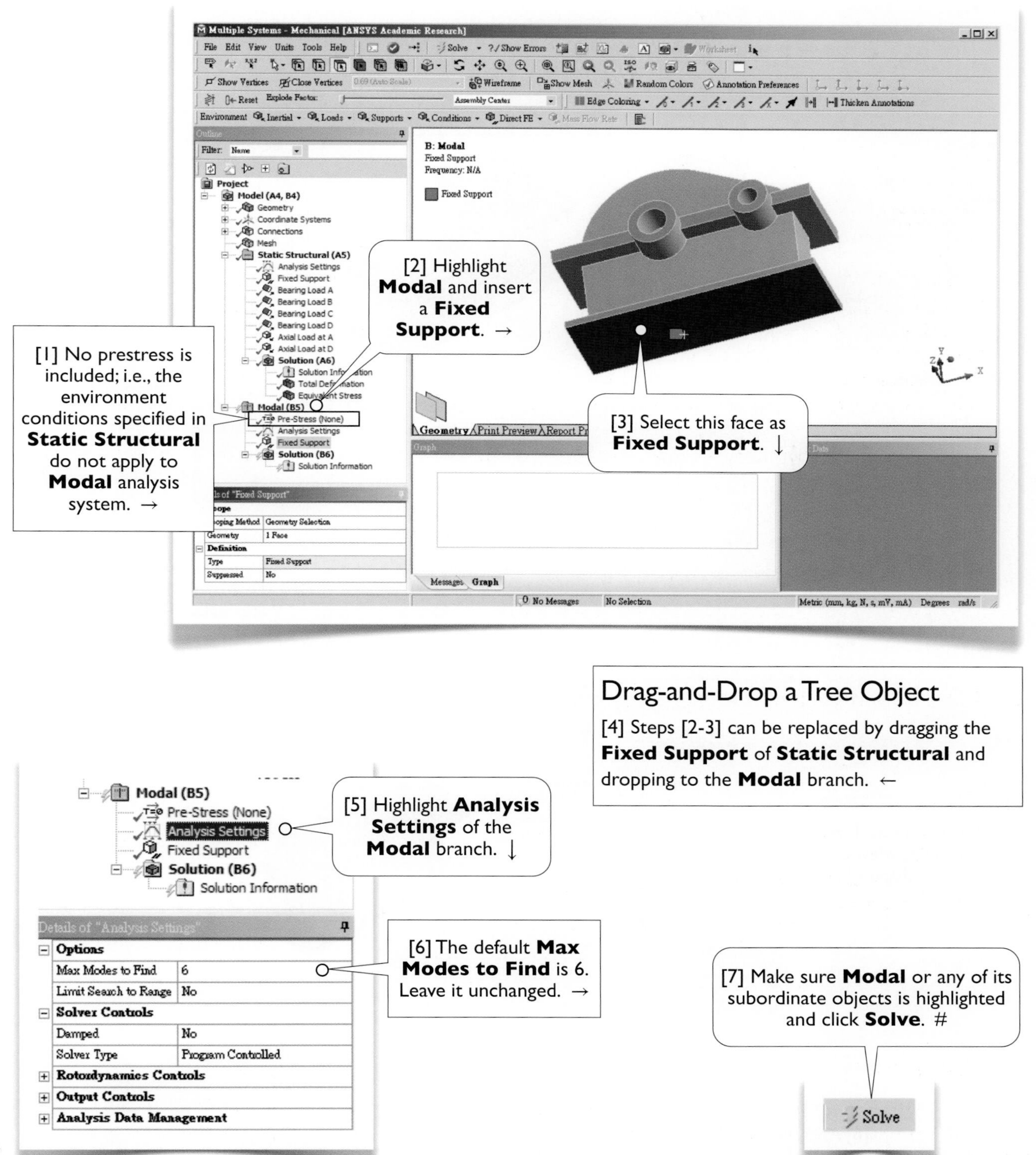

11.1.4 View the Results

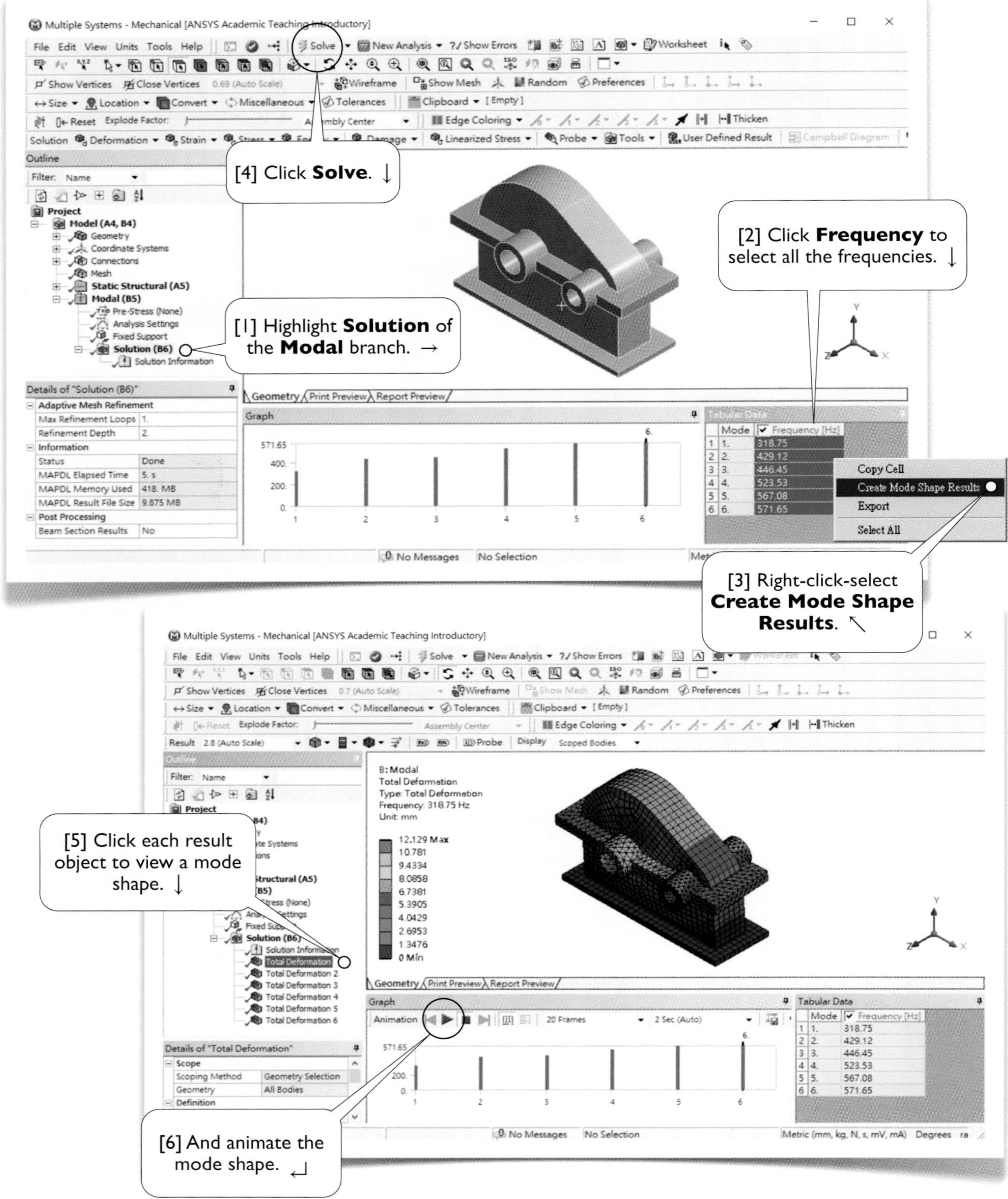

Free Vibration Mode Shapes

[7] The vibrations you observed in [5-6] (last page) are called *free vibrations* since no external forces are applied on the structure. Like the buckling mode shapes discussed in the last chapter, values of deformation (and the corresponding stress, strain) have no physical significance; it is the shape of a vibration mode that is meaningful.

Natural Frequencies

The frequencies corresponding to the free vibrations are called *natural frequencies*. The lowest natural frequency is called the *fundamental natural frequency*, or simply *fundamental frequency*. In this case, the fundamental frequency is 318.89 Hz (19,133 rpm), far beyond the operational frequency (630 rpm). Before we jump to conclude that the resonance is not an issue, let's make sure that the prestress is negligible.

Rigid Body Modes

When performing a modal analysis, you should provide enough supports to avoid any rigid body motions. If you didn't, then rigid body modes would be included in the results. Rigid body modes have infinite period, or, equivalently, zero frequencies. Rigid body modes are superfluous, and should not be present (see 10.1.4[5], page 370). #

11.1.5 Perform Prestressed Modal Analysis

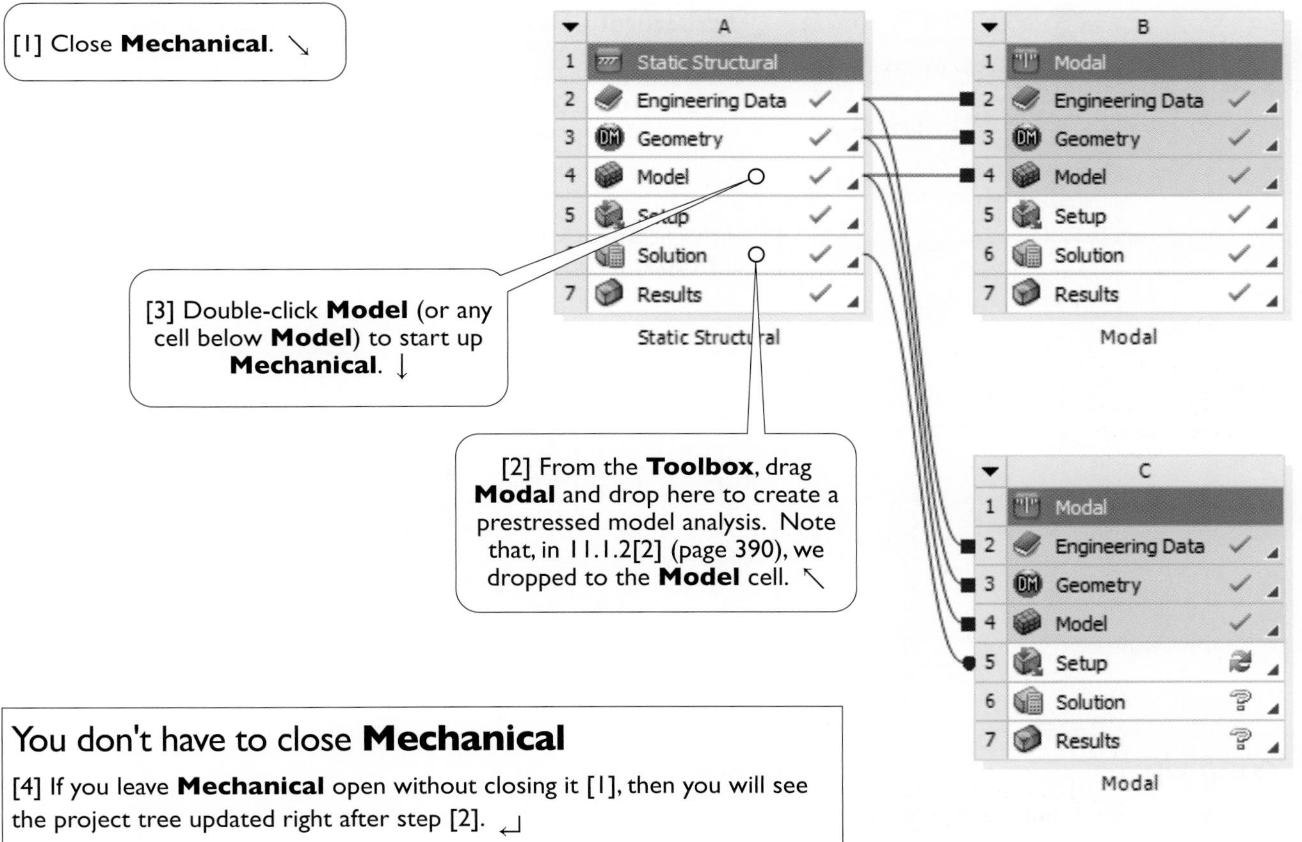

You don't have to close **Mechanical**

[4] If you leave **Mechanical** open without closing it [1], then you will see the project tree updated right after step [2]. ↵

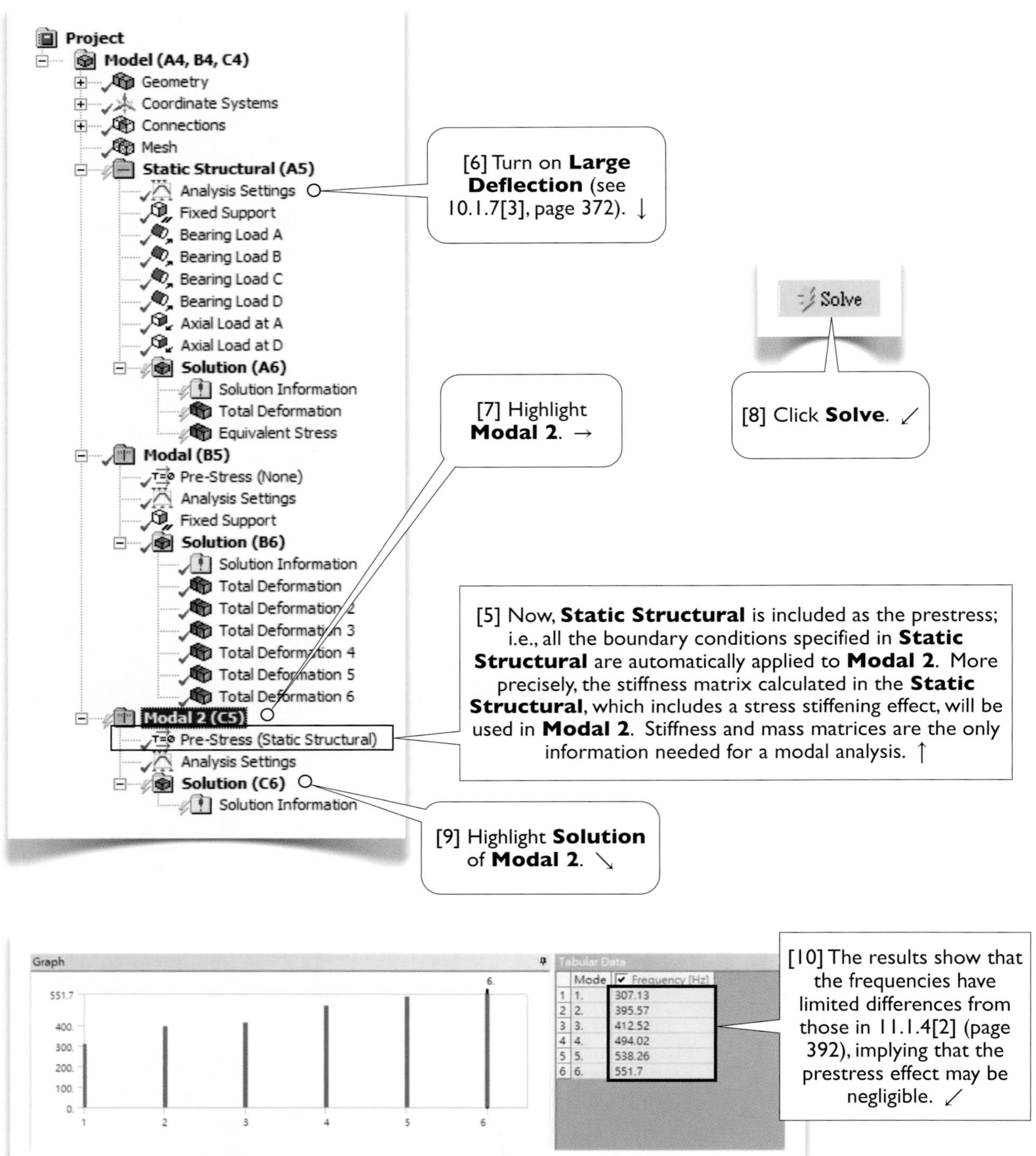

Wrap Up

[11] Save the project and exit Workbench. #

Section 11.2

Two-Story Building

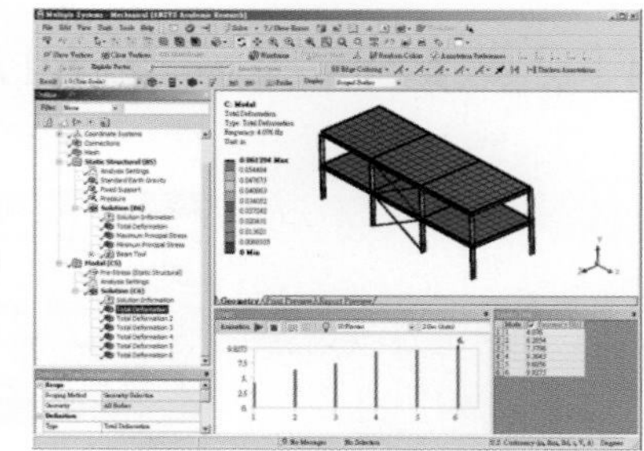

11.2.1 About the Two-Story Building

[1] In Section 7.3, we performed a static structural simulation for a two-story building. Under the static loads, the deformation and stresses are within the safety margin. Dynamic behavior, however, should be investigated for a case like this.

A structure's fundamental natural frequency is proportional to the structure's stiffness: the higher the frequency, the stiffer the structure. The stiffnesses of two structures can be compared using their fundamental frequencies. For a structure, stiffnesses in different directions can be compared using the lowest frequencies in the respective directions. When we want to reinforce a structure, we should first reinforce the direction which has the lowest frequency.

Local building codes usually require a minimum frequency so that a building would not be too soft (i.e., not stiff enough), both for comfort and safety concerns. In this section, we will perform modal analyses using the two-story building model. Prestress is included in the simulation. We will find that the fundamental frequency of the building is too low. We then propose a simple solution to improve the stiffness of the building. #

11.2.2 Resume the Project **Building**

[1] Launch Workbench. Open the project **Building**, which was saved in Section 7.3. ↙

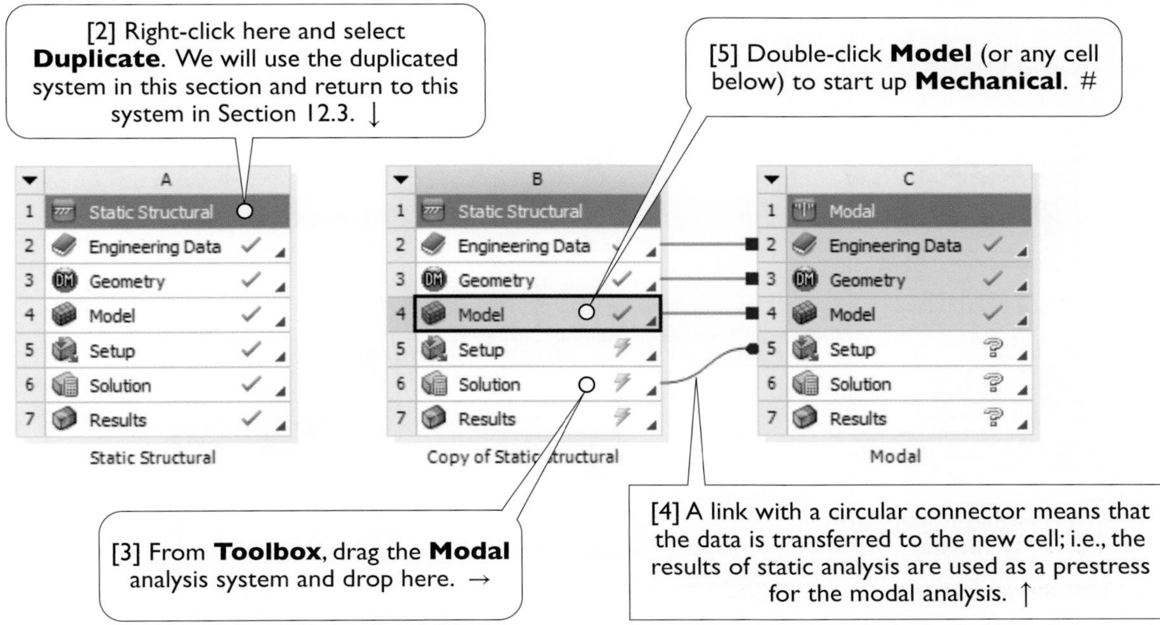

11.2.3 Perform Modal Analysis

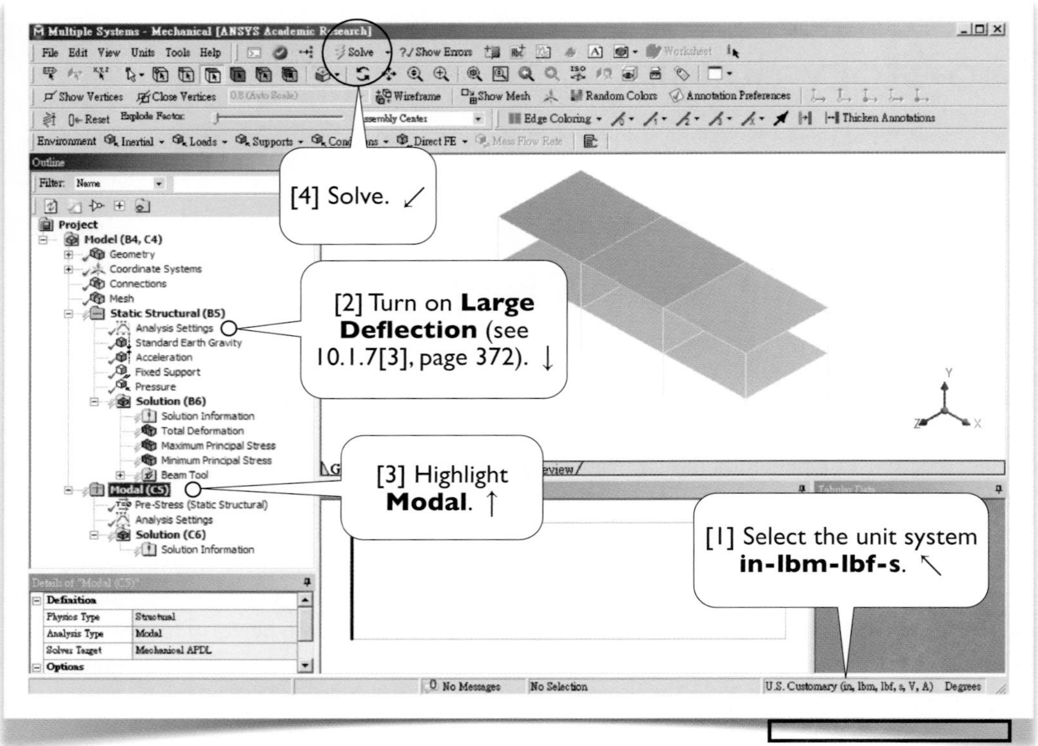

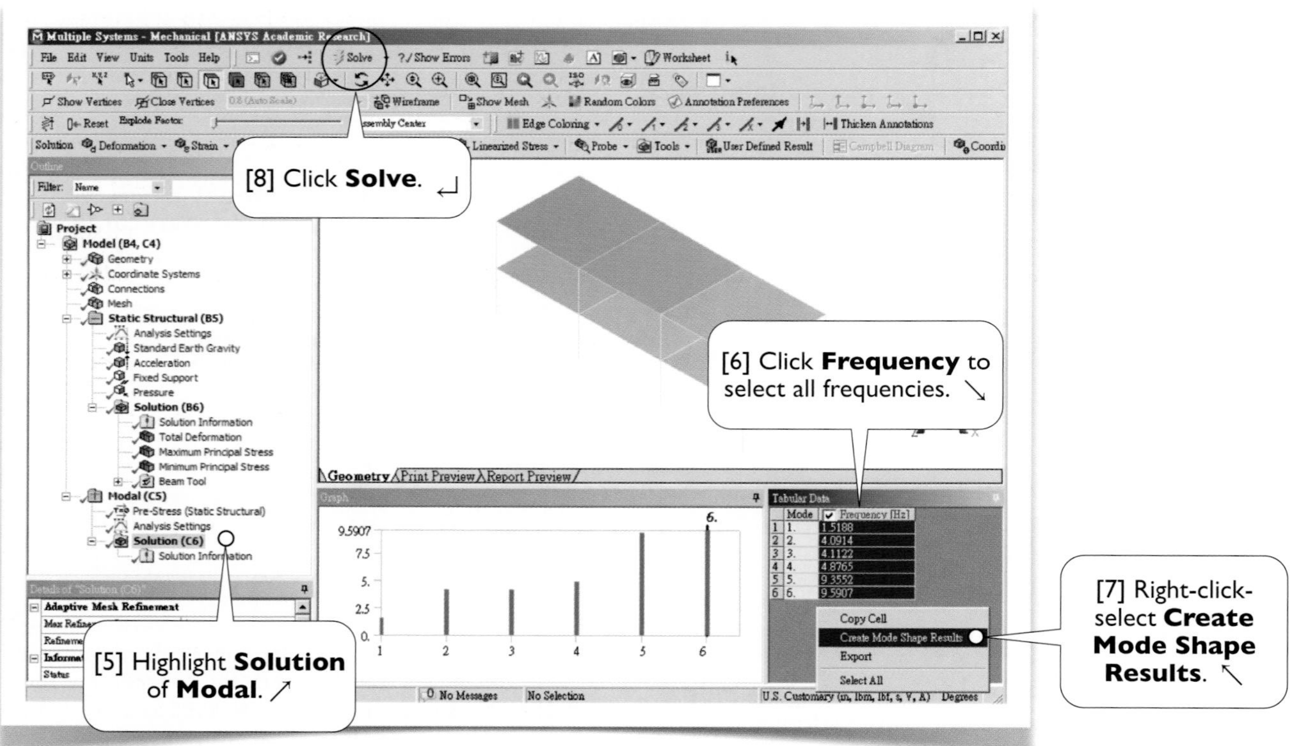

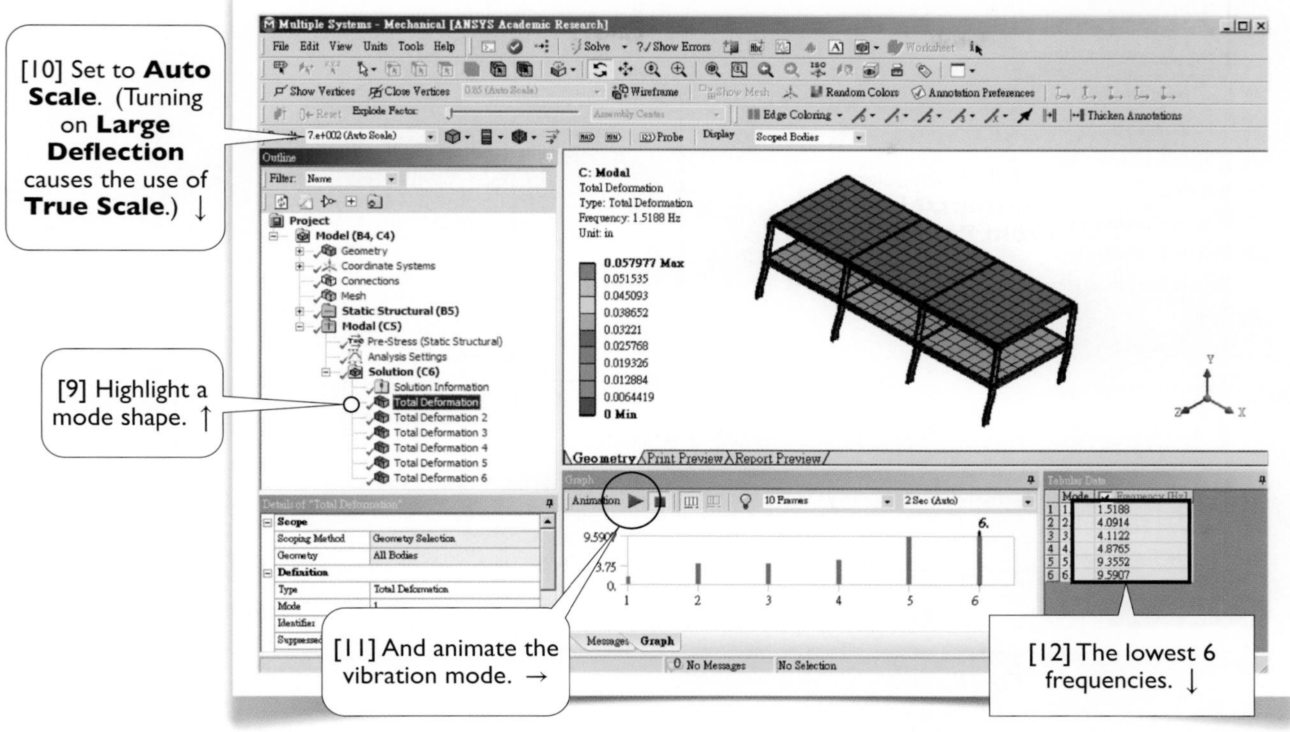

Discussion of the Vibration Modes

[13] The fundamental frequency is 1.52 Hz, a vibration in X-direction (this can be observed from the animation). The second lowest frequency is 4.09 Hz, a vibration in Z-direction. The third lowest frequency mode is also in X-direction. The fourth lowest frequency mode is a torsional vibration in XZ-plane. The fifth and sixth modes are vertical (Y-direction) vibrations of the floors.

A frequency of 1.52 Hz is not only uncomfortable but also unsafe. Imagine a group of young people dancing on the building's floor. The rhythmic loading of the floor may cause a safety issue, since the tempo of the music is possibly close to the building's fundamental frequency. A harmonic response analysis will be conducted to clear up this safety issue in Section 12.3. Some local building codes require a structure's natural frequency be larger than 5 Hz if the structure is to be used as a dance floor in a venue.

A simple remedy for the building is to add diagonal members to the weakest direction. Location of the diagonal members should be carefully chosen to avoid conflicting with the building's architectural functionalities. #

11.2.4 Modify the Geometry

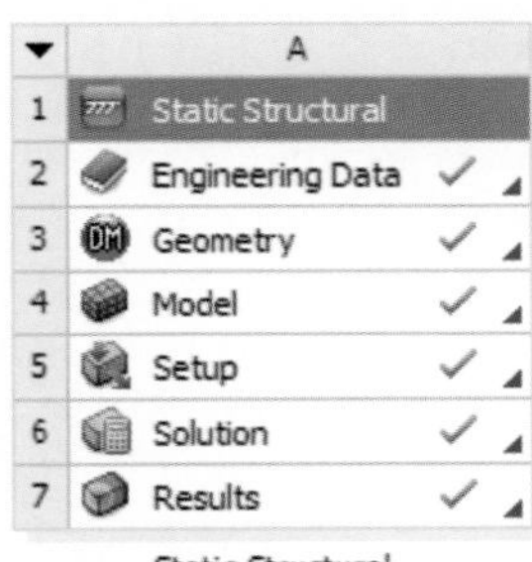

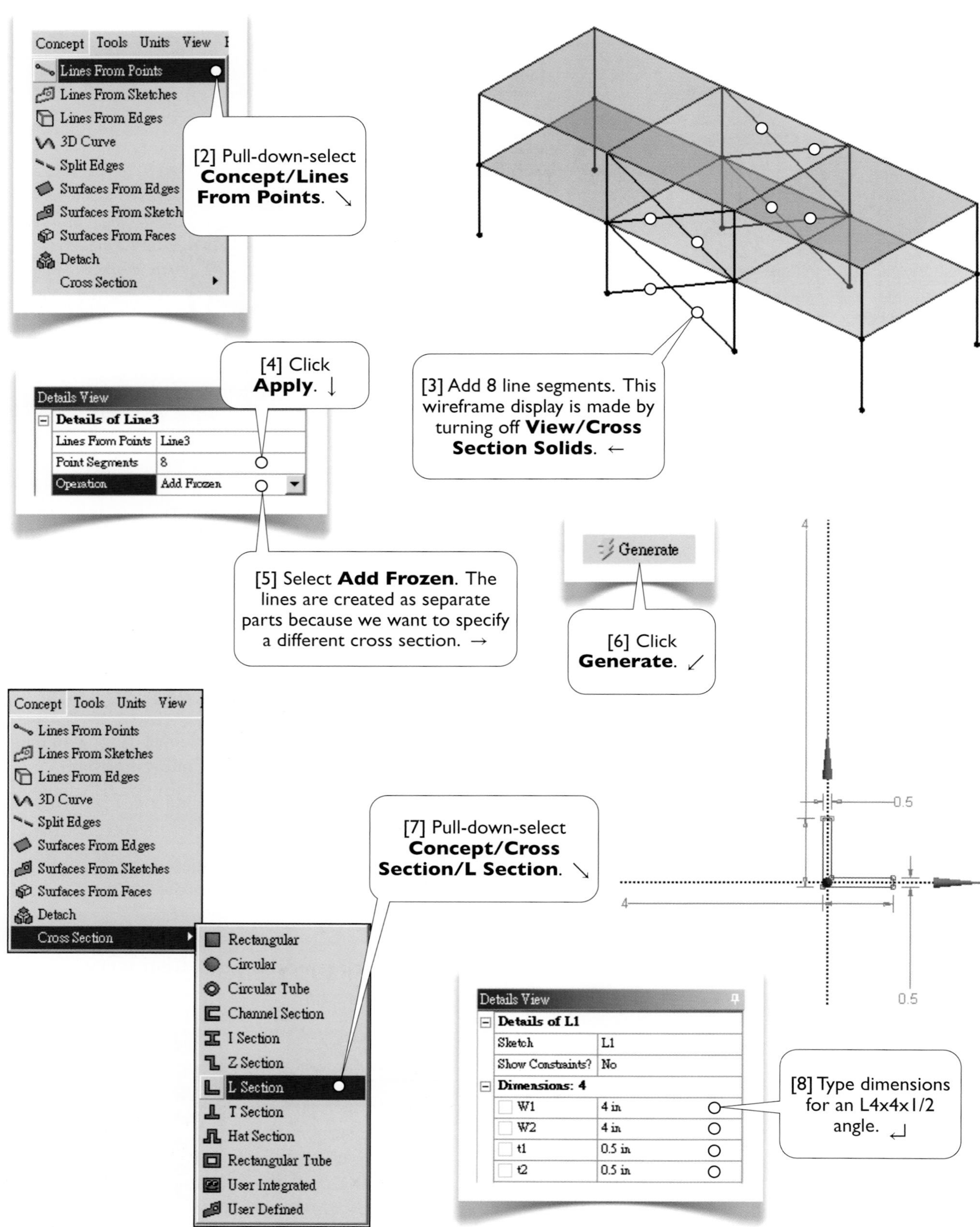

Concept Tools Units View
Lines From Points
Lines From Sketches
Lines From Edges
3D Curve
Split Edges
Surfaces From Edges
Surfaces From Sketches
Surfaces From Faces
Detach
Cross Section
[2] Pull-down-select **Concept/Lines From Points**. ↘
[3] Add 8 line segments. This wireframe display is made by turning off **View/Cross Section Solids**. ←
[4] Click **Apply**. ↓
Details View
Details of Line3
Lines From Points | Line3
Point Segments | 8
Operation | Add Frozen
[5] Select **Add Frozen**. The lines are created as separate parts because we want to specify a different cross section. →
Generate
[6] Click **Generate**. ↙
Rectangular
Circular
Circular Tube
Channel Section
I Section
Z Section
L Section
T Section
Hat Section
Rectangular Tube
User Integrated
User Defined
[7] Pull-down-select **Concept/Cross Section/L Section**. ↘
4
0.5
4
0.5
Details View
Details of L1
Sketch | L1
Show Constraints? | No
Dimensions: 4
W1 | 4 in
W2 | 4 in
t1 | 0.5 in
t2 | 0.5 in
[8] Type dimensions for an L4x4x1/2 angle. ↲

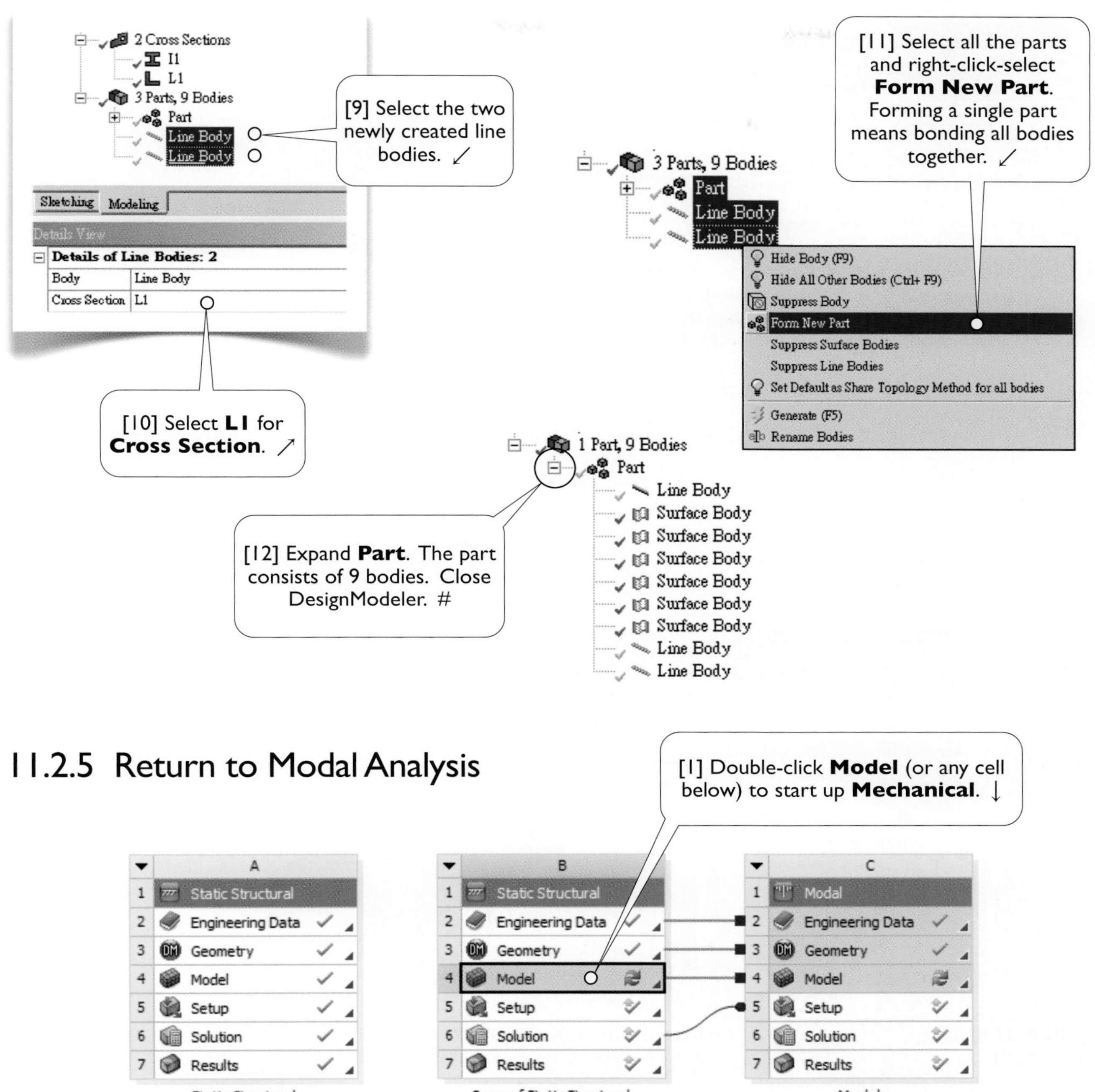

11.2.5 Return to Modal Analysis

Check the environment conditions every time you modify geometry

[2] After you modify the geometry, always check environment conditions such as loads and supports. In this case, some of the fixed supports are lost; we need to fix them (see [3], next page). We'll delete **Acceleration** in **Static Structural** branch [4], which represents an earthquake load. We usually don't consider earthquake load as a "prestress." We do consider earth gravity as prestress. However, the effect of the prestress on natural frequencies is very limited. This is left as an exercise (11.5.2, page 418).

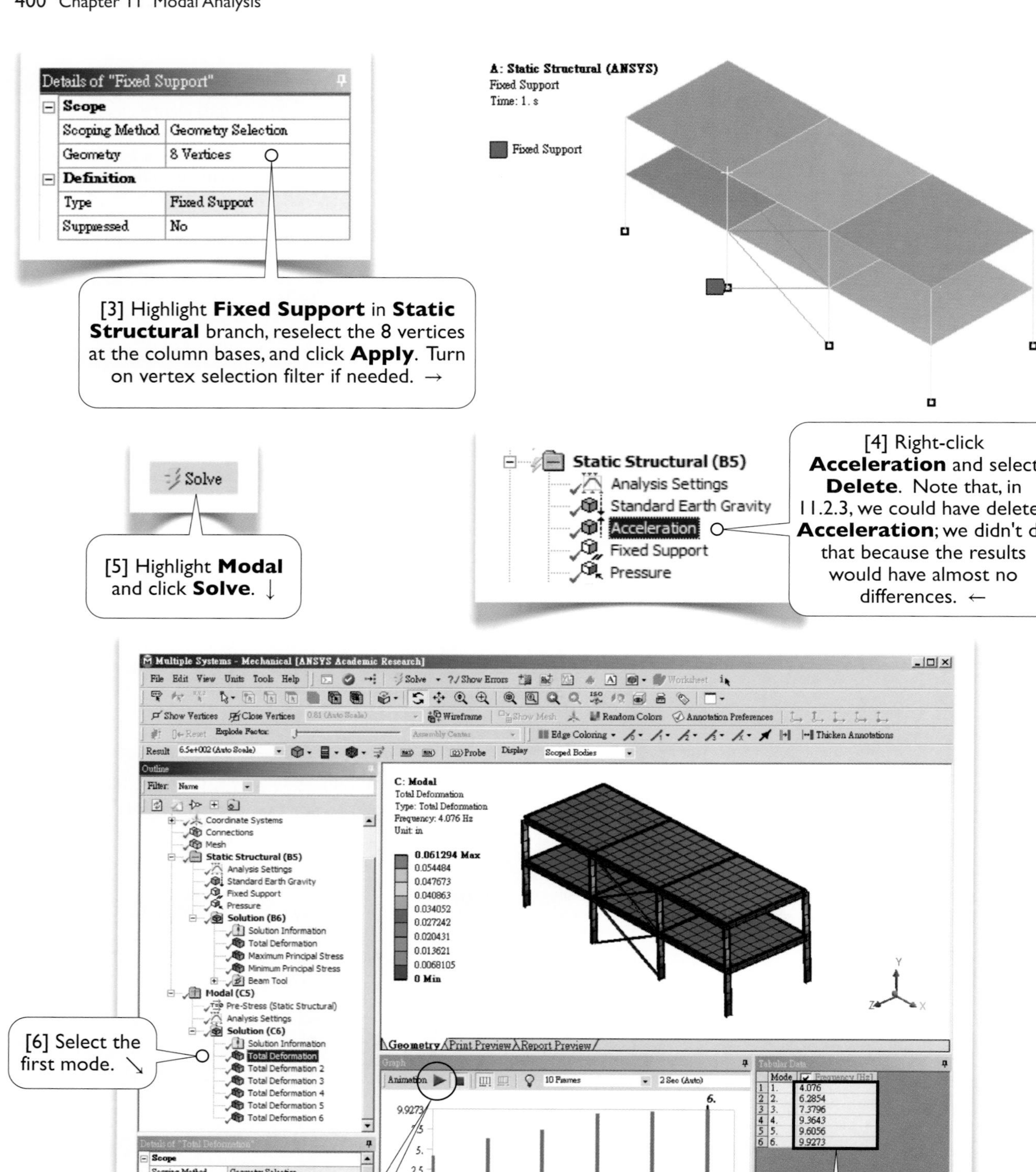

[3] Highlight **Fixed Support** in **Static Structural** branch, reselect the 8 vertices at the column bases, and click **Apply**. Turn on vertex selection filter if needed. →

[4] Right-click **Acceleration** and select **Delete**. Note that, in 11.2.3, we could have deleted **Acceleration**; we didn't do that because the results would have almost no differences. ←

[5] Highlight **Modal** and click **Solve**. ↓

[6] Select the first mode. ↘

[7] The 1.52 Hz mode disappears. The structure's stiffness is improved. The building now has a fundamental frequency of 4.1 Hz. ←

[8] Click **Play** to animate the first mode. Now, Z-direction is the weakest direction. ↵

Discussion

[9] After the reinforcement of X-direction, Z-direction now becomes the weakest direction ([8], last page). If we want to further improve the structure's stiffness, we may add bracing members in the Z-direction. This is left as an exercise (11.5.2, page 418). ↓

Wrap Up

[10] Save the project and exit Workbench. #

Section 11.3

Compact Disk

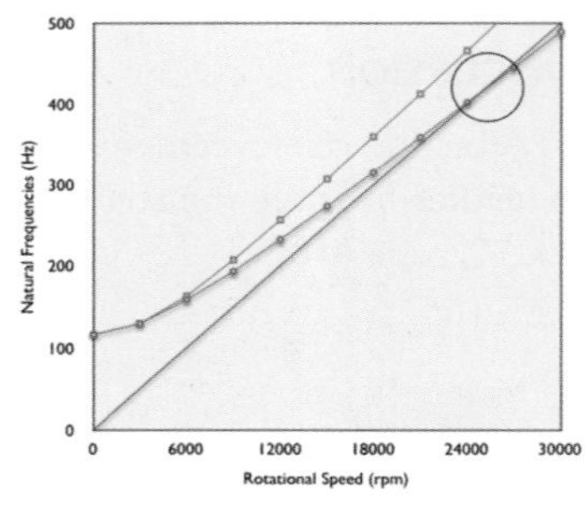

11.3.1 About the Compact Disk

[1] A CD is made of a polycarbonate (PC) [2], with density 1200 kg/m^3, Young's modulus 2.2 GPa, Poisson's ratio 0.37, and tensile strength 65 MPa. The outer diameter is 120 mm, the inner (hole) diameter is 15 mm, and the thickness is 1.2 mm. For a 52x CD drive, the maximum rotational speed reaches 27,500 rpm (458 Hz) when reading the inner tracks[Ref 1].

The television series *MythBusters* conducted experiments[Refs 2-4] in which they succeeded in shattering CDs at speeds of 23,000 rpm. When conducting the experiments, they press the CD between two nuts, which are 27 mm in diameter[Ref 5].

In this section, we first want to find out the maximum stress in the CD due to the centrifugal force when rotating in 27,500 rpm, to justify that the shattering may not be due to the centrifugal stress.

Second, we want to find the natural frequencies of the CD to investigate the possibility of resonant vibrations. We will conclude that the CD shattering may be due to sustaining vibrations rather than centrifugal stress. Also, we want to demonstrate that the natural frequencies increase with increasing rotational speed. →

11.3.2 Start Up a New Project

[1] Launch Workbench. Save the project as **CD**. Create a **Static Structural** analysis system [2]. Use **SI** as the project units. Double-click **Engineering Data** [3] and create a new material with the name **PC** ([4], next page). Input the material properties: a density (1200 kg/m^3), a Young's modulus (2.2 GPa), and a Poisson's ratio (0.37) [5-7]. Click **Project** to return to **Project Schematic** [8]. →

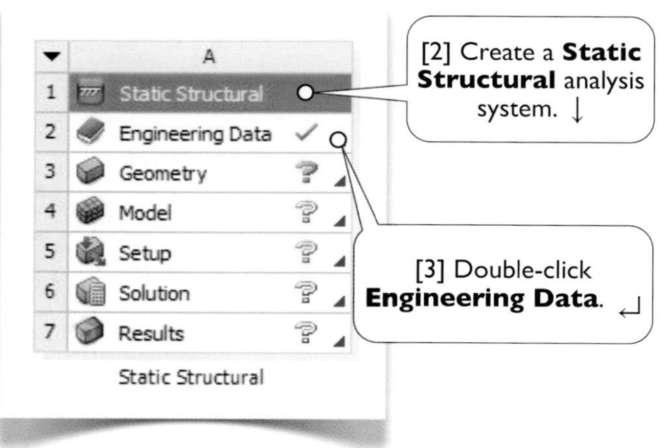

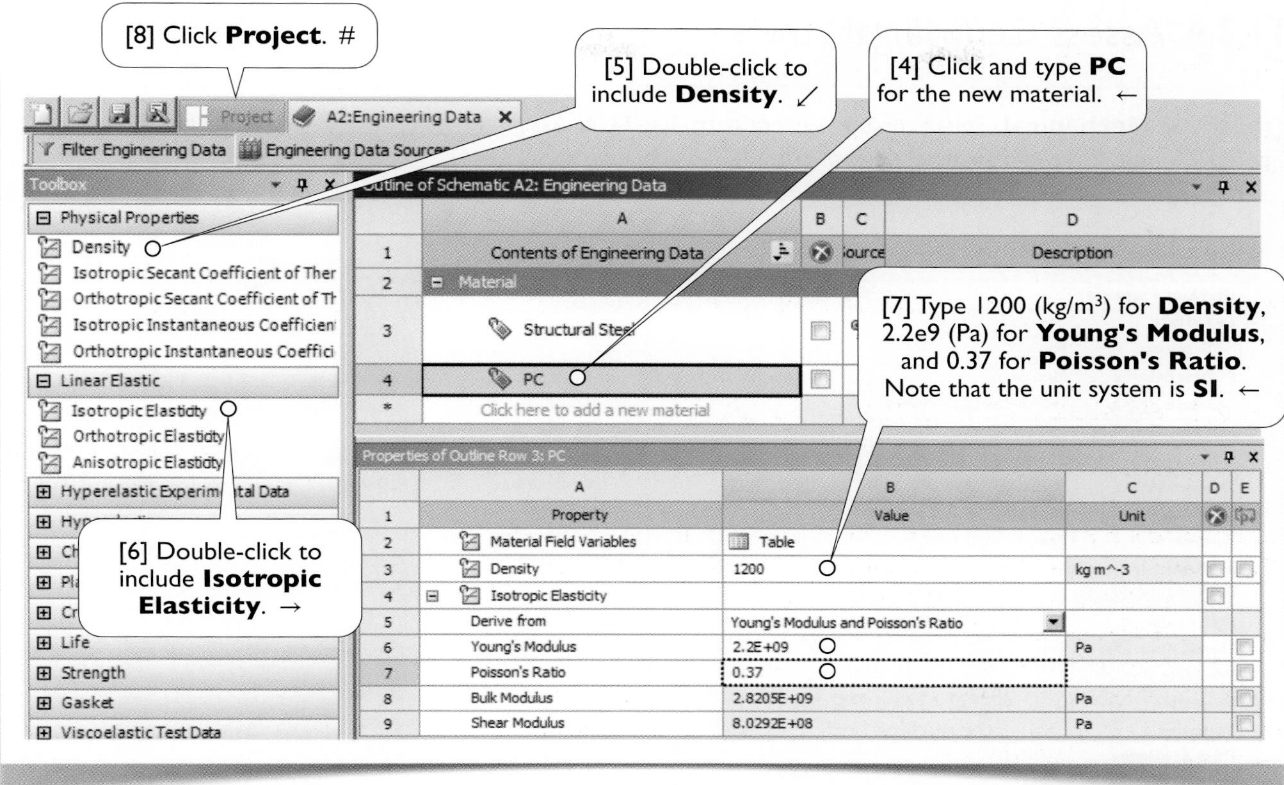

11.3.3 Create Geometry in DesignModeler

[1] Start up DesignModeler. Select **Millimeter** as the length unit. Draw a sketch on **XYPlane** as shown in [2]. Pull-down-select **Concept/Surfaces From Sketches** to create a surface body from the sketch [3]. Close DesignModeler. ↓

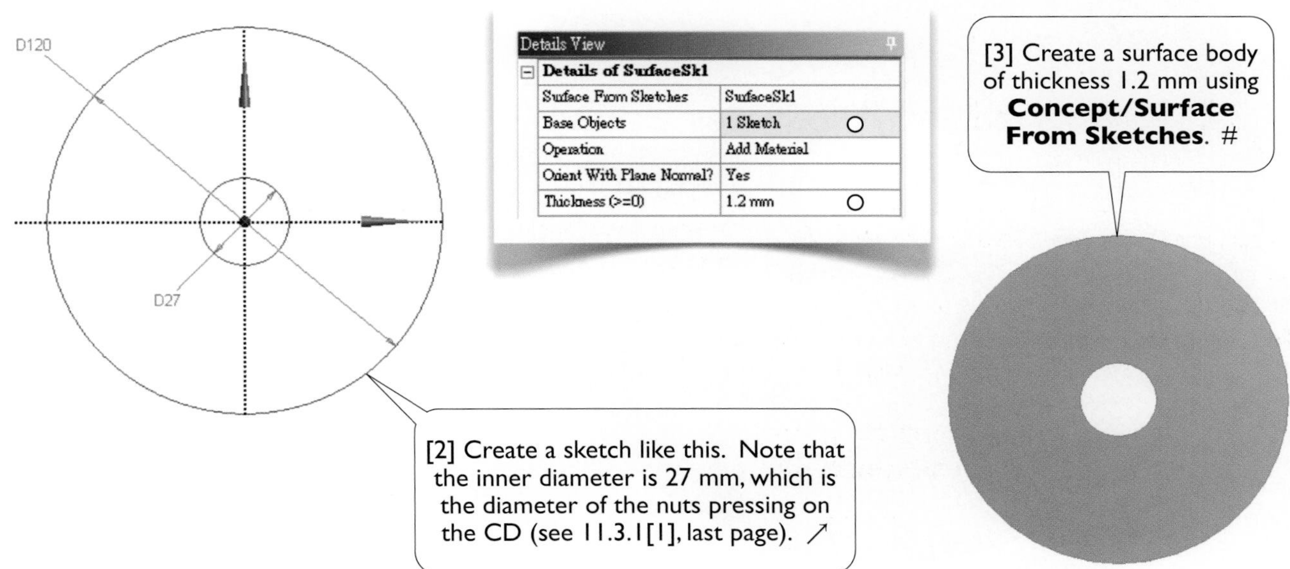

11.3.4 Assess Centrifugal Stress

[1] Start up **Mechanical**. Select the unit system **mm-kg-N-s**. Assign the material **PC** for the surface body [2]. Specify 2.0 (mm) for the maximum element size [3] and generate mesh. ↙

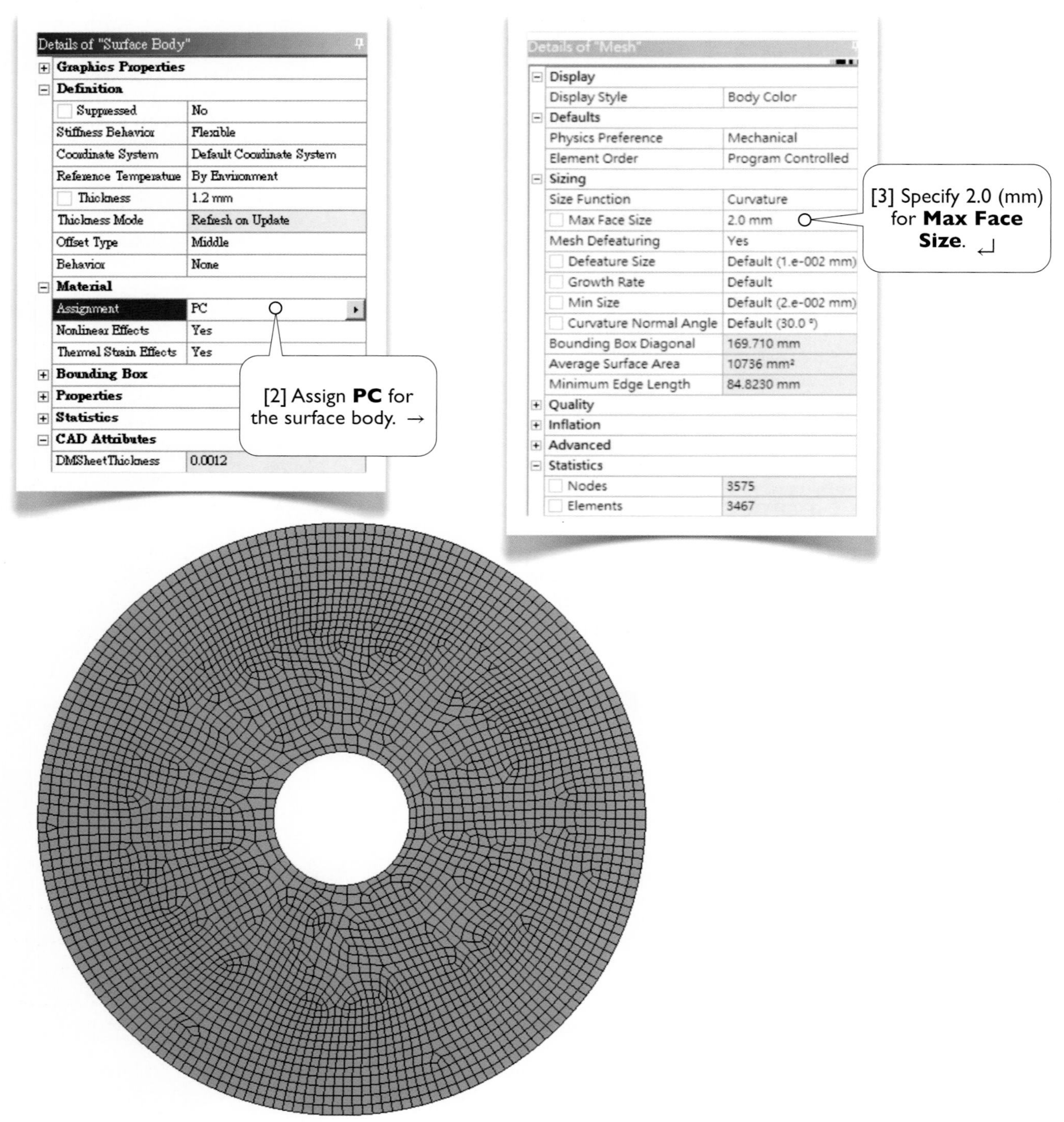

[4] Select the inner rim and insert a **Fixed Support** [5]. Insert an **Inertial/Rotational Velocity** load and type 27,500 (RPM) for **Z Component** [6-7]. Insert a **Maximum Principal Stress** [8] and solve the model [9]. ↙

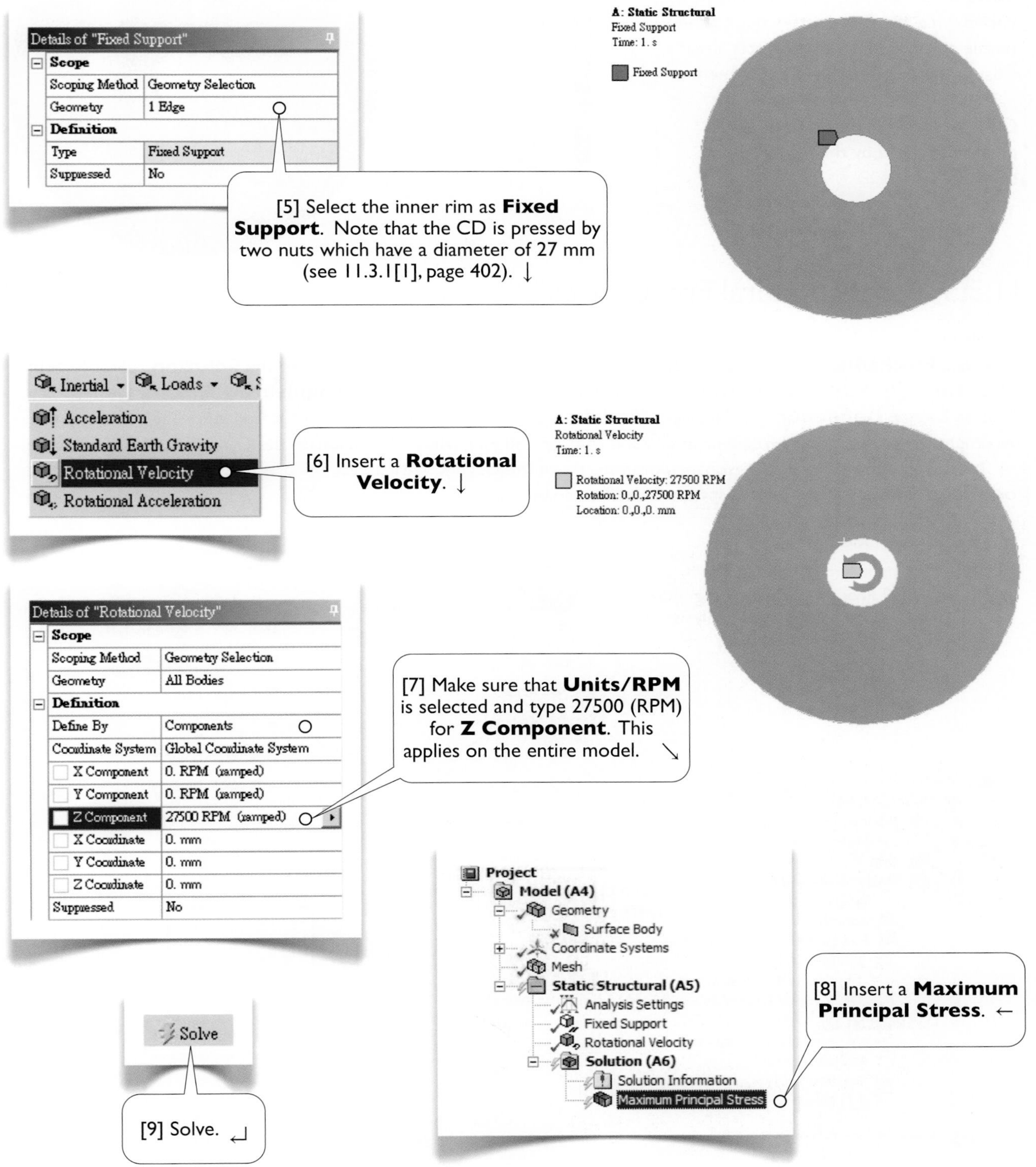

[10] The maximum tensile stress is about 20 MPa [11], which occurs at the inner rim as expected and is far less than the material's tensile strength (65 MPa). It is unlikely that the CD shatters due to the tensile stress, unless some defects already exist in the CD before the experiments.

As the next step, let's assess the CD's natural frequencies under the prestress caused by the high-speed spinning. →

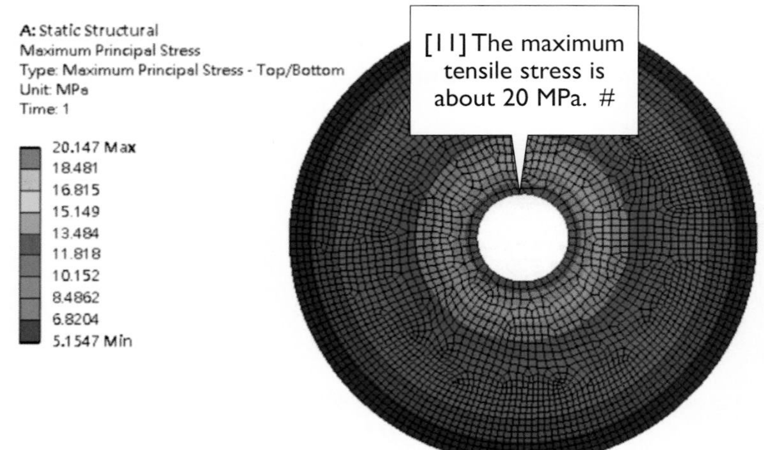

11.3.5 Assess Natural Frequencies

[1] Close **Mechanical**. From **Toolbox**, drag-and-drop the **Modal** analysis system to the **Solution** cell of **Static Structural** [2]. Start up **Mechanical** by double-clicking any cells other than **Engineering Data** or **Geometry**. Turn on **Large Deflection** [3]. Highlight **Modal** and solve the model [4]. The lowest frequency [5] is close to the maximum of the operational frequencies (2,7500 rpm, or 458 Hz), which may resonate the CD and may cause damage.

Let's look into this issue of resonant vibrations more thoroughly. We'll assess the prestressed natural frequencies for a range of possible operational speeds, from 0 to 30,000 rpm. ↙

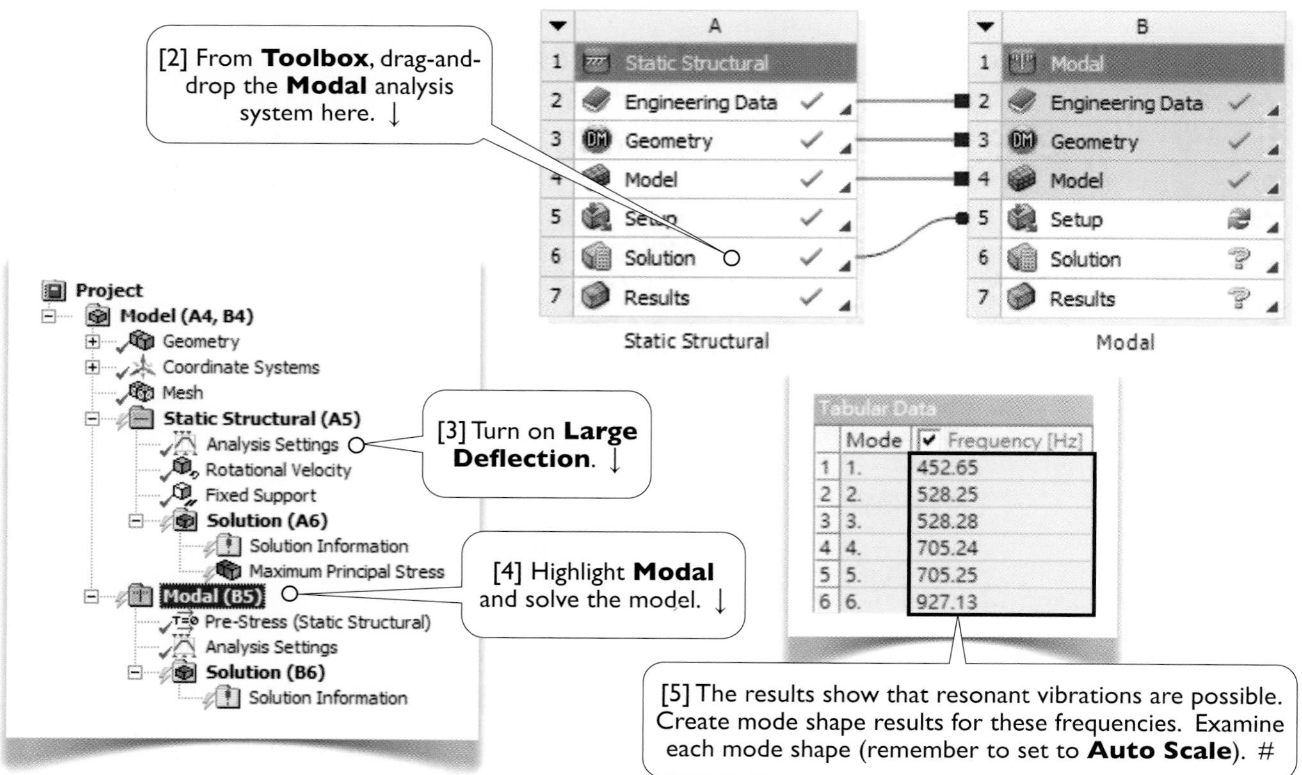

11.3.6 Assess Natural Frequencies over the Range of Rotational Speeds

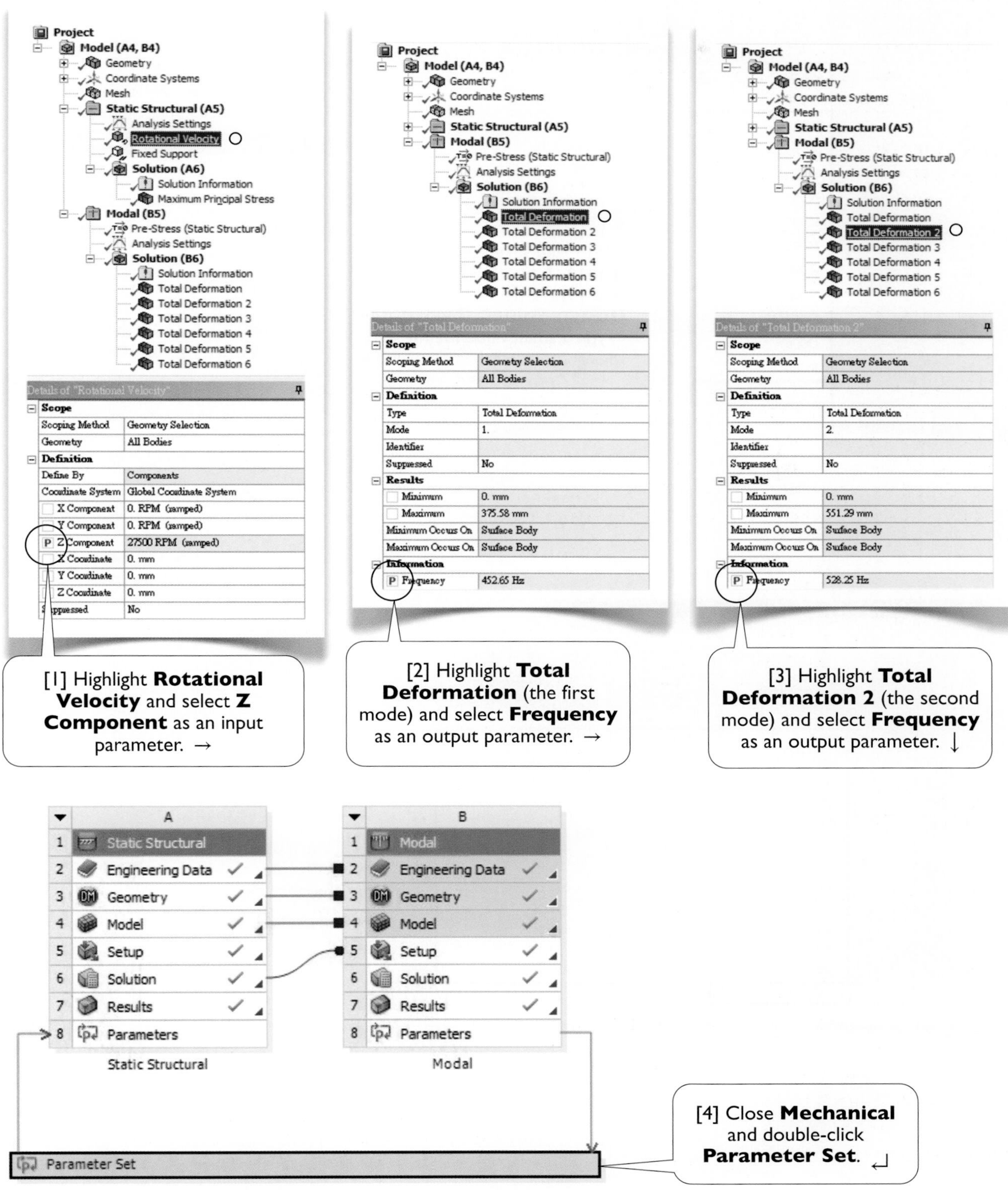

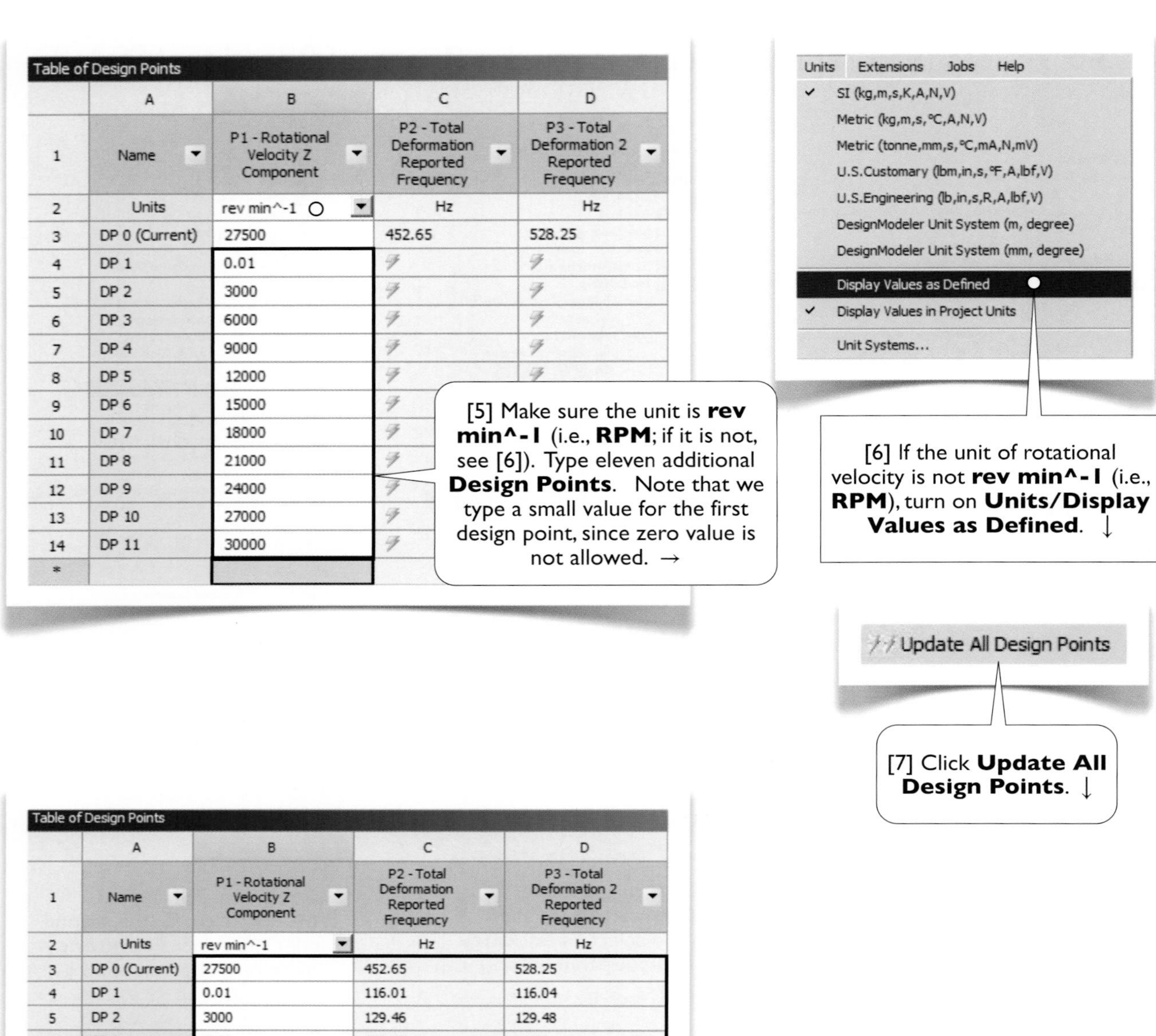

Table of Design Points

	A	B	C	D
1	Name	P1 - Rotational Velocity Z Component	P2 - Total Deformation Reported Frequency	P3 - Total Deformation 2 Reported Frequency
2	Units	rev min^-1	Hz	Hz
3	DP 0 (Current)	27500	452.65	528.25
4	DP 1	0.01		
5	DP 2	3000		
6	DP 3	6000		
7	DP 4	9000		
8	DP 5	12000		
9	DP 6	15000		
10	DP 7	18000		
11	DP 8	21000		
12	DP 9	24000		
13	DP 10	27000		
14	DP 11	30000		
*				

Table of Design Points

	A	B	C	D
1	Name	P1 - Rotational Velocity Z Component	P2 - Total Deformation Reported Frequency	P3 - Total Deformation 2 Reported Frequency
2	Units	rev min^-1	Hz	Hz
3	DP 0 (Current)	27500	452.65	528.25
4	DP 1	0.01	116.01	116.04
5	DP 2	3000	129.46	129.48
6	DP 3	6000	158.92	163.13
7	DP 4	9000	193.16	207.11
8	DP 5	12000	231.99	255.88
9	DP 6	15000	273.11	306.97
10	DP 7	18000	315.44	359.27
11	DP 8	21000	358.42	412.26
12	DP 9	24000	401.78	465.65
13	DP 10	27000	445.37	519.29
14	DP 11	30000	489.11	573.1
*				

[8] We'll use these results to plot a chart on the next page. ↵

[9] The data [8] (last page) are plotted in a chart as shown [10-11], which shows that the CD may be resonated when the rotational speed reaches about 25000 rpm [12-13]. The excitation may in turn cause damage.

According to these studies, we may conclude that the CD shattering in the experiments conducted by the TV series *Mythbusters* is probably due to resonant vibration effects, rather than the centrifugal stress.

A real CD drive usually provides supports for the CD, to reduce the vibrations as well as the radial deformations; therefore, the vibrations may not occur in a real CD drive. ↓

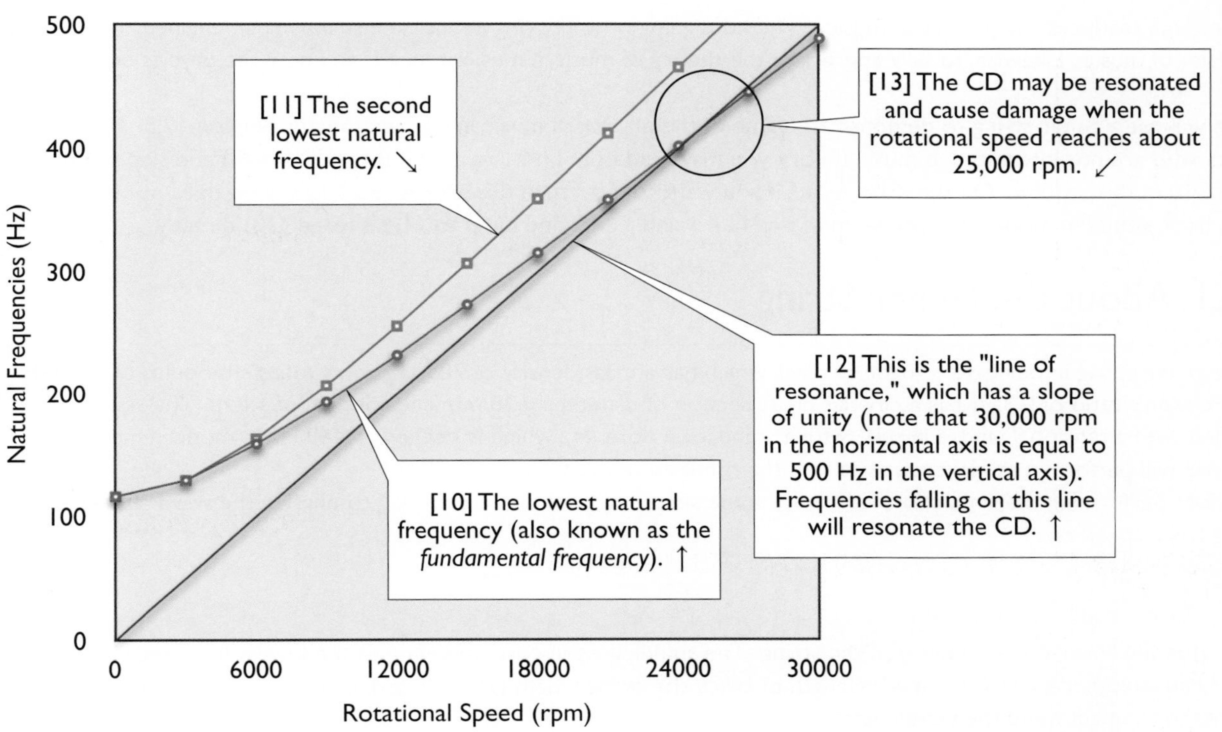

Wrap Up

[14] Save the project and exit Workbench. #

References

1. Wikipedia>Optical disc drive
2. https://www.youtube.com/watch?v=g7puuZlh-oM
3. Wikipedia>MythBusters (2003 season)>Episode 2.3 Exploding CDs.
4. https://www.youtube.com/watch?v=zs7x1Hu29Wc
5. Acknowledgement: Thanks to Professor Per Blomqvist of the University of Gavle in Sweden, who pointed out this fact and suggested a more realistic simulation of this exercise. In the earlier versions of the book, I assumed that the CD was fixed in the inner rim, which has a diameter of 15 mm.

Section 11.4

Guitar String

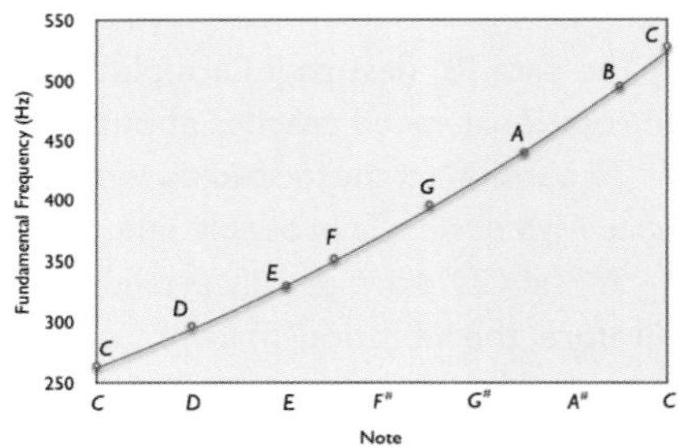

This section introduces the physics of music. When designing or improving a musical instrument, an engineer must know the physics of music. Likewise, to fully appreciate the theory of music, a musician needs to know the physics behind the music.

We will use a guitar string to demonstrate some of the physics of music in this section and Section 12.5. For those students who are not interested in music theory, you may read only 11.4.1 to 11.4.3 (pages 410-413) and skip the rest of the material in this section. On the other hand, if you want to introduce this article to a friend who does not have enough background in modal analyses, he may skip 11.4.1 and 11.4.2 and jump to 11.4.3 (page 413) directly.

11.4.1 About the Guitar String

[1] The guitar string in our case is made of steel, which has a mass density of 7850 kg/m^3, a Young's modulus of 200 GPa, and a Poisson's ratio of 0.3. It has a circular cross section of diameter 0.28 mm and a length of 1.0 m. The string is stretched with a tension T, and is in tune with a standard A note (*la*), which is defined as 440 Hz in modern music. In 11.4.2, we will perform a modal analysis to find the required tension T.

Before performing the simulation, let's make some simple calculations. According to physics, the wave traveling on a string has a speed of

$$v = \sqrt{\frac{T}{\mu}} \tag{1}$$

Where μ is the linear density (kg/m) of the string. The standing wave corresponding to the lowest frequency is called the *first harmonic mode*, which has a wavelength of twice the string length ($2L$). According to the relation between the velocity, the frequency, and the wavelength,

$$f = \frac{v}{\lambda} = \frac{v}{2L} \tag{2}$$

According to (1) and (2), we can estimate the required tension,

$$T = \mu\left(2fL\right)^2 = 7850 \times \frac{\pi(0.00028)^2}{4}\left(2 \times 440 \times 1.0\right)^2 = 374.32 \text{ N} \tag{3}$$ #

11.4.2 Perform Modal Analysis

[1] Launch Workbench. Save the project as **String**. Create a **Static Structural** system. Drag-and-drop the **Modal** analysis system to the **Solution** cell of the **Static Structural** system. In **Engineering Data**, make sure the material properties for **Structural Steel** are consistent with those of the guitar string given in 11.4.1 (a mass density of 7850 kg/m^3, a Young's modulus of 200 GPa, and a Poisson's ratio of 0.3).

Start up DesignModeler. Select **Millimeter** as the length unit. On **XYPlane**, draw a line of 1000 mm on X-axis. Create a line body from the sketch. Create a circular cross section of radius 0.14 mm, and assign the cross section to the line body. Close DesignModeler. ↵

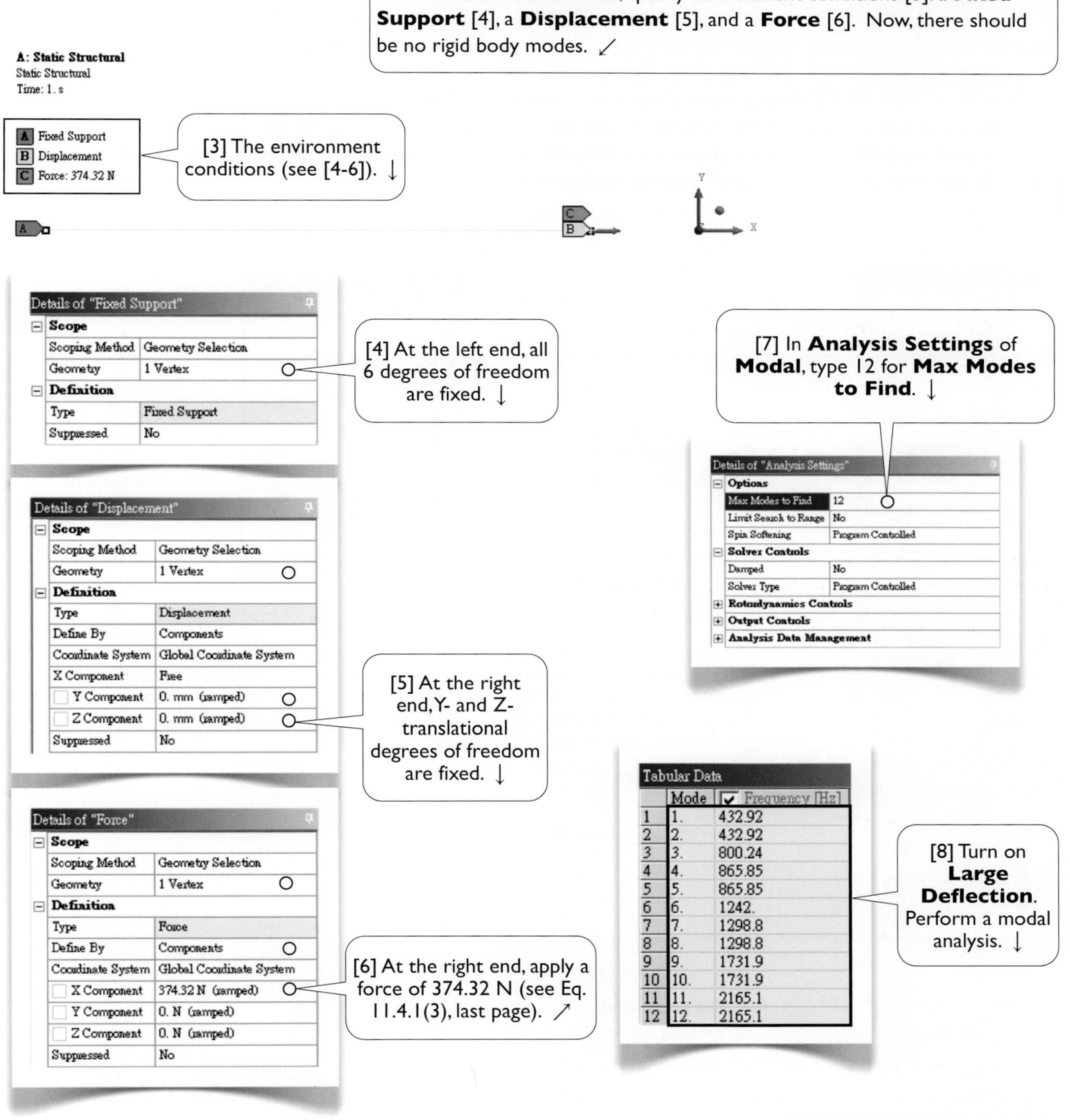

	Mode	Frequency [Hz]
1	1.	432.92
2	2.	432.92
3	3.	800.24
4	4.	865.85
5	5.	865.85
6	6.	1242.
7	7.	1298.8
8	8.	1298.8
9	9.	1731.9
10	10.	1731.9
11	11.	2165.1
12	12.	2165.1

[9] The results of the modal analysis [8] show that the lowest frequency is not exactly 440 Hz. It slightly deviates from what we've predicted using Eq. 11.4.1(3), last page. The main reason is that when applying the tension, the length of the string increases (about 30 mm). Another reason is due to the use of beam model, which is slightly different from the pure tension-only model used in our hand-calculation.

[10] After several trial-and-errors (try this on your own), we come up with a tension of 387.16 N [11] that exactly produces a fundamental frequency of 440 Hz [12].

As expected, the frequencies [12] include 440 Hz, 880 Hz, 1320 Hz, 1760 Hz, 2200 Hz (which are all integral multiplications of the fundamental frequency, 440 Hz), with negligible numerical errors. Before we discuss these "harmonic modes" in 11.4.3 (next page), let's take a look at those non-harmonic modes first.

The third mode (800.8 Hz) is a rotation mode in X-direction [13-14]. In the real-world, it exists, but its magnitude is usually very small, and our ears hardly sense it since the air pressure is hard to be excited in this way.

The sixth mode (1241.3 Hz) is a stretching mode [15], visible in an animation. In the real-world, it exists too, but again, we hardly sense it since the air pressure is hard to be excited in this way. ↙

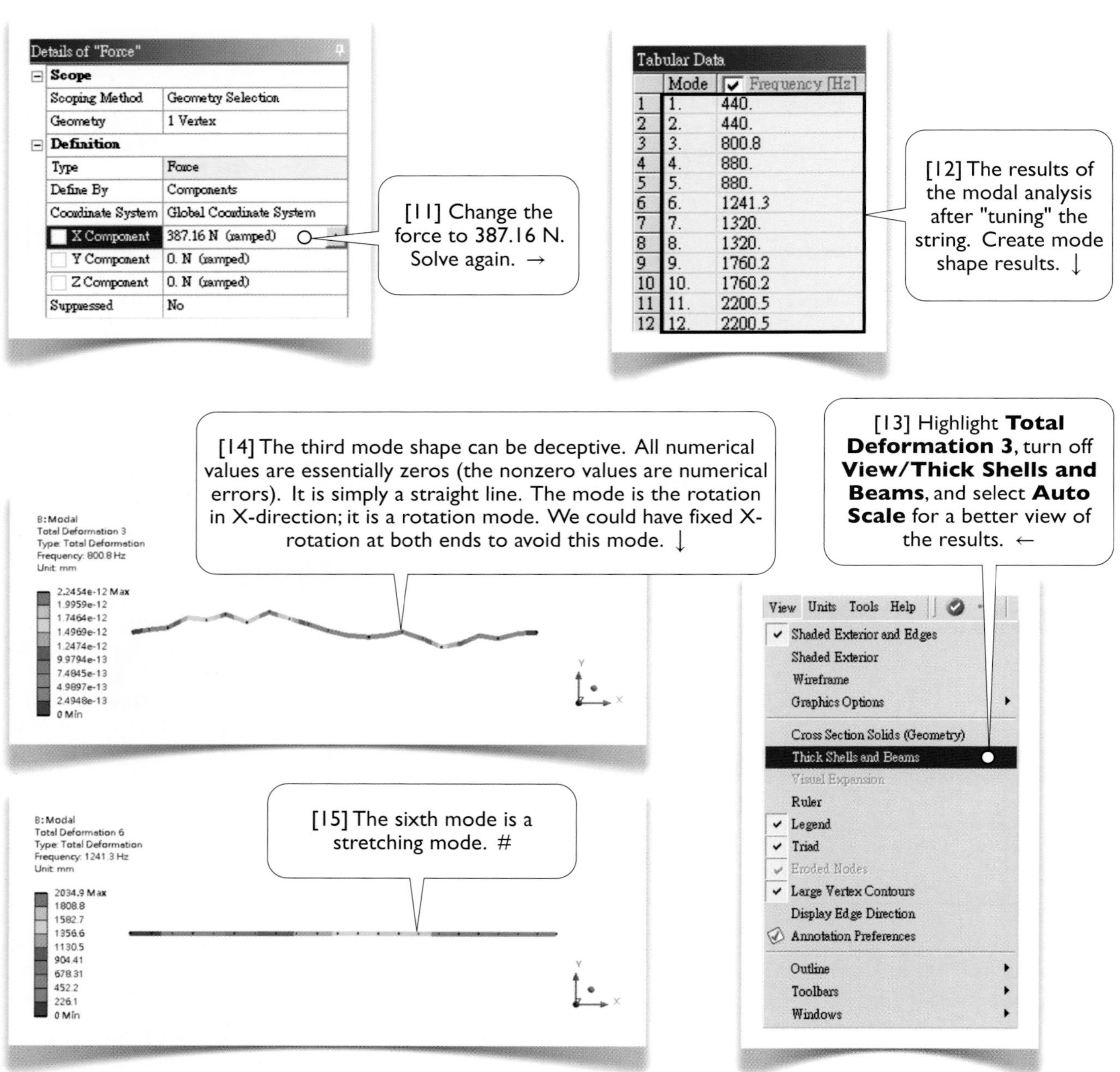

11.4.3 Harmonic Series

[1] A *harmonic mode* has a frequency that is an integral multiplication of the fundamental frequency. The first 5 harmonic modes of the guitar string are shown below [2-6].

If you pluck a string, you will produce a tone made up of all harmonic modes. Although all the plucks produce the same note (in this case, note *A*), the *harmonic mixes* determine the quality of the note. If you pluck the string near the midpoint of the string, you will produce a tone dominated by the first harmonic [2]. If you pluck the string near the quarter point (in a guitar, that is near the sound hole), you will produce a tone dominated by the second harmonic [3]. And so forth. You can produce an "overtone" (a tone made up of harmonics that are all above the fundamental mode) by touching the string lightly at the midpoint and, at the same time, plucking the string; the first harmonic mode will be suppressed. You can produce other overtones in a similar way.

Different musical instruments generate tones that have different harmonic mixes. A trumpet usually produces much higher harmonics in its frequency spectrum. This gives the trumpet a "brassy" sound. A flute usually produces a tone dominated by the first harmonic with almost no higher harmonics. This gives the flute a unique "pure" sound.

Knowing these physics, an engineer should be able to produce sounds of any musical instruments (including human voices) using a frequency-generating device. ↘

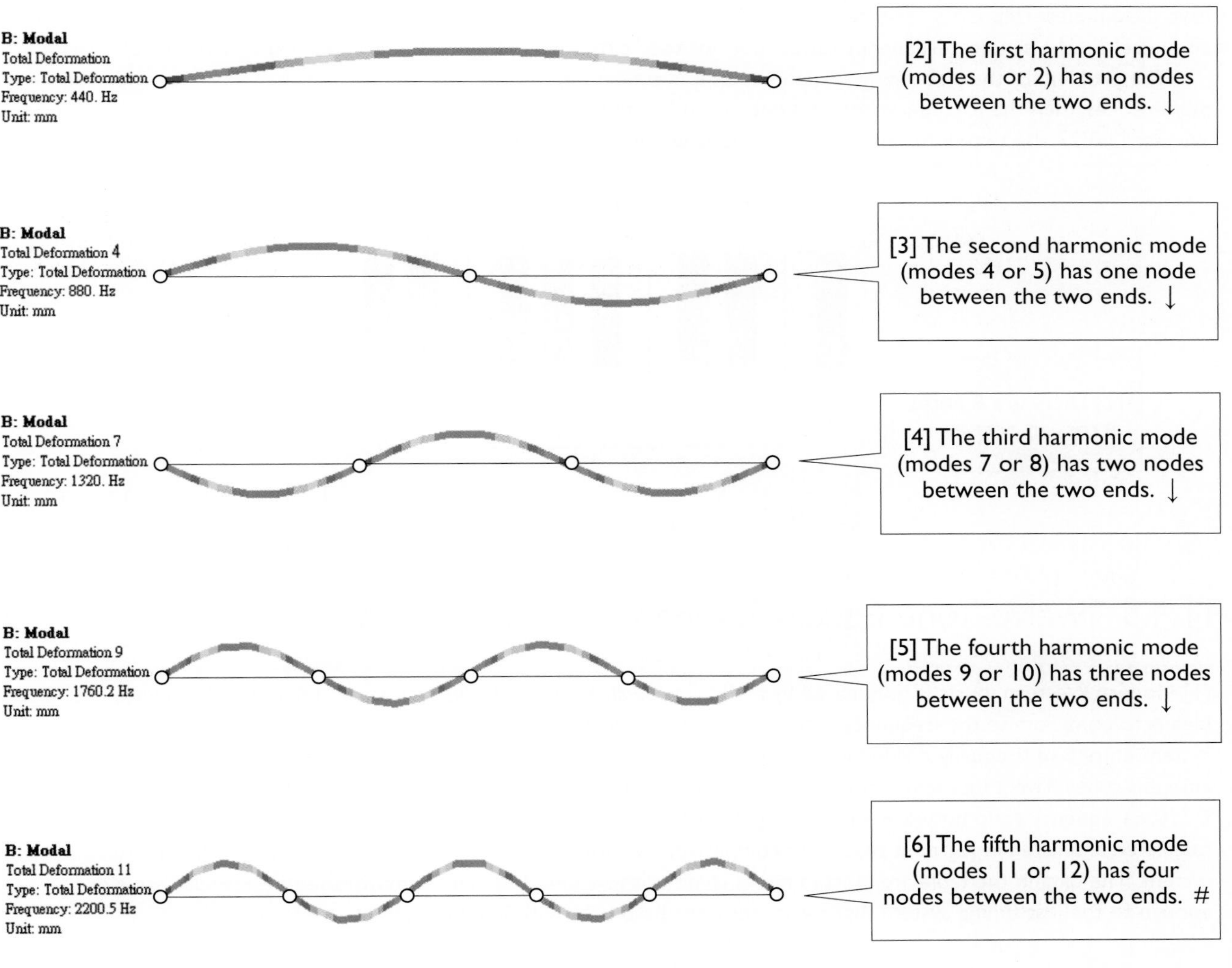

[2] The first harmonic mode (modes 1 or 2) has no nodes between the two ends. ↓

[3] The second harmonic mode (modes 4 or 5) has one node between the two ends. ↓

[4] The third harmonic mode (modes 7 or 8) has two nodes between the two ends. ↓

[5] The fourth harmonic mode (modes 9 or 10) has three nodes between the two ends. ↓

[6] The fifth harmonic mode (modes 11 or 12) has four nodes between the two ends. #

11.4.4 Just Tuning System[Ref 1]

[1] Why do some notes sound pleasing to our ears when played together, while others do not? We know from the experience that when two notes have a simple frequency ratio, they sound harmonious with each other. The simpler the ratio, the more harmonious it sounds; we'll explain this in 11.4.6 (page 416).

In Western music, an 8-tone musical scale has traditionally been used. When learning to sing, we identify the eight tones in the scale by the syllables *do, re, mi, fa, sol, la, ti, do*. For a *C*-major scale in a piano, there are 8 white keys from a *C* to the higher pitch of *C* [2]. The two *C*'s have a frequency ratio of 2:1, and are said to be an *octave* apart. If we play two notes an octave apart, they sound very similar. In fact, we often have difficulty telling the difference between two notes an octave apart. This is because, except for the fundamental harmonic of the lower note, two notes have most of the same higher harmonics.

For the following discussion, let's arbitrarily assume the frequency of the lower pitch *C* as 1. (In a modern piano, the middle *C* has a frequency of 261.63 Hz; see 11.4.5[2], next page.) Then the frequency of the higher pitch *C* is 2. Before being replaced by the "equal temperament" (11.4.5) in the early 20th century, the "just tuning" systems prevailed in the music world. In a just tuning music system, the frequencies of the notes between the 2 *C*'s are chosen according to the "simple ratio" rule, in order to be harmonious to each other. They are summarized in [3]. Note that we didn't show the frequency ratios for the black keys (the semitones) to simplify our discussion.

Now, you can appreciate that if we play the notes *do* and *sol* together, the sound is pleasing to our ears, since they have the simplest frequency ratio between 1 and 2. You also can appreciate that the major cord *C* consists of the notes *do, me, sol, do*, the simplest frequency ratios (but not too "close," to avoid *beats*; see 11.4.6, page 416) between 1 and 2.

The problem of the just tuning system is that it is almost impossible to play in another key. For example, when we play in *D* key, then the frequency ratio between *D* and its *fifth* (*A*) is no longer 3/2. Instead, the frequency ratio is an awkward 40/27; the two notes are not harmonious enough any more. ↙

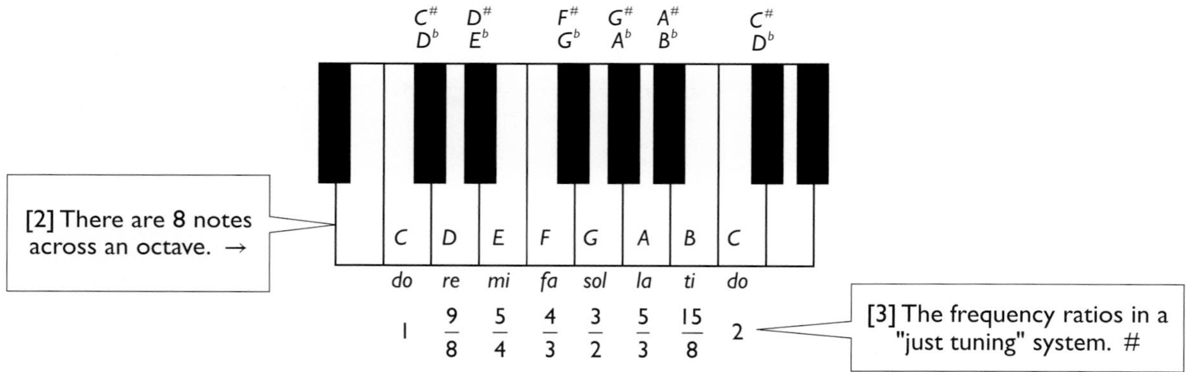

11.4.5 Twelve-Tone Equally Tempered Tuning System[Ref 2]

[1] Modern Western music is dominated by a *12-tone equally tempered tuning system*, or simply *equal temperament*. The idea is to compromise the frequency ratios between the notes, so that they can be played in different keys. In this system, an octave is equally divided into 12 tones (including semitones) in logarithmic scale. In other words, the adjacent tones have a frequency ratio of $2^{1/12}$, or 1.05946. For example, the frequency ratio between the $C^{\#}$ and the *C* is $2^{1/12}$; the frequency ratio between an *A* and the lower *C* is $2^{9/12}$. According to this idea, frequencies of the notes can be calculated and listed in [2], next page. For comparison, we also list the frequencies of the notes in the just tuning system. The data in the table are plotted into a chart as shown in [3-4]. The compromised frequencies are close enough to the just tuning system that most musicians have been satisfied with this system for centuries. ↵

Note	Just Tuning		Equal Temperament	
	Frequency Ratio	Frequency	Frequency Ratio	Frequency
C	1	264.00	1	261.63
$C^{\#}$ (D^b)			$2^{1/12}$	277.18
D	9/8	297.00	$2^{2/12}$	293.66
$D^{\#}$ (E^b)			$2^{3/12}$	311.13
E	5/4	330.00	$2^{4/12}$	329.63
F	4/3	352.00	$2^{5/12}$	349.23
$F^{\#}$ (G^b)			$2^{6/12}$	369.99
G	3/2	396.00	$2^{7/12}$	392.00
$G^{\#}$ (A^b)			$2^{8/12}$	415.30
A	5/3	440.00	$2^{9/12}$	440.00
$A^{\#}$ (B^b)			$2^{10/12}$	466.16
B	15/8	495.00	$2^{11/12}$	493.88
C	2	528.00	2	523.25

[2] The frequencies of the notes. ↙

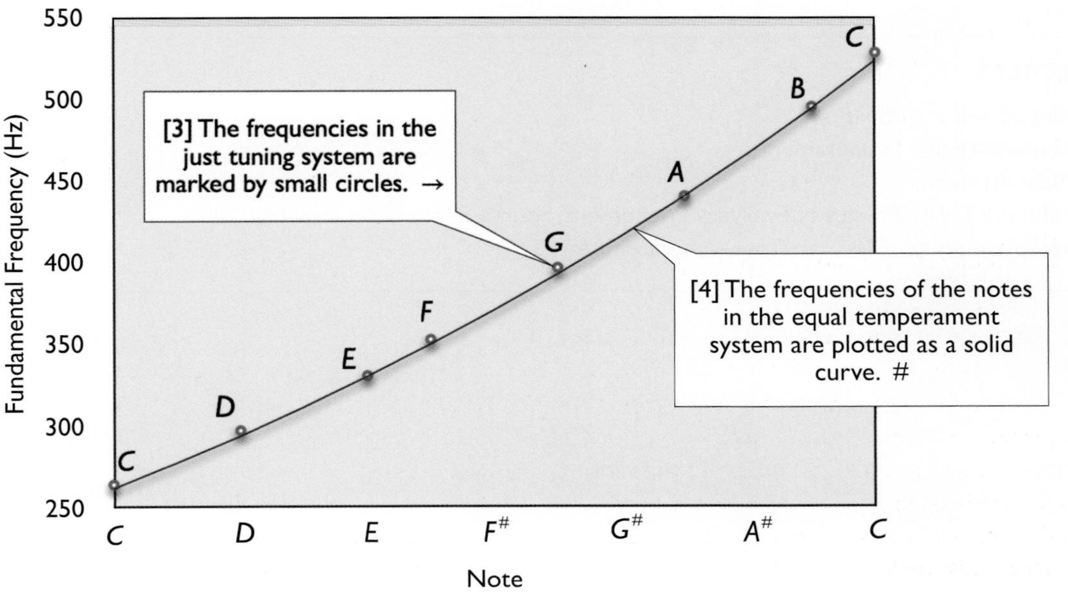

11.4.6 Beat Frequency[Refs 3, 4]

[1] Back to the question in the beginning of 11.4.4[1], page 414: Why do some notes sound pleasing to our ears when played together, while others do not? Why does a simple frequency ratio imply a harmonious sound? To explain this, another physical phenomenon called *beats* is at work.

When two waves of different frequencies are combined, they interfere with each other. When they are in phase, the combined wave has a large amplitude. When they are out of phase, the amplitude becomes smaller. This fluctuation in amplitude of the combined wave is called *beats*, and the frequency is called the *beat frequency*.

The beat frequency is equal to the frequency difference of the two waves. If the two notes are very close in frequency, the beat frequency is slow enough to be heard as a variation in amplitude, that is, you can hear the sound getting louder and softer in a repetitive pattern. This effect can be useful in tuning a musical instrument, since the beats disappear when two frequencies are in tune.

What happens when we play *C* and *D* together? The beat frequency is 32.03 Hz (293.66 - 261.63, see 11.4.5[2], last page), which is large enough that we hear a harsh buzz, which is unpleasant for our ears. In music, however, a dissonant sound is sometimes used to produce desired effects, for example, a sad mood.

What happens when we play *C* and *G* together? The beat frequency is 130.37 Hz (392.00 - 261.63, see 11.4.5[2], last page), which is very close to one half of the middle *C* (261.63 Hz). In other words, the beat frequency is an octave below middle *C*. Therefore, it fits nicely within a chord containing *C* and *G*. Thus, beats can explain the harmony when we play major chords. ↓

Wrap Up

[2] Save the project and exit Workbench. #

References

1. Wikipedia>Just Intonation.
2. Wikipedia>Equal Temperament.
3. Wikipedia>Beat.
4. Griffith, W.T., The Physics of Everyday Phenomena, Fourth Edition, McGraw-Hill, 2004.
5. http://plasticity.szynalski.com/tone-generator.htm

Section 11.5

Review

11.5.1 Keywords

Choose a letter for each keyword from the list of descriptions

1. ( ) Dynamic Behavior
2. ( ) Free Vibration
3. ( ) Fundamental Natural Frequency
4. ( ) Harmonic Mode
5. ( ) Modal Analysis
6. ( ) Natural Frequencies
7. ( ) Rigid Body Modes
8. ( ) Structural Dynamic Analysis
9. ( ) Transient Structural Analysis

Answers:

1. (B) 2. (E) 3. (G) 4. (I) 5. (C) 6. (F) 7. (H) 8. (A) 9. (D)

List of Descriptions

(A) Technique used to determine the dynamic behavior of a structure.

(B) Includes vibration characteristics, effect of harmonic loads, and effect of general time-varying loads.

(C) A dynamic analysis to investigate the vibration characteristics, specifically the frequencies and shapes, of a structure without any time-varying loads. The frequencies are called natural frequencies.

(D) A dynamic analysis to investigate the response of a structure under general time-varying loads.

(E) The vibration of a structure without any external forces.

(F) The frequencies corresponding to free vibrations.

(G) The lowest frequency of free vibration.

(H) They have infinite period, or, equivalently, zero frequencies. In general, these modes should not be present.

(I) It has a frequency that is an integer multiple of the fundamental frequency.

11.5.2 Additional Workbench Exercises

Gravity Is Negligible When Evaluating Frequencies of the Two-Story Building

In 11.2.5[2] (page 399), we mentioned that the earth's gravity affects natural frequencies of the two-story building very little. Verify this.

Improving the Stiffness of the Two-Story Building

In 11.2.5[9] (page 401), we mentioned that if we want to further improve the structure's solidity, we may add bracing members in the Z-direction. Implement this idea.

Model Airplane Wing

A wing of a model airplane is detailed in the ANSYS Help System[Ref 1]. Carry out the modal analysis for the airplane wing. How do you stiffen the wing further without increasing the weight?

Reference

1. ANSYS Help//Mechanical APDL//Mechanical APDL Introductory Tutorials//8. Modal Tutorial

Chapter 12

Transient Structural Simulations

In the real world, all loads are time-varying, so are the structural responses. For example, imagine that you hang a block on a spring and slowly release it. The force on the spring increases gradually from zero until it reaches the weight of the block, and then the block moves up and down for a while, and finally steadies at a certain position, due to the damping of the system. To know the whole process, you need to perform a dynamic simulation. If you are concerned about only the final state (final position), called the *steady state*, of the response, a static simulation is adequate. The response before the steady state is called a *transient state*. A situation where a dynamic simulation can be replaced by a series of static simulations is that if the structure displaces so slowly that the dynamic effects (inertia and damping effects) are negligible. A series of time-varying static simulations is called a *quasistatic simulation*.

Other than these two categories of cases (cases of finding a steady state solution or cases that the structure displaces slowly), dynamic effects must be taken into account and dynamic simulations are needed.

Purpose of This Chapter

The purpose of this chapter is to provide background knowledge as well as practical examples for students to master techniques for dynamics simulations. It is a sequel of the last chapter, modal analysis, in which no external forces are involved. Except Section 12.3, in which **Harmonic Response** analyses are carried out, all the other exercises are performed with **Transient Structural** analysis system, which uses an implicit integration method to calculate the response. **Explicit Dynamics** analysis system, on the other hand, uses an explicit integration method. The explicit dynamics will be introduced in Chapter 15, where differences between implicit and explicit methods will be discussed.

About Each Section

Section 12.1 intends to equip students with background knowledge of structural dynamics. Again, it is the concepts, rather than the mathematics details, that we want to emphasize. We first use a single-degree-of-freedom structural system to illustrate some ideas, and then conceptually generalize these ideas for multiple-degrees-of-freedom structural systems.

Section 12.2 provides a practical example to demonstrate the application of dynamic loads and other considerations such as integration time steps and damping. Section 12.3 uses the two-story building (7.3, 11.2) as an example, demonstrating the procedure of harmonic response analysis. Section 12.4 performs an impact simulation. One of its purposes is to demonstrate how to specify a simple initial condition, namely uniform velocity. Another purpose is to show the limitation of implicit integration methods and the necessity of explicit methods for high-speed impact simulations.

Section 12.5 is a sequel of Section 11.4. As mentioned in 11.4.3 (page 413), if you pluck a string, you will produce a tone made up of all harmonic modes. We will strum different locations on the string to observe the responses. Another purpose of Section 12.5 is to demonstrate how to specify a more general initial condition. Some initial conditions need a static simulation themselves. In this section, we will use the results of a static simulation as an initial condition for a transient simulation.

Section 12.1

Basics of Structural Dynamics

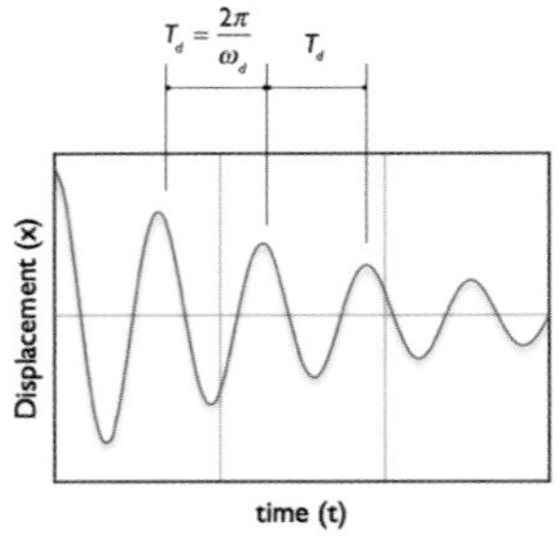

The purpose of this section is to provide basics of structural dynamics, so that students can understand the exercises in this chapter. The concepts and terminology introduced in this section will be used throughout this chapter and the rest of the book. We first use a single-degree-of-freedom lumped mass model to explain some basic behavior of dynamic response. The results will be conceptually extended to multiple-degrees-of-freedom cases (and not limited to lumped mass models).

12.1.1 Lumped Mass Model

[1] In the old days, many dynamics problems were simplified as lumped mass systems of a few degrees of freedom. As an example, to find the lateral displacements, the two-story building (7.3, 11.2) is modeled as a two-degrees-of-freedom system as shown [2-8]. A lumped mass model like this can be solved with a circuit (made of inductors, resistors, capacitors, etc.) that has the same form of governing equations as the lumped mass model. This simple device, used to solve engineering problems long before digital computers became available, is essentially an analog computer.

The parameters ($m_1, m_2, k_1, k_2, c_1, c_2$) must be reasonably evaluated to obtain an acceptable solution. Each of the masses m_1 and m_2 [3-4] may include the floor mass, part of the columns mass, and the equivalent mass of the loads on the floor. Each of the spring constants k_1 and k_2 [5-6] may be calculated according to the bending stiffnesses of the columns and beams at the floor. Each of the damping coefficients c_1 and c_2 [7-8] represents all energy dissipating mechanisms of the floor.

The energy dissipating mechanisms include frictions between the building and the surrounding air (viscous damping), material's internal frictions (material damping), and the frictions in the joints connecting structural members (Coulomb damping). Evaluating damping is one of the most challenging tasks of engineering practice. Fortunately, damping in most real-world structures is usually very small. Consequently, in many cases, it may not be crucial and we may choose a reasonable value according to engineering experiences. In other cases, conducting experiments to evaluate damping values may be needed. We'll discuss damping in 12.1.3, pages 423-425. ↓

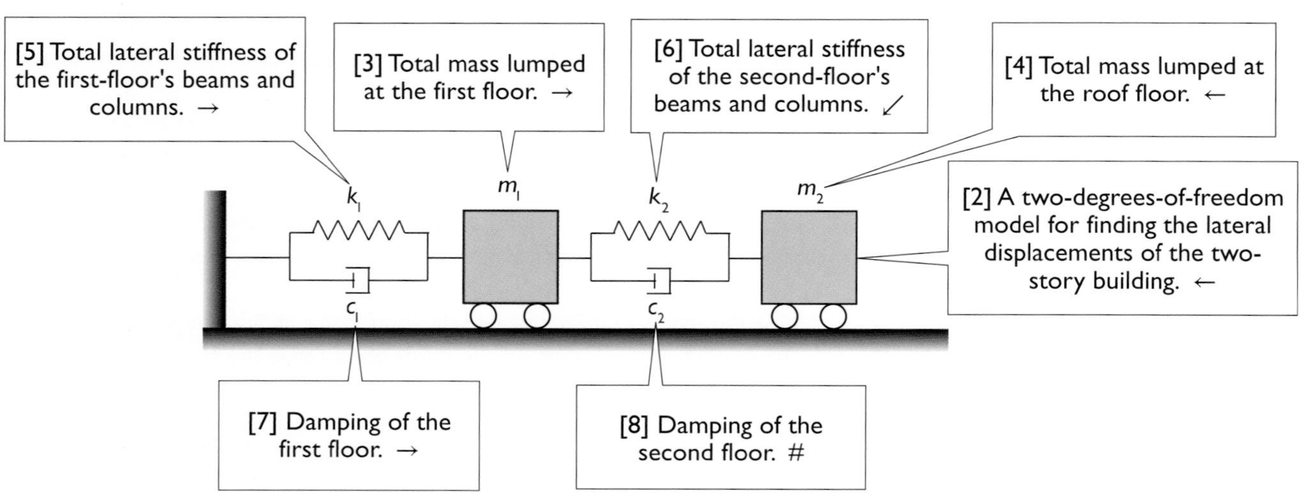

12.1.2 Single Degree of Freedom Model

[1] Consider a single-degree-of-freedom (SDOF) model [2]. Applying Newton's law of motion on the block of mass *m*, we have

$$\sum F = ma$$

$$p - kx - c\dot{x} = m\ddot{x}$$

or

$$m\ddot{x} + c\dot{x} + kx = p \qquad (1)$$

Eq. (1) is the governing equation of the SDOF model. We'd assumed that the spring force is linearly proportional to the displacement, and the damping force is linearly proportional to the velocity. This kind of damping is called a *viscous damping*. We will discuss this assumption of damping in 12.1.3, pages 423-425.

If no external forces exist, Eq. (1) becomes

$$m\ddot{x} + c\dot{x} + kx = 0 \qquad (2)$$

Eq. (2) represents a *free vibration* system.

If the damping is negligible, then the equation becomes

$$m\ddot{x} + kx = 0 \qquad (3)$$

Eq. (3) represents an *undamped free vibration* system.

Undamped Free Vibration

The general solution of Eq. (3) is

$$x = A\sin\left(\omega t + B\right) \qquad (4)$$

where *A* and *B* are arbitrary real numbers and

$$\omega = \sqrt{\frac{k}{m}} \qquad (5)$$

Verification of this solution can be done by substituting Eqs. (4-5) into Eq. (3). A typical plot of Eq. (4) is shown in [3], where we arbitrarily assume *B* = 0, since it is irrelevant. The vibration can be generated by holding up the mass of an undamped SDOF system [2], displacing an arbitrary amount *A*, releasing it, and letting it vibrate freely. The meaning of the *natural angular frequency* ω is also shown in the plot [3]. Relations between the natural angular frequency ω (rad/s), *natural frequency* *f* (Hz), and *natural period* *T* (s) are

$$f = \frac{\omega}{2\pi} \qquad (6)$$

$$T = \frac{1}{f} \qquad (7)$$

↗

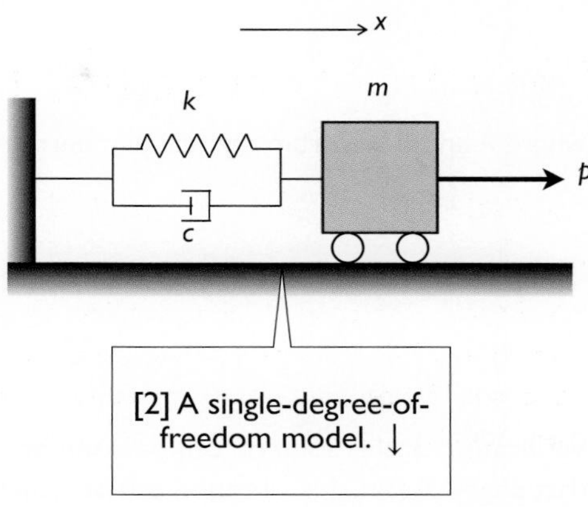

[2] A single-degree-of-freedom model. ↓

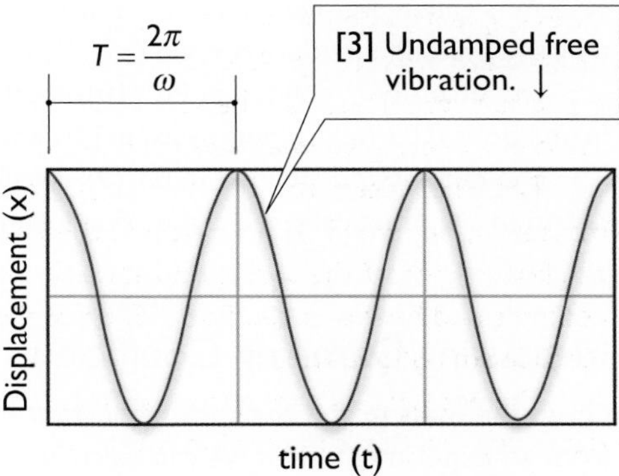

[3] Undamped free vibration. ↓

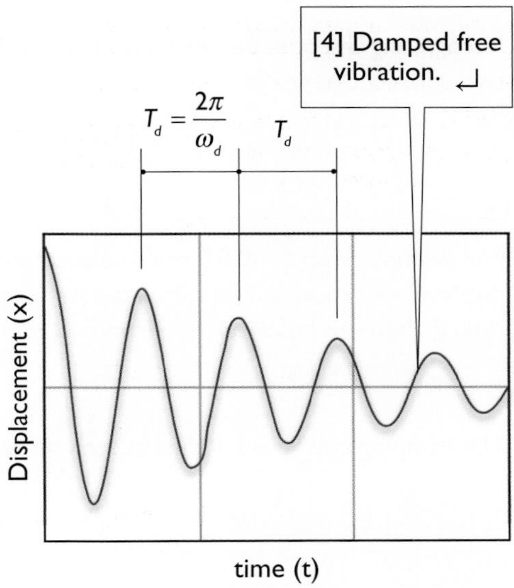

[4] Damped free vibration. ↵

Damped Free Vibration

[5] Assume the damping c is smaller than the critical damping, defined in Eq. (11), then the general solution of Eq. (2) is

$$x = Ae^{-\xi\omega t}\sin\left(\omega_d t + B\right) \tag{8}$$

where A and B are arbitrary real numbers and

$$\omega_d = \omega\sqrt{1-\xi^2} \tag{9}$$

$$\xi = \frac{c}{c_c} \tag{10}$$

$$c_c = 2m\omega \tag{11}$$

Verification of this solution can be done by substituting Eqs. (8-11) into Eq. (2). A typical plot of Eq. (8) is shown in [4] (last page). Again, this vibration can be generated by holding up the mass of a SDOF system [2], displacing an arbitrary amount A, releasing it, and letting it vibrate freely.

The quantity c_c, defined in Eq. (11), is called the *critical damping coefficient.* It can be shown that if the damping c is larger than or equal to c_c, then the Eq. (8) is no longer valid. Actually, the motion is no longer oscillatory, and the system is called *over-damped* (if $c > c_c$) or *critically damped* (if $c = c_c$). For most structures, we may reasonably assume that the system is under-damped ($c < c_c$), and the Eq. (8) is always valid.

The quantity ξ, defined in Eq. (10), is called the *damping ratio.* Values of damping ratio for typical structures range from about 0.02 (e.g., piping systems) to about 0.07 (e.g., bolted structures, reinforced concrete)[Ref 1].

The quantity ω_d, defined in Eq. (9), is called *damped natural angular frequency*. Note that, for a small damping ratio, ω_d and ω are practically the same. For example, with $\xi = 0.07, \omega_d = 0.9975\omega$.

Measuring Damping Coefficient

Given the mass m and the spring constant k of a SDOF system, and if we can obtain a damped free vibration curve [4] from an experiment, then we can calculate the damping coefficient from this information, as follows. From Eq. (8), the peak displacement is

$$x_{peak} = Ae^{-\xi\omega t}$$

Recognizing the time between two peaks is $T = 2\pi/\omega_d \approx 2\pi/\omega$, we can write down the displacement ratio R between two consecutive peaks

$$R = \frac{x_{peak2}}{x_{peak1}} = \frac{Ae^{-\xi\omega(t+\frac{2\pi}{\omega})}}{Ae^{-\xi\omega t}} = e^{-2\pi\xi} \tag{12}$$

For example, with $\xi = 0.07$, the displacement ratio between two consecutive peaks is $R = 0.64$. In practice, the displacement ratio can be calculated by averaging the displacement ratios of *several* cycles of vibrations. The damping ratio ξ then can be calculated from Eq. (12)

$$\xi = \frac{-\ln R}{2\pi} \tag{13}$$

The damping coefficient, if desired, can be calculated using Eqs. (10) and (11),

$$c = \frac{-\ln R}{\pi} m\omega \tag{14}$$

Viscous damping coefficient c is not an intrinsic property of a material

[6] Imagine that the SDOF system [2] represents a cantilever beam made of a material and you want to characterize the damping property for the material. You gather a specimen, generate a free vibration curve [4], and calculate the damping coefficient according to Eq. (14). It seems easy. The problem is that, in this way, the damping coefficient depends on the geometry of the specimen: different geometries have different damping coefficients. To characterize damping for a material, we need more knowledge of damping mechanisms. #

12.1.3 Damping

Damping Mechanisms

[1] As mentioned, damping includes all energy dissipating mechanisms. In a structural system, all energy dissipating mechanisms boil down to one word: friction. In a structure, three categories of frictions can be identified: First, friction between the structure and its surrounding fluid, called *viscous damping*. Second, internal friction in the material, called *material damping*, *solid damping*, or *elastic hysteresis*. Third, friction in the connection between structural members, called *dry friction* or *Coulomb friction*. When the structure is surrounded by the air, the viscous damping is usually very small, and the major sources of damping are material damping and Coulomb friction.

Viscous Damping

Viscous damping is the friction between a structure and its surrounding fluid. If small, viscous damping force can reasonably assume to be proportional to the velocity of the structural displacement

$$F_D = c\dot{x} \tag{1}$$

The viscous damping coefficient c can be input directly as an element parameter, such as a spring element [2]. For each material, you can include a constant damping ratio as a material property (see [13], page 425). In some analysis systems, the viscous damping can be specified using a global damping ratio ξ [3]. The damping coefficient c is then calculated according to Eqs. 12.1.2(10-11) (last page),

$$c = 2m\omega\xi = 2\xi\sqrt{mk} \tag{2}$$

In general, to introduce viscous damping to a structure, you may add individual elements involving viscous damping, such as spring elements [2]. In **Harmonic Response** analysis (Section 12.3), which calculates the response under various frequencies, viscous damping can be specified using a global damping ratio [3]. ↙

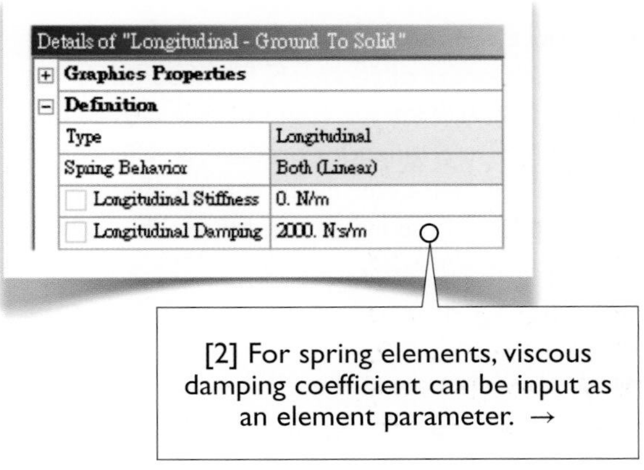

[2] For spring elements, viscous damping coefficient can be input as an element parameter. →

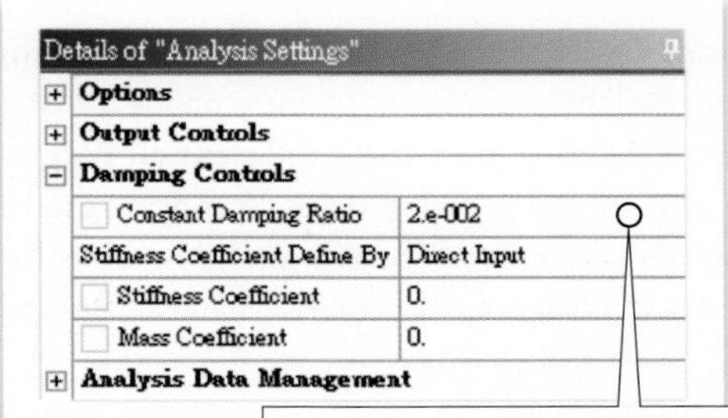

[3] In Harmonic Response analysis, viscous damping can be specified using a global damping ratio. ↵

Material Damping

[4] Material damping is the internal friction between molecules or grains of the material. Other names for material damping include *solid damping* and *elastic hysteresis*. It is often the major sources of damping in a real-world structure.

To understand the material damping, consider a typical stress-strain relation in a uniaxial material test [5-7]. Unlike the tests shown in 1.4.1[2-6] (page 40), here we repeatedly increase the stress [5] and then release the stress [6]. If the plastic deformation (permanent deformation) is not present, the strain will return to the original state when releasing the stress, and the curve of releasing stress is likely to be a straight line. A material in which the strain state returns to the original state when the stress is released is called an *elastic material*.

The area enclosed by the curves [7] represents the energy dissipation (to the environment) due to the internal friction of the material. This area is typically very small; the plot [5-7] is exaggerated for instructional purposes.

The plot [5-7] is analogous to a *B-H* curve of a magnetic material [8], where *B* is the *magnetic flux density* and *H* is the *magnetic field intensity*. A magnetic field *H* is applied on a magnetic material to create a *B* field within the material (i.e., the material is magnetized). The area enclosed by the *B-H* curve represents an energy dissipation, and is called the *magnetic hysteresis*. Because of the analogy, the energy dissipation in a stress-strain curve is called the *elastic hysteresis*. ↗

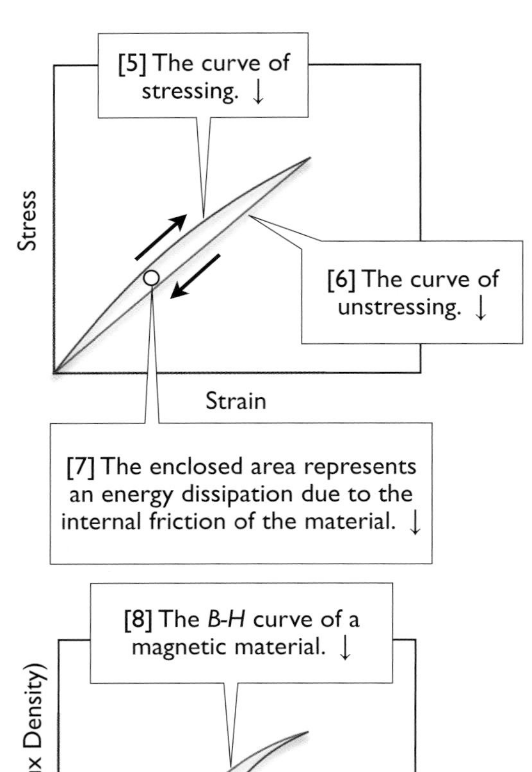

[9] Recognizing that the damping is small for a structure and the global behavior is similar regardless of the sources of damping, Workbench assumes, as a simplification, that the material damping force is proportional to the structural velocity (this somehow deviates from reality, in which the material damping force is more likely proportional to the structural displacement rather than the structural velocity), the same as the viscous damping,

$$F_D = c\dot{x} \tag{3}$$

However, we cannot characterize a material using a damping coefficient *c*, since, as mentioned in 12.1.2[6] (page 423), the damping coefficient *c* is not an intrinsic property of a material. To filter out factor of geometry, we need more elaboration. Eq. (2) shows how the coefficient *c* relates to the mass *m* and stiffness *k* for the case of single degree of freedom; for cases of multiple degrees of freedom, the relation is not so simple. In engineering practice, an efficient way to characterize a material is proposing a mathematics form with parameters and then determining the parameters using data fitting. With this idea, the coefficient *c* is assumed a linear combination of the mass *m* and the stiffness *k*[Ref 1],

$$c = \alpha m + \beta k \tag{4}$$

Now, the parameters α and β are used to characterize the damping property of a material. Eq. (4) is based on an observation that damping is related to the mass and the stiffness of the structure. ↵

[10] Using $c = 2m\omega\xi$ (Eq. (2)) and $k = m\omega^2$ (Eq. 12.1.2(5), page 421), we may rewrite Eq. (4) in terms of frequency and damping ratio,

$$2\omega\xi = \alpha + \beta\omega^2 \tag{5}$$

If we can make a single material specimen and measure the damping ratios ξ_i under different excitation frequencies ω_i, or make several material specimens of different sizes, and measure the damping ratios ξ_i under their respective fundamental frequencies ω_i or, even better, a combination of the above ideas, then we can evaluate the material parameters α and β by a standard data fitting procedure.

Workbench allows you to input the α and β values for each material as material properties [11]. A **Transient Structural** analysis system also allows you to input a global β value (**Stiffness Coefficient**) and a global α value (**Mass Coefficient**), in a details view of **Analysis Settings** [12]. ↓

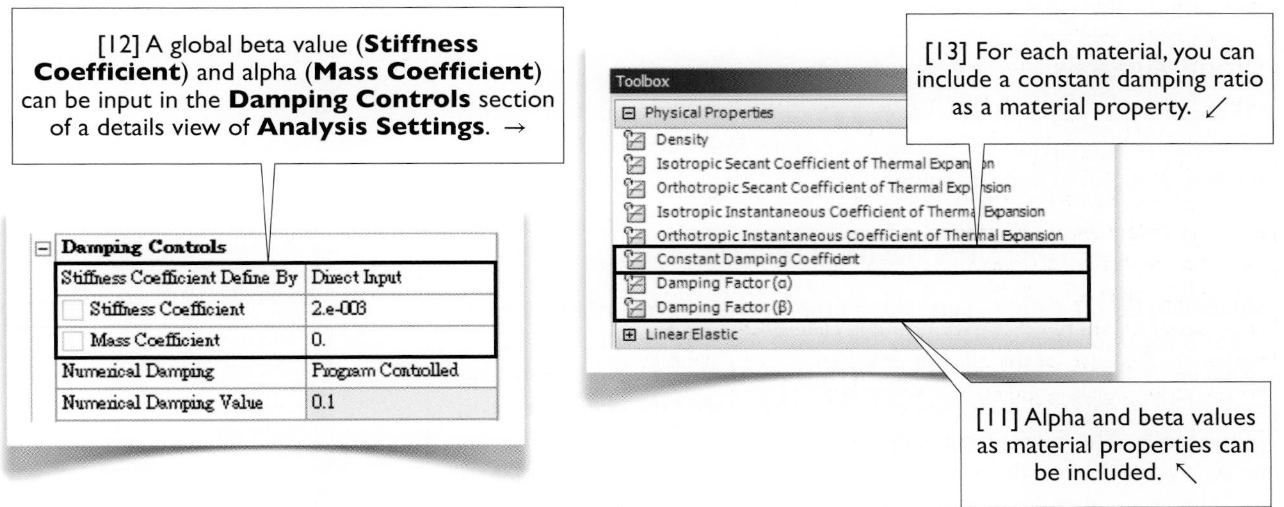

[14] If we assume $\alpha = 0$, Eq. (5) becomes

$$\beta = \frac{2\xi}{\omega} \tag{6}$$

Eq. (6) is a simple relation between the β value and the damping ratio ξ. It can be used to estimate one value when knowing the other one, if the frequency is also known.

Although the damping ratio is meant to be used for the viscous damping ([3], page 423) rather than the material damping, in practice, the engineers often use a damping ratio to simplify the overall damping effect, when the damping is not critical to the response. As mentioned in 12.1.2[5] (page 422), values of damping ratio for typical structures range from 0.02 to 0.07.

Coulomb Friction

Another major source of damping is the friction in the connection between structural members. It is called the *dry friction* or *Coulomb friction*. In Workbench, it is implemented as *frictional contacts*. To include the Coulomb friction, you have to specify frictional contacts between parts, which will be discussed in Chapter 13. #

12.1.4 Analysis Systems

[1] Generally, we are dealing with multiple-degrees-of-freedom systems. The foregoing concepts may be extended for general cases. Specifically, Eq. 12.1.2(1) (page 421) can be generalized for multiple-degrees-of-freedom cases,

$$[M]\{\ddot{D}\}+[C]\{\dot{D}\}+[K]\{D\}=\{F\} \tag{1}$$

Where $\{D\}$ is the nodal displacements vector, $\{F\}$ is the nodal external forces vector, $[M]$ is the *mass matrix*, $[C]$ is the *damping matrix*, and $[K]$ is the *stiffness matrix*. Eq. (1) also can be viewed as a generalization of Eq. 1.3.1(1) (page 35).

Eq. (1) represents the governing equation of a transient structural simulation. It can be viewed as a force equilibrium relation. On the right hand side of the equation is the external force $\{F\}$. On the left hand side, the first item $[M]\{\ddot{D}\}$ is the *inertia force*, the second item $[C]\{\dot{D}\}$ is the *damping force*, and the third item $[K]\{D\}$ is the *elastic force*. Combination of the inertia force, damping force, and the elastic force balances with the external force.

Let's look at some specialized cases of Eq. (1) (also see [2]). ↘

Modal Analysis

[3] Imagine that you displace a structure a certain amount and then release. There is no external force involved; it is called a free vibration. Eq. (1), since there is no external force, becomes

$$[M]\{\ddot{D}\}+[C]\{\dot{D}\}+[K]\{D\}=0 \tag{2}$$

Solution of Eq. (2) is not unique. For a problem of n degrees of freedom, it has at most n solutions, denoted by $\{D_i\}, i=1,2,...,n$. These solutions are called *mode shapes* of the structure. Each mode shape $\{D_i\}$ can be resonantly excited by an external excitation of frequency ω_i, called the *natural frequency* of the mode. The lowest frequency is called the *fundamental natural frequency*, or simply *fundamental frequency*. Finding all or some of the mode shapes and their corresponding natural frequencies is called a *modal analysis*.

In a modal analysis, since we are usually interested only in the natural frequencies and the relative shapes of the vibration modes (the absolute values of deformation depend on the energy that excites the structure), the damping effect is usually neglected (see Eq. 12.1.2(9), page 422) to simplify the calculation; Eq. (2) becomes

$$[M]\{\ddot{D}\}+[K]\{D\}=0 \tag{3}$$

It is Eq. (3) that Workbench solves in a **Modal** analysis system. Note that modal analysis is a linear analysis; all nonlinearities are ignored.

We've performed some modal analyses in Chapter 11. ↓

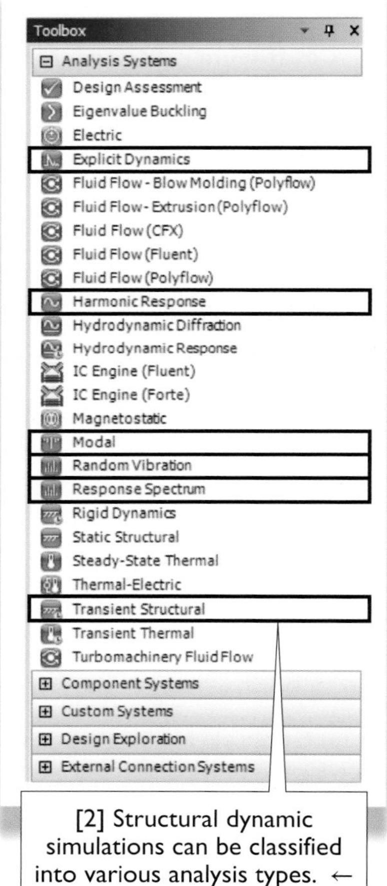

[2] Structural dynamic simulations can be classified into various analysis types. ←

Transient Structural Analysis

[4] **Transient Structural** analysis solves the general form of Eq. (1). External force $\{F\}$ can be time-dependent forces. All nonlinearities can be included. It uses a *direct integration method* to calculate the dynamic response. In this chapter, all exercises, except Section 12.3, are carried out with **Transient Structural** system [2]. Through these exercises, we will demonstrate how to specify initial conditions, dynamic loads, and set up **Analysis Settings**. ↵

Harmonic Response Analysis

[5] Imagine that a rotatory machine is installed on the floor of the two-story building (Sections 7.3, 11.2), its operational speed 3000 rpm. Due to an inevitable eccentricity of the rotation, the machine generates an up-and-down harmonic force on the floor. After it is started up, the machine's speed increases from zero up to 3000 rpm. Any natural frequencies of the building ranging from zero to 3000 rpm may be excited by the harmonic force. While vibrations of the building or its structural components are unavoidable, the question is how large the amplitude of the vibrations will be. Do the vibrations cause any safety concern or psychological annoyance? A **Harmonic Response** analysis can be carried out to answer that question.

Harmonic Response analysis solves a special form of Eq. (1), in which the external force on *i*th degree of freedom is of the form

$$F_i = A_i \sin(\Omega t + \phi_i) \tag{4}$$

where A_i is the amplitude of the force, ϕ_i is the phase angle of the force, and Ω is the angular frequency of the external force. Due to the special form of the external forces, the calculation is much more efficient than a general transient response analysis. The steady-state solution of the equation will be of the form

$$D_i = B_i \sin(\Omega t + \varphi_i) \tag{5}$$

The goal of the harmonic response analysis is to find the magnitude B_i and the phase angle φ_i of the response for each degree of freedom, under a range of frequencies of the external forces. Note that harmonic response analysis is a linear analysis; all nonlinearities are ignored.

In Section 12.3, we will use the two-story building to demonstrate the procedure of harmonic response analysis.

Explicit Dynamics

Similar to **Transient Structural**, **Explicit Dynamics** also solves the general form of Eq. (1). The external force $\{F\}$ can be time-dependent forces. All nonlinearities can be included. It also uses a direct integration method to calculate the dynamic response. The difference is described as follows.

The direct integration method used in **Transient Structural** analysis is called an *implicit integration method*. The implicit method works fine for most applications except for high-speed impact simulations.

In high-speed impact simulations, the duration of impact time is so short that the integration time needs to be extremely small (e.g., micro to nano seconds) to catch the details of the behavior. If the implicit integration method is used, the total number of time steps becomes so huge that the computational time is unbearable. That calls for an *explicit integration method*, implemented in **Explicit Dynamics** analysis system.

For many transient dynamic simulations, the explicit method is not popular for one reason: it requires very small integration time steps to achieve an accurate solution. A small integration time is exactly what a high-speed impact simulation needs; therefore, it is not a disadvantage any more. The advantage, on the other hand, is that the calculation is very efficient in each time step. Overall, the high-speed impact simulations are possible only with the explicit integration method.

There are cases, other than high-speed impact simulations, that benefit by using the explicit method. Highly nonlinear simulations usually require very small time steps to overcome the convergence difficulties. In such cases, explicit method may be used.

In this chapter, we will restrict the discussion to implicit dynamics only (i.e., using **Transient Structural** analysis system). The applications of **Explicit Dynamics** will be postponed until Chapter 15. The reason is that, since explicit dynamics usually involves nonlinearities, we need more background on nonlinear simulations, which will be covered in Chapters 13 and 14.

For the beginners, the origination of the names "implicit" or "explicit" may not be important. You may regard them as code names. For those students with strong curiosity, you will learn them in Section 15.1. ↵

Response Spectrum Analysis

[6] We often design an engineering object such that it can withstand oscillatory or repeated loadings. For example, a building must withstand the strikes of earthquakes. When designing a building, we may use a well-recorded earthquake as a "design earthquake." The history of the earthquake (typically a history of acceleration versus time) can be input as loads and a transient structural simulation is then carried out. After the simulation, the maximum stress (or any other responses) at each location of the structure is collected. The members of the structure are designed according to these maximum values of responses. If the maximum response is the sole purpose of the simulation, a much more inexpensive way of simulation is available: response spectrum analysis.

A time history of earthquake can be transformed to a response spectrum, a maximum response versus frequency (or period) plot. The response spectrum is then input to a **Response Spectrum** analysis system. The output is the maximum response (e.g., maximum stress) at each location of the structure.

Random Vibration Analysis

There exists a design methodology that considers probabilistic loads, instead of deterministic loads. The probabilistic loads are described by a spectrum representing probability distribution of excitation at varying frequencies in known directions. The technique is used to design structures withstanding probabilistic loadings, for example, a space vehicle subject to probabilistic strikes of meteorites or asteroids. #

References

1. Cook, R.D., Milkus, D. S., Plesha, M. E., and Witt, R. J., *Concepts and Applications of Finite Element Analysis, Fourth Edition*, John Wiley & Sons, Inc., 2002; Section 11.5 Damping.
2. ANSYS Help//Mechanical APDL//Mechanical APDL Theory Reference//14.3. Damping Matrices

Section 12.2

Lifting Fork

12.2.1 About the Lifting Fork

[1] In Section 4.3, we built a model for a lifting fork and glass, in which the fork was modeled as solid body and the glass as surface body. The lifting fork [2] is used in an LCD factory to handle a glass panel [3], which is so large and thin that the engineers are concerned about its vertical deflections during dynamic handling.

The fork is made of steel with a density of 7850 kg/m^3, Young's modulus of 200 GPa, and Poisson's ratio of 0.3. The glass has a density of 2370 kg/m^3, Young's modulus of 70 GPa, and Poisson's ratio of 0.22.

In this section we will perform a static structural simulation first, to evaluate the vertical deflection of the glass panel under the gravitational force. This is critical when determining the clearance of the processing machine [4]. During a dynamic handling, the fork accelerates upward from rest to a velocity of 1.5 m/s in 0.3 seconds, and then decelerates to a full stop in another 0.3 seconds, causing the glass panel to vibrate [5]. We want to know the time duration when the vibration is settled to a certain amount so that the glass can be moved into the processing machine [4]. We also want to know the maximum stress during the handling. ↘

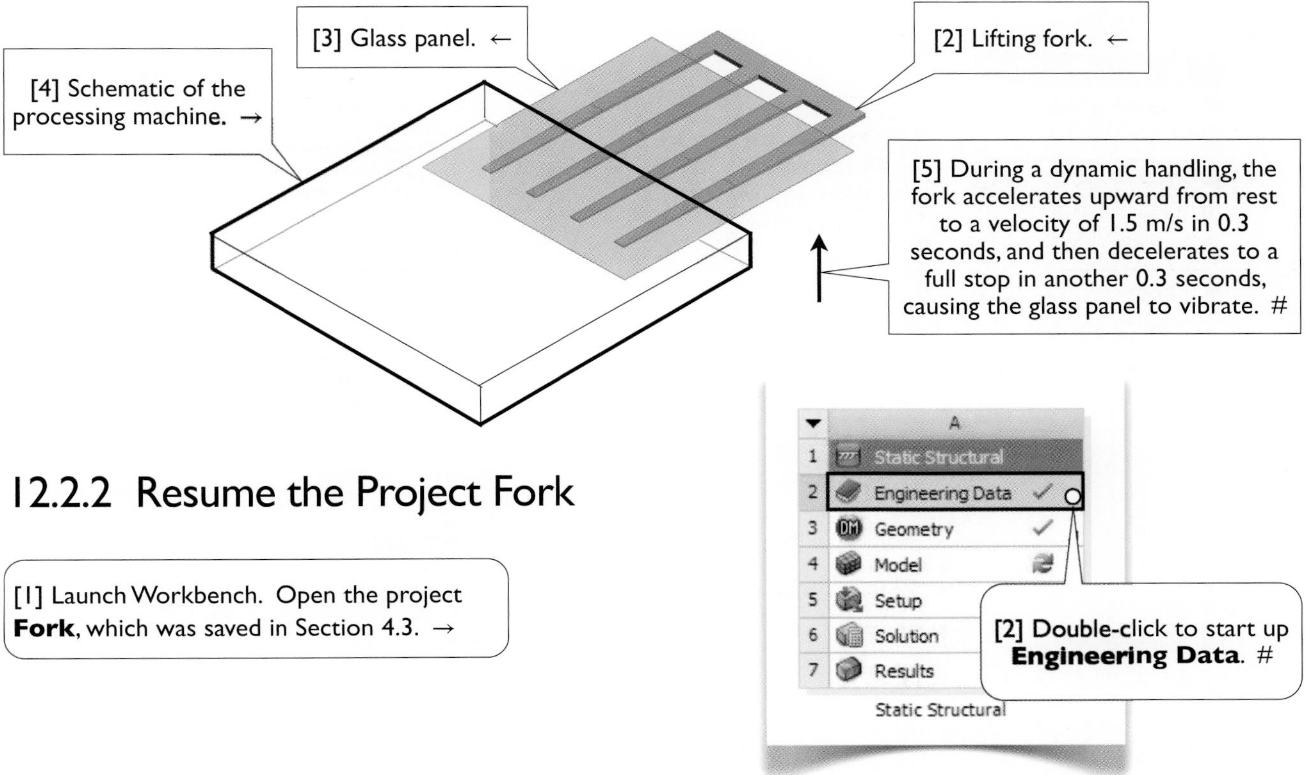

12.2.2 Resume the Project Fork

[1] Launch Workbench. Open the project **Fork**, which was saved in Section 4.3. →

	A
1	Static Structural
2	Engineering Data
3	Geometry
4	Model
5	Setup
6	Solution
7	Results

Static Structural

[2] Double-click to start up **Engineering Data**. #

12.2.3 Set Up the Model

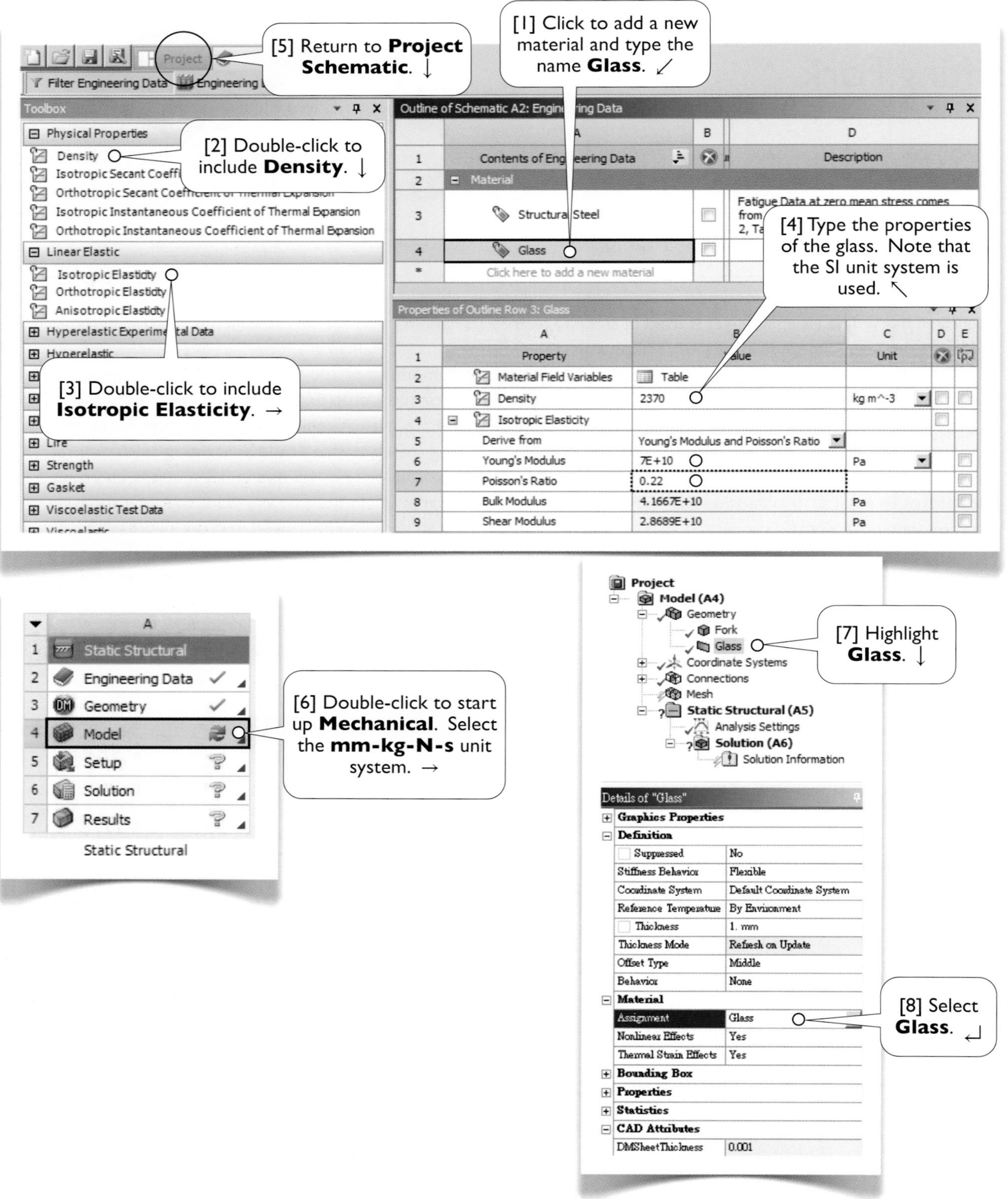

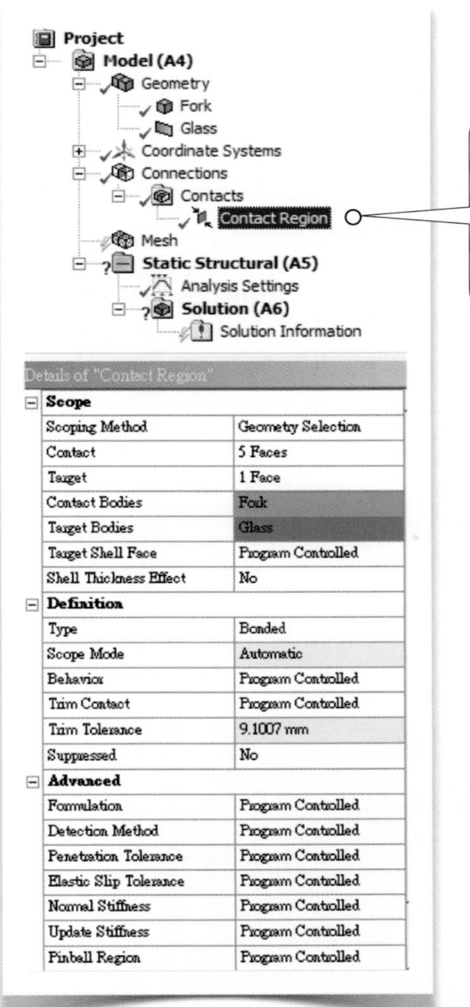

[9] Highlight **Contact Region**. Workbench correctly established contact between the glass and fork. We don't need to change anything for this case. ↓

About the Contact Region

[10] Note that, in this case, we use **Bonded** (default) rather than more realistic contact types such as **Frictionless** or **Frictional**, because **Bonded** should be accurate enough for this case; it avoids introducing nonlinearity into the simulation system.

The contact types **Bonded** and **No Separation** are the only two contact types that do not introduce nonlinearity. **No Separation** contact condition between two surfaces prohibits separation in their normal direction, but allows sliding relative to each other. The sliding is assumed to be very small such that the small-deformation theory can apply and the simulation remains linear. ↙

[11] With **Mesh** highlighted, select **Mesh Control/Face Meshing**. →

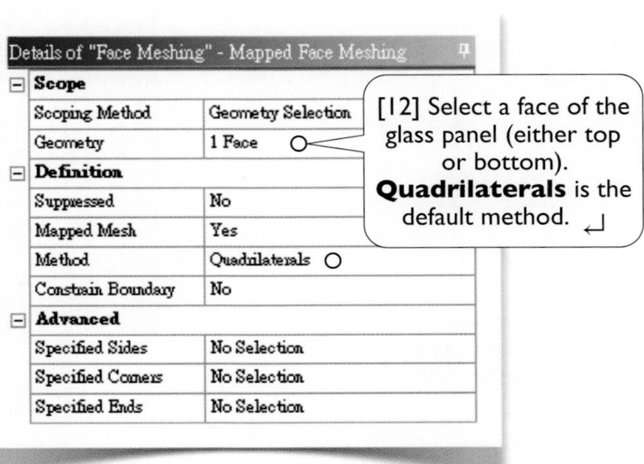

[12] Select a face of the glass panel (either top or bottom). **Quadrilaterals** is the default method. ↲

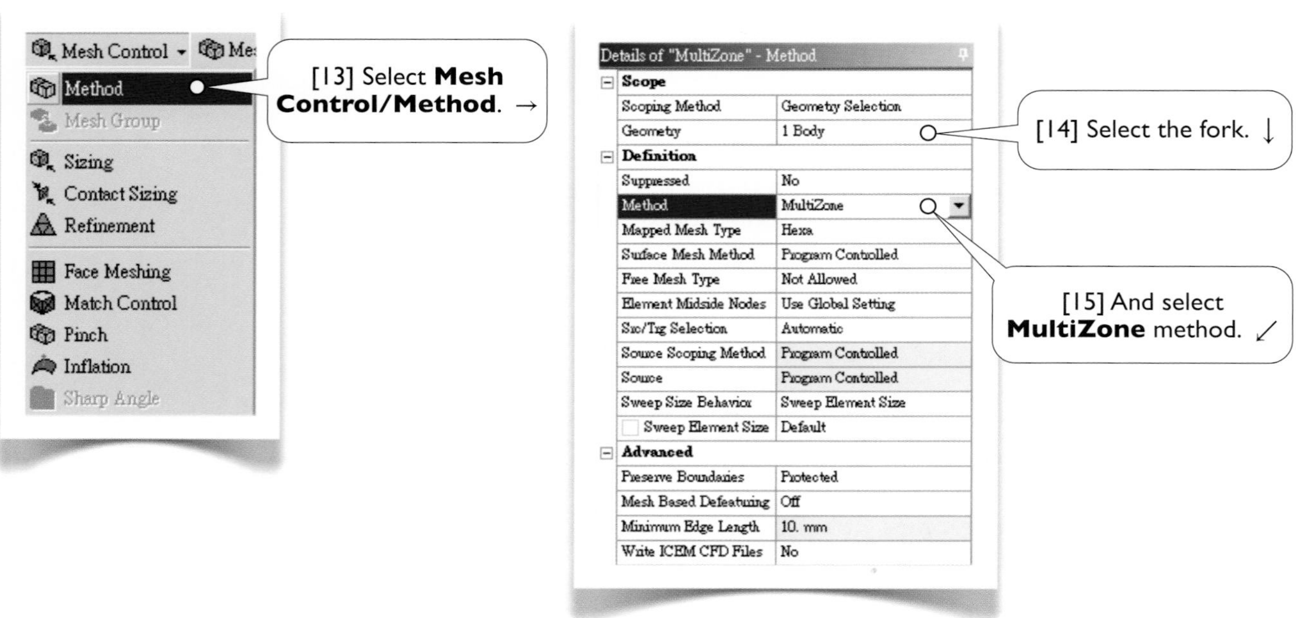

Mesh | Mesh Control | Mes

Generate Mesh

Preview Surface Mesh

Preview Source and Target Mesh

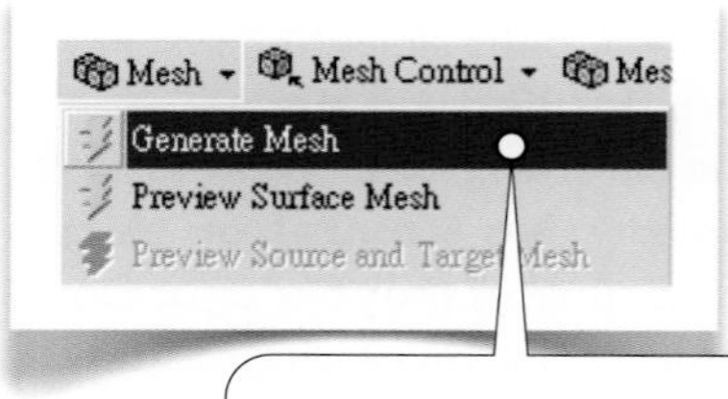

[16] Generate mesh. ↓

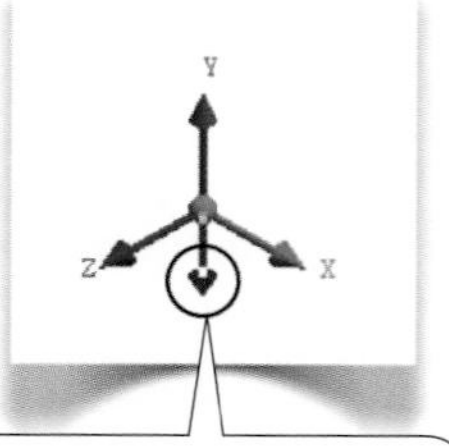

[17] To view the mesh from bottom, click **-Y** of the triad. #

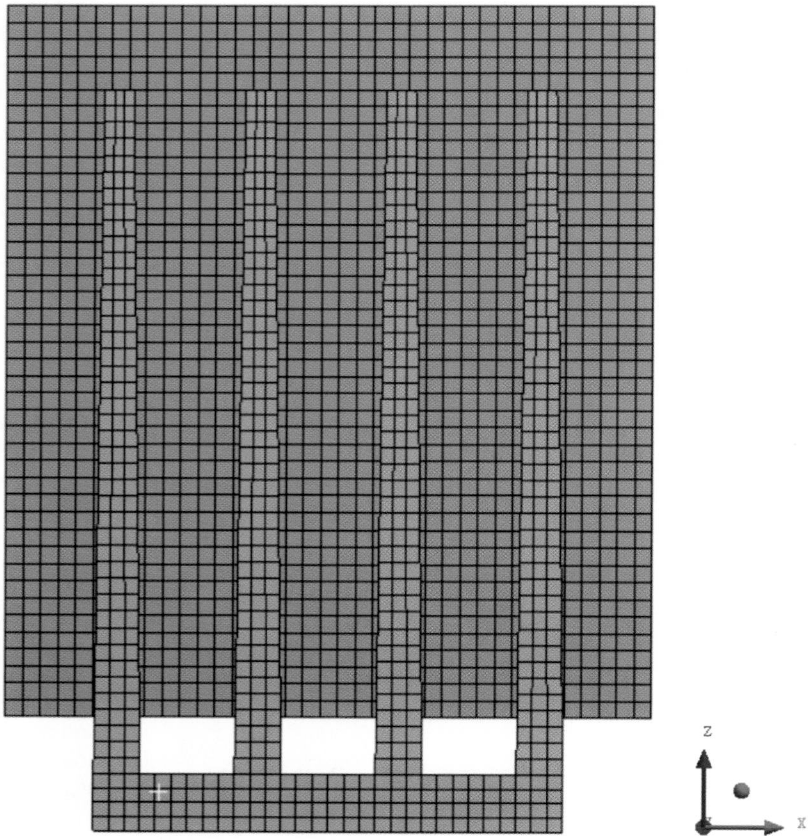

12.2.4 Evaluate Deflection under Gravity

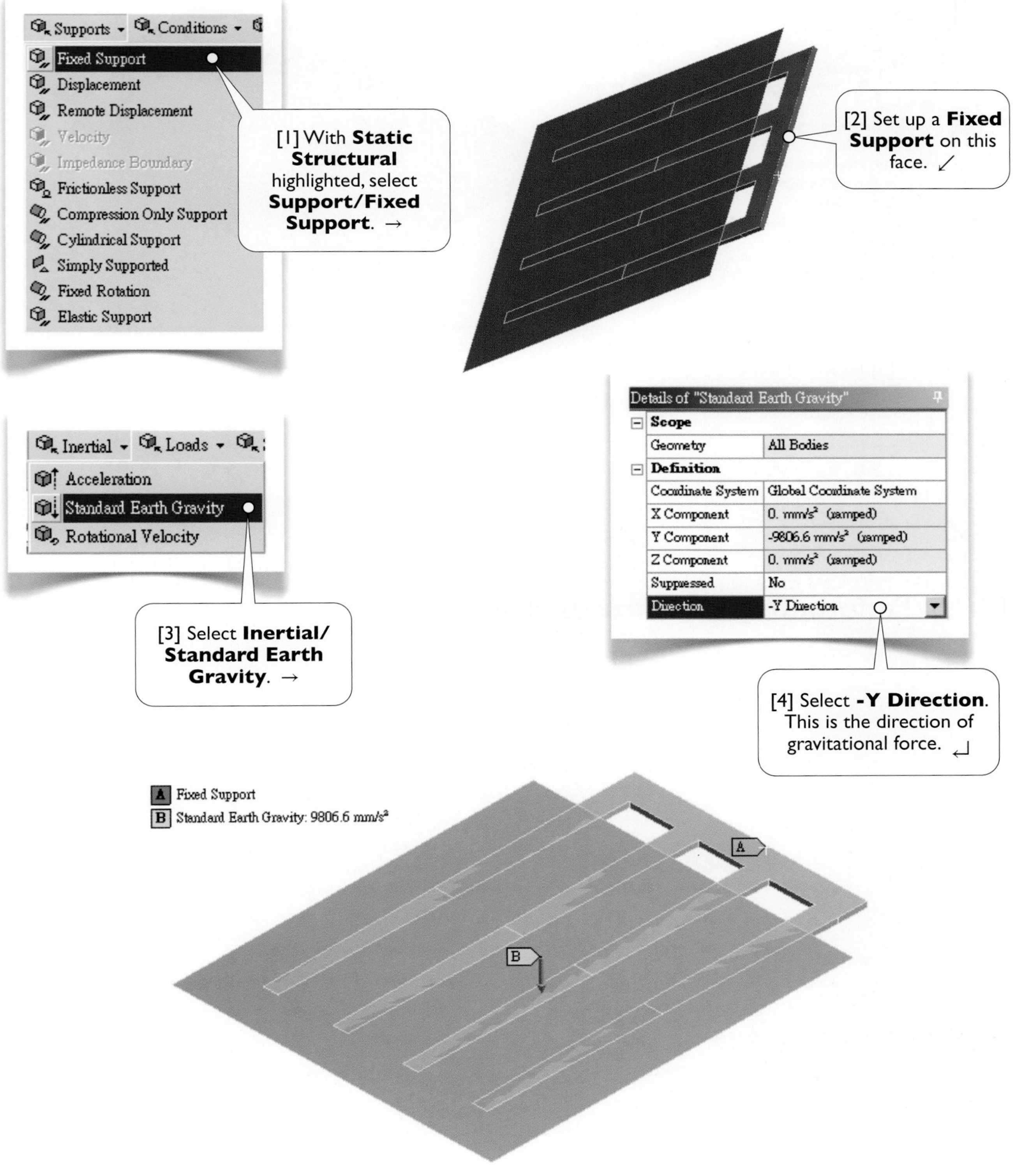

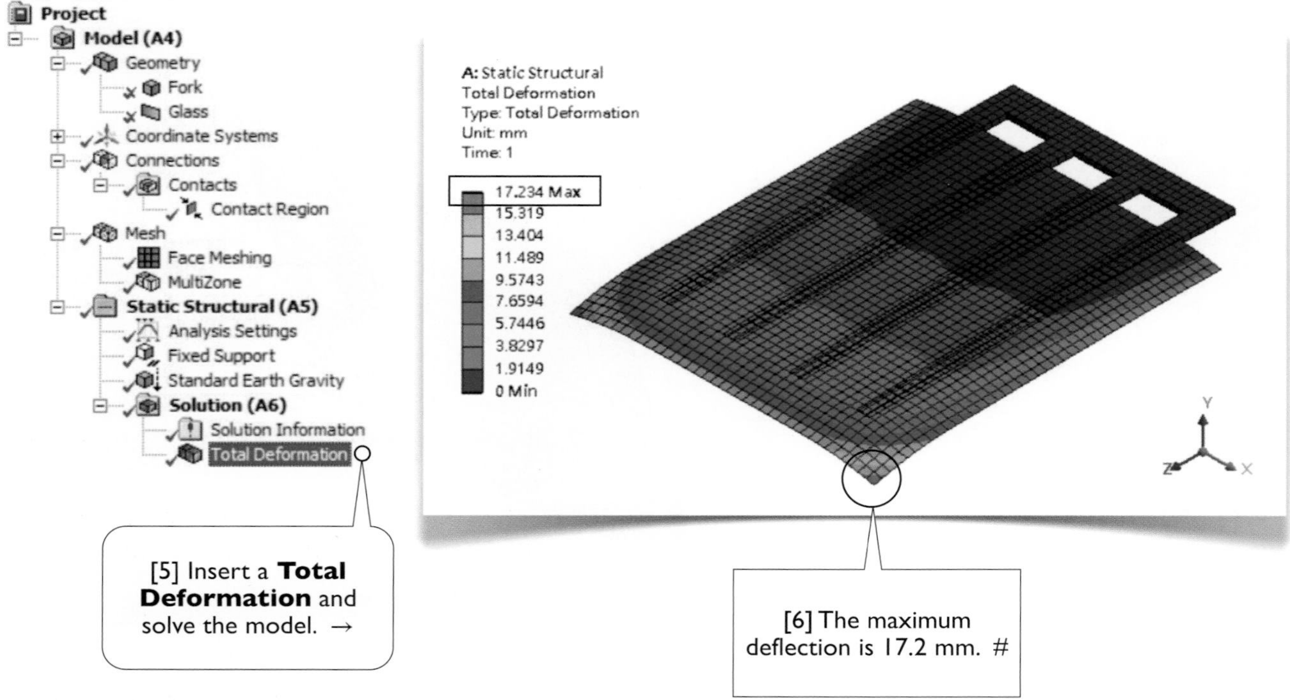

12.2.5 Perform Transient Structural Simulation

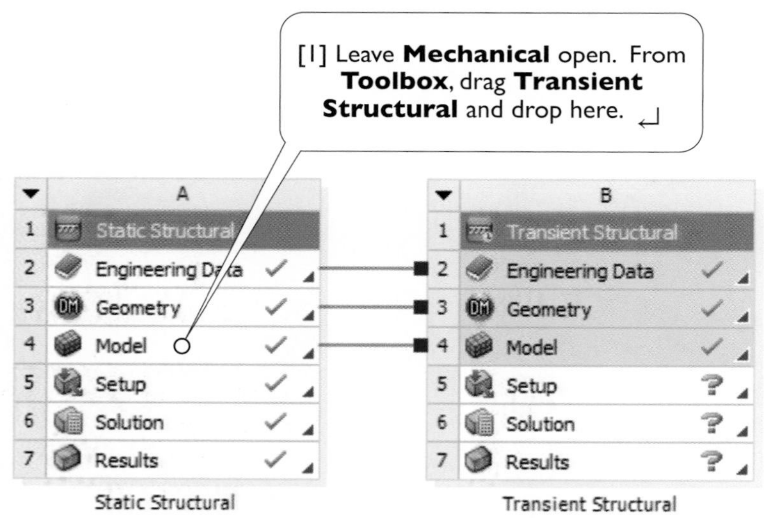

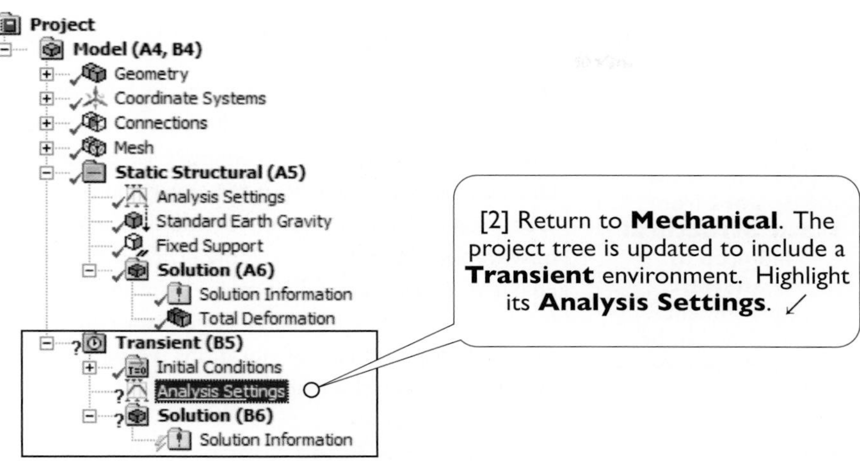

Details of "Analysis Settings"

Step Controls	
Number Of Steps	1.
Current Step Number	1.
Step End Time	1.5 s
Auto Time Stepping	On
Define By	Time
Initial Time Step	5.e-003 s
Minimum Time Step	1.e-003 s
Maximum Time Step	2.e-002 s
Time Integration	On
Solver Controls	
Solver Type	Program Controlled
Weak Springs	Off
Large Deflection	Off
Restart Controls	
Nonlinear Controls	
Output Controls	
Stress	Yes
Strain	No
Nodal Forces	No
Contact Miscellaneous	No
General Miscellaneous	No
Store Results At	Equally Spaced Points
— Value	100.
Damping Controls	
Stiffness Coefficient Define By	Direct Input
Stiffness Coefficient	1.e-003
Mass Coefficient	0.
Numerical Damping	Program Controlled
Numerical Damping Value	0.1
Analysis Data Management	

[3] Set up **Analysis Settings** like this. →

Step End Time

[4] This is the total simulation time. According to engineers' experience, the vibration should settle to a negligible amount in 1.5 sec.

Initial Time Step

As a guideline, it is suggested that the integration time step be about 1/20 of the response period, to catch the detail behavior of the structural response. The response frequency is estimated to be 10 Hz (a simulation shows that the response frequency is about 6-10 Hz; see [19], page 437). According to this guideline, the integration time step is about

$$ITS = \frac{1}{20f} = \frac{1}{20(10)} = 0.005 \text{ sec}$$

Output Controls

Transient dynamic simulations usually generate huge amounts of data. **Output Controls** allows users to cut down data storage space and computing time.

Beta Damping Value

The beta damping value (stiffness coefficient) is reported to be 0.001 from a lab test (see Eq. 12.1.3(4), page 424).

We could specify this value as a material property. Since the vibration is dominated by the glass, here, we choose to input the beta value as a global damping value, that is, neglecting the difference caused by the steel material. ↵

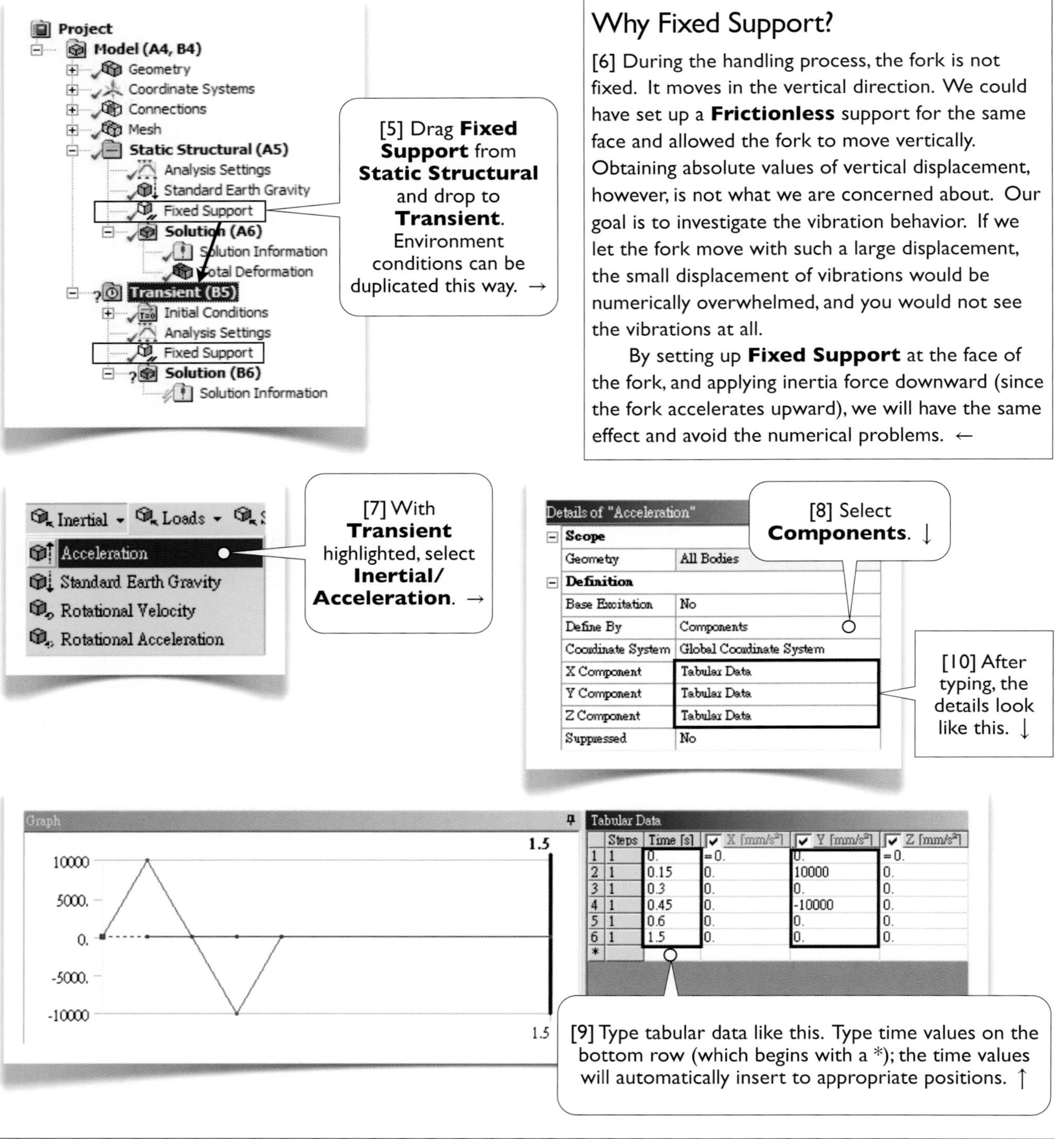

	Steps	Time [s]	X [mm/s²]	Y [mm/s²]	Z [mm/s²]
1	1	0.	= 0.	0.	= 0.
2	1	0.15	0.	10000	0.
3	1	0.3	0.	0.	0.
4	1	0.45	0.	-10000	0.
5	1	0.6	0.	0.	0.
6	1	1.5	0.	0.	0.
*					

Why Fixed Support?

[6] During the handling process, the fork is not fixed. It moves in the vertical direction. We could have set up a **Frictionless** support for the same face and allowed the fork to move vertically. Obtaining absolute values of vertical displacement, however, is not what we are concerned about. Our goal is to investigate the vibration behavior. If we let the fork move with such a large displacement, the small displacement of vibrations would be numerically overwhelmed, and you would not see the vibrations at all.

By setting up **Fixed Support** at the face of the fork, and applying inertia force downward (since the fork accelerates upward), we will have the same effect and avoid the numerical problems. ←

How are the acceleration data calculated?

[11] During the handling, the fork accelerates upward to a speed of 1.5 m/s in 0.3 sec and then decelerates to a full stop in another 0.3 sec. The average acceleration is thus 5 m/s^2. In order to reduce the amplitude of vibrations, the system controls the acceleration such that it increases linearly to 10 m/s^2 (note that this is about the same as a gravitational acceleration) and then decreases to zero linearly. The same idea applies during the deceleration. ↵

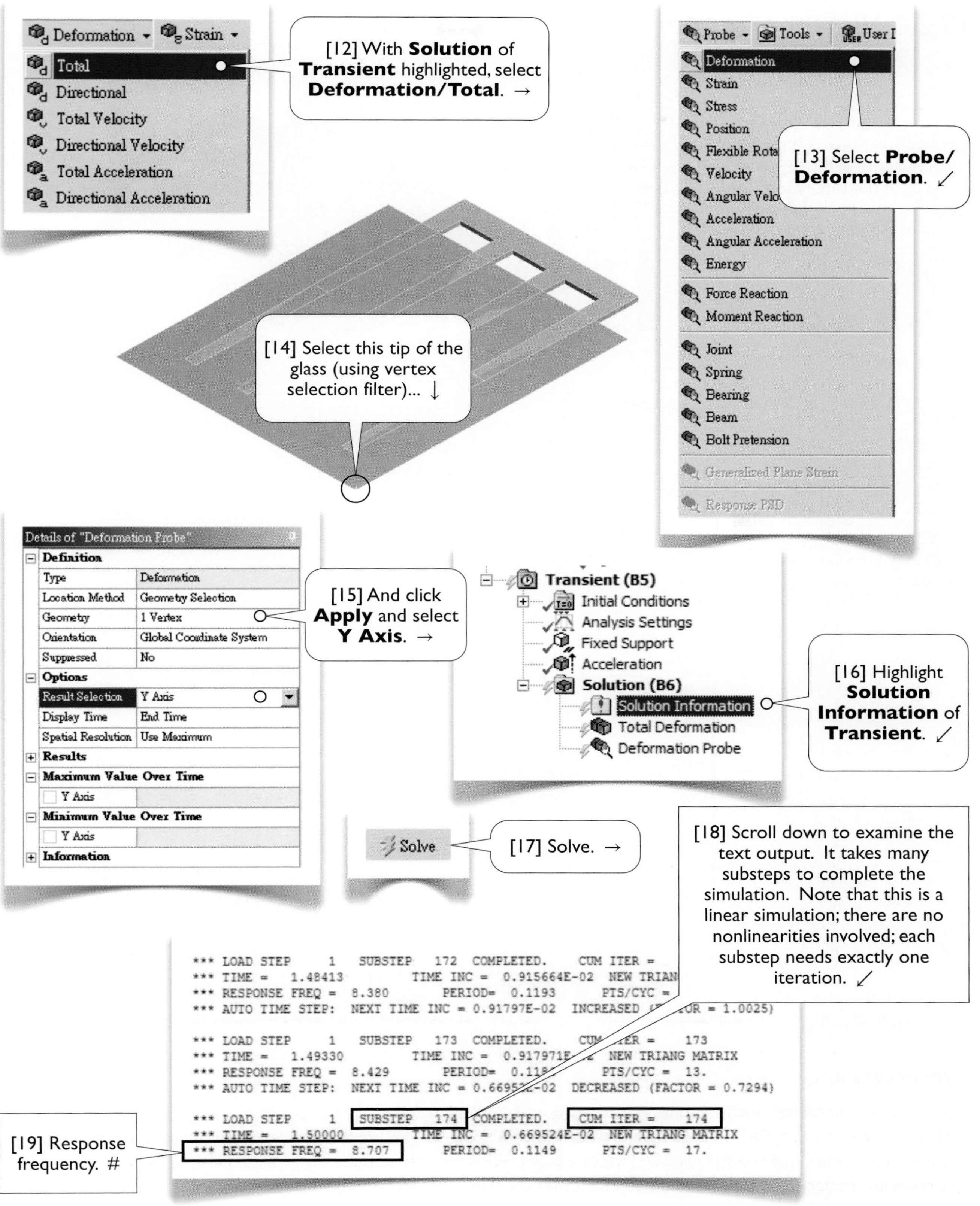

Deformation
Strain
Total
Directional
Total Velocity
Directional Velocity
Total Acceleration
Directional Acceleration
[12] With **Solution** of **Transient** highlighted, select **Deformation/Total**. →
Probe
Tools
User
Deformation
Strain
Stress
Position
Flexible Rota
Velocity
Angular Velo
Acceleration
Angular Acceleration
Energy
Force Reaction
Moment Reaction
Joint
Spring
Bearing
Beam
Bolt Pretension
Generalized Plane Strain
Response PSD
[13] Select **Probe/ Deformation**. ↙
[14] Select this tip of the glass (using vertex selection filter)... ↓
Details of "Deformation Probe"
Definition
Type Deformation
Location Method Geometry Selection
Geometry 1 Vertex
Orientation Global Coordinate System
Suppressed No
Options
Result Selection Y Axis
Display Time End Time
Spatial Resolution Use Maximum
Results
Maximum Value Over Time
Y Axis
Minimum Value Over Time
Y Axis
Information
[15] And click **Apply** and select **Y Axis**. →
Transient (B5)
Initial Conditions
Analysis Settings
Fixed Support
Acceleration
Solution (B6)
Solution Information
Total Deformation
Deformation Probe
[16] Highlight **Solution Information** of **Transient**. ↙
Solve
[17] Solve. →
[18] Scroll down to examine the text output. It takes many substeps to complete the simulation. Note that this is a linear simulation; there are no nonlinearities involved; each substep needs exactly one iteration. ↙
*** LOAD STEP 1 SUBSTEP 172 COMPLETED. CUM ITER =
*** TIME = 1.48413 TIME INC = 0.915664E-02 NEW TRIAN
*** RESPONSE FREQ = 8.380 PERIOD= 0.1193 PTS/CYC =
*** AUTO TIME STEP: NEXT TIME INC = 0.91797E-02 INCREASED (FACTOR = 1.0025)
*** LOAD STEP 1 SUBSTEP 173 COMPLETED. CUM ITER = 173
*** TIME = 1.49330 TIME INC = 0.917971E-02 NEW TRIANG MATRIX
*** RESPONSE FREQ = 8.429 PERIOD= 0.1186 PTS/CYC = 13.
*** AUTO TIME STEP: NEXT TIME INC = 0.66952E-02 DECREASED (FACTOR = 0.7294)
*** LOAD STEP 1 SUBSTEP 174 COMPLETED. CUM ITER = 174
*** TIME = 1.50000 TIME INC = 0.669524E-02 NEW TRIANG MATRIX
*** RESPONSE FREQ = 8.707 PERIOD= 0.1149 PTS/CYC = 17.
[19] Response frequency. #

12.2.6 View the Results

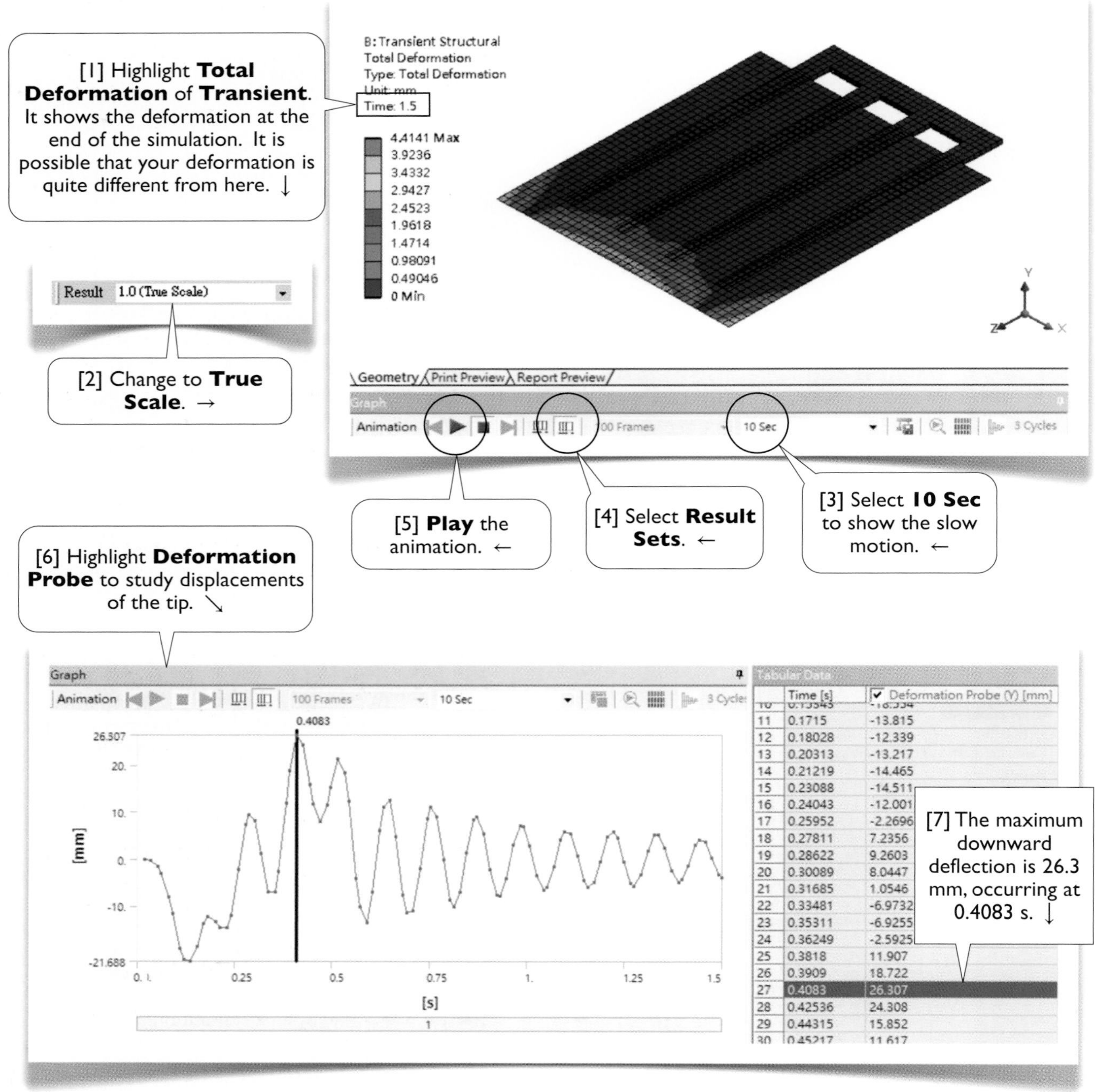

Observation

[8] The maximum deflection 26.3 mm [7], occurring at 0.4083 sec, is about 50% more than that of the static deflection, which is 17.2 mm (12.2.4[6], page 434). The vibration damps out fast and reduces to less than 5 mm in 1.5 sec. The response frequency is about 9 Hz (12.2.5[19], last page). Workbench automatically adjusts the time step according to the response frequency. Next, we want to investigate the maximum stress and its location. ↵

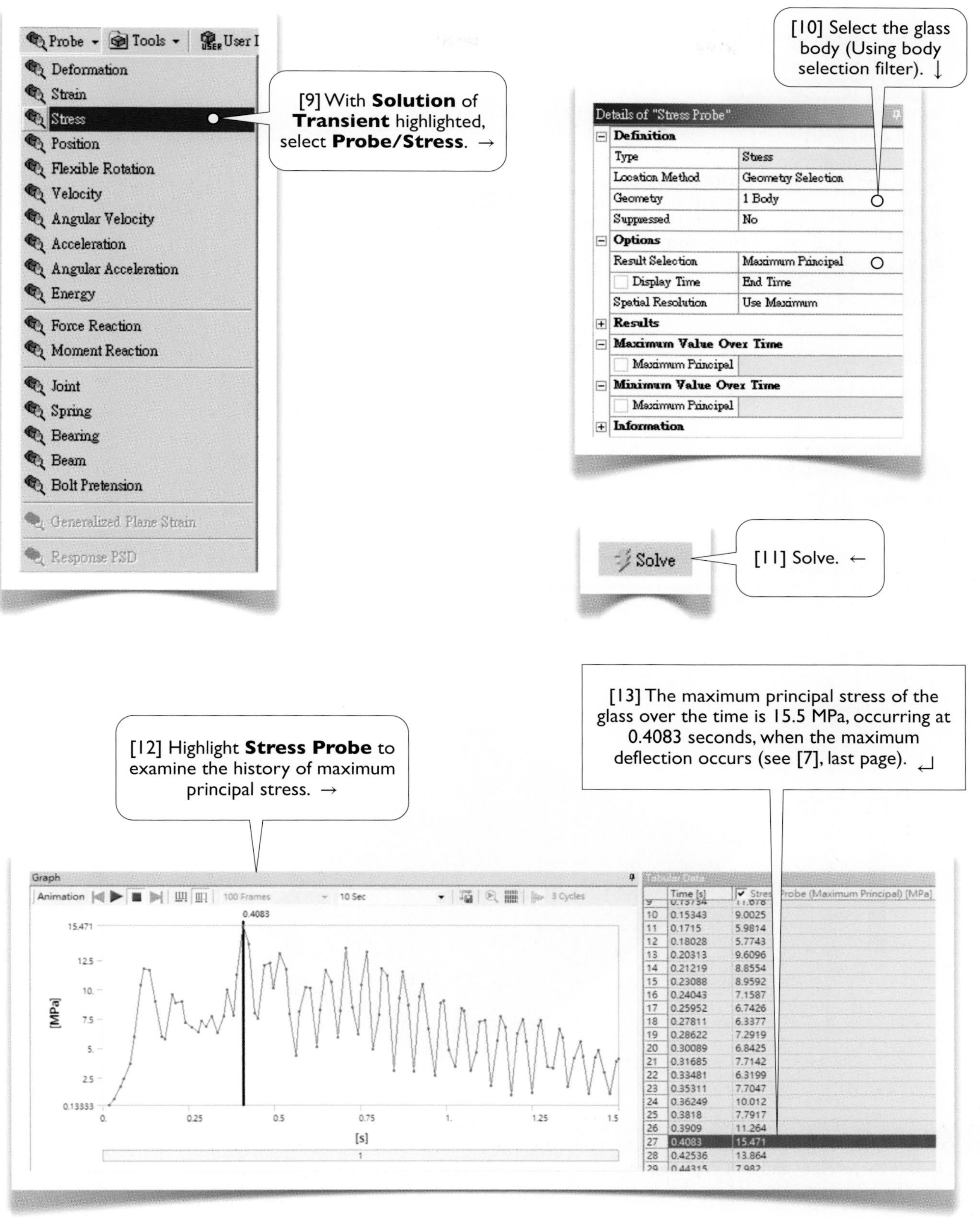
Probe
Tools
User I
Deformation
Strain
Stress
Position
Flexible Rotation
Velocity
Angular Velocity
Acceleration
Angular Acceleration
Energy
Force Reaction
Moment Reaction
Joint
Spring
Bearing
Beam
Bolt Pretension
Generalized Plane Strain
Response PSD
[9] With **Solution** of **Transient** highlighted, select **Probe/Stress**. →
[10] Select the glass body (Using body selection filter). ↓
Details of "Stress Probe"
Definition
Type Stress
Location Method Geometry Selection
Geometry 1 Body
Suppressed No
Options
Result Selection Maximum Principal
Display Time End Time
Spatial Resolution Use Maximum
Results
Maximum Value Over Time
Maximum Principal
Minimum Value Over Time
Maximum Principal
Information
Solve
[11] Solve. ←
[12] Highlight **Stress Probe** to examine the history of maximum principal stress. →
[13] The maximum principal stress of the glass over the time is 15.5 MPa, occurring at 0.4083 seconds, when the maximum deflection occurs (see [7], last page). ↵
Graph
Animation
100 Frames
10 Sec
3 Cycles
0.4083
15.471
12.5
10.
7.5
5.
2.5
0.13333
[MPa]
0.
0.25
0.5
0.75
1.
1.25
1.5
[s]
1
Tabular Data
Time [s]
Stress Probe (Maximum Principal) [MPa]
10 0.15343 9.0025
11 0.1715 5.9814
12 0.18028 5.7743
13 0.20313 9.6096
14 0.21219 8.8554
15 0.23088 8.9592
16 0.24043 7.1587
17 0.25952 6.7426
18 0.27811 6.3377
19 0.28622 7.2919
20 0.30089 6.8425
21 0.31685 7.7142
22 0.33481 6.3199
23 0.35311 7.7047
24 0.36249 10.012
25 0.3818 7.7917
26 0.3909 11.264
27 0.4083 15.471
28 0.42536 13.864

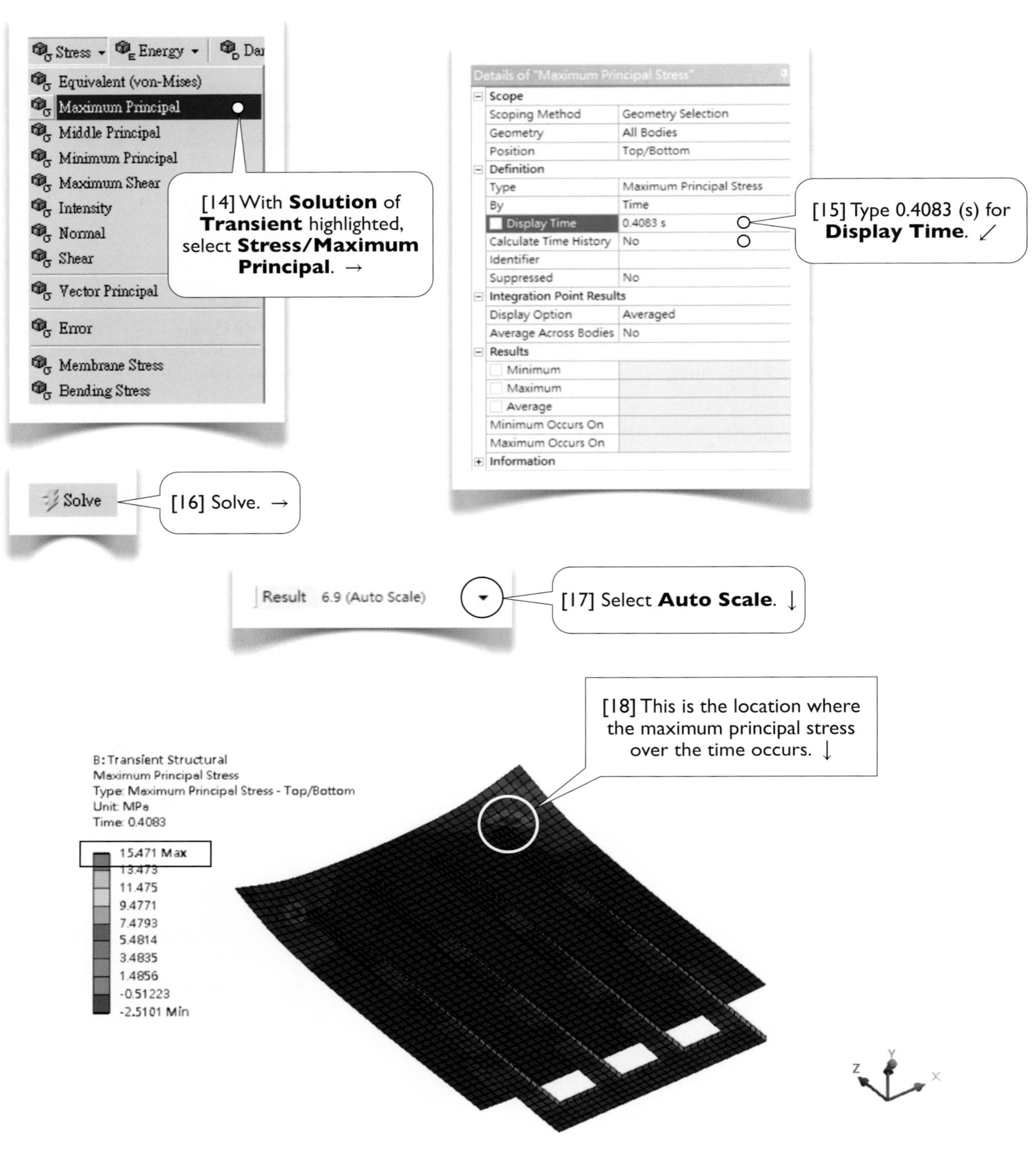

Wrap Up

[19] Save the project and exit Workbench. #

Section 12.3

Harmonic Response Analysis: Two-Story Building

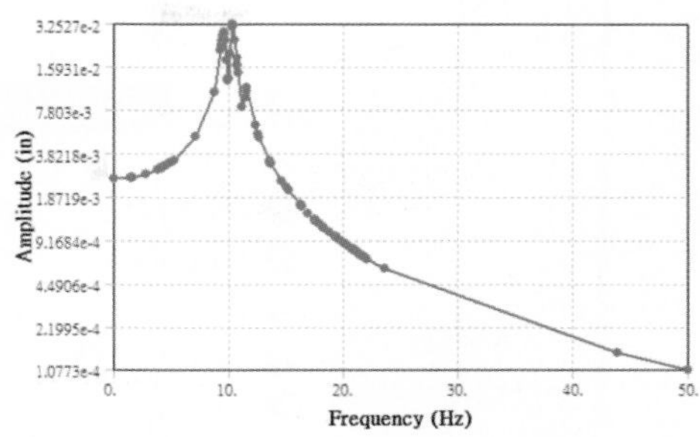

12.3.1 About the Two-Story Building

[1] In this section, we will demonstrate the procedure of a harmonic response analysis (12.1.4[5], page 427). The two-story building (7.3, 11.2) is used again to demonstrate the procedures.

Harmonic Response Analysis

In 11.2.3[13] (page 397), we mentioned that the rhythmic loading on the floor may cause a safety issue. Is "dancing on the floor" really an issue? Since the building is designed to withstand a live load of 50 psf (lb/ft^2), we will assume that a group of young people of 50 psf is dancing on a side-span floor deck [2] to simulate an asymmetric loading that will cause the building to side sway. The dancing is so hard that the young people generate a vertical periodical force of 10 psf, that is, the loading fluctuates from 40 psf to 60 psf.

Engineers usually don't consider "dancing" as a serious issue. Let's look at a more realistic engineering consideration. Imagine that an electric motor (or any rotatory machine) is installed on the floor deck [2]. The operational speed of the machine is 3000 rpm. When started up, the machine's speed increases from zero up to 3000 rpm. Is the vibration caused by the rotatory machine an issue?

In this section, we will perform a harmonic response analysis to answer these questions. →

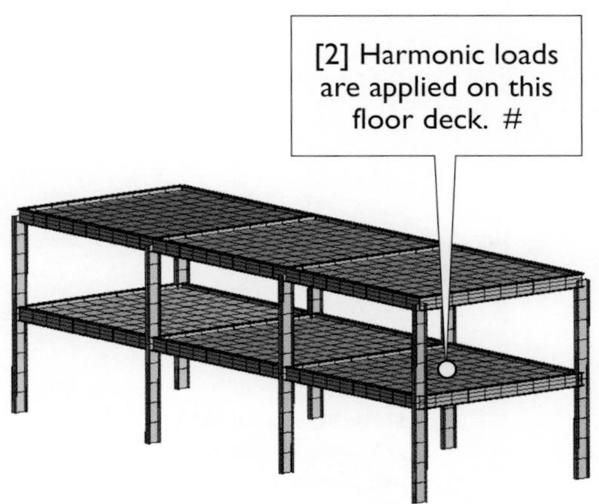

12.3.2 Perform Modal Analysis

[1] Launch Workbench. Open the project **Building**, which was saved in Section 11.2. ↙

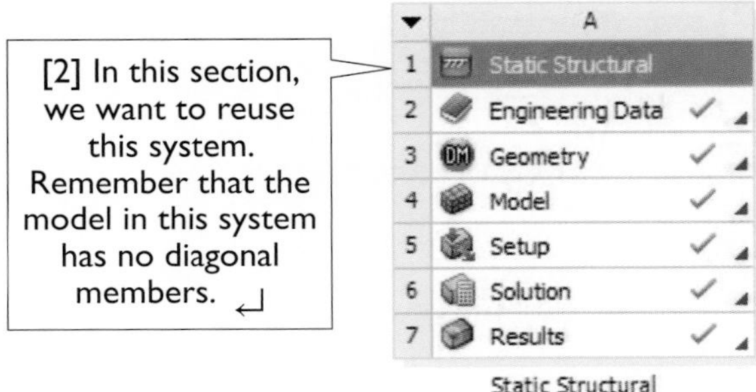

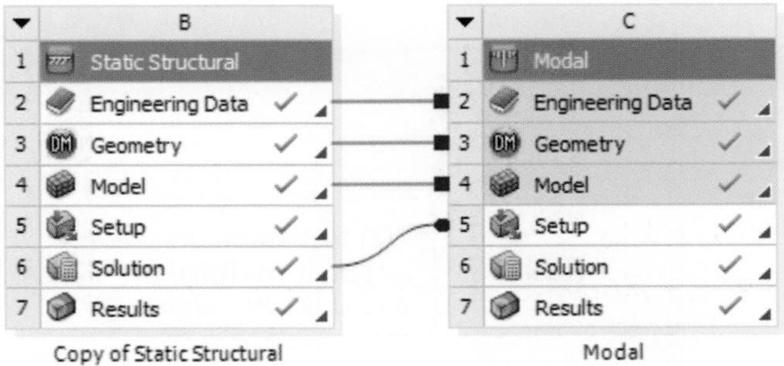

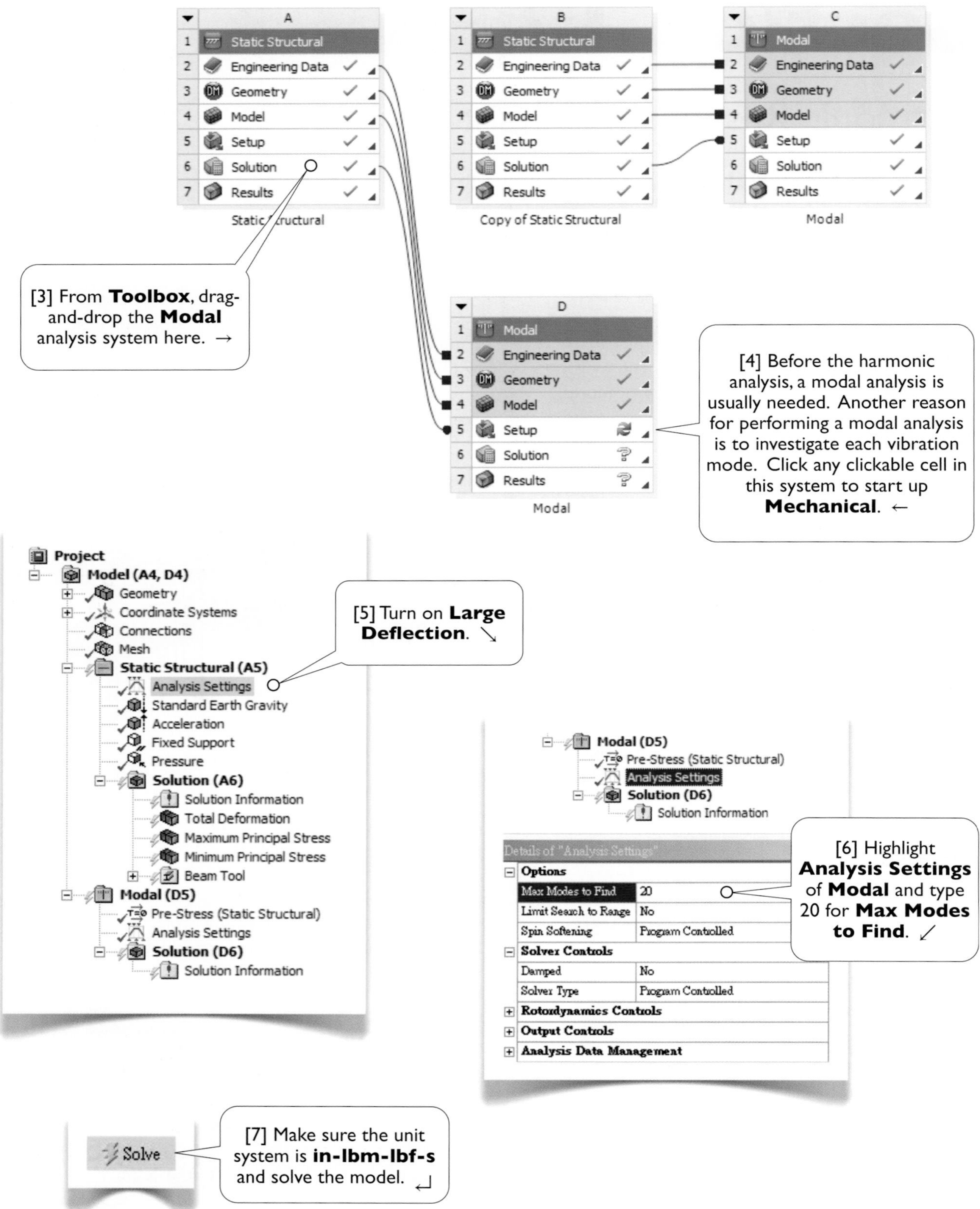
A
1 Static Structural
2 Engineering Data
3 Geometry
4 Model
5 Setup
6 Solution
7 Results
Static Structural
B
1 Static Structural
2 Engineering Data
3 Geometry
4 Model
5 Setup
6 Solution
7 Results
Copy of Static Structural
C
1 Modal
2 Engineering Data
3 Geometry
4 Model
5 Setup
6 Solution
7 Results
Modal
[3] From **Toolbox**, drag-and-drop the **Modal** analysis system here. →
D
1 Modal
2 Engineering Data
3 Geometry
4 Model
5 Setup
6 Solution
7 Results
Modal
[4] Before the harmonic analysis, a modal analysis is usually needed. Another reason for performing a modal analysis is to investigate each vibration mode. Click any clickable cell in this system to start up **Mechanical**. ←
Project
Model (A4, D4)
Geometry
Coordinate Systems
Connections
Mesh
Static Structural (A5)
Analysis Settings
Standard Earth Gravity
Acceleration
Fixed Support
Pressure
Solution (A6)
Solution Information
Total Deformation
Maximum Principal Stress
Minimum Principal Stress
Beam Tool
Modal (D5)
Pre-Stress (Static Structural)
Analysis Settings
Solution (D6)
Solution Information
[5] Turn on **Large Deflection**. ↘
Modal (D5)
Pre-Stress (Static Structural)
Analysis Settings
Solution (D6)
Solution Information
Details of "Analysis Settings"
Options
Max Modes to Find 20
Limit Search to Range No
Spin Softening Program Controlled
Solver Controls
Damped No
Solver Type Program Controlled
Rotordynamics Controls
Output Controls
Analysis Data Management
[6] Highlight **Analysis Settings** of **Modal** and type 20 for **Max Modes to Find**. ↙
Solve
[7] Make sure the unit system is **in-lbm-lbf-s** and solve the model. ↵

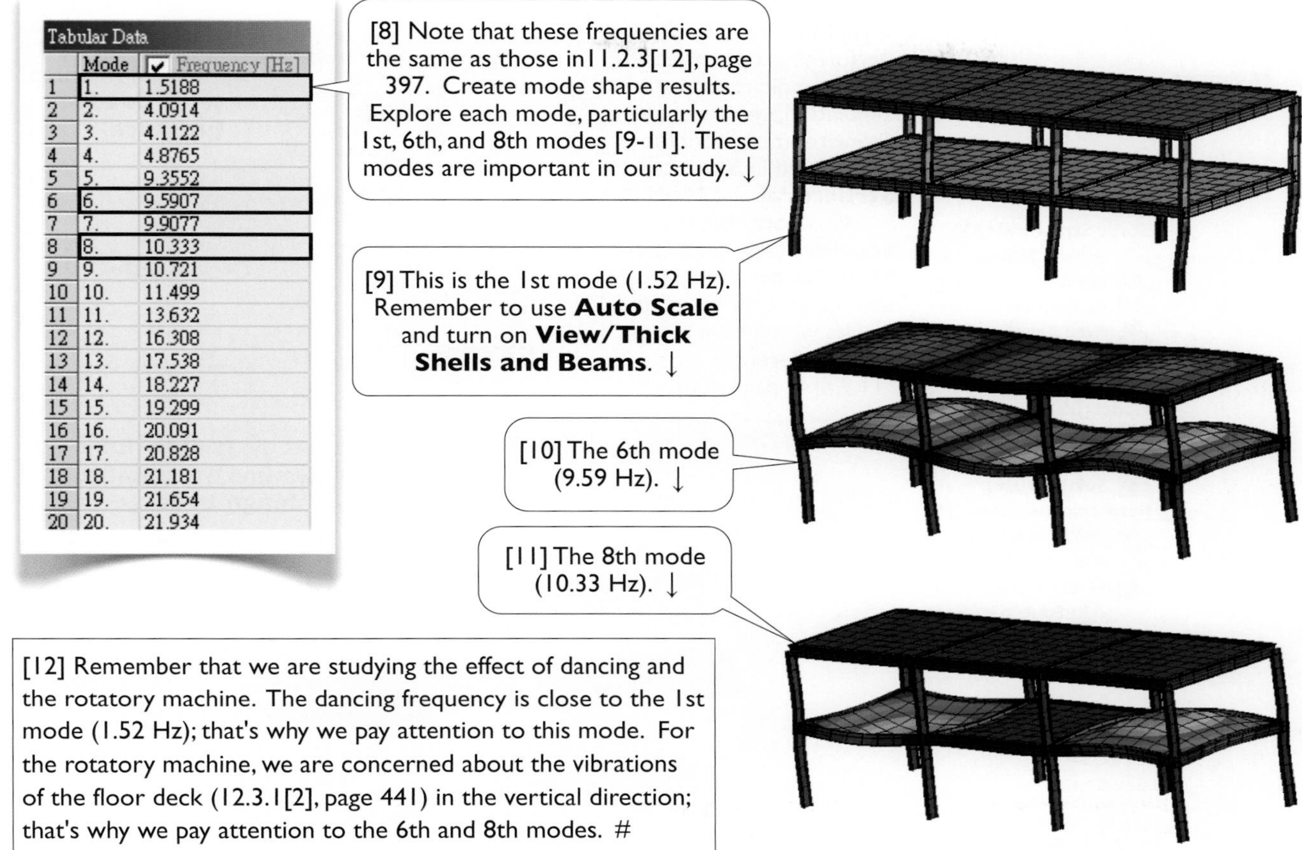

[12] Remember that we are studying the effect of dancing and the rotatory machine. The dancing frequency is close to the 1st mode (1.52 Hz); that's why we pay attention to this mode. For the rotatory machine, we are concerned about the vibrations of the floor deck (12.3.1[2], page 441) in the vertical direction; that's why we pay attention to the 6th and 8th modes. #

12.3.3 Perform Harmonic Response Analysis

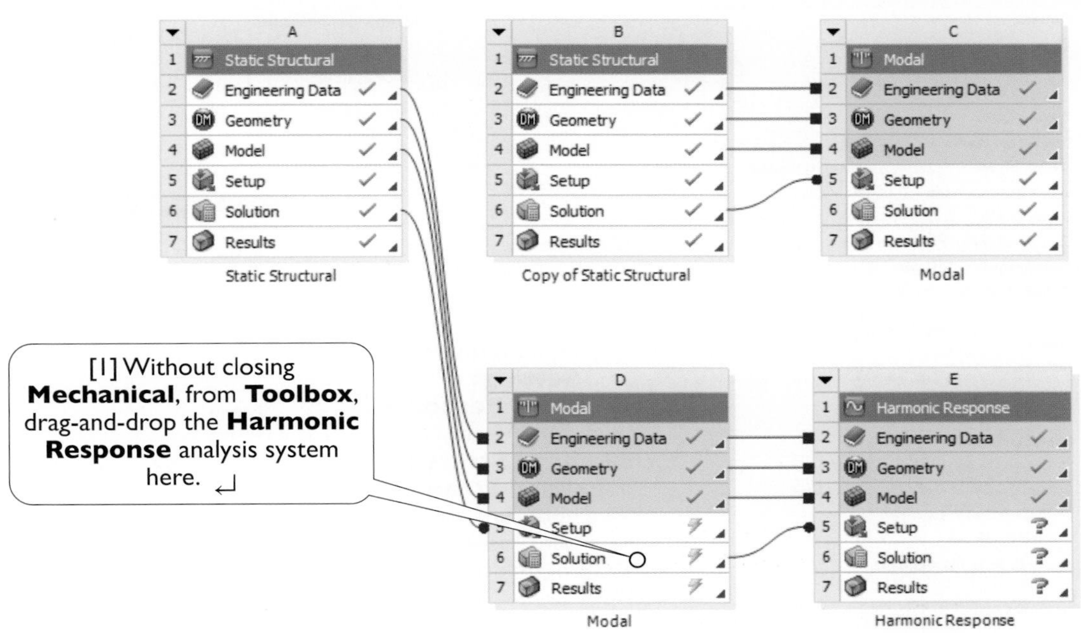

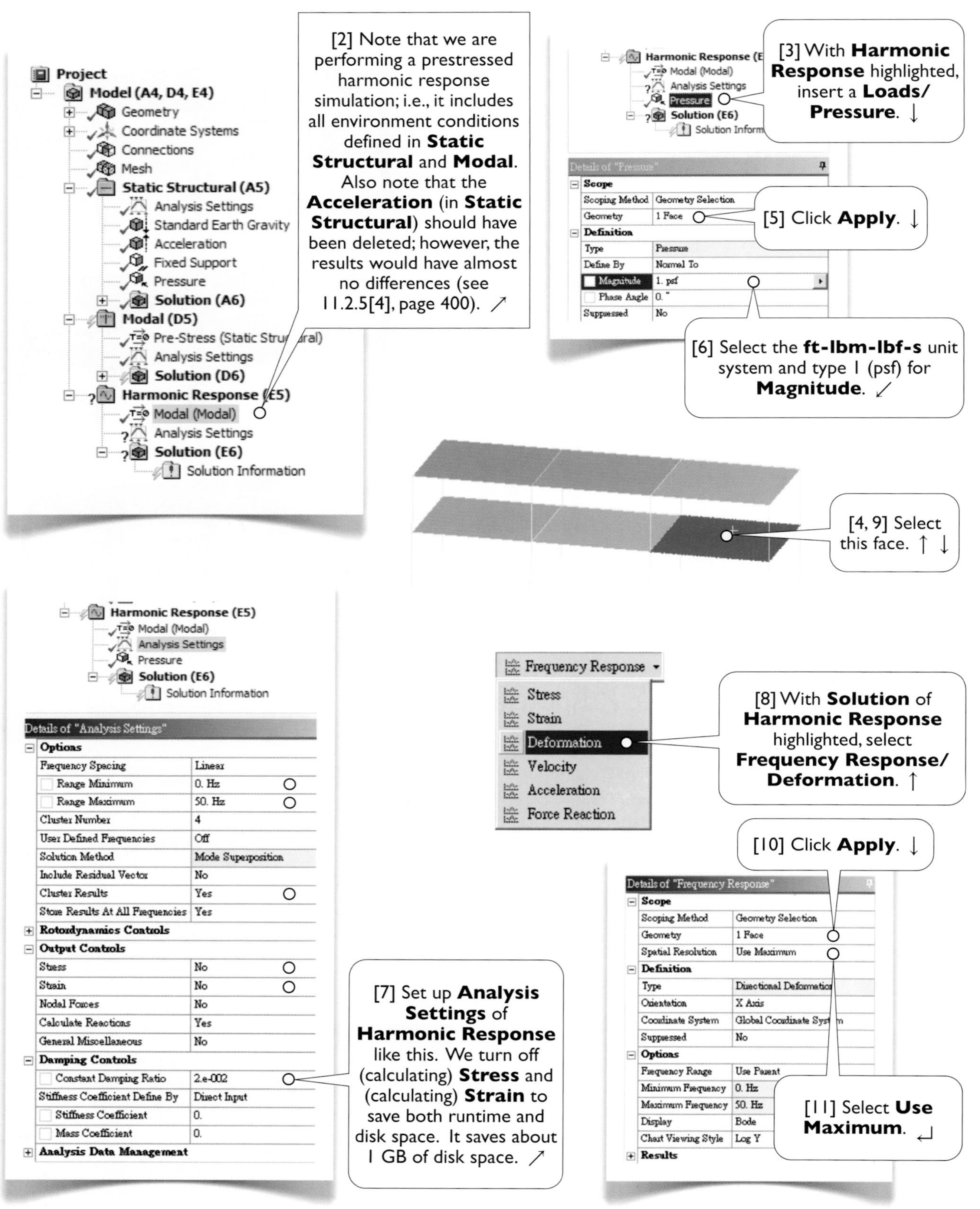

[2] Note that we are performing a prestressed harmonic response simulation; i.e., it includes all environment conditions defined in Static Structural and Modal. Also note that the Acceleration (in Static Structural) should have been deleted; however, the results would have almost no differences (see 11.2.5[4], page 400).
[3] With Harmonic Response highlighted, insert a Loads/Pressure.
[5] Click Apply.
[6] Select the ft-lbm-lbf-s unit system and type 1 (psf) for Magnitude.
[4, 9] Select this face.
[8] With Solution of Harmonic Response highlighted, select Frequency Response/Deformation.
[10] Click Apply.
[7] Set up Analysis Settings of Harmonic Response like this. We turn off (calculating) Stress and (calculating) Strain to save both runtime and disk space. It saves about 1 GB of disk space.
[11] Select Use Maximum.
Project
Model (A4, D4, E4)
Geometry
Coordinate Systems
Connections
Mesh
Static Structural (A5)
Analysis Settings
Standard Earth Gravity
Acceleration
Fixed Support
Pressure
Solution (A6)
Modal (D5)
Pre-Stress (Static Structural)
Analysis Settings
Solution (D6)
Harmonic Response (E5)
Modal (Modal)
Analysis Settings
Solution (E6)
Solution Information
Details of "Pressure"
Scope
Scoping Method Geometry Selection
Geometry 1 Face
Definition
Type Pressure
Define By Normal To
Magnitude 1. psf
Phase Angle 0. °
Suppressed No
Frequency Response
Stress
Strain
Deformation
Velocity
Acceleration
Force Reaction
Details of "Analysis Settings"
Options
Frequency Spacing Linear
Range Minimum 0. Hz
Range Maximum 50. Hz
Cluster Number 4
User Defined Frequencies Off
Solution Method Mode Superposition
Include Residual Vector No
Cluster Results Yes
Store Results At All Frequencies Yes
Rotordynamics Controls
Output Controls
Stress No
Strain No
Nodal Forces No
Calculate Reactions Yes
General Miscellaneous No
Damping Controls
Constant Damping Ratio 2.e-002
Stiffness Coefficient Define By Direct Input
Stiffness Coefficient 0.
Mass Coefficient 0.
Analysis Data Management
Details of "Frequency Response"
Scope
Scoping Method Geometry Selection
Geometry 1 Face
Spatial Resolution Use Maximum
Definition
Type Directional Deformation
Orientation X Axis
Coordinate System Global Coordinate System
Suppressed No
Options
Frequency Range Use Parent
Minimum Frequency 0. Hz
Maximum Frequency 50. Hz
Display Bode
Chart Viewing Style Log Y
Results

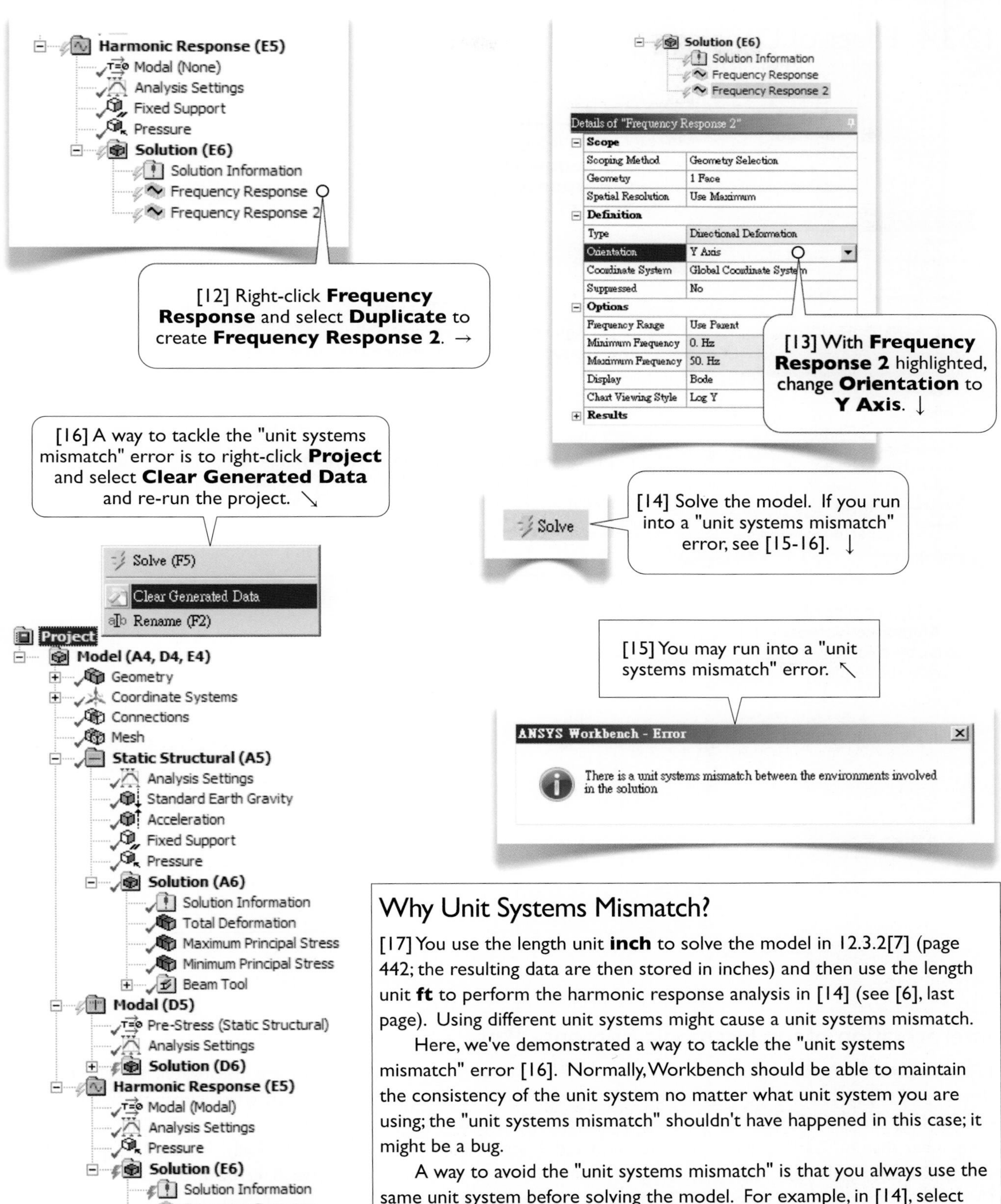

Why Unit Systems Mismatch?

[17] You use the length unit **inch** to solve the model in 12.3.2[7] (page 442; the resulting data are then stored in inches) and then use the length unit **ft** to perform the harmonic response analysis in [14] (see [6], last page). Using different unit systems might cause a unit systems mismatch.

Here, we've demonstrated a way to tackle the "unit systems mismatch" error [16]. Normally, Workbench should be able to maintain the consistency of the unit system no matter what unit system you are using; the "unit systems mismatch" shouldn't have happened in this case; it might be a bug.

A way to avoid the "unit systems mismatch" is that you always use the same unit system before solving the model. For example, in [14], select **in-lbm-lbf-s** unit system before clicking **Solve**. #

12.3.4 Effect of Dancing

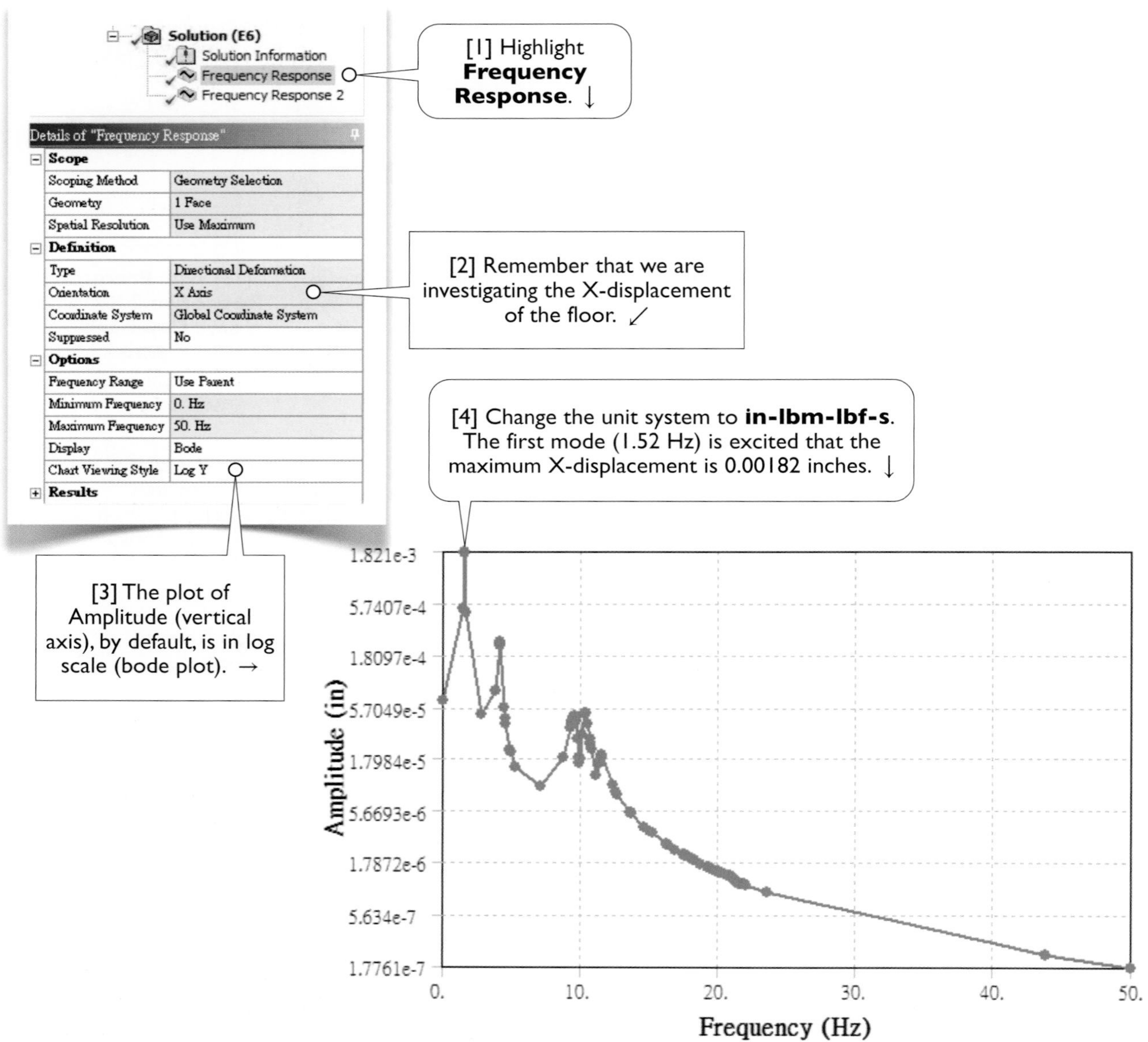

Interpretation of the Harmonic Response Plot

[5] The harmonic response plot [4] is an amplitude versus frequency plot. Because we are investigating the effect of the dancing, we should look at the frequencies that are less than 3 or 4 Hz (people don't dance faster than that). At dancing frequency of 1.52 Hz, the structure is excited such that the maximum X-displacement is 0.00182 in. Remember that in 12.3.1[1] (page 441), the estimated dancing load is about 10 psf periodically. In 12.3.3[6] (page 444), we input 1 psf of harmonic load; therefore, the estimated response should be 10 times 0.00182 in, that is 0.0182 in. (0.46 mm). Obviously, this value is too small to be worried about.

We conclude that dancing is not an issue for this building. #

12.3.5 Effect of Rotatory Machine

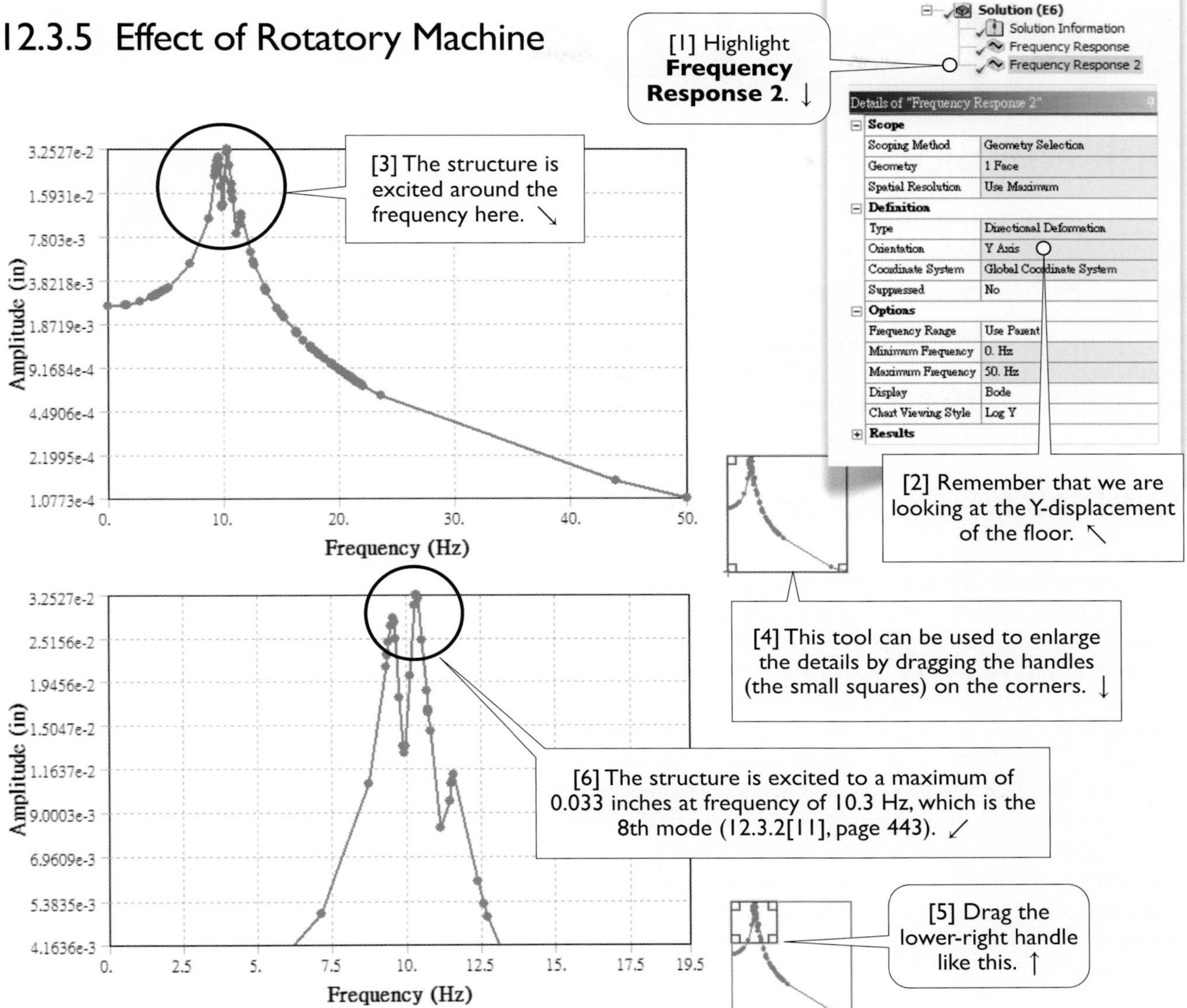

Interpretation of the Harmonic Response Plot

[7] We were investigating the effect of the rotatory machine from 0 to 3000 rpm (50 Hz). We estimated that the amplitude of the harmonic load (of the electric motor) should be no more than 0.1 psf distributing on the floor (that totals to 40 lb). In 12.3.3[6] (page 444), we input 1 psf of harmonic load; therefore, the estimated response should be 0.1 times of the response shown in [3] (or [6]).

Although high frequencies do excite the floor, the values are very small. At frequency of 10.3 Hz, the excitation reaches a maximum of 0.0033 in (0.1 times of 0.033 in), or 0.084 mm, that is even smaller than the dancing effect. Therefore, the value is too small to cause an issue.

We conclude that the rotatory machine is safe for this building. ↓

Wrap Up

[8] Save the project and exit Workbench. #

Section 12.4

Disk and Block

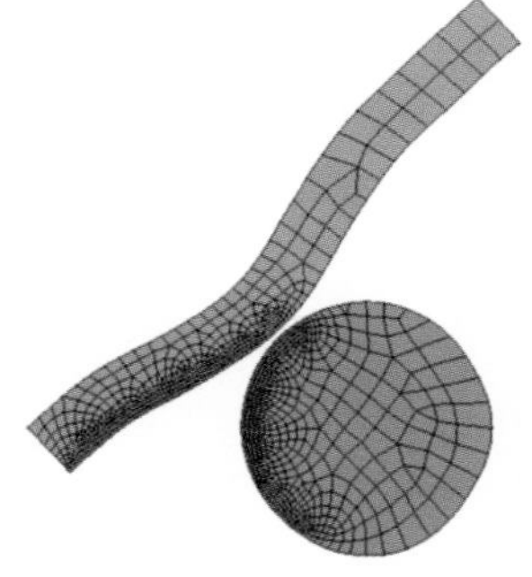

This exercise has two purposes: First, to demonstrate how to apply a simple initial condition, namely uniform velocity, on a body. Second, to show a limitation of **Transient Structural** analysis system for impact simulations and to motivate the students learning **Explicit Dynamics** in Chapter 15.

12.4.1 About the Disk and Block

[1] Consider a disk of radius of 40 mm and a block of 200 mm x 20 mm on a frictionless horizontal surface [2-3]; both have a thickness of 10 mm. Both are made of a very soft polymer of Young's modulus of 10 kPa, Poisson's ratio of 0.4, and mass density of 1000 kg/m^3.

Right before the impact, the disk moves toward the block with a velocity of 0.5 m/s, and the positions of the disk and the block are as shown.

We purposely consider a very soft material (Young's modulus of 10 kPa) and a very slow-speed impact (velocity of 0.5 m/s) to relieve numerical difficulty. Increasing either of them would make the impact duration shorter, and in turn require a shorter integration time step (to find a solution). This is left as an exercise for you at the end of this chapter. →

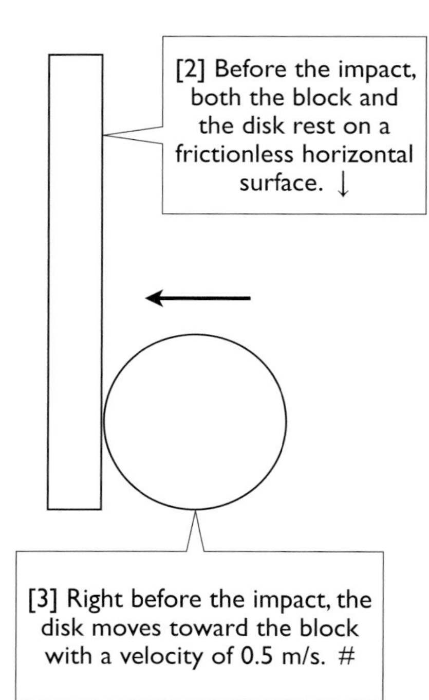

[2] Before the impact, both the block and the disk rest on a frictionless horizontal surface. ↓

[3] Right before the impact, the disk moves toward the block with a velocity of 0.5 m/s. #

12.4.2 Start Up

[1] Launch Workbench. Create a **Transient Structural** analysis system [2] by double-clicking it in **Toolbox**. Save the project as **Disk**.

Double-click **Engineering Data** to prepare material data [3]. →

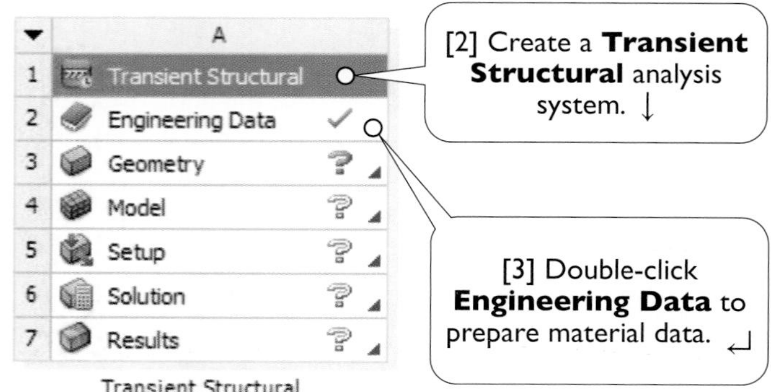

[2] Create a **Transient Structural** analysis system. ↓

[3] Double-click **Engineering Data** to prepare material data. ↵

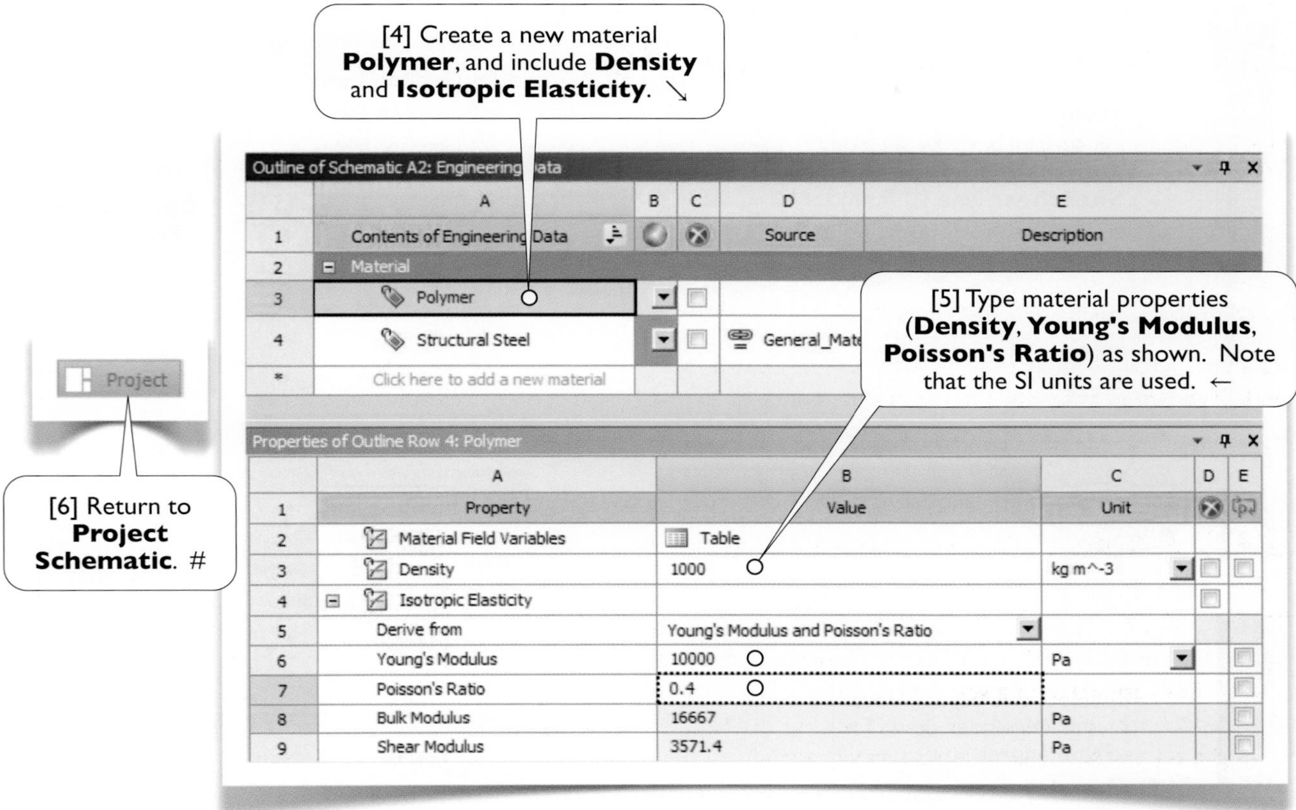

12.4.3 Create Geometry in DesignModeler

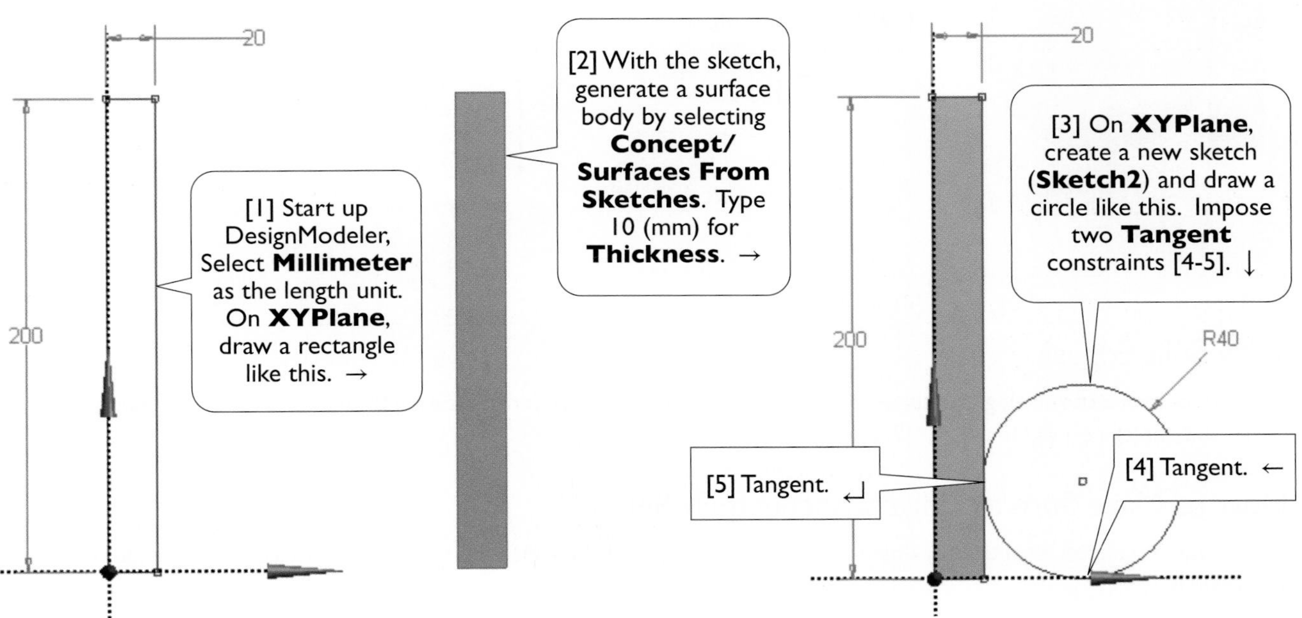

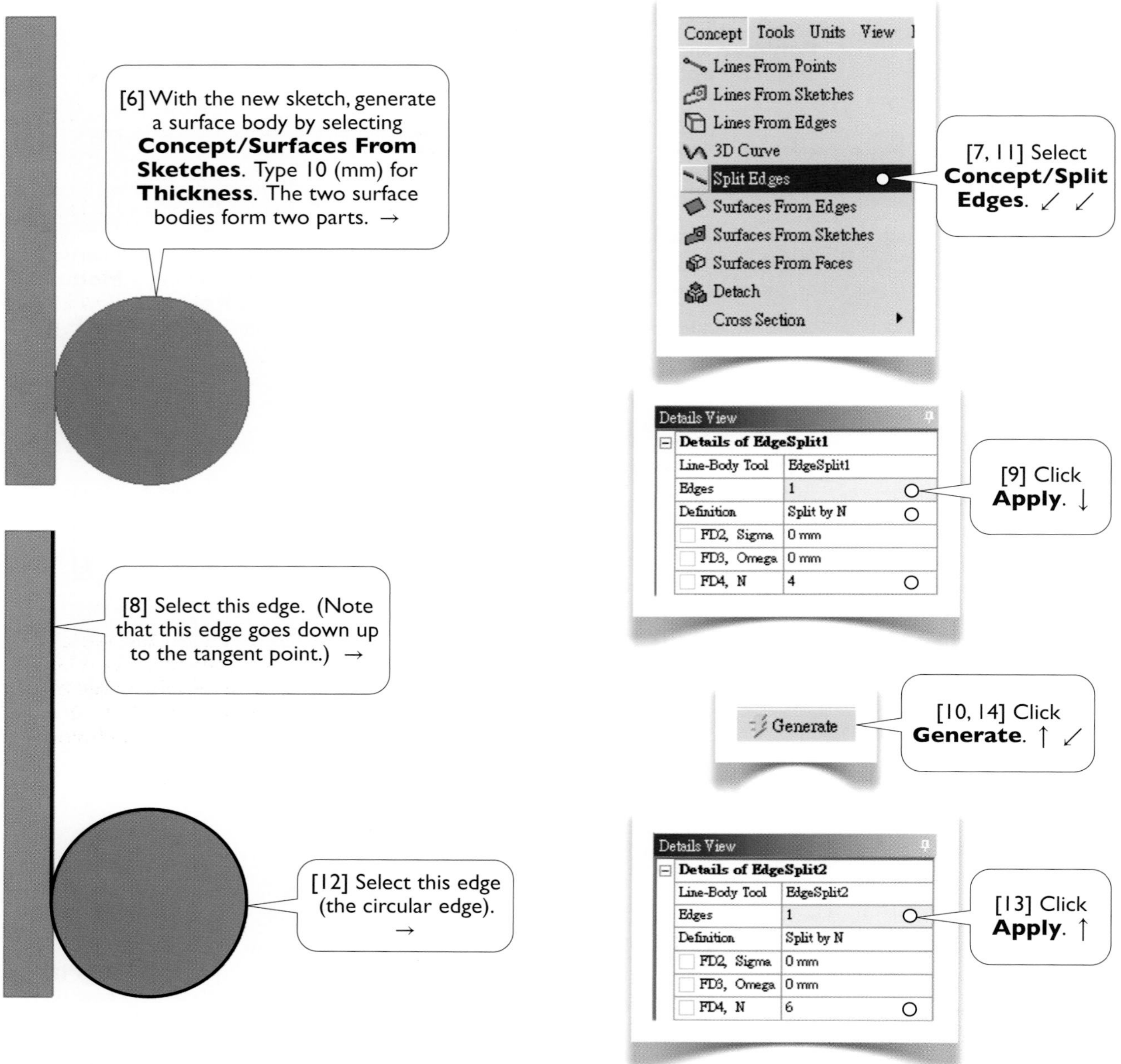

Why Split Edges?

[15] The purpose of splitting the edges is that we may specify a mesh density for each segment. We need finer mesh near the contact region (see 12.4.4[7-9], page 452).

How to Make Sure an Edge is Successfully Split?

Move your mouse on a split edge, which is then highlighted. Go through each split edge to make sure an edge is successfully split. ↵

A: Transient Structural
XYPlane
Sketch1
Sketch2
ZXPlane
YZPlane
SurfaceSk1
SurfaceSk2
EdgeSplit1
EdgeSplit2
EdgeSplit3
2 Parts, 2 Bodies
Block
Disk

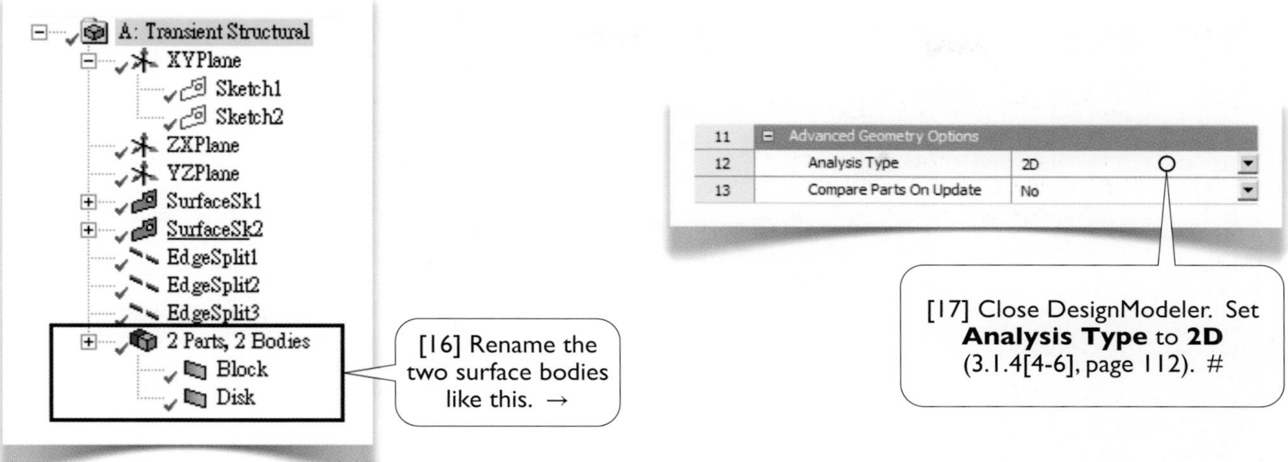

[16] Rename the two surface bodies like this. →

11 Advanced Geometry Options
12 Analysis Type 2D
13 Compare Parts On Update No

[17] Close DesignModeler. Set **Analysis Type** to **2D** (3.1.4[4-6], page 112). #

12.4.4 Simulation in **Mechanical**

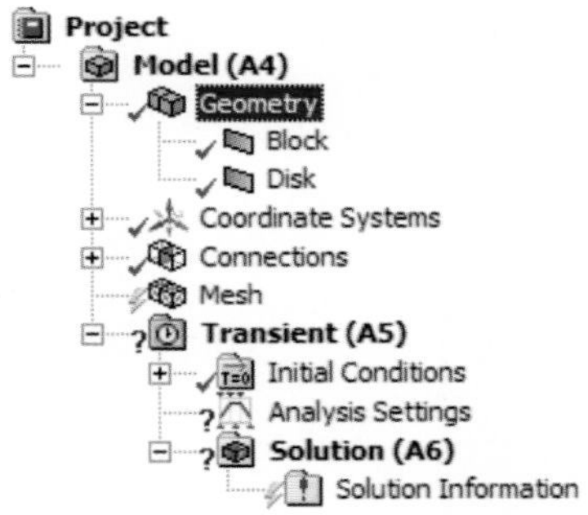

Project
Model (A4)
Geometry
Block
Disk
Coordinate Systems
Connections
Mesh
Transient (A5)
Initial Conditions
Analysis Settings
Solution (A6)
Solution Information

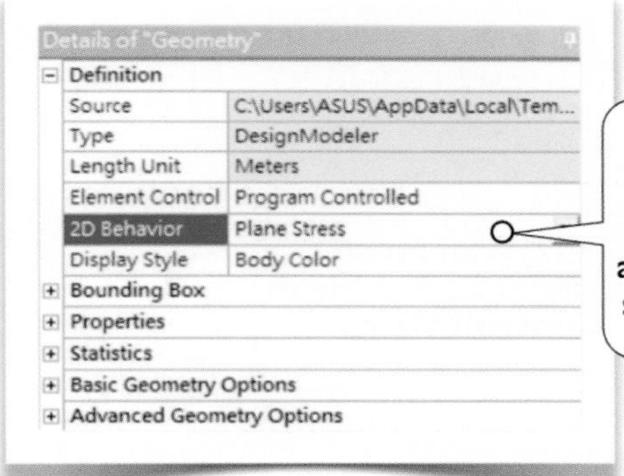

Details of "Geometry"
Definition
Source C:\Users\ASUS\AppData\Local\Tem...
Type DesignModeler
Length Unit Meters
Element Control Program Controlled
2D Behavior Plane Stress
Display Style Body Color
Bounding Box
Properties
Statistics
Basic Geometry Options
Advanced Geometry Options

[1] Start up **Mechanical** and select the **mm-kg-N-s** unit system. Highlight **Geometry**, and make sure **Plane Stress** is selected for **2D Behavior**. ↙

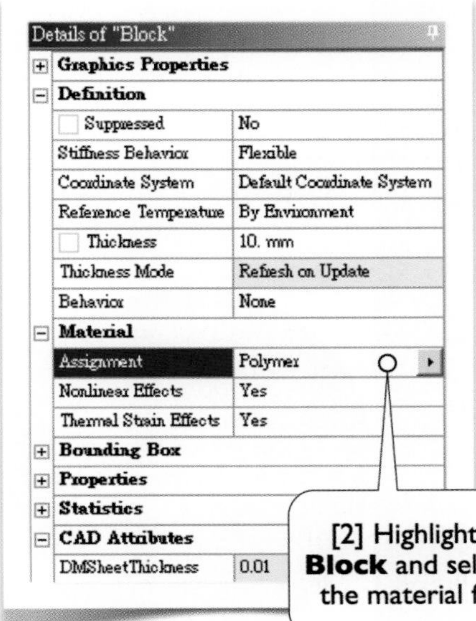

Details of "Block"
Graphics Properties
Definition
Suppressed No
Stiffness Behavior Flexible
Coordinate System Default Coordinate System
Reference Temperature By Environment
Thickness 10. mm
Thickness Mode Refresh on Update
Behavior None
Material
Assignment Polymer
Nonlinear Effects Yes
Thermal Strain Effects Yes
Bounding Box
Properties
Statistics
CAD Attributes
DMSheetThickness 0.01

[2] Highlight **Geometry/Block** and select **Polymer** as the material for the body. →

Details of "Disk"
Graphics Properties
Definition
Suppressed No
Stiffness Behavior Flexible
Coordinate System Default Coordinate System
Reference Temperature By Environment
Thickness 10. mm
Thickness Mode Refresh on Update
Behavior None
Material
Assignment Polymer
Nonlinear Effects Yes
Thermal Strain Effects Yes
Bounding Box
Properties
Statistics
CAD Attributes
DMSheetThickness

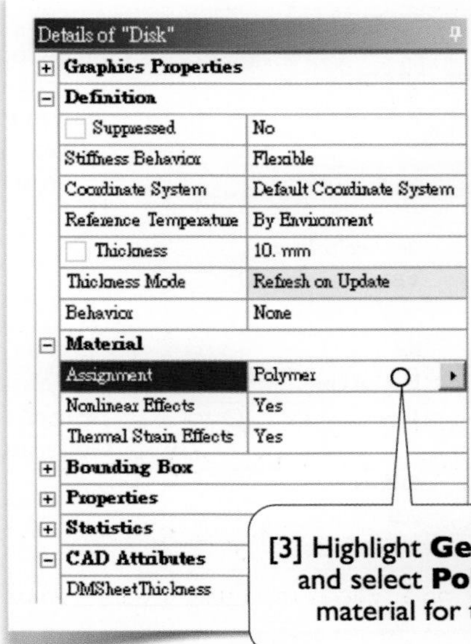

[3] Highlight **Geometry/Disk** and select **Polymer** as the material for the body. ↵

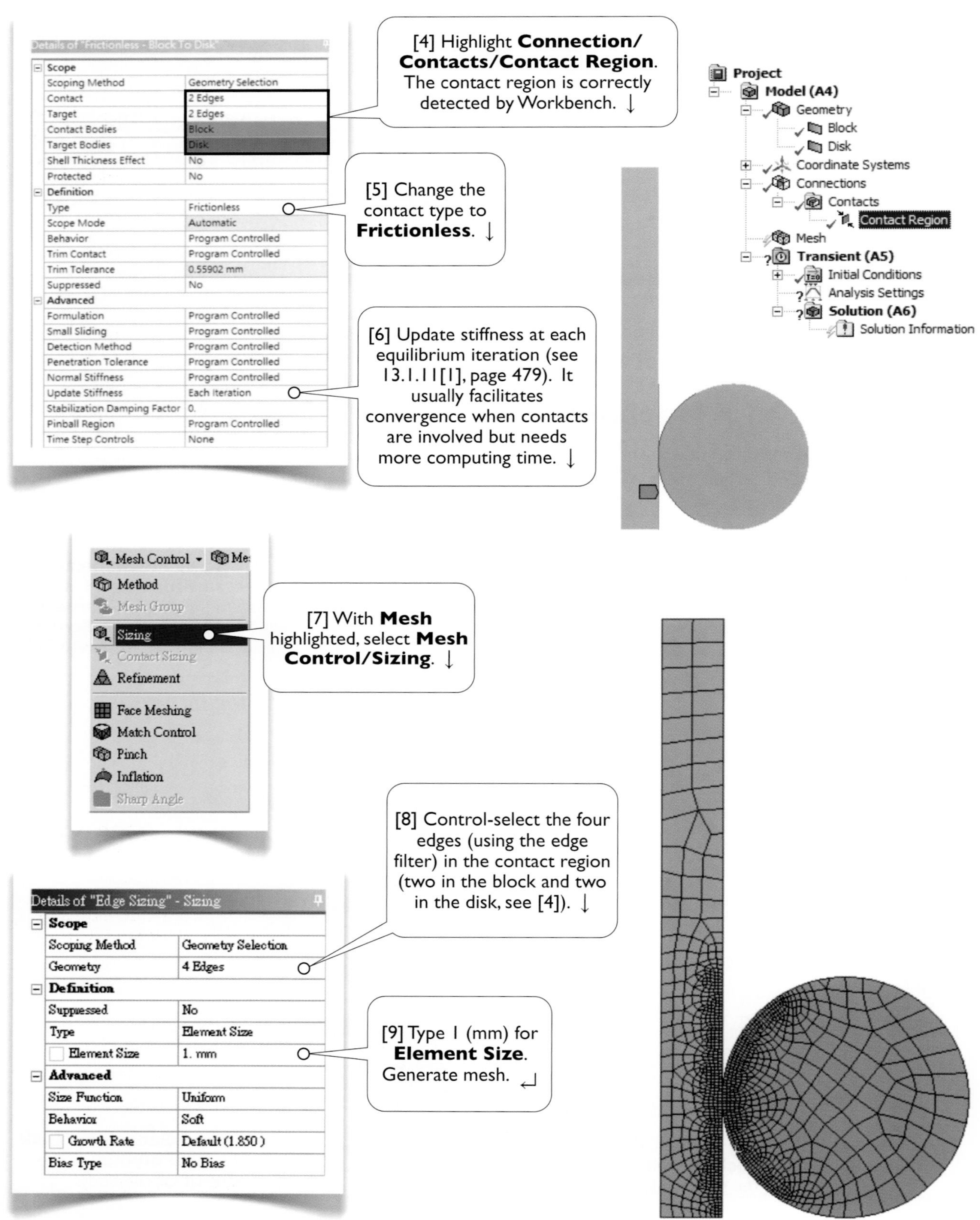
[4] Highlight **Connection/ Contacts/Contact Region**. The contact region is correctly detected by Workbench.
[5] Change the contact type to **Frictionless**.
[6] Update stiffness at each equilibrium iteration (see 13.1.11[1], page 479). It usually facilitates convergence when contacts are involved but needs more computing time.
[7] With **Mesh** highlighted, select **Mesh Control/Sizing**.
[8] Control-select the four edges (using the edge filter) in the contact region (two in the block and two in the disk, see [4]).
[9] Type 1 (mm) for **Element Size**. Generate mesh.
Scope
Scoping Method
Geometry Selection
Contact
2 Edges
Target
2 Edges
Contact Bodies
Block
Target Bodies
Disk
Shell Thickness Effect
No
Protected
No
Definition
Type
Frictionless
Scope Mode
Automatic
Behavior
Program Controlled
Trim Contact
Program Controlled
Trim Tolerance
0.55902 mm
Suppressed
No
Advanced
Formulation
Program Controlled
Small Sliding
Program Controlled
Detection Method
Program Controlled
Penetration Tolerance
Program Controlled
Normal Stiffness
Program Controlled
Update Stiffness
Each Iteration
Stabilization Damping Factor
0.
Pinball Region
Program Controlled
Time Step Controls
None
Project
Model (A4)
Geometry
Block
Disk
Coordinate Systems
Connections
Contacts
Contact Region
Mesh
Transient (A5)
Initial Conditions
Analysis Settings
Solution (A6)
Solution Information
Mesh Control
Method
Mesh Group
Sizing
Contact Sizing
Refinement
Face Meshing
Match Control
Pinch
Inflation
Sharp Angle
Details of "Edge Sizing" - Sizing
Scope
Scoping Method
Geometry Selection
Geometry
4 Edges
Definition
Suppressed
No
Type
Element Size
Element Size
1. mm
Advanced
Size Function
Uniform
Behavior
Soft
Growth Rate
Default (1.850)
Bias Type
No Bias

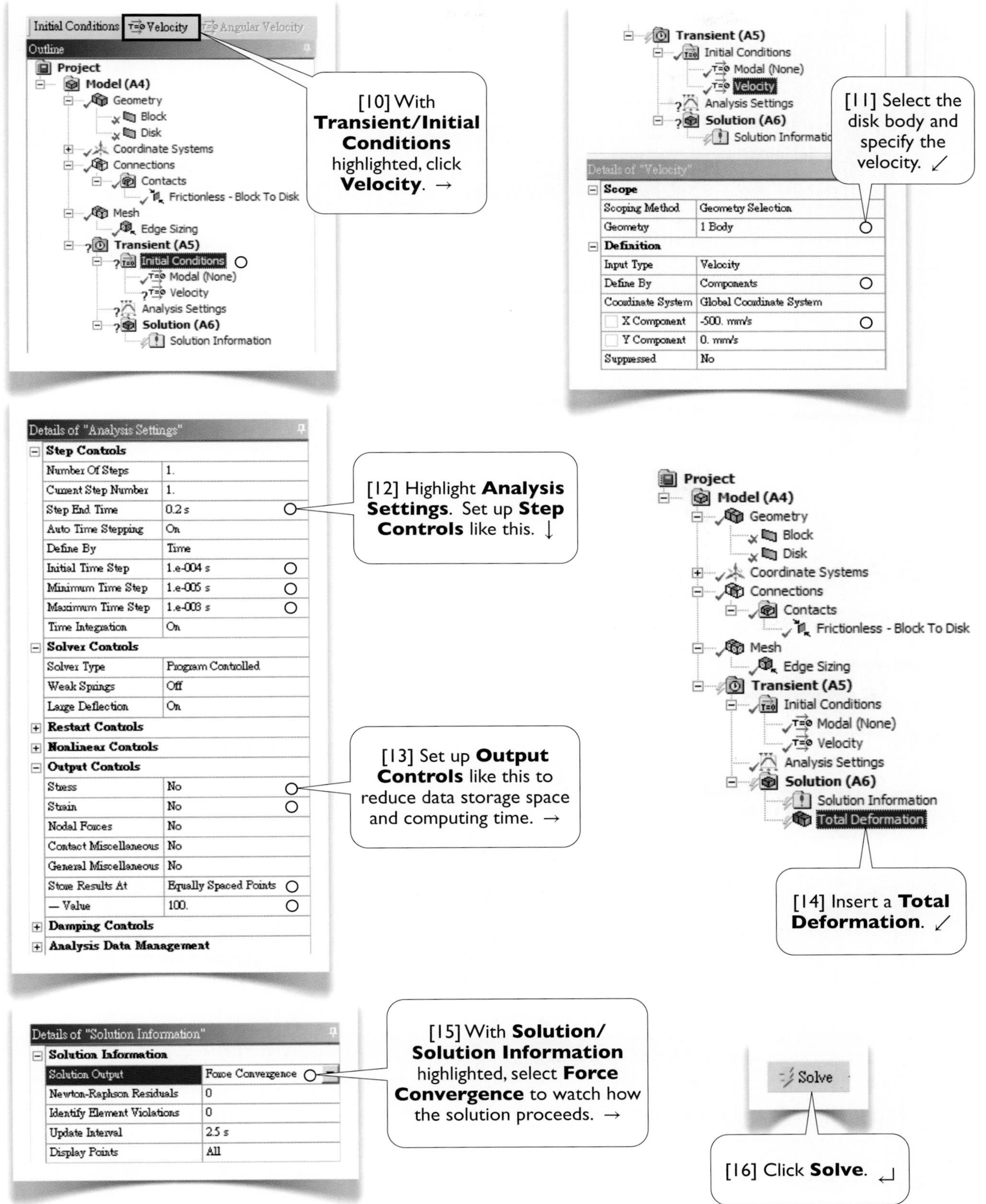
[10] With **Transient/Initial Conditions** highlighted, click **Velocity**. →
[11] Select the disk body and specify the velocity.
[12] Highlight **Analysis Settings**. Set up **Step Controls** like this. ↓
[13] Set up **Output Controls** like this to reduce data storage space and computing time. →
[14] Insert a **Total Deformation**.
[15] With **Solution/Solution Information** highlighted, select **Force Convergence** to watch how the solution proceeds. →
[16] Click **Solve**.
Initial Conditions
Velocity
Angular Velocity
Outline
Project
Model (A4)
Geometry
Block
Disk
Coordinate Systems
Connections
Contacts
Frictionless - Block To Disk
Mesh
Edge Sizing
Transient (A5)
Initial Conditions
Modal (None)
Velocity
Analysis Settings
Solution (A6)
Solution Information
Total Deformation
Details of "Velocity"
Scope
Scoping Method
Geometry Selection
Geometry
1 Body
Definition
Input Type
Velocity
Define By
Components
Coordinate System
Global Coordinate System
X Component
-500. mm/s
Y Component
0. mm/s
Suppressed
No
Details of "Analysis Settings"
Step Controls
Number Of Steps
1.
Current Step Number
1.
Step End Time
0.2 s
Auto Time Stepping
On
Define By
Time
Initial Time Step
1.e-004 s
Minimum Time Step
1.e-005 s
Maximum Time Step
1.e-003 s
Time Integration
On
Solver Controls
Solver Type
Program Controlled
Weak Springs
Off
Large Deflection
On
Restart Controls
Nonlinear Controls
Output Controls
Stress
No
Strain
No
Nodal Forces
No
Contact Miscellaneous
No
General Miscellaneous
No
Store Results At
Equally Spaced Points
Value
100.
Damping Controls
Analysis Data Management
Details of "Solution Information"
Solution Information
Solution Output
Force Convergence
Newton-Raphson Residuals
0
Identify Element Violations
0
Update Interval
2.5 s
Display Points
All
Solve

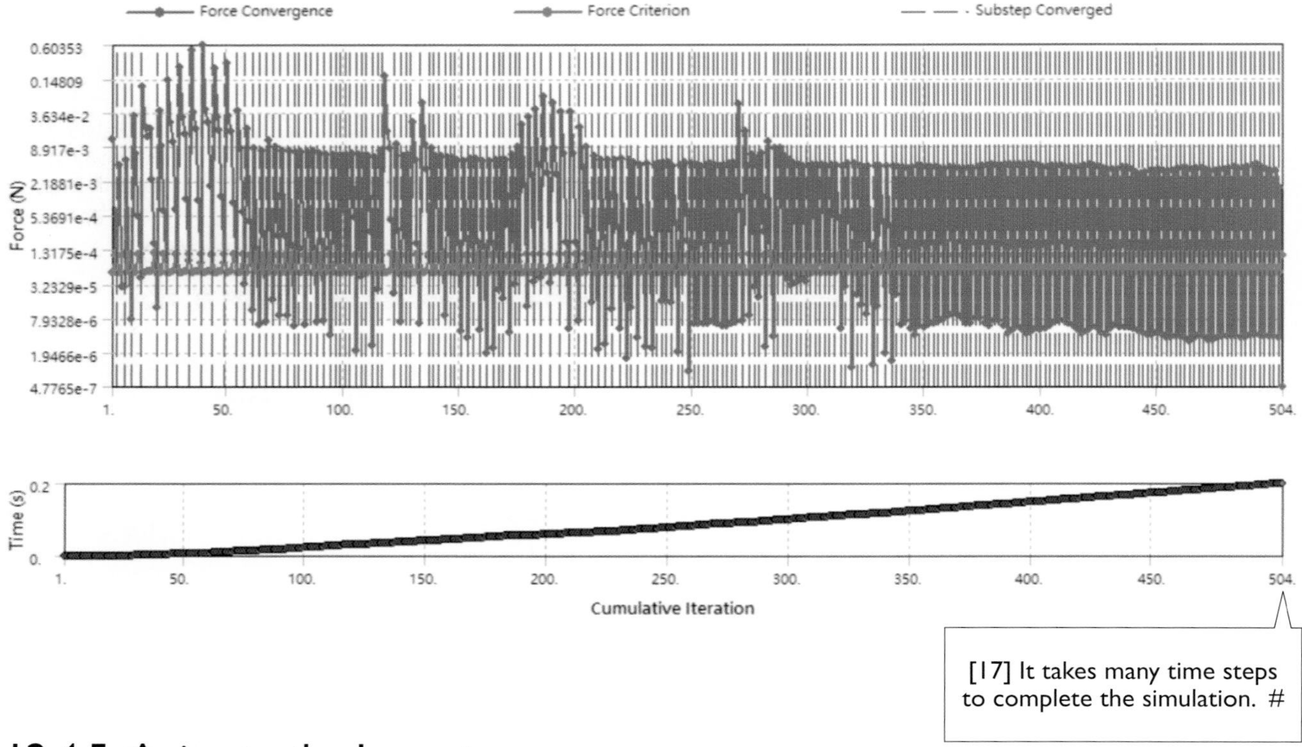

12.4.5 Animate the Impact

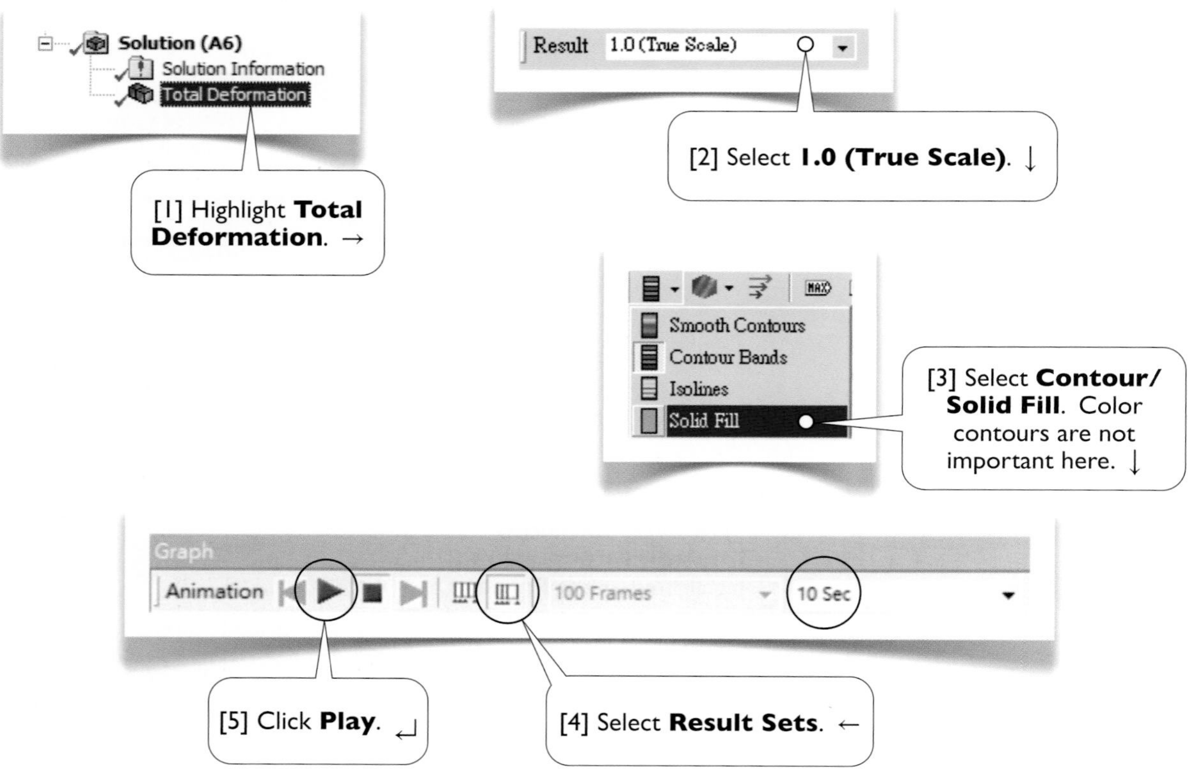

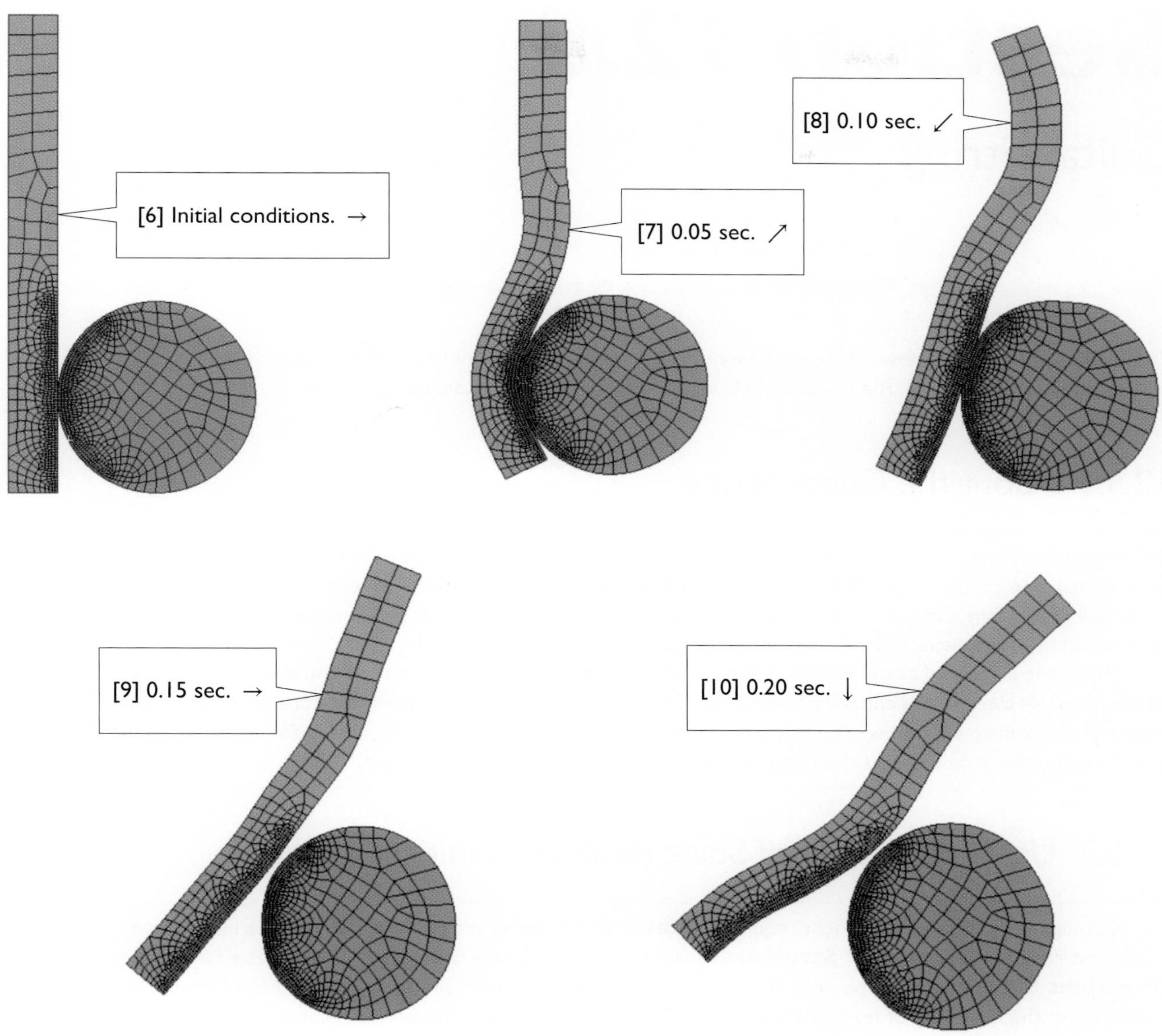

How to Obtain a Snapshot?

[11] Highlight **Total Deformation**, type the time for **Display Time**, and click **Solve**. ↓

Wrap Up

[12] Save the project and exit Workbench. #

Section 12.5

Guitar String

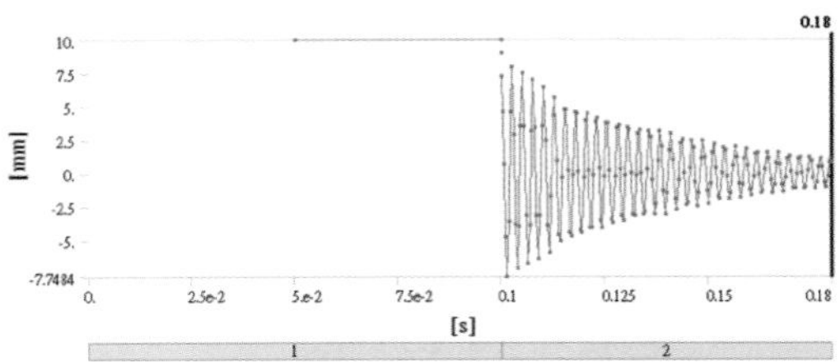

This exercise has two purposes: First, to demonstrate how to set up an initial condition using the result of a static simulation. Second, to show the transient behavior of a guitar string, as a sequel of Section 11.4.

12.5.1 About the Guitar String

[1] As mentioned, in 11.4.3[1] (page 413), if you pluck a string, you will produce a tone made up of all harmonic modes. It is the harmonic mixes that differentiate the sound of one music instrument from others. A good guitar player knows very well that he/she can control the quality of sound by plucking the string at different locations. We will explore this phenomenon in this section, using the guitar string introduced in Section 11.4.

The string is plucked at the middle, quarter, and eighth points respectively, and its transient responses are observed. More precisely, the string is applied a vertical displacement of 10 mm at respective locations, and then instantaneously released to produce vibrations. The vibrations will eventually reach a steady state, that is, free vibrations, after a certain duration of time. It, however, is the transient vibrations that impress our ears most. #

12.5.2 Using the Result of Static Analysis as Initial Condition

[1] Applying the vertical displacement requires a **Static Structural** analysis. Releasing the string to produce vibrations requires a **Transient Structural** analysis using the results of the static analysis as an initial condition. In a **Transient Structural**, it is possible to specify the first step as a static simulation. The results become the initial condition of the next step of transient simulation. This two-step method has become a standard procedure, when the simulation requires a static simulation as an initial condition.

Before we demonstrate the two-step method, let's clear up some important concepts. In transient dynamic simulation, each time step needs an initial condition to carry it on. The results of the last time step become the initial condition of the next time step. The initial condition, more specifically, is the position and velocity of each node.

In a **Transient Structural** simulation, Workbench allows you to turn off/on **Time Integration** ([2], next page). Turning **Time Integration** off means turning the dynamic effects (1.1.10[1], page 21) off, and Eq. 12.1.4(1) (page 426) reduces to Eq. 1.3.1(1) (page 35), i.e., a static simulation. You can turn on/off **Time Integration** in any load step.

Make sure you don't confuse load steps (or simply called steps), substeps, and equilibrium iterations. Multiple load steps can be created when you need to specify different **Analysis Settings** for each step. Each step is further divided into substeps (also called time steps) for two reasons: First, in a transient dynamic simulation, each substep is a time integration step. Second, in a nonlinear simulation, a substep must be small enough to achieve convergence within that substep. For nonlinear simulations (either dynamic or static), each substep may need multiple equilibrium iterations to achieve convergence. ↵

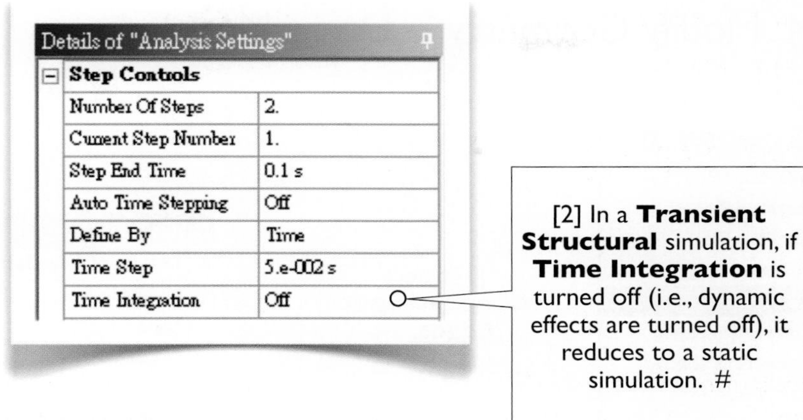

Details of "Analysis Settings"	
Step Controls	
Number Of Steps	2.
Current Step Number	1.
Step End Time	0.1 s
Auto Time Stepping	Off
Define By	Time
Time Step	5.e-002 s
Time Integration	Off

[2] In a **Transient Structural** simulation, if **Time Integration** is turned off (i.e., dynamic effects are turned off), it reduces to a static simulation. #

12.5.3 Resume the Project **String**

[1] Launch Workbench. Open the project **String**, which was saved in Section 11.4. →

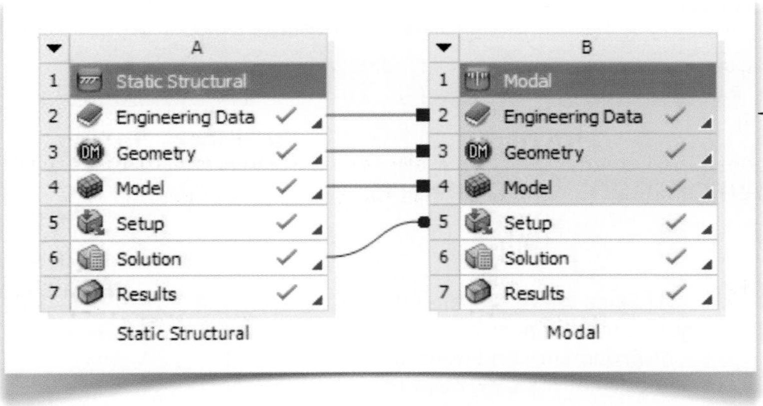

	A		B
1	Static Structural	1	Modal
2	Engineering Data ✓	2	Engineering Data ✓
3	Geometry ✓	3	Geometry ✓
4	Model ✓	4	Model ✓
5	Setup ✓	5	Setup ✓
6	Solution ✓	6	Solution ✓
7	Results ✓	7	Results ✓
	Static Structural		Modal

[2] This is the project schematic at the end of Section 11.4. ↓

[3] From **Toolbox**, drag-and-drop **Transient Structural** here. You might instead drop to **Static Structural**; the schematic would be different, but the simulation results would be the same. ←

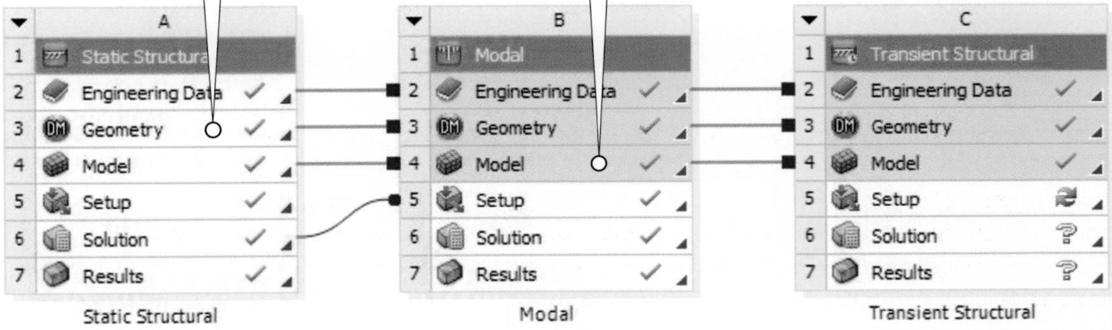

[4] Double-click **Geometry** to start up DesignModeler. #

	A		B		C
1	Static Structural	1	Modal	1	Transient Structural
2	Engineering Data ✓	2	Engineering Data ✓	2	Engineering Data ✓
3	Geometry ✓	3	Geometry ✓	3	Geometry ✓
4	Model ✓	4	Model ✓	4	Model ✓
5	Setup ✓	5	Setup ✓	5	Setup
6	Solution ✓	6	Solution ✓	6	Solution
7	Results ✓	7	Results ✓	7	Results
	Static Structural		Modal		Transient Structural

12.5.4 Modify Geometry in DesignModeler

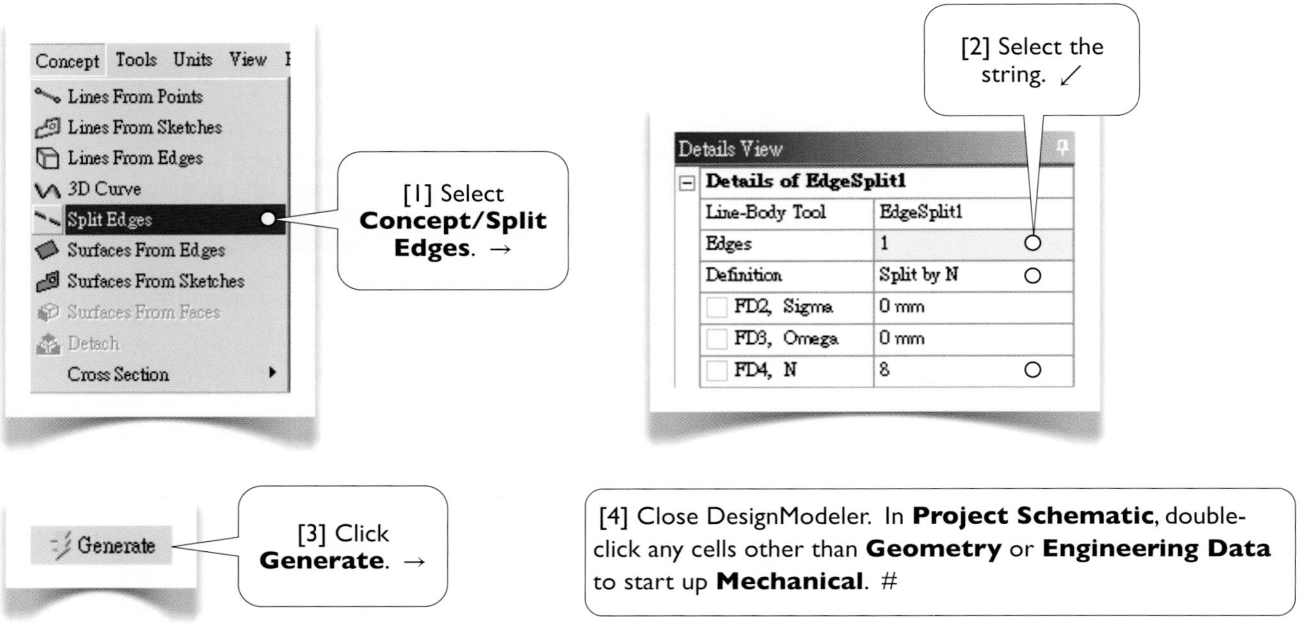

12.5.5 Set Up Environment Conditions

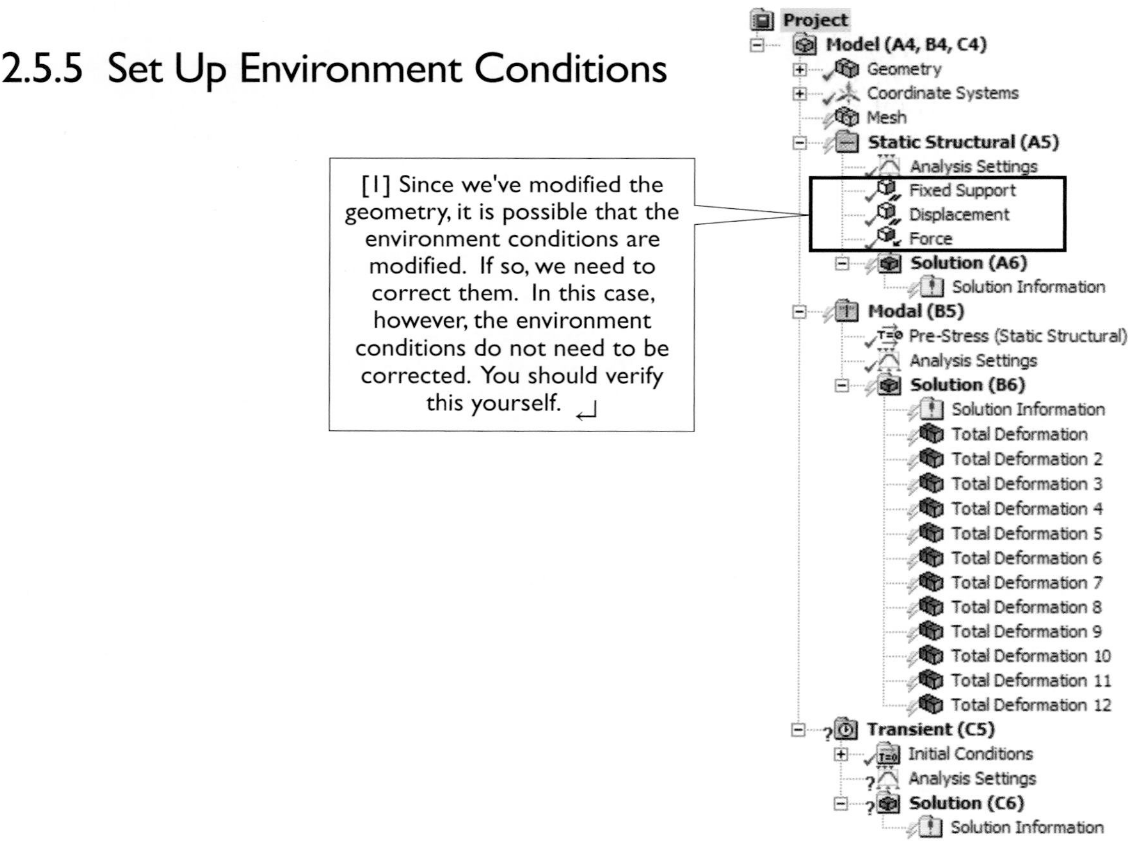

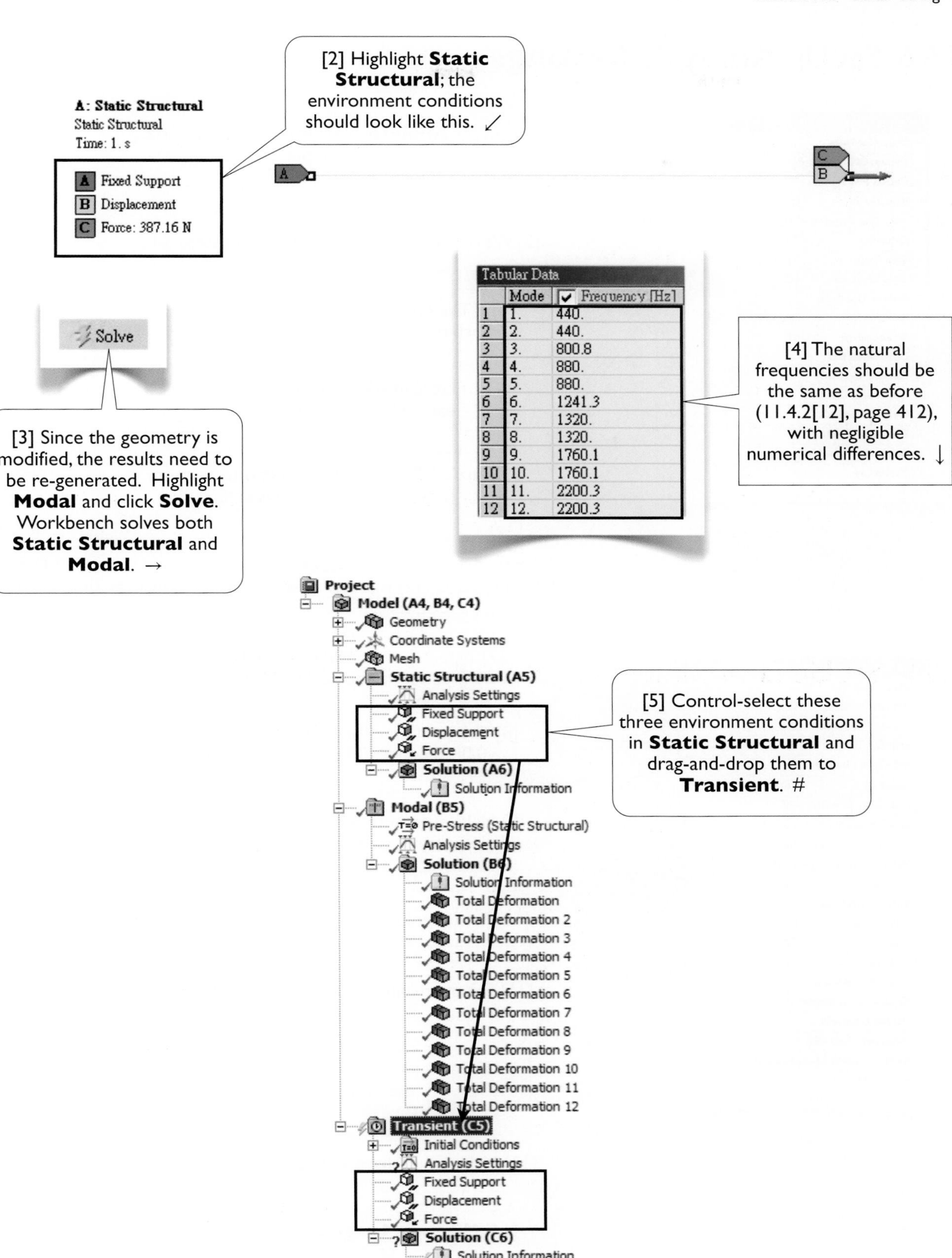

	Mode	Frequency [Hz]
1	1.	440.
2	2.	440.
3	3.	800.8
4	4.	880.
5	5.	880.
6	6.	1241.3
7	7.	1320.
8	8.	1320.
9	9.	1760.1
10	10.	1760.1
11	11.	2200.3
12	12.	2200.3

12.5.6 Set Up **Analysis Settings**

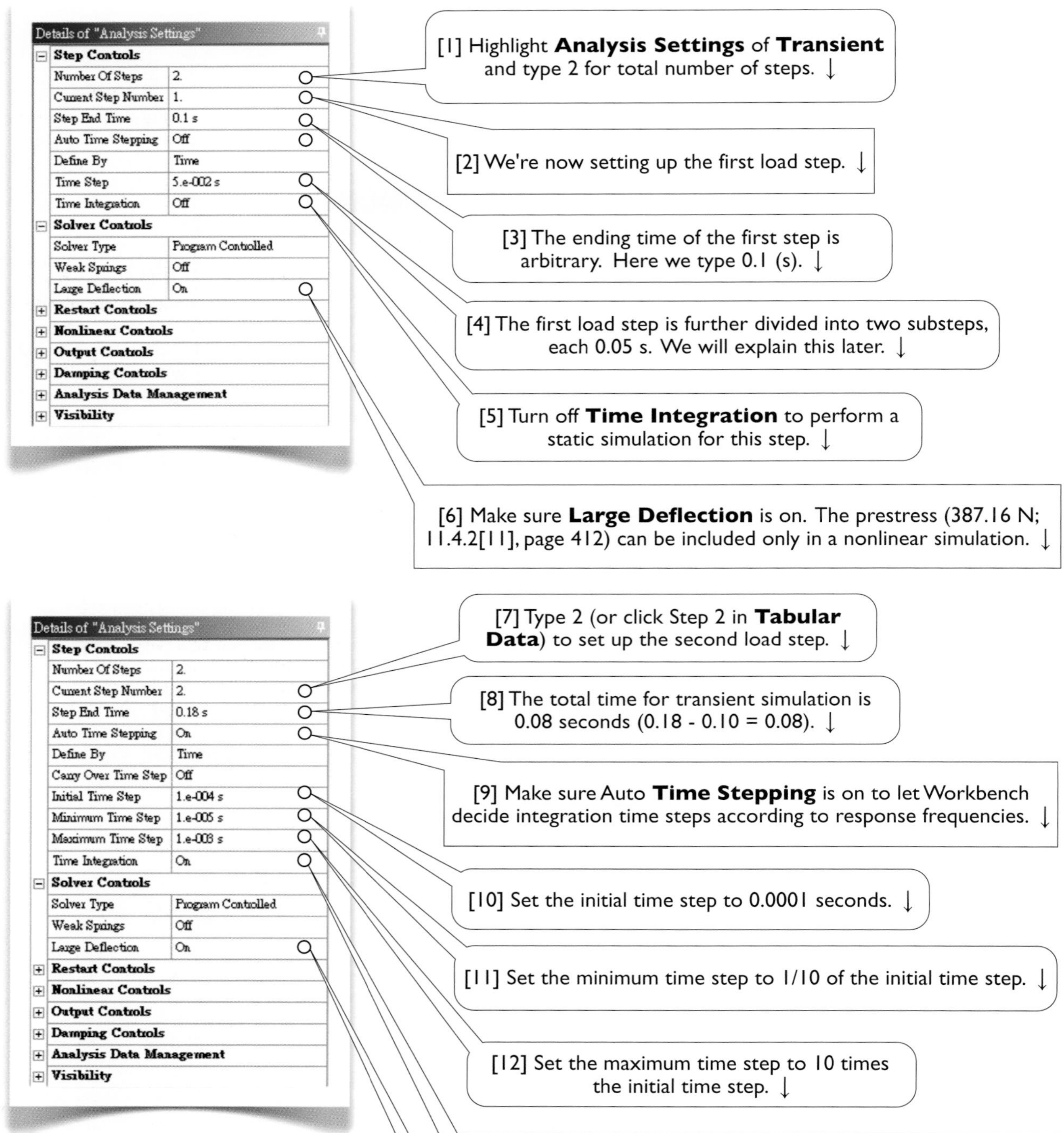

12.5.7 Set Up Initial Condition

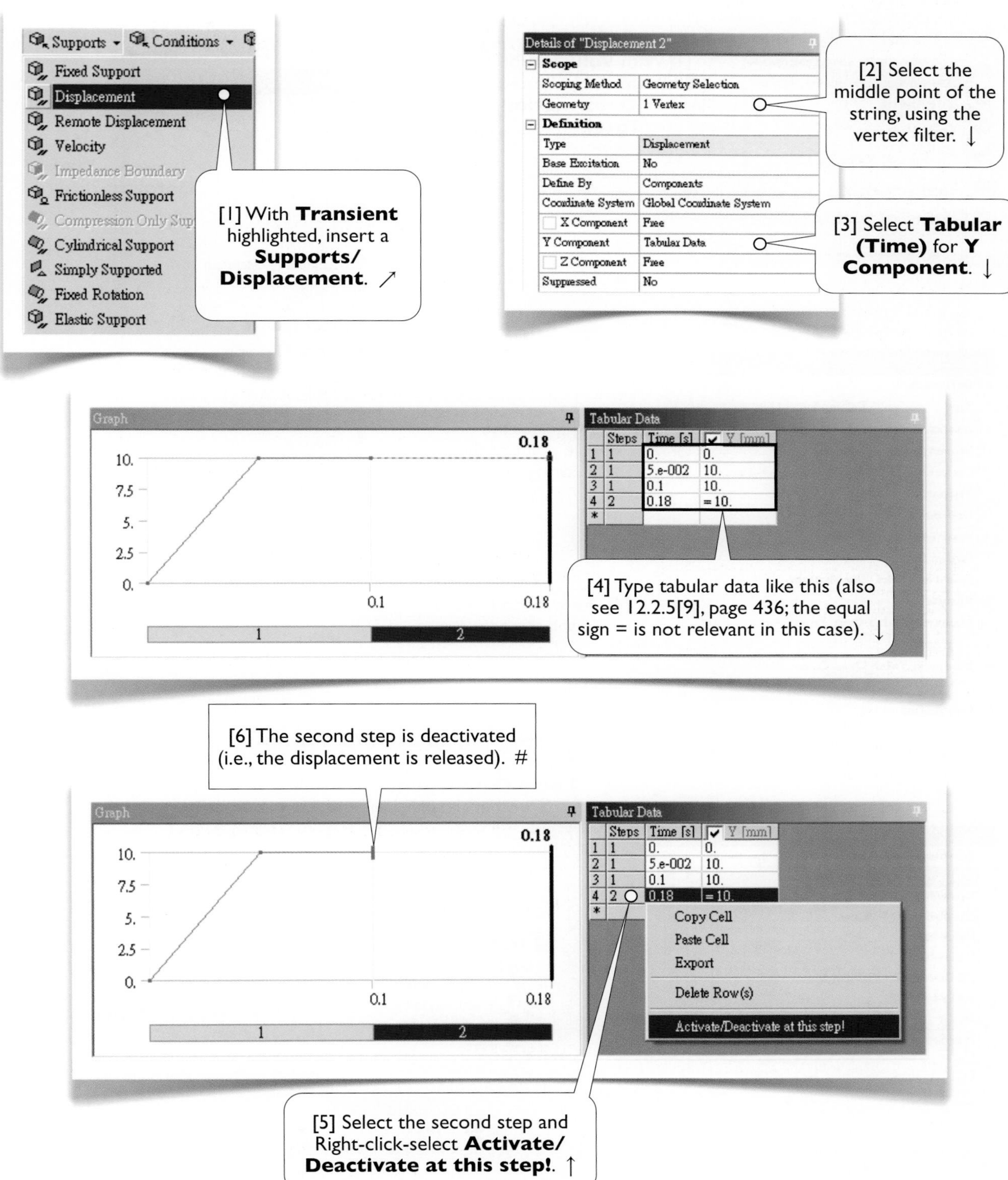

12.5.8 Insert Result Objects and Solve the Model

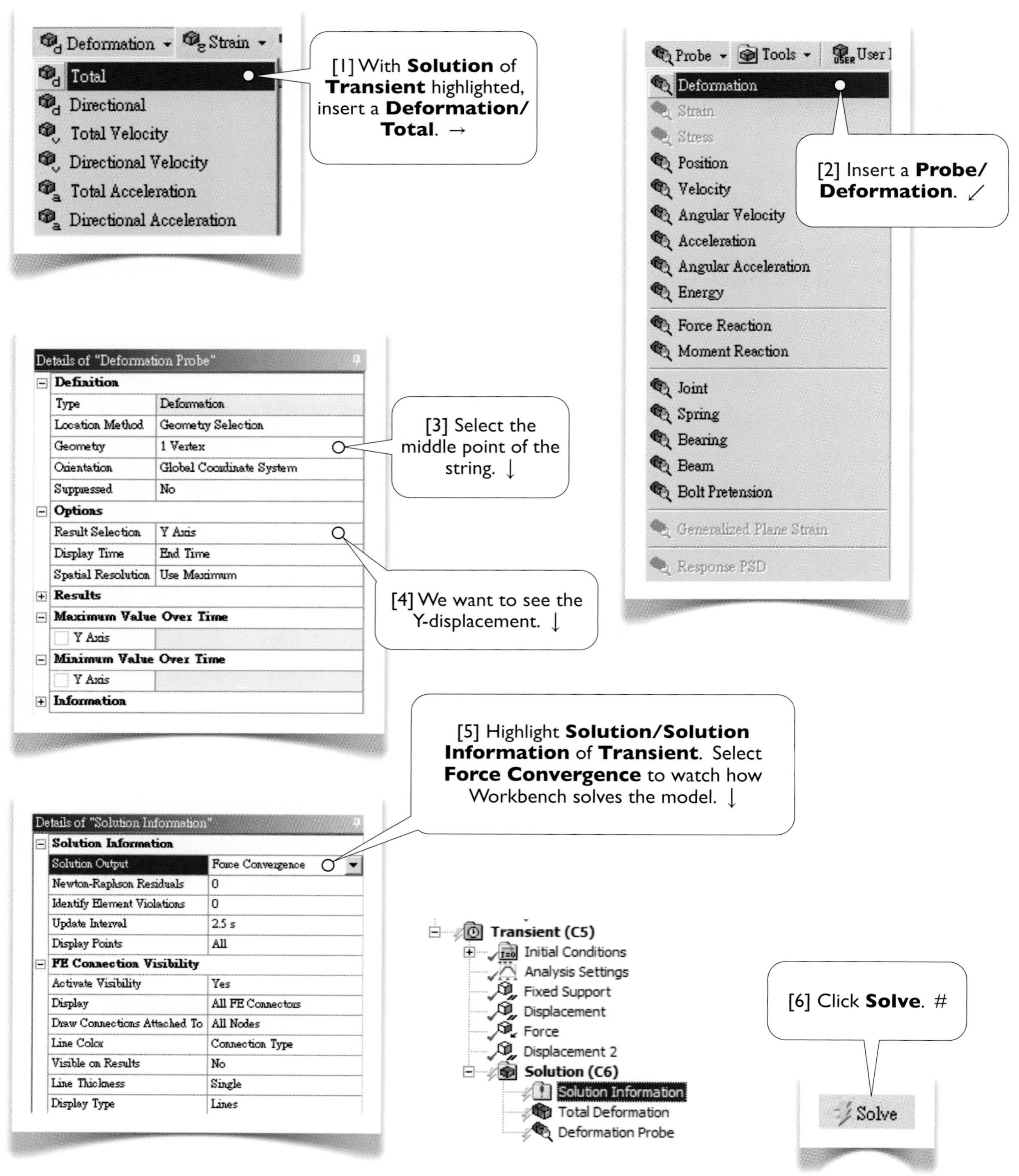

12.5.9 View the Results

[1] Highlight **Deformation Probe**; it shows a deflection-versus-time plot for the middle point. ↓

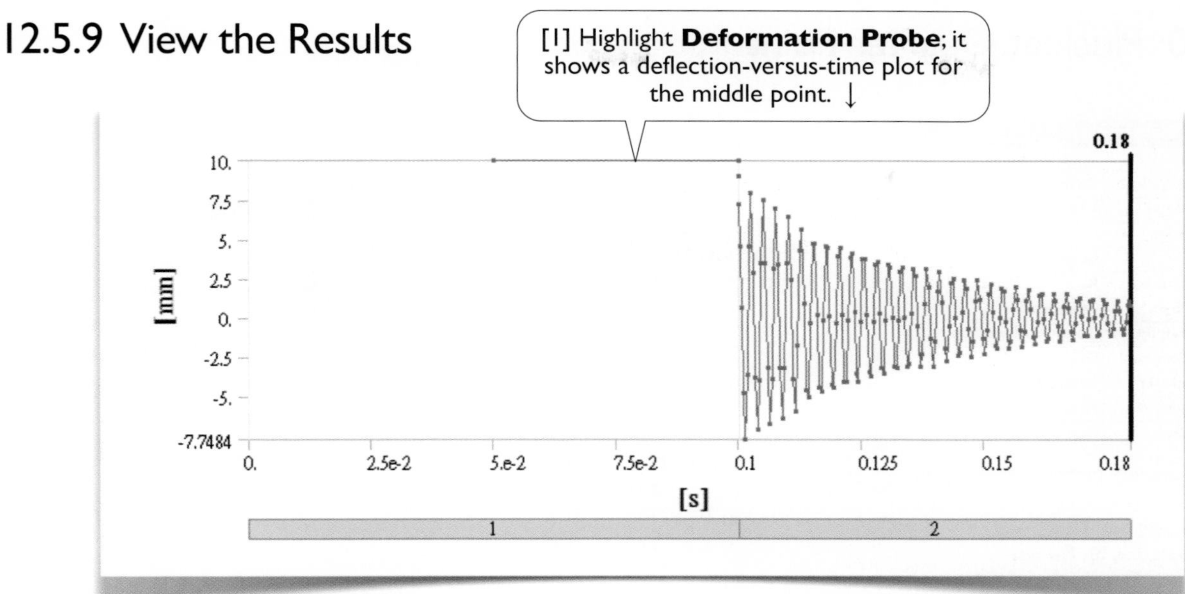

[2] We are now preparing an animation. Highlight **Solution/Total Deformation** of **Transient**. Make sure **View/Thick Shells and Beams** is off. →

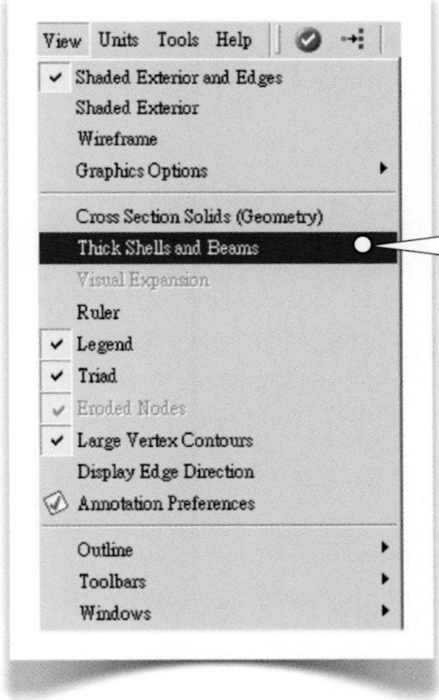

[3] Select **Solid Fill**. Color contours are not important here. ↙

[6] Click **Play**. ←

[4] Select **Result Sets**. →

[5] Select slowest animation speed. ←

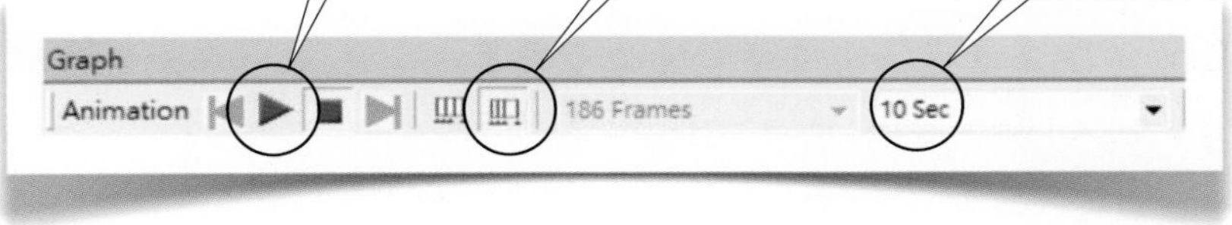

[7] Initial condition. #

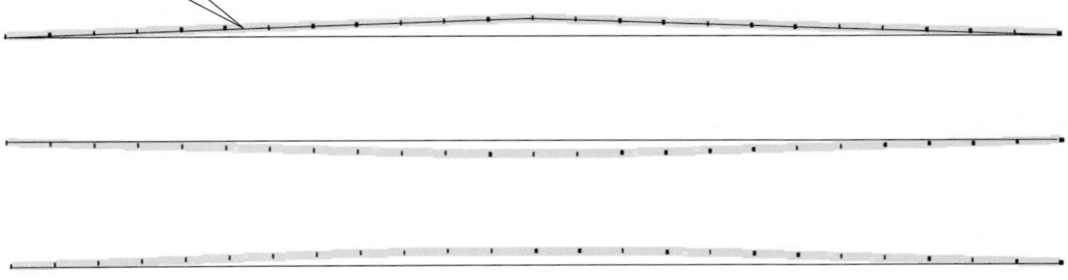

12.5.10 Pluck at Quarter Point

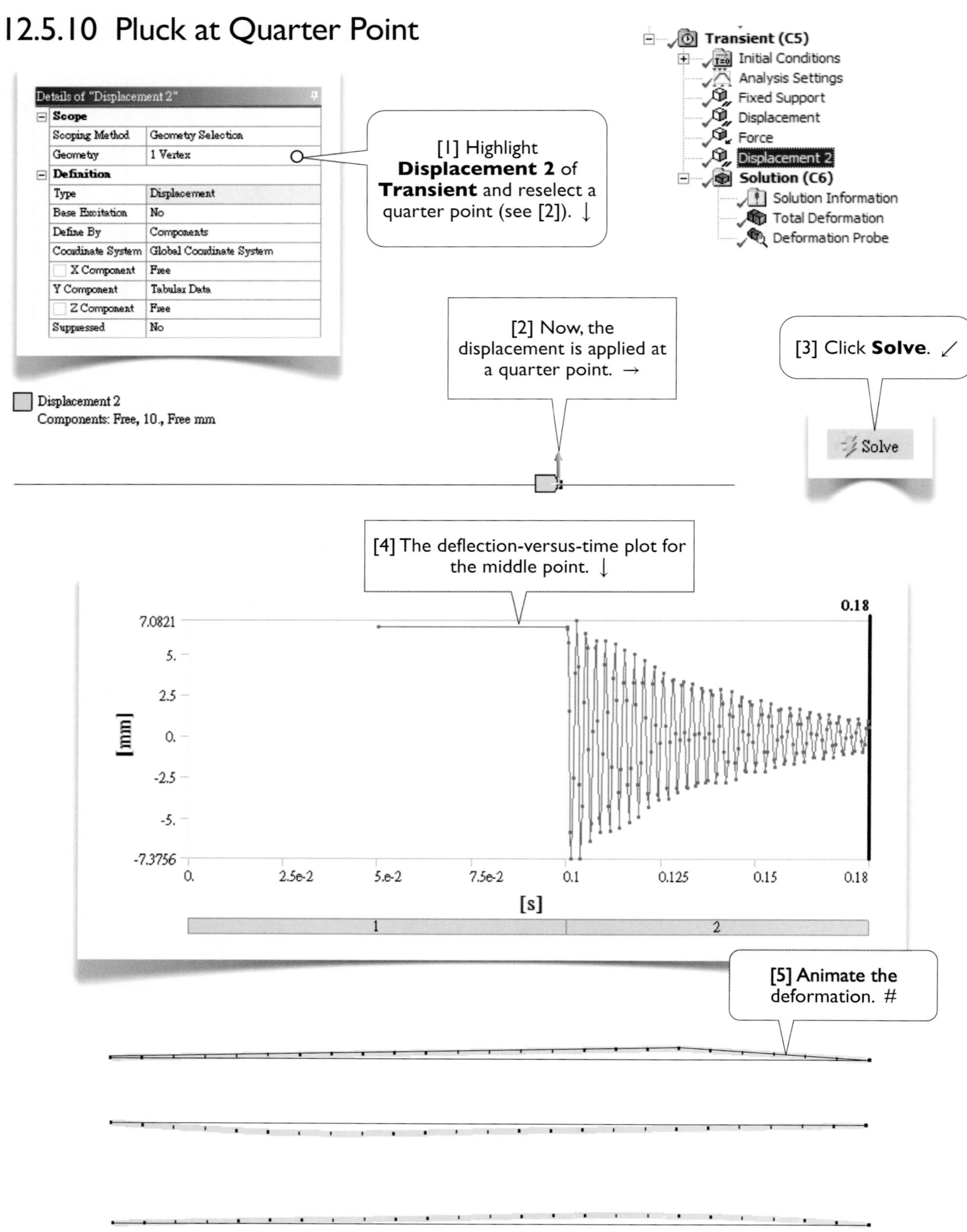

12.5.11 Pluck at Eighth Point

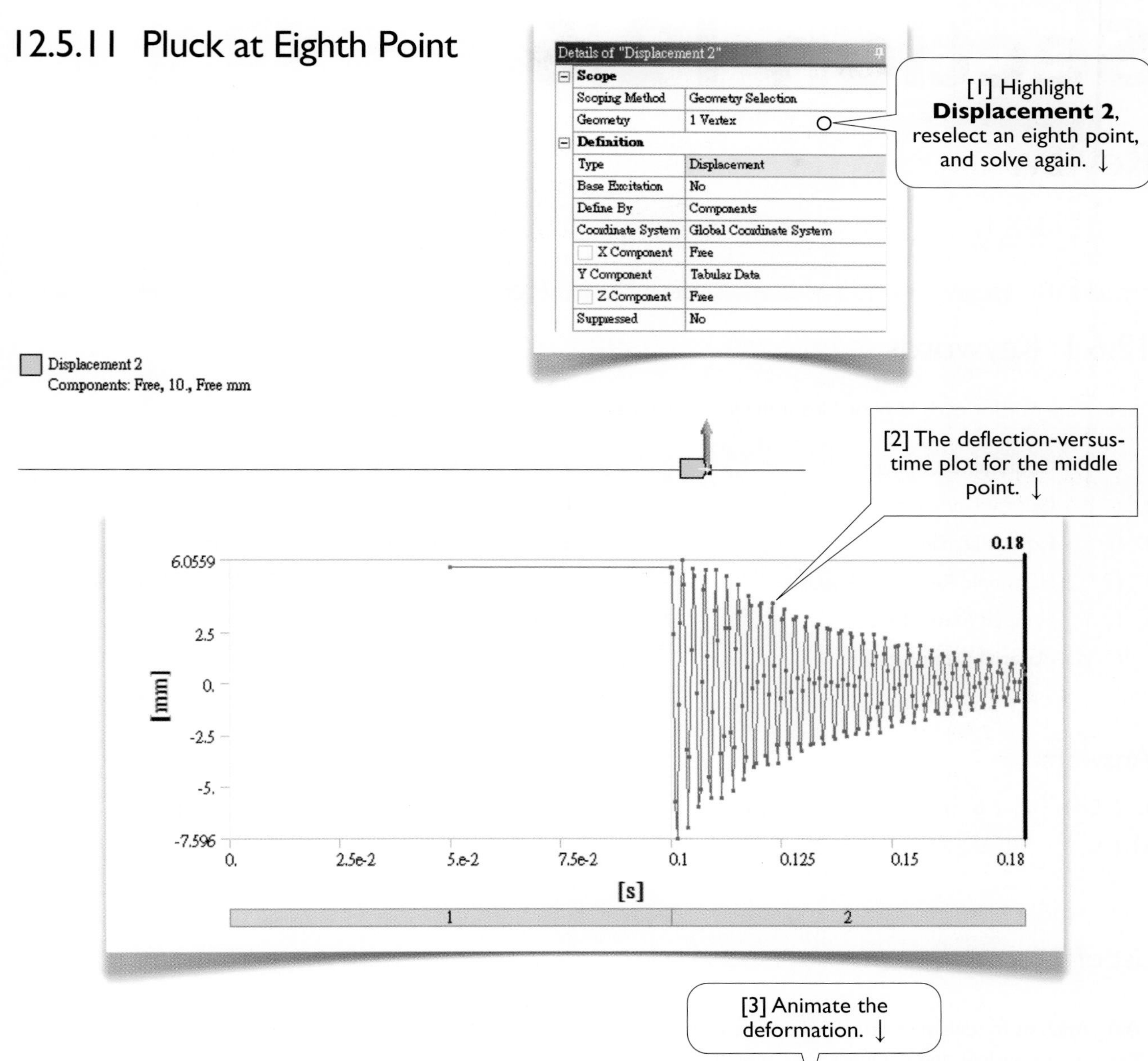

[3] Animate the deformation. ↓

Wrap Up

[4] Save the project and exit Workbench. #

Section 12.6

Review

12.6.1 Keywords

Choose a letter for each keyword from the list of descriptions

1. ( ) Coulomb Damping
2. ( ) Critical Damping
3. ( ) Explicit Dynamics
4. ( ) Harmonic Response Analysis
5. ( ) Lumped Mass Model
6. ( ) Material Damping
7. ( ) No Separation Contact
8. ( ) Random Vibration Analysis
9. ( ) Response Spectrum Analysis
10. ( ) Steps, Substeps, and Equilibrium Iterations
11. ( ) Viscous Damping

Answers:

1. (E) 2. (B) 3. (I) 4. (F) 5. (A) 6. (D) 7. (K) 8. (H) 9. (G) 10. (J) 11. (C)

List of Descriptions

(A) Analysis models that assume the mass concentrated at certain locations. The discrete masses are typically connected by springs and dampers. The models are often used to study dynamic systems.

(B) When the damping of a dynamic system is smaller than a critical value, its free vibration is oscillatory. On the other hand, if the damping is larger than the critical value, the motion is not oscillatory.

(C) The damping due to the friction between the structure and its surrounding fluid. The damping force typically assumes to be proportional to the velocity of the structural displacement.

(D) Also called solid damping or elastic hysteresis. The damping due to the internal friction in the material. The behavior is still an open research topic. In the Workbench, assuming small material damping, we may express it by an equivalent viscous damping. Further, the viscous damping coefficient (of this equivalent viscous damping) is assumed to be a linear combination of the stiffness matrix and the mass matrix. The form of linear combination (alpha and beta) is then determined by lab experiments and data fittings.

(E) Also called Coulomb friction or dry friction. The damping is due to the friction in the connection between structural members. In Workbench, it can be modeled using frictional contact.

(F) A dynamic analysis to investigate the maximum response of a structure under steady harmonic (sinusoidal) loads.

(G) A dynamic analysis to evaluate the maximum response of a structure under loading conditions described by a spectrum representing the maximum response at varying frequencies in known directions to a specific time history. The technique is often used to design structures withstanding multiple short-duration loadings, such as earthquakes.

(H) A dynamic analysis to evaluate the probabilistic response of a structure under probabilistic loads described by a spectrum representing probability distribution of excitation at varying frequencies in known directions. The technique is used to design structures withstanding random loadings.

(I) A technique for transient dynamic analysis. Explicit integration method is used, which requires a very small integration time step to achieve solution accuracy. For a single time step, it is much more efficient than an implicit method, which is used in a **Transient Structural** analysis system. These features make it very efficient for a high-speed impact simulation, or highly nonlinear simulations.

(J) We may divide the whole loading history into steps, to specify different **Analysis Settings** for each step. Each step can be further divided into substeps. For transient dynamic simulation, each substep is an integration time step. For nonlinear simulation, dividing into substeps is to expedite convergence. In nonlinear simulations, each substep may need several equilibrium iterations to find its solution.

(K) Two surfaces with **No Separation** contact condition prohibit separation in their normal direction, but allow small sliding relative to each other. Both **Bonded** and **No Separation** contact types do not introduce contact nonlinearity.

12.6.2. Additional Workbench Exercises

High-Speed Impact Simulation

The impact simulation in the Section 12.4 is actually not useful, since the material (of Young's Modulus 10 kPa) is not realistic and the impact speed (0.5 m/s) is slow. Try more realistic simulations by yourself using the same geometric model. Gradually increase the Young's modulus and the impact speed. Each time you may need to decrease the integration time step. This exercise is to experience the limitation of the implicit method, which is used in a **Transient Structural** analysis system. High-speed impact simulations are more suitable by using **Explicit Dynamics** analysis system (Chapter 15).

Chapter 13

Nonlinear Simulations

When the relationship between the response and the load of a structure is linear, the structure is a *linear structure*, and the simulation is a *linear simulation*. Otherwise, the structure is a *nonlinear structure* and the simulation is a *nonlinear simulation*. In the real world, all structures are more or less nonlinear. In many cases, however, when nonlinearities are negligible, we may predict their behavior using linear simulations. For other cases, when nonlinearities are not negligible, nonlinear simulations are needed.

Structural nonlinearities come from three sources: large deformation, change of connectivity, and nonlinear stress-strain relations. Nonlinearity due to large deformation is called *geometry nonlinearity*. Nonlinearity due to the change of connectivity is called *topology nonlinearity*, which includes failure of structural components and change of contact status. In this chapter, we will discuss the change of contact status only, which can be termed *contact nonlinearity*. Nonlinearity due to nonlinear stress-strain relations is called *material nonlinearity* and will be covered in Chapter 14.

In general, nonlinear simulations are much more challenging than linear simulations. They not only take much computing time, but sometimes fail to find a solution. Solution behaviors of nonlinear simulations highly depend on settings of solution parameters. A thorough comprehension of these solution parameters becomes critical when you want to adjust these parameters to reduce the computing time, or try to successfully find a solution.

Purpose of This Chapter

This chapter discusses nonlinear solution algorithms and focuses on geometry nonlinearity and contact nonlinearity. Material nonlinearity will be covered in Chapter 14. This chapter provides some basics of nonlinear simulations process, so that you can understand and use various solution parameters.

About Each Section

Section 13.1 provides some basics of nonlinear simulations, including Newton-Raphson method, the solution method used in Workbench. Concepts of convergence follow the introduction. Section 13.2 provides a step-by-step example of geometric nonlinearity. Section 13.3 provides a step-by-step example of contact nonlinearity. Section 13.4 provides an additional exercise of contact nonlinearity.

Section 13.1

Basics of Nonlinear Simulations

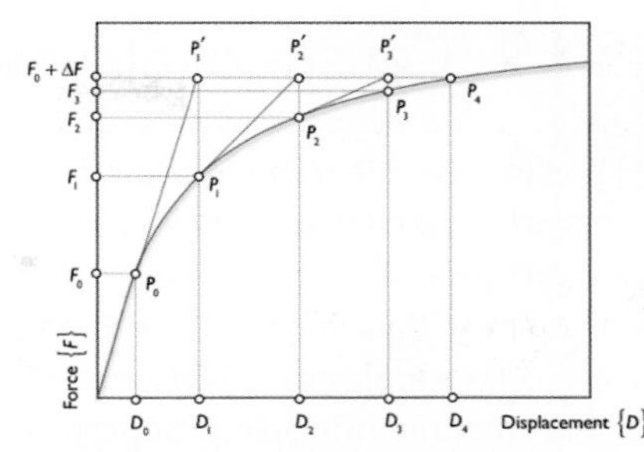

PART A. NONLINEAR SOLUTION METHODS

13.1.1 What Are Nonlinear Simulations

[1] In a linear simulation, the response is linearly proportional to the load. More specifically, let's use Eq. 1.3.1(1) for the upcoming discussion

$$[K]\{D\}=\{F\} \qquad \text{Copy of Eq. 1.3.1(1), page 35}$$

The nodal force $\{F\}$ may represent the load; the nodal displacement $\{D\}$ may represent the response (stress and strain can be computed from the displacement); the stiffness matrix $[K]$ contains the proportionality coefficients between the force and the displacement. Eq. 1.3.1(1) can be conceptually plotted as shown in [2]. Note that both the horizontal axis and the vertical axis are actually multi-dimensional (i.e., vectors), and $[K]$ is actually the gradient of $\{F\}$ with respect to $\{D\}$. $[K]$ is a matrix of dimension n by n, where n is the degrees of freedom of the system. In cases of single degree of freedom, $[K]$ is a scalar and is the "spring constant" of the structure.

Performing a linear static simulation means to solve Eq. 1.3.1(1), once and for all. For linear transient dynamic simulations, Eq. 12.1.4(1) (page 426) is solved instead; it involves integrations over time domain. In each integration time step, the equation is solved exactly once. ↓

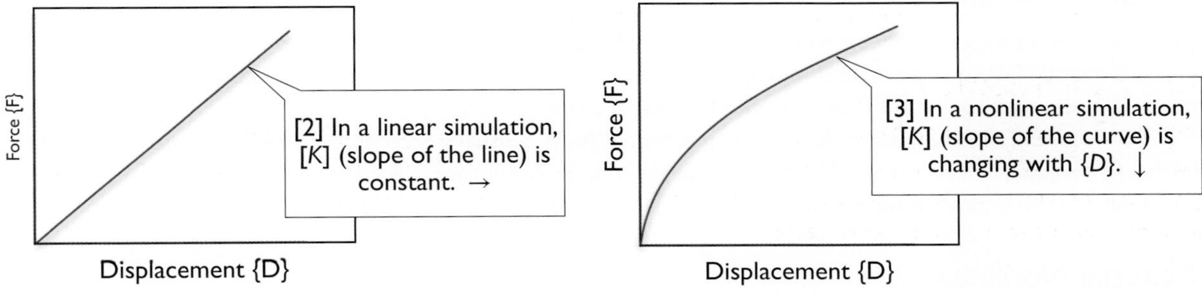

[4] In a nonlinear simulation, the relation between $\{F\}$ and $\{D\}$ is nonlinear, as shown in [3]. Note that, in [3], we've shown a "concave down" curve, but it may also be a "concave up" curve, or even a curve with inflection points. In nonlinear cases, $[K]$ matrix in Eq. 1.3.1(1) is no longer a constant matrix, it changes with $\{D\}$; that is, $[K]$ is a function of $\{D\}$. To emphasize this, we may rewrite

$$[K(D)]\{D\}=\{F\} \qquad (1)$$

Challenges of nonlinear simulations come from the difficulties of solving Eq. (1). #

13.1.2 Causes of Structural Nonlinearities

[1] As mentioned in the opening of this chapter, sources of structural nonlinearities can be classified into three categories: geometry nonlinearity, topology nonlinearity, and material nonlinearity. A problem may include more than one category of nonlinearities.

A linear simulation implies that no nonlinearities are present; it in turn implies (a) the deformation is very small, (b) there is no topological change, and (c) the stress-strain relation is linear; i.e., it can be described by Hooke's law. It also implies that the principle of superposition is applicable and the solution is independent of loading history. On the other hand, in nonlinear problems, the principle of superposition is not applicable and the solution may depend on loading history. ↓

Geometry Nonlinearity

[2] Geometry nonlinearity is due to large deformation of structures. The stiffness matrix $[K]$ is composed by element stiffness matrices, and each element stiffness is a function of the element's material properties as well as geometry. When the deformation of structure is so large that the stiffness matrix $[K]$ is changed substantially, geometry nonlinearity must be considered. To include geometry nonlinearity, simply turn on **Large Deflection** in **Analysis Settings** [3]. Among the three sources of nonlinearities, geometry nonlinearity is usually the easiest to tackle: reducing time steps is usually enough to improve the convergence. Some exceptions (e.g., the pneumatic finger of Section 9.1) need special treatments. In this book, so far, we've experienced many simulations involving geometry nonlinearity. Section 13.2 provides one more exercise for geometry nonlinearity. →

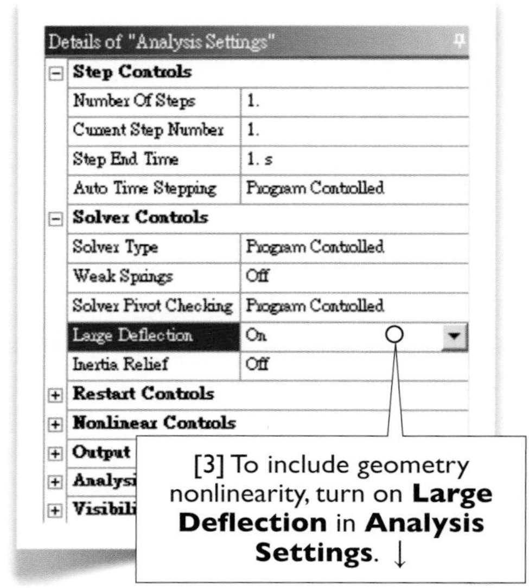

[3] To include geometry nonlinearity, turn on **Large Deflection** in **Analysis Settings**. ↓

Topology Nonlinearity

[4] When the topology (connectivity) of a structure changes, its stiffness matrix also changes. Possible topology changes include failure of structural members or materials, and the changes of contact status. In this chapter, we will cover only contact nonlinearity; i.e., change of contact status.

Contact nonlinearity itself is challenging and, moreover, it is usually accompanied by large deformation. In this book, so far, we've experienced many simulations involving contact nonlinearity. Sections 13.3 and 13.4 are two more exercises for the contact nonlinearity.

Material Nonlinearity

When its material's stress-strain relation is not linear (i.e., it cannot be described by Hooke's law), the problem involves material nonlinearity. In these cases, we need other ways of describing the relation between stress and strain (or stress and strain rate, when dealing with viscous materials). A mathematical model used to describe a stress-strain relationship is called a material model. Eqs. 1.2.8(3-4) (page 32), are two examples of material models. A material model is usually a mathematic form with some parameters; these parameters are usually determined by data fitting using material test data.

Besides Hooke's law, Workbench provides many other material models. Use of these material models is one of the most challenging tasks in the finite element simulations. We will discuss a few material models in Chapter 14. #

13.1.3 Load Steps, Substeps, and Equilibrium Iterations

Steps (Load Steps)

[1] You can divide the entire loading history into one or more *load steps*, or simply called *steps*. The number of steps can be specified in the details view of **Analysis Settings** [2]. To switch between steps, you can type a step number in the details view [3], click a step number in **Graph** [4], or click a step number in **Tabular Data** [5]. Each step can have its own analysis settings [6].

Time Steps (Substeps)

Each load step is further divided into substeps, or time steps. In dynamic simulations, time step is used for integration over time domain (Chapter 12); main consideration of the time step size is to capture the response characteristics. In static simulation, a load step can be divided into substeps to achieve or enhance convergence. Smaller time step size usually converges easier, but, of course, needs a greater number of time steps to complete a load step.

Iterations (Equilibrium Iterations)

For nonlinear problems, each time step itself needs several iterations to solve Eq. 13.1.1(1) (page 469). Each iteration involves solving a subproblem, Eq. 1.3.1(1) (page 35), the linearized equilibrium equation. Solving Eq. 1.3.1(1) is called an equilibrium iteration, or simply iteration. We will introduce the process of equilibrium iterations, known as the Newton-Raphson method, in the next subsection. ↗

[2] Total number of steps can be specified here. ↓

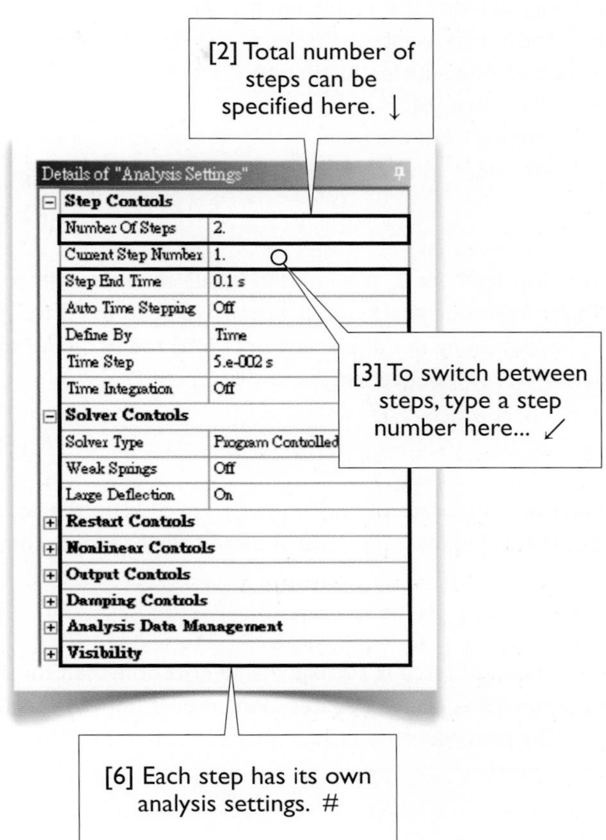

[3] To switch between steps, type a step number here... ↙

[6] Each step has its own analysis settings. #

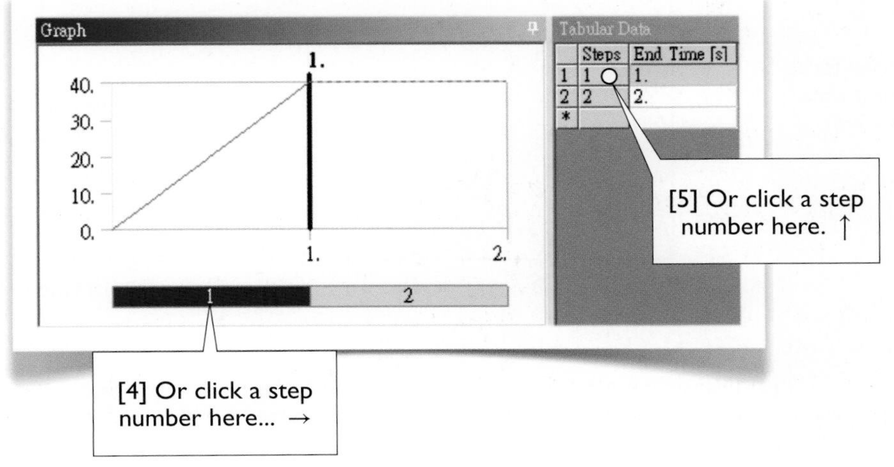

[5] Or click a step number here. ↑

[4] Or click a step number here... →

13.1.4 Newton-Raphson Method[Ref 1]

[1] Suppose that the simulation proceeds at a certain time step [2-4], where the displacement is D_0, the external force is F_0, and P_0 represents the point at the response curve described by Eq. 13.1.1(1) (page 469). Now, the time is increased one substep further, so the external force is increased to $F_0 + \Delta F$ [5], and we want to find the displacement at next time step [6].

Starting from the point P_0, Workbench calculates a *tangent stiffness* $\left[K(D_0)\right]$, the linearized stiffness, and solves the following equation

$$\left[K(D_0)\right]\left\{\Delta D\right\} = \left\{\Delta F\right\} \qquad (1)$$

The displacement D_0 is increased by ΔD and advances to D_1. Now, in the *D-F* space, we are at $(D_1, F_0 + \Delta F)$ (i.e., the point P_1'), far from our goal P_4. To proceed, we need to "drive" the point P_1' back to the actual curve (i.e., P_1).

Substituting the displacement D_1 into the left-hand side of the governing equation, Eq. 13.1.1(1), we can calculate the actual force F_1 needed for the displacement D_1,

$$\left[K(D_1)\right]\left\{D_1\right\} = \left\{F_1\right\}$$

Now we can locate the point (D_1, F_1), which is on the actual force-displacement curve. The difference between the external force (here, $F_0 + \Delta F$) and the balanced force (here, F_1) is called the *residual force* of that equilibrium iteration,

$$F_1^R = (F_0 + \Delta F) - F_1$$

If the residual force is smaller than a criterion, then the substep is said to be converged, otherwise, another equilibrium iteration takes place. The iterations repeat until the *convergence criterion* satisfies.

The procedure described above is called the *Newton-Raphson Method.* ↓

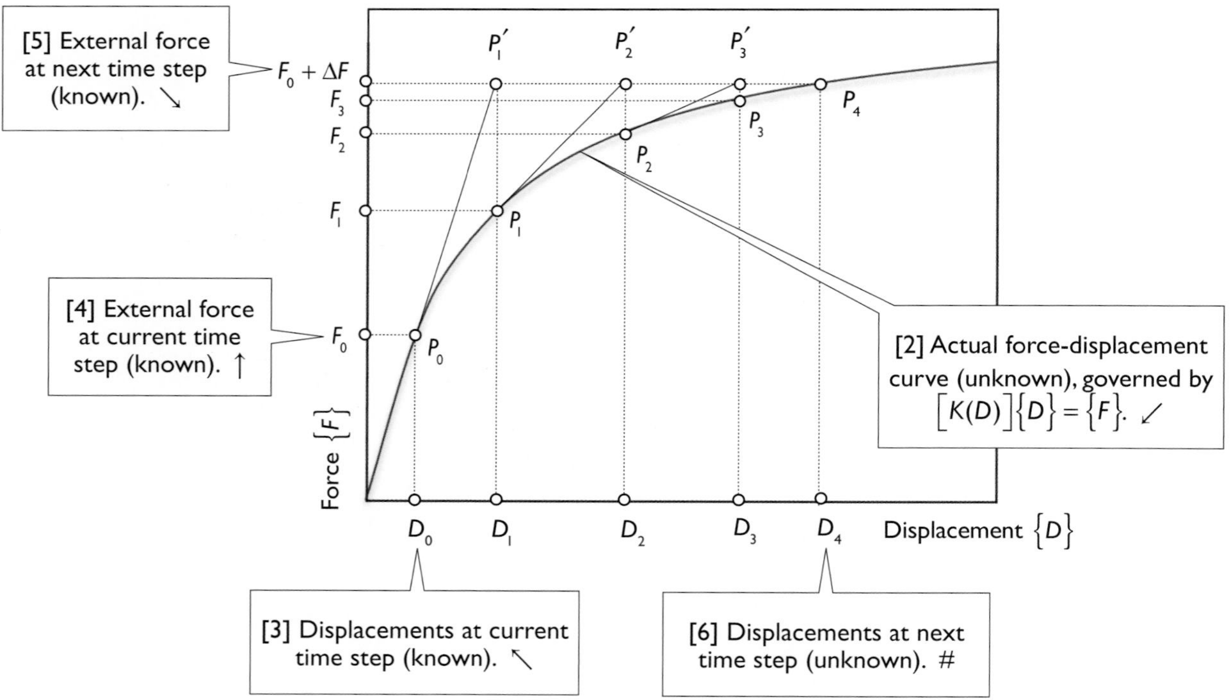

13.1.5 Convergence Criteria[Ref 2]

[1] In the last subsection, we stated that when the residual force F^R is smaller than a criterion, then the substep is converged. This statement is not strictly correct. There are at most four convergence criteria that can be activated under your control, namely, **force convergence** [2], **displacement convergence** [3], **moment convergence** [4], and **rotation convergence** [5]. The moment convergence and rotation convergence can be activated only when shell elements or beam elements are used. These convergence monitoring methods are all defaulted to **Program Controlled**, that is, Workbench automatically turns on any of them when it is appropriate. You may manually turn off or turn on any of them.

When you turn on any of them, you may specify a **Value**, a **Tolerance**, and a **Minimum Reference**. The criterion is then

$$\text{Criterion} = \text{Tolerance} \times \text{maximum}(\text{Value},\ \text{Minimum Reference})$$

The force (or moment) convergence satisfies when

$$\left\| F^R \right\| < \text{Criterion} \tag{1}$$

The displacement (or rotation) convergence satisfies when

$$\left\| \Delta D \right\| < \text{Criterion} \tag{2}$$

where $\left\| \cdot \right\|$ denotes the norm of the underlying vector, and is called a **Convergence Value**. Value defaults to ANSYS Calculated, which usually means the current maximum value. In 13.1.4[2-6] (last page), as an example, the current maximum force value is $\left\| F_0 \right\|$, and the current maximum displacement value is $\left\| D_0 \right\|$. Tolerances default to 0.5%. Note that setting up a Minimum Reference is to avoid a never-convergent situation when Value is near zero. ↓

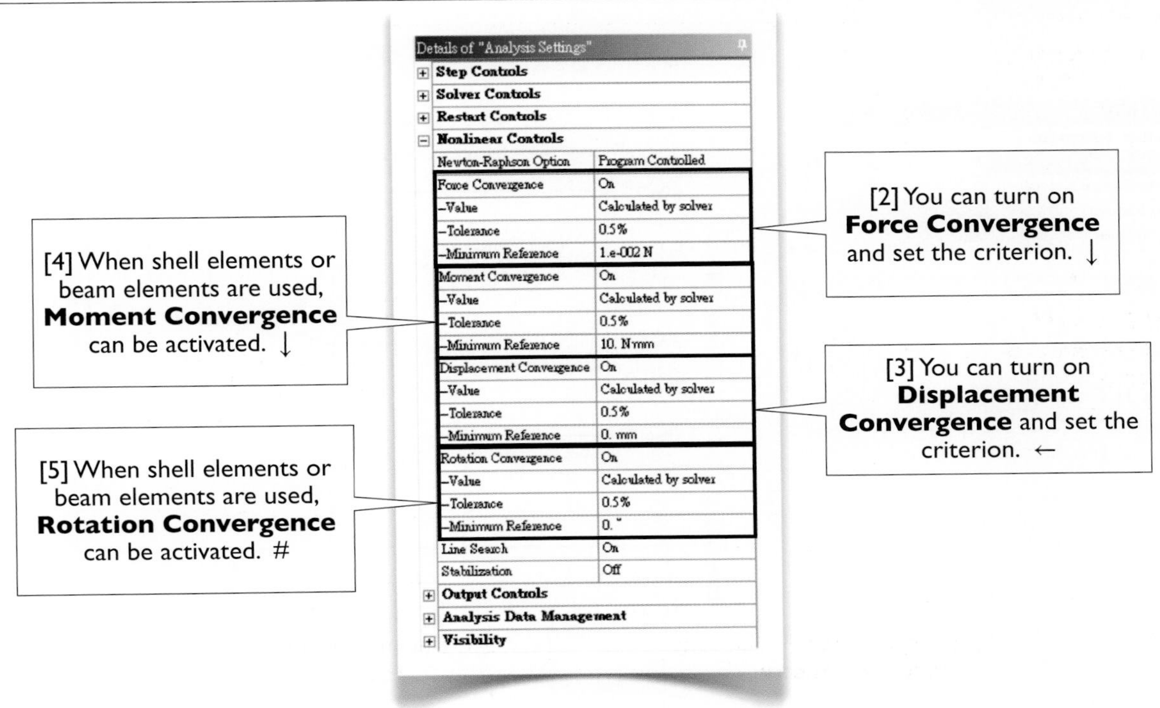

[2] You can turn on **Force Convergence** and set the criterion. ↓

[3] You can turn on **Displacement Convergence** and set the criterion. ←

[4] When shell elements or beam elements are used, **Moment Convergence** can be activated. ↓

[5] When shell elements or beam elements are used, **Rotation Convergence** can be activated. #

13.1.6 Solution Information

[1] A text form of convergence values and criteria for each iteration is available in **Solution Information** [2]. A graphics form of this information is also available [3]. A key to tackle nonlinear problems is the ability to interpret and take measures using this information. ↘

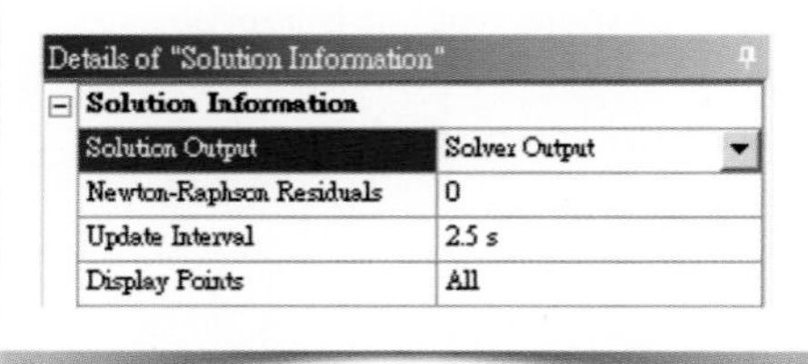

[2] In **Solver Output**, convergence values and criteria are listed in a text form. ↙

```
    FORCE CONVERGENCE VALUE  = 0.9949E-01  CRITERION= 0.2227E-01
   EQUIL ITER   1 COMPLETED.  NEW TRIANG MATRIX.  MAX DOF INC=  0.9229E-02
    LINE SEARCH PARAMETER =   1.000      SCALED MAX DOF INC =  0.9229E-02
    FORCE CONVERGENCE VALUE  = 0.6719E-02  CRITERION= 0.2173E-01 <<< CONVERGED
   >>> SOLUTION CONVERGED AFTER EQUILIBRIUM ITERATION   1
*** LOAD STEP     1   SUBSTEP     9  COMPLETED.    CUM ITER =     20
*** TIME =   0.900000         TIME INC =  0.100000

    FORCE CONVERGENCE VALUE  = 0.9542E-01  CRITERION= 0.2455E-01
   EQUIL ITER   1 COMPLETED.  NEW TRIANG MATRIX.  MAX DOF INC=  0.9888E-02
    LINE SEARCH PARAMETER =   1.000      SCALED MAX DOF INC =  0.9888E-02
    FORCE CONVERGENCE VALUE  = 0.4753E-02  CRITERION= 0.2415E-01 <<< CONVERGED
   >>> SOLUTION CONVERGED AFTER EQUILIBRIUM ITERATION   1
*** LOAD STEP     1   SUBSTEP    10  COMPLETED.    CUM ITER =     21
*** TIME =   1.00000          TIME INC =  0.100000
```

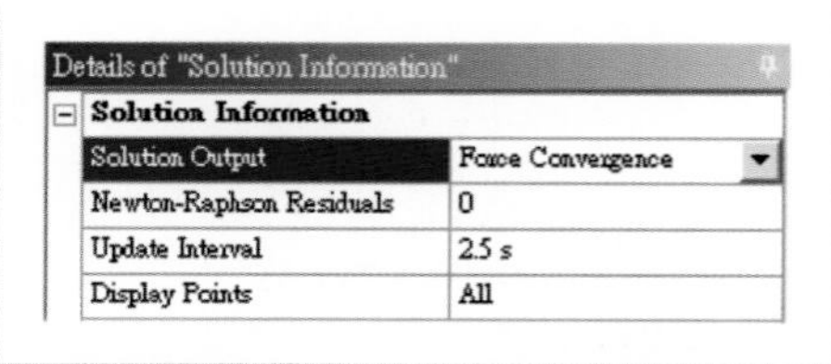

[3] Convergence values and criteria are displayed in graphics form. #

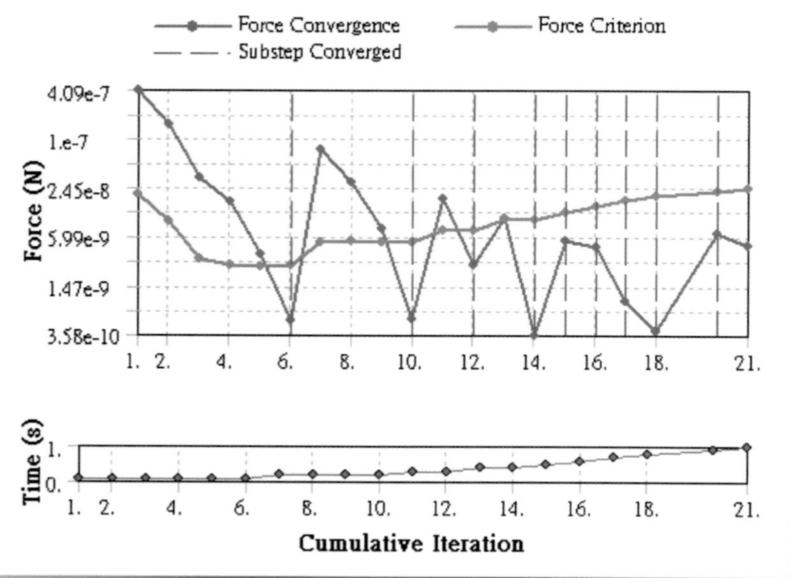

13.1.7 Line Search[Ref 3]

[1] In the example of 13.1.4[2-6] (page 472), each equilibrium iteration (Eq. 13.1.4(1), page 472) calculates a displacement ΔD, that is smaller than the goal, therefore several iterations are needed to reach the goal. That is true in cases when the *F-D* curves are monotonically "concave down," such as 13.1.4[2].

In cases when *F-D* curves are highly nonlinear or "concave up," the calculated displacement ΔD in a single iteration may "overshoot" the goal [2]. In such cases, a numerical technique called **Line Search** can be activated to "scale down" the incremental displacement [3-4]. In these cases, **Line Search** is helpful, but takes extra computing time. →

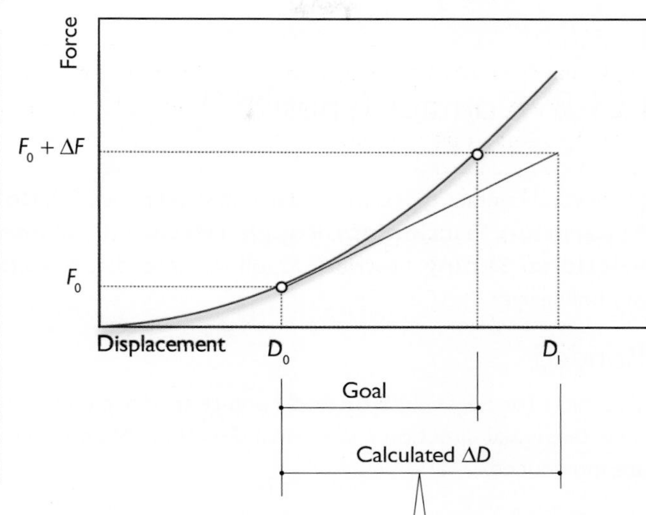

[2] In cases when the *F-D* curve is highly nonlinear or concave up, the calculated ΔD in a single iteration may overshoot the goal. ↙

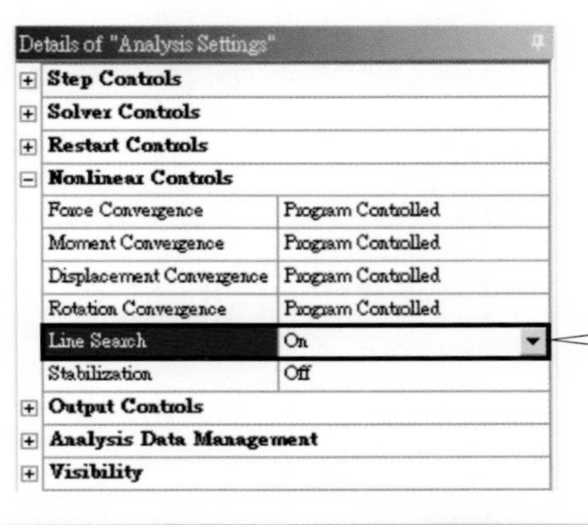

[3] **Line Search** can be turned on to scale down the incremental displacement. By default, it is **Program Controlled**. →

[4] **Line Search Parameter** scales down the incremental displacement. #

```
  FORCE CONVERGENCE VALUE  =  327.2       CRITERION= 0.2962
 EQUIL ITER   1 COMPLETED.  NEW TRIANG MATRIX.  MAX DOF INC= -0.1956
  LINE SEARCH PARAMETER =  0.5779      SCALED MAX DOF INC = -0.1131
  FORCE CONVERGENCE VALUE  =  179.4       CRITERION= 0.2133
 EQUIL ITER   2 COMPLETED.  NEW TRIANG MATRIX.  MAX DOF INC=  0.1267E-01
  LINE SEARCH PARAMETER =   1.000      SCALED MAX DOF INC =  0.1267E-01
  FORCE CONVERGENCE VALUE  =  34.92       CRITERION= 0.2253
 EQUIL ITER   3 COMPLETED.  NEW TRIANG MATRIX.  MAX DOF INC=  0.1004
  LINE SEARCH PARAMETER =  0.6958      SCALED MAX DOF INC =  0.6985E-01
  FORCE CONVERGENCE VALUE  =  30.58       CRITERION= 0.2916
 EQUIL ITER   4 COMPLETED.  NEW TRIANG MATRIX.  MAX DOF INC=  0.1047E-01
  LINE SEARCH PARAMETER =   1.000      SCALED MAX DOF INC =  0.1047E-01
  FORCE CONVERGENCE VALUE  =  5.867       CRITERION= 0.3076
 EQUIL ITER   5 COMPLETED.  NEW TRIANG MATRIX.  MAX DOF INC=  0.1734E-01
  LINE SEARCH PARAMETER =  0.8577      SCALED MAX DOF INC =  0.1487E-01
  FORCE CONVERGENCE VALUE  =  2.724       CRITERION= 0.3313
 EQUIL ITER   6 COMPLETED.  NEW TRIANG MATRIX.  MAX DOF INC=  0.3472E-03
  LINE SEARCH PARAMETER =   1.000      SCALED MAX DOF INC =  0.3472E-03
  FORCE CONVERGENCE VALUE  = 0.2257       CRITERION= 0.3382     <<< CONVERGED
  >>> SOLUTION CONVERGED AFTER EQUILIBRIUM ITERATION   6
 *** LOAD STEP     2   SUBSTEP   114  COMPLETED.    CUM ITER =    1889
 *** TIME =    1.89395         TIME INC =  0.288325E-01
 *** AUTO TIME STEP:  NEXT TIME INC = 0.43249E-01  INCREASED (FACTOR = 1.5000)
```

PART B. CONTACT NONLINEARITY

13.1.8 Contact Types[Ref 4]

[1] Several contact types are available in Workbench [2]: **Bonded**, **No Separation**, **Frictionless**, **Rough**, **Frictional**, and **Forced Frictional Sliding**, described as follows according to increasing degree of nonlinearity.

Bonded

Two faces (or edges) in **Bonded** contact are coupled together both in their tangential direction and normal direction. No contact nonlinearities are introduced.

No Separation

Two faces (or edges) in **No Separation** contact are coupled in their normal direction only. The tangential direction allows a small sliding on each other. No contact nonlinearities are introduced since small displacement theory is assumed for the sliding.

Rough

Two faces (or edges) in **Rough** contact are free to separate. But, when in contact, they cannot slide in tangential direction, due to large friction between them. This contact type introduces contact nonlinearities. No small displacement theory is assumed.

Frictionless

Two faces (or edges) in **Frictionless** contact are free to separate. And, when in contact, they may slide in tangential direction without any friction force. This contact type introduces contact nonlinearities. No small displacement theory is assumed.

Frictional

Two faces (or edges) in **Frictional** contact are free to separate. And, when in contact, they may slide only when the shear stress between them exceeds a critical value, calculated by multiplying the normal stress by a friction coefficient, which is input as a contact property. This contact type introduces contact nonlinearities. No small displacement theory is assumed.

Forced Frictional Sliding

Similar to **Frictional** except that there is no "sticking" state; i.e., two faces (or edges) slide even when the shear stress is below the critical value. ↗

[2] Several types of contact are available. ↓

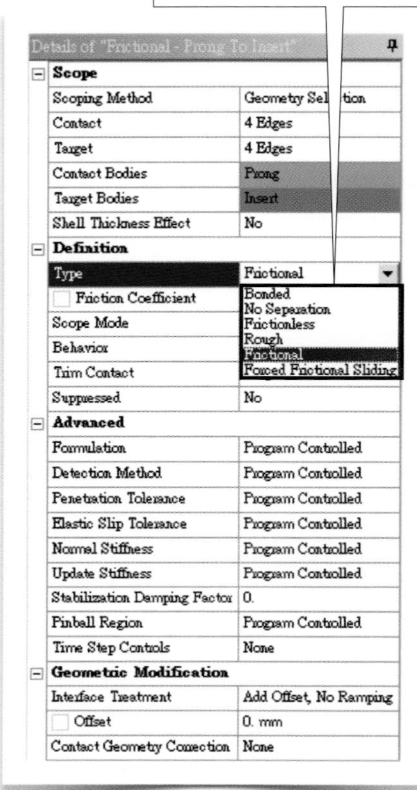

Linear vs. Nonlinear Contacts

[3] The contact types **Bonded** and **No Separation** are called linear contacts because they assume small deformation and involve no nonlinearities. The other contact types will introduce contact nonlinearities. #

13.1.9 Contact versus Target

[1] To specify a contact region, you need to select a set of **Contact** faces (or edges) and a set of **Target** faces (or edges) [2]. During the solution, Workbench checks the contact status for each node (or integration point) on the **Contact** faces against the **Target** faces.

If **Behavior** is set to **Symmetric**, the roles of **Contact** and **Target** will be symmetric [3]; i.e., Workbench checks each point on the **Contact** against the **Target**, as well as each point on the **Target** against the **Contact**. If **Behavior** is set to **Asymmetric**, the checking is only one-sided.

Consider a point, on a **Contact** face, which is approaching a **Target** face. The point is called a contacting point. Workbench keeps tracing the contacting point so that it won't penetrate into the target face. When the point is in contact with the surface, Workbench starts to enforce contact compatibility (i.e., preventing penetration).

In some cases, when **Symmetric** behavior is not available, or when you want to set to **Asymmetric**, to save the run time, selection of **Contact** and **Target** becomes important. You may select **Target** faces using the following guidelines: the faces belong to a body with fixed supports, the faces belong to a body with more rigid material (higher Young's modulus), the faces of less curvature, etc.

For **Symmetric**, results are reported for both contact and target sides. When **Asymmetric** is used, all result data is on the contact side. ↗

[2] To specify a contact region, you need to select a set of **Contact** faces (or edges) and a set of **Target** faces (or edges). ↓

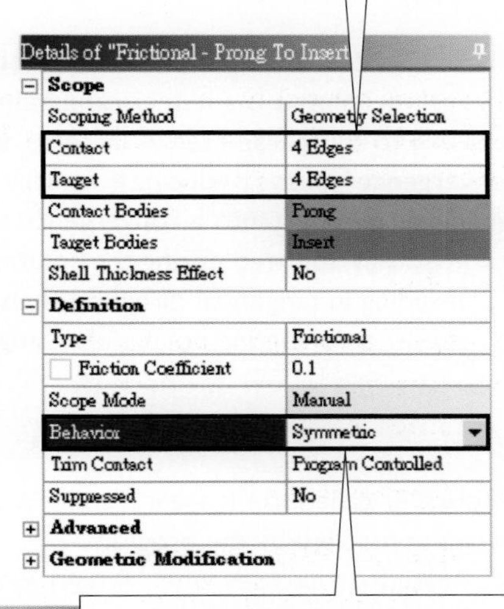

[3] If **Behavior** is set to **Symmetric**, the roles of **Contact** and **Target** will be symmetric. #

13.1.10 Contact Formulations[Ref 5]

[1] Workbench offers several **Formulation** options [2] to enforce contact compatibility at the contact interface; i.e., preventing penetration of the contacting point into the target faces.

MPC (Multi-Point Constraint)

For linear contact types (**Bonded** and **No Separation**), a multi-point constraint (**MPC**) formulation is available. **MPC** internally adds constraint equations to couple the displacements between contacting faces. Although you can use other formulations for **Bonded** and **No Separation** contact types, **MPC** is recommended for these linear contact types, since it is a direct, efficient formulation. ↗

[2] Workbench offers several **Formulation** options to enforce contact compatibility. ↓

Contact	4 Edges
Target	4 Edges
Contact Bodies	Prong
Target Bodies	Insert
Shell Thickness Effect	No
Definition	
Advanced	
Formulation	Pure Penalty (Program Controlled / Augmented Lagrange / Pure Penalty / MPC / Normal Lagrange)
Detection Method	
Penetration Tolerance	
Elastic Slip Tolerance	
Normal Stiffness	Program Controlled
Update Stiffness	Program Controlled
Stabilization Damping Factor	0.
Pinball Region	Program Controlled
Time Step Controls	None
Geometric Modification	

[3] **Normal Stiffness** (see [4], next page) can be input here. The input value is used to multiply a stiffness value calculated by the program. ↵

Pure Penalty

[4] Whenever a contacting point penetrates normally by an amount x_n into a target face, it will be pushed back by a normal force F_n,

$$F_n = k_n x_n \quad (1)$$

where k_n is called the **Normal Stiffness** ([3], last page) of the contact region. A **Normal Stiffness** has no real physical meaning; it is a numerical parameter of the penalty algorithm. Solution convergence behavior is usually sensitive to this parameter. A larger k_n usually gives a more accurate solution (less penetration), but may raise convergence issues. Reducing k_n usually helps convergence, but results in increasing penetration. As a simple rule, whenever bumping into a convergence problem, try reducing k_n first. The **Normal Stiffness** can also be automatically adjusted during the solution (see **Update Stiffness** in 13.1.11[1], next page).

If sliding in tangential direction is also prohibited (e.g., **Rough** contact), a similar treatment can be implemented. Whenever a contacting point slides tangentially by an amount x_t, it will be pushed back by a tangential force F_t,

$$F_t = k_t x_t \quad (2)$$

where k_t is called the *tangential stiffness* of the contact region. Unlike **Normal Stiffness**, the tangential stiffness is always controlled by the program.

Note that, in cases when **Augmented Lagrange** or **Normal Lagrange** (to be discussed) formulation is used, the formulations apply on the normal direction only. The tangential direction still uses **Pure Penalty** formulation, Eq. (2). When **MPC** is used, both normal and tangential directions use **MPC** formulation.

Normal Lagrange

Normal Lagrange formulation adds an extra degree of freedom, namely, contact pressure, to satisfy compatibility. Whenever a contacting point is in touch with the target face, the contact pressure is explicitly calculated. It is the contact pressure that prevents further penetration,

$$F_n = \lambda \quad (3)$$

where λ is the contact pressure, and is traditionally called a *Lagrange multiplier*. Note that this formulation is used in the normal direction only. In the tangential direction, **Pure Penalty** formulation, Eq. (2), is used. This is where the name **Normal Lagrange** comes from.

Normal Lagrange formulation does not require a normal stiffness, and, theoretically, it can enforce zero penetration. However, since no penetration is allowed, the contact status is either open or closed (a step function). This can sometimes make convergence difficult because contact points may oscillate between open and closed status. This behavior is called *chattering*.

Augmented Lagrange

The idea is to combine **Pure Penalty** and **Normal Lagrange**: the push-back normal force is

$$F_n = k_n x_n + \lambda \quad (4)$$

Because of the contact pressure λ, **Augmented Lagrange** formulation is less sensitive to **Normal Stiffness** k_n. Note that again, in the tangential direction, **Pure Penalty** formulation, Eq. (2), is used.

Although **Pure Penalty** is the default setting, **Augmented Lagrange** is recommended for general frictional or frictionless contact in large deformation problems. Since Lagrange method adds an extra degree of freedom, it also takes extra computing time. #

13.1.11 Advanced Contact Settings[Refs 6, 7]

Pinball Region

[1] The pinball is a sphere region; its radius can be defined in **Pinball Region**. Consider again that a contacting point approaches a target face. If a target node is within the pinball region centered at a contacting node, the contacting node is considered to be in "near" contact with the target node and will be monitored. Target nodes outside of the pinball region will not be monitored.

If **Bonded** type is specified, surfaces that have a gap smaller than the pinball radius are treated as bonded.

Interface Treatment

For **Bonded** contact type, a large enough pinball radius may allow any gap between contacting faces to be ignored. For **Frictional** or **Frictionless** contact types, an initial gap is not automatically ignored, no matter how large the pinball is, since the gap may represent the real geometry.

If an initial gap is present [2] and a force is applied, one part may "fly away" relative to another part [3] if the initial contact is not established right at the end of the time step.

To alleviate situations where a gap (clearance) is modeled but needs to be ignored to establish initial contact for **Frictional** or **Frictionless** contact types, **Interface Treatment** can internally offset the contact surfaces by a specified amount. Note that this treatment is intended for small gaps. Don't apply it in a large gap.

Time Step Controls

Time Step Controls tries to enhance convergence by allowing adjustments of time step size based on contact behavior.

By default, contact behavior does not affect auto time stepping, since adjustment of time step based on contact behavior may increase computing time too much.

With **Time Step Controls** turned on, Workbench adjusts the time step size based on contact behavior.

Update Stiffness

The **Normal Stiffness** can be automatically adjusted during the solution. Whenever convergence difficulties arise, **Normal Stiffness** will be reduced automatically. →

Details of "Frictional - Prong To Insert"

Scope		
Scoping Method	Geometry Selection	
Contact	4 Edges	
Target	4 Edges	
Contact Bodies	Prong	
Target Bodies	Insert	
Shell Thickness Effect	No	
Definition		
Advanced		
Formulation	Pure Penalty	
Detection Method	Program Controlled	
Penetration Tolerance	Program Controlled	
Elastic Slip Tolerance	Program Controlled	
Normal Stiffness	Program Controlled	
Update Stiffness	Each Iteration	○
Stabilization Damping Factor	0.	
Pinball Region	Program Controlled	○
Time Step Controls	None	○
Geometric Modification		
Interface Treatment	Add Offset, No Ramping	○
Offset	0. mm	○
Contact Geometry Correction	None	

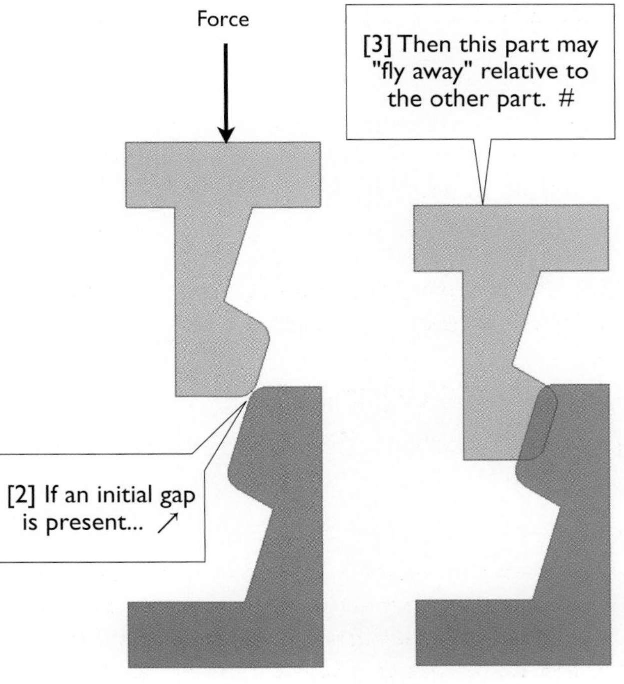

References

1. ANSYS Help//Mechanical APDL//Mechanical APDL Theory Reference//14.12. Newton-Raphson Procedure
2. ANSYS Help//Mechanical APDL//Mechanical APDL Theory Reference//14.12.2. Convergence
3. ANSYS Help//Mechanical APDL//Mechanical APDL Theory Reference//14.12.5. Line Search
4. ANSYS Help//Mechanical Application//Mechanical Users' Guide//Setting Connections//Contact//Contact Settings//Definition Settings
5. ANSYS Documentation//Mechanical Applications//ANSYS Mechanical Users' Guide//Setting Connections//Contact//Contact Formulation Theory
6. ANSYS Documentation//Mechanical Applications//ANSYS Mechanical Users' Guide//Setting Connections//Contact//Contact Settings//Advanced Settings
7. ANSYS Documentation//Mechanical Applications//ANSYS Mechanical Users' Guide//Setting Connections//Contact//Contact Settings//Geometric Modification

Section 13.2

Translational Joint[Ref 1]

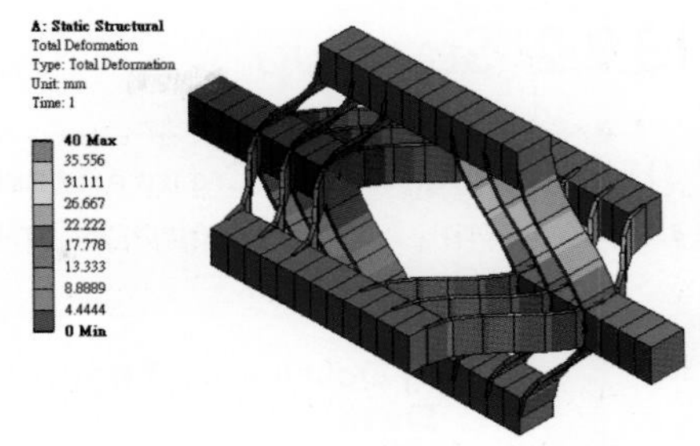

13.2.1 About the Translational Joint

[1] A translational joint is used to connect two machine components, so that the relative motion of the two components is restricted to translation in a specific direction. Conventionally, translational joints are designed as mechanisms, composed by parts, between which the clearance or interference is inevitable; they either decrease precision or increase friction. The translational joint in this section is not a mechanism; rather, it is a unitary flexible structure, in which no clearance or interference exist.

The translational joint [2-5] is made of POM (polyoxymethylene, a plastic), which has a Young's modulus of 2 GPa and a Poisson's ratio of 0.35. The most important design consideration is that the rigidity of translational direction should be much less than all other directions, so that the motion can be restricted in that direction only.

Here, we want to explore the geometric nonlinearity of the structure: how the applied force increases nonlinearly with the translational displacement. For this purpose, we will model the structure using line bodies entirely. The goal of the simulation is to plot a force-versus-displacement chart. The unit system used is **mm-kg-N-s**. ↙

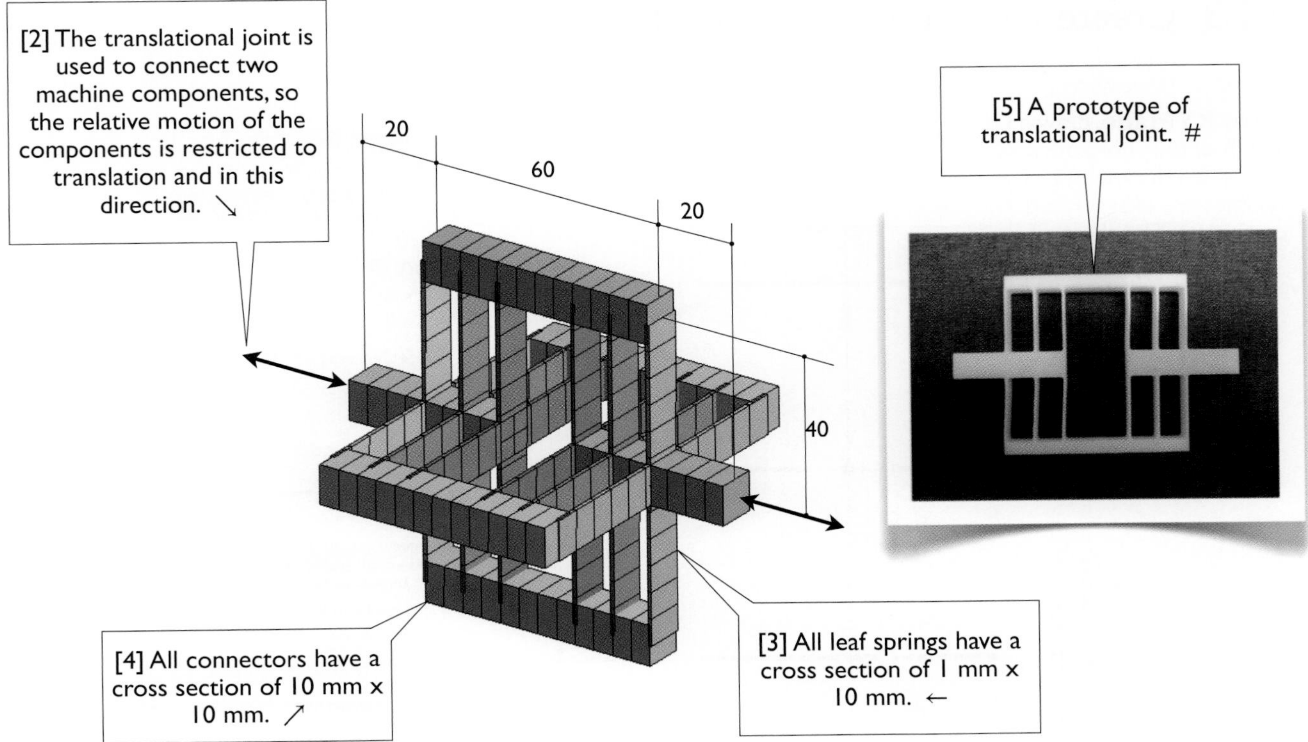

13.2.2 Start Up

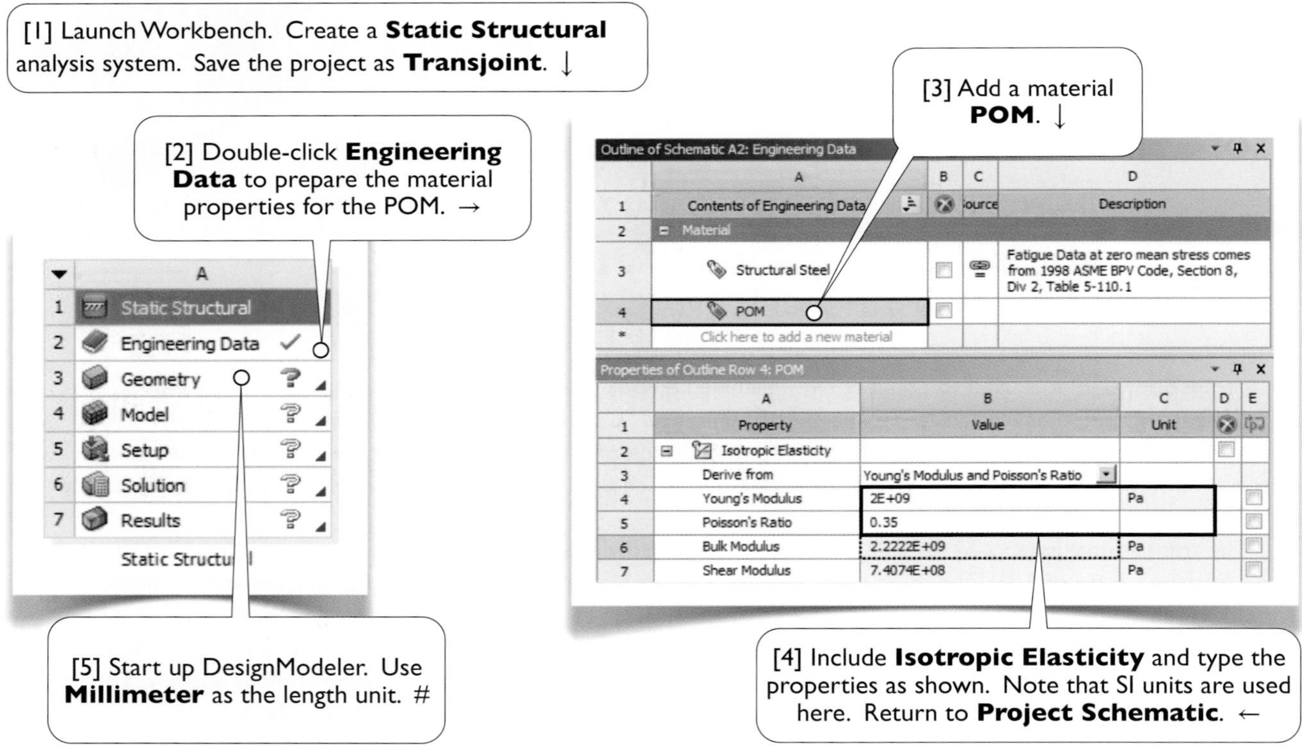

13.2.3 Create Geometry in DesignModeler

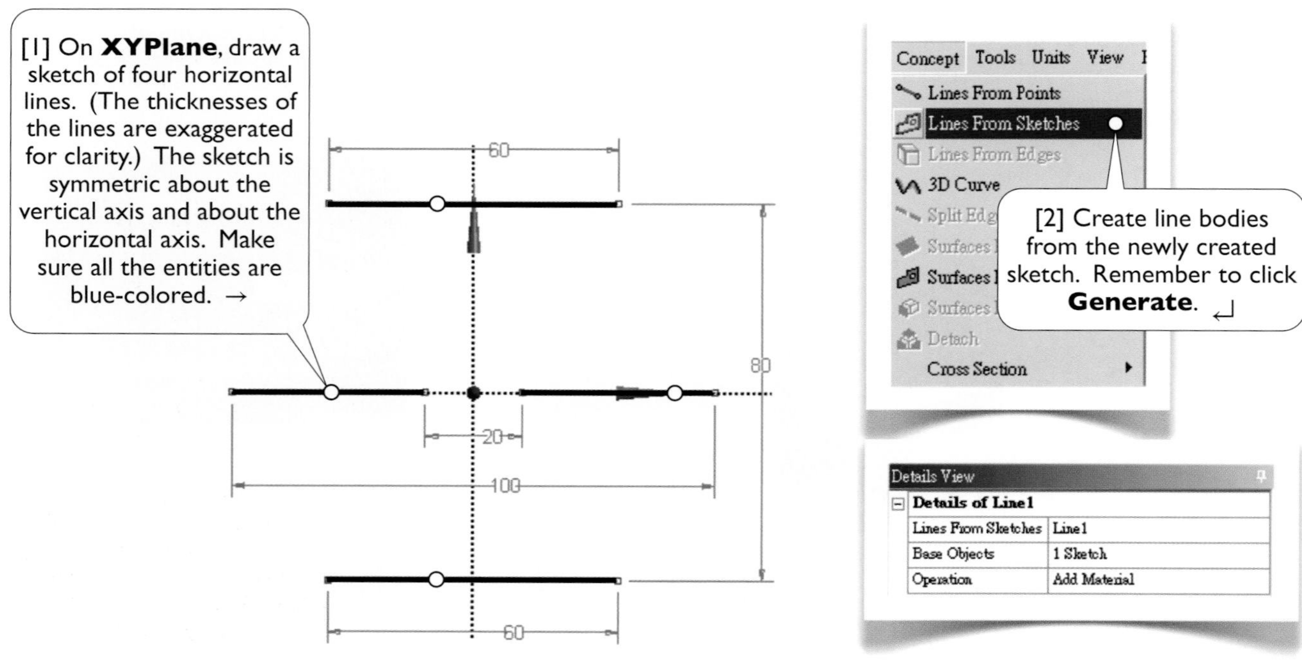

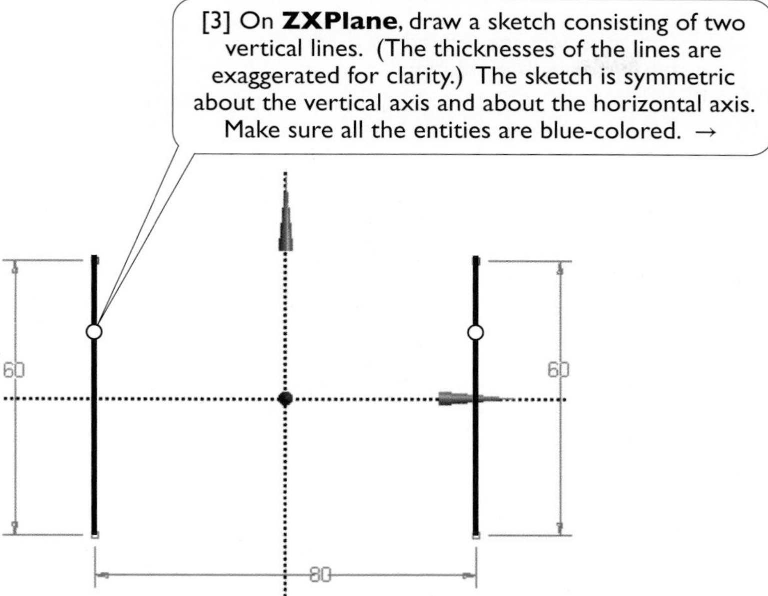
[3] On **ZXPlane**, draw a sketch consisting of two vertical lines. (The thicknesses of the lines are exaggerated for clarity.) The sketch is symmetric about the vertical axis and about the horizontal axis. Make sure all the entities are blue-colored. →
60
60
80

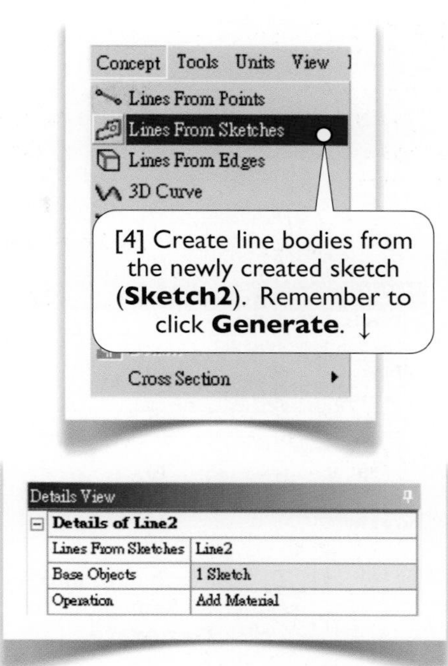
Concept
Tools
Units
View
Lines From Points
Lines From Sketches
Lines From Edges
3D Curve
[4] Create line bodies from the newly created sketch (**Sketch2**). Remember to click **Generate**. ↓
Cross Section

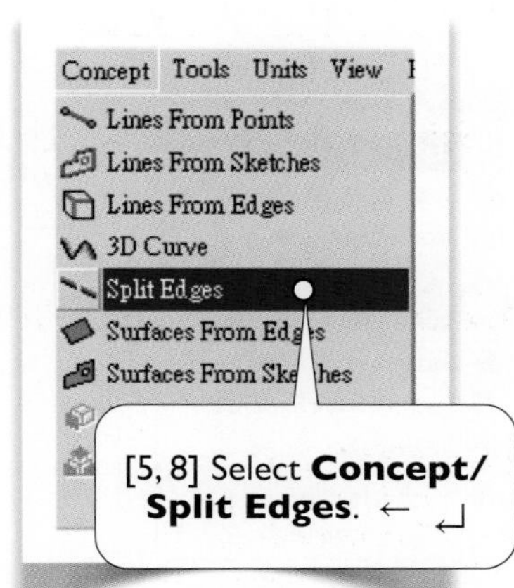
Details View
Details of Line2
Lines From Sketches | Line2
Base Objects | 1 Sketch
Operation | Add Material

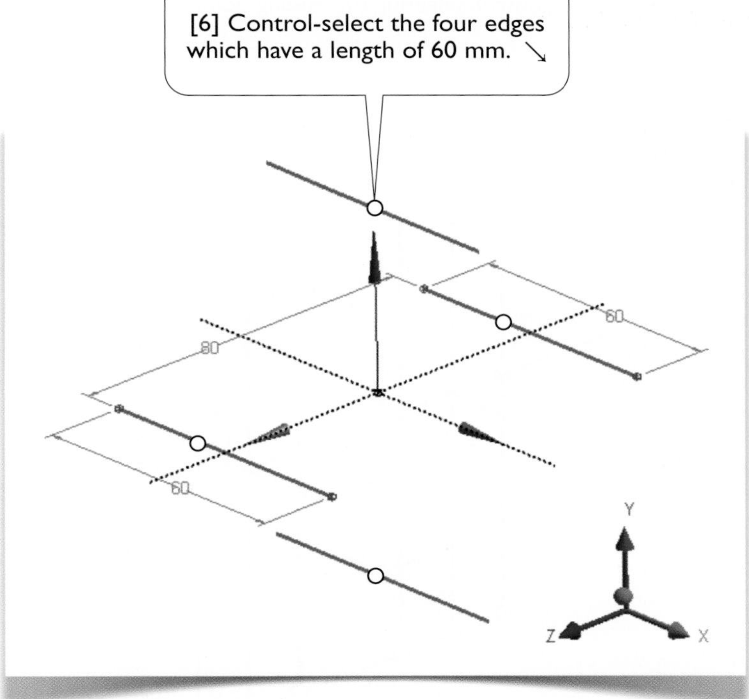
[6] Control-select the four edges which have a length of 60 mm. ↘
80
60
60
Y
Z
X

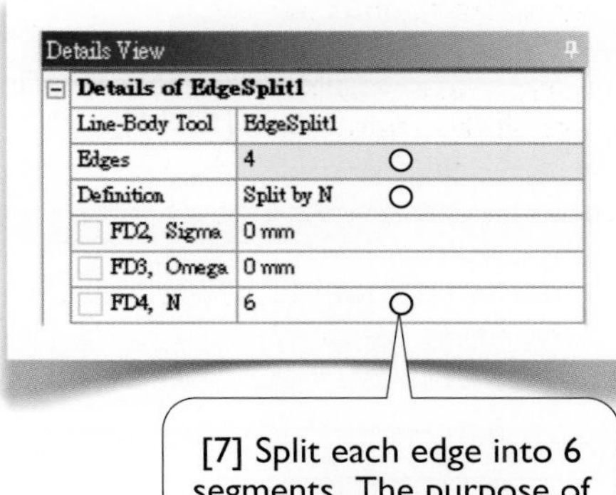
Concept
Tools
Units
View
Lines From Points
Lines From Sketches
Lines From Edges
3D Curve
Split Edges
Surfaces From Edges
Surfaces From Sketches
[5, 8] Select **Concept/ Split Edges**. ← ↵
Details View
Details of EdgeSplit1
Line-Body Tool | EdgeSplit1
Edges | 4
Definition | Split by N
FD2, Sigma | 0 mm
FD3, Omega | 0 mm
FD4, N | 6
[7] Split each edge into 6 segments. The purpose of splitting is to create construction points on the edges. Remember to click **Generate**. ↑

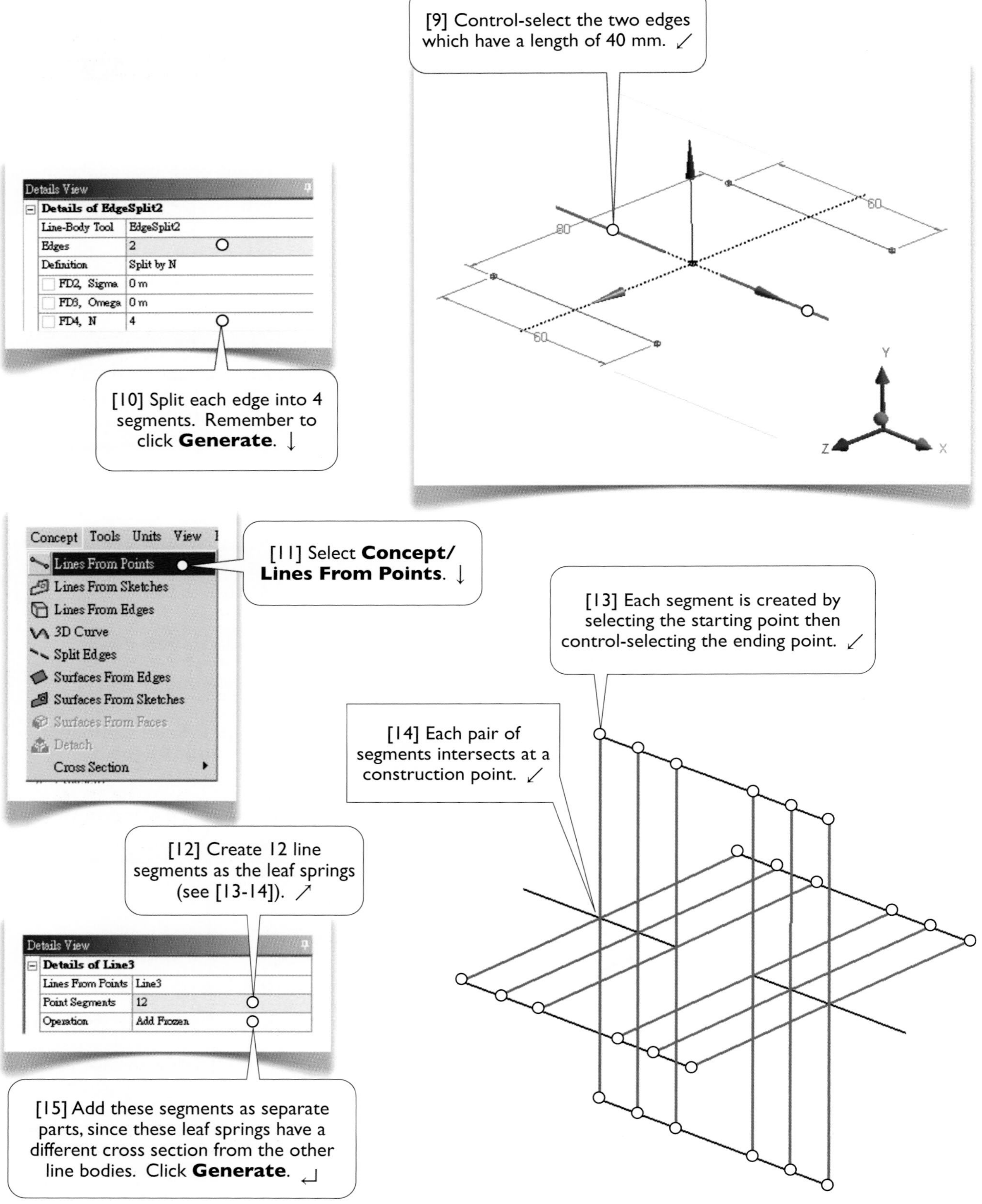

[9] Control-select the two edges which have a length of 40 mm.
80
60
60
Y
Z
X
Details View
Details of EdgeSplit2
Line-Body Tool
EdgeSplit2
Edges
2
Definition
Split by N
FD2, Sigma
0 m
FD3, Omega
0 m
FD4, N
4
[10] Split each edge into 4 segments. Remember to click **Generate**.
Concept Tools Units View
Lines From Points
Lines From Sketches
Lines From Edges
3D Curve
Split Edges
Surfaces From Edges
Surfaces From Sketches
Surfaces From Faces
Detach
Cross Section
[11] Select **Concept/ Lines From Points**.
[13] Each segment is created by selecting the starting point then control-selecting the ending point.
[14] Each pair of segments intersects at a construction point.
[12] Create 12 line segments as the leaf springs (see [13-14]).
Details View
Details of Line3
Lines From Points
Line3
Point Segments
12
Operation
Add Frozen
[15] Add these segments as separate parts, since these leaf springs have a different cross section from the other line bodies. Click **Generate**.

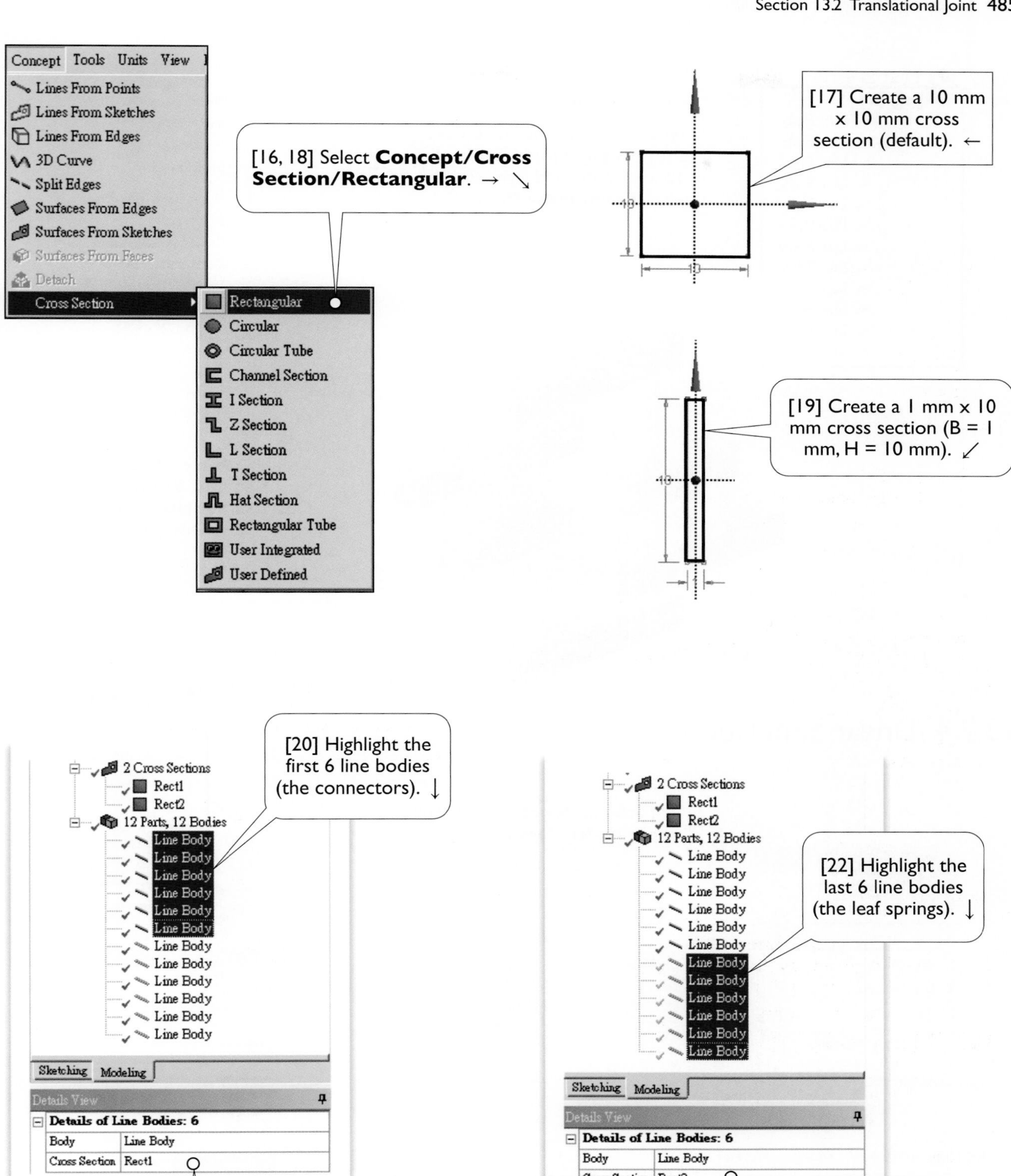
Concept Tools Units View
Lines From Points
Lines From Sketches
Lines From Edges
3D Curve
Split Edges
Surfaces From Edges
Surfaces From Sketches
Surfaces From Faces
Detach
Cross Section
Rectangular
Circular
Circular Tube
Channel Section
I Section
Z Section
L Section
T Section
Hat Section
Rectangular Tube
User Integrated
User Defined
[16, 18] Select **Concept/Cross Section/Rectangular**. → ↘
[17] Create a 10 mm x 10 mm cross section (default). ←
[19] Create a 1 mm x 10 mm cross section (B = 1 mm, H = 10 mm). ↙
2 Cross Sections
Rect1
Rect2
12 Parts, 12 Bodies
Line Body
[20] Highlight the first 6 line bodies (the connectors). ↓
Sketching
Modeling
Details View
Details of Line Bodies: 6
Body
Line Body
Cross Section
Rect1
[21] Assign **Rect1**, which has a cross section of 10 mm x 10 mm, to the connectors. ↗
[22] Highlight the last 6 line bodies (the leaf springs). ↓
Rect2
[23] Assign **Rect2**, which has a cross section of 1 mm x 10 mm, to the leaf springs. ↵

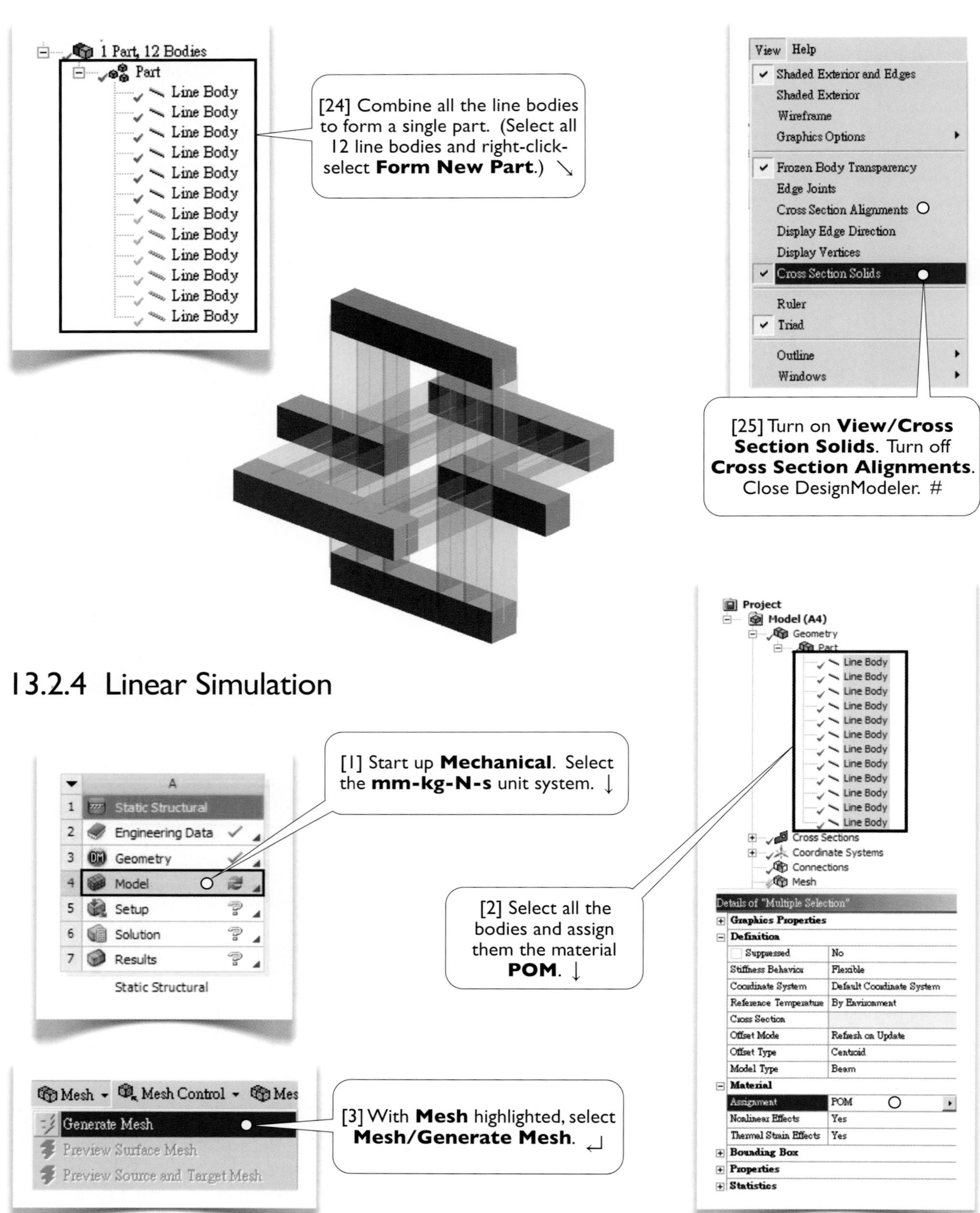

13.2.4 Linear Simulation

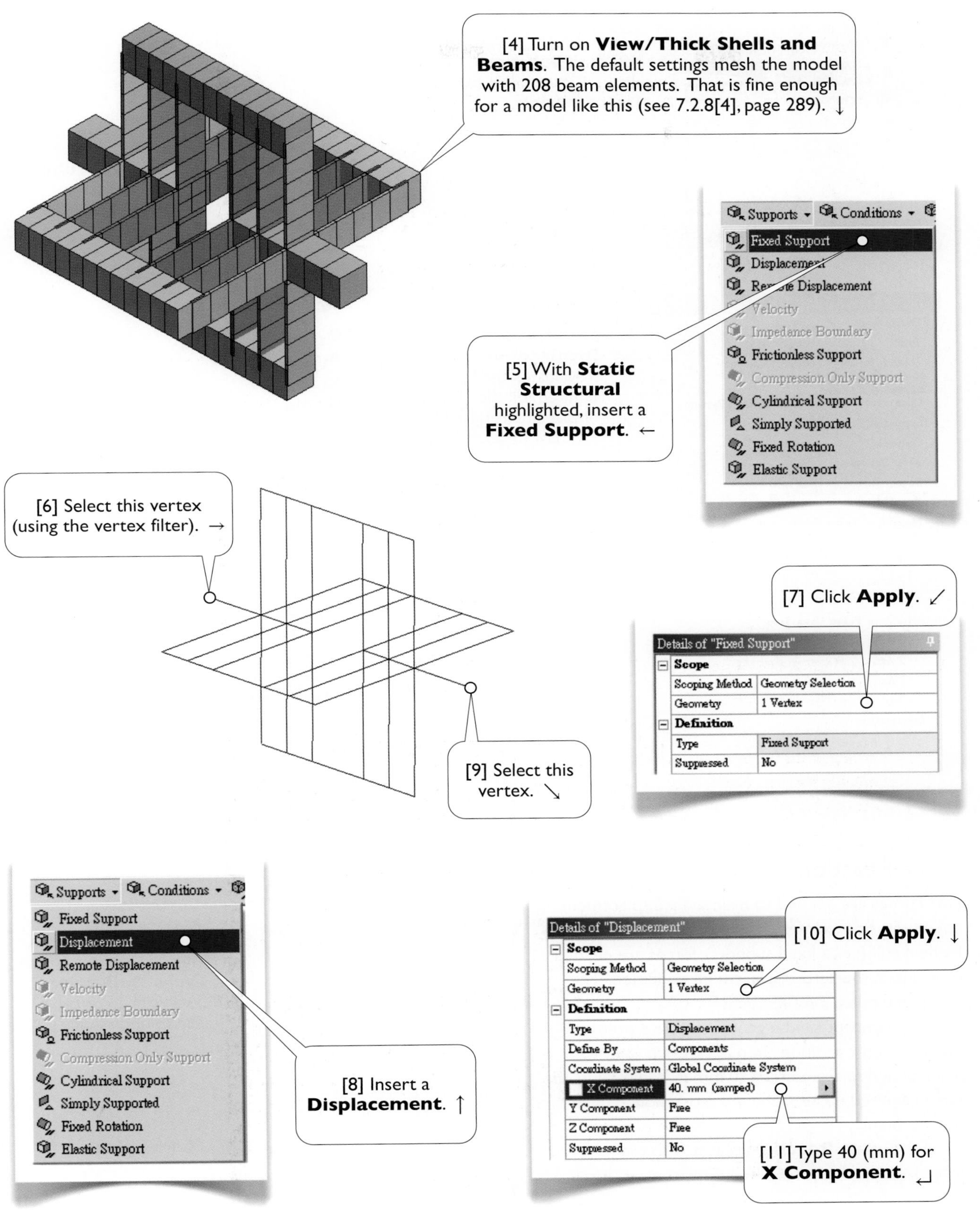

[4] Turn on **View/Thick Shells and Beams**. The default settings mesh the model with 208 beam elements. That is fine enough for a model like this (see 7.2.8[4], page 289). ↓
Supports
Conditions
Fixed Support
Displacement
Remote Displacement
Velocity
Impedance Boundary
Frictionless Support
Compression Only Support
Cylindrical Support
Simply Supported
Fixed Rotation
Elastic Support
[5] With **Static Structural** highlighted, insert a **Fixed Support**. ←
[6] Select this vertex (using the vertex filter). →
[7] Click **Apply**. ↙
Details of "Fixed Support"
Scope
Scoping Method
Geometry Selection
Geometry
1 Vertex
Definition
Type
Fixed Support
Suppressed
No
[9] Select this vertex. ↘
Supports
Conditions
Fixed Support
Displacement
Remote Displacement
Velocity
Impedance Boundary
Frictionless Support
Compression Only Support
Cylindrical Support
Simply Supported
Fixed Rotation
Elastic Support
[8] Insert a **Displacement**. ↑
Details of "Displacement"
Scope
Scoping Method
Geometry Selection
Geometry
1 Vertex
Definition
Type
Displacement
Define By
Components
Coordinate System
Global Coordinate System
X Component
40. mm (ramped)
Y Component
Free
Z Component
Free
Suppressed
No
[10] Click **Apply**. ↓
[11] Type 40 (mm) for **X Component**. ↵

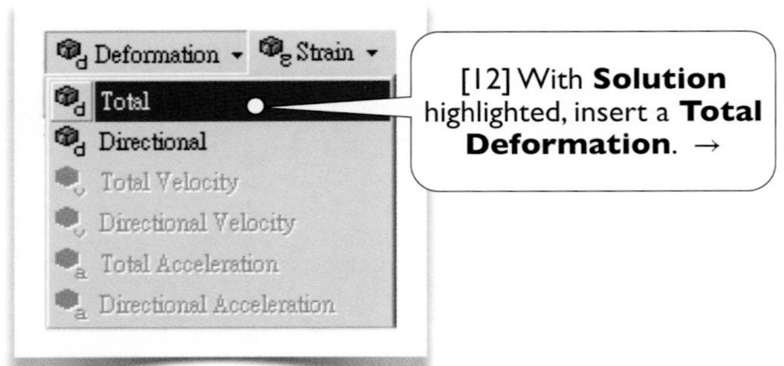

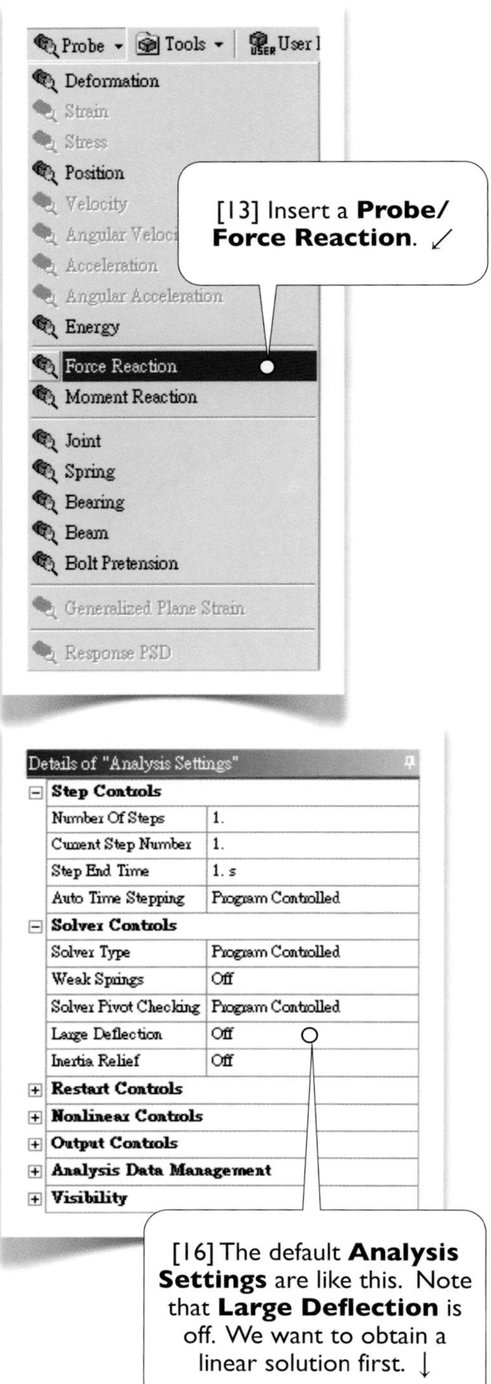

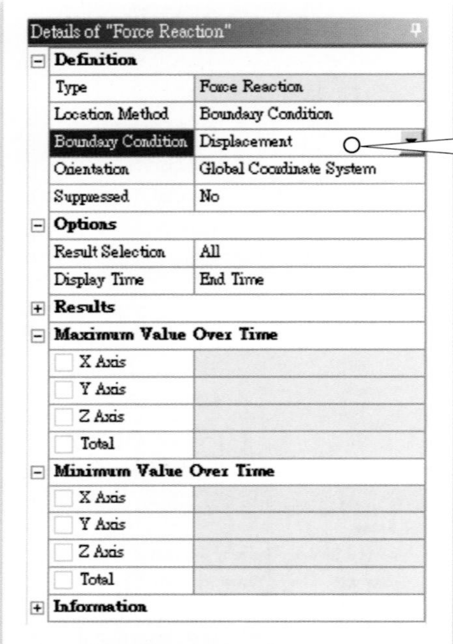

Details of "Force Reaction"

Definition	
Type	Force Reaction
Location Method	Boundary Condition
Boundary Condition	Displacement
Orientation	Global Coordinate System
Suppressed	No
Options	
Result Selection	All
Display Time	End Time
Results	
Maximum Value Over Time	
X Axis	
Y Axis	
Z Axis	
Total	
Minimum Value Over Time	
X Axis	
Y Axis	
Z Axis	
Total	
Information	

Details of "Analysis Settings"

Step Controls	
Number Of Steps	1.
Current Step Number	1.
Step End Time	1. s
Auto Time Stepping	Program Controlled
Solver Controls	
Solver Type	Program Controlled
Weak Springs	Off
Solver Pivot Checking	Program Controlled
Large Deflection	Off
Inertia Relief	Off
Restart Controls	
Nonlinear Controls	
Output Controls	
Analysis Data Management	
Visibility	

[16] The default **Analysis Settings** are like this. Note that **Large Deflection** is off. We want to obtain a linear solution first. ↓

Force Reaction

[15] Forces are needed to maintain a displacement condition. For example, if a point of a structure is to be held still (i.e., a fixed support), forces are needed to maintain the zero-displacement condition. The forces are called *force reactions*.

This concept can be generalized to a nonzero displacement condition. In our case, we want to maintain a displacement of 40 mm in X-direction; the force required is called the force reaction for that displacement.

The **Force Reaction** object is to evaluate the force required for the 40-mm displacement. →

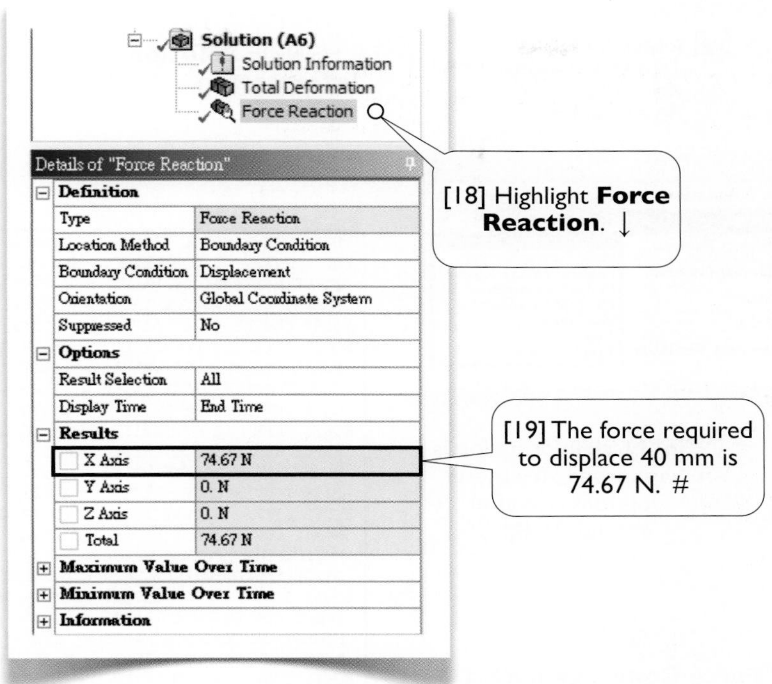

13.2.5 Nonlinear Simulation

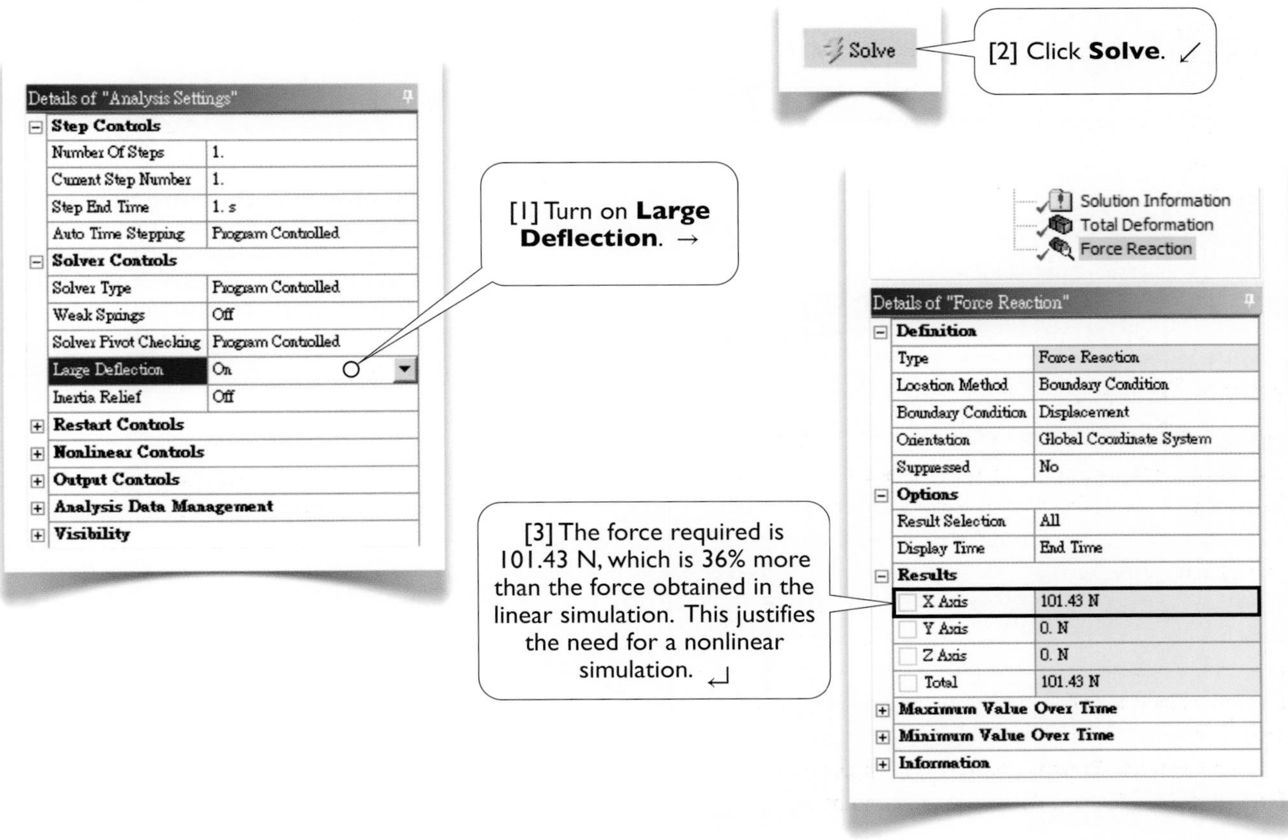

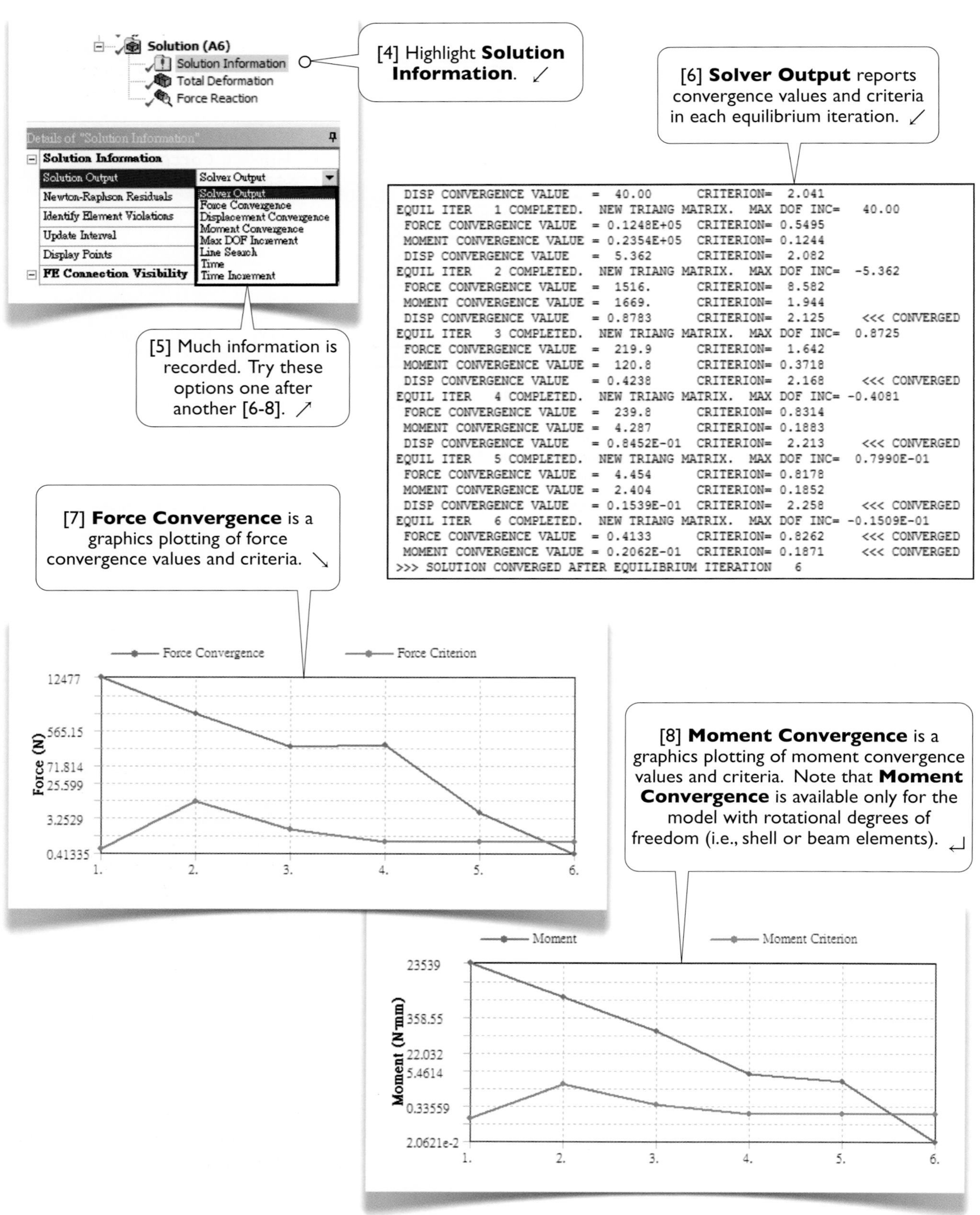

Solution (A6)
Solution Information
Total Deformation
Force Reaction
Details of "Solution Information"
Solution Information
Solution Output
Solver Output
Newton-Raphson Residuals
Identify Element Violations
Update Interval
Display Points
FE Connection Visibility
Solver Output
Force Convergence
Displacement Convergence
Moment Convergence
Max DOF Increment
Line Search
Time
Time Increment
[4] Highlight **Solution Information**.
[5] Much information is recorded. Try these options one after another [6-8].
[6] **Solver Output** reports convergence values and criteria in each equilibrium iteration.
DISP CONVERGENCE VALUE = 40.00 CRITERION= 2.041
EQUIL ITER 1 COMPLETED. NEW TRIANG MATRIX. MAX DOF INC= 40.00
FORCE CONVERGENCE VALUE = 0.1248E+05 CRITERION= 0.5495
MOMENT CONVERGENCE VALUE = 0.2354E+05 CRITERION= 0.1244
DISP CONVERGENCE VALUE = 5.362 CRITERION= 2.082
EQUIL ITER 2 COMPLETED. NEW TRIANG MATRIX. MAX DOF INC= -5.362
FORCE CONVERGENCE VALUE = 1516. CRITERION= 8.582
MOMENT CONVERGENCE VALUE = 1669. CRITERION= 1.944
DISP CONVERGENCE VALUE = 0.8783 CRITERION= 2.125 <<< CONVERGED
EQUIL ITER 3 COMPLETED. NEW TRIANG MATRIX. MAX DOF INC= 0.8725
FORCE CONVERGENCE VALUE = 219.9 CRITERION= 1.642
MOMENT CONVERGENCE VALUE = 120.8 CRITERION= 0.3718
DISP CONVERGENCE VALUE = 0.4238 CRITERION= 2.168 <<< CONVERGED
EQUIL ITER 4 COMPLETED. NEW TRIANG MATRIX. MAX DOF INC= -0.4081
FORCE CONVERGENCE VALUE = 239.8 CRITERION= 0.8314
MOMENT CONVERGENCE VALUE = 4.287 CRITERION= 0.1883
DISP CONVERGENCE VALUE = 0.8452E-01 CRITERION= 2.213 <<< CONVERGED
EQUIL ITER 5 COMPLETED. NEW TRIANG MATRIX. MAX DOF INC= 0.7990E-01
FORCE CONVERGENCE VALUE = 4.454 CRITERION= 0.8178
MOMENT CONVERGENCE VALUE = 2.404 CRITERION= 0.1852
DISP CONVERGENCE VALUE = 0.1539E-01 CRITERION= 2.258 <<< CONVERGED
EQUIL ITER 6 COMPLETED. NEW TRIANG MATRIX. MAX DOF INC= -0.1509E-01
FORCE CONVERGENCE VALUE = 0.4133 CRITERION= 0.8262 <<< CONVERGED
MOMENT CONVERGENCE VALUE = 0.2062E-01 CRITERION= 0.1871 <<< CONVERGED
>>> SOLUTION CONVERGED AFTER EQUILIBRIUM ITERATION 6
[7] **Force Convergence** is a graphics plotting of force convergence values and criteria.
Force Convergence
Force Criterion
Force (N)
12477
565.15
71.814
25.599
3.2529
0.41335
1. 2. 3. 4. 5. 6.
[8] **Moment Convergence** is a graphics plotting of moment convergence values and criteria. Note that **Moment Convergence** is available only for the model with rotational degrees of freedom (i.e., shell or beam elements).
Moment
Moment Criterion
Moment (N·mm)
23539
358.55
22.032
5.4614
0.33559
2.0621e-2
1. 2. 3. 4. 5. 6.

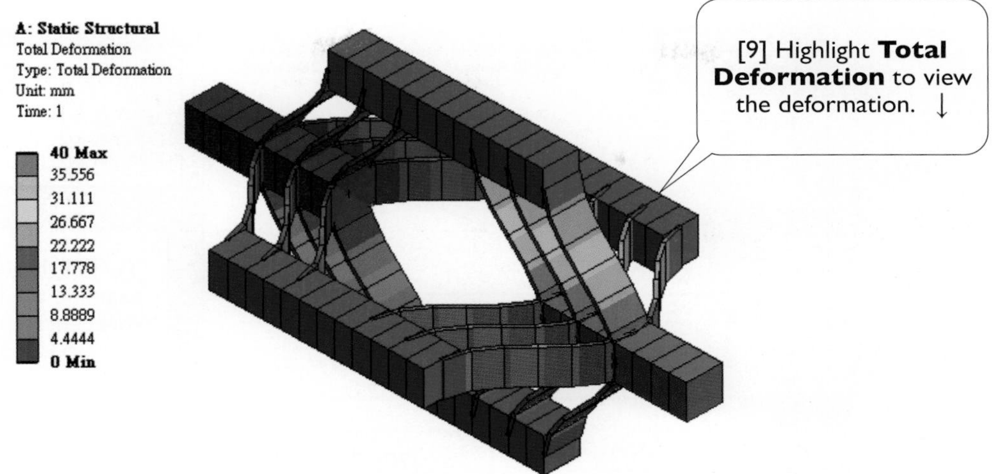

[10] Workbench solves this case easily in only one substep, and the substep takes 6 equilibrium iterations to converge (see [7-8], last page). To plot a force-versus-displacement chart, let's use a smaller initial substep value and let Workbench automatically adjust the substep values for the subsequent substeps. This feature is called **Auto Time Stepping**. ↓

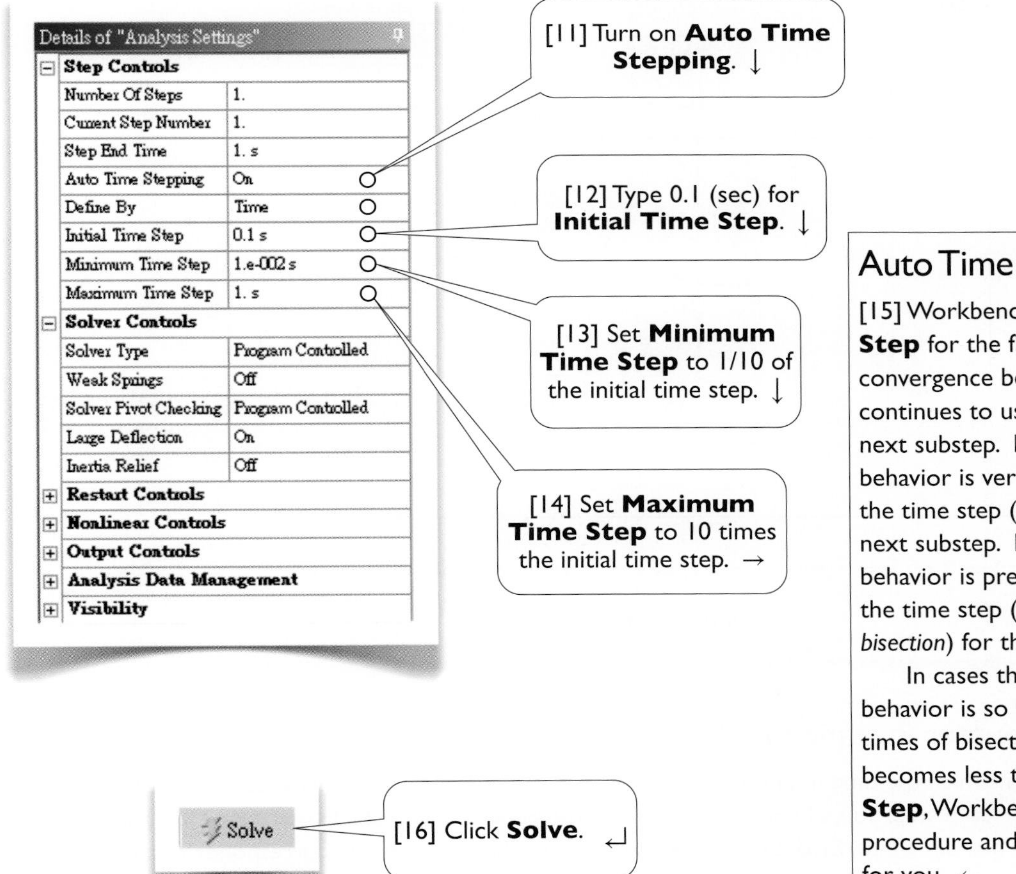

Auto Time Stepping

[15] Workbench uses **Initial Time Step** for the first substep. If the convergence behavior is merely okay, it continues to use that time step for the next substep. If the convergence behavior is very good, it may increase the time step (typically 50%) for the next substep. If the convergence behavior is pretty bad, it may decrease the time step (typically 50%; called *bisection*) for the next substep.

In cases that the convergence behavior is so bad that, after several times of bisections, the time step becomes less than **Minimum Time Step**, Workbench will stop the solution procedure and report an error message for you. ←

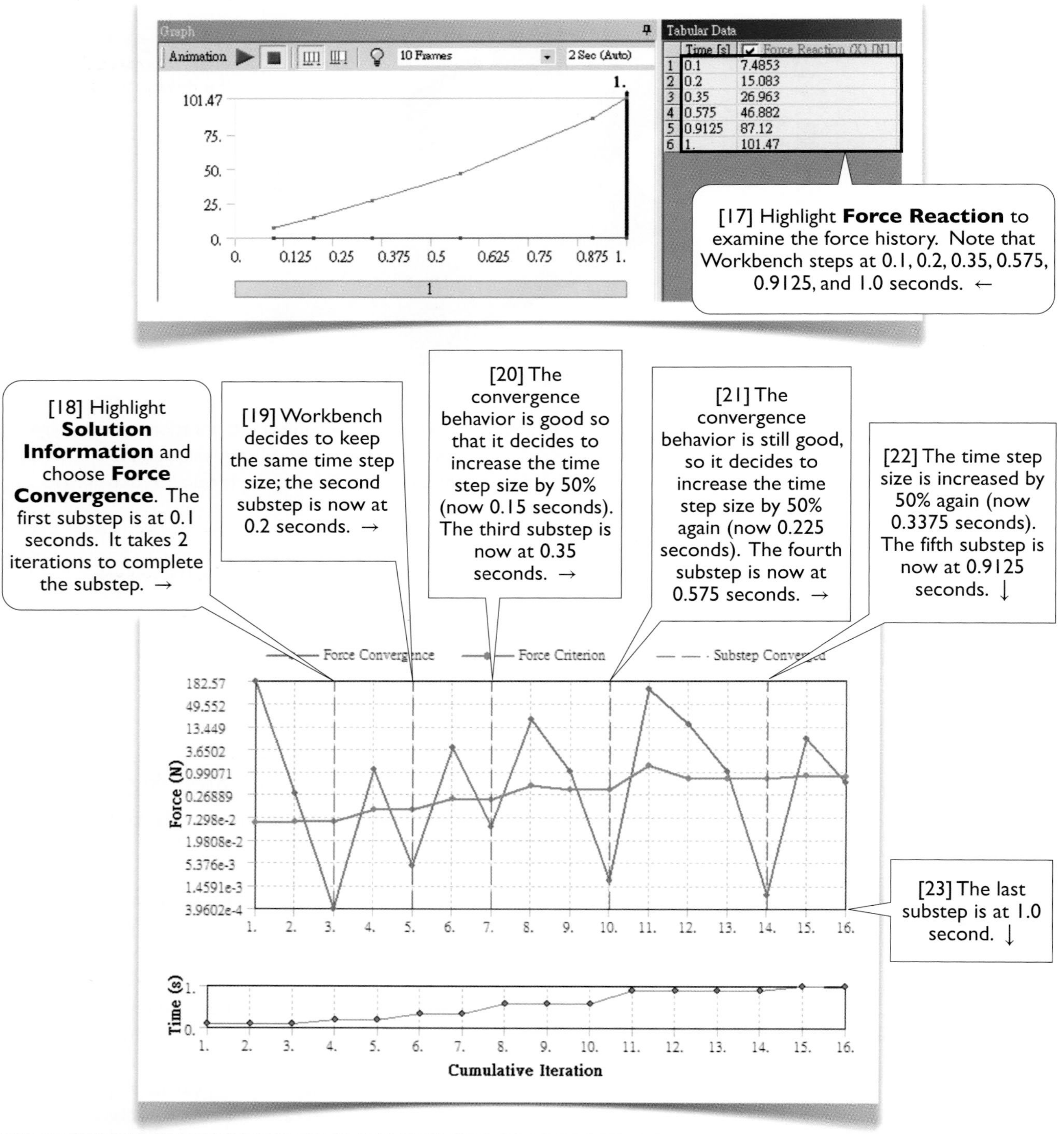

	Time [s]	Force Reaction (X) [N]
1	0.1	7.4853
2	0.2	15.083
3	0.35	26.963
4	0.575	46.882
5	0.9125	87.12
6	1.	101.47

Time is used as a counter in static simulations

[24] In a static simulation, in which the time has no real-world meaning, Workbench uses **Time** as a counter. By default, it set 1.0 second for each step. You, of course, can change that value as you wish. In our case, since the total displacement is 40 mm, we may set **Step End Time** to 40 seconds, so that each second corresponds to a millimeter of displacement. ↵

More data points; more uniform data points

[25] In the foregoing simulation, there are only 6 data points calculated ([17], last page), and they are distributed unevenly. To plot a more accurate force-versus-displacement relation, we need more time steps which preferably distribute evenly. Let's modify **Analysis Settings** further. ↓

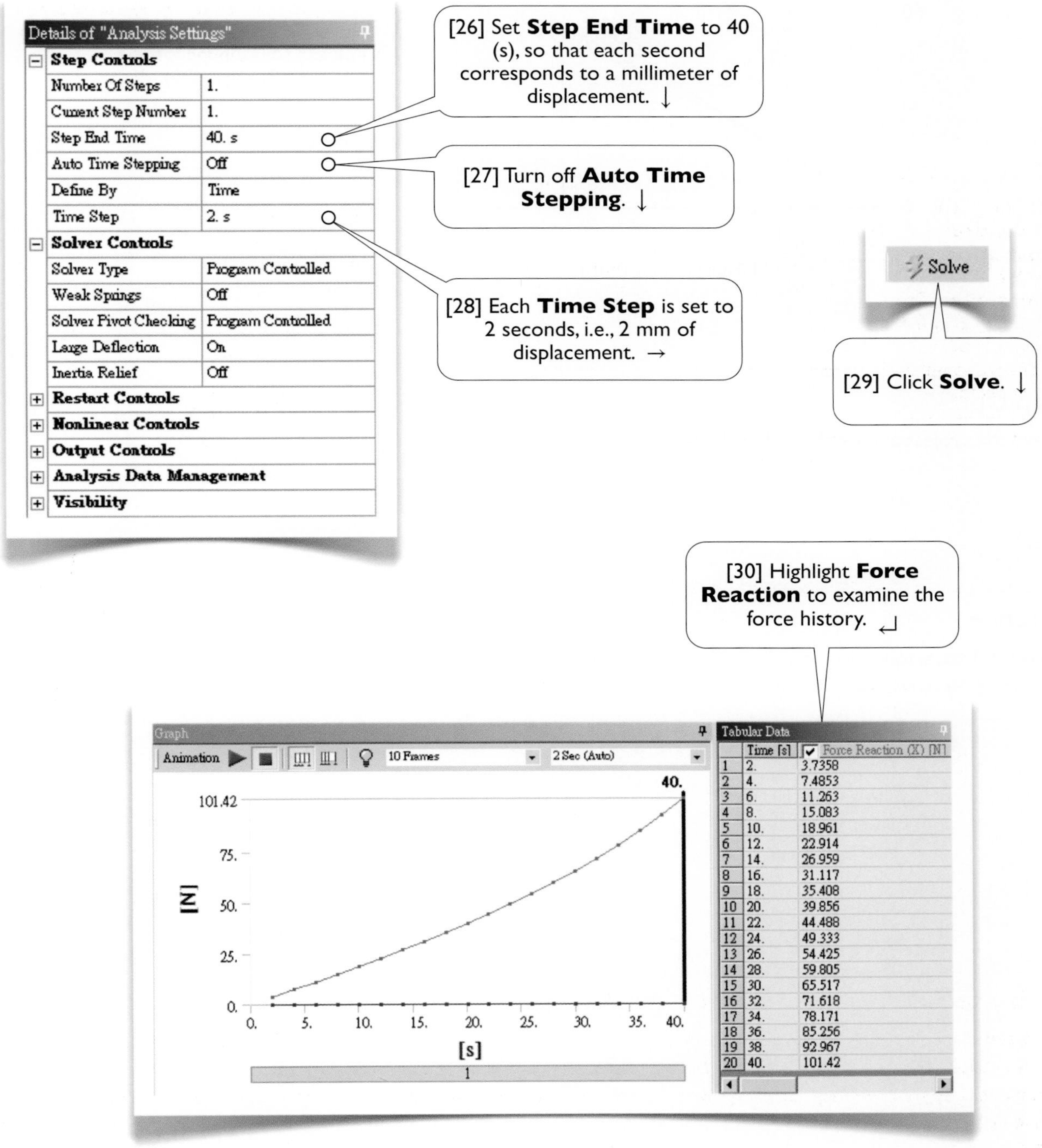

	Time [s]	Force Reaction (X) [N]
1	2.	3.7358
2	4.	7.4853
3	6.	11.263
4	8.	15.083
5	10.	18.961
6	12.	22.914
7	14.	26.959
8	16.	31.117
9	18.	35.408
10	20.	39.856
11	22.	44.488
12	24.	49.333
13	26.	54.425
14	28.	59.805
15	30.	65.517
16	32.	71.618
17	34.	78.171
18	36.	85.256
19	38.	92.967
20	40.	101.42

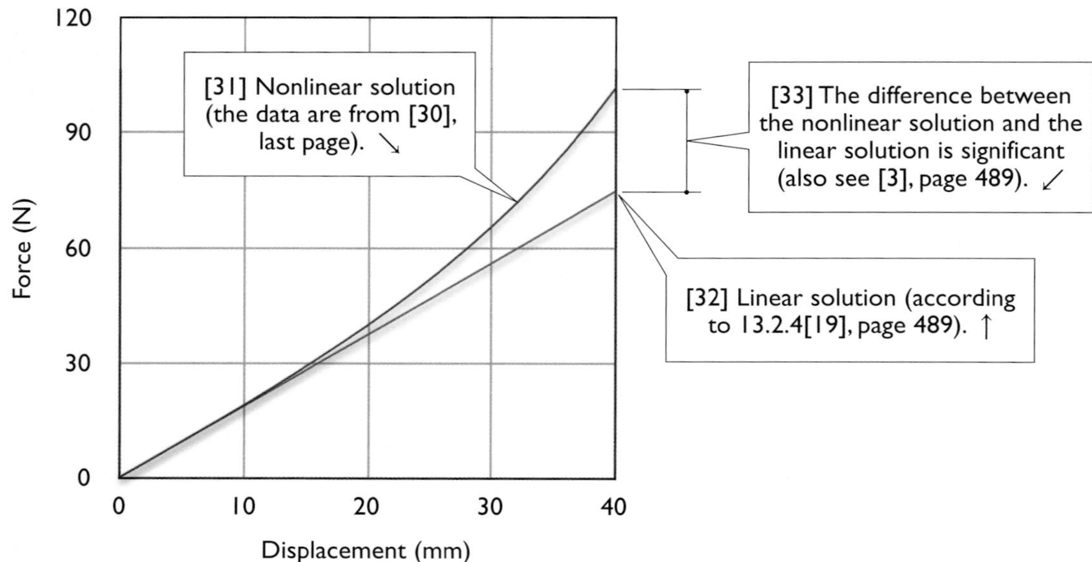

Wrap Up

[34] Save the project and exit Workbench. #

Reference

1. Brian P. Trease, Yong-Mo Moon, and Sridhar Kota, 2005, "Design of Large-Displacement Compliant Joints," *ASME Journal of Mechanical Design*, Vol. 127, pp. 788-798.

Section 13.3

Microgripper[Refs 1, 2]

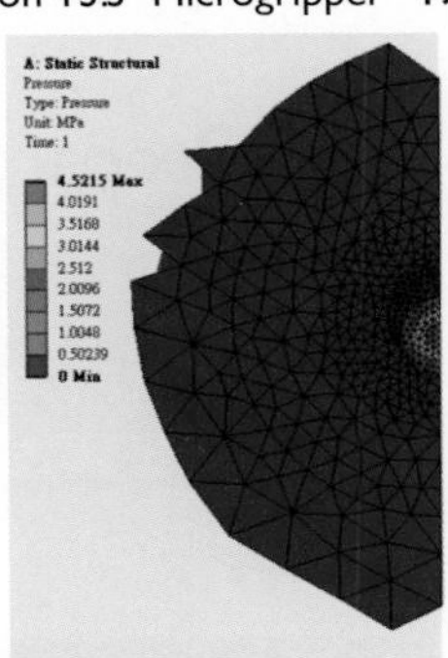

13.3.1 About the Microgripper

[1] In Section 2.6, we introduced the microgripper and created a solid model for it. The microgripper is made of a rubber-like polymer material and actuated by a shape memory alloy (SMA) actuator; it is tested by gripping a steel bead in a lab. In this section, we want to assess the gripping pressure on the bead under an actuation force of 40 μN exerted by the SMA. The polymer material has a Young's modulus of 200 MPa and a poisson's ration of 0.48. Since the steel bead is much more rigid than the polymer material, we will model the steel bead as a rigid body. This will ease some of the computation (convergence) difficulties. By considering the symmetries, we will model only one quarter of the microgripper. #

13.3.2 Resume the Project **Microgripper**

[1] Launch Workbench. Open the project **Microgripper**, which was saved in Section 2.6. ↓

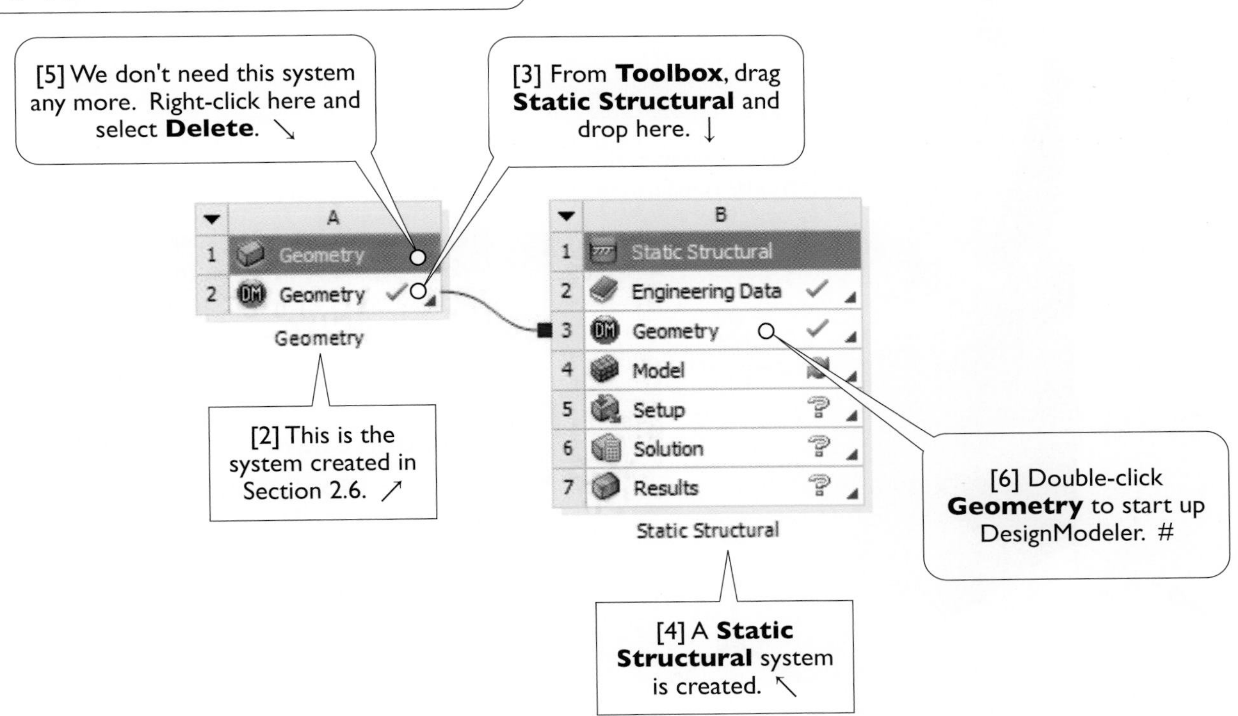

13.3.3 Prepare Geometry in DesignModeler

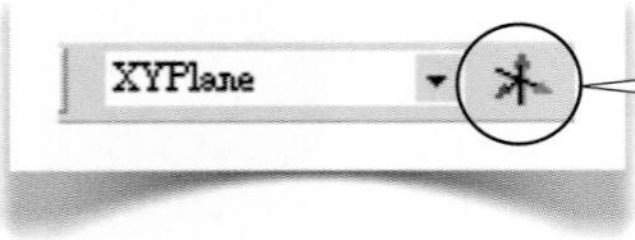

XYPlane

[1] Click **New Plane**. ↓

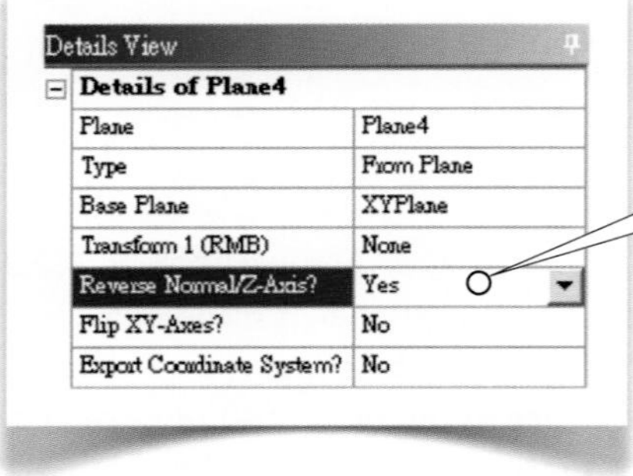

Details View	
Details of Plane4	
Plane	Plane4
Type	From Plane
Base Plane	XYPlane
Transform 1 (RMB)	None
Reverse Normal/Z-Axis?	Yes
Flip XY-Axes?	No
Export Coordinate System?	No

[2] The new plane is simply a copy of **XYPlane** with its Z-axis reversed. Click **Generate**. →

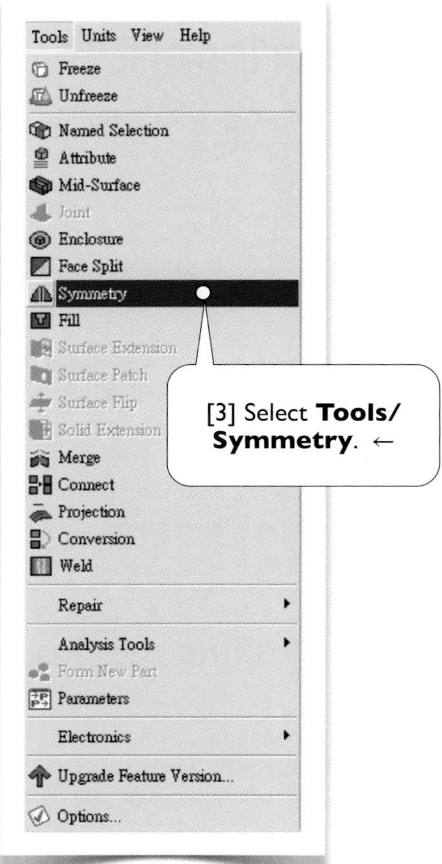

[3] Select **Tools/ Symmetry**. ←

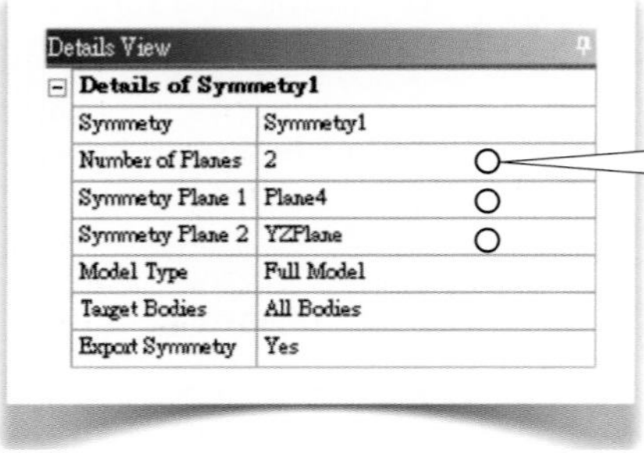

Details View	
Details of Symmetry1	
Symmetry	Symmetry1
Number of Planes	2
Symmetry Plane 1	Plane4
Symmetry Plane 2	YZPlane
Model Type	Full Model
Target Bodies	All Bodies
Export Symmetry	Yes

[4] Select **2** for **Number of Planes**. Select **Plane4** and **YZPlane** as the planes of symmetry. Click **Generate**. ↓

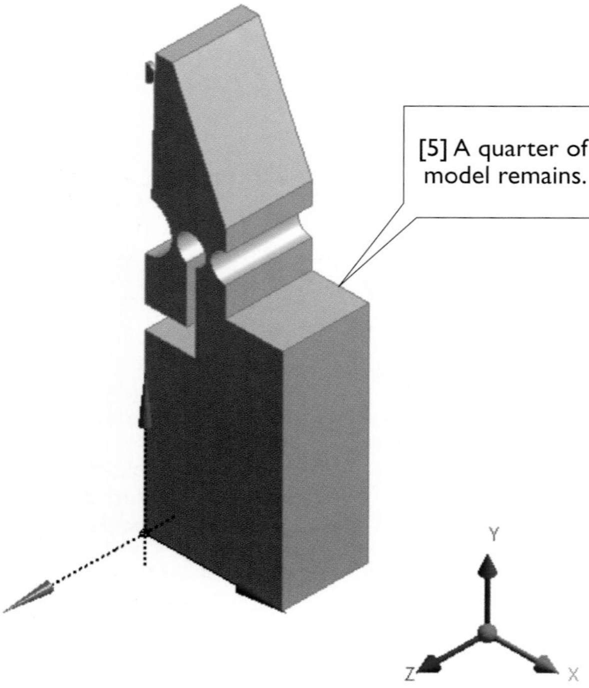

[5] A quarter of the model remains. →

Y

Z X

Why use Plane4 instead of XYPlane?

[6] Because we want to keep the half of the model behind **XYPlane**, so that it has better visual effect. When you specify a plane of symmetry, DesignModeler keeps the portion that is in the +Z side of the plane coordinate system.

You don't have to build a full model

To create a **Tools/Symmetry** feature, you don't have to build a full model and then slice it to a half. You may just build a half model and use **Tools/Symmetry** to specify a plane of symmetry. The information that Workbench needs is the planes of symmetry, not the other half of the model. ↵

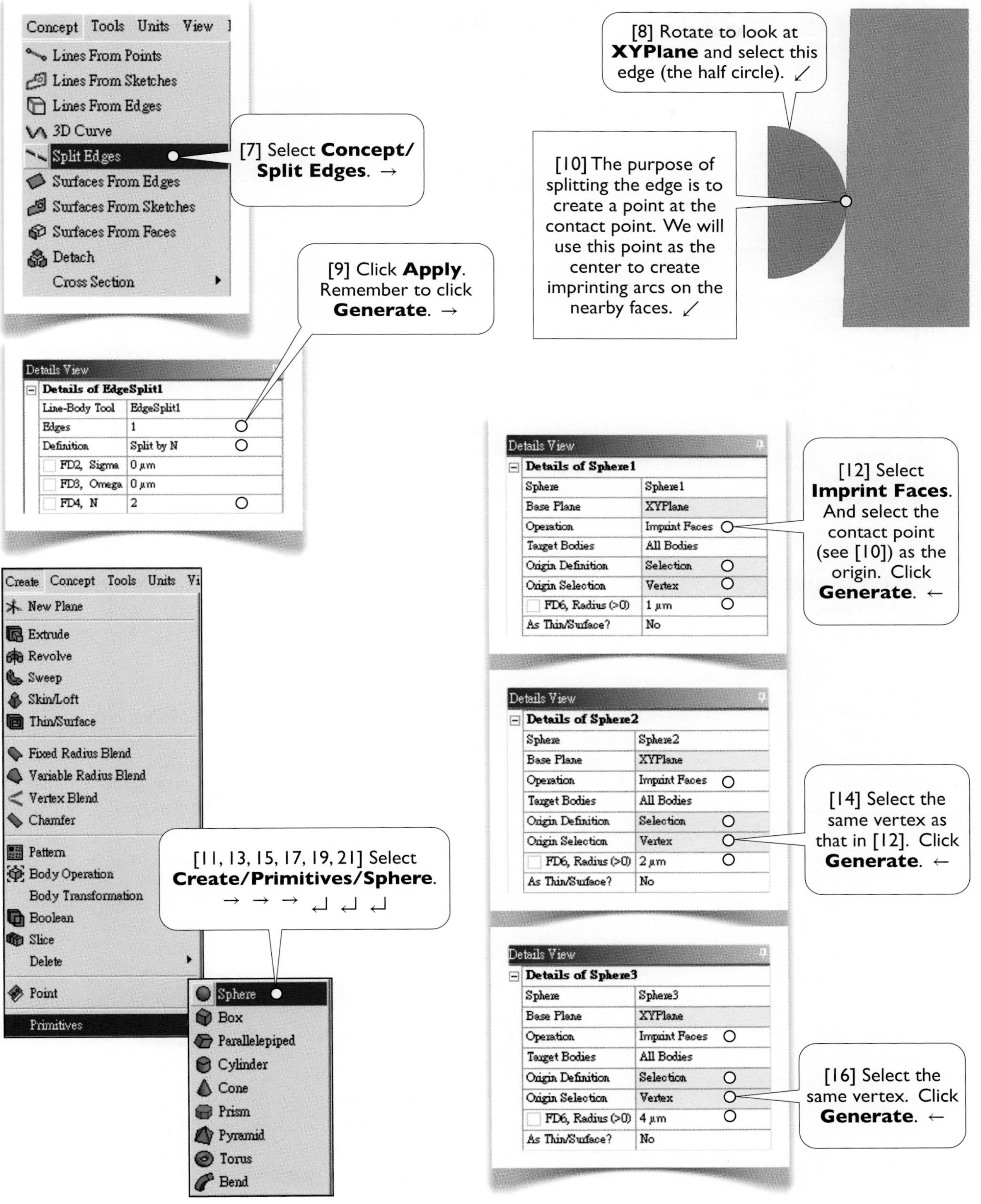

Concept Tools Units View
Lines From Points
Lines From Sketches
Lines From Edges
3D Curve
Split Edges
Surfaces From Edges
Surfaces From Sketches
Surfaces From Faces
Detach
Cross Section
[7] Select **Concept/ Split Edges**. →
[8] Rotate to look at **XYPlane** and select this edge (the half circle).
[10] The purpose of splitting the edge is to create a point at the contact point. We will use this point as the center to create imprinting arcs on the nearby faces.
[9] Click **Apply**. Remember to click **Generate**. →
Details View
Details of EdgeSplit1
Line-Body Tool EdgeSplit1
Edges 1
Definition Split by N
FD2, Sigma 0 µm
FD3, Omega 0 µm
FD4, N 2
Details View
Details of Sphere1
Sphere Sphere1
Base Plane XYPlane
Operation Imprint Faces
Target Bodies All Bodies
Origin Definition Selection
Origin Selection Vertex
FD6, Radius (>0) 1 µm
As Thin/Surface? No
[12] Select **Imprint Faces**. And select the contact point (see [10]) as the origin. Click **Generate**. ←
Create Concept Tools Units
New Plane
Extrude
Revolve
Sweep
Skin/Loft
Thin/Surface
Fixed Radius Blend
Variable Radius Blend
Vertex Blend
Chamfer
Pattern
Body Operation
Body Transformation
Boolean
Slice
Delete
Point
Primitives
[11, 13, 15, 17, 19, 21] Select **Create/Primitives/Sphere**.
Sphere
Box
Parallelepiped
Cylinder
Cone
Prism
Pyramid
Torus
Bend
Details View
Details of Sphere2
Sphere Sphere2
Base Plane XYPlane
Operation Imprint Faces
Target Bodies All Bodies
Origin Definition Selection
Origin Selection Vertex
FD6, Radius (>0) 2 µm
As Thin/Surface? No
[14] Select the same vertex as that in [12]. Click **Generate**. ←
Details View
Details of Sphere3
Sphere Sphere3
Base Plane XYPlane
Operation Imprint Faces
Target Bodies All Bodies
Origin Definition Selection
Origin Selection Vertex
FD6, Radius (>0) 4 µm
As Thin/Surface? No
[16] Select the same vertex. Click **Generate**. ←

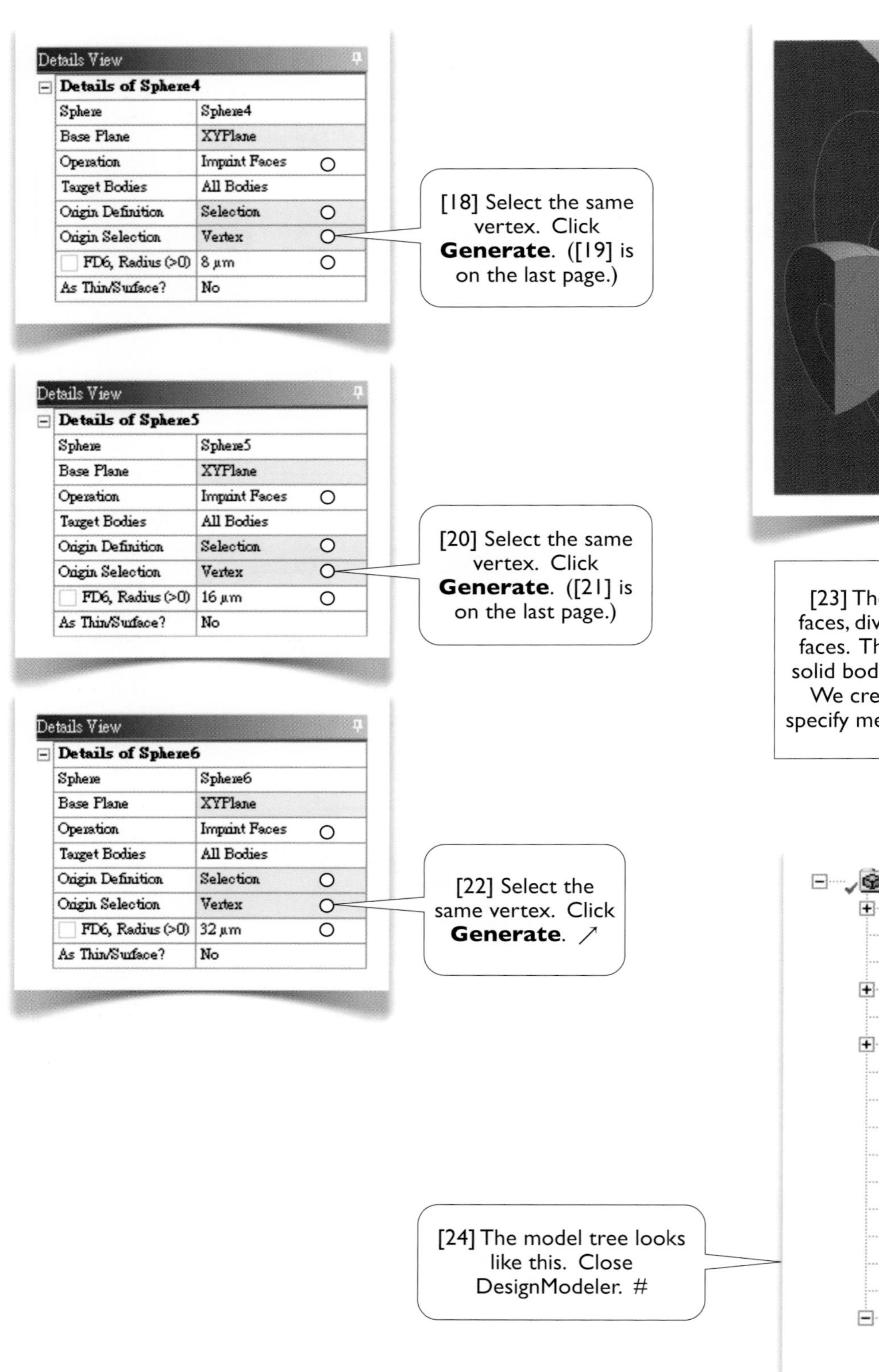

[18] Select the same vertex. Click **Generate**. ([19] is on the last page.)

[20] Select the same vertex. Click **Generate**. ([21] is on the last page.)

[22] Select the same vertex. Click **Generate**. ↗

[23] These spheres **imprint** the faces, dividing the faces into several faces. The spheres don't "slice" the solid bodies; they slice the faces only. We create these faces so we can specify mesh density for each face. ↙

[24] The model tree looks like this. Close DesignModeler. #

13.3.4 Set Up Material in **Engineering Data**

Model the Bead as a Rigid Body

[1] The Young's modulus of steel (200 GPa) is 1,000 times that of the polymer material (200 MPa). Whenever two bodies are so different in rigidity, you should model the more rigid one as a rigid body. This not only reduces the problem size (rigid bodies won't deform, they don't need to be meshed), but also eases numerical difficulties during finite element simulations.

A rigid bodies won't deform; therefore, its Young's modulus and Poisson's ratio are not relevant. For a static simulation, we can assign any material to a rigid body. For dynamic simulation, the mass density should be specified. ↗

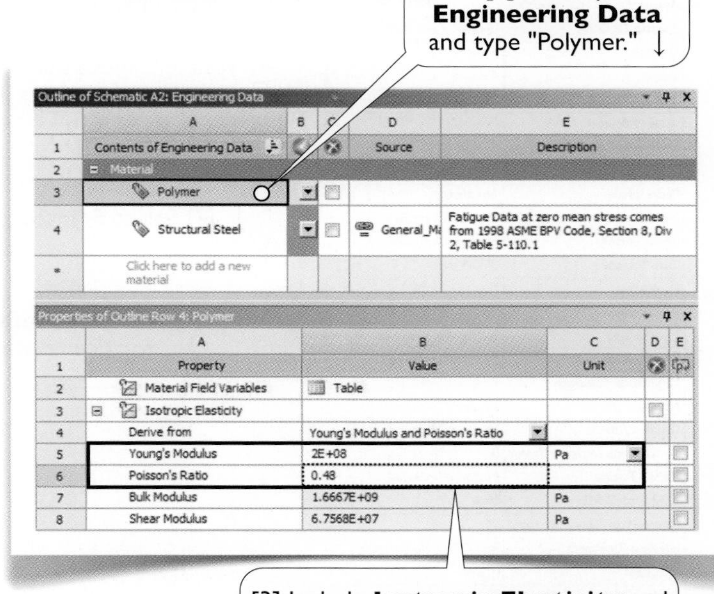

13.3.5 Set Up Model

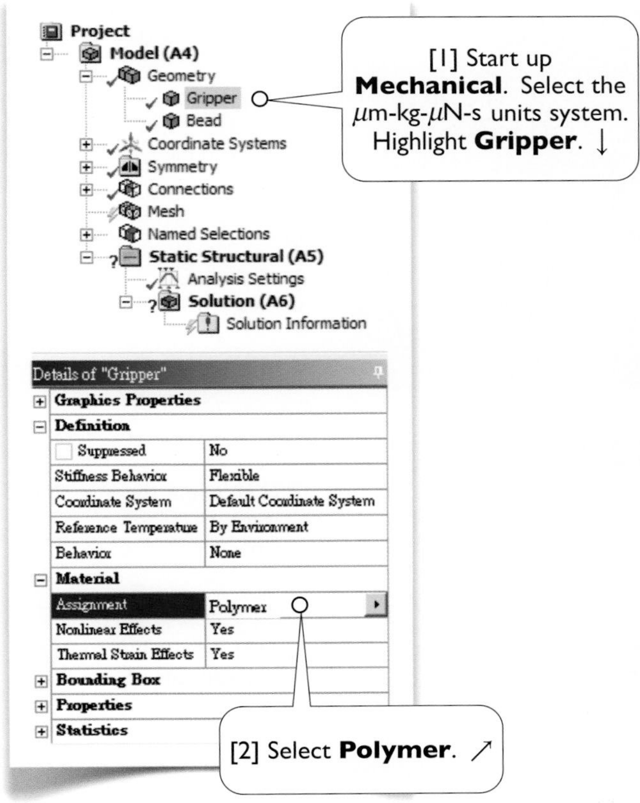

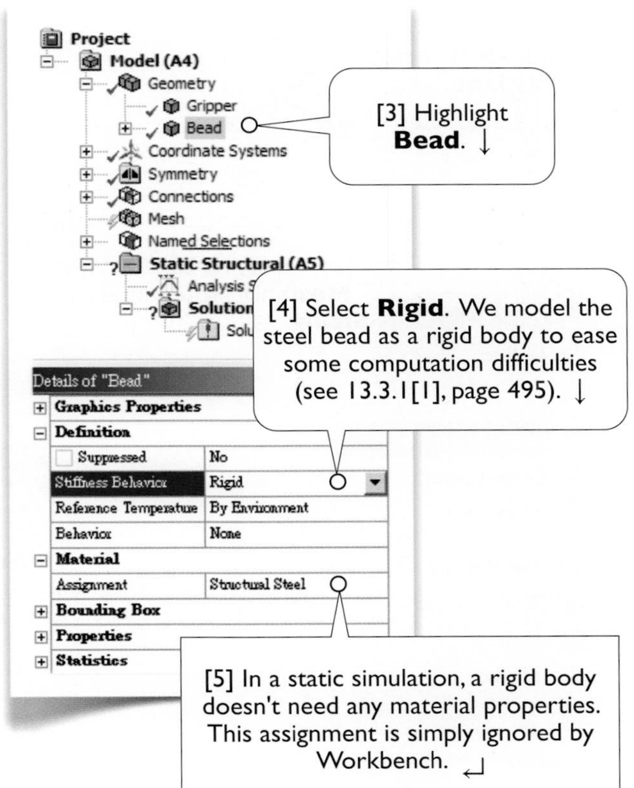

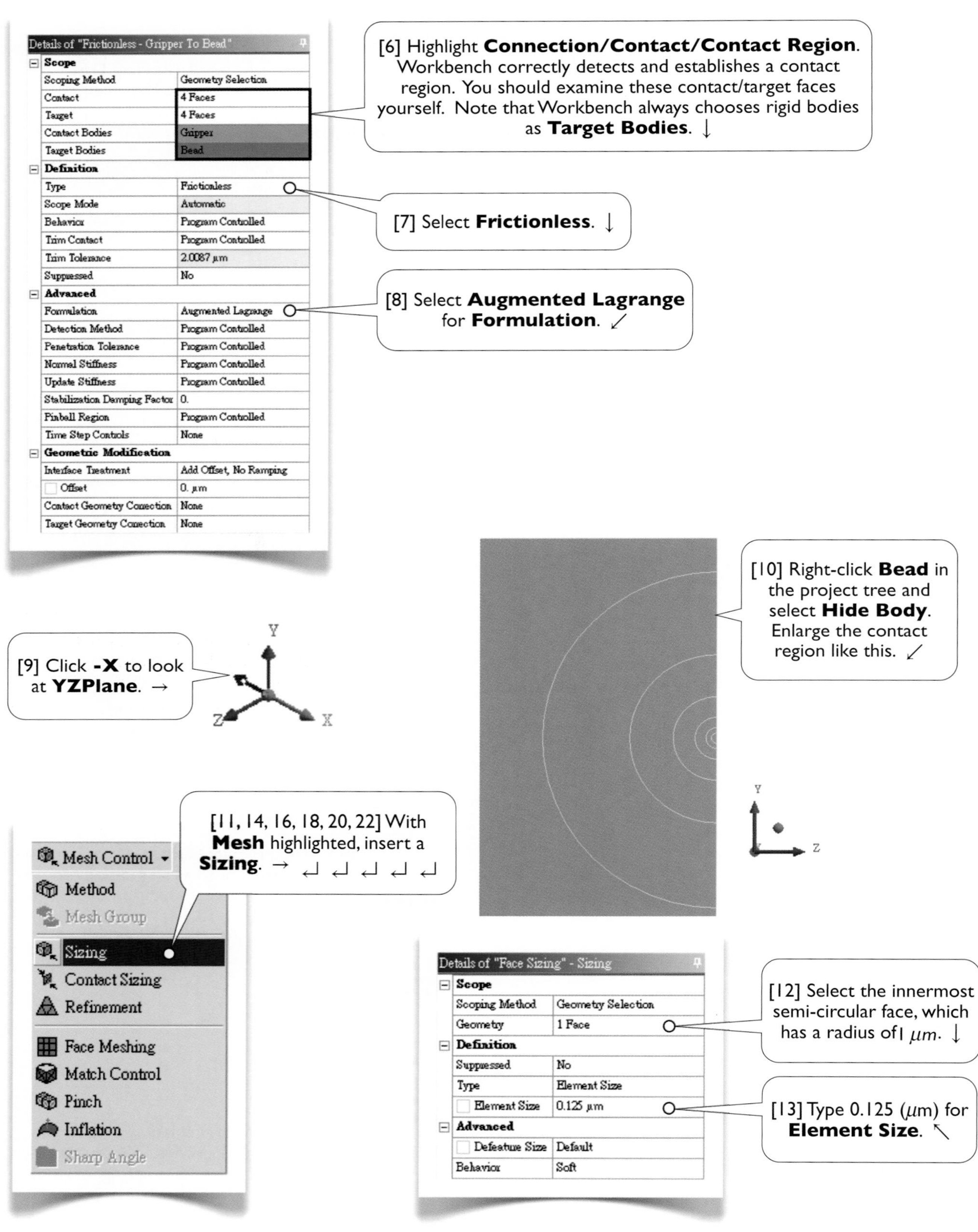

Details of "Frictionless - Gripper To Bead"
Scope
Scoping Method | Geometry Selection
Contact | 4 Faces
Target | 4 Faces
Contact Bodies | Gripper
Target Bodies | Bead
Definition
Type | Frictionless
Scope Mode | Automatic
Behavior | Program Controlled
Trim Contact | Program Controlled
Trim Tolerance | 2.0087 µm
Suppressed | No
Advanced
Formulation | Augmented Lagrange
Detection Method | Program Controlled
Penetration Tolerance | Program Controlled
Normal Stiffness | Program Controlled
Update Stiffness | Program Controlled
Stabilization Damping Factor | 0.
Pinball Region | Program Controlled
Time Step Controls | None
Geometric Modification
Interface Treatment | Add Offset, No Ramping
Offset | 0. µm
Contact Geometry Correction | None
Target Geometry Correction | None
[6] Highlight **Connection/Contact/Contact Region**. Workbench correctly detects and establishes a contact region. You should examine these contact/target faces yourself. Note that Workbench always chooses rigid bodies as **Target Bodies**. ↓
[7] Select **Frictionless**. ↓
[8] Select **Augmented Lagrange** for **Formulation**. ↙
[9] Click **-X** to look at **YZPlane**. →
Y
Z
X
[10] Right-click **Bead** in the project tree and select **Hide Body**. Enlarge the contact region like this. ↙
Y
Z
Mesh Control
Method
Mesh Group
Sizing
Contact Sizing
Refinement
Face Meshing
Match Control
Pinch
Inflation
Sharp Angle
[11, 14, 16, 18, 20, 22] With **Mesh** highlighted, insert a **Sizing**. → ↵ ↵ ↵ ↵ ↵
Details of "Face Sizing" - Sizing
Scope
Scoping Method | Geometry Selection
Geometry | 1 Face
Definition
Suppressed | No
Type | Element Size
Element Size | 0.125 µm
Advanced
Defeature Size | Default
Behavior | Soft
[12] Select the innermost semi-circular face, which has a radius of 1 μm. ↓
[13] Type 0.125 (μm) for **Element Size**. ↖

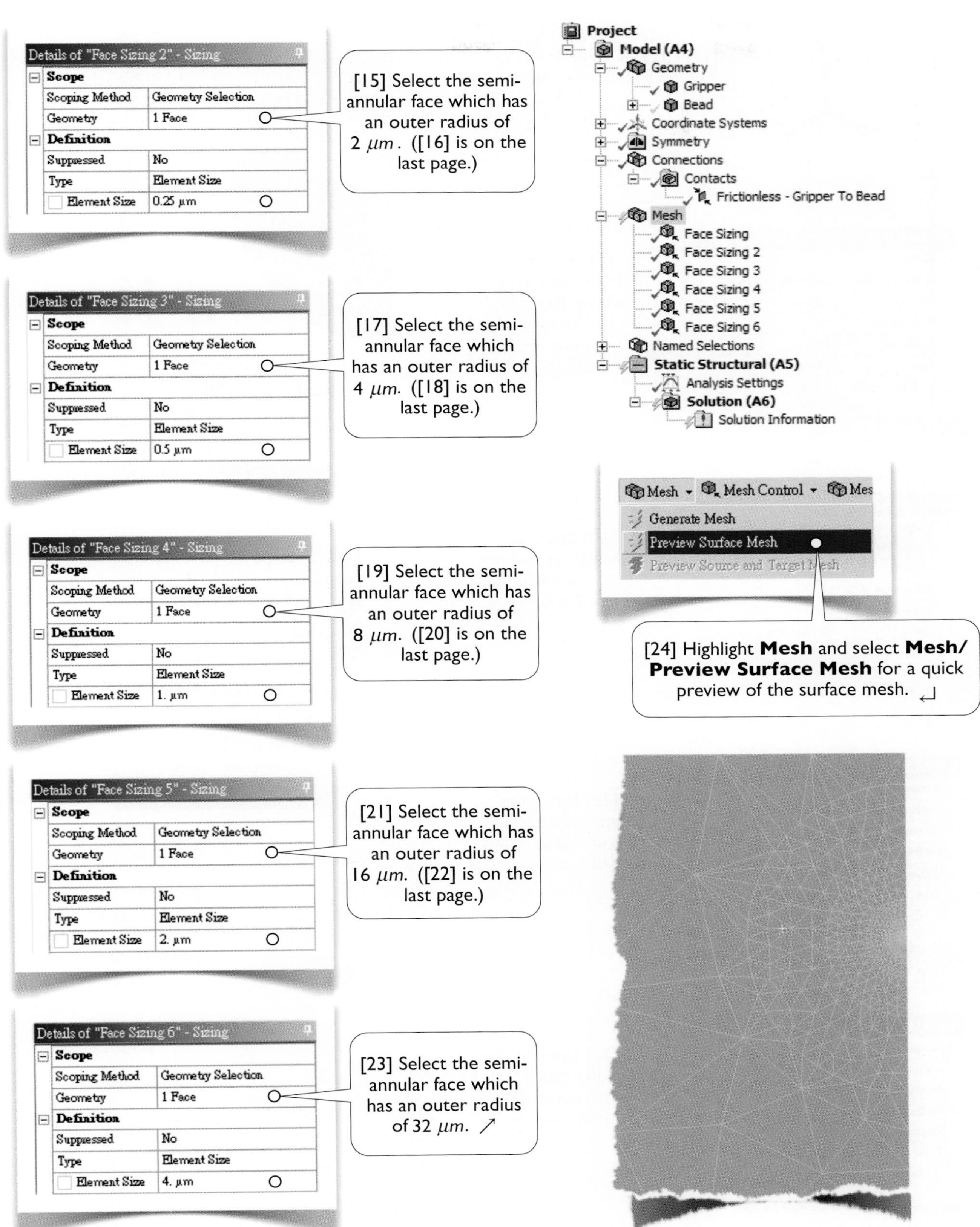

Details of "Face Sizing 2" - Sizing
Scope
Scoping Method | Geometry Selection
Geometry | 1 Face
Definition
Suppressed | No
Type | Element Size
Element Size | 0.25 µm
[15] Select the semi-annular face which has an outer radius of 2 µm. ([16] is on the last page.)
Details of "Face Sizing 3" - Sizing
Scope
Scoping Method | Geometry Selection
Geometry | 1 Face
Definition
Suppressed | No
Type | Element Size
Element Size | 0.5 µm
[17] Select the semi-annular face which has an outer radius of 4 µm. ([18] is on the last page.)
Details of "Face Sizing 4" - Sizing
Scope
Scoping Method | Geometry Selection
Geometry | 1 Face
Definition
Suppressed | No
Type | Element Size
Element Size | 1. µm
[19] Select the semi-annular face which has an outer radius of 8 µm. ([20] is on the last page.)
Details of "Face Sizing 5" - Sizing
Scope
Scoping Method | Geometry Selection
Geometry | 1 Face
Definition
Suppressed | No
Type | Element Size
Element Size | 2. µm
[21] Select the semi-annular face which has an outer radius of 16 µm. ([22] is on the last page.)
Details of "Face Sizing 6" - Sizing
Scope
Scoping Method | Geometry Selection
Geometry | 1 Face
Definition
Suppressed | No
Type | Element Size
Element Size | 4. µm
[23] Select the semi-annular face which has an outer radius of 32 µm.
Project
Model (A4)
Geometry
Gripper
Bead
Coordinate Systems
Symmetry
Connections
Contacts
Frictionless - Gripper To Bead
Mesh
Face Sizing
Face Sizing 2
Face Sizing 3
Face Sizing 4
Face Sizing 5
Face Sizing 6
Named Selections
Static Structural (A5)
Analysis Settings
Solution (A6)
Solution Information
Mesh
Mesh Control
Generate Mesh
Preview Surface Mesh
Preview Source and Target Mesh
[24] Highlight Mesh and select Mesh/Preview Surface Mesh for a quick preview of the surface mesh.

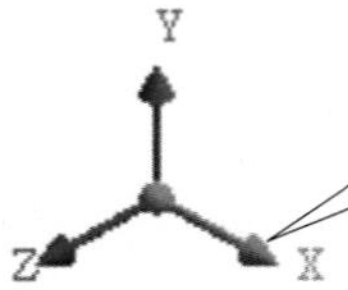

[25] Click **X** to look at **YZPlane**. →

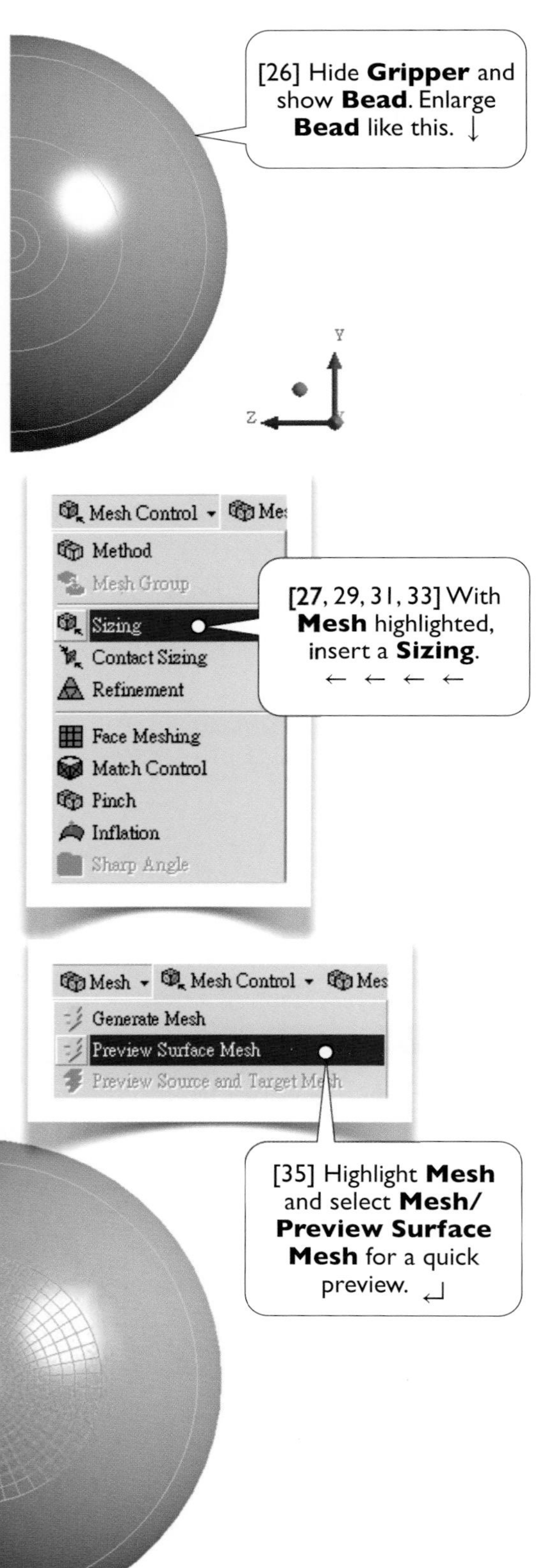

Details of "Face Sizing 7" - Sizing

Scope	
Scoping Method	Geometry Selection
Geometry	1 Face
Definition	
Suppressed	No
Type	Element Size
Element Size	0.125 µm

[28] Select the innermost semi-circular face, which has a radius of 1 μm. →

Details of "Face Sizing 8" - Sizing

Scope	
Scoping Method	Geometry Selection
Geometry	1 Face
Definition	
Suppressed	No
Type	Element Size
Element Size	0.25 µm

[30] Select the annular face which has an outer radius of 2 μm. →

Details of "Face Sizing 9" - Sizing

Scope	
Scoping Method	Geometry Selection
Geometry	1 Face
Definition	
Suppressed	No
Type	Element Size
Element Size	0.5 µm

[32] Select the annular face which has an outer radius of 4 μm. →

Details of "Face Sizing 10" - Sizing

Scope	
Scoping Method	Geometry Selection
Geometry	1 Face
Definition	
Suppressed	No
Type	Element Size
Element Size	1. µm

[34] Select the annular face which has an outer radius of 8 μm. →

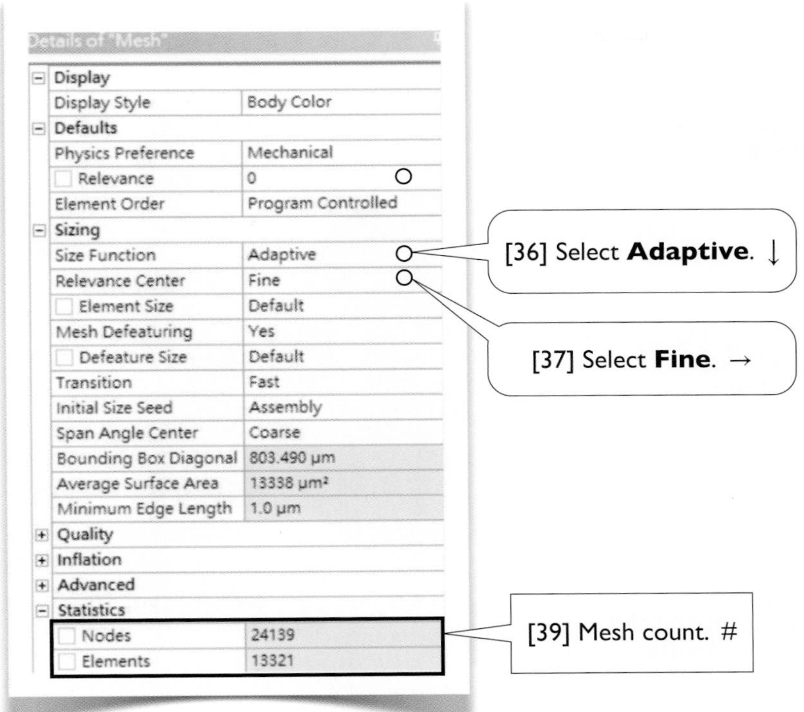

Details of "Mesh"

Display	
Display Style	Body Color
Defaults	
Physics Preference	Mechanical
Relevance	0
Element Order	Program Controlled
Sizing	
Size Function	Adaptive
Relevance Center	Fine
Element Size	Default
Mesh Defeaturing	Yes
Defeature Size	Default
Transition	Fast
Initial Size Seed	Assembly
Span Angle Center	Coarse
Bounding Box Diagonal	803.490 µm
Average Surface Area	13338 µm²
Minimum Edge Length	1.0 µm
Quality	
Inflation	
Advanced	
Statistics	
Nodes	24139
Elements	13321

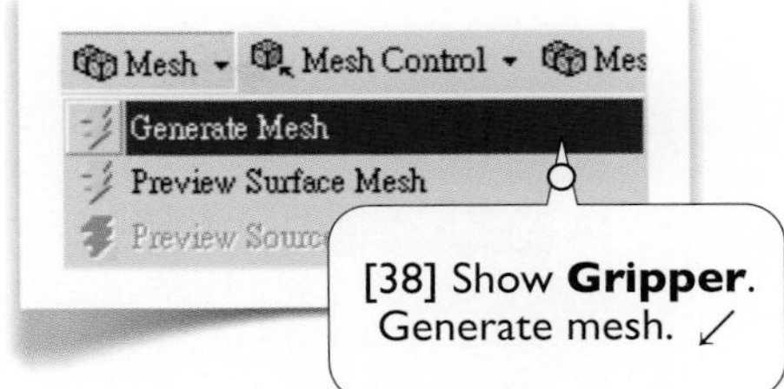

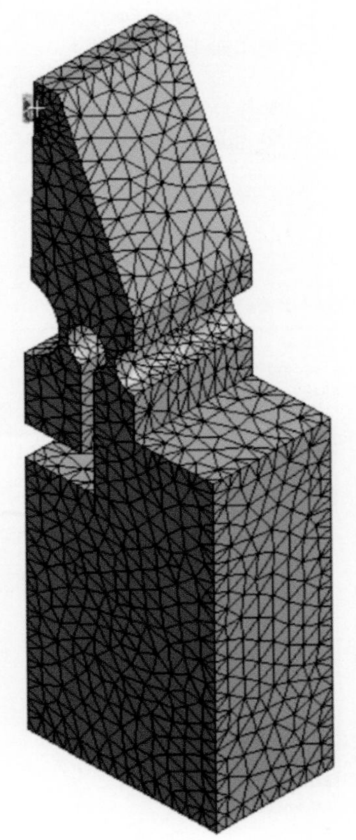

13.3.6 Set Up Environment Conditions

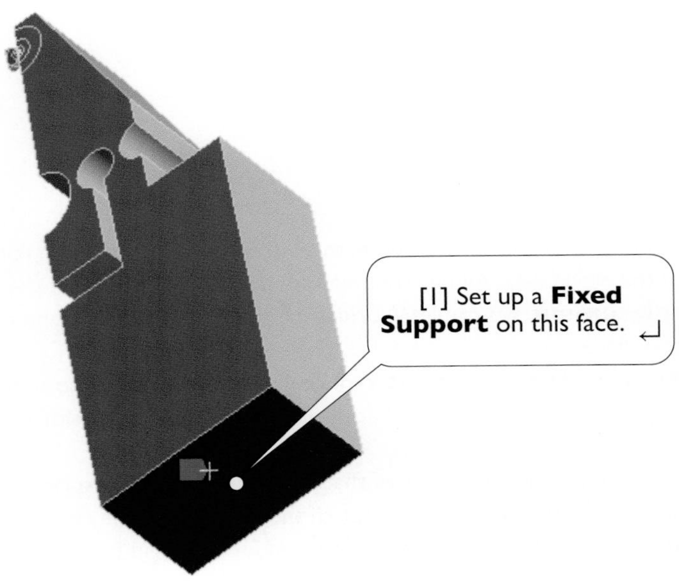

Details of "Fixed Support"

Scope	
Scoping Method	Geometry Selection
Geometry	1 Face
Definition	
Type	Fixed Support
Suppressed	No

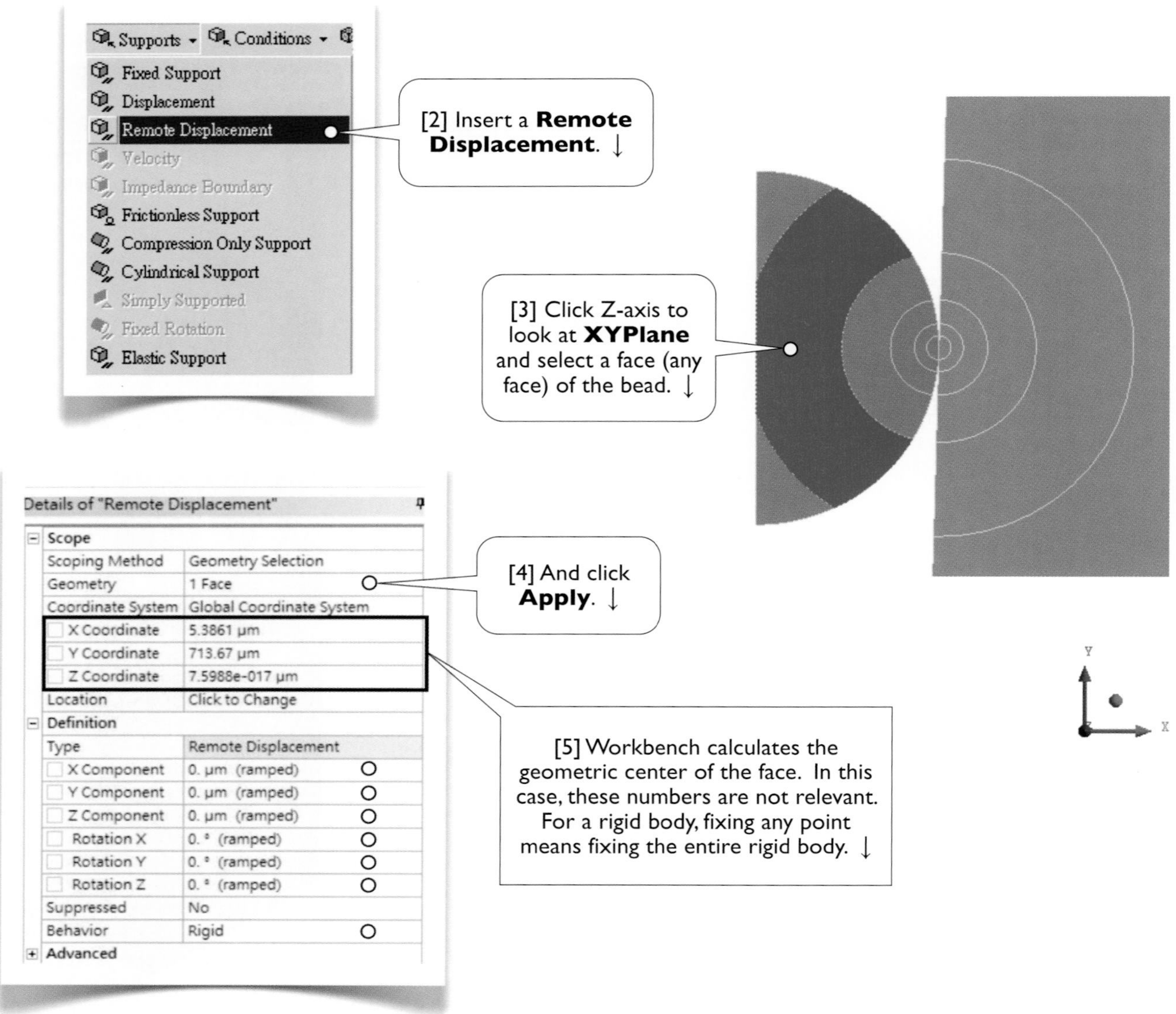

Remote Displacement

[6] A rigid body is internally represented as a node of 6 degrees of freedom, three translational and three rotational. The node is located at the geometric center of the rigid body. All environment conditions are applied at the geometric center of the rigid body. Using **Remote Displacement**, we can let Workbench transform the environment conditions to the node.

When we select a face [3-4] and apply a **Remote Displacement**, Workbench calculates the geometry center for that face, and the specified displacement is applied at that location. That displacement is then transformed to the geometry center of the rigid body.

In our case, we arbitrarily select a face [3-4]. The geometric center of that face will have zero displacements and zero rotations. These conditions will be transformed to the geometry center of the rigid body. The result is that the entire rigid body is fixed, both translationally and rotationally. That's what we intend for this rigid body.

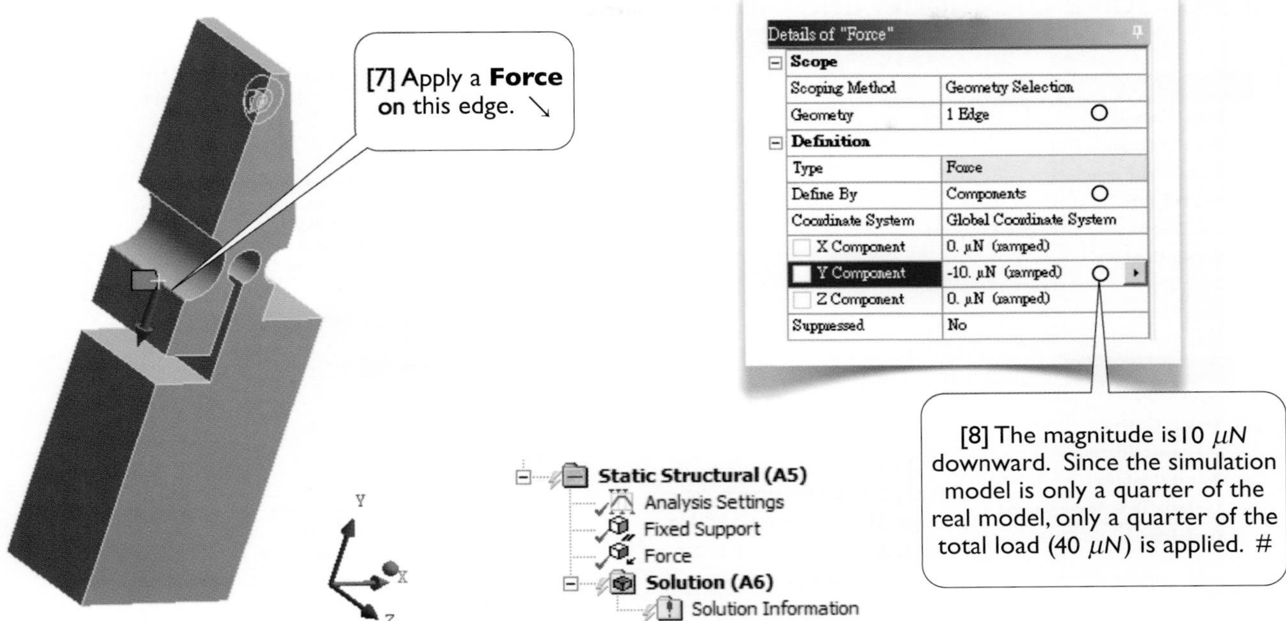

13.3.7 Set Up Result Objects

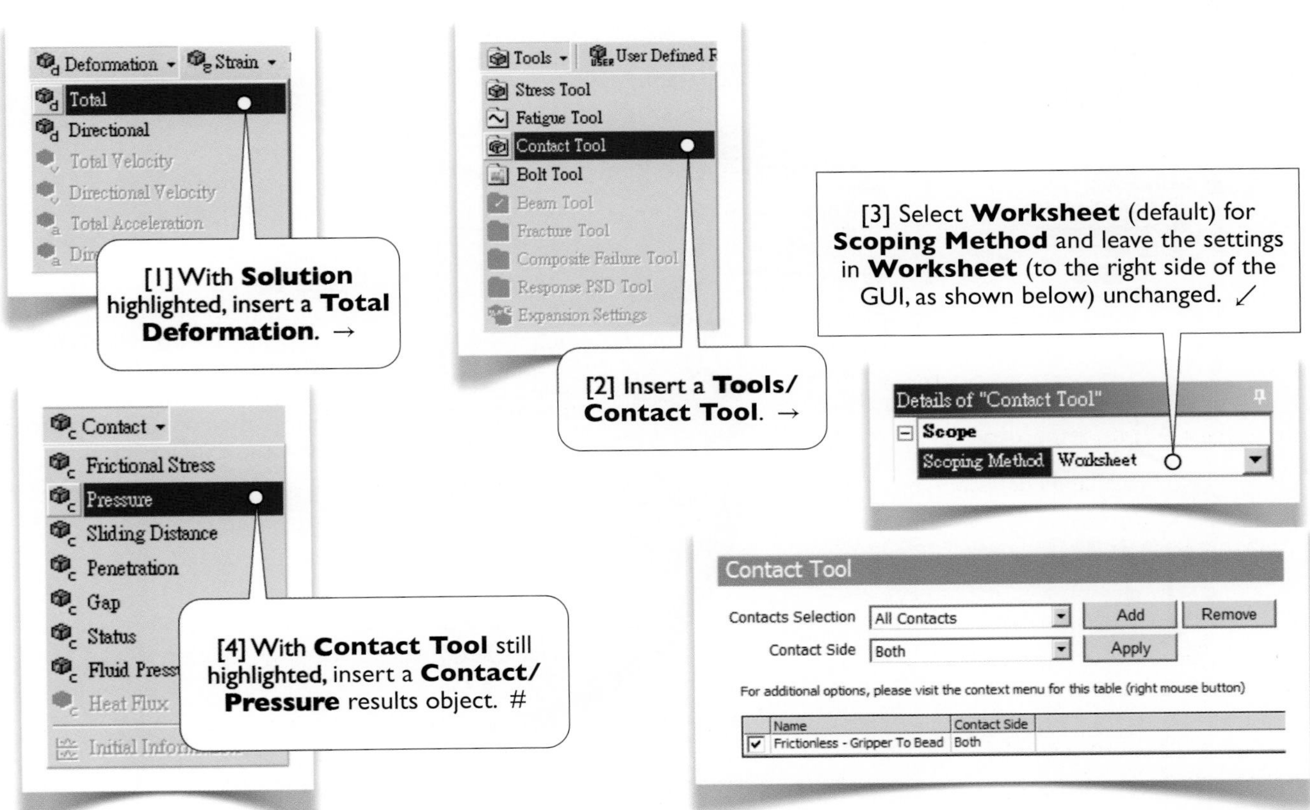

13.3.8 Solve the Model and View the Results

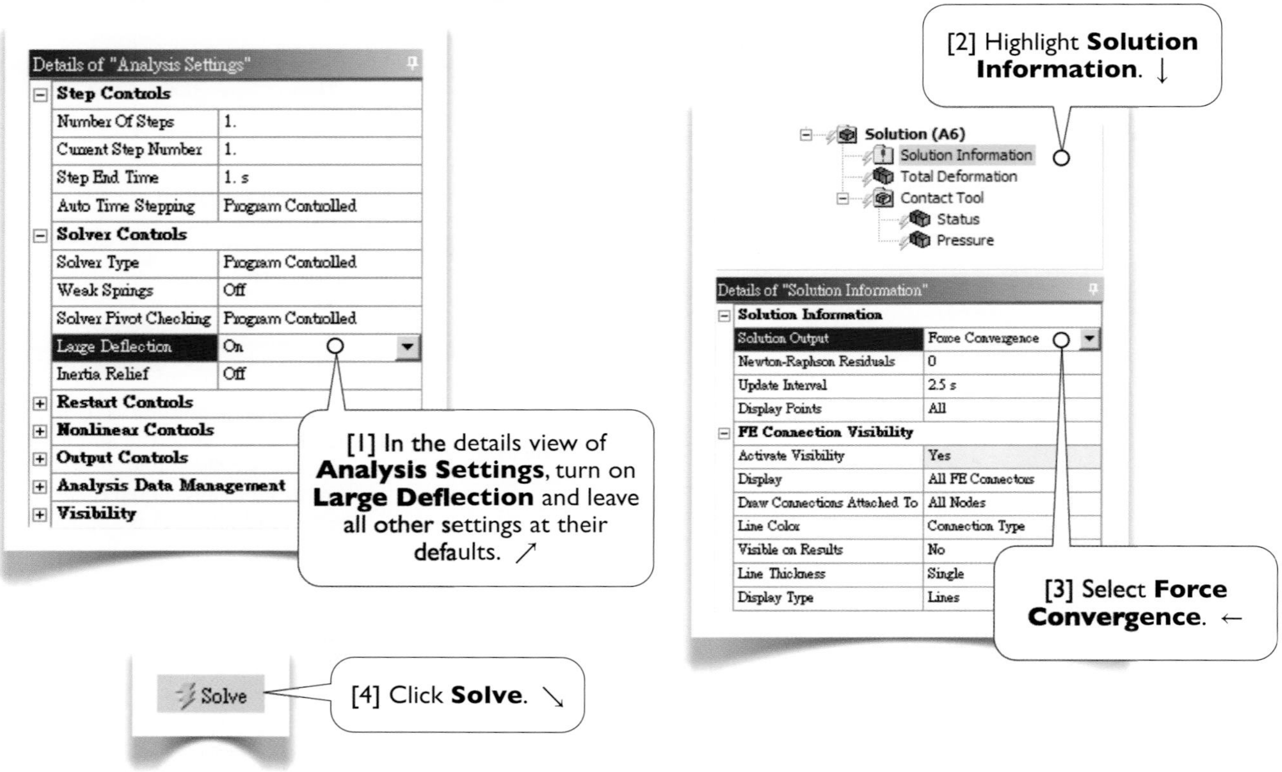

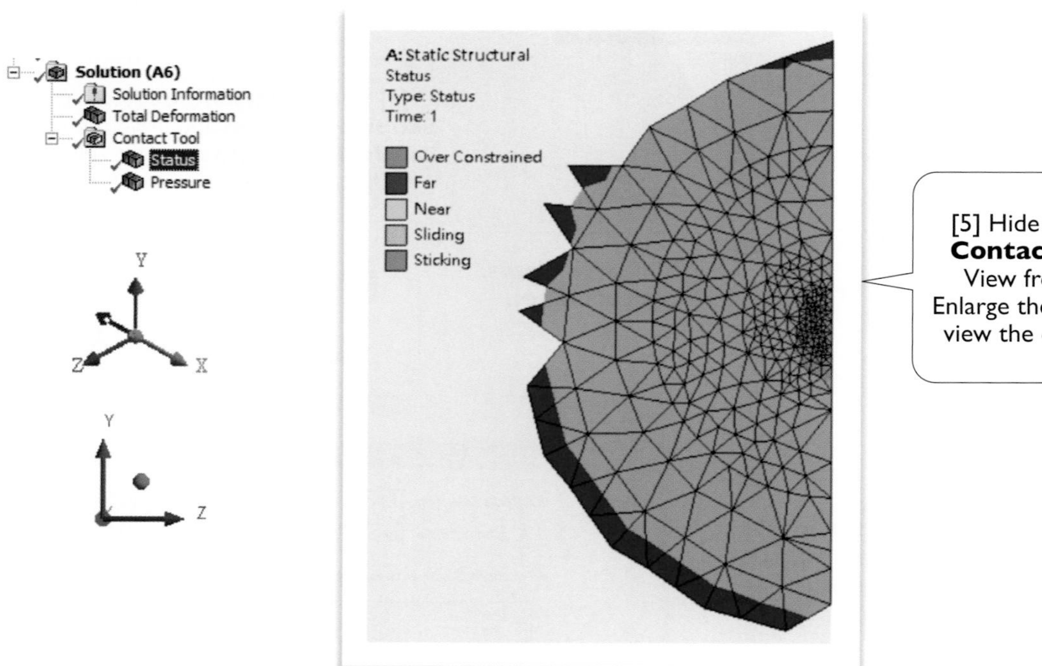

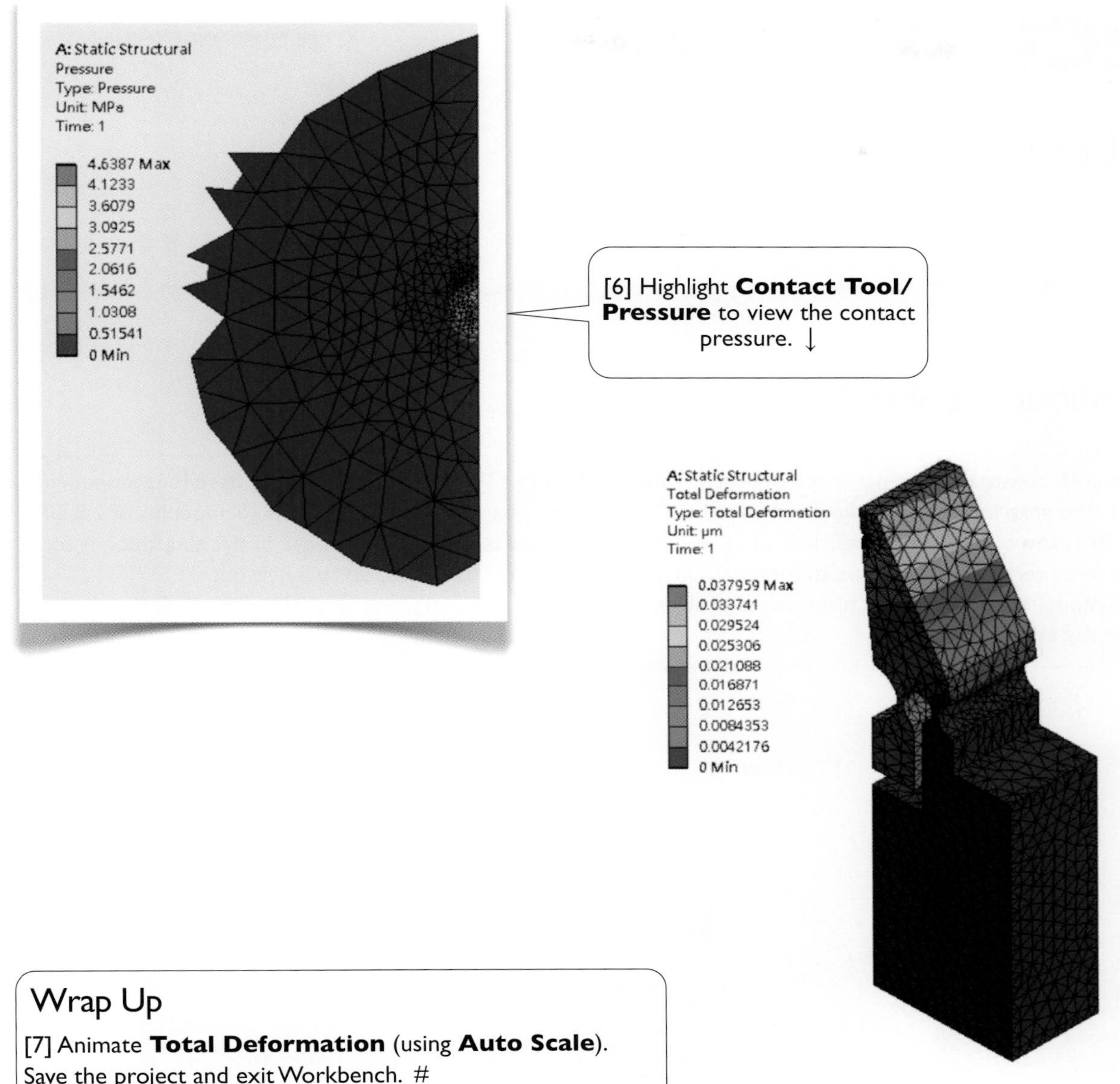

Wrap Up

[7] Animate **Total Deformation** (using **Auto Scale**). Save the project and exit Workbench. #

References

1. Chang, R. J., Lin ,Y. C., Shiu, C. C., and Hsieh,Y. T., "Development of SMA-Actuated Microgripper in Micro Assembly Applications," IECON, IEEE,Taiwan, 2007.
2. Shih, P.W., *Applications of SMA on Driving Micro-gripper*, MS Thesis, NCKU, ME,Taiwan, 2005.

Section 13.4

Snap Lock

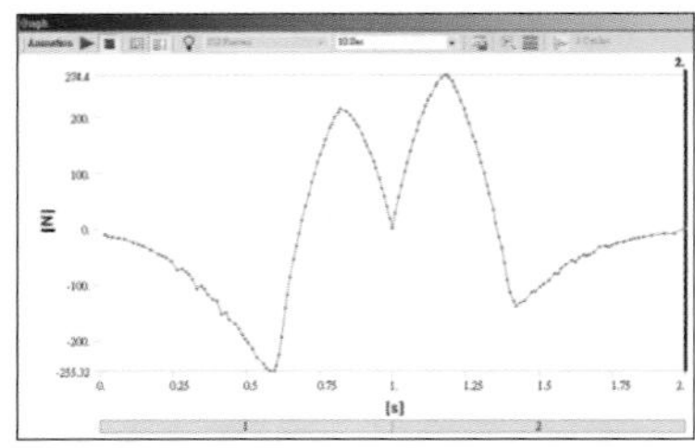

13.4.1 About the Snap Lock

[1] The snap lock consists of two parts: the **insert** [2] and the **prong** [3]. It is locked when the **insert** is pushed into position [4]. The snap lock has a thickness of 5 mm and is made of a plastic material with a Young's modulus of 2.8 GPa and a Poisson's ratio of 0.35. The coefficient of friction between the parts is 0.1. The purpose of this simulation is to find out the force required to push the **insert** into the position and the force required to pull it out.

We will model the problem as a plane stress problem. Due to the symmetry, only one half of the snap lock is modeled for the simulation. ↓

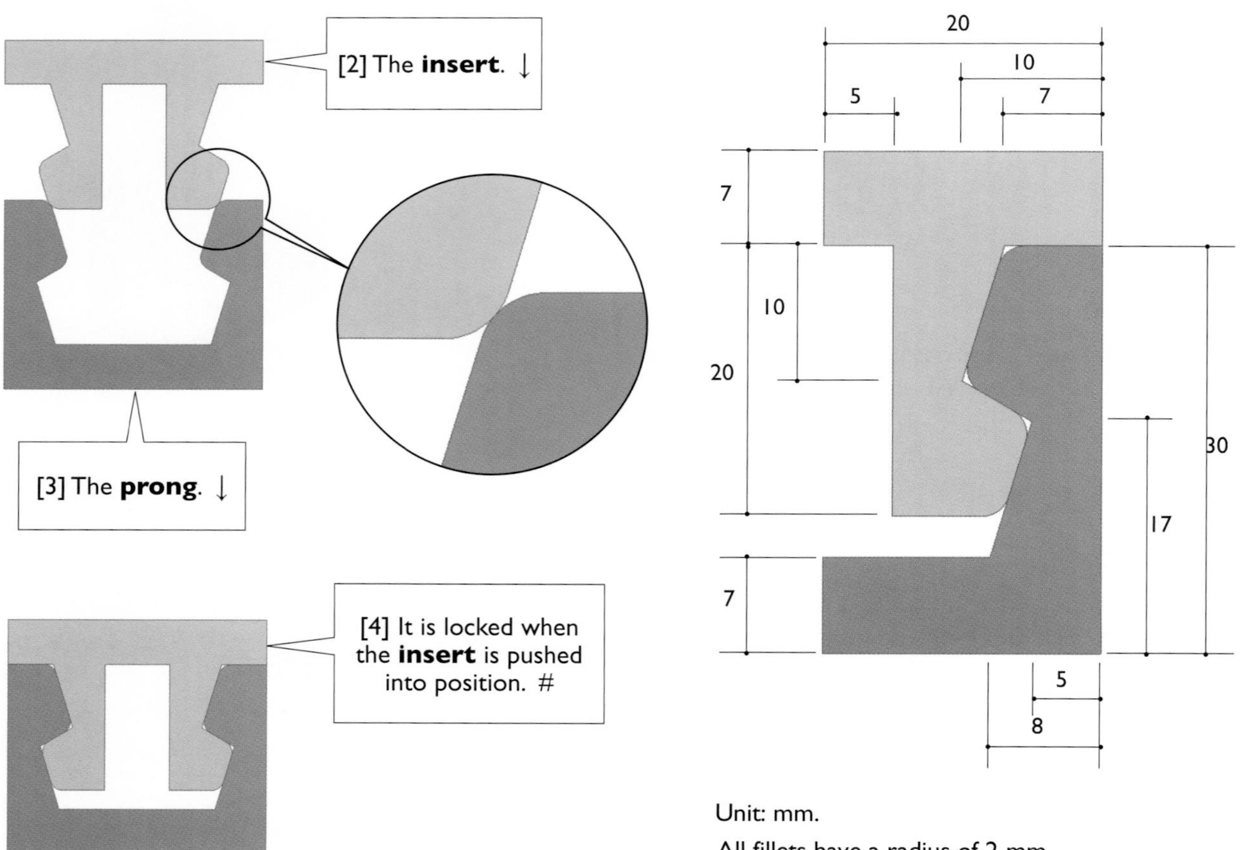

13.4.2 Start Up

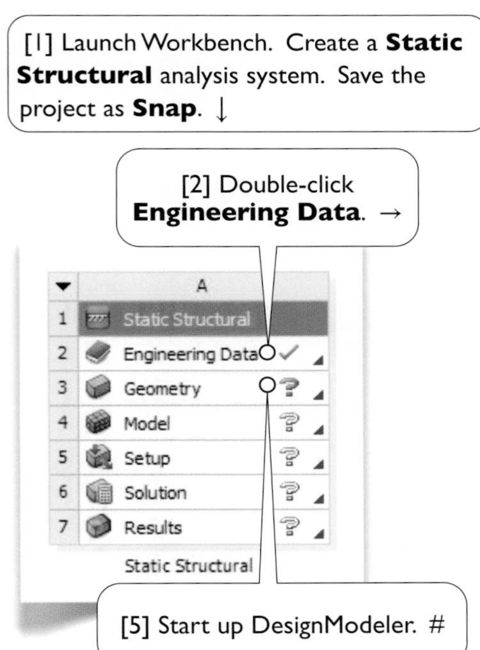

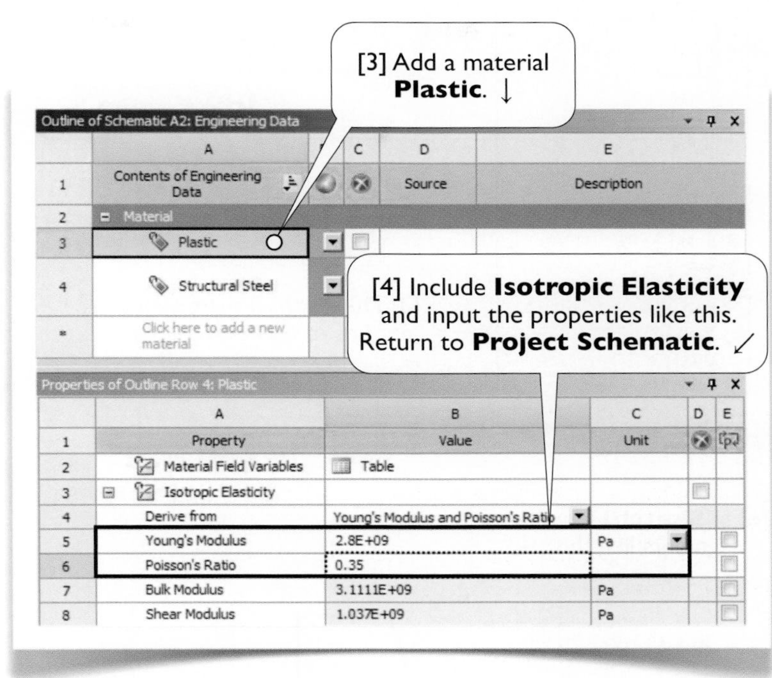

13.4.3 Create Geometry in DesignModeler

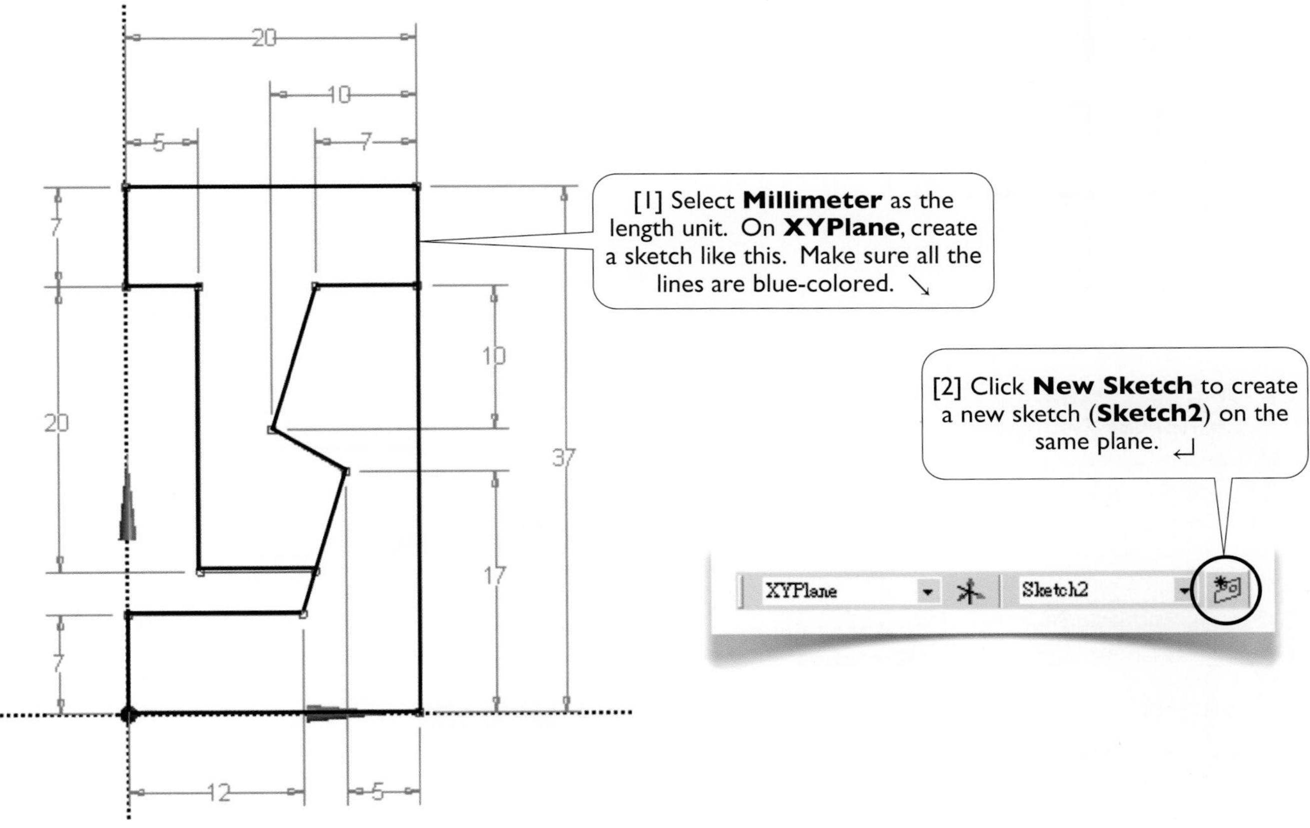

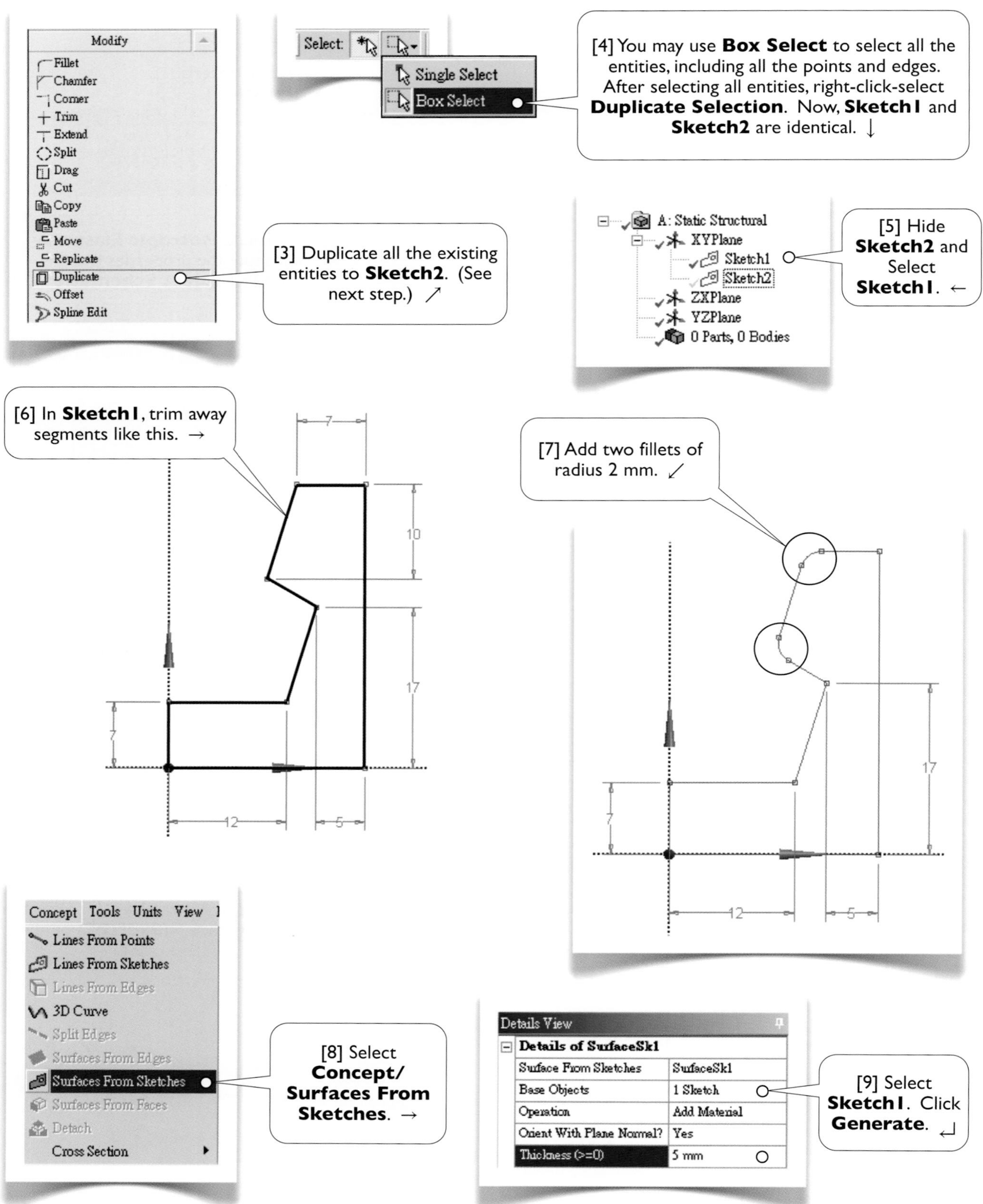

Modify
Fillet
Chamfer
Corner
Trim
Extend
Split
Drag
Cut
Copy
Paste
Move
Replicate
Duplicate
Offset
Spline Edit
[3] Duplicate all the existing entities to **Sketch2**. (See next step.) ↗
Select:
Single Select
Box Select
[4] You may use **Box Select** to select all the entities, including all the points and edges. After selecting all entities, right-click-select **Duplicate Selection**. Now, **Sketch1** and **Sketch2** are identical. ↓
A: Static Structural
XYPlane
Sketch1
Sketch2
ZXPlane
YZPlane
0 Parts, 0 Bodies
[5] Hide **Sketch2** and Select **Sketch1**. ←
[6] In **Sketch1**, trim away segments like this. →
7
10
17
7
12
5
[7] Add two fillets of radius 2 mm. ↙
17
7
12
5
Concept Tools Units View
Lines From Points
Lines From Sketches
Lines From Edges
3D Curve
Split Edges
Surfaces From Edges
Surfaces From Sketches
Surfaces From Faces
Detach
Cross Section
[8] Select **Concept/ Surfaces From Sketches**. →
Details View
Details of SurfaceSk1
Surface From Sketches SurfaceSk1
Base Objects 1 Sketch
Operation Add Material
Orient With Plane Normal? Yes
Thickness (>=0) 5 mm
[9] Select **Sketch1**. Click **Generate**. ↵

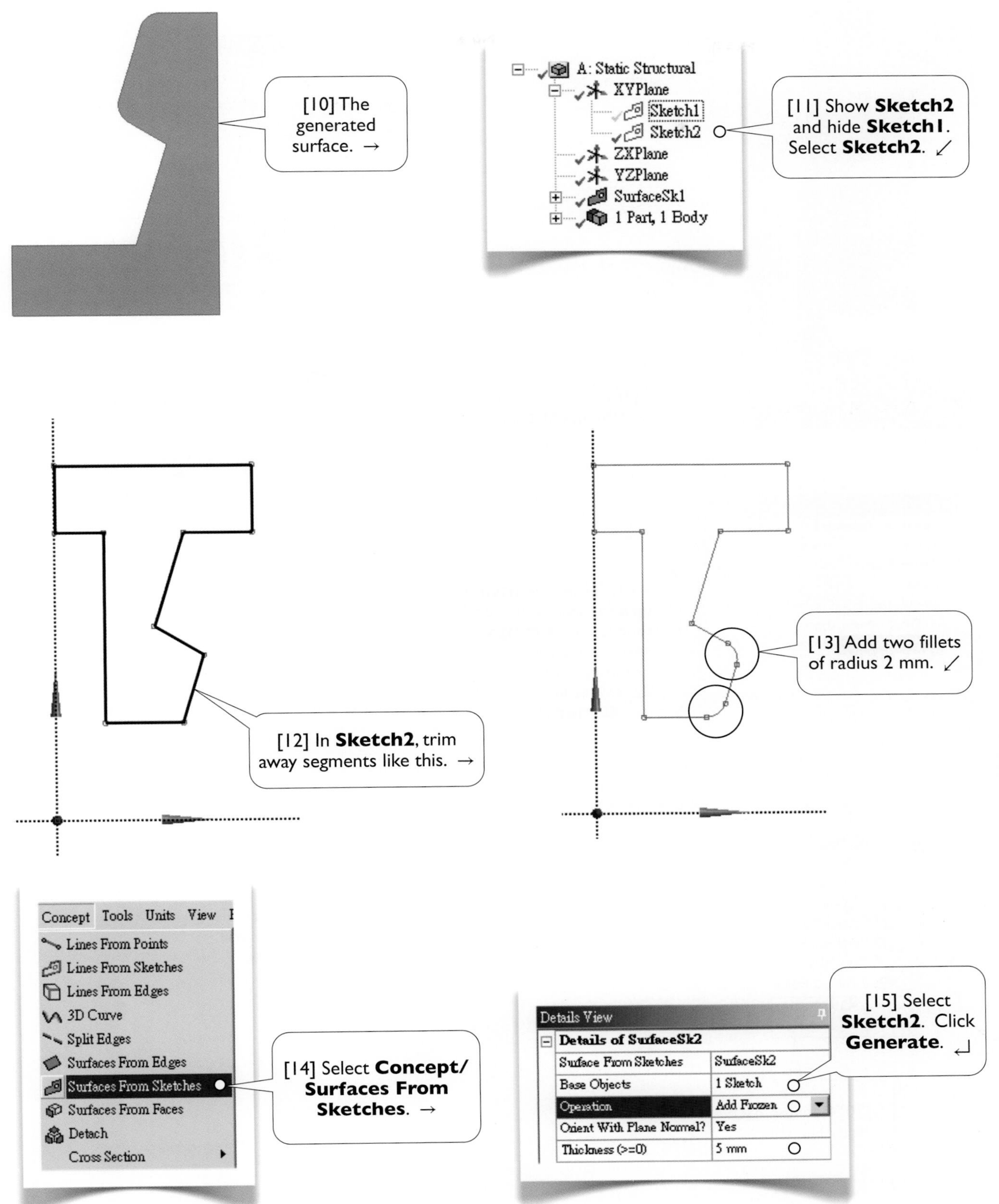
[10] The generated surface. →
A: Static Structural
XYPlane
Sketch1
Sketch2
ZXPlane
YZPlane
SurfaceSk1
1 Part, 1 Body
[11] Show **Sketch2** and hide **Sketch1**. Select **Sketch2**.
[12] In **Sketch2**, trim away segments like this. →
[13] Add two fillets of radius 2 mm.
Concept Tools Units View
Lines From Points
Lines From Sketches
Lines From Edges
3D Curve
Split Edges
Surfaces From Edges
Surfaces From Sketches
Surfaces From Faces
Detach
Cross Section
[14] Select **Concept/ Surfaces From Sketches**. →
Details View
Details of SurfaceSk2
Surface From Sketches SurfaceSk2
Base Objects 1 Sketch
Operation Add Frozen
Orient With Plane Normal? Yes
Thickness (>=0) 5 mm
[15] Select **Sketch2**. Click **Generate**.

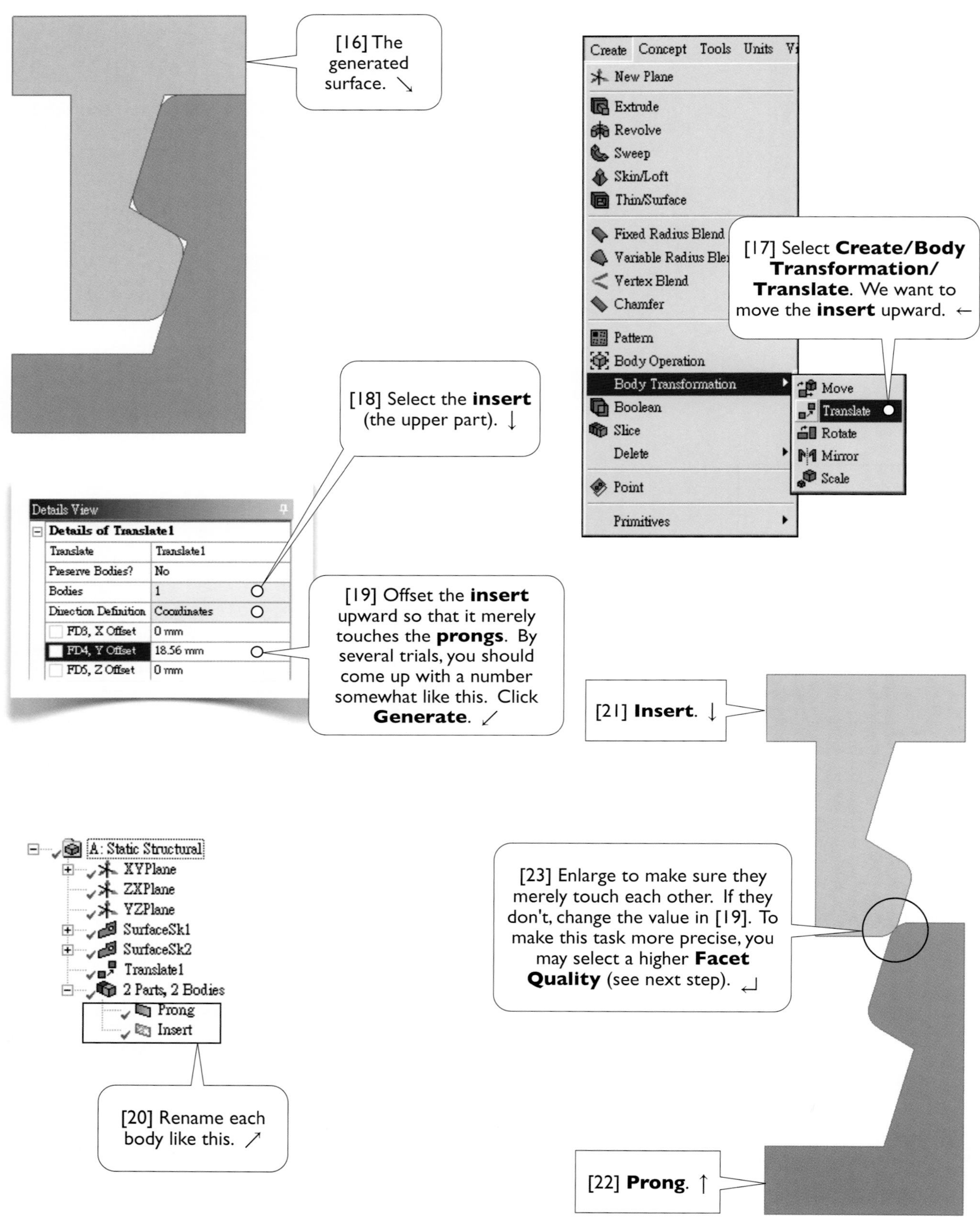

[16] The generated surface. ↘
Create Concept Tools Units
New Plane
Extrude
Revolve
Sweep
Skin/Loft
Thin/Surface
Fixed Radius Blend
Variable Radius Blend
Vertex Blend
Chamfer
Pattern
Body Operation
Body Transformation
Boolean
Slice
Delete
Point
Primitives
Move
Translate
Rotate
Mirror
Scale
[17] Select **Create/Body Transformation/Translate**. We want to move the **insert** upward. ←
[18] Select the **insert** (the upper part). ↓
Details View
Details of Translate1
Translate | Translate1
Preserve Bodies? | No
Bodies | 1
Direction Definition | Coordinates
FD3, X Offset | 0 mm
FD4, Y Offset | 18.56 mm
FD5, Z Offset | 0 mm
[19] Offset the **insert** upward so that it merely touches the **prongs**. By several trials, you should come up with a number somewhat like this. Click **Generate**. ↙
[21] **Insert**. ↓
A: Static Structural
XYPlane
ZXPlane
YZPlane
SurfaceSk1
SurfaceSk2
Translate1
2 Parts, 2 Bodies
Prong
Insert
[23] Enlarge to make sure they merely touch each other. If they don't, change the value in [19]. To make this task more precise, you may select a higher **Facet Quality** (see next step). ↵
[20] Rename each body like this. ↗
[22] **Prong**. ↑

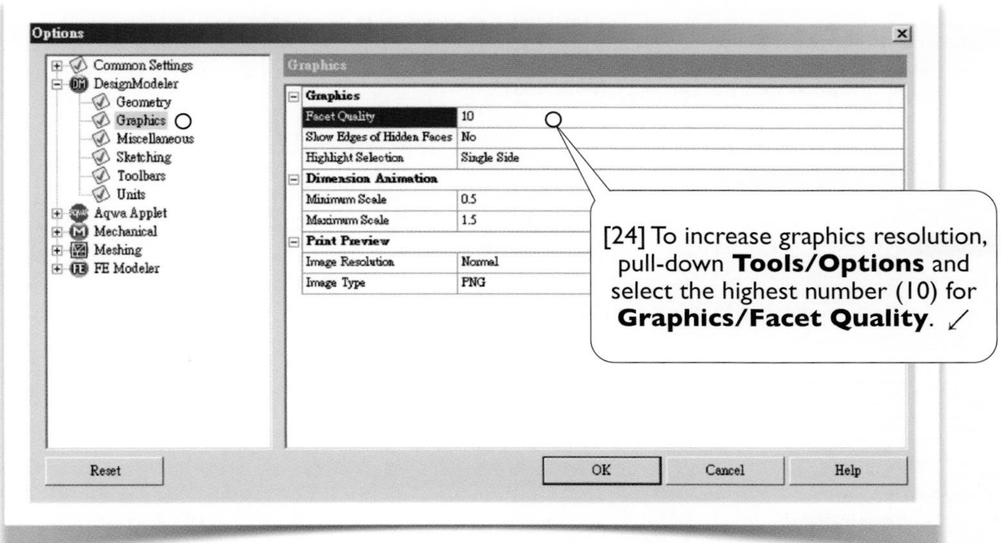

[24] To increase graphics resolution, pull-down **Tools/Options** and select the highest number (10) for **Graphics/Facet Quality**. ↙

[25] Close DesignModeler. Before attaching the geometry to **Mechanical**, remember to select **2D** option (3.1.4[4-6], page 112). #

13.4.4 Set Up Model

[1] Start up **Mechanical**. Select the unit system **mm-kg-N-s**. ↓

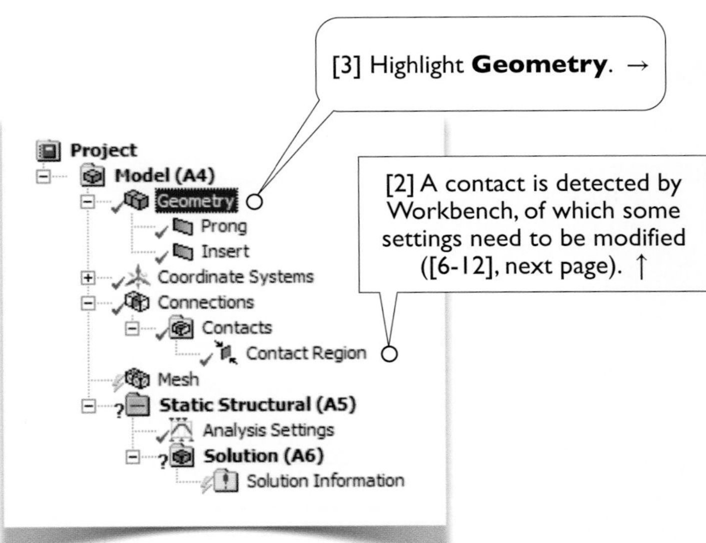

[2] A contact is detected by Workbench, of which some settings need to be modified ([6-12], next page). ↑

[3] Highlight **Geometry**. →

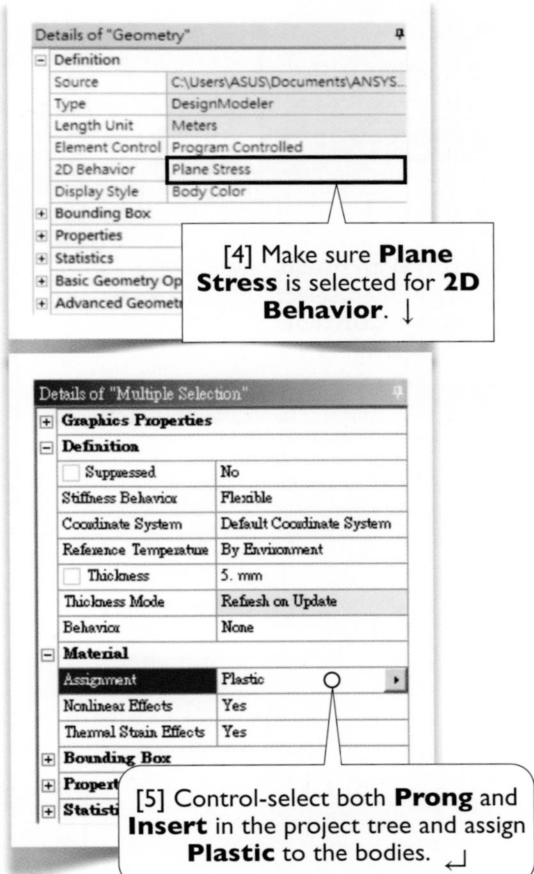

[4] Make sure **Plane Stress** is selected for **2D Behavior**. ↓

[5] Control-select both **Prong** and **Insert** in the project tree and assign **Plastic** to the bodies. ↵

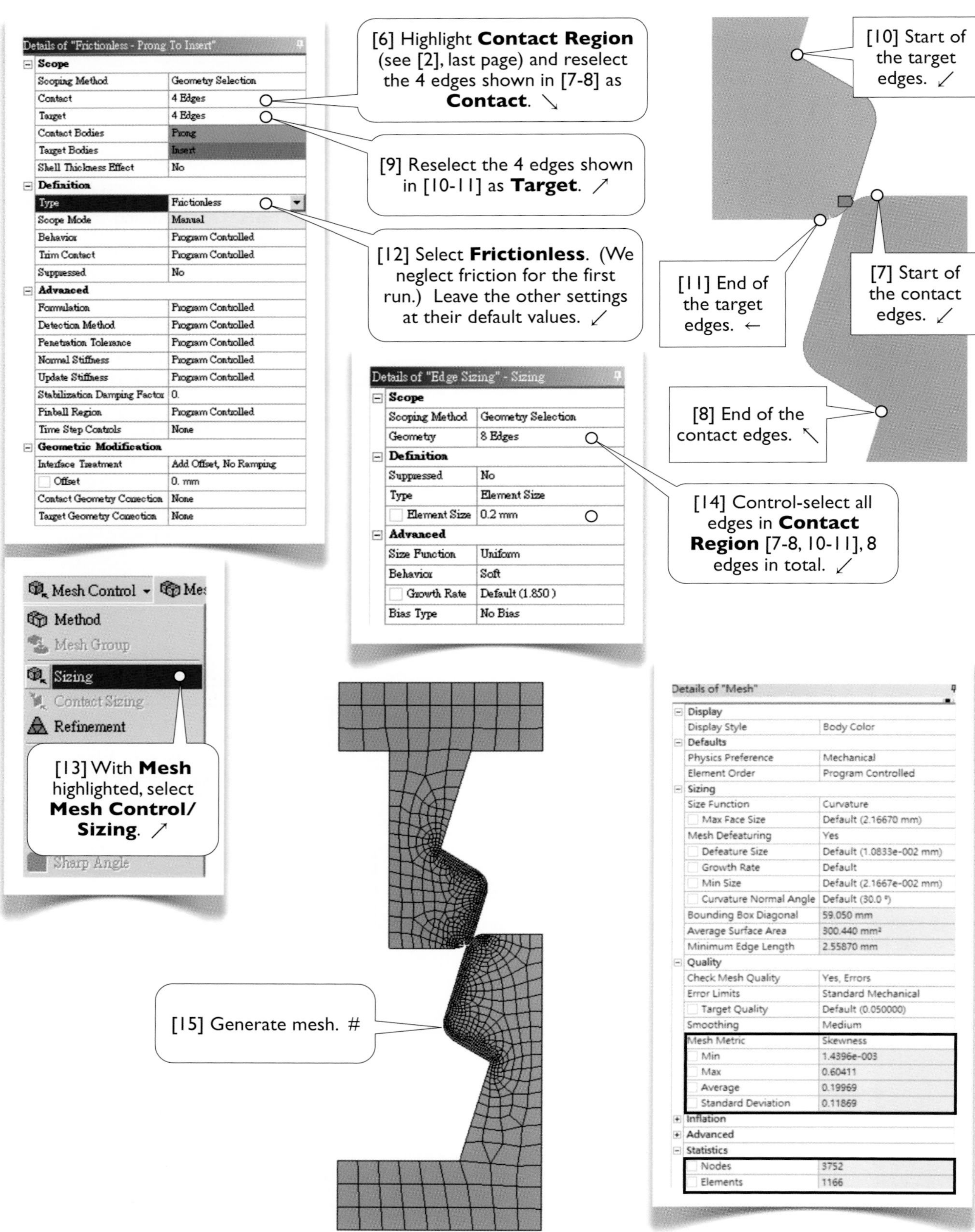
Details of "Frictionless - Prong To Insert"
Scope
Scoping Method | Geometry Selection
Contact | 4 Edges
Target | 4 Edges
Contact Bodies | Prong
Target Bodies | Insert
Shell Thickness Effect | No
Definition
Type | Frictionless
Scope Mode | Manual
Behavior | Program Controlled
Trim Contact | Program Controlled
Suppressed | No
Advanced
Formulation | Program Controlled
Detection Method | Program Controlled
Penetration Tolerance | Program Controlled
Normal Stiffness | Program Controlled
Update Stiffness | Program Controlled
Stabilization Damping Factor | 0.
Pinball Region | Program Controlled
Time Step Controls | None
Geometric Modification
Interface Treatment | Add Offset, No Ramping
Offset | 0. mm
Contact Geometry Correction | None
Target Geometry Correction | None
[6] Highlight **Contact Region** (see [2], last page) and reselect the 4 edges shown in [7-8] as **Contact**. ↘
[9] Reselect the 4 edges shown in [10-11] as **Target**. ↗
[12] Select **Frictionless**. (We neglect friction for the first run.) Leave the other settings at their default values. ↙
[10] Start of the target edges. ↙
[11] End of the target edges. ←
[7] Start of the contact edges. ↙
[8] End of the contact edges. ↖
Details of "Edge Sizing" - Sizing
Scope
Scoping Method | Geometry Selection
Geometry | 8 Edges
Definition
Suppressed | No
Type | Element Size
Element Size | 0.2 mm
Advanced
Size Function | Uniform
Behavior | Soft
Growth Rate | Default (1.850)
Bias Type | No Bias
[14] Control-select all edges in **Contact Region** [7-8, 10-11], 8 edges in total. ↙
Mesh Control | Mes
Method
Mesh Group
Sizing
Contact Sizing
Refinement
Sharp Angle
[13] With **Mesh** highlighted, select **Mesh Control/ Sizing**. ↗
[15] Generate mesh. #
Details of "Mesh"
Display
Display Style | Body Color
Defaults
Physics Preference | Mechanical
Element Order | Program Controlled
Sizing
Size Function | Curvature
Max Face Size | Default (2.16670 mm)
Mesh Defeaturing | Yes
Defeature Size | Default (1.0833e-002 mm)
Growth Rate | Default
Min Size | Default (2.1667e-002 mm)
Curvature Normal Angle | Default (30.0 °)
Bounding Box Diagonal | 59.050 mm
Average Surface Area | 300.440 mm²
Minimum Edge Length | 2.55870 mm
Quality
Check Mesh Quality | Yes, Errors
Error Limits | Standard Mechanical
Target Quality | Default (0.050000)
Smoothing | Medium
Mesh Metric | Skewness
Min | 1.4396e-003
Max | 0.60411
Average | 0.19969
Standard Deviation | 0.11869
Inflation
Advanced
Statistics
Nodes | 3752
Elements | 1166

13.4.5 Set Up Analysis Settings

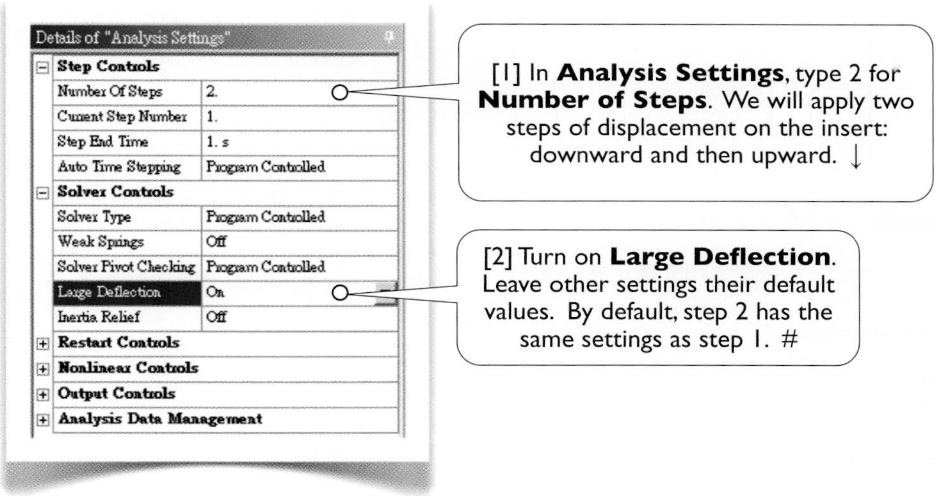

13.4.6 Set Up Environment Conditions

[1] Set up a **Fixed Support** on this edge. ↑

Details of "Fixed Support"	
Scope	
Scoping Method	Geometry Selection
Geometry	1 Edge
Definition	
Type	Fixed Support
Suppressed	No

[2] Set up a **Frictionless Support** on this edge and the edge in the next step. ↓

Details of "Frictionless Support"	
Scope	
Scoping Method	Geometry Selection
Geometry	2 Edges
Definition	
Type	Frictionless Support
Suppressed	No

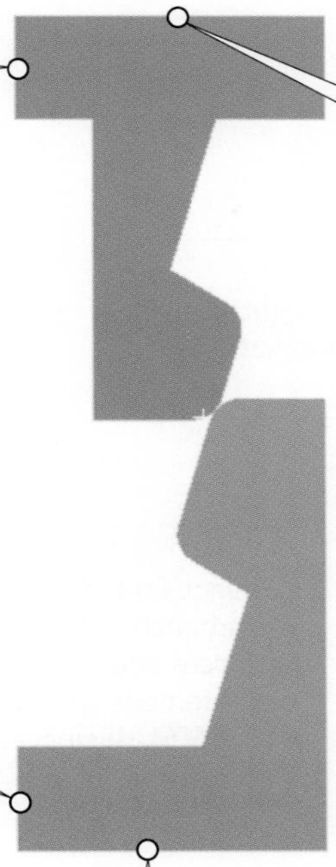

[3] And this edge. For solid elements, a **Frictionless Support** is equivalent to a symmetry condition. ↗

[4] Set up a **Displacement** on this edge (see [5-6]). ↓

Details of "Displacement"	
Scope	
Scoping Method	Geometry Selection
Geometry	1 Edge
Definition	
Type	Displacement
Define By	Components
Coordinate System	Global Coordinate System
X Component	Free
Y Component	Tabular Data
Suppressed	No
Tabular Data	
Independent Variable	Time

[5] Select **Tabular** for Y **Component**. ↵

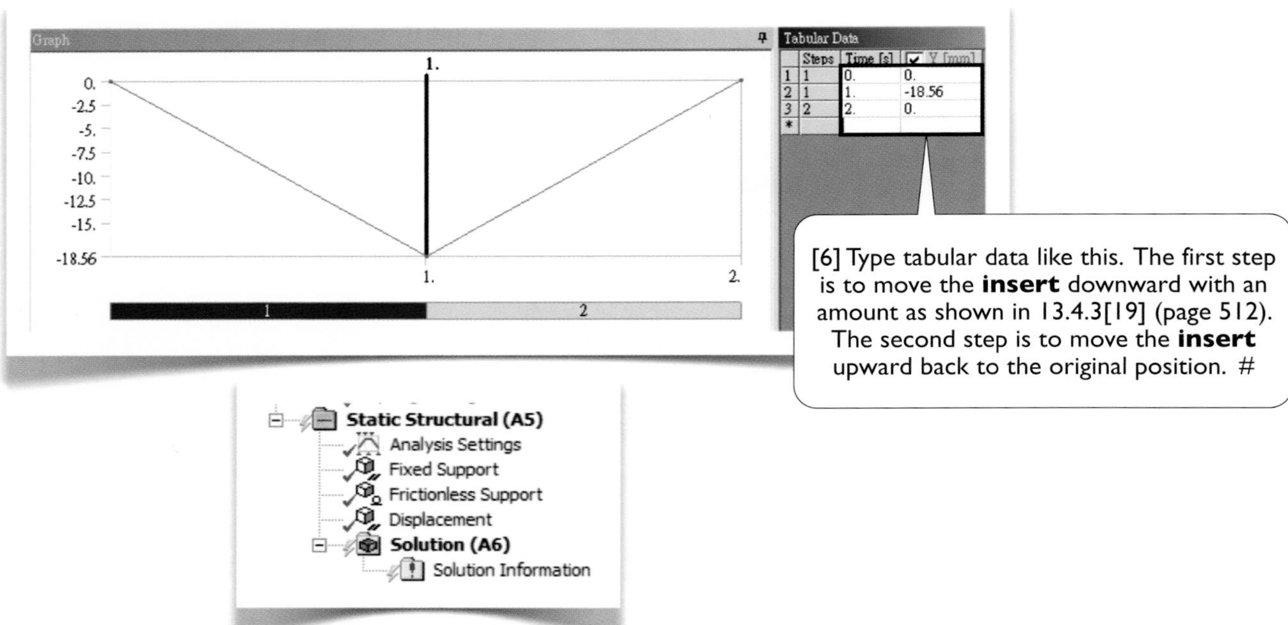

13.4.7 Set Up Result Objects

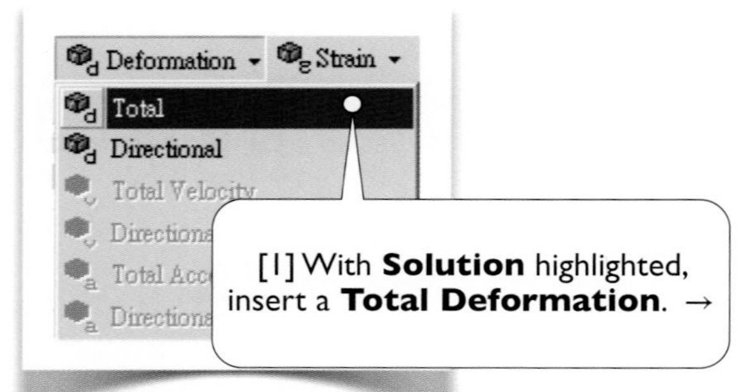

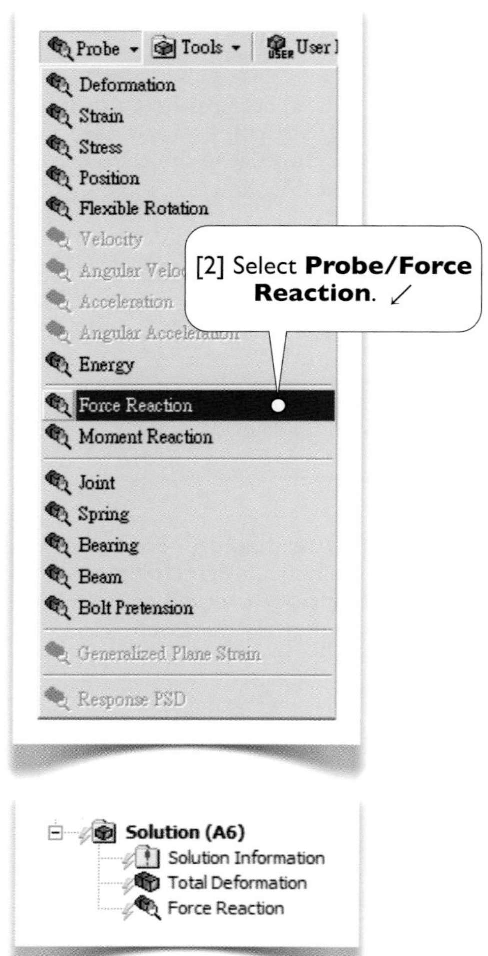

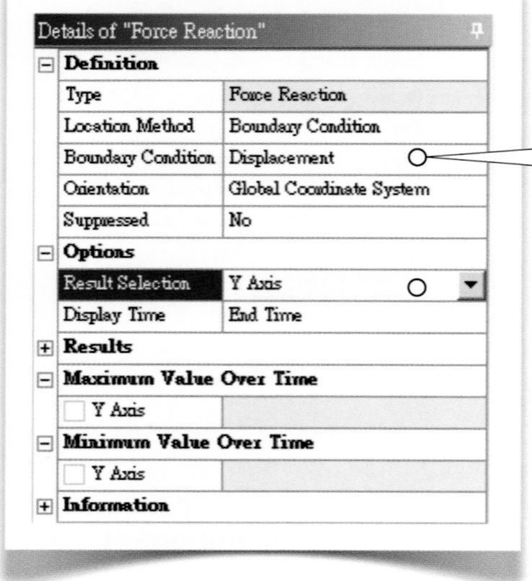

[3] Select **Displacement**. Workbench will calculate the forces required for the displacement specified in 13.4.6[6], this page. #

13.4.8 Solve the Model and View the Results

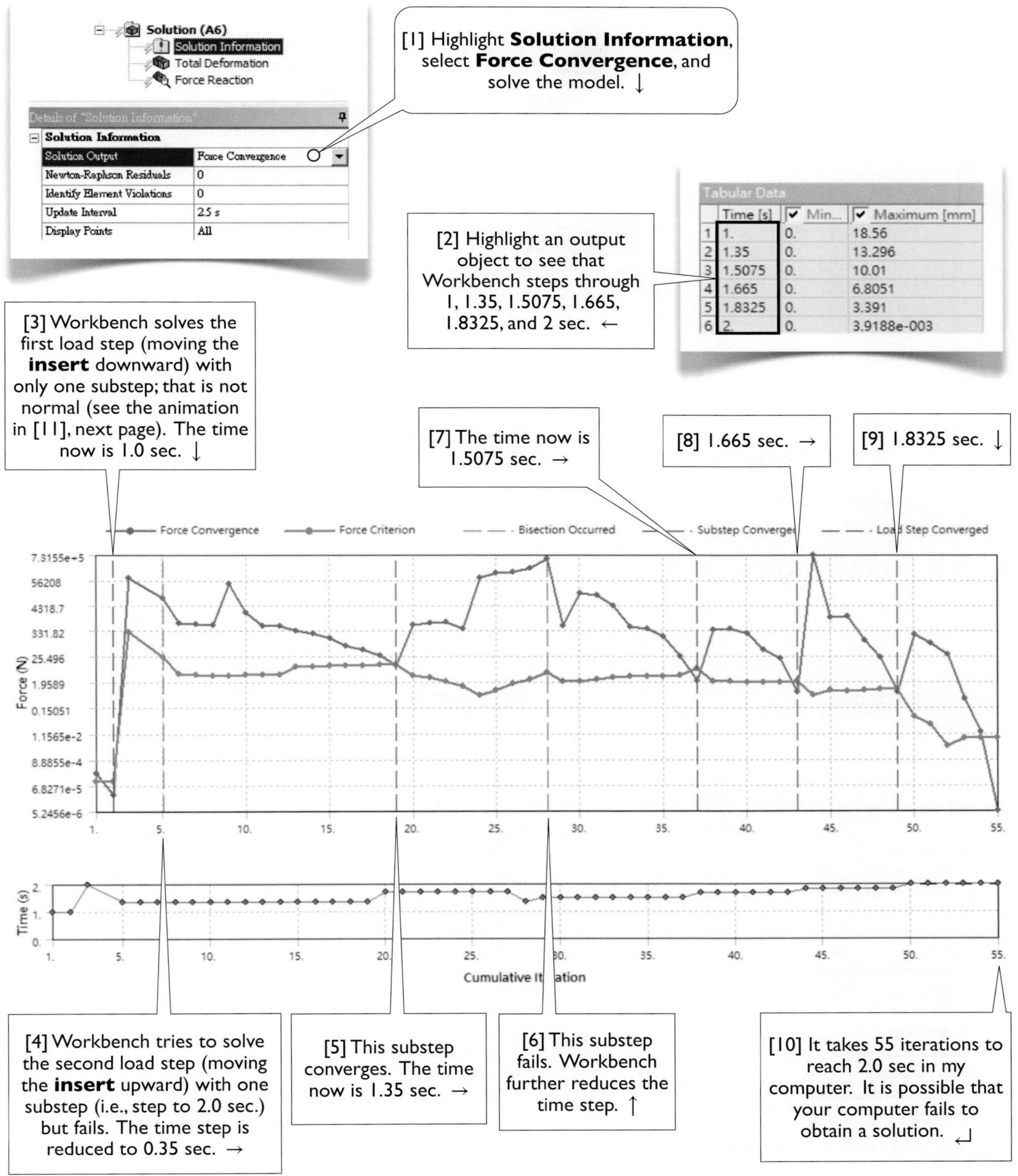

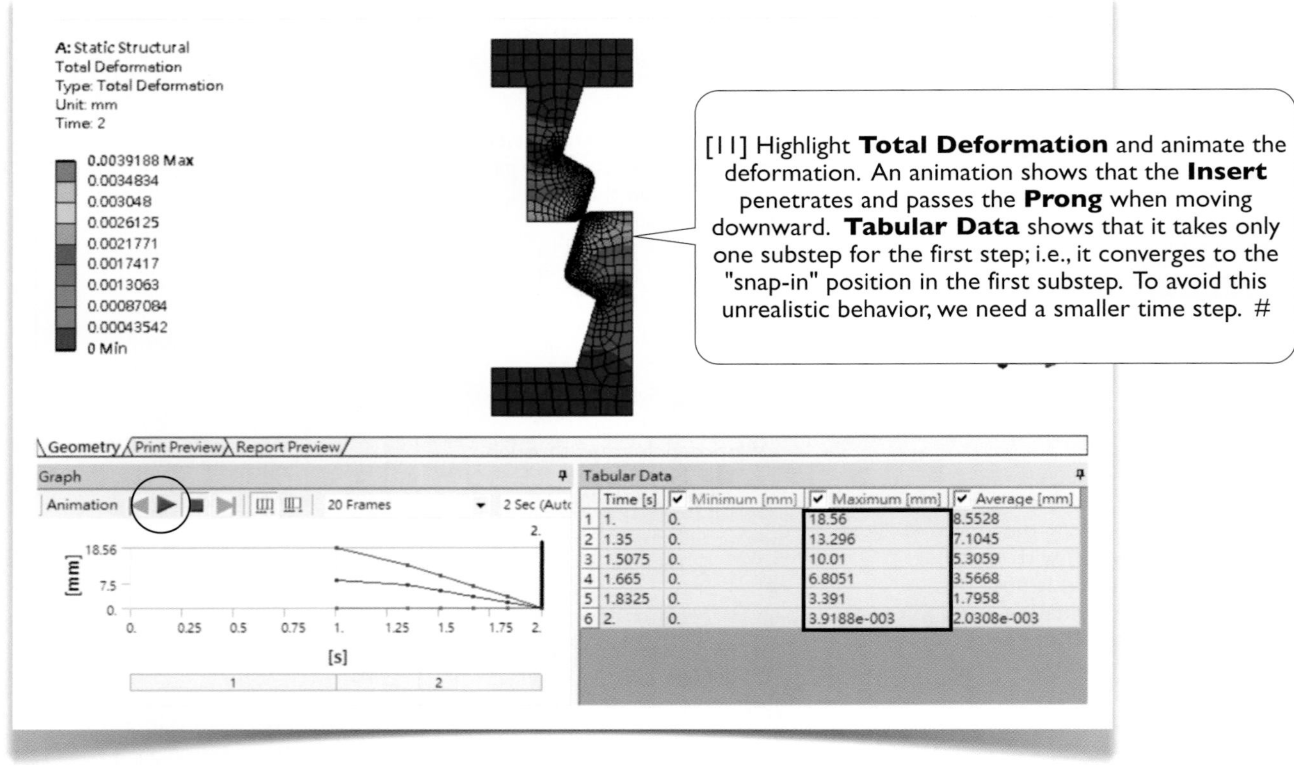

	Time [s]	Minimum [mm]	Maximum [mm]	Average [mm]
1	1.	0.	18.56	8.5528
2	1.35	0.	13.296	7.1045
3	1.5075	0.	10.01	5.3059
4	1.665	0.	6.8051	3.5668
5	1.8325	0.	3.391	1.7958
6	2.	0.	3.9188e-003	2.0308e-003

13.4.9 Reduce Time Steps and Solve Again

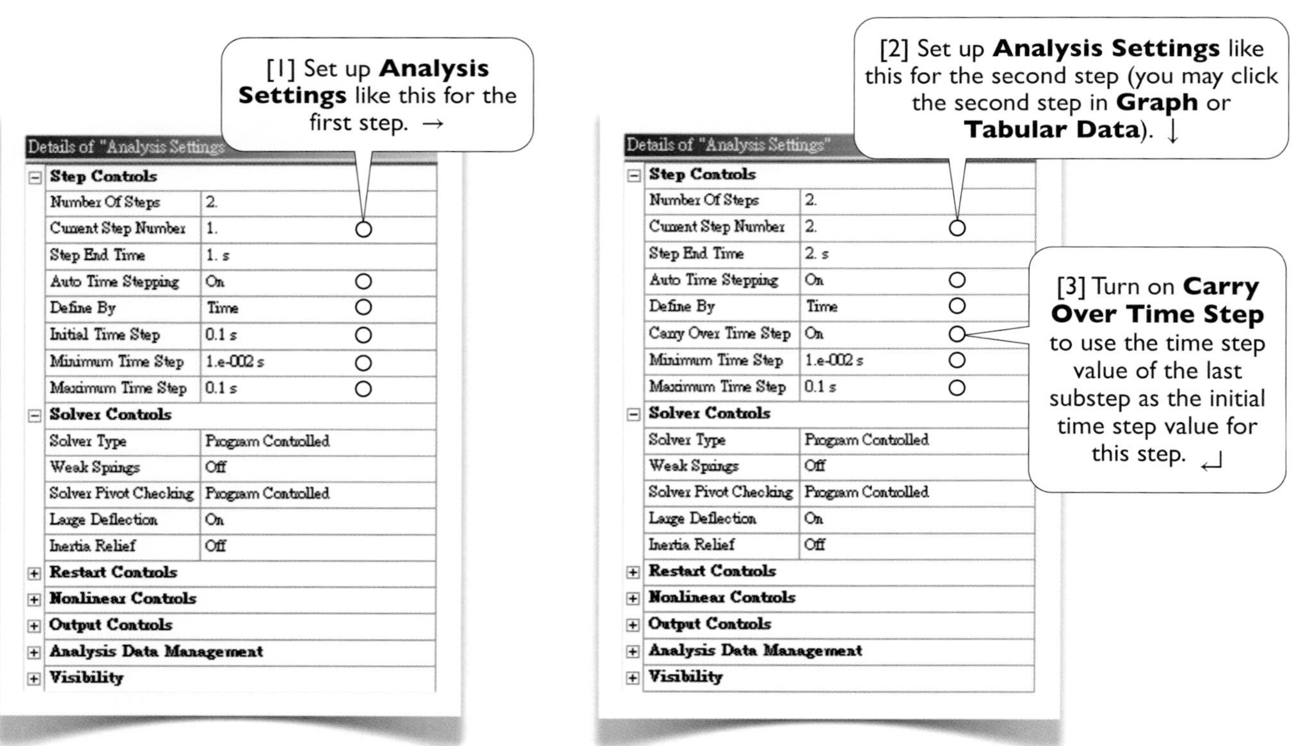

Details of "Analysis Settings" (first step)

Step Controls	
Number Of Steps	2.
Current Step Number	1.
Step End Time	1. s
Auto Time Stepping	On
Define By	Time
Initial Time Step	0.1 s
Minimum Time Step	1.e-002 s
Maximum Time Step	0.1 s
Solver Controls	
Solver Type	Program Controlled
Weak Springs	Off
Solver Pivot Checking	Program Controlled
Large Deflection	On
Inertia Relief	Off
Restart Controls	
Nonlinear Controls	
Output Controls	
Analysis Data Management	
Visibility	

Details of "Analysis Settings" (second step)

Step Controls	
Number Of Steps	2.
Current Step Number	2.
Step End Time	2. s
Auto Time Stepping	On
Define By	Time
Carry Over Time Step	On
Minimum Time Step	1.e-002 s
Maximum Time Step	0.1 s
Solver Controls	
Solver Type	Program Controlled
Weak Springs	Off
Solver Pivot Checking	Program Controlled
Large Deflection	On
Inertia Relief	Off
Restart Controls	
Nonlinear Controls	
Output Controls	
Analysis Data Management	
Visibility	

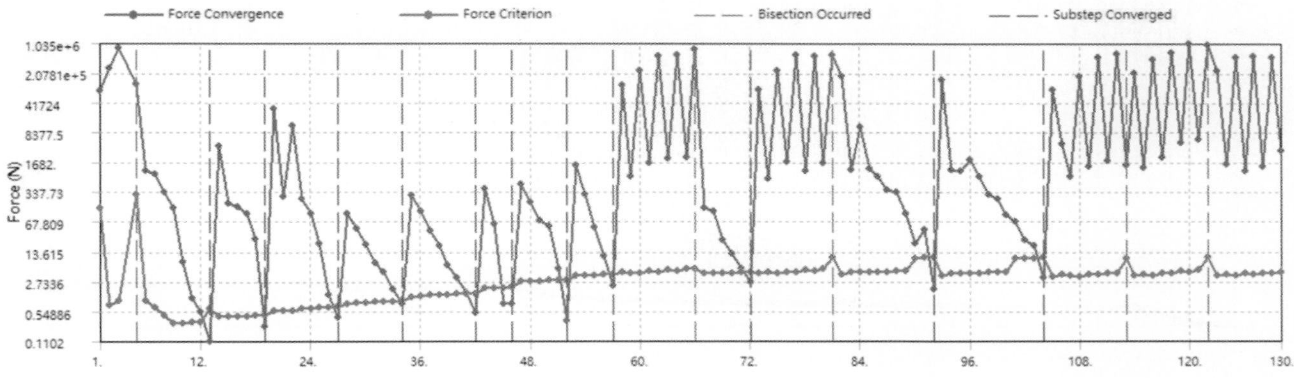

[5] Workbench fails to obtain a solution. ↓

Solution (A6)
Solution Information
Total Deformation
Force Reaction

Details of "Analysis Settings"

Restart Analysis	
Restart Type	Program Controlled
Load Step	Initial
Substep	Initial
Time	Initial
Step Controls	
Number Of Steps	2.
Current Step Number	1.
Step End Time	1. s
Auto Time Stepping	On
Define By	Time
Initial Time Step	0.1 s
Minimum Time Step	1.e-003 s
Maximum Time Step	0.1 s
Solver Controls	
Solver Type	Program Controlled
Weak Springs	Off
Solver Pivot Checking	Program Controlled
Large Deflection	On
Inertia Relief	Off
Rotordynamics Controls	
Restart Controls	
Nonlinear Controls	
Output Controls	
Stress	No
Strain	No
Nodal Forces	No
Contact Miscellaneous	No
General Miscellaneous	No
Store Results At	Equally Spaced Points
— Value	100.
Analysis Data Management	
Visibility	

[6] Reduce **Minimum Time Step** to 0.001 sec. ↓

[7] To save disk storage and computing time, set up **Output Controls** like this. ↵

Details of "Analysis Settings"

Restart Analysis	
Restart Type	Program Controlled
Load Step	Initial
Substep	Initial
Time	Initial
Step Controls	
Number Of Steps	2.
Current Step Number	2.
Step End Time	2. s
Auto Time Stepping	On
Define By	Time
Carry Over Time Step	On
Minimum Time Step	1.e-003 s
Maximum Time Step	0.1 s
Solver Controls	
Solver Type	Program Controlled
Weak Springs	Off
Solver Pivot Checking	Program Controlled
Large Deflection	On
Inertia Relief	Off
Rotordynamics Controls	
Restart Controls	
Nonlinear Controls	
Output Controls	
Stress	No
Strain	No
Nodal Forces	No
Contact Miscellaneous	No
General Miscellaneous	No
Store Results At	Equally Spaced Points
— Value	100.
Analysis Data Management	
Visibility	

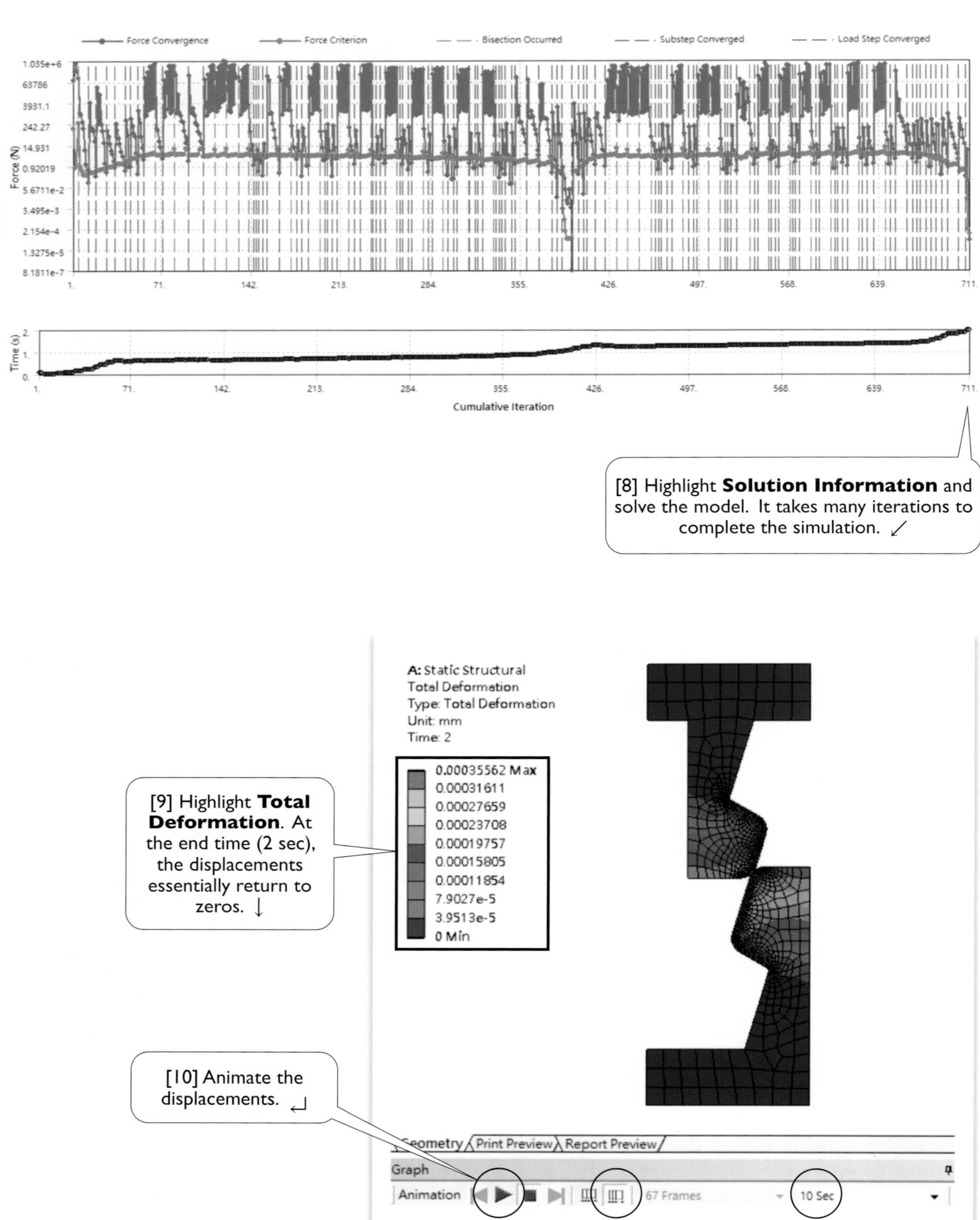

Force Convergence
Force Criterion
Bisection Occurred
Substep Converged
Load Step Converged
Force (N)
Time (s)
Cumulative Iteration
[8] Highlight **Solution Information** and solve the model. It takes many iterations to complete the simulation.
A: Static Structural
Total Deformation
Type: Total Deformation
Unit: mm
Time: 2
0.00035562 Max
0.00031611
0.00027659
0.00023708
0.00019757
0.00015805
0.00011854
7.9027e-5
3.9513e-5
0 Min
[9] Highlight **Total Deformation**. At the end time (2 sec), the displacements essentially return to zeros.
[10] Animate the displacements.
Geometry
Print Preview
Report Preview
Graph
Animation
67 Frames
10 Sec

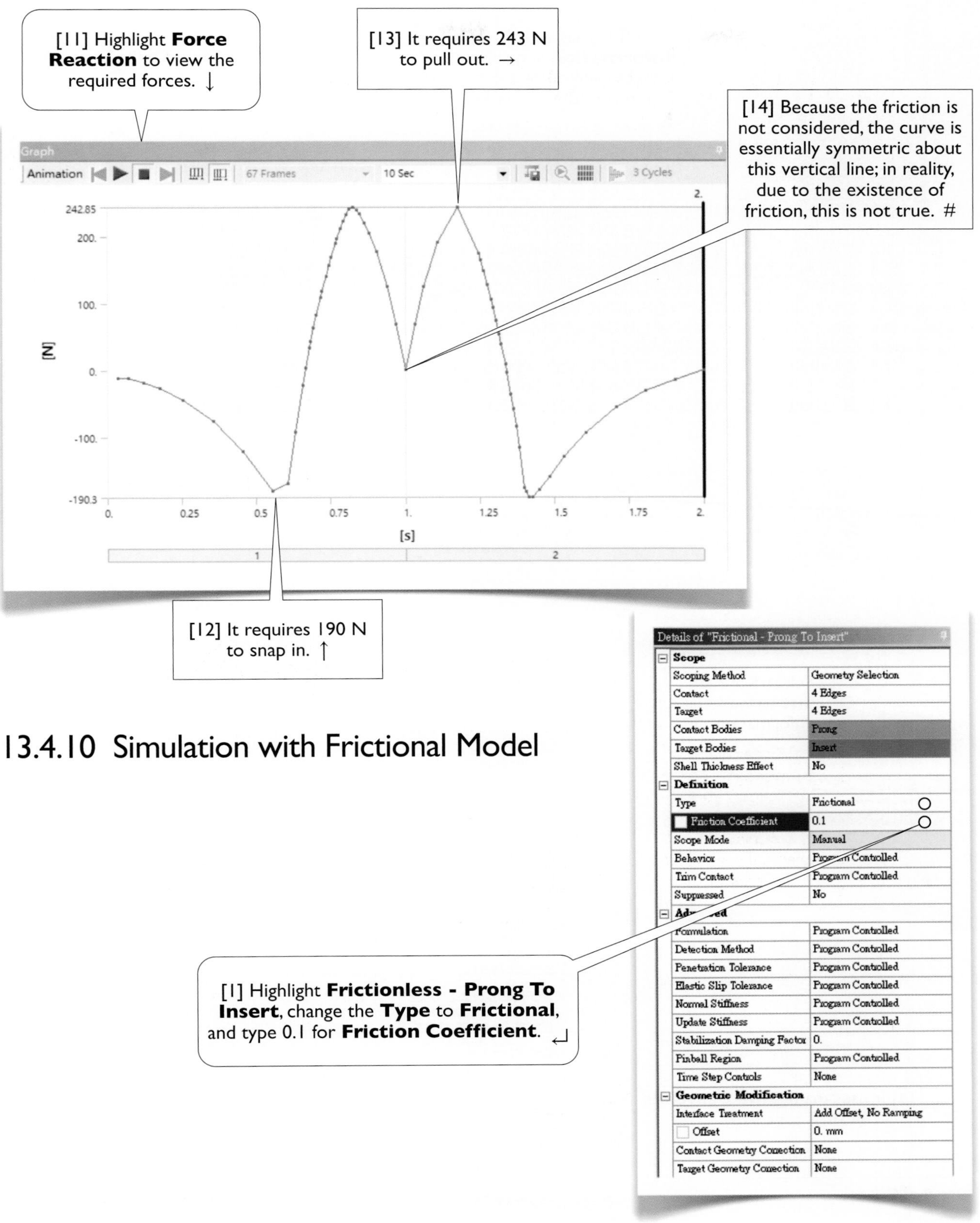

13.4.10 Simulation with Frictional Model

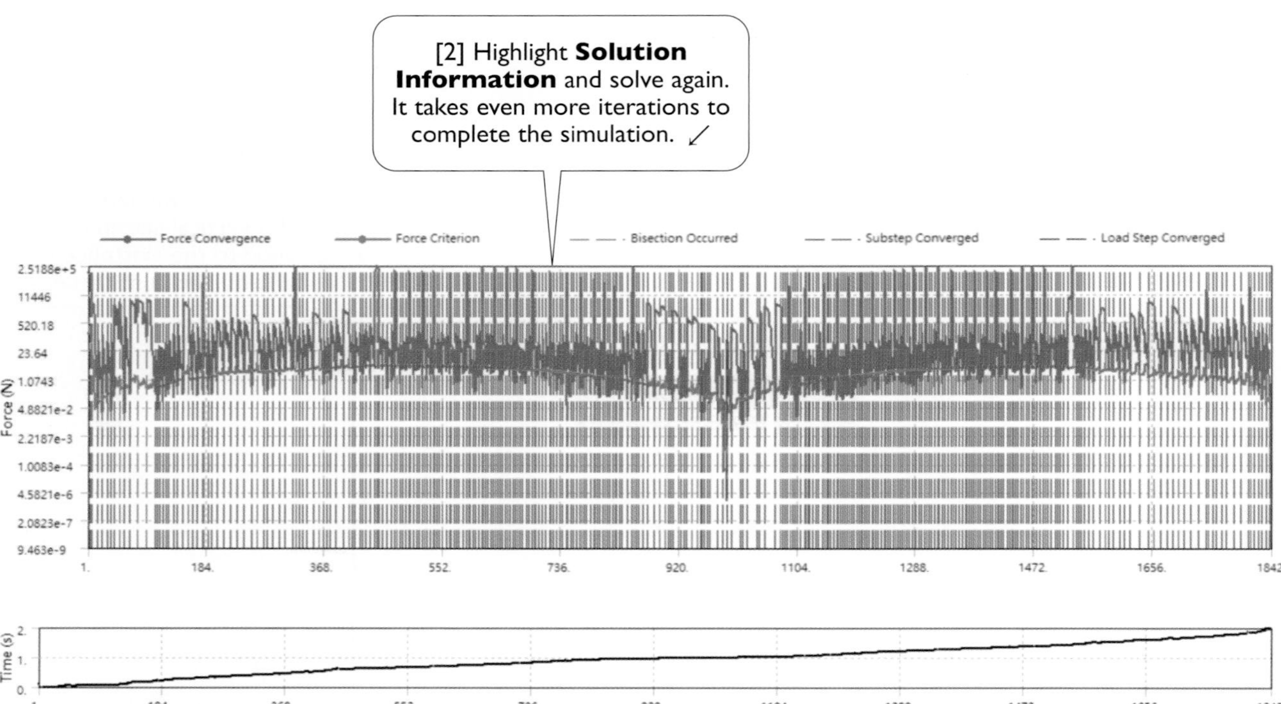

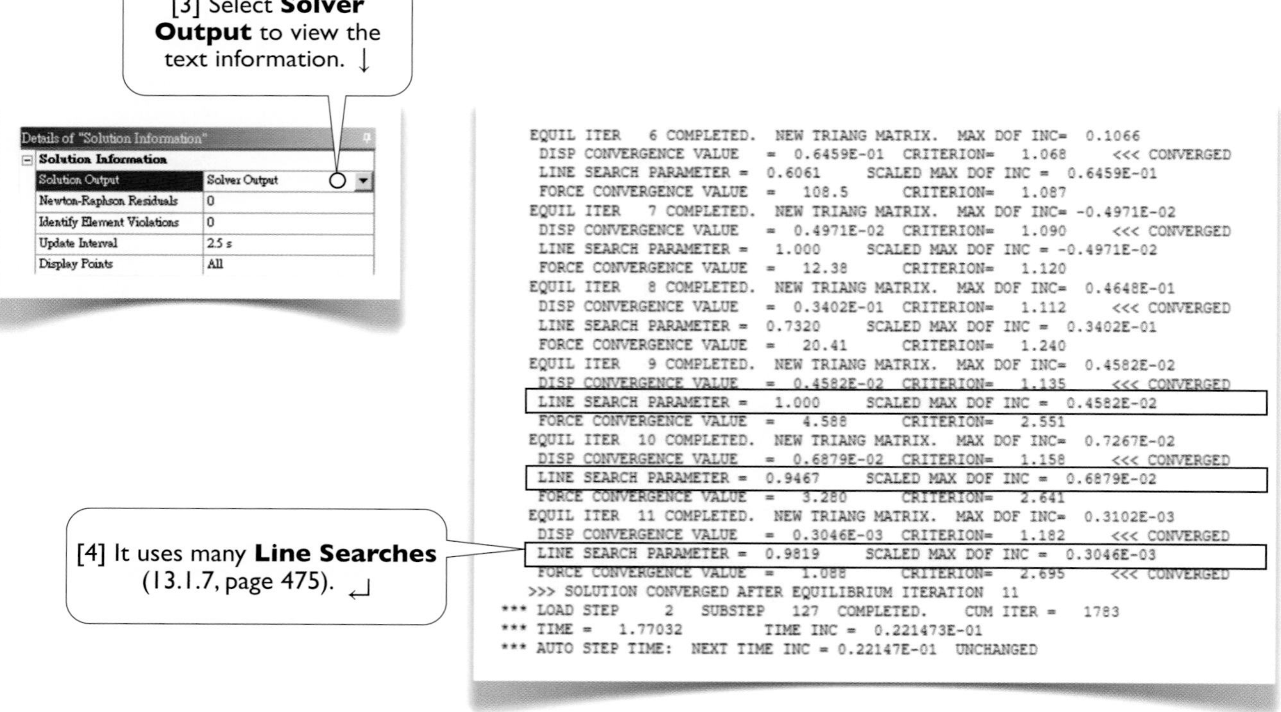


```
 EQUIL ITER   6 COMPLETED.  NEW TRIANG MATRIX.  MAX DOF INC=  0.1066
  DISP CONVERGENCE VALUE    =  0.6459E-01  CRITERION=   1.068      <<< CONVERGED
  LINE SEARCH PARAMETER =  0.6061      SCALED MAX DOF INC =  0.6459E-01
  FORCE CONVERGENCE VALUE   =   108.5      CRITERION=   1.087
 EQUIL ITER   7 COMPLETED.  NEW TRIANG MATRIX.  MAX DOF INC= -0.4971E-02
  DISP CONVERGENCE VALUE    =  0.4971E-02  CRITERION=   1.090      <<< CONVERGED
  LINE SEARCH PARAMETER =   1.000      SCALED MAX DOF INC = -0.4971E-02
  FORCE CONVERGENCE VALUE   =   12.38      CRITERION=   1.120
 EQUIL ITER   8 COMPLETED.  NEW TRIANG MATRIX.  MAX DOF INC=  0.4648E-01
  DISP CONVERGENCE VALUE    =  0.3402E-01  CRITERION=   1.112      <<< CONVERGED
  LINE SEARCH PARAMETER =  0.7320      SCALED MAX DOF INC =  0.3402E-01
  FORCE CONVERGENCE VALUE   =   20.41      CRITERION=   1.240
 EQUIL ITER   9 COMPLETED.  NEW TRIANG MATRIX.  MAX DOF INC=  0.4582E-02
  DISP CONVERGENCE VALUE    =  0.4582E-02  CRITERION=   1.135      <<< CONVERGED
  LINE SEARCH PARAMETER =   1.000      SCALED MAX DOF INC =  0.4582E-02
  FORCE CONVERGENCE VALUE   =   4.588      CRITERION=   2.551
 EQUIL ITER  10 COMPLETED.  NEW TRIANG MATRIX.  MAX DOF INC=  0.7267E-02
  DISP CONVERGENCE VALUE    =  0.6879E-02  CRITERION=   1.158      <<< CONVERGED
  LINE SEARCH PARAMETER =  0.9467      SCALED MAX DOF INC =  0.6879E-02
  FORCE CONVERGENCE VALUE   =   3.280      CRITERION=   2.641
 EQUIL ITER  11 COMPLETED.  NEW TRIANG MATRIX.  MAX DOF INC=  0.3102E-03
  DISP CONVERGENCE VALUE    =  0.3046E-03  CRITERION=   1.182      <<< CONVERGED
  LINE SEARCH PARAMETER =  0.9819      SCALED MAX DOF INC =  0.3046E-03
  FORCE CONVERGENCE VALUE   =   1.088      CRITERION=   2.695      <<< CONVERGED
 >>> SOLUTION CONVERGED AFTER EQUILIBRIUM ITERATION  11
*** LOAD STEP     2   SUBSTEP   127  COMPLETED.    CUM ITER =   1783
*** TIME =   1.77032         TIME INC =  0.221473E-01
*** AUTO STEP TIME:  NEXT TIME INC = 0.22147E-01  UNCHANGED
```

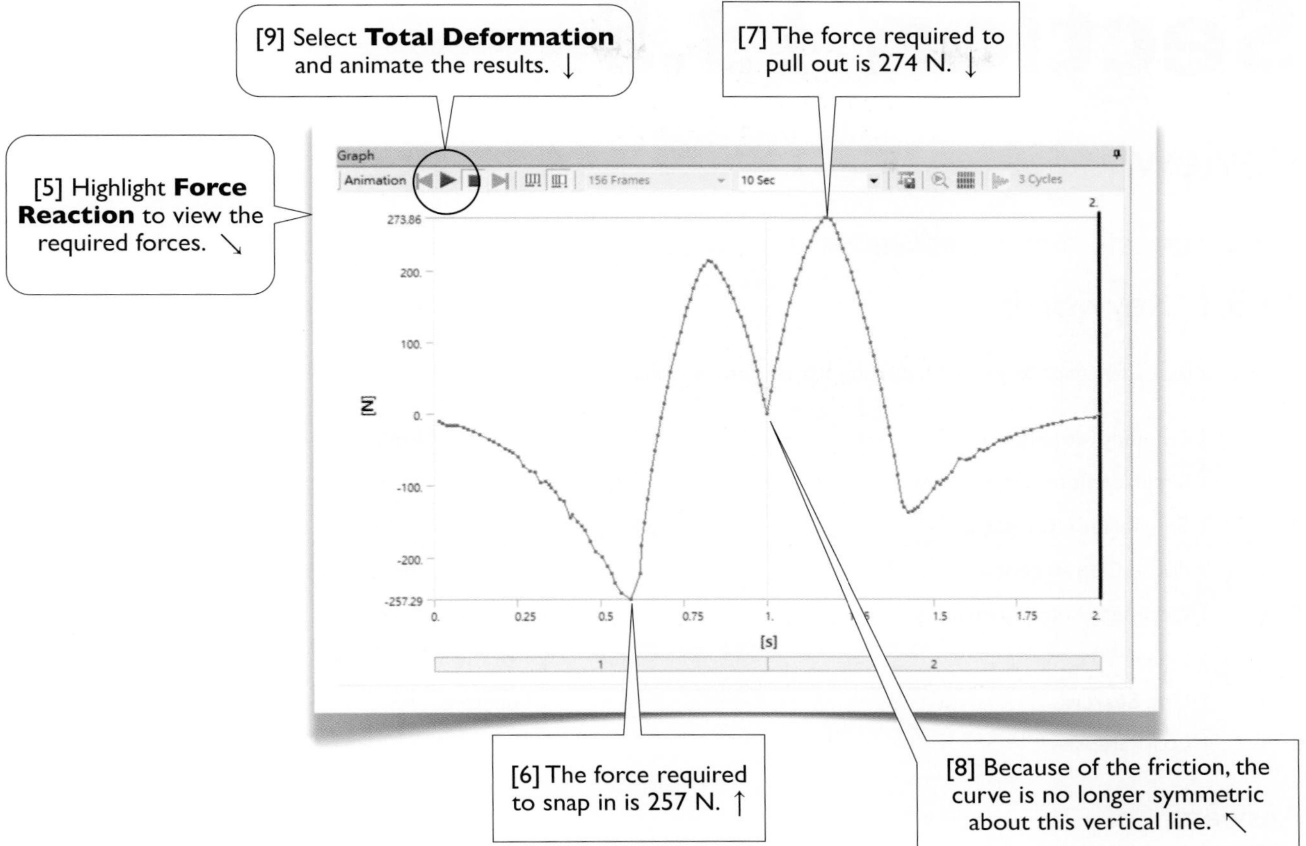

Wrap Up

[10] Save the project and exit Workbench. #

Section 13.5

Review

13.5.1 Keywords

Choose a letter for each keyword from the list of descriptions

1. ( ) Contact Nonlinearity
2. ( ) Displacement Convergence
3. ( ) Equilibrium Iterations
4. ( ) Force Convergence
5. ( ) Geometry Nonlinearity
6. ( ) Linear Structures
7. ( ) Line Search
8. ( ) Load Steps
9. ( ) Material Nonlinearity
10. ( ) Moment Convergence
11. ( ) Newton-Raphson Method
12. ( ) Nonlinear Structures
13. ( ) Residual Force
14. ( ) Rotation Convergence
15. ( ) Substeps

Answers:

1. (D) 2. (L) 3. (H) 4. (K) 5. (C) 6. (A) 7. (O) 8. (F) 9. (E) 10. (M)
11. (I) 12. (B) 13. (J) 14. (N) 15. (G)

List of Descriptions

(A) The structures in which the relation between the responses and the loads is linear.

(B) The structures in which the relation between the responses and the loads is not linear.

(C) Nonlinearity due to large deformation.

(D) Nonlinearity due to the change of contact status.

(E) Nonlinearity due to the presence of nonlinear stress-strain relation.

(F) Also called steps. In Workbench simulations, the entire loading history can be divided into one or more load steps, so that different analysis settings can be specified for each load step.

(G) Also called time steps. In dynamic simulations, time step size is used for integration over the time domain; the time step size must be small enough to capture the response characteristics. In static simulation, a load step can be divided into substeps small enough to achieve or enhance convergence.

(H) Also called iterations. For nonlinear problems, each time step needs several iterations to complete. For linear problems, each time step requires exactly one equilibrium iteration.

(I) The method used by Workbench to solve a substep in a nonlinear simulation. External force of that substep is applied. Equilibrium equation is solved for the displacement using the tangent stiffness. Internal force is calculated using updated displacement and stiffness. This completes an iteration. The process iterates until all the active convergence criteria are satisfied.

(J) During the Newton-Raphson equilibrium iterations, the difference between external force and the calculated internal force is called a residual force.

(K) A substep is said to satisfy the force convergence when the norm of the residual force is less than a criterion.

(L) A substep is said to satisfy the displacement convergence when the norm of the difference of displacements between two iterations is less than a criterion.

(M) When shell or beam elements are used in a model, moment convergence can be activated. A substep is said to satisfy the moment convergence when the norm of the residual moment is less than a criterion.

(N) When shell or beam elements are used in a model, rotation convergence can be activated. A substep is said to satisfy the rotation convergence when the norm of the difference of rotations between two iterations is less than a criterion.

(O) In some cases when a force-displacement relation is highly nonlinear or "concave up," during the Newton-Raphson iterations of a substep, the calculated displacement in a single iteration may overshoot the goal. In such cases, a numerical technique called line search can be activated to "scale down" the incremental displacement. Line search often helps convergence, but takes extra computing time.

13.5.2 Additional Workbench Exercises

Contact Stiffness Study

In 13.1.10[4] (page 478), when introducing **Pure Penalty** formulation, we mentioned that, in many cases, solution convergence behavior may be sensitive to **Normal Stiffness**. VM63[Ref 1] of **Verification Manual for the Mechanical APDL Application** is a good exercise to study this behavior. First, complete a simulation using all the parameters given in the verification manual, and verify the solution with the verification manual. Next, increase the Young's modulus from the original value (1 GPa) to 200 GPa (which is the Young's modulus of steel). This should create a convergence difficulty. When you run into a convergence difficulty, try to ease the difficulty by decreasing **Normal Stiffness**. This will introduce more penetration into the solution. Tabulate data (or draw curve) to show how **Normal Stiffness** affects penetration and other results.

Reference

1. ANSYS Help//Verification Manuals//ANSYS Mechanical APDL Verification Manual//I. Verification Test Case Descriptions//VM63: Static Hertz Contact Problem

Chapter 14

Nonlinear Materials

When the stress-strain relation of a material is linear, it is called a *linear material*. On the other hand, if the stress-strain relation is not linear, it is called a *nonlinear material*. Stress-strain relation of an isotropic linear material can be expressed by Hooke's law, Eq. 1.2.8(1) (page 31), in which two independent material parameters are needed to define the stress-strain relation. So far, in foregoing simulations, we assumed the materials are linear.

In reality, most of the materials exhibit nonlinearities to some degree. In many cases, the nonlinearities are negligible, and we use Hooke's law to describe the stress-strain relation. In some other cases, when the material nonlinearities are not negligible, *nonlinear material models* must be used to define stress-strain relations. A material model is usually a mathematics form with some parameters, called *material parameters*. To assign a material to a body, you select a material model from Workbench and provide its material parameters. The material parameters are usually obtained by data-fitting using the results of a set of material testings.

Purpose of This Chapter

Workbench provides a variety of nonlinear material models. In this chapter, we will introduce two categories of nonlinear material models: *plasticity* and *hyperelasticity*. Background knowledge will be introduced first, and two step-by-step exercises will follow to demonstrate their applications.

About Each Section

Section 14.1 gives basics of plasticity and hyperelasticity. Section 14.2 provides a step-by-step example to demonstrate the use of a plastic material model. Section 14.3 uses another step-by-step example to demonstrate the application of a hyperelastic material model.

Section 14.1

Basics of Nonlinear Materials

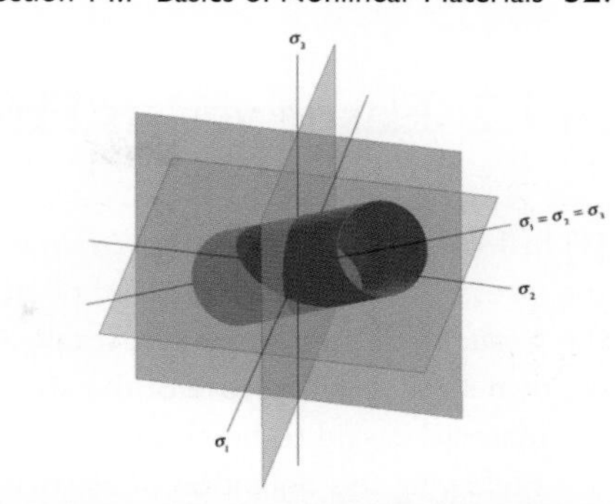

PART A. INTRODUCTION

14.1.1 Linear versus Nonlinear Materials

[1] When the stress-strain relation of a material is linear, it is called a *linear material* [2], otherwise the material is called a *nonlinear material*. For an isotropic linear elastic material, the stress-strain relation can be expressed by Eq. 1.2.8(1) (page 31), in which two independent material parameters are needed to define the material. Note that Eq. 1.2.8(1) assumes an *isotropic material;* i.e., the Young's modulus and the Poisson's ratio are independent of directions. Orthotropic (Eq. 1.2.8(4), page 32) and anisotropic linear elasticity are also available in Workbench [3].

Besides a linear relation, Hooke's law also assumes that the stress-strain relation is elastic, time-independent, and rate-independent. Materials that violate any of these behaviors cannot be described by Hooke's law, and are categorized as nonlinear materials, even though the stress-strain relation is linear for both stressing and unstressing [4].

Workbench provides non-elastic material models, called plastic material models.

Time-Dependent Stress-Strain Relations

When you apply a stress on steel, the strain occurs instantaneously and remains the same forever. This statement is not strictly true. A more rigorous statement should be like this, "The strain occurs ALMOST instantaneously and ALMOST remains the same forever."

Consider a big water tank made of plastic material. Starting from the moment right after you fill up the water, the plastic tank begins to deform: the diameter expands slowly. It may take months until the deformation stops. This gives us a lesson that the deformation may take time. This time-dependent stress-strain behavior is called *creeping*, or *viscosity*.

For the steel under room temperature, the creeping behavior is practically negligible. In engineering practice, we often neglect creeping for solid materials under a fairly lower temperature, for example, below half of the melting point.

We will not discuss time-dependent behavior (although creep models are available in Workbench) for the rest of this book. ↗

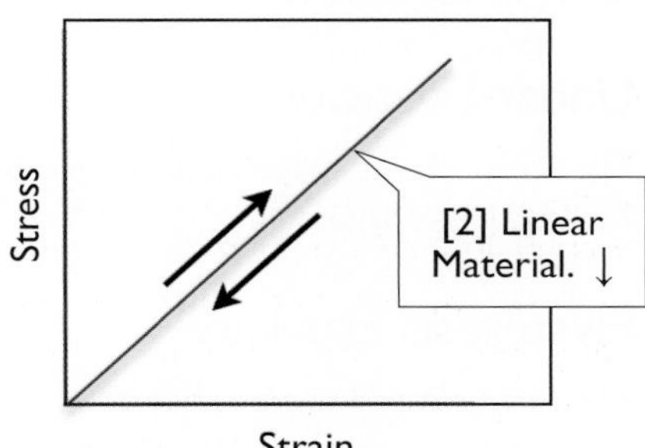

[2] Linear Material. ↓

Toolbox
Physical Properties
Linear Elastic
Isotropic Elasticity
Orthotropic Elasticity
Anisotropic Elasticity
Hyperelastic Experimental Data
Hyperelastic
Chabo
Plastic
Creep
Life

[3] **Orthotropic** and **anisotropic** linear material models are also available in Workbench. ↓

[4] This is not a linear material, even though the stress-strain relation is linear for both stressing and unstressing. #

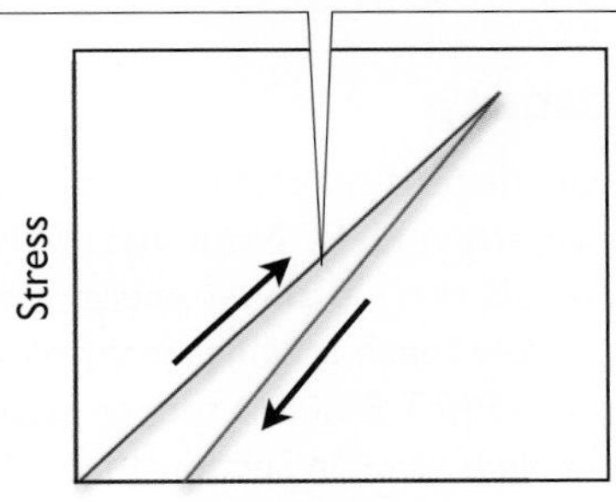

14.1.2 Elastic versus Plastic Materials

[1] In 14.1.1[4] (last page), the strain is not totally recovered after release of the stress. This behavior is called plasticity, and the residual strain is called the *plastic strain*. If the strain is totally recoverable, that is, if there is no residual strain after release of the stress, the behavior is called elasticity, and the material is said to be elastic.

Following this definition of elasticity, we may classify the elastic materials into three categories: (1) linear elastic, (2) nonlinear hysteresis elastic, and (3) nonlinear non-hysteresis elastic, or simply called nonlinear elastic.

Linear Elasticity

The linear elastic material is defined by Hooke's law and depicted in 14.1.1[2], last page.

Hysteresis Elasticity

The term *elastic hysteresis* has been introduced in 12.1.3[4-7] (page 424). The current version of ANSYS (even with APDL) doesn't directly provide a material model to include the hysteresis behavior. However, you may include the hysteresis behavior in terms of material damping (12.1.3, page 423-425).

Most materials have hysteresis behavior to some extent. However, as long as it is small enough, we may neglect the hysteresis behavior.

Nonlinear Elasticity

Nonlinear non-hysteresis elastic materials are characterized by the fact that the stressing curve and the unstressing curve are coincident [2]: the energy is conserved in stressing-unstressing cycles.

The challenge of implementing nonlinear elastic material models is that the strain may be as large as 100% or even 200%, such as rubber under stretching or compression. Under such large strains, the stretching and compression behaviors may not be described by the same parameters. This kind of super-large deformation elasticity is given a special name: *hyperelasticity*.

Workbench provides many hyperelastic material models and we will discuss them in PART C of this section (pages 532-535). A step-by-step example is given in Section 14.3.

Plasticity

Plastic materials are characterized by the presence of the residual strain, or plastic strain [3-4]. Note that the hysteresis is always present in plastic materials: there is always energy loss in stressing-unstressing cycles.

Workbench provides many plastic material models and we will discuss them in PART B of this section (starting from next page). A step-by-step example is given in Section 14.2. ↗

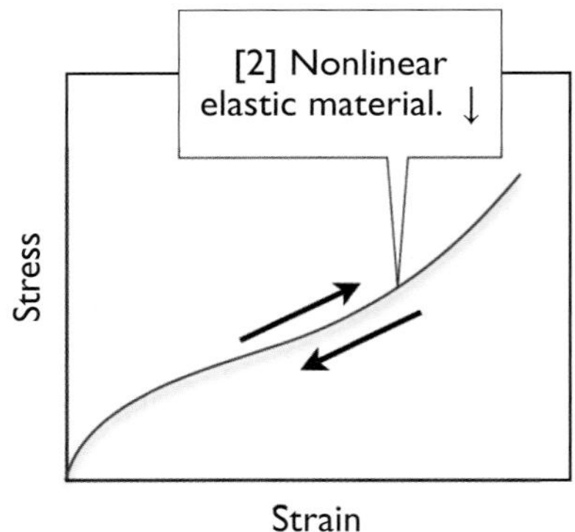

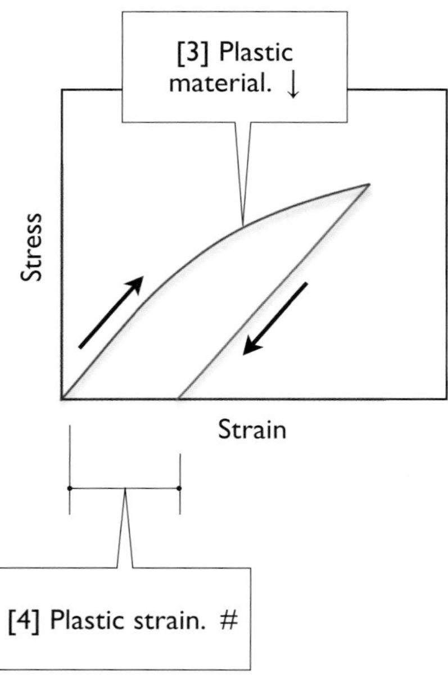

PART B. PLASTICITY

14.1.3 Idealized Stress-Strain Curve for Plasticity

[1] Plasticity behavior typically occurs in ductile metals subject to large deformation. Plastic deformation is a result of slips between grains due to shear stresses. It is essentially a rearrangement of atoms in the crystal structure.

In Workbench, a typical stress-strain relation, such as 14.1.2[3] (last page), is idealized as shown in [2-5]. The stress-strain curve is composed of several straight segments. The slope of the first segment is the Young's modulus [4]. When the stress is released, the strain decreases with a slope equal to the Young's modulus [5]. This implies that if the stress/strain state is on the first segment, the behavior is elastic and no plastic strain remains after releasing the stress. The point at the end of the first segment is called *elastic limit*, or *initial yield point* [3]. All points higher than the initial yield point are called *subsequent yield points*, since they all represent *yield states*.

A uniaxial stress-strain relation such as [2-5] is not sufficient to fully define a plasticity behavior. There are other characteristics that must be described for general multiaxial cases: (a) What is the yield criterion? (b) What is the hardening rule? ↗

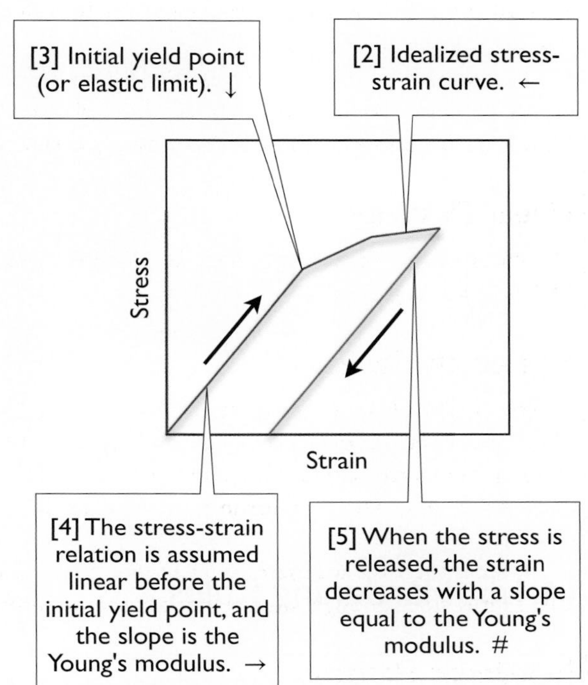

14.1.4 Yield Criteria

[1] A stress-strain curve such as 14.1.3[2-5] is usually obtained by a uniaxial tensile test. It provides an initial yield strength σ_y of the uniaxial tensile test. In three-dimensional cases, the stress state is multiaxial. According to what criteria can we say that a stress state reaches a yield state? Workbench uses von Mises criterion (1.4.5, pages 43-46) as the yield criterion; i.e., a stress state reaches yield state when the von Mises stress σ_e is equal to the CURRENT *uniaxial yield strength* σ_y', or

$$\sqrt{\frac{1}{2}\left[(\sigma_1-\sigma_2)^2+(\sigma_2-\sigma_3)^2+(\sigma_3-\sigma_1)^2\right]}=\sigma_y' \qquad (1)$$

The yielding initially occurs when $\sigma_y' = \sigma_y$, and the "current" uniaxial yield strength σ_y' may change subsequently. As mentioned in 1.4.5[4] (page 46), when plotted in the σ_1-σ_2-σ_3 space, Eq. (1) is a cylindrical surface aligned with the axis $\sigma_1 = \sigma_2 = \sigma_3$ and with a radius of $\sqrt{2}\sigma_y'$. It is called a *von Mises yield surface* ([2], next page). If the stress state is inside the cylinder, no yielding occurs. If the stress state is on the surface, yielding occurs. No stress state can be outside the yield surface. If the stress state is on the surface and the loads continue to "push" the yield surface outward, the size (radius) or the location of the yield surface will change. The rule that describes how the yield surface changes its size or location is called a *hardening rule* (14.1.5, next page).

Note that, in a uniaxial test, we are talking about "yield points" in the stress axis. In a biaxial case, the yielding states form a "yield line" in a stress plane, while in 3D cases, the yielding states become a "yield surface" in a stress space ([2], next page). ↵

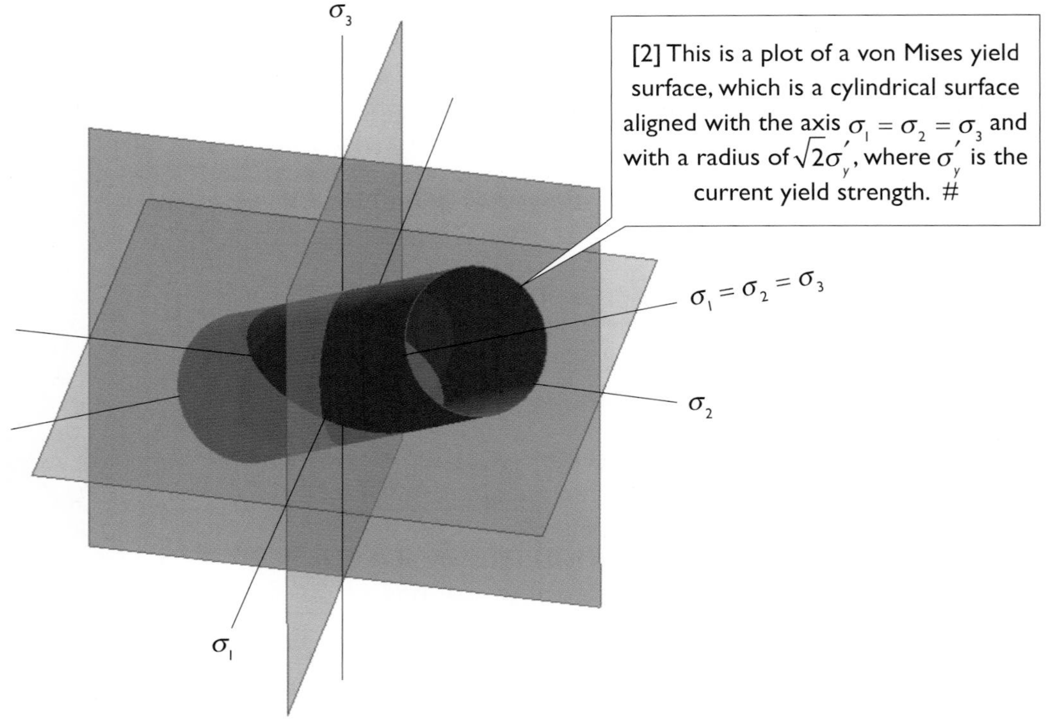

14.1.5 Hardening Rules

[1] Workbench implements two hardening rules: (a) kinematic hardening, and (b) isotropic hardening. In metal plasticity, hardening behavior is often a mix-up of kinematic and isotropic.

Kinematic Hardening

Kinematic hardening assumes that, if a stress state is on the yield surface and the loads continue to "push" a yield surface outward, the yield surface will change its location, according to the "pushing direction," but preserve the size of the yield surface. In a uniaxial test, it is equivalent to say that the difference between the tensile yield strength and the compressive yield strength remains a constant of $2\sigma_y$ [2].

Kinematic hardening is generally used for small strain, cyclic loading applications, especially metals. ↓

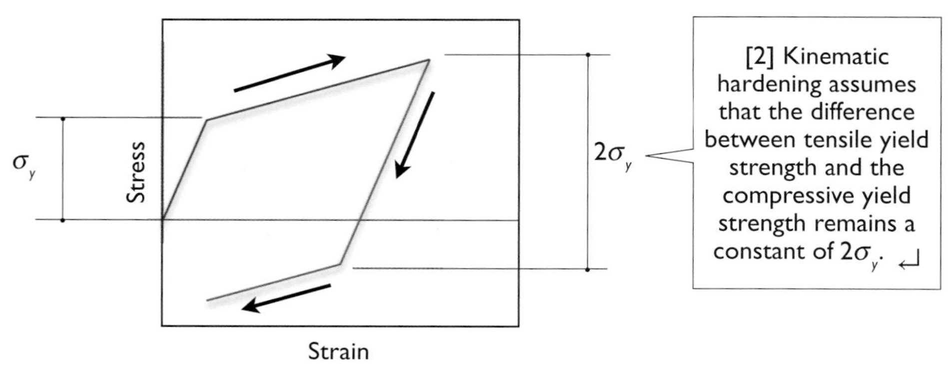

Isotropic Hardening

[3] Isotropic hardening assumes that, when the loads continue to "push" a yield surface, the yield surface will expand its size, but preserve the axis of the yield surface. In a uniaxial test, it is equivalent to say that the "current" tensile yield strength and the compressive yield strength remain equal in magnitude [4].

Isotropic hardening is often used for large strain simulations. It is usually not applicable for cyclic loading applications. ↘

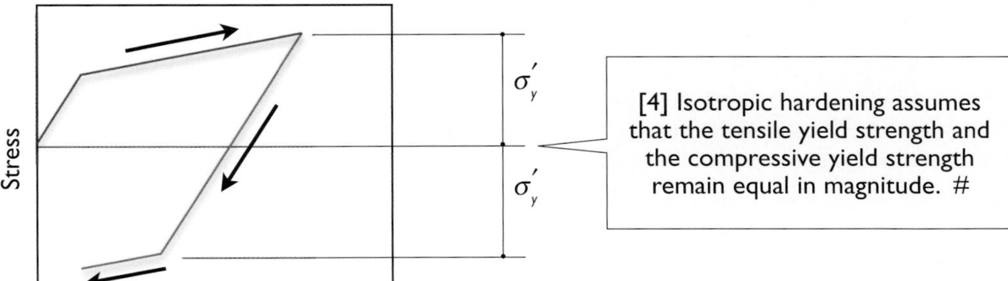

[4] Isotropic hardening assumes that the tensile yield strength and the compressive yield strength remain equal in magnitude. #

14.1.6 Workbench Plasticity Models

[1] Workbench provides many plasticity models [2]. Besides choosing from either of the hardening rules, you can choose from either type of stress-strain curves: bilinear or multilinear.

Linear Elastic Properties Must be Included

To describe a material model for plasticity, you must include a set of linear elastic properties (e.g., Young's modulus and Poisson's ratio in cases of isotropy). The Young's modulus is used as the initial slope of the stress-strain curve, in which the material initially behaves as linear elasticity.

Bilinear Stress-Strain Curve

Examples of bilinear models are shown in 14.1.5[2, 4] (last and this pages); the stress-strain curve is composed of two straight segments. Besides the Young's modulus and Poisson's ratio, you need to supply an *initial yield stress* and a *tangent modulus* (the slope of the second segment).

Multilinear Stress-Strain Curve

An example of multilinear models was shown in 14.1.3[2-5] (page 529); the stress-strain curve is composed of several straight segments. Besides the Young's modulus and Poisson's ratio, you need to supply a tabular form of data describing the *subsequent yield stresses* and the corresponding *plastic strains*. An example is given in 14.2.6[6-10], pages 548-549. →

[3] To complete a description of plasticity model, you must include its linear elastic properties (e.g., Young's modulus and Poisson's ratio for isotropic materials). #

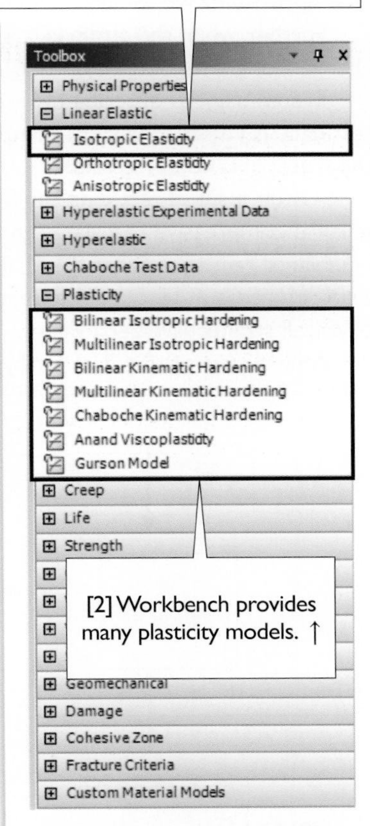

[2] Workbench provides many plasticity models. ↑

PART C. HYPERELASTICITY

14.1.7 Test Data Needed for Hyperelasticity

[1] As mentioned in 14.1.2[1] (page 528), the challenge of implementing nonlinear elastic models is that the strain may be as large as 100% (or even 200%), such as rubber under stretching.

In plasticity or linear elasticity, we use a stress-strain curve to describe its behavior, and the stress-strain curve is usually obtained by a uniaxial tensile test. Since only tension behavior is investigated, other behaviors (e.g., compressive, shearing) must be drawn from the tensile test data. In plasticity or linear elasticity, we implicitly made the following assumptions: (a) The compressive behavior is symmetric to the tension behavior; i.e., they have the same Young's modulus, and the same Poisson's ratio. The symmetry may not be true when the strain is large. We may need to conduct a compressive test to assess the compressive behavior. (b) The shear modulus G is related to the Young's modulus and the Poisson's ratio by Eq. 1.2.8(2) (page 31). Again, this assumption may not be true when the strain is large. We may need to conduct a shear test to assess the shearing behavior. (c) We also assume that the bulk modulus B is related to the Young's modulus and the Poisson's ratio by

$$B = \frac{E}{3(1-2\nu)} \quad (1)$$

Again, this assumption may not be true when the strain is large. We may need to conduct a volumetric test to assess the volumetric behavior. However, in many cases, when the bulk modulus is almost infinitely large (i.e., the material is incompressible), we usually assume incompressibility without conducting a volumetric test.

Further, when the strain is large, all the moduli (tensile, compressive, shear, and bulk) are no longer constant; they change along stress-strain curves. Nonlinear elasticity with large strain is also called *hyperelasticity*.

In summary, to describe hyperelasticity behavior, we may need the following test data: (a) a set of uniaxial tensile test data, (b) a set of uniaxial compressive test data, (c) a set of shear test data, and (d) a set of volumetric test data if the material is compressible.

Often, a set of test data can be obtained by superposing two sets of other test data. For example, the set of uniaxial compressive test data can be obtained by adding a set of hydrostatic compressive test data to a set of equibiaxial tensile test data [2-4]. There are two reasons for doing this: (a) Biaxial tensile test may be easier to conduct than compressive test in some testing devices; (b) For incompressible materials, hydrostatic compressive test data are trivial: all strains have zero values.

An example of test data for hyperelasticity is shown in [5-7] (next page), which will be used in Section 14.3. ↙

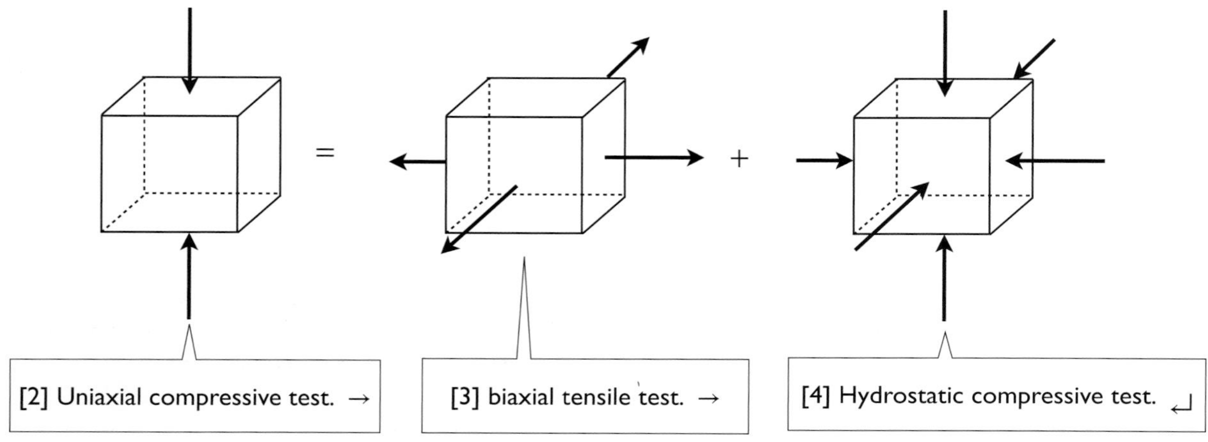

[2] Uniaxial compressive test. →

[3] biaxial tensile test. →

[4] Hydrostatic compressive test. ↲

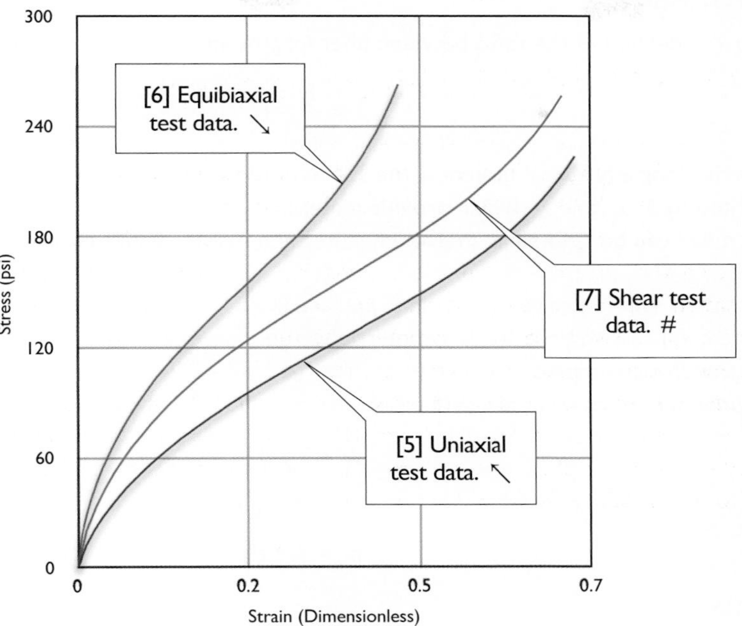

14.1.8 Strain Energy Function

[1] Workbench provides a material model, called **Response Function** (see 14.1.9[2], page 535), which uses experimental data (such as 14.1.7[5-7], this page) directly. A drawback of using **Response Function** model is that it may not be efficient enough--it may cost too many iterations, even if convergence is eventually achieved.

A better idea is described as follows.

As mentioned in 14.1.2[1] (page 528), hyperelasticity is characterized by the fact that the stressing curve and the unstressing curve are coincident (14.1.2[2], page 528). During the stressing and unstressing, the energy is conserved, or, in other words, the stressing and unstressing are path independent. The stress state depends only on the strain state, and vice versa. They are independent of the stressing/unstressing history. This implies that there exists a *potential energy function* that depends on the state of the stress or strain. It reminds us of the strain energy density function, which does depend only on the state of stress or strain. With these in mind, we propose a mathematical form for the strain energy

$$W = W(\varepsilon_{ij}) \tag{1}$$

And the stress can be calculated from the strain energy using

$$\sigma_{ij} = \frac{\partial W}{\partial \varepsilon_{ij}} \tag{2}$$

The strain state ε_{ij} consists of 6 strain components (Eq. 1.2.4(4), page 28). To further simplify the strain energy function and develop a coordinate-independent expression, we may replace the 6 strain components (which are coordinate-dependent) with 3 strain invariants (which are coordinate-independent). Before going further, we need more background. Let's introduce some terms in solid mechanics. ↵

Principal Stretch Ratios

[2] The *stretch ratio* λ is defined as the ratio between fiber lengths after and before deformation,

$$\lambda = \frac{L}{L_0} = 1 + \varepsilon \tag{3}$$

When the direction is along a principal direction, the stretch ratio is called a *principal stretch ratio*. There are 3 principal stretch ratios, denoted by λ_1, λ_2, and λ_3, which provide a measure of the deformation.

The *volumetric ratio* J can be defined as the volume after and before deformation,

$$J = \frac{V}{V_0} = \lambda_1 \lambda_2 \lambda_3 \tag{4}$$

Note that, if the material is incompressible, $J = 1$.

The *deviatoric principal stretch ratios* are defined as

$$\begin{aligned} \bar{\lambda}_1 &= \lambda_1 / \sqrt[3]{J} \\ \bar{\lambda}_2 &= \lambda_2 / \sqrt[3]{J} \\ \bar{\lambda}_3 &= \lambda_3 / \sqrt[3]{J} \end{aligned} \tag{5}$$

Strain Invariants

Let I_1, I_2, and I_3 be the characteristic values (eigenvalues) of the strain state; they are also called *strain invariants*. It can be proved that

$$\begin{aligned} I_1 &= \lambda_1^2 + \lambda_2^2 + \lambda_3^2 \\ I_2 &= \lambda_1^2 \lambda_2^2 + \lambda_2^2 \lambda_3^2 + \lambda_3^2 \lambda_1^2 \\ I_3 &= \lambda_1^2 \lambda_2^2 \lambda_3^2 \end{aligned} \tag{6}$$

The *deviatoric strain invariants* are defined as

$$\begin{aligned} \bar{I}_1 &= \lambda_1 / \sqrt[3]{J^2} \\ \bar{I}_2 &= \lambda_2 / \sqrt[3]{J^2} \\ \bar{I}_3 &= \lambda_3 / \sqrt[3]{J^2} \end{aligned} \tag{7}$$

Strain Energy Functions

We can replace the 6 strain components in Eq. (1) with either strain invariants or principal stretch ratios; i.e.,

$$W = W(I_1, I_2, I_3)$$

or

$$W = W(\lambda_1, \lambda_2, \lambda_3)$$

Or, we can split the strain energy into deviatoric part and volumetric part, and write

$$W = W_d(\bar{I}_1, \bar{I}_2) + W_b(J) \tag{8}$$

or

$$W = W_d(\bar{\lambda}_1, \bar{\lambda}_2, \bar{\lambda}_3) + W_b(J) \tag{9}$$

Note that $I_3 = J^2$, so $\bar{I}_3$ is not used in the definition of W. #

14.1.9 Workbench Hyperelasticity Models

[1] Workbench provides many hyperelasticity models [2]; they are based on either Eq. 14.1.8(8) or Eq. 14.1.8(9), last page. ↘

Polynomial Form

[3] The polynomial form is based on Eq. 14.1.8(8), last page,

$$W = \sum_{i+j=1}^{N} c_{ij}(\bar{I}_1 - 3)^i(\bar{I}_2 - 3)^j + \sum_{k=1}^{N} \frac{1}{d_k}(J-1)^{2k} \quad (1)$$

For example, **Polynomial 1st Order** (N = 1) is

$$W = c_{10}(\bar{I}_1 - 3) + c_{01}(\bar{I}_2 - 3)^1 + \frac{1}{d_1}(J-1)^2 \quad (2)$$

It has three parameters, c_{10}, c_{01}, and d_1. Note that, for incompressible materials, $J = 1, d_1 = 0$, and the last term is dropped.

Ogden Form

The Ogden form is based on Eq. 14.1.8(9), last page,

$$W = \sum_{i=1}^{N} \frac{\mu_i}{\alpha_i}(\bar{\lambda}_1^{\alpha_i} + \bar{\lambda}_2^{\alpha_i} + \bar{\lambda}_3^{\alpha_i} - 3) + \sum_{i=1}^{N} \frac{1}{d_i}(J-1)^{2i} \quad (3)$$

For example, **Ogden 1st Order** (N = 1) is

$$W = \frac{\mu_1}{\alpha_1}(\bar{\lambda}_1^{\alpha_1} + \bar{\lambda}_2^{\alpha_1} + \bar{\lambda}_3^{\alpha_1} - 3) + \frac{1}{d_1}(J-1)^2 \quad (4)$$

It has three parameters, μ_1, α_1, and d_1.

Mooney-Rivlin, Yeoh, and Neo-Hookean

These are reduced forms of the generalized polynomial.

Which Form to Use?

The choice depends on type of material, maximum strain, and the test data available. In general, the best form of strain energy density function is the one that produces the closest curve fit of the test data.

Section 14.3 provides an example to demonstrate some details. #

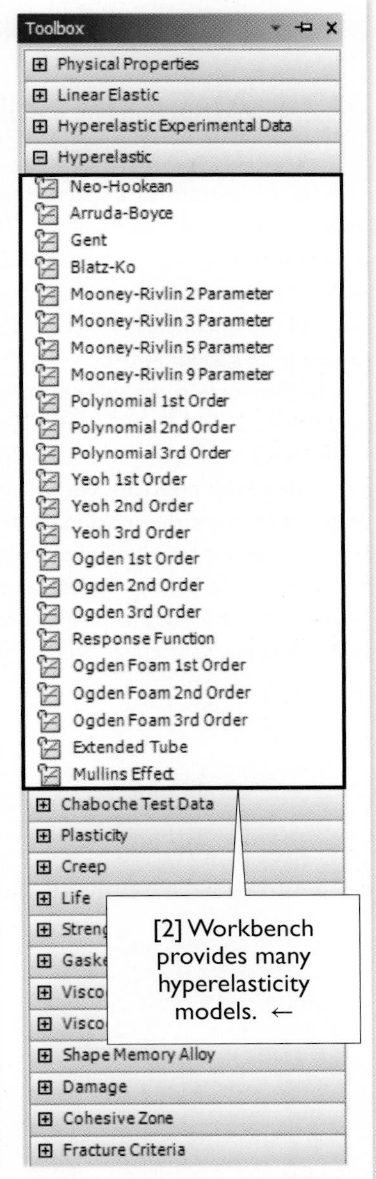

Section 14.2

Belleville Washer

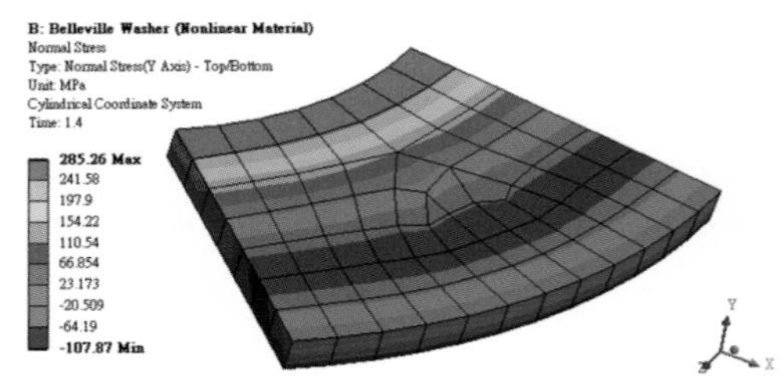

14.2.1 About the Belleville Washer

[1] The dimensions of a *Belleville washer* [2], also called a *Belleville spring*, are shown in [3]. The washer is made of a steel with a Young's modulus of 200 GPa, a Poisson's ratio of 0.3, and an initial yield stress of 250 MPa. Beyond the initial yield stress, it displays plasticity behavior as shown in [4].

In this section, we will compress the Belleville spring by 1.0 mm and then release the displacement completely. As a preliminary study, before performing a simulation with plasticity, we will assume a linear material to see if the maximum stress exceeds the yield stress. If so, then we will explore the plasticity behavior of the Belleville spring. Specifically, we will examine the residual stress after the spring is completely released. A force-displacement curve will be plotted. ↗

[2] A Belleville washer[Ref 1]. ↙

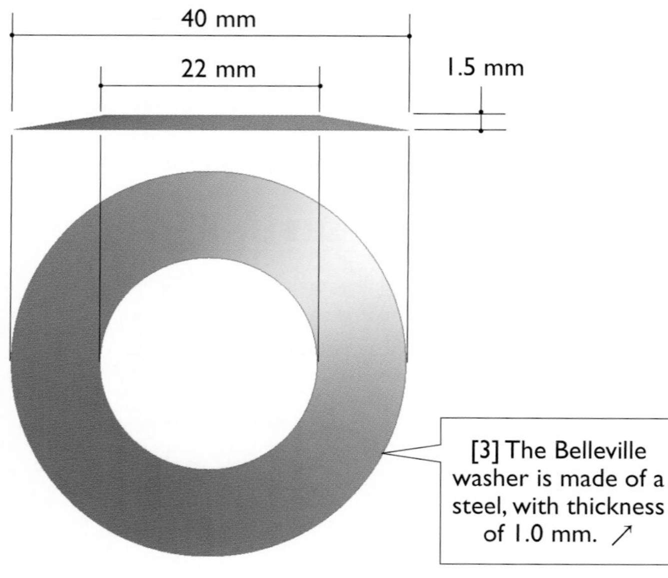

[3] The Belleville washer is made of a steel, with thickness of 1.0 mm. ↗

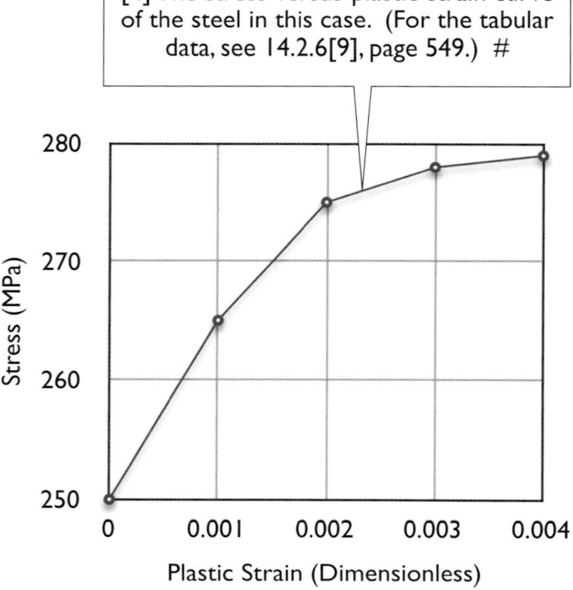

[4] The stress-versus-plastic-strain curve of the steel in this case. (For the tabular data, see 14.2.6[9], page 549.) #

14.2.2 Start Up

[1] Launch Workbench. Create a **Static Structural** analysis system. Save the project as **Belleville**. ↓

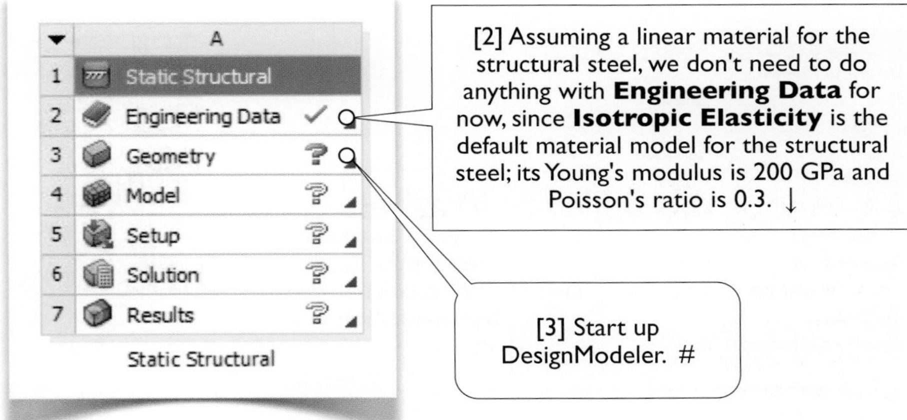

14.2.3 Create Geometry

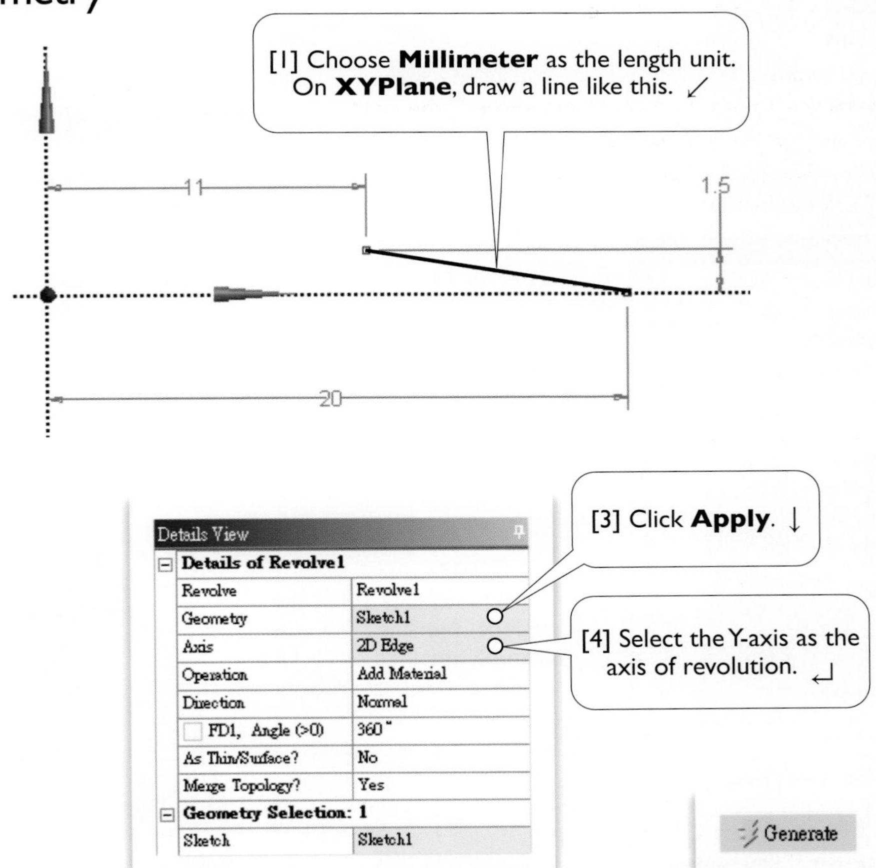

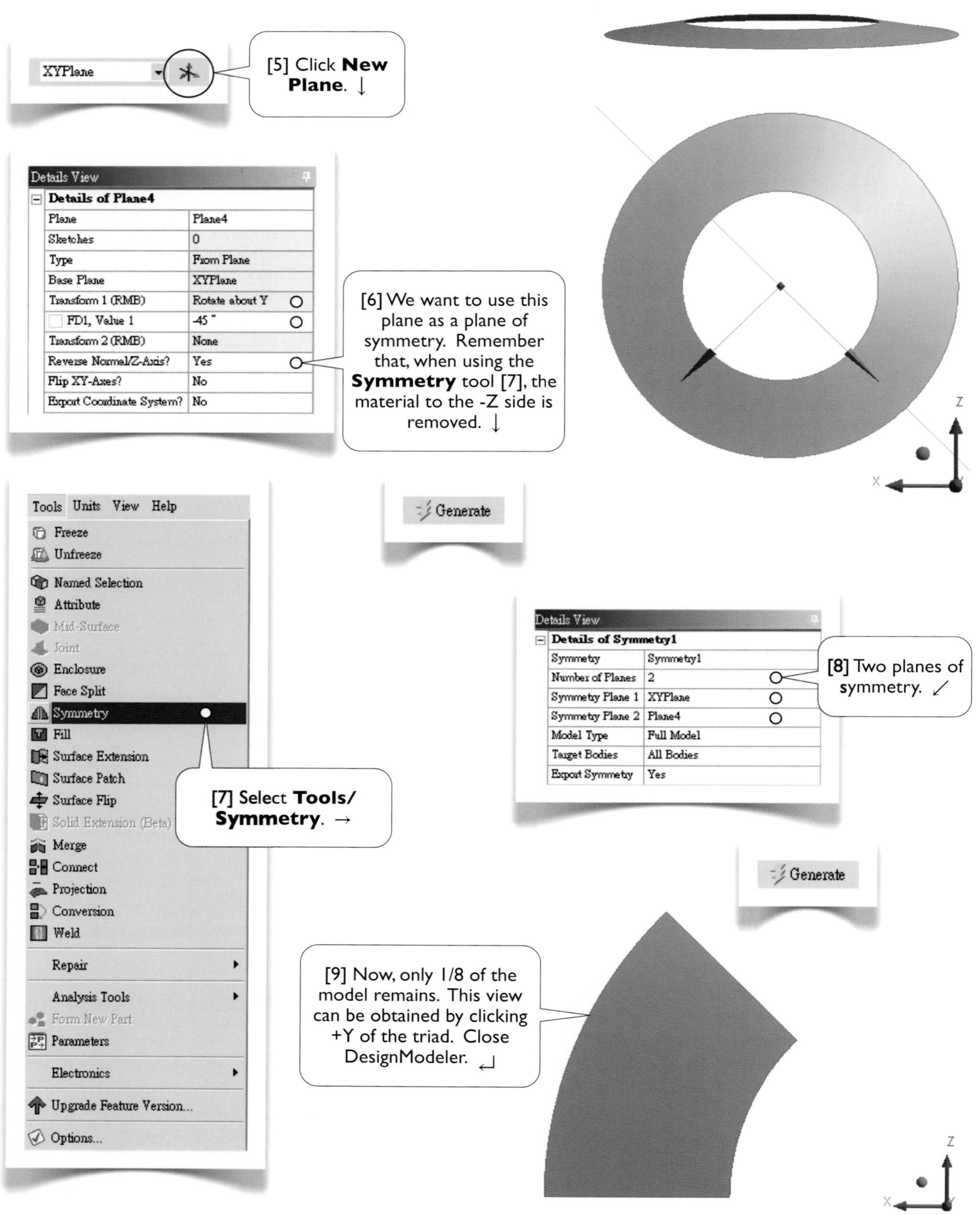
XYPlane
[5] Click **New Plane**. ↓
Details View
Details of Plane4
Plane | Plane4
Sketches | 0
Type | From Plane
Base Plane | XYPlane
Transform 1 (RMB) | Rotate about Y
FD1, Value 1 | -45 °
Transform 2 (RMB) | None
Reverse Normal/Z-Axis? | Yes
Flip XY-Axes? | No
Export Coordinate System? | No
[6] We want to use this plane as a plane of symmetry. Remember that, when using the **Symmetry** tool [7], the material to the -Z side is removed. ↓
Z
X
Tools Units View Help
Freeze
Unfreeze
Named Selection
Attribute
Mid-Surface
Joint
Enclosure
Face Split
Symmetry
Fill
Surface Extension
Surface Patch
Surface Flip
Solid Extension (Beta)
Merge
Connect
Projection
Conversion
Weld
Repair
Analysis Tools
Form New Part
Parameters
Electronics
Upgrade Feature Version...
Options...
[7] Select **Tools/ Symmetry**. →
Generate
Details View
Details of Symmetry1
Symmetry | Symmetry1
Number of Planes | 2
Symmetry Plane 1 | XYPlane
Symmetry Plane 2 | Plane4
Model Type | Full Model
Target Bodies | All Bodies
Export Symmetry | Yes
[8] Two planes of symmetry. ↙
Generate
[9] Now, only 1/8 of the model remains. This view can be obtained by clicking +Y of the triad. Close DesignModeler. ↵
Z
X

[10] There are actually infinite planes of symmetry for this mode (any plane derived from rotating **XYPlane** about Y-axis could be a plane of symmetry for this model). In general, an axisymmetric geometry has an infinite number of planes of symmetry; in 2D, we can use any two of them. In this case, we decide to use 1/8 of the model. In practice, we often choose a 45° sector of the model. Alternatively, we may model the problem as an axisymmetric problem using 2D option. This is left as an exercise (14.4.2, page 569). #

14.2.4 Set Up for Simulation with Linear Material

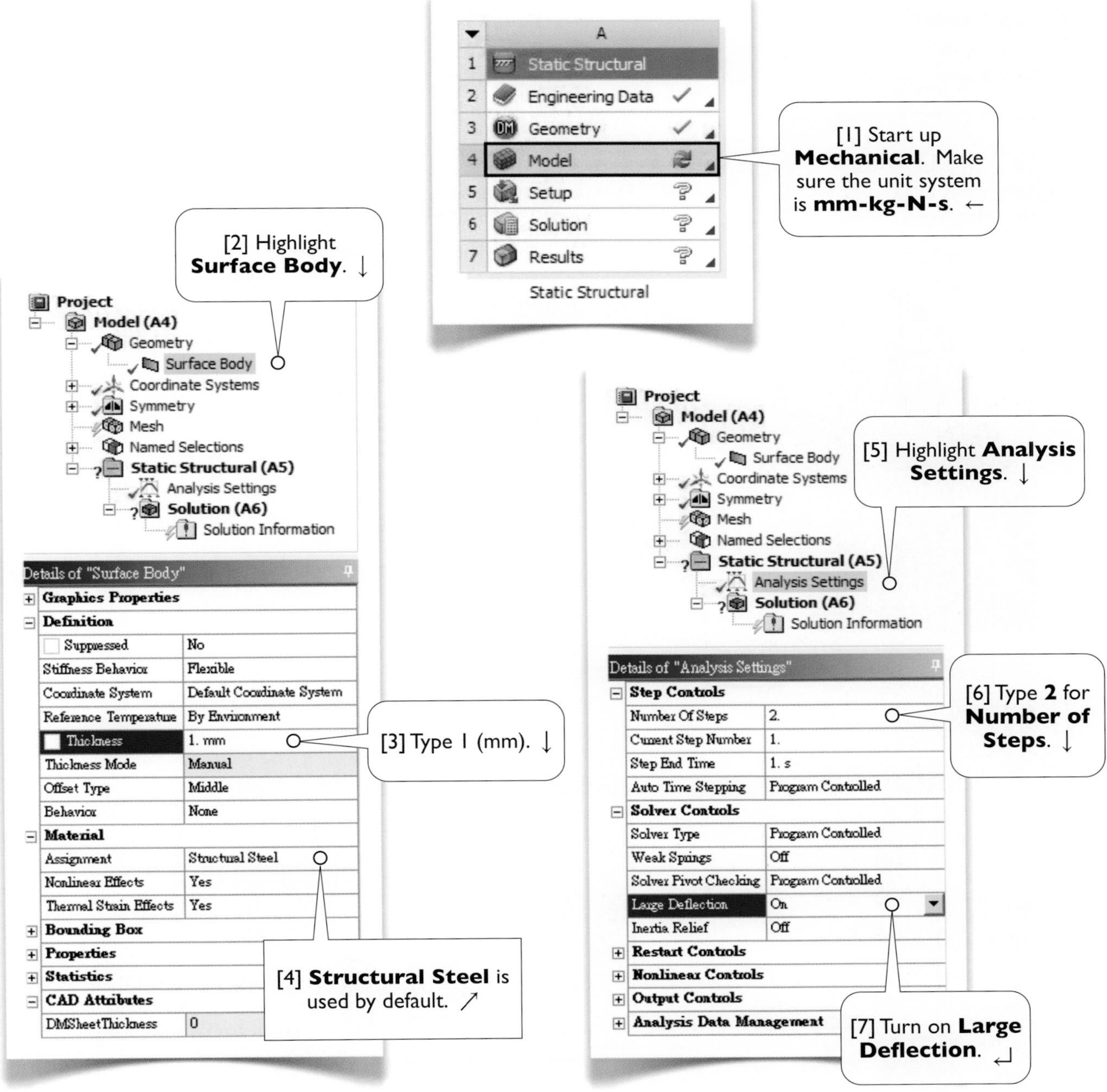

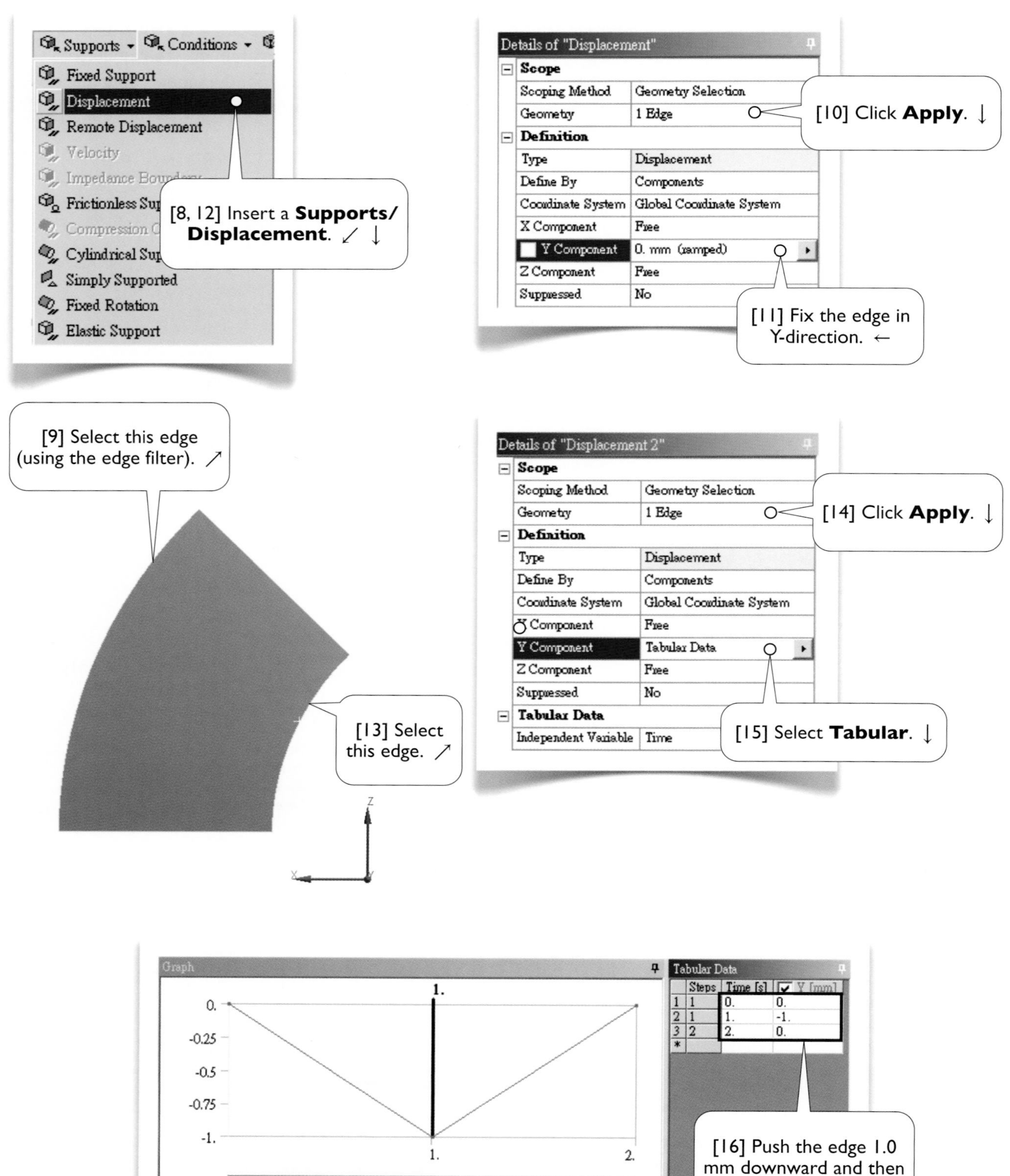
Supports
Conditions
Fixed Support
Displacement
Remote Displacement
Velocity
Impedance Boundary
Frictionless Support
Compression Only Support
Cylindrical Support
Simply Supported
Fixed Rotation
Elastic Support
[8, 12] Insert a **Supports/ Displacement**.
Details of "Displacement"
Scope
Scoping Method
Geometry Selection
Geometry
1 Edge
[10] Click **Apply**.
Definition
Type
Displacement
Define By
Components
Coordinate System
Global Coordinate System
X Component
Free
Y Component
0. mm (ramped)
Z Component
Free
Suppressed
No
[11] Fix the edge in Y-direction.
[9] Select this edge (using the edge filter).
[13] Select this edge.
Details of "Displacement 2"
Scope
Scoping Method
Geometry Selection
Geometry
1 Edge
[14] Click **Apply**.
Definition
Type
Displacement
Define By
Components
Coordinate System
Global Coordinate System
X Component
Free
Y Component
Tabular Data
Z Component
Free
Suppressed
No
Tabular Data
Independent Variable
Time
[15] Select **Tabular**.
Graph
Tabular Data
Steps
Time [s]
Y [mm]
[16] Push the edge 1.0 mm downward and then pull it back to the original position.

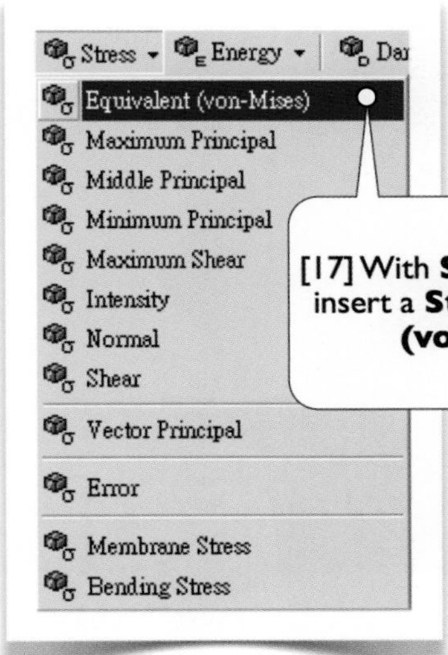

[17] With **Solution** highlighted, insert a **Stress/Equivalent (von-Mises)**. ↓

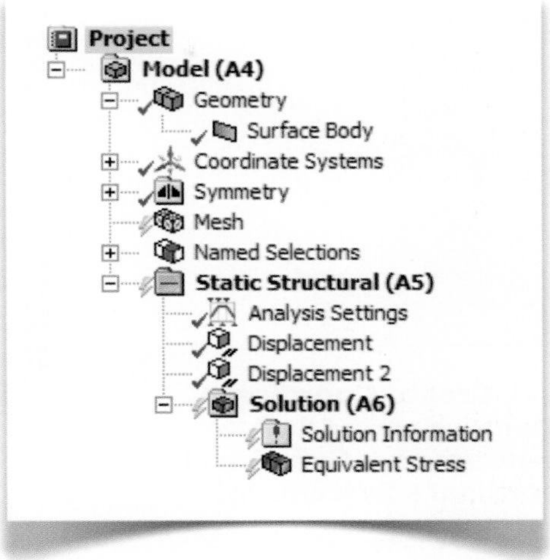

[18] Solve. Workbench completes the solution with only two substeps, at 1.0 second and 2.0 seconds. #

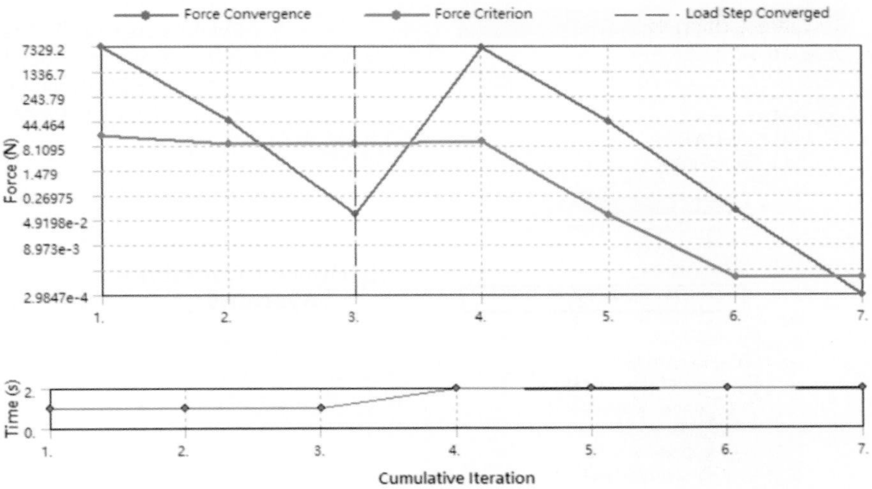

14.2.5 Results of the Linear Material Simulation

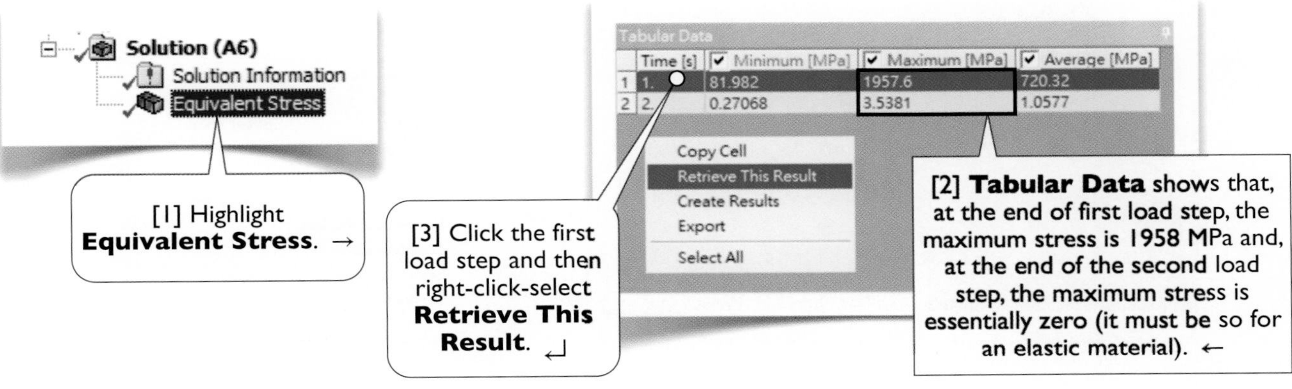

[1] Highlight **Equivalent Stress**. →

[3] Click the first load step and then right-click-select **Retrieve This Result**. ↵

[2] **Tabular Data** shows that, at the end of first load step, the maximum stress is 1958 MPa and, at the end of the second load step, the maximum stress is essentially zero (it must be so for an elastic material). ←

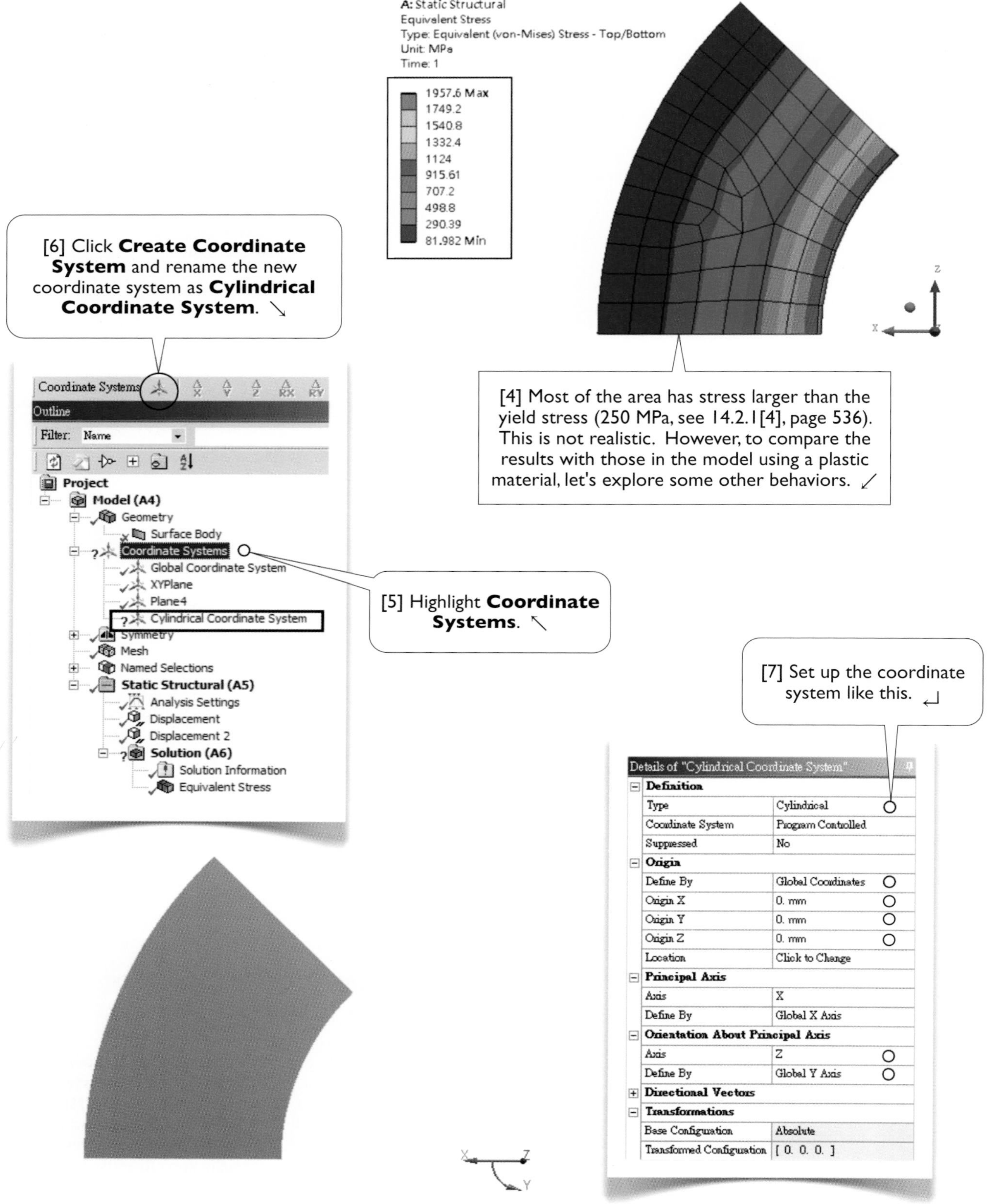
A: Static Structural
Equivalent Stress
Type: Equivalent (von-Mises) Stress - Top/Bottom
Unit: MPa
Time: 1
1957.6 Max
1749.2
1540.8
1332.4
1124
915.61
707.2
498.8
290.39
81.982 Min
[6] Click **Create Coordinate System** and rename the new coordinate system as **Cylindrical Coordinate System**.
[4] Most of the area has stress larger than the yield stress (250 MPa, see 14.2.1[4], page 536). This is not realistic. However, to compare the results with those in the model using a plastic material, let's explore some other behaviors.
Coordinate Systems
Outline
Filter: Name
Project
Model (A4)
Geometry
Surface Body
Coordinate Systems
Global Coordinate System
XYPlane
Plane4
Cylindrical Coordinate System
Symmetry
Mesh
Named Selections
Static Structural (A5)
Analysis Settings
Displacement
Displacement 2
Solution (A6)
Solution Information
Equivalent Stress
[5] Highlight **Coordinate Systems**.
[7] Set up the coordinate system like this.
Details of "Cylindrical Coordinate System"
Definition
Type | Cylindrical
Coordinate System | Program Controlled
Suppressed | No
Origin
Define By | Global Coordinates
Origin X | 0. mm
Origin Y | 0. mm
Origin Z | 0. mm
Location | Click to Change
Principal Axis
Axis | X
Define By | Global X Axis
Orientation About Principal Axis
Axis | Z
Define By | Global Y Axis
Directional Vectors
Transformations
Base Configuration | Absolute
Transformed Configuration | [0. 0. 0.]

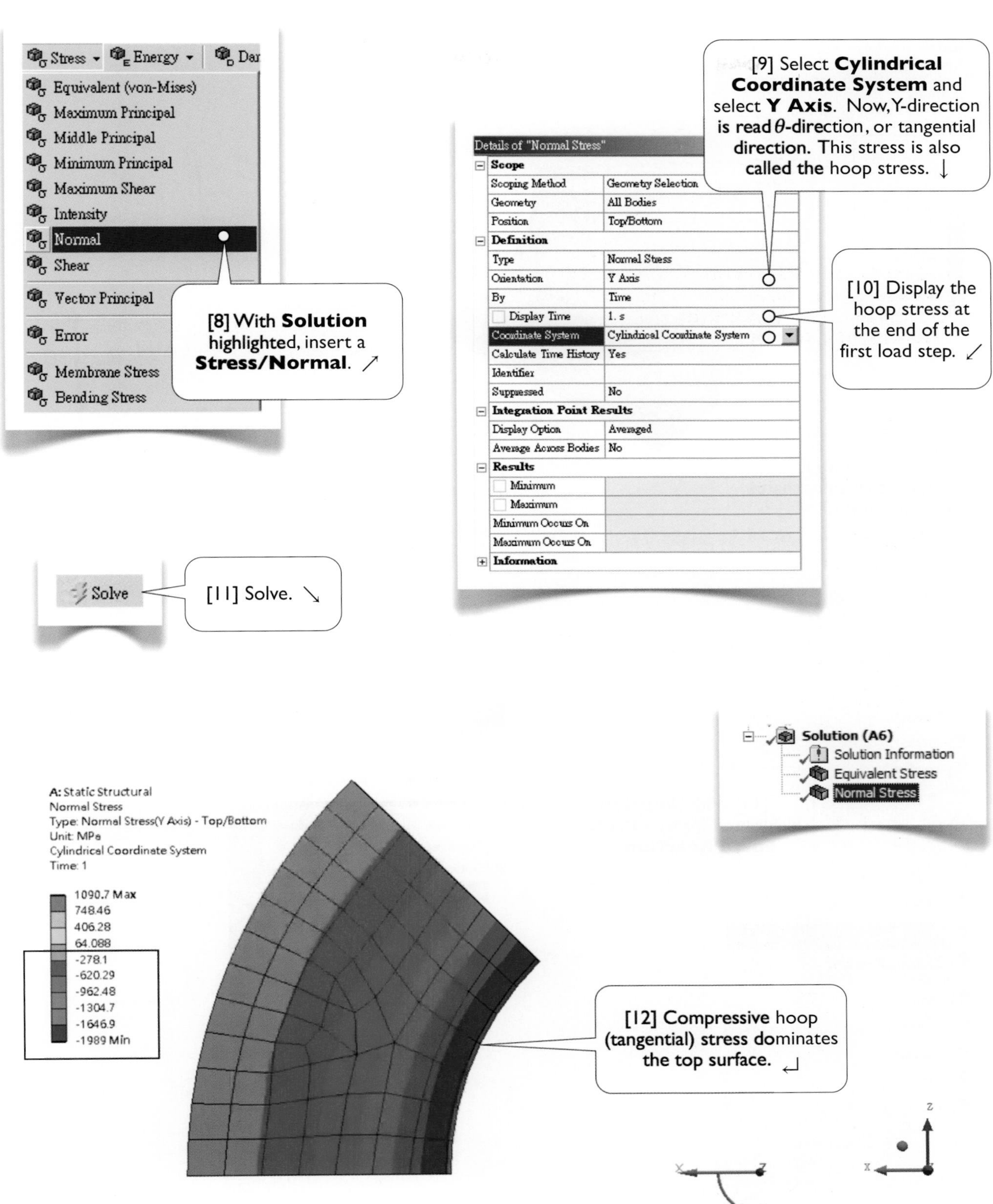
Stress
Energy
Equivalent (von-Mises)
Maximum Principal
Middle Principal
Minimum Principal
Maximum Shear
Intensity
Normal
Shear
Vector Principal
Error
Membrane Stress
Bending Stress
[8] With **Solution** highlighted, insert a **Stress/Normal**.
[9] Select **Cylindrical Coordinate System** and select **Y Axis**. Now, Y-direction is read θ-direction, or tangential direction. This stress is also called the hoop stress.
Details of "Normal Stress"
Scope
Scoping Method Geometry Selection
Geometry All Bodies
Position Top/Bottom
Definition
Type Normal Stress
Orientation Y Axis
By Time
Display Time 1. s
Coordinate System Cylindrical Coordinate System
Calculate Time History Yes
Identifier
Suppressed No
Integration Point Results
Display Option Averaged
Average Across Bodies No
Results
Minimum
Maximum
Minimum Occurs On
Maximum Occurs On
Information
[10] Display the hoop stress at the end of the first load step.
Solve
[11] Solve.
Solution (A6)
Solution Information
Equivalent Stress
Normal Stress
A: Static Structural
Normal Stress
Type: Normal Stress(Y Axis) - Top/Bottom
Unit: MPa
Cylindrical Coordinate System
Time: 1
1090.7 Max
748.46
406.28
64.088
-278.1
-620.29
-962.48
-1304.7
-1646.9
-1989 Min
[12] Compressive hoop (tangential) stress dominates the top surface.

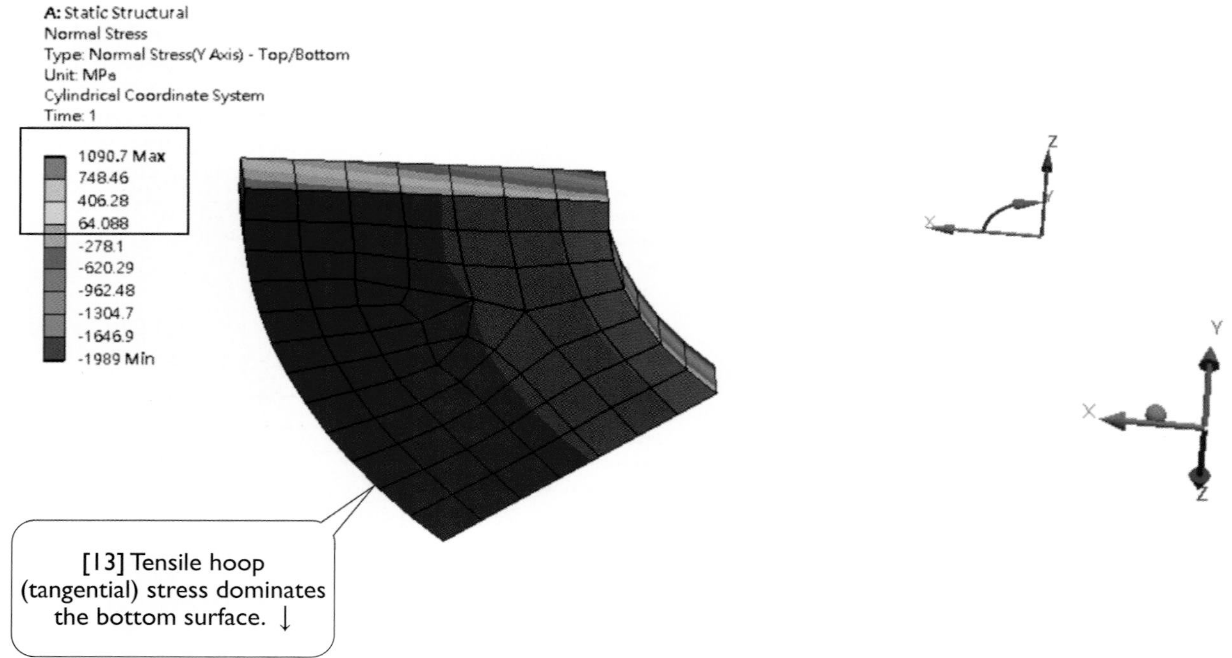
A: Static Structural
Normal Stress
Type: Normal Stress(Y Axis) - Top/Bottom
Unit: MPa
Cylindrical Coordinate System
Time: 1
1090.7 Max
748.46
406.28
64.088
-278.1
-620.29
-962.48
-1304.7
-1646.9
-1989 Min
[13] Tensile hoop (tangential) stress dominates the bottom surface. ↓

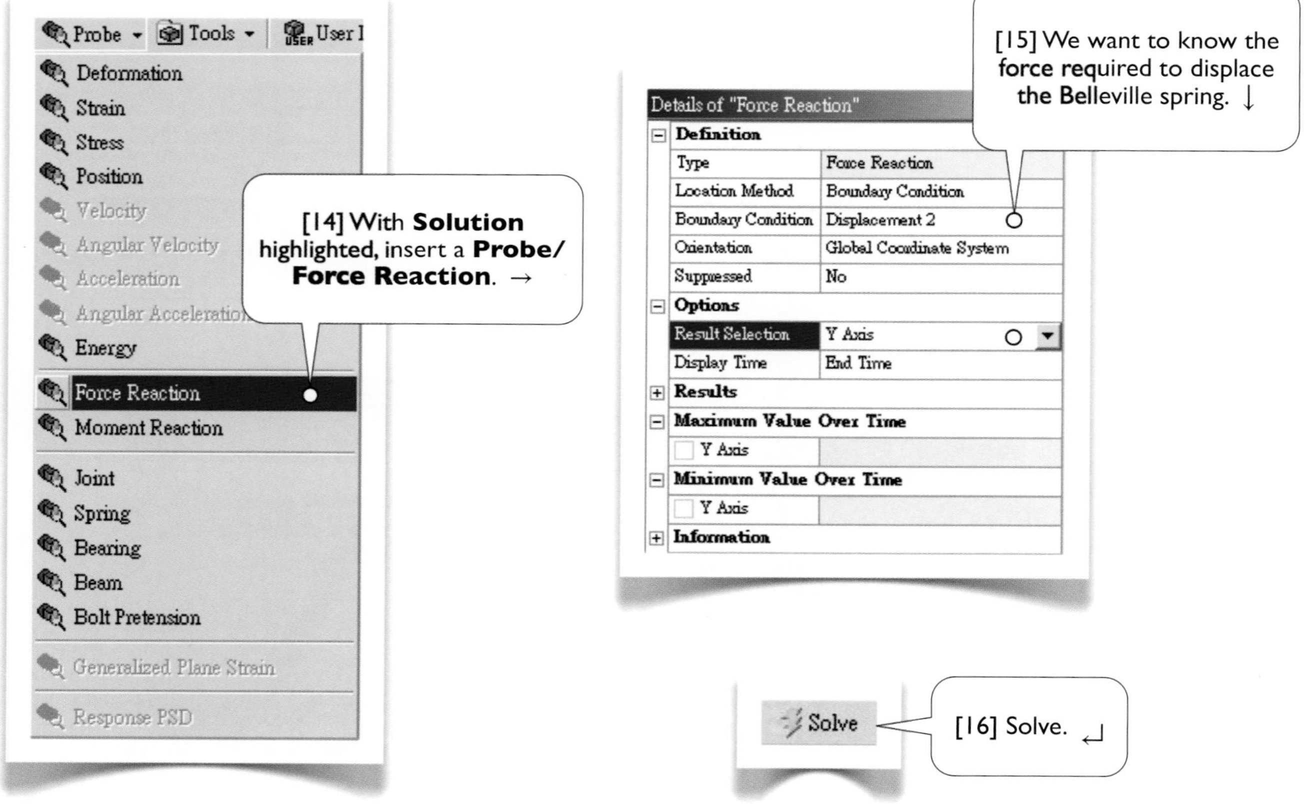
Probe
Tools
User
Deformation
Strain
Stress
Position
Velocity
Angular Velocity
Acceleration
Angular Acceleration
Energy
Force Reaction
Moment Reaction
Joint
Spring
Bearing
Beam
Bolt Pretension
Generalized Plane Strain
Response PSD
[14] With **Solution** highlighted, insert a **Probe/ Force Reaction**. →
[15] We want to know the **force required** to displace **the Belleville** spring. ↓
Details of "Force Reaction"
Definition
Type | Force Reaction
Location Method | Boundary Condition
Boundary Condition | Displacement 2
Orientation | Global Coordinate System
Suppressed | No
Options
Result Selection | Y Axis
Display Time | End Time
Results
Maximum Value Over Time
Y Axis
Minimum Value Over Time
Y Axis
Information
Solve
[16] Solve. ↵

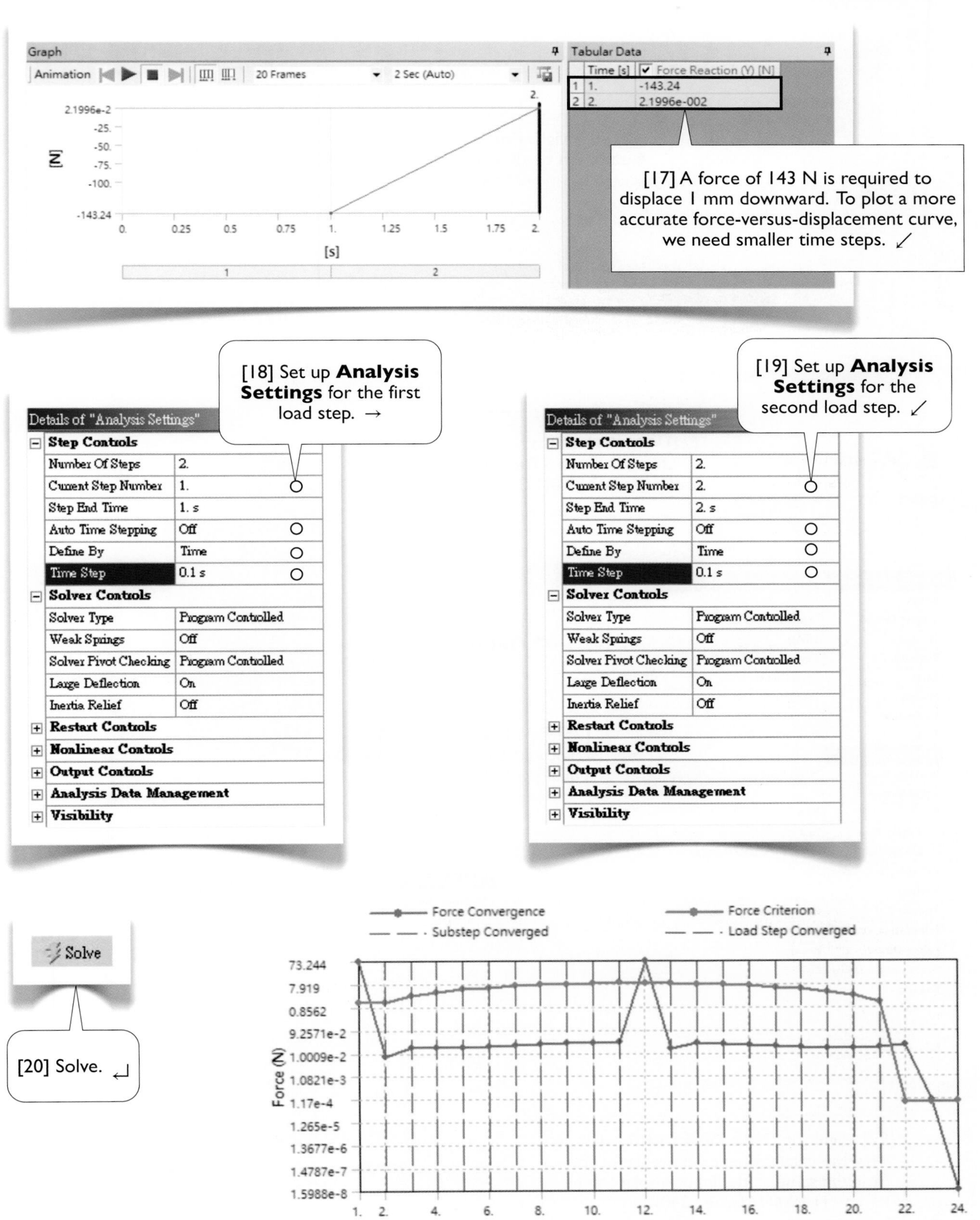

Graph
Animation
20 Frames
2 Sec (Auto)
2.1996e-2
-25.
-50.
-75.
-100.
-143.24
[N]
0. 0.25 0.5 0.75 1. 1.25 1.5 1.75 2.
[s]
1
2
Tabular Data
Time [s] Force Reaction (Y) [N]
1 1. -143.24
2 2. 2.1996e-002
[17] A force of 143 N is required to displace 1 mm downward. To plot a more accurate force-versus-displacement curve, we need smaller time steps.
[18] Set up **Analysis Settings** for the first load step.
Details of "Analysis Settings"
Step Controls
Number Of Steps 2.
Current Step Number 1.
Step End Time 1. s
Auto Time Stepping Off
Define By Time
Time Step 0.1 s
Solver Controls
Solver Type Program Controlled
Weak Springs Off
Solver Pivot Checking Program Controlled
Large Deflection On
Inertia Relief Off
Restart Controls
Nonlinear Controls
Output Controls
Analysis Data Management
Visibility
[19] Set up **Analysis Settings** for the second load step.
Details of "Analysis Settings"
Step Controls
Number Of Steps 2.
Current Step Number 2.
Step End Time 2. s
Auto Time Stepping Off
Define By Time
Time Step 0.1 s
Solver Controls
Solver Type Program Controlled
Weak Springs Off
Solver Pivot Checking Program Controlled
Large Deflection On
Inertia Relief Off
Restart Controls
Nonlinear Controls
Output Controls
Analysis Data Management
Visibility
Solve
[20] Solve.
Force Convergence
Substep Converged
Force Criterion
Load Step Converged
73.244
7.919
0.8562
9.2571e-2
1.0009e-2
1.0821e-3
1.17e-4
1.265e-5
1.3677e-6
1.4787e-7
1.5988e-8
Force (N)
1. 2. 4. 6. 8. 10. 12. 14. 16. 18. 20. 22. 24.

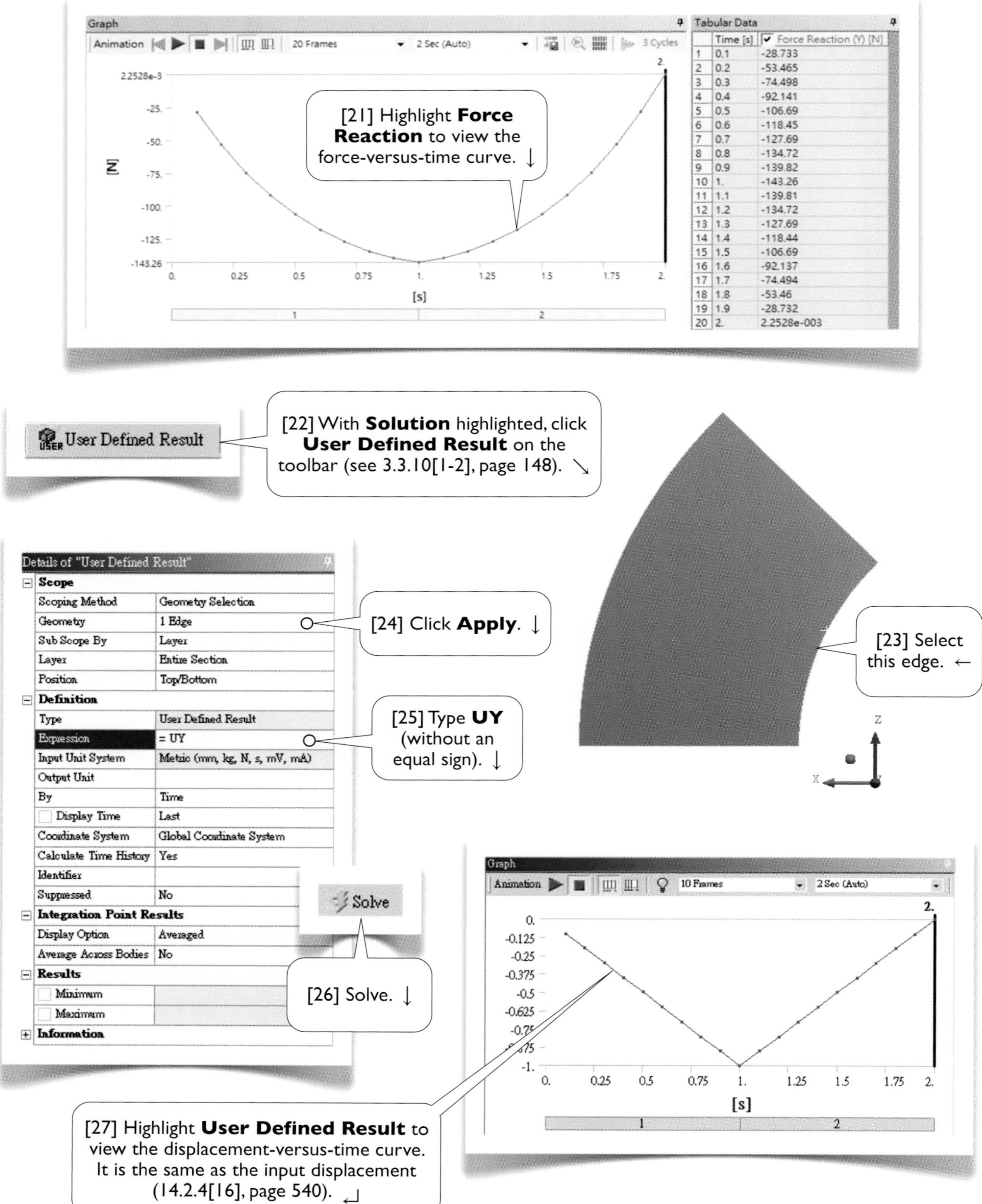

	Time [s]	Force Reaction (Y) [N]
1	0.1	-28.733
2	0.2	-53.465
3	0.3	-74.498
4	0.4	-92.141
5	0.5	-106.69
6	0.6	-118.45
7	0.7	-127.69
8	0.8	-134.72
9	0.9	-139.82
10	1.	-143.26
11	1.1	-139.81
12	1.2	-134.72
13	1.3	-127.69
14	1.4	-118.44
15	1.5	-106.69
16	1.6	-92.137
17	1.7	-74.494
18	1.8	-53.46
19	1.9	-28.732
20	2.	2.2528e-003

Create a Force-versus-Displacement Curve

[28] In step [21] (last page), we have a force-versus-time curve, while in step [27] we have a displacement-versus-time curve. Now, we want to combine them and create a force-versus-displacement curve. →

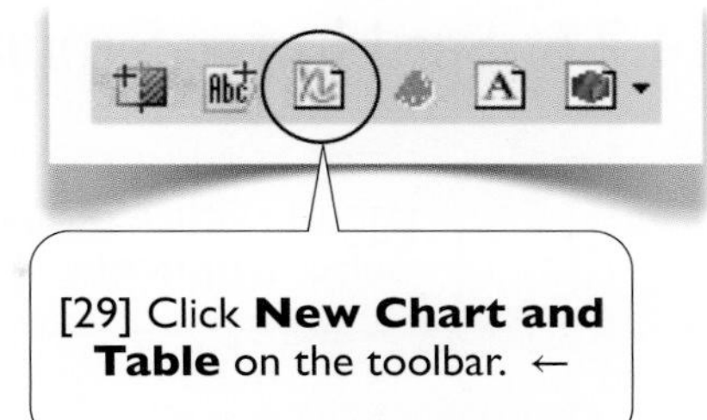

[29] Click **New Chart and Table** on the toolbar. ←

[30] Click to bring up **Apply/Cancel** buttons and control-select **Force Reaction** and **User Defined Result** in the project tree and click **Apply**. ↓

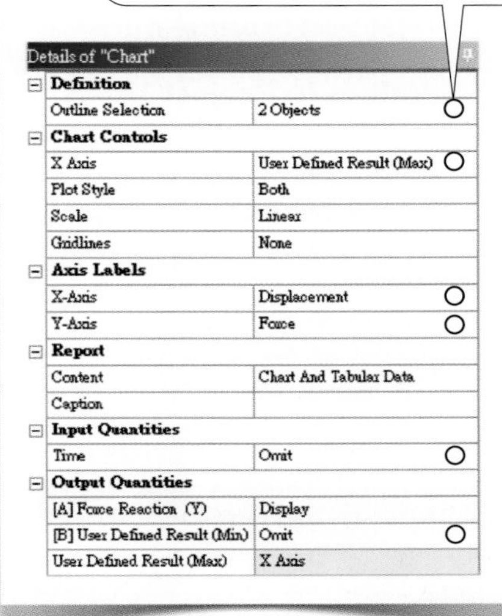

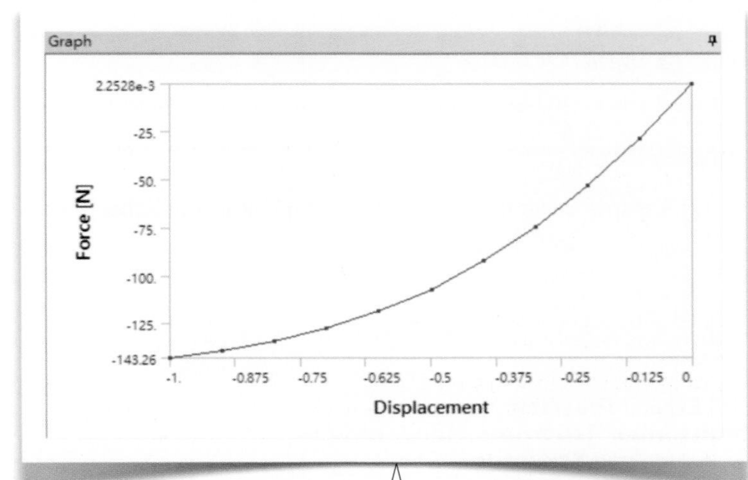

[31] A force-versus-displacement curve. ↓

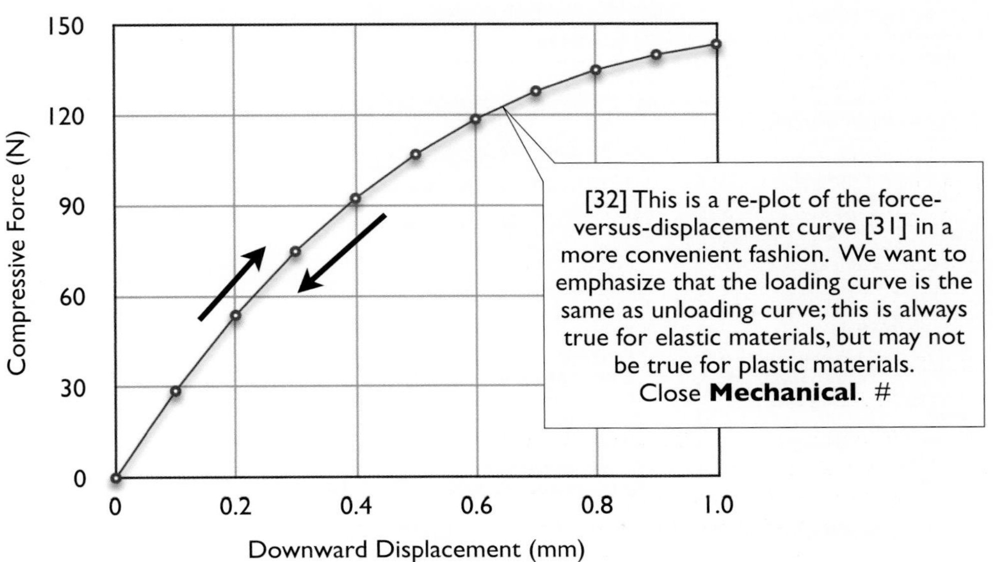

[32] This is a re-plot of the force-versus-displacement curve [31] in a more convenient fashion. We want to emphasize that the loading curve is the same as unloading curve; this is always true for elastic materials, but may not be true for plastic materials. Close **Mechanical**. #

14.2.6 Set Up for Simulation with Plastic Material Model

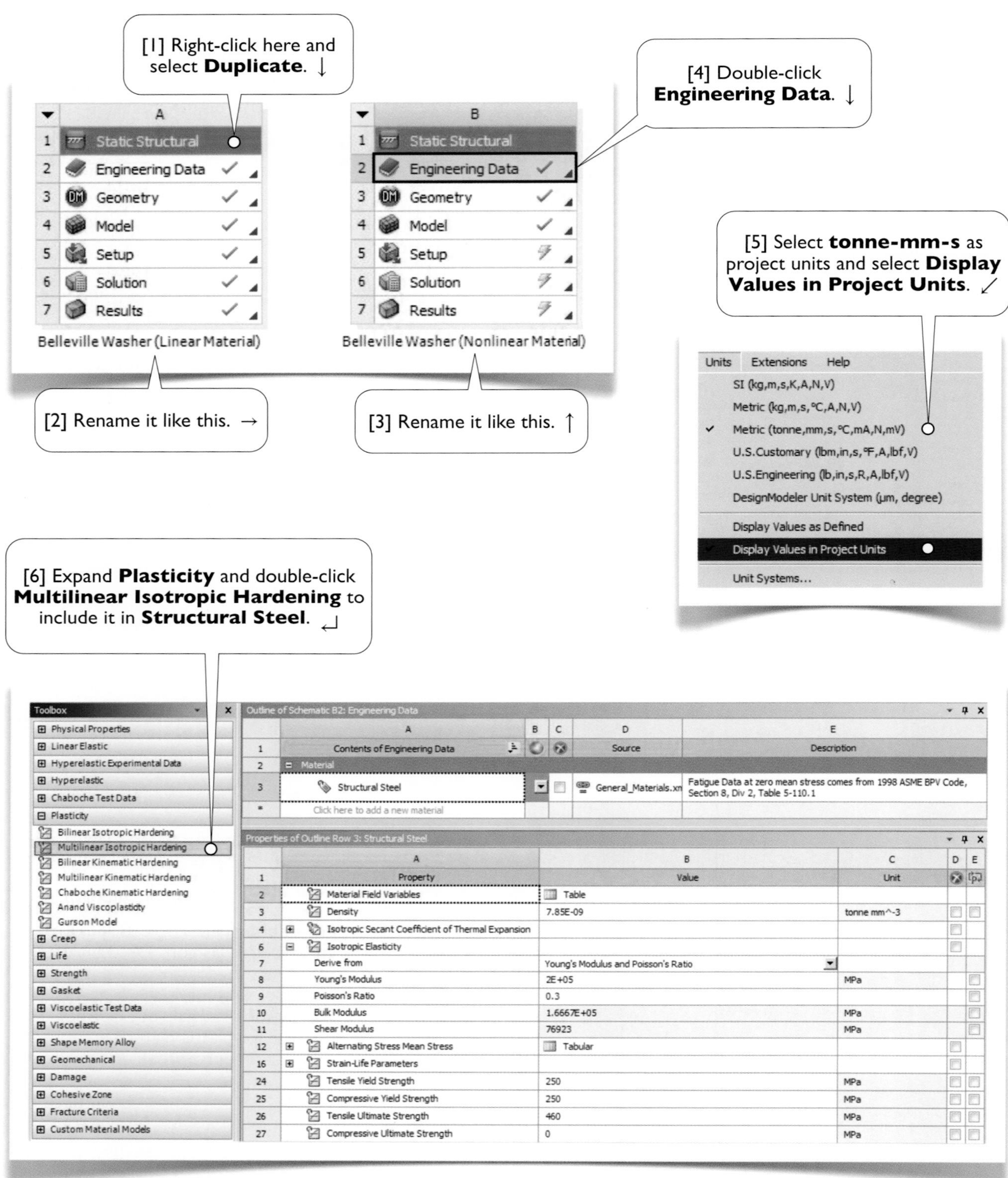

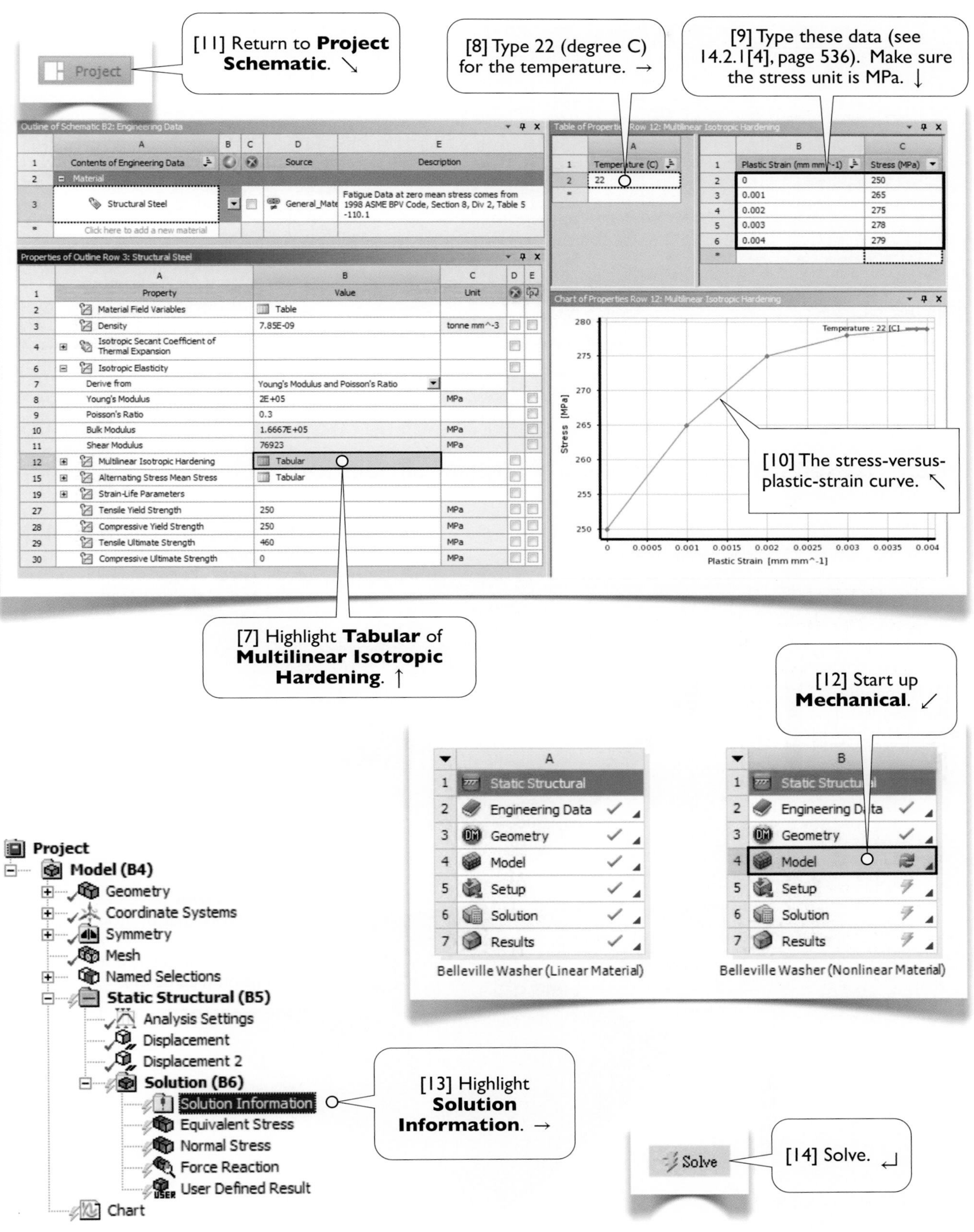
[11] Return to **Project Schematic**.
Project
[8] Type 22 (degree C) for the temperature.
[9] Type these data (see 14.2.1[4], page 536). Make sure the stress unit is MPa.
Outline of Schematic B2: Engineering Data
Contents of Engineering Data
Source
Description
Material
Structural Steel
General_Mate
Fatigue Data at zero mean stress comes from 1998 ASME BPV Code, Section 8, Div 2, Table 5 -110.1
Click here to add a new material
Table of Properties Row 12: Multilinear Isotropic Hardening
Temperature (C)
22
Plastic Strain (mm mm^-1)
Stress (MPa)
0 250
0.001 265
0.002 275
0.003 278
0.004 279
Properties of Outline Row 3: Structural Steel
Property
Value
Unit
Material Field Variables
Table
Density
7.85E-09
tonne mm^-3
Isotropic Secant Coefficient of Thermal Expansion
Isotropic Elasticity
Derive from
Young's Modulus and Poisson's Ratio
Young's Modulus
2E+05
MPa
Poisson's Ratio
0.3
Bulk Modulus
1.6667E+05
MPa
Shear Modulus
76923
MPa
Multilinear Isotropic Hardening
Tabular
Alternating Stress Mean Stress
Tabular
Strain-Life Parameters
Tensile Yield Strength
250
MPa
Compressive Yield Strength
250
MPa
Tensile Ultimate Strength
460
MPa
Compressive Ultimate Strength
0
MPa
Chart of Properties Row 12: Multilinear Isotropic Hardening
Temperature : 22 [C]
Stress [MPa]
Plastic Strain [mm mm^-1]
[10] The stress-versus-plastic-strain curve.
[7] Highlight **Tabular** of **Multilinear Isotropic Hardening**.
[12] Start up **Mechanical**.
Static Structural
Engineering Data
Geometry
Model
Setup
Solution
Results
Belleville Washer (Linear Material)
Belleville Washer (Nonlinear Material)
Project
Model (B4)
Geometry
Coordinate Systems
Symmetry
Mesh
Named Selections
Static Structural (B5)
Analysis Settings
Displacement
Displacement 2
Solution (B6)
Solution Information
Equivalent Stress
Normal Stress
Force Reaction
User Defined Result
Chart
[13] Highlight **Solution Information**.
Solve
[14] Solve.

Force Convergence
Force Criterion
Substep Converged
Load Step Converged
Force (N)
73.244
23.117
7.2964
2.3029
0.72685
0.22941
7.2407e-2
2.2853e-2
7.213e-3
2.2766e-3
7.1855e-4
Time (s)
Cumulative Iteration

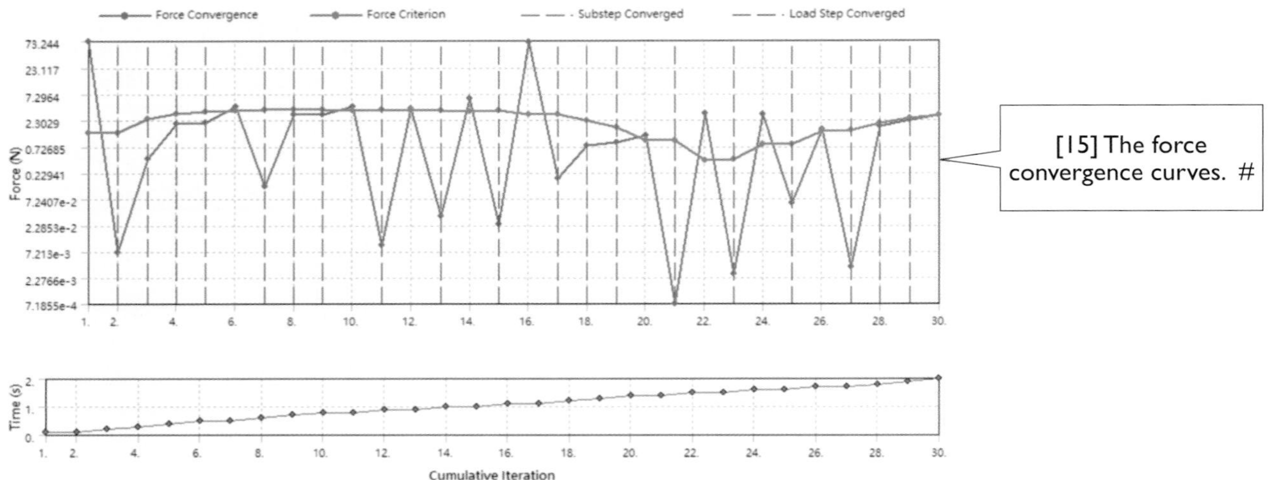

[15] The force convergence curves. #

14.2.7 Results of the Nonlinear Material Simulation

[1] Highlight **Equivalent Stress**. A majority of the surface area is plastic (stress larger than 250 MPa) at the end of the first load step. ←

B: Belleville Washer (Nonlinear Material)
Equivalent Stress
Type: Equivalent (von-Mises) Stress - Top/Bottom
Unit: MPa
Time: 1.
279 Max
249.14
219.29
189.43
159.58
129.72
99.87
70.014
40.159
10.304 Min

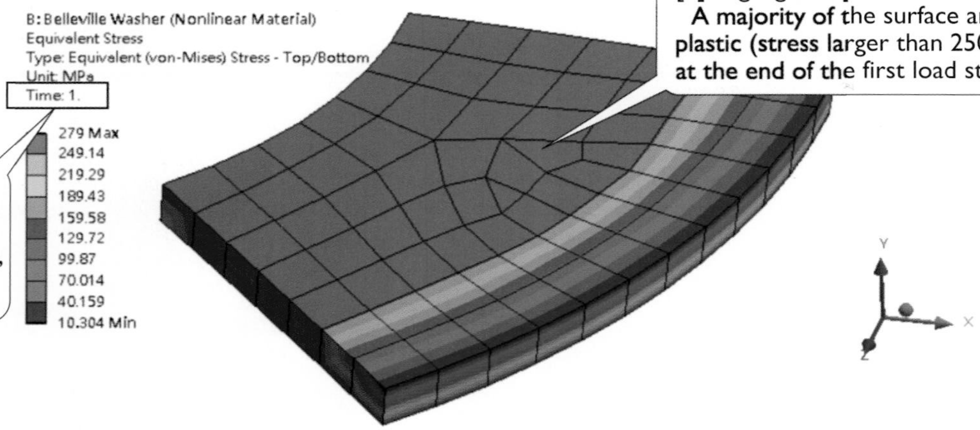

[2] Remember the time is 1.0 second (14.2.5[3], page 541). →

[3] Highlight **Normal Stress** to view the hoop stress. Note that the top surface is dominated by compression (blue color), while the bottom surface is dominated by tension (red color). ↵

B: Belleville Washer (Nonlinear Material)
Normal Stress
Type: Normal Stress(Y Axis) - Top/Bottom
Unit: MPa
Cylindrical Coordinate System
Time: 1.
279.51 Max
216.09
152.67
89.244
25.823
-37.598
-101.02
-164.44
-227.86
-291.28 Min

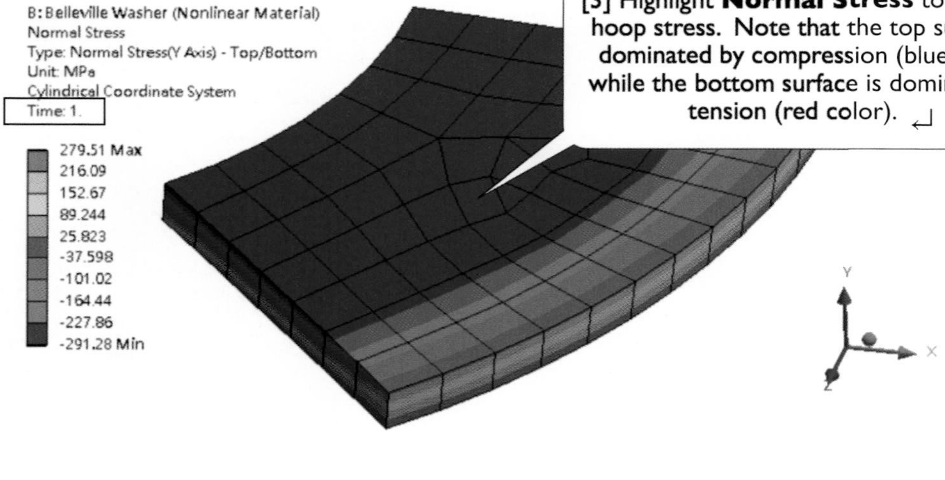

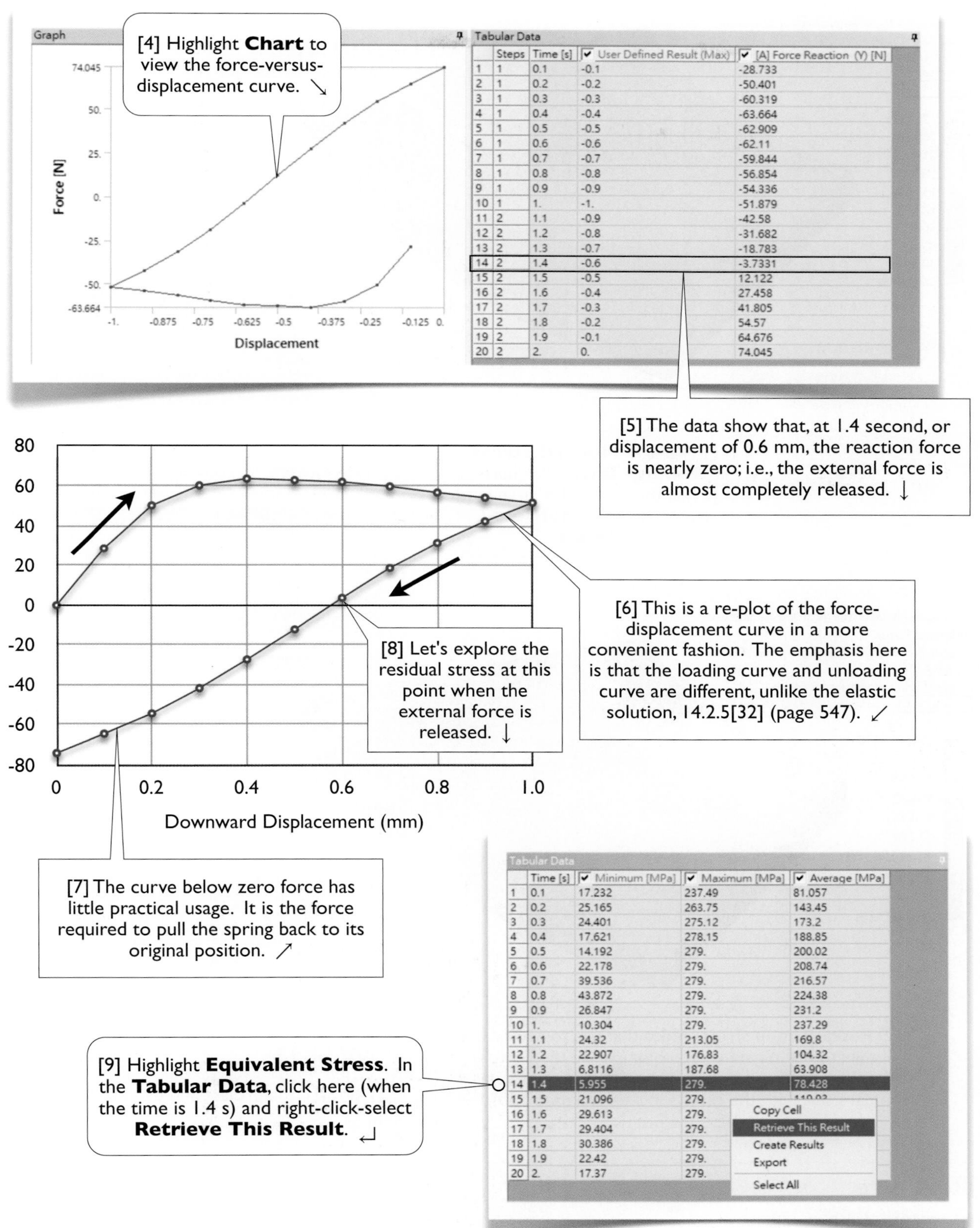

	Steps	Time [s]	User Defined Result (Max)	[A] Force Reaction (Y) [N]
1	1	0.1	-0.1	-28.733
2	1	0.2	-0.2	-50.401
3	1	0.3	-0.3	-60.319
4	1	0.4	-0.4	-63.664
5	1	0.5	-0.5	-62.909
6	1	0.6	-0.6	-62.11
7	1	0.7	-0.7	-59.844
8	1	0.8	-0.8	-56.854
9	1	0.9	-0.9	-54.336
10	1	1.	-1.	-51.879
11	2	1.1	-0.9	-42.58
12	2	1.2	-0.8	-31.682
13	2	1.3	-0.7	-18.783
14	2	1.4	-0.6	-3.7331
15	2	1.5	-0.5	12.122
16	2	1.6	-0.4	27.458
17	2	1.7	-0.3	41.805
18	2	1.8	-0.2	54.57
19	2	1.9	-0.1	64.676
20	2	2.	0.	74.045

	Time [s]	Minimum [MPa]	Maximum [MPa]	Average [MPa]
1	0.1	17.232	237.49	81.057
2	0.2	25.165	263.75	143.45
3	0.3	24.401	275.12	173.2
4	0.4	17.621	278.15	188.85
5	0.5	14.192	279.	200.02
6	0.6	22.178	279.	208.74
7	0.7	39.536	279.	216.57
8	0.8	43.872	279.	224.38
9	0.9	26.847	279.	231.2
10	1.	10.304	279.	237.29
11	1.1	24.32	213.05	169.8
12	1.2	22.907	176.83	104.32
13	1.3	6.8116	187.68	63.908
14	1.4	5.955	279.	78.428
15	1.5	21.096	279.	
16	1.6	29.613	279.	
17	1.7	29.404	279.	
18	1.8	30.386	279.	
19	1.9	22.42	279.	
20	2.	17.37	279.	

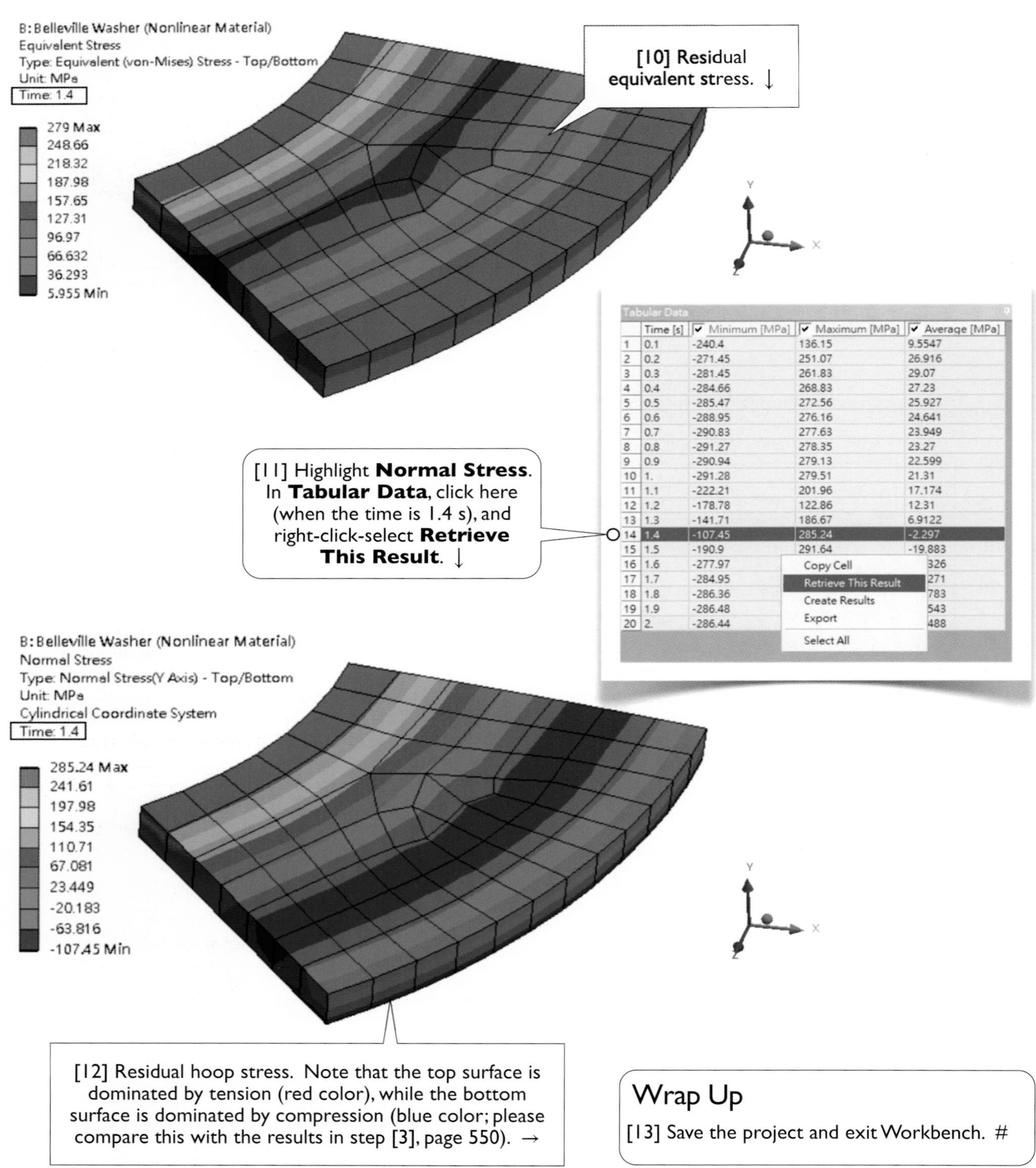

	Time [s]	Minimum [MPa]	Maximum [MPa]	Average [MPa]
1	0.1	-240.4	136.15	9.5547
2	0.2	-271.45	251.07	26.916
3	0.3	-281.45	261.83	29.07
4	0.4	-284.66	268.83	27.23
5	0.5	-285.47	272.56	25.927
6	0.6	-288.95	276.16	24.641
7	0.7	-290.83	277.63	23.949
8	0.8	-291.27	278.35	23.27
9	0.9	-290.94	279.13	22.599
10	1.	-291.28	279.51	21.31
11	1.1	-222.21	201.96	17.174
12	1.2	-178.78	122.86	12.31
13	1.3	-141.71	186.67	6.9122
14	1.4	-107.45	285.24	-2.297
15	1.5	-190.9	291.64	-19.883
16	1.6	-277.97		326
17	1.7	-284.95		271
18	1.8	-286.36		783
19	1.9	-286.48		543
20	2.	-286.44		488

[10] Residual equivalent stress. ↓

[11] Highlight **Normal Stress**. In **Tabular Data**, click here (when the time is 1.4 s), and right-click-select **Retrieve This Result**. ↓

[12] Residual hoop stress. Note that the top surface is dominated by tension (red color), while the bottom surface is dominated by compression (blue color; please compare this with the results in step [3], page 550). →

Wrap Up

[13] Save the project and exit Workbench. #

Reference

1. Wikipedia>Belleville Washer.

Section 14.3

Planar Seal

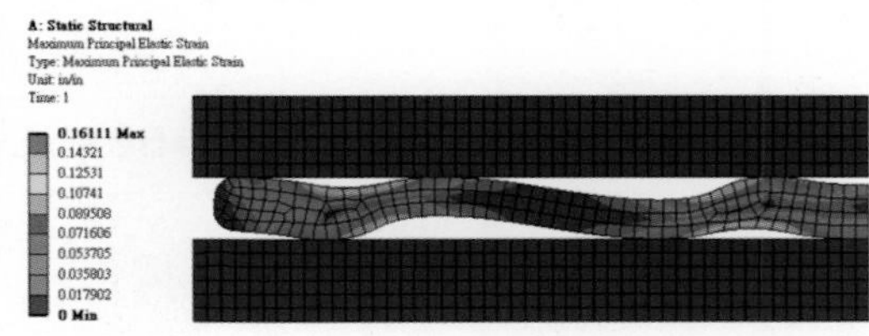

14.3.1 About the Planar Seal

[1] The seal shown in [2-5] is used in the door of a refrigerator. The seal is a long strip of rubber, and we will model it as a plane strain problem (3.3.2, page 141). A series of material tests has been conducted, including a uniaxial tensile test, a biaxial tensile test, and a shear test [6-8].

The strain range of the original test data covers much more than the data shown in [6-8]. However, a preliminary study of the problem shows that the maximum strain does not exceed 0.3. Therefore, we decided to use the portion of data up to a strain of 0.3. The data fitting will be better if we use only the relevant data. A series of trials of data fitting shows that, for these material testing data, the two-parameter Mooney-Rivlin hyperelastic model fits the data better than other models. We decide to use two-parameter Mooney-Rivlin model.

The unit system used in this section is **in-lbm-lbf-s**. ↘

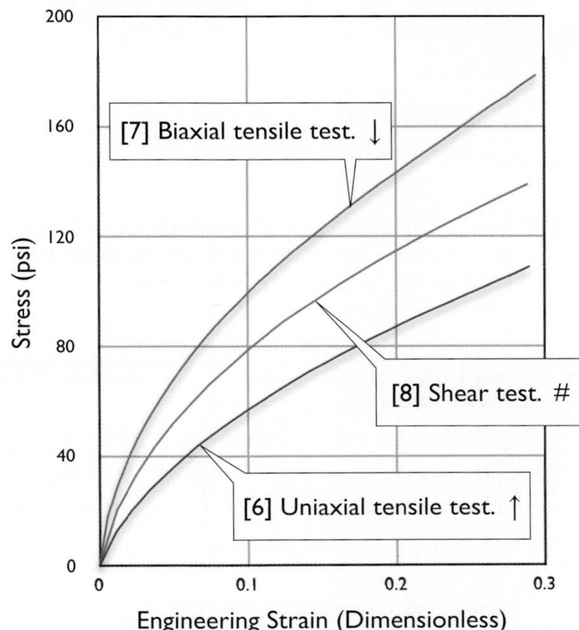

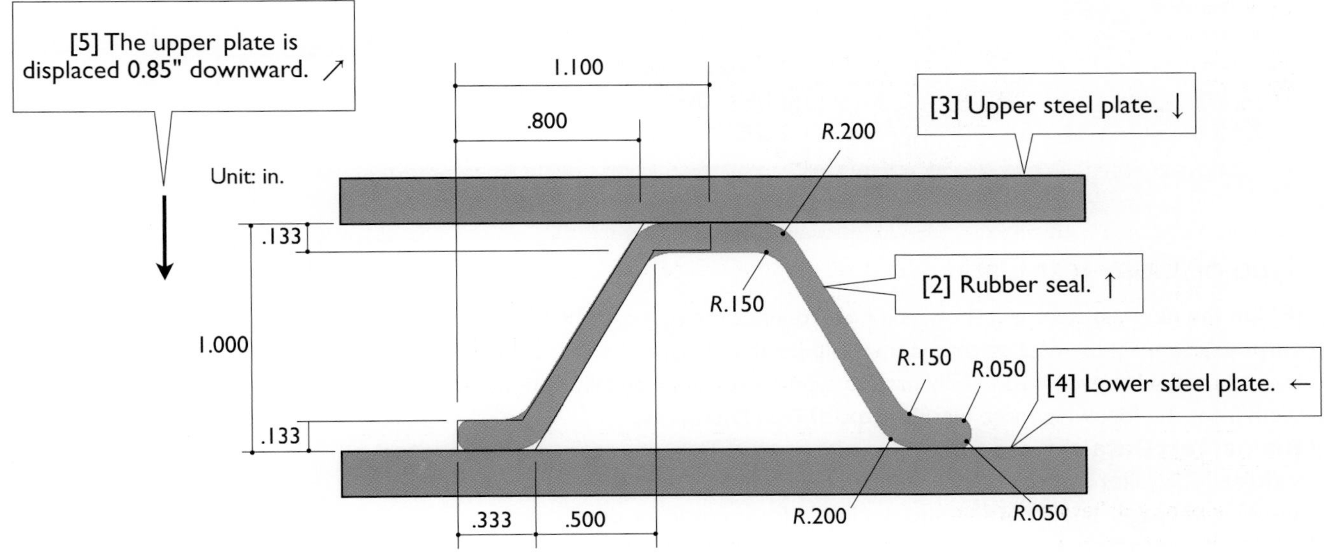

14.3.2 Prepare Material Properties for the Rubber

[1] Launch Workbench. Create a **Static Structural** analysis system. Save the project as **Seal**. ↙

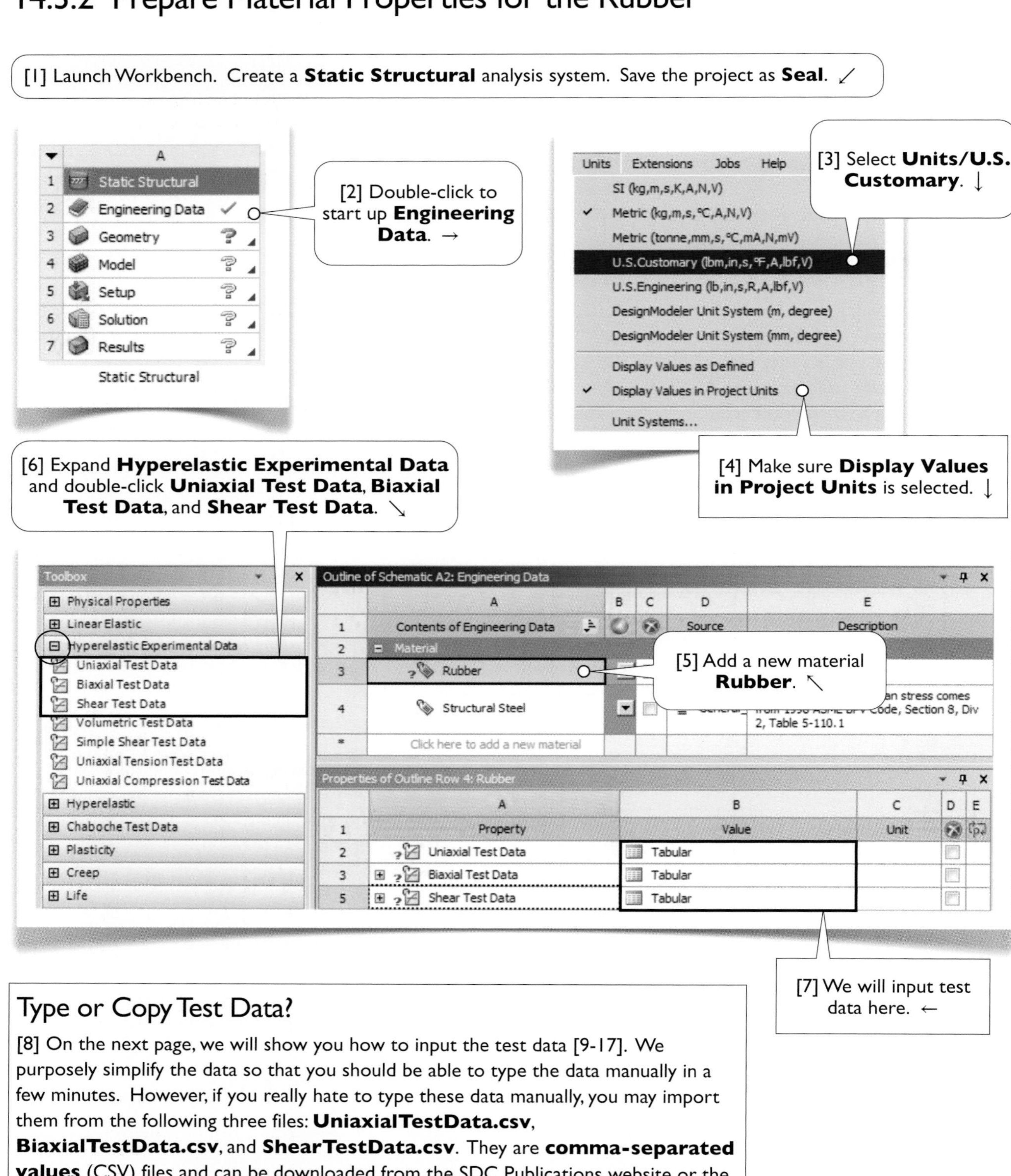

Type or Copy Test Data?

[8] On the next page, we will show you how to input the test data [9-17]. We purposely simplify the data so that you should be able to type the data manually in a few minutes. However, if you really hate to type these data manually, you may import them from the following three files: **UniaxialTestData.csv**, **BiaxialTestData.csv**, and **ShearTestData.csv**. They are **comma-separated values** (CSV) files and can be downloaded from the SDC Publications website or the author's webpage. See **Preface** and the inside front cover for details. ↵

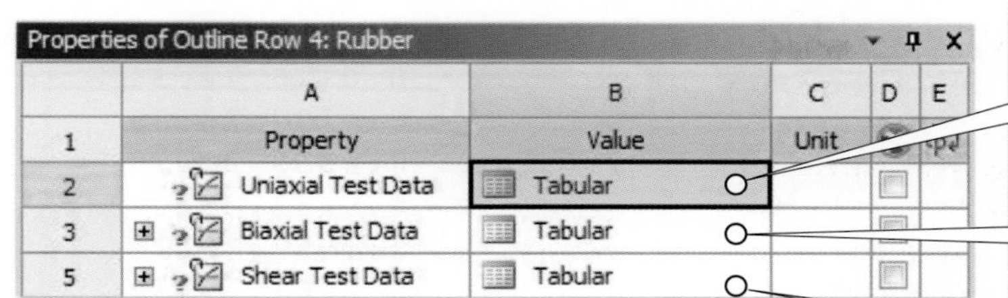

[9] Click **Tabular** of **Uniaxial Test Data**. ↙

[12] Click **Tabular** of **Biaxial Test Data**. ↓

[15] Click **Tabular** of **Shear Test Data**. ↓

[10] Type data (or import the data from **UniaxialTestData.csv**, as demonstrated in [19-24], next page) like this. ↓

	B	C
1	Strain (in in^-1)	Stress (psi)
2	0	0
3	0.0116	12.34
4	0.0227	20.35
5	0.0339	27.28
6	0.0451	33.06
7	0.0557	38.29
8	0.0664	43.45
9	0.0773	47.86
10	0.088	52.12
11	0.0989	56.3
12	0.1095	60.06
13	0.1203	63.74
14	0.131	67.28
15	0.1415	70.67
16	0.1524	73.77
17	0.1628	76.84
18	0.1737	79.87
19	0.184	82.8
20	0.1947	85.61
21	0.2051	88.5
22	0.2154	91.13
23	0.2259	93.77
24	0.2366	96.37
25	0.2475	99.07
26	0.2584	101.6
27	0.2687	104.09
28	0.2797	106.58
29	0.2903	109.29
*		

[13] Type data (or import the data from **BiaxialTestData.csv**) like this. ↓

	B	C
1	Strain (in in^-1)	Stress (psi)
2	0	0
3	0.0056	17.94
4	0.0124	29.12
5	0.0192	38.22
6	0.0264	46.45
7	0.0337	53.81
8	0.041	60.37
9	0.0481	66.18
10	0.0555	71.76
11	0.0628	77.06
12	0.0702	81.97
13	0.0775	86.67
14	0.0851	91.21
15	0.0924	95.29
16	0.1	99.56
17	0.1078	103.8
18	0.1154	107.33
19	0.1231	110.94
20	0.1312	114.72
21	0.1394	118.19
22	0.1475	121.68
23	0.1558	125.28
24	0.1642	128.86
25	0.1726	132.26
26	0.1813	135.78
27	0.19	139.2
28	0.1992	142.49
29	0.2083	146.07
30	0.2174	149.37
31	0.2266	152.75
32	0.2361	156.28
33	0.2458	160.03
34	0.2551	163.66
35	0.2649	167.41
36	0.2748	170.67
37	0.2847	174.77
38	0.2948	178.44
*		

[16] Type data (or import the data from **ShearTestData.csv**) like this. ↓

	B	C
1	Strain (in in^-1)	Stress (psi)
2	0	0
3	0.012	20.22
4	0.0252	33.06
5	0.0376	43.07
6	0.0499	51.64
7	0.0624	59.27
8	0.0744	66.15
9	0.0874	72.3
10	0.0999	78.36
11	0.1124	83.77
12	0.1243	88.93
13	0.1371	93.76
14	0.1498	98.36
15	0.1619	102.75
16	0.1744	106.86
17	0.187	110.9
18	0.1998	114.76
19	0.2126	118.48
20	0.2252	122.09
21	0.238	125.61
22	0.2511	129.05
23	0.2633	132.32
24	0.2763	135.64
25	0.2892	138.91
*		

	A
1	Temperature (F)
2	70
*	

[11, 14, 17] In **Table of Properties**, type 70 (degree F) for **Temperature**. ↑ ↑ →

[18] Go to [25], page 557. ↵

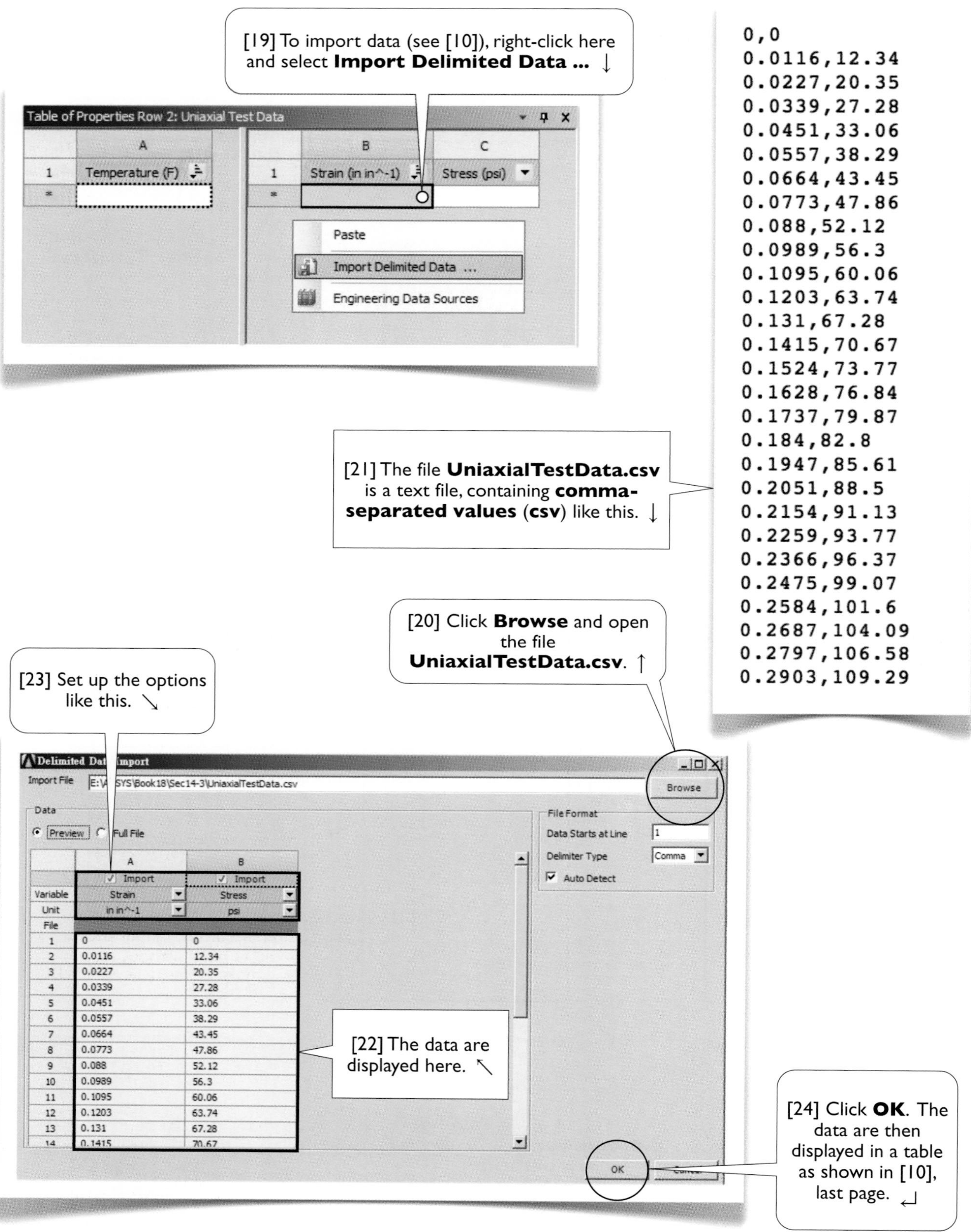
[19] To import data (see [10]), right-click here and select **Import Delimited Data ...** ↓
Table of Properties Row 2: Uniaxial Test Data
A
B
C
Temperature (F)
Strain (in in^-1)
Stress (psi)
Paste
Import Delimited Data ...
Engineering Data Sources
0,0
0.0116,12.34
0.0227,20.35
0.0339,27.28
0.0451,33.06
0.0557,38.29
0.0664,43.45
0.0773,47.86
0.088,52.12
0.0989,56.3
0.1095,60.06
0.1203,63.74
0.131,67.28
0.1415,70.67
0.1524,73.77
0.1628,76.84
0.1737,79.87
0.184,82.8
0.1947,85.61
0.2051,88.5
0.2154,91.13
0.2259,93.77
0.2366,96.37
0.2475,99.07
0.2584,101.6
0.2687,104.09
0.2797,106.58
0.2903,109.29
[21] The file **UniaxialTestData.csv** is a text file, containing **comma-separated values** (**csv**) like this. ↓
[20] Click **Browse** and open the file **UniaxialTestData.csv**. ↑
[23] Set up the options like this. ↘
Delimited Data Import
Import File
Browse
Data
Preview
Full File
File Format
Data Starts at Line
1
Delimiter Type
Comma
Auto Detect
Import
Variable
Strain
Stress
Unit
in in^-1
psi
File
[22] The data are displayed here. ↖
OK
[24] Click **OK**. The data are then displayed in a table as shown in [10], last page. ↵

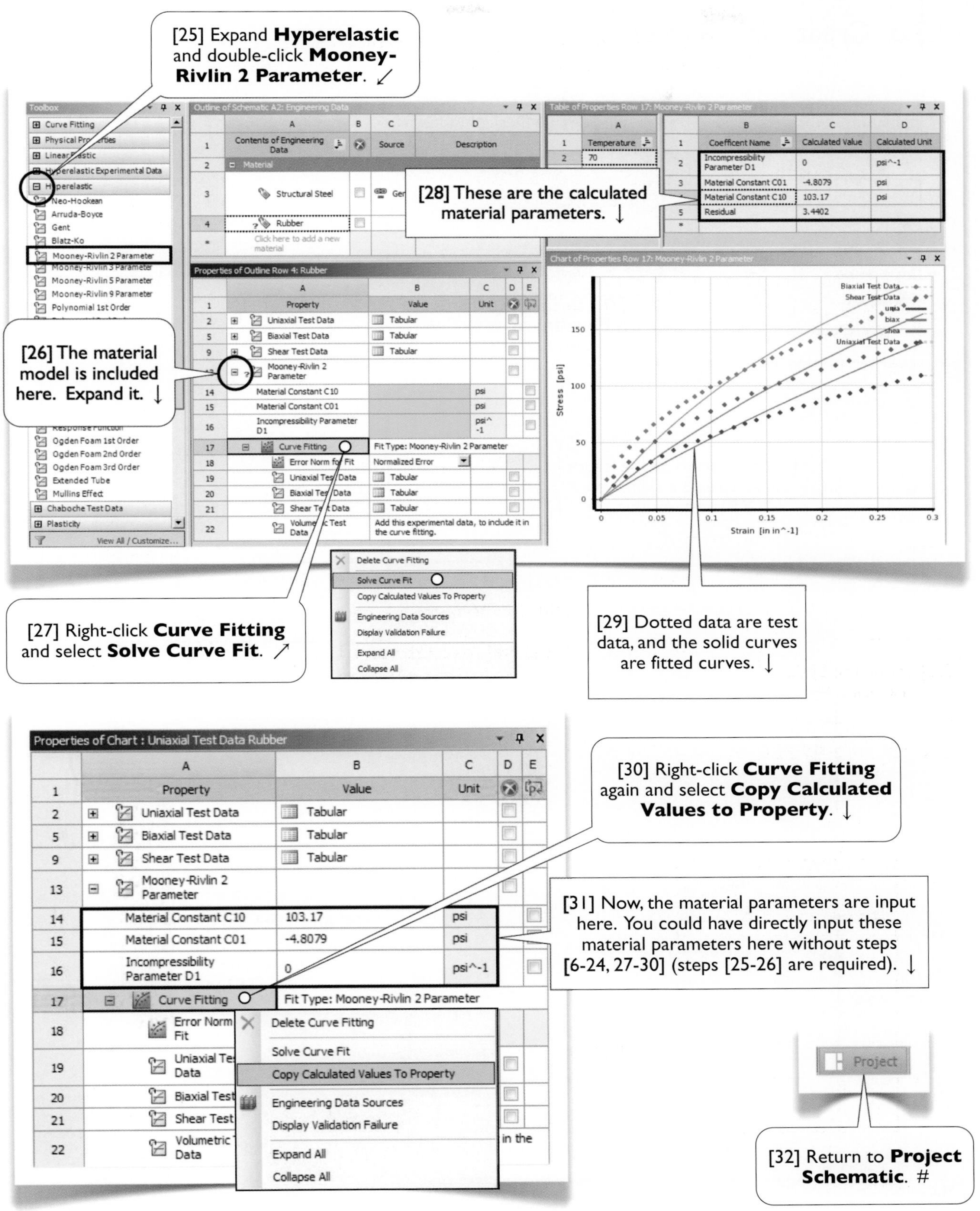
[25] Expand **Hyperelastic** and double-click **Mooney-Rivlin 2 Parameter**.
[26] The material model is included here. Expand it.
[27] Right-click **Curve Fitting** and select **Solve Curve Fit**.
[28] These are the calculated material parameters.
[29] Dotted data are test data, and the solid curves are fitted curves.
[30] Right-click **Curve Fitting** again and select **Copy Calculated Values to Property**.
[31] Now, the material parameters are input here. You could have directly input these material parameters here without steps [6-24, 27-30] (steps [25-26] are required).
[32] Return to **Project Schematic**. #
Properties of Chart : Uniaxial Test Data Rubber
Material Constant C10
103.17
Material Constant C01
-4.8079
Incompressibility Parameter D1
0
Curve Fitting
Fit Type: Mooney-Rivlin 2 Parameter
Delete Curve Fitting
Solve Curve Fit
Copy Calculated Values To Property
Engineering Data Sources
Display Validation Failure
Expand All
Collapse All
Project

14.3.3 Create 2D Geometry

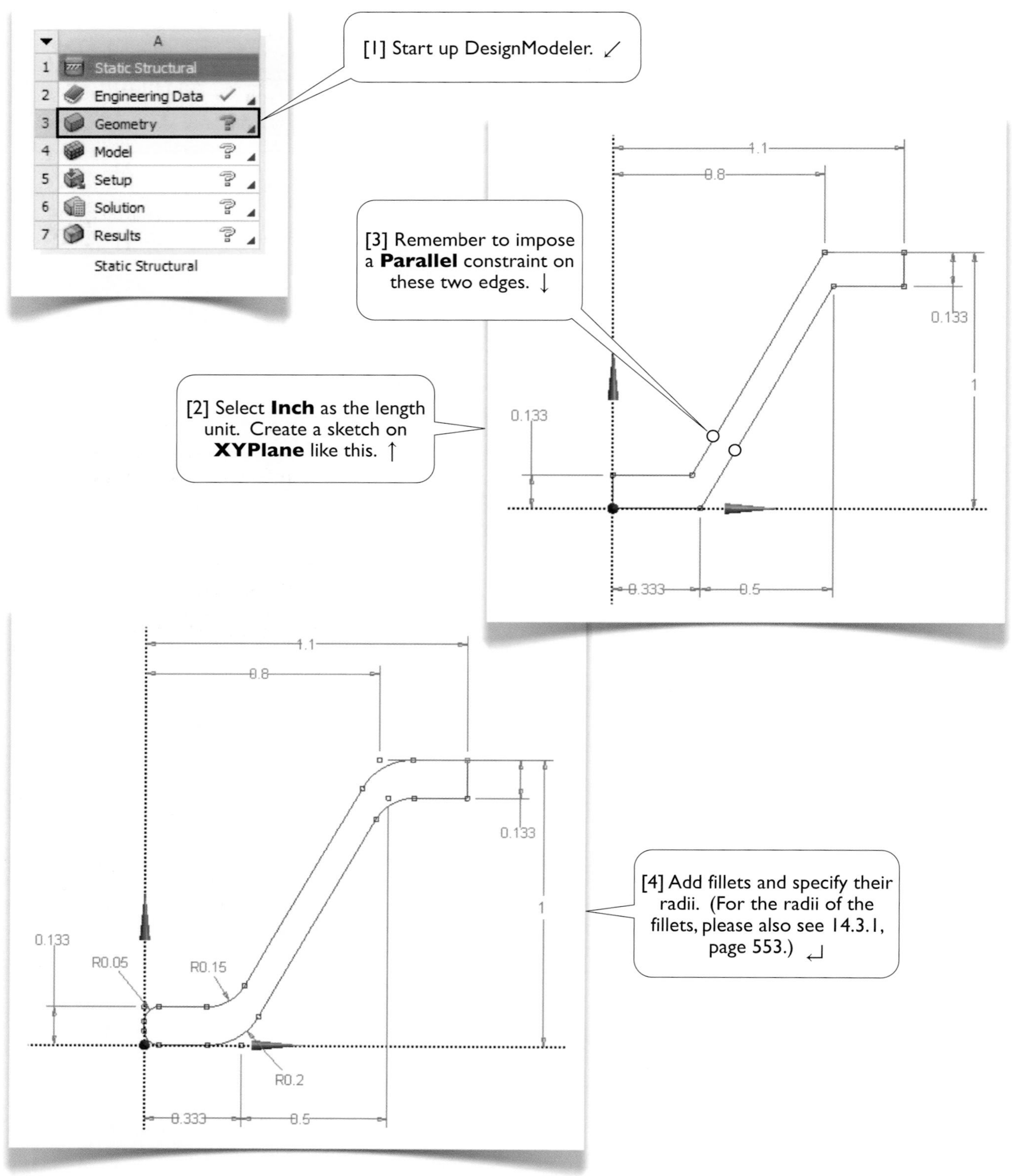

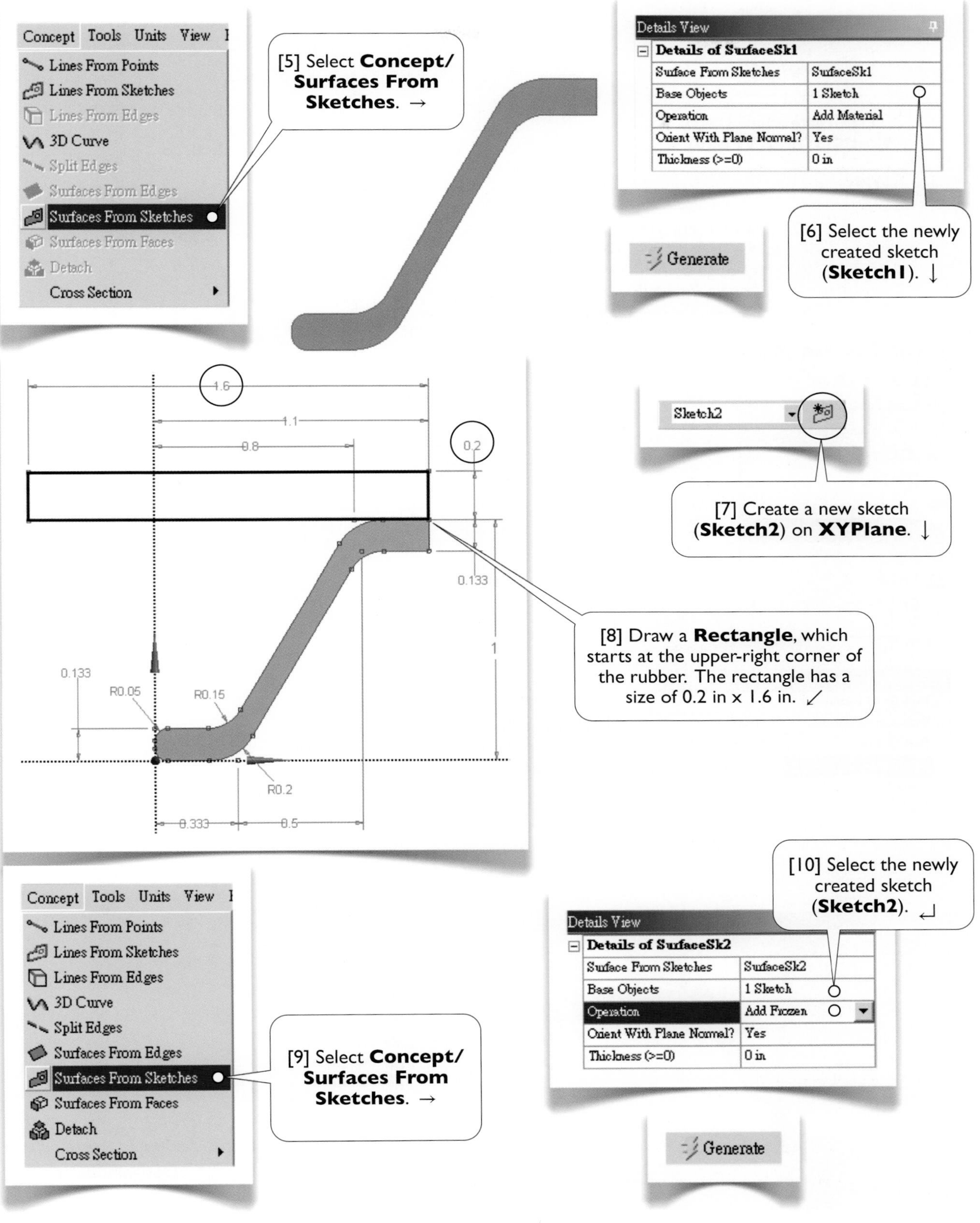
Concept
Tools
Units
View
Lines From Points
Lines From Sketches
Lines From Edges
3D Curve
Split Edges
Surfaces From Edges
Surfaces From Sketches
Surfaces From Faces
Detach
Cross Section
[5] Select **Concept/ Surfaces From Sketches**. →
Details View
Details of SurfaceSk1
Surface From Sketches
SurfaceSk1
Base Objects
1 Sketch
Operation
Add Material
Orient With Plane Normal?
Yes
Thickness (>=0)
0 in
Generate
[6] Select the newly created sketch (**Sketch1**). ↓
1.6
1.1
0.8
0.2
0.133
1
0.133
R0.05
R0.15
R0.2
0.333
0.5
Sketch2
[7] Create a new sketch (**Sketch2**) on **XYPlane**. ↓
[8] Draw a **Rectangle**, which starts at the upper-right corner of the rubber. The rectangle has a size of 0.2 in x 1.6 in. ↙
[9] Select **Concept/ Surfaces From Sketches**. →
[10] Select the newly created sketch (**Sketch2**). ↲
Details View
Details of SurfaceSk2
Surface From Sketches
SurfaceSk2
Base Objects
1 Sketch
Operation
Add Frozen
Orient With Plane Normal?
Yes
Thickness (>=0)
0 in
Generate

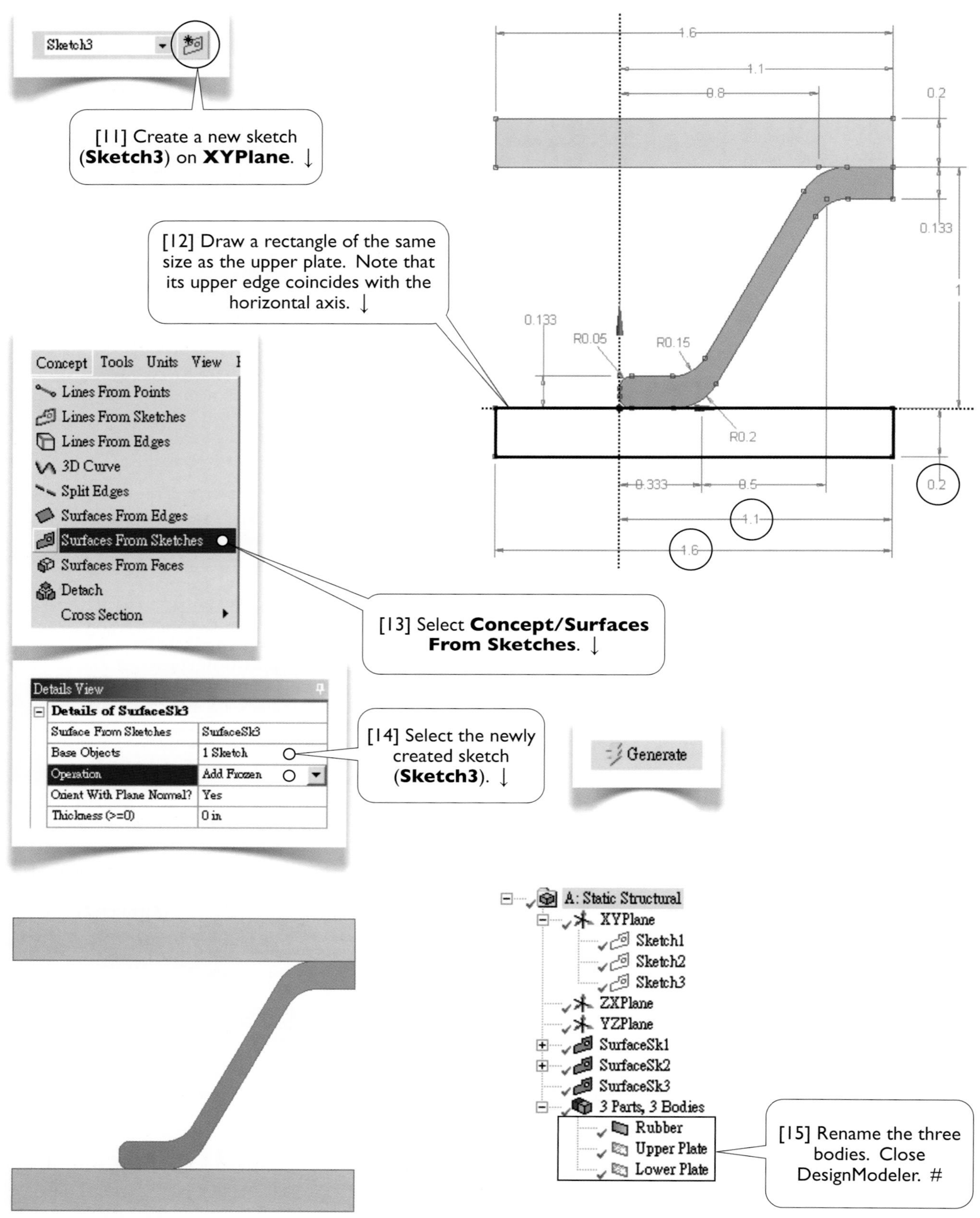

Sketch3
[11] Create a new sketch (**Sketch3**) on **XYPlane**. ↓
1.6
1.1
0.8
0.2
0.133
1
[12] Draw a rectangle of the same size as the upper plate. Note that its upper edge coincides with the horizontal axis. ↓
0.133
R0.05
R0.15
R0.2
0.333
0.5
0.2
1.1
1.6
Concept
Tools
Units
View
Lines From Points
Lines From Sketches
Lines From Edges
3D Curve
Split Edges
Surfaces From Edges
Surfaces From Sketches
Surfaces From Faces
Detach
Cross Section
[13] Select **Concept/Surfaces From Sketches**. ↓
Details View
Details of SurfaceSk3
Surface From Sketches
SurfaceSk3
Base Objects
1 Sketch
Operation
Add Frozen
Orient With Plane Normal?
Yes
Thickness (>=0)
0 in
[14] Select the newly created sketch (**Sketch3**). ↓
Generate
A: Static Structural
XYPlane
Sketch1
Sketch2
Sketch3
ZXPlane
YZPlane
SurfaceSk1
SurfaceSk2
SurfaceSk3
3 Parts, 3 Bodies
Rubber
Upper Plate
Lower Plate
[15] Rename the three bodies. Close DesignModeler. #

14.3.4 Set Up for Simulation

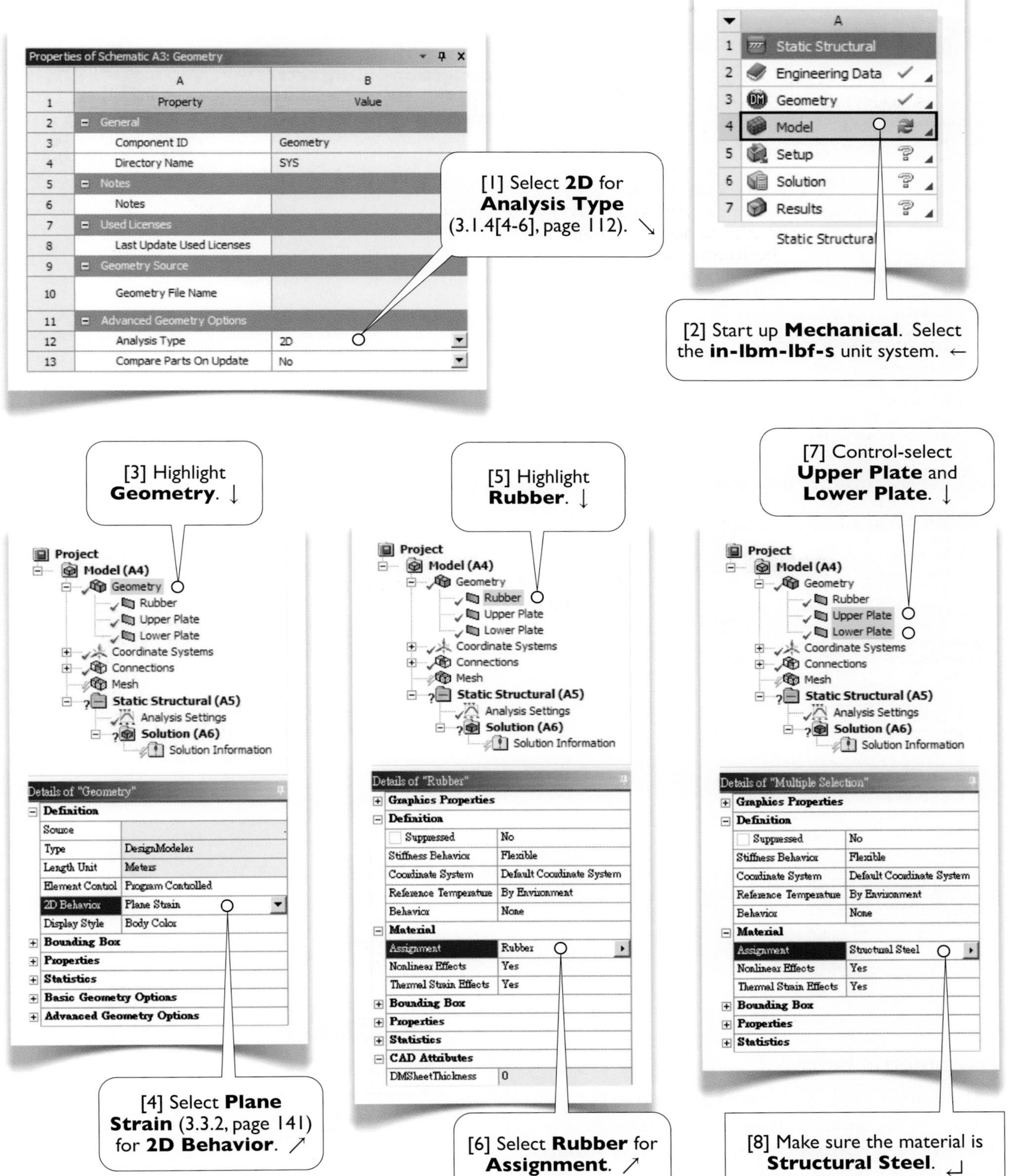

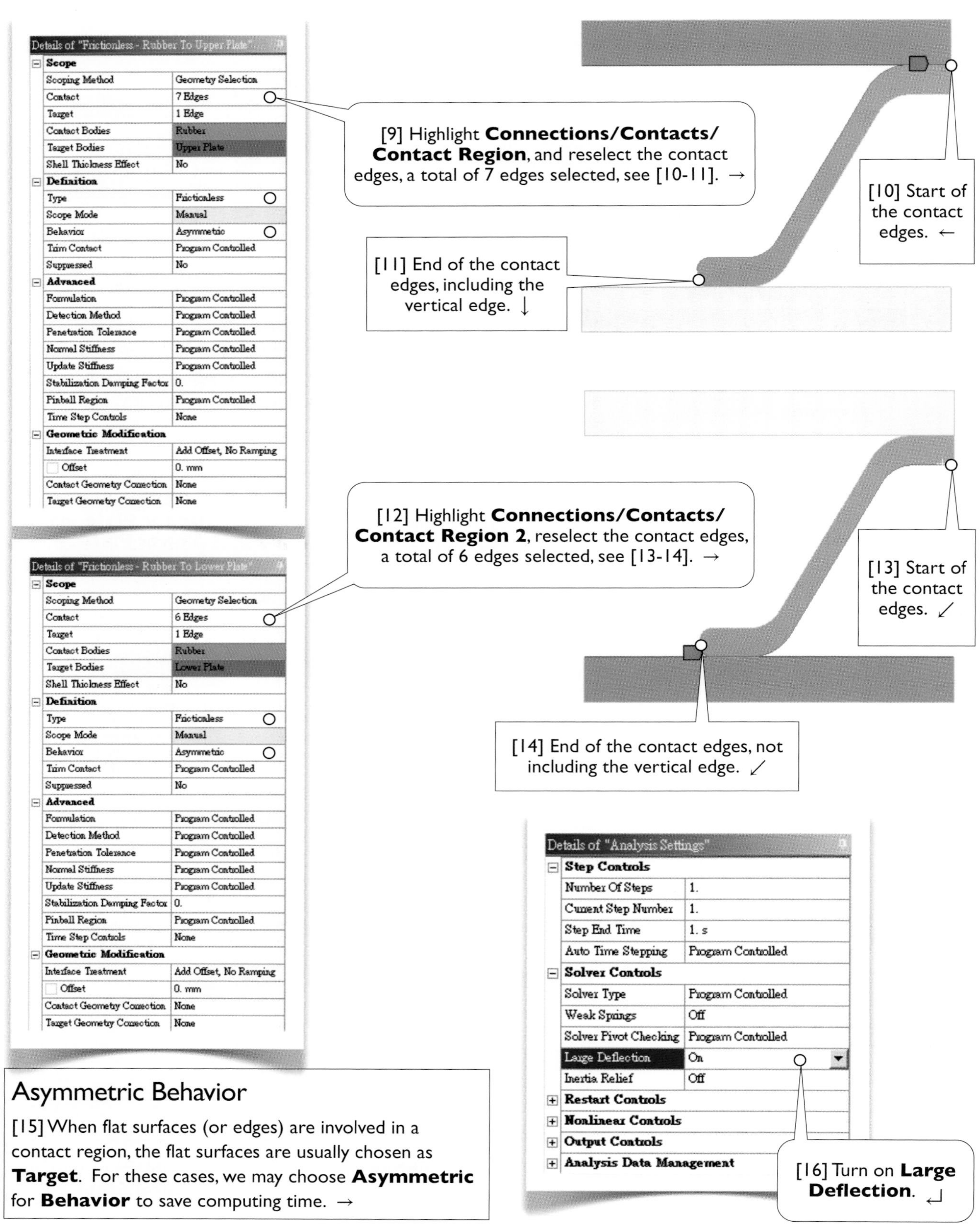

Details of "Frictionless - Rubber To Upper Plate"

Scope	
Scoping Method	Geometry Selection
Contact	7 Edges
Target	1 Edge
Contact Bodies	Rubber
Target Bodies	Upper Plate
Shell Thickness Effect	No
Definition	
Type	Frictionless
Scope Mode	Manual
Behavior	Asymmetric
Trim Contact	Program Controlled
Suppressed	No
Advanced	
Formulation	Program Controlled
Detection Method	Program Controlled
Penetration Tolerance	Program Controlled
Normal Stiffness	Program Controlled
Update Stiffness	Program Controlled
Stabilization Damping Factor	0.
Pinball Region	Program Controlled
Time Step Controls	None
Geometric Modification	
Interface Treatment	Add Offset, No Ramping
Offset	0. mm
Contact Geometry Correction	None
Target Geometry Correction	None

Details of "Frictionless - Rubber To Lower Plate"

Scope	
Scoping Method	Geometry Selection
Contact	6 Edges
Target	1 Edge
Contact Bodies	Rubber
Target Bodies	Lower Plate
Shell Thickness Effect	No
Definition	
Type	Frictionless
Scope Mode	Manual
Behavior	Asymmetric
Trim Contact	Program Controlled
Suppressed	No
Advanced	
Formulation	Program Controlled
Detection Method	Program Controlled
Penetration Tolerance	Program Controlled
Normal Stiffness	Program Controlled
Update Stiffness	Program Controlled
Stabilization Damping Factor	0.
Pinball Region	Program Controlled
Time Step Controls	None
Geometric Modification	
Interface Treatment	Add Offset, No Ramping
Offset	0. mm
Contact Geometry Correction	None
Target Geometry Correction	None

Details of "Analysis Settings"

Step Controls	
Number Of Steps	1.
Current Step Number	1.
Step End Time	1. s
Auto Time Stepping	Program Controlled
Solver Controls	
Solver Type	Program Controlled
Weak Springs	Off
Solver Pivot Checking	Program Controlled
Large Deflection	On
Inertia Relief	Off
Restart Controls	
Nonlinear Controls	
Output Controls	
Analysis Data Management	

Asymmetric Behavior

[15] When flat surfaces (or edges) are involved in a contact region, the flat surfaces are usually chosen as **Target**. For these cases, we may choose **Asymmetric** for **Behavior** to save computing time. →

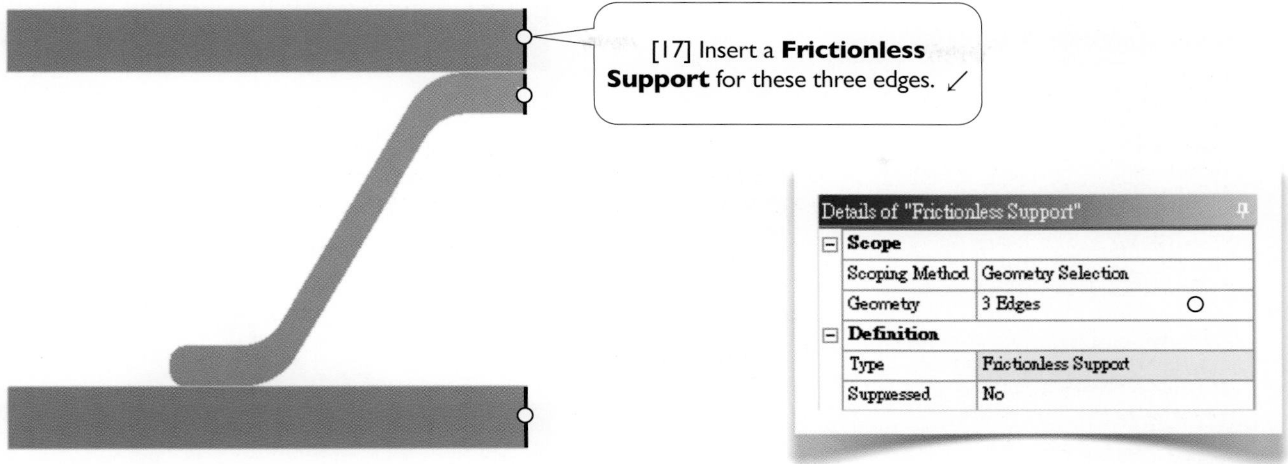

Details of "Frictionless Support"

Scope	
Scoping Method	Geometry Selection
Geometry	3 Edges
Definition	
Type	Frictionless Support
Suppressed	No

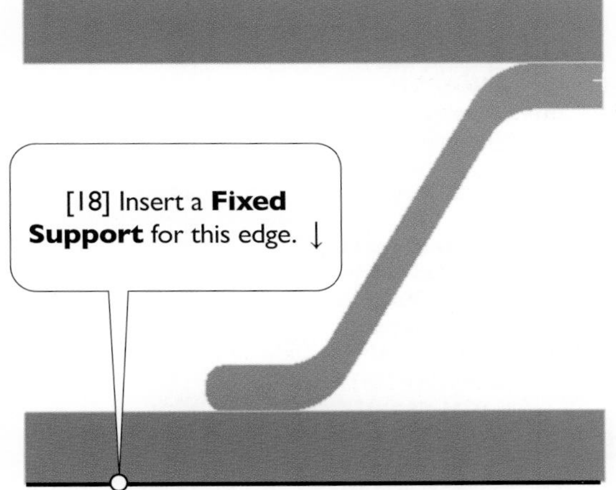

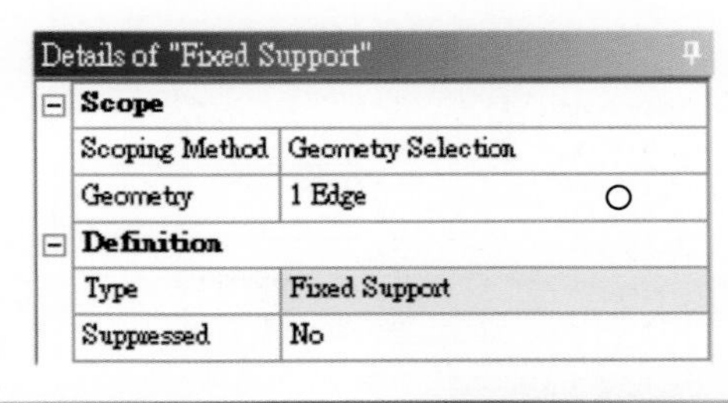

Details of "Fixed Support"

Scope	
Scoping Method	Geometry Selection
Geometry	1 Edge
Definition	
Type	Fixed Support
Suppressed	No

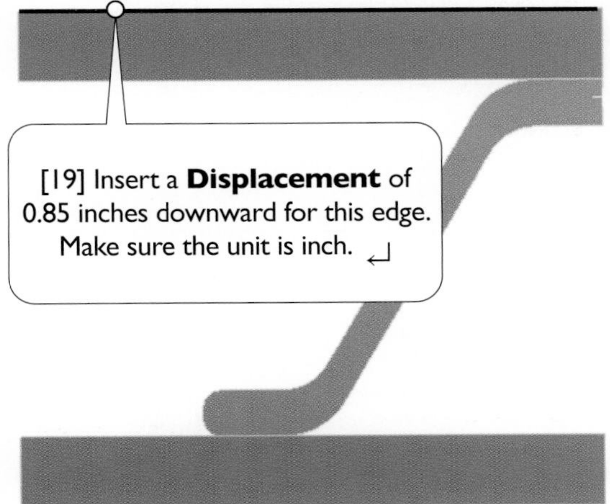

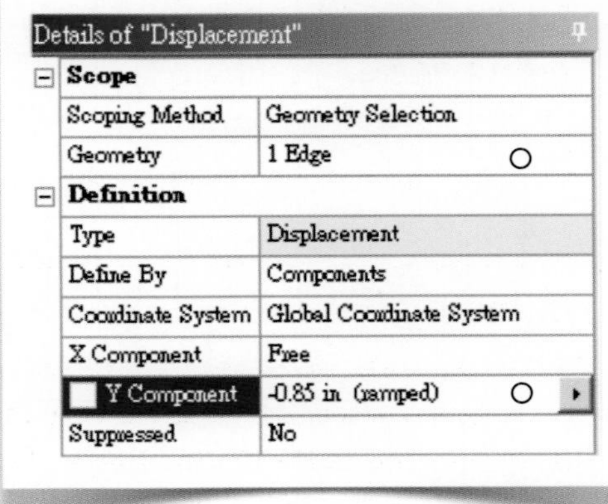

Details of "Displacement"

Scope	
Scoping Method	Geometry Selection
Geometry	1 Edge
Definition	
Type	Displacement
Define By	Components
Coordinate System	Global Coordinate System
X Component	Free
Y Component	-0.85 in (ramped)
Suppressed	No

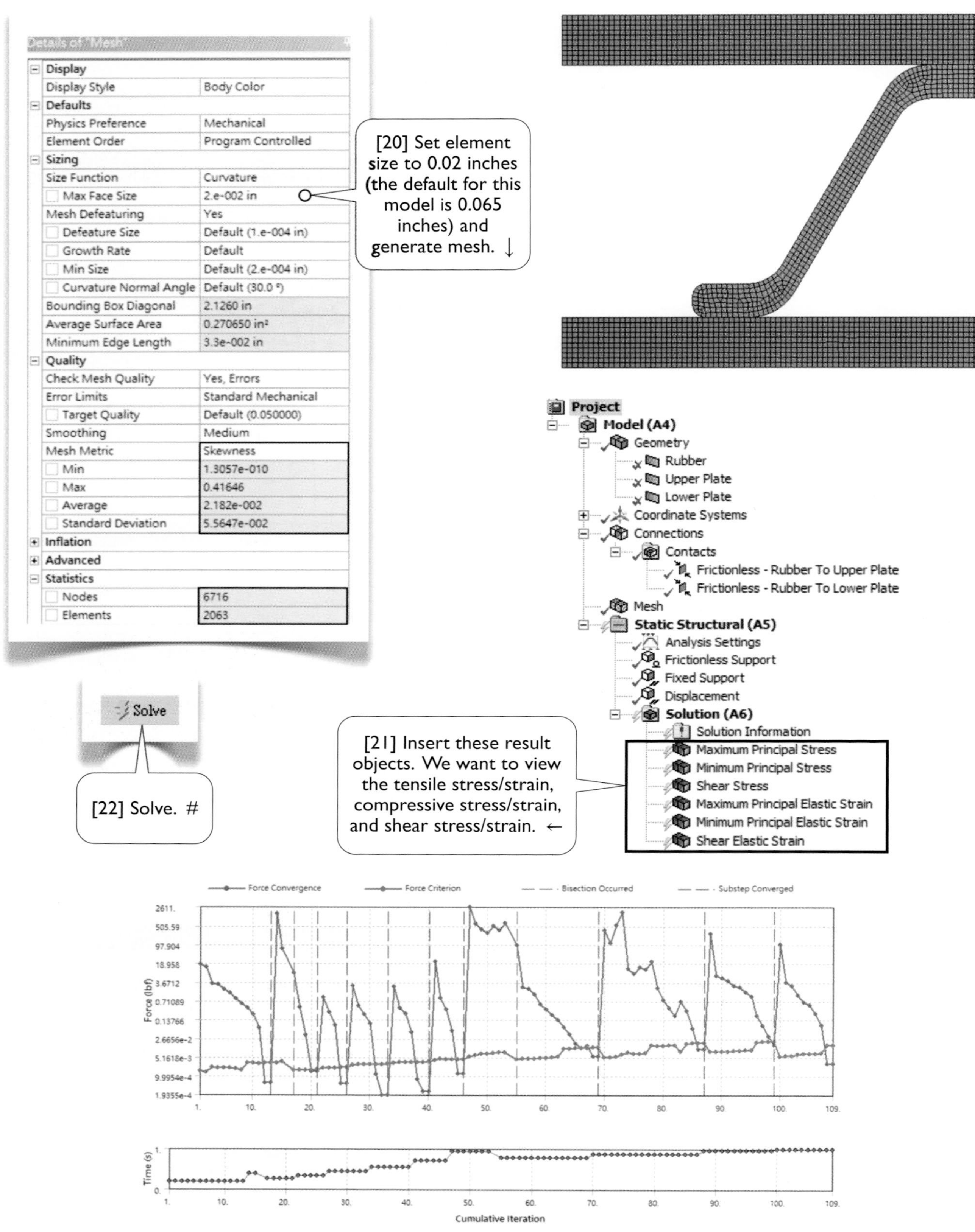

Details of "Mesh"
Display
Display Style | Body Color
Defaults
Physics Preference | Mechanical
Element Order | Program Controlled
Sizing
Size Function | Curvature
Max Face Size | 2.e-002 in
Mesh Defeaturing | Yes
Defeature Size | Default (1.e-004 in)
Growth Rate | Default
Min Size | Default (2.e-004 in)
Curvature Normal Angle | Default (30.0 °)
Bounding Box Diagonal | 2.1260 in
Average Surface Area | 0.270650 in²
Minimum Edge Length | 3.3e-002 in
Quality
Check Mesh Quality | Yes, Errors
Error Limits | Standard Mechanical
Target Quality | Default (0.050000)
Smoothing | Medium
Mesh Metric | Skewness
Min | 1.3057e-010
Max | 0.41646
Average | 2.182e-002
Standard Deviation | 5.5647e-002
Inflation
Advanced
Statistics
Nodes | 6716
Elements | 2063
[20] Set element size to 0.02 inches (the default for this model is 0.065 inches) and generate mesh. ↓
Project
Model (A4)
Geometry
Rubber
Upper Plate
Lower Plate
Coordinate Systems
Connections
Contacts
Frictionless - Rubber To Upper Plate
Frictionless - Rubber To Lower Plate
Mesh
Static Structural (A5)
Analysis Settings
Frictionless Support
Fixed Support
Displacement
Solution (A6)
Solution Information
Maximum Principal Stress
Minimum Principal Stress
Shear Stress
Maximum Principal Elastic Strain
Minimum Principal Elastic Strain
Shear Elastic Strain
Solve
[22] Solve. #
[21] Insert these result objects. We want to view the tensile stress/strain, compressive stress/strain, and shear stress/strain. ←
Force Convergence
Force Criterion
Bisection Occurred
Substep Converged
Force (lbf)
2611.
505.59
97.904
18.958
3.6712
0.71089
0.13766
2.6656e-2
5.1618e-3
9.9954e-4
1.9355e-4
1. 10. 20. 30. 40. 50. 60. 70. 80. 90. 100. 109.
Time (s)
1.
0.
Cumulative Iteration

14.3.5 View the Results

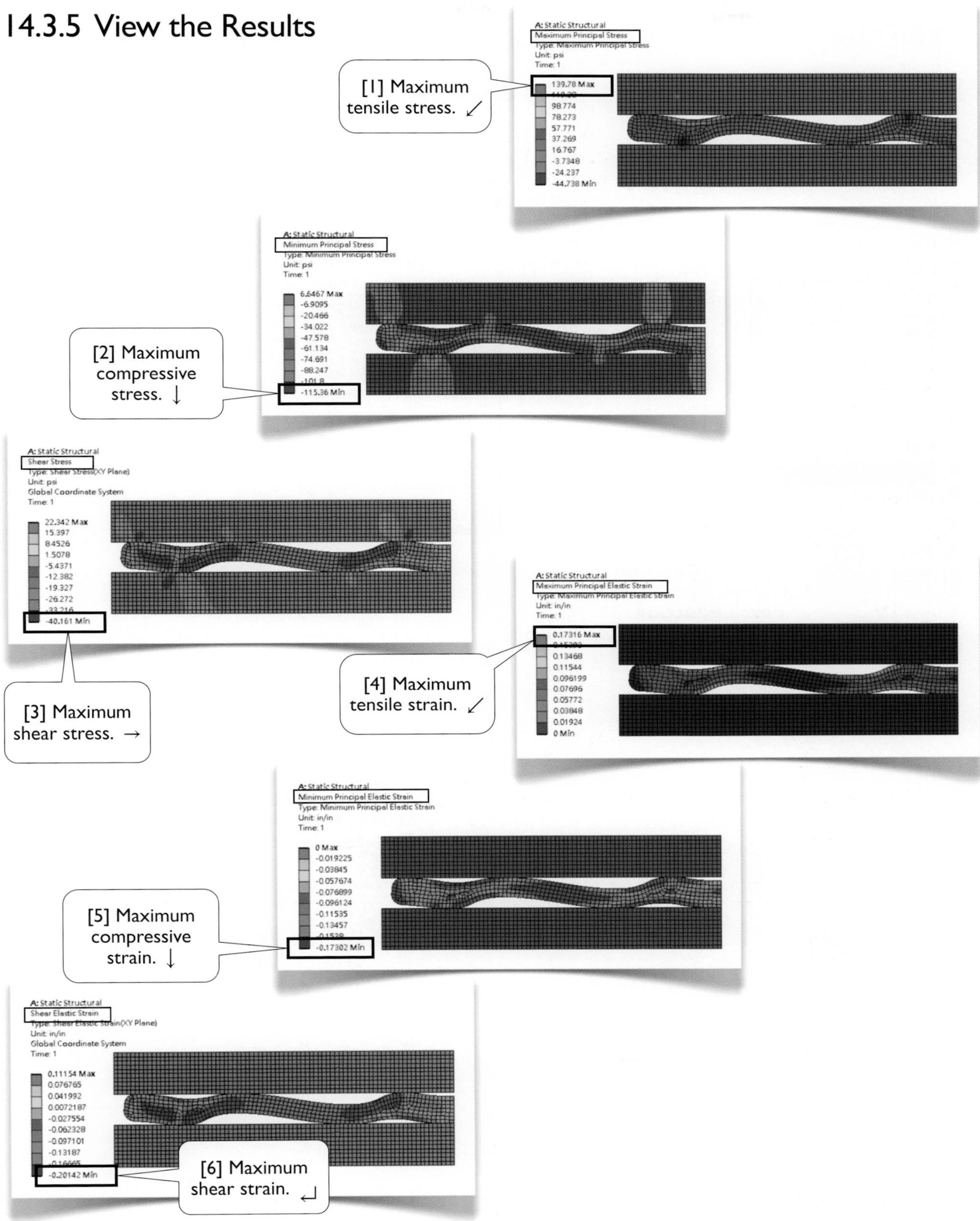

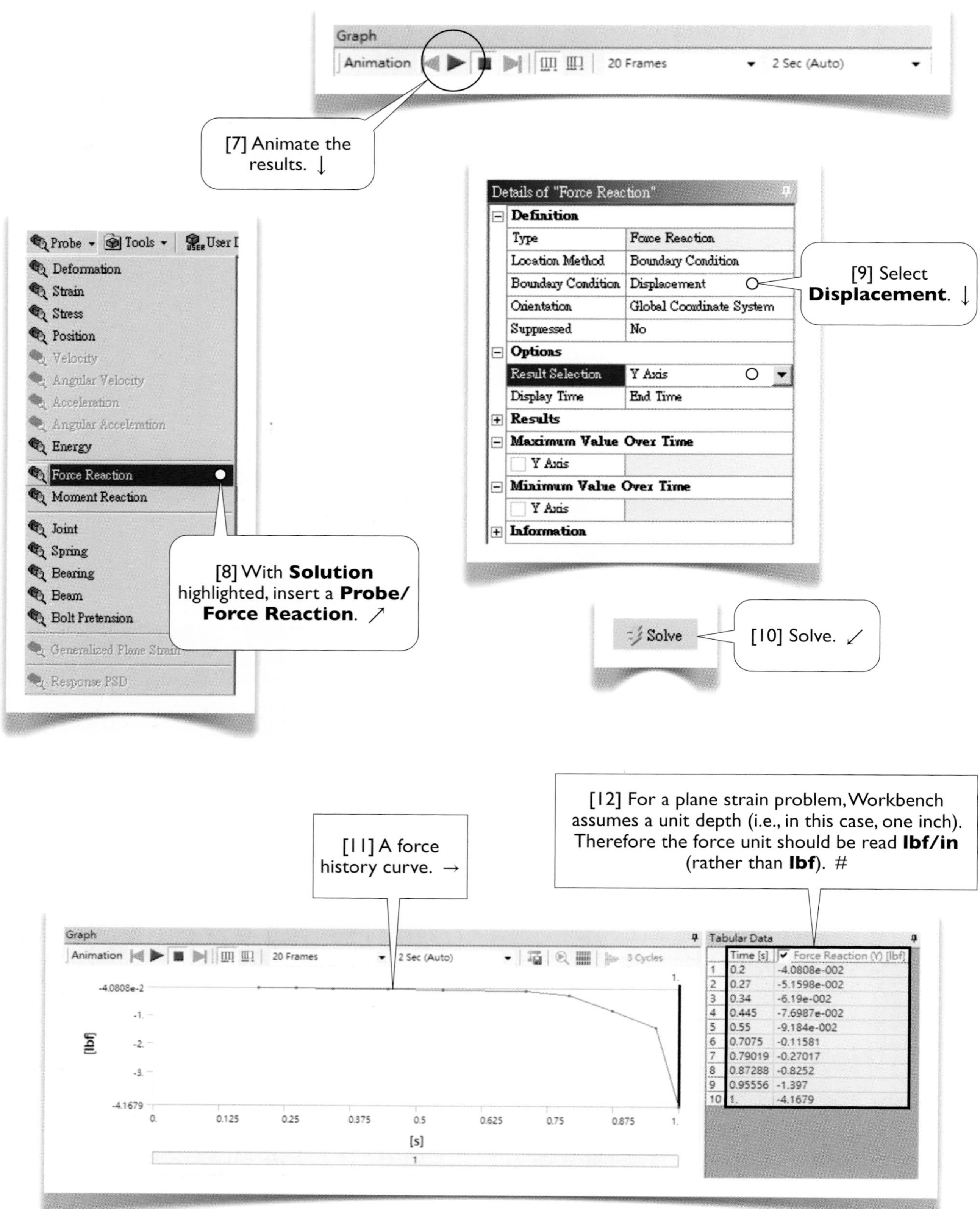

	Time [s]	Force Reaction (Y) [lbf]
1	0.2	-4.0808e-002
2	0.27	-5.1598e-002
3	0.34	-6.19e-002
4	0.445	-7.6987e-002
5	0.55	-9.184e-002
6	0.7075	-0.11581
7	0.79019	-0.27017
8	0.87288	-0.8252
9	0.95556	-1.397
10	1.	-4.1679

14.3.6 Create a Force-versus-Displacement Chart

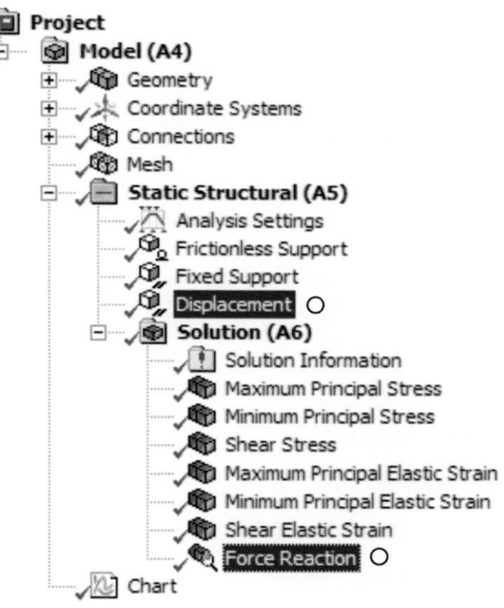

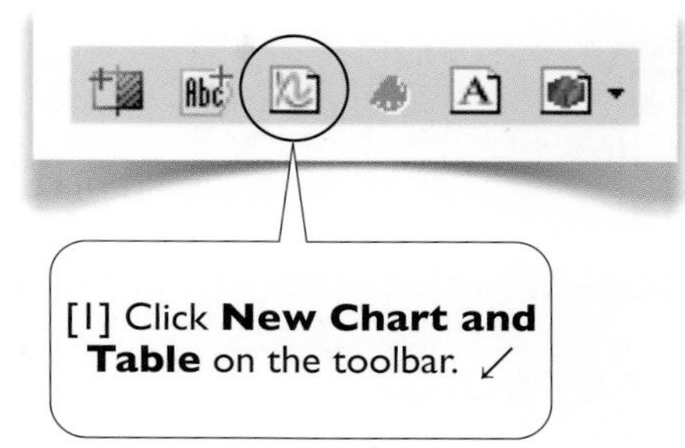

[1] Click **New Chart and Table** on the toolbar. ↙

[2] Click to bring up **Apply/Cancel** buttons and control-select **Static Structural/Displacement** and **Solution/Force Reaction**. →

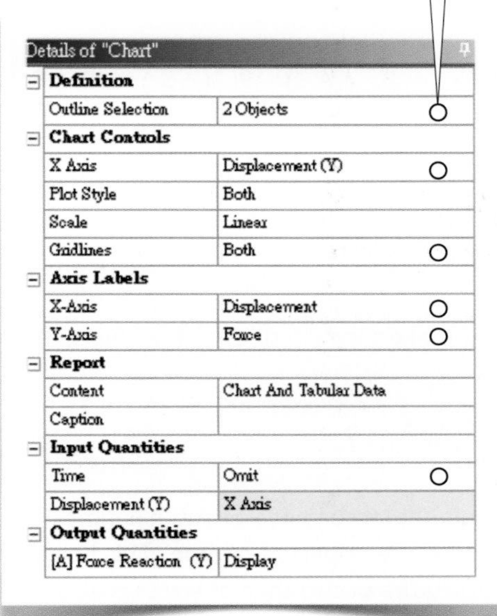

Details of "Chart"

Definition	
Outline Selection	2 Objects
Chart Controls	
X Axis	Displacement (Y)
Plot Style	Both
Scale	Linear
Gridlines	Both
Axis Labels	
X-Axis	Displacement
Y-Axis	Force
Report	
Content	Chart And Tabular Data
Caption	
Input Quantities	
Time	Omit
Displacement (Y)	X Axis
Output Quantities	
[A] Force Reaction (Y)	Display

[3] A force-versus-displacement curve. Note that the force unit should be read **lbf/in** (instead of **lbf**; also see 14.3.5[12], last page). ↓

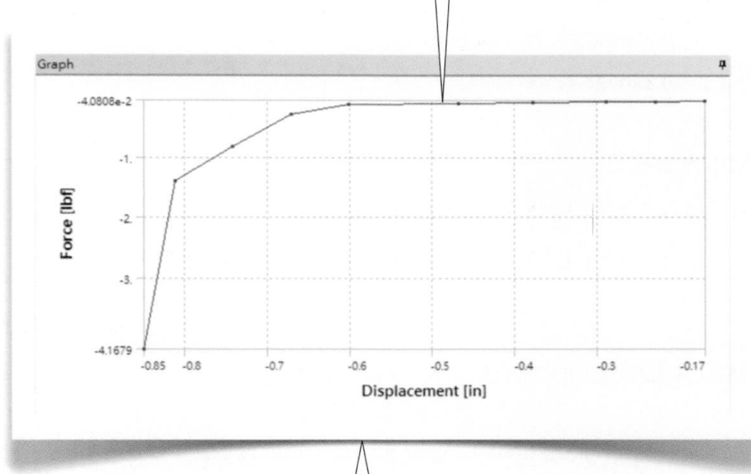

[4] This curve is plotted using only 10 points since the model is solved with 10 substeps (see 14.3.5[12], last page). To obtain more data points for a more accurate curve, you need to change the settings in **Analysis Settings**. ←

Wrap Up

[5] Save the project and exit Workbench. #

Section 14.4

Review

14.4.1 Keywords

Choose a letter for each keyword from the list of descriptions

1. ( ) Elastic Material
2. ( ) Hardening Rules
3. ( ) Hyperelastic Materials
4. ( ) Kinematic Hardening
5. ( ) Initial Yield Point
6. ( ) Initial Yield Surface
7. ( ) Isometric Hardening
8. ( ) Linear Materials
9. ( ) Nonlinear Materials
10. ( ) Plastic Materials
11. ( ) Rate-Dependent Materials
12. ( ) Subsequent Yield Surface
13. ( ) Time-Dependent Materials
14. ( ) Von Mises Yield Surface
15. ( ) Yield Surface

Answers:

1. (B) 2. (M) 3. (G) 4. (N) 5. (H) 6. (K) 7. (O) 8. (A) 9. (E) 10. (F)
11. (D) 12. (L) 13. (C) 14. (J) 15. (I)

List of Descriptions

(A) A material that has a linear stress-strain relation and can be described by Hooke's law. In this book, a linear material also implies elasticity, time-independent, and rate-independent.

(B) Materials in which the strain returns to its original state when the stress is removed.

(C) Also called viscous materials. If you apply a constant stress on the materials, the occurrence of strain is time-dependent. The strain typically increases with time and stabilizes to a constant value after a period of time. This behavior is called creeping. On the other hand, if you apply a constant strain on the materials, the occurrence of stress is also time-dependent. The stress typically decreases with time and stabilizes to a constant value after a period of time. This behavior is called stress-relaxation.

(D) Materials in which the stress-strain relation is such that the magnitude of stress depends on not only the magnitude of strain but also the rate of the strain, and vice versa.

(E) In this book, all materials that are not described by Hooke's law are categorized as nonlinear materials.

(F) Materials that exhibit plasticity behavior, which is characterized by strain that does not return to its original state after the stress is removed. Rather, residual strain, or plastic strain, remains.

(G) Materials that remain elastic under a very large strain.

(H) In plasticity, yield point may change. Stress at the elastic limit is called the initial yield point.

(I) In a uniaxial test, yielding occurs at certain stress values; they are called yield points. In 3D cases, yielding occurs at certain stress states; these stress states form a "surface" (or hyper-surface) in the multiaxial stress space. The surface is called a yield surface. During stressing, the yield surface may change in size as well as location.

(J) Different yield criterion results in different yield surface. If von Mises yield criterion is used, the yield surface is a cylindrical surface in the σ_1-σ_2-σ_3 space.

(K) In plasticity, yield surface may change in size as well as location during stressing. The original yield surface before any yielding occurs (i.e., the behavior is still elastic) is called the initial yield surface.

(L) In plasticity, during stressing, yield surface may change in size as well as location. All possible yield surfaces except the initial yield surface are called the subsequent yield surfaces.

(M) Rules that describe how yield surface changes its size and location.

(N) The yield surface changes only the location but not the size.

(O) The yield surface changes only the size but not the location.

14.4.2 Additional Workbench Exercises

Model the Belleville Washer as a 2D Problem

In Section 14.2, we modeled the Belleville washer as a surface body and meshed it with shell elements. It is possible to model it as a 2D problem and mesh it with 2D axisymmetric solid elements. Do it and draw the pros and cons for both modeling methods.

Pneumatic Finger

Replace the material of the pneumatic finger, simulated in Section 9.1, with the rubber used in the Section 14.3. Redo the simulation for the pneumatic finger.

Chapter 15
Explicit Dynamics

Many transient dynamic simulations require extremely small integration time steps, for example, high-speed impact, drop test, or highly nonlinear problems. In such cases, the time steps may be as small as a few nanoseconds, and the use of **Transient Structural** analysis system becomes impractical, since the run time would be too enormous. The integration method used in **Transient Structural** analysis system is an *implicit method.*

Explicit Dynamics analysis system also deals with transient structural dynamics; however, it uses an *explicit integration method.* The *explicit method* is very efficient for each time step. It thus allows a large number of time steps to calculate within an acceptable time. A characteristic of explicit methods is that the integration time steps must be very small (e.g., microseconds or nanoseconds) in order to achieve stable solutions. If the dynamic behavior must be observed within a long duration (say, several seconds), then many millions of time steps are required to complete the simulation. In these situations, high-performance computing facilities are usually used to facilitate the computations.

Purpose of This Chapter

This chapter provides the basics of explicit dynamics so that the students have enough background knowledge to perform some simple explicit dynamics simulations. A high-speed impact simulation and a drop test simulation are used to demonstrate the applications of **Explicit Dynamics**.

About Each Section

Section 15.1 provides the basics of explicit dynamics. Section 15.2 presents a high-speed impact example and Section 15.3 presents a drop test simulation. Both examples are simple yet instructional. We use default settings as much as possible, so that the students can appreciate the built-in capabilities of **Explicit Dynamics**.

Section 15.1

Basics of Explicit Dynamics

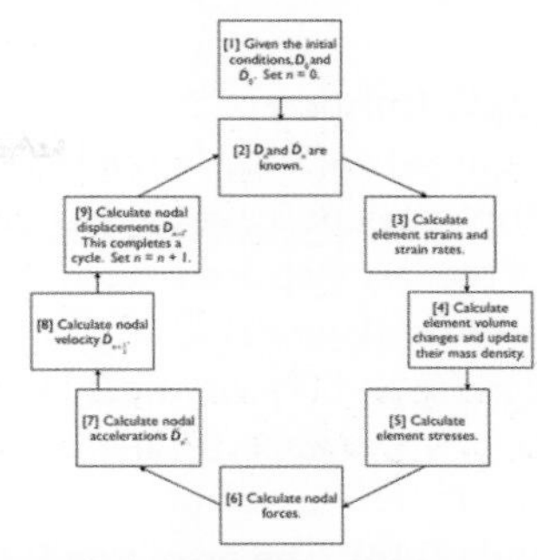

15.1.1 Implicit Integration Methods

[1] As mentioned, transient dynamic simulations involve solving the equation,

$$[M]\{\ddot{D}\}+[C]\{\dot{D}\}+[K]\{D\}=\{F\} \qquad \text{Copy of Eq. 12.1.4(1), page 426}$$

Consider a typical time step at t_n. Let $D_n, \dot{D}_n$, and $\ddot{D}_n$ be the displacement, velocity, and acceleration at t_n, and $D_{n+1}, \dot{D}_{n+1}$, and $\ddot{D}_{n+1}$ at t_{n+1}. Also, let $\Delta t = t_{n+1} - t_n$. We temporarily assume that the acceleration is linear over the time step (i.e., $\ddot{\ddot{D}}_n = \ddot{\ddot{D}}_{n+1} = 0$), then, by Taylor series expansions at t_n,

$$\dot{D}_{n+1} = \dot{D}_n + \Delta t \ddot{D}_n + \frac{\Delta t^2}{2}\dddot{D}_n \tag{1}$$

$$D_{n+1} = D_n + \Delta t \dot{D}_n + \frac{\Delta t^2}{2}\ddot{D}_n + \frac{\Delta t^3}{6}\dddot{D}_n \tag{2}$$

The quantity $\dddot{D}_n$ can be approximated by

$$\dddot{D}_n = \frac{\ddot{D}_{n+1} - \ddot{D}_n}{\Delta t} \tag{3}$$

Substitution of Eq. (3) into Eqs. (1) and (2) respectively yields

$$\dot{D}_{n+1} = \dot{D}_n + \frac{\Delta t}{2}\left(\ddot{D}_{n+1} + \ddot{D}_n\right) \tag{4}$$

$$D_{n+1} = D_n + \Delta t \dot{D}_n + \Delta t^2\left(\frac{1}{6}\ddot{D}_{n+1} + \frac{1}{3}\ddot{D}_n\right) \tag{5}$$

Eqs. (4) and (5) can be regarded as a special case of Newmark methods,

$$\dot{D}_{n+1} = \dot{D}_n + \Delta t\left[\gamma \ddot{D}_{n+1} + (1-\gamma)\ddot{D}_n\right] \tag{6}$$

$$D_{n+1} = D_n + \Delta t \dot{D}_n + \frac{1}{2}\Delta t^2\left[2\beta \ddot{D}_{n+1} + (1-2\beta)\ddot{D}_n\right] \tag{7}$$

If you substitute $\gamma = 1/2$ and $\beta = 1/6$ into Eqs. (6) and (7) respectively, you will come up with Eqs. (4) and (5).

Eqs. (6) and (7) are used in **Transient Structural** analysis system. The parameters γ and β are chosen to control characteristics of the algorithm such as accuracy, numerical stability, etc. It is called an *implicit method* because the calculation of $\dot{D}_{n+1}$ and D_{n+1} requires knowledge of $\ddot{D}_{n+1}$. That is, the response at the current time step depends on not only the historical information but also the current information; therefore, solving Eqs. (6) and (7) involves iterative process. ↵

[2] Calculation of the response at time t_{n+1} is depicted in [3-8]. In the beginning [3], the displacement D_n, velocity $\dot{D}_n$, and acceleration $\ddot{D}_n$ of the last step are already known (For $n = 0$, we may assume $\ddot{D}_0 = 0$). Since $\ddot{D}_{n+1}$ is needed in Eqs. (6) and (7), we use $\ddot{D}_n$ as an initial guess of $\ddot{D}_{n+1}$. Knowing D_n, $\dot{D}_n$, and $\ddot{D}_{n+1}$, the quantities $\dot{D}_{n+1}$ and D_{n+1} can be calculated according to Eqs. (6) and (7) [4]. The next step [5] is to substitute $\ddot{D}_{n+1}$, $\dot{D}_{n+1}$, and D_{n+1} into Eq. 12.1.4(1). If Eq. 12.1.4(1) is satisfied [6], then the calculation of the response at time t_{n+1} is complete [7], otherwise, $\ddot{D}_{n+1}$ is updated and another iteration is initiated [8]. Update of $\ddot{D}_{n+1}$ [8] is similar to the Newton-Raphson method described in 13.1.4, page 472.

With implicit methods, integration time step is typically about milliseconds; a typical simulation time is about 0.1 to 10 seconds, which requires hundreds to ten-thousands of integration time steps.

Implicit methods can be used for most transient structural simulations. However, for highly nonlinear problems, it often fails due to convergence issues; for high-speed impact problems, the integration time is so small that the computing time becomes intolerable. In such cases, explicit methods are more applicable. ↗

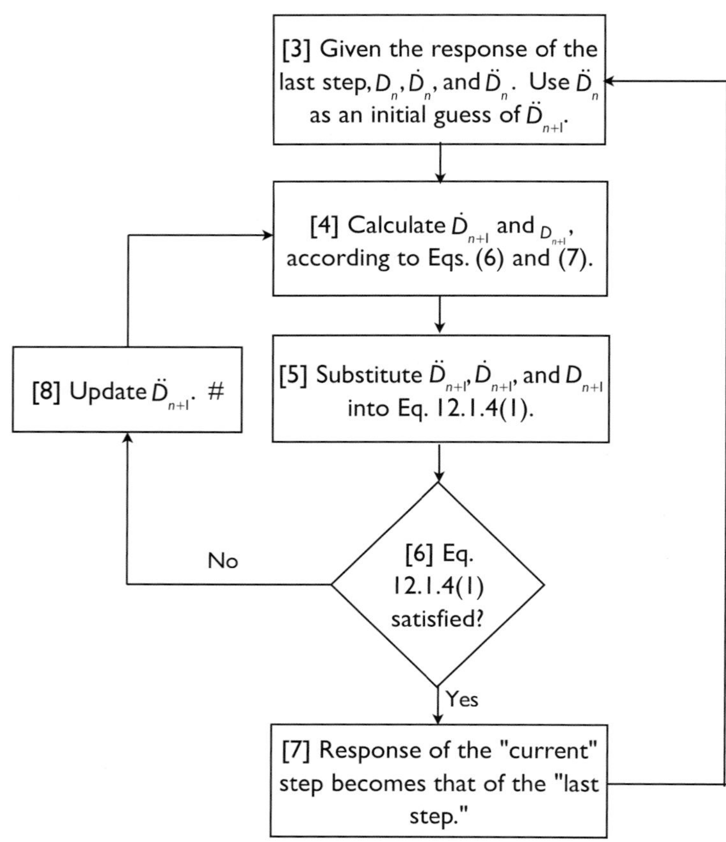

15.1.2 Explicit Integration Methods

[1] The explicit method used in **Explicit Dynamics** analysis system is based on half-step central differences

$$\ddot{D}_n = \frac{\dot{D}_{n+\frac{1}{2}} - \dot{D}_{n-\frac{1}{2}}}{\Delta t}, \text{ or } \dot{D}_{n+\frac{1}{2}} = \dot{D}_{n-\frac{1}{2}} + \ddot{D}_n \Delta t \qquad (1)$$

$$\dot{D}_{n+\frac{1}{2}} = \frac{D_{n+1} - D_n}{\Delta t}, \text{ or } D_{n+1} = D_n + \dot{D}_{n+\frac{1}{2}} \Delta t \qquad (2)$$

Eqs. (1) and (2) are called *explicit methods* because the calculation of $\dot{D}_{n+\frac{1}{2}}$ and D_{n+1} requires knowledge of historical information only. That is, the response at the current time can be calculated explicitly; no iterations within a time step are needed. Therefore, it is very efficient to complete a time step, also called a *cycle*. One of the distinct characteristics of the explicit method is that its integration time step needs to be very small to achieve a stable solution. ↵

[2] The procedure used in the **Explicit Dynamics** analysis system is illustrated in [3-11]. In the beginning of a cycle [3-4], the displacement D_n and velocity $\dot{D}_n$ of the last cycle are already known. With this information, we can calculate the strain and strain rate for each element [5], using the relations such as Eqs. 1.3.2(2) (page 36) and 1.2.7(1) (page 31). The volume change for each element is then calculated, according to the equations of state, and the mass density is updated [6]. The volumetric information is needed for the calculation of stresses. With this information, the element stresses can be calculated [7] according to a relation between stresses and strains/strain rates, such as Eq. 1.2.8(1), page 31. The stresses are integrated over the elements, and the external loads are added to form the nodal forces F_n [8]. The nodal accelerations are then calculated [9] using

$$\ddot{D}_n = \frac{F_n}{m} + \frac{b}{\rho} \tag{3}$$

where b is the body force (Eq. 1.2.6(2), page 30), m is the nodal mass, and ρ is the mass density. The nodal velocities at $t_{n+\frac{1}{2}}$ are calculated [10] using Eq. (1) and the nodal displacements at t_{n+1} are calculated [11] using Eq. (2).

With explicit methods, a typical integration time step is about nanoseconds to microseconds; a typical simulation time is about 1 millisecond to 1 second, which will need many thousands or millions of cycles.

Explicit methods are useful for high-speed impact problems and highly nonlinear problems. For low-speed problems, where the durations are usually long, using explicit methods becomes impractical due to an enormous computing time, since it requires very small integration time steps. ↓

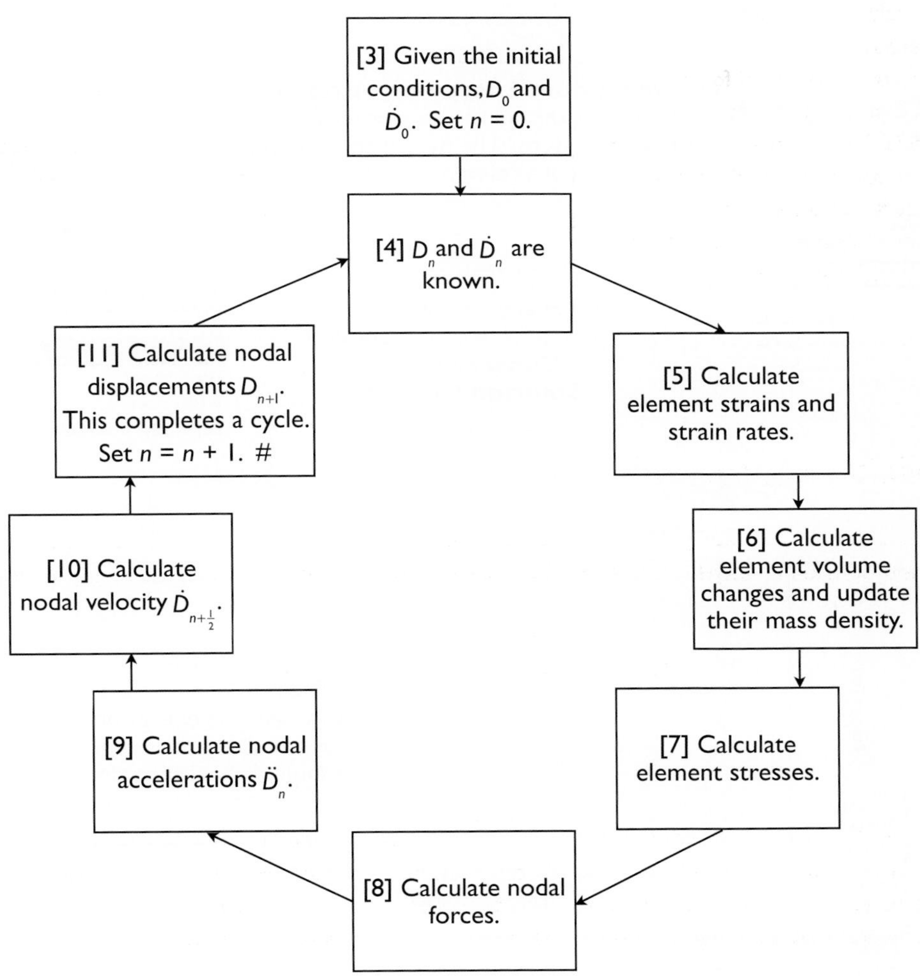

15.1.3 Solution Accuracy

[1] In **Transient Structural**, in which an implicit method is used, convergence criteria are used to control the solution accuracy, similar to the Newton-Raphson method described in 13.1.4, page 472. Equilibrium iterations imply that force balance must be satisfied. In **Explicit Dynamics**, since no equilibrium iterations are involved, solution accuracy is not controlled with convergence criteria. Instead, it uses the *principle of conservation of energy* to monitor the solution accuracy. It calculates overall energy at each cycle. If the energy error (to be defined) reaches a threshold, the solution is regarded as unstable and stops. The default threshold is 10% of a reference energy [2]. Energy statistics can be viewed by selecting **Energy Conservation** in **Solution Output** [3-4].

At any time, *Current Energy* of the system can be calculated, including its kinetic energy and strain energy. The *principle of work and energy*, a form of the principle of conservation of energy, states

$$(\mathit{Reference\ Energy}) + (\mathit{Work\ Done})_{\mathrm{Reference}\rightarrow\mathrm{Current}} = (\mathit{Current\ Energy}) \tag{1}$$

Where *Reference Energy* is the total energy of a reference time, default to the initial time. *Energy Error* is defined by

$$\mathit{Energy\ Error} = \frac{\left|(\mathit{Current\ Energy})\text{-}(\mathit{Reference\ Energy}) - (\mathit{Work\ Done})_{\mathrm{Reference}\rightarrow\mathrm{Current}}\right|}{\max\left(\left|\mathit{Current\ Energy}\right|, \left|\mathit{Reference\ Energy}\right|, \left|\mathit{Kinetic\ Energy}\right|\right)} \tag{2}$$

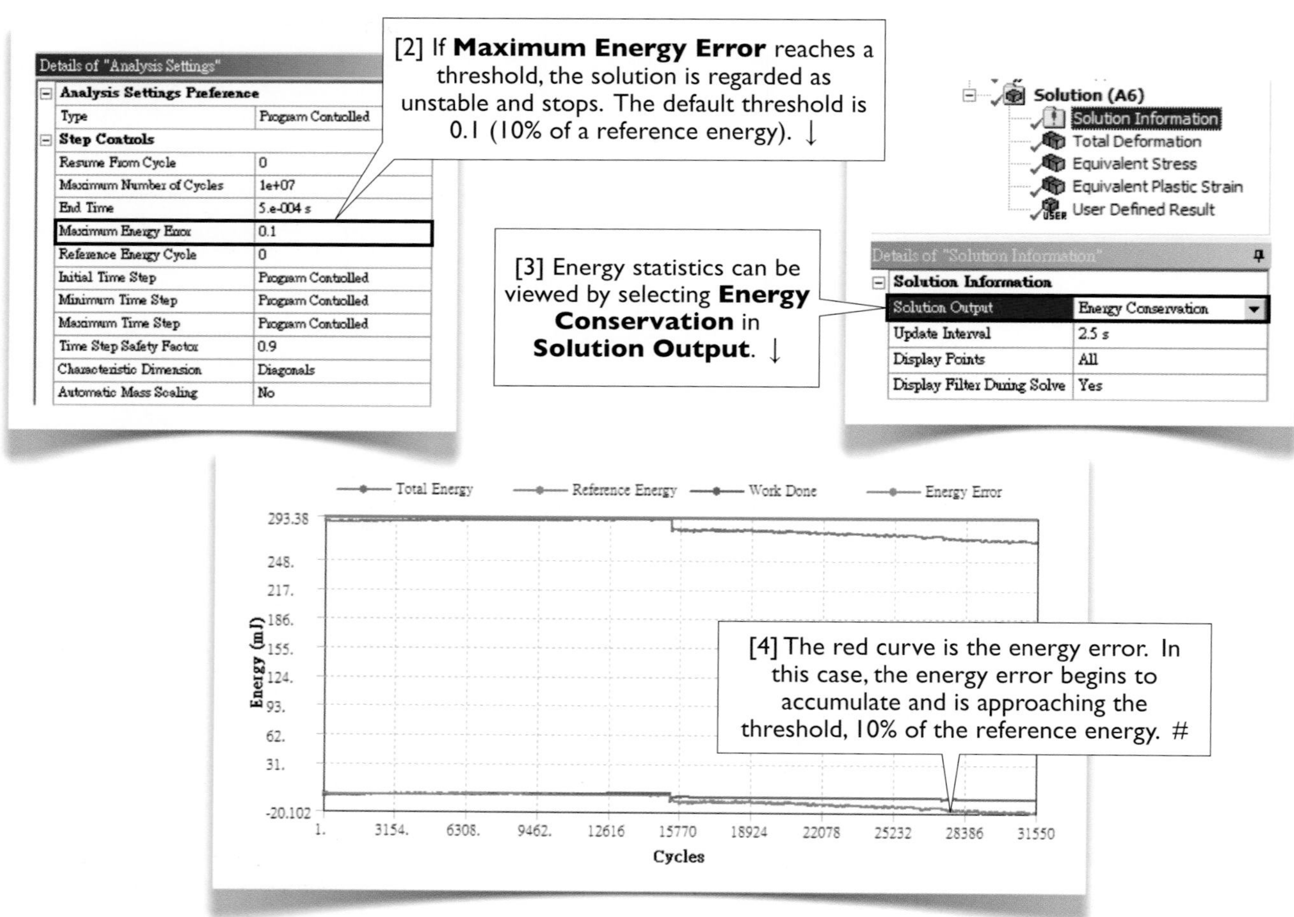

15.1.4 Integration Time Steps

[1] With explicit methods, the integration time step needs to be small enough to ensure stability and accuracy of the solution. How small should the time step be? The German mathematicians, Courant, Friedrichs, and Lewy[Refs 1, 2], suggested that, in a single time step Δt, a wave should not travel further than the smallest element size; i.e.,

$$\Delta t \le \frac{h}{c} \qquad (1)$$

where h is the smallest element size, c is the wave speed in the element. Eq. (1) is called the *CFL condition*. In **Explicit Dynamics**, a safety factor f is used to further ensure the solution stability [2-3]; i.e.,

$$\Delta t \le f\frac{h}{c} \qquad (2)$$

When generating meshes for **Explicit Dynamics**, you should make sure that a few very small elements do not control the time step, which is calculated according to the CFL condition. In general, a uniform mesh size is desirable in **Explicit Dynamics** simulations. ↘

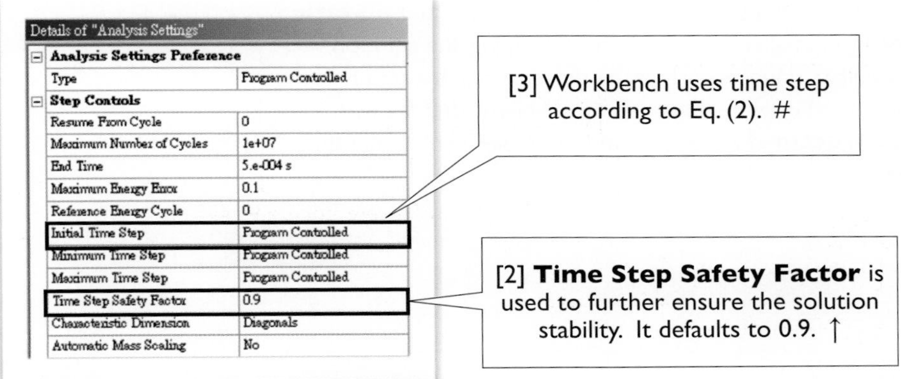

[3] Workbench uses time step according to Eq. (2). #

[2] **Time Step Safety Factor** is used to further ensure the solution stability. It defaults to 0.9. ↑

15.1.5 Automatic Mass Scaling

[1] The wave speed in a material is $c = \sqrt{E/\rho}$, where E is the Young's modulus and ρ is the mass density of the material. Further, $\rho = m/V$, where m is the mass and V is the volume of an element. Substitution of these into Eq. 15.1.4(2) yields

$$\Delta t \le fh\sqrt{\frac{m}{VE}} \qquad (1)$$

The idea of mass scaling is to artificially increase the mass of small elements, so that the stability time step can be increased. Mass scaling is applied only to those elements which have a calculated stability time step less than a specified value, default to 1e-20 sec [2], which is to ensure that no mass scaling takes place. If a mesh contains very few small elements, this idea can be useful. Note that mass scaling changes the inertial properties of the model. Be careful to ensure that the model remains valid for the physical problem. →

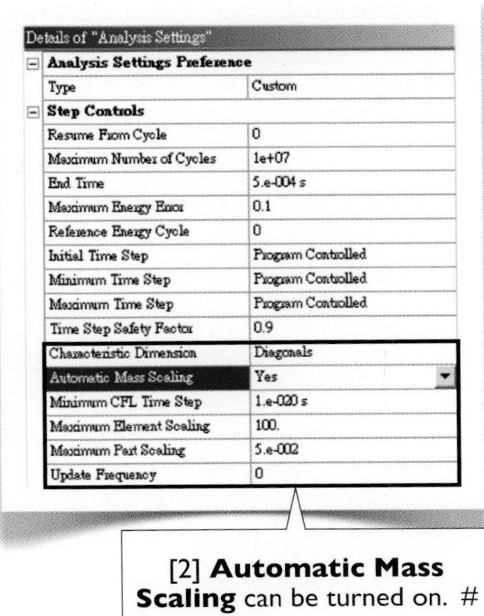

[2] **Automatic Mass Scaling** can be turned on. #

15.1.6 Static Damping

[1] **Explicit Dynamics** is designed for solving transient dynamic problems. To solve a static problem, we may perform a transient dynamic analysis and find the steady-state solution. **Static Damping** option [2] is to facilitate the finding of the steady-state solution. The idea is to introduce a damping force, to critically damp the lowest mode of oscillation.

The value of **Static Damping** for critical damping of the lowest mode of vibration is

$$\frac{2f\Delta t}{1+2\pi f\Delta t} \quad (1)$$

where f is the lowest frequency of the system.

Using the critical damping may minimize simulation run time. If **Static Damping** is larger than the critical damping, the solution would take unnecessarily longer time (than critical damping) to reach a steady state. On the other hand, if **Static Damping** is smaller than the critical damping, the solution would oscillate unnecessarily many times to reach a steady state.

For some highly-nonlinear static problems that fail with **Static Structural** analysis system, you may want to try this idea. →

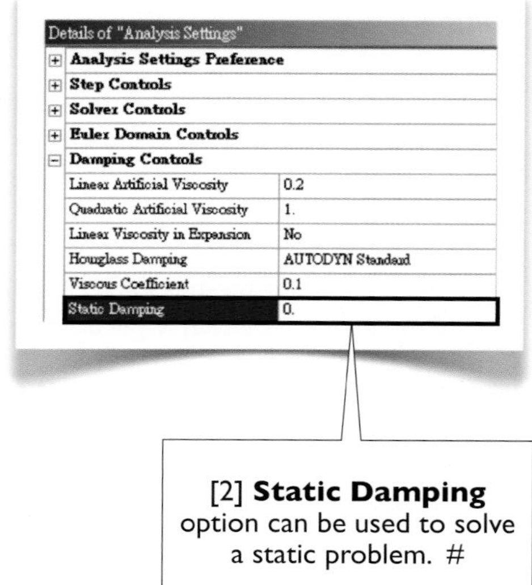

[2] **Static Damping** option can be used to solve a static problem. #

References

1. Wikipedia>Courant-Friedrichs-Lewy condition. (An English translation of the original paper can be downloaded from the webpage.)
2. Cook, R.D., Milkus, D. S., Plesha, M. E., and Witt, R. J., *Concepts and Applications of Finite Element Analysis, Fourth Edition*, John Wiley & Sons, Inc., 2002.

Section 15.2

High-Speed Impact

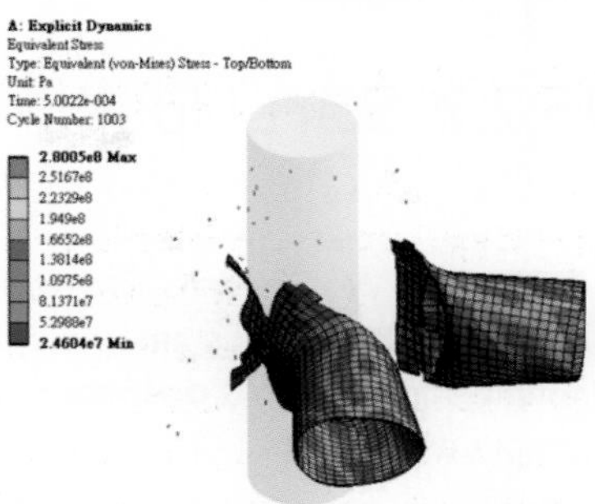

15.2.1 About the High-Speed Impact Simulation

[1] Imagine that, during an explosion, an aluminum pipe blasts away under the explosive pressure, hits a solid steel column, deforms, and is finally torn to fragments due to excessive strain (see snapshots below). In this section, we will simulate this scenario. We will use the default settings as much as possible to demonstrate that a complicated simulation like this can be done in **Explicit Dynamic** analysis system with just a few input data.

Both the aluminum pipe and the steel solid column have a diameter of 50 mm and a length of 200 mm. The steel column is modeled as a rigid body fixed in space. The aluminum pipe has a thickness of 1 mm and, right before hitting the pipe, has a speed of 300 m/s, about the speed of sound in the air. The aluminum is modeled as a bilinear isotropic plasticity material (Section 14.1) using the material parameters stored in **Engineering Data** with a modification that the tangent modulus is set to zero; i.e., the aluminum is modeled as a perfectly elastic-plastic material. It is assumed that the aluminum will be torn apart when the plastic strain is larger than 75%.

Millimeter will be used to create the geometry and the **SI** unit systems will be used in the simulation. #

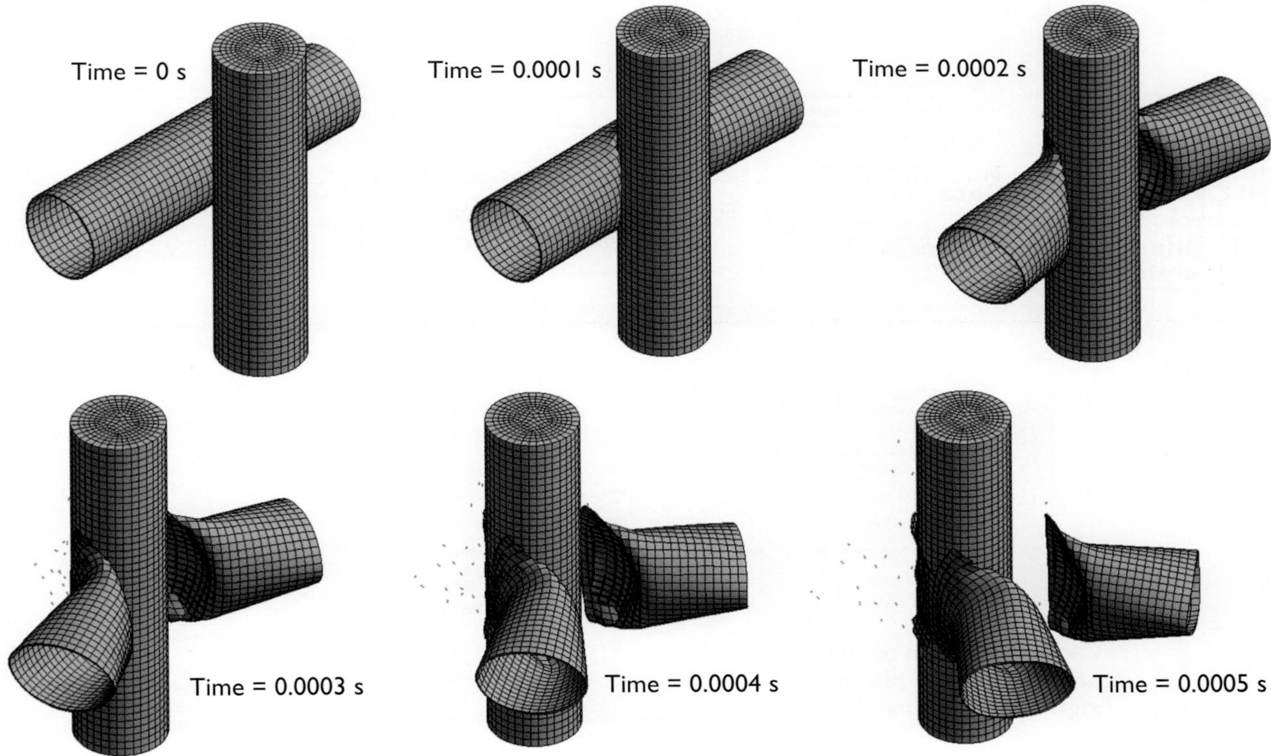

15.2.2 Start Up

[1] Launch Workbench. Create an **Explicit Dynamics** analysis system by double-clicking it in **Toolbox**. Save the project as **Impact**. →

[2] Double-click **Engineering Data** to prepare material properties. #

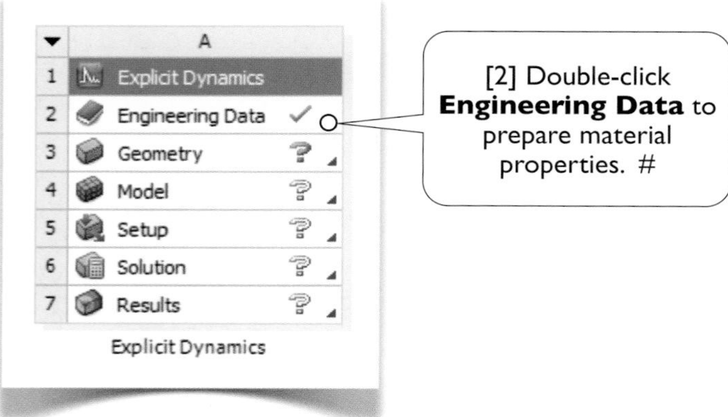

15.2.3 Prepare Material Properties for Aluminum

[1] Click to switch to **Engineering Data Sources**. ↓

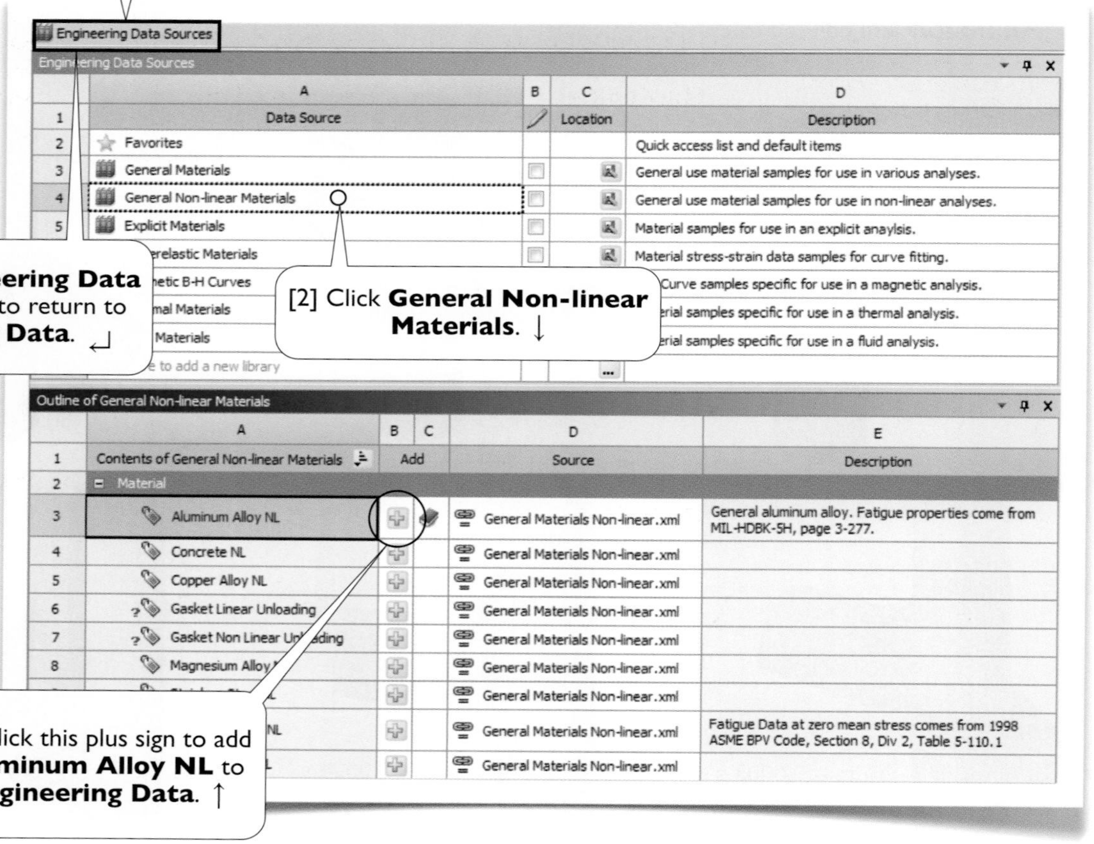

[2] Click **General Non-linear Materials**. ↓

[3] Click this plus sign to add **Aluminum Alloy NL** to **Engineering Data**. ↑

[4] Click **Engineering Data Sources** again to return to **Engineering Data**. ↵

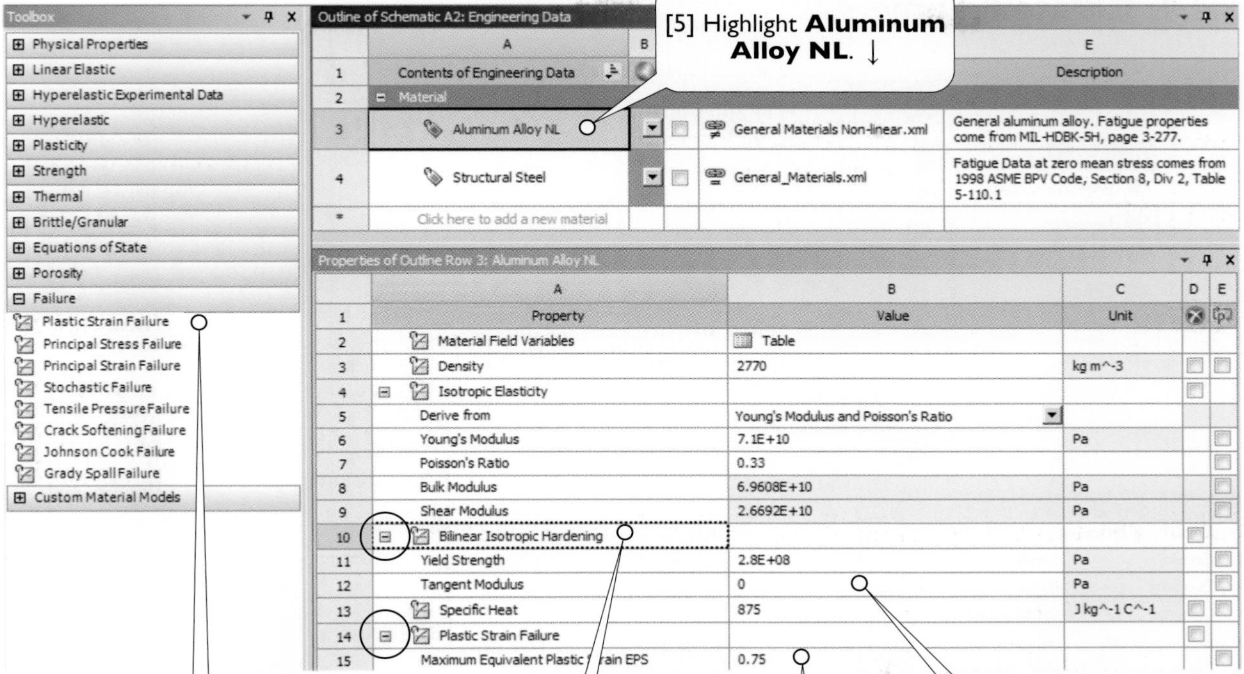

Toolbox
Physical Properties
Linear Elastic
Hyperelastic Experimental Data
Hyperelastic
Plasticity
Strength
Thermal
Brittle/Granular
Equations of State
Porosity
Failure
Plastic Strain Failure
Principal Stress Failure
Principal Strain Failure
Stochastic Failure
Tensile Pressure Failure
Crack Softening Failure
Johnson Cook Failure
Grady Spall Failure
Custom Material Models
Outline of Schematic A2: Engineering Data
Contents of Engineering Data
Description
Material
Aluminum Alloy NL
General Materials Non-linear.xml
General aluminum alloy. Fatigue properties come from MIL-HDBK-5H, page 3-277.
Structural Steel
General_Materials.xml
Fatigue Data at zero mean stress comes from 1998 ASME BPV Code, Section 8, Div 2, Table 5-110.1
Click here to add a new material
Properties of Outline Row 3: Aluminum Alloy NL
Property
Value
Unit
Material Field Variables
Table
Density
2770
kg m^-3
Isotropic Elasticity
Derive from
Young's Modulus and Poisson's Ratio
Young's Modulus
7.1E+10
Pa
Poisson's Ratio
0.33
Bulk Modulus
6.9608E+10
Pa
Shear Modulus
2.6692E+10
Pa
Bilinear Isotropic Hardening
Yield Strength
2.8E+08
Pa
Tangent Modulus
0
Pa
Specific Heat
875
J kg^-1 C^-1
Plastic Strain Failure
Maximum Equivalent Plastic Strain EPS
0.75

[5] Highlight **Aluminum Alloy NL**. ↓

[6] Expand **Bilinear Isotropic Hardening** and type 0 for **Tangent Modulus**. ←

[7] Expand **Failure** and double-click **Plastic Strain Failure** to include this failure criterion. →

[8] Expand **Plastic Strain Failure** and type 0.75 for **Maximum Equivalent Plastic Strain EPS**. ←

[9] Highlight **Bilinear Isotropic Hardening**. ↓

Chart of Properties Row 9: Bilinear Isotropic Hardeni
Bilinear Isotropic Hardening
Stress (x10⁸) [Pa]
Strain [m m^-1]
0
0.5
1
1.5
2
2.5
0.005
0.01
0.015
0.02

[10] The stress-strain curve for **Aluminum Alloy NL**. →

Project

[11] Return to **Project Schematic**. #

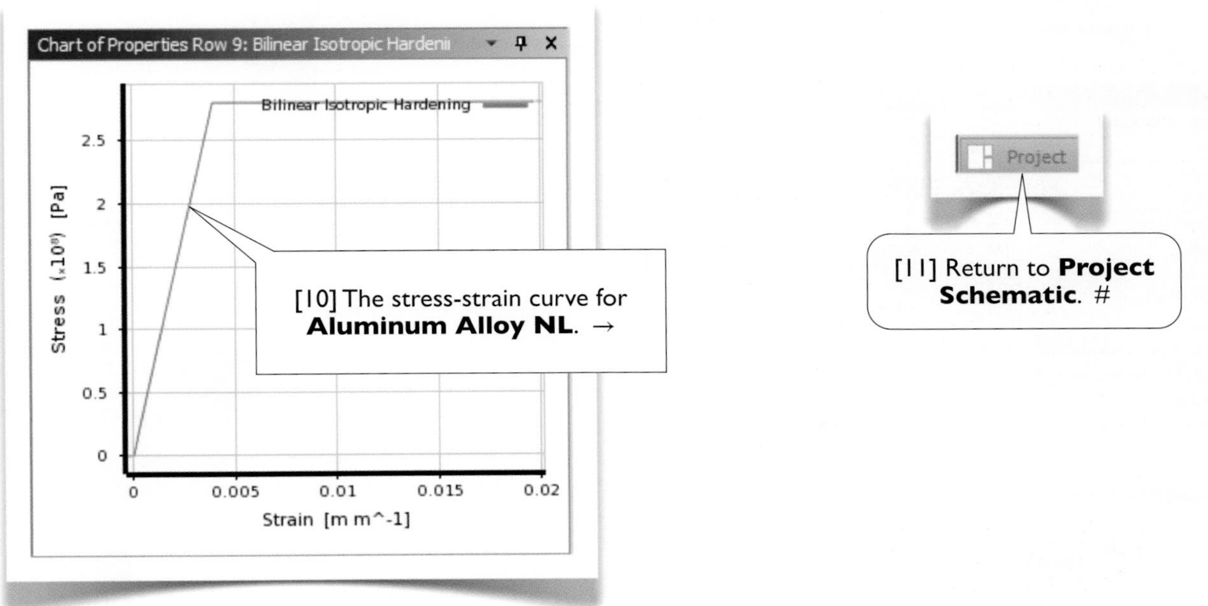

15.2.4 Create Geometry

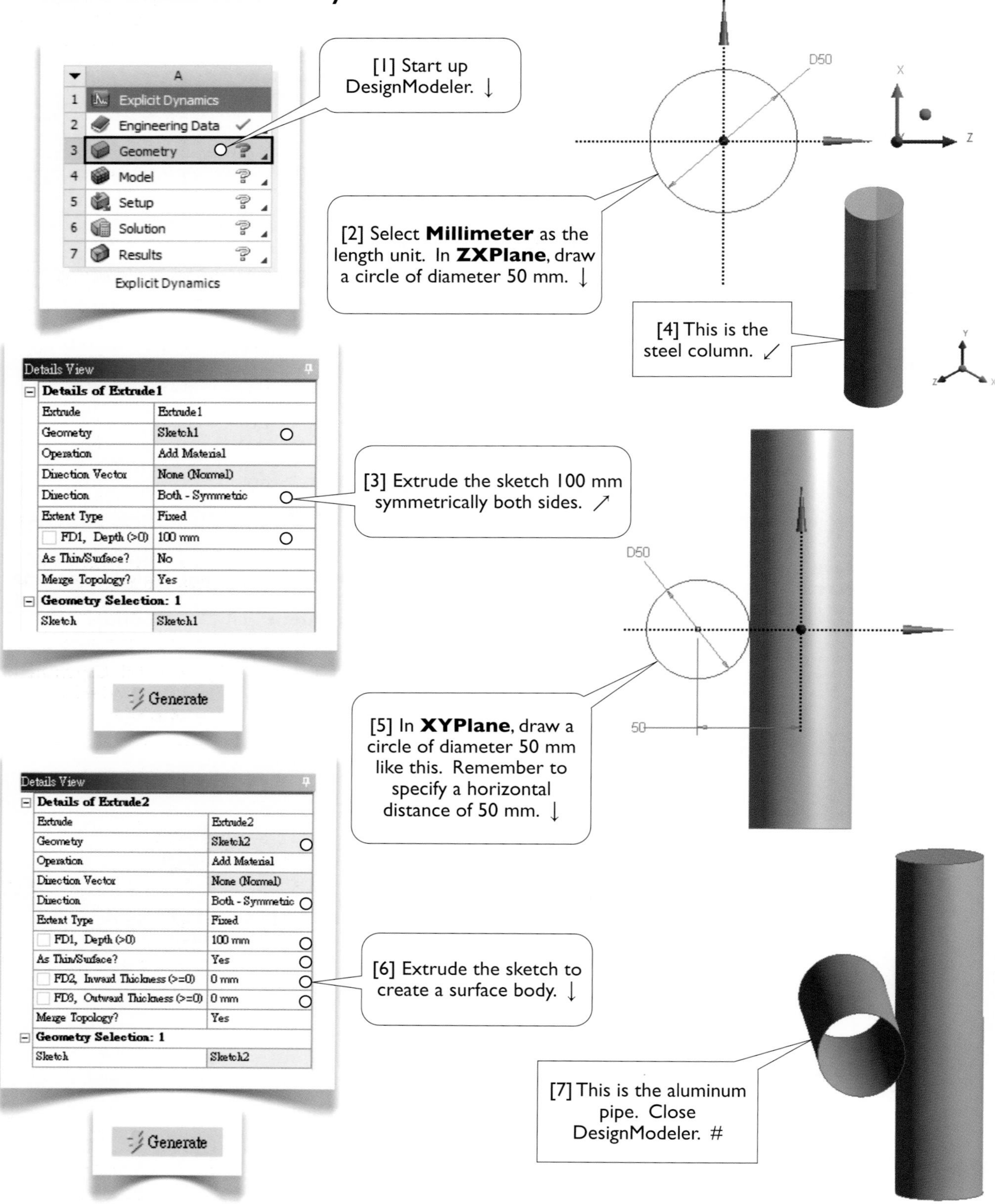

15.2.5 Set Up for Simulation

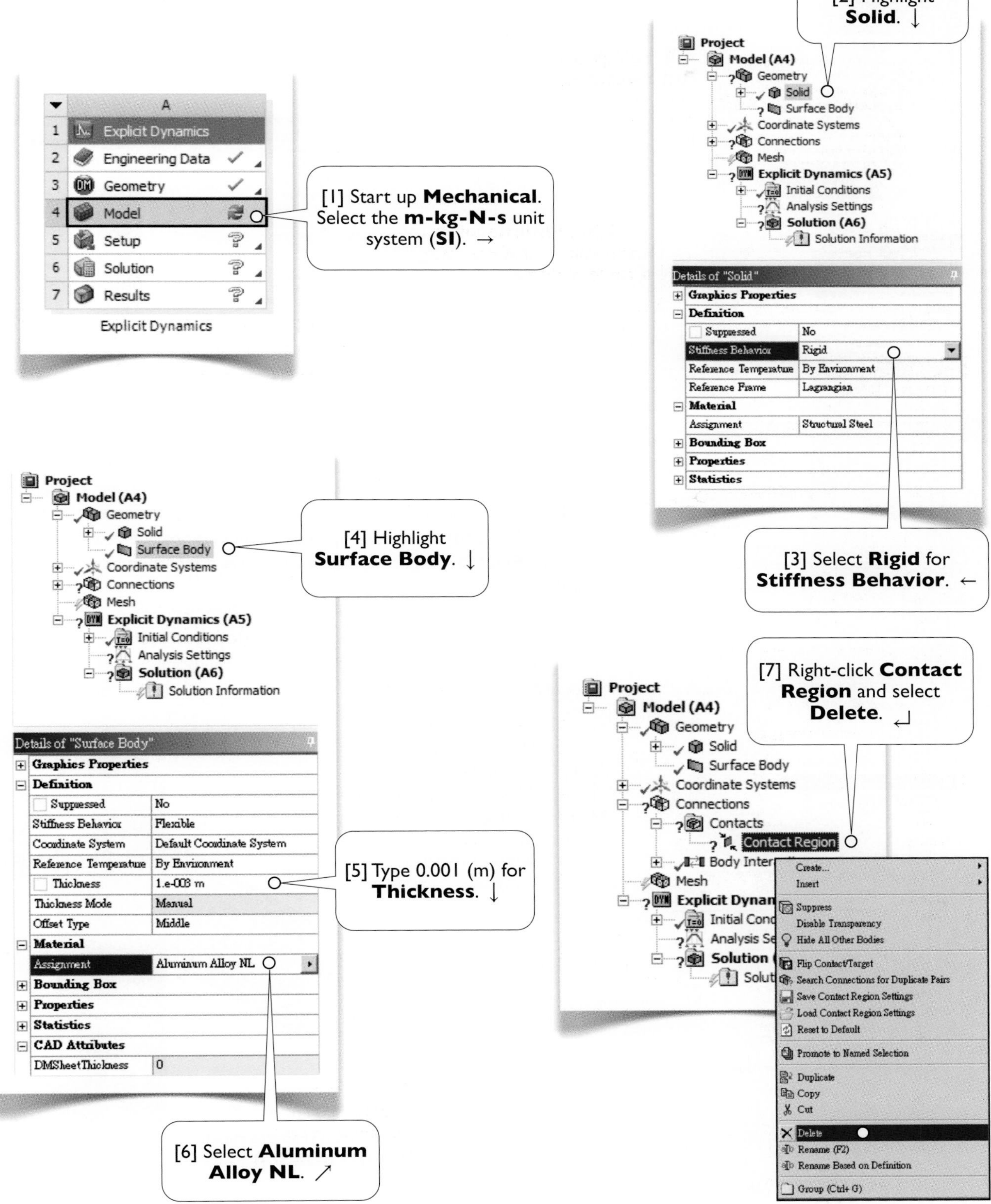

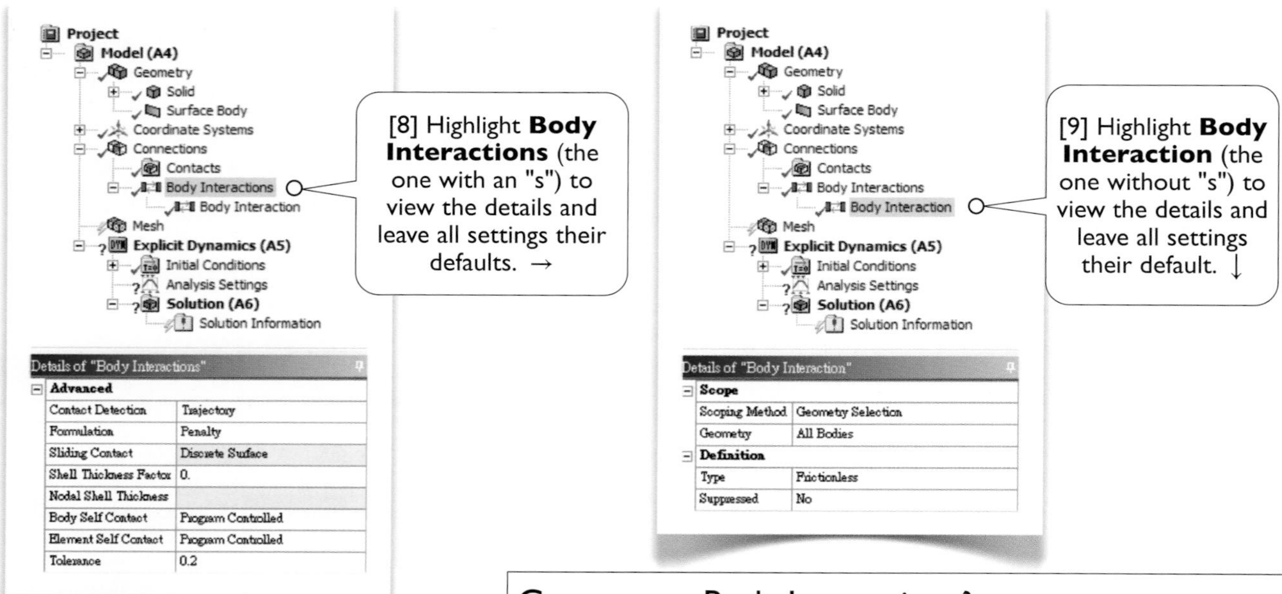

[8] Highlight **Body Interactions** (the one with an "s") to view the details and leave all settings their defaults. →

[9] Highlight **Body Interaction** (the one without "s") to view the details and leave all settings their default. ↓

Contacts vs. Body Interactions?

[10] **Body Interactions** is to specify contacts between bodies while **Contacts** is to specify contacts between surfaces. You can choose either way to specify the contact relations. **Body Interactions** is simpler, but **Contacts** may be more computationally efficient. By default, **Frictionless** body interactions are established among all bodies.

A feature of **Body Interactions** is that two bodies can be specified as both **Bonded** and **Frictionless** (or **Frictional**). In that case, two bodies are bonded initially. After the bond breaks during the simulation, the frictionless (or frictional) contact will take place. ↙

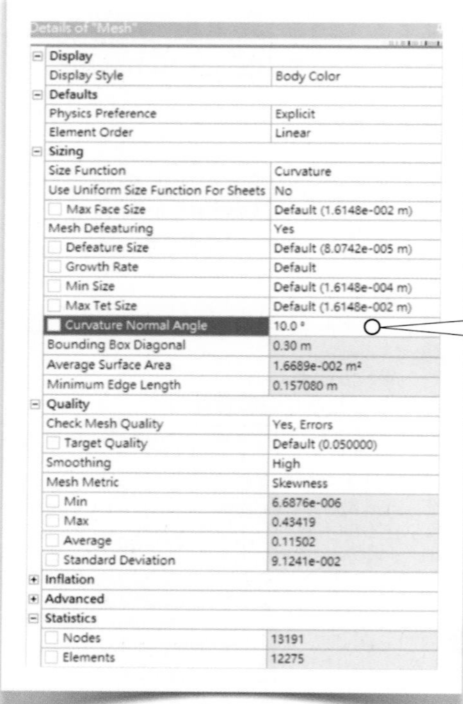

[11] Highlight **Mesh** and type 10 (degrees) for **Curvature Normal Angle**. This divides a circle into 36 elements. Generate mesh. ↓

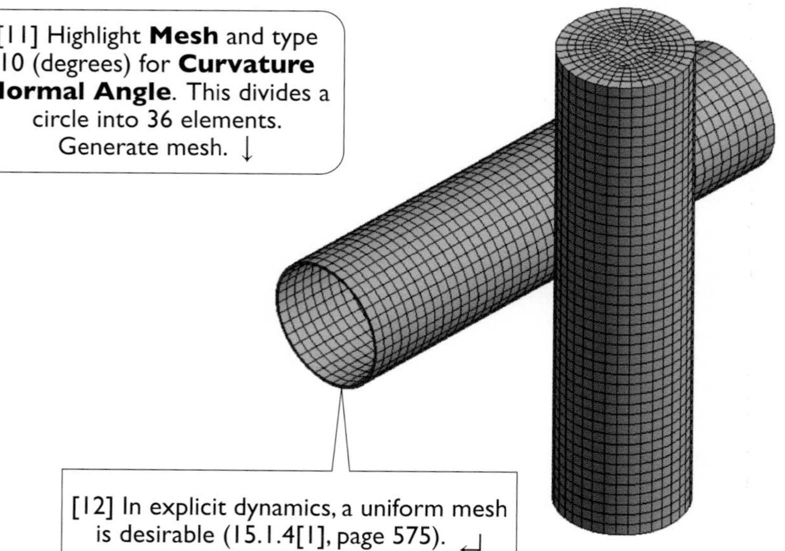

[12] In explicit dynamics, a uniform mesh is desirable (15.1.4[1], page 575). ↵

Initial Conditions
Velocity
Angular Velocity
Outline
Filter: Name
Project
Model (A4)
Geometry
Solid
Surface Body
Coordinate Systems
Connections
Contacts
Body Interactions
Body Interaction
Mesh
Explicit Dynamics (A5)
Initial Conditions
Pre-Stress (None)
Velocity
Analysis Settings
Solution (A6)
Solution Information
Details of "Velocity"
Scope
Scoping Method | Geometry Selection
Geometry | 1 Body
Definition
Input Type | Velocity
Define By | Components
Coordinate System | Global Coordinate System
X Component | 300. m/s
Y Component | 0. m/s
Z Component | 0. m/s
Suppressed | No

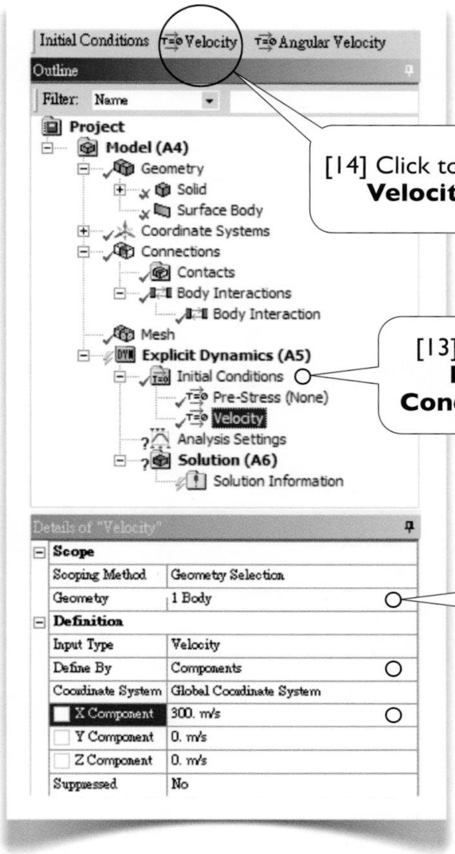

[14] Click to insert a **Velocity**. ↓

[13] Highlight **Initial Conditions**. ↑

[15] Select the surface body (aluminum pipe). ↗

[16] Highlight **Analysis Settings.** Type 0.0005 (s) for **End Time**. ↓

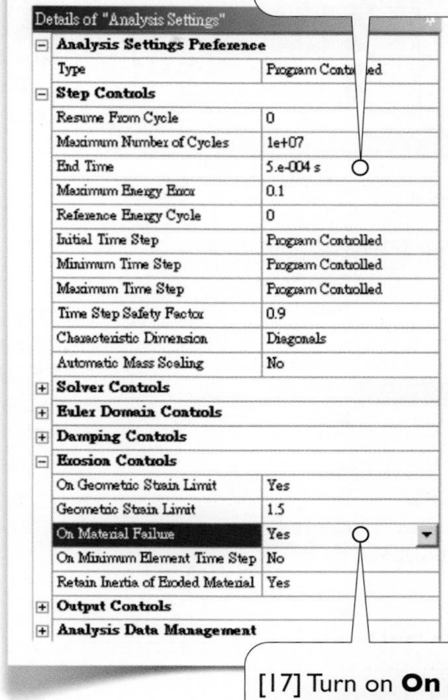

Details of "Analysis Settings"
Analysis Settings Preference
Type | Program Controlled
Step Controls
Resume From Cycle | 0
Maximum Number of Cycles | 1e+07
End Time | 5.e-004 s
Maximum Energy Error | 0.1
Reference Energy Cycle | 0
Initial Time Step | Program Controlled
Minimum Time Step | Program Controlled
Maximum Time Step | Program Controlled
Time Step Safety Factor | 0.9
Characteristic Dimension | Diagonals
Automatic Mass Scaling | No
Solver Controls
Euler Domain Controls
Damping Controls
Erosion Controls
On Geometric Strain Limit | Yes
Geometric Strain Limit | 1.5
On Material Failure | Yes
On Minimum Element Time Step | No
Retain Inertia of Eroded Material | Yes
Output Controls
Analysis Data Management

[17] Turn on **On Material Failure** (see [18]). ↓

Erosion Controls[Ref 1]

[18] **Erosion Controls** in **Analysis Settings** determines the conditions under which an element will be removed. The default condition is that an element is removed when its *geometric strain*, or *effective strain*[Ref 1], exceeds a limit of 150%. This value is large enough to assure that no elements are removed by default.

In this case, we add another failure condition: an element is removed when its plastic strain exceeds 75% (see [17] and 15.2.3[8], page 579). ↓

Supports
Conditions
Fixed Support
Displacement
Remote Displacement
Velocity
Impedance Boundary
Frictionless Support
Compression Only Support
Cylindrical Support
Simply Supported
Fixed Rotation
Elastic Support

[19] Insert a **Fixed Support**. →

[20] Select the solid body (the steel column). ↲

Details of "Fixed Support"
Scope
Scoping Method | Geometry Selection
Geometry | 1 Body
Definition
Type | Fixed Support
Suppressed | No

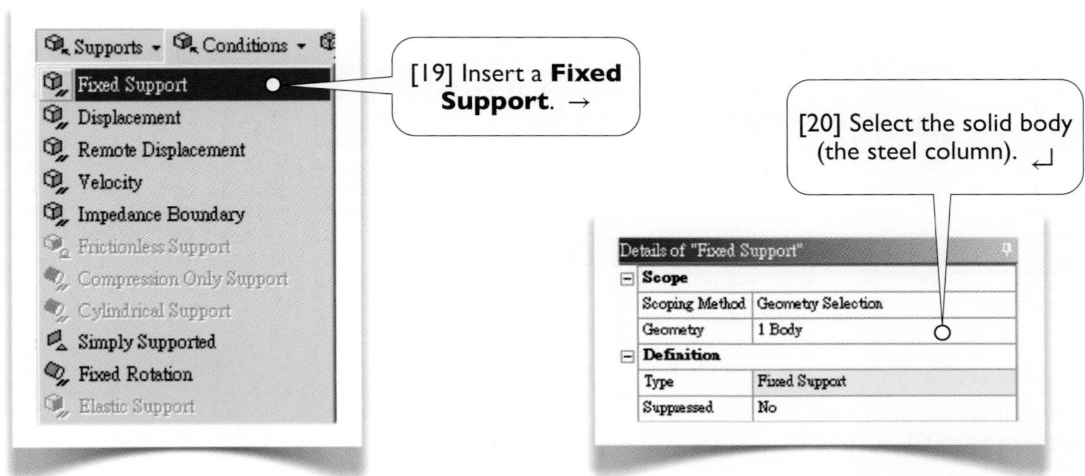

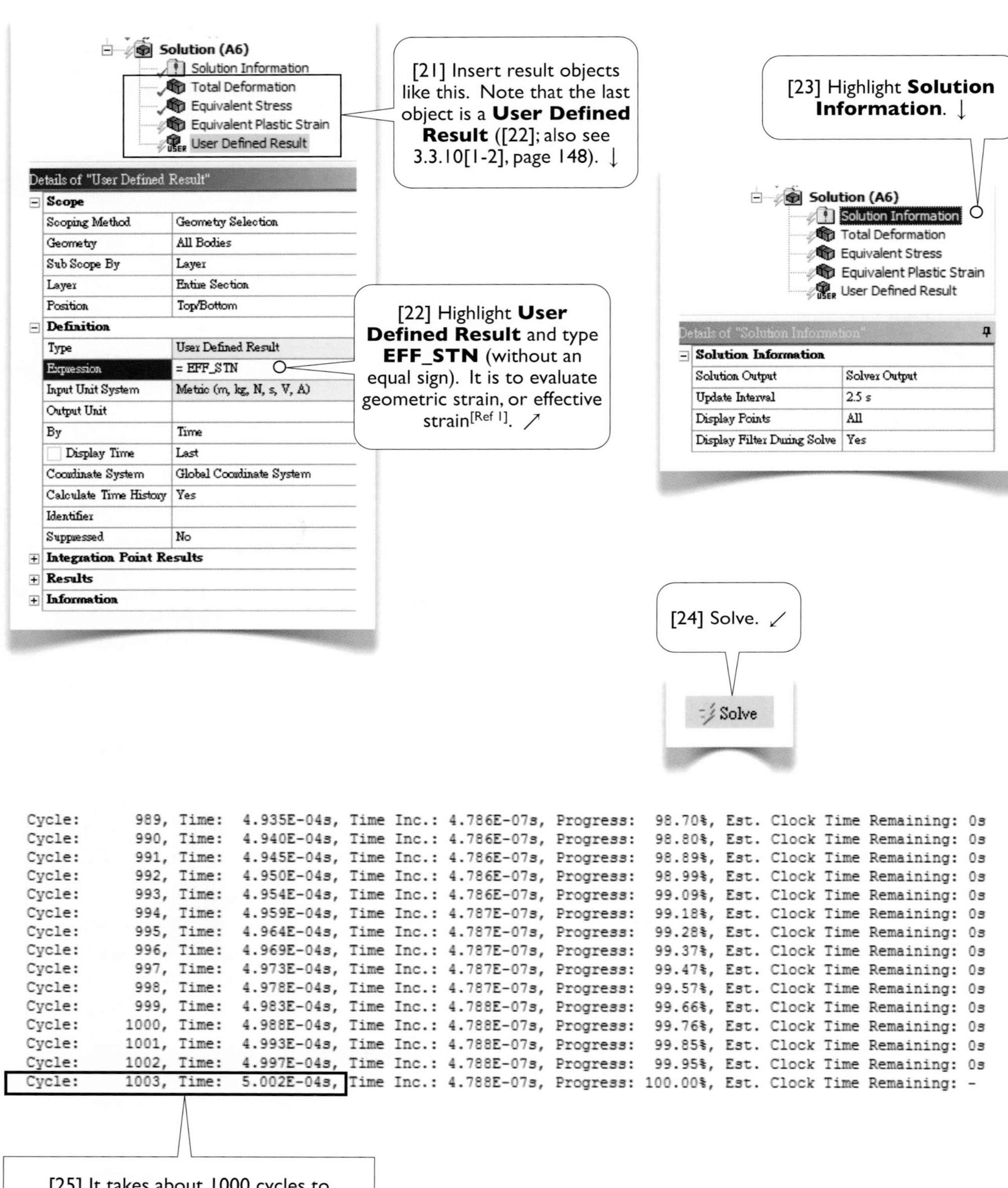
Solution (A6)
Solution Information
Total Deformation
Equivalent Stress
Equivalent Plastic Strain
User Defined Result
[21] Insert result objects like this. Note that the last object is a User Defined Result ([22]; also see 3.3.10[1-2], page 148).
Details of "User Defined Result"
Scope
Scoping Method Geometry Selection
Geometry All Bodies
Sub Scope By Layer
Layer Entire Section
Position Top/Bottom
Definition
Type User Defined Result
Expression = EFF_STN
Input Unit System Metric (m, kg, N, s, V, A)
Output Unit
By Time
Display Time Last
Coordinate System Global Coordinate System
Calculate Time History Yes
Identifier
Suppressed No
Integration Point Results
Results
Information
[22] Highlight User Defined Result and type EFF_STN (without an equal sign). It is to evaluate geometric strain, or effective strain[Ref 1].
[23] Highlight Solution Information.
Details of "Solution Information"
Solution Information
Solution Output Solver Output
Update Interval 2.5 s
Display Points All
Display Filter During Solve Yes
[24] Solve.
Solve
Cycle: 989, Time: 4.935E-04s, Time Inc.: 4.786E-07s, Progress: 98.70%, Est. Clock Time Remaining: 0s
Cycle: 990, Time: 4.940E-04s, Time Inc.: 4.786E-07s, Progress: 98.80%, Est. Clock Time Remaining: 0s
Cycle: 991, Time: 4.945E-04s, Time Inc.: 4.786E-07s, Progress: 98.89%, Est. Clock Time Remaining: 0s
Cycle: 992, Time: 4.950E-04s, Time Inc.: 4.786E-07s, Progress: 98.99%, Est. Clock Time Remaining: 0s
Cycle: 993, Time: 4.954E-04s, Time Inc.: 4.786E-07s, Progress: 99.09%, Est. Clock Time Remaining: 0s
Cycle: 994, Time: 4.959E-04s, Time Inc.: 4.787E-07s, Progress: 99.18%, Est. Clock Time Remaining: 0s
Cycle: 995, Time: 4.964E-04s, Time Inc.: 4.787E-07s, Progress: 99.28%, Est. Clock Time Remaining: 0s
Cycle: 996, Time: 4.969E-04s, Time Inc.: 4.787E-07s, Progress: 99.37%, Est. Clock Time Remaining: 0s
Cycle: 997, Time: 4.973E-04s, Time Inc.: 4.787E-07s, Progress: 99.47%, Est. Clock Time Remaining: 0s
Cycle: 998, Time: 4.978E-04s, Time Inc.: 4.787E-07s, Progress: 99.57%, Est. Clock Time Remaining: 0s
Cycle: 999, Time: 4.983E-04s, Time Inc.: 4.788E-07s, Progress: 99.66%, Est. Clock Time Remaining: 0s
Cycle: 1000, Time: 4.988E-04s, Time Inc.: 4.788E-07s, Progress: 99.76%, Est. Clock Time Remaining: 0s
Cycle: 1001, Time: 4.993E-04s, Time Inc.: 4.788E-07s, Progress: 99.85%, Est. Clock Time Remaining: 0s
Cycle: 1002, Time: 4.997E-04s, Time Inc.: 4.788E-07s, Progress: 99.95%, Est. Clock Time Remaining: 0s
Cycle: 1003, Time: 5.002E-04s, Time Inc.: 4.788E-07s, Progress: 100.00%, Est. Clock Time Remaining: -
[25] It takes about 1000 cycles to complete a 0.0005-second simulation. #

15.2.6 Animate the Deformation

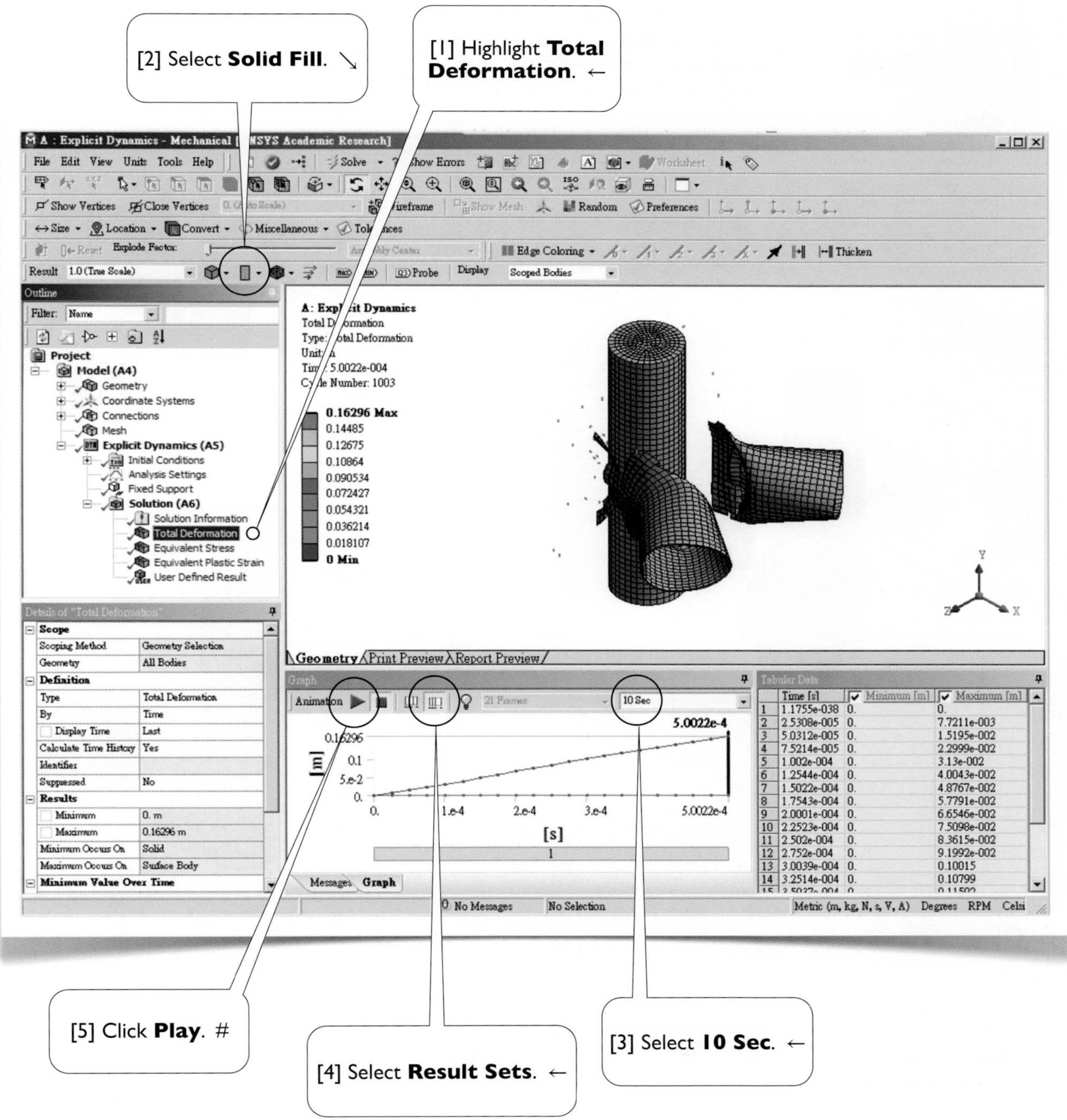

15.2.7 View Numerical Results

[1] Select **Smooth Contours** and examine the results. ↓

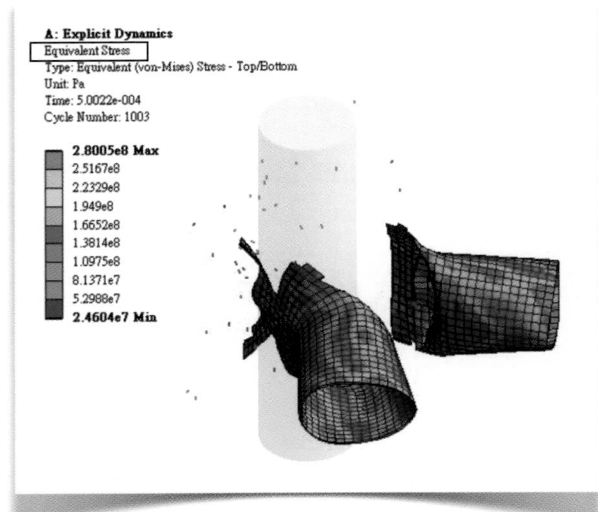

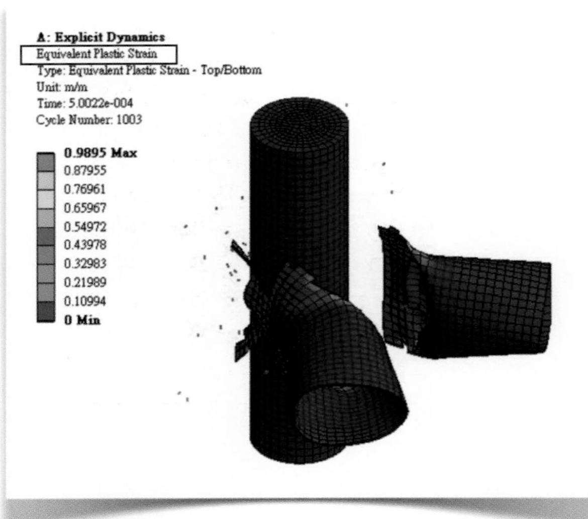

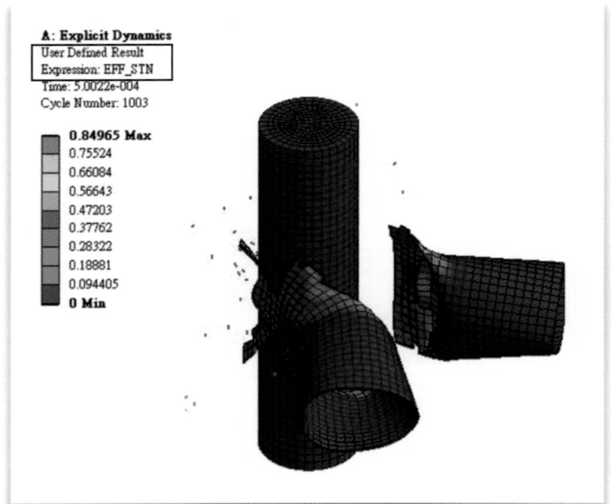

Wrap Up

[2] Save the project and exit Workbench. #

Reference

1. ANSYS Help//Mechanical Application//ANSYS Mechanical User's Guide//Appendix G. Explicit Dynamics Theory Guide// Analysis Settings//Erosion Controls

Section 15.3

Drop Test

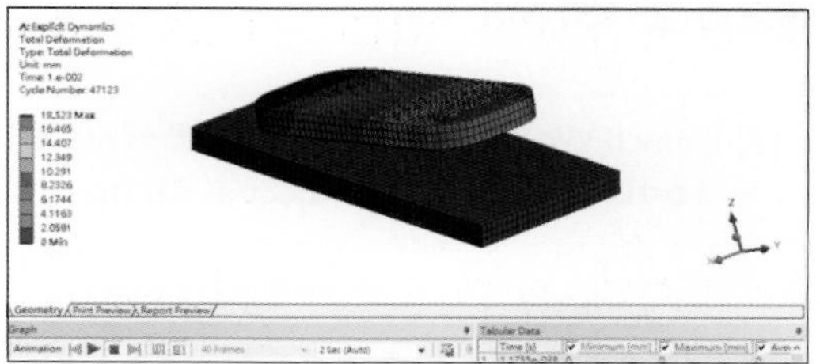

15.3.1 About the Drop Test Simulation

[1] Drop test simulation is a special case of impact simulation, in which one of the impacting objects is a stationary floor, typically made of concrete, steel, or stone. In this section, we consider a scenario that a mobile phone falls out of your pocket and drops on a concrete floor. This kind of simulation typically takes hours of computing time. We learned from Section 15.2 that a typical integration time step in **Explicit Dynamics** is 10^{-7} to 10^{-8} seconds. It would take about 100,000 to 1,000,000 cycles to complete a 0.01 seconds drop test. In this section, we will simplify the model to minimize the run time. A more realistic model will be suggested and left as an exercise (15.4.2, page 600).

The phone body is a shell of thickness 0.5 mm and made of an aluminum alloy [2]. The concrete floor is modeled as an 160 mm x 80 mm x 10 mm block [3]. When the phone hits the floor, its velocity is 5 m/s, which is equivalent to a free fall from a height of 1.25 m. We will assume that the phone body forms an angle of 20° with the horizon when it hits the floor.

We will use **mm-kg-N-s** unit system in the simulation. ↓

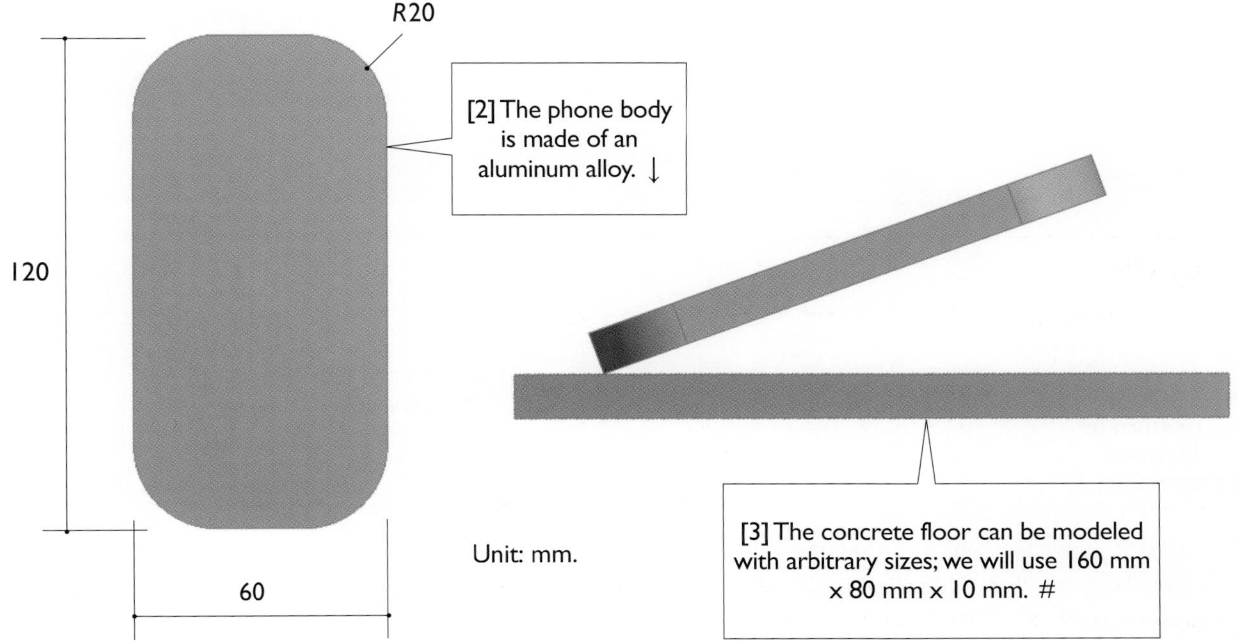

[2] The phone body is made of an aluminum alloy. ↓

[3] The concrete floor can be modeled with arbitrary sizes; we will use 160 mm x 80 mm x 10 mm. #

15.3.2 Start Up

[1] Launch Workbench. Create an **Explicit Dynamics** analysis system by double-clicking it in **Toolbox**. Save the project as **Drop**. ↓

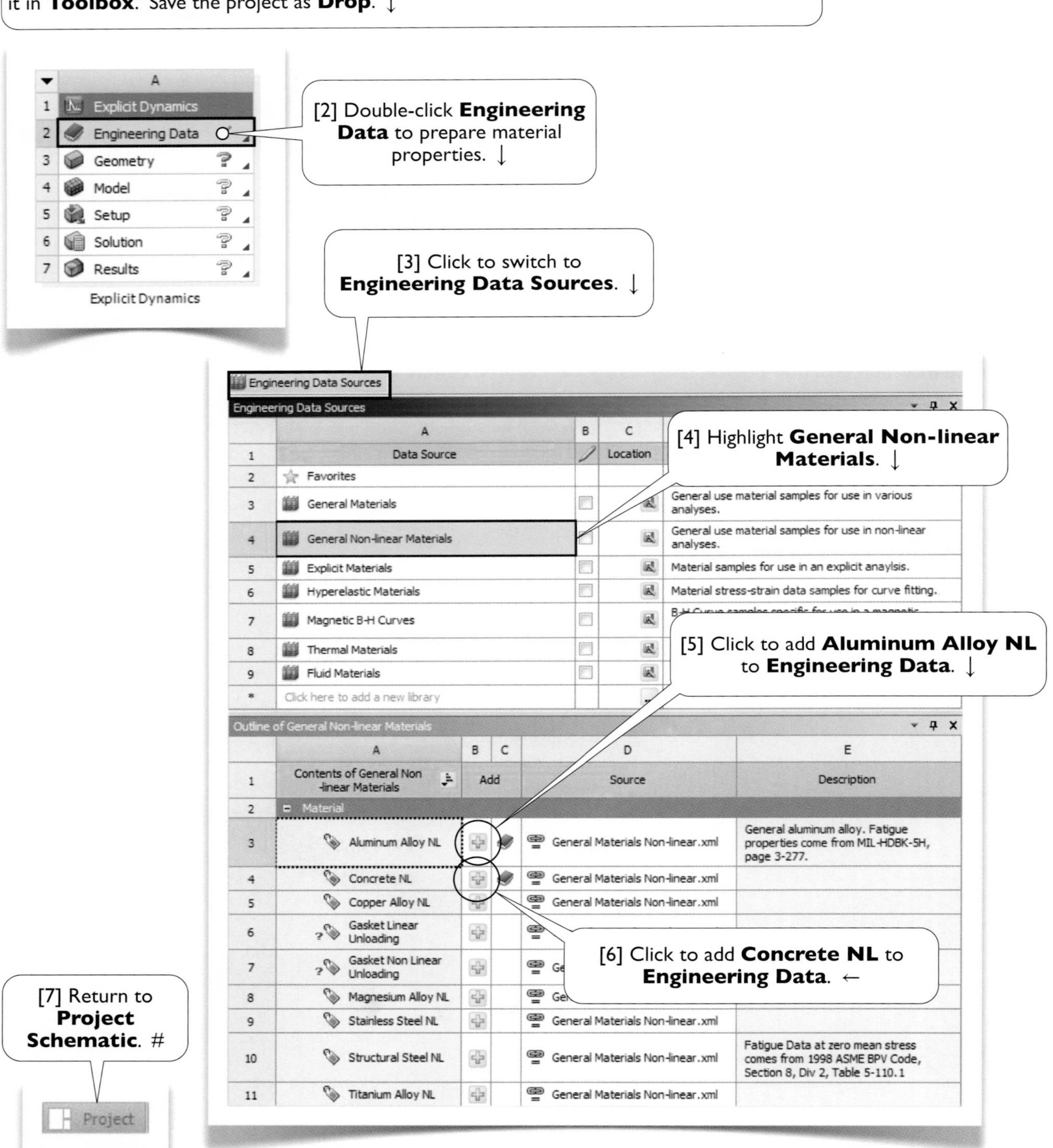

15.3.3 Create Geometry

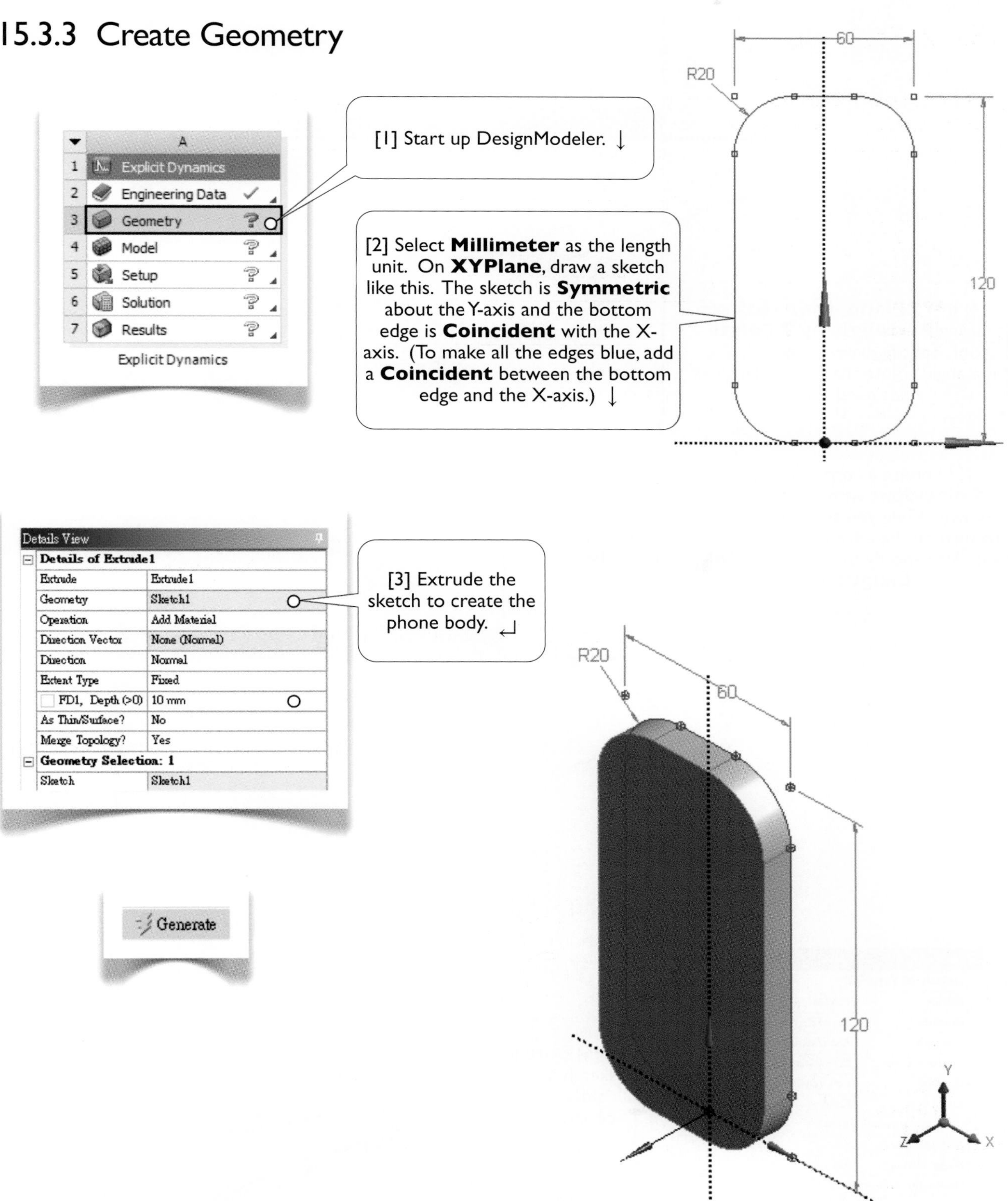

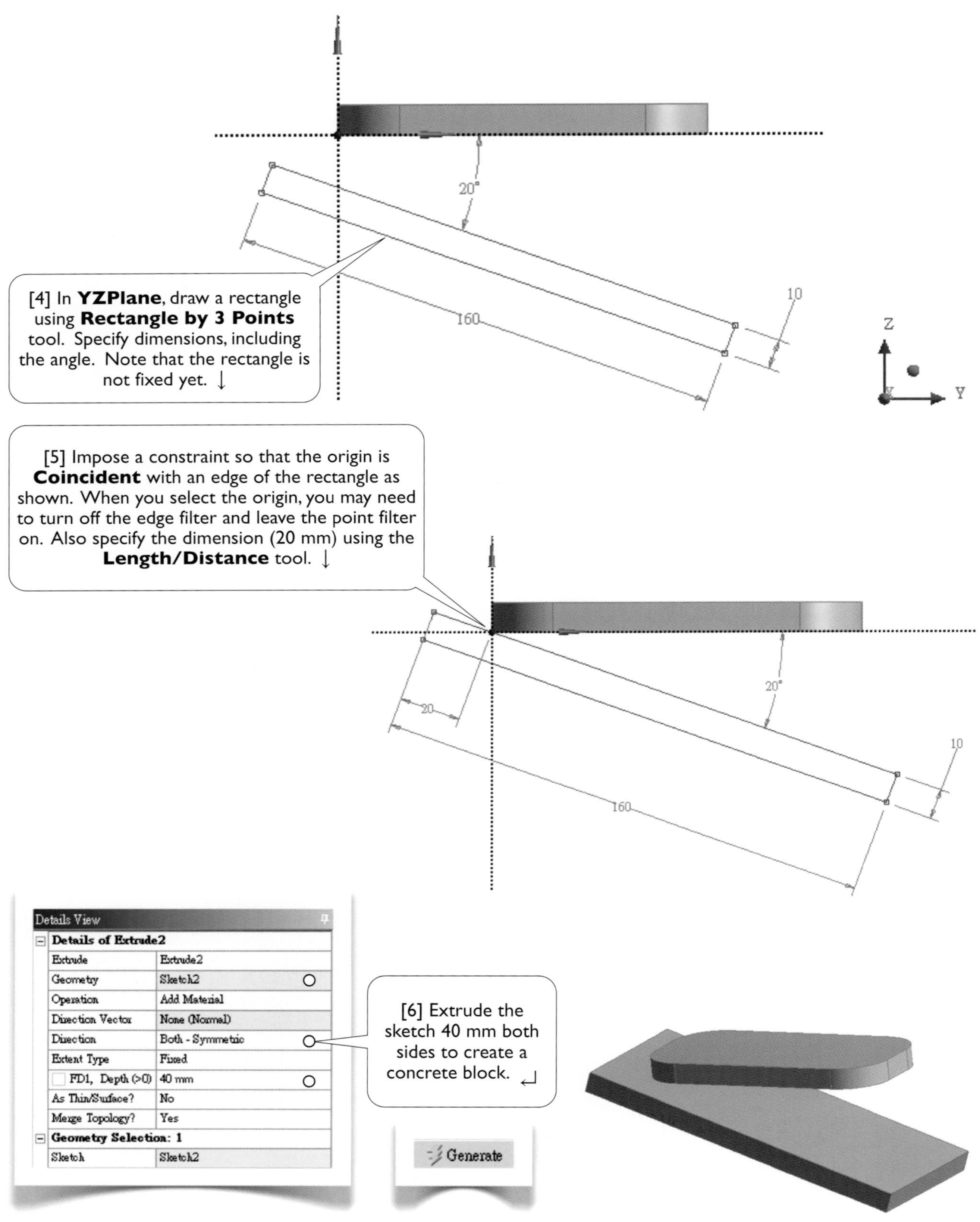

20°
160
10
Z
Y
[4] In **YZPlane**, draw a rectangle using **Rectangle by 3 Points** tool. Specify dimensions, including the angle. Note that the rectangle is not fixed yet. ↓
[5] Impose a constraint so that the origin is **Coincident** with an edge of the rectangle as shown. When you select the origin, you may need to turn off the edge filter and leave the point filter on. Also specify the dimension (20 mm) using the **Length/Distance** tool. ↓
20
20°
160
10
Details View
Details of Extrude2
Extrude | Extrude2
Geometry | Sketch2
Operation | Add Material
Direction Vector | None (Normal)
Direction | Both - Symmetric
Extent Type | Fixed
FD1, Depth (>0) | 40 mm
As Thin/Surface? | No
Merge Topology? | Yes
Geometry Selection: 1
Sketch | Sketch2
[6] Extrude the sketch 40 mm both sides to create a concrete block. ↵
Generate

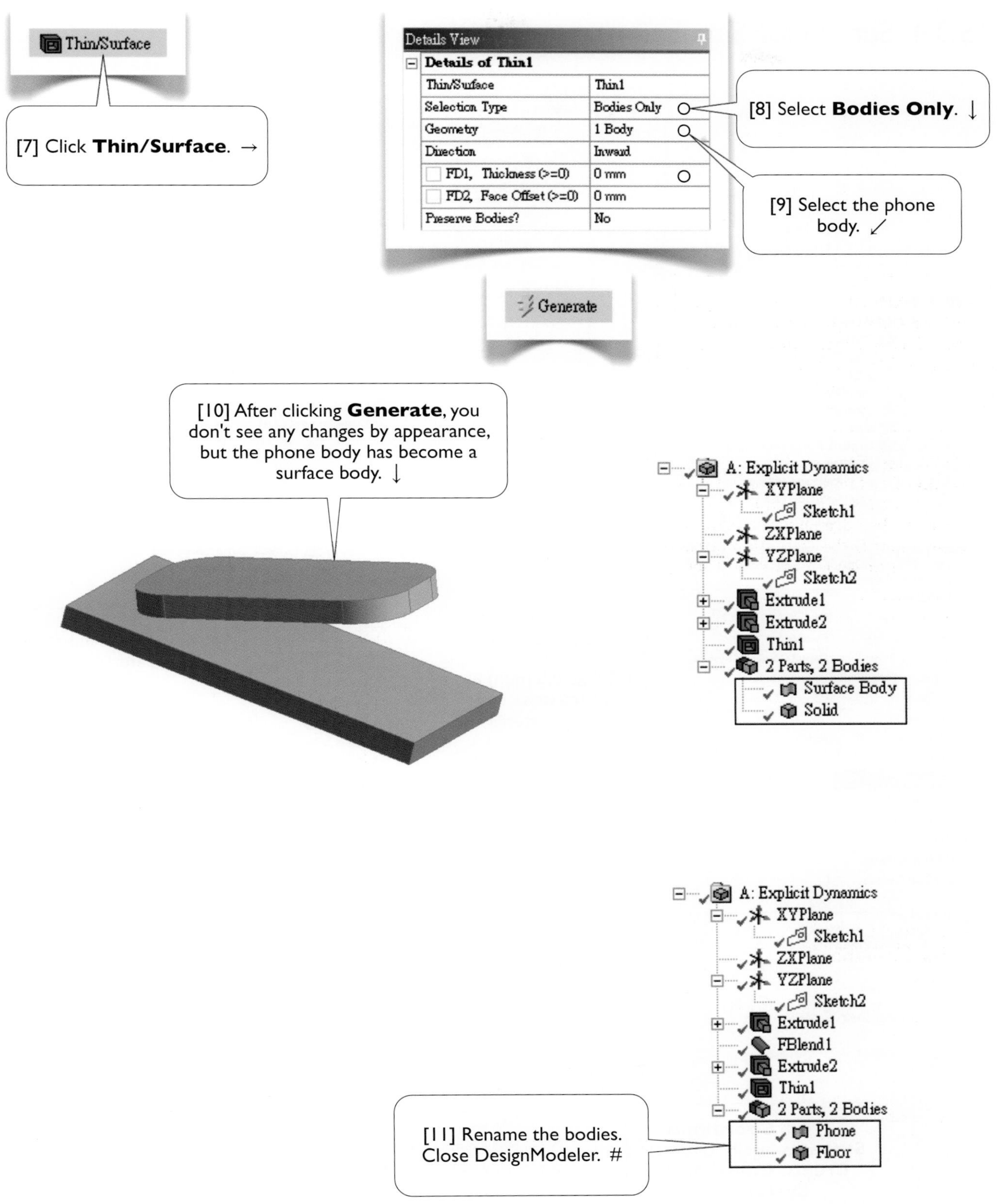
Thin/Surface
[7] Click **Thin/Surface**. →
Details View
Details of Thin1
Thin/Surface
Thin1
Selection Type
Bodies Only
Geometry
1 Body
Direction
Inward
FD1, Thickness (>=0)
0 mm
FD2, Face Offset (>=0)
0 mm
Preserve Bodies?
No
[8] Select **Bodies Only**. ↓
[9] Select the phone body. ↙
Generate
[10] After clicking **Generate**, you don't see any changes by appearance, but the phone body has become a surface body. ↓
A: Explicit Dynamics
XYPlane
Sketch1
ZXPlane
YZPlane
Sketch2
Extrude1
Extrude2
Thin1
2 Parts, 2 Bodies
Surface Body
Solid
A: Explicit Dynamics
XYPlane
Sketch1
ZXPlane
YZPlane
Sketch2
Extrude1
FBlend1
Extrude2
Thin1
2 Parts, 2 Bodies
Phone
Floor
[11] Rename the bodies. Close DesignModeler. #

15.3.4 Set Up for Simulation

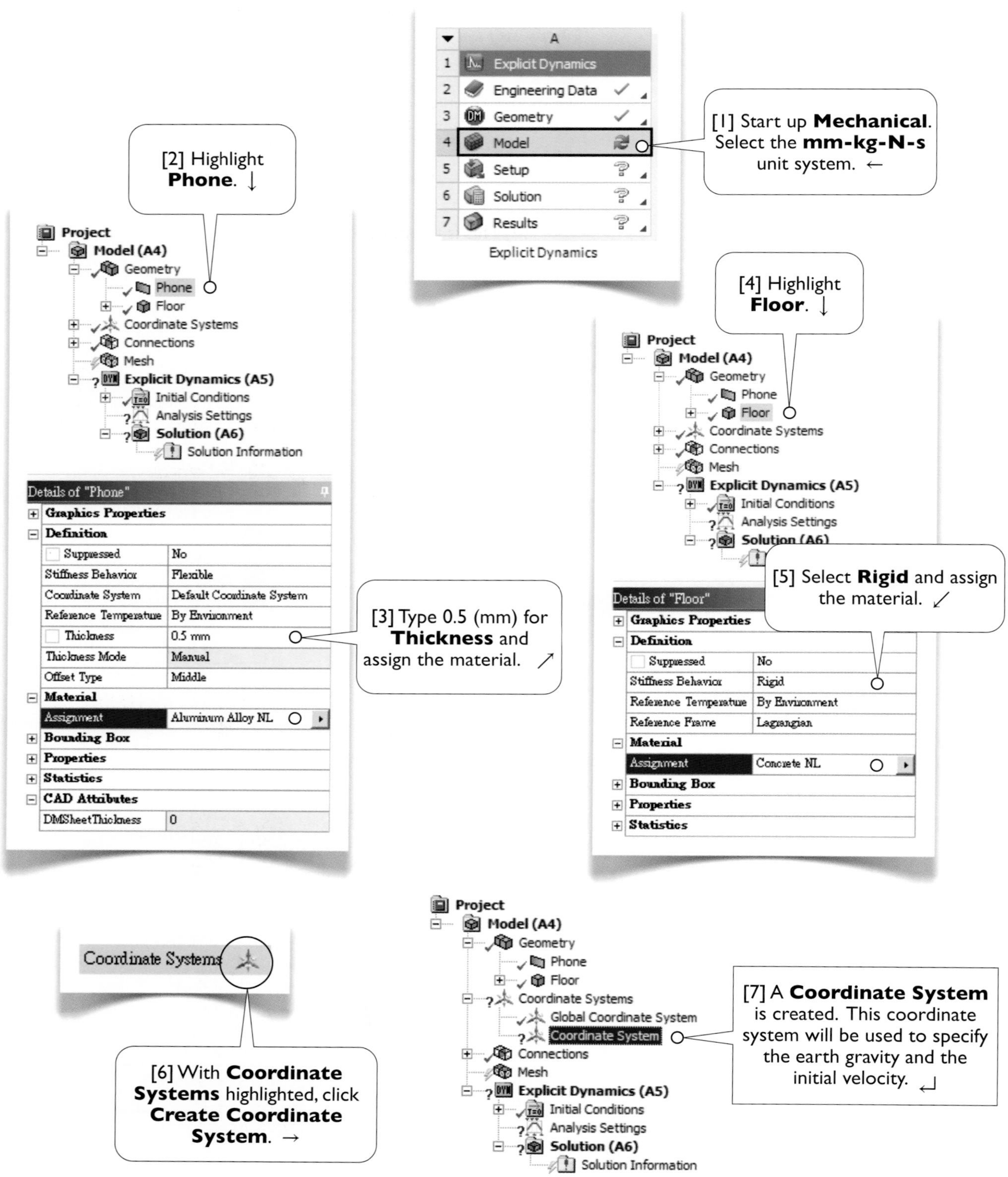

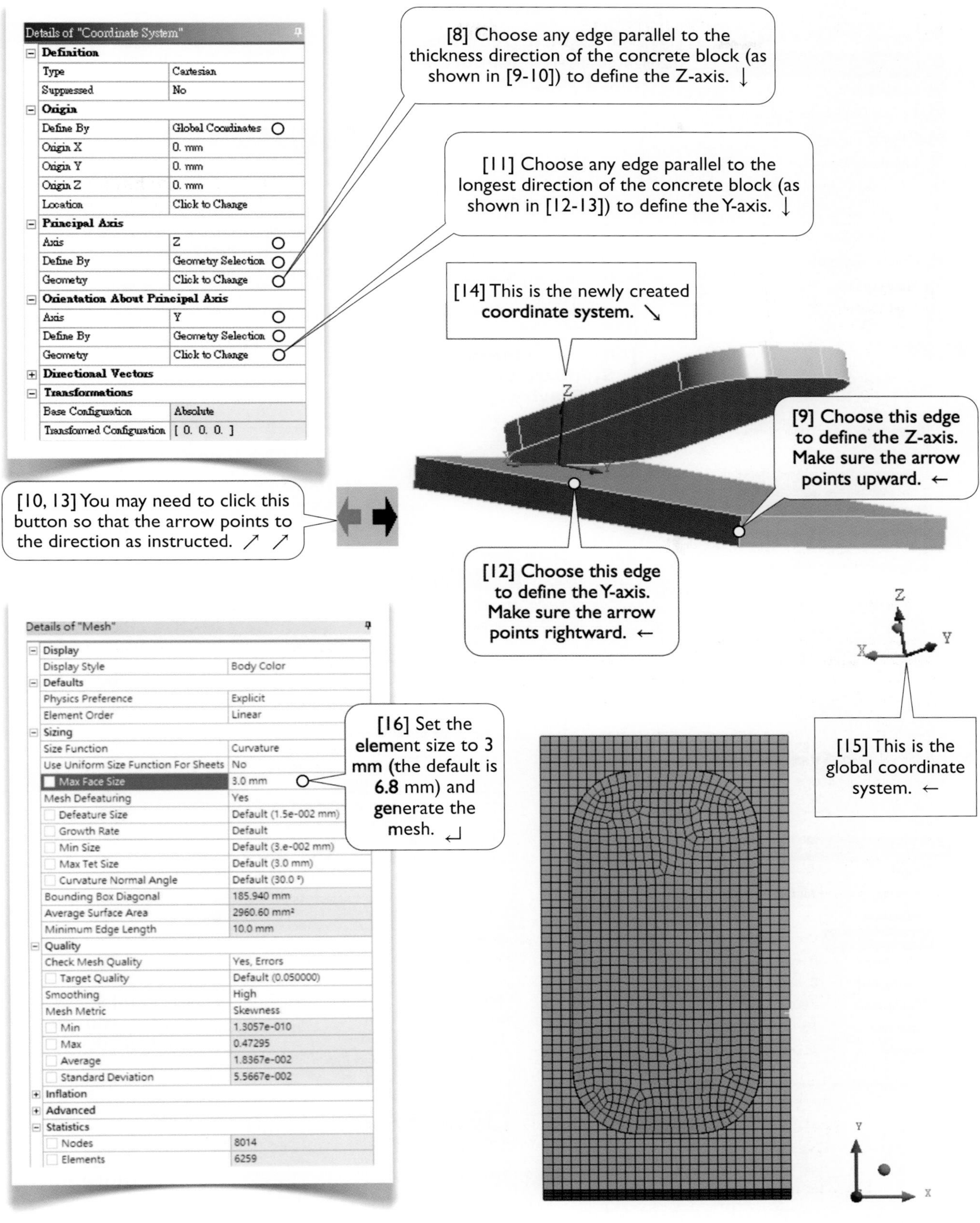
Details of "Coordinate System"
Definition
Type | Cartesian
Suppressed | No
Origin
Define By | Global Coordinates
Origin X | 0. mm
Origin Y | 0. mm
Origin Z | 0. mm
Location | Click to Change
Principal Axis
Axis | Z
Define By | Geometry Selection
Geometry | Click to Change
Orientation About Principal Axis
Axis | Y
Define By | Geometry Selection
Geometry | Click to Change
Directional Vectors
Transformations
Base Configuration | Absolute
Transformed Configuration | [0. 0. 0.]
[8] Choose any edge parallel to the thickness direction of the concrete block (as shown in [9-10]) to define the Z-axis.
[11] Choose any edge parallel to the longest direction of the concrete block (as shown in [12-13]) to define the Y-axis.
[14] This is the newly created coordinate system.
Z
[9] Choose this edge to define the Z-axis. Make sure the arrow points upward.
[10, 13] You may need to click this button so that the arrow points to the direction as instructed.
[12] Choose this edge to define the Y-axis. Make sure the arrow points rightward.
Z
X
Y
[15] This is the global coordinate system.
Details of "Mesh"
Display
Display Style | Body Color
Defaults
Physics Preference | Explicit
Element Order | Linear
Sizing
Size Function | Curvature
Use Uniform Size Function For Sheets | No
Max Face Size | 3.0 mm
Mesh Defeaturing | Yes
Defeature Size | Default (1.5e-002 mm)
Growth Rate | Default
Min Size | Default (3.e-002 mm)
Max Tet Size | Default (3.0 mm)
Curvature Normal Angle | Default (30.0 °)
Bounding Box Diagonal | 185.940 mm
Average Surface Area | 2960.60 mm²
Minimum Edge Length | 10.0 mm
Quality
Check Mesh Quality | Yes, Errors
Target Quality | Default (0.050000)
Smoothing | High
Mesh Metric | Skewness
Min | 1.3057e-010
Max | 0.47295
Average | 1.8367e-002
Standard Deviation | 5.5667e-002
Inflation
Advanced
Statistics
Nodes | 8014
Elements | 6259
[16] Set the element size to 3 mm (the default is 6.8 mm) and generate the mesh.
Y
X

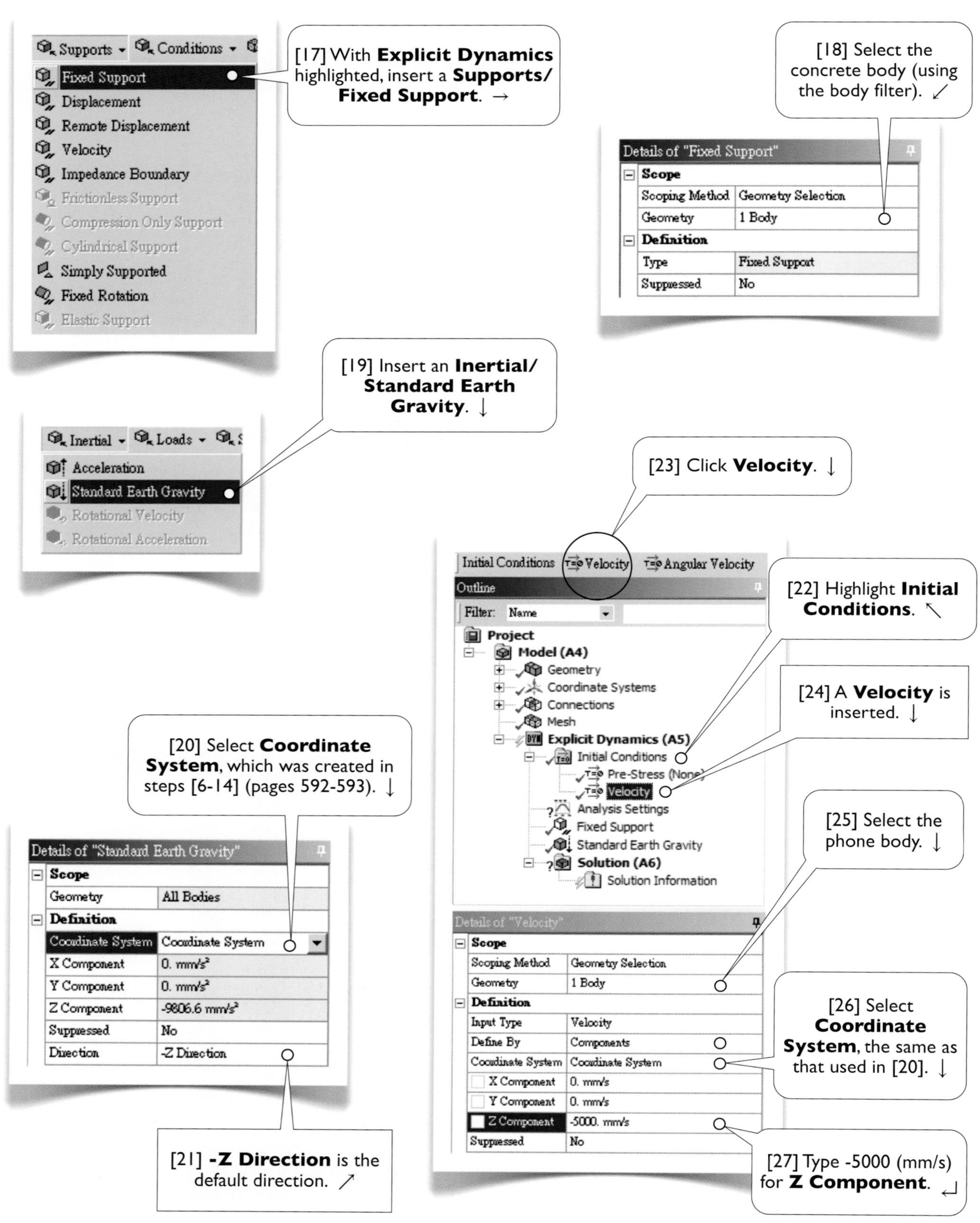

Supports
Conditions
Fixed Support
Displacement
Remote Displacement
Velocity
Impedance Boundary
Frictionless Support
Compression Only Support
Cylindrical Support
Simply Supported
Fixed Rotation
Elastic Support
[17] With **Explicit Dynamics** highlighted, insert a **Supports/ Fixed Support**. →
[18] Select the concrete body (using the body filter). ↙
Details of "Fixed Support"
Scope
Scoping Method
Geometry Selection
Geometry
1 Body
Definition
Type
Fixed Support
Suppressed
No
[19] Insert an **Inertial/ Standard Earth Gravity**. ↓
Inertial
Loads
Acceleration
Standard Earth Gravity
Rotational Velocity
Rotational Acceleration
[23] Click **Velocity**. ↓
Initial Conditions
Velocity
Angular Velocity
Outline
Filter:
Name
Project
Model (A4)
Geometry
Coordinate Systems
Connections
Mesh
Explicit Dynamics (A5)
Initial Conditions
Pre-Stress (None)
Velocity
Analysis Settings
Fixed Support
Standard Earth Gravity
Solution (A6)
Solution Information
[22] Highlight **Initial Conditions**. ↖
[24] A **Velocity** is inserted. ↓
[25] Select the phone body. ↓
[20] Select **Coordinate System**, which was created in steps [6-14] (pages 592-593). ↓
Details of "Standard Earth Gravity"
Scope
Geometry
All Bodies
Definition
Coordinate System
Coordinate System
X Component
0. mm/s²
Y Component
0. mm/s²
Z Component
-9806.6 mm/s²
Suppressed
No
Direction
-Z Direction
[21] **-Z Direction** is the default direction. ↗
Details of "Velocity"
Scope
Scoping Method
Geometry Selection
Geometry
1 Body
Definition
Input Type
Velocity
Define By
Components
Coordinate System
Coordinate System
X Component
0. mm/s
Y Component
0. mm/s
Z Component
-5000. mm/s
Suppressed
No
[26] Select **Coordinate System**, the same as that used in [20]. ↓
[27] Type -5000 (mm/s) for **Z Component**. ↵

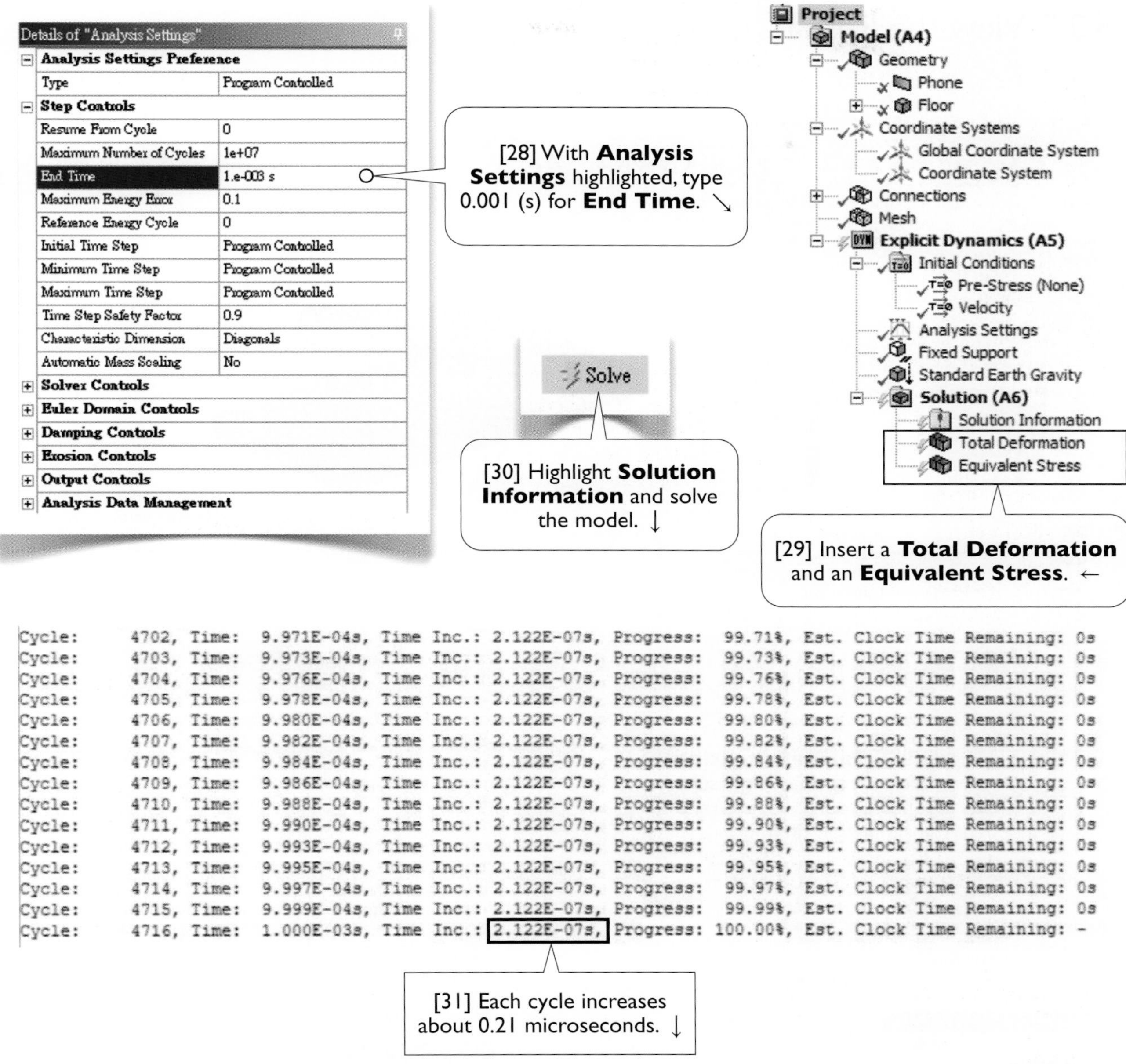

Integration Time Steps

[32] In this case, the integration time step is about 0.21 microseconds. This number, proportional to the size of the smallest element, controls the overall run time (15.1.4, page 575). In **Explicit Dynamics**, a mesh of uniform element size is the most efficient mesh.

Perform Simulation Incrementally

A drop test simulation typically takes hours of computing time. If you make a mistake, it will waste a significant amount of time. As a good practice, always perform the simulation incrementally. In this case, we perform a 0.001 seconds simulation first to see if anything goes wrong. If everything is okay, then we can continue the simulation starting from the last cycle. This feature is called the **Restart** of the simulation. #

15.3.5 View the Results

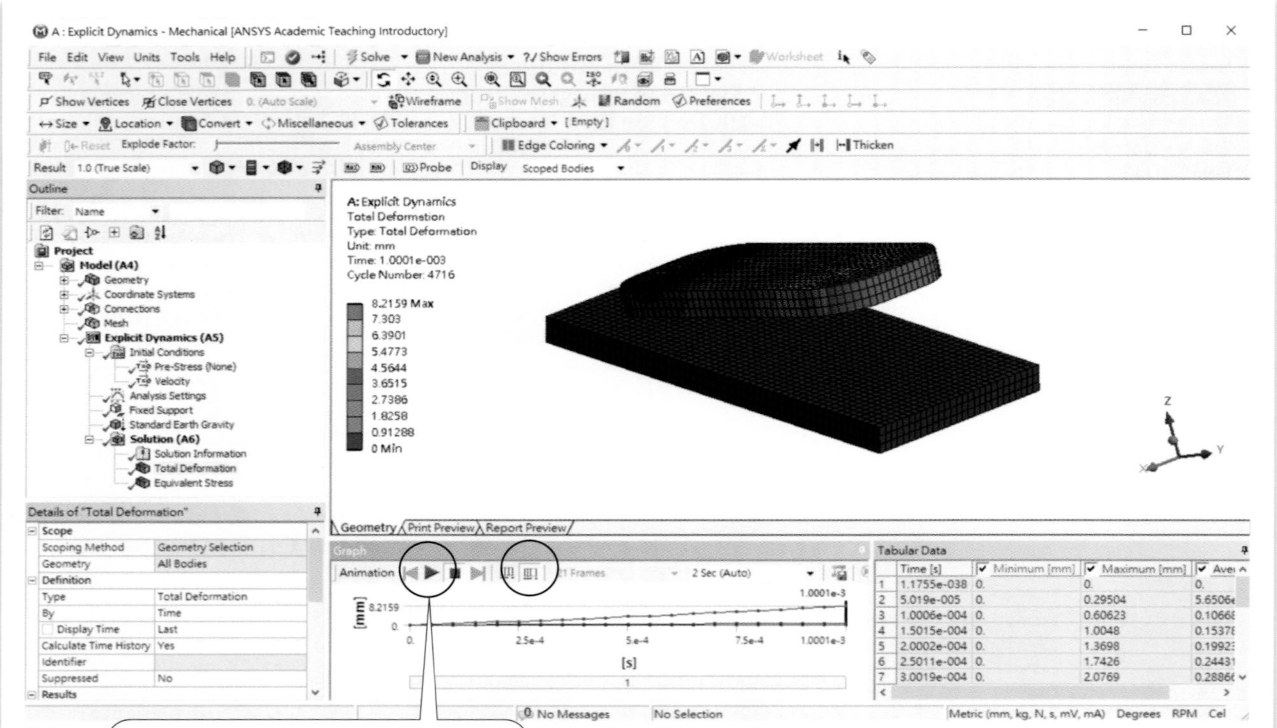

[1] Highlight **Total Deformation** and animate the results. The results look reasonable so far. Let's extend the simulation time. ↓

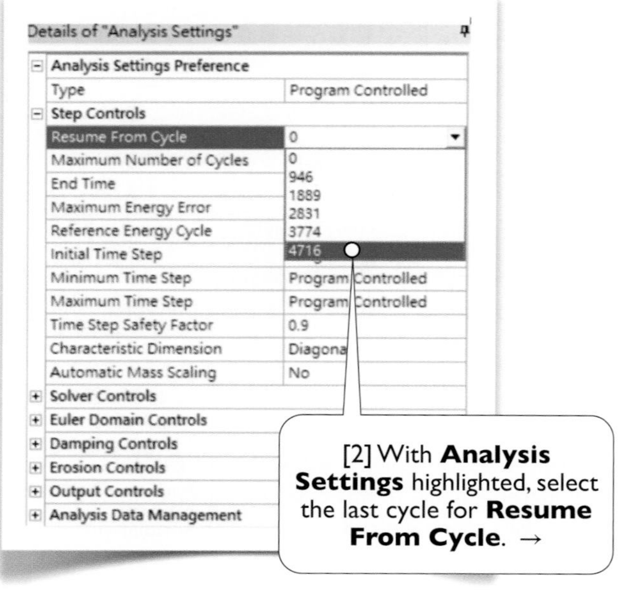

[2] With **Analysis Settings** highlighted, select the last cycle for **Resume From Cycle**. →

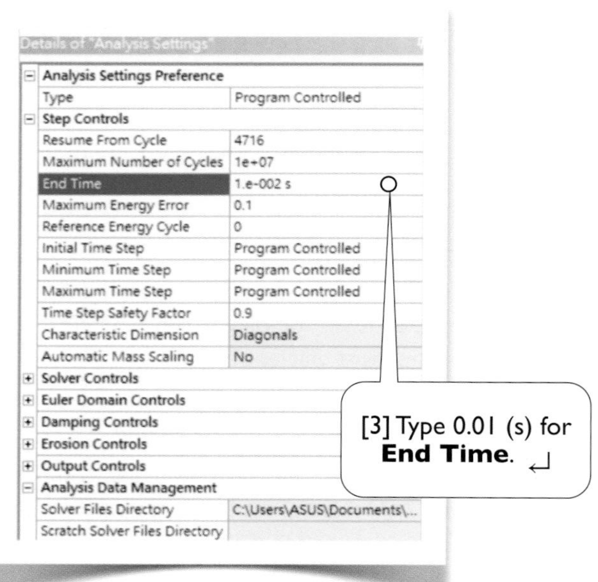

[3] Type 0.01 (s) for **End Time**. ↵

```
Cycle:    47109, Time:  9.997E-03s, Time Inc.: 2.123E-07s, Progress:  99.97%, Est. Clock Time Remaining: 0s
Cycle:    47110, Time:  9.997E-03s, Time Inc.: 2.123E-07s, Progress:  99.97%, Est. Clock Time Remaining: 0s
Cycle:    47111, Time:  9.998E-03s, Time Inc.: 2.123E-07s, Progress:  99.97%, Est. Clock Time Remaining: 0s
Cycle:    47112, Time:  9.998E-03s, Time Inc.: 2.123E-07s, Progress:  99.97%, Est. Clock Time Remaining: 0s
Cycle:    47113, Time:  9.998E-03s, Time Inc.: 2.123E-07s, Progress:  99.98%, Est. Clock Time Remaining: 0s
Cycle:    47114, Time:  9.998E-03s, Time Inc.: 2.123E-07s, Progress:  99.98%, Est. Clock Time Remaining: 0s
Cycle:    47115, Time:  9.998E-03s, Time Inc.: 2.123E-07s, Progress:  99.98%, Est. Clock Time Remaining: 0s
Cycle:    47116, Time:  9.999E-03s, Time Inc.: 2.123E-07s, Progress:  99.98%, Est. Clock Time Remaining: 0s
Cycle:    47117, Time:  9.999E-03s, Time Inc.: 2.123E-07s, Progress:  99.99%, Est. Clock Time Remaining: 0s
Cycle:    47118, Time:  9.999E-03s, Time Inc.: 2.123E-07s, Progress:  99.99%, Est. Clock Time Remaining: 0s
Cycle:    47119, Time:  9.999E-03s, Time Inc.: 2.123E-07s, Progress:  99.99%, Est. Clock Time Remaining: 0s
Cycle:    47120, Time:  9.999E-03s, Time Inc.: 2.123E-07s, Progress:  99.99%, Est. Clock Time Remaining: 0s
Cycle:    47121, Time:  1.000E-02s, Time Inc.: 2.123E-07s, Progress: 100.00%, Est. Clock Time Remaining: 0s
Cycle:    47122, Time:  1.000E-02s, Time Inc.: 2.123E-07s, Progress: 100.00%, Est. Clock Time Remaining: 0s
Cycle:    47123, Time:  1.000E-02s, Time Inc.: 2.123E-07s, Progress: 100.00%, Est. Clock Time Remaining: -
```

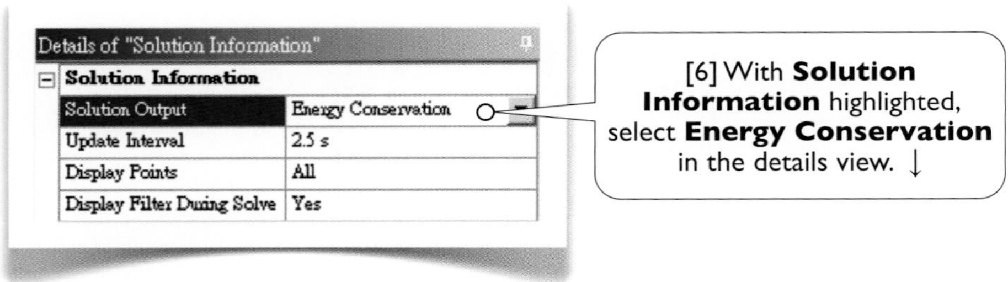

[6] With **Solution Information** highlighted, select **Energy Conservation** in the details view. ↓

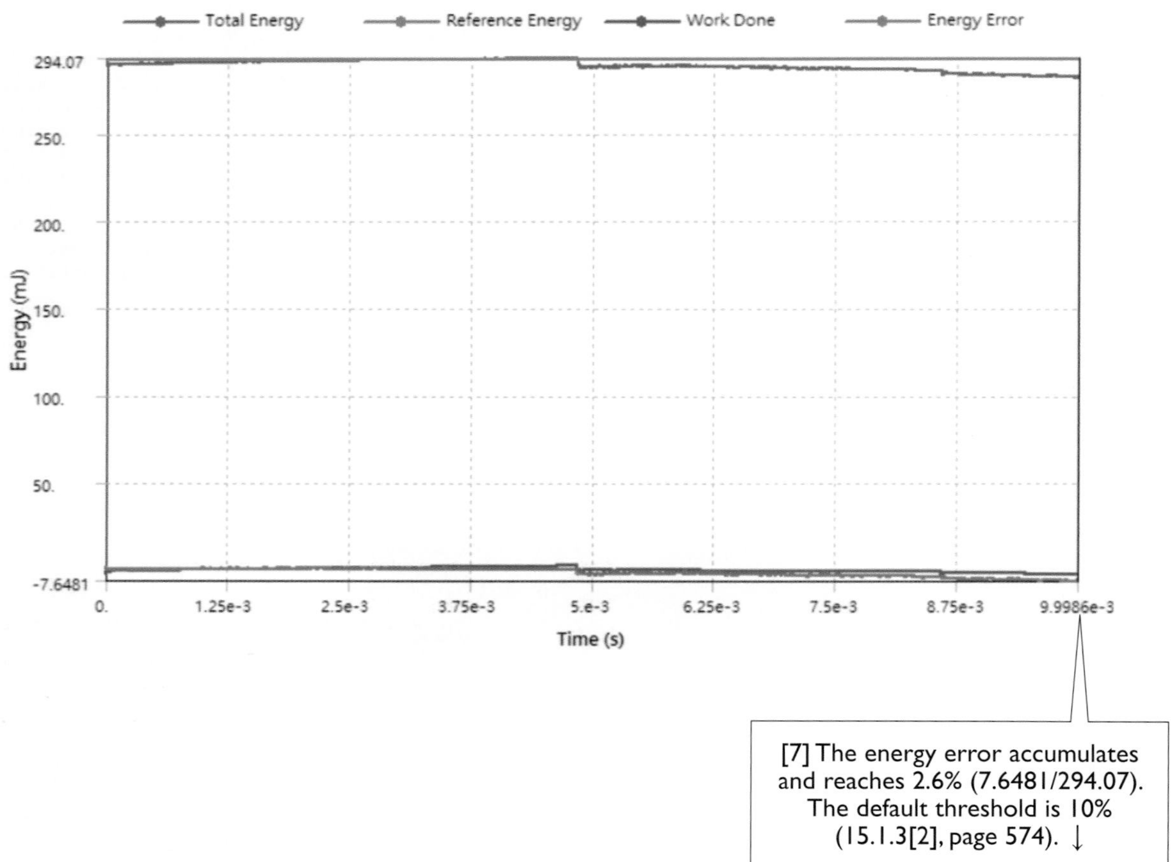

[7] The energy error accumulates and reaches 2.6% (7.6481/294.07). The default threshold is 10% (15.1.3[2], page 574). ↓

Wrap Up

[8] Save the project and exit Workbench. #

Section 15.4

Review

15.4.1 Keywords

Choose a letter for each keyword from the list of descriptions

1. (   ) Automatic Mass Scaling
2. (   ) CFL Condition
3. (   ) Energy Error
4. (   ) Explicit Method
5. (   ) Implicit Method
6. (   ) Principle of Work and Energy
7. (   ) Static Damping

Answers:

1. (F) 2. (E) 3. (D) 4. (B) 5. (A) 6. (C) 7. (G)

List of Descriptions

(A) A time integration method used in **Transient Structural** analysis system. It is so named because the method calculates the response in the present time using implicit information. It thus requires iterations for a time step, implying an expensive runtime for a time step. It, however, allows relatively large time steps. Overall, it is suitable for most transient simulations except high-speed or highly-nonlinear simulations.

(B) A time integration method used in **Explicit Dynamics** analysis system. It is so named because the method calculates the response in the present time using explicit information. It thus doesn't require iterations for a time step, implying an efficient runtime for a time step. It, however, requires very small time steps. Overall, it is suitable for high-speed or highly-nonlinear simulations.

(C) The principle of work and energy states that the energy at a reference time plus the work done from the reference time to a specific time is equal to the energy at that specific time.

(D) **Explicit Dynamics** uses this value to monitor the solution stability. If the energy error reaches a threshold, the solution is regarded as unstable and the computation stops. According to the principle of work and energy, the energy at a reference time plus the work done from the reference time to the present time is equal to the energy at the current time. If not equal, the difference is called an energy error. The energy error is further transformed into a dimensionless value by dividing using the maximum energy.

(E) In a single time step, a wave should not travel further than the smallest element size. This condition is used by **Explicit Dynamics** to determine the integration time step.

(F) The idea of mass scaling is to artificially increase the mass of some small elements, so that the stability time step is increased, to reduce the overall runtime.

(G) **Explicit Dynamics** is primarily for solving transient dynamic problems. However, a steady-state solution may also be obtained by introducing a damping force to critically damp the lowest mode of oscillation.

15.4.2 Additional Workbench Exercises

Performing a More Realistic Drop Test

As mentioned in 15.3.1[1] (page 587), to reduce the runtime, the model is over-simplified. In reality, the housing of a mobile phone may include a battery compartment. Include a slide-in battery in the model and perform a drop test to see if the battery will fall off upon impact with the floor.

Index

2D, 112
2D behavior, **113**, 130, 451
2D body, **111**, 126
2D graphics control, 72
2D model, 15, **138**, 140
2D simulation, 15, **109**, 112, 569
2D solid body, 38
2D solid model, **15**, 272
3D 20-node elements, **34**
3D body, **111**, 126
3D feature, 197
3D geometric modeling, 15
3D line modeling, 15, **272**
3D solid body, **37**, 68
3D solid element, 353
3D surface body, 38, **238**
3D solid modeling, 172
3D surface modeling, 15
3D truss, **283**, 379
3D view, 68
3D view sketching, 175
abort a tool, 85
ABS, 203
ABS (PA-757), 231
absolute displacement, 29
Acceleration, 147
acrylonitrile-butadiene-styrene, 203
Activate/Deactivate at this step, 461
active plane, 82
active sketch, 68, **82**
Adaptive meshing, **229**, 236
Adaptive Multiple-Objective, 318
Adaptive Single-Objective, **318**, 325
Add Frozen, 128, **129**, 193
Add Material, 129
addendum, 97
Adjust to Touch, 153
advanced contact settings, 479
All Quad, 162
All-Hexa mesh, 228
Allowable Change, 230
Alternate Angle, 88
Amplitude, 446
analysis, 12
Analysis Settings, 143, **460**
Analysis Systems, 12, **112**, 426
Analysis Type, 112
Angle, 88
angle unit, 71
animate, **18**, 88
anisotropic, 527
ANSYS Parametric Design Language, 149
APDL, 37, 47, **149**, 170, 239, 248
APDL verification manual, 171
Apply, 68
Arc by 3 Points, 85
Arc by Center, 73, **85**
Arc by Tangent, 85
arrow, 198
Asymmetric, 562
Augmented Lagrange, 131, 153, **478**
Auto Constraints, **84**, 90, 107
Auto Time Stepping, 279, 338, **491**
automatic contact detection, 269
Automatic Mass Scaling, **575**, 599
averaged stress, **160**, 291
AVI, 216
axial load, 267
axial stress, 142
axis of symmetry, 125, **136**
Axisymmetric, 130
axisymmetric body, 270
axisymmetric condition, 142
axisymmetric problem, 142
Bar, 157
base circle, 98
base feature, **200**, 201
Base Plane, 120
beam, 39
beam bracket, **173**, 210, 249, 383
beam element, 271, 278, **282**, 283, 310
Beam Results/Axial Force, 292
Beam Results/Bending Moment, 293
Beam Results/Shear Force, 293
Beam Results/Torsional Moment, 293
Beam Tool, 148

BEAM188, **39**, 283
Bearing Load, **146**, 232, 236, 261, 265
bearing support, 256
beat, 414
beat frequency, 416
Behavior, **131**, 477
Belleville, 537
Belleville spring, 536
Belleville washer, **536**, 569
bellows joint, 239
bending stress, 151
beta damping, 435
biaxial tensile test, **532**, 553, 554, 555
Bilinear Isotropic Hardening, 579
bisection, 481
black arrow, 198
Blend, 201
Blend/Fixed Radius, 177
blue color, **60**, 66
bode plot, 446
body force, 30
Body Interactions, 582
body types, **37**, 196
Body View, 130
Bonded, 476
Boolean, 195
boundary condition, **17**, 47, 136
boundary-value problems, **33**, 34
Box Select, 77
Box Zoom, 64, 72, **83**
Bracket, **174**, 210, 249, 384
branch, 107
brittle material, **40**, 42, 47
buckling, **20**, 47, 367, 368
buckling analysis, **380**, 384
buckling load, 365, **373**, 374, 379
buckling mode, 367
buckling mode shape, 378
Building, **296**, 395, 441
built-in unit systems, 144
bulk modulus, 220
calculate strain, 444
calculate stress, 444
Candidate Points, **320**, 321, 327
Cantilever, 353
Carry Over Time Step, 518
cast iron, 40
causes of structural nonlinearities, 470
CD, 402
cell, 12
centrifugal stress, 404
ceramics, 40
CFL condition, **575**, 599
Chamfer, **86**, 201
change dimension name, 88
change dimension value, 88
Chart, 551
chattering, 478
Circle, 85
Circle by 3 Tangents, 85
circularity, **219**, 225
click, 64
click-sweep, 64
Closed End, 71, **85**
coefficient of thermal expansion, **32**, 283
Coincident, 63, **89**
compact disk, 402
Component Systems, 57
Components, 135
Compression Only Support, **147**, 232, 265
Concentric, **90**, 206
Concept/Cross Section/I Section, 299
Concept/Cross Section/L Section, **287**, 398
Concept/Cross Section/Rectangular, **275**, 485
Concept/Lines From Points, **286**, 297, 298, 398, 484
Concept/Split Edges, **450**, 458, 483
Concept/Surfaces From Edges, 301
Concept/Surfaces From Sketches, **111**, 126
conceptual model, 15
consistent unit system, 144
constraint status, 60, **76**, 107
Constraints, 60, 65, 83, **89**, 319
Construction Geometry, 164
Construction Point at Intersection, 85
Contact, 131, 153, **477**
contact body, 133
Contact Body View, 132
contact element, 133
contact formulation, 477
contact nonlinearity, 468, 470, **476**, 524
Contact Region, **130**, 153
contact stiffness, 525
contact stress, 151
Contact Tool, 148
Contact Tool/Pressure, 507
Contact Tool/Status, 506
contact type, 476
Contact/Pressure, 505
context menu, **63**, 64, 74, 83, 107
continuous selection, 64
contour band, 229
contour display, 228
control-middle-click-drag, 64
convergence criteria, 137, 338, **473**
convergence study, 163, 244, 277, 282, 310, **353**, 366
convergence value, 473
coordinate system, **119**, 181, 224, 228, 236
Coordinates, 152
Coordinates File, 285
Copies, 194

Copy, 63, **86**
Copy Calculated Values to Property, 557
Copy inputs to Current, **321**, 328
copy test data, 554
Corner, 86
Coulomb damping, 466
Coulomb friction, 425
Cover, **180**, 219, 346
Create Coordinate System, **224**, 592
Create Mode Shape Results, **377**, 381, 396
Create/Body Transformation/Mirror, 105
Create/Body Transformation/Translate, **152**, 298, 512
Create/Boolean, 195
Create/Pattern, **102**, 194
Create/Primitives/Sphere, 497
creeping, 527
critical damping, 466
critical damping coefficient, 422
critical load, 373
critically damped, 422
cross section, 275
Cross Section Alignments, **276**, 288, 300
current design, **316**, 321, 322
current energy, 574
current sketching plane, 59
Cursor, 84
Curvature Normal Angle, 582
Curve Fitting, 557
Cut, 85, **87**
Cut Material, **185**, 186
cylinder cover, **179**, 219, 346
Cylindrical, 224
cylindrical coordinate system, **224**, 246, 542
Cylindrical Support, 147
D, 314
damped free vibration, 421, **422**
damped natural angular frequency, 422
damping, 423
damping coefficient, 422, 423
Damping Controls, 425
damping effect, 21
damping force, 426
damping matrix, 426
damping mechanism, 423
damping ratio, 422
database, 18
Date and Time, 229
dedendum, 97
Define By, 135
definition of stress, 26
deformed shape, 18
degenerated element, 47
degree of freedom, **35**, 36, 47
Delete, 85, 111
delete constraint, 90
delete dimension, 89
delete edge, 85
Density, 403
Depth, 68
Design Exploration, 11, **317**
design point, **316**, 329, 408
design process, 311
design space, **316**, 329
design variable, 312, **316**
DesignModeler, 11, **14**, 47, 57, 70
DesignModeler GUI, 58, **81**, 197
DesignXplorer, 312
Details of Mesh, 16
Details View, **60**, 66, 81
deviatoric energy, **44**, 45
deviatoric strain energy, 44
deviatoric stress, **43**, 44
Diameter, 88
Dimension Display, 61, **62**
dimension name, 61
dimension value, 61, **88**
Dimensions, 60, 66, 73, 83, **87**
direct integration method, 426
Direct Optimization, 317, 325
Direction Definition, 152
Direction of surface body, 302
Disk, 448
disk and block, 448
displacement, **24**, 47
Displacement, 146
Displacement Convergence, **338**, 473, 524
displacement field, **24**, 36, 160
displacement scaling, 228
Display, **61**, 88
Display Model, **127**, 182
Display Option, **158**, 159, 291
Display Plane, **68**, 79, 100
Display Values in Project Units, **313**, 548
divider, 229
Drag, 86
Draw, 63, 83, **84**
driving parameter, 314
Drop, 588
drop test, **587**, 600
dry friction, 425
ductile material, **40**, 42, 43, 44, 47
Duplicate, **87**, 111, 118, 182
Duplicate Selection, **182**, 510
dynamic behavior, 417
dynamic effect, 21
dynamic simulation, **21**, 47
earth gravity, 306
edge, **82**, 107
edge selection, 302
edge selection filter, 300

edge-to-surface contact, 264
edges display, 228
Edit, 88
EFF_STN, 584
effective strain, **583**, 584
effective stress, 46
Eigenvalue Buckling, **375**, 380, 384, 387
elastic force, 426
elastic hysteresis, 424
elastic limit, 529
elastic material, 424, **528**, 568
Elastic Support, 147
elasticity, 528
elastomer, **10**, 11
element, 16, **35**
element convergence study, 170
Element Order, **159**, 168, 355
Element Size, **134**, 154, 289, 330, 356
element stress, 291
element type, 37
Ellipse, 85
End, 64
End Release, 283, **310**
End selection/Place Offset, 77
End/Set Paste Handle, 74
End/Use Plane Origin as Handle, 63
Energy Conservation, **574**, 598
energy error, 574, 599
Engineering Data, 11, **12**, 47, 211
Engineering Data Sources, **262**, 578, 588
Engineering Plastic, 220
environment branch, 143
environment condition, 16, 17, 24, 47, **145**, 170
Equal Distance, 90
Equal Length, 72, 76, **90**
Equal Radius, 90
equal temperament, 414
equilibrium equation, **30**, 33, 34
equilibrium iteration, 456, 466, **471**, 524
equivalent strain, 46
equivalent stress, **46**, 118
Erosion Controls, 583
ESC, **64**, 85
Explicit Dynamics, 427, 466, **571**
explicit integration method, 427, **572**
explicit method, 21, 570, **573**, 599
Export Video File, 216
Extend, 86
extend selection, 197, **199**
Extend to Adjacent, 199
Extend to Limits, **199**, 245
Extrude, 14, 15, 68, 79, **201**
face pair, 251
Facet Quality, 512
Failure, 579
failure criterion, 19, **40**, 42, 47
failure mode, 40
failure point, 40
Fatigue Tool, 148
FE convergence, 157, **161**
feature-based 3D modeling, 200
FEM procedure, 35
File/Close DesignModeler, 69
File/Exit, **69**, 80
File/Start Over, 83, **85**
Fillet, 67, **86**, 177
filleted bar, 157
Find Face Pairs, 250
Finger, 331
finite element, **16**, 47
finite element mesh, 12, **16**, 17, 47
finite element method, 16, 18, 33, **35**
finite element model, 12, **16**, 17, 18, 47
first harmonic mode, 410
first-order element, **37**, 47
Fix Endpoints, 89
Fix Guide Line, 193
Fixed, 60, 66, **89**
Fixed Support, 17, **146**
Flexible, 99
flexible gripper, **272**, 312
flexible spline, 87
Flip Horizontal, 64
Flip Surface Normal, **301**, 302
Flood Area, 199
Flood Blends, 199
Force, 146
Force Convergence, 137, **473**, 490, 524
force distribution, 135
force intensity, 25
Force Reaction, **488**, 521
Forced Frictional Sliding, 476
Fork, **191**, 429
Form New Part, **303**, 399
Formulation, 131, 153, **477**
fracture point, 40
fracture strength, 42
free boundary, 47
Free Face Mesh Type, 162
free rotation, 198
Free vibration, 22, 389, 417, **421**
Frequency, **392**, 396
Frequency Response/Deformation, 444
Friction Coefficient, **131**, 521
Frictional, 131, **476**, 521
Frictionless, 476
Frictionless Support, 125, **147**
From Coordinates File, 285
From Face, **185**, 188
fundamental frequency, **393**, 426

fundamental natural frequency, **393**, 417, 426
GA, **272**, 281, 312
Gear, **99**, 151
Gearbox, 256, **257**, 390
General, 60, 66, **87**
General Materials, **262**, 303
General Non-linear Materials, **578**, 588
Generate, 68
Generate Mesh, **114**, 133
geometric advantage, **272**, 312
geometric model, 13
geometric modeling, 15
geometric strain, **583**, 584
Geometry, **57**, 68
geometry display, 228
geometry modeler, 57
geometry nonlinearity, 23, 448, **470**, 524
Global, 84
global coordinate system, 173
Global Coordinates, 224
global mesh control, **227**, 330, 347
governing equation, 16, 18, **30**, 34, 47
Graph, 143
graphic user interface, 11
graphics window, **81**, 143
gravitational acceleration, 25
gravitational force, 17
greenish-blue color, 67
Grid, 91
Gripper, **273**, 313
guitar string, **410**, 456
hardening rule, 529, **530**, 568
harmonic mix, 413
harmonic mode, **413**, 417
harmonic response analysis, **427**, 441, 443, 466
harmonic response plot, **446**, 447
harmonic series, 413
Henri Tresca, 42
Hex Dominant, **340**, 351
hexahedron, **38**, 362
Hide Progress, 320
Hide Sketch, 127
high-speed impact, 467, **577**
higher-order 2D element, 366
higher-order element, **37**, 361
higher-order hexahedron, 359, 361, 362, **364**
higher-order parallel prism, **360**, 361, 363
higher-order perpendicular prism, **360**, 361, 363
higher-order tetrahedron, 359, 361, 362, **364**
history of ANSYS, 51
homogeneous, 24
Hooke's law, 24, **31**, 32
hoop direction, 247
hoop stress, **142**, 286, 543, 552
Horizontal, 63, 66, **88**, 89
hydrostatic compressive test, 532
Hydrostatic Pressure, 146
hydrostatic stress, **43**, 44
Hyperelastic Experimental Data, 554
hyperelasticity, 19, 528, **532**, 568
hyperelasticity model, 535
hysteresis elasticity, 528
idealized stress-strain curve, 529
Ignore Axis, 65
Impact, 578
implicit integration method, 427, **571**
implicit method, 21, 570, **599**
imprint face, 498
in-plane rotation, 278
inconsistency, 76
inconsistent unit system, **144**, 170
inertia effect, 21
inertia force, 24, **426**
Inertial, 145, **147**
Inertial/Acceleration, 306
Inertial/Rotational Velocity, 405
Inertial/Standard Earth Gravity, **306**, 594
Initial Condition, 456, 461, **594**
Initial Time Step, 338, **435**, 491
initial yield point, **529**, 568
initial yield surface, 568
input parameter, **316**, 319, 324, 329
insert a new object, 82
insert APDL command, 149
Insert as Design Point, **321**, 327
Insert Commands, 149
Insert/Convergence, 229
Insert/Deformation/Total, **117**, 136
Insert/Force, 135
Insert/Frictionless Support, 136
Insert/Sizing, 134
Insert/Stress/Equivalent (von-Mises), 117
Insert/Stress/Normal, 136
integration time step, 575, **595**
Interface Treatment, 153, **479**
interpolating function, 36
involute curve, **98**, 99
isometric view, 68, 79, 197, **198**
isotropic, 24
Isotropic Elasticity, **12**, 31, 47, 211, 220
isotropic hardening, **531**, 568
isotropic material, 527
iteration, 471
Joint Load, 146
just tuning system, 414
kinematic hardening, **530**, 568
Label, 148
Lagrange multiplier, 478
Large Deflection, **23**, 279, 338, 339, 470
LCD, 190

LCD display support, **203**, 231
legend control, 229
length unit, **58**, 71
Length/Distance, 88
lifting fork, 190, 429
Line, 84
line body, **15**, 271, 295
Line by 2 Tangents, 84
line model, **271**, 289
line of action, **97**, 98
line of centers, **97**, 98
line of resonance, 409
line search, **475**, 522, 524
Linear, **159**, 355
linear buckling analysis, 367
linear buckling analysis with constant load, 388
linear contact, 476
linear elasticity, 528
linear element, 37
linear material, 526, **527**, 568
linear simulation, **49**, 468, 486
linear solution, **337**, 494
linear structure, 468, **524**
Linearized Stress, 148
linearly elastic, **24**, 31
liquid crystal display, 190
Load Multiplier, **377**, 382, 386
load step, 456, **471**, 524
Loads, 17, 145, **146**
Loads/Bearing Load, 265
Loads/Force, 213
Loads/Moment, 155
Loads/Pressure, 305
local coordinate system, **175**, 274
lofting guide line, 193
Look At, 59, 64, **72**, 182
Lower Bound, **319**, 326
lower-order element, **37**, 361
lower-order hexahedron, 355, 361, 362, **364**
lower-order parallel prism, **361**, 363
lower-order perpendicular prism, **361**, 363
lower-order prism, 357, 358, **364**
lower-order tetrahedron, 356, 361, 362, **364**
lower-order triangular element, 159
lower/upper bound, 319
lumped mass model, **420**, 466
M20x2.5, 92
magnitude, 145
major diameter, 92
Major Grid Spacing, 91
Manual Input, 296
Manual Source, 357
Mass, 324
Mass Coefficient, 425
mass density, 25
mass matrix, 426
material assignment, 113, **212**
material damping, **424**, 466
material model, 19, **31**
material nonlinearity, 468, **470**, 524
material parameter, **31**, 526
material property, 220
material symmetry, 32
Max Modes to Find, **376**, 380, 391
Max Refinement Loops, 230
Maximum Energy Error, 574
Maximum Equivalent Plastic Strain EPS, 579
maximum normal stress, 41
maximum principal stress, 41
maximum shear stress, 41
maximum shear stress criterion, 42
Maximum Time Step, **338**, 491
measuring damping coefficient, 422
Mechanical, 11, **16**, 49
Mechanical GUI, 16, 18, 113, **142**
mechanical properties of PDMS, 11
medium principal stress, **41**, 43
member force, 293
member moment, 293
Mesh, 114
mesh control, 159
Mesh Control/Face Meshing, 431
Mesh Control/Method, 217
Mesh Control/Sizing, **167**, 514
mesh count, 227
mesh density, **218**, 347
Mesh Metric, **335**, 347
mesh quality, 217, 218, **335**, 336
Mesh/Generate Mesh, **133**, 135
Mesh/Preview Surface Mesh, **501**, 502
meshing, 330
Method, 159
Method Name, 325
Microgripper, 103, **104**, 495
Micrometer, 104
Mid-Surface, 249, **251**
middle mouse button, 198
middle-click-drag, 64
Midpoint, 89
midside node, 37
mild steel, 40
minimum normal stress, 41
minimum principal stress, **41**, 383
Minimum Reference, 473
Minimum Size Limit, 349
Minimum Time Step, **338**, 491, 519
Minor-Steps per Major, 91
Mirror, 105
MISQP, 318
Modal, 390

modal analysis, 22, 49, 389, 381, **426**
Mode, **377**, 381
mode shape, 426
mode tab, 81
Model, 16
model airplane wing, 418
model tree, 68, 79, **82**, 107
Modeling mode, **68**, 81, 107
Modify, 63, 65, 67, 83, **86**, 118
MOGA, 318
Mohr's circle, 27, **41**, 42
Moment, 146
Moment Convergence, 473, **490**, 524
moment equilibrium, 27
Mooney-Rivlin, **535**, 553
Mooney-Rivlin 2 Parameter, 557
mouse cursor, 198, **199**
mouse operation, 64
Move, **87**, 88
move dimension, 67
MPC, 477
multi-point constraint, 477
Multilinear Isotropic Hardening, 548
MultiZone, 217, **227**
MultiZone method, 236, **341**, 352
MythBusters, 402
Name, 61
natural angular frequency, 421
natural frequency, 22, 389, **393**, 406, 407, 417, 421, 426
natural period, 421
negative-X-face, 26, **27**
negative-Y-face, 26, **27**
negative-Z-face, 26, **27**
Neo-Hookean, 535
New Chart and Table, **547**, 567
New Plane, **82**, 120, 181
New Section Plane, 343
New Sketch, **82**, 106, 127, 186
Newton-Raphson method, **472**, 524
Next View, 72
NLPQL, **318**, 329
No Separation, 466, **476**
nodal displacement, 36
nodal stress, 291
node, 16, 35, **49**
nonlinear buckling analysis, 367, **387**
nonlinear contact, 476
Nonlinear Effects, 212
nonlinear elastic material, 528
nonlinear elasticity, 528
nonlinear material, 526, **527**, 568
nonlinear simulation, 49, 137, 337, 468, **469**, 489
nonlinear solution, 338, 344, **494**
nonlinear structure, 468, **524**
nonlinearity, 264
Normal Lagrange, 478
Normal Stiffness, 477, **478**, 525
normal strain, 28, **29**
normal stress, **26**, 27
number of buckling modes, 376
number of digits, 229
Number of Steps, **471**, 515
Number Of Substeps, 279
numerical method, 33
object, 107
objective, 319
Objectives and Constraints, **319**, 325
octave, 414
Offset, 77, **87**, 183, 194
Ogden form, 535
Open End, 63, **85**, 99
Open End with Fit Points, 99
Operation, 128
optimal design, **320**, 321, 327
Optimization, 311, **317**, 318, 320, 325
optimization method, 318
optimization problem, 325
order of element, 36
Orientation, 136
Orientation About Principal Axis, 228
Origin, 228
orthotropic, 527
Orthotropic Elasticity, 32, 527
out-of-plane translation, 278
Outline, **143**, 316
Outline Plane, 181, **182**
Output Controls, **435**, 453, 519
output parameter, 316, 324, **329**
Oval, 85
over-constrained, 76
over-damped, 422
overtone, 413
P, 315
paint-select, 77
Pan, 64, **72**, 83
Parallel, **90**, 558
parallel prism, **357**, 363
Parameter Set, **314**, 315
parameter study, 373
part, 82
pascal, 31
Paste, 64, **87**
Paste at Plane Origin, **64**, 75
paste handle, **75**, 107, 127
patch, 348
Patch Conforming, 348
Patch Independent, **348**, 349
Path, **164**, 165
PC, 402
PDMS, 10, **11**

perfectly elastic-plastic material, 577
Perpendicular, 89
perpendicular prism, **358**, 363
Physical Properties, 220
physics of music, 410
pin-jointed, 310
Pinball Region, 479
Pitch, 198
pitch circle, **97**, 98
pitch point, **97**, 98, 99
pitch radius, 97, **98**
placed feature, 200, **201**
planar seal, 553
plane of symmetry, **17**, 125
plane outline, 182
Plane Strain, 561
Plane Stress, 110, **113**, 513, 451
plane view sketching, 175
plane-strain condition, 141
plane-strain problem, **141**, 170, 553
plane-stress condition, 140
plane-stress problem, **140**, 170
PLANE182, 38
PLANE183, 38
plastic material, **528**, 548, 568
plastic strain, **528**, 577, 583
Plastic Strain Failure, 579
plasticity, 528, **529**, 536
plasticity model, 531
Play, **18**, 216
pneumatic finger, **10**, 35, 331, 569
Point, **73**, 74, 82
Point, 296
Poisson's ratio, 11, **31**, 32
Poisson's Ratio, 12, **220**
polycarbonate, 402
polydimethylsiloxane, 10
Polygon, 84
Polyline, 63, **84**
polynomial form, 535
polyoxymethylene, **272**, 481
POM, **272**, 481
Preserve Bodies, 152
Pressure, 25, **146**
pressure angle, 97
pressure cylinder, **179**, 219, 346
prestressed modal analysis, 390, **393**
pretension, 125
Previous View, 72
Principal Axis, 228
principal direction, 41
principal strain, 19
principal stress, **41**, 49
principal stretch ratio, 534
principal view, 197
principle of conservation of energy, 574
principle of work and energy, **574**, 599
Probe, 148
Probe/Deformation, **437**, 462
Probe/Force Reaction, **488**, 516, 544, 566
Probe/Stress, 439
problem domain, 24
procedure of FEM, 49
Progress, 320
Project, 211
project name, **69**, 80
Project Schematic, 11, 12, 49
project tree, **143**, 170
project unit system, 144
Properties, 112
Pull-down menu, **81**, 143
Pure Penalty, 478
quadratic element, 37
quadrilateral, 38
Quadrilateral Dominant, 162
quadrilateral element, **37**, 162, 163
quadrilateral-based pyramid, 38
quality of mesh, 16
quasistatic simulation, 419
quadrilateral shell, 38
radial deformation, 247
radial direction, 246
radial stress, 142
Radius, 67, 86, **88**
random vibration analysis, **428**, 466
rate-dependent material, 568
Re-Fit Spline, 99
reaction force, 488
Rectangle, 59
Rectangle by 3 Points, 85
Redo, 72
reference energy, 574
Refresh Geometry, 123
Relevance, **114**, 227, 330
Relevance Center, **227**, 330
Remote Displacement, 147, **504**
Remote Force, 146
Rename, 137
Rename Based on Definition, 385
Replicate, 74, **87**
Reset, 112, **346**
reset legend, 229
residual force, **472**, 524
residual strain, 528
response, **24**, 30
response frequency, 435, **437**
response spectrum analysis, **428**, 466
restart, 595
Result Sets, **280**, 438
results, 18

results object, 117, **148**
results tool, 148
results toolbar, 148
results view control, 228
Resume From Cycle, 596
Retrieve This Result, **541**, 551, 552
Reverse Normal/Z-Axis, 120
revolute joint, 283
Revolve, 96, **201**, 242
Richard von Mises, 44
right-click, 64
right-click-drag, 64
Rigid, **581**, 592
rigid body, 499
rigid body mode, 370, **393**, 417
rigid body motion, 29, **117**, 118
rigid-Jointed, 310
roll, 198
rotate, 64, 197, **198**
Rotate about Global Z, 120
Rotate by r, **74**, 75
Rotation Convergence, **473**, 524
rotation mode, 412
Rotational Velocity, **147**, 405
Rough, 476
round, 177
round-cornered textbox, 58
RPM, 408
rubber, 553
ruler, **61**, 113
Safety Factor, 215
Save Project, **69**, 80
Scoping Method, **165**, 505
Screening, 318
scroll-wheel, 64
SDOF, 421
Seal, 554
second-order element, **37**, 49
section plane, 343
section view, 125, **343**
Select All, 245
select mode, 77
Select new symmetry axis, **60**, 65
selection, 199
Selection Filter, 74, 107, 197, **199**
selection pane, 197, **199**, 301, 302
Semi-Automatic, 88
separator, **81**, 143, 211
Settings, 83, **91**
shape function, **36**, 49
shape memory alloy, **103**, 495
sharp-cornered textbox, 58
shear failure, **40**, 42
Shear Modulus, 31, 32, **220**
shear strain, **28**, 29
shear strength, 42
shear stress, **26**, 27
Shear Test Data, 532, 553, **554**, 555
shell element, 238, 269, **270**, 278
SHELL181, 38
SHELL208, 248
Shift-middle-click-drag, 64
Show Constraints, 90
Show Elements, 307
Show in 2D, 91
Show Progress, 320
Show Sketch, 128
Show Whole Elements, 343
Show/Sweepable Bodies, 341
SimpleBeam, 369
simply supported beam, 368
simulation, 12
single degree of freedom model, 421
singular point, 169
singular stress, 166
sizing, **154**, 330
sketch, **82**, 107
sketched feature, 200
Sketching, 15, 56, **62**
Sketching mode, 59, **81**, 107
sketching option, 61
sketching plane, **82**, 107
sketching toolboxes, 81, **83**
Skewness, **335**, 347
Skin/Loft, **192**, 201
SMA, **103**, 495
small cube, 26
Small deformation, 24, **28**, 29, 31
Smooth Contours, 586
Snap, 91, **509**
Snap lock, 508
Snaps per Minor, 91
solid body, 15
solid damping, 424
solid element, 270
Solid Fill, 463
solid model, 255
SOLID186, 37
SOLID187, 37
Solution, 117, **143**
solution accuracy, 575
Solution Information, 137, 143, 338, **474**, 490
Solution Output, 137
Solve, **18**, 116
Solve Curve Fit, 557
Solver Output, **474**, 490
source face, **228**, 341
Sphere Radius, **167**, 168
Spline, **85**, 99
Spline Edit, **87**, 99

Spline Edit option, 87
Split, 86
Split Edges, 450
Split option, 86
spur gear, **97**, 151
square bracket, 10
stability analysis, 367
Standard Earth Gravity, 147
standing wave, 410
Static Damping, **576**, 599
static simulation, 21
Static Structural, 11, 112
statically indeterminate structure, 379
Statistics, 16, **114**
status bar, **81**, 83, 143
status symbol, 149
steady state, 419
Step End Time, 435
Steps, 456, 466, **471**
stiffness, 49
Stiffness Behavior, 581
Stiffness Coefficient, **425**, 435
stiffness matrix, **36**, 49, 426
still water, 25
Stop, 216
strain, **28**, 49
strain component, 28
strain energy, **533**, 534
strain field, 36
strain invariant, 534
strain state, 49, **141**, 142
strain-displacement relation, **31**, 33, 36
stress, **25**, 49
stress component, **26**, 27
stress concentration, 163
stress discontinuity, **157**, 170
stress field, 36, **160**
stress intensity, **42**, 49
stress singularity, 157, **166**, 169, 170
stress state, **27**, 41, 42, 46, 49, 140, 142
stress stiffening, 20, 49, 367, **368**
stress stiffening effect, **373**, 387
Stress Tool, 148
stress-strain curve of PDMS, 11
stress-strain relation, **31**, 33, 36
Stress/Error, 161
Stress/Maximum Bending Stress, 291
Stress/Minimum Bending Stress, 291
stretching mode, 412
String, **410**, 457
strong dimension, 79
structural analysis problem, 24
structural dynamic analysis, 417
structural dynamics, 420
Structural Error, 157, **161**, 170, 216
structural mechanics, 24
structural nonlinearity, 22
structural steel, 113
SU316, 239
subsequent yield point, **529**, 568
Substeps, 456, 466, **471**, 524
Subtype, 182
Support, 17, 136, 145, **146**, 204, 231
support condition, 265
supporting file, 69
Supports/Fixed Support, 212
surface body, 15, 113, 242, 257, **269**, 295, 301
surface force, 30
surface model, 238, **255**, 289
surgical parallel robot system, 10
Sweep, 192, **201**
Sweep method, 236, **341**
sweep-select, 77
sweepable, 228
sweepable body, 341
switch between steps, 471
Symmetric, **131**, 477
Symmetry, **60**, 65, 76, 89, 118, 270
symmetry condition, 278
symmetry of shear stress, 27
Table of Design Points, **316**, 317
Table of Outline, 326
Table of Schematic, **319**, 320
Tabular Data, **143**, 461, 518, 540
tag, 228
Tangent, **89**, 106
Tangent Line, 84
Tangent Modulus, 579
tangent stiffness, 472
tangential stiffness, 478
Target, 131, 153, **477**
Target Body View, 132
target face, **228**, 341
Temperature, 555
temperature change, **17**, 24
tensile failure, **40**, 42
test data, 532
tetrahedron, **38**, 362
Tetrahedrons method, 356
textbox, 10
Thermal Condition, 146
thermal effects, 32
thermal strain, 32
Thermal Strain Effects, 212
Thickness, **113**, 244
Thickness assignment, 253
thin body, 206
thin solid body, 257, **269**
Thin/Surface, **201**, 206, 256, 257

threaded bolt, 92
threaded bolt-and-nut, 125
Threads, **93**, 126
threshold, 250
thunderbolt, 121
Time, 338
Time Integration, 457
Time Step, 471
Time Step Controls, 479
Time Step Safety Factor, 575
time-dependent stress-strain relation, **527**, 568
Tolerance, 473
toolbar, **81**, 143
toolbar menu, 115
Toolbox, **11**, 12, 316
Tools, 148
Tools/Beam Tool, **279**, 291
Tools/Contact Tool, 505
Tools/Mid-Surface, **250**, 251
Tools/Options, 62
Tools/Stress Tool, 213
Tools/Symmetry, 119, 120
topology nonlinearity, 468, **470**
torsional vibration, 397
total strain energy, 44
Transfer Data To New/Eigenvalue Buckling, 374
transformation, 181
transient state, 419
transient structural simulation, 21, 417, **419**, 426, 434
Transient/Initial Conditions, 453
Transjoint, 482
translational joint, 481
Tree Outline, 81
Tresca criterion, 42
Triad, 197
triangle, 38, **159**, 364
triangle-based prism, 38
triangular element, **162**, 163
triangular shell, 38
Trim, 65, **86**
Triplate, **80**, 110, 323
True Scale, **156**, 247, 438
Truss, **284**, 380
twelve-tone equally tempered tuning system, 414
two-step method, 456
two-story building, **295**, 395, 418, 441
Type, 521
type of coordinate system, 228
U.S. customary unit system, 151
Unaveraged, **158**, 159
unaveraged stress, **160**, 291
undamped free vibration, 421
undeformed shape, 10
under-constrained, **67**, 76
Undo, 72
uniaxial compressive test, 532
uniaxial tensile test, 19, **40**, 42, 532, 553, 554, 555
uniform velocity, 448
unit system, 144
Unit Systems Mismatch, 445
Unite, 195
Units, 58, 71, 113, **144**
Units/Display Values as Defined, 408
Units/RPM, 405
unprestressed modal analysis, 390
Update, 320
Update All Design Points, 317
Update Selected Design Points, **317**, 321, 328
Update Stiffness, 479
Upper Bound, **319**, 326
Use Manufacturable Values, 326
Use Maximum, 444
Use Plane Origin as Handle, 75
User Defined Result, **148**, 546, 584
UY, 546
Value, 61, **62**, 473
vector display, 228
Velocity, 453
verification manual, 525
verification manual for Workbench, 237
Vertical, 63, **88**, 89
vibration mode, 22
vibration mode shape, 389, **393**, 397
view results, 148
view rotation, 198
View/Cross Section Alignments, **274**, 286, 288
View/Cross Section Solids, **275**, 288, 300, 486
View/Reset Workspace, 211, **315**
View/Ruler, **61**, 113
View/Shaded Exterior, 178
View/Thick Shells and Beams, **254**, 412, 487
View/Windows/Reset Layout, 142
View/Windows/Section Planes, 343
viscosity, 527
viscous damping, **423**, 466
VM25, 171
volumetric test, 532
von Mises strain, 46
von Mises stress, 19, 20, **46**, 49
von Mises yield criterion, **43**, 44, 46, 529
von Mises yield surface, **46**, 568
W16x50, 69
W16x50 beam, **57**, 295, 299
wave speed, 410
weak constraint, 65
weak dimension, **67**, 78, 86
Weak Springs, **116**, 170
well-defined, **60**, 66, 76
Workbench GUI, **11**, 49, 57
Worksheet, **148**, 505

X-face, 26
XYPlane, **14**, 59
Y-face, 26
Yeoh, 535
yield criterion, 529
yield point, 40
yield state, 529
yield strength, 211
yield surface, 568
young's modulus, **11**, 31, 32, 529
Young's Modulus, **12**, 220
Z-face, 26
Zoom, 72
zoom in/out, 64
Zoom to Fit, 60, 64, 67, 68, **72**